S0-CBI-139

THE

INTERNATIONAL SERIES

OF

MONOGRAPHS ON PHYSICS

COSMICAL MAGNETIC FIELDS

THEIR ORIGIN AND THEIR ACTIVITY

BY

E. N. PARKER

CLARENDON PRESS · OXFORD
1979

Oxford University Press, Walton Street, Oxford OX2 6DP

OXFORD LONDON GLASGOW
NEW YORK TORONTO MELBOURNE WELLINGTON
KUALA LUMPUR SINGAPORE JAKARTA HONG KONG TOKYO
DELHI BOMBAY CALCUTTA MADRAS KARACHI
IBADAN NAIROBI DAR ES SALAAM CAPE TOWN

Published in the United States by
Oxford University Press, New York.

British Library Cataloguing in Publication Data

Parker, E N
Cosmical magnetic fields.
1. Magnetic fields (Cosmic physics)
I. Title
523.01′8′8 QC809.M25 78–40488

ISBN 0-19-851290-2

Printed in Great Britain by J. W. Arrowsmith Ltd., Bristol

PREFACE

THIS monograph treats the physics of large-scale magnetic fields in fluids of high electrical conductivity. The terrestrial laboratory is too small to demonstrate more than the simplest magnetic effects, whereas the astronomical universe abounds with the complicated activity of the fields. Consequently, the theoretical investigation is guided by the observed behaviour of the magnetic fields in the universe about us, particularly by the detailed and comprehensive observations of the magnetic fields of Earth and of the sun.

In the last two decades a remarkable variety of exotic active objects in the astronomical universe, such as pulsars, active supernovae remnants, x-ray stars, x-ray bursters, γ-ray sources, active galaxies, and quasars, have been discovered all of which appear to involve magnetic fields in either a primary or strongly supporting role. The discovery of these remarkable creatures has enormously expanded our horizons. Clearly there are more things in the heavens than anyone anticipated. The familiar planets, stars, and star clouds are only the more prosaic inhabitants of the universe. The extraordinary energy of many of the active objects suggests gravitational collapse, which has inspired theoretical work on the dynamics of fields and fluids in close proximity to massive, compact, spinning objects. These circumstances have also inspired development of the quantum electrodynamics of particle decay and of pair creation and annihilation in the Kerr metric to understand more fully the capture and escape of particles and radiation from black holes. Indeed, the theory of the creation and annihilation of black holes themselves has been explored, with very interesting results. But for all the fascination of these fundamental physical effects, and their suggested association with the active exotic objects, the effects in the objects lie beyond the resolution of our finest instruments. Hence, their exploration is limited to inference based on the information in the total integrated signal from each object.

On the other hand, we know from detailed studies of the activity in the terrestrial magnetosphere and in the solar atmosphere that the basic concepts form only a naïve beginning. Close inspection always shows major unanticipated cooperative effects. Nature is much more imaginative than we are. Indeed, were the creativity of nature as restricted as our own, there would be no pulsars, x-ray stars, γ-ray bursts, and quasars, for we certainly did not anticipate any of them from our 'basic concepts'. The fundamental equations of physics may contain all knowledge, but they are close-mouthed and do not volunteer that knowledge.

It is amusing to look back at the first ideas on the activity of the magnetic fields of Earth and the sun developed prior to the detailed observations of the last two decades. Basic principles and naïve extrapolation to detailed models were all mixed together in the literature. Observations have since shown the validity of the basic principles and the *naïveté* and inadequacy of much of their early superficial application. It seems not unlikely that the present popular explanations of x-ray stars and quasars form a similar witch's brew. It is to be expected that the active objects represent cooperative effects largely unanticipated by the separate application of basic concepts. Detailed observations are required to light the way for the advance of theory, and such observations lie in the future, if they exist anywhere at all. Hence, we attempt far less in this writing, in the expectation of accomplishing more, presenting only the general idea of magnetic activity and then the separate elementary physical properties of magnetic fields, insofar as they are known, with particular attention to the generation and active decay of magnetic fields. The detailed observations of Earth, the sun, and the galaxy are the guideposts. We expect that, even for Earth and the sun, there is much that we do not yet know or understand. The theory of the distant active objects is left to others with greater physical insight and computational ability.

The present exposition is built on the direct application of physical principles, combined with formal mathematical examples to illustrate the nature and immediate consequences of the effects. Consequently, the applied mathematician may feel the work to be heuristic, because it makes use of rigorous physical arguments, and the observer may feel it to be formalistic, because it makes use of formal calculation. However, that is the nature of physics, as distinct from mathematics on the one hand and observation on the other.

As a final comment, anyone assaulting this book by continuous reading from the beginning will soon discover that individual points and principles may be stated several times. Such redundancy is necessary for two reasons. The first is to avoid having to read all the pages prior to the nth page to understand what the nth page has to say; each section should be intelligible in its own right with a minimum of references to earlier sections.

The second reason is that redundancy is an essential part of the presentation of a topic as complex as the web of interconnecting magnetic effects. The printed or spoken word forms a single continuous thread, and if it is to describe a web, it can do so only by describing each link in the web and then noting in linear sequence how the links fit together to form the web. So the exposition notes each separate magnetic effect and then points out the interconnections, while continually reminding the reader of the overall pattern of the web. Hence, there is repetition of principles and concepts at appropriate places in the text. We do not intend to hand the reader a bushel of potsherds with the remark that it contains the wisdom of ancient Sumer. Facts do not make knowledge until they are assembled in systematic form, the major interrelations noted, and some general impressions and conclusions formed. Needless to say, the description of properly assembled potsherds is a tedious business. We hope that the results are worth the effort.

E. N. P.

Chicago
December 1977

ACKNOWLEDGEMENTS

It is a pleasure to acknowledge the help and assistance rendered to me by others in the preparation of this monograph. First of all, I am indebted to the University of Chicago for maintaining the ancient custom of scholarship and research in the face of mounting bureaucratic pressure to do otherwise. In particular, I am grateful for the University sending me from administrative duties at Chicago to work for six months at the Sterrewacht of the University of Utrecht at the kind invitation of Professor C. de Jager, director of the Sterrewacht. The gracious hospitality and the freedom from official responsibility enjoyed by a visitor of the Sterrewacht were essential for a timely finishing of the manuscript.

I wish to express, too, my debt to all of my colleagues at Chicago, Utrecht, Sacramento Peak, Boulder, Cambridge, St. Andrews, Göttingen, Tuscon, Pasadena, Stanford, Greenbelt, New York, and elsewhere who have contributed so much to the development of the ideas through general discussions over the years. Even the perpetually skeptical critics have contributed because they stimulated development of a sharper presentation of the concepts. The students who have endured my lectures on the properties of magnetic fields have contributed in no small way too. I have occasionally remarked to them—and they steadfastly refuse to believe—that the lecturer gains at least as much in the preparation and delivery of a lecture as they do in listening to it.

The published works of other researchers are fundamental building blocks in the present construction and I am grateful for permission to reproduce figures from some of those works. Thus, Figs 16.6 and 16.7 are taken directly from a published paper of Dr. N. O. Weiss at the University of Cambridge. Figure 17.4 is taken from the published numerical experiments of Dr. R. H. Kraichnan of Dublin, New Hampshire. Figures 19.17–19.20 are from a paper by Dr. E. H. Levy, presently at the University of Arizona. The numerical experiments on dynamo models published by Dr. M. Stix, presently at the Fraunhofer Institute in Freiburg, contributed Figs 19.21 and 19.22, while Figs 21.2 and 21.3 represent the work of Dr. H. Köhler of Göttingen, and Figs 21.4 and 21.5 are from Dr. H. Yoshimura at the University of Tokyo. Finally, Figs 22.1 and 22.2 are from the published work of Dr. W. A. Hiltner, presently at the University of Michigan.While on the subject of figures, I should like to express my gratitude to Mr. Rudolph Banovich and Mr. Edward Pool of the Graphic Arts Department of the University of Chicago for converting my numerous tremulous sketches into respectable drawings.

The enormous task of converting manuscript to typescript was carried out by Mrs. E. C. Hutton, Mrs. Gail Wagner, Mrs. Sylvia Berry, Mrs. Jo Muster, and Mrs. Yvette McLean. I am particularly grateful to my wife Niesje for typing the two longest chapters while we were in Utrecht and for all the help and moral support in the final months of correcting and editing, without which many more errors would have gone unnoticed and the finished product would have been much delayed.

CONTENTS

1

THE ROLE OF MAGNETIC FIELDS

In many respects the astronomical universe has reached the stage of middle-age, with its violent youth behind it and its final stages of senility still safely in the future. Observations indicate that the universe started on its present course of expansion and degradation from a hot compact state some 10–20 billion (10^9) years ago. The matter in the universe has evolved into galaxies and stars, losing all outward trace of the initial compact state. The only recognizable surviving remnants of the early moments are the deuterium and helium in the early (and by now very old) faint stars (Danziger 1970; Pagel 1973; Hirschberg 1973), and the universal 4° K microwave radiation in space (Gamow 1949, Alpher and Herman 1948; Alpher *et al.* 1953; Penzias and Wilson 1965; Dicke *et al.* 1965: see review by Harrison 1973 and references therein; Hoyle 1975). Our galaxy formed about 10 billion years ago, and the sun 5 billion years ago when the galaxy was already well along in years. The sun is halfway through its 10 billion years of normal life on the main sequence. The synthesis of the heavy elements from hydrogen in massive, short-lived stars is well advanced (Wagoner *et al.* 1967; Pagel 1973; Arnett 1973; Iben 1974). The heavy elements carbon, oxygen, iron, and silicon, together with the general hydrogen background, have formed complex molecules in interstellar space (Zuckerman and Palmer 1974; Litvak 1974), and then planets together with new stars.

These effects continue today, but the initial excitement has passed and so far as one can tell, the universe is looking forward to unlimited expansion into ultimate oblivion (Gott *et al.* 1974). It is with some surprise, therefore, that examination of the universe on a small-scale shows so much activity. After 10 billion years new stars are still being formed. After 5 billion years, the sun is still popping and boiling, unable to settle down into the decadent middle age that simple theoretical considerations would suggest. Perhaps the most singular property of the general activity is the continual acceleration of a small fraction of the atoms to high speeds. As a result one finds throughout the galaxy, and particularly in the neighbourhood of active stars, electrons, protons, and helium and heavier nuclei, with nearly the speed of light. These *cosmic rays* are the outcasts from the normal thermal activity of the stars, catapulted (Fermi 1949, 1954; Parker 1958, 1965; Gunn and Ostriker 1969; Kadomtsev and Tsytovich 1970; Evans 1974; Papadopoulos and

Coffey 1974; Scott and Chevalier 1975; McIlwain 1975) into the far corners of phase space, where they reside for a time before collisions restore them to the thermal state.

It appears that the radical element responsible for the continuing thread of cosmic unrest is the magnetic field. Magnetic fields are familiar in the laboratory, and indeed in the household, where their properties are well known; they are easily controlled, and they serve at our beck and call. In the large dimensions of the astronomical universe, however, the magnetic field takes on a role of its own, quite unlike anything in the laboratory. The magnetic field exists in the universe as an 'organism', feeding on the general energy flow from stars and galaxies. The presence of a weak field causes a small amount of energy to be diverted into generating more magnetic field, and that small diversion is responsible for the restless activity in the solar system, in the galaxy, and in the universe. Over astronomical dimensions the magnetic field takes on qualitative characteristics that are unknown in the terrestrial laboratory. The cosmos becomes the laboratory, then, in which to discover and understand the magnetic field and to apprehend its consequences.

The liberation of nuclear and gravitational energy in stars and the occasional liberation of gravitational energy in collections of stars feeds the universe, causing the stars to shine and their gases to churn. Most of the energy is degraded directly and passively as it seeps out from the hot interior to the cool surface of each star. From the surface the energy is radiated away into space where it is diluted and ultimately consumed by the expansion of the universe. But there is a small fraction of the energy which follows various complex side-paths to oblivion and we shall concern ourselves with one of those paths as it produces magnetic fields and the strange effects that follow.

There are other paths with strange consequences, of course. For instance, we should not fail to note that a microscopic portion of the energy from stars falls on the surface of planets where a small fraction is used by plants to synthesize the complex organic molecules that are the basis of life—perhaps somewhere intelligent life. In the final analysis, psychology and sociology, not to mention the writing of books, are no more than one of the exotic degradations of sunlight, an insignificant branch of the main stream of energy flow. A byproduct of the energy release is the previously mentioned synthesis of heavy elements which makes possible the planets and their life forms[1]. Some of the radioactive heavy elements contribute to the high temperatures in the interior of

[1] We should not fail to recognize the mythological implications of the ethyl alcohol fumes in interstellar space (Zuckerman *et al.* 1975) as a consequence of some of the heavier elements and a little of the energy from stars. Nor should we overlook the irony of the interstellar formaldehyde (Zuckerman *et al.* 1970).

Earth, and presumably, therefore, to the magnetic field of Earth. We could go on endlessly identifying diverse trickles of energy degradation. It is, after all, the sum total of these paths that makes up the subject of astrophysics. But to return to the purpose of our narrative, we are interested in the energy that is diverted temporarily into magnetic fields.

Magnetic fields are produced by fluid motions. We note that the liquid core of Earth appears to be in a state of turbulent convection. The outer layers of many stars, including the sun, are subject to convection. The interstellar gas making up the gaseous disc of the galaxy is continually churned by heat from the luminous stars that are created within it. In all of these cases the convection—the churning of the fluid—gives rise to magnetic fields, and in all cases the magnetic fields give rise to a complicated activity that would otherwise not occur. Once the energy is invested in fields, the possibilities for further complexity become almost limitless. Sunspots, solar flares, and solar eruptions are but a few of the phenomena produced on the sun. (See review by Tandberg-Hanssen 1973; Smith and Gottlieb 1974.) The magnetic field may refrigerate its surroundings, as in the sunspot where the temperature is reduced to two-thirds the surrounding 6000 K. Or it may cause intense heating, as in the enhanced chromosphere and corona above regions where magnetic field emerges through the surface of the sun. The temperature may rise to 3 000 000 K or more in the enhanced corona, some 500 times greater than the surface temperature. The entire sunspot cycle with its quasi-periodic variation of 11 and 22 years is a magnetic phenomenon, evidently produced in the convection that is driven by the outflow and degradation of energy from the core of the sun. The magnetic fields on the sun represent a diversion of less than 10^{-4} of the total solar output, yet they are responsible for qualitative effects on the enivironment in space and on Earth.

The occasional intense x-rays and ultraviolet radiation from the sun are a product of solar activity and, hence, of magnetic fields. The bursts of x-rays (McKenzie *et al.* 1973; Datlowe and Lin 1973; Kane *et al.* 1974; Datlowe *et al.* 1974; Goldberg 1974) and the occasional γ-ray burst (Lingenfelter and Ramaty 1967; Audouze *et al.* 1967; Chupp *et al.* 1973; see also Ramaty and Stone 1973; de Feiter 1974) are caused by the collision of fast electrons and atomic nuclei with thermal nuclei, after acceleration to high speeds by the magnetic field activity that makes up the solar flare. The enormous outbursts of fast electrons and protons from solar flares sometimes fill the inner solar system to levels that are a problem for human life in spacecraft (Datlowe 1971; Mogro-Campero and Simpson 1972; Bertsch *et al.* 1972; Anglin *et al.* 1973; Sakurai 1973; Simnett 1974; Pomerantz and Duggal 1974) and affect the terrestrial ionosphere (Simpson 1960; Reid 1961; Ney 1962; Davis 1962; Krimigis and Van Allen 1967; Scholer *et al.* 1972; Aarsnes and Amundsen 1972). The solar wind is powerfully affected by magnetic events on the sun, as is the effect of the solar wind on Earth. The magnetic sector structure of the wind, the turbulence, the standing shock waves (Wilcox and Ness 1965, 1967, 1968), and the occasional blast wave from a flare all owe their existence to magnetic conditions in the sun (Parker 1963; Hundhausen 1972; Jokipii 1973). The large cosmic ray variations at Earth are the result of magnetic variations in the wind (Parker 1963; Quenby 1967; Jokipii 1971). The effects of the solar wind on Earth are transmitted through the magnetic field of Earth where the aurora and the Van Allen radiation belts are added to the magnetic creations of the sun (Dessler 1967; Williams and Mead 1968; Akasofu and Chapman 1972). The remarkable conditions surrounding the planet Jupiter, with intense belts of fast particles trapped in a distended magnetic field emitting bursts of protons and relativistic electrons into space with each rotation of the planet, were recently observed by direct passage of the Pioneer X and XI spacecrafts (see, for example, Mead 1974). The fast protons trapped in the magnetic field of Jupiter are so numerous as to reduce the life of electronic circuits to only a few hours. Pioneer X was nearly incapacitated during its swift passage through the most intense part of the Jupiter radiation belts. Strong activity at Jupiter had been anticipated for many years form the occasional bursts of radio waves, that could be produced only by electrons moving near the speed of light in a magnetic field. It had been deduced from the frequency, intensity, and distribution of radio emission around the planet (Warwick 1961) that Jupiter must have a magnetic field of the same dipole form as Earth but some twenty times stronger. But, of course, the precise exotic nature of the activity could be determined only by observations *in situ.* The activity is a direct consequence of the rapid rotation (11-h period) of Jupiter and its strong magnetic field in the solar wind.

The sun is a star in middle age, of a class of no particular distinction. Other stars, although far away and difficult to observe, are presumably as active as the sun, and it is an observed fact that some are very much more so. The flare star, for instance, is a star one-thousandth as bright as the sun, with frequent flares a thousand times larger than any exhibited by the sun (Luyten 1949; Joy and Humason 1949; Lovell 1971; Webber *et al.* 1973; Kunkel 1973; Moffet 1973, 1974; Rodono 1974; Worden 1974). The whole flare star brightens with the flare. Indeed, were this not so, the flare could not be observed from our great distance. Some of these stars show light variations that can be explained only as a large cool 'sunspot' on the surface of the star, rotating around with the star so that when the spot is facing us the star is noticeably dimmer (Bopp and Evans 1973; Torres and Mello 1973; Lacy, 1977).

Over 200 magnetic stars, with fields in excess of 100 G and ranging up to 34 000 G, have come to light since the development of the Babcock magnetograph and related instruments (Babcock 1958, 1960, 1963; Sargent 1964; Ledoux and Renson 1966; Preston 1967, 1969; Severny 1970; Borra and Landstreet 1973). Some of them show peculiar variations with time. Unfortunately it is only possible to observe the magnetic field of a distant star if the algebraic mean of the field in the line of sight over the visible hemisphere of the star exceeds 100–200 G, and then only under favourable circumstances. The magnetic field of the sun (some few G) could not be detected from the distance of its nearest neighbour, α-Centauri. Nor can one detect the magnetic field of any star whose spectral lines are faint, or are broadened by temperature, rotation, or 'turbulence'. The fact that more than 200 magnetic stars are known is testimony to the common occurrence of strong magnetic fields.

The extensive observations of Wilson (1963, 1966) and Wilson and Skumanich (1964) show that youthful stars have strong chromospheric lines, which disappear by middle age, indicating enhanced magnetic activity for the first few hundred million years of life. What is more Wilson (1971) has shown that the activity of many of these youthful stars varies with a period of several years, suggesting a cycle of activity much like that of the sun.

In connection with the question of stellar winds, the observations of a variety of classes of stars (Parker 1960, 1963) suggest that there is no major class of star for which one cannot infer activity (i.e. turbulence or other fluctuation) either from spectral line widths, from the theoretical expectation of a convective zone, or from a massive outflow of gas. On the basis of this broad but fragmentary information we are inclined to the hypothesis that most stars are at least as active magnetically as the sun, and some very much more so.

There are some collapsed degenerate stars whose intense magnetic fields put them in a class by themselves. Magnetic fields of 10^6–10^7 G have been detected in white dwarf stars (Kemp 1970; Kemp *et al.* 1970; Angel and Landstreet 1970, 1971, 1972; Kemp *et al.* 1971; Landstreet and Angel 1971; Angel *et al.* 1972). There is some evidence that magnetic fields are involved in x-ray stars (Landstreet and Angel 1972; Kemp and Wolstencroft 1973a, 1973b). It is generally believed that the pulsar is a spinning neutron star with strong pulsed emission of radiation as a consequence of magnetic fields up to 10^{12} G (Pacini 1968; Gunn and Ostriker 1969). It is presumed that the intense magnetic fields of the white dwarf and the pulsar are the direct result of collapse of the original star, which has a radius of some 10^6 km, like the present sun. Supposing the original star had a general magnetic field of 10^2 G, the collapse to 10^4 km to form the white dwarf involves a decrease in surface area, and an associated compression of magnetic field, by a factor of 10^4, producing 10^6 G from the original 10^2 G. The collapse to 10 km to form a neutron star involves a decrease of surface area by a factor of 10^{10}, compressing the original field to 10^{12} G.

When we look at the universe on a larger scale, there is a magnetic field in interstellar space, evidently extending around the disc of the galaxy (Hiltner 1949, 1951, 1956; Chandrasekhar and Fermi 1953; Wentzel 1963; van de Hulst 1967; Berge and Seielstad 1967; Gardner *et al.* 1969; Fujimoto *et al.* 1971; Manchester 1974). The field is embedded in the interstellar gas and confines the cosmic rays produced by the stars within it. The field pressure is comparable to the pressure of the gas and of the cosmic rays, and plays a role in shaping the dynamics and the accumulation of the interstellar gas into cloud complexes (Parker 1966, 1969; Mouschovias 1974, 1975; Mouschovias *et al.* 1974). There is direct evidence of magnetic fields in the Magellanic clouds (Mathewson and Ford 1970) and in a number of active galaxies and radio sources based on the polarization of starlight and radio emission received from those objects.

Magnetic fields (and their inevitable offspring the fast particles) are found everywhere in the universe where we have the means to look for them. The magnetic fields do not alter or impede the overall course of the cosmos in its expansion and degradation, but they greatly complicate the local evolution of stars and galaxies. The continual activity of the magnetic field contrasts sharply with the direct passive degradation of thermal energy. The magnetic field is responsible in large part for the existence of massive, luminous (and hence short-lived) stars at the present late date in cosmology, by controlling the condensation of gas into the large interstellar cloud complexes in which stars are formed. The magnetic field is responsible for the present activity of the sun and its effects upon the terrestrial environment. Earth is illuminated by the aurora, and its weather patterns change in puzzling ways that bear no evident connection to the supposedly steady heat output of the sun. Cosmic rays are trapped in the magnetic field of the galaxy so that their pressure accumulates and rivals that of starlight and of the interstellar gas. Supernova remnants—the debris of stellar explosions—remain active, filled with relativistic particles and producing intense radio emission, for millenia after the initial explosion has died away (Lundmark 1921; Duyvandak 1942; Shklovsky 1960).

What then, is a magnetic field and how does it operate in the astronomical universe to cause all the 'trouble' that we have attributed to it? What is this fascinating entity that, like a biological form, is able to reproduce itself and carry on an active life in the general outflow of starlight, and from there alter the behaviour of stars and galaxies?

References

AARSNES, K. and AMUNDSEN, R. (1972). *Planet. Space Sci.* **20,** 1835.
AKASOFU, S. -I. and CHAPMAN, S. (1972). *Solar terrestrial physics.* Clarendon Press, Oxford.
ALPHER, R. A. and HERMAN, R. (1948). *Nature* **162,** 774.
ALPHER, R. A., FOLLIN, J. W., and HERMAN, R. C. (1953). *Phys. Rev.* **92,** 1347.
ANGEL, J. R. P., ILLING, R. M. E., and LANDSTREET, J. D. (1972). *Astrophys. J. Lett.* **175,** L85.
ANGEL, J. R. P. and LANDSTREET, J. D. (1970). *Astrophys. J. Lett.* **160,** L147.
—— (1971). ibid. **164,** L15; **165,** L71.
—— (1972). ibid. **178,** L21.
ANGLIN, J. D., DIETRICH, W. F., and SIMPSON, J. A. (1973). *Astrophys. J. Lett.* **186,** L46.
ARNETT, W. D. (1973). *Annl. Rev. Astron. Astrophys.* **11,** 73.
AUDOUZE, J., EPHERE, M., and REEVES, H. (1967). *High energy nuclear reactions in astrophysics* (ed. B. S. P. SHEN), p. 255. W. A. Benjamin, New York.
BABCOCK, H. W. (1958). *Astrophys. J. Suppl.* **3** (30) 141.
—— (1960). Stars and stellar systems, *Stellar atmospheres,* (ed. J. L. GREENSTEIN) Vol. 6, p. 282. University of Chicago Press, Chicago.
—— (1963). *Annl. Rev. Astron. Astrophys.* **1,** 41. (Palo Alto: Annual Reviews, Inc.)
BERGE, G. L. and SEIELSTAD, G. A. (1967). *Astrophys. J.* **148,** 367.
BERTSCH, D. L., FICHTEL, C. E., and REAMES, D. V. (1972) *Astrophys. J.* **171,** 169.
BOPP, B. W. and EVANS, D. S. (1973). *Mon. Not. Roy. Astron. Soc.* **164,** 343.
BORRA, E. F. and LANDSTREET, J. D. (1973). *Astrophys. J. Lett.* **185,** L139.
CHANDRASEKHAR, S. and FERMI, E. (1953). *Astrophys. J.* **118,** 113.
CHUPP, E. L., FORREST, D. J., HIGBIE, P. R., SURI, A. N., TSAI, C., and DUNPHY, P. P. (1973). *Nature* **241,** 333.
DANZIGER, I. J. (1970). *Annl. Rev. Astron. Astrophys.* **8,** 161.
DATLOWE, D. (1971). *Solar Phys.* **17,** 436.
—— HUDSON, H. S., and PETERSON, L. E. (1974). *Solar Phys.* **35,** 193.
—— and LIN, R. P. (1973). *Solar Phys.* **32,** 459.
DAVIS, L. R. (1962). *J. Geophys. Res.* **67,** 1711.
DESSLER, A. J. (1967). *Rev. Geophys.* **5,** 1.
DICKE, R. H., PEEBLES, P. J. E., ROLL, P. G., and WILKINSON, D. T. (1965). *Astrophys. J.* **142,** 414.
DUYVANDAK, J. J. L. (1942). *Pub. Astron. Soc. Pacific* **54,** 91.
EVANS, D. S. (1974). *J. Geophys. Res.* **79,** 2853.
de FEITER, L. D. (1974). *Space Sci. Rev.* **16,** 3.
FERMI, E. (1949). *Phys. Rev.* **75,** 1169.
—— (1954). *Astrophys. J.* **119,** 1.
FUJIMOTO, M., KAWABATA, K., and YOSHIAKI, S. (1971). *Prog. Theor. Phys. Suppl.* **49,** 181.
GAMOW, G. (1949). *Rev. Mod. Phys.* **21,** 367.
GARDNER, F. F., MORRIS, D., and WHITEOAK, J. B. (1969). *Austral. J. Phys.* **22,** 813.
GOLDBERG, L. (1974). *Astrophys. J.* **191,** 1.
GOTT, J. R., GUNN, J. E., SCHRAMM, D. N., and TINSLEY, B. M. (1974). *Astrophys. J.* **194,** 543.

GUNN, J. and OSTRIKER, J. (1969). *Phys. Rev. Lett.* **22,** 728.
HARRISON, E. R. (1973). *Annl. Rev. Astron. Astrophys.* **11,** 155.
HILTNER, W. A. (1949). *Astrophys. J.* **109,** 471.
—— (1951). ibid. **114,** 241.
—— (1956). *Astrophys. J. Suppl.* **2,** 389.
HIRSHBERG, J. (1973). *Rev. Geophys. Space Phys.* **11,** 115.
HOYLE, F. (1975). *Astrophys. J.* **196,** 661.
VAN DE HULST, H. C. (1967). *Annl. Rev. Astron. Astrophys.* **5,** 167.
HUNDHAUSEN, A. J. (1972). *Coronal expansion and solar wind.* Springer-Verlag, New York.
IBEN, I. (1974). *Annl. Rev. Astron. Astrophys.* **12,** 215.
JOKIPII, J. R. (1971). *Rev. Geophys. Space Phys.* **9,** 27.
—— (1973). *Annl. Rev. Astron. Astrophys.* **11,** 1.
JOY, A. H. and HUMASON, M. L. (1949). *Publ. Astron. Soc. Pacific* **61,** 133.
KADOMTSEV, B. B. and TSYTOVICH, V. N. (1970). *Interstellar gas dynamics. IAU Symposium No.* 39 (ed. HABING, H. J.), p. 108. Reidel, Dordrecht.
KANE, S. R., KREPLIN, R. W., MARTRES, M. J., PICK, M., and SORU–ESCAUT, I. (1974). *Solar Phys.* **38,** 483.
KEMP, J. C. (1970). *Astrophys. J.* **162,** 169.
—— SWEDLUND, J. B., LANDSTREET. J. D., and ANGEL, J. R. P. (1970). *Astrophys. J. Lett.* **161,** L77.
—— —— and WOLSTENCROFT, R. D. (1971). *Astrophys. J. Lett.* **164,** L17.
—— and WOLSTENCROFT, R. D. (1973a). *Astrophys. J. Lett.* **182,** L43.
—— —— (1973b). ibid. **185,** 21.
KRIMIGIS, S. M. and VAN ALLEN, J. A. (1967). *J. Geophys. Res.* **72,** 4471.
KUNKEL, W. E. (1973). *Astrophys. J. Suppl.* **25,** 1.
LACY, C. H. (1977), *Astrophys. J.* **218,** 444.
LANDSTREET, J. D. and ANGEL, J. R. P. (1971). *Astrophys. J. Lett.* **165,** L67.
—— —— (1972). *Astrophys. J.* **172,** 443.
LEDOUX, P. and RENSON, P. (1966). *Annl. Rev. Astron. Astrophys.* **4,** 293.
LINGENFELTER, R. E. and RAMATY, R. (1967). High energy nuclear reactions in astrophysics (ed. B. S. SHEN), p. 99. W. A. BENJAMIN, New York.
LITVAK, M. M. (1974). *Annl. Rev. Astron. Astrophys.* **12,** 97.
LOVELL, B. (1971). *Quart. J. Roy. Astron. Soc.* **12,** 98.
LUNDMARK, K. (1921). *Publ. Astron. Soc. Pacific* **33,** 225.
LUYTEN, W. J. (1949). *Astrophys. J.* **109,** 532.
MANCHESTER, R. N. (1974). *Astrophys. J.* **188,** 637.
MATHEWSON, D. S. and FORD, V. L. (1970). *Astrophys. J. Lett.* **160,** L43.
MCILWAIN, C. E. (1975). *Proceedings of the Nobel symposium* (Kiruna, Sweden) Plenum Press, London.
MCKENZIE, D. L., DATLOWE, D. W., and PETERSON, L. E. (1973). *Solar Phys.* **28,** 175.
MEAD, G. D. (1974). *J. Geophys. Res.* **79,** (25) September 1. *Pioneer* 10 *Mission, Jupiter Encounter* (ed. G. D. MEAD). See also *Science* **183,** 301–324 (1974) and *Science* May 2 (1975).
MOFFETT, T. L. (1973). *Mon. Not. Roy. Astron. Soc.* **164,** 11.
—— (1974). *Astrophys. J. Suppl.* **29,** (273) 1.
MOGRO–CAMPERO, A. and SIMPSON, J. A. (1972). *Astrophys. J. Lett.* **177,** L37 L43.
MOUSCHOVIAS, T. Ch. (1974). *Astrophys. J.* **192,** 37.
—— (1975). Ph.D. thesis, Department of Astronomy, University of California at Berkeley.

—— Shu, F. H., and Woodward, P. R. (1974). *Astron. Astrophys.* **33,** 73.
Ney, E. P. (1962). *J. Geophys. Res.* **62,** 2087.
Pacini, F. (1968). *Nature,* **219,** 145.
Pagel, B. E. J. (1973). *Space Sci. Rev.* **15,** 1.
Papadopoulos, K. and Coffey, T. (1974). *J. Geophys. Res.* **79,** 1558.
Parker, E. N. (1958). *Phys. Rev.* **109,** 1328.
—— (1960). *Astrophys. J.* **132,** 821.
—— (1963). *Interplanetary dynamical processes* Chap. XV. Wiley-Interscience, New York.
—— (1965). *Phys. Rev. Lett.* **14,** 55.
—— (1966). *Astrophys. J.* **145,** 811.
—— (1969). *Space Sci. Rev.* **9,** 651.
Penzias, A. A. and Wilson, R. W. (1965). *Astrophys. J.* **142,** 419.
Pomerantz, M. A. and Duggal, S. P. (1974). *Rev. Geophys. Space Sci.* **12,** 343.
Preston, G. W. (1967). *The magnetic and related stars* (ed. R. C. Cameron), p. 3. Mono Book Corp., Baltimore.
—— (1969). *Astrophys. J.* **158,** 243.
Quenby, J. J. (1967). *Handbuch d. Physik XLVI/2 Cosmic Rays,* Springer–Verlag, Berlin.
Ramaty, R. and Stone, R. G. (1973). Symposium Proceedings: *High energy phenomena on the sun,* Goddard Space Flight Center, Greenbelt, Maryland.
Rodono, M. (1974). *Astron. Astrophys.* **32,** 337.
Reid, G. C. (1961). *J. Geophys. Res.* **66,** 4071.
Sakurai, K. (1973). *Planet. Space Sci.* **21,** 793.
Sargent, W. L. W. (1964). *Annl. Rev. Astron. Astrophys.* **2,** 297.
Scholer, M., Hausler, B., and Hovestadt, D. (1972). *Planet. Space Sci.* **20,** 271.
Scott, J. S. and Chevalier, R. A. (1975). *Astrophys. J. Lett.* **197,** L5.
Severny, A. (1970). *Astrophys. J. Lett.* **159,** L73.
Shklovsky, I. S. (1960). *Cosmic Radio Waves,* Harvard University Press, Cambridge, Massachusetts.
Simnett, G. M. (1974). *Space Sci. Rev.* **16,** 257.
Simpson, J. A. (1960). *J. Geophys. Res.* **65,** 1615.
Smith, E. V. P. and Gottlieb, D. M. (1974). *Space Sci. Rev.* **16,** 771.
Tandberg–Hanssen, E. (1973). *Rev. Geophys. Space Phys.* **11,** 469.
Torres, C. A. O. and Mello, F. S. (1973). *Astron. Astrophys.* **27,** 231.
Wagoner, R. V., Fowler,, W. A., and Hoyle, F. (1967). *Astrophys. J.* **148,** 3.
Warwick, J. (1961). *Astrophys. J.* **137,** 41.
Webber, J. C., Yoss, K. M., Deming, D., and Yang, K. S. (1973). *Publ. Astron. Soc. Pacific* **85,** 739.
Wentzel, D. G. (1963). *Annl. Rev. Astron. Astrophys.* **1,** 195.
Wilcox, J. M. and Ness, N. F. (1965). *J. Geophys. Res.* **70,** 5793.
—— —— (1967). *Solar Phys.* **1,** 437.
—— —— (1968). *Space Sci. Rev.* **8,** 258.
Williams, D. J. and Mead, G. D. (1968). *Rev. Geophys.* **7,** (1, 2).
Wilson, O. C. (1963). *Astrophys. J.* **138,** 832.
—— (1966). ibid. **144,** 695.
—— (1971). *Proc. Asilomar Solar Wind Conf.,* 21–26 April, Pacific Grove, California.
—— and Skumanich, A. (1964). *Astrophys. J.* **140,** 1401.
Worden, S. P. (1974). *Publ. Astron. Soc. Pacific* **86,** 595.

Zuckerman, B., Buhl, D., Palmer, P., and Snyder, L. E. (1970). *Astrophys. J.* **160,** 485.
—— and Palmer, P. (1974). *Annl. Rev. Astron. Astrophys.* **12,** 279.
—— Turner, B. E., Johnson, D. R., Clark, F. O., Lovas, F. J., Fourikis, N., Palmer, P., Morris, M., Lilley, A. E., Ball, J. A., Gottlieb, C. A., Litvak, M. M., and Penfield, H. (1975). *Astrophys. J. Lett.* **196,** L99.

2

THE NATURE OF MAGNETIC FIELDS

THE magnetic field is a syndrome, composed of a number of effects that occur in combination. Some of the effects have been known since ancient times. Thales of Miletus (*circa* 500 B.C.) mentioned the mutual attraction of pieces of lodestone or magnetite (the iron oxide Fe_3O_4) and Plato mentioned the attraction of pieces of iron to the lodestone. The Chinese appear to be the first to have used the lodestone for surveying and for navigation, and are therefore to be credited with discovering and using, if not comprehending, the general magnetic field of Earth (*circa* 1100 A.D.). The simplest effect by which we may identify a magnetic field is, then, the magnetic needle—the compass. A needle of iron, or of magnetite, when freely suspended turns so as to take up a unique alignment relative to Earth. The Chinese found that a lodestone (magnetite) suspended on a fibre, or placed on a free-floating cork in a pan of water, turned to align itself north and south. The alignment occurred wherever one happened to be, without regard for political or religious affiliation, personality, altitude, time, or weather. It was a fundamental natural phenomenon transcending the local environmental conditions. What concept the Chinese attached to the universal alignment of lodestones is not known today. There is no indication that they realized the alignment to be caused by a property of Earth itself. It is an interesting fact that the ancient walls of Peking were lined up with magnetic north rather than geographic north, a difference at that time of about 1°. We may presume that the surveyor found it easier to work with his compass needle by day than to sight on the pole star by night. A concise review of the history of magnetism has been given by Chapman and Bartels (1940) based in large part on the earlier extensive researches of Mitchell (1932, 1937, 1939).

It remained for Gilbert (1958), physician to Queen Elizabeth I of England, to collect information from the worldwide navigation of his day to formulate a concise picture of terrestrial magnetism, pointing out, in the year 1600, that Earth itself behaves like a large lodestone. The small lodestone of the navigator points in directions relative to Earth in the same way that Gilbert's small lodestone behaved in the neighbourhood of a large lodestone, pointing in at the north pole and out at the south pole, and generally north and south at latitudes between. Figure 2.1 is a sketch of the directions pointed to by the lodestone, or compass needle, in the vicinity of Earth. The directions pointed to by the compass needle vary

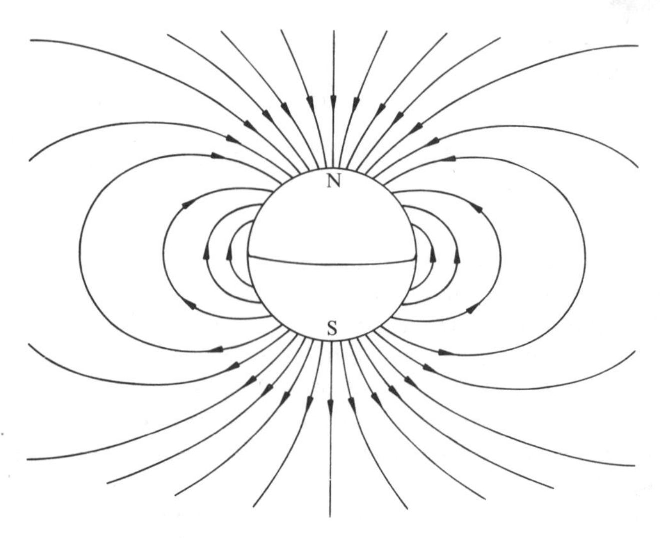

FIG. 2.1. Map of the lines of force of a dipole field such as the external magnetic field of earth. The arrows on the lines of force indicate the direction pointed by the north-seeking pole of a compass needle. The closed circle represents the surface of Earth with geographic north and south indicated at the top and bottom, respectively.

continuously from one position to another and so can be used to define a continuous family of lines, called the *magnetic lines of force*, illustrated in Fig. 2.1. The lines of force are a graphic and useful concept invented by Michael Faraday in the nineteenth century for representing the configuration of the magnetic field. We shall use the concept extensively in later discussion of the dynamical properties of magnetic fields.

The strength B of a magnetic field is a measure of the force that it exerts to align the compass needle along the lines of force. The strength and direction together determine the vector field $\mathbf{B}$, which may be thought of, then, as an arrow of length B pointing in the direction of the aligned compass needle. Using the unit of field strength called a gauss (G), the strength of the magnetic field of Earth is approximately 0·3 G at the equator and 0·6 G at the north and south poles. Fields of this magnitude are weak, by human standards. We discover to our sorrow how little damage to the needle bearing of a pocket compass prevents free alignment.

A more absolute measure of the gauss, expressed in terms of electric currents, can be stated in terms of the force exerted on a wire carrying an electric current across the magnetic field. An electric current of 1 A, approximately the current drawn by a 110-V, 100-W lamp, flowing along a wire for one metre across a magnetic field of 100 G is pushed sidewise by the magnetic field with a force of 10^3 dyn, or about the weight of one gram. This is, of course, just the effect on which electric motors are based, except that they may use one thousand gauss, tens of amperes, and many metres of wire.

Compared to the 0·6-G polar field of Earth, the planet Jupiter has a relatively strong field, of some 10 G, sufficient to exert forces on a piece of iron that are sensible to the human hand. A small permanent magnet may have a field of 100 G and be suitable for the magnetic latch of a cupboard door. Fields above a few hundred gauss are able to wrench a piece of iron from the hand. The field in the gap between the pole pieces of a powerful electromagnet may be 10^4 G. Steady magnetic fields of 10^5 G can be produced in electromagnets wound with superconducting wires, and transient fields of up to 10^6 G can be produced using chemical explosives to compress a magnetic field trapped momentarily inside a metal tube. As we mentioned earlier, magnetic fields of as little as 100–200 G have been found in stars where special conditions permit easy detection. The star HD215441 exhibits a mean field of 34 000 G. Three white dwarfs are known with fields of the order of 10^7 G, exceeding the maximum field strengths that can be produced in the laboratory.

With these brief remarks on the strength of magnetic fields, let us move on to the second basic property of the phenomenon, or syndrome, that we call the magnetic field. The second property concerns electric fields. On the one hand, we must understand that magnetic fields and electric fields are distinct effects, the magnetic field aligning the iron needle and the electric field pushing on electric charges. Thus, for instance, an *electric* charge placed in the magnetic field experiences no force whatsoever[1]. On the other hand, although electric and magnetic fields are distinct, they are intimately related for if we move relative to one, we find the other. Imagine being in a metal building in which all electric fields (the atmosphere is normally charged to some $100\,\mathrm{V\,m^{-1}}$) are neutralized by the electrically conducting structure of the building. Place a small permanent magnet on a table. There are no electric fields present in the room. A small electric charge experiences no force anywhere in the interior of the building[2]. But if we now move with velocity v across the field of the magnet on the table, we observe an electric field! An electric charge q moved with velocity v across the magnetic field B experiences a force

[1] A magnetic needle of iron placed in an electric field will tend to line-up with the electric field because of the electric charges induced on it by the electric field. But the effect is readily distinguishable from magnetic alignment because it is independent of whether the needle is magnetized. A copper (non-magnetic) needle is aligned in an electric field to the same degree as a magnetized iron needle.

[2] If the small test charge q is a small distance h from a metal wall, it will induce an equal and opposite charge on the wall, giving an attractive force $q^2/4h^2$. But this force is second order in q and so can be neglected for small values of q. We are concerned only with forces first order in q.

proportional to qvB. The strength of the electric field is $10^{-4}\,vB$ V m^{-1} if B is measured in G and v in m s^{-1}. The electric field vector **E** is perpendicular to both the velocity **v** and the magnetic field **B**. Thus we may stand motionless in front of the magnet in the room free of electric fields and reflect that there are electric fields all about in the room, in other moving frames of reference. Clearly, then, there is a semantic difficulty involved in the question whether there are electric fields in a given space. Provided that there is a magnetic field, there is an electric field in all moving frames but one. In that one frame in which the electric field is zero, we say we are at rest with respect to the magnetic field.

It is instructive to consider for a moment the magnitude of the electric fields and the total potential differences, produced by various motions. If, for example, an observer moves at a speed of 300 m s^{-1} (approximately the speed of sound) across a magnetic field of 1000 G, he sees an electric field of 30 V m^{-1}. Thus an electromotive force of 120 V (necessary to power a standard light bulb) arises along 4 m of wire carried at 300 m s^{-1} across a 1000-G field on the spinning armature of a generator. One can move either the magnetic field or the wire. It is only the relative motion that counts[3].

In astrophysical circumstances the calculated potential difference, or voltage, can be extremely large because of the large dimensions. On the other hand, the potential differences have no physical meaning unless we can show that the frame of reference in which they are computed has some particular physical significance. One example will suffice. The solar wind is composed of ionized electrically conducting gas from the solar corona. It streams past Earth with a velocity that is typically 400 km s^{-1}, carrying a magnetic field with a component of 3×10^{-5} G perpendicular to the motion. Hence we calculate that in the frame of Earth there is an electric field of $1{\cdot}2\times10^{-3}$ V m^{-1}. Over dimensions of one astronomical unit (1 a.u. $=1{\cdot}5\times10^{13}$ cm $=$ sun–Earth distance) we calculate $1{\cdot}8\times10^{8}$ V. This is a very impressive potential difference, but by itself it is without significance. We can have any electric field we wish, depending upon the frame of reference in which we carry out the calculation. In the frame moving with the solar wind the electric field is zero. In a frame of reference moving inward toward the sun at 400 km s^{-1} the potential difference is $3{\cdot}6\times10^{8}$ V. We are presented with the same semantic dilemma as in the question whether there is an electric field in the metal shielded building containing a permanent magnet. There is, and there is

[3] We should not fail to emphasize that the complementary effect occurs, too. In a room free of magnetic field, but containing an electric field **E**, an observer moving with velocity v across E sees a magnetic field B of strength $vE/c\times10^{12}$ G where v is in m s^{-1} and E is in V m^{-1}. If the room is free of magnetic field in its own frame of reference, there is a magnetic field in every moving frame of reference.

not, depending upon the motion of the frame of reference in which the question is asked. One must, therefore, frame the complete context of the question before talking about potential differences. They may, or may not, be relevant. There is almost always a frame of reference in which the local electric field is zero. It is usually in, or near, that frame of reference in which the thermal particles reside.

The electric field that we detect because of motion relative to a magnetic field is said to be *induced*. The electric field induced from magnetic fields (the induced electromotive force) was discovered independently by Michael Faraday at the Royal Institution, London and Joseph Henry at Albany Academy, Albany, New York in the period 1830–1833. Faraday published his results first and the effect that the total voltage (electromotive force) around any closed path is proportional to the time rate of change of the number of lines of force enclosed by the path, is called Faraday's Law of induction. Joseph Henry was recognized by having the unit of self-inductance named after him. At a demonstration of induction, Faraday was asked by Gladstone, Chancellor of the Exchequer, 'But after all, what use is it?' Faraday retorted, 'Why, Sir, there is every probability that you will soon be able to tax it.' Indeed, there was rapid development of the idea so that within fifty years of the first discovery commercial power plants and power transmission systems were put into operation (about 1885) in Europe and America. In contrast to such fundamental developments of the last century, it is curious that in our present complex technological state 100 years later, the RANN program (Research Applied to National Needs) of the National Science Foundation operates on the policy that only those ideas that can reach final application in *five* years are worthy of support.

Consider, then, the final basic property of magnetic fields, and that is their intimate association with electric currents, discovered by Oersted in 1820. There is no magnetic field without an associated electric current[4]. Magnetic fields for laboratory and industrial purposes are usually produced by forcing an electric current through a coil of wire, i.e. an electromagnet. The effect is often enhanced by placing iron inside the coil, to take advantage of the electric currents that naturally flow inside iron atoms and are easily aligned to supplement the currents forced through the coil of wire. With a suitably designed iron circuit around the coil of wire the strength of the field obtained from a current in a wire can be enhanced a thousandfold. As a consequence of our familiar means of making magnetic field, we have developed the habit of saying that the

[4] In electromagnetic waves, such as radio waves, light waves, x-rays, etc. the associated rapid variation of the electric field, called the displacement current by Maxwell, plays the role of the electric current. We are not concerned here with such rapidly varying magnetic and electric fields.

electric current *causes* the magnetic field. The energy is introduced via the electric current and gives rise to the magnetic field.

But now suppose that we have built a large electromagnet with a large iron core so that it operates efficiently. We have closed the switch connecting the coil to a d.c. generator, building up a substantial electric current through the coil and producing a magnetic field of, say, 10^4 G in the iron core. Having accomplished this, we open the switch in order to shut off the power. We discover that the current does not immediately cease to flow. Indeed, if we cut off the power by opening a switch, the electric current will arc across the open switch for several seconds, perhaps doing considerable damage to the switch, before dying away. The reason is that the current cannot stop until the magnetic field is gone, and the magnetic field cannot disappear until its energy is dispersed in some way. As the magnetic field decays, it disappears by shrinking back into the coil, crossing over the many turns of wire and inducing an enormous electromotive force[5] which keeps the current going, regardless of whether we have opened the switch. Forcing the current through the coil (and arcing across the open switch) consumes the energy of the magnetic field so that the field soon dies away[6]. During this phase of the operation it is meaningless to say that the current is causing the field. Indeed, the field is causing—inducing—the current. The energy is flowing from the field into the wires and our open switch may be burned and melted by the energy delivered from the *field*.

In the astrophysical universe there are no wires which carry electric current and energy from a region of generation into a distant active site. Hence there are few general circumstances of which we are aware where one can say unambiguously that the current *causes* the magnetic field. In nature, in the absence of insulating materials to channel and direct the flow of electric currents, the generator supplying the power, and the electromagnet producing the magnetic field, are one and the same, a single amorphous mass of gas.

A magnetic field, then, is a collection of effects to which we have attached a name. The term 'field' indicates that the effects extend over space. The field aligns the compass needle and has electric currents associated with it. Motion relative to the magnetic field sees an electric

[5] One may say alternatively that the magnetic flux through the coil is decreasing and hence there is an induced e.m.f. Whatever words we choose, the fact is that the current and field die away only as the energy is dissipated by the electric currents.

[6] If we contrive to make an electromagnet using superconducting wire, there is no dissipation of energy whatsoever. The magnetic field persists in the coil without diminishing for as long as the low temperature necessary for the superconductivity of the wire is maintained.

field. In 1864 Maxwell unified electricity and magnetism in his celebrated electromagnetic equations, and 40 years later Lorentz extended the ideas to relatively moving frames of reference, which were incorporated into Einstein's special theory of relativity in 1905.

We are concerned here with motions that are slow compared to the speed of light c so that if we neglect terms that are as small as v^2/c^2 compared to the others, Maxwell's equations reduce to the induction equation

$$c\nabla \times \mathbf{E} = -\partial \mathbf{B}/\partial t \tag{2.1}$$

and Ampere's law

$$\nabla \times \mathbf{B} = 4\pi \mathbf{j}/c \tag{2.2}$$

where $\mathbf{E}$ is the electric field, $\mathbf{B}$ is the magnetic field, and $\mathbf{j}$ the electric current density.

This is perhaps the appropriate place to remark that we shall employ the centimetre–gram–second (cgs) system of units, expressing electric charges in electrostatic units (so that the electronic charge is $e = 4{\cdot}8 \times 10^{-10}$ e.s.u.), electric fields in statvolts cm^{-1} (1 statvolt = 300 V) and magnetic fields in gauss. Those with a strong preference for other systems of units are free to interpolate the equations into their favourite form, of course.

The Lorentz transformation of fields between frames of reference with relative velocity are

$$\mathbf{E}' = \mathbf{E} + \mathbf{v} \times \mathbf{B}/c, \qquad \mathbf{B}' = \mathbf{B} \tag{2.3}$$

where the prime denotes fields in the frame moving with velocity $\mathbf{v}$ relative to the frame in which the fields are $\mathbf{E}$ and $\mathbf{B}$.

These relations are based on laboratory experiments, testing them in all conceivable ways under a wide variety of circumstances. The general equations, from which the above reduced forms were obtained, have shown themselves to be accurate under the most extreme conditions of small scale, rapid variation, and intense field, found in high-speed collisions of electrons and/or nuclear particles, giving rise to x-rays, γ-rays, and pair production and annihilation. Thus quantum electrodynamics is the most severe test of Maxwell's field equations. However, we must not forget the Michelson–Morley experiment, demonstrating that light in a vacuum is seen by all observers, regardless of their own motion, to pass by with the same speed. Although the results of the experiment were baffling until Einstein's special theory of relativity—and for some 'scientists' they remained baffling long after that—they were a fundamental test of Maxwell's equations. The equations, being Lorentz invariant, passed the test with flying colours, having anticipated special relativity by several decades.

The equations represent, then, the sum total of our knowledge of the magnetic field **B** in terms of the sum of the particle motions **j**. Fortunately the complicated question of the particle motions does not usually enter into the structure of the final field equations for **B** because the universe is an excellent conductor of electricity almost everywhere.

Maxwell's equations have not been critically tested over large astrophysical dimensions, but they are applied without hesitation[7] to the magnetic field of Earth, to the sun, and to the galaxy. We are compelled to accept the general validity of the basic equations until evidence arises from observation that the equations are inadequate[8]. Indeed it would be our fondest hope to discover that corrections must be made to the equations for fields extending over large dimensions. But we do not count on the fascinating possibility of the equations being found wanting. Rather it is the curious behaviour of the magnetic syndrome in the universe that draws us into the arena, not the expectation of corrections to already existing laws of physics.

To return, then, to a point made earlier, we are familiar with the magnetic field in the laboratory. There we control it with electric currents. The behaviour of the field in space is quite different. There the fields are a power unto themselves, producing effects that cannot be duplicated in the laboratory, although we think the strange and unanticipated effects in space are describable by the basic equations learned in the laboratory. We believe that the difference, pointed out by Cowling (1933) Alfven (1942, 1950) Walen (1946), and Elsasser (1946, 1954), is merely a matter of scale. The nature of the mathematical solutions of the basic equations depends upon the relative magnitudes of the time and distance scales. So we cannot look to the laboratory for guidance in understanding the sunspot, or the origin of the magnetic field of Earth. We must look at the sunspot itself. The basic point emphasized by Cowling and Alfven is that a magnetic field in an electrically conducting body of dimension l and electrical conductivity of σ (e.s.u.) decays away in a characteristic time $t = l^2\sigma/c^2$; that is to say, the field decays in a time proportional to the current carrying capacity (area times conductivity) of the body. Thus the magnetic field through a cubic metre of copper, for which $\sigma = 5\times10^{17}$ e.s.u. ($l = 10^2$ cm and $c = 3\times10^{10}$ cm s^{-1}) dissipates in about 5 s. It is the same situation that arose in switching off the electromagnet, discussed earlier. The electric current must accompany the magnetic field, and the current is dissipated by the electrical resistivity of the medium. So

[7] Except in very strong gravitational fields where modification of the geometry of space requires a restatement of the electromagnetic equations.

[8] The frequent success of such sweeping extrapolation from laboratory experiment to the universe has been remarked on by Wigner (1967) in his essay *The unreasonable effectiveness of mathematics in the natural sciences.*

the field exhausts itself driving the current. Over astronomical dimensions this may require long periods of time.

Now a cubic metre of copper is very large, expensive, and clumsy, and even so a magnetic field through it persists for only a few seconds. There is little that can be done to manipulate the block of copper in that time. Ionized gases can be manipulated in the laboratory in microseconds, but their electrical conductivity is much inferior to copper. For ionized hydrogen, the electrical conductivity at temperature T is (Cowling 1953, 1957; Spitzer 1956)

$$\sigma \cong 10^7 T^{\frac{3}{2}} \text{ e.s.u.} \tag{2.4}$$

Thus for $T = 10^5$ K (10 eV) we have $\sigma = 3 \times 10^{14}$ e.s.u. One can produce such a temperature across a gas column 10 cm in diameter, yielding a relaxation time of the magnetic field of 3×10^{-5} s. In this period of 30 μs it is possible to turn the gas over once to distort the magnetic field significantly. But this is nothing to compare with the situation in the cosmos.

Consider the molten (and hence non-magnetic) metallic core of Earth with a radius of $l \cong 3 \times 10^8$ cm and electrical conductivity of $\sigma \cong 10^{16}$ e.s.u. The magnetic relaxation time is 10^{12} s or 3×10^4 years. Earth is much older than 3×10^4 years, having been formed more or less simultaneously with the sun nearly 5×10^9 years ago. Therefore, in view of the decay time of 3×10^4 years we conclude that, if left to itself, the magnetic field of Earth would have disappeared long ago. Hence the present magnetic field must be produced by contemporary effects. With this problem in mind, then, we return to our consideration of the manipulation of the magnetic field by the fluid (Elsasser 1946). The convective motions in the core are typically $v = 0{\cdot}1 - 1$ mm s^{-1}, with a characteristic turnover time of a few hundred years. Thus the fluid in the core of Earth mixes around perhaps 100 times in the period over which the magnetic fields decays. The magnetic field, unable to escape from the fluid, is carried into grotesque contortions, beyond anything that can be accomplished in the laboratory.

Consider the sun. The central temperature of the sun is in excess of 10^7 K, producing a gaseous core with an electrical conductivity $\sigma \cong 3 \times 10^{17}$ e.s.u., comparable with aluminium, with a radius of 2×10^5 km ($l = 2 \times 10^{10}$ cm). On this basis the decay time of a magnetic field caught up in the core of the sun at the time it was formed is 4×10^9 years, comparable to the present age of the sun. On this basis Cowling (1953) suggested that any primordial field in the sun should still be there, having decayed no more than by a factor of two or three[9]. The point is that the

[9] As a matter of fact we believe now that there are ways by which a primordial magnetic field of any significant strength can wiggle out of a fluid more rapidly than anticipated by $t = l^2\sigma/c^2$. This will be one of the points developed in the chapters that follow.

breadth, and consequent current-carrying capacity, of the sun is so enormous that the magnetic field is trapped in its fluid embrace for an extended period of time while the fluid circulates and convects into wholly new configurations. Even a very small part of the sun, such as a sunspot, has the magnetic field firmly in its grip. The escape time of a magnetic field from a sunspot, with dimensions of 10^4 km ($l = 10^9$ cm) and with $T = 10^4$ K beneath the surface so that $\sigma = 10^{13}$ e.s.u., is 300 years. The fluid initially in the area may disperse, carrying the field with it, of course, but the two cannot be separated in a time shorter than 300 years unless broken up into smaller dimensions. It is this trapping of the magnetic field at each locale in a fluid of broad extent that led to Alfven's (1950) statement of the basic concept that the magnetic lines of force are 'frozen' into the fluid and carried bodily with it.

As a final example, consider the gaseous disc of the galaxy, with a thickness of the order of $l \cong 200$ pc $= 6 \times 10^{20}$ cm and an electrical conductivity[10] evidently not less than 10^{11} e.s.u. The magnetic relaxation time, $l^2\sigma/c^2$, is then calculated to be 4×10^{31} s or about 10^{24} years, to be compared with the estimated age of the galaxy (10^{10} years) and compared to the periods of $1\text{–}5 \times 10^8$ years over which the differential rotation and turbulence mixes and circulates the gas through the disc. It is easy to argue that the present galactic field is an artefact, trapped since the formation of the galaxy a mere 10^{10} years ago.

Piddington (1970, 1972a, b) has eloquently expounded the view that the magnetic fields observed today in both the sun and the galaxy are the primordial fields, trapped since the formation. We will have more to say on this question later.

The indisputable conclusion is that the magnetic fields in the cosmos behave in strange ways because the dimensions of the gases and fields are so large that the field is trapped and carried around with the fluid motions. Larmor (1919) was the first to suggest that swirling fluid motions might produce magnetic fields, although, as Cowling (1933) showed later, the suggestion was posed in the wrong context. Elsasser (1941, 1946, 1950, 1955, 1956) was the first to make positive progress along these lines in his inquiry into the origin of the magnetic field of Earth in the liquid core of Earth. Alfven (1950) developed the ideas of hydromagnetism, as it is now called, in a variety of ways.

References

ALFVEN, H. (1942). *Ark. Matematik, Astron. Fys.* **29A,** (11); **29B,** (2).
—— (1950). *Cosmical electrodynamics.* Clarendon Press, Oxford.

[10] The electrical conductivity of the interstellar gas may be very low in some dense, cold clouds, but the intercloud medium temperature is presumed to be generally 10^3 K or more, for which $\sigma \gtrsim 10^{11}$ e.s.u.

CHAPMAN, S. and BARTELS, J. (1940). *Geomagnetism*, Vol. I, II. Clarendon Press, Oxford.
COWLING, T. G. (1933) *Mon. Not. Roy. Astron. Soc.* **94,** 39
—— (1953). *The sun* (ed. G. P. KUIPER), chap. 8. University of Chicago Press, Chicago.
—— (1957). *Magnetohydrodynamics.* Wiley-Interscience, New York.
ELSASSER, W. M. (1941). *Phys. Rev.* **60,** 876.
—— (1946). ibid. **69,** 106; **70,** 202.
—— (1950). *Rev. Mod. Phys.* **22,** 4.
—— (1954). *Phys. Rev.* **95,** 1.
—— (1955). *Amer. J. Phys.* **23,** 590.
—— (1956). ibid. **24,** 85.
GILBERT, W. (1958). *On the magnet* (ed. D. J. PRICE). Basic Books, New York.
LARMOR, J. (1919). *Brit. Ass. Rep.* p. 150.
MITCHELL, A. C. (1932). *Terr. Magn.* **37,** 105.
—— (1937). ibid. **42,** 241.
—— (1939). ibid. **44,** 77.
PIDDINGTON, J. H. (1970). *Aust. J. Phys.* **23,** 731.
—— (1972a). *Solar Phys.* **22,** 3.
—— (1972b). *Cosmic Electrodyn.* **3,** 60; **5,** 731.
SPITZER, L. (1956). *The physics of fully ionized gases.* Wiley-Interscience, New York.
WALEN, C. (1946). *Ark. Matematik, Astron. Fys.* **33A,** (18).
WIGNER, E. P. (1967). *Symmetries and reflections*, p. 222. Indiana University Press, Bloomington.

3

ELECTRIC AND MAGNETIC FIELDS

BEFORE going on to develop the properties of large-scale magnetic fields it is worthwhile pausing for a moment to reflect on the radically different roles played by electric and magnetic fields in the cosmos. Maxwell's field equations

$$\partial \mathbf{B}/\partial t = -c\nabla \times \mathbf{E}, \qquad \partial \mathbf{E}/\partial t = +c\nabla \times \mathbf{B}$$

are symmetric in **E** and **B**. Why is the symmetry not reflected in the universe? The answer lies, of course, in the microscopic structure of the universe. Matter is composed of electrically charged particles. All known particles are electrically charged, in units of $e = 4{\cdot}8 \times 10^{-10}$ e.s.u. What is more, it appears that positively- and negatively-charged particles are precisely equal in number so that there is no net electric charge in the universe[1]. On the other hand, no magnetically charged particles—magnetic monopoles—have ever been found, although considerable effort and ingenuity have been employed in searching for them. Hence the general existence of large-scale magnetic fields and the general weakness of large-scale electric fields (in the frame of reference of the gaseous matter) are consequences of the fundamental electrical nature of matter.

To examine the connection between the large-scale fields and the particulate matter, we note first that the apparent precise large-scale electrical neutrality of the universe implies that there are no electrostatic fields on a cosmological scale. There are, then, only local electrostatic fields caused by the separation of equal and opposite charges. Writing $\mathbf{E} = -\nabla\phi$ and denoting the charge density by δ, we have

$$\nabla^2\phi = -4\pi\delta.$$

But it appears that everywhere in the universe there are enough x-rays, ultraviolet radiation, cosmic rays, and thermal excitation to guarantee that at least a few electrons everywhere will be dislodged from the local atoms at any given time. Certainly this is the rule in the sun, where even in the cool photosphere the carbon and the alkali metals are at least partially ionized. Elsewhere in the sun the gases are fully ionized and

[1] The possibility of a net electric charge in the universe was once suggested by Lyttleton and Bondi (1959, 1960) (see comments by Hoyle (1960)). The experimental results of Zorn *et al.* (1963) rule out the effect as a consequence of an electron–proton charge difference.

saturated with free electrons. In interstellar space the electron density appears to be 10^{-3} cm^{-3} or more (Hobbs 1974). One exception to this rule might be the central region of some cold and very dense interstellar gas cloud so thick as to shield its interior from all external ionizing radiation. The other exception is the rare, but familiar, space occupied by cold planetary atmospheres, in which electrostatic phenomena—the 100 V m^{-1} electrostatic field in the atmosphere and the thunderstorm phenomenon—are commonplace. But apart from these obscure nooks, space is universally populated with free electrons. The free electrons quickly neutralize any separated charges. The characteristic time to carry out the neutralization is the plasma period $(\pi m/N_e e^2)^{\frac{1}{2}}$ s where m is the electron mass, e the charge, and N_e the number of free electrons per cm^3. Numerically this is $10^{-4}/N_e^{\frac{1}{2}}$ s. Even in intergalactic space where N_e may be 10^{-10}–10^{-8} cm^{-3}, the neutralization time is only 1–10 s. Elsewhere it is much less[2].

It follows, then, that any large-scale electrostatic field is quickly and effectively neutralized by the universal free electrons. The general occurrence of electrostatic fields is restricted to microscopic scales, inside atoms or within a Debye radius of an ion. The large-scale electric field in the frame of the gas containing the free thermal electrons is essentially zero. It is motion relative to the thermal gas that sees an induced electric field. Thus, for instance, a particle, with velocity $\mathbf{w}$ relative to the gas, sees an electric field

$$\mathbf{E}' = \mathbf{w} \times \mathbf{B}/c$$

as a consequence of its motion across the large-scale magnetic field $\mathbf{B}$. If the particle has an electric charge q, it experiences a force

$$\mathbf{F} = q\mathbf{E}' = q\mathbf{w} \times \mathbf{B}/c,$$

causing it to move in a circle about the magnetic field $\mathbf{B}$. The radius of the circle (the so-called cyclotron radius) is

$$R = Mw_\perp c/qB$$

where $w_\perp$ is the component of the particle velocity $\mathbf{w}$ perpendicular to $\mathbf{B}$. The angular frequency of the motion is the cyclotron frequency $\Omega = w_\perp/R = qB/Mc$. The motion perpendicular to $\mathbf{B}$ is, in this approximation, a stationary circle relative to the background gas. The velocity $\mathbf{w}_\parallel$ parallel to $\mathbf{B}$ is constant. If the gas has a velocity $\mathbf{v}$ relative to some fixed frame of

[2] In the very high electric current densities localized in the extreme conditions of the solar flare, in the neutral sheet in the magnetic tail of Earth or Jupiter, or in auroral sheets, the electrical resistivity may be enormously enhanced by the excitation of plasma turbulence (discussed in Chapter 4), to the extent that local electric fields may occur. This interesting and important exception to the general rule will be discussed at length in later chapters.

reference then, of course, the particle velocity seen in the fixed frame is $\mathbf{w}_{\parallel}+\mathbf{w}_{\perp}+\mathbf{v}$, where $\mathbf{w}_{\perp}$ is the cyclotron motion around the magnetic field. The mean motion of the particle has the steady value $\mathbf{w}_{\parallel}+\mathbf{v}$, so that in the direction perpendicular to $\mathbf{B}$ it moves with the gas.

Consider now the implications of the universal occurrence of large-scale magnetic fields. They could not exist if magnetic monopoles were abundant. They would be neutralized in exactly the same way that electric fields are neutralized. In that case there could be no induced electric fields either. What, then, is the upper limit on the abundance of magnetic monopoles in the universe? Indeed, why should we expect any?

The answer to the latter question, as to why we might expect magnetic monopoles, is that there is no theoretical basis for expecting either their presence or absence. We have only the philosophical possibility, suggested by the symmetry between electric and magnetic fields in Maxwell's equations, that if there are electric charges, then why not magnetic charges? The concept of global symmetry has proved remarkably fruitful in understanding the fundamental properties of the elementary particles. Particles and their anti-particles are a particular favourite, leading to such speculation as the idea that half of the stars in the universe are composed of anti-matter (Alfven and Klein 1962; Alfven 1965; Omnes 1971, 1972). It is philosophically repugnant, is it not, to think that the universe should be prejudiced in favour of particles and against anti-particles? If, then, we entertain the idea that the universe is symmetric in particles and anti-particles, we are philosophically obliged to consider the possibility that the universe is symmetric in electric charges and magnetic charges. Consequently the existence of monopoles has been pursued in the laboratory for decades, with one or two spectacular false alarms (Ehrenhaft 1945, 1949; Price *et al.* 1975) to complicate the search.

The idea of electric-magnetic charge symmetry would suggest that the magnetic charge g is just equal to the electric charge e, and the masses of the magnetic particles should equal their electric counterparts, etc. If this were the case, then, a distant star composed of magnetic material would be indistinguishable in every respect from an ordinary star of the same total mass. What is more, if $g=e$, then g would be so small as to make detection of a monopole very difficult. So there appears to be no direct objection to the possibility of magnetically charged particles. Nor, of course, is there any need of such an hypothesis.

However, it was pointed out by Dirac (1931, 1948) and Schwinger (1968) that one might expect that g is equal to 137/2 or 137 times e, for if g does not have some such value as $137e/2$, then there arise ambiguities in the phase differences of wave functions around closed paths, so that a Hamiltonian formulation of quantum mechanics is not possible. It is interesting to note that if $g = 137e$, then there is a bound state of an electron in the field of a magnetic monopole. On the other hand, it has been argued by Cabibbo and Ferrari (1962) that $g = e$ can be rendered consistent with quantum electrodynamics (see discussion by Wentzel 1966). The remarkable success of quantum electrodynamics in treating the electromagnetic interactions of particles suggests that, in the absence of any experimental information on monopoles, we should not be wholly blind to its suggestions[3]. In any case, the suggestion that $g \geqslant 137e/2$ is welcome because it is a magnetic charge sufficiently large to make detection feasible.

Now magnetic monopoles, if they exist at all, are rare. There would be no large-scale magnetic fields in astrophysical bodies if they were not rare. If there are magnetic monopoles in space, they are accelerated by the force $g\mathbf{B}$ exerted on them by the large-scale magnetic field and so should be travelling with nearly the speed of light. The magnetic field of Earth would accelerate a magnetic charge $g = e$ to an energy of about 3×10^{10} eV and a charge of $g = 137e/2$ to 2×10^{12} eV. A sunspot field of 3×10^3 G over 10^4 km would accelerate a monopole $g = e$ to 10^{15} eV. A galactic field of 3×10^{-6} G would accelerate a monopole to 3×10^{18} eV over a distance of one kiloparsec (3×10^{21} cm). Therefore we should look among cosmic rays for magnetic monopoles (see for instance Porter 1960). If monopoles strike the atmosphere of Earth, or the surface of the moon, they may be embedded there. So there has been an extensive search in terrestrial rocks, ocean bottom sediments, meteorites, and material from the moon (Goto *et al.* 1963; Fleischer *et al.* 1969a,b,c, 1970; Kolm *et al.* 1971). Even if there are no monopoles in space one may conjecture that they are produced as secondaries in high-energy collisions of cosmic ray particles with Earth, the moon, or meteorites. Carithers *et al.* (1966), and others since, have searched for fast monopoles in the terrestrial atmosphere (see review by Eberhard *et al.* 1971).

A low upper limit on the abundance of magnetic monopoles follows from the work of Ross *et al.* (1973) in which they processed about 12 kg of lunar material (from Apollo 11, 12, and 14) with a detector sensitive to any net magnetic charge in excess of $30e$. They found no monopoles with $g \geqslant 30e$, implying less than $1{\cdot}7 \times 10^{-4}$ monopoles per g of material, or less than one monopole per $3 \times 10^{+27}$ nucleons. This places the monopole production cross-section in nucleon–nucleon collisions at less than 10^{-18} barns if the monopole mass is comparable to the mass of the proton (see also Gurevich *et al.* 1972).

[3] Although there is no reason to think that a theory founded explicitly on the absence of monopoles, ($\nabla \cdot \mathbf{B} = 0$, $\mathbf{B} = \nabla \times \mathbf{A}$) should necessarily correctly describe monopoles.

There is no information available so far on the abundance of monopoles for which g is as small as e. The quantum interferometer (Vant Hull 1968) is the only device to date that, in principle, is sufficiently sensitive to detect $g=e$.

Consider, then, the theoretical upper limit on the abundance of magnetic monopoles placed by the general occurrence of large-scale magnetic fields (Parker 1970). We have already noted that magnetic monopoles would be accelerated to enormous energies in the existing magnetic fields. The monopole velocities would approach the speed of light c. Hence, if there were N_g monopoles per unit volume, the magnetic current density[4] would be $J=N_g gc$. The rate at which the magnetic field does work on the monopoles is $\mathbf{J}\cdot\mathbf{B}$ erg cm^{-3} s^{-1}. In view of the acceleration of the monopoles along $\mathbf{B}$, we expect that $\mathbf{J}$ will generally have a significant component along $\mathbf{B}$, so that the rate of dissipation of magnetic field energy is of the order of JB. The magnetic energy density $B^2/8\pi$ declines, therefore, at an approximate rate

$$\frac{\mathrm{d}}{\mathrm{d}t}\frac{B^2}{8\pi}\cong BN_g gc$$

$$\frac{1}{B^2}\frac{\mathrm{d}B^2}{\mathrm{d}t}=\frac{8\pi N_g gc}{B}$$

The characteristic decay time τ is

$$\tau=B/8\pi N_g gc.$$

Applying this result to the galaxy, in which $B\cong 3\times10^{-6}$ G, suppose that the present galactic field is a remnant of a primordial field. Then presumably the dissipation time τ is not less than the age of the galaxy, 10^{10} years. Hence $\tau\gtrsim 3\times10^{17}$ s. If $g=e$, then it follows that $N_g<3\times10^{-26}$ cm^{-3}. If there is one nucleon per cm^3 in interstellar space, this is equivalent to the statement that there is less than one monopole per 3×10^{25} nucleons. If we suppose that $g=137e$, then we require less than one monopole per 4×10^{27} nucleons.

It appears, however, that the galactic field is not primordial, for reasons that will be discussed in a later chapter. It is generated by the fluid motions in the galaxy on a characteristic time scale of the order of 3×10^8 years $=10^{16}$ s. Hence, we can assert only that $\tau\gtrsim10^{16}$ s, requiring that there be fewer than one monopole ($g=137e$) per 10^{26} nucleons. The present laboratory results, that there are fewer than one monopole per 3×10^{27} nucleons, gives a comfortable margin. There should be little or

[4] Maxwell's equations become $4\pi\mathbf{j}+\partial\mathbf{E}/\partial t=c\nabla\times\mathbf{B}$ and $4\pi\mathbf{J}+\partial\mathbf{B}/\partial t=-c\nabla\times\mathbf{E}$, together with $\nabla\cdot\mathbf{E}=4\pi\delta_e$ and $\nabla\cdot\mathbf{B}=4\pi\delta_g$. It is no longer possible to write merely $\mathbf{B}=\nabla\times\mathbf{A}$.

no dissipation of the galactic magnetic field by monopoles. Other, usually less stringent, restrictions can be placed on monopole abundances in the sun and in Earth (Parker 1970). It appears, then, in view of the recent laboratory results, that the existence of large-scale magnetic fields is entirely consistent with the measured abundance of monopoles. But it must be remembered that the laboratory results refer only to monopoles for which $g \gtrsim 30e$. If, in fact, there should be a true symmetry between electric and magnetic particles, then $g = e$ and the laboratory stands mute. In that case we have only the general existence of large-scale magnetic fields in the universe to indicate the upper limit on the number of monopoles. As we have seen, that upper limit is of the order of one monopole per 10^{24} nucleons, based on estimates from the galactic magnetic field. The upper limit defines the laboratory problem of searching for magnetic monopoles with $g = e$ in samples of matter or among cosmic rays. The general astrophysical prevalence of large-scale magnetic fields places a severe limit on the abundance of monopoles and hence a severe limit on their stability or their production cross-section in high-energy (cosmic ray) particle interactions.

In the absence of any affirmative evidence we can only assume that the universe contains no magnetically charged particles and proceed to develop the theoretical properties of the magnetic field $\mathbf{B}$ as though $\nabla \cdot \mathbf{B} = 0$ and there is no direct dissipation $\mathbf{J} \cdot \mathbf{B}$ by magnetic currents. Perhaps someday a monopole will be discovered and we will revise this simple hypothesis. For the time being the fact is that no monopoles have been discovered in the terrestrial environment. Nor have any monopoles been discovered among the cosmic rays, which come to us from distances probably of the order of a kiloparsec.

We should note that the same situation prevails with regard to stars composed of anti-matter. Anti-protons are absent among the primary cosmic rays arriving at Earth (Evenson 1972; Smoot *et al.* 1975). It is generally believed that the cosmic rays are produced by active stars—supernovae, flare stars, etc. within some distance of the order of 1 kpc. Hence we conclude that there are no stars composed of anti-matter in our neighbourhood of the galaxy. In view of the non-uniform rotation and general mixing of stars in the disc of the galaxy, we take this to mean that anti-matter stars occur nowhere in the galaxy. The same statements apply to magnetic matter.

Altogether, the evidence at hand shows a strong prejudice in the universe for *electric matter* and against anti-matter (Steigman 1973, 1974) or magnetic matter.

It is interesting to note that the theory developed in this book is insensitive to whether a locality is populated solely by electric or by magnetic particles. Maxwell's equations are symmetric in $\mathbf{E}$ and $\mathbf{B}$, so that the fields can be interchanged if the charge on particles is interchanged. A difference appears only in a region where *both* electric and magnetic charges are present. In that case there would be neither large-scale electric nor large-scale magnetic fields. There would be no large-scale electromagnetic activity and the locale would be a dull place indeed.

References

ALFVEN, H. (1965). *Rev. Mod. Phys.* **37,** 652.
—— and KLEIN, O. (1962). *Ark. Fys.* **23,** 187.
CABIBBO, N. and FERRARI, E. (1962). *Nuovo Cim.* **23,** 1147.
CARITHERS, W. C., STEFANSKI, R., and ADAIR, R. K. (1966). *Phys. Rev.* **149,** 1970.
DIRAC, P. A. M. (1931). *Proc. Roy. Soc.* A **133,** 60.
—— (1948). *Phys. Rev.* **74,** 817.
EBERHARD, P. H., ROSS, R. R., ALVAREZ, L. W. and WATT, R. D. (1971). *Phys. Rev.* D **4,** 3260.
EHRENHAFT, F. (1945). *Phys. Rev.* **67,** 63, 201.
—— (1949). ibid. **75,** 1334.
EVENSON, P. (1972). *Astrophys. J.* **176,** 797.
FLEISCHER, R. L., JACOBS, I. S., SCHWARZ, W. M., and PRICE, P. B. (1969a). *Phys. Rev.* **177,** 2029.
—— PRICE, P. B., and WOODS, R. T. (1969b). *Phys. Rev.* **184,** 1398.
—— HART, H. R., JACOBS, I. S., PRICE, P. B., SCHWARZ, W. M., and AUMENTO, F. (1969c). *Phys. Rev.* **184,** 1393.
—— —— —— —— —— and WOODS, R. T. (1970). *J. Appl. Phys.* **41,** 958.
GOTO, E., KOLM, H. H., and FORD, K. W. (1963). *Phys. Rev.* **132,** 387.
GUREVICH, I. I., KHAKIMOV, S. KH., MARTEMIANAV, V. P., MISHAKOVA, A. P., MAKAR'INA, L. A., ORGURTZOV, V. V., TARASENKOV, V. G., CHERNISHOVA, L. A., BARKOV, L. M., ZOLOTOREV, M. S., OHAPKIN, V. S., and TARAKANOV, N. M. (1972). *Phys. Lett.* B **38,** 549.
HOBBS, L. M. (1974). *Astrophys. J. Lett.* **188,** L107.
HOYLE, F. (1960). *Proc. Roy. Soc.* A **257,** 431.
KOLM, H. H., VILLA, F., and ODIAN, A. (1971). *Phys. Rev.* D **4,** 1285.
LYTTLETON, R. A. and BONDI, H. (1959). *Proc. Roy. Soc.* A **252,** 313.
—— —— (1960). ibid. **257,** 442.
OMNES, R. (1971). *Astron. Astrophys.* **10,** 228; **11,** 450; **15,** 275.
—— (1972). *Phys. Rep. Phys. Lett.* **3C** (no. 1), 1.
PARKER, E. N. (1970). *Astrophys. J.* **160,** 383.
PORTER, N. A. (1960). *Nuovo Cim.* **16,** 958.
PRICE, P. B., SHIRK, E. K., OSBORNE, W. Z., and PINSKY, L. S. (1975). *Phys. Rev. Lett.* **35,** 487.
ROSS, R. R., EBERHARD, P. H., and ALVAREZ, L. W. (1973). *Phys. Rev.* D **8,** 698.
SCHWINGER, J. (1968). *Phys. Rev.* **173,** 1536.
SMOOT, G. F., BUFFINGTON, A., and ORTH, C. D. (1975). *Phys. Rev. Lett.* **35,** 258.
STEIGMAN, G. (1973). *Cargese lectures in physics* 6 (ed. E. SCHATZMAN). Gordon and Breach, New York.

—— (1974). *Confrontation of cosmological theories with observational data, IAU Symposium No.* 63 (*Copernicus Symposium II*) D. Reidel, Dordrecht–Holland.
VANT HULL, L. (1968). *Phys. Rev.* **173,** 1412.
WENTZEL, G. (1966). *Prog. Theor. Phys. Suppl.* **35,** 163.
ZORN, J. C., CHAMBERLAIN, G. E., and HUGHES, V. W. (1963). *Phys. Rev.* **129,** 2566.

4

THE BASIC EQUATIONS

4.1. The reduction of Maxwell's equations

THE task before us is to develop the theoretical behaviour of large-scale magnetic fields from Maxwell's equations. The theory is already well known, with a number of excellent texts on the subject (see, for instance, Cowling 1957; Landau and Lifschitz 1960; Chandrasekhar, 1961; Alfven and Falthammer 1963; Kendall and Plumpton 1964; Ferraro and Plumpton 1966; Jeffrey 1966; in particular, see Roberts 1967). We will not try to duplicate what has already been so well expounded, but, after the necessary preliminaries, will proceed directly to the development of the theory in the form most appropriate to understanding the activity in the universe. The two principal goals will be to explain (a) the general origin of magnetic fields in convecting rotating bodies and (b) the general absence of hydrostatic equilibrium when magnetic fields are present. The origin has been treated in earlier monographs, but only mathematically. The physical questions, that have gone largely unmentioned, will be the main concern here. The absence of equilibrium has not been treated at all, and lies at the heart of the unceasing activity in magnetic fields. Observations of the active magnetic fields in the universe tell us that the behaviour is complex so that we approach the theoretical problem expecting to find it extensive, and quite different from the conventional field theories developed for isolated microscopic phenomena. Our progress will be limited only by our energy and imagination in recognizing the basic physical effects among the enormous observational and theoretical complexity. In view of the hints and suggestions to be gleaned from observations of magnetic activity in the universe, it is only natural that we should respond with suggestions as to the role played in nature by the theoretical effects. But it is the basic theoretical properties of the magnetic field, rather than the tentative suggestions of their role in nature, that is our primary goal. In most cases the observations do not define the circumstances well enough to confirm the tentative suggestions. In the meantime, we concentrate on the basic physics.

Maxwell's equations for an electric field **E** (statvolts/cm) and magnetic field **B** (G) in any inertial coordinate system in the presence of a charge density δ (e.s.u.) and current density $\mathbf{j}$ are

$$4\pi\mathbf{j}+\partial\mathbf{E}/\partial t=c\nabla\times\mathbf{B}, \tag{4.1}$$

$$\partial\mathbf{B}/\partial t=-c\nabla\times\mathbf{E}, \tag{4.2}$$

$$\nabla\cdot\mathbf{E}=4\pi\delta, \tag{4.3}$$

$$\nabla\cdot\mathbf{B}=0, \tag{4.4}$$

where c (3×10^{10} cm s^{-1}) is the speed of light[1].

The electric and magnetic fields $\mathbf{E}'$ and $\mathbf{B}'$ respectively, in a frame of reference moving with velocity $\mathbf{u}$ relative to the coordinate system are

$$\mathbf{E}'=(\mathbf{E}+\mathbf{u}\times\mathbf{B}/c)/(1-u^2/c^2)^{\frac{1}{2}} \tag{4.5}$$

$$\mathbf{B}'=(\mathbf{B}-\mathbf{u}\times\mathbf{E}/c)/(1-u^2/c^2)^{\frac{1}{2}} \tag{4.6}$$

Imagine, then, a magnetic field $\mathbf{B}(\mathbf{r},t)$ with characteristic dimension l embedded in an electrically conducting medium with fluid velocity $\mathbf{v}$. The characteristic time of variation of the field is l/v. In order of magnitude, then, $\partial/\partial t\sim v/l$ and $|\nabla|\sim1/l$. We are concerned with velocities small compared to the speed of light and will develop the theory neglecting all terms second order in v/c, essentially along the lines outlined by Elsasser (1954). For simplicity we suppose that the background fluid is a classical fluid with a simple scalar electrical conductivity σ, so that in the frame of reference of the fluid the electric current density $\mathbf{j}'$ is related to the electric field by Ohm's law

$$\mathbf{j}'=\sigma\mathbf{E}' \tag{4.7}$$

The electric field $\mathbf{E}'$ in the frame of the fluid follows from (4.5) as

$$\mathbf{E}'=\mathbf{E}+\mathbf{v}\times\mathbf{B}/c, \tag{4.8}$$

neglecting terms second order in v/c.

Now in view of the generally large electrical conductivity, the only electric fields present are those that are induced, of the order of vB/c. Hence $\partial\mathbf{E}/\partial t$ has a magnitude of the order of $(v/l)vB/c=(cB/l)v^2/c^2$. The term $c\nabla\times\mathbf{B}$ on the right hand side of (4.1) is of the order of cB/l, larger

[1] We define δ and $\mathbf{j}$ to be the entire charge and current density, including both free and bound charges, so that the total electric field is $\mathbf{E}$ and the magnetic field is $\mathbf{B}$. It is convenient sometimes to use the more conventional approach, in which the dielectric polarization $\mathbf{P}$ (dipole moment per unit volume) is added to the electric field to give the total electric displacement vector $\mathbf{D}=\mathbf{E}+4\pi\mathbf{P}$, and the magnetic field $\mathbf{B}$ is divided into the two parts $\mathbf{H}$ and $4\pi\mathbf{M}$ with $\mathbf{H}$ denoting that part associated with the conduction currents and $\mathbf{M}$ the magnetization associated with the electric currents circulating within individual atoms. For most astrophysical fields these considerations do not arise and so are merely confusing. The reader who desires to do so can easily translate our discussion into his own favourite form.

$\frac{\partial B}{\partial t} = \eta \nabla^2 B$

$G(r-r', t) = \frac{1}{(4\pi\eta t)^{\frac{3}{2}}} \exp\left[-\frac{(r-r')^2}{4\eta t}\right]$

$B_i(r, t) = \iiint d^3r' \, G(r-r', t) B_i(r', 0)$

lacement current $\partial \mathbf{E}/\partial t$ from
form

(4.9)

of cB/l. We must not forget
nplies a net electric charge
plies a contribution $\mathbf{v}\delta$ to the
der of vB/c, it follows from
nd hence negligible. There-
the current densities in the
follows from (4.7) and (4.8)

(4.10)

olving for $\mathbf{E}$, there results

$\times \mathbf{B}$. (4.11)

equation (4.2) yields the familiar hydromagnetic equation

$$\partial \mathbf{B}/\partial t = \nabla \times (\mathbf{v} \times \mathbf{B}) - \nabla \times (\eta \nabla \times \mathbf{B}), \tag{4.12}$$

where for convenience we have defined the resistive diffusion coefficient

$$\eta = c^2/4\pi\sigma.$$

If η is independent of position, then (4.12) reduces to

$$\partial \mathbf{B}/\partial t = \nabla \times (\mathbf{v} \times \mathbf{B}) + \eta \nabla^2 \mathbf{B}, \tag{4.13}$$

and otherwise

$$\partial \mathbf{B}/\partial t = \nabla \times (\mathbf{v} \times \mathbf{B}) + \eta \nabla^2 \mathbf{B} - \nabla \eta \times (\nabla \times \mathbf{B}). \tag{4.14}$$

4.2. Physical interpretation of the hydromagnetic equation

The physical interpretation of eqn (4.12) is, of course, that the term $\nabla \times (\mathbf{v} \times \mathbf{B})$ is the convection term, carrying the magnetic lines of force bodily with the fluid. The term $\nabla \times (\eta \nabla \times \mathbf{B})$ represents diffusion and resistive dissipation. For uniform resistivity the magnetic field satisfies the familiar heat flow equation

$$\partial \mathbf{B}/\partial t = \eta \nabla^2 \mathbf{B} \tag{4.15}$$

in the absence of any fluid motions (see §6.8 for the general case). Each cartesian component of the magnetic field diffuses with time from its initial configuration $B_i(\mathbf{r}, 0)$ according to the general Green's function (for an infinite space)

$$G(\mathbf{r}-\mathbf{r}', t) = \frac{1}{(4\pi\eta t)^{\frac{3}{2}}} \exp\left[-\frac{(\mathbf{r}-\mathbf{r}')^2}{4\eta t}\right]$$

showing how the field at each initial point $\mathbf{r}'$ spreads out with a Gaussian profile of width $(4\eta t)^{\frac{1}{2}}$ in a time t. The field at a time t (>0) is formally related to the initial field ($t=0$) by the familiar integral expression

$$B_i(\mathbf{r}, t) = \iiint \mathrm{d}^3\mathbf{r}' G(\mathbf{r}-\mathbf{r}', t) B_i(\mathbf{r}', 0).$$

Field gradients soften with time and opposite fields merge and cancel.

Now consider the convection term. In the limit of large electrical conductivity, we have $\eta \ll lv$, and (4.12) reduces to[2]

$$\partial \mathbf{B}/\partial t = \nabla \times (\mathbf{v} \times \mathbf{B}). \tag{4.16}$$

There are a number of ways in which convection of the magnetic field can be demonstrated. The concept of the magnetic lines of force comes into prominence here. We indicated in Chapter 2 that a magnetic field is represented by a continuous vector function of position, $\mathbf{B}(\mathbf{r})$. The lines of force are defined, in terms of $\mathbf{B}(\mathbf{r})$, as the two-parameter family of solutions of the two equations

$$\mathrm{d}x/B_x = \mathrm{d}y/B_y = \mathrm{d}z/B_z \tag{4.17}$$

at *each instant* of time t. The association of lines of force at *different* times is arbitrary, and we shall, therefore, choose the association that is most convenient.

The first point, then, is to note that application of Gauss's theorem to (4.4) shows that the total magnetic flux across any closed surface S is zero,

$$\int_S \mathrm{d}\mathbf{S} \,.\, \mathbf{B} = 0 \tag{4.18}$$

where $\mathrm{d}\mathbf{S}$ is the outward drawn normal to the element of area $\mathrm{d}S$ in S. Lines of force can terminate only on charges. There are no magnetic charges, so every line of force entering S must also leave. It follows then that the total number of lines of force Φ through any closed contour C can be written

$$\Phi = \int_C \mathrm{d}\mathbf{S} \,.\, \mathbf{B} \tag{4.19}$$

where the integration is over *any* surface enclosed by C.

[2] This equation can also be written $\mathrm{d}\mathbf{B}/\mathrm{d}t \equiv \partial\mathbf{B}/\partial t + (\mathbf{v}\,.\,\nabla)\mathbf{B} = (\mathbf{B}\,.\,\nabla)\mathbf{v} - \mathbf{B}\nabla\,.\,\mathbf{v}$. If the fluid density ρ is variable, then $\mathrm{d}\rho/\mathrm{d}t + \rho\nabla\,.\,\mathbf{v} = 0$ and $(\mathrm{d}/\mathrm{d}t)(\mathbf{B}/\rho) = [(\mathbf{B}/\rho)\,.\,\nabla]\mathbf{v}$. If the fluid motion is incompressible, then $\nabla\,.\,\mathbf{v} = 0$ so that

$$\partial B_i/\partial t = (\partial/\partial x_j)(v_i B_j - v_j B_i)$$

and $\mathrm{d}B_i/\mathrm{d}t = (\partial/\partial x_j) v_i B_j$.

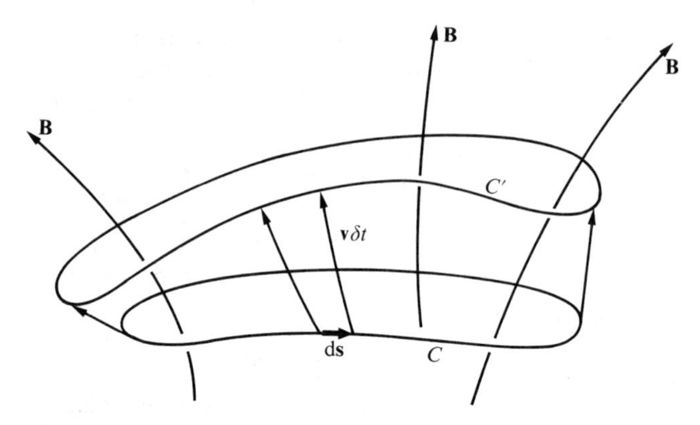

FIG. 4.1. A sketch of the contour C and the position C' to which the fluid would carry C in a time δt.

Now consider the change with time of the magnetic flux through a closed contour moving with the fluid. Its initial position at time t is the closed contour C and a short time δt later it has undergone the displacement $\mathbf{v}\delta t$, as a consequence of the fluid velocity $\mathbf{v}(\mathbf{r}, t)$, to a new position C', sketched in Fig. 4.1. Denote by $\mathrm{d}\mathbf{S}_C$ an element of area on any surface bounded by C, and by $\mathrm{d}\mathbf{S}'_C$ the corresponding element on any surface bounded by C'. The element of area (with outward normal) swept out by the element of arc length $\mathrm{d}\mathbf{s}$ of the contour moving a distance $\mathbf{v}\delta t$ is $\mathrm{d}\mathbf{s}\times\mathbf{v}\delta t$. It follows from (4.18) that the total flux through the closed surface shown in Fig. 4.1 is

$$\int_C \mathrm{d}\mathbf{S}\,.\,\mathbf{B}(t+\delta t)-\int_{C'}\mathrm{d}\mathbf{S}'_C\,.\,\mathbf{B}(t+\delta t)-\oint_C \mathbf{B}(t+\delta t)\,.\,\mathrm{d}\mathbf{s}\times\mathbf{v}\delta t=0$$

at time $t+\delta t$. Hence the change in flux through the moving contour from time t to $t+\delta t$ is

$$\begin{aligned}\delta\Phi &\equiv \int_{C'}\mathrm{d}\mathbf{S}'_C\,.\,\mathbf{B}(t+\delta t)-\int_C \mathrm{d}\mathbf{S}_C\,.\,\mathbf{B}(t)\\ &= \int_C \mathrm{d}\mathbf{S}_C\,.\,[\mathbf{B}(t+\delta t)-\mathbf{B}(t)]-\delta t\oint \mathbf{B}\,.\,\mathrm{d}\mathbf{s}\times\mathbf{v}\\ &= \delta t\left\{\int_C \mathrm{d}\mathbf{S}_C\,.\,\frac{\partial\mathbf{B}}{\partial t}-\oint_C \mathbf{B}\,.\,\mathrm{d}\mathbf{s}\times\mathbf{v}\right\}.\end{aligned} \tag{4.20}$$

Then using (4.16) and Stokes' theorem,

$$\begin{aligned}\delta\Phi &= \delta t\left\{\int_C \mathrm{d}\mathbf{S}_C\,.\,\nabla\times(v\times\mathbf{B})-\oint_C \mathbf{B}\,.\,\mathrm{d}\mathbf{s}\times\mathbf{v}\right\}\\ &= \delta t\oint\{\mathrm{d}\mathbf{s}\,.\,\mathbf{v}\times\mathbf{B}-\mathbf{B}\,.\,\mathrm{d}\mathbf{s}\times\mathbf{v}\}.\end{aligned}$$

Using the well known vector identity for combinations of vector and scalar products, we have $d\mathbf{s} \cdot \mathbf{v} \times \mathbf{B} = \mathbf{B} \cdot d\mathbf{s} \times \mathbf{v}$, so that

$$\delta\Phi = 0. \tag{4.21}$$

The magnetic flux through any contour moving with the fluid is constant in time, not matter what the field configuration **B** or the velocity field **v**. This is true for all contours, located anywhere relative to the local inhomogeneities in the field[3]. Imagine, then, a magnetic field composed of separate, isolated tubes of magnetic flux. It is clear that the individual tubes of flux move with the fluid, for if there were any motion of a tube relative to the fluid, there would be closed contours moving with the fluid that would experience changes in magnetic flux, contrary to (4.21). Thus it is awkward to adopt any view other than that the magnetic lines of force move exactly with the fluid.

Now any continuous field distribution $\mathbf{B}(\mathbf{r}, t)$ can be made up of the superposition of two or more configurations of discrete and isolated flux tubes. (The hydromagnetic equation, (4.19) or (4.16), is linear in **B**.) Therefore, if we are compelled to the view that the individual lines of force of the discontinuous fields move with the fluid, we arrive at the position that the lines of force of the continuous field—the superposition of several discontinuous fields—move with the fluid. Any other view is inconvenient, to say the least. The concept is vividly phrased by Alfven's (1950) statement that the magnetic lines of force are 'frozen' into the fluid. The reader will find Lundquist's early treatment (Lundquist 1952) particularly interesting.

It follows that the magnetic lines of force move in the frame of reference in which the electric field is zero. The electric field is held to zero by the high conductivity of the fluid. To put it another way, note that in the limit of large electrical conductivity, $lv\sigma \gg c^2$ (i.e. $vl \gg \eta$), (4.10) reduces to

$$\mathbf{E} = -\mathbf{v} \times \mathbf{B}/c. \tag{4.22}$$

Hence the field in the local frame of the fluid is, according to (4.5)

$$\mathbf{E}' = \mathbf{E} + \mathbf{v} \times \mathbf{B}/c = 0. \tag{4.23}$$

The point is that the reference frame of the magnetic field is the frame in which the electric field is zero. This is the same convention as adopted in

[3] It should be understood that any quantity such as **v** and **B** representing a physical entity is automatically finite and continuous, with derivatives to all orders. Mathematical discontinuities cannot be established by experiment, even if they existed, so the only discontinuities that can appear are idealizations that we ourselves arbitrarily, or inadvertently, introduce for computational convenience, such as shock fronts, etc.

Chapter 3 to discuss magnetic fields and the electric fields induced from them.

Another demonstration of the motion of the magnetic lines of force follows from the representation of the magnetic field in terms of the Euler potentials, $\alpha(\mathbf{r})$ and $\beta(\mathbf{r})$ (Sweet 1950; Lundquist 1952; Dungey 1953; see discussion in Stern 1966, 1976; and Roberts 1967). One writes

$$\mathbf{B}=\nabla\alpha\times\nabla\beta=\nabla\times(\alpha\nabla\beta),$$

which automatically satisfies the divergence condition, $\nabla\cdot\mathbf{B}=0$. The magnetic lines of force are given by the intersections of the two families of surfaces, $\alpha=constant$, $\beta=constant$, as sketched in Fig. 4.2. This is obvious from the fact that the vector $\nabla\alpha$ is perpendicular to the surface $\alpha=constant$ so that any line perpendicular to $\nabla\alpha$ lies in the surface $\alpha=constant$. The same is true for β. The magnetic field is perpendicular to both $\nabla\alpha$ and $\nabla\beta$. Hence it lies in both surfaces $\alpha=constant$, $\beta=constant$. Hence $\mathbf{B}$ must lie in the intersection of the two surfaces.

Now (4.16) becomes

$$\begin{aligned}\nabla\alpha\times\nabla\frac{\partial\beta}{\partial t}-\nabla\beta\times\nabla\frac{\partial\alpha}{\partial t}&=\nabla\times[\mathbf{v}\times(\nabla\alpha\times\nabla\beta)],\\&=\nabla\times[\nabla\alpha(\mathbf{v}\cdot\nabla\beta)-\nabla\beta(\mathbf{v}\cdot\nabla\alpha)],\\&=\nabla(\mathbf{v}\cdot\nabla\beta)\times\nabla\alpha-\nabla(\mathbf{v}\cdot\nabla\alpha)\times\nabla\beta.\end{aligned}$$

Hence

$$\nabla\alpha\times\nabla\frac{\mathrm{d}\beta}{\mathrm{d}t}=\nabla\beta\times\nabla\frac{\mathrm{d}\alpha}{\mathrm{d}t} \tag{4.24}$$

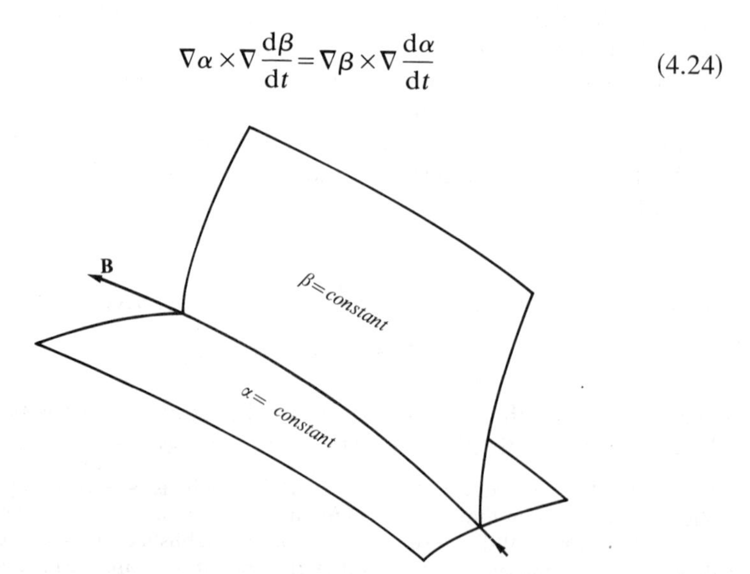

FIG. 4.2. A sketch of the intersection of $\alpha=constant$, $\beta=constant$ defining a line of force of the magnetic field $\mathbf{B}=\nabla\alpha\times\nabla\beta$.

where

$$\frac{d\alpha}{dt} \equiv \frac{\partial \alpha}{\partial t} + \mathbf{v} \cdot \nabla \alpha, \qquad \frac{d\beta}{dt} = \frac{\partial \beta}{\partial t} + \mathbf{v} \cdot \nabla \beta.$$

There is a whole family of solutions to (4.24), but the simplest is obviously $d\alpha/dt = d\beta/dt = 0$. That is to say, we are permitted to suppose that the Euler potentials, the intersection of the equipotential surfaces, and the lines of force, move with the fluid.

An equivalent calculation without the ambiguities of the Euler potentials (see discussion of ambiguities in Stern 1966) can be constructed working directly with the magnetic lines of force. Consider a fluid element with position $x_{1i}(t)$ at which $B_i(x_{1j}) \neq 0$. Through the point x_{1i} there passes one line of force, i.e. a single solution of

$$dx_i/ds = B_i/B$$

where ds is an element of arc length and B is the magnitude of the field, $B^2 = B_i B_i$. Then consider another fluid element at position $x_{2i}(t)$ somewhere on the same line of force at time t so that

$$x_{2i}(t) = x_{1i} + \int_{x_1}^{x_2} ds\, n_i \tag{4.25}$$

where $n_i = B_i/B$ is the unit vector along B_i and the integration is carried along the magnetic line of force from x_{1i} to x_{2i}. It is a simple matter to show that the motion of the magnetic line of force (4.25) with the fluid is consistent with the hydromagnetic equation (4.16). Differentiation of (4.25) yields

$$\frac{dx_{2i}}{dt} = \frac{dx_{1i}}{dt} + \int_{x_1}^{x_2} ds \frac{dn_i}{dt} + \int_{x_1}^{x_2} n_i \frac{d}{dt} ds \tag{4.26}$$

where the last term on the right-hand side symbolizes the time rate of change of the path of integration.

Now if the field line moves with the fluid, then the velocity difference between x_{2i} and x_{1i} must be just the integral of the velocity gradient along the field line,

$$\frac{dx_{2i}}{dt} = \frac{dx_{1i}}{dt} + \int_{x_1}^{x_2} ds\, n_j \frac{\partial v_i}{\partial x_j}. \tag{4.27}$$

To show that (4.26) reduces to this expression, note that the time rate of change of the element of arc length ds moving with the fluid is just equal to the difference in velocity of the two ends of ds,

$$\frac{d}{dt} ds = ds\, n_i n_j \frac{\partial v_i}{\partial x_j}. \tag{4.28}$$

The quantity dn_i/dt must be evaluated from the hydromagnetic equation (4.16)[2]. To do this we note first that direct differentiation of $n_i = B_i/B$ yields

$$dn_i/dt = (\delta_{ij} - n_i n_j)(1/B)\, dB_j/dt.$$

Then since (4.16) can be written

$$(1/B)\, dB_j/dt = n_k \partial v_j/\partial x_k - n_j \partial v_k/\partial x_k,$$

it follows that

$$dn_i/dt = (\delta_{ij} - n_i n_j) n_k \partial v_j/\partial x_k. \tag{4.29}$$

Substituting (4.28) and (4.29) into (4.26) leads immediately to (4.27), q.e.d. It follows that any two points initially connected by the same magnetic line of force will always be connected by a line of force. We choose, then, to define the lines of force by the material points that lie upon them, stating that in the perfectly conducting fluid the magnetic lines of force move with the fluid.

4.3. Mathematical solutions for field convection:

The hydromagnetic equation (4.16) is, by itself, a kinematical statement of the evolution of the magnetic field $\mathbf{B}(\mathbf{r}, t)$ for a given velocity field $\mathbf{v}(\mathbf{r}, t)$. It is identical in form to the dynamical equation for the vorticity $\boldsymbol{\omega} = \nabla \times \mathbf{v}$ of a perfect fluid,

$$\partial \boldsymbol{\omega}/\partial t = \nabla \times (\mathbf{v} \times \boldsymbol{\omega}). \tag{4.30}$$

The vortex lines move exactly with the fluid. The Cauchy solution to (4.30) is applicable to (4.16) (Lundquist 1952), leading to a formal relation between the field configurations at any two times t_1 and t_2 in terms of the Lagrangian displacement of the fluid. The result is easily derived.

As already noted[2] the field convection equation can be written

$$dF_i/dt = F_j \partial v_i/\partial x_j \tag{4.31}$$

where $F_i(\mathbf{r}, t) = B_i(\mathbf{r}, t)/\rho$. Consider, then, the field and fluid at position $x_i(t)$ at time t. A short time δt later the field and fluid have moved to the position $x_i(t + \delta t) = x_i(t) + v_i \delta t$. The field is

$$\begin{aligned} F_i[x_k(t+\delta t), t+\delta t] &= F_i[x_k(t), t] + \delta t\, dF_i/dt, \\ &= F_i[x_k(t), t] + F_j[x_k(t), t] \frac{\partial v_i}{\partial x_j} \delta t, \\ &= F_j[x_k(t), t] \frac{\partial}{\partial x_j(t)} [x_i(t) + v_i(t)\delta t], \\ &= F_j[x_k(t), t] \frac{\partial x_i(t+\delta t)}{\partial x_j(t)}. \end{aligned}$$

Thus the strain $\partial x_i(t+\delta t)/\partial x_j(t)$ projects the field F_i forward in time. Repeating the forward projection n times yields

$$F_i[x_s(t+n\delta t), t+n\delta t] = F_j[x_s(t), t]\frac{\partial x_k(t+\delta t)}{\partial x_j(t)}\frac{\partial x_i(t+2\delta t)}{\partial x_k(t+\delta t)}\cdots,$$

$$= F_j[x_s(t), t]\frac{\partial x_i(t+n\delta t)}{\partial x_j(t)}.$$

That is to say, if the fluid element at position X_i at time t_1 is at position $x_i = x_i(X_k, t_2)$ at time t_2, then the field $F_i(x_k, t_2)$ at x_i at time t_2 has evolved from its initial value $F_i(X_k, t_1)$ at X_i according to the total accumulated strain

$$F_i(x_k, t_2) = F_j(X_k, t_1)\frac{\partial x_i}{\partial X_j}. \tag{4.32}$$

It is necessary, then, only to compute the unfolding of the Lagrangian coordinates $x_i = x_i(X_k, t)$ with time in order to work out the evolution of the magnetic field.

There are other methods of solving (4.16), of course. The equation is linear in B_i and may be approached as an ordinary initial value problem, to be solved by separation of variables, etc. We shall adopt whatever method of solution is most convenient for the particular problem at hand.

It is worth noting the solution of (4.16) in terms of the vector potential. Let $\mathbf{B} = \nabla \times \mathbf{A}$ ($B_i = \epsilon_{ijk}\partial A_k/\partial x_j$ where ϵ_{ijk} is the usual permutation tensor, equal to zero if any of the three indices are equal, and otherwise equal to ± 1 depending upon whether ijk is an even or odd permutation on 123). Then (4.16) becomes

$$\nabla \times [\partial \mathbf{A}/\partial t - \mathbf{v} \times (\nabla \times \mathbf{A})] = 0.$$

Integration yields

$$\partial \mathbf{A}/\partial t = \mathbf{v} \times (\nabla \times \mathbf{A}) - \nabla \phi$$

where ϕ is an arbitrary function of position and time. The equation is more conveniently cast in the form

$$\frac{\mathrm{d}A_i}{\mathrm{d}t} \equiv \frac{\partial A_i}{\partial t} + v_j\frac{\partial A_i}{\partial x_j} = v_j\frac{\partial A_j}{\partial x_i} - \frac{\partial \phi}{\partial x_i}.$$

Since ϕ is arbitrary, contributing nothing to B_i, let $\phi = v_jA_j$. The result is

$$\frac{\mathrm{d}A_i}{\mathrm{d}t} = -A_j\frac{\partial v_j}{\partial x_i}, \tag{4.33}$$

which is sufficiently like (4.31) that a similar method of integration can be employed.

Consider the element of fluid and field at x_i at time $t+\delta t$. A short time δt earlier the same field and fluid were at $x_i(t+\delta t)-v_i\delta t$. Then

$$A_i[x_s(t+\delta t),\, t+\delta t]=A_i[x_s(t),\, t]+\frac{\mathrm{d}A_i}{\mathrm{d}t}\delta t+O[(\delta t)^2],$$

$$=A_i[x_s(t),\, t]-A_j[x_s(t),\, t]\frac{\partial v_j}{\partial x_i}\delta t,$$

$$=A_j[x_s(t),\, t]\frac{\partial[x_j(t+\delta t)-v_j\delta t]}{\partial x_i(t+\delta t)}+O[(\delta t)^2],$$

$$=A_j[x_s(t),\, t]\partial x_j(t)/\partial x_i(t+\delta t).$$

Repeated application of this infinitesimal projection yields

$$A_i(x_k,\, t_2)=A_j(X_k,\, t_1)\partial X_j/\partial x_i. \tag{4.34}$$

This is the solution for A_i comparable to (4.31) for the field B_i/ρ. Note that if we multiply both sides of the equation by $\partial x_i/\partial X_k$ and sum over i, we have

$$\frac{\partial x_i}{\partial X_k}\frac{\partial X_j}{\partial x_i}=\delta_{jk}$$

and

$$A_j(X_k,\, t_1)=A_i(x_k,\, t_2)\partial x_i/\partial X_j \tag{4.35}$$

for the projection backward in time[4].

It should be noted that, with the gauge used to obtain (4.33), the divergence of A_i is generally not zero. The gauge in which $\partial A_j/\partial x_j=0$ is treated in the next section.

It is instructive to work backwards from (4.34) to obtain (4.33). The total Lagrangian time derivative of (4.34) follows as

$$\frac{\mathrm{d}A_i(x_k,\, t)}{\mathrm{d}t}=A_j(X_k,\, t_1)\frac{\mathrm{d}}{\mathrm{d}t}\frac{\partial X_j}{\partial x_i},$$

since the initial position X_k of the Lagrangian coordinate x_k is fixed in time, $\mathrm{d}X_k/\mathrm{d}t=0$. Consider the two moving points $(x_k,\, x_k+\delta x_k)$ with initial positions $(X_k,\, X_k+\delta X_k)$. The initial vector separation is δX_k, and at time t the separation is δx_k with

$$\delta X_i=(\partial X_i/\partial x_j)\delta x_j.$$

Differentiate with respect to time and note that δX_i does not change with time. It follows that

$$0=\delta x_j\frac{\mathrm{d}}{\mathrm{d}t}\frac{\partial X_i}{\partial x_j}+\frac{\partial X_i}{\partial x_j}\frac{\mathrm{d}\delta x_j}{\mathrm{d}t}.$$

[4] The discussion of the mathematical properties of the strain tensor $\partial x_i/\partial X_i$ given in Roberts (1967) is particularly useful and interesting.

The two points move with velocities $v_i(x_k, t)$ and $v_i(x_k+\delta x_k, t)$. Hence the rate of separation is

$$\frac{\mathrm{d}\delta x_j}{\mathrm{d}t}=\frac{\partial v_j}{\partial x_k}\,\delta x_k.$$

With this expression for $\mathrm{d}\delta x_j/\mathrm{d}t$ it follows that

$$\delta x_j\left[\frac{\mathrm{d}}{\mathrm{d}t}\frac{\partial X_i}{\partial x_j}+\frac{\partial X_i}{\partial x_k}\frac{\partial v_k}{\partial x_j}\right]=0.$$

This is a general expression for any separation δx_j. Hence the quantity in brackets must be identically zero, so that

$$\frac{\mathrm{d}}{\mathrm{d}t}\frac{\partial X_j}{\partial x_i}=-\frac{\partial X_j}{\partial x_k}\frac{\partial v_k}{\partial x_i}.$$

It follows then that

$$\begin{aligned}\frac{\mathrm{d}A_i(x_s, t)}{\mathrm{d}t}&=-A_j(X_s, t)\frac{\partial X_j}{\partial x_k}\frac{\partial v_k}{\partial x_i}\\&=-A_k(x_s, t)\frac{\partial v_k}{\partial x_i},\end{aligned}$$

upon using (4.34). The result is just (4.33), of course q.e.d.

It is an interesting fact, then, that the vector potential in an element of fluid at a position x_k at time t_2 is related solely to the vector potential in the same element of fluid at any earlier time t_1. The vector potential, like the magnetic field, is not affected by its value in other neighbouring elements of fluid (so long as the diffusion term is neglected, of course). The difference between the behaviours of B_i and A_i is that while the total flux of B_i is conserved, A_i is not. The conservation theorem for A_i follows from the conservation of magnetic flux through any contour moving with the fluid,

$$\begin{aligned}0&=\frac{\mathrm{d}}{\mathrm{d}t}\int_c \mathrm{d}\mathbf{S}\,.\,\mathbf{B}\\&=\frac{\mathrm{d}}{\mathrm{d}t}\int_c \mathrm{d}\mathbf{S}\,.\,\nabla\times\mathbf{A}\\&=\frac{\mathrm{d}}{\mathrm{d}t}\oint_c \mathrm{d}\mathbf{s}\,.\,\mathbf{A}.\end{aligned}$$

The line integral of **A** around any closed contour moving with the fluid is invariant over time.

4.4. Mathematical solution of the complete hydromagnetic equations

Mathematical solution of the complete hydromagnetic equation (4.12) for $\mathbf{B}$ is possible through standard methods employing separation of variables. The equation is linear and homogeneous in $\mathbf{B}$, although the coefficients $\mathbf{v}$ and η may be variable. Usually one chooses η to be uniform over the region of solution since its variability is difficult to handle and does not, in most cases, introduce anything of physical interest. The equation can be rewritten in terms of the vector potential $\mathbf{B}=\nabla\times\mathbf{A}$ if desired. Integrating once then yields

$$\partial\mathbf{A}/\partial t=\mathbf{v}\times\nabla\times\mathbf{A}-\eta\nabla\times\nabla\times\mathbf{A}-c\nabla\phi$$

where ϕ is an arbitrary scalar function of space and time, which does not affect the magnetic field in any way. Hence, it is usually put equal to zero so that $\mathbf{A}$ is constant in time in any region where $\mathbf{v}=0$ and $\mathbf{j}=0$ $(\nabla\times\nabla\times\mathbf{A}=0)$. With the choice $c\nabla^2\phi=\mathbf{B}\cdot\nabla\times\mathbf{v}-\mathbf{v}\cdot\nabla\times\mathbf{B}$, it follows that $\nabla\cdot\mathbf{A}=0$ (there is no general advantage to the Lorentz gauge in non-relativistic hydromagnetic problems) and

$$\partial\mathbf{A}/\partial t=\mathbf{v}\times\nabla\times\mathbf{A}+\eta\nabla^2\mathbf{A}. \tag{4.36}$$

Under steady conditions $(\partial/\partial t=0)$ eqn (4.12) can be integrated once to give the first order equation, equivalent to (4.36),

$$\mathbf{v}\times\mathbf{B}=\eta\nabla\times\mathbf{B}+c\nabla\psi$$

where ψ is an arbitrary function of position and time and $-\nabla\psi$ represents the electric field in the region. If we can pick our coordinate system to be at rest with respect to the fluid at infinity, then $\nabla\psi=0$ and

$$\mathbf{v}\times\mathbf{B}=\eta\nabla\times\mathbf{B}. \tag{4.37}$$

Finally we should point out that, if η is uniform and there is no fluid motion $(\mathbf{v}=0)$, the hydromagnetic equation (4.12) reduces to the vector heat flow equation (4.15). The Cartesian components of the field each diffuse separately, and the wealth of mathematical techniques developed to treat heat flow (see, for instance, Carslaw and Jaeger 1959) can be employed. The Green's function solution already mentioned is only one of several methods.

4.5. Resistive diffusion and magnetic lines of force

The convection and diffusion terms making up the right-hand side of (4.14) are in competition with each other. The convection term has a magnitude of the order of vB/l, if l is the smallest characteristic scale of the structure of $\mathbf{v}$ and $\mathbf{B}$. The diffusion term has a magnitude of the order of $\eta B/l^2$. The ratio of the magnitude of the convection term to the diffusion term is called the *magnetic Reynolds number*

$$R_m = vl/\eta. \qquad (4.38)$$

When R_m is large compared to one, the convection term dominates and the magnetic field is carried bodily with the fluid. The diffusion term is small and may be neglected unless it plays some crucial role in the discussion (as it does, for instance, in the hydromagnetic dynamo and turbulent diffusion).

The magnetic Reynolds number is very large compared to one in most astrophysical circumstances. Even over the small dimensions of a solar granule (10^3 km) in the low electrical conductivity of the solar photosphere ($\sigma = 10^{11}$ e.s.u., $\eta = 10^9\ \text{cm}^2\ \text{s}^{-1}$) the convective velocities of $1\ \text{km s}^{-1}$ yield $R_m = 10^4$. The solar wind ($v = 400\ \text{km s}^{-1}$, $\sigma \gtrsim 10^{14}$ e.s.u., $\eta \lesssim 10^6\ \text{cm}^2\ \text{s}^{-1}$) yields $R_m = 6 \times 10^{14}$ across a sector of width 1 a.u. = $1{\cdot}5 \times 10^{13}$ cm. The gaseous disc of the galaxy, of characteristic thickness $l = 100\ \text{pc} = 3 \times 10^{20}$ cm, gas velocity $v = 10\ \text{km s}^{-1}$, and electrical conductivity of 10^{11} e.s.u. or more ($\eta \lesssim 10^9\ \text{cm}^2\ \text{s}^{-1}$) yields $R_m \gtrsim 3 \times 10^{17}$.

In the presence of resistive diffusion it is generally not possible to attach a permanent identity to the individual magnetic lines of force since the field pattern changes and the connection of magnetic lines of force between material points varies with time. A simple example is a twisted rope of magnetic field in a motionless medium. Resistive diffusion causes the rope to unwind at a rate depending upon the distance ϖ from the axis of the rope, while the material particles making up the fluid do not move at all.

In those special cases wherein the field pattern remains fixed in time, because $\nabla \times (\mathbf{v} \times \mathbf{B})$ exactly balances the diffusion term $\nabla \times (\eta \nabla \times \mathbf{B})$, the field lines may be considered fixed, of course, with the fluid slipping across them with velocity $\mathbf{v}$ (see example in Roberts 1967). But only in the simplest cases is the geometrical concept of persistent magnetic lines useful.

4.6. The electric current and the electric field

This is the appropriate time to remark on our use of the simple scalar Ohm's law (4.7). In a solid or liquid, or in a dense gas in which the electronic collision frequency is larger than the electron cyclotron frequency, the electron drift is dominated by collisions and (4.7) is the correct description, with

$$\sigma = 10^7 T^{\frac{3}{2}}\ \text{s}^{-1}, \qquad \eta = 10^{13} T^{-\frac{3}{2}}\ \text{cm}^2\ \text{s}^{-1} \qquad (4.39)$$

(Cowling 1953, 1957; Spitzer 1956; Chapman and Cowling 1958). On the other hand, if the gas is so tenuous that the electron and ion collision frequencies are small compared to their cyclotron frequencies Ω, then their motion is approximately that of a free particle moving with velocity w in the large-scale fields $\mathbf{E}$ and $\mathbf{B}$. In the large-scale magnetic field $\mathbf{B}(l \gg w/\Omega)$ the particle motion is readily calculated in terms of the 'guiding centre approximation', yielding the result that the particles move approximately in the frame of reference with velocity $\mathbf{u}$ where (Alfven 1950)

$$\mathbf{u} = c\mathbf{E} \times \mathbf{B}/B^2. \tag{4.40}$$

The particles move across the magnetic lines of force with the 'electric' drift velocity $\mathbf{u}$ such that they see no induced electric field. This motion is, then, identical with the motion of a highly conducting fluid, in whose local frame of reference the electric field is zero.

In the tenuous collisionless plasma, an electric field *parallel* to the magnetic field is possible, in principle. It depends upon the anisotropy of the electron and ion distribution functions and may not be negligible in the presence of violent activity, but quickly decays through excitation of plasma turbulence when the activity ceases. Hence we shall consider the parallel component of $\mathbf{E}$ only in connection with special localized intense magnetic gradients (see below). For the present we disregard it.

To continue with the problem at hand, relative to the frame with velocity $\mathbf{u}$, the electrons and ions drift slightly as a consequence of the curvature in the magnetic lines of force and the gradient of the field density. The electric current resulting from this slow drift automatically satisfies Ampere's law (4.9) (see, for instance, Parker 1957). The vector product of $\mathbf{B}$ with (4.40) yields

$$\mathbf{E} = -\mathbf{u} \times \mathbf{B}/c + \mathbf{n}(\mathbf{n} \cdot \mathbf{E}),$$

where again $\mathbf{n}$ is the unit vector in the direction of $\mathbf{B}$. Since $\mathbf{n} \cdot \mathbf{E}$ is generally small, we neglect it. Then $\mathbf{E} = -\mathbf{u} \times \mathbf{B}/c$. Substitution into (4.2) yields

$$\partial \mathbf{B}/\partial t = \nabla \times (\mathbf{u} \times \mathbf{B}). \tag{4.41}$$

This is the same form as (4.16) indicating that the magnetic lines of force move with the velocity $\mathbf{u}$. There is no dissipation or diffusion and the system behaves in the same way as in a highly conducting fluid to which Ohm's law (4.7) applies. Thus the hydromagnetic equations describing large-scale $(l \gg w/\Omega)$, slowly varying $(l \gg u/\Omega)$ magnetic fields are the same whether the collision frequency is large or small compared to the cyclotron frequency (see discussion in Schlüter 1950, 1952; Chew *et al.* 1956; Watson 1956; Brueckner and Watson 1956; Parker 1957; Northrop 1963).

It does *not* follow, however, that (4.16) or (4.41) can be used in all intermediate cases. When the gas is both tenuous and contains a large portion of neutral atoms, as in the terrestrial ionosphere, the deviations from (4.16) are important (see, for instance, Chapman 1964; Hines *et al.* 1965; Paulikas 1974; Farley 1974; Akasofu 1974). We will restrict attention here to circumstances where (4.12), or its special cases (4.15) or (4.16) are applicable.

While we are discussing the validity of Ohm's law, it is important to note that in gases with a very low level of ionization, such as the solar photosphere, or the interior of a dense, cold interstellar cloud, the electrical conductivity may be very much reduced from the value (4.39) for a gas that is significantly, if not fully, ionized. The reduction is due to the lack of free charge carriers and may be considerable. The conductivity in the solar photosphere evidently dips as low as $10^9\ s^{-1}$ (see discussion in Nagasawa 1955; Kopecky 1957, 1958; Kopecky and Kuklin 1966; Oster 1968; Kopecky and Obridko 1968). In such cases we must be sure to include the large diffusion η.

It is interesting to note the important complications that may arise from plasma turbulence when the field gradient is very steep (high current density). Unstable resistive tearing modes (Furth *et al.* 1963; Jaggi 1963; Biskamp and Schindler 1971; Coppi and Friedland 1971) and unstable ion acoustic modes (Linhart 1960; Alfvén and Carlquist 1967; Hamberger and Friedman 1968; Hamberger and Jancarik 1970; Coppi and Mazzucato 1971; Burchenko *et al.* 1971; Coppi 1975) may be excited, enormously increasing the effective large-scale resistive diffusion coefficient η. Such effects probably occur in the aurora and the geomagnetic tail, and in solar flares, contributing to the violent dissipation of magnetic fields and the acceleration of particles. These interesting and important effects lie outside the scope of the hydromagnetic equations, except insofar as we introduce them *ad hoc* through an increased value of the effective resistive diffusion coefficient η.

We should note, too, the dissipative effect called *ambipolar diffusion* (Schlüter and Biermann 1950; Piddington 1954; Cowling 1956, 1957; Parker 1963), arising when there are relatively few ions and electrons embedded in a background of neutral gas. The magnetic field is tied to the electrons and ions, whose motion through the neutral gas is impeded by their collisions with neutral atoms. A force **F** per cm^3 exerted on the ions by the magnetic field causes them to drift through the neutral gas at a rate $\Delta\mathbf{v}$ such that

$$\mathbf{F} = M_i N_i \nu \Delta\mathbf{v}$$

where M_i is the ion mass, N_i the number of ions per cm^3, and ν is the rate at which the individual ion collides with the neutral atoms. The electrons, by virtue of their small mass, may be neglected here. If the ion–neutral collision cross-section is A and the mean thermal velocity is w, with N neutral atoms/cm^3, then $\nu = NAw$. Hence

$$\Delta\mathbf{v} = \mathbf{F}/M_i N_i N A w.$$

Osterbrock (1961) has pointed out that for the slow collisions of ions with neutrals, $Aw \cong 2{\cdot}5 \times 10^{-9}\ \text{cm}^3\ \text{s}^{-1}$ under most circumstances of interest. If a gradient (with scale l) in the magnetic pressure $B^2/8\pi$ is the cause of the force $\mathbf{F}$ exerted on the ions, we have $F = B^2/8\pi l$ and

$$\Delta v = B^2/8\pi l M_i N_i N A w. \tag{4.42}$$

In interstellar space, where $B^2/8\pi \simeq 0{\cdot}5 \times 10^{-12}$ dyn cm^{-2} and $l \cong 10^2$ pc $=$ 3×10^{20} cm, together with $M_i = 10^{-23}$ g, $N_i = 10^{-3}$, and $N = 1$, there results a drift of the order of 10^2 cm s^{-1}, traversing only 10 pc in 10^{10} years. In the solar photosphere where, say, $B = 10^3$ G, $l = 10^3$ km, $M_i = 10^{-23}$ g, and $N_i = 10^{-4} N$, with $N \cong 10^{17}$ cm^{-3}, we have $\Delta v = 10^{-2}$ cm s^{-1}. Ambipolar diffusion can be ignored under most astrophysical circumstances (see examples in Cowling 1957).

Let us return, then, to the electric field in a highly conducting fluid in which the electric current and electric field are related by the simple Ohm's law (4.7). The electric field in the frame of reference of the fluid is, according to (4.7) and (4.9)

$$\mathbf{E}' = (\eta/c)\nabla \times \mathbf{B}$$

Hence in a magnetic field of magnitude $\mathbf{B}$ and scale l, the electric field is of the order of $(\eta B/cl) = (vB/c)/R_m$. Thus, $\mathbf{E}'$ is smaller than the induced field (of order vB/c) in the fixed frame of reference by a factor comparable to the magnetic Reynolds number $R_m = vl/\eta$. The enormous magnitude of R_m reduces the electric field $\mathbf{E}'$ in the fluid to 'insignificant' levels in large-scale magnetic fields.

4.7. The charge density and the electric field

The electric field in the fixed frame of reference is given by (4.22). It satisfies the divergence condition (4.3). Hence there is an electrostatic charge density

$$\delta = -\nabla \cdot (\mathbf{v} \times \mathbf{B})/4\pi c \tag{4.43}$$

in the fluid. In terms of the electron cyclotron frequency $\Omega_e = eB/mc$ and the plasma frequency $\omega_p = (4\pi N_e e^2/m)^{\frac{1}{2}}$ where N_e is the number of electrons per cm^3, the charge density in a field variation of scale l is of the order of $vB/4\pi cl$, and the fractional change $\Delta N_e/N_e$ in electron density relative to the ion density is

$$\Delta N_e/N_e = (v/l)\Omega_e/\omega_p^2.$$

With $\Omega_e \cong 1{\cdot}8 \times 10^7 B$ rad s^{-1} and $\omega_p = 5 \times 10^4 N_e^{\frac{1}{2}}$ rad s^{-1} the fractional change is slight. Even in the extreme case of the solar wind impacting the magnetic field of Earth ($v \cong 400$ km s^{-1}, $l \cong 10^2$ km, $N_e \cong 1/\text{cm}^3$, $B \cong 10^{-4}$ G) we have $\Delta N_e/N_e \cong 3 \times 10^{-6}$. The charge on the individual electrons is so large that a slight rearrangement takes care of the induced electric field. The universe is saturated with electric charge, as we emphasized earlier.

The variation of the charge density from one frame of reference to another raises an interesting question that is worthy of brief notice. The charge density (4.43) can be written

$$\delta = (\mathbf{v} \cdot \nabla \times \mathbf{B} - \mathbf{B} \cdot \nabla \times \mathbf{v})/4\pi c.$$

In terms of the current density $4\pi\mathbf{j} = c\nabla \times \mathbf{B}$ and the vorticity $\boldsymbol{\omega} = \nabla \times \mathbf{v}$ this is

$$\delta = \frac{\mathbf{v}}{c} \cdot \frac{\mathbf{j}}{c} - \frac{\mathbf{B} \cdot \boldsymbol{\omega}}{4\pi c}. \tag{4.44}$$

If we view the fluid from a frame of reference moving with uniform velocity $\mathbf{u}$ relative to the fixed frame, the velocity of the fluid is $\mathbf{v}' = \mathbf{v} - \mathbf{u}$. The current density is unaffected, $\mathbf{j} = \mathbf{j}'[1 + O(v^2/c^2)]$. Replacing $\mathbf{v}$ in (4.44) by $\mathbf{v}'$ we calculate the charge density in the moving frame to be

$$\delta' = \delta + \mathbf{u} \cdot \mathbf{j}/c^2.$$

The two charge densities are not equal. How is it then that two observers with relative motion observe charge densities that differ by an amount first order in u/c? The answer is, of course, that the electric current implies a conduction velocity $\mathbf{w}$ of the electrons relative to the ions, which gives a Lorentz contraction of the electron density and automatically produces that part of the charge density given by $\mathbf{u} \cdot \mathbf{j}/c^2$. The effect is easily calculated in elementary terms. Denote by L_o the mean spacing of both electrons and ions (singly-charged) in the frame of the fluid. Then in the fixed frame of reference, relative to which the fluid velocity is $\mathbf{v}$, the mean ion spacing in the direction of fluid motion $\mathbf{v}$ is $L_i = L_o(1 - v^2/c^2)^{\frac{1}{2}}$. The mean spacing of the electrons is $L_e = L_o[1 - (\mathbf{v} + \mathbf{w})^2/c^2]^{\frac{1}{2}}$. Hence

$$L_e/L_i = 1 - \mathbf{v} \cdot \mathbf{w}/c^2 + O(w^2/c^2).$$

The spacing in the two directions perpendicular to $\mathbf{v}$ is unchanged. Hence the electron density is enhanced relative to the ion density by the factor

$$N_e/N_i = L_i/L_o \cong 1 + \mathbf{v} \cdot \mathbf{w}/c^2$$

so that there is a net charge

$$e(N_i - N_e) = -N_e e \mathbf{v} \cdot \mathbf{w}/c^2$$

in the fixed frame of reference. This net charge can be written $-\mathbf{v} \cdot \mathbf{j}/c^2$ since the current density is $\mathbf{j} = -N_e e w$. This is the origin of the first term on the right-hand side of (4.44). It follows that in the frame with velocity $\mathbf{u}$ the contribution is $-(\mathbf{v} - \mathbf{u}) \cdot \mathbf{j}/c^2$, giving the difference $\mathbf{u} \cdot \mathbf{j}/c^2$ that aroused our curiosity in the first place.

References

Akasofu, S. E. (1974). *Planet. Space Sci.* **22,** 885.
Alfvén, H. (1950) *Cosmical Electrodynamics* 1st. edn., Clarendon Press, Oxford.
Alfvén, H. and Carlquist, P. (1967). *Solar Phys.* **1,** 220.
—— and Falthammar, C.-G. (1963). *Cosmical electrodynamics*, 2nd edn. Clarendon Press, Oxford.
Biskamp, D. and Schindler, K. (1971). *Plasma Phys.* **13,** 1013.
Brueckner, K. A. and Watson, K. M. (1956). *Phys. Rev.* **102,** 19.
Burchenko, P. Y., Volkov, E. D., Rudakov, V. A., Sizanenko, V. L., and Stepanov, K. N. (1971). paper, C.N. 28/H-9, *Conference on plasma physics and controlled thermonuclear research, Madison, Wisconsin*, I.A.E.A. Vienna.
Carslaw, H. S. and Jaeger, J. C. (1959). *Conduction of heat in solids*, 2nd edn. Clarendon Press, Oxford.
Chandrasekhar, S. (1961). *Hydrodynamic and hydromagnetic stability*. Clarendon Press, Oxford.
Chapman, S. (1964). *Solar plasma, geomagnetism, and aurora*, Gordon and Breach, New York.
—— and Cowling, T. G. (1958). *Mathematical theory of nonuniform gases*, 2nd edn., p. 337. Cambridge University Press.
Chew, G. F., Goldberger, M. L., Low, F. E. (1956). *Proc. Roy. Soc. A* **236,** 112.
Coppi, B. (1975). *Astrophys. J.* **195,** 545.
—— and Friedland, A. B. (1971). *Astrophys. J.* **169,** 379.
—— and Mazzucato, E. (1971). *Phys. Fluids* **64,** 134.
Cowling, T. G. (1953). *The sun* (ed. G. P. Kuiper), chap. 8. University of Chicago Press.
—— (1956). *Mon. Notic. Roy. Astron. Soc.* **116,** 114.
—— (1957). *Magnetohydrodynamics.* Wiley-Interscience, New York.
Dungey, J. W. (1953). *Mon. Notic. Roy. Astron. Soc.* **113,** 679.
Elsasser, W. M. (1954). *Phys. Rev.* **95,** 1.
Farley, D. T. (1974). *Rev. Geophys. Space Phys.* **12,** 285.
Ferraro, V. C. A. and Plumpton, C. (1966). *An introduction to magneto-fluid mechanics.* Clarendon Press, Oxford.
Furth, H. P., Killeen, J. and Rosenbluth, M. N. (1963). *Phys. Fluids* **6,** 459.
Hamberger, S. M. and Friedman, M. (1968). *Phys. Rev. Lett.* **21,** 674.
Hamberger, S. M. and Jancarik, J. (1970). *Phys. Rev. Lett.* **25,** 999.
Hines, C. O., Paghis, I., Hartz, T. R., and Fejer, J. A. (1965). *Physics of the Earth's upper atmosphere.* Prentice-Hall, Englewood Cliffs, New Jersey.
Jaggi, R. K. (1963). *J. Geophys. Res.* **68,** 4429.
Jeffrey, A. (1966). *Magnetohydrodynamics.* Wiley-Interscience, New York.
Kendall, P. C. and Plumpton, C. (1964). *Magnetohydrodynamics.* Macmillan, New York.
Kopecky, M. (1957). *Bull. Astron. Inst. Czech.* **8,** 71.
—— (1958). *Electromagnetic phenomena in cosmical physics* (IAU Symposium No. 6) Cambridge, p. 513.
—— and Kuklin, C. V. (1966). *Bull. Astron. Inst. Czech.* **17,** 45.
—— and Obridko, V. (1968). *Solar Phys.* **5,** 354.
Landau, L. D. and Lifshitz, E. M. (1960). *Electrodynamics of continuous media.* Pergamon Press, New York.
Linhart, J. G. (1960). *Plasma physics* North-Holland, Amsterdam.
Lundquist, S. (1952). *Ark. Fys.* **5,** (15) 297.
Nagasawa, S. (1955). *Publ. Astron. Soc. Japan* **7,** 9.

NORTHROP, T. G. (1963). *The adiabatic motion of charged particles.* Wiley-Interscience, New York.
OSTER, L. (1968). *Solar Phys.* **3,** 593.
OSTERBROCK, D. E. (1961). *Astrophys. J.* **134,** 270.
PARKER, E. N. (1957). *Phys. Rev.* **107,** 924.
—— (1963). *Astrophys. J. Suppl.* **8,** 177.
PAULIKAS, G. A. (1974). *Rev. Geophys. Space Phys.* **12,** 117.
PIDDINGTON, J. H. (1954). *Mon. Notic. Roy. Astron. Soc.* **94,** 638, 651.
ROBERTS, P. H. (1967). *An introduction to magnetohydrodynamics.* American Elsevier, New York.
SCHLÜTER, A. (1950). *Z. Naturforsch.* A**5,** 72.
—— (1952). *Ann. Phys.* **10,** 422.
—— and BIERMANN, L. (1950). *Z. Naturforsch.* **52,** 237.
SPITZER, L. (1956). *The physics of fully ionized gases.* Wiley-Interscience, New York.
STERN, D. P. (1966). *Space Sci. Rev.* **6,** 147.
—— (1976). *Rev. Geophys. Space Phys.* **14,** 199.
SWEET, P. A. (1950). *Mon. Notic. Roy. Astron. Soc.* **110,** 69.
WATSON, K. M. (1956). *Phys. Rev.* **102,** 12.

5

MAGNETIC FIELD STRESS AND ENERGY

5.1. Magnetic field stresses

THE magnetic field transmits stresses between regions of material particles and fluids. It is these stresses that are responsible for the non-equilibrium and the continual activity of fluids in astrophysical bodies. The stresses are derivable directly from the expression

$$\mathbf{f}' = q\mathbf{E}' \tag{5.1}$$

for the force transmitted to an electric charge q by the electromagnetic fields $\mathbf{E}'$ and $\mathbf{B}'$ in the frame of the charge. In terms of the electric field $\mathbf{E}$ and magnetic field $\mathbf{B}$ in the fixed frame, relative to which the charge q has a velocity $\mathbf{w}$, we have from (4.5) that

$$\mathbf{f}' = q\gamma(\mathbf{E}+\mathbf{w}\times\mathbf{B}/c) \tag{5.2}$$

where for convenience we have written $\gamma=(1-w^2/c^2)^{-\frac{1}{2}}$. The force $\mathbf{f}'$ is the momentum imparted per unit time in the frame of q. The momentum imparted per unit time in the fixed frame is denoted by $\mathbf{f}$. It is smaller than $\mathbf{f}'$ because of the time dilatation $\Delta t=\gamma\Delta t'$ between the two frames. Hence

$$\mathbf{f}=\mathbf{f}'/\gamma \tag{5.3}$$

$$=q(\mathbf{E}+\mathbf{w}\times\mathbf{B}/c) \tag{5.4}$$

This is, of course, the familiar expression for the Lorentz force on the charge q.

The fluid is made up of many particles with various individual charges q, together giving a net charge density δ per unit volume[1]. In the same way the net current density $\mathbf{j}(\mathbf{r}, t)$ is the net density of $q\mathbf{w}$ per unit volume. The net force $\mathbf{F}$ per unit volume is, then,

$$\mathbf{F}=\delta\mathbf{E}+\mathbf{j}\times\mathbf{B}/c \tag{5.5}$$

[1] If there are random variations in the charge density, as a consequence of small-scale fluctuations in particle density, we imagine an ensemble of systems, all statistically identical on the macroscopic scales with which we are dealing here, but statistically independent on the microscopic scale. The mean charge density $\delta(\mathbf{r}, t)$ is then taken to be the ensemble average charge density at the position $\mathbf{r}$ and time t.

on the fluid. The force is readily transformed into the divergence of a stress tensor, M_{ij}, representing the stresses carried in the field. We use (4.3) to eliminate δ and (4.1) to eliminate $\mathbf{j}$, so that

$$4\pi\mathbf{F}=\mathbf{E}\nabla\cdot\mathbf{E}+(\nabla\times\mathbf{B})\times\mathbf{B}-(\partial\mathbf{E}/\partial t)\times\mathbf{B}/c.$$

In view of (4.4) we may add the term $\mathbf{B}\nabla\cdot\mathbf{B}$, to keep the expression symmetric in $\mathbf{E}$ and $\mathbf{B}$. We may write the last term as

$$(\partial\mathbf{E}/\partial t)\times\mathbf{B}=\partial(\mathbf{E}\times\mathbf{B})/\partial t-\mathbf{E}\times\partial\mathbf{B}/\partial t$$

and then use (4.2) to write $\partial\mathbf{B}/\partial t$ in terms of $\mathbf{E}$. The result is

$$\begin{aligned}4\pi\mathbf{F}&=\mathbf{E}\nabla\cdot\mathbf{E}+\mathbf{B}\nabla\cdot\mathbf{B}+(\nabla\times\mathbf{B})\times\mathbf{B}+(\nabla\times\mathbf{E})\times\mathbf{E}-\partial(\mathbf{E}\times\mathbf{B}/c)/\partial t,\\&=\mathbf{E}\nabla\cdot\mathbf{E}+(\mathbf{E}\cdot\nabla)\mathbf{E}-\nabla\tfrac{1}{2}E^2+\mathbf{B}\nabla\cdot\mathbf{B}+(\mathbf{B}\cdot\nabla)\mathbf{B}-\nabla\tfrac{1}{2}B^2-\partial(\mathbf{E}\times\mathbf{B}/c)/\partial t,\end{aligned}$$

upon using a vector identity for $(\nabla\times\mathbf{B})\times\mathbf{B}$ and $(\nabla\times\mathbf{E})\times\mathbf{E}$. In terms of the Poynting vector

$$\mathbf{Q}=c\mathbf{E}\times\mathbf{B}/4\pi \tag{5.6}$$

and the Maxwell stress tensor

$$M_{ij}=-\delta_{ij}(E^2+B^2)/8\pi+(E_iE_j+B_iB_j)/4\pi \tag{5.7}$$

the expression can be written, using index notation exclusively[2]

$$F_i+\frac{\partial}{\partial t}\frac{Q_i}{c^2}=\frac{\partial M_{ij}}{\partial x_j}. \tag{5.8}$$

The Maxwell stress tensor represents the stress transmitted through the electromagnetic field. We use the convention that M_{ij} represents the force (per unit area) in the i-direction. The force is exerted *by* the field on the *positive* side of an element of area (with normal in the j-direction) *on* the field on the *negative* side. Hence pressure is negative and tension is positive. The stress across an element of area $\mathrm{d}S_j$ is $M_{ij}\,\mathrm{d}S_j$, and it is readily shown from Gauss's theorem that the force per unit volume diverted from the field to the material particles is the divergence $\partial M_{ij}/\partial x_j$.

[2] Had we restricted $\mathbf{j}$ to be only the electric conduction current, writing $4\pi\mathbf{j}+\partial\mathbf{D}/\partial t=c\nabla\times\mathbf{H}$ and $\partial\mathbf{B}/\partial t=-c\nabla\times\mathbf{E}$, as is the convention of many authors, with $\mathbf{H}$ limited to that part of $\mathbf{B}$ associated with the conduction current $\mathbf{j}$ (see footnote 1, Chapter 4), the expression for $4\pi\mathbf{F}$ becomes $\mathbf{E}\nabla\cdot\mathbf{D}+\mathbf{H}\nabla\cdot\mathbf{B}+(\nabla\times\mathbf{H})\times\mathbf{B}+(\nabla\times\mathbf{E})\times\mathbf{D}-(\partial/\partial t)\mathbf{D}\times\mathbf{B}/c$, and further progress requires knowing the relations between $\mathbf{E}$ and $\mathbf{D}$, $\mathbf{H}$ and $\mathbf{B}$. With the simple scalar relations $\mathbf{D}=\epsilon\mathbf{E}$ and $\mathbf{B}=\mu\mathbf{H}$ the field pressure becomes $(\epsilon E^2+\mu H^2)/8\pi$. This differs from the $(E^2+B^2)/8\pi$ worked out above because it includes the stresses within the matter. We prefer in our formulation to separate the stresses in the macroscopic fields from the elastic stresses (microscopic fields) in the material medium.

The stress tensor M_{ij} is necessarily symmetric in its two indices, because any non-vanishing anti-symmetric part represents an unbalanced torque on each infinitesimal element of volume.

The momentum density in the electromagnetic field is Q_i/c^2.

The electromagnetic stress tensor M_{ij}, (5.7), is symmetric in E_i and B_i because the field equations are symmetric in E_i and B_i. The first term on the right-hand side of (5.7) represents an isotropic pressure. The second term represents a tension $E^2/4\pi$ along the electric lines of force and a tension $B^2/4\pi$ along the magnetic lines of force. To demonstrate that $B_iB_j/4\pi$ represents a tension along B_i, consider an element of area $n_i\,\mathrm{d}S$ with its normal along the magnetic field B_i, where again n_i is the unit vector B_i/B along B_i. Then the stress exerted across $\mathrm{d}S_i$ is

$$\begin{aligned}\mathrm{d}S\, n_j B_i B_j/4\pi &= B_i B_j B_j\, \mathrm{d}S/4\pi B \\ &= n_i\, \mathrm{d}S\, B^2/4\pi\end{aligned}$$

which is a tension (because it is positive) of density $B^2/4\pi$ in the direction of the local field n_i. The force of $B_iB_j/4\pi$ across an element of area with its normal perpendicular to B_i is zero. A similar analysis holds for the tension $E_iE_j/4\pi$ along the electric field.

Now in a highly conducting fluid with a velocity v_i (everywhere small compared to c) we have E_i smaller than B_i by the ratio v/c. Hence, neglecting all terms second order in v/c, the Maxwell stress tensor reduces to

$$M_{ij} \cong -\delta_{ij}B^2/8\pi + B_iB_j/4\pi. \tag{5.9}$$

The magnetic field exerts an isotropic pressure $B^2/8\pi$ in all three directions, and carries a tension $B^2/4\pi$ along the magnetic lines of force. Once again we see the prominent role played by the magnetic lines of force. Each small flux tube is like a rubber band, under tension, and infinitely elastic. Neighbouring flux tubes expand against each other with a pressure $B^2/8\pi$. Equilibrium exists only where it is possible to balance the tension and pressure against each other.

The electromagnetic momentum density Q_i/c^2 can be neglected compared to the momentum density ρv_i of the fluid because their ratio is of the order of

$$\frac{|Q_i/c^2|}{|\rho v_i|} \cong \frac{EB}{4\pi\rho v c} = \frac{B^2}{4\pi\rho c^2}$$

and is equal to the field energy density divided by the rest energy density of the fluid. The characteristic dynamical pressures of fluid motions resulting from unbalanced stresses in B_i are comparable to the magnetic pressure. So $\frac{1}{2}\rho v^2 \approx B^2/8\pi$ and $v \approx B/(4\pi\rho)^{\frac{1}{2}}$. Hence the ratio is of the order of v^2/c^2, which we neglect. Another way of looking at the problem is to note that the magnetic field pressure $B^2/8\pi$ is often not greater than the gas pressure p written ρw^2 in terms of the mean square thermal velocity w in any one direction. The ratio is then of the order of w^2/c^2, which we neglect.

The equation of motion for the fluid can be written directly from (5.8). A fluid with velocity v_i, density ρ, and pressure p subject to a gravitational force $-\rho\partial\theta/\partial x_i$ and the magnetic force $F_i = \partial M_{ij}/\partial x_j$ accelerates at a rate

$$\rho\frac{dv_i}{dt} \equiv \rho\left(\frac{\partial v_i}{\partial t} + v_j\frac{\partial v_i}{\partial x_j}\right)$$

$$= -\frac{\partial p}{\partial x_i} - \rho\frac{\partial \theta}{\partial x_i} + \frac{\partial M_{ij}}{\partial x_j}. \tag{5.10}$$

The reader can add the standard viscous stresses to this equation when the situation demands it (see, for instance, Landau and Lifshitz 1959; Roberts 1967).

The magnetic force F_i per unit volume can be written in a variety of ways,

$$\mathbf{F} = -\nabla B^2/8\pi + (\mathbf{B}\cdot\nabla)\mathbf{B}/4\pi \tag{5.11}$$

$$= (\nabla\times\mathbf{B})\times\mathbf{B}/4\pi \tag{5.12}$$

$$= \mathbf{j}\times\mathbf{B}/c, \tag{5.13}$$

the last form to be compared to (5.5). The electrostatic forces $\delta\mathbf{E}$ are negligible, being second order in v/c compared to $\mathbf{j}\times\mathbf{B}/c$.

It is evident from (5.12) or (5.13) that the magnetic field exerts no forces on the fluid in the direction parallel to the field. The tension in the field is always in equilibrium along the magnetic lines of force because there is no coupling between field and fluid in that direction. Only if there were magnetic monopoles could there be a component of $\mathbf{F}$ along $\mathbf{B}$.

The magnetic stresses are carried through the region entirely within the field, and not transmitted to the fluid, whenever $\nabla\times\mathbf{B} = 0$ (i.e. $\mathbf{B} = -\nabla\psi$) or $\nabla\times\mathbf{B} = h(\mathbf{r})\mathbf{B}$. Such a field is said to be *force-free*, i.e. $\mathbf{F} = 0$.

The field exerts a force on the fluid in a direction perpendicular to the field whenever the field pattern is distorted from the condition of local equilibrium of the Maxwell stresses, $(\nabla\times\mathbf{B})\times\mathbf{B} = 0$. Then if the magnetic lines of force have a local radius of curvature R, their tension $B^2/4\pi$ exerts a transverse force $B^2/4\pi R$ per unit volume. Insofar as this curvature stress is not balanced by the gradient of the magnetic pressure $B^2/8\pi$, there is a force exerted on the fluid.

It is the field stress M_{ij} that tugs at the magnets held in our hand, that holds the refrigerator door firmly closed, and that pushes the armature of an electric motor around on its shaft. It is the same field stress that propels the eruptive prominence on the sun, and confines the fast particles of the Van Allen radiation belts around Earth, and restricts the motions of the interstellar gas.

Generally speaking the tension along the magnetic lines of force of most field line topologies cannot be balanced by gas pressure, so there can be no equilibrium. The fluid acceleration dv_i/dt vanishes only with the vanishing of the magnetic field B_i. Hydrostatic equilibrium with non-vanishing B_i is a figment of the highly symmetric, and hence unlikely, idealized examples commonly employed by theoreticians.

5.2. Magnetic field energy

The rate at which the electromagnetic field does work on a charge q with velocity $\mathbf{w}$ is

$$\mathbf{w}\cdot\mathbf{f}=q\mathbf{w}\cdot\mathbf{E}$$

where $\mathbf{f}$ is the Lorentz force (5.4). The sum (ensemble average) over all particles gives a mean local rate per unit volume

$$\mathrm{d}W/\mathrm{d}t=\mathbf{j}\cdot\mathbf{E}(\mathrm{erg\ cm^{-3}\ s^{-1}}). \tag{5.14}$$

With the help of (4.1) and then (4.2) this can be transformed into

$$\begin{aligned}4\pi\,\mathrm{d}W/\mathrm{d}t&=c\mathbf{E}\cdot\nabla\times\mathbf{B}-\mathbf{E}\cdot\partial\mathbf{E}/\partial t\\&=-\nabla\cdot(c\mathbf{E}\times\mathbf{B})+c\mathbf{B}\cdot\nabla\times\mathbf{E}-\partial\tfrac{1}{2}E^2/\partial t\\&=-\nabla\cdot(c\mathbf{E}\times\mathbf{B})-\partial\tfrac{1}{2}(E^2+B^2)/\partial t.\end{aligned} \tag{5.15}$$

In terms of the Poynting vector Q_i, defined in (5.6), we have, then, the relation

$$\frac{\mathrm{d}W}{\mathrm{d}t}+\frac{\partial}{\partial t}\frac{E^2+B^2}{8\pi}+\frac{\partial Q_i}{\partial x_i}=0, \tag{5.16}$$

expressing the interchange of energy between particles and field. The Poynting vector represents the energy flux in the electromagnetic field so that its divergence is the local rate of energy deposition. The energy density of the field is $(E^2+B^2)/8\pi$, and $\mathrm{d}W/\mathrm{d}t$ is the rate at which energy is transferred to the particles[3].

In the presence of a highly conducting fluid the electric field energy is second order in w/c and can be neglected compared to the magnetic

[3] Again had we restricted $\mathbf{j}$ to be only the electric conduction current[2] (see footnote 1, Chapter 4), then the Poynting vector becomes $c\mathbf{E}\times\mathbf{H}/4\pi$ and the energy densities of the fields are $\mathbf{E}\cdot\mathbf{D}/8\pi$ and $\mathbf{H}\cdot\mathbf{B}/8\pi$. The potential energy of the electric polarization and the atomic magnetization are then included in the field energy, whereas in our formulation, where $\mathbf{j}$ represents the total of all electric currents, these energies are placed with the particles.

energy. Then with **E** given by (4.22) it follows that the energy flux in the field is

$$\begin{aligned} \mathbf{Q} &= \mathbf{B}\times(\mathbf{v}\times\mathbf{B})/4\pi \\ &= [\mathbf{v}B^2-\mathbf{B}(\mathbf{v}\,.\,\mathbf{B})]/4\pi \\ &= B^2[\mathbf{v}-\mathbf{n}(\mathbf{n}\,.\,\mathbf{v})]/4\pi \\ &= \mathbf{v}_\perp B^2/4\pi \end{aligned} \tag{5.17}$$

where again $\mathbf{n}=\mathbf{B}/B$ is the unit vector in the direction of the magnetic field and $\mathbf{v}_\perp$ is the component of fluid velocity perpendicular to **B**. The Poynting vector is, then, merely the convection of magnetic enthalpy in the conducting fluid: The convection of magnetic energy contributes $\mathbf{v}_\perp B^2/8\pi$, and the magnetic pressure $B^2/8\pi$ of the field behind any surface moving with the fluid does work on the fluid ahead at a rate $\mathbf{v}_\perp B^2/8\pi$ erg cm^{-2} s^{-1}, so that the total transfer of energy is the sum, (5.17)[4].

The magnetic energy density $B^2/8\pi$ is directly related to the stress density of the field, of course. If, for instance, the fluid is displaced in opposition to the magnetic stresses, then the work done by the fluid against the stresses is just equal to the increase in field energy. This is easily demonstrated from (4.16), which can be written

$$\partial B_i/\partial t+v_j\partial B_i/\partial x_j=B_j\partial v_i/\partial x_j-B_i\partial v_j/\partial x_j.$$

Multiply by $B_i/4\pi$ and sum on i. The terms can be rearranged to give the energy equation

$$\begin{aligned} \frac{\partial}{\partial t}\frac{B^2}{8\pi}+\frac{\partial}{\partial x_i}\left(v_i\frac{B^2}{8\pi}\right) &= \frac{B_iB_j}{4\pi}\frac{\partial v_i}{\partial x_j}-\frac{B^2}{8\pi}\frac{\partial v_i}{\partial x_j} \\ &= M_{ij}\partial v_i/\partial x_j. \end{aligned} \tag{5.18}$$

The right-hand side is the rate of production of magnetic field energy $B^2/8\pi$ per unit volume. The first term represents the energy input due to stretching of the magnetic lines of force (in opposition to the tension $B^2/4\pi$) and the second term represents compression of the field (in opposition to the isotropic pressure $B^2/8\pi$). If we integrate (5.18) over the entire system, enclosed by the surface S on which v_i and/or B_i vanish, then with the aid of Gauss's theorem and the vanishing surface integrals, we have

$$\frac{\mathrm{d}}{\mathrm{d}t}\int_S \mathrm{d}^3\mathbf{r}\,\frac{B^2}{8\pi}=-\int \mathrm{d}^3\mathbf{r}\,v_i\frac{\partial M_{ij}}{\partial x_j} \tag{5.19}$$

[4] A detailed discussion of the energy, including the energy of the particles, may be found in Landau and Lifshitz (1959) and Roberts (1967).

which states that the rate of increase of magnetic energy is precisely equal to the rate at which the fluid velocity v_i works against the magnetic forces on the fluid.

Altogether, then, it is the magnetic energy density $B^2/8\pi$ that powers the magnetic forces. It is the magnetic energy in the electromagnet that must be dissipated before the field can disappear and the electric current cease to flow. It is the magnetic energy of an active region on the sun that fuels the solar flare. The magnetic stresses are of such a form, involving tension as well as an isotropic pressure, that magnetic energy can be continually generated and released.

5.3. Global properties of magnetic stress

The magnetic stress M_{ij} is the cause of so much activity and non-equilibrium in the universe because a magnetic field in an astrophysical body cannot help but exert strong forces on the fluid. We show in a later chapter that the forces cannot generally be balanced by fluid pressure and gravity. To demonstrate the necessity for strong forces between the magnetic field and the materials, consider the net stress exerted on the material particles in a volume V containing a magnetic field $B_i(\mathbf{r})$. The most effective way to get at the dynamical problem is through the first moment of the momentum equation, (5.10), the so-called virial equations (Chandrasekhar and Fermi 1953; Parker 1953, 1954, 1969; Chandrasekhar 1961, 1969). Imagine, then, a fluid with density ρ and pressure p, subject to the magnetic force $\partial M_{ij}/\partial x_j$ and a general force (gravity, etc.) $\mathscr{F}_i$ per unit volume. The equation of motion is

$$\rho\left(\frac{\partial v_i}{\partial t}+v_j\frac{\partial v_i}{\partial x_j}\right)=-\frac{\partial p}{\partial x_i}+\frac{\partial M_{ij}}{\partial x_j}+\mathscr{F}_i.$$

The equation for conservation of mass is

$$\frac{\partial \rho}{\partial t}+\frac{\partial}{\partial x_i}\rho v_i=0.$$

Multiply this equation by v_i and add it to the momentum equation, obtaining the familiar result

$$\frac{\partial}{\partial t}\rho v_i+\frac{\partial}{\partial x_j}\rho v_i v_j=-\frac{\partial p}{\partial x_i}+\frac{\partial M_{ij}}{\partial x_j}+\mathscr{F}_i \tag{5.20}$$

for the momentum density ρv_i. Multiply this equation by x_k and add it to the same equation with i and k interchanged to give the symmetric part[5].

[5] The anti-symmetric part, obtained by subtracting the two equations with i and k interchanged, represents the angular momentum equation and is of little interest in the present context.

Then integrate over the fixed volume V. The result can be written

$$\frac{\mathrm{d}}{\mathrm{d}t}\int_V \mathrm{d}^3\mathbf{r}\rho(x_i v_k + x_k v_i) + \int_V \mathrm{d}^3\mathbf{r}\left[x_k \frac{\partial}{\partial x_j}\rho v_i v_j + x_i \frac{\partial}{\partial x_j}\rho v_k v_j\right]$$
$$= -\int_V \mathrm{d}^3\mathbf{r}\left(x_k \frac{\partial p}{\partial x_i} + x_i \frac{\partial p}{\partial x_k}\right) + \int_V \mathrm{d}^3\mathbf{r}\left(x_k \frac{\partial M_{ij}}{\partial x_j} + x_i \frac{\partial M_{kj}}{\partial x_j}\right)$$
$$+ \int_V \mathrm{d}^3\mathbf{r}(x_k \mathscr{F}_i + x_i \mathscr{F}_k). \quad (5.21)$$

To manipulate this equation into the desired form[6] we note that the time derivative of the moment of inertia tensor

$$I_{ij} \equiv \int_V \mathrm{d}^3\mathbf{r}\rho x_i x_j \quad (5.22)$$

is

$$\frac{\mathrm{d}I_{ik}}{\mathrm{d}t} = \int_V \mathrm{d}^3\mathbf{r} x_i x_k \frac{\partial \rho}{\partial t},$$
$$= -\int_V \mathrm{d}^3\mathbf{r} x_i x_k \frac{\partial}{\partial x_j}\rho v_j,$$
$$= -\int_V \mathrm{d}^3\mathbf{r} \frac{\partial}{\partial x_j} x_i x_k \rho v_j + \int_V \mathrm{d}^3\mathbf{r}\rho(x_k v_i + x_i v_k).$$

We use Gauss's theorem to reduce the first volume integral to a surface integral and then assume that there is no fluid motion across the surface of V. It follows that

$$\frac{\mathrm{d}I_{ik}}{\mathrm{d}t} = \int_V \mathrm{d}^3\mathbf{r}\rho(x_k v_i + x_i v_k). \quad (5.23)$$

Hence the first integral on the left-hand side of (5.21) is just $\mathrm{d}^2 I_{ik}/\mathrm{d}t^2$.

The second integral on the left-hand side of (5.21) can be rewritten, after an integration by parts and application of Gauss's theorem, in terms of the kinetic tensor T_{ij}, defined as the volume integral of the Reynolds stresses $\rho v_i v_j$,

$$T_{ik} = \int_V \mathrm{d}^3\mathbf{r}\, \tfrac{1}{2}\rho v_i v_j, \quad (5.24)$$

[6] It is mathematically easier to consider the fluid as an aggregate of individual particles, summing the left-hand side over the individual particles instead of integrating over a continuous volume. But we must remember in that case that the kinetic pressure p appears in the Reynolds stress on the left-hand side, instead of being inserted on the right.

Its trace is the total kinetic energy T and the diagonal components T_{11} etc. represent the kinetic energy of the motions in the various directions. The result is

$$\int_V \mathrm{d}^3\mathbf{r}\left(x_k \frac{\partial}{\partial x_j}\rho v_i v_j + x_i \frac{\partial}{\partial x_j}\rho v_i v_k\right) = -4T_{ik} \tag{5.25}$$

The right-hand side can be rewritten after integrating by parts and applying Gauss's theorem to give

$$\frac{1}{2}\frac{\mathrm{d}^2 I_{ik}}{\mathrm{d}t^2} - 2T_{ik} = \int_V \mathrm{d}^3\mathbf{r}\{\delta_{ik}p - M_{ik} + \tfrac{1}{2}(x_k\mathcal{F}_i + x_i\mathcal{F}_k)\}$$
$$+\frac{1}{2}\int_S \mathrm{d}S_j\{-p(\delta_{ij}x_k + \delta_{kj}x_i) + x_k M_{ij} + x_i M_{kj}\} \tag{5.26}$$

where the integral $\int \mathrm{d}S_j$, is over the surface enclosing V. This is the desired tensor virial equation (Parker 1969).

As a first example consider the trace of this equation for an isolated system for which all motions and fields vanish on the enclosing surface S. Put $i = k$ and sum on i,

$$\frac{1}{2}\frac{\mathrm{d}^2 I}{\mathrm{d}t^2} = 2T + 3\int_V \mathrm{d}^3\mathbf{r}p + \int_V \mathrm{d}^3\mathbf{r}\frac{B^2}{8\pi} + \int_V \mathrm{d}^3\mathbf{r}x_i\mathcal{F}_i \tag{5.27}$$

where $I = I_{ii}$, $T = T_{ii}$, and $M_{ii} = -B^2/8\pi$. The volume integral of the fluid pressure is positive, and, for an ideal gas, can be written in terms of the kinetic energy of the mean square thermal velocity $\langle u^2\rangle$ as $p = \frac{1}{3}\rho\langle u^2\rangle$. Hence in terms of the total thermal kinetic energy T_t, we have

$$\begin{aligned} 3\int_V \mathrm{d}^3\mathbf{r}p &= \int_V \mathrm{d}^3\mathbf{r}\rho\langle u^2\rangle, \\ &= 2T_t. \end{aligned} \tag{5.28}$$

The volume integral of $B^2/8\pi$ is just the total magnetic energy $\mathcal{M}$ of the system,

$$\mathcal{M} \equiv \int_V \mathrm{d}^3\mathbf{r}B^2/8\pi. \tag{5.29}$$

Altogether, then,

$$\frac{1}{2}\frac{\mathrm{d}^2 I}{\mathrm{d}t^2} = 2(T + T_t) + \mathcal{M} + \int_V \mathrm{d}^3\mathbf{r}x_i\mathcal{F}_i. \tag{5.30}$$

It is clear that T, T_t, and $\mathcal{M}$ are all positive quantities because they depend upon the square of the velocity and magnetic fields. Hence the fluid motion T, the gas pressure T_t, and the magnetic field $\mathcal{M}$ all contribute to expansion ($\mathrm{d}^2 I/\mathrm{d}t^2 > 0$). The magnetic field with its tension $B^2/4\pi$ in the one direction along the magnetic line of force, and its isotropic pressure $B^2/8\pi$ in all three dimensions, is basically expansive, with the net effect equal to the total over the three dimensions,

$$1\times\frac{B^2}{4\pi}-3\times\frac{B^2}{8\pi}=-\frac{B^2}{8\pi},$$

representing a net pressure (Parker 1957). It follows, then, that there must be inward forces exerted on the system if it is to be maintained in equilibrium, $\mathrm{d}^2I/\mathrm{d}t^2=0$ (Chandrasekhar and Fermi 1953; Parker 1958, Biermann and Davis 1960). The inward forces could be external in origin, i.e. an external pressure exerted on the surface of the system. In many cases of astrophysical interest the external pressures are negligible and the confining force is internal and gravitational. The fundamental point here is that a magnetic field necessarily exerts outward forces on the body in which it is embedded; the magnetic field must have inward forces exerted on it by the material in the body. The magnetic field cannot contain its own stresses. The stress tensor M_{ij} must somewhere in the body, have non-vanishing divergence where inward forces in the amount

$$\int_V \mathrm{d}^3\mathbf{r}\,x_iF_i=-\mathcal{M} \tag{5.31}$$

are exerted upon the magnetic field. The nature and consequences of the strong forces between field and matter are the subject of the remainder of this monograph. One important aspect of the forces is that they are anisotropic as a consequence of the tension along the lines of force. Consider, then the individual diagonal components of (5.26).

5.4. Magnetic stress in one and two dimensions

The component of (5.26) for $i=k=3$ is

$$\frac{1}{2}\frac{\mathrm{d}^2I_{33}}{\mathrm{d}t^2}=T_{33}+\int_V \mathrm{d}^3\mathbf{r}\left(p+\frac{B_1^2+B_2^2-B_3^2}{8\pi}+x_3\mathcal{F}_3\right)$$
$$-\int_S \mathrm{d}S_3x_3\left(p+\frac{B^2}{8\pi}\right)+\int_S \mathrm{d}S_jB_jx_3\frac{B_3}{4\pi}. \tag{5.32}$$

If the normal component of magnetic field vanishes at the surface, then $\mathrm{d}S_jB_j=0$ and the last term vanishes. Denote by angular brackets with a subscript three the mean value in the direction $i=3$ across the volume for fixed x_1 and x_2. Thus,

$$\langle p\rangle_3=\frac{1}{x_3^{(2)}-x_3^{(1)}}\int_{x_3^{(1)}}^{x_3^{(2)}}\mathrm{d}x_3p \tag{5.33}$$

where $x_3^{(1)}$ and $x_3^{(2)}(x_3^{(2)} > x_3^{(1)})$ denote the x_3 coordinates of the enclosing surface at fixed (x_1, x_2). It follows that

$$\begin{aligned}\int_V \mathrm{d}^3\mathbf{r}p &\equiv \int \mathrm{d}x_1 \int \mathrm{d}x_2 \int \mathrm{d}x_3 p, \\ &= \int \mathrm{d}x_1 \int \mathrm{d}x_2 \langle p\rangle_3 (x_3^{(2)} - x_3^{(1)}), \\ &= \int \mathrm{d}S_3 x_3 \langle p\rangle_3. \end{aligned} \tag{5.34}$$

A similar result follows for B_1^2 etc.

Then if we denote by the subscript s the value of a quantity at the surface, (5.32) can be written

$$\frac{1}{2}\frac{\mathrm{d}^2 I_{33}}{\mathrm{d}t^2} = T_{33} + \int_S \mathrm{d}S_3 x_3 \left[\left\{\langle p\rangle_3 + \frac{\langle B_1^2\rangle_3 + \langle B_2^2\rangle_3}{8\pi}\right\} - \left\{p_s + \frac{(B_1^2)_s + (B_2^2)_s}{8\pi}\right\} - \frac{\langle B_3^2\rangle_3 - (B_3^2)_s}{8\pi}\right] + \int_V \mathrm{d}^3\mathbf{r} x_3 \mathscr{F}_3. \tag{5.35}$$

It follows that the internal excess of $p + (B_1^2 + B_2^2)/8\pi$ over its value at the surface causes expansion, in keeping with the fact that the magnetic field components perpendicular to $i = 3$ exert pressure. The excess tension $B_3^2/8\pi$ over pressure in the $i = 3$ direction causes contraction. Thus the two opposite effects of magnetic pressure and tension can be brought into balance in any one direction, but, as was shown in §5.3, not in all three.

If we now imagine a field confined to a right cylinder extending parallel to the $i = 3$ axis, and suppose that the normal component of field vanishes on the lateral sides but not on the plane ends, then $\mathrm{d}S_3 B_3 \neq 0$ across the ends. The surface integral in (5.32) is

$$\int \mathrm{d}S_j B_j x_3 B_3 / 4\pi = \int \mathrm{d}S_3 x_3 B_3^2 / 4\pi$$

and we have

$$\frac{1}{2}\frac{\mathrm{d}^2 I_{33}}{\mathrm{d}t^2} = T_{33} + \int_S \mathrm{d}S_3 x_3 \left[\left\{\langle p\rangle_3 + \frac{\langle B_1^2\rangle_3 + \langle B_2^2\rangle_3 - \langle B_3^2\rangle_3}{8\pi}\right\} - \left\{p_s + \frac{(B_1^2)_s + (B_2^2)_s - (B_3^2)_s}{8\pi}\right\}\right] + \int \mathrm{d}^3\mathbf{r} x_3 \mathscr{F}_3. \tag{5.36}$$

We see that the internal tension and the tension across the surface are in direct competition with one another. The net force exerted on, or by, the magnetic field is the difference between the average internal value of $p - M_{33} = p + (B_1^2 + B_2^2 - B_3^2)/8\pi$ and its value at the surface. Again, we can arrange for the net force to be zero in the $i = 3$ direction.

Consider, then, the net effect in the other two dimensions, $i = 1, 2$. From (5.26)

$$\begin{aligned}\frac{1}{2}\frac{\mathrm{d}^2}{\mathrm{d}t^2}(I_{11}+I_{22}) = T_{11} + T_{22} &+ 2\int_V \mathrm{d}^3\mathbf{r}\left(p + \frac{B_3^2}{8\pi}\right) \\ &+ \int_V \mathrm{d}^3\mathbf{r}(x_1\mathscr{F}_1 + x_2\mathscr{F}_2) \\ &+ \int \mathrm{d}S_j B_j \frac{x_1 B_1 + x_2 B_2}{4\pi} \\ &- \int_S \mathrm{d}S_1 x_1\left(p + \frac{B^2}{8\pi}\right) - \int \mathrm{d}S_2 x_2\left(p + \frac{B^2}{8\pi}\right).\end{aligned} \tag{5.37}$$

In the absence of external pressures, the surface integrals vanish and

$$\begin{aligned}\frac{1}{2}\frac{\mathrm{d}^2}{\mathrm{d}t^2}(I_{11}+I_{22}) = T_{11} + T_{22} &+ \int_V \mathrm{d}^3\mathbf{r}(x_1\mathscr{F}_1 + x_2\mathscr{F}_2) \\ &+ 2\int_V \mathrm{d}^3\mathbf{r}\left(p + \frac{B_3^2}{8\pi}\right).\end{aligned} \tag{5.38}$$

The net effect of the magnetic field is the pressure of the component B_3 perpendicular to the two dimensions $i = 1, 2$. The internal stresses and tensions of B_1 and B_2 automatically balance over the two dimensions. Thus, if $B_3 = 0$, the magnetic field does not contribute at all to the virial for the combined $i = 1, 2$ directions. However, the field still contributes strongly for $i = 3$, with (5.35) becoming

$$\frac{1}{2}\frac{\mathrm{d}^2 I_{33}}{\mathrm{d}t^2} = T_{33} + \int_V \mathrm{d}^3\mathbf{r}\left(p + \frac{B^2}{8\pi}\right) \tag{5.39}$$

where now $B^2 = B_1^2 + B_2^2$. Thus the magnetic stresses can be made to vanish in any two dimensions. But then the full pressure $B^2/8\pi$ of the field is brought to bear on the third, and the system becomes unstable. So confinement in two dimensions does not occur in nature. To elaborate, suppose that $B_3 = 0$ and the system is self-confining in the $i = 1, 2$ directions. Then pressure in the third dimension causes the system to lengthen in the third dimension, and if constrained from doing so, to buckle. Therefore, it is not possible, without strong external constraints, to devise a magnetic field configuration that is self-confined in two dimensions (see discussion in §6.2). An extensive study of the anisotropic dynamical properties of a magnetic field in an isolated gas cloud is available in the literature (Chandrasekhar and Limber 1954; Parker 1957) for the reader interested in the magnetic cloud as a real entity in the universe.

References

Biermann, L. and Davis, L. (1960). *Z. Astrophys.* **51,** 19.

Chandrasekhar, S. (1961). *Hydrodynamic and hydromagnetic stability*, pp. 577–598. Clarendon Press, Oxford.

—— (1969). *Ellipsoidal figures of equilibrium*, chap. 2. Yale University Press, New Haven.

—— and Fermi, E. (1953). *Astrophys. J.* **118,** 116; **122,** 208.

—— and Limber, D. N. (1954). *Astrophys. J.* **119,** 10.

Landau, L. D. and Lifshitz, E. M. (1959). *Fluid mechanics.* Addison-Wesley, Reading, Massachussetts.

Parker, E. N. (1953). *Astrophys. J.* **117,** 169.

—— (1954). *Phys. Rev.* **96,** 1686.

—— (1957). *Astrophys. J. Suppl.* **3,** 51.

—— (1958). *Phys. Rev.* **109,** 1440.

—— (1969). *Space Sci. Rev.* **9,** 651.

Roberts, P. H. (1967). *An introduction to magnetohydrodynamics.* American Elsevier, New York.

6

MAGNETIC EQUILIBRIUM

6.1. The problem of equilibrium

IF the purpose of this monograph is to explore the general non-equilibrium of magnetic fields, then we should begin with the special conditions for which equilibria exist. There has been a vast theoretical effort exploring the possibilities for hydrostatic equilibrium of a magnetic field **B** embedded in a conducting fluid of density ρ and pressure p, perhaps in the presence of a gravitational potential ψ, or otherwise confined by rigid boundaries. For hydrostatic equilibrium the magnetic field must satisfy the momentum balance

$$0 = -\nabla p + (\nabla \times \mathbf{B}) \times \mathbf{B}/4\pi - \rho \nabla \psi \tag{6.1}$$

There exist equilibria for fields with a sufficiently high degree of symmetry. The equilibria may be stable against perturbations constrained to the same symmetry, but if the fluid is a real gas, i.e. not incompressible, then a gravitational acceleration introduces instability through the magnetic buoyancy of the magnetic field (the outward pressure of the field noted in Chapter 5 reduces the fluid pressure and density within the field, causing the region of field to rise, Parker, 1955).

We do not expect perfect symmetry in nature, of course. It can be shown that in the absence of symmetry there is generally no equilibrium, even in the absence of gravity (Parker 1965, 1972), which is a topic of succeeding chapters. The present chapter gives a brief review of known equilibrium configurations and their stability in fluid bodies, beginning with the most highly constrained and symmetrical forms. There are two general configurations that interest us. One is the magnetic field within a self-gravitating fluid object, of the nature of a magnetic star. The other is the magnetic field above a fixed surface, essentially the external field of a planet, or a star, such as the sun.

In the circumstances where an equilibrium exists, the question is whether the equilibrium is stable. Although general energy principles exist for attacking the stability question (Bernstein *et al.* 1958; Newcomb 1960; Chandrasekhar 1961) it is usually not easy to show that a system is stable, for that requires a general proof that there is no perturbation of any form under which the system is unstable. On the other hand, instability is established by demonstrating a single unstable mode. If an equilibrium is unstable, with a significant growth rate, then we do not expect to find that equilibrium form in nature.

We will find that there are many field topologies for which there is no equilibrium. They will be demonstrated in later chapters. In the absence of equilibrium the question of stability (to small perturbations about an equilibrium) does not arise. The field configuration evolves continuously, the rate of evolution controlled by the inertia of the fluid and the dissipative reconnection of the topology.

This is perhaps the appropriate place to make some preliminary remarks on the topological properties of the magnetic lines of force, although we will take up the subject at length again in later chapters. First of all, the condition $\nabla \cdot \mathbf{B} = 0$ means that the individual lines of force do not end. Hence they either extend to infinity or, if confined to a finite volume of space, they are almost always ergodic within that volume. If constrained to any two-dimensional surface, the lines of force either extend to infinity or form closed curves. Any field whose lines of force are confined to parallel planes, say $[B_x(x, y, z), B_y(x, y, z), 0]$, can be written in terms of a single scalar function $A(x, y, z)$ according to

$$B_x = +\partial A/\partial y, \qquad B_y = -\partial A/\partial x.$$

The magnetic lines of force are given by (4.17), which becomes

$$\frac{\partial A}{\partial x}\,dx + \frac{\partial A}{\partial y}\,dy = 0,$$

with the integral

$$A(x, y, z) = constant$$

on any plane $z = constant$. The contours of constant A are the lines of force. This representation is easily generalized to planes which are not parallel, as in the poloidal field with axial symmetry, represented in (6.17) in terms of a scalar function P, for which the lines of force are $\varpi^2 P = constant$. Extension to more general surfaces is possible, but not often interesting or useful.

It is the tension along the lines of force, together with the closure of the lines of force in surfaces, that gives equilibrium and stability to highly-symmetric magnetic field configurations. The self-confinement in two dimensions was discussed in §5.4, where it was noted that the magnetic stresses can be brought into a balance in a two-dimensional space, with the field in the third dimension appearing only as a pressure.

6.2. Magnetic fields in infinite spaces

Beginning with the simplest case first, consider a magnetic field constrained to vary only with the coordinate x (i.e. $\partial/\partial y = \partial/\partial z = 0$). Then the field can have no varying x-component, for if it did, $\nabla . \mathbf{B}=0$ would require it to have a y- and/or a z-component depending upon y and z. Supposing that the x-component is zero, the lines of force are confined to planes perpendicular to the x-axis, and the field in each plane is independent of the y, z-coordinates in the plane. Hence the lines of force are straight lines, and we may as well rotate the coordinate system so that the z-axis lies along the field, and the field has components $\{0, 0, B_z(x)\}$. The lines of force extend from $z=-\infty$ to $z=+\infty$. Substituting into (6.1) yields the equilibrium condition

$$p(x)+B_z^2(x)/8\pi = constant \tag{6.2}$$

if there are no gravitational forces. Gravitational forces, such as $\mathbf{g}=(g, 0, 0)$ add the familiar barometric variation of pressure

$$p(x)+B_z^2(x)/8\pi = constant - \int dx\, g\rho(x)$$

and play no essential role in the character of the equilibrium (provided $\partial T/\partial y=\partial T/\partial z=0$). Any perturbations within the constraints $\partial/\partial y=\partial/\partial z=0$ are stable, representing magneto–acoustic waves (fast-mode hydromagnetic waves) and are discussed in Chapter 7.

Now consider what happens when the constraints are removed one by one. If the field B_z depends upon both x and y, so that the only remaining constraint is $\partial/\partial z=0$, there are many more possibilities. Equilibrium leads to

$$p(x, y)+B_z^2(x, y)/8\pi = constant \tag{6.3}$$

in the absence of gravity. The system has neutral stability to interchange of tubes of magnetic flux. If we introduce a gravitational acceleration with a component perpendicular to the field, then equilibrium disappears because of magnetic buoyancy (Parker 1955).

The intense flux tubes rise and the weak ones sink. The effect is considered in Chapter 8. Equilibrium (and neutral stability) is restored with the constraint of incompressibility $\rho = constant$, which eliminates the buoyant forces.

Suppose now that in addition to $B_z(x, y)$ we add $B_x(x, y)$, $B_y(x, y)$. Then since $\nabla . \mathbf{B}=0$, it follows that $\partial B_x/\partial x+\partial B_y/\partial y=0$. Hence B_x and B_y can be expressed in terms of a vector potential $A(x, y)$ as

$$B_x=+\partial A/\partial y, \qquad B_y=-\partial A/\partial x. \tag{6.4}$$

The projection of the magnetic lines of force on the xy-plane is given by $A = constant$, as noted above. The three components of the equilibrium equation (6.1) are ($\nabla_{xy}^2 \equiv \partial^2/\partial x^2 + \partial^2/\partial y^2$)

$$0 = \frac{\partial}{\partial x}\left(p + \frac{B^2}{8\pi}\right) + \frac{\nabla_{xy}^2 A}{4\pi}\frac{\partial A}{\partial x}, \tag{6.5}$$

$$0 = \frac{\partial}{\partial y}\left(p + \frac{B^2}{8\pi}\right) + \frac{\nabla_{xy}^2 A}{4\pi}\frac{\partial A}{\partial y}, \tag{6.6}$$

$$0 = \frac{\partial A}{\partial x}\frac{\partial B_z}{\partial y} - \frac{\partial A}{\partial y}\frac{\partial B_z}{\partial x} \equiv \frac{\partial(A, B_z)}{\partial(x, y)}, \tag{6.7}$$

in the absence of gravity (Chandrasekhar 1961). Multiply (6.5) by $\partial A/\partial y$ and (6.6) by $\partial A/\partial x$ and subtract. The result is

$$\frac{\partial(A, p + B_z^2/8\pi)}{\partial(x, y)} = 0, \tag{6.8}$$

The general solution to (6.7) is $B_z = B_z(A)$ and the general solution to (6.8) is

$$p + B_z^2/8\pi = F(A)$$

where B_z and F are arbitrary functions of A. Hence $p = p(A)$, which tells us that the gas pressure extends freely along the magnetic lines of force $A = constant$ because the magnetic field exerts no force ($\mathbf{F} = (\nabla \times \mathbf{B}) \times \mathbf{B}/4\pi$) on the gas in that direction.

Equations (6.5) and (6.6) together become

$$[\nabla_{xy}^2 A + 4\pi F'(A)]\nabla A = 0.$$

Since $\nabla A \neq 0$, we require that

$$\nabla_{xy}^2 A + 4\pi F'(A) = 0. \tag{6.9}$$

The form of the fluid pressure $p(A)$ and the longitudinal field $B_z(A)$ determine the function $F(A)$, which determines the form of eqn (6.9) for $A(x, y)$. If $F(A)$ is equal to a constant or is a linear or quadratic function of A, then (6.9) is a linear equation and general methods of solution are available. Otherwise special methods of solution must be devised.

The addition of the transverse fields, B_x and B_y, to the longitudinal field $B_z(x, y)$ eliminates the neutral stability of the longitudinal field alone. A modest combination of B_z and A appears to yield stability, with B_z stabilizing the sausage and kink instabilities of B_x and B_y. The upper limit on B_x and B_y for stability is discussed in later sections of this chapter and in Chapter 9. Introduction of a gravitational field may destroy equilibrium through magnetic buoyancy, if the fluid is not incompressible. The magnetic buoyancy can be resisted by the stresses in the field only if the field lines can be suitably anchored at a 'rigid' boundary (see discussion in §6.5 and in Chapters 7 and 9.

6.3. The twisted rope of magnetic field

One of the simplest examples of a magnetic field confined to a fluid body is the axi-symmetric field in an infinitely long circular cylinder. Place the z-axis of the coordinate system along the axis of the cylinder and denote distance from the axis by $\varpi=(x^2+y^2)^{\frac{1}{2}}$. Angular position about the z-axis is denoted by ϕ, measured from the x-axis, so that $x=\varpi\cos\phi$, $y=\varpi\sin\phi$, and (ϖ, ϕ, z) represents a right-handed cylindrical coordinate system.

For an axi-symmetric field, we have $\partial B_i/\partial\phi=0$. If the field is also uniform along the cylinder, then $\partial B_i/\partial z=0$. It then follows from (4.4) that the radial component B_ϖ is zero. The field is made up of longitudinal and azimuthal components, B_z and B_ϕ. If the fluid pressure is denoted by $p(\varpi)$, the density by $\rho(\varpi)$, and the gravitational potential by $\psi(\varpi)$, then for hydrostatic equilibrium (6.1) or (5.10) reduces to

$$0=\frac{\mathrm{d}}{\mathrm{d}\varpi}\left(p+\frac{B_\phi^2+B_z^2}{8\pi}\right)+\frac{B_\phi^2}{4\pi\varpi}+\rho\frac{\mathrm{d}\psi}{\mathrm{d}\varpi} \tag{6.10}$$

where $B_\phi^2/4\pi\varpi$ represents the confining force of the tension along the lines of force. Following the scheme first proposed by Lüst and Schlüter (1954) the magnetic field satisfying this equation can be expressed in terms of p and ρ through a generating function $F(\varpi)$ defined as the total pressure,

$$F(\varpi)=p+(B_\phi^2+B_z^2)/8\pi. \tag{6.11}$$

It follows from (6.10) that

$$\frac{B_\phi^2}{8\pi}=-\tfrac{1}{2}\varpi\left(\frac{\mathrm{d}F}{\mathrm{d}\varpi}+\rho\frac{\mathrm{d}\psi}{\mathrm{d}\varpi}\right) \tag{6.12}$$

and from (6.11) that

$$\frac{B_z^2}{8\pi}=F+\tfrac{1}{2}\varpi\frac{\mathrm{d}F}{\mathrm{d}\varpi}-p+\tfrac{1}{2}\varpi\rho\frac{\mathrm{d}\psi}{\mathrm{d}\varpi}. \tag{6.13}$$

The generating function $F(\varpi)$ is arbitrary except that for a given $p(\varpi)$ and $\rho(\varpi)$ it must vary in such a way as to assure that B_ϕ and B_z are real. Thus we specify $p(\varpi)$ and $\rho(\varpi)$, deducing ψ from ρ with the gravitational field equation[1]

$$\frac{1}{\varpi}\frac{\mathrm{d}}{\mathrm{d}\varpi}\varpi\frac{\mathrm{d}\psi}{\mathrm{d}\varpi}=4\pi G\rho(\varpi)$$

[1] We must not forget that the right-hand side of the gravitational equation includes *all* of the matter present. Thus in the gaseous disc of the galaxy the interstellar gas density ρ, in which the galactic field is embedded, must be augmented by the (generally larger) mean density of stars, to which the magnetic field in (6.10) is not directly coupled.

and obtaining all possible field configurations from (6.12) and 6.13). It is important to note that for a given distribution ρ and p, the field distribution is not wholly constrained. The arbitrary generating function $F(\varpi)$ represents the freedom of variation of B_ϕ and B_z that is self-supporting and does not involve forces exerted directly on the fluid. Indeed, Lüst and Schlüter first proposed $F(\varpi)$ for the force-free field, in which there is no force exerted on the fluid by the field throughout the interior of the cylinder. Then the gas supports itself, $\mathrm{d}p/\mathrm{d}\varpi = -\rho\, \mathrm{d}\psi/\mathrm{d}\varpi$, and (6.10) reduces to

$$0 = \frac{\mathrm{d}}{\mathrm{d}\varpi}\left(\frac{B_\phi^2 + B_z^2}{8\pi}\right) + \frac{B_\phi^2}{4\pi\varpi}.$$

Then write $F = (B_z^2 + B_\phi^2/8\pi$, with B_ϕ and B_z given by

$$B_z^2 = F + \tfrac{1}{2}\varpi\, \mathrm{d}F/\mathrm{d}\varpi,\ B_\phi^2 = -\tfrac{1}{2}\varpi\, \mathrm{d}F/\mathrm{d}\varpi. \tag{6.14}$$

The field is entirely self-supporting throughout the interior. In order that B_ϕ and B_z are real we require only that F should decline monotonically with ϖ, but not faster than $1/\varpi^2$, i.e. $0 \geqslant \mathrm{d}F/\mathrm{d}\varpi \geqslant -2F/\varpi$.

The stability of a longitudinal magnetic field in a self-gravitating cylinder has been treated by Chandrasekhar and Fermi (1953) employing the virial equation. They show (see also Simon 1958) that the magnetic field helps to stabilize Jean's gravitational instability, although in an infinite medium this is not the case (Chandrasekhar 1961). They did not show that the magnetic field was itself stable. A number of authors (Roberts 1956; Callebaut and Voslander 1962; Anzer 1968) have considered the stability of a twisted magnetic field in a cylinder without gravitation and with and without external confining pressure or boundaries i.e. $F = 0$ at the surface. The calculations show instability to spiral kinking in all cases. We can understand the instability in simple physical terms directly from the virial equation (5.36). In the absence of external pressure the magnetic fields must vanish on the surface, so (5.36) reduces to

$$\frac{1}{2}\frac{\mathrm{d}^2 I_{33}}{\mathrm{d}t^2} = T_{33} + \int_V \mathrm{d}^3\mathbf{r}\left\{p + \frac{B_\phi^2 - B_z^2}{8\pi}\right\} + \int \mathrm{d}^3\mathbf{r}\, z\mathscr{F}_z$$

with the second term on the right-hand side denoting the contribution of the gas pressure and the magnetic stresses. In the absence of gravitational forces this term may be reduced with the aid of (6.13) to

$$\begin{aligned} Z &\equiv \int_V \mathrm{d}^3\mathbf{r}\left\{p + \frac{B_\phi^2 - B_z^2}{8\pi}\right\}, \\ &= \int \mathrm{d}z 2\pi \int \mathrm{d}\varpi\varpi\left\{2p - \frac{\mathrm{d}}{\mathrm{d}\varpi}\,\varpi F\right\}, \\ &= \int \mathrm{d}z 2\pi \int \mathrm{d}\varpi\left\{\varpi(2p + F) - \frac{\mathrm{d}}{\mathrm{d}\varpi}\,\varpi^2 F\right\}, \end{aligned}$$

where the integration of z is over the length of the tube and the integration of ϖ is from the axis $\varpi=0$ to the surface, where the total pressure F vanishes. Hence

$$Z=2\pi\int \mathrm{d}z\int \mathrm{d}\varpi\varpi(2p+F) \tag{6.15}$$

and is positive. The tube is under compression and therefore subject to buckling. This is the cause of the universal instability found by Roberts. The tube of flux must be twisted so tightly to be self-confining that the mean square value of B_ϕ greatly exceeds the mean square of B_z. Hence there is no possibility for the equilibrium to be realized in nature.

6.4. The axi-symmetric magnetic field

It is possible to treat the axi-symmetric field without the constraint that it be uniform along the axis. The mathematical treatment is much more difficult than in the uniform twisted rope ($\partial B_i/\partial z=0$), but the elegance of the formulation, and the physical perspective which it provides, merit a discussion. Following Lüst and Schlüter (1954), and Chandrasekhar (1956a), Chandrasekhar and Prendergast (1956) and Chandrasekhar and Kendall (1957) we decompose the magnetic field into an azimuthal, or toroidal, component

$$B_\phi(\varpi, z)=\varpi T(\varpi, z) \tag{6.16}$$

and a meridional, or poloidal, component

$$B_\varpi(\varpi, z)=-\varpi\partial P/\partial z, \qquad B_\phi(\varpi, z)=(1/\varpi)\partial(\varpi^2 P)/\partial\varpi \tag{6.17}$$

where $P=P(\varpi, z)$. The projection of the lines of force on the meridional planes gives $\varpi^2 P=$ *constant.* It is readily shown that

$$\nabla\times\mathbf{B}=-\mathbf{e}_\varpi\varpi\frac{\partial T}{\partial z}-\mathbf{e}_\phi\varpi\Delta_5 P+\mathbf{e}_z\frac{1}{\varpi}\frac{\partial}{\partial\varpi}(\varpi^2 T) \tag{6.18}$$

where Δ_5 is the Laplacian operator in a space of five dimensions,

$$\Delta_5=\frac{\partial^2}{\partial\varpi^2}+\frac{3}{\varpi}\frac{\partial}{\partial\varpi}+\frac{\partial^2}{\partial z^2}.$$

The force per unit volume $F_i=\partial M_{ij}/\partial x_j$ is then

$$\begin{aligned}\mathbf{F}&=(\nabla\times\mathbf{B})\times\mathbf{B}/4\pi\\ &=-\left\{\mathbf{e}_\varpi\Delta_5 P\frac{\partial}{\partial\varpi}\varpi^2 P+T\frac{\partial}{\partial\varpi}\varpi^2 T\right\}\\ &\quad-\mathbf{e}_\phi\left\{\frac{\partial P}{\partial z}\frac{\partial}{\partial\varpi}\varpi^2 T-\frac{\partial T}{\partial z}\frac{\partial}{\partial\varpi}\varpi^2 P\right\}\\ &\quad-\mathbf{e}_z\varpi^2\left\{T\frac{\partial T}{\partial z}+\frac{\partial P}{\partial z}\Delta_5 P\right\}.\end{aligned} \tag{6.19}$$

Finally

$$\nabla\times\mathbf{F}=-\mathbf{e}_\varpi\frac{\partial F_\phi}{\partial z}+\mathbf{e}_\phi\left(\frac{\partial F_\varpi}{\partial z}-\frac{\partial F_z}{\partial\varpi}\right)+\mathbf{e}_z\frac{1}{\varpi}\frac{\partial}{\partial\varpi}\varpi F_\phi. \tag{6.20}$$

The equation for hydrostatic equilibrium is (6.1). Then if the density ρ is uniform on the level surfaces $\psi=$ *constant*, if not uniform throughout the entire body, we have $\rho=\rho(\psi)$ and $\rho\nabla\psi=\nabla\pi(\psi)$ where $\pi(\psi)$ is defined as $\rho=\mathrm{d}\pi/\mathrm{d}\psi$. The curl of (6.1) yields $\nabla\times\mathbf{F}=0$, so that the requirement that F_ϕ be finite and well-behaved at $\varpi=0$ leads through the ϖ- and z-components of (6.20) to $F_\phi=0$. Hence, from (6.19) the Jacobian of ϖ^2P and ϖ^2T must vanish,

$$\frac{\partial(\varpi^2P,\varpi^2T)}{\partial(\varpi,z)}=0. \tag{6.21}$$

This determinant of derivatives can be made to vanish generally only if

$$\varpi^2T=\Psi(\varpi^2P), \tag{6.22}$$

where Ψ is an arbitrary function of its argument ϖ^2P.

The vanishing of the ϕ-component of (6.20) is the condition

$$\frac{\partial(\Delta_5P,\varpi^2P)}{\partial(\varpi,z)}=\varpi\frac{\partial T^2}{\partial z}. \tag{6.23}$$

This mathematical relationship is easily reduced by noting that the right-hand side can be written as a determinant, in the same form as the left. To show this consider the form

$$\frac{\partial[\varpi^2P,G(\varpi^2P)/\varpi^2]}{\partial(\varpi,z)}=\varpi\frac{\partial T^2}{\partial z} \tag{6.24}$$

where G is a function of ϖ^2P. This expression reduces directly to

$$2G\frac{\partial}{\partial z}\varpi^2P=\frac{\partial}{\partial z}\varpi^4T^2$$

and hence

$$G=\frac{1}{2}\frac{\mathrm{d}\varpi^4T^2}{\mathrm{d}\varpi^2P}$$
$$=\Psi\Psi' \tag{6.25}$$

with the aid of (6.22), where the prime denotes differentiation with respect to the argument, ϖ^2P. Using (6.24) to replace the right-hand side of (6.23), it follows that

$$\frac{\partial(\Delta_5P+G/\varpi^2,\varpi^2P)}{\partial(\varpi,z)}=0$$

and hence $\Delta_5 P + G/\varpi^2$ must be a function only of $\varpi^2 P$,

$$\Delta_5 P + \Psi(\varpi^2 P)\Psi'(\varpi^2 P)/\varpi^2 = \Phi(\varpi^2 P), \tag{6.26}$$

where we have used (6.25) to write G in terms of Ψ, and Φ is an arbitrary function of $\varpi^2 P$. Equation (6.26) is the differential equation for P. The equation is second order in the derivatives with respect to ϖ and z, but the form of the equation depends upon the choice of the functions Ψ and Φ.

The function $\Psi(\varpi^2 P)$ is the relation between the toroidal field function T and the poloidal vector potential P. If, for instance, there is no azimuthal field, then $F = 0$ and (6.26) reduces to

$$\Delta_5 P = \Phi(\varpi^2 P), \tag{6.27}$$

which has been treated by Ferraro (1954) and Roberts (1955).

The special case of a force-free field, in which the field exerts no force $\mathbf{F} = (\nabla \times \mathbf{B}) \times \mathbf{B}/4\pi$ on the fluid throughout some specified region of space, requires that

$$\nabla \times \mathbf{B} = h(\mathbf{r})\mathbf{B} \tag{6.30}$$

throughout the region. Since $\nabla \cdot \mathbf{B} = 0$ and the divergence of a curl is identically zero, the divergence of (6.30) yields

$$\mathbf{B} \cdot \nabla h = 0 \tag{6.31}$$

so that h is constant along each magnetic line of force. This statement is equivalent to the statement of conservation of electric current, since it follows from (4.9) that (6.30) is just $4\pi\mathbf{j} = ch\mathbf{B}$. The current must be parallel to $\mathbf{B}$ in order that $\mathbf{F} = 0$, so that the fluxes of both $\mathbf{B}$ and $\mathbf{j}$ are conserved if, and only if, h is constant along each tube of flux. The special case $h = constant$ throughout the field leads to the maximum magnetic energy density for a given total dissipation (Chandrasekhar and Woltjer 1958).

For an axi-symmetric field we use (6.17) and (6.18). Then since $\partial h/\partial\phi = 0$, (6.31) becomes

$$\frac{\partial(h, \varpi^2 P)}{\partial(\varpi, z)} = 0$$

so that h is an arbitrary function of $\varpi^2 P$,

$$h = h(\varpi^2 P).$$

The three components of (6.30) reduce to

$$\Delta_5 P + hT = 0, \tag{6.32}$$

$$\nabla\varpi^2 T - h\nabla\varpi^2 P = 0. \tag{6.33}$$

In terms of the function β whose derivative is h, (6.33) can be written

$$\nabla\{\varpi^2 T - \beta(\varpi^2 P)\} = 0.$$

Integration yields

$$\varpi^2 T = \beta(\varpi^2 P) \tag{6.34}$$

where the integration constant has been absorbed into β. Equation (6.32) now becomes

$$\varpi^2 \Delta_5 P + \beta\beta' = 0. \tag{6.35}$$

Comparing this result with (6.22) it is evident that the force free field corresponds to the special choice $\Psi' = h$ (so that $T = \alpha P$) and $\Phi = 0$.

Analytical solutions to (6.35) are readily available for special forms of β. In particular, if the curl or current density is everywhere in the same proportion to the field density itself, then $h = constant$ and (6.35) reduces to

$$\Delta_5 P + h^2 P = 0. \tag{6.36}$$

In spherical coordinates this is

$$\left(\frac{\partial^2}{\partial r^2} + \frac{4}{r}\frac{\partial}{\partial r} + \frac{1-\mu^2}{r^2}\frac{\partial^2}{\partial \mu^2} - \frac{4\mu}{r^2}\frac{\partial}{\partial \mu} + h^2\right) P = 0$$

where $\mu = \cos\theta$ and θ is the polar angle. The solutions are represented by the modes (Chandrasekhar 1956a)

$$P_n = \frac{C_n^{3/2}(\mu)}{r^{3/2}}\left[A_n J_{n+\frac{3}{2}}(hr) + B_n J_{-n-\frac{3}{2}}(hr)\right] \tag{6.37}$$

where $C_n^{\frac{3}{2}}(\mu)$ represents the Gegenbauer polynomial (Morse and Feshbach 1953) and A_n and B_n are arbitrary constants.

The magnetic field corresponding to the *nth* mode has (r, θ, ϕ) components

$$-\frac{1}{r^2}\frac{\partial}{\partial \mu} P_n r^2(1-\mu^2),\ -\frac{1}{r(1-\mu)^{\frac{1}{2}}}\frac{\partial}{\partial r}\{P_n r^2(1-\mu^2)\},\ hr(1-\mu^2)^{\frac{1}{2}} P_n. \tag{6.38}$$

The poloidal function P varies asymptotically as $\sin(hr+q)/r^2$ for $hr \gg 1$, where q is an arbitrary constant. Hence the amplitude of the radial component diminishes as $1/r^2$, and the θ- and ϕ-components as $1/r$. The solution may be cut off at any spherical surface $r = R$ on which the radial component vanishes,

$$A_n J_{n+\frac{3}{2}}(hR) + B_n J_{-n-\frac{3}{2}}(hR) = 0. \tag{6.39}$$

Note that the magnetic pressure of the θ- and ϕ-components on the confining surface $r = R$ diminishes as $1/r^2$, so that the net force per steradian is undiminished as R becomes large. This is a direct manifestation of the necessity for physical confinement of a magnetic field.

It is a simple matter to show the consistency of the field (6.38) with the virial equation. The trace of (5.26) for static equilibrium ($T_{ij}=0$, $d^2I_{ij}/dt^2=0$) and vanishing non-magnetic forces ($\mathcal{F}_i=0$) yields

$$3\int_V d^3\mathbf{r}p+\mathcal{M}=\int_S dS_i x_i p_S$$

where p_S is the external pressure exerted on the surface S. The total magnetic energy $\mathcal{M}$ increases as R. The magnetic pressure at the surface decreases as $1/R^2$ so that the inward pressure differential at the surface, which must balance the magnetic pressure there, also decreases as $1/R^2$. Hence the surface integral on the right-hand side is proportional to R and is able to balance $\mathcal{M}$ on the left-hand side. In this way the field within $r=R$ is confined by inward forces on the surface.

The field throughout the interior of $r=R$ is made up of spherical shells of helical coils of field, the coils extending azimuthally around the shells. The bounding radii of the individual shells are given by the successive zeros of (6.39). The individual coils of field in each shell extend over θ between successive zeros of the Gegenbauer polynomial $C_n^{\frac{3}{2}}(\mu)$.

6.5. The axi-symmetric magnetic field in a finite volume

Prendergast (1956) worked out the solution for an axi-symmetric magnetic field confined inside a self-gravitating sphere of incompressible fluid a liquid magnetic 'star'). His field vanished at the surface of the star (but see Monaghan 1976) and was confined by internal forces $-\mathbf{F}$ exerted by the fluid on the field. He solved (6.26) with $\Phi=constant$ and $\Psi\Psi'=\varpi^2P$. The solution to the homogeneous part of the equation is again just (6.37). The solution to the inhomogeneous equation is $P=constant$, representing the force exerted by the fluid on the field. Prendergast showed that the mode $n=0$ represents a magnetic equilibrium configuration. A particularly interesting feature of the calculation is the demonstration that the magnetic field can be wholly confined within the fluid sphere. The magnetic 'star' is in complete hydrostatic equilibrium and the field-free surface is not distorted from a perfect sphere. Chandrasekhar and Fermi (1953) had pointed out earlier that a *uniform* field through a liquid 'star', producing an exterior dipole field, distorts the otherwise spherical star into an oblate spheroid. Gjellestad (1954a) (see also Ferraro 1954; Roberts 1955; Gjellestad 1957) determined the eccentricity of the star for a given magnetic field[2]. Unfortunately the configuration is unstable and so cannot represent the basic dynamical circumstances of a magnetic star.

[2] Gjellestad (1954b) treated the complementary problem of a field-free self-gravitating incompressible body immersed in an external magnetic field.

Sykes (1957) extended Prendergast's equilibrium field to a rotating spheroidal star. The formal treatment of an axi-symmetric magnetic field with axi-symmetric fluid motions has been given by Chandrasekhar (1956b, 1958), Woltjer (1958b, c), and Lilliequist *et al.* (1971). The general subject of rotating self-gravitating liquid spheres without magnetic field has been summarized by Chandrasekhar (1969) (see also Trehan and Singh 1974).

A fundamental question is the stability of the axi-symmetric magnetic field confined to a self-gravitating sphere, i.e. the magnetic star. Instability has not been found in the cases of moderate field strength (Prendergast 1957) where the fluid is constrained to incompressibility and the perturbations are constrained to axial symmetry (Lundquist 1951; Chandrasekhar and Fermi 1953; Chandrasekhar and Limber 1954; Lyttkens 1954; Jenson 1955; Prendergast 1957; Hain *et al.* 1957; Trehan 1957; Woltjer 1958a). Indeed the magnetic field, as mentioned earlier, serves to suppress gravitational instability in a long self-gravitating cylinder (Simon 1958). But as soon as the constraints of incompressibility and axial symmetry are relaxed, instability is the rule rather than the exception, as a consequence of magnetic buoyancy.

The real magnetic star, with rotation and radiative equilibrium, is much more complicated to deal with than the 'liquid' models discussed so far. Real magnetic stars, with fields in excess of 10^2 G (Babcock 1958, 1960a, b; Preston 1967a, 1969a, b, 1971) have received considerable attention in recent years. Their periodic variations can be understood largely in terms of rotation of the star with an oblique dipole field (Deutsch 1958; Monaghan 1966; Preston 1967b, 1972; Davies 1968; Wright 1969; Wolff and Wolff 1970, 1972; Monaghan and Robson 1971; Monaghan 1973; Bath *et al.* 1974; Stift 1974; Mestel and Moss 1977) or an offset dipole field (Landstreet 1970; Borra 1974; Stift 1974), although the pattern of anomalous abundances of elements rotating with some of these stars is a puzzle under any circumstances. The internal structure of the fields within the body of the magnetic star is not known, of course. There has been an extensive discussion of simple models (Monaghan 1966; Davies 1968; Wright 1969; Monaghan and Robson 1971; Moss 1973) with particular attention to the effects on the form and brightness of the star. The stellar magnetic field may have a strong effect on the rotation period of a star through its braking effect in connection with the stellar wind of the star (Weber and Davis 1967; Mestel 1968; Pneuman and Kopp 1971; Okamoto 1974).

The theoretical question of the stability of the observed stellar magnetic fields is a difficult question, particularly in view of our ignorance of the internal fields. Magnetic fields in degenerate stars evidently cause convective instabilities (Vandakurov 1972; Tayler 1973a, b; Chanmugam 1974).

It has been demonstrated that almost any perturbing forces give a breakdown in radiative equilibrium, causing large-scale convection in normal stars (Eddington 1929; Sweet 1950; Opik 1951; Baker and Kippenhahn 1959; Mestel 1965a, b, 1966). It appears that all poloidal configurations are unstable but that a toroidal (azimuthal) field may stabilize the principal instabilities of the poloidal field (Markey and Tayler 1973, 1974; Wright 1973). Lilliequist *et al.* (1971) study, by numerical methods, the dynamics of a toroidal field alone, and also as it might interact with a poloidal field, showing the general activity of such a system. One way or another, it is clear that the main-sequence star has lost most of its primeval field presumably through some convective instability (Mestel 1965a, b, 1971; Wright 1974). Although there are general energy principles, etc. available for treating the stability of a hydromagnetic system (Bernstein *et al.* 1958; Newcomb 1960; Chandrasekhar 1961) it remains yet to carry out a systematic and comprehensive dynamical study of self-gravitating axi-symmetric magnetic star models incorporating general (non-axi-symmetric) perturbations in the presence of radiative equilibrium and the enormous density variations between the centre and the surface of the star. The equilibrium of a self-gravitating *polytrope* containing a magnetic field has been considered (Chandrasekhar and Lebovitz 1964; Trehan and Uberoi 1972) and the virial equations applied to study various modes of oscillation (Chandrasekhar and Lebovitz 1964; Trehan and Billings 1971; Sood and Trehan 1972). Kochhar and Trehan (1974) have applied the third order virial equations to the stability of a rotating, self-gravitating polytropic body with an internal magnetic field, working out the effects of the magnetic field on the characteristic frequencies of the various oscillatory modes.

We conjecture that there are no stable equilibrium magnetic configurations confined to self-gravitating gaseous sphere, i.e. that in every case there is at least one mode which is unstable. This conjecture is based on the general fact of magnetic buoyancy (Parker 1955). The fluid must push inward on the magnetic field, so the pressure, and presumably the density, are reduced inside the field. The field tends to rise relative to the surrounding fluid. A perturbation which breaks the symmetry of the initial axi-symmetric field configuration sets things in motion. The field comes to the surface as the fluid redistributes itself along the lines of force. It is nothing more than a magnetic Rayleigh–Taylor instability, with the heavy (non-magnetic) fluid streaming downward through the light (magnetized) fluid, thereby uncovering and releasing the field from the centre of the confining body. The characteristic growth rate is comparable to the free-fall time or the Alfven transit time (Chandrasekhar 1961; Parker 1966, 1969). Many illustrations of magnetic buoyancy are given in

§8.7 and Chapter 13. They form the physical basis for the conjecture of general instability. It would be interesting to see if some special case can be found that is a counter example to our conjecture.

The onset of the instability soon puts the field in a non-equilibrium state, rising up through the body at a rate determined by the local physical conditions. If the fluid is convectively unstable, the field rises and escapes at a rate comparable to the Alfven speed. If the fluid is stable against convection, the rate of rise is much slower, and is limited then by the rate of heat transfer (Parker 1974, 1975). We suggest that in general the magnetic fields embedded in stars and in the galaxy are in non-equilibrium. The extremely strong primordial fields, that might have been compressed into the body during its initial condensation from a large gas cloud (10^7 G or more for a star), have long since departed as a consequence of the magnetic buoyancy (Parker 1974). Hence the surviving fields are generally weak[3], as remarked above, and represent a balance between the rate of loss through buoyancy (and turbulent diffusion) and the rate of generation in the fluid motions within the parent body.

6.6. The magnetic field extending from a fixed surface

Only the external magnetic fields of stars and planets can be observed, the internal fields being generally inaccessible below the visible surface. For that reason it is the theory principally of the external field, extending outward into space from the surface of the object, that comes to grips with observations. It is here that nature shows us her card tricks openly and directly, and challenges us to understand them. It is worthwhile, therefore, to pause briefly in the exposition of the formal mathematical methods for treating magnetostatic equilibrium to review some of the equilibria presented by the external magnetic fields of stars and planets for our serious consideration.

6.6.1. *Physical problems suggested by the sun*

The magnetic fields observed on the sun emerge from the interior through the visible surface and extend outward into space. These external fields are held and controlled by the dense fluid beneath the photosphere, which disgorges them for observation while continually contorting their configuration. The portion of the field above the surface extends through gases so tenuous as to have little effect upon the field, i.e. $\mathbf{F} = (\nabla \times \mathbf{B}) \times \mathbf{B}/4\pi \cong 0$ and the fields are essentially force-free. The magnetic stresses above the surface form their own equilibrium. It is particularly fruitful to study the magnetic configurations above the surface of the sun because it

[3] Some white dwarfs and the pulsars (Neutron star) evidently retain their compressed fields for a time after collapse to their final degenerate state.

affords one of the few cases where the field configuration can be observed in some detail.

Sunspot fields, and the fields of active regions in general, are the strongest and most conspicuous examples of solar magnetic field. Given that they are firmly anchored below the photosphere, their most curious property is the observed fact (Zwaan 1968; Newkirk *et al.* 1968; Kai 1969; Wild 1970; Zirin 1972; Sheeley *et al.* 1975) that no matter how complicated the evolution of the magnetic pattern at the surface may be, the field configuration above is nearly a potential field (i.e. current-free, $\nabla \times \mathbf{B} = 0$, so that $\mathbf{B} = -\nabla \phi$). Maintenance of the simple magnetic connections of the potential field in the face of rapid change (in a day) of the pattern of field at the visible surface implies that the magnetic lines of force have at their disposal some means for rapid reconnection in spite of the high electrical conductivity of the gas. We will have more to say on this in later chapters. We note that the potential field is the lowest energy state given the distribution of the magnetic lines of force at the surface of the sun. Hence the potential field represents a stable equilibrium.

On those occasions when the magnetic field of a sunspot group is observed to depart significantly from the simple potential field, the departure is in the form of a twisting or shearing, so that the magnetic lines of force emerging from the sunspot spiral as they extend outward across the penumbra and into the surrounding photosphere. Evidently there has been a net rotation of the field in the umbra relative to the surrounding photosphere at some point in the history of the field (perhaps before emergence). Nakagawa *et al.* (1971) have worked out a variety of force-free field configurations (see §6.7.2 (below) for the simple case $h = constant$ (in (6.30)) and have shown that they fit the observed spiral patterns very well (see further discussion and elaboration in Dicke 1970; Raadu and Nakagawa 1971; Nakagawa and Raadu 1972; Nakagawa 1973; Nakagawa *et al.* 1973). Coronal fields were studied along the same lines a little earlier by Altschuler and Newkirk (1969), Newkirk *et al.* (1970), Raadu (1972b). The extension of coronal magnetic fields into space by the hydrodynamic expansion of the solar corona leads to interesting magnetic forms and shapes (Parker 1963; Sturrock and Smith 1968; Pneuman and Kopp 1971). The coronal streamers are one result, where the expanding corona stretches a re-entrant loop of field out into space, sketched in Fig. 6.1. There is a neutral sheet ($\mathbf{B} = 0$) between the lines of force extending out on one side and back on the other. The magnetic field observed in interplanetary space is merely the extension of the solar magnetic field by the solar wind, blowing outward through the solar system.

A more complex problem is the storage of energy in the twisted field of sunspots and active regions to produce the solar flare. It is generally

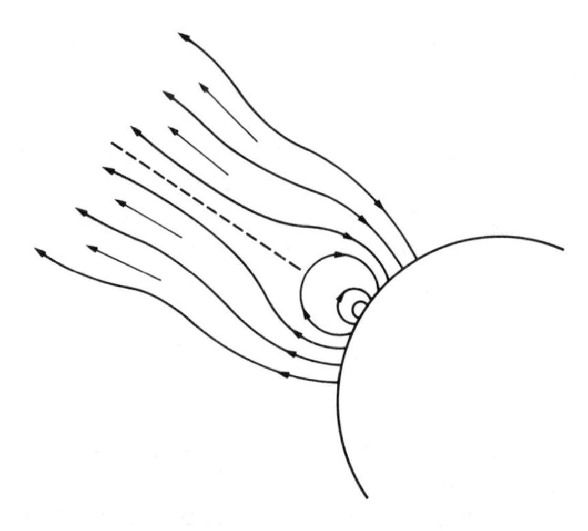

FIG. 6.1. Sketch of the magnetic lines of force over an active region of the sun. The short arrows indicate the outward expansion of the corona that extends the outer lines of force to infinity.

believed that the only adequate energy source for the solar flare is the magnetic field of the active region in which the flare occurs (Lundquist 1950; Lüst and Schlüter 1954; Parker 1957; Dungey 1958). The general idea is that in some way the sunspot fields are twisted, compressed, or stretched into a configuration of elevated energy (see, for instance, Gold and Hoyle 1960; Sturrock and Coppi 1966; Sturrock 1972; Lilliequist *et al.* 1971). The onset of a sudden and violent instability, involving rapid reconnection of magnetic lines of force to relieve the enhanced field stresses, produces the flare. Many possible contorted equilibria have been considered. A number of authors (Barnes and Sturrock 1972; Raadu 1972a; Sturrock 1972; Tanaka and Nakagawa 1973) have considered models of the twisted force-free fields as the flare energy source. Twisted force-free configurations of sufficiently short length appear to be stable until a critical value in the twisting is reached, whereupon they become unstable. They are, therefore, of possible interest for the flare. The long twisted axi-symmetric rope of flux (Gold and Hoyle 1960), on the other hand, is unstable for any small amount of twisting (Anzer 1968) and does not appear promising for the flare. Evidently the equilibrium configuration of lowest energy for the long rope takes up spiral, rather than axial, symmetry as it is twisted. It has yet to be shown whether the spiral equilibrium develops instabilities. Observations of the magnetic changes during flares (Moreton and Severny 1968) indicate relaxation and the release of magnetic energy in quantities comparable to the estimated flare energy. Tanaka and Nakagawa (1973) propose that the observed flare loops represent lines of force in the strained pre-flare field. The observed rotation of the loops during the flare then indicates the relaxation of the field and the release of magnetic energy.

The quiescent prominence above the surface of the sun presents another challenging equilibrium problem. The prominence is a long-lived, extended (10^5 km), thin (5×10^3 km), vertical (5×10^4 km) equilibrium cloud of dense cool gas (10^{11} atoms/cm at 5–10×10^3 K) at a height of 10^4–10^5 km in the corona where the normal gas is 10^9 atoms/cm^3 at 10^6 K or more. Clearly the dense cold prominence cannot be supported by the surrounding coronal gas. The only possible explanation for its suspension is the magnetic field extending outward from the active region at the surface with which the prominence is associated (Dungey 1953; Kippenhahn and Schlüter 1957; Brown 1958). The problem is to deduce a magnetic configuration that gives *stable* support. The basic idea is that the thermal instability of the gas (i.e. the tendency to radiate more rapidly if its temperature is reduced; see Parker 1953; Whitaker 1963; Nakagawa 1970) causes the temperature to plunge from 10^6 to 10^4 K, with the density increasing as the gas is compressed by the pressure of the surrounding 10^6 K corona. Thermal equilibrium is possible only when the temperature drops so low that the radiative losses no longer increase and thermal conduction from the surrounding corona into the prominence can balance the radiative losses (see, for instance, Orrall and Zirker 1961). The weight of the cool condensation is then presumed to be supported by horizontal magnetic fields through the region (Kippenhahn and Schlüter 1957). Insofar as the field is horizontal and its direction changes at a suitable rate with height, the equilibrium has no evident instabilities (Nakagawa 1970). But of course there are no truly horizontal fields in the solar corona. The magnetic lines of force all extend up into the corona from the surface of the sun, so that the magnetic field passes only through the horizontal at the apex of magnetic arches (Kippenhahn and Schlüter 1957) as sketched in Fig. 6.2(a). The weight of the prominence presumably depresses the apex, as sketched in Fig. 6.2(b), suggesting a precarious equilibrium at best. Some authors (de Jager 1959; Kuperus and Tandberg-Hanssen 1967; Raadu and Kuperus 1973; Kuperus and Raadu 1974) now suggest that the prominence occurs in the neutral sheet (current sheet where $\mathbf{B}=0$ but $\nabla\times\mathbf{B}\neq0$) in the extended field above the arch, as sketched in Fig. 6.2(c). It is an observed fact (Veeder and Zirin 1970; Zirin 1972) that the quiescent prominences lie above the photosphere along the places (paths) where the field in the photosphere and chromosphere is horizontal (i.e. where the magnetic field in the line of sight passes through zero, sometimes called neutral lines upon forgetting that the unobserved component of field perpendicular to the line of sight may not be zero). Rust (1967) infers horizontal magnetic fields of 5–50 G in quiescent prominences, perhaps increasing with height (1 G per

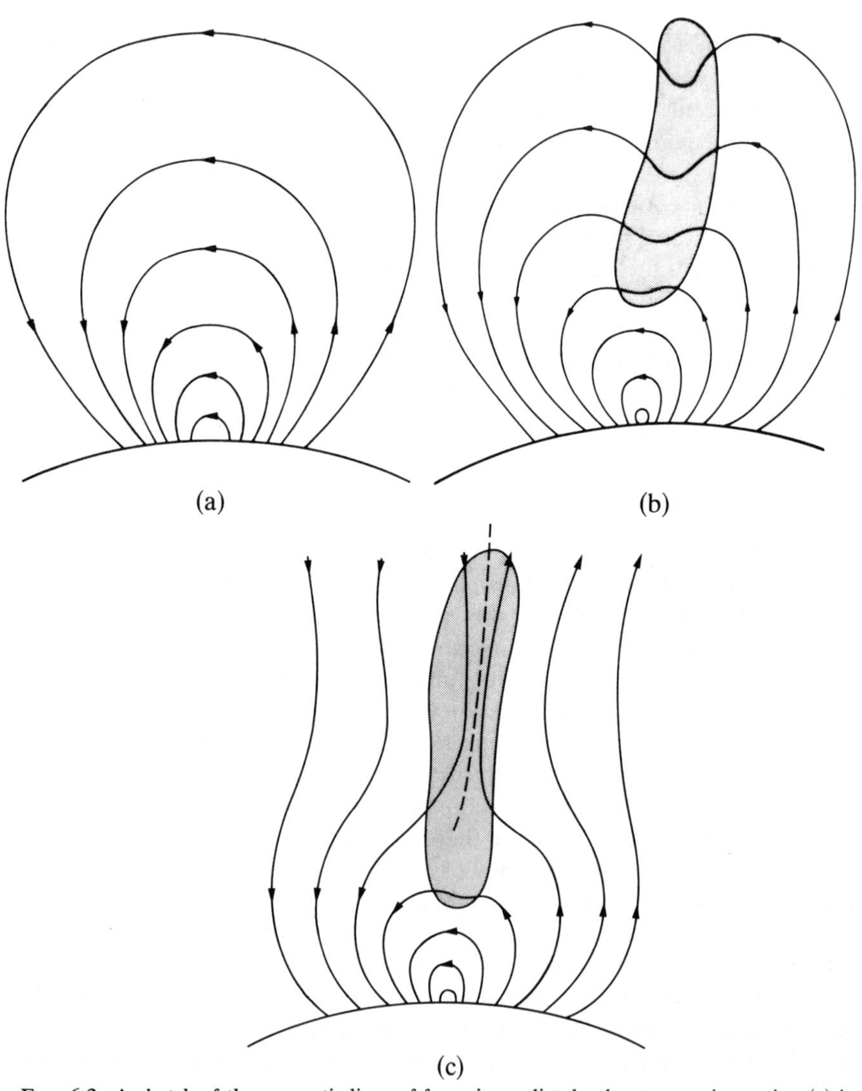

FIG. 6.2. A sketch of the magnetic lines of force immediately above an active region (a) in the absence of a quiescent prominence or coronal expansion; (b) supporting the weight of a quiescent prominence at the apex of the arched lines of force; (c) supporting a quiescent prominence formed in the neutral sheet caused by coronal expansion.

10^4 km). Altogether, then, it appears that the quiescent prominence is indeed a creation of thermal instability in and above the magnetic arches in the corona. It remains to demonstrate the precise form of the equilibrium and its apparent thermal and dynamical stability. Quite apart from the perplexing question of the stability of the configuration sketched in Fig. 6.2(c) (which has been suggested (Carmichael 1964; Sturrock 1972) as the site of the magnetic instability causing flares) there is the basic question of the equilibrium support of the dense prominence gas. Since the force $\mathbf{F} = (\nabla \times \mathbf{B}) \times \mathbf{B}/4\pi$ exerted by the field on the gas has no component parallel to $\mathbf{B}$, what is to prevent the gas from sliding downward along the lines of force into the photosphere. Figure 6.2(b) may perhaps avoid this problem, but its stability is problematical.

6.6.2. *Physical problems suggested by the planets and pulsars*

The magnetic dipole field of Earth, (with a strength of 0.6 G at each pole) has been mapped extensively on the ground and in space, and it poses a number of challenging problems. Within a few radii of Earth the field approximates closely to a potential field (a dipole plus higher harmonics) with the sources within the body of the planet. It was discovered that the field fluctuates occasionally, which we know now is caused by motions in the ionosphere, by the variable pressure and drag of the solar wind, and by the pressure of fast particles trapped in the field. The historical development of knowledge of these fluctuations, as recorded and studied from laboratories on the surface of Earth, is the subject of Chapman and Bartels (1940). The subject has advanced rapidly and qualitatively since it has been possible to study conditions directly in space (Kennel 1969; Willis 1971; Nagata 1971; Parker and Ferraro 1971; Rostoker 1972; Akasofu and Chapman 1972; Roederer 1972; Lanzerotti 1972; Sckopke 1972; Vasyliunas and Wolf 1973; Carovillano and Siscoe 1973; Hoch 1973; Burch 1974; Hill 1974). The basic hydromagnetic effect is the distortion and confinement of magnetic field to within some ten or twenty radii of Earth by the solar wind. The turbulent coupling of the wind to the magnetic field at the outer boundary (the magnetopause) drags the lines of force in the anti-solar direction to form the geomagnetic tail, with a (nearly) neutral sheet down the middle, as sketched in Fig. 6.3. The 24-h rotation of Earth and its field inside the eccentric boundary of the field, together with the drag of the wind causes the magnetic field to convect (Axford and Hines 1961; Walbridge 1967; Axford 1969; Mozer 1973). The planets are unique in having a layer of cold, electrically insulating lower atmosphere between the ionized gases of the outer atmosphere in space and the conducting solids and fluids in the interior. The layer of insulation breaks the tying of magnetic lines of force between material particles above and below it, permitting convection of the external gases and field lines. The convection produces potential differences in the frame of reference of Earth of the order of 5×10^4 V. These potential differences appear between the magnetic lines of force at various positions in the field and play a basic role in producing the aurora. A number of interesting ideas have been developed along these lines (see, for instance, the original paper by Axford and Hines

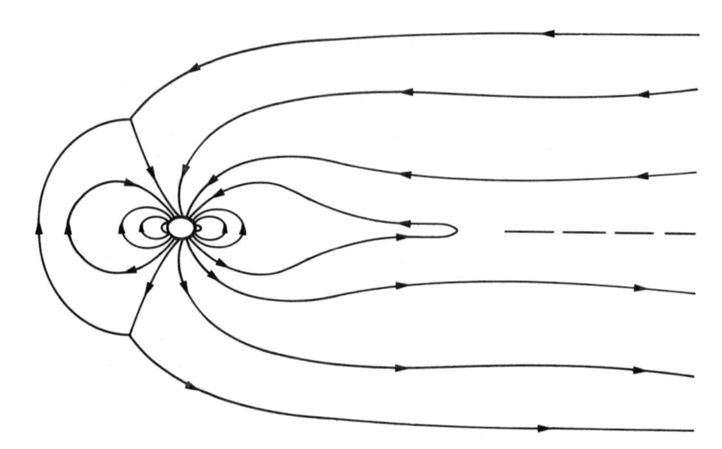

FIG. 6.3. Sketch of the magnetic lines of force of the geomagnetic field as deformed by the solar wind.

(1961); Axford (1969)). There is some suggestion that there may be a strong electric field *parallel* to the magnetic field that produces the fast particles responsible for the aurora (Alfven and Carlquist 1967). The electric current densities in some auroral arcs are so high as to suggest that they excite plasma turbulence, enormously enhancing the effective resistivity (Linhart 1960; Mozer and Bruston 1966, 1967; Kindel and Kennel 1971; Block 1972; Papadopoulos and Coffey 1974; Evans 1974). These ideas are interesting and it is to be hoped that they can be put together into a model that will illustrate the functioning of at least the basic principles.

Jupiter has a dipole field, of some 5–10 G at the poles (Carr and Gulkis 1969) so that the interaction with the solar wind would be similar to that at Earth were it not for the rapid rotation of Jupiter, and the enormous scale of the Jovian magnetosphere. The basic comparison with Earth is easily calculated. The dynamical pressure (Reynolds stress) of the solar wind is ρv^2. The solar wind velocity is approximately constant but, because the wind is radial, the density declines approximately as the inverse square of the distance R from the sun. Hence the dynamical pressure of the wind declines as $1/R^2$. The magnetic field of a planet extends out from the planet to about the point where the magnetic stress density $B^2/8\pi$ declines to the value of the dynamical pressure of the wind. Beyond that point, the wind dominates and sweeps away any lines of force from the planet. Hence, if the field of a planet has a value B_0 at the surface $r = r_0$, than it has a value $B \cong B_0(r_0/r)^3$ at a distance r. The field extends out a distance r_1, to where $B^2/8\pi \cong \rho v^2$. The characteristic dimensions of the geomagnetic field are $r_1 \cong 15r_0 \cong 10^5$ km (the field extends to 10 Earth radii in the sunward direction, about 18 at the dawn–dusk line, and essentially to infinity in the anti-solar direction). The algebra is then straightforward and it is readily shown (designating Earth and Jupiter by the subscripts E and J, respectively) that the characteristic dimension of Jupiter's field is $r_{1J} \cong 50 r_{0J} \simeq 500 r_{0E} \simeq 3\times10^6$ km $\cong 0{\cdot}02$ a.u. It follows that the rotational velocity of the outer regions of the Jovian magnetic field is of the order of $\Omega_J r_{1J} \cong 600$ km s^{-1} (compared to only 7 km s^{-1} for Earth). This exceeds the usual 300–400 km s^{-1} velocity of the solar wind. The outer boundary of the magnetic field on the dawn side of Jupiter experiences a solar wind with a relative velocity of 10^3 km s^{-1}, while the field on the dusk side experiences a wind with a relative velocity of 300 km s^{-1} *toward* the sun. More important for the activity of the magnetic field is the fact that the centrifugal force $\Omega_J^2 r$ of the rotation exceeds the gravitational attraction of Jupiter beyond only about $2r_{0J}$ ($g_{0J} = 2{\cdot}5\times10^3$ cm s^{-2}) where $\Omega_J^2 r = 6\times10^2$ cm s^{-1} (the corresponding figure for Earth is approximately $6r_{0E}$). Hence everywhere beyond two Jovian radii the outer atmosphere of Jupiter is thrown outward by the centrifugal force. Direct observations show that the field is enormously distended in the radial direction at the equatorial plane, presumably as a consequence of the centrifugal force. The reader is referred to the various published compendia of current knowledge on the fascinating properties of the magnetic field of Jupiter (Carr and Gulkis 1969; Mead 1975). Again, the outstanding feature is the copious production of fast particles.

High angular velocity suggests the neutron star or pulsar, with rotation periods as small as 0·07 s (the Crab pulsar, $\Omega = 1\text{–}2\times10^2$ rad s^{-1}) and presumably much shorter when younger and newly formed from the collapsed core of a supernova. It is inferred that the pulsar has a magnetic field with a dipole component at the surface ($r_0 = 5$ km) of the general order of 10^{12} G. Such a field must dominate space for a distance of at least many astronomical units (1 a.u. = Earth–sun distance, $1{\cdot}5\times10^{13}$ cm). Rotation with the field at the angular velocity $\Omega = 10^{+2}$ rad s^{-1} of the neutron star leads to centrifugal forces that exceed gravity at $r = 50r_0 =$ 250 km. The rotational velocity reaches the speed of light at 3×10^3 km. The properties of the magnetic field of the pulsar are a problem in relativistic plasma physics and, for all their fascination, lie beyond the scope of this work. The reader is referred to the literature (Pacini 1968; Goldreich and Julian 1969; Ostriker and Gunn 1969; Hewish 1970; Gunn and Ostriker 1970, 1971; Lerche 1970; Ruderman 1972) for a review of the interesting physics and copious speculation motivated by the discovery of the pulsar (Hewish *et al.* 1968) in 1968. The complex internal workings of the Crab nebula and perhaps other supernova remnants (Woltjer 1972; Scott and Chevalier 1975; Chevalier 1975) are presumed to be the direct consequences of the machinations of the pulsar at its centre.

6.7. Formal mathematical treatment of extended magnetic fields

6.7.1. *General remarks*

The formal theoretical treatment of the magnetic fields extending outward from the surface of an object may be treated by the methods outlined in the earlier sections of this chapter. For a potential field ($\mathbf{B}=-\nabla W$, $\nabla^2 W=0$) then, of course, all of the classical harmonic analysis is available. For axial symmetry ($\partial/\partial\phi=0$) the general method of §6.4 is available, which has as a special case the force-free field ($\mathbf{F}=0$). If the field is independent of a linear coordinate (say $\partial/\partial z=0$) then the method of §6.2 can be employed. Noting that the fluid pressure p is communicated freely along each magnetic line of force ($\mathbf{F}\cdot\mathbf{B}=0$), it follows that the equilibrium calculation of the field leading to (6.9) is unaffected by a gravitational field in any direction if the fluid is incompressible, i.e. if $\rho=$ *constant.* For in that case we add $-\rho\psi$ to the fluid pressure $p(A)$. This addition to p cancels the effect $-\rho\nabla\psi$ of the gravitational field. It is also apparent that a force-free field (in which $p(A)=p_0$) described by $B_z(A)$ and (6.4) is unaffected by a gravitational field in the z-direction provided that the scale height (temperature) $\Lambda(z)$ of the gas is independent of the vector potential A. For if the scale height is independent of A, then the pressure varies as

$$p(A,z)=p_0\exp\left\{-\int_0^z\frac{\mathrm{d}z'}{\Lambda(z')}\right\}$$

along every line of force, $A=$ *constant.* Hence p is independent of A, and does not interfere with the field.

A more general statement of this physical principle is that the gas pressure is communicated freely along the magnetic lines of force (i.e. $\mathbf{B}\cdot(\nabla p+\rho\nabla\psi)=0$). Hence if a reference surface S lies across the magnetic field so that each line of force crosses S at least once, the pressure conditions on S are conveyed to all parts of the magnetic field. Hence if the gas is in hydrostatic equilibrium across S (i.e. if $\nabla p+\rho\nabla\psi=0$ in the surface S) the gas is in hydrostatic equilibrium everywhere throughout the field and the field is force-free.

6.7.2. *Force-free magnetic fields*

Nakagawa *et al.* (1971), Raadu and Nakagawa (1971) and Nakagawa and Raadu (1972) have given an extensive exposition of force-free magnetic fields with $h=$ *constant* (see (6.30)), with particular attention to the configurations outlined by the filaments and chromospheric fibrils around single sunspots and around bipolar regions. They show that the simple solutions for $h=$ *constant* conform remarkably well in many cases to the observed patterns. In particular they consider solutions of (6.36) in cylindrical coordinates,

$$P_n = \frac{1}{\varpi}\frac{\partial^n}{\partial q^n}\{J_1(k\varpi)\exp(\pm qz)\}, \qquad T_n = hP_n$$

where $k=(h^2+q^2)^{\frac{1}{2}}$ and n is any positive integer. For the field extending upward from a surface $z=0$ we choose the negative exponent so that the solution vanishes at $z=+\infty$.

The essential difference between solutions of order $n=0, 1, 2, \ldots$ is illustrated by the spiral angle $\theta^{(n)}$, between the radial direction and the projection of the lines of force on the planes $z=$ *constant.* Then

$$\tan\theta^{(n)} \equiv B_\phi/B_\varpi = -hP_n/(\partial P_n/\partial z).$$

For $n=0$ the spiral is constant, $\tan\theta^{(0)}=h/\beta$, while for $n=1$,

$$\tan\theta^{(1)} = h\frac{kzJ_1(k\varpi)-q\varpi J_0(k\varpi)}{k(qz-1)J_1(k\varpi)-q^2\varpi J_0(k\varpi)}.$$

Near the axis of symmetry ($k\varpi \ll 1$) this is

$$\tan\theta^{(1)} \cong \frac{h}{q}\frac{kz-2q/k}{kz-2q/k-k/q}$$

which is positive for $kz=0$, vanishes at $kz=2q/k$, is negative for $2q/k< kz<2q/k+k/q$, with $\theta^{(1)}$ going to $-\pi/2$ as kz approaches $2q/k+k/q$. For $kz>2q/k+k/q$, $\theta^{(1)}$ is positive and declines monotonically from $\pi/2$. Thus the spiral angle at small $k\varpi$ rotates through nearly 180° with increasing z. The higher order solutions rotate correspondingly more.

Nakagawa *et al.* (1971) present a variety of solutions for $n=0, 1$ viewed from various angles, and show the comparison with photographs of large circular stable sunspots. It is evident from the comparison that the special case $h=$ *constant* captures the principle features of the twisted field of the sunspot.

Raadu and Nakagawa (1971) go on to generate mathematical forms for fields without axial symmetry by noting that, if $\mathbf{B}$ is an axi-symmetric solution to (6.30), then so is $(\mathbf{I}\cdot\nabla)\mathbf{B}$ a solution, where $\mathbf{I}$ is the unit vector $(\cos\phi, -\sin\phi, 0)$ and

$$\mathbf{I}\cdot\nabla = \cos\phi\frac{\partial}{\partial\varpi} - \sin\phi\frac{1}{\varpi}\frac{\partial}{\partial\phi}$$

That $(\mathbf{I}\cdot\nabla)\mathbf{B}$ is a solution of (6.30) is easily verified by direct differentiation and substitution into (6.30), noting that differentiation of the unit vectors $\mathbf{e}_\varpi$ and $\mathbf{e}_\phi$ gives zero except for

$$\partial\mathbf{e}_\varpi/\partial\phi = \mathbf{e}_\phi, \qquad \partial\mathbf{e}_\phi/\partial\phi = -\mathbf{e}_\varpi.$$

The method is applied to the axi-symmetric solutions P_n just given.

They go on to work out a second class of non-axi-symmetric fields based on the general representation of a magnetic field (Chandrasekhar 1961, Appendix III).

$$\mathbf{B}(\mathbf{r}) = \nabla\times\nabla\times(R\mathbf{r}) + \nabla\times(S\mathbf{r}),$$

where $\mathbf{r}$ is the position vector from the origin, and R and S are functions of (ϖ, ϕ, z). With this representation the special ease $h = constant$ reduces (6.30) to

$$\nabla\times[\nabla\times\nabla\times R\mathbf{r} - hS\mathbf{r} + \nabla\times\{(S-hR)\mathbf{r}\}] = 0.$$

Consider, then, the special solution represented by

$$S = hR$$

and

$$\nabla\times\nabla\times R\mathbf{r} - \alpha S\mathbf{r} = \nabla G$$

where G is an arbitrary function. Note that

$$\begin{aligned}\nabla\times\nabla\times R\mathbf{r} &= \nabla\times(\nabla R\times\mathbf{r}) \\ &= 3\nabla R - \mathbf{r}\nabla^2 R + (\mathbf{r}\cdot\nabla)\nabla R - (\nabla R\cdot\nabla)\mathbf{r} \\ &= -\mathbf{r}\nabla^2 R + \nabla\{R + (\mathbf{r}\cdot\nabla)R\}.\end{aligned}$$

the last step is most easily carried out using index notation. It follows that

$$\mathbf{r}\{\nabla^2 R + h^2 R\} = \nabla\{R + (\mathbf{r}\cdot\nabla)R - G\}.$$

Then, with the choice $G = R + (\mathbf{r}\cdot\nabla)R$, we have

$$\nabla^2 R + h^2 R = 0.$$

In rectangular coordinates there are solutions of the form

$$R = \exp(ik_x x + ik_y y - qz)$$

where the wave numbers are related by the dispersion relation $k_x^2 + k_y^2 = q^2 + h^2$. In cylindrical coordinates there are solutions

$$R = J_m(k\varpi)\exp(\pm im\phi)\exp(-qz)$$

where $k^2 = q^2 + h^2$. Raadu and Nakagawa (1971) show how this second class of non-axi-symmetric field modes is related to the first type, the nth mode of the first being expressible in terms of modes of the second up to order $n-1$. They plot the lines of force of a number of examples of both types of field, showing the resemblance to the observed field structure around twisted bipolar sunspots. The question of stability is not considered. We can imagine that the configuration is unstable when the total twist is in excess of one-half a revolution.

The general method developed in these calculations can be extended to more complex structures, with $m \geqslant 2$. It would be particularly interesting to extend the method to highly twisted forms with convoluted flux connections to explore the possibilities for instability.

6.7.3. *Magnetic fields subject to pressure*

Low (1975a) has given a method for calculating the hydrostatic equilibrium of the field that is independent of z but subject to the weight of a gas in a uniform transverse gravitational acceleration, g, say in the y-direction (see also Parker 1968). Then (6.5) and (6.7) are unaffected while (6.6) becomes

$$0=\frac{\partial}{\partial y}\left(p+\frac{B_z^2}{8\pi}\right)+\frac{\nabla_{xy}^2 A}{4\pi}\frac{\partial A}{\partial y}+\frac{P}{\Lambda} \tag{6.42}$$

where $\Lambda=\Lambda(x, y)$ is the scale height $kT(x, y)/Mg$. We know that the pressure extends freely along the lines of force $A=$ *constant*. Hence if T or Λ, is a function only of A and y, we know from the barometric law that

$$p(A, y)=p(A, y_0)\exp\left\{-\int_{y_0}^{y}\frac{dy'}{\Lambda(A, y')}\right\} \tag{6.43}$$

where the integration is along a magnetic line of force $A=$ *constant*. With this expression for p we proceed as before.

To obtain (6.43) by formal means, note from (6.7) that $B_z=B_z(A)$ again. Then multiply (6.5) by $\partial A/\partial y$ and (6.42) by $\partial A/\partial x$ and subtract. The result is

$$\frac{\partial(A, p)}{\partial(x, y)}+\frac{p}{\Lambda}\frac{\partial A}{\partial x}=0 \tag{6.44}$$

which is, in fact, just the scalar product of $\mathbf{B}$ with (6.1). Now $p=p(x, y)$, $\Lambda=\Lambda(x, y)$ can always be written as $p[A(x, y), y]$ and $\Lambda[A(x, y), y]$, giving p and Λ as a function of height y along any line of force, $A=$ *constant*. With this form for p and Λ, (6.44) becomes

$$\left(\frac{\partial p}{\partial y}+\frac{p}{\Lambda}\right)\frac{\partial A}{\partial x}=0. \tag{6.45}$$

Since $\partial A/\partial x\neq 0$, we have the standard barometric law whose integral over y is just (6.43), with the other variable, A, held constant. With (6.45) it is readily shown that (6.5) and (6.42) reduce to

$$\frac{\nabla_{xy}^2 A}{4\pi}+\frac{\partial}{\partial A}\left\{p(A, y)+\frac{B_z^2(A)}{8\pi}\right\}=0. \tag{6.46}$$

Thus, we choose the functions $B_z(A)$, $p(A, y_0)$, and the scale height, or temperature, $\Lambda(A, y)$. The pressure follows from (6.43), and the differential equation (6.46) for $A(x, y)$ is determined. The method is readily adopted to axial symmetry (see Appendix, Low 1975a). The isothermal case $\Lambda = constant$ without a longitudinal field ($B_z = 0$) has been treated by Bhatnager *et al.* (1951), Dungey (1953), Kippenhahn and Schlüter (1957), and Brown (1958) in connection with the structure of quiescent prominences. Observations (Nakagawa and Malville 1969; Tandberg-Hanssen and Anzer 1970) indicate that the longitudinal field B_z is not zero. Presumably the presence of a longitudinal field would enhance the precarious stability of the prominence model (Brown 1958; Nakagawa 1970). More conspicuous, however, is the restriction to the isothermal case when representing a prominence whose existence is the result of a temperature less than 10^{-2} of the ambient value. Consequently Low (1975b) gives two examples to illustrate the equilibrium of a cold vertical sheet of gas supported by a transverse magnetic field and surrounded by a hot plasma. Put $B_z = 0$. In the first example choose the vector potential to be of the form

$$A(x, y) = B_0\{\lambda U(x) \pm y\}, \tag{6.47}$$

where $U(x)$ is a function to be determined, λ is a constant length, and B_0 is a field strength. Then write

$$\Lambda(x, y) = \Lambda(A, y) = \Lambda\{U(x)\} \tag{6.48}$$

and note that along a line of force $A = constant$ (6.47) yields $\mathrm{d}y = \mp\lambda\,\mathrm{d}U$, so that (6.43) becomes

$$p(A, y) = p_0 \exp\left\{\pm\lambda \int_C^U \frac{\mathrm{d}\xi}{\Lambda(\xi)}\right\} \tag{6.49}$$

where C is a constant. The differential equation (6.46) becomes

$$\frac{\mathrm{d}^2 U}{\mathrm{d}\zeta^2} + \gamma \frac{\partial}{\partial U} \exp\left\{\pm\lambda \int_C^U \frac{\mathrm{d}\xi}{\Lambda(\xi)}\right\} = 0 \tag{6.50}$$

where $\zeta = x/\lambda$ and $\gamma = 4\pi p_0/B_0^2$.

This equation is reducible by quadratures. Multiply by $\mathrm{d}U/\mathrm{d}\zeta$ and integrate,

$$\left(\frac{\mathrm{d}U}{\mathrm{d}\zeta}\right)^2 + 2\gamma \exp\left\{\pm\lambda \int_C^U \frac{\mathrm{d}\xi}{\Lambda(\xi)}\right\} = 2\gamma D \tag{6.51}$$

where D is a constant. Then separating variables and integrating,

$$(2r)^{\frac{1}{2}}\zeta = \int \mathrm{d}U\left[D - \exp\left\{\pm\lambda \int_C^U \frac{\mathrm{d}\xi}{\Lambda(\xi)}\right\}\right]^{-\frac{1}{2}}. \tag{6.52}$$

It is obvious from the absence of the vertical coordinate y in (6.48), (6.49), and (6.52) that the solution represents an infinite vertical uniform sheet across which the temperature varies as $\Lambda[U(x)]$ and the density as p/Λ. The first example, $A = B_0[\lambda U(x) - y]$ and $\Lambda = \lambda(U/U_0)^{\frac{2}{7}}$ is equivalent to a vertical layer of plasma with an internal heat sink (presumably radiative cooling) proportional to the density, into which heat is conducted from both sides. The choice $\Lambda \propto U^{\frac{2}{7}}$ is equivalent to heat conduction with a thermal conduction coefficient proportional to $T^{\frac{5}{2}}$, appropriate for an ionized gas (Chapman 1954). The pressure, density, and temperature, and the form of the field are given from numerical evaluation of (6.52). The magnetic lines of force sag under the weight of the vertical slab of cold gas. Outside the slab, with surface mass density $d(\mathrm{g\ cm^{-2}})$ the vertical and horizontal components of the field are given by

$$B_x B_y/4\pi = gd/2,$$

in order that the tension in the lines of force on each side support the weight of the gas. As in all models which do not employ an inhibition of thermal conduction into the sides of the layer of cool gas (presumably as a consequence of a longitudinal field B_z) the characteristic thickness of the model layer proves to be several times thicker than in the real prominence.

The second example

$$A = B_0\{\lambda U(x) + y\},$$
$$C = 0, \qquad D = 1,$$
$$\Lambda/\lambda = 1 + E\exp(-U),$$

is designed to give analytic solutions which have as a subclass the isothermal solutions given earlier by Kippenhahn and Schlüter (1957). Then it is readily shown that (6.51) reduces to

$$(\mathrm{d}U/\mathrm{d}\zeta)^2 = 2\Gamma(1 - \exp U)$$

where, for convenience, we have written $\Gamma = \gamma/(1+E) = 4\pi p_0/B_0^2(1+E)$. Integration yields

$$\psi = 2\ln\operatorname{sech}(\Gamma/2)^{\frac{1}{2}}\zeta,$$

where the origin of the coordinates is placed so that $U = 0$ where $\zeta = 0$. Then

$$A = B_0[2\lambda\ \ln\{\operatorname{sech}(\Gamma/2)^{\frac{1}{2}}\zeta\} + y]$$
$$\Lambda = 1 + E\cosh^2(\Gamma/2)^{\frac{1}{2}}\zeta$$
$$p = p_0[E + \operatorname{sech}^2(\Gamma/2)^{\frac{1}{2}}\zeta]/(E+1).$$

The temperature increases exponentially away from the dense cold plasma sheet supported in the neighbourhood of $\zeta = 0$ by the magnetic field. Putting $E = 0$ gives the isothermal case. The pressure has the value

p_0 in the centre of the plasma sheet, falling to the fraction $E/(E+1)$ at $x = \pm\infty$. The density $\rho = p/g\Lambda$ reduces to

$$\rho = \{p_0/g\lambda(E+1)\}\text{sech}^2\{(\Gamma/2)^{\frac{1}{2}}x/\lambda\}.$$

which declines exponentially to zero with distance from the plasma sheet, thereby compensating for the exponential increase of temperature to give $p \sim$ *constant* as $x \to \pm\infty$. The total mass per unit area of the prominence sheet is $\int dx\rho = 2^{\frac{3}{2}}p_0/g\Gamma^{\frac{1}{2}}(E+1)$, and its weight is supported by the magnetic field

$$B_x = B_0$$

$$B_y = B_0(2\Gamma^{\frac{1}{2}} \tanh\{(\Gamma/2)^{\frac{1}{2}}x/\lambda\}.$$

The supporting field outside the sheet of plasma makes an angle arc $\tan(2\Gamma)^{\frac{1}{2}}$ with the horizontal.

This simple example provides an illustration of the confinement and support of a gas in a gravitational field. It is clear that not many examples can be worked out exactly by analytical methods, but until the stability of the solutions can be explored, the example given by Low should suffice. The requirement of long term stability may introduce further requirements that are, so far, unanticipated.

6.8. Diffusion of an equilibrium configuration

This is, perhaps the appropriate place to discuss the problem of the passive diffusion of a magnetic field in a resistive medium. The diffusion of a magnetic field in a medium constrained to rest is described by the linear heat flow equation (4.15), the solutions of which are well known. Thus, the diffusion of a *weak* field, which leaves the ambient fluid relatively undisturbed, is a straightforward problem, with results that are familiar from classical heat flow. But if the field is not weak, so that its varying pressure causes the fluid to move, the problem is not so simple. A simultaneous solution of (4.12) and (5.10), with the equation of continuity and the equation of state, is required. Such problems have received but little attention, although they are an integral part of basic hydromagnetic theory. We present a simple example here to illustrate the general principles.

Consider the diffusion of a slab of isothermal gas of density $\rho(x, t)$ and fixed temperature T embedded in a magnetic field $B(x, t)$ in the y-direction. The pressure $p(x, t)$ of the gas is $u^2\rho(x, t)$ where u is the thermal velocity $(kT/m)^{\frac{1}{2}}$. The diffusion of the field is accompanied by the fluid velocity $v(x, t)$ in the x-direction. The hydromagnetic induction equation is

$$\partial B/\partial t + \partial vB/\partial x = \eta\partial^2 B/\partial x^2 \qquad (6.53)$$

for a uniform resistive diffusion coefficient η. Conservation of matter requires

$$\partial p/\partial t+\partial vp/\partial x=0. \tag{6.54}$$

The momentum equation is

$$\rho(\partial v/\partial t+v\partial v/\partial x)=-(\partial/\partial x)(p+B^2/8\pi). \tag{6.55}$$

Simultaneous solution of these three equations yields the diffusion of the gas pressure $p(x, t)$ with time.

To display the 'diffusion' equation resulting from simultaneous solution of (6.53)–(6.55), suppose that the diffusion coefficient η is not so large that the diffusion velocity v has significant inertial effects. Then the left-hand side of (6.55) can be neglected and the result integrated to give

$$p+B^2/8\pi=P_0 \tag{6.56}$$

where the constant P_0 represents the total pressure. Then use (6.56) to express p in terms of B in (6.54). Solve for $\partial B/\partial t$ and substitute into (6.53). The result is the expression

$$(P_0+B^2/8\pi)\partial v/\partial x=(\eta B/4\pi)\partial^2 B/\partial x^2$$

for $\partial v/\partial x$ in terms of B. Then write (6.53) in the form

$$v=-\left\{\left(\frac{\partial}{\partial t}-\eta\frac{\partial^2}{\partial x^2}\right)B+B\frac{\partial v}{\partial x}\right\}\bigg/\frac{\partial B}{\partial x}.$$

Differentiate with respect to x and replace $\partial v/\partial x$ with the expression above. The final result can be written

$$\frac{B}{4\pi}\frac{\partial^2 B}{\partial x^2}+\left(P_0+\frac{B^2}{8\pi}\right)\frac{\partial}{\partial x}\left\{\frac{\partial B/\partial t}{\eta\partial B/\partial x}\right.$$
$$\left.-\frac{\partial^2 B/\partial x^2}{\partial B/\partial x}\frac{P_0-B^2/8\pi}{P_0+B^2/8\pi}\right\}=0. \tag{6.57}$$

This is the form of the diffusion equation, in place of (4.15). The equation is third order, instead of second, and highly non-linear.

To obtain a solution of (6.57) to illustrate the basic physical effects, suppose that the magnetic field is strong, in the sense that any fluctuations δp in the gas pressure are small compared to the magnetic pressure, $\delta p\ll B^2/8\pi$. Then suppose that p_0 is the minimum gas pressure anywhere in the system (this may be zero or may be very large compared to $B^2/8\pi$) and B_0 is the maximum magnetic field (occurring where $p=p_0$) so that $P_0=p_0+B_0^2/8\pi$. Then write $B=B_0+\delta B$ and note that $\delta B\leqslant 0$ and $|\delta B|\ll B_0$ because $\delta p\ll B_0^2/8\pi$. For convenience write $p=p_0+\delta p$, but note that δp is not necessarily small compared to p_0. Then neglecting terms second order in $\delta B/B_0$, (6.56) reduces to the linear relation

$$\delta p + (B_0^2/8\pi)\delta B/B_0 \cong 0 \tag{6.58}$$

between the variation of the field and the gas pressure. We proceed now to an exact solution of (6.43), (6.54), and (6.58).

Following the same procedure as before, use (6.58) to write δp in terms of δB in (6.54), solve for $\partial\delta B/\partial t$, and use the result to eliminate $\partial\delta B/\partial t$ from (6.53). The result is

$$(1+\tfrac{1}{2}\beta)\partial v/\partial x = \eta(\partial^2/\partial x^2)\delta B/B_0$$

where β is the ratio of gas to field pressure $8\pi p_0/B_0^2$. Integrate once and transform to the frame of reference in which the gas velocity is zero where the gas pressure is p_0,

$$v = \{\eta/(1+\tfrac{1}{2}\beta)\}(\partial/\partial x)\delta B/B_0. \tag{6.59}$$

Then (6.54) becomes

$$\frac{\partial}{\partial t}\frac{\delta B}{B_0} = \frac{\eta}{1+\tfrac{1}{2}\beta}\frac{\partial}{\partial x}\left\{\left(\tfrac{1}{2}\beta - \frac{\delta B}{B_0}\right)\frac{\partial}{\partial x}\frac{\delta B}{B_0}\right\}. \tag{6.60}$$

This is a diffusion equation with both a linear diffusion term and a non-linear term in which the effective diffusion coefficient is proportional to the perturbation of the field. Going back to (6.58) eqn (6.60) can be written in terms of the gas perturbation

$$\frac{\partial}{\partial t}\frac{\delta p}{p_0} = \frac{\tfrac{1}{2}\eta\beta}{1+\tfrac{1}{2}\beta}\frac{\partial}{\partial x}\left\{\left(1+\frac{\delta p}{p_0}\right)\frac{\partial}{\partial x}\frac{\delta p}{p_0}\right\}, \tag{6.61}$$

wherein $\delta p/p_0$ is not necessarily small compared to one. Note, however, that the total gas pressure $p_0+\delta p$ is non-negative, so the effective diffusion coefficient is positive and the gas distribution, and hence the field perturbation, spread out with the passage of time.

Now to solve either (6.60) or (6.61) let $S = \tfrac{1}{2}\beta - \delta B/B_0$ so that (6.60) becomes

$$\partial S/\partial t = \{\tfrac{1}{2}\eta/(1+\tfrac{1}{2}\beta)\}\partial^2 S^2/\partial x^2 \tag{6.62}$$

and let $\Pi = 1 + \delta p/p_0 = p/p_0$, so that (6.61) becomes

$$\partial\Pi/\partial t = \{\tfrac{1}{4}\beta\eta/(1+\tfrac{1}{2}\beta)\}\partial^2\Pi^2/\partial x^2. \tag{6.63}$$

The two equations are of the same form, so consider the solution of (6.63).

There are a variety of solutions that can be constructed to illustrate the nature of the diffusion. In particular, consider the diffusion of an isolated slab of gas that is initially very thin and located at $x=0$ in an infinite space $-\infty < x < +\infty$. Then initially $\rho(x,0) = \mu\delta(x)$, where μ is the mass per unit area. Then the background gas pressure p_0 is zero and (6.63) reduces to

$$\partial p/\partial t = (2\pi\eta/B_0^2)\partial^2 p^2/\partial x^2. \tag{6.64}$$

This equation is easily solved using the similarity form

$$p = t^{-\alpha} f(\xi), \qquad \xi = xt^{-\alpha},$$

for which

$$u^2\mu = \int_{-\infty}^{+\infty} \mathrm{d}x p(x, t) = \mathit{constant}.$$

Then, substituting this form for p into (6.64), it is seen that if $\alpha = \frac{1}{3}$, (6.63) can be expressed entirely in terms of ξ, becoming

$$(6\pi\eta/B_0^2)\, \mathrm{d}^2 f^2/\mathrm{d}\xi^2 + \mathrm{d}\xi f/\mathrm{d}\xi = 0.$$

Integrating once yields

$$(6\pi\eta/B_0^2)\, \mathrm{d}f^2/\mathrm{d}\xi + \xi f = C,$$

where C is the integration constant. The gas distribution is symmetric about $x = 0$, so the desired solution is an even function of x. Hence $C = 0$ and

$$f\{(12\pi\eta/B_0^2)\, \mathrm{d}f/\mathrm{d}\xi + \xi\} = 0.$$

There are, then, two solutions,

$$f_1(\xi) = 0$$

and

$$f_2(\xi) = f(0) - (B_0^2/24\pi\eta)\xi^2$$

where $f(0)$ is the integration constant. Since $f(\xi)$ represents pressure or density, it must be positive. Hence the physical solution is the inverted parabola $f_2(\xi)$ for $\xi^2 \leqslant \xi_1^2 = 24\pi\eta f(0)/B_0^2$ and $f_1(\xi)$ for $\xi^2 > \xi_1^2$. The gas spreads out with time, the thickness of the gas distribution increasing as $t^{\frac{1}{3}}$. The gas density distribution is an inverted parabola, falling to zero at $x = \pm\xi_1 t^{\frac{1}{3}}$, and zero everywhere beyond. There is, then, no precursor, as with linear diffusion, because the effective diffusion coefficient falls to zero along with the density.

The integration constant $f(0)$ is a measure of the mass density μ in the initial sheet of gas,

$$u^2\mu = \int_{-\xi_1}^{+\xi_1} \mathrm{d}\xi\{f(0) - \xi^2 B_0^2/24\pi\eta\},$$

from which it follows that

$$f(0) = (3u^2\mu/4)^{\frac{2}{3}}(B_0^2/24\pi\eta)^{\frac{1}{3}}.$$

The gas distribution evolves from the initial $\mu\delta(x)$ in the form of the inverted parabola

$$p(x, t) = (B_0^2/24\pi\eta t)^{\frac{1}{3}}\{(3u^2\mu)^{\frac{2}{3}} - x^2(B_0^2/24\pi\eta t)^{\frac{2}{3}}\} \quad (6.65)$$

of width

$$x_1(t) = 2(9\pi u^2\mu\eta t/B_0^2)^{\frac{1}{3}}. \quad (6.66)$$

It is interesting to note that, although the diffusion equation (6.64) is non-linear, separate initial thin sheets of gas diffuse independently of each other until such time as neighbouring solutions meet. It should be noted too that the inverted parabola (6.60) is the asymptotic form at large time for any initial distribution confined to a finite interval h.

The advancing edge of the gas distribution can be illustrated directly by solutions of the form $p = p(\zeta)$ with $\zeta = x - wt$. Then (6.64) becomes

$$(\mathrm{d}/\mathrm{d}\zeta)[\{w + (4\pi\eta/B_0^2)\,\mathrm{d}p/\mathrm{d}\zeta\}p] = 0.$$

Integrating once yields

$$\{w + (4\pi\eta/B_0^2)\,\mathrm{d}p/\mathrm{d}\zeta\}p = wp_1$$

where p_1 is the constant of integration. For the present case where p falls to zero while $\mathrm{d}p/\mathrm{d}\zeta$ remains finite, put $p_1 = 0$. Then there are the two solutions

$$p = 0, \qquad p = -(B_0^2 w/4\pi\eta)(\zeta - \zeta_0).$$

The second solution obtains for $\zeta < \zeta_0$ and the first for $\zeta > \zeta_0$. We have an advancing front of gas with a linear decline to zero at the leading edge $x = \zeta_0 + wt$. Beyond the edge there is no gas. The velocity of advance w is proportional to the rate of decline toward the edge, $w = -(4\pi\eta/B_0^2)\,\mathrm{d}p/\mathrm{d}\zeta$.

More generally, write $\tau = \frac{1}{4}t\beta\eta/(1 + \frac{1}{2}\beta)$ and put $\Pi = \Pi(\zeta)$ in (6.63). The result is

$$\Pi(2\,\mathrm{d}\Pi/\mathrm{d}\zeta + w) = w\Pi_1.$$

Integrating again, we obtain

$$\Pi + \Pi_1 \ln(\Pi - \Pi_1) = \tfrac{1}{2}w(\zeta_1 - \zeta),$$

where ζ_1 is the second constant of integration. If $\Pi_1 > 0$, note that for $\zeta \to -\infty$ the linearly increasing gas density

$$\Pi \sim -\tfrac{1}{2}w(\zeta - \zeta_1)$$

advances with the velocity $\mathrm{d}x/\mathrm{d}\tau = w$. The front has an exponential tail extending ahead of it to $\zeta = +\infty$,

$$\Pi - \Pi_1 \cong \exp\{-\tfrac{1}{2}w(\zeta - \zeta_1)/\Pi_1\}.$$

Clearly, the validity of these solutions is limited to values of Π that are not so large as to violate $\delta B/B_0 \ll 1$.

6.9. Laboratory fields and plasmas

We should not fail to note the extensive experimental and theoretical work on the equilibrium of magnetic fields and plasmas carried on in connection with the worldwide effort at plasma confinement and, hopefully, controlled thermonuclear power. The laboratory geometry is complementary to that encountered in the astrophysical body, so there is little direct application of the laboratory results to the astrophysical problem. But there are some general implications that should certainly not be overlooked. To contrast and compare the two, note first that in the astrophysical body it is the plasma, bound by gravity, which grips and holds the magnetic field. On the other hand, in the laboratory the magnetic field is produced and held by electric currents forced through rigid metal conductors, and the purpose is to trap a small amount of plasma at some particular place in the magnetic field. Working from the foundation provided by the rigid conductors it is possible to construct an equilibrium configuration for confinement of an ideal fluid in the laboratory magnetic field. Indeed, it is possible to construct equilibria which, in the hydromagnetic (ideal fluid) approximation, have most of their large-scale instabilities strongly suppressed, relying heavily, of course, on the surrounding rigid conductors, end plates, etc. Unfortunately, the system is generally subject to small-scale plasma instabilities—microturbulence—as a consequence of the plasma not being an ideal fluid but rather a gas, composed of electrons and ions, each with a finite cyclotron radius in the magnetic field, and an anisotropic velocity distribution in space, as well as overall density and temperature gradients. The microturbulence leads to anomalous diffusion of the plasma across the magnetic field and out of the system. Only with the greatest ingenuity has it been possible in recent years to design systems which minimize the anomalous diffusion and approach the question of the hydromagnetic stability problems. The difficulty of achieving stable magnetohydrostatic equilibrium is clearly impressed upon the mind of anyone making a serious study of the laboratory plasma confinement problem (See, for instance, Artsimovich 1964; Kunkel 1966; Jeffrey and Taniuti 1966; Kadomtsev 1968; International Atomic Energy Commission 1969; Solovev and Shafranov 1970; Dyachenko and Imshennik 1970, and references therein).

It is interesting to note, then, that in the astrophysical problem, wherein the field is trapped by the gas, rather than by rigid metal conductors, the basic confining force of gravity unavoidably introduces magnetic buoyancy. The buoyancy is one of the major causes of non-equilibrium and instability, so that continual dynamical activity rather than stable equilibrium appears to be the rule. Microturbulence and anomalous diffusion appear in astrophysics, too, at least in the extreme field gradients across magnetic neutral sheets, believed to play a fundamental role in the solar flare and in the geomagnetic tail.

References

AKASOFU, S. I. and CHAPMAN, S. (1972). *Solar–terrestrial physics.* Clarendon Press, Oxford.
ALFVEN, H. and CARLQUIST, P. (1967). *Solar Phys.* **1,** 220.
ALTSCHULER, M. D. and NEWKIRK, G. (1969). *Solar Phys.* **9,** 131.
ANZER, U. (1968). *Solar Phys.* **3,** 298; **4,** 101.
ARTSIMOVICH, L. A. (1964). *Controlled thermonuclear reactions.* Gordon and Breach, New York.
AXFORD, W. I. (1969). *Rev. Geophys. Space Phys.* **7,** 421.
—— and HINES, C. O. (1961). *Can. J. Phys.* **39,** 1433.
BABCOCK, H. W. (1958). *Astrophys. J. Suppl.* **3,** 141.
—— (1960a) in *Stars and stellar systems, Stellar atmospheres* (ed. J. L. GREENSTEIN), vol. 6, p. 282. University of Chicago Press, Chicago.
—— (1960b). *Astrophys. J.* **132,** 521.
BAKER, N. and KIPPENHAHN, R. (1959). *Z. Astrophys.* **48,** 140.
BARNES, C. W. and STURROCK, P. A. (1972). *Astrophys. J.* **174,** 659.
BATH, G. T., EVANS, W. D., and PRINGLE, J. E. (1974). *Mon. Notic. Roy. Astron. Soc.* **166,** 113.
BERNSTEIN, I. B., FRIEDMAN, E. A., KRUSKAL, M. D., and KULSRUD, R.M. (1958). *Proc. Roy. Soc., Ser. A.* **244,** 17.
BHATNAGER, P. L., KROOK, M., and MENZEL, D. H. (1951). *Report of conference on dynamics of ionized media,* University College, London
BLOCK, L. (1972). *Cosmic Electrodyn.* **3,** 349.
BORRA, E. F. (1974). *Astrophys. J.* **187,** 271.
BROWN, A. (1958). *Astrophys. J.* **128,** 646.
BURCH, J. L. (1974). *Rev. Geophys. Space Phys.* **12,** 363.
CALLEBAUT, D. K. and VOSLANDER, D. (1962). *Phys. Rev.* **127,** 1857.
CARMICHAEL, H. (1964). *The physics of solar flares,* NASA SP-50 (ed. W. N. HESS) p. 451. U.S. Government Printing Office, Washington D.C..
CAROVILLANO, R. L. and SISCOE, G. L. (1973). *Rev. Geophys. Space Phys.* **11,** 289.
CARR, T. D. and GULKIS, S. (1969). *Annl. Rev. Astron. Astrophys.* **7,** 577.
CHANDRASEKHAR, S. (1956a). *Proc. Nat. Acad. Sci.* **42,** 1.
—— (1956b). *Astrophys. J.* **124,** 232.
—— (1958). *Proc. Nat. Acad. Sci.* **44,** 843.
—— (1961). *Hydrodynamic and hydromagetic stability.* Clarendon Press, Oxford.
—— (1969). *Ellipsoidal figures of equilibrium.* Yale University Press. New Haven.
—— and FERMI, E. (1953). *Astrophys. J.* **118,** 116.
—— and KENDALL, P. C. (1957). *Astrophys. J.* **126,** 457.
—— and LEBOVITZ, N. R. (1964). *Astrophys. J.* **140,** 1517.
—— and LIMBER, D. N. (1954). *Astrophys. J.* **11** , 10.
—— and PRENDERGAST, D. N. (1954). *Astrophys. J.* **119,** 10.
—— and WOLTJER, L. (1958). *Proc. Nat. Acad. Sci.* **44,** 285.
CHANMUGAM, G. (1974). *Mon. Notic. Roy. Astron. Soc.* **169,** 353.
CHAPMAN, S. (1954). *Astrophys. J.* **120,** 151.

—— and BARTELS, J. (1940). *Geomagnetism*, vol. I, II. Clarendon Press, Oxford.
CHEVALIER, R. (1975). *Astrophys. J.* **198,** 355; **200,** 698.
DAVIES, G. F. (1968). *Austl. J. Phys.* **21,** 294.
DEUTSCH, A. J. (1958). *I.A.U. Symposium* No. 6 (ed. B. LEHNERT), p. 209. Cambridge University Press, Cambridge.
DICKE, R. H. (1970). *Astrophys. J.* **159,** 25.
DUNGEY, J. W. (1953). *Mon. Notic. Roy. Astron. Soc.* **113,** 180.
—— (1958). *Cosmic electrodynamics.* Cambridge University Press.
DYACHENKO, V. F. and IMSHENNIK, V. S. (1970). *Rev. Plasma Phys.* **5,** 447.
EDDINGTON, A. S. (1929). *Mon. Notic. Roy. Astron. Soc.* **90,** 54.
EVANS, D. S. (1974). *J. Geophys. Res.* **79,** 2853.
FERRARO, V. C. A. (1954). *Astrophys. J.* **119,** 407.
GJELLESTAD, G. (1954a). *Astrophys. J.* **119,** 14.
—— (1954b). ibid. **120,** 172.
—— (1957) ibid. **126,** 565.
GOLD, T. and HOYLE, F. (1960). *Mon. Notic. Roy. Astron. Soc.* **120,** 89.
GOLDREICH, P. and JULIAN, W. (1969). *Astrophys. J.* **157,** 869.
GUNN, J. E. and OSTRIKER, J. P. (1970). *Astrophys. J.* **160,** 979.
—— —— (1971). ibid. **165,** 523.
HAIN, K., LÜST, R., and SCHLÜTER, A. (1957). *Z. Naturforsch. A* **12,** 833.
HEWISH, A. (1970). *Ann. Rev. Astron. Astrophys.* **8,** 265.
—— BELL, S. J., PILKINGTON, J. D., SCOTT, J. D., and COLLINS, R. A. (1968). *Nature* **217,** 709.
HILL, T. W. (1974). *Rev. Geophys. Space Phys.* **12,** 379.
HOCH, R. J. (1973). *Rev. Geophys. Space Phys.* **11,** 935.
International Atomic Energy Agency (1969). *Plasma physics and controlled nuclear fusion research,* Proc. Third Intern. Conf. Plasma Phys. and Controlled Nuclear Fusion Res. Novosibirsk, August 1968 Vol. I, II. (IAEC, Vienna).
DE JAGER, C. (1959). *Handbuch der Physik* (ed. S. FLUGGE), vol. 52, p. 151. Springer-Verlag, Berlin.
JEFFREY, A. and TANIUTI, T. (1966). *Magnetohydrodynamic stability and thermonuclear containment.* Academic Press, New York.
JENSON, E. (1955). *Astrophys. J. Suppl.* **2,** (16).
KADOMTSEV, B. B. (1968). *Rev. Plasma Phys.* **2,** 153.
KAI, K. (1969). *Proc. Astron. Soc. Aust.* **1,** 186.
KENNEL, C. F. (1969), *Rev. Geophys. Space Phys.* **7,** 379.
KINDEL, J. M. and KENNEL, C. F. (1971). *J. Geophys. Res.* **76,** 3055.
KIPPENHAHN, R. and SCHLÜTER, A. (1957), *Z. Astrophys.* **43,** 36.
KOCHHAR, R. K. and TREHAN, S. K. (1974). *Astrophys. Space Sci.* **26,** 271.
KUNKEL, W. B. (1966). *Plasma physics in theory and application.* McGraw-Hill, New York.
KUPERUS, M. and RAADU, M. A. (1974). *Astron. Astrophys.* **31,** 189.
KUPERUS, M. and TANDBERG-HANSSON, E. (1967). *Solar Phys.* **2,** 39.
LANDSTREET, J. D. (1970). *Astrophys. J.* **159,** 1001.
LANZEROTTI, L. J. (1972). *Rev. Geophys. Space Phys.* **10,** 379.
LERCHE, I. (1970). *Astrophys. J.* **159,** 229; **160,** 1003; **162,** 153.
LILLIEQUIST, C. G., ALTSCHULER, M. D., and NAKAGAWA, Y. (1971). *Solar Phys.* **20,** 348.
LINHART, J. G. (1960). *Plasma physics,* p. 173. North-Holland, Amsterdam.
LOW, B. C. (1975a). *Astrophys. J.* **197,** 251.
—— (1975b), ibid. **198,** 211.

LUNDQUIST, S. (1950). *Ark. Fys.* **2,** 361.
—— (1951). *Phys. Rev.* **83,** 307.
LÜST, R. and SCHLÜTER, A. (1954). *Z. Astrophys.* **34,** 263, 365.
LYTTKENS, E. (1954). *Astrophys. J.* **119,** 413.
MARKEY, P. and TAYLER, R. J. (1973). *Mon. Notic. Roy. astron. Soc.* **163,** 77.
—— —— (1974). ibid. **168,** 505.
MEAD, G. D. (1974). *J. Geophys. Res.* 79, No. 25, September 1. *Pioneer* 10 *Mission* Jupiter encounter, (ed. G. D. MEAD). See also *Science* **183,** 301–324 (1974) and *Science* May 2 (1975).
MESTEL, L. (1965a). *Stars and stellar systems* (ed. L. ALLER and D. MCLAUGHLIN), vol. 8, p. 465. University of Chicago Press, Chicago.
—— (1965b). *Quart. J. Roy. Astron. Soc.* **6,** 161, 265.
—— (1966). *Z. Astrophys.* **63,** 196.
—— (1968). *Mon. Notic. Roy. Astron. Soc.* **138,** 359.
—— (1971). *Quart. J. Roy. Astron. Soc.* **12,** 402.
—— *and* MOSS, D. L. (1977). *Mon. Notic. Roy. Astron. Soc.* **178,** 27.
MONAGHAN, J. J. (1966). *Mon. Notic. Roy. Astron. Soc.* **132,** 1.
—— (1973). ibid. **163,** 423.
—— (1976). *Astrophys. Space Sci.* **40,** 385.
MONAGHAN, J. J. and ROBSON, K. W. (1971). *Mon. Notic. Roy. Astron. Soc.* **155,** 231.
MORETON, G. E. and SEVERNY, A. B. (1968). *Solar Phys.* **3,** 282.
MORSE, P. M. and FESHBACH, H. (1953). *Methods of theoretical physics,* p. 782. McGraw-Hill, New York.
MOSS, M. L. (1973). *Mon. Notic. Roy. Astron. Soc.* **164,** 33.
MOZER, F. S. (1973). *Rev. Geophys. Space Phys.* **111,** 755.
—— and P. BRUSTON (1966). *J. Geophys. Res.* **71,** 4461.
—— —— (1967). ibid. **72,** 1109.
NAGATA, T. (1971). *Handbuch d. Physik,* vol. 49(3), p. 1. Springer-Verlag, Berlin.
NAKAGAWA, Y. (1970). *Solar Phys.* **12,** 419.
—— (1973). *Astron. Astrophys.* **27,** 95.
—— and MALVILLE, J. M. (1969). *Solar Phys.* **9,** 102.
—— and RAADU, M. A. (1972). *Solar Phys.* **25,** 127.
—— ——, BILLINGS, D. E., and MCNAMARA, D. (1971). *Solar Phys.* **19,** 72.
—— ——, and J. HARVEY 1973, *Solar Phys.* **30,** 421.
NEWCOMB, W. A. (1960). *Ann. Phys.* **10,** 232.
NEWKIRK, G., ALTSCHULER, M. D., and HARVEY, J. (1968). in *Structure and development of active regions,* I.A.U. Symposium 35 (ed. K. O. KIEPENHEUER), p. 379.
—— DUPREE, R. G., and SCHMAHL, E. J. (1970). *Solar Phys.* **13,** 131; **15,** 15.
OKAMOTO, I. (1974). *Mon. Notic. Roy. Astron. Soc.* **166,** 683.
OPIK, E. J. (1951). *Mon. Notic. Roy. Astron. Soc.* **111,** 278.
ORRALL, F. Q. and ZIRKER, J. B. (1961). *Astrophys. J.* **134,** 72.
OSTRIKER, J. P. and GUNN, J. E. (1969). *Astrophys. J.* **157,** 1395.
PACINI, F. (1968). *Nature* **219,** 145.
PAPADOPOULOS, K. and COFFEY, T. (1974). *J. Geophys. Res.* **79,** 674, 1558.
PARKER, E. N. (1953). *Astrophys. J.* **117,** 431.
—— (1955), ibid. **121,** 491.
—— (1957). *Phys. Rev.* **107,** 830.
—— (1963). *Interplanetary dynamical processes.* Wiley-Interscience, New York.

—— (1965). *Astrophys. J.* **142,** 584.
—— (1966). ibid. **145,** 811.
—— (1968). ibid. **154,** 57 (Appendix A).
—— (1969). *Space Sci. Rev.* **9,** 651.
—— (1972). *Astrophys. J.* **174,** 499.
—— (1974). *Astrophys. Space Sci.* **31,** 261.
—— (1975). *Astrophys. J.* **198,** 205.
—— and FERRARO, V. C. A. (1971). *Handbuch d. Physik*, vol. 49(3), p. 131. Springer-Verlag, Berlin.
PNEUMAN, G. W. and KOPP, R. A. (1971). *Solar Phys.* **18,** 258.
PRENDERGAST, K. H. (1956). *Astrophys. J.* **123,** 498.
——(1957). ibid. **128,** 361.
PRESTON, G. W. (1967a) *The magnetic and related stars* (ed. R. C. CAMERON), p. 3. Mono Book Corp., Baltimore.
—— (1967b). *Astrophys. J.* **150,** 547.
—— (1969a). ibid. **156,** 967.
—— (1969b). ibid. **158,** 243, 1081.
—— (1971). ibid **164,** 309.
—— (1972). ibid. **175,** 465.
RAADU, M. A. (1972a). *Solar Phys.* **22,** 425.
—— (1972b). ibid. 443.
—— and KUPERUS, M. (1973). *Solar Phys.* **28,** 77.
—— and NAKAGAWA, Y. (1971). *Solar Phys.* **20,** 64.
ROBERTS, P. H. (1955). *Astrophys. J.* **122,** 508.
—— (1956). ibid. **124,** 430.
ROEDERER, J. G. (1972). *Rev. Geophys. Space Phys.* **10,** 599.
ROSTOKER, G. (1972). *Rev. Geophys. Space Phys.* **10,** 157.
RUDERMAN, M. (1972). *Ann. Rev. Astron. Astrophys.* **10,** 427.
RUST, D. M. (1967). *Astrophys. J.* **150,** 313.
SCKOPKE, N. (1972). *Cosmic Electrodyn.* **3,** 330.
SCOTT, J. S. and CHEVALIER, R. A. (1975). *Astrophys. J. Lett.* **197,** L5.
SHEELEY, N. R., BOHLIN, J. D., BRUECKNER, G. E., PURCELL, J. D., SCHERRER, V., and TOUSEY, R. (1975). *Solar Phys.* **40,** 103.
SIMON, R. (1958). *Astrophys. J.* **128,** 375.
SOLOVEV, L. S. and SHAFRANOV, V. D. (1970). *Rev. Plasma Phys.* **5,** 1.
SOOD, N. K. and TREHAN, S. K. (1972). *Astrophys. Space Sci.* **16,** 451; **19,** 441.
STIFT, M. J. (1974). *Mon. Notic. Roy. Astron. Soc.* **169,** 471.
STURROCK, P. A. (1972). *Solar Phys.* **23,** 438.
—— and COPPI, B. (1966). *Astrophys. J.* **143,** 3.
—— and SMITH, S. M. (1968). *Solar Phys.* **5,** 87.
SWEET, P. A. (1950). *Mon. Notic. Roy. Astron. Soc.* **110,** 548.
SYKES, J. (1957). *Astrophys. J.* **125,** 615.
TANAKA, K. and NAKAGAWA, Y. (1973). *Solar Phys.* **33,** 187.
TANDBERG-HANSSEN, E. and ANZER, U. (1970). *Solar Phys.* **15,** 158.
TAYLER, R. J. (1973a). *Mon. Notic. Roy. Astron. Soc.* **161,** 365.
—— (1973b). ibid. **162,** 17.
TREHAN, S. K. (1957). *Astrophys. J.* **126,** 429.
—— and BILLINGS, D. F. (1971). *Astrophys. J.* **169,** 567.
—— and SINGH, M. (1974). *Astrophys. Space Sci.* **26,** 167.
—— and UBEROI, M. S. (1972). *Astrophys. J.* **175,** 161.
VANDAKUROV, YU. V. (1972). *Sov. Astron.* **16,** 265.

VASYLIUNAS, V. M. and WOLF, R. A. (1973). *Rev. Geophys. Space Phys.* **11,** 181.
VEEDER, G. J. and ZIRIN, H. (1970). *Solar Phys.* **12,** 391.
WALBRIDGE, E. (1967). *J. Geophys. Res.* **21,** 5213.
WEBER, E. J. and DAVIS, L. (1967). *Astrophys. J.* **148,** 217.
WHITAKER, W. A. (1963). *Astrophys. J.* **137,** 914.
WILD, J. P. (1970). *Proc. Astron. Soc. Aust.* **1,** 365.
WILLIS, D. M. (1971). *Rev. Geophys. Space Phys.* **9,** 953.
WOLFF, S. C. and WOLFF, R. J. (1970). *Astrophys. J.* **160,** 1049.
—— —— (1972). ibid. **176,** 433.
WOLTJER, L. (1958a). *Astrophys. J.* **128,** 384.
—— (1958*b*). *Bull. Astron. Inst. Netherlands* **14,** 39.
—— (1958c). *Proc. Nat. Acad. Sci.* **44,** 489, 833.
—— (1972). *Ann. Rev. Astron. Astrophys.* **10,** 129.
WRIGHT, G. A. E. (1969). *Mon. Notic. Roy. Astron. Soc.* **146,** 197.
—— (1973). ibid. **162,** 339.
—— (1974). ibid. **167,** 527.
ZIRIN, H. (1972). *Solar Phys.* **22,** 34.
ZWAAN, C. (1968). *Ann. Rev. Astron. Astrophys.* **6,** 135.

7

PROPAGATION OF DISTURBANCES

7.1. General discussion

A small disturbance introduced into a magnetic field in a conducting fluid is propagated away by the stresses in the field and fluid. If the field and fluid system is in a stable equilibrium state, the propagation is in the form of hydromagnetic waves, which are the subject of this chapter. If the system is not stable, or possesses no equilibrium, then the situation is more complicated and no simple generalization is possible.

The stable modes of an equilibrium system confined to a finite region of space are essentially standing waves and have been treated at length in the references given in Chapter 6. A system is unstable, and unlikely to exist in nature, if there is at least one disturbance, or eigenmode, that grows exponentially with time so that it cannot be described as a wave. The present discussion is limited to disturbances of a characteristic scale l in a system of large-scale $\lambda(\gg l)$ so that the magnetic field at the site of the disturbance may be taken as locally uniform. We also suppose that, as a consequence of $l \ll \lambda$ any instability of the large system grows sufficiently slowly compared to the frequency of the waves of the disturbance that the local field may be considered constant in time as well as in space. With this idealization of small-scale disturbances it is possible to characterize the basic wave properties of the magnetic fluid medium by the propagation of disturbances of small amplitude. We begin the discussion with a treatment of an incompressible ideal fluid (viscosity μ and resistivity η are zero, the density uniform, and the sound speed $V_S = (\mathrm{d}p/\mathrm{d}\rho)^{\frac{1}{2}}$ large compared to the Alfven speed $V_A = B/(4\pi\rho)^{\frac{1}{2}}$ and fluid velocity v). The equations are then particularly simple (Lundquist 1952) and yield to general solution of arbitrary form and amplitude.

7.2. Disturbances in an ideal magnetic fluid

The equations for the velocity $\mathbf{v}(\mathbf{r}, t)$ and magnetic field $\mathbf{B}(\mathbf{r}, t)$ of an ideal, incompressible, infinitely conducting uniform fluid with pressure $p(\mathbf{r}, t)$ are given by (5.10) and (4.16),

$$\left(\frac{\partial}{\partial t}+\mathbf{v}\cdot\nabla\right)\mathbf{v} = -\nabla\left(\frac{p}{\rho}+\frac{B^2}{8\pi\rho}\right)+\frac{(\mathbf{B}\cdot\nabla)\mathbf{B}}{4\pi\rho}, \qquad \nabla\cdot\mathbf{v}=0, \tag{7.1}$$

$$\left(\frac{\partial}{\partial t}+\mathbf{v}\cdot\nabla\right)\mathbf{B} = (\mathbf{B}\cdot\nabla)\mathbf{v}, \qquad \nabla\cdot\mathbf{B}=0, \tag{7.2}$$

respectively. The first of these equations is mechanical, and the second electromagnetic, in origin. Yet the symmetry of the roles of the velocity and magnetic fields in both equations is conspicuous. Writing the magnetic field **B** in terms of its characteristic Alfven velocity $\mathbf{u}=\mathbf{B}/(4\pi\rho)^{\frac{1}{2}}$, (7.1) and (7.2) can be expressed as

$$\left(\frac{\partial}{\partial t}+\mathbf{v}\cdot\nabla\right)\mathbf{v}-\left(\mathbf{u}\cdot\nabla\right)\mathbf{u}=-\nabla(p/\rho+\tfrac{1}{2}u^2), \tag{7.3}$$

$$\left(\frac{\partial}{\partial t}+\mathbf{v}\cdot\nabla\right)\mathbf{u}-(\mathbf{u}\cdot\nabla)\mathbf{v}=0, \tag{7.4}$$

$$\nabla\cdot\mathbf{v}=\nabla\cdot\mathbf{u}=0. \tag{7.5}$$

An obvious solution is the steady dynamical equilibrium

$$p=constant-\tfrac{1}{2}\rho u^2, \tag{7.6}$$

with

$$\mathbf{v}=\pm\mathbf{u}, \qquad \partial\mathbf{v}/\partial t=0. \tag{7.7}$$

Chandrasekhar (1961) has shown that this dynamical equilibrium solution is stable. Thus any complicated entangled field pattern is possible if it is accompanied by a fluid velocity **v** along the field at the local Alfven speed.

Of particular interest to the present question of wave propagation is the case in which the magnetic field **B** is uniform everywhere throughout space, with a constant value $\mathbf{B}_0$, except in a finite domain $\mathscr{V}$ in the neighbourhood of the origin where the field has some arbitrary and complicated form $\mathbf{B}_0+\mathbf{B}_1(\mathbf{r})$, as sketched in Fig. 7.1. Suppose that the

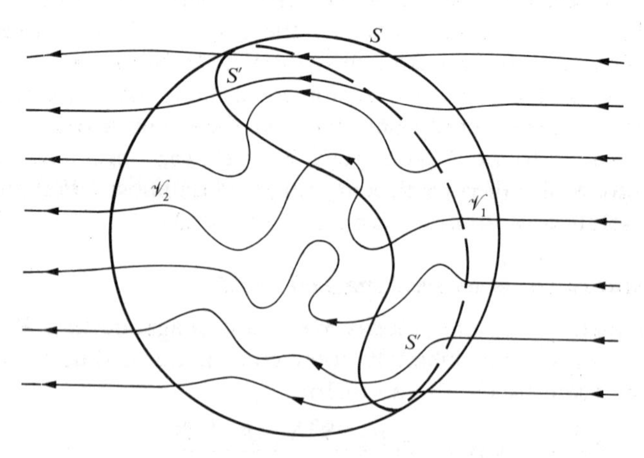

FIG. 7.1. Sketch of a localized disturbance in a large-scale magnetic field $\mathbf{B}_0$. The disturbance occupies the volume $\mathscr{V}$ and vanishes on the enclosing surface S. The volume is divided into two parts $\mathscr{V}_1$ and $\mathscr{V}_2$ by the intersecting surface S'.

fluid velocity has the uniform value $\mathbf{v}=-\mathbf{V}_A=-\mathbf{B}_0/(4\pi\rho)^{\frac{1}{2}}$ everywhere outside $\mathcal{V}$, and the value $-[\mathbf{B}_0+\mathbf{B}_1(\mathbf{r})]/(4\pi\rho)^{\frac{1}{2}}$ within, so that (7.6) and (7.7) apply throughout the entire space. Then transform to the frame of reference moving with velocity $-\mathbf{V}_A$. The fluid, in that frame, is at rest except for the volume $\mathcal{V}$, which now moves with speed $+\mathbf{V}_A$. The field inside $\mathcal{V}$ is now $\mathbf{B}_0+\mathbf{B}_1(\mathbf{r}-\mathbf{V}_A t)$, and the fluid velocity $\mathbf{v}'$ within $\mathcal{V}$ is

$$\mathbf{v}'=-\mathbf{B}_1(\mathbf{r}-\mathbf{V}_A t)/(4\pi\rho)^{\frac{1}{2}} \tag{7.8}$$

The solution in this frame of reference is clearly to be interpreted as a wave packet of arbitrary form and amplitude[1] $\mathbf{B}_1(\mathbf{r}-\mathbf{V}_A t)$ propagating along the uniform field $\mathbf{B}_0$ with the Alfven speed $V_A=B_0/(4\pi\rho)^{\frac{1}{2}}$.

There is no net momentum transported by the wave packet, nor any net displacement of fluid,

$$\int d^3\mathbf{r}\rho\mathbf{v}=0. \tag{7.9}$$

This follows directly from the two facts that $\nabla\cdot\mathbf{v}=0$, and $\mathbf{v}$ vanishes on the surface S of the volume $\mathcal{V}$ to which the packet is restricted. Consider the simple surface S' dividing $\mathcal{V}$ into two parts (Fig. 7.1) $\mathcal{V}_1$ and $\mathcal{V}_2$, and the enclosing surface S into two parts S_1 and S_2, respectively. Then $S'+S_1$ and $S'+S_2$ each form a closed surface. Since $\nabla\cdot\mathbf{v}=0$, we have

$$\begin{aligned} 0&=\int_{\mathcal{V}_1} d^3\mathbf{r}\nabla\cdot\mathbf{v},\\ &=\int_{S'+S_1} d\mathbf{S}\cdot\mathbf{v}. \end{aligned}$$

But $\mathbf{v}$ vanishes everywhere on S, so

$$\int_{S'} d\mathbf{S}\cdot\mathbf{v}=0. \tag{7.10}$$

Thus the net flux of fluid is zero across any surface which divides the wave packet into two separate volumes. Integrating over the dimension perpendicular to the surface then yields (7.9). The wave packet carries no momentum.

[1] The dynamical balance of the wave is readily understood from the fact that the pressure variation $-\frac{1}{2}\rho v^2$ of the Bernoulli effect exactly compensates for the magnetic pressure $B^2/8\pi=\frac{1}{2}\rho u^2$, while the centrifugal force $K\rho v^2$ of the fluid streaming along the field lines with curvature K exactly cancels the force $KB^2/4\pi=K\rho u^2$ of the tension of those same curved lines of force.

The total energy $\mathscr{E}$ carried by the wave packet is the volume integral of the excess energy density above the background $B_0^2/8\pi$,

$$\mathscr{E} = \int_{\mathscr{V}} d^3\mathbf{r}\left\{\frac{(\mathbf{B}_0+\mathbf{B}_1)^2 - B_0^2}{8\pi} + \tfrac{1}{2}\rho v'^2\right\},$$

$$= \rho\int d^3\mathbf{r}\,\mathbf{v}' \cdot (\mathbf{V}_\mathrm{A} + \mathbf{v}'),$$

$$= \rho\mathbf{V}_\mathrm{A} \cdot \int d^3\mathbf{r}\,\mathbf{v}' + \rho\int d^3\mathbf{r}\,v'^2.$$

It follows from (7.9) that the first item is identically zero, so that

$$\mathscr{E} = \int_v d^3\mathbf{r}\rho v'^2,$$

$$= \int_v d^3\mathbf{r}\left(\frac{B_1^2}{8\pi} + \tfrac{1}{2}\rho v'^2\right). \tag{7.11}$$

The energy $\mathscr{E}$ moves along the field $\mathbf{B}_0$ with the Alfven velocity $\mathbf{B}_0/(4\pi\rho)^{\frac{1}{2}}$.

Now, it is apparent from (7.8) that the wave packet propagates without dispersion. Any initial wave form is preserved. Hence in this one case of an ideal incompressible fluid the wave behaves much as though the equations were linear (which, of course, they are not). The waves are non-dispersive and the principle of superposition applies to waves *propagating in the same direction.* On the other hand, it should be noted that while waves of arbitrary form propagate without dispersion in any *one* direction along the field, waves with *opposite* directions of propagation cannot be superimposed without strong non-linear coupling causing distortion and dispersion. The dynamical equations are non-linear, notwithstanding the remarkable fact that their solution is the simple linear relationship (7.7). Thus, for instance, (7.7) avails us little in treating hydromagnetic turbulence. The separate solutions $\mathbf{v}_1 = +\mathbf{u}$ and $\mathbf{v}_2 = -\mathbf{u}$ interact as strongly when they meet as separate Fourier components, and the result is just as inscrutable (see for instance, Kraichnan and Nagarajan 1967).

There is, however, one fundamental theorem on turbulence that can be stated on the basis of (7.7)–(7.10). Suppose that for $t < t_0$ the space is permeated by the uniform magnetic field $\mathbf{B}_0$ embedded in the ideal fluid of uniform density ρ. Then suppose that at time $t = t_0$ external forces transmit some arbitrary complicated velocity distribution $\mathbf{v}_0(\mathbf{r})$ to the fluid within some finite region $\mathscr{V}$ (the necessary impulse is $\delta(t-t_0)\rho\mathbf{v}_0(\mathbf{r})$) setting the fluid in $\mathscr{V}$ into some complicated motion with zero total momentum $\int d^3\mathbf{r}\,\mathbf{v}_0(\mathbf{r}, t) = 0$. The subsequent turbulent velocity and magnetic field variations are described by (7.1) and (7.2). Assuming that $\mathbf{v}_0$ is not small compared to $\mathbf{V}_A$, the magnetic field will become significantly distorted, and the situation soon might look something like Fig. 7.1. We assert that in the limit of large time the turbulence in $\mathscr{V}$ converts itself into two separate hydromagnetic wave trains (Alfven waves) propagating without dispersion in opposite directions away from $\mathscr{V}$ along $\mathbf{B}_0$. The assertion is obvious if we recall that the radiation of Alfven waves along $\mathbf{B}_0$ from either side of $\mathscr{V}$ is irreversible. Any perturbation of the lines of force at the surface of $\mathscr{V}$ means an outgoing Alfven wave, and, once emitted, a wave does not return. Then note that for an ideal fluid, no other means of energy loss is available. The turbulent energy may spread out in space, as a consequence of eddy diffusion, and the turbulent energy may migrate up and down the wave number scale[2], but there is no escape of the turbulent energy from the region of non-linear interaction except as an Alfven wave emitted from the surface. Only in the family of modes $\omega = \pm k V_A$ is there the possibility of a final resting place, and, once there, the energy leaves the region, travelling without dispersion along $\mathbf{B}_0$. Alfven waves are the only leak in the system, from which it follows that the turbulence must eventually disentangle itself into two trains of Alfven waves, going their separate ways along $\mathbf{B}_0$ away from the region $\mathscr{V}$ of initial turbulence. A particular simple example is given in §8.5 of the separation of a pulse into two separate wave packets in an isolated flux tube.

In a sense, then, it is not incorrect to say that all motions of an ideal incompressible fluid permeated by a magnetic field are Alfven waves. The statement is generally useful if the magnetic field is strong ($V_A \gg v$), for in that instance the non-linear term $(\mathbf{v} \cdot \nabla)\mathbf{v}$ is small compared to the interaction terms $(\mathbf{v} \cdot \nabla)\mathbf{B}$ and $(\mathbf{B} \cdot \nabla)\mathbf{v}$, and the principal effect is propagation as Alfven waves. If v is *not* small compared to V_A, then the Alfven wave characterization of the turbulence becomes useful only after sufficient time for the oppositely-moving waves to disentangle themselves from each other.

7.3. Alfven waves in an ideal fluid

In view of the fundamental importance of the Alfven waves in redistributing stress and momentum in a magnetic field embedded in an ideal fluid, we recast eqns (7.1) and (7.2) in a form suitable for treating the energy transport by the waves. Once again write $\mathbf{B} = \mathbf{B}_0 + \mathbf{B}_1(\mathbf{r}, t)$ where $\mathbf{B}_0$ is a uniform field of magnitude B_0 in the direction of the unit vector $\mathbf{n}_0 = \mathbf{B}_0/B_0$. The wave field $\mathbf{B}_1(\mathbf{r}, t)$ is arbitrary in form and not necessarily small compared to $\mathbf{B}_0$. Represent $\mathbf{B}_0$ by its Alfven velocity $\mathbf{V}_A = \mathbf{B}_0/(4\pi\rho)^{\frac{1}{2}}$

[2] There is no characteristic dimension in the problem apart from the scales introduced by the initial excitation of the turbulence.

and the wave field $\mathbf{B}_1$ by its Alfven velocity $\mathbf{w}=\mathbf{B}_1/(4\pi\rho)^{\frac{1}{2}}$. Then with $p=constant-\frac{1}{2}\rho(\mathbf{V}_A+\mathbf{w})^2$ again, (7.1) and (7.2) reduce to

$$\left(\frac{\partial}{\partial t}+\mathbf{v}\cdot\nabla\right)\mathbf{v}-(\mathbf{V}_A\cdot\nabla+\mathbf{w}\cdot\nabla)\mathbf{w}=0, \tag{7.12}$$

$$\left(\frac{\partial}{\partial t}+\mathbf{v}\cdot\nabla\right)\mathbf{w}-(\mathbf{V}_A\cdot\nabla+\mathbf{w}\cdot\nabla)\mathbf{v}=0. \tag{7.13}$$

If we let $\mathbf{v}=\pm\mathbf{w}$, then both equations reduce to

$$\frac{\partial\mathbf{v}}{\partial t}\mp(\mathbf{V}_A\cdot\nabla)\mathbf{v}=0 \tag{7.14}$$

which has as its general solution any arbitrary function of $\mathbf{r}\pm\mathbf{V}_A t$,

$$\mathbf{v}=\mathbf{v}(\mathbf{r}\pm\mathbf{V}_A t). \tag{7.15}$$

It is instructive to compute the energy flux transported by the waves again. The total energy flux is made up of the Poynting flux $c\mathbf{E}\times\mathbf{B}/4\pi=\mathbf{B}\times(\mathbf{v}\times\mathbf{B})/4\pi$ and the mechanical energy flux $\mathbf{v}(p+U+\frac{1}{2}\rho v^2)$ convected by the fluid (Landau and Lifshitz 1959, 1960; Roberts 1967) where U is the internal energy per unit volume of fluid and $p+U=\gamma NkT/(\gamma-1)$ is the enthalpy per unit volume. Thus, altogether the total energy flux $\mathbf{Q}$(erg/cm^2 s) is

$$\mathbf{Q}=\mathbf{v}B^2/4\pi-\mathbf{B}(\mathbf{v}\cdot\mathbf{B})/4\pi+\mathbf{v}(p+U+\tfrac{1}{2}\rho v^2). \tag{7.16}$$

In the present case of an incompressible fluid, the internal degrees of freedom are absent and $U=0$. Further, we have from (7.6) that $p+B^2/8\pi=constant$. Then since $\mathbf{B}=(4\pi\rho)^{\frac{1}{2}}(\mathbf{V}_A\pm\mathbf{v})$, it is possible to rearrange the terms to give

$$\mathbf{Q}=\mathbf{v}(p+B^2/8\pi+\tfrac{1}{2}\rho V_A^2)-\rho\mathbf{V}_A(\mathbf{v}\cdot\mathbf{V}_A)\mp\rho v^2\mathbf{V}_A. \tag{7.17}$$

The first two terms on the right-hand side represent circulation of energy around inside the packet, with no net transport. This is readily apparent from the fact that $p+B^2/8\pi+\frac{1}{2}\rho V_A^2=constant$, so that according to (7.10) the integral of the first term over any surface S' extending across the packet is zero. The same is true of the second term, for if we denote by $dS'_\parallel$ and $v_\parallel$ the projection of the element of area dS' and the velocity v, respectively, onto the direction of $\mathbf{B}_0$ we have $\mathbf{V}_A\cdot d\mathbf{S}'=V_A\,dS'_\parallel$ and $\mathbf{v}\cdot\mathbf{V}_A=V_A v_\parallel$ so that

$$\int_{S'} d\mathbf{S}'\cdot\mathbf{V}_A(\mathbf{v}\cdot\mathbf{V}_A)=V_A^2\int dS_\parallel v_\parallel$$

$$=0$$

from (7.10). The third term on the right-hand side, then, represents the net energy transport of the wave, which is just the transport of the kinetic and magnetic energy densities $\frac{1}{2}\rho v^2 + B_1^2/8\pi = \rho v^2(\mathbf{r} \pm \mathbf{V}_A t)$ of the wave at the Alfven speed $\mathbf{V}_A$ along the large-scale field $\mathbf{B}_0$. This is, of course, just the result (7.11) worked out in §7.2 from a slightly different approach. For sinusoidal waves, the mean value over one cycle yields $\frac{1}{2}\rho v_1^2 V_A$, where v_1 is the velocity amplitude of the wave. Note that half of the energy transport is mechanical and half is electromagnetic or potential.

7.4. Hydromagnetic waves, a compressible medium

The next stage of generalization is to include the compressibility of the fluid into the dynamical equations, supposing that the sound speed V_s is not necessarily large compared to the Alfven speed. To keep the problem tractable we restrict the fluid velocity $\mathbf{v}$ to be small compared to both the Alfven speed V_A and sound speed V_S. Then if the magnetic field is $\mathbf{B}(\mathbf{r}, t) = \mathbf{B}_0 + \delta\mathbf{B}(\mathbf{r}, t)$, where $\mathbf{B}_0$ is again a uniform field, we have $|\delta\mathbf{B}| \ll B_0$ so that the dynamical equations can be linearized. The Alfven velocity is again defined as $\mathbf{V}_A = \mathbf{B}_0/(4\pi\rho)^{\frac{1}{2}}$. The pressure and density fluctuations, δp and $\delta\rho$, accompanying the wave are small, and related by the definition of the sound speed,

$$\delta p = V_S^2 \delta\rho, \tag{7.18}$$

where

$$V_S^2 = \gamma kT/M$$

with the coefficient γ chosen appropriately for the number of degrees of freedom involved. We suppose that both V_A and V_S are uniform, so that the linearized dynamical eqns (7.1) and (7.2) reduce to

$$\rho\, \partial\mathbf{v}/\partial t = -\nabla(\delta p + \mathbf{B}_0 \cdot \delta\mathbf{B}/4\pi) + (\mathbf{B}_0 \cdot \nabla)\delta\mathbf{B}/4\pi, \tag{7.19}$$

$$\partial\delta\,\mathbf{B}/\partial t = (\mathbf{B}_0 \cdot \nabla)\mathbf{v} - \mathbf{B}_0 \nabla \cdot \mathbf{v}. \tag{7.20}$$

Linearization of the condition for conservation of matter yields

$$\partial\delta\rho/\partial t + \rho\nabla \cdot \mathbf{v} = 0, \tag{7.21}$$

and, of course,

$$\nabla \cdot \delta\mathbf{B} = 0. \tag{7.22}$$

Use (7.18) to express δp in (7.19) in terms of $\delta\rho$. Then differentiate (7.19) with respect to t and use (7.20) and (7.21) to eliminate the time derivatives of $\delta\mathbf{B}$ and $\delta\rho$. The result can be written

$$\partial^2\mathbf{v}/\partial t^2 = (V_S^2 + V_A^2)\nabla\nabla \cdot \mathbf{v} + (\mathbf{V}_A \cdot \nabla)[(\mathbf{V}_A \cdot \nabla)\mathbf{v} - \nabla(\mathbf{V}_A \cdot \mathbf{v}) - \mathbf{V}_A \nabla \cdot \mathbf{v}]. \tag{7.23}$$

The coefficients are all constants so that the solution is made up of plane waves $\exp i(\omega t-\mathbf{k}\cdot\mathbf{r})$. Without loss of generality we may rotate the coordinate system so that the z-axis lies along the wave vector $\mathbf{k}$ and $\mathbf{B}_0$ lies somewhere in the yz-plane, making an angle θ with the wave vector $\mathbf{k}$,

$$\mathbf{V}_A = V_A(\mathbf{e}_y \sin\theta + \mathbf{e}_z \cos\theta). \tag{7.24}$$

Then the solutions are of the form

$$\mathbf{v} = \mathbf{V} \exp i(\omega t - kz) \tag{7.25}$$

where $\mathbf{V}$ is a constant vector. Then (7.23) reduces to the three components

$$V_x(\omega^2 - k^2 V_A^2 \cos^2\theta) = 0, \tag{7.26}$$

$$V_y(\omega^2 - k^2 V_A^2 \cos^2\theta) + V_z k^2 V_A^2 \sin\theta \cos\theta = 0, \tag{7.27}$$

$$V_y k^2 V_A^2 \sin\theta \cos\theta + V_z(\omega^2 - k^2 V_S^2 - k^2 V_A^2 \sin^2\theta) = 0. \tag{7.28}$$

The dispersion relation follows from equating the determinant of the coefficients equal to zero. In particular, if $v_x \neq 0$, it follows from (7.26) that

$$\omega^2 - k^2 V_A^2 \cos^2\theta = 0 \tag{7.29}$$

representing Alfven waves propagating along $\mathbf{B}_0$ at the Alfven speed V_A so that their phase velocity (in the z-direction) is $V_A \cos\theta$. The Alfven waves involve no motions or fields in the y- and z-directions. It is readily shown from (7.20) that $\delta B_x = \mp B_0 v_x / V_A$. The propagation of Alfven waves was treated in §7.2 and 7.3, and need not be repeated here.

Suppose, then that $v_x = 0$ so that (7.29) is not required. For waves with $v_y, v_z \neq 0$, (7.27) and (7.28) yield the determinant

$$(\Omega^2 - \cos^2\theta)(\Omega^2 - \beta^2 - \sin^2\theta) - \sin^2\theta \cos^2\theta = 0 \tag{7.30}$$

where $\Omega = \omega / k V_A$ represents the phase velocity and $\beta = V_S / V_A$ is the sound velocity, all in units of the Alfven speed V_A. There are, then, the two modes

$$\Omega^2(\beta, \theta) = \tfrac{1}{2}[\beta^2 + 1 \pm \{(\beta^2 - 1)^2 + 4\beta^2 \sin^2\theta\}^{\frac{1}{2}}]$$
$$= \tfrac{1}{2}[\beta^2 + 1 \pm \{(\beta + 1)^2 - 4\beta^2 \cos^2\theta\}^{\frac{1}{2}}]. \tag{7.31}$$

Since Ω represents the phase velocity, the waves given by the upper sign are called the *fast* mode and by the lower sign the *slow* mode. The fast and slow modes involve fluid motions confined to the plane of $\mathbf{k}$ and $\mathbf{B}_0$. The phase velocity is independent of the magnitude of the wave number (although not of the direction) so that the hydromagnetic waves of small amplitude are non-dispersive.

The relation between the two velocity components follows from (7.27) as

$$V_z = -V_y(\Omega^2 - \cos^2\theta)/\sin\theta\cos\theta. \tag{7.32}$$

It is readily shown from (7.20) that $\delta B_x = \delta B_z = 0$ and

$$\begin{aligned}\delta B_y &= B_0(v_z \sin\theta - v_y \cos\theta)/V_A\Omega,\\ &= -B_0 \frac{\Omega}{\cos\theta}\frac{v_y}{V_A}.\end{aligned} \tag{7.33}$$

It follows from (7.21) that

$$\delta\rho = \rho v_z/\Omega V_A, \tag{7.34}$$

with δp following directly through (7.18).

A few words are in order describing the physical nature of the fast and slow modes, and the variation of the waves with angle of propagation θ with respect to the magnetic field. First of all note from (7.31) that

$$\Omega(\beta, \theta) = \beta\Omega(1/\beta, \theta), \tag{7.35}$$

showing that the form of the angular dependence of phase velocity $\Omega(\beta, \theta)$ depends only on the relative proportions of V_S and V_A without regard for which is the larger. Note further that $\Omega(\beta, \theta)/\beta^{\frac{1}{2}}$ is invariant to inverting V_S/V_A.

The phase velocity $\Omega(\beta, \theta)$ in units of the Alfven speed V_A is plotted as a polar diagram in Fig. 7.2 for the special value $\beta = 1$ ($V_S = V_A$). The inner loop represents the slow mode and the outer loop the fast mode. The fast mode is also plotted for $\beta = 2$ ($V_S = 2V_A$) to show its nearly circular form. The phase diagram for other values of β is readily illustrated by noting that in the limit $\beta \to \infty$ ($V_S \gg V_A$) the inner loop changes but little, becoming the circle $\Omega = \cos\theta$ shown in Fig. 7.2 (representing Alfven waves, as indicated by (7.29)), while the outer loop becomes the circle $\Omega = \beta + \sin^2\theta/2\beta + \ldots$ of increasing radius β representing the isotropic propagation of sound waves. Except for the special value $\beta = 1$, the inner and outer loops do not connect together. The limiting forms for $\beta = 1$ and $\beta \gg 1$ together illustrate the general form of $\Omega(\beta, \theta)$ for $\beta \geqslant 1$. The limiting case $\beta \to 0$ reduces to the Alfven waves treated in §6.3. For $\beta < 1$ it follows from (7.35) that $\Omega(\beta, \theta)$ has the same form, but with smaller magnitude (because it is expressed in units of V_A). In the limit as $\beta \to 0$ the slow mode vanishes and the fast mode becomes a purely hydromagnetic wave with velocity V_A in all directions.

To examine the nature of the waves in more detail, note that when the wave vector lies along the magnetic field ($\theta = 0$), one mode is a pure sound wave and the other a pure Alfven wave, with the larger speed V_S

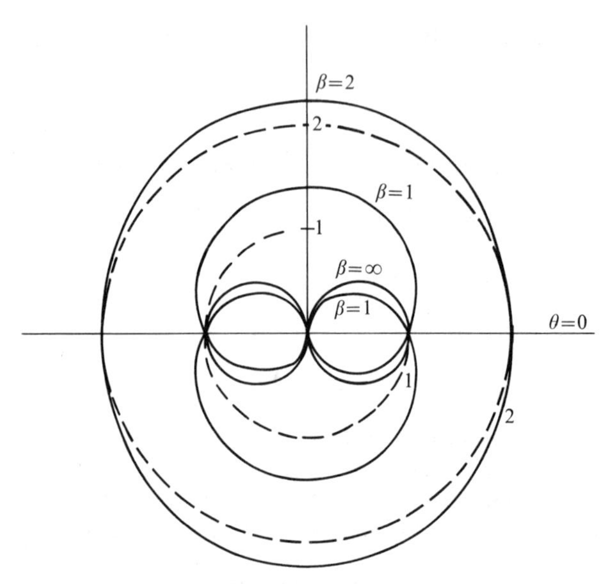

FIG. 7.2. A polar diagram of the phase velocity $\Omega \equiv \omega/kV_A$ as a function of the angle θ between the magnetic field and the wave vector. The heavy line is for the intermediate value $\beta = 1$ ($V_S = V_A$) with the fast mode shown also for $\beta = 2$ and the slow mode for $\beta = \infty$.

or V_A identified with the fast mode and the smaller with the slow mode. When the wave vector is perpendicular to the magnetic field ($\theta = \pi/2$) the fast mode is the longitudinal wave—the magnetosonic wave—with phase velocity $(V_A^2 + V_S^2)^{\frac{1}{2}}$. The fast mode compresses both the gas and the field, giving a velocity of propagation larger than in either alone.

This all follows formally from (7.31) and (7.32) upon recalling that a wave is principally longitudinal (compressional) if the fluid motion is more nearly parallel than perpendicular to the wave vector ($V_z^2 > V_y^2$), and transverse if *vice versa* ($V_z^2 < V_y^2$). Thus, for instance for $\beta > 1$ ($V_S > V_A$) we have $\{(\beta^2 - 1)^2\}^{\frac{1}{2}} = \beta^2 - 1$ in (7.31), and it follows for propagation parallel to the field ($\theta \to 0$) that the fast mode $\Omega^2 = \beta^2\{1 + O(\theta^2)\}$ yields $V_z = -V_y(\beta^2 - 1)/\theta \to \infty$, from (7.32), indicating the purely longitudinal sound wave already mentioned. The slow mode $\Omega^2 \cong 1 - \beta^2\theta^2/(\beta^2 - 1)$ yields $V_z = V_y\theta/(\beta^2 - 1) \to 0$, indicating the purely transverse Alfven wave. For propagation perpendicular to the field ($\pi/2 - \theta \to 0$) the fast mode $\Omega^2 = 1 + \beta^2 + O\{(\pi/2 - \theta)^2\}$ yields the longitudinal form $V_z = -V_y(1 + \beta^2)/(\pi/2 - \theta) \to \infty$, while the slow mode $\Omega^2 = \beta^2(\pi/2 - \theta)^2/(1 + \beta^2)$ yields the transverse form $V_z = V_y(\pi/2 - \theta)/(1 + \beta^2) \to 0$ as the velocity of propagation goes to zero. Altogether, then, if $\beta^2 > 1$ the fast wave is longitudinal for all θ, while the slow mode is transverse.

If, on the other hand, $\beta < 1$ ($V_S < V_A$) then $\{(\beta^2 - 1)^2\}^{\frac{1}{2}} = 1 - \beta^2$ in (7.31) and the fast mode is transverse for propagation along the field and longitudinal for propagation across. The slow mode is the opposite—longitudinal along the field and transverse across. The transition between longitudinal and transverse occurs for both the fast and the slow modes at $\sin^2\theta = \frac{1}{2}(1-\beta^2)$, for which case $2\Omega^2 = (1+\beta^2) \pm (1-\beta^4)^{\frac{1}{2}}$ and $V_z = \mp V_y$.

The limitations of space prevent us from going further into the general properties of hydromagnetic waves. Nor in view of the excellent literature available today is there any virtue in our doing so.

The reader interested in pursuing further the properties of hydromagnetic waves will find it profitable to consult such basic references as Allis, *et al.* (1963) where the hydromagnetic waves are placed in perspective within the general spectrum of waves that occur in an isotropic plasma. The extensive general discussions given by Jeffrey (1966) and Ferraro and Plumpton (1966) of wave propagation of both small and large amplitude, and the wave fronts that develop from a point source, are particularly useful and instructive.

7.5. Energy transport by hydromagnetic waves

Consider the transport of energy by the hydromagnetic waves of small amplitude in an ideal, infinitely conducting, compressible medium. The Alfven waves, as noted in §7.3 propagate exactly along the large-scale field, transporting an energy density ρv^2 at the Alfven speed V_A regardless of the orientation of the wave vector **k**. The other modes are more complicated.

We recall that the z-axis of the coordinate system of §7.4 is oriented along the wave vector with the magnetic field at an angle θ. Hence the phase velocity is in the z-direction, at the speed ΩV_A, given by (7.30) or (7.31). In as much as Ω is independent of the magnitude of the wave number k, the propagation is without dispersion, although the phase velocity varies with the direction θ of the wave vector relative to the magnetic field. The group velocity $d\omega/dk$ is equal to the phase velocity ω/k for any fixed θ.

Consider, then, the direction and rate of energy flow, **Q**, given formally by (7.16). It is evident that we cannot use (7.16) to compute **Q** from the solutions of the linearized equations (7.19)–(7.21) because **Q** is second order in the wave amplitude, requiring retention of all second order terms in (7.16)[3]. Instead we note that the energy transported by the wave is just the rate at which the field and fluid on one side of a surface S' do work on the field and fluid on the other side. In terms of the Maxwell stresses of the wave, $\delta M_{ij} = -\delta_{ij} B_k \delta B_k / 4\pi + (B_i \delta B_j + B_j \delta B_i)/4\pi$, the Reynold's stresses, $\delta R_{ij} = -\rho v_i v_j$, and the gas pressure $-\delta_{ij} p$ the rate at which work is

[3] The problem arises in the convection of internal energy $\mathbf{v}\{(\mathbf{B}+\delta\mathbf{B})^2/8\pi + U + \delta U\}$ wherein the second order terms in the displacement of $B^2/8\pi + U$ cancel exactly the product $\mathbf{v}\{\mathbf{B}\,.\,\delta\mathbf{B}/8\pi + \delta U\}$ of the first order terms.

done across S' is

$$Q_i = v_j\{-\delta M_{ij} - \delta R_{ij} + \delta_{ij}p\} \text{ (erg/cm}^2\text{ s)}. \tag{7.36}$$

With this formulation of the energy flow Q_i, which contains no zero order factors, it is sufficient to compute the stresses to lowest order and then multiply by the velocity. Correct to second order, then, the Reynolds stresses drop out in the present circumstance and

$$\mathbf{Q} = \mathbf{v}(\mathbf{B} \cdot \delta\mathbf{B}/4\pi + \delta p) - \delta\mathbf{B}(\mathbf{v} \cdot \mathbf{B})/4\pi - \mathbf{B}(\mathbf{v} \cdot \delta\mathbf{B})/4\pi, \tag{7.37}$$

which is to be compared with (7.16). The convection of internal energy $B^2/8\pi + U + \delta U + \frac{1}{2}\rho v^2$ is absent.

Since $\delta B_x = \delta B_z = 0$ and $v_x = 0$ for the waves (7.30), it follows that $Q_x = 0$ and

$$Q_y = \frac{\frac{1}{2}\rho V_y^2 V_A \Omega}{\sin\theta\cos\theta}\{1 - \beta^2 - \Omega^2 + (\beta^2/\Omega^2)\cos^2\theta\} \tag{7.38}$$

$$Q_z = \frac{\frac{1}{2}\rho V_y^2 V_A}{\Omega\sin^2\theta\cos^2\theta}\{\Omega^4\sin^2\theta + \beta^2(\Omega^2 - \cos^2\theta)^2\}, \tag{7.39}$$

in terms of the amplitude V_y of the y-component of the velocity. The factor $\frac{1}{2}$ is the result of averaging over one cycle. The energy flow vector $\mathbf{Q}$ makes an angle ϑ with the wave vector, where

$$\begin{aligned}\tan\vartheta &= Q_y/Q_z \\ &= \sin\theta\cos\theta\frac{\beta^2\cos^2\theta + \Omega^2(1-\beta^2) - \Omega^4}{\Omega^4\sin^2\theta + \beta^2(\Omega^2 - \cos^2\theta)^2}\end{aligned} \tag{7.40}$$

For $\beta = 1$ we have $\Omega^2 = 1 \pm \sin\theta$ for the fast and slow modes, yielding, then,

$$\tan\vartheta = \mp\cos\theta/(1 \pm \sin\theta),$$

or

$$\vartheta = \tfrac{1}{2}(\theta \mp \pi/2), \tag{7.41}$$

where the upper sign applies to the fast mode and the lower sign to the slow mode. This case ($\beta = 1$) is singular at $\theta = 0$, yielding $\vartheta = \pm\pi/4$ (and incidentally $V_z^2 = V_y^2$ from (7.32)) when, in fact we know that ϑ must equal zero because the modes are a sound wave and an Alfven wave, both with an energy flow in the direction of the wave vector ($\vartheta = 0$). To avoid the singular value suppose that $\beta \neq 1$ and $\theta \ll (\beta^2 - 1)$. Then it is readily shown that

$$\tan\vartheta = -\frac{2\theta}{\beta^2 - 1}[1 + O\{\theta^2/(\beta^2 - 1)^2\}] \tag{7.42}$$

and ϑ goes to zero with θ. If β is either large or small compared to one, it is readily shown that the direction of energy flow in the slow mode is given by

$$\tan\vartheta = 2\tan\theta, \tag{7.43}$$

neglecting terms second order in $1/\beta$ or β, respectively. It is evident that the energy flow in the slow mode is exactly parallel to the magnetic field for $\theta = 0$, $\pi/2$ and approximately along the magnetic field for intermediate values θ. For instance, for $\theta = 45°$ we have $\vartheta - \theta = 18{\cdot}4°$.

If β is large compared to one, the fast mode yields

$$\tan\vartheta = -(1/\beta^2)\sin 2\theta \tag{7.44}$$

while if β is small compared to one,

$$\tan\vartheta = -\beta^2 \sin 2\theta, \tag{7.45}$$

telling us that the energy flow is always approximately along the wave vector, irrespective of the magnetic field direction. It will be recalled that, even in the extreme case $\beta = 1$ (7.41) gives a maximum deviation ϑ of $\pi/4$.

Altogether, then, hydromagnetic waves propagate with a group velocity equal to the phase velocity ΩV_{A} in any given direction. We must be careful, however, in identifying energy transport and signal velocity (see discussion in Sommerfeld 1952, 1964) with the group velocity, because we cannot construct wave packets of finite length and breadth and permanent form. Only wave packets of infinite transverse width have permanent form, and in such waves the energy transport in the transverse x- and y-directions is unrestricted and undefined by the boundaries (which are only $z = constant$). We note, then, that (7.16) and (7.40) show that the Alfven mode transports energy exactly along the magnetic field, the slow mode transports energy very approximately along the magnetic field, and the fast mode transports it approximately along the wave vector. Thus the direction of departure of the energy of a localized disturbance depends very much upon which wave modes are excited by the disturbance. The final asymptotic deposition of that energy throughout the surrounding space, then, depends further upon the dissipation of the waves. Dissipation is the subject of the next section.

It would be interesting to go on here to introduce into (7.19) the effects of the gravitational acceleration $\mathbf{g}$, and the associated buoyancy $\mathbf{g}\delta\rho$ (indicated in (5.10)). Unfortunately the complications introduced by the variation of the background pressure and density ($\nabla p = -\mathbf{g}\rho$) with height place the problem beyond the scope of the present work. The problem is so complex that only the most rudimentary results are available in the literature (see review by Stein and Leibacher 1974 and references therein, as well as the recent work by Nye and Thomas 1974 on running penumbral waves). The problem is particularly important for understanding the oscillations and motions observed (Giovanelli 1972) above an active region on the sun. The problem has been considered in the absence of a magnetic field in connection with the general 300-s oscillations observed in the sun (see, for instance, Stein and Leibacher 1974; Chiuderi and Giovanardi 1975) and in connection with the vertical motions of the terrestrial ionosphere (Hines 1960, 1963, 1964).

7.6. Dissipation of hydromagnetic waves in a fluid

In a real fluid the finite viscosity and resistivity cause hydromagnetic waves to decay. The viscosity μ leads to the Navier–Stokes equation (Landau and Lifschitz 1959)

$$\rho\left(\frac{\partial}{\partial t}-\mu\frac{\partial^2}{\partial x_j\partial x_j}\right)v_i+\rho v_j\frac{\partial v_i}{\partial x_j}=-\frac{\partial}{\partial x_i}\left(p+\frac{B^2}{8\pi}\right)+\frac{B_i}{4\pi}\frac{\partial B_i}{\partial x_j}$$
$$+\frac{\mu}{3}\frac{\partial}{\partial x_i}\frac{\partial v_j}{\partial x_j}+\frac{\partial\mu}{\partial x_j}\left(\frac{\partial v_i}{\partial x_j}+\frac{\partial v_j}{\partial x_i}-\tfrac{2}{3}\delta_{ij}\frac{\partial v_k}{\partial x_k}\right),\tag{7.46}$$

in place of (7.1), and the resistivity η leads to (4.12), which we choose to write as,

$$\left(\frac{\partial}{\partial t}-\eta\frac{\partial^2}{\partial x_j\partial x_j}\right)B_i+\frac{\partial}{\partial x_j}(v_jB_i-v_iB_j)=\frac{\partial\eta}{\partial x_j}\left(\frac{\partial B_i}{\partial x_j}-\frac{\partial B_j}{\partial x_i}\right),\tag{7.47}$$

in place of (7.2). Then, in place of the simple adiabatic relation (7.18) we should employ the heat flow equation

$$\frac{\partial U}{\partial t}-\frac{\partial}{\partial x_j}\left(\kappa\frac{\partial T}{\partial x_j}\right)+\frac{\partial}{\partial x_j}v_jU+p\frac{\partial v_j}{\partial x_j}=0\tag{7.48}$$

where κ is the thermal conduction coefficient and U is the thermal energy $NkT/(\gamma-1)$ per unit volume. We have neglected to write in the heat source from the viscous dissipation, from radiative transfer, from recombination and dissociation, etc. (see for instance, Landau and Lifshitz 1959).

Suppose for simplicity that ρ, μ, η, and κ are uniform over space and constant in time. Then the kinematic viscosity $\nu=\mu/\rho$ and the thermometric conductivity $K=\kappa(\gamma-1)/Nk$ are constant. It is evident from the appearance of the operator $(\partial/\partial t-q\nabla^2)$ on the left-hand sides of (7.46)–(7.48) that ν, η, and κ cause a wave mode $\exp(i\mathbf{k}\cdot\mathbf{r})$ to decay exponentially with time. Viscosity contributes a decay factor $\exp(-\alpha_1\nu k^2t)$ where α_1 is a number of the order of unity. For the resistivity, we note that except for sound waves propagating parallel to the magnetic field ($\theta=0$), the magnetic field is significantly perturbed and the resistivity contributes a decay factor $\exp(-\alpha_2\eta k^2t)$, where α_2 is a number of the order of unity. Insofar as the wave is longitudinal there will be density, and hence temperature, fluctuations, and the thermal conductivity contributes a decay factor $\exp(-\alpha_3Kk^2t)$, where α_3 is a number of the order of unity. Consider then, the magnitudes of these three dissipative contributions.

The viscosity μ of fully-ionized hydrogen is (Chapman 1954)

$$\nu = 1{\cdot}2\times 10^{-16}T^{\frac{5}{2}}/\rho \text{ (cm}^2\text{/s)}. \tag{7.49}$$

The resistive diffusion coefficient η is given by (4.39) as

$$\eta = 10^{13}T^{-\frac{3}{2}} \text{ (cm}^2\text{/s)} \tag{7.50}$$

and the thermal conduction coefficient is (Chapman 1954; Spitzer 1956)

$$\kappa = 6\times 10^{-7}T^{\frac{5}{2}} \text{ erg/cm s K},$$

so that the thermometric conductivity is

$$K = 5\times 10^{-15}T^{\frac{5}{2}}/\rho \text{ (cm}^2\text{ s)}. \tag{7.51}$$

Comparing (7.49) and (7.51) it is evident that $K \cong 40\nu$, so that the thermal conductivity is more important than the viscosity (Prandtl number $K/\nu \gg 1$) in damping any wave with a significant longitudinal component. Only the transverse Alfven wave of small amplitude, or with circular polarization, is sufficiently free of temperature fluctuations to escape the damping effect of thermal conductivity. The viscosity affects all waves, of course[4]. It is evident, therefore, that such viscous, conducting atmospheres as the stellar corona act as a strong filter, passing Alfven waves and perpendicular magnetosonic waves to the exclusion of others.

In the hot tenuous gases that occupy all of space outside stellar photospheres, and outside the cool interstellar neutral hydrogen clouds, the viscous and thermal damping are both large compared to the resistive dissipation. On the other hand, in the high densities of the stellar photosphere, and everywhere inside any non-degenerate star, resistive decay dominates the viscosity and thermal conductivity. Resistive diffusion exceeds viscosity wherever

$$\rho > 10^{-29}T^4 \text{ (g/cm}^3\text{)}$$

Thus in the typical stellar photosphere ($\rho \cong 10^{-7}$ g/cm^3, $T \cong 10^4$ K) and in the typical stellar interior ($\rho \cong 10^2$ g/cm^3, $T \cong 10^7$ K) the inequality is amply satisfied. The effects of viscosity and thermal conductivity can be ignored compared to resistivity. Resisitivity is the dominant source of

[4] In the hot tenuous 'collisionless' plasma the viscosity is suppressed by the magnetic field, of course, but in that case other wave–particle interactions enter the picture and the waves are strongly damped anyway. See discussion in §7.7.

dissipation but may be insignificant for many purposes. Only in the *photospheres* of the coolest stars is resistivity a strong effect. For instance in the solar photosphere where η may be as large as $10^9\ \mathrm{cm^2/s^{-1}}$ the characteristic decay time of a wave with length λ is $\lambda^2/4\times10^{10}$ s. So for waves longer than, say, 100 km, the decay period is an hour or more and generally longer than the transit time of the wave. Elsewhere within the sun, resistive damping is completely negligible, and viscosity and thermal conductivity even less important.

However, as soon as we move out from the photosphere to the low densities of the chromosphere ($\rho\cong10^{-12}\ \mathrm{g/cm^3}$, $T\cong2\times10^4$ K), viscous and thermal conductive damping are strong and are important for waves of small length. In this case $\nu\cong7\times10^6\ \mathrm{cm^2\,s^{-1}}$, and the viscous decay time is $\lambda^2/3\times10^8$ s. A wavelength of 100 km is unaffected by resistivity, damped only slightly by viscosity, and decays in 10^4 s by thermal conduction. The low corona, ($\rho\cong10^{-16}\ \mathrm{g/cm^3}$, $T\cong10^6$ K) because of its high temperature and low density, is strongly dissipative through viscosity and thermal conductivity. The kinematic viscosity is $\nu\cong1{\cdot}2\times10^{15}\ \mathrm{cm^2\,s^{-1}}$ so that viscous damping occurs in $\lambda^2/5\times10^{16}$ s. Any wavelength less than 10^5 km is strongly damped by viscosity, and all but the Alfven waves, or magnetosonic waves propagating perpendicular to the magnetic field, are quenched even more rapidly by thermal conduction. The stellar corona has a deadening effect then, filtering out much of the high-frequency noise from the star.

In the *outer* corona and stellar wind regions of a star the particle mean free path becomes longer than most wavelengths and the formal concept of the coefficients of viscosity, resistivity, and thermal conductivity is no longer applicable. The energy of a wave cannot be dissipated by collisions in less than one collision time. But, as it turns out wave–particle interactions become important in the absence of collisions (taken up in the next section) so that dissipation is not absent.

In the interstellar regions of ionized hydrogen (the H_{II} regions where $\rho\cong10^{-24}\ \mathrm{g\,cm^{-3}}$ and $T\cong10^4$ K) we have $\nu\cong10^{18}\ \mathrm{cm^2\,s^{-1}}$ and a viscous damping time of $\lambda^2/2\times10^{20}$ s. With typical Alfven and sound speeds of $10\ \mathrm{km\,s^{-1}}$, any wave with a length greater than 10^{16} cm propagates several times its own length before being damped by viscosity and thermal conduction.

In summary, then, hydromagnetic waves with $\lambda>10^2$ km propagate without significant damping in stellar photospheres and stellar interiors[5], while only the longest wavelengths can escape out through the stellar

[5] Clearly the reader can confound this simple generalization by thinking of a magnetic field sufficiently weak, or distances sufficiently great, that the propagation time becomes longer than the computed decay time. Hence, it is advisable to assess each new situation afresh, without relying on general statements.

corona. For most purposes the damping in interstellar H_{II} regions may be neglected unless for some reason we are interested in very short wavelengths or very long times.

In the cool and only slightly ionized regions of interstellar space (the H_I regions, where $T \cong 10^2$ K, $N \sim 1$–10 H-atoms cm^{-3} and $N_e \sim 10^{-3}$ cm^{-3}) the viscosity dominates over thermal conduction and resistivity, while ambipolar diffusion, discussed in §4.6, is most important of all (Cowling 1957). The ions and electrons are tied to the magnetic field and participate in the wave motion. The massive neutral component of the gas participates in the wave motion only to the degree that it is compelled by collisions with the ions and electrons. We note then, that an ion collides with the N neutral atoms cm^{-3} at a rate $f = NAw$ where A is the collision cross-section and w the relative velocity. Since A is approximately inversely proportional to w, it follows that Aw has a nearly constant value, which proves to be of the order of $2{\cdot}5 \times 10^{-9}$ cm^3 s^{-1} (Osterbrock 1961). Thus, for $N = 1$ atom cm^{-3} we have $f^{-1} = 4 \times 10^8$ s between collisions. A roughly comparable number applies to electrons, but their momentum is so small that their effect may be neglected. It is clear then that if the wave period is small compared to the collision period 4×10^8 s (corresponding to the very small wavelength of 4×10^{14} cm $\cong 10^{-4}$ pc for the typical value $V_A = 10$ km s^{-1}) the oscillations of the ions die out in one collision time, of 4×10^8 s $\cong 10$ years. More interesting is the circumstance where the wave period is long compared to the ion–neutral collision time, and both the neutral and ionized component participate in the wave motion. The rate of dissipation is the frictional force **F** between the ionized and neutral component, multiplied by the relative motion $\Delta\mathbf{v}$ of the two components (see §4.6 and (4.42)). The characteristic dissipation time τ in a wave of amplitude B and scale l is, then, of the order of $\mathbf{F} \cdot \Delta\mathbf{v}$ divided by the energy density in the wave, which for simplicity we take to be just $B^2/8\pi$. Hence with $|\mathbf{F}|$ of the order of $B^2/8\pi l$ it follows that

$$
\begin{aligned}
1/\tau &\cong \mathbf{F} \cdot \Delta\mathbf{v}/B^2/8\pi, \\
&\cong \Delta v/l, \\
&= \frac{B^2/8\pi}{l^2 M_i N_i N A w}.
\end{aligned}
$$

With $M_i \cong 10^{-23}$ g, $N \cong 1$ cm^{-3}, $N_i \cong 10^{-3}$ cm^{-3}, and with $B \cong 3 \times 10^{-6}$ G, typical of the cool interstellar H_I region, a wave with the small scale of $0{\cdot}1$ pc decays in about 3×10^{12} s, during which time it propagates 1 pc or 10 wavelengths (if $V_A \cong 10$ km s^{-1}). A wave of 1 pc propagates some 10^2 pc while decaying. Thus, the ambipolar diffusion strongly affects the short wavelengths, but not those in excess of 1 pc. A more sophisticated discussion of the problem may be found in Cowling (1957).

Altogether, it would appear that the longer Alfven waves of small amplitude travel unhindered nearly everywhere. The magnetosonic waves of small amplitude are more heavily damped, but even there the longer wavelengths are significantly dissipated only in the high thermal conductivity of stellar coronas. In interstellar space the regions of hot, tenuous, and fully ionized gas transmit waves longer than about 10^{16} cm pretty well, while the cool neutral regions transmit only those waves longer than about 1 pc. It would appear the muffling of interstellar noise is sufficient to suppress any significant reverberation.

7.7. Dissipation of hydromagnetic waves in a collisionless plasma

In the hot tenuous plasmas of the stellar winds from most stars, the collisional mean free path is so long as to preclude any significant collisional damping of hydromagnetic waves. The plasma may be considered to be essentially without collisions and the degradation of energy, insofar as it exists, must depend upon wave–particle interactions in lieu of particle–particle collisions. The fascinating and endless complications of the collisionless plasma, with its inexhaustible supply of internal degrees of freedom, are beyond the scope of this work, but the damping by wave–particle interactions is a large effect, and we are obliged to note its effect on the large-scale fluid properties of the plasma. First of all, we point out that the large-scale behaviour of the collisionless plasma is, apart from the dissipation, described correctly and fully by the fluid equations, relating the bulk velocity $\mathbf{v}(\mathbf{r}, t)$, the pressure, and the magnetic field. The pressure may be anisotropic in principle, and the degradation of the fluid motions cannot be described by simple diffusion coefficients. And, of course, the conventional elementary derivation of the hydrodynamic equations is no longer valid. But it is no accident that the basic hydromagnetic relations (4.16) and the fluid equation (5.10) both remain valid if we replace the scalar pressure $\delta_{ij}p$ by the pressure tensor p_{ij}. The fluid equation is nothing more than the statement of local conservation of momentum, on the basis of which the Maxwell stress tensor was defined in the first place (see the formal discussion in Brueckner and Watson 1956; Chew *et al.* 1956; Parker 1957; Montgomery and Tidman 1964, Chap. 13; Frieman *et al.* 1966). The computation of the pressure tensor p_{ij} is much more difficult than for the classical fluid, of course, because of the possible anisotropy and the small-scale internal instabilities that an anisotropy excites, not to mention the other wave–particle effects that contribute to dissipation of waves. Fortunately the large-scale fluid motions in astrophysical circumstances have characteristic periods that are very large compared to the small-scale internal degrees of freedom (such as the ion cyclotron motion and the electron and ion plasma oscillation). Consequently it is sometimes a workable approximation to compute the large-scale motion as for a fluid, taking into account no more than the simple anisotropy of the thermal motions relative to the magnetic field direction. Then the modifications as a consequence of the accumulated small-scale wave–particle interactions are introduced in a qualitative manner through modified coefficients of resistivity, etc. Sometimes more direct methods are required.

Consider, then, the wave particle interactions pertinent to the question of the degradation of hydromagnetic waves (defined to have frequencies small compared to the ion cyclotron frequency). The effects are sometimes referred to under the general term of Landau damping, following Landau's first identification of a collisionless damping effect (Landau 1946) in plasma oscillations. There are a variety of wave–particle interactions involving transit time, or resonant effects (Bohm and Gross 1949; Gross 1951; Van Kampen 1955, 1957; Stix 1962; Montgomery and Tidman 1964; Tanenbaum 1967). They can be deduced only from detailed simultaneous solution of the collisionless Boltzmann equation (the Vlasov equation) and Maxwell's equations, or some equivalent self-consistent treatment of the fields and the individual particle motions. Until 1966 Landau damping was known only in the small-scale waves with frequencies above the ion cyclotron frequency. Then Barnes (1966, 1967, 1968) showed that all hydromagnetic waves are subject to Landau damping of one form or another, with the sole exception of those waves propagating *precisely* at right angles to the ambient magnetic field ($\theta = \pi/2$) and the Alfven mode. Barnes showed that when the magnetic pressure $B^2/8\pi$ is comparable to, or less than, the gas pressure p (i.e. $\beta \gtrsim 1$, $V_S > V_A$), the damping is large. Indeed the slow mode is essentially obliterated, the waves decaying in only a fraction of their period. It is this situation ($B^2/8\pi \lesssim p$) with which we are usually concerned in astrophysical circumstances (the solar corona immediately above an active region, and the magnetosphere of a pulsar being the only two exceptions that come immediately to mind). So there is a major dissipation effect in any region such as the outer corona and solar wind, or a supernova remnant where the hot tenuous gas is effectively collisionless.

As a consequence of wave–particle interactions, the fast mode magnetosonic wave has a characteristic decay time of only 2–20 wave periods (a characteristic decay length of 2–20 wavelengths). Only when the wave vector is within about half a degree of 90° from the large-scale field does the damping become negligible. The dissipation for propagation near the direction of the field ($\theta < 40°$) is associated with ion heating, while the sharp peak of dissipation from $\theta = 80°$ to $89°$ is associated mainly with electrons. The wave heats those particles that move along with it, so that when $V_S \cong V_A$, some of the ions can keep up with the wave and are accelerated (heated) thereby. On the other hand, the electrons move very fast compared to the wave speed and keep pace with the wave only when travelling along the ambient magnetic field at nearly right angles to the direction of wave propagation. Altogether the damping of hydromagnetic waves in a collisionless plasma is stronger than in circumstances where there are collisions, for there the collisions serve to suppress the wave–particle interactions that damp the wave so strongly in the collisionless plasma.

What then of the Alfven wave of small amplitude, which avoids these sordid effects? Further theoretical investigation of hydromagnetic waves in a collisionless plasma has shown that the pure *undamped* Alfven mode of *small* amplitude is, in fact, an exact simultaneous solution of the Vlasov equation and Maxwell's equations, even in the relativistic case (Barnes and Suffolk 1971; Barnes and Hollweg 1974). But it has been shown that there are other pitfalls, which the Alfven mode with *finite* amplitude cannot escape. As a consequence of the non-linear effects, neglected in the treatment of waves of small amplitude, it has been shown that all hydromagnetic waves, except the Alfven wave with precise circular polarization, steepen as they propagate (Parker 1958; Montgomery 1959; Kulikovsky and Lyubimov, 1965; Barnes and Hollweg 1974) evolving into other modes or into a collisionless shock (Tidman and Krall 1971). In particular, Sagdeev and Galeev (1969), Galeev and Oraevskii (1963) have shown that the linearly polarized Alfven wave is unstable to decay into a magnetosonic wave and a back-scattered Alfven wave. Hence the initial wave evolves into another mode that is not immune to Landau damping (Cohen and Kulsrud 1974; Cohen and Dewar 1974). What is more, the non-linear interaction of two circularly polarized Alfven waves leads to a form of Landau damping (Lee and Völk 1973) i.e. the thermal particles gain energy at the expense of the waves. Finally we should note the demonstration by Hollweg (1971) of *non-linear* Landau damping of Alfven waves (Barnes and Hung 1972), and the non-linear damping of magnetosonic waves by a two-stream instability (Papadopolous 1973). Altogether it is evident that all hydromagnetic waves in a hot tenuous (collisionless) plasma are subject to damping in either first or second order in their amplitudes.

The question of the propagation and dissipation of hydromagnetic waves in a hot plasma has been studied principally in connection with the waves in the solar wind, generated by activity at the sun and by the interaction of streams of wind with different velocities. The waves are often observed to be of large amplitude at the orbit of Earth. It is inferred that they contribute substantially to the momentum (Belcher and Davis 1971) as well as the energy (temperature) of the wind (Barnes 1971; Hollweg 1972, 1973, 1974). The solar wind is a particularly effective laboratory for studying hydromagnetic waves because the density, velocity, thermal velocity distribution, and magnetic field of the passing wave can be observed and measured directly with instruments carried on spacecraft (see review and discussion by Barnes 1975, and Hollweg 1975; see also Barnes and Scargle 1973, for a discussion of wave propagation in the collisionless plasma of the Crab Nebula). The solar wind is the ideal laboratory for study of the propagation of hydromagnetic waves, something that is very difficult in the small dimensions of the terrestrial laboratory.

For present purposes we note the various forms of damping of hydromagnetic disturbances and pass on to the question of the non-equilibrium and instability of magnetic fields that are likely to produce those disturbances.

References

Allis, W. P., Buchsbaum, S. J. and Abraham, B. (1963). *Waves in anisotropic plasmas.* MIT Press, Cambridge, Massachusetts.
Barnes, A. (1966). *Phys. Fluids* **9,** 1483.
—— (1967), ibid. **10,** 2427.
—— (1968). ibid. **11,** 2644.
—— (1971). *J. Geophys. Res.* **76,** 7522.
—— (1975). *Rev. Geophys. and Space Phys.* **13,** No. 3, 1049.
—— and Hollweg, J. V. (1974). *J. Geophys. Res.* **79,** 2302.
—— and Hung, R. J. (1972). *J. Plasma Phys.* **8,** 197.
—— and Scargle, J. D. (1973). *Astrophys. J.* **184,** 251.
—— and Suffolk, G. C. J. (1971). *J. Plasma Phys.* **5,** 315.
Belcher, J. W. and Davis, L. (1971). *J. Geophys. Res.* **76,** 3534.
Bohm, D. and Gross, E. P. (1949). *Phys. Rev.* **75,** 1851.
Brueckner, K. A. and Watson, K. M. (1956). *Phys. Rev.* **102,** 19.
Chandrasekhar, S. (1961). *Hydrodynamic and hydromagnetic stability*, p. 551. Clarendon Press, Oxford.
Chapman, S. (1954). *Astrophys. J.* **120,** 151.
Chew, G. F., Goldberger, M. L., and Low, F. E. (1956). *Proc. Roy. Soc. Ser. A* **236,** 112.
Chiuderi, C. and Giovanardi, C. (1975). *Astrophys. J. Lett.* **200,** L165.
Cohen, R. H. and Dewar, R. L. (1974). *J. Geophys. Res.* **79,** 4174.
—— and Kulsrud, R. M. (1974). *Phys. Fluids* **17,** 2215.
Cowling, T. G. (1957). *Magnetohydrodynamics.* Wiley-Interscience, New York.
Ferraro, V. C. A. and Plumpton, C. (1966). *An introduction to magneto-fluid mechanics.* Clarendon Press, Oxford.
Frieman, E., Davidson, R., and Langdon, B. (1966). *Phys. Fluids* **9,** 1475.
Galeev, A. A. and Oraevskii, V. N. (1963). *Sov. Phys. Dokl.* **7,** 988.
Giovanelli, R. G. (1972). *Solar Phys.* **27,** 71.
Gross, E. P. (1951). *Phys. Rev.* **82,** 232.
Hines, C. O. (1960). *Can J. Phys.* **38,** 1441.
—— (1963). *Quart. J. Roy. Met. Soc.* **89,** 1.
—— (1964). *Research in geophysics* (ed. H. Odishaw), chap. 12. MIT Press, Cambridge, Massachusetts.
Hollweg, J. V. (1971). *Phys. Rev. Lett.* **27,** 1349.

HOLLWEG, J. V. (1972). *Cosmical Electrodyn.* **2,** 423.
—— (1973). *Astrophys. J.* **181,** 547.
—— (1974). *J. Geophys. Res.* **28,** 3643.
—— (1975). *Rev. Geophys. Space Phys.* **13,** 263.
JEFFREY, A. (1966). *Magnetohydrodynamics.* Oliver and Boyd, Edinburgh; Wiley-Interscience, New York.
KRAICHNAN, R. H. and NAGARAJAN, S. (1967). *Phys. Fluids* **10,** 859.
KULIKOVSKY, A. G. and LYUBIMOV, G. A. (1965). *Magnetohydrodynamics.* Addison Wesley, Reading, Massachusetts.
LANDAU, L. (1946). *J. Phys. USSR* **10,** 25.
—— and LIFSHITZ, E. M. (1959). *Fluid mechanics.* pp. 49, 184. Pergamon Press, London.
—— —— (1960). *Electrodynamics of continuous media,* p. 215. Pergamon Press, New York.
LEE, M. A. and VÖLK H. J. (1973). *Astrophys. Space Sci.* **24,** 31.
LUNDQUIST, S. (1952). *Ark. Fys.* **5,** 297.
MONTGOMERY, D. C. (1959). *Phys. Rev. Lett.* **2,** 36.
—— and TIDMAN, D. A. (1964). *Plasma kinetic theory.* McGraw-Hill, New York.
NYE, A. H. and THOMAS, J. H. (1974). *Solar Phys.* **38,** 399.
OSTERBROCK, D. E. (1961). *Astrophys. J.* **134,** 270.
PAPADOPOULOS, K. (1973). *Astrophys. J.* **179,** 939.
PARKER, E. N. (1957). *Phys. Rev.* **107,** 924.
—— (1958). ibid. **109,** 1328.
ROBERTS, P. H. (1967). *An introduction to magnetohydrodynamics,* p. 20. American Elsevier, New York.
SAGDEEV, R. Z. and GALEEV, A. A. (1969). *Nonlinear plasma theory* (ed. T. O'NEIL and D. BOOK), p. 8. W. A. Benjamin, New York.
SOMMERFELD, A. (1952). *Electrodynamics,* p. 231. Academic Press, New York.
—— (1964). *Optics,* p. 22. Academic Press, New York.
SPITZER, L. (1956). *The physics of fully ionized gases.* Wiley-Interscience, New York.
STEIN, R. F. and LEIBACHER, J. (1974). *Ann. Rev. Astron. Astrophys.* **12,** 407.
STIX, T. H. (1962). *The theory of plasma waves.* McGraw-Hill, New York.
TANENBAUM, B. S. (1967). *Plasma physics.* McGraw-Hill, New York.
TIDMAN, D. A. and KRALL, N. A. (1971). *Shock waves in collisionless plasmas.* Wiley-Interscience, New York.
VAN KAMPEN, N. G. (1955). *Physica* **21,** 949.
—— (1957). ibid. **23,** 641.

8

THE ISOLATED FLUX TUBE

8.1. Basic physical properties

THIS chapter begins the development of the properties of the individual tube of flux. A continuous field distribution $\mathbf{B}(\mathbf{r})$ is a juxtaposition of elemental flux tubes, each tube crowding against its neighbours with pressure $B^2/8\pi$ and striving to shorten its length with tension $B^2/4\pi$. The eqn (6.1) for magnetohydrostatic equilibrium is the statement of balance of these forces. In many circumstances in nature, such as the solar photosphere, the magnetic fields are observed to split into separated flux tubes of small cross-section. Thus the dynamical properties of the individual tubes provide an understanding of large-scale continuous field distributions, as well as the phenomena of separate flux tubes in the sun and other bodies. Curiously, even the most elementary properties of the individual flux tube have been largely ignored in the conventional development of hydromagnetic theory, most theoretical work being confined to analytical or numerical solution of the hydromagnetic equations for continuous field distributions $\mathbf{B}(\mathbf{r})$. The conventional approach has provided a wealth of knowledge on equilibrium fields (see Chapter 6). But the high degree of symmetry of conventional analytical solutions completely misses the general non-equilibrium of the less symmetric, and hence more probable, field topologies. The non-equilibrium of almost all magnetic fields is associated with the broken symmetry of the topology of their lines of force, or elemental flux tubes. Hence, this chapter initiates the development leading to the general non-equilibrium of magnetic fields. We begin with the equilibrium properties of the elemental tube of force, with obvious immediate application of the general principles to the isolated flux tubes observed in the sun.

To fix ideas, consider a slender flux tube of small cross-section[1] A embedded in an ideal fluid of infinite electrical conductivity, as sketched in Fig. 8.1. The statement that the tube is slender means that the characteristic transverse dimension, or diameter, $(4A/\pi)^{\frac{1}{2}}$ is small compared to the scale of variation along the tube, such as the radius of curvature, the scale height of the surrounding gas, etc. The field strength B is presumed uniform across A. Hence in hydrostatic equilibrium the

[1] We may as well imagine a tube of circular cross-section, of radius $R=(A/\pi)^{\frac{1}{2}}$, although the calculations and conclusions are generally correct for any shape.

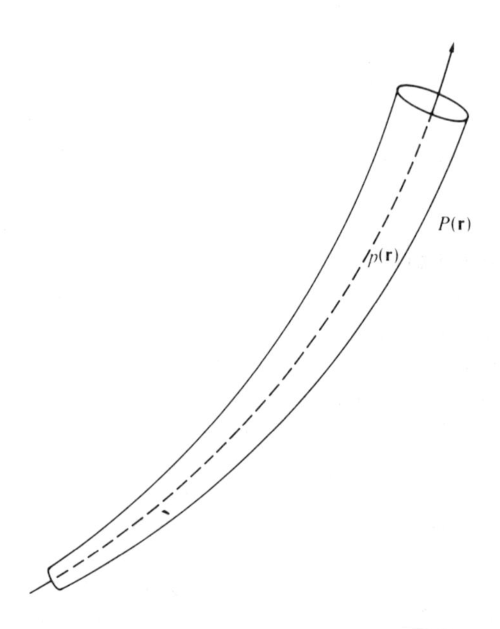

FIG. 8.1. A sketch of the slender flux tube with internal field $\mathbf{B}(\mathbf{r})$, confined by the local external pressure $P(\mathbf{r})$.

fluid pressure p_i within the tube is also uniform across A. Conservation of magnetic flux requires that the product AB be independent of distance s along the tube,

$$A(s)B(s) = \Phi. \tag{8.1}$$

where the constant Φ represents the total flux in the tube. The fluid moves freely along the tube of flux, communicating its pressure as it would along any open flexible tube. Hence the internal pressure p_i varies with longitudinal distance s in the manner described by the longitudinal component of the hydrostatic equilibrium eqn (6.1),

$$\partial p_i/\partial s + \rho_i \partial\psi/\partial s = 0 \tag{8.2}$$

where ρ_i is the fluid density inside the tube and ψ is the gravitational potential (the gravitational acceleration is $\mathbf{g} = -\nabla\psi$). The external pressure P exerted on the surface of the tube may be made up of an external fluid pressure p_e plus the magnetic pressure of neighbouring flux tubes. For the present discussion suppose that the external pressure is solely the pressure p_e of an external fluid in hydrostatic equilibrium in the gravitational potential ψ,

$$\nabla p_e + \rho_e \nabla\psi = 0. \tag{8.3}$$

Suppose further that the fluid has the properties of an ideal gas, so that both inside and outside the tube

$$p = \rho kT/M$$

where M is the mass per particle and T is the temperature. In order to solve (8.3) to obtain p_e as a function of height z, note that the curl of (8.3) yields

$$\nabla \rho_e \times \nabla \psi = 0$$

while the vector product with $\nabla \psi$ yields

$$\nabla p_e \times \nabla \psi = 0.$$

It follows that ρ_e and p_e, and hence T_e, must be functions only of ψ

$$p_e = p_e(\psi), \qquad \rho_e = \rho_e(\psi), \qquad T_e = T_e(\psi)$$

for hydrostatic equilibrium. The hydrostatic equilibrium condition (8.3) can be rewritten in terms of p_e and T_e as

$$\nabla \ln p_e + \{M/kT_e(\psi)\}\nabla \psi = 0.$$

Integrating this equation yields

$$p_e(\psi) = p_e(\psi_0)\exp\left\{-\int_{\psi_0}^{\psi} d\psi M/kT_e(\psi)\right\}. \tag{8.4}$$

The internal pressure p_i is defined along the tube so that (8.2) can be integrated to give

$$p_i(\psi) = p_i(\psi_0)\exp\left\{-\int_{\psi_0}^{\psi} d\psi M/kT_i(\psi)\right\}. \tag{8.5}$$

Now consider the question of local hydrostatic equilibrium across the diameter of the tube. As already noted, we are considering an elemental tube whose diameter $A^{\frac{1}{2}}$ is small compared to the scale of variation L of conditions along the tube. Hence the transverse component of (6.1) is dominated by the term $\partial(p_i + B^2/8\pi)/\partial n$ where n denotes distance in the direction perpendicular to the tube axis. The other terms are separately all smaller by a factor $A^{\frac{1}{2}}/L$ or more. Hence

$$\frac{\partial}{\partial n}\left(p_i + \frac{B^2}{8\pi}\right) \cong 0.$$

This equation states that the total pressure $p_i + B^2/8\pi$ does not vary significantly across the tube and, hence, must be equal to the total pressure immediately outside. It follows that

$$p_i + B^2/8\pi = p_e \tag{8.6}$$

at every location along the tube. Since p_i and p_e are both determined by the gravitational potential ψ, it follows that $B = B(\psi)$ as well.

Altogether the basic conditions (8.4)–(8.6) yield

$$\frac{B^2(\psi)}{8\pi} = p_e(\psi_0)\exp\left\{-\int_{\psi_0}^{\psi} \mathrm{d}\psi M/kT_e(\psi)\right\}$$

$$-p_i(\psi_0)\exp\left\{-\int_{\psi_0}^{\psi} \mathrm{d}\psi M/kT_i(\psi)\right\}. \quad (8.7)$$

Given the temperatures $T_i(\psi)$ and $T_e(\psi)$, the field strength depends only upon the gravitational potential ψ, and is independent of the path of the flux tube through the medium. It should be noted that, if by some artificial means we were to set up a situation in which (8.7) were not satisfied, then upon releasing the system the first thing to happen would be to establish the local condition (8.6) of transverse hydrostatic equilibrium, in approximately the Alfven transit time $A^{\frac{1}{2}}/V_A$ across the diameter of the tube. Longitudinal equilibrium takes longer, of the order of the Alfven transit time L/V_A along the length L of the tube. Thus, for instance, if we had artificially compressed the tube in the neighbourhood of some position $s = s_1$, making the magnetic field abnormally strong there, then, upon releasing the tube, the immediate establishment of lateral equilibrium (8.6) implies a reduced internal gas pressure p_i. The reduced gas pressure in the region 'pulls' fluid along the tube into the region, inflating the tube and restoring the tube to equilibrium (8.7). Two formal examples of the disturbance, and the return to equilibrium, are presented in §8.5. The point to be emphasized here is that *the specification of the hydrostatic fluid pressure on any surface which intersects the flux tube, say, the surface $\psi = \psi_0$, determines the field strength everywhere along the tube.*

If we suppose, for instance, that, as a consequence of low electrical conductivity somewhere along the tube, it is possible for fluid to leak into the tube, gradually neutralizing the pressure difference $p_e - p_i$, it is clear from (8.6) that the field strength must decline with time and the tube expand *everywhere* along the length of the tube. By the same token, if in some way the interior is evacuated, thereby enhancing $p_e - p_i$, the field increases and the tube radius shrinks *everywhere* along the tube. This will occur, for instance, when the tube is lengthened by external forces, thereby increasing its internal volume without supplying additional gas to the interior. Presumably it is an evacuation of the interior that is responsible for the very concentrated flux tubes (1000–4000 G) that appear at the solar photosphere. Some static or dynamical effect must evacuate the interior of those tubes.

8.2. A flux tube in thermal equilibrium with its surroundings

One or two examples are in order to show the control of the field density by the gas pressure. Suppose, for instance, that the temperature is the same inside and outside the flux tube, so that the scale heights are equal, $\Lambda_i(z)=\Lambda_e(z)=kT(z)/Mg\equiv\Lambda(z)$, where, it will be recalled, z is the vertical dimension and g is the local gravitational acceleration in that direction. Define

$$m(z)=\int_0^z dz'/\Lambda(z')$$

to be the number of scale heights above $z=0$. It follows from (8.4) and (8.5) that

$$\frac{p_i(z)}{p_i(0)}=\frac{p_e(z)}{p_e(0)}=\exp\{-m(z)\}.$$

It follows from (8.6) that the field pressure varies in direct proportion to the gas pressure, $B^2(z)/8\pi p_e(z)=constant$, and the field strength varies as

$$B(z)=B(0)\exp\{-\tfrac{1}{2}m(z)\}. \tag{8.8}$$

The ratio of the Alfven speed to the sound speed is independent of z. It follows from (8.1) that the characteristic transverse dimension $A^{\frac{1}{2}}(z)$ of the tube varies as

$$A^{\frac{1}{2}}(z)=A^{\frac{1}{2}}(0)\exp\{-\tfrac{1}{4}m(z)\}. \tag{8.9}$$

Thus the tube expands and the field declines with height. The tube diameter and field strength are a function only of height z, regardless of the path followed by the flux tube. The radius of the tube declines upward with a scale height equal to four times the pressure scale height in the surrounding atmosphere. This upward relaxation is the expected nature of a slender flux tube beneath the surface of a star when the tube is in thermal and hydrostatic equilibrium with its surroundings. On the other hand, the flux tubes observed in the sun show the remarkable property of intense concentration where they come up through the photosphere, indicating that something has driven them far from local thermal equilibrium, and perhaps also hydro*static* equilibrium.

8.3. Flux tube with cooled interior

Suppose, then, that the temperature $T_i(z)$ within the tube differs from the temperature $T_e(z)$ outside by a small amount, ΔT. Then

$$T_i(z)=T_e(z)-\Delta T(z).$$

with $\Delta\Lambda(z) \equiv k\Delta T(z)/Mg$ the scale height within the tube is

$$\frac{1}{\Lambda_i(z)} = \frac{1}{\Lambda_e(z)}\left\{1 + \frac{\Delta\Lambda(z)}{\Lambda_e(z)} + \ldots\right\}.$$

It follows from (8.7) that

$$\begin{aligned}\frac{B^2(z)}{8\pi} &= p_e(z)\left[1 - \frac{p_i(0)}{p_e(0)}\exp\left\{-\int_0^z dz' \frac{\Delta\Lambda(z')}{\Lambda_e^2(z')}\right\}\right] \\ &= p_e(z)\left[1 - \left\{1 - \frac{B^2(0)}{8\pi p_e(0)}\right\}\exp\left\{-\int_0^z \frac{dz'\Delta\Lambda(z')}{\Lambda_e^2(z')}\right\}\right]. \end{aligned} \quad (8.10)$$

The ratio of magnetic pressure $B^2/8\pi$ to gas pressure $p_e(z)$ increases monotonically, asymptotically approaching the value of unity at large heights.

$$B^2(z)/8\pi \sim p_e(z) \quad (8.11)$$

On the other hand, note that the magnetic pressure itself reaches a maximum at the level $z = z_1$, at which $dB^2/dz = 0$, where

$$\left\{1 - \frac{B^2(0)}{8\pi p_e(0)}\right\}\exp\left\{-\int_0^{z_1} dz' \frac{\Delta\Lambda(z')}{\Lambda_e^2(z')}\right\} = 1 - \frac{\Delta\Lambda(z_1)}{\Lambda_e(z_1)}. \quad (8.12)$$

It is apparent that $z_1 > 0$ if $B^2(0)/8\pi p_e(0)$ is less than $\Delta\Lambda(z_1)/\Lambda_e(z_1)$, and otherwise negative. The field density declines above $z = z_1$, although not as fast as the gas pressure $p_e(z)$. The maximum field strength, at z_1, has the value

$$B_{max}^2/8\pi \cong p_e(z_1)\Delta\Lambda(z_1)/\Lambda_e(z_1). \quad (8.13)$$

The maximum magnetic pressure is the fraction $\Delta\Lambda/\Lambda_e$ of the gas pressure at $z = z_1$.

As a simple special case suppose that Λ_e and $\Delta\Lambda$ are constant and that $B^2(0)/8\pi p_e(0)$ is so small compared to $\Delta\Lambda/\Lambda$ as to be negligible. Then it follows from (8.12) that $z_1 = \Lambda_e$, where the pressure is

$$\begin{aligned} p_e(z_1) &= p_e(0)/e \\ B_{max}^2/8\pi &= \{p_e(0)/e\}\Delta\Lambda/\Lambda_e \end{aligned} \quad (8.14)$$

(with $e = 2{\cdot}718$). The maximum magnetic pressure is reached within one scale height of the reference level $z = 0$ and is the fraction $\Delta\Lambda/e\Lambda_e$ of the gas pressure at the base reference level. The magnetic pressure relative to the gas pressure continues to increase above $z = z_1$

$$B^2(z)/8\pi p_e(z) = 1 - \exp(-z\Delta\Lambda/\Lambda_e^2) \quad (8.15)$$

being essentially equal to the ambient gas pressure above $\Lambda_e/\Delta\Lambda$ scale heights.

In this example the cool interior of the flux tube is evacuated by gravity. The reduced temperature allows the interior gas to drop down the tube, evacuating the upper portion so that the field strength is compressed to the limiting value (8.11) by the external gas pressure. The important point is that a very small temperature reduction ΔT leads to nearly complete evacuation of the tube if the reduction extends over many scale heights.

In view of the usually large characteristic scale of the variation of temperature compared to the pressure scale height in most stellar photospheres, it follows that any ΔT will generally extend over many scale heights. Thus even a very modest ΔT leads to the condition that $B^2/8\pi$ is of the same order as the gas pressure p_e. Presumably it is this effect that permits the modest temperature reduction in the sunspot or starspot ($\Delta T/T = \Delta\Lambda/\Lambda \cong 0{\cdot}3$) to compress the magnetic field to 3000 G or more (Parker 1955, 1975b; Schlüter and Temesvary 1958).

8.4. Flux tube without confining pressures

If a magnetic flux tube is not confined by a pressure differential $p_e - p_i$, then it expands to fill all the available volume, with a corresponding decrease in the mean energy density. As an example imagine that the axis of a flux tube with total flux Φ coincides with the z-axis, and the tube is confined by external pressures to the small radius a everywhere in $z^2 > L^2$. The field strength within this thin tube is $B_0 = \Phi/\pi a^2$. On the other hand, where it extends between the planes $z = \pm L$, the field is without confinement. The field expands to fill the volume $-L < z < +L$ as shown in Fig. 8.2. The field is easily calculated, writing

$$\mathbf{B} = -\nabla W, \qquad \nabla^2 W = 0 \tag{8.16}$$

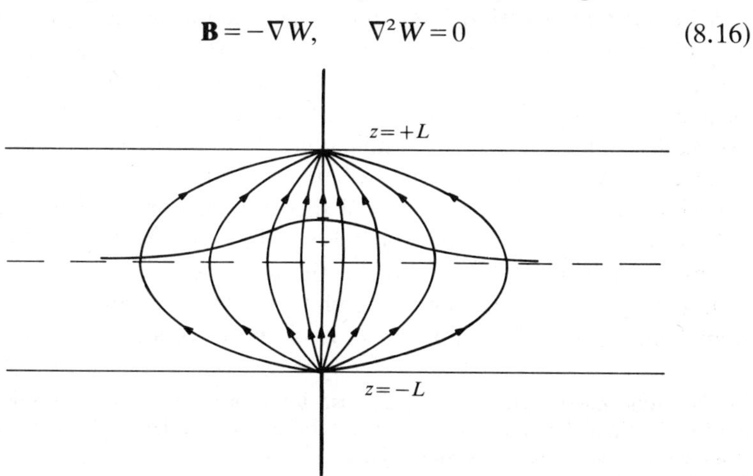

FIG. 8.2. The magnetic lines of force of a flux tube without confining pressure between the planes $z = \pm L$. The field density is plotted along the central plane $z = 0$ in units of $\Phi/2\pi L^2$.

in the region between the planes. The boundary conditions are that **B** vanishes at large radial distance $\varpi=(x^2+y^2)^{\frac{1}{2}}$ from the z-axis, the normal component $B_z=-\partial W/\partial z$ vanishes on $z=\pm L$, and the total flux across $z=L-\epsilon$ and $z=-L+\epsilon$ is Φ, as $\epsilon\to 0$. The flux is confined to the small neighbourhood a of $\varpi=0$. In terms of the infinite sequence of positive images at $z=(4n-1)L$ and negative images at $z=(4n+1)L$, the potential is

$$W=\frac{\Phi}{2\pi}\sum_{n=-\infty}^{+\infty}\{[\varpi^2+\{z-(4n-1)\}^2]^{-\frac{1}{2}}-[\varpi^2+\{z-(4n+1)L\}^2]^{-\frac{1}{2}}\}\tag{8.17}$$

so that

$$B_z=\frac{\Phi}{2\pi}\sum_{n=-\infty}^{+\infty}\left\{\frac{z-(4n-1)L}{[\varpi^2+\{z-(4n-1)L\}^2]^{\frac{3}{2}}}-\frac{z-(4n+1)L}{[\varpi^2+\{z-(4n+1)L\}^2]^{\frac{3}{2}}}\right\}.\tag{8.18}$$

Where the lines of force issue from the confining cylinder $\varpi=a$ in $z^2\geqslant L^2$, the field is radial and falls off inversely with the square of the distance, which follows directly from (8.17),

$$B=\Phi/2\pi r^2=\tfrac{1}{2}B_0a^2/r^2.$$

The field on the z-axis falls to a minimum of $0{\cdot}94\ B_0a^2/L^2$ at the midpoint $z=0$ between the bounding planes $z=\pm L$. The field is plotted as a function of ϖ on the midplane $z=0$ in Fig. 8.2.

8.5. Stability of a magnetic flux tube

An important question in studying the equilibrium of a magnetic flux tube is the dynamical behaviour of the tube when deformed from equilibrium, so that (8.2), (8.3), and (8.6) are not satisfied. We have already asserted the physically obvious fact in §8.1, that the flux tube is stable and eventually returns to the longitudinal equilibrium[2] (8.7). It is interesting however, to work out the formal mathematical theory of the dynamical behaviour and illustrate the general theory with some simple examples.

Consider, then, the slender flux tube[3] with total flux Φ embedded in an incompressible fluid ($B^2/8\pi\ll p_e$) in the absence of a gravitational field, so that the external pressure p_e is uniform along the tube. The equilibrium configuration is a uniform cross-section A_0 and field strength B_0

[2] A flux tube has neutral stability against deformation of its cross-section. Since the longitudinal dynamics of a slender tube is independent of the shape of the cross-section, we ignore the shape here.

[3] It is not necessary to assume that the tube is straight when it is slender, although that is the simplest case conceptually.

along the length of the tube, with $A_0B_0=\Phi$, and an internal fluid pressure

$$p_i=p_e-B_0^2/8\pi$$

as a consequence of the magnetic pressure $B_0^2/8\pi$.

Suppose that the tube is placed in a non-equilibrium state with a cross-sectional area varying as $A(s)$ with distance along its length, and perhaps with some longitudinal fluid velocity $v(s)$. Suppose that the tube is slender, i.e. the characteristic scale of variation of A and v along the tube, *viz.* $(\partial \ln A/\partial s)^{-1}$ and $(\partial \ln v/\partial s)^{-1}$, is large compared to the characteristic transverse dimension $A^{\frac{1}{2}}$. Then in the ensuing motion the transverse velocities, characterized by $\partial A^{\frac{1}{2}}/\partial t$, will be small to the same order compared to the longitudinal fluid velocity v. Hence, the equation of motion for the longitudinal component of the momentum equation reduces to

$$\frac{\partial v}{\partial t}+v\frac{\partial v}{\partial s}+\frac{1}{\rho}\frac{\partial p_i}{\partial s}=0 \tag{8.19}$$

neglecting terms $O[A^{\frac{1}{2}}/\partial \ln A/\partial s]$ compared to one. In view of the smallness of the transverse fluid velocity both inside and outside the tube, the condition for transverse equilibrium is

$$p_i(s)+B^2(s)/8\pi=p_e(s) \tag{8.20}$$

to the same order. Conservation of fluid requires that

$$\partial A/\partial t+\partial(vA)/\partial s=0. \tag{8.21}$$

It is a straightforward procedure to solve (8.1) and (8.19)–(8.21). Use (8.1) to eliminate B from (8.20). Then remembering that p_e is uniform it follows that

$$\frac{\partial p_i}{\partial s}=-\frac{\Phi^2}{8\pi}\frac{\partial}{\partial s}\left(\frac{1}{A^2}\right). \tag{8.22}$$

Using this relation to eliminate p_i from (8.19) the equation of motion may be written in terms of v and A, which must be solved simultaneously with (8.21). To do this we introduce the Lagrangian coordinate $\xi(S, t)$ of the element of fluid initially at the position $s=S$ at time $t=0$. Then conservation of fluid (8.21) requires that the element of volume dV within the tube between the two elements of fluid at ξ and $\xi+d\xi$ be constant. Writing $d\xi=(\partial\xi/\partial S)\,dS$, it follows that

$$dV=A(\xi, t)(\partial\xi/\partial S)\,dS=A(S, 0)\,dS \tag{8.23}$$

where $A(S, 0)=A(S)$ is the initial cross-section. This relation is the integral of (8.21). There remains, then, the equation of motion (8.19), with the pressure gradient given by (8.22). It is convenient to introduce the accumulated volume along the tube,

$$V(S) = \int^{S} \mathrm{d}SA(S, 0) \tag{8.24}$$

as the coordinate of position. If we choose the lower limit of the integration to lie at some point where the fluid is motionless over the time interval with which we are concerned, then it follows from (8.23) that

$$V(S) = \int^{\xi(S,t)} \mathrm{d}\zeta A(\zeta, t), \qquad \partial V/\partial t = 0$$

We also have

$$\begin{aligned}\partial\xi/\partial V &= (\partial\xi/\partial S)/A(S, 0),\\ &= 1/A(\xi, t).\end{aligned}$$

Noting that the fluid velocity $v(\xi, t)$ is just $\partial\xi/\partial t$, the equation of motion can now be written

$$\begin{aligned}\frac{\partial^2\xi}{\partial t^2} &= \frac{\Phi^2}{8\pi\rho}\frac{\partial}{\partial\xi}\frac{1}{A^2},\\ &= \frac{\Phi^2}{8\pi\rho}\frac{1}{\partial\xi/\partial S}\frac{\partial}{\partial S}\left(\frac{\partial\xi}{\partial V}\right)^2,\\ &= \frac{\Phi^2}{8\pi\rho}\frac{1}{\partial\xi/\partial V}\frac{\partial}{\partial V}\left(\frac{\partial\xi}{\partial V}\right)^2,\\ &= \frac{\Phi^2}{8\pi\rho}\frac{\partial^2\xi}{\partial V^2}.\end{aligned} \tag{8.25}$$

The general solution is

$$\xi(S, t) = \tfrac{1}{2}F\{V(S) + \alpha t\} + \tfrac{1}{2}G\{V(S) - \alpha t\}, \tag{8.26}$$

where $\alpha \equiv \Phi/(4\pi\rho)^{\frac{1}{2}}$ and the functions F and G are arbitrary. The fluid velocity at the position $\xi(S, t)$ is

$$v(\xi, t) = \frac{\alpha}{2}[F'\{V(S) + \alpha t\} - G'\{V(S) - \alpha t\}]. \tag{8.27}$$

The magnetic field is

$$\begin{aligned}B(\xi, t) &= \Phi/A(\xi, t)\\ &= \Phi\partial\xi/\partial V\\ &= \tfrac{1}{2}\Phi[F'\{V(S) + \alpha t\} + G'\{V(S) - \alpha t\}].\end{aligned} \tag{8.28}$$

The solution (8.26) represents waves propagating without dispersion along the coordinate V with constant velocity $\mathrm{d}V/\mathrm{d}t = \pm\alpha$. The coordinate velocity is the Alfvén speed $\partial S/\partial t = \pm\alpha/A(S, 0)$. The displacement of an element of fluid from its initial position S is $\xi(S, t) - S$.

The functions F and G are determined by the initial conditions. Since S is the initial linear coordinate of the fluid that is later at $\xi(S, t)$ at time t, it follows that $S = \xi(S, 0)$. Then (8.26) becomes

$$2S = F\{V(S)\} + G\{V(S)\}. \tag{8.29}$$

The initial fluid velocity follows from (8.27),

$$v(S, 0) = \tfrac{1}{2}\alpha[F'\{V(S)\} - G'\{V(S)\}], \tag{8.30}$$

and the initial field follows from (8.28)

$$B(S, 0) = \tfrac{1}{2}\Phi[F'\{V(S)\} + G'\{V(S)\}], \tag{8.31}$$

which is equivalent, through (8.1), to the reciprocal of the initial cross-sectional area $A(S, 0)$. Prescription of the initial velocity $v(S, 0)$ and initial cross-section $A(S, 0)$ together with (8.25) serves to determine F, G, and $V(S)$. Note that differentiation of V with respect to S gives A, while differentiation of S with respect to V converts (8.29) into (8.31).

Suppose, as an example, that the fluid is initially at rest. Then (8.30) requires

$$F'(V) = G'(V)$$

and

$$F(V) = G(V) + \textit{constant}. \tag{8.32}$$

It follows from (8.29) that

$$S = F\{V(S)\}, \tag{8.33}$$

where the integration constant in (8.32) has been put equal to zero by suitable location of the origin of the coordinates S. It follows that F is the inverse function of V,

$$F = V^{-1}. \tag{8.34}$$

The function $V(S)$ now follows from specification of $A(S, 0)$, and F follows from (8.34). The subsequent displacement of an element of fluid from its initial position S is, then

$$\xi(S, t) - S = \tfrac{1}{2}F\{V(S) + \alpha t\} + \tfrac{1}{2}F\{V(S) - \alpha t\} - F\{V(S)\} \tag{8.35}$$

with $v(\xi, t)$ and $B(\xi, t)$ given by (8.27) and (8.28) with $G = F$.

Two examples will suffice to illustrate the effects. Suppose that the tube is compressed over a length L to a cross section $1 - \nu$, and expanded over an adjacent length L to a cross section $1 + \nu$, where ν is a number

$0<\nu<1$. It is convenient to place the origin of the coordinates $S=0$ at the point between the two sections so that

$$A(S,0)=\begin{cases}1 & \text{for}\quad S^2>L^2\\ 1+\nu & \text{for}\quad 0<S<L\\ 1-\nu & \text{for}\quad -L<S<0.\end{cases}\tag{8.36}$$

The total volume of the tube is conserved in this way. It is readily shown that $V(S)$ has the piecewise linear form

$$\begin{aligned}&V(S)=\nu L+S, && \partial S/\partial V=1 \quad \text{for}\quad S^2>L^2\\ &V(S)=(1+\nu)S, && \partial S/\partial V=1/(1+\nu) \quad \text{for}\quad 0<S<L\\ &V(S)=(1-\nu)S, && \partial S/\partial V=1/(1-\nu) \quad \text{for}\quad -L<S<0.\end{aligned}$$

The inverse function $F=V^{-1}$ is then piecewise linear as well. The disturbance separates into two symmetric waves propagating in opposite directions, with velocities $\pm\alpha$. After separation the pulse travelling in the $+s$-direction has a fluid velocity $v(\xi,t)$ and cross-section $A(\xi,t)$

$$\begin{aligned}&v=+\alpha\nu/2(1+\nu), && A=(1+\nu)/(1+\tfrac{1}{2}\nu) \quad \text{for}\quad -\tfrac{1}{2}\nu L<\xi-\alpha t<L\\ &v=-\alpha\nu/2(1-\nu), && A=(1-\nu)/(1-\tfrac{1}{2}\nu) \quad \text{for}\quad -L<\xi-\alpha t<-\tfrac{1}{2}\nu L\end{aligned}$$

within the pulse, and the undisturbed values $v=0$, $A=1$, $(\xi-\alpha t)^2>L^2$ ahead and behind. The pulse is plotted in Fig. 8.3 for the special case $\nu=\frac{2}{3}$. Note the asymmetry of the pulse, created by the non-linear interaction with the other pulse prior to separation.

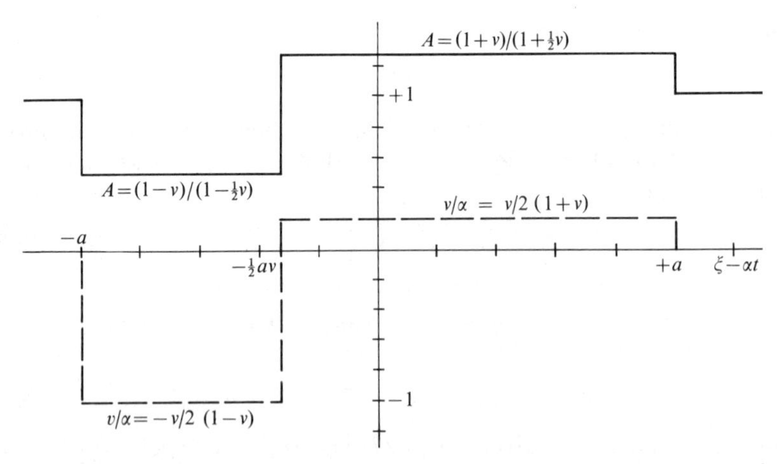

FIG. 8.3. The solid curve is a plot of the cross-sectional area and the dashed curve is the longitudinal fluid velocity in the pulse propagating in the positive z-direction away from the initial disturbance (8.36). The curves are plotted for $\nu=\frac{2}{3}$, although the algebraic values are indicated for general ν.

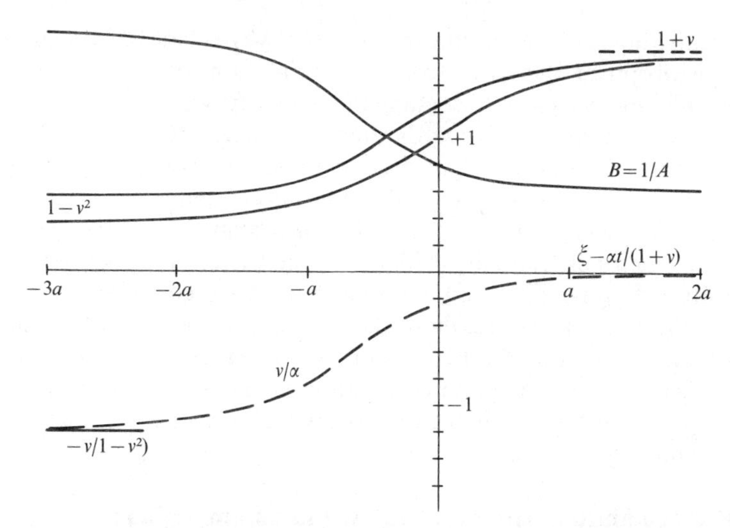

FIG. 8.4. The solid curve is a plot of the cross-sectional area and the dashed curve is the longitudinal fluid velocity in the pulse propagating in the positive z-direction away from the initial disturbance (8.37). The curves are plotted for $\nu=\frac{2}{3}$, although the algebraic values of the asymptotes are given for general ν.

For the second example suppose that the initial cross-section is

$$A(S)=1+\nu S/(L^2+S^2)^{\frac{1}{2}}$$

and (8.37)

$$V(S)=S+\nu\{(L^2+S^2)-L\}$$

so that the cross-section is $1+\nu$ for $S \gg L$ and $1-\nu$ for $S \ll -L$ with a smooth transition across $S=0$, plotted in Fig. 8.4 for comparison with the pulse moving in the $+S$-direction. There is a net displacement of fluid toward $+S$, with the total integrated volume in the interval $(-S, +S)$ undisturbed. The inverse function is

$$S(V)=(1-\nu^2)^{-1}\{V+\nu L-\nu[L^2+2\nu LV+V^2]^{\frac{1}{2}}\}.$$

The disturbance separates into two wave trains, of course. The wave propagating toward positive S (into the cross-section $1+\nu$, and field strength $B_0/(1+\nu)$) moves with the local Alfven speed $\alpha/(1+\nu)$ and is located in the neighbourhood of $s=\alpha t/(1+\nu)$. The velocity profile is plotted in Fig. 8.4 for $\nu=\frac{2}{3}$. The velocity takes up the constant value $-\alpha\nu/(1+\nu)$ behind the pulse, where the cross-section goes asymptotically to $1-\nu^2$ and the field to $B_0/(1-\nu^2)$, which, incidentally, is the mean of the values $B_0/(1\pm\nu)$ at $S=\pm\infty$. The pulse is a 'rarefaction' wave. The other pulse, located in the neighbourhood of $s=-\alpha t/(1-\nu)$ is a 'shock' wave, with the field increasing through the wave. Note again the non-linear distortion of the wave from the initial form of the cross-section, arising while the waves are disentangling themselves prior to separation.

These two examples illustrate the stable character of the flux tube, with the departure from equilibrium producing waves propagating along the tube. Any such waves interact with oppositely moving disturbances and their form is altered thereby, but they are stable and they are non-dispersive when not 'colliding' with oncoming waves. They leave the tube behind in a state nearer to equilibrium than the initial state: In the first example the tube is left exactly in equilibrium; in the second the field is the arithmetic mean of the two extremes, so that further reverberations are necessary before complete equilibrium is established. It is clear that any dissipation in the system, causing the waves to damp, eventually leads to hydrostatic equilibrium.

8.6. The equilibrium path of a flux tube in an atmosphere

The previous sections have dealt with the local transverse hydrostatic equilibrium of the slender magnetic flux tube, showing how the field strength is determined at every location along the length of the tube by the ambient gas pressure. For a given temperature field $T_i(\psi)$ and $T_e(\psi)$ the field strength $B(\psi)$ is a function only of ψ and is independent of the path of the tube. We now consider the net transverse and longitudinal stress in the flux tube, and the consequent path of the flux tube through the atmosphere.

The net tension, or longitudinal stress, along a flux tube is readily computed from the complete stress tensor

$$M_{ij}-\delta_{ij}p=-\delta_{ij}(p+B^2/8\pi)+B_iB_j/4\pi.$$

The total pressure $p+B^2/8\pi$ is, according to (8.6), continuous across the tube and into the surrounding medium. Hence the only extra stress contributed by the flux tube above the level of the surrounding medium is the tension $B^2/4\pi$ along the tube. The total tension $\mathcal{T}$ along the tube of cross-section A is, then,

$$\begin{aligned}\mathcal{T}&=AB^2/4\pi\\&=\Phi B/4\pi\end{aligned}\tag{8.38}$$

in terms of the total flux $\Phi=AB$ and the field intensity B. Thus the net tension varies directly with the field strength B and declines with height, as discussed in the special cases of §§8.2 and 8.3.

Now in addition to the tension along the tube there is a buoyant force $+A(\rho_e-\rho_i)\nabla\psi$ per unit length as a consequence of the different density ρ_i inside the tube. Noting that $\rho=pM/kT$, it follows that the buoyant force

$\mathscr{F}$ per unit length is

$$\mathscr{F} = Ag(\rho_e - \rho_i)$$

$$= A(p_e/\Lambda_e - p_i/\Lambda_i) \tag{8.39}$$

$$= A\left\{\frac{1}{\Lambda_i}\frac{B^2}{8\pi} + p_e\left(\frac{1}{\Lambda_e} - \frac{1}{\Lambda_i}\right)\right\} \tag{8.40}$$

upon application of (8.6) and the scale heights defined as kT/Mg in §8.2. Note that $\mathscr{T}$ and $\mathscr{F}$ are functions only of ψ irrespective of the path followed by the flux tube.

It is evident from (8.40) that if the temperatures inside and outside the flux tube are the same, the density reduction and the attendant buoyant force per unit length are simply proportional to the pressure of the magnetic field,

$$\mathscr{F} = AB^2/8\pi\Lambda. \tag{8.41}$$

In the general case differentiate (8.7) with respect to z. Then

$$\frac{\mathrm{d}}{\mathrm{d}z}\frac{B^2}{8\pi} = \frac{p_i}{\Lambda_i} - \frac{p_e}{\Lambda_e}. \tag{8.42}$$

Note from (8.38) that

$$\frac{\mathrm{d}\mathscr{T}}{\mathrm{d}z} = A\frac{\mathrm{d}}{\mathrm{d}z}\frac{B^2}{8\pi}$$

$$= A(p_i/\Lambda_i - p_e/\Lambda_e)$$

$$= -\mathscr{F} \tag{8.43}$$

upon employing (8.39). This relation applies to any flux tube regardless of its orientation relative to the vertical. The relation follows from the condition (8.6) that the internal and external pressures balance and are each described by the barometric relation. For a vertical flux tube the relation clearly must be satisfied because the buoyant force per unit length must be balanced by the upward decline of the tension. The fact that (8.43) also follows from the lateral pressure equilibrium and barometric relations for the gas merely illustrates the internal consistency of the system which makes possible the simple hydrostatic equilibrium of the tube.

For a straight horizontal flux tube there is no component of the tension to oppose the buoyant force, with the result that the tube rises steadily upward through the atmosphere. The rate of rise is worked out in §§8.7 and 8.8 and proves to be rapid in the convective zones of stars and in the galaxy, and often important, if not rapid, in the stable interior of a star.

A flux tube that is not entirely horizontal may be held in equilibrium by the tension $\mathcal{T}$ over horizontal distances of the order of the scale height as a consequence of the curvature of the tube (Parker 1975b). If the slender flux tube makes an angle $\theta(z)$ with the horizontal direction at the height z, where the tension in the tube is $\mathcal{T}(z)$, then equilibrium requires that the horizontal component of the tension be constant,

$$\frac{\mathrm{d}}{\mathrm{d}z}\mathcal{T}(z)\cos\theta(z)=0. \tag{8.44}$$

What is more, the vertical component of the tension must just balance the buoyant force

$$\mathrm{d}\mathcal{T}(z)\sin\theta(z)=-\mathcal{F}(z)\,\mathrm{d}s.$$

Since $\mathrm{d}z=\mathrm{d}s\sin\theta(z)$, this is

$$\sin\theta\frac{\mathrm{d}}{\mathrm{d}z}\mathcal{T}\sin\theta=-\mathcal{F}. \tag{8.45}$$

The tension is given as a function of height by (8.7) and (8.38) in terms of the temperature of the gas, so the only unknown function is $\theta(z)$. Both (8.44) and (8.45) must be satisfied by a single $\theta(z)$, but they are readily shown to be equivalent. Use (8.43) to eliminate $\mathrm{d}\mathcal{T}/\mathrm{d}z$ from (8.44), yielding

$$\mathcal{T}\sin\theta\frac{\mathrm{d}\theta}{\mathrm{d}z}+\mathcal{F}\cos\theta=0. \tag{8.46}$$

Use (8.43) to eliminate $\mathrm{d}\mathcal{T}/\mathrm{d}z$ from (8.45), combine terms in $\mathcal{F}$, and divide by $\cos\theta$. The result is (8.46) again, the same as was obtained from (8.44).

To calculate the path of a slender flux tube, rotate the coordinate system about the vertical z-axis so that the tube lies in the yz-plane. Then if $y=y(z)$ denotes the path, it follows that

$$\mathrm{d}z/\mathrm{d}y=\tan\theta(z). \tag{8.47}$$

Integrate (8.44),

$$\mathcal{T}(z)\cos\theta(z)=\mathcal{T}(z_0)\cos\theta(z_0),$$

and solve for $\cos\theta(z)$ in terms of the tension $\mathcal{T}(z)$, and eliminate $\theta(z)$ from (8.47). Separation of variables in (8.47) and integration yields

$$y(z)=y(z_0)-\int_{z_0}^{z}\frac{\mathrm{d}\zeta}{\{\mathcal{T}^2(\zeta)/\mathcal{T}^2(z_0)\cos^2\theta(z_0)-1\}^{\frac{1}{2}}} \tag{8.48}$$

for the path from the foot point $y=y(z_0)$. The tension $\mathcal{T}(z)$ follows from (8.38) and (8.7), so that the problem of computing the path is reduced to a quadrature.

Since the flux tube is of the nature of a buoyant rubber band, it is obvious that the path will be an arch of some form or other. It is easier, therefore, to see the consequences of (8.48) if the integration is carried downward from the apex of the arch (where $\theta = 0$) at some height z_2. Without loss of generality the coordinates may be moved so that the apex lies at $y = 0$, midway between the foot points of the arch at $z = z_0$ ($< z_2$) and $y = \pm y(z_0)$. Then

$$y(z) = \pm \int_z^{z_2} \frac{d\zeta}{\{\mathcal{T}^2(\zeta)/\mathcal{T}^2(z_0) - 1\}^{\frac{1}{2}}}. \tag{8.49}$$

To explore the nature of this path suppose that the gas temperatures internal and external to the tube are equal, $T_i(z) = T_e(z) \equiv T(z)$,as we would expect in thermal equilibrium, so that (8.7) reduces to (8.8) and, through (8.39), (8.49) becomes

$$y(z) = \pm \int_z^{z_2} \frac{d\zeta}{[\exp\{-m(\zeta)\} - 1]^{\frac{1}{2}}} \tag{8.50}$$

where, now,

$$m(z) = \int_z^{z_2} d\zeta/\Lambda(\zeta) \tag{8.51}$$

in terms of the scale height $\Lambda(z) = kT(z)/Mg$.

The obvious question is how far apart can the footpoints be, where the tube is anchored in some suitably dense restraining medium. Noting that the arch gets wider and wider the farther down it extends from its apex, it is obvious that the maximum width is $y(-\infty)$. This is finite in any circumstance in which the integral on the right-hand side of (8.50) converges. It is necessary and sufficient for convergence that $|m(z)|$ increase faster than $2 \ln |z|$ as $z \to -\infty$, i.e. that the temperature $T(z)$ not increase as fast as $|z|$. In view of our belief that the internal temperatures of astrophysical bodies are bounded, it follows that the convergence condition is universally satisfied and the arch formed by a flux tube in hydrostatic equilibrium is limited to finite width. If the foot points are separated more widely than the finite distance $2y(-\infty)$, then the tension in the tube is not sufficient to overcome the buoyant forces, there is no equilibrium, and the tube floats upward toward $z = +\infty$.

Two examples are sufficient to illustrate the limitations. The simplest is the isothermal case in a uniform gravitational field, for which $\Lambda(z)$ is independent of depth. Then (8.50) yields the path

$$\tan^2(y/2\Lambda) = \exp[(z_2 - z)/\Lambda] - 1, \tag{8.52}$$

so that the maximum value of y is $\pi\Lambda$. In an isothermal atmosphere, any flux tube anchored at points separated by more than $2\pi\Lambda$ cannot be held in equilibrium against the buoyant forces.

In a polytropic atmosphere, extending downward from its top at $z=0$, we have a linear variation of temperature

$$T = T_0(-z/\lambda)$$

and

$$N = N_0(-z/\lambda)^\alpha, \qquad p = p_0(-z/\lambda)^{\alpha+1},$$

where α is a free parameter[4] and the temperature scale λ is $(\alpha+1)kT_0/Mg$. Note that

$$\mathrm{d}(\ln p)/\mathrm{d}(\ln T) = \alpha + 1.$$

Then (8.50) reduces to

$$y(z) = |z_2| \int_1^{z/z_2} \mathrm{d}u/(u^{\alpha+1}-1)^{\frac{1}{2}} \tag{8.53}$$

where now both z and z_2 are negative, and $|z_2| < |z|$. The width at $z=-\infty$ is finite for all $\alpha > 1$. For $\alpha > 1$, the maximum width of the arch is

$$2y(-\infty) = 2\,|z_2| \frac{\Gamma(\frac{1}{2})\Gamma\{(\alpha-1)/2(\alpha+1)\}}{(\alpha+1)\Gamma\{\alpha/(\alpha+1)\}}. \tag{8.54}$$

The maximum width declines as $(\alpha-1)^{-1}$ as α increases from unity, reaching $2y(-\infty)/|z_2| = 9{\cdot}08$ at the adiabatic value $\alpha = \frac{3}{2}$ ($\gamma = \frac{5}{3}$), 5·06 at $\alpha = 2$, 2·42 at $\alpha = 3$, and declining asymptotically as $2\pi/\alpha$ for $\alpha \gg 1$. The special case $\alpha = 1$ yields

$$z = z_2 \cosh(y/z_2). \tag{8.55}$$

For $\alpha = 2$,

$$y(z) = |z_2| 3^{-\frac{1}{4}} F(\psi, k), \tag{8.56}$$

where $F(\psi, k)$ is an elliptic integral of the first kind, with

$$\cos\psi = \frac{\sqrt{3}+1-z/z_2}{\sqrt{3}-1+z/z_2},$$

$$k^2 = (2-\sqrt{3})/4 = \sin^2(\pi/12).$$

For $\alpha = 3$,

$$y(z) = |z_2|\, 2^{-\frac{1}{2}} F\{\arccos(z_1/z), 2^{-\frac{1}{2}}\}. \tag{8.57}$$

[4] In the solar convective zone, for instance, the effective value of α is one or greater, with $\alpha = 1$ at a depth of 800 km (where $T = 9 \times 10^3$ K), rising to a maximum of about 8 at 10^3 km, and falling slowly to 2 at 10^4 km, and finally to the adiabatic value 1·5 at 2×10^5 km (Spruit 1974).

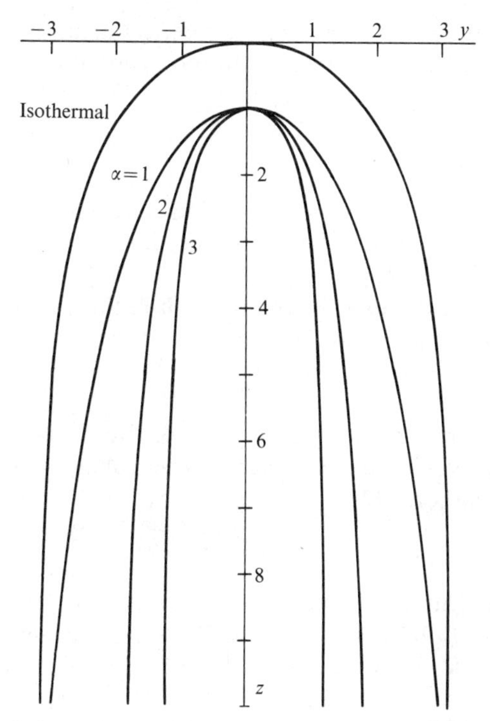

FIG. 8.5. A plot of the path $y = y(z)$ of a flux tube in hydrostatic equilibrium in an isothermal atmosphere, given by (8.52) in units of the scale height Λ, and in the polytrope atmospheres $\alpha = 1, 2, 3$, given by (8.55) in units of the depth z_1, of the apex below the upper surface of the atmosphere.

The equilibrium paths (8.52), (8.55), (8.56), and (8.57) are plotted in Fig. 8.5. Figure 8.5 emphasizes again the limited width of the arches (except 8.55), and we conclude that in general magnetic buoyancy cannot be suppressed over horizontal widths greater than a few scale heights. Hence, any magnetic fields in the interior of an astrophysical body must eventually appear at the surface. This brings us to the question of the rate of rise of a buoyant magnetic field and the question of whether, indeed, it is possible to anchor a magnetic field at all, as we have tacitly assumed in our discussion of the arched equilibrium path.

8.7. Rate of rise of magnetic field in thermal equilibrium

Consider the rate of rise of a slender horizontal flux tube of small cross-section A, extending across a stratified atmosphere of pressure $p_e(z)$ and temperature $T_e(z)$, all subject to a uniform gravitational acceleration g in the negative z-direction. The buoyant force per unit length is given by (8.40). Suppose in the first instance that the gas inside the tube is in thermal equilibrium with the gas outside, so that the temperature T_i within is equal to the temperature outside. Then (8.40) reduces to

$$\mathscr{F} = \frac{A}{\Lambda}\frac{B^2}{8\pi}\ \text{dyn cm}^{-1}. \tag{8.58}$$

This upward force causes the tube to rise at a rate u, which is opposed by the aerodynamic drag. The drag on various circular, elliptical, and other cylinders has been studied extensively (see, for instance, Goldstein 1938) both theoretically and experimentally. If we suppose, for instance, that the cross-section of the tube is circular with radius R, then $A = \pi R^2$ and the aerodynamic drag associated with the rate of rise u is

$$\mathscr{F}_D = \tfrac{1}{2}\rho u^2 R C_D, \tag{8.59}$$

where C_D is the drag coefficient. The rate of rise of the flux tube reaches its limiting value when the aerodynamic drag balances the buoyant force, in which case (Parker 1975a)

$$u = V_A(\pi/C_D)^{\frac{1}{2}}(R/\Lambda)^{\frac{1}{2}}. \tag{8.60}$$

The drag coefficient C_D is well known from experiment for Reynolds numbers $N_R = Ru/\nu$ between 10^{-2} and 10^6 (cf. Goldstein 1938). It is large compared to one at small Reynolds numbers, declining to 30 at $N_R = 1$, 4 at $N_R = 10$, and 1 at $N_R = 10^2$. For $N_R > 10^2$ the drag coefficient lies between 0·1 and 1·0. The molecular viscosity of the gases in stellar envelopes is so small that for any magnetic field and tube radius in which we are interested, the rate of rise is sufficiently large that $N_R \gg 1$. Hence $(\pi/C_D)^{\frac{1}{2}}$ typically has values of 2–5. For R of the same order as Λ, then, the rate of rise is comparable to the Alfven speed, $u = O(V_A)$. It is not possible to anchor flux tubes for long in thermodynamic equilibrium with the exterior gas ($T_i = T_e$). Thus, for instance, a magnetic field of 10^2 G at a depth of 10^5 km in the solar convective zone, where $\rho = 5 \times 10^{-2}$ g cm^{-3}, produces an Alfven speed of 1·2 m s^{-1}. A buoyant rise of 1 m s^{-1} carries the tube the distance of 10^5 km in 10^8 s, or 3 years. Generally speaking, it is not possible to anchor any but very weak fields in the convective zone of the sun, or other stars, or in the gaseous disc of the galaxy, for extended periods of time. It is clear that the dynamo in a star or a galaxy can produce fields only up to the strength where the buoyant rise time is comparable to the time for regeneration and amplification of that field. The regeneration time for the azimuthal field of the sun is of the order of the sunspot period of 11 years. Hence, it would appear, from the simple estimate given above, that it is not possible to produce an azimuthal field in the sun in excess of a few hundred gauss, and any such fields must be produced at depths of 10^5 km or more (Parker 1977).

Now it should be noted that the calculation based on the simple drag formula (8.59) involves a number of idealizations. First of all it is based on the idealization of a circular equilibrium cross-section, whereas in fact the cross-section of a slender flux tube has neutral stability and may be seriously deformed. Twisting the tube helps to stabilize the cross-section and is taken up in §9.8. Generally speaking the cross-section tends to widen across the direction of motion, becoming a flat ribbon of great width (because $A = constant$) as a consequence of the stagnation pressure $\frac{1}{2}\rho u^2$ at the top and bottom of the cross-section, and the reduced pressure at the sides. Indeed, we anticipate that a large cross-section tends to flatten and then break up into several smaller tubes, which in turn flatten, etc. The rate of rise will be reduced below the idealization (8.60) for the circular cross-section. Altogether the question is an interesting non-linear theoretical problem in fluid mechanics. It is essentially the dynamics of a two-dimensional free bubble without surface tension, and it is particularly difficult because of the free and variable boundary. It has not, so far as we are aware, received attention in the literature on fluid mechanics. Observations of the flux tubes coming up through the surface of the sun show no evidence of flattening, either on the large scale (10^4–10^5 km) of the active region or on the small scale (10^3 km) of the individual magnetic knots (cf. Beckers and Schröter 1968). Hence we suggest that a circular cross-section is a useful approximation to the truth (see Goldstein 1938 for the modifications of the drag on cylinders with elliptic cross-section).

More serious corrections to (8.60), for application to such real problems as the flux tubes in the solar convective zone, arise from the general turbulence of the fluid through which the flux tube is rising and from the convective instability of the fluid. The drag formula (8.59) is based on motion through a passive and quiescent medium, with the turbulence limited to the wake, if the Reynolds number Ru/ν is large enough to produce any turbulence at all. If the fluid is turbulent in its natural state, the question arises as to the modification of the drag formula (8.59). There is little or no appropriate wind tunnel data. The only theoretical tool available to treat the problem is the conventional mixing length theory of Prandtl (cf. Goldstein 1938; Dryden *et al.* 1956). In its most primitive form the mixing length theory describes the effect of small-scale turbulence on the large-scale flow of the fluid entirely in terms of the product of the characteristic dimension l and r.m.s. velocity v of the dominant eddies in the small-scale turbulence. The assertion is that the dynamical effect on the large-scale flow is effectively an enhanced kinematic viscosity, $\nu_e \cong \frac{1}{3}vl$ referred to as the equivalent *eddy viscosity*. One then proceeds to work out the large-scale flow from the Navier–Stokes equations with the molecular viscosity ν augmented by the eddy viscosity, so that ν is replaced by $\nu + \nu_e$. This simple approach, characterized by a single constant ν_e, ignores the boundary layer effects, the variation of the mixing length, and the diffusion of turbulence in gradients in the large-scale velocity. Although commonly used in treating astrophysical problems, it must be regarded as an order-of-magnitude estimate of the effect of turbulence, at best. More sophisticated forms of the mixing length theory are available, in which the mixing length l is modified by the presence of the boundaries of the large-scale flow, but they are cumbersome and inspire no great confidence in any special case until checked by direct experiment. Consequently, we proceed using the simplest form of the theory to estimate the order of magnitude of the rate of rise of a buoyant tube through a turbulent fluid, being fully aware of the lack of precision in the result. Indeed, the fact should not be overlooked that the scale l and velocity v of the turbulence in the convective zone of the sun, or other star, to which we might apply the results, are known only from application of the same mixing length theory to the problem of convective heat transport. Consequently l and v are themselves both uncertain in any real circumstance.

To continue, then, an eddy viscosity $\nu_e = \frac{1}{3}vl$ yields an effective Reynolds number

$$N_R \equiv uR/(\nu + \nu_e) \cong 3(R/l)(u/v).$$

The drag force and the rate of rise now follow from (8.59) and (8.60) with the drag coefficient C_D appropriate to the Reynolds number N_R computed from the eddy viscosity ν_e.

Now, for the tube radius R comparable to the eddy size l and $N_R \cong 3u/v$, it is evident that N_R is of the order of one only if the magnetic field is as large as the equipartition value so that $V_A \cong v$ and $u = O(V_A)$. This is the case already discussed, with $(\pi/C_D)^{\frac{1}{2}}$ of the order of one, for which $u \sim V_A$. If, on the other hand, the field is much weaker than the equipartition value, then $V_A \ll v$ so that $u \ll v$ and $N_R \ll 1$. Hence, for fields below the equipartition value, the Reynolds number is small and the drag coefficient C_D is large, so that the factor $(\pi/C_D)^{\frac{1}{2}}$ in (8.60) decreases from 0·3 at $N_R = 1$ to 0·1 at $N_R = 0{\cdot}1$. The rate of rise is significantly below the Alfven speed for small Reynolds numbers. Wind tunnel data is available for small Reynolds numbers (cf. Goldstein 1938). Fortunately, when $N_R \lesssim 1$ the hydrodynamic flow of a viscous fluid around a cylinder can be treated by formal methods (cf. Lamb 1932). The drag per unit length of tube proves to be

$$\mathcal{F}_D = 4\pi\rho u\nu/Q, \quad \text{i.e.} \quad C_D = 8\pi\nu/QuR.$$

The factor Q is the number $\ln(4/N_R) + \frac{1}{2} - \gamma$, whose γ is Euler's constant, equal to 0·577. This theoretical result for the drag agrees very well with the drag measured experimentally, $N_R < 1$. The terminal velocity of rise is, then,

$$u = QR^2 V_A^2 / 8\Lambda\nu$$

$$= V_A(3Q/8)(R^2/\Lambda l)V_A/v.$$

In the application of the mixing length theory to convective zones of stars it is customary to suppose that $l = \Lambda$, for lack of a better value. Hence, for weak fields, the rate of rise is

$$u/V_A = (3Q/8)(R/\Lambda)^2 V_A/v. \tag{8.61}$$

The Reynolds number is

$$N_R = (9Q/8)(R/\Lambda)^3 (V_A/v)^2.$$

The factor $3Q/8$ is of the order of one. If R/Λ is also of the general order of one, then N_R is as small as $(V_A/v)^2$. The rate of rise is the small fraction V_A/v of the Alfven speed.

To illustrate the application of these theoretical results to the rise of flux tubes in the convective zone of the sun, consider the broad fields that emerge through the surface of the sun to produce the large bipolar magnetic regions. The total flux may be as high as 3×10^{21} Mx, in the form of 40 G in a region with radius $R = 5\times10^4$ km. Suppose that the field is 10^2 G at a depth of 10^5 km, halfway to the bottom of the convective zone. The tube radius is then about 3×10^4 km. The scale height at a depth of 10^5 km is 4×10^4 km and the gas density is 5×10^{-2} g cm^{-3}, so that the Alfven speed is 1·2 m s^{-1}. The standard mixing length theory of the convective zone (cf. Spruit 1974) suggests a turbulent convective velocity $v = 30$ m s^{-1}. Then $v/V_A = 25$, $Q \cong 7$, $R/\Lambda = \frac{3}{4}$, and it follows that $N_R = 0{\cdot}004$. The rate of rise is $u = 0{\cdot}06\ V_A = 7$ cm s^{-1}, so that the characteristic rise time over one scale height is $\Lambda/u = 4{\cdot}3\times10^8$ s $= 14$ years. The rise time exceeds the 11-year half period of the magnetic cycle of the sun, in which the field is amplified by the solar dynamo (see Chapter 21). Hence, it seems possible to amplify the azimuthal field of the sun up to 10^2 G deep in the convective zone without undue loss of field by the buoyant rise. But note that a field of 200 G yields $V_A =$ 2·4 m s^{-1}, so that $N_R = 0{\cdot}02$, $Q \cong 7$, and the rate of rise is $v = 0{\cdot}12\ V_A =$ 29 cm s^{-1}. The characteristic buoyant rise time Λ/u is then $1{\cdot}4\times10^8$ s, or 4·4 years. The field, upon reaching 200 G, rises rapidly toward the surface. Even in the lower convective zone where ρ is as large as 0·1 g cm^{-3} and the scale height is somewhat larger, field strengths cannot be pushed beyond about 300 G before escape. Altogether it would seem that the buoyant rise limits the amplification of the general azimuthal field in the solar convective zone to no more than a few hundred G[5,6].

It must be remembered, of course, that this result carries with it all the uncertainties of the simple mixing length theory. In addition, it omits the possibility that the temperature within the rising flux tube may be different from the temperature outside ($T_i \neq T_e$). That is to say, contrary to the special condition of thermodynamic equilibrium $T_i = T_e$ on which this section is based, the gas within the tube may have a positive or negative buoyancy of its own. The next section takes up the problem in an atmosphere that is stable against convection, so that T_i within a rising tube is less than the temperature T_e outside. The point to be noted here is that the general convective instability of the ionization zone beneath the surface of the sun, and other stars, implies that the temperature within a rising flux tube is *higher* than the temperature outside, accelerating the buoyant rise. The effect is small for weak fields, but for strong fields it may contribute significantly to the rate of rise (see §§10.10 and 10.11). In that case (8.60) provides only a lower limit on the rate of rise. The very intense flux tubes, for which $V_A^2 \gg v^2$, near the surface of the sun may rise significantly faster than suggested by (8.60).

Zwaan (1978) has pointed out that if a flux tube is somehow confined solely by the dynamical pressure of external turbulence, then it may have less buoyancy than given by (8.58). For suppose that $T_i = T_e$ and $p_i = p_e$ so that the external pressure is larger than the internal pressure only because of the external turbulence, exerting a mean dynamical pressure $\frac{1}{2}\rho_e v^2$. Then, in place of (8.6) we have $B^2/8\pi = \frac{1}{2}\rho_e v^2$. It follows from $T_e = T_i$ and $p_e = p_i$ that $\rho_e = \rho_i$, so there is no magnetic buoyancy. As a matter of fact the external turbulence deforms the flux tube and produces internal velocity fluctuations with their own dynamical pressure, $\frac{1}{2}\rho_i v_i^2$. The special case of a thin flux tube squeezed periodically and equally on both sides is worked out in §10.4. There is an enhancement of the mean field $\langle\langle B \rangle\rangle$ along the tube given by (10.38). For the same internal and external gas pressures, the dynamical enhancement of the mean field is by the fraction $\frac{1}{4}n^2$ of the external dynamical pressure $\frac{1}{2}\rho_e v^2$, where n is a number less than one. Thus a modest reduction of buoyancy might be accomplished, but it does not significantly alter the estimated limiting field strengths.

[5] Other authors have proposed rates of rise considerably smaller than those estimated here. Unno and Ribes (1976) employ a formula equivalent to (8.61) without the factor Q. Schüssler (1977) uses values of the eddy viscosity that are about 30 times larger than we have used.

[6] The equipartition field computed from Spruit's model (Spruit 1974) of the convective zone is of the order of 2×10^3 G in the lower convective zone, so that a few hundred G is clearly a weak field.

8.8. Rate of rise of magnetic field in a stable atmosphere

If we take into account the possibility that the gas within the flux tube is not precisely in thermodynamic equilibrium with the outside, there are two possibilities. In an atmosphere that is unstable to convection the rising gas finds itself warmer than its surroundings, $T_i > T_e$, and the buoyancy (8.40) is enhanced; in a stable atmosphere the rising gas finds itself cooler than its surroundings, $T_i < T_e$, and the rise is suppressed. In the former case, then, in the convective atmosphere (such as the convective zone of most stars), the simple estimate (8.60) is still to be considered correct as a lower bound and the rate of rise is of the order of the Alfven speed. The only exception to this would be for the weakest fields where the flux tube is completely at the mercy of the turbulent convection, being wafted about to such a degree that (8.60) becomes irrelevant.

In the second case, of a stable atmosphere in which the rising tube becomes cooler than its surroundings, the rate of rise may be enormously reduced below the Alven speed, and the problem needs reconsideration (Parker 1974, 1975a). The slender horizontal tube rises until the internal temperature falls so far below the external temperature that the internal and external gas densities are equal and the buoyancy force (8.40) vanishes. Thereafter the rate of rise is controlled simply by the rate of heat transfer into the flux tube, warming the interior and permitting further upward motion of the tube.

Given the basic condition (8.6) for pressure equilibrium, it follows that the tube rises until the interior is cooler than the exterior by the amount $\Delta T = T_e - T_i$ at which $\rho_i = \rho_e$ and the buoyancy vanishes. Then since $T = M_p / k\rho$, it follows that

$$\begin{aligned} \Delta T &= \frac{M}{k}\frac{p_e - p_i}{\rho_e}, \\ &= \frac{M}{k}\frac{B^2}{8\pi\rho_e}, \\ &= T_e\frac{B^2}{8\pi p_e}, \\ &= \frac{B^2}{8\pi k N_e}. \end{aligned} \tag{8.62}$$

This temperature reduction leads to an inflow of heat, say at a rate $\mathrm{d}Q/\mathrm{d}t$ per unit length. In keeping with the idealization that the conditions within the slender tube are uniform across the diameter, we use Newton's law of cooling, that the rate of heat loss $\mathrm{d}Q/\mathrm{d}t$ is simply proportional to the difference ΔT between the internal and external temperatures,

$$\mathrm{d}Q/\mathrm{d}t = \alpha\kappa\Delta T \tag{8.63}$$

where α is a numerical constant of the order of unity[7] and κ is the total effective heat transfer coefficient of the medium (erg/s cm K).

Now in terms of the first law of thermodynamics it follows that the inflowing heat will increase the thermal energy as well as do work in expanding the system. Thus

$$\frac{dQ}{dt}=\frac{kAN_i}{\gamma-1}\frac{dT_i}{dt}+p_i\frac{dA}{dt} \tag{8.64}$$

where $k/(\gamma-1)$ is the specific heat per particle at constant volume. This equality relates the rate of rise dT_i/dt and dA/dt to ΔT and ultimately $B^2/8\pi$. If the velocity of rise is u, write

$$\frac{dT_i}{dt}=u\frac{dT_i}{dz}, \qquad \frac{dA}{dt}=u\frac{dA}{dz},$$

and consider how T_i and A vary with the elevation z of the tube. Conservation of magnetic flux requires that $AB=\Phi=\textit{constant}$ and conservation of matter requires $N_iA=M=\textit{constant}$. But as already noted, the tube is always very close to the height at which $N_i=N_e$, so $N_eA=\mathcal{M}$ and $B/N_e=\Phi/\mathcal{M}$. Hence, it follows from $T_i=T_e-\Delta T$ and (8.62) that

$$\begin{aligned}\frac{dT_i}{dz}&=\frac{dT_e}{dz}-\frac{d}{dz}\left(\frac{B^2}{8\pi kN_e}\right)\\&=\frac{dT_e}{dz}-\frac{B^2}{8\pi kN_e^2}\frac{dN_e}{dz}.\end{aligned}$$

Hence (8.64) can be written

$$\frac{dQ}{dt}=uAp_e\left[\frac{1}{\gamma-1}\frac{1}{T_e}\frac{dT_e}{dz}-\left\{1+\frac{2-\gamma}{\gamma-1}\frac{B^2}{8\pi p_e}\right\}\frac{1}{N_e}\frac{dN_e}{dz}\right]. \tag{8.65}$$

Use (8.62) to write dQ/dt in (8.63) in terms of T_e, equate the result to (8.65), and solve for u. The result is

$$u=\frac{\alpha\kappa T_eB^2}{8\pi Ap_e^2}\left(-\frac{1}{T_e}\frac{dT_e}{dz}\right)^{-1}\left(\delta+\frac{B^2}{8\pi p_e}\frac{n(2-\gamma)}{\gamma-1}\right)^{-1} \tag{8.66}$$

where we have defined the indices $n\equiv d(\ln N_e)/d(\ln T_e)$, and $\delta\equiv n-1/(\gamma-1)$. In an adiabatic atmosphere, $n=1/(\gamma-1)$ and $\delta=0$. In a

[7] It is a simple matter to show from the heat flow equation $C\partial T/\partial t=\kappa\nabla^2T$ that the internal temperature across a uniform circular cylinder of fixed radius R, specific heat per unit volume C, and thermal transport coefficient κ, subject to the steadily declining external temperature $T_e(t)=T_0(1-t/\tau)$, is $T_i(\varpi,t)=T_e(t)-T_0C(R^2-\varpi^2)/4\kappa\tau$ in terms of the distance ϖ from the axis. In terms of the difference ΔT between the mean temperature inside and outside, $\Delta T=T_0CR^2/2\kappa\tau$, the rate of heat loss per unit length is then $dQ/dt=\pi R^2CT_0/\tau=2\pi\kappa\Delta T$, corresponding to $\alpha=2\pi$.

stable atmosphere, the temperature declines with height less rapidly than in the adiabatic atmosphere, so that $\delta > 0$. This formula for the rate of rise can be simplified, noting that generally $B^2/8\pi p_e \ll 1$, while $\gamma - 1 = O(1)$. Then, in terms of the total heat flux I upward through the atmosphere

$$I = -\kappa\, dT_e/dz \text{ erg cm}^{-2}\text{ s}^{-1},$$

and the temperature scale height λ,

$$1/\lambda = d(\ln T_e)/dz,$$

(8.66) reduces to

$$u \cong \frac{\alpha}{\pi\delta} \frac{\lambda^2}{R^2} \frac{B^2}{8\pi p_e} \frac{I}{p_e}, \tag{8.67}$$

where we have written the cross-sectional area as $A = \pi R^2$. In terms of the mean upward velocity of flow u_T of the thermal energy,

$$u_T \equiv (\gamma - 1) I/p_e \tag{8.68}$$

the rate of rise of the flux tube is (Parker 1974, 1975a)

$$u = u_T \frac{\alpha}{\pi\delta(\gamma-1)} \frac{\lambda^2}{R^2} \frac{B^2}{8\pi p_e}. \tag{8.69}$$

For a slender flux tube $R \ll \Lambda$, λ the rate of rise can be very rapid; even for weak fields, $B^2/8\pi p_e \ll 1$. A number of examples will serve to illustrate the rates under various circumstances. Gurm and Wentzel (1967) treat the rise of spherical cells of magnetic field, with comparable results.

In the sun the convective zone is believed to begin at a depth h of about 2×10^5 km beneath the surface (Weymann 1957; Schwarzschild 1958; Baker and Temesvary 1966). Above that level (8.60) is applicable, yielding high rates of rise. The density in the lower convective zone is of the order of $0{\cdot}1 \text{ g cm}^{-3}$ (Spruit 1974), so that for a field of 10^2 G, $V_A \cong 1 \text{ m s}^{-1}$ and, from (8.60), $u = O(V_A)$. The characteristic rise time is $h/u \cong 2 \times 10^8 \text{ s} \cong 7$ years. In the stable region beneath the convective zone, (8.67) or (8.69), is applicable and the rate of rise is very much slower. At the top of the stable interior, immediately below the base of the convective zone, the density is of the order of $0{\cdot}2 \text{ g cm}^{-3}$ and the temperature is 2×10^6 K. Since $I \cong 10^{11} \text{ erg cm}^{-2}\text{ s}^{-1}$, and $\gamma = \frac{5}{3}$, it follows that the upward transport velocity of thermal energy is given by (8.68) to be about $2 \times 10^{-3} \text{ cm s}^{-1}$. The temperature scale height is $\lambda \cong 10^5$ km while the pressure scale height Λ is half as much. The index n increases rapidly from its adiabatic value of $\frac{3}{2}$ in the convective zone to 3 or more, so that $\delta \gtrsim \frac{3}{2}$ in the stable interior. Put $\alpha/\pi\delta(\gamma - 1) = O(1)$. Then for 10^2 G, $B^2/8\pi p_e \cong 10^{-11}$, and a flux tube with radius $R = 10^3$ km ($\lambda/R \cong 10^2$) rises at a rate of $u = 10^{-7}\ u_T = 2 \times 10^{-10}\ \mathrm{cm\ s^{-1}}$. In the 4×10^9-year life of the sun, the distance traversed is only 300 km. A field of 3000 G is required to raise the flux tube of radius 10^3 km a distance of one scale height (5×10^4 km) in the 4×10^9 year age of the sun. A tube of larger radius rises more slowly, of course, with the requirement $B \propto R$ to achieve a given velocity u/u_T. Altogether, then, magnetic buoyancy is important for ejecting 10^2-G fields from the convective zone, but it has negligible effect in the stable radiative interior of the sun. A flux tube of 10^2 G in the stable interior may be considered to be permanently anchored.

It is interesting, then, to consider what primordial fields might possibly be buried in the stable radiative core of the sun. Some authors (Bartenwerfer 1973; Chitre *et al.* 1973) have suggested that there may be fields of sufficient strength to modify the gas pressure and, hence, the neutrino emission, to some significant degree. To make the best possible case for this, suppose that the field is coherent over a radius as large as $R = \Lambda$. Then (8.69) reduces to

$$u = u_T B^2/8\pi p_e$$

and the rate of rise is the same fraction of u_T as the magnetic pressure is of the total pressure. Thus if $B^2/8\pi p_e$ is as large as 0·1, so as to have a significant effect on the nuclear reaction rates, then u is as large as 0·1 u_T, and the field rises rapidly. To treat a specific example, consider a position at a radial distance of $r = 10^{10}$ cm (0·14 $R_\odot$) from the centre of the sun, where $p_e \cong 10^{17}\ \mathrm{dyn\ cm^{-2}}$, $I = 2 \times 10^{12}\ \mathrm{erg\ cm^{-2}\ s^{-1}}$, and $u_T \cong 10^{-5}\ \mathrm{cm\ s^{-1}}$. Then if the gas pressure is reduced 10 per cent by the magnetic field ($B^2/8\pi p_e = 0{\cdot}1$), the rate of rise of the field is $u = 10^{-6}\ \mathrm{cm\ s^{-1}}$. The characteristic rise time if $t = r/u = 10^{16}\ \mathrm{s} = 3 \times 10^8$ years. Any such field (5×10^8 G) in the early sun would have departed long ago. Hence we doubt that magnetic fields trapped in the interior of the sun since its formation have any significant effect on the thermodynamic state and the thermonuclear rates in the interior of the sun (Parker 1974). Note, however, that a much weaker field of 10^7 G ($B^2/8\pi p_e = 4 \times 10^{-5}$) has a characteristic rise time of 10^{19} s, or 3×10^{11} years, and so might conceivably be trapped in the interior of the sun without our being aware of it. Such a field has no sensible effect on the interior structure of the sun, of course, but the total flux might be as large as $r^2 B \cong 10^{27}$ Mx. If some portion of this were to appear at the surface of the sun, it might contribute significantly to the normal fluxes of 10^{23} Mx observed there. Any such scenario must be worked out carefully, however, before it can be seriously considered. There is no evidence from the mapping of the surface field of the sun that there is a permanent field extending up from the interior. The fields that appear at the surface seem to be part of the general solar dynamo (see discussion in §§21.1 and 21.2). On the other hand, in the magnetic A-stars, exhibiting rigid rotation of oblique field patterns of 10^3 G, we may be observing a primordial magnetic field embedded in the star and preserved to the present time. Many of the magnetic stars can be of no great age (say, not more than 10^8 years) so we should be surprised if they had lost all their primordial field in so short a time.

8.9. The mutual attraction of rising flux tubes

8.9.1. The problem posed by observations

It is an observed fact (Beckers and Schröter 1968) that the magnetic knots, appearing in such great numbers in bipolar magnetic regions on the surface of the sun, exhibit a mutual attraction for each other during the formative stage of the sunspots. The attraction is between magnetic knots of like polarity, whose magnetic fields repel each other. The growth of the sunspots appears to be a consequence of the mutual attraction and coalescence of knots to form pores and to join existing pores and spots. No presentation of the theory of individual flux tubes can pretend to be complete without taking up the challenge of this remarkable behaviour.

The magnetic knots represent the intersection of magnetic flux tubes with the surface of the sun, with fields of 1100–1400 G (Beckers and Schröter 1968). Their mutual magnetic repulsion, then, is strong. An important clue to the cause of the *attraction* is that it extends only to magnetic flux tubes newly arriving at the surface. Once the development stage is passed, and there are no new flux tubes arriving at the surface, the knots lose their mutual attraction and begin to move apart, as one would expect them to do. Indeed, the decay of sunspots is nothing more than their decomposition into smaller spots, and into magnetic knots which then disperse over the surface of the sun.

The explanation for the mutual attraction during the formative stages appears to be the upward motion of the tubes so that they are subject to the hydrodynamic forces of the rising tubes located nearby (Parker 1978b). Two bubbles of the same size rising up through water attract each other whether they are rising side by side, or one behind the other. If they are side by side, then they are attracted by the Bernoulli effect. The fluid streams by faster on the sides facing each other, thereby reducing the pressure and sucking the rising objects toward each other. On the other hand, if one is nearly behind the other, then it is boosted along in the wake of the leading bubble. The rate of rise of flux tubes in a convective zone is comparable to the Alfven speed, with the result that the hydrodynamic forces of attraction are strong.

The basic facts on magnetic knots are fields of 1100–1400 G over a diameter of about 10^3 km, giving a total flux Φ of about 10^{19} Mx. Beckers

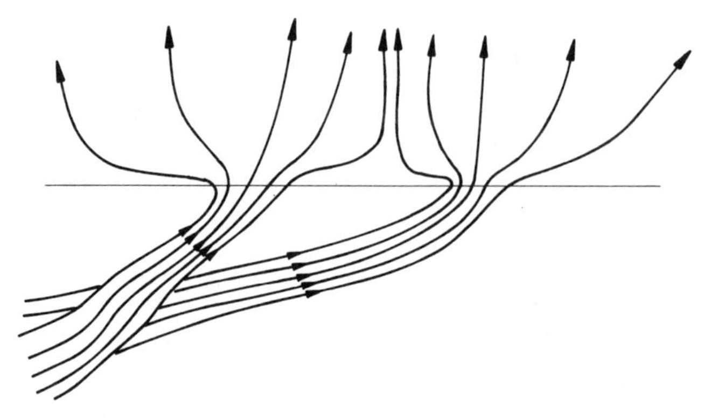

FIG. 8.6. A sketch of two neighbouring flux tubes rising through the surface of the sun with the field above the surface expanded to fill the available volume.

and Schröter (1968) found about 2×10^3 knots within $1{\cdot}4\times10^5$ km of a unipolar spot. About one-third of the knots had the same polarity, and two-thirds had the opposite polarity, of the sunspot, so that the algebraic sum of the magnetic flux in the knots was comparable in magnitude, and opposite in sign, to the spot.

Now the repulsion of two knots of like polarity is readily calculated. The field of a single knot with flux Φ spreads out radially above the surface of the sun in the same manner as the field above the surface of a magnetic pole $\mathcal{M}=\Phi/2\pi$ lying on the surface. The repulsion of two poles separated by a distance $2h$ (large compared to the radius a of the individual knots) is $(\mathcal{M}/2h)^2$. But, of course, in the real case the field exists only in the half-space above the surface so that the repulsion is half this amount. Altogether, then, magnetic knots of radius a and field strength B_0, so that $\Phi=\pi a^2B_0$, repel each other with a force

$$F_r=\tfrac{1}{8}B_0^2a^2(a/2h)^2$$

above the surface of the sun. Below the surface of the sun the fields are compressed into isolated flux tubes so there is no magnetic interaction. There are only the hydrodynamic forces caused by the rapid rise of the tubes. Figure 8.6 is a sketch of the field configuration of two neighbouring flux tubes. Consider first the hydrodynamic attraction of two horizontal flux tubes rising side by side with a velocity u.

8.9.2. Transverse attraction of rising tubes

The physical principles involved in the mutual hydrodynamic attraction of two rising flux tubes separated initially by a horizontal distance $2h_0$ are readily illustrated by the example of two circular cylinders of equal radius set in parallel motion in an ideal fluid. Consider, then, two cylinders, each of radius a with their axes parallel to the z-axis and lying symmetrically at $(x, y = \pm h_0)$, as shown in Fig. 8.7. At time $t = 0$ suppose that both cylinders are set in motion with a velocity u_0 in the x-direction. At the same time they are given velocities $\pm v_0$ in the y-direction, so that in the subsequent motion their distance $h(t)$ on either side of the xz-plane varies with time $\mathrm{d}h/\mathrm{d}t = v(t)$ as they move in the x-direction with velocity $u(t)$. The fluid in the neighbourhood of the cylinders is set in motion by the cylinders. The subsequent motion of the fluid and cylinders was worked out first by Hicks (1879), and in a different form by Greenhill (1882), and is presently available in the republication of Basset (1888), whose exposition can be taken over directly for the present purposes. The fluid velocity is determined from the fact that the motion is irrotational so that the velocity potential and the stream function are the conjugate parts of an analytic complex function. The total kinetic energy of the two cylinders plus the fluid is represented by $2T$ and can be written

$$T = \tfrac{1}{2}R(u^2 + v^2). \tag{8.70}$$

The quantity R is the mass of each cylinder plus the effective mass of the moving fluid associated with it. It is a function of the separation $2h$ of the cylinders and can be shown to be

$$R(h) = \pi a^2\sigma + \pi a^2\rho\left\{1 + 2\sum_{n=1}^{\infty}\frac{(1-q)^2 q^n}{(1-q^{n+1})^2}\right\}. \tag{8.71}$$

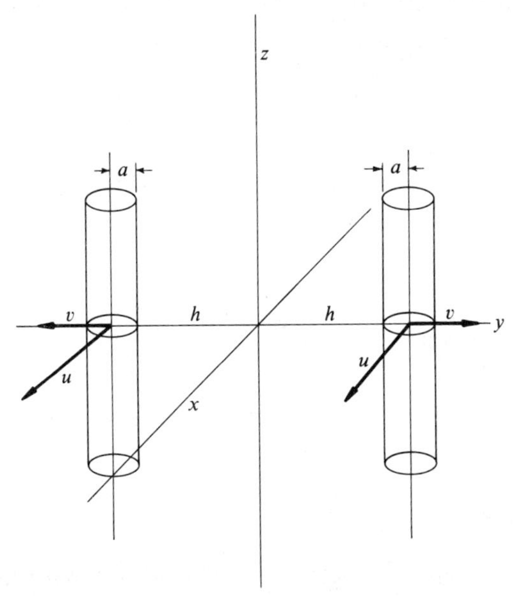

FIG. 8.7. The geometry for considering the mutual hydrodynamic attraction of parallel flux tubes with circular cross-sections moving together in the x-direction.

where ρ is the density of the fluid and σ is the density of the cylinder. The quantity $q = q(t)$ is defined as $\exp(-2\alpha)$ where α is the parameter relating the separation $h(t)$ and the cylinder radius a,

$$h(t) = a \cosh \alpha(t). \tag{8.72}$$

With the understanding that the subscript zero denotes the initial value, and u, v, h, q, α without the subscript imply the value at time t, note that

$$\alpha = \ln\{h/a + (h^2/a^2 - 1)^{\frac{1}{2}}\}. \tag{8.73}$$

The mass function R is positive and declines monotonically with increasing separation h. For $h = \infty$ it is obvious that $q = 0$ and R has the limiting value $\pi a^2(\rho + \sigma)$, which tells us that, so far as total kinetic energy is concerned, an isolated circular cylinder moves in the company of an equivalent mass of fluid equal to the mass of the fluid it displaces.

At the other end of the scale ($h \to a$), where the cylinders come into physical contact, the velocity of the fluid streaming between the cylinders becomes large without bound. We expect therefore, that R remains well defined (because T is well defined) as $h \to a$ but that the derivatives d^nR/dh^n may show some unusual behaviour. It is evident that q increases to one as h/a decreases to one. Let $q = 1 - \epsilon$, then, so that

$$h/a = (1 - \tfrac{1}{2}\epsilon)/(1-\epsilon)^{\frac{1}{2}} \cong 1 + \tfrac{1}{8}\epsilon^2 + \tfrac{1}{8}\epsilon^3 + \ldots,$$

Then note that the nth term Y_n in the sum defining R can be written

$$\begin{aligned} Y_n &= q^n/(1 + q + q^2 + \ldots + q^n)^2 \\ &\cong \{1 - \tfrac{1}{6}n(1 + \tfrac{1}{2}n)\epsilon^2 + O(\epsilon^4)\}/(n+1)^2 \end{aligned}$$

with the aid of the sums

$$1 + 2 + 3 + \ldots + n = \tfrac{1}{2}n(n+1), \qquad 1 + 3 + 6 + \ldots + \tfrac{1}{2}n(n-1) = \tfrac{1}{6}n(n-1)(n+1).$$

Hence

$$R(h) = \pi a^2(\sigma + \rho) + 2\pi a^2 \rho \left\{ \sum_{n=1}^{\infty} \frac{1}{(n+1)^2} - \frac{\epsilon^2}{6} \sum_{n=1}^{\infty} \frac{n(1 + \frac{1}{2}n)}{(n+1)^2} + \ldots \right\}.$$

The first sum yields $\frac{1}{6}\pi^2 - 1$. The second sum is divergent, indicating that the second derivative d^2R/dq^2 is singular at $q = 1$. At $h = a$ $(q = 1)$ the value of R is

$$R(a) = \pi a^2\{\sigma + \rho(\tfrac{1}{3}\pi^2 - 1)\} = \pi a^2(\sigma + 2{\cdot}290\rho). \tag{8.74}$$

Thus $R(a)$ is larger than $R(\infty)$ by $(\sigma + 2{\cdot}290\rho)/(\sigma + \rho)$, which equals 1·645 for the case that $\sigma = \rho$. The equivalent mass of fluid carried with each cylinder as they come into physical contact ($h = a$) is 2·290 times the mass $\pi a^2\rho$ of the fluid displaced by the cylinder. It follows from conservation of mass and energy that the velocity of the cylinders increases with increasing separation and decreases as they approach each other.

The motion of the cylinders is readily deduced from the expression (8.70) for the kinetic energy. With no external forces applied to the cylinders, the kinetic energy T is a constant of the motion. It is obvious, too, that the momentum of the system in the x-direction along the plane of symmetry is a constant of the motion. The conjugate momentum in the x-direction is $\partial T/\partial u = Ru$. In terms of the initial values, then,

$$Ru = R_0 u_0, \tag{8.75}$$

$$R(u^2+v^2) = R_0(u_0^2+v_0^2), \tag{8.76}$$

with

$$u = \mathrm{d}x/\mathrm{d}t, \qquad v = \mathrm{d}h/\mathrm{d}t, \qquad R = R(h). \tag{8.77}$$

Using (8.75) to eliminate u from (8.76) it is readily shown that

$$v^2 = (u_0^2+v_0^2)(R_0/R) - u_0^2(R_0/R)^2,$$

from which the quadrature

$$t = \pm\int_{h_0}^{h} \frac{\mathrm{d}h}{h_0\{(u_0^2+v_0^2)R_0/R - u_0^2R_0^2/R^2\}^{\frac{1}{2}}}$$

follows directly. The path $(x, \pm h)$ of the cylinders also follows directly, from the fact that $\mathrm{d}h/\mathrm{d}x = v/u$, so that

$$\mathrm{d}h/\mathrm{d}x = \pm\{(1+v_0^2/u_0^2)R/R_0 - 1\}^{\frac{1}{2}}. \tag{8.78}$$

Now suppose that initially the cylinders are moving in the positive x-direction ($u_0 > 0$) while moving apart ($v_0 > 0$) so that h increases with time and R/R_0 declines from its initial value of one. Initially $\mathrm{d}h/\mathrm{d}x$ is positive. If $1+v_0^2/u_0^2$ is so large that $(1+v_0^2/u_0^2)R/R_0$ remains larger than one as $h\to\infty$ (and R declines to R_∞), then it is obvious from (8.78) that $\mathrm{d}h/\mathrm{d}x$ remains positive as h becomes large without limit. The cylinders forever move apart on a path with asymptotic slope

$$\mathrm{d}h/\mathrm{d}x \sim \{(1+v_0^2/u_0^2)R_\infty/R_0 - 1\}^{\frac{1}{2}}.$$

On the other hand, if $1+v_0^2/u_0^2$ is not so large, and $(1+v_0^2/u_0^2)R_\infty/R_0$ is less than one, then when h increases to some finite positive value $h_{\max}$, $\mathrm{d}h/\mathrm{d}x$ declines to zero. At that point in time the cylinders are moving along parallel courses and the hydrodynamic attraction subsequently causes them to move together. The slope $\mathrm{d}h/\mathrm{d}x$ passes through zero and becomes negative. The cylinders coast on until they collide. It follows that the necessary and sufficient condition that the moving cylinders come together is

$$(1+v_0^2/u_0^2)R_\infty < R_0.$$

That is to say, their initial velocity of separation v_0 must be less than

$$v_0 < u_0(R_0/R_\infty - 1)^{\frac{1}{2}}. \tag{8.79}$$

If v_0 is larger than this, the cylinders escape from each other. Hence it follows that two cylinders moving on parallel courses ($v_0 = 0$) with the same initial velocity u_0 will eventually come together, as a consequence of their mutual hydrodynamic attraction, *no matter how large their initial separation may be.*

To examine the mutual attraction in more detail, consider two cylinders that are widely separated, $h \gg a$. Then $q \sim (a/2h)^2$ and is very small. It follows from (8.71) that the mass function is

$$R = \pi a^2\{\sigma + \rho(1 + a^2/2h^2 + \ldots)\} \tag{8.80}$$

to lowest order. To compute how far the cylinders move along together before colliding, consider the case where $t = 0$ when they are moving parallel to each other with velocity u_0 and large separation $2h_0$. Then the distance Δx over which the separation declines from $2h_0$ to $2h$ is given by (8.78) as

$$\begin{aligned}\frac{\Delta x}{h_0} &= \int_{h/h_0}^{1} \frac{d(h/h_0)}{(R/R_0 - 1)^{\frac{1}{2}}}\\ &\cong \left\{\frac{2(\sigma+\rho)}{\rho}\right\}^{\frac{1}{2}} \frac{h_0}{a} \int_{h/h_0}^{1} \frac{d(h/h_0)(h/h_0)}{(1 - h^2/h_0^2)^{\frac{1}{2}}}\\ &= \left\{\frac{2(\sigma+\rho)}{\rho}\right\}^{\frac{1}{2}} \frac{h_0}{a}\left(1 - \frac{h^2}{h_0^2}\right)^{\frac{1}{2}}.\end{aligned}$$

The distance to collision ($h = a \ll h_0$) is, then, Δx_c given by

$$\frac{\Delta x_c}{h_0} = \left\{\frac{2(\sigma+\rho)}{\rho}\right\}^{\frac{1}{2}} \frac{h_0}{a}. \tag{8.81}$$

For $\sigma = \rho$ this is

$$\Delta x_c/h_0 = 2h_0/a \tag{8.82}$$

and is large compared to the initial separation because of the initial weak attraction at large separation. The path followed by the cylinder is the ellipse

$$\left(\frac{\Delta x}{h_0}\right)^2 \left(\frac{a}{2h_0}\right)^2 + \left(\frac{h}{h_0}\right)^2 = 1$$

in the coordinates Δx and h.

The final direction of the path at the time of collision ($h=a$) is readily worked out from (8.70) and (8.71). The value of R at $h=a$ is given by (8.74), while the initial value follows from (8.80) (for $h_0 \gg a$). From (8.75) the final value of u is

$$u_a = u_0 \frac{\sigma+\rho}{\sigma+\rho(\frac{1}{3}\pi^2-1)}$$
$$= u_0 6/\pi^2 = 0{\cdot}607 u_0$$

for $\sigma=\rho$. It follows from (8.76) that the final value of v i.e. the velocity of approach of each cylinder to the plane, is

$$v_a^2 = u_0^2(R_0/R_a)(1-R_0/R_a)$$
$$= u_0^2 \frac{\rho(\sigma+\rho)(\frac{1}{3}\pi^2-2)}{\{\sigma+\rho(\frac{1}{3}\pi^2-1)\}^2}.$$

Hence, for $\sigma=\rho$, $v_a=0{\cdot}49u_0$, so that $v_a/u_a=0{\cdot}81$. The path makes an angle of about 39° with the x-axis, so the collision is moderately hard.

It remains to give a direct result for the attractive forces exerted on each cylinder. Consider an external force describable by a potential function $V(x, y^2)$. Then the Lagrangian for the cylinder at (x, y) is

$$L = \tfrac{1}{2}R(u^2+v^2) - V.$$

Hence the equations of motions are

$$\frac{\mathrm{d}}{\mathrm{d}t} Ru + \frac{\partial V}{\partial x} = 0$$

$$\frac{\mathrm{d}}{\mathrm{d}t} Rv - \tfrac{1}{2}(u^2+v^2)\frac{\mathrm{d}R}{\mathrm{d}y} + \frac{\partial V}{\partial y} = 0.$$

Suppose that the external $V(x, y)$ is constructed so that u and v are both constant, with the cylinder moving along a straight path with constant speed. Then we require that

$$\partial V/\partial x = -uv\, \mathrm{d}R/\mathrm{d}h, \qquad \partial V/\partial y = \tfrac{1}{2}(u^2-v^2)\, \mathrm{d}R/\mathrm{d}h$$

with $h=h_0+vt$. At large separations, then, the force required to maintain the constant motion is

$$F_x = -\partial V/\partial x = -\pi a^2 \rho(uv/a)(a/h)^3, \tag{8.83}$$

$$F_y = -\partial V/\partial y = -\pi a^2 \rho\{(u^2-v^2)/2a\}(a/h)^3. \tag{8.84}$$

For parallel motion ($v=0$), the force of attraction is, then,

$$F_y = -\pi a^2 \rho(u^2/2a)(a/h)^3 \tag{8.85}$$

per unit length of cylinder. The force per unit mass is of the same order as

u^2/a and varies inversely with the cube of the separation. The criterion (8.79) for capture becomes

$$v_0 < u_0 \frac{\rho}{2(\sigma+\rho)} \left(\frac{a}{h}\right)^2$$

and declines as h^{-2}. The inverse cube relation for the attractive force is readily understood from the fact that the velocity perturbation produced by one cylinder at the position of the other varies inversely with the separation h. Hence the Bernoulli effect Δp varies as h^{-2}. It is the gradient of the Bernoulli pressure Δp that causes the attraction, leading to h^{-3}.

It was pointed out that, in the region above the surface of the sun, two flux tubes repel each other with a force

$$F_r = \tfrac{1}{8}\pi a^2 \rho V_A^2 (a/h)^2.$$

Beneath the surface of the sun the tubes attract each other with the force (8.85) for the motion u perpendicular to their axes (see Fig. 8.6). The attraction over any length of cylinder in excess of L overcomes the repulsion, where $LF = F_r$, or

$$L = \tfrac{1}{4} h V_A^2 / u^2. \tag{8.86}$$

With u comparable to V_A the mutual hydrodynamic attraction over a length equal to one-eighth of the separation $2h$ of two tubes is sufficient to overcome the repulsion and bring the tubes together. It would seem, then, that the mutual magnetic repulsion of magnetic knots of like polarity is readily overcome by the mutual hydrodynamic attraction of rising flux tubes. The observed mutual attraction and coalescence of magnetic knots appears to result from hydrodynamic attraction. The effect continues so long as the flux tubes are rising. Once the flux tubes approach a vertical posture and cease to rise, there is no longer a mutual hydrodynamic attraction. Separate magnetic knots repel each other and flux tubes already bundled together to form a sunspot are inclined to separate again into individual magnetic knots (unless other forces come into play to hold them together; Meyer *et al.* 1974).

The mutual hydrodynamic attraction of two rising flux tubes separated vertically as well as horizontally can be worked out along the same lines as the simple case of two tubes on the same level. However, the algebraic manipulations are formidable because of the loss of the principal symmetry of the problem. The example of two tubes rising side by side is sufficient to illustrate the principles. We turn to the other symmetric situation, in which one tube follows directly behind the other.

8.9.3. The longitudinal attraction of rising tubes

The physical principles involved in the mutual hydrodynamic attraction of two rising flux tubes following one behind the other are readily illustrated by the properties of the wake behind a horizontal circular cylinder rising upward through a passive fluid. The passage of an object, such as a flux tube, through any but an ideal inviscid fluid is accompanied by a drag force F_D on the object. After a time t the total momentum imparted to the fluid is

$$\textit{Momentum} = \int_0^t dt' F_D(t').$$

The momentum spreads out from the path of the body so that the motion of the fluid is slower than that of the body, with the result that the moving fluid is soon left behind and is termed the *wake* of the body. The intimate relation between the drag and the wake of a moving body has led to an extensive study of wakes by aerodynamicists (cf. Goldstein 1938; Coutanceau and Bouard 1977). The properties of wakes are remarkable, their vortex structure exhibiting beautiful symmetric patterns behind the moving body. Mathematical formulations of comparable beauty have been developed to describe the wake (cf. review in Goldstein 1938). The elementary mathematical treatment of the long thin wake of a cylinder moving transversely with constant velocity U through a homogeneous, and initially static, fluid is well known. If the cylinder is moving along the x-axis, leaving behind a narrow wake that spreads out in the y-direction, then, in the frame of reference of the wake, the Navier–Stokes equations

$$\rho(\mathbf{v}\cdot\nabla)\mathbf{v} = -\nabla p + \mu\nabla^2\mathbf{v}$$

reduce to a simple form. Far down stream the fluid in the wake has nearly the background velocity U in the x-direction, so write $v_x = U + \delta v_x$ where $\delta v_x \ll U$. The principal velocity gradients are across, rather than along, the wake ($\partial/\partial y \gg \partial/\partial x$) so that the incompressibility,

$$\partial\delta v_x/\partial x + \partial\delta v_y/\partial y = 0,$$

leads to a transverse velocity δv_y that is small compared to the longitudinal velocity δv_x. Neglecting $\partial/\partial x$ except where it is multiplied by U, and neglecting δv_y compared to δv_x, the x-component of the Navier–Stokes equations reduces to

$$\rho U \partial\delta v_x/\partial x \cong \mu\partial^2\delta v_x/\partial y^2.$$

This equation is nothing more than the familiar heat flow equation with t replaced by x. The desired solution is

$$\delta v_x = \frac{C}{x^{\frac{1}{2}}}\exp\left(-\frac{\rho U y^2}{4\mu x}\right)$$

for which δv_x is a maximum on the path of the body along the x-axis and falls to zero at $x=+\infty$ and $y=\pm\infty$. It is apparent that the characteristic width of the wake is $(4\mu/\rho U)^{\frac{1}{2}}x^{\frac{1}{2}}$, increasing in proportion to the square root of the distance x behind the body. If a is the radius of the cylinder producing the wake, then the Reynolds number is $R\equiv\rho Ua/\mu$ and the width of the wake can be written simply as $(x/a)^{\frac{1}{2}}/R^{\frac{1}{2}}$. The forward velocity δv_x in the wake declines only very slowly, $x^{-\frac{1}{2}}$, with distance x behind the body, from a value at high Reynolds number that is comparable to U immediately behind the body. It is the extended upward motion of the fluid in the wake of a rising tube that hastens the upward progress of a tube back in the wake, causing it to catch up.

Now any attempt at a *quantitative* assessment of the upward motion in the wake of a rising flux tube in the sun must include the effect of the pre-existing turbulence and the effect of the convective forces that boost the wake along in its upward motion. The effect of the pre-existing turbulence may be included in a conventional, if not precise, manner using the familiar mixing length theory and the concept of eddy viscosity. On the other hand, the effect of the convective forces, which tend to accelerate the rising flux tube and its wake, is beyond the scope of the present introductory treatment and will be discussed but not treated quantitatively.

Consider, then, the rate of rise U of the flux tube producing the wake. As a starting point, note that the rate of rise of a tube, with circular cross-section of radius a, through an atmosphere with neutral convective stability (i.e. a temperature gradient equal to the adiabatic gradient) is given by (8.60), where the drag coefficient C_D is equal to 30 for a Reynolds number $N_R=1$, and to 4 at $N_R=10$. The analytic form (8.61) applies (as well as 8.60) if the eddy viscosity of the background turbulence of the convective core is strong. Putting the mixing length l equal to the tube radius R, the Reynolds number is

$$N_R=(9\pi/C_D)^{\frac{1}{2}}(R/\Lambda)^{\frac{1}{2}}V_A/v.$$

The numerical factor $(9\pi/C_D)^{\frac{1}{2}}$ is of the order of unity for $N_R=1$. The tube radius near the surface of the sun is presumably comparable to the scale height, so $(R/\Lambda)^{\frac{1}{2}}=O(1)$. We are concerned with intense fields, considerably in excess of the equipartition value (see §§10.10 and 10.11). With $V_A^2\cong 10v^2$ immediately below the surface, and probably larger at depth, it follows that $N_R\gtrsim 3$ and $C_D\lesssim 10$. Hence, it follows from (8.60) that the rate of rise is $0{\cdot}5\ V_A$ or more. With such large velocities the Bernoulli effect is strong, so that the horizontal attraction of rising tubes is effective to large distances. The characteristic Reynolds stresses $\rho u^2\cong \rho V_A^2$ in the wake exceed the Reynolds stresses ρv^2 of the ambient turbulence, dying out inversely with distance back along the wake. Hence the wake effect is strong.

We do not attempt to calculate here the enhancement of the wake of a rising flux tube by the convective forces. It is possible that the wake initiated by a passing tube may go on to develop an identity of its own, feeding on the convective energy available from the superadiabatic energy gradient. Thus the first tube to rise through the region may initiate a column of upwelling fluid that persists for some time afterward. Subsequent tubes are caught up in the rising fluid so that they are boosted along the trail initiated by the first tube. The passage of subsequent tubes serves to redefine the rising column of fluid and reaffirm the established trail.

Altogether, the observed mutual attraction of magnetic knots on the surface of the sun seems to follow in a straightforward way from the hydrodynamic effects of rising flux tubes. Without such effects the mutual repulsion of knots of like polarity, and the tension along the lines of force, would dominate, leading to the breakup of magnetic knots and pores, rather than their coalescence. Breakup and decay is exactly what occurs when the tubes cease to rise through the surface, i.e. when no new flux appears. It is only the rising flux that has the attractive effects associated with it.

8.10. Transverse oscillation

The transverse oscillation of a slender magnetic flux tube about the equilibrium path merits brief attention. To treat the two-dimensional problem first, consider the magnetic field B in the z-direction confined in equilibrium to the slab $-a<y<+a$, whose centre plane coincides with the xz-plane. The whole is immersed in an ideal, incompressible fluid of uniform density ρ. Imagine a small transverse sinusoidal displacement of the slab in the y-direction by the amount $\xi(z, t)=\epsilon \exp i(\omega t-kz)$. In keeping with the idealization of a slender flux tube, assume that ϵ, $ak \ll 1$. The fluid outside the tube is initially at rest, so that it has no vorticity. The displacement of the slab sets the fluid in motion, and the velocity $\mathbf{v}(y, z, t)$ can be written as $-\nabla\Phi(y, z, t)$. The velocity satisfies Euler's equation

$$\partial\mathbf{v}/\partial t+(\mathbf{v}\cdot\nabla)\mathbf{v}+\nabla p/\rho=0.$$

Noting the vector identity $(\mathbf{v}\cdot\nabla)\mathbf{v}=(\nabla\times\mathbf{v})\times\mathbf{v}+\nabla v^2/2$, the equation can be written

$$\nabla\{-\partial\Phi/\partial t+\tfrac{1}{2}(\nabla\Phi)^2+p/\rho\}=0,$$

leading to the well known result

$$\partial\Phi/\partial t=p/\rho+\tfrac{1}{2}(\nabla\Phi)^2+\textit{constant}. \tag{8.87}$$

Now since $\nabla \cdot \mathbf{v}=0$, it follows that $\nabla^2\Phi=0$. The transverse component (the y-component) of the displacement is continuous across the boundaries $y=\pm a$. Hence, for the space $y \geqslant a$ to the right of the slab, write

$$\Phi=+(\mathrm{i}\omega\epsilon/k)\exp\{-k(y-a)\}\exp \mathrm{i}(\varpi t-kz),$$

with a similar expression to the left ($y \leqslant -a$) in which the y-dependence is $-\exp\{+k(y+a)\}$. Then neglecting terms second order in ϵ, the associated pressure fluctuations follow from (8.87) as

$$\begin{aligned}p(y, z, t) &\cong \rho\partial\Phi/\partial t\\ &=-\rho\epsilon(\omega^2/k)\exp\{-k(y-a)\}\exp \mathrm{i}(\omega t-kz)\end{aligned}$$

in $y>a$, with a similar expression for $y \leqslant -a$.

To write the equation of motion for the slab of field and fluid, denote by $\mathcal{T}$ the tension per unit area in the field, equal to $B^2/4\pi$ for a uniform field in the z-direction. Then, to first order in the displacement $\xi(y, z, t)$, the curvature of the slab is $\partial^2\xi/\partial z^2$, around which the total tension $2a\mathcal{T}$ produces a restoring force $2a\mathcal{T}\partial^2\xi/\partial z^2$. The pressure of the fluid outside exerts a combined force $-p(a, z, t)+p(-a, z, t)$ on the surface. Altogether, Newton's equation of motion for the slab is

$$2a\rho\frac{\partial^2\xi}{\partial t^2}=2a\mathcal{T}\frac{\partial^2\xi}{\partial z^2}-p(a, z, t)+p(-a, z, t).$$

Substituting the expressions for ξ and p, and cancelling all common factors, yields the dispersion relation

$$\omega^2(1+1/ak)-k^2\mathcal{T}/\rho=0$$

or a phase velocity

$$\omega/k=\pm(\mathcal{T}/\rho)^{\frac{1}{2}}(ak)^{\frac{1}{2}}/(1+ak)^{\frac{1}{2}}$$

We recognize $(\mathcal{T}/\rho)^{\frac{1}{2}}$ as the expression for the velocity of propagation of a transverse wave in a string with tension $\mathcal{T}$ and density ρ, equal to the Alfven speed $V_{\mathrm{A}}=B/(4\pi\rho)^{\frac{1}{2}}$ in a uniform field B. The reduction of phase velocity by $\{ak/(1+ak)\}^{\frac{1}{2}}$ is the direct result of the inertia of the external fluid, which is set in motion out to a characteristic distance $y=\pm 1/k$ from the slab no matter how thin the slab ($ak \ll 1$). Hence, the longer wave lengths propagate slower. The waves are dispersive, so that wave forms are not preserved during propagation. The group velocity $\mathrm{d}\omega/\mathrm{d}k$ is larger than the dispersive phase velocity by the factor $(3+2ak)/2(1+ak)$.

A flux tube limited to a finite cross-section, in place of the infinitely broad slab, is not so heavily burdened by the external fluid. The tube slips back and forth without forcing more than the local fluid into motion. The phase velocity is corresponding larger, equal to $V_{\mathrm{A}}/\sqrt{2}$ for a tube with circular cross-section. To demonstrate this consider a flux tube with radius a extending along the z-axis and undergoing a periodic displacement $\xi = \epsilon \exp i(\omega t - kz)$ in the y-direction, where $\epsilon \ll a \ll 1/k$. In view of the long wavelength $2\pi/k$ in the z-direction, the motion of the external field is principally in the x- and y-directions. If ϖ denotes distance $(x^2+y^2)^{\frac{1}{2}}$ from the z-axis and ϕ the polar angle measured from the x-axis, then the velocity potential of the external fluid is

$$\Phi = (a^2 \epsilon i\omega/\varpi)\sin\phi \exp i(\omega t - kz),$$

satisfying the boundary condition that the radial component of velocity is continuous across $\varpi = a$, on which $v_\varpi = \sin\phi\, \partial\xi/\partial t$. It follows that the associated pressure fluctuation is

$$p(\varpi, \phi, t) = -\rho a^2 \omega^2 (\epsilon/\varpi)\sin\phi \exp i(\omega t - kz).$$

Consider the equation of motion in the y-direction for the thin slice of the flux tube contained in the interval $(x, x+dx)$. The width of the slice is $2a \sin\phi$ in the y-direction so that Newton's equation yields

$$2\rho a \sin\phi\, dx \frac{\partial^2\xi}{\partial t^2} = 2a\mathcal{T} \sin\phi\, dx \frac{\partial^2\xi}{\partial z^2} - p(a, \phi, t)\, dx + p(a, -\phi, t)\, dx.$$

Substituting the expressions for ξ and p, and cancelling all common factors, yields the dispersion relation

$$\omega^2/k^2 = \mathcal{T}/2\rho,$$

independent of x. All parts of the cross-section oscillate together, so that the circular cross-section is preserved. It should, however, be noted again that the cross-section of the untwisted tube has only neutral stability.

For the tension $\mathcal{T} = B^2/4\pi$ of a uniform field B along the tube, ω/k is just the phase velocity $\pm V_A/\sqrt{2}$. The phase velocity is only slightly impeded by the external fluid. The waves are non-dispersive so that forms are preserved during propagation.

It is a simple exercise in elliptic coordinates to work out the phase velocity for a tube with an elliptic cross-section of arbitary orientation relative to the direction of oscillation. Detailed calculations for surface waves in a slab of field, and for the oscillations of a slab for which ka is of the order of unity, or greater, have been given elsewhere (Parker 1974). It is also a simple matter to treat the oscillation of flux tubes with an internal density ρ_i different from the surrounding fluid. But the consequences are neither surprising nor of general interest, so we do not burden the reader further.

It is evident that signals can be transmitted along the slender flux tube at speeds only a little less than the Alfven speed. In the real world the waves are damped by viscosity, so that their range of propagation is not, of course, unlimited. In the convective zones of stars, where the isolated flux tube is expected, the eddy viscosity of the turbulent convection is the dominant effect. The convection produces the principal signal and is also the principal agent for the destruction of that signal. Consequently it is not evident that one can learn much about subsurface conditions from the twitching of the flux tubes appearing at the surface of a star.

Twisted magnetic flux tubes are taken up in the next chapter. Their circular cross-section is stable, rather than neutral. The mean tension is less than $B^2/4\pi$, so that the phase velocity of transverse waves is reduced, all the way to zero in the limiting case of strong twisting.

8.11. Summary

There are two points that emerge from the basic theory. The first is that in hydrostatic equilibrium the field density at any place along a slender flux tube uniquely determines the field density everywhere else along the flux tube, in terms of the temperature of the gas in which the tube is immersed. The result was worked out for the simple case of a tube of small cross-section across which the field is uniform. It is clear from the nature of the problem, however, that a large number of elemental tubes of different field strength can be placed side by side to form a tube whose field varies over the cross-section. So the statement of unique determination of B along the tube applies to any slender tube whether its field is uniform across the diameter or not.

There are a number of obvious conclusions concerning the detailed behaviour of the magnetic fields observed, for instance, in the solar photosphere. Their most outstanding feature is their remarkable concentration, to fields of 10^3 G or more in the thin tubes in the supergranule boundaries, and to $3–4\times10^3$ G in sunspots. Their existence over periods of more than a minute implies transverse hydrostatic equilibrium (8.6), from which we conclude that the pressure within is reduced, by some mechanism somewhere along the tube, by the amount $B^2/8\pi$ below the ambient gas pressure. Indeed, the longevity of the concentrated fields suggests in many cases an approach to *longitudinal* hydrostatic equilibrium along some extended length of the tube. In such cases, the concentration of field at the surface reflects conditions along the length of the tube, and not merely at the surface. We conclude that the concentrated tubes cannot be in thermal equilibrium with their surroundings everywhere along their length if they are also in hydrostatic equilibrium, because, if they were, then, as noted in §8.2, the ratio of the Alfven speed V_A in the tube to the sound speed V_S in the gas is independent of depth below the surface. But $B^2/8\pi$ is comparable to p_e at the surface and hence *everywhere* below, indicating that $V_A \cong V_S$ *everywhere* below the surface. According to the results of §8.7, the tubes would rise at the sound speed and escape from the sun in an hour. One might save the situation by anchoring the tubes in the deep stable interior below the convective zone, but with such permanent anchorage, it is then difficult to understand their origin and their variation with the sunspot cycle. Their intensity would reach 2×10^7 G at the base of the convective zone! It seems more conservative to suppose that the tubes are *not* everywhere in thermal equilibrium with their environment (see §10.10).

The cool sunspot is clearly not in thermodynamic equilibrium with its surroundings, of course, and, as pointed out at the end of §8.3, its intense field appears to be a direct consequence of this fact. The accumulated effect of the cooling over many scale heights produces the concentrated field at the surface, with the magnetic pressure comparable to the gas pressure. The small flux tubes in supergranule boundaries may be similarly accounted for (Parker 1976, 1978a; Roberts 1976) but there are still many unanswered questions on their stability, etc., just as with sunspots. It may be that there are dynamical effects, in addition to the hydrostatic forces discussed so far, but that is a topic for discussion in Chapter 10.

The second point is the magnetic buoyancy of the magnetic flux tube which prevents an overall hydrostatic equilibrium. We worked out in §8.6 the arched equilibrium path of a slender flux tube in a stratified atmosphere when the two ends of the tube are 'clamped' in some suitable way. We found that if the tube is held in clamps separated horizontally by more that a few scale heights, they cannot hold down the tube between. That part of the tube floats up through the atmosphere and escapes into space. The rates of rise worked out in §§8.7 and 8.8 indicate that there is no possibility for 'clamping' flux tubes in the solar convective zone, but in the stable solar interior below, the rate of rise of fields of 10^2–10^5 G is sufficiently slow that they may be considered clamped for many purposes. From these facts we conclude that the continual emergence of magnetic fields through the solar photosphere is an inevitable consequence of the presence (and generation) of magnetic fields in the convective zone.

Altogether it would appear that in no stellar interior is it possible to store primordial fields of such pressure as to affect the internal structure of the star. In no stellar convective zone is it possible for there to be fields of 10^2 G or more that do not make their appearance at the surface of the star. Indeed, the rapid rate of rise of such modest fields restricts their generation by the hydromagnetic dynamo effects of the convection and circulation to the lower parts of the convective zone, which is a point of concern in Chapter 22. We may safely say that if no fields appear at the surface of a star, then there are none of any significance in the convective zone. We will take up the matter again in Chapter 13 where the question of buoyancy and the escape of magnetic field is treated from the point of view of the Rayleigh–Taylor instability.

References

BAKER, N. and TEMESVARY, S. (1966). *Tables of convective stellar atmospheres*, 2nd edn. NASA Institute for Space Studies, New York.

BARTENWERFER, D. (1973). *Astron. Astrophys.* **25,** 455.

BASSET, A. B. (1888). *A treatise on hydrodynamics.* Deighton, Bell and Co. Republished 1961, pp. 220, 221. Dover, New York.

BECKERS, J. M. and SCHRÖTER, E. H. (1968). *Solar Phys.* **4,** 142, 165.

CHITRE, S. M., EZER, E. and STOTHERS, R. (1973). *Astrophys. Lett.* **14,** 37.

COUTANCEAU, M. and BOUARD, R. (1977). *J. Fluid. Mech.* **79,** 231, 257.

DRYDEN, H. L., MURNAGHAN, F. D. and BATEMAN, H. (1956). *Report of Committee on Hydrodynamics* (Dover Publications: New York).

GOLDSTEIN, S. (1938). *Modern developments in fluid dynamics*, vol. I, pp. 15, 16, 24; vol. II, chap. XIII. Clarendon Press, Oxford.

GREENHILL, A. G. (1882). *Quart. J. pure appl. Math.* **18,** 231, 346.

GURM, W. S. and WENTZEL, D. G. (1967). *Astrophys. J.* **149,** 139.

HICKS, W. M. (1879). *Quart. J. pure appl. Math.* **16,** 113, 193.

LAMB, H. (1932). *Hydrodynamics*, 6th edn. Cambridge University Press. Republished 1945, Dover, New York.

MEYER, F., SCHMIDT, H. U., WEISS, N. O. and WILSON, P. R. (1974). *Mon. Not. Roy. Astron. Soc.* **169,** 35.

PARKER, E. N. (1955). *Astrophys. J.* **121,** 491.

—— (1974). *Astrophys. Space Sci.* **31,** 261.

—— (1975a). *Astrophys. J.* **198,** 205.

—— (1975b). ibid. **201,** 494.

—— (1976). ibid. **204,** 259.

—— (1977). ibid. **215,** 370.

—— 1978a, ibid. **221,** 368.

—— 1978b, ibid. **222,** 357.

ROBERTS, B. (1976). *Astrophys. J.* **204,** 268.

SCHLÜTER, A. and TEMESVARY, S. (1958). *Electromagnetic phenomena in cosmic physics, IAU Symposium No.* 6 (ed. B. LEHMERT), p. 263. Cambridge University Press.

SCHÜSSLER, M. (1977). *Astron. Astrophys.* **56,** 439.

SCHWARZSCHILD, M. (1958). *Structure and evolution of the stars.* Princeton University Press.

SPRUIT, H. C. (1974). *Solar Phys.* **34,** 277.

UNNO, W. and RIBES, E. (1976). *Astrophys. J.* **208,** 222.

WEYMANN, R. (1957). *Astrophys. J.* **126,** 208.

ZWAAN, C. (1978). *Solar Phys.*, in press.

9

THE TWISTED FLUX TUBE

9.1. Basic properties of a twisted tube

THE magnetic flux tube appearing in the turbulent astrophysical environment cannot be entirely free of twisting, and is often referred to as a *flux rope.* Only a few revolutions in a long tube are enough to produce the qualitative dynamical effects of instability and non-equilibrium. The torsion concentrates in the upper expanded portions of the tube, reducing the total tension in the flux rope to zero and negative values, causing buckling of the tube into a corkscrew form. Hence, if there are two or three revolutions anywhere in a long tube, they will be concentrated at the apex of the tube where the tube emerges through the surface of the star, causing dynamical non-equilibrium, dissipation, and general activity (Parker 1974, 1975a). Thus it is no coincidence that the flux ropes observed emerging through the surface of the sun are often active. Their activity reflects the concentration at their apex of some modest degree of torsion produced by the convection and circulation at depth. Indeed, the accumulated torsion at depth is severely limited by escape to the apex and immediate destruction through line cutting, etc. The non-equilibrium of the flux rope under most astrophysical circumstances is a major source of stellar, and perhaps galactic, activity, and merits examination in some detail.

The equilibrium magnetic configuration of a uniform axi-symmetric twisted flux tube of radius R with field components $b_z(\varpi)$ and $b_\phi(\varpi)$ is given by (6.11) as

$$p(\varpi)+b_z^2(\varpi)/8\pi = f(\varpi)+\tfrac{1}{2}\varpi\,\mathrm{d}f/\mathrm{d}\varpi, \qquad b_\phi^2(\varpi)/8\pi = -\tfrac{1}{2}\varpi\,\mathrm{d}f/\mathrm{d}\varpi \quad (9.1)$$

where ϖ denotes distance from the axis of the tube (the z-axis) and $p(\varpi)$ is the fluid pressure. The generating function $f(\varpi)$ is the total pressure $p+(b_z^2+b_\phi^2)/8\pi$ and is arbitrary except that $f\geq 0$, $\mathrm{d}f/\mathrm{d}\varpi\leq 0$, and $\mathrm{d}(\varpi^2 f)/\mathrm{d}\varpi\geq 0$ in order that the field components be real and the pressure p positive. The equality signs in the first two conditions are achieved only for the tube without twisting, and in the third only for a purely azimuthal field (complete twisting). The total pressure is continuous across the lateral boundary $\varpi=R$ of the tube, so that $f(R)$ is equal to the ambient pressure outside. For the present we suppose that the tube is surrounded by a gas of uniform pressure $P=f(R)$, although pressure variations with height, and the possibility of external magnetic fields, are taken up later.

Note that b_z^2 and p are entirely interchangeable within the freedom provided by (9.1) for radial stress balance. Thus if p is zero, it is replaced by a suitable magnetic pressure $b_z^2/8\pi$. In general if $p(\varpi)$ is constant across the diameter of the tube, it may be dropped from (9.1) by the simple expedient of replacing $f-p$ by a new function, call it f again, representing the magnetic pressure alone. Then $f(R)$ is equal to the fluid pressure *increment* $P-p$ across the boundary rather than the total pressure outside. In the discussion that follows we frequently consider the simple force-free case $p(\varpi)=$ *constant*, using (9.1) without the fluid pressure reckoned into f.

The individual magnetic lines of force are helices of fixed radius ϖ. The angle $\vartheta(\varpi)$ between a line of force and the z-direction—the pitch angle—is

$$\tan\vartheta(\varpi)=\varpi\,\mathrm{d}\phi/\mathrm{d}z=b_\phi(\varpi)/b_z(\varpi)$$
$$=\left(\frac{\varpi^2\,\mathrm{d}f/\mathrm{d}\varpi}{2\varpi p-d(\varpi^2 f/\mathrm{d}\varpi)}\right)^{\frac{1}{2}}. \tag{9.2}$$

where ϕ is the azimuthal angle measured around the axis of the tube. The equation for the line of force crossing $z=0$ at (ϖ_0, ϕ_0) is

$$\varpi=\varpi_0,\ z=\varpi_0\cot\vartheta_0(\phi-\phi_0), \tag{9.3}$$

the line making one revolution around the tube in a distance

$$\lambda(\varpi)=2\pi\,\mathrm{d}z/\mathrm{d}\phi$$
$$=2\pi\varpi\cot\vartheta(\varpi). \tag{9.4}$$

To understand fully the representation (9.1) in terms of the total pressure $f(\varpi)$ note that $f(\varpi)=$ *constant* represents a flux tube without twisting ($b_\phi=0$), for which $p+b_z^2/8\pi=$ *constant* across the interior. In the other extreme, the most rapid variation of $f(\varpi)$ is $1/\varpi^2$, for which $p+b_z^2/8\pi=0$ and the magnetic field is the familiar azimuthal field of a line current, $b_\phi\propto\varpi^{-1}$. The tube in this case is 'completely twisted' with $b_z^2=0$. All possibilities between these two extremes are possible.

A few remarks are in order on the form of $f(\varpi)$. To avoid the physical impossibility of infinite Ohmic dissipation in any small resistivity in the fluid, we suppose that the electric current density $(c/4\pi)\nabla\times\mathbf{b}$ is finite everywhere within the tube. It follows that b_ϕ must vanish at least as fast as ϖ toward the axis of the tube. Hence $\mathrm{d}f/\mathrm{d}\varpi$ vanishes at least as fast as ϖ and, in the neighbourhood of $\varpi=0$, $f(\varpi)$ can be expanded as $f(\varpi)=f(0)+\frac{1}{2}f''(0)\varpi^2+\ldots$ where $f''(0)<0$. Hence b_ϕ drops out and $p+b_z^2/8\pi=f(0)$ on the axis of the tube. The theoretical maximum value of $p+b_z^2/8\pi$ at any radius ϖ occurs if $b_\phi(\varpi)=0$, so that the fluid and the longitudinal field must bear the full pressure. This follows formally from

(9.1) noting that $f(\varpi)<f(0)$ and $df/d\varpi \leq 0$. Hence the theoretical maximum in the longitudinal field at a radius ϖ occurs if there is a level place (a ledge) in the generally declining $f(\varpi)$, in which case $p+b_z^2/8\pi=f(\varpi)<f(0)$. This upper limit on the longitudinal component of the field declines outward with $f(\varpi)$. Generally $b_z^2/8\pi$ will be less than the theoretical maximum, of course. Altogether, then, $p(\varpi)+b_z^2(\varpi)/8\pi<p(0)+b_z^2(0)/8\pi$ because it must balance the accumulation of inward force from the tension in b_ϕ. It will be shown in a moment that for fixed P, the mean value $\langle p\rangle+\langle b_z^2\rangle/8\pi$ across the tube is unaffected by the twisting. Hence $p(0)+b_z^2(0)/8\pi$ must increase, and $p(R)+b_z^2(R)/8\pi$ must decrease with the twisting of the tube,

$$p(0)+b_z^2(0)/8\pi>\langle p\rangle+\langle b_z^2\rangle/8\pi>p(R)+b_z^2(R)/8\pi.$$

This is perhaps the proper place to remark that we do not expect to find in nature flux ropes more complex than a more or less uniform twist across the radius of the tube, so that the azimuthal field $b_\phi(\varpi)$ increases monotonically with ϖ while $p(\varpi)+b_z^2(\varpi)/8\pi$ declines monotonically. We expect neither local horizontal ledges in $f(\varpi)$, yielding layers of purely longitudinal field, nor local sharp drops, with $f(\varpi)\propto\varpi^{-2}$, yielding layers of purely azimuthal field. Above all we expect no layers of reverse twist in the flux tube; b_ϕ has the same sign for all ϖ. The mathematics that we employ in developing the properties of flux tubes is quite general, of course, permitting all manner of field variation across the radius of the tube. But there is little reason at present to clutter our conceptual picture with more than the simple twisted tube. The exposition that follows concentrates on the simple case in which $b_z^2(\varpi)$ declines monotonically with increasing radius, $b_z^2(0)>b_z^2(\varpi)>b_z^2(R)$. The more exotic cases produce amusing field configurations upon radial expansion, but they seem to be of little practical application in astrophysics, and in any case the reader can see at once the consequences once he is familiar with the basic effects presented here.

Now in the absence of a gravitational field, the pressure $p(\varpi)$ is uniform along each line of force, and in most cases is determined by external conditions at the 'ends' of the tube. Hence, with $p(\varpi)$ given, one thinks of the equilibrium in terms of an adjustment of $b_z(\varpi)$ and $b_\phi(\varpi)$, through local expansion or contraction, so that the radial gradient of $(b_z^2+b_\phi^2)/8\pi$ balances $-dp/d\varpi$ and the inward force $b_\phi^2/4\pi\varpi$ of the tension in the lines of force wrapping around the tube. If the external conditions are such as to produce $p(\varpi)=$ *constant* across the radius of the tube, then p drops out of f and the equilibrium adjusts itself into a force-free field with $(b_z^2+b_\phi^2)/8\pi$ balanced solely against the tension $b_\phi^2/4\pi$ in the lines of force wrapped around the tube. It is the tension in

b_ϕ that produces the concentration of field toward the axis of the flux tube, but with the limitations cited in §§5.4 and 6.3. In the present context note that the mean value of the pressure $p+b_z^2/8\pi$ is

$$\begin{aligned}\langle p\rangle+\langle b_z^2\rangle/8\pi &= \frac{2}{R^2}\int_0^R \mathrm{d}\varpi\varpi\left(f+\frac{1}{2}\varpi\frac{\mathrm{d}f}{\mathrm{d}\varpi}\right)\\ &= \frac{1}{R^2}\int_0^R \mathrm{d}\varpi\frac{\mathrm{d}(\varpi^2 f)}{\mathrm{d}\varpi}\\ &= f(R). \end{aligned} \tag{9.5}$$

That is to say, the mean pressure of the fluid and longitudinal field within the tube is just equal to the external pressure exerted on the boundary of the tube. This relation for the force-free twisted tube is identical with the untwisted tube, as indicated in §8.1 by (8.6).

For the force-free tube the condition reduces to $\langle b_z^2\rangle/8\pi=f(R)$ where now $f(R)$ represents the increment in fluid pressure at the boundary. Hence the mean square longitudinal field in a force-free tube is unaffected by the twisting. We may go on to show, then, that the mean longitudinal field declines with increased twisting: In the untwisted tube there is only a longitudinal field, and it is uniform across the tube; the square of the longitudinal field is equal to the mean square and to the square of the mean. When the tube is twisted, the longitudinal flux Φ along the tube is conserved

$$\Phi=2\pi\int_0^R \mathrm{d}\varpi\varpi\left(f+\frac{1}{2}\varpi\frac{\mathrm{d}f}{\mathrm{d}\varpi}\right)^{\frac{1}{2}},$$

while an azimuthal field appears and the tube radius R increases in association with the decline in the mean longitudinal field. The generating function—i.e. the total magnetic pressure—which was initially uniform across the tube, now declines monotonically from the axis to the boundary, so that b_z is no longer uniform[1]. But if the longitudinal field is not uniform then the mean square exceeds the square of the mean,

$$\langle b_z^2\rangle>\langle b_z\rangle^2.$$

The inequality becomes more pronounced as the twisting increases. But $\langle b_z^2\rangle$ is not affected, so for a fixed confining pressure increment $P-p$, the *mean longitudinal field declines as the twisting is increased.* The twisting

[1] Only for the special case $f(\varpi)=C_1+C_2/\varpi^2$, for which $b_z^2/8\pi=C_1$, does the longitudinal field remain uniform. But such a field has a singularity at $\varpi=0$ where an enormous fluid pressure gradient is required to balance the azimuthal field $b_\phi^2/8\pi=C_2/\varpi^2$, contrary to our assertion that we are treating a force-free field.

increases the total field pressure $f(\varpi)$ within the tube because it increases the azimuthal field $b_\phi^2(\varpi) = -\frac{1}{2}\varpi\, \mathrm{d}f/\mathrm{d}\varpi$, but it does so at the expense of the mean longitudinal field. It is important to keep these facts in mind when we come to the question of the very intense flux tubes ($>10^3$ G) that appear in, say, the solar photosphere. Contrary to the conventional folklore, twisted flux ropes cannot provide the effect. Only the peak field, in the neighbourhood of the axis of the tube, increases, while the rest of the field spreads out so much as to reduce the mean.

To illustrate these effects consider what happens when a force-free flux tube is twisted uniformly along its length. Suppose that initially the flux tube is not twisted, with radius R_0 and a uniform field B_0 along its length. The field is confined by the fluid pressure increment $P_0 = B_0^2/8\pi$ at the surface of the tube. The total flux along the tube is $\Phi = \pi R_0^2 B_0$. Now suppose that the tube is twisted by one revolution in each $2\pi a$ of length. The fluid pressure is held fixed along each line of force so that the field is force-free across the radius R and is confined by the same pressure differential P_0 across the surface. The generating function is

$$f(\varpi) = P_0(1 + R^2/a^2)/(1 + \varpi^2/a^2). \tag{9.6}$$

Initially, in the absence of twisting, a was infinite and $f(\varpi) = P_0$. It is readily shown that

$$\frac{b_z^2}{8\pi} = \frac{P_0(1+R^2/a^2)}{(1+\varpi^2/a^2)^2}, \qquad \frac{b_\phi^2}{8\pi} = \frac{P_0(1+R^2/a^2)}{(1+\varpi^2/a^2)^2}\frac{\varpi^2}{a^2}.$$

Hence $b_\phi/b_z = \varpi/a$, and it follows from (9.2) and (9.4) that the distance along the tube per revolution is $2\pi a$, independent of the radial distance ϖ.

The mean square fields are

$$\left\langle\frac{b_z^2}{8\pi}\right\rangle = P_0, \qquad \left\langle\frac{b_\phi^2}{8\pi}\right\rangle = P_0\left\{\left(\frac{a^2}{R^2}+1\right)\ln\left(1+\frac{R^2}{a^2}\right)-1\right\},$$

while the mean longitudinal field is

$$\langle b_z\rangle = B_0(1+R^2/a^2)^{\frac{1}{2}}(a^2/R^2)\ln(1+R^2/a^2).$$

Then the total flux is

$$\begin{aligned}\Phi &= \pi R^2\langle b_z\rangle \\ &= \pi a^2 B_0(1+R^2/a^2)^{\frac{1}{2}}\ln(1+R^2/a^2)\end{aligned}$$

and is fixed, of course. Hence, as a decreases with increasing twisting, this relation gives R/a. The decline of the mean field with increasing twisting (declining a) is shown in Fig. 9.1, together with the increasing radius of the tube and the increasing peak field on the axis of the tube. The total tension (eqn (9.10)) in the tube becomes negative when a/R falls below

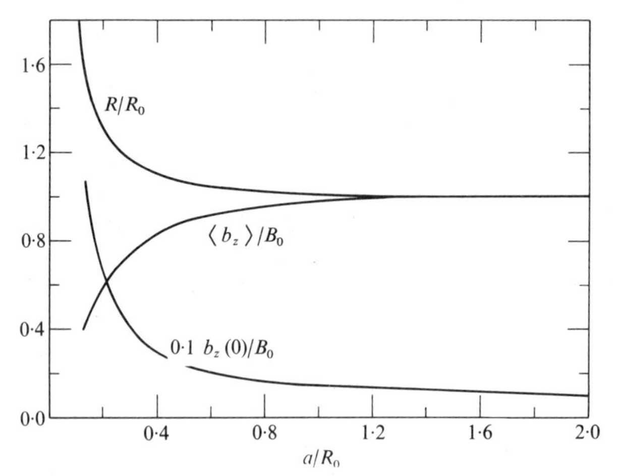

FIG. 9.1. A plot of the mean longitudinal field $\langle b_z \rangle$, in units of the field B_0 in the absence of twisting, as a function of the length a in which the lines of force make one revolution around the tube. The increase of the tube radius R and the peak field $b_z(0)/B_0$ are also shown.

0·252, at which point the peak field has risen to 4·09 B_0 and the mean longitudinal field has fallen to 0·73 B_0. Further twisting only causes the tube to buckle.

9.2. Total tension and stability of a flux rope

With the above facts in mind, consider the total tension $\mathcal{T}$ in the flux tube. For the uniform flux tube without twisting the total pressure $p + b_z^2/8\pi$ is uniform across the tube, and is equal to the external pressure, so that the only outstanding stress is the tension $b_z^2/4\pi$. It follows that for a tube of radius R,

$$\mathcal{T} = \pi R^2(b_z^2/4\pi).$$

For the twisted tube, described by (9.1) in terms of the total pressure $f(\varpi)$, the net tension $\mathcal{T}$ in the tube above background is the integral over the cross-section of the total stress within the tube minus the uniform background pressure $f(R)$. With the convention that tension is positive and pressure negative, it follows that

$$\begin{aligned}
\mathcal{T} &= 2\pi \int_0^R \mathrm{d}\varpi\varpi \left[\frac{b_z^2}{4\pi} - \{f(\varpi) - f(R)\}\right], \\
&= 2\pi \int_0^R \mathrm{d}\varpi\varpi \left\{\frac{\mathrm{d}\varpi f}{\mathrm{d}\varpi} - 2p(\varpi) + f(R)\right\}, \\
&= \pi R^2\{3f(R) - \langle f \rangle - 2\langle p \rangle\}, \\
&= \pi R^2[3\{f(R) - \langle p \rangle\} + \langle p \rangle - \langle f \rangle], \qquad (9.7)
\end{aligned}$$

etc. where the angular brackets denote the average over the cross-section,

$$\langle f\rangle = \frac{2}{R^2}\int_0^R \mathrm{d}\varpi\varpi f(\varpi)$$

etc. For a force-free field, in which the gas pressure $p(\varpi)$ is uniform across the radius of the tube, and so can be dropped from (9.1), the tension can be written

$$\begin{aligned}\mathcal{T} &= \pi R^2\{3f(R) - \langle f\rangle\}\\ &= \pi R^2\{3f(R) - \langle b_z^2\rangle/8\pi - \langle b_\phi^2\rangle/8\pi\}.\end{aligned} \tag{9.8}$$

If we suppose that the confining pressure increment $f(R)$ remains unchanged, then $\langle b_z^2\rangle$ is unchanged and the only effect of the twisting is to increase $\langle b_\phi^2\rangle$. The pressure of $b_\phi^2/8\pi$ reduces the tension, and with sufficient twisting, the total tension becomes negative. The tube is under longitudinal compression rather than tension, causing it to buckle, going over into a corkscrew form.

As an illustrative example, consider again the uniformly twisted, force-free tube (9.6). The tension is readily shown from (9.7) or (9.8) to be

$$\mathcal{T} = \pi R^2 P_0\{3 - (1 + a^2/R^2)\ln(1 + R^2/a^2\}. \tag{9.9}$$

Twisting the tube (decreasing a) decreases the total tension, from the maximum value

$$\mathcal{T}_0 = \pi R^2 B_0^2/4\pi = 2\pi R^2 P_0 \tag{9.10}$$

in the untwisted tube, to zero when R/a reaches 3·97, and to negative values, indicating a net compression, for $R/a > 3{\cdot}97$.

In the development that follows we will have frequent recourse to (9.7) and/or (9.8) as an indication of the stability of the cylindrical flux tube. The slender flux tube without twisting is stable, being stretched taut (and stable) by the tension $\pi R^2(B^2/4\pi)$ along its length (see §8.5). The introduction of twisting diminishes the tension until the simple cylindrical equilibrium form of the tube becomes unstable to buckling and kinking, which clearly must occur by the time the tension goes negative (if not before), placing the tube under longitudinal compression. The strongly twisted tube then slips into a corkscrew form for partial relief of torsion. Unfortunately the corkscrew equilibrium is difficult to treat. Further twisting brings adjacent loops of the corkscrew into contact with one another, and it will be shown in Chapter 14 that there is then no equilibrium at all. There is dynamical line cutting, and hence magnetic activity, at the contact areas, preventing the accumulation of further torsion. Altogether, then, active magnetic dissipation dominates and prevents torsion accumulating beyond the approximate upper limit indicated by $\mathcal{T} \cong 0$.

The complexity of the conditions for the onset of buckling may be seen from the fact that the tension is not uniformly distributed over the cross-section of the tube. The longitudinal magnetic stress density is

$$(b_z^2 - b_\phi^2)/8\pi = P_0(1+R^2/a^2)(1-\varpi^2/a^2)/(1+\varpi^2/a^2)^2.$$

It represents a tensile stress of $P_0(1+R^2/a^2)$ dyn cm^{-2} on the axis of the tube, which increases as the twisting (measured by R/a) increases, but it falls to zero at the radius $\varpi = a$. Beyond $\varpi = a$ the stress is compressive, rising to a maximum $P_0(R^2/a^2-1)/(R^2/a^2+1)$ at the surface of the tube. The compression at the surface increases asymptotically to P_0 for large R/a, while the tension along the axis becomes increasingly peaked there. It is the outer portions of the tube that impose the buckling.

In this first exploratory exposition we will be no more quantitative than the approximate criterion $\mathcal{T} = 0$ for the onset of buckling and kinking, although it would be desirable, certainly, to work out more precise criteria in the future. To give an idea of the validity of this rough criterion consider the uniformly twisted rope

$$b_z = B_0, \qquad b_\phi = B_0\varpi/q$$

so that

$$p(\varpi) = p(0) - (B_0^2/4\pi)\varpi^2/q^2,$$
$$f(\varpi) = p(0) + (B_0^2/8\pi)(1-\varpi^2/q^2)$$

out to a radius $\varpi = R$, beyond which there is a uniform pressure $P = f(R)$. The rope is untwisted and stable for $R/q = 0$. It is readily shown from (9.7) that

$$\mathcal{T} = \pi R^2 \frac{B_0^2}{8\pi}\left(1 - \frac{R^2}{4q^2}\right).$$

Hence $\mathcal{T} = 0$ when q decreases to $0{\cdot}5R$. Roberts (1956) treated the stability to infinitesimal perturbations directly from the hydromagnetic equations, finding the precise result

$$q^2 = 4/(k^2 + j_{mn}^2/R^2) - m^2/k^2$$

for marginal stability of the spiral mode $\exp \mathrm{i}(m\phi + kz)$ where j_{mn} is the nth zero of the mth order Bessel function. We presume that with the onset of instability the equilibrium form of the flux tube changes from a cylinder to a corkscrew form, the degree of spiralling of the corkscrew increasing with further twisting. The first instability to appear as the twisting increases (q decreases) occurs for $m = 0$, and then for $k = 0$, choosing the smallest root $j_{01} = 2{\cdot}405$. Roberts predicts instability when q falls below $0{\cdot}83R$, somewhat before the total tension falls to zero (at $0{\cdot}5R$).

More recent work shows the situation to be somewhat more complicated, based on the fact that when the ratio of the mean values of the transverse and longitudinal fields b_ϕ and b_z exceed the ratio of the circumference $2\pi R$ to the length λ of the flux tube, then the equilibrium is subject to the kink instability (Shafranov 1957, 1970; Kruskal and Tuck 1958; Kadomtsev 1966; Ware and Haas 1966; Raadu 1972; Furth *et al.* 1973; Goedbloed and Sakanaka 1974; Sakanaka and Goedbloed 1974; Molodensky 1974, 1975, 1976). Thus a long straight tube is subject to the kink instability for very small amounts of twisting $\langle b_\phi \rangle / \langle b_z \rangle \cong 2\pi R/\lambda$. However, it appears that for small $\langle b_\phi \rangle / \langle b_z \rangle$ there is a nearby equilibrium, of a slightly corkscrew form, that is stable. The larger is $\langle b_\phi \rangle / \langle b_z \rangle$, the more extreme becomes the corkscrew shape of the flux tube, of course. The criterion that the twisting be small enough for $\mathcal{T} > 0$ in the straight tube is equivalent to the statement that the straight tube relapses into only a moderate corkscrew form. Further twisting of the tube increases the looping of the corkscrew, until adjacent loops touch and rapid reconnection occurs (see Chapters 14 and 15). Recent observations of a flux tube in a solar flare (Cheng 1977) show the increasing corkscrew form of the flux tube as b_ϕ increases relative to b_z.

9.3. Dilatation and stretching of a flux rope

As a first example of the behaviour of a flux rope consider the equilibrium of a twisted tube of uniform radius $R = c$, described by (9.1), when the tube expands (or compresses) to a new radius C as a consequence of a change in the fluid pressure on its boundary, at the same time that the tube is stretched (or contracted) longitudinally. It is obvious from the topological invariance of the lines of force in the infinitely conducting fluid that radial expansion and/or longitudinal contraction increase the pitch angle of the helical lines of force, i.e. on a given line of force b_ϕ / b_z increases. The tension $\mathcal{T}$ along the rope is reduced, so that buckling occurs if the radial expansion, or longitudinal contraction, goes too far. Thus a flux tube can be made unstable by radial expansion or longitudinal contraction.

To illustrate the basic effects as directly as possible (Parker 1974) suppose that the fluid pressure extending along each line of force within the tube is the same, so that $p(\varpi)$ is uniform throughout the tube ($\varpi < c$) and the magnetic field is force-free, with

$$b_z^2(\varpi)/8\pi = f(\varpi) + \tfrac{1}{2}\varpi \, \mathrm{d}f/\mathrm{d}\varpi, \qquad b_\phi^2(\varpi) = -\tfrac{1}{2}\varpi \, \mathrm{d}f/\mathrm{d}\varpi. \tag{9.11}$$

The tube is confined by an external pressure excess

$$\Delta p \equiv P - p = f(c) \tag{9.12}$$

at the boundary $\varpi = c$.

Then suppose that the tube is stretched lengthwise by a factor α, at the same time that the confining pressure excess Δp takes up the new value ΔP, which might be larger or smaller than Δp. The helical lines of force initially with radius ϖ ultimately find themselves at some new radius $\Pi = \Pi(\varpi)$, and the radius of the tube is $C = \Pi(c)$. The field components $b_z(\varpi)$ and $b_\phi(\varpi)$ are expanded or compressed to $B_z(\Pi)$ and $B_\phi(\Pi)$. Assuming that in equilibrium the fluid pressure is again constant across the radius[2], it follows that B_z and B_ϕ can be expressed in terms of the generating function $F(\Pi)$,

$$B_z^2/8\pi = F + \tfrac{1}{2}\Pi \, \mathrm{d}F/\mathrm{d}\Pi, \qquad B_\phi^2/8\pi = -\tfrac{1}{2}\Pi \, \mathrm{d}F/\mathrm{d}\Pi. \tag{9.13}$$

The boundary condition at the surface is

$$\Delta P = F(C). \tag{9.14}$$

Conservation of the longitudinal flux initially through the annulus $(\varpi, \varpi + \mathrm{d}\varpi)$, and ultimately through $(\Pi, \Pi + \mathrm{d}\Pi)$, requires that

$$b_z(\varpi)\varpi \, \mathrm{d}\varpi = B_z(\Pi)\Pi \, \mathrm{d}\Pi. \tag{9.15}$$

Conservation of azimuthal flux initially through $\mathrm{d}\varpi \, \mathrm{d}z$ at radius ϖ, and ultimately through $\mathrm{d}\Pi \, \mathrm{d}Z$ at Π, requires that

$$b_\phi(\varpi) \, \mathrm{d}\varpi = \alpha B_\phi(\Pi) \, \mathrm{d}\Pi. \tag{9.16}$$

Square (9.15) and (9.16) and use (9.11) and (9.13) to write the field components in terms of the generating functions f and F. Then introduce the new radial coordinates $u = \varpi^2/L^2$, $U = \Pi^2/L^2$ where L is some suitable length scale, to be determined later to suit our convenience. The conditions (9.15) and (9.16) can be written

$$\mathrm{d}(UF)/\mathrm{d}U = (\mathrm{d}u/\mathrm{d}U)^2 \, \mathrm{d}(uf)/\mathrm{d}u, \tag{9.17}$$

$$\alpha^2 \, \mathrm{d}F/\mathrm{d}U = (\mathrm{d}u/\mathrm{d}U)^2 \, \mathrm{d}f/\mathrm{d}u. \tag{9.18}$$

To obtain the differential equation for the mapping $u(U)$, multiply (9.18) by U/α^2 and subtract from (9.17), obtaining

$$F = (\mathrm{d}u/\mathrm{d}U)^2 \{\mathrm{d}uf/\mathrm{d}u - (U/\alpha^2) \, \mathrm{d}f/\mathrm{d}u\}. \tag{9.19}$$

Differentiate with respect to U and eliminate $\mathrm{d}F/\mathrm{d}U$ with the aid of

[2] The mathematical problem is tractable for the more general case (9.1) wherein the field is not force-free, but it is necessary in that case to specify the new pressure distribution in terms of the initial $p(\varpi)$. The extra specifications then obscure the basic points of the illustration.

(9.18). The result is

$$0=\frac{\mathrm{d}u}{\mathrm{d}U}\left\{\frac{\mathrm{d}^2u}{\mathrm{d}U^2}\left(\frac{\mathrm{d}(uf)}{\mathrm{d}u}-\frac{U}{\alpha^2}\frac{\mathrm{d}f}{\mathrm{d}u}\right)\right.$$

$$\left.+\frac{1}{2}\left(\frac{\mathrm{d}u}{\mathrm{d}U}\right)^2\left(\frac{\mathrm{d}^2(uf)}{\mathrm{d}u^2}-\frac{U}{\alpha^2}\frac{\mathrm{d}^2f}{\mathrm{d}u^2}\right)-\frac{1}{\alpha^2}\frac{\mathrm{d}f}{\mathrm{d}u}\frac{\mathrm{d}u}{\mathrm{d}U}\right\}. \tag{9.20}$$

Now before plunging into the formal solution of this equation, it is well to have an idea of what we wish to extract from it. We know from simple physical considerations that the degree of twisting is increased by radial expansion and reduced by compression. We would like to work out one or two illustrative quantitative examples, to show, for instance at what point the total tension $\mathcal{T}$ falls to zero and equilibrium ceases to exist. We will find that, in the limiting case of large expansion or contraction, the profile of the field across the radius of the tube approaches a simple asymptotic form, regardless of the initial profiles. For instance, for extreme contraction the field becomes uniform without significant torsion ($B_\phi/B_z \to 0$, $B_z = constant$), the vestiges of the initial twisting then being confined to a surface layer of vanishing thickness. For extreme expansion the field becomes essentially azimuthal ($B_z/B_\phi \to 0$, $B_\phi \propto 1/\varpi$) with the vestiges of the longitudinal field confined to a small neighbourhood of the axis. A property of both of these asymptotic states is that the form of each is an invariant to both radial and longitudinal expansion or contraction. Thus, obviously, a flux tube without twisting (the asymptotic limit of compression) remains untwisted for any expansion, no matter how large. It is equally obvious that a flux tube composed of an azimuthal field (the asymptotic limit of expansion) remains azimuthal for any degree of compression. As we shall soon see, there are intermediate forms of twisting which are also invariant. With this picture in mind, then, let us begin the tedious algebraic journey that leads ultimately to formal illustration of the effects.

9.4. An example of dilatation and stretching

Noting that (9.20) is homogeneous in f, it is clear that, without loss of generality, $f(u)$ can be normalized to one at $u=0$. The generating function $f(u)$ can be expanded in a series of ascending powers of u,

$$f(u)=1-u+a_2u^2+a_3u^3+\ldots, \tag{9.21}$$

normalizing $\mathrm{d}f/\mathrm{d}u$ to -1 at $u=0$ by suitable choice of L. Solution of (9.20) involves inverting this expansion[3] so, to consider the simplest

[3] Let $\theta \equiv \mathrm{d}(uf)/\mathrm{d}u-\alpha U\,\mathrm{d}f/\mathrm{d}u$ so that (9.17) can be written $\mathrm{d}(\ln(\mathrm{d}u/\mathrm{d}U))/\mathrm{d}U+(\mathrm{d}\theta/\mathrm{d}U-\alpha^2\,\mathrm{d}f/\mathrm{d}u)/2\theta$. To proceed with the integration it is necessary to express u in terms of f and θ.

example, suppose that $f(u)$ is expressed by the first two terms of the series $f(u)=1-u$ with the restriction $u<\frac{1}{2}$ so that the fields

$$b_z^2/8\pi=1-2u, \qquad b_\phi^2/8\pi=u$$

are real. The azimuthal field grows linearly with radius, while the longitudinal field is uniform near the axis, dropping with increasing steepness to zero at $\varpi=L/\sqrt{2}$. The fields may cut off at any radius less than $L/\sqrt{2}$, of course, with the boundary condition (9.12).

If $\mathrm{d}u/\mathrm{d}U\neq 0$, (9.20) becomes

$$\frac{\mathrm{d}^2u}{\mathrm{d}U^2}(1-2u+U/\alpha^2)-\left(\frac{\mathrm{d}u}{\mathrm{d}U}\right)^2+\frac{1}{\alpha^2}\frac{\mathrm{d}u}{\mathrm{d}U}=0. \tag{9.22}$$

To solve this equation, introduce the scale change $U=\alpha^2\mu$. Then write $\psi=1-2u+\mu$. The result is

$$2\psi\frac{\mathrm{d}^2\psi}{\mathrm{d}\mu^2}=1-\left(\frac{\mathrm{d}\psi}{\mathrm{d}\mu}\right)^2$$

which can be rewritten

$$\frac{2\mathrm{d}\psi/\mathrm{d}\mu\ \mathrm{d}^2\psi/\mathrm{d}\mu^2}{1-(\mathrm{d}\psi/\mathrm{d}\mu)^2}=\frac{1}{\psi}\frac{\mathrm{d}\psi}{\mathrm{d}\mu}.$$

The integral is obviously

$$\psi[1-(\mathrm{d}\psi/\mathrm{d}\mu)^2]=\mathit{constant}$$

or

$$(\mathrm{d}\psi/\mathrm{d}\mu)^2=1+K/\psi. \tag{9.23}$$

Hence

$$\mu=\pm\int_1^{1-2u+\mu}\mathrm{d}\psi/(1+K/\psi)^{\frac{1}{2}}, \tag{9.24}$$

where the arbitrary constant of integration has been evaluated from the fact that u and U both vanish on the axis of the tube.[4]

Before proceeding further, the sign of K must be determined. Noting that $0\leq u\leq\frac{1}{2}$ and $\mu>0$ it is clear that $\psi>0$. It follows, from (9.23) and the requirement that $\mathrm{d}\psi/\mathrm{d}\mu$ be real, that K is positive or negative depending upon whether $(\mathrm{d}\psi/\mathrm{d}\mu)^2$ is larger or smaller than one. Suppose, then, that the twisted flux rope is sufficiently expanded radially in proportion to the

[4] In the circumstance that fluid has been injected along the axis of the tube, producing an internal column of fluid of radius μ_2, the axis $u=0$ maps into μ_2 and the lower limit on the integral is $1+\mu_2$.

stretching α that μ is larger than the corresponding u, so that $0 < \mathrm{d}u/\mathrm{d}\mu < 1$. Then since

$$\mathrm{d}\psi/\mathrm{d}\mu = 1 - 2\,\mathrm{d}u/\mathrm{d}\mu,$$

it follows that $-1 < \mathrm{d}\psi/\mathrm{d}\mu < +1$, requiring that K be negative. If, on the other hand, the tube is compressed and/or stretched so that $\mathrm{d}u/\mathrm{d}\mu > 1$, then $\mathrm{d}\psi/\mathrm{d}\mu < -1$ and K must be positive.

9.4.1. Compression of the flux tube

Suppose that the rope is compressed sufficiently so that $\mathrm{d}u/\mathrm{d}\mu > 1$. Then $\mathrm{d}\psi/\mathrm{d}\mu < -1$ and (9.24) yields

$$-\mu = (1-2u+\mu)^{\frac{1}{2}}(K+1-2u+\mu)^{\frac{1}{2}} - (K+1)^{\frac{1}{2}} + K\sinh^{-1}(1/K^{\frac{1}{2}}) - K\sinh^{-1}\{(1-2u+\mu)/K\}^{\frac{1}{2}} \quad (9.25)$$

where the negative sign is chosen because ψ diminishes outward from the axis of the tube. The constant K is to be determined from the fact that c maps into C (with $C < c$). Altogether, then, the mapping $u = u(U)$ proves to be a transcendental relationship. The generating function $F(U)$ follows from (9.19) and (9.23) upon noting that $\mathrm{d}u/\mathrm{d}\mu > 1$ so that $0 > \mathrm{d}\psi/\mathrm{d}\mu = -(1+K/\psi)^{\frac{1}{2}}$. Then

$$\begin{aligned} F &= (\mathrm{d}u/\mathrm{d}\mu)^2\psi/\alpha^4 \\ &= (\psi/4\alpha^4)(1-\mathrm{d}\psi/\mathrm{d}\mu)^2 \\ &= (\psi/4\alpha^4)\{1+(1+K/\psi)^{\frac{1}{2}}\}^2. \end{aligned} \quad (9.26)$$

Given that the tube terminates where F has the value given by (9.14) at $\Pi = C$ (say at $\mu = \mu_s = \alpha^2C^2/L^2$ and $u = u_s = c^2/L^2$) it follows that

$$K = \psi_s[\{(4\alpha^4\Delta P/\psi_s)^{\frac{1}{2}} - 1\}^2 - 1] \quad (9.27)$$

where ψ_s is the value of $1 - 2u_s + \mu_s$ at the surface. This equation must now be solved simultaneously with (9.25) to obtain K and C/L for a given ΔP and α.

In order to demonstrate the full effect of compression, suppose that the initial field $f(\varpi)$ extends all the way out to the limiting radius $c = L/\sqrt{2}$ ($u_s = \frac{1}{2}$), where b_z falls to zero and the total flux along the tube is $(8\pi)^{\frac{1}{2}}\pi L^2/3$. Then $\psi_s = \mu_s$ in (9.25) and

$$0 = \mu_s + \mu_s^{\frac{1}{2}}(K+\mu_s)^{\frac{1}{2}} - (K+1)^{\frac{1}{2}} + K\sinh^{-1}(1/K^{\frac{1}{2}}) - K\sinh^{-1}(\mu_s/K)^{\frac{1}{2}}. \quad (9.28)$$

A strong compression ($\mu_s \ll u_s$) reduces this to

$$K^{\frac{1}{2}} = (2/3\mu_s)(1 - \mu_s - (\tfrac{27}{40})\mu_s^2 + \ldots) \quad (9.29)$$

with K related to ΔP by (9.27). It then follows from (9.25), neglecting terms $O(\mu^2)$ compared to one, that

$$\psi \cong \{1-(1-\mu_s^{\frac{3}{2}})\mu/\mu_s\}^{\frac{2}{3}} \cong (1-\mu/\mu_s)^{\frac{2}{3}} \quad (9.30)$$

$$u \cong \tfrac{1}{2}[1+\mu-\{1-(1-\mu_s^{\frac{3}{2}})\mu/\mu_s\}^{\frac{2}{3}}]$$

$$\cong \tfrac{1}{2}\{1-(1-\mu/\mu_s)^{\frac{2}{3}}\}. \quad (9.31)$$

It follows from (9.19) or (9.26) that

$$F(\mu) \cong \frac{1-2\mu_s^{\frac{3}{2}}}{9\mu_s^2\alpha^4}[1+3\mu_s\{1-(1-\mu_s^{\frac{3}{2}})\mu/\mu_s\}^{\frac{1}{3}}] \quad (9.32)$$

$$\cong \frac{1}{9\mu_s^2\alpha^4}\{1+3\mu_s(1-\mu/\mu_s)^{\frac{1}{3}}\}. \quad (9.33)$$

Then, from (9.13), the longitudinal field is

$$B_z^2/8\pi = F+\mu\, dF/d\mu$$

$$= \frac{1}{9\mu_s^2\alpha^4}\left[1+\frac{3\mu_s-4\mu}{\{1-(1-\mu_s^{\frac{3}{2}})\mu/\mu_s\}^{\frac{2}{3}}}\right]. \quad (9.34)$$

This field is uniform, with the value

$$B_z^2/8\pi \cong 1/9\mu_s^2\alpha^4$$

except in the very near neighbourhood ($1-\mu/\mu_s<\frac{3}{2}\mu_s^{\frac{3}{2}}$) of the surface U_s, where the field plunges precipitously to zero

$$B_z^2/8\pi \cong (2/27\mu_s^{\frac{7}{2}}\alpha^4)(1-\mu/\mu_s). \quad (9.35)$$

The azimuthal field is

$$B_\phi^2/8\pi = -\mu\, dF/d\mu$$

$$= (\mu/9\mu_s^2\alpha^4)\{1-(1-\mu_s^{\frac{3}{2}})\mu/\mu_s\}^{\frac{2}{3}} \quad (9.36)$$

and is very weak compared to $B_z^2/8\pi$ except in the thin region at the surface $1-\mu/\mu_s<3\mu_s^{\frac{3}{2}}/2$ where $B_\phi^2/8\pi$ jumps abruptly to its value $\frac{1}{9}\mu_s^2\alpha^4$ at the boundary. The total magnetic pressure is the sum of (9.34) and (9.36), giving the essentially uniform pressure

$$F(\mu) = 1/9\mu_s^2\alpha^4 \quad (9.37)$$

across the entire tube.

It is evident that, if the initial tube did not extend out to $u_s=\frac{1}{2}$, the outermost layers would be missing, and the abrupt drop of B_z and rise of B_ϕ, would both be missing from the compressed tube. The compression would leave the tube essentially straight and untwisted, with the pitch angle ϑ, given by (9.3), close to zero.

It is obvious by inspection of (9.20) that the uniform untwisted flux tube, for which $f=$ *constant*, is an invariant field form. For (9.20) integrates to $du/dU=$ *constant* so that $F=$ *constant*, from (9.19).

9.4.2. Expansion of the flux tube

Suppose that the tube is expanded sufficiently that $du/d\mu < 1$. Then $K < 0$, and it is convenient to write $Q \equiv -K > 0$, so that (9.24) yields

$$\mu = \psi^{\frac{1}{2}}(\psi - Q)^{\frac{1}{2}} - (1 - Q)^{\frac{1}{2}} + Q \cosh^{-1}(\psi/Q)^{\frac{1}{2}} - Q \cosh^{-1}(1/Q)^{\frac{1}{2}}. \tag{9.38}$$

If again we consider the complete flux tube, out to $u_s = \frac{1}{2}$, then again $\psi_s = \mu_s$, and substitution into (9.38) yields the relation between μ_s and Q. For large expansion, $\mu_s \gg 1$, u, the result reduces to

$$Q \cong \frac{2}{\ln(\mu_s)} \left\{ 1 - \frac{\frac{3}{2}}{\ln^2 \mu_s} + O\left(\frac{1}{\ln^3(\mu_s)} \right) \right\} \tag{9.39}$$

which is related to ΔP through (9.24). To this order $\psi \cong 1 + \mu$ and

$$\begin{aligned} u &\cong \frac{1}{2} \left\{ \frac{\ln(1+\mu)}{\ln \mu_s} \left(1 - \frac{\frac{3}{2}}{\ln^2 \mu_s} \right) + \frac{3\mu}{2(1+\mu)\ln^2 \mu_s} \right\} \\ &\quad \times \left\{ 1 + \frac{1 - \frac{3}{2}\ln^2 \mu_s}{(1+\mu_s)\ln \mu_s} \right\}^{-1} + O\left(\frac{1}{\ln^4 \mu_s} \right) \\ &\cong \frac{\ln(1+\mu)}{2\ln(\mu_s)}. \end{aligned} \tag{9.40}$$

It follows from (9.19) or (9.26) that, to the same order,

$$F(\mu) = 1/\{4\alpha^4(1+\mu)\ln^2(\mu_s)\}. \tag{9.41}$$

Hence

$$B_\phi^2/8\pi\mu = B_z^2/8\pi = 1/\{4\alpha^4(1+\mu)^2\ln^2(\mu_s)\}. \tag{9.42}$$

The longitudinal field is negligible compared to the azimuthal field throughout the outer portion of the radius $1 \leqslant \mu \leqslant \mu_s$. The magnetic field is principally azimuthal. Generally speaking, longitudinal stretching of the flux rope has the opposite effect of radial expansion, reducing the azimuthal field in proportion to the longitudinal field and reducing the pitch angle of the spiralling lines of force.

9.4.3. Invariant field configuration

It was pointed out in §9.4.1 that the limiting form $F = \textit{constant}$ for compression is an invariant form of the field. So too is (9.41). To show this, suppose that $f(u)$ is given *exactly* by[5]

$$f(u) = 1/(1+u) \tag{9.43}$$

[5] This is the same form as (9.6), representing a helical field in which all lines of force make one revolution in the distance $2\pi L$.

for $0 \leqslant u \leqslant +\infty$. Then (9.20) reduces to

$$\frac{d^2u}{d\mu^2}-\frac{1}{1+u}\left(\frac{du}{d\mu}\right)^2+\frac{1}{1+\mu}\frac{du}{d\mu}=0 \tag{9.44}$$

where again $U=\alpha^2\mu$. Then divide through by $du/d\alpha$ and integrate, obtaining

$$\frac{du}{d\mu}=M\frac{1+u}{1+\mu} \tag{9.45}$$

where M is the constant of integration. Integrating again,

$$1+u=N(1+\mu)^M \tag{9.46}$$

where N is the integration constant. If u and μ are to vanish simultaneously, we must put $N=1$, so that

$$1+u=(1+\mu)^M. \tag{9.47}$$

It is obvious that $M>1$ yields $du/d\mu>1$, etc. If u_s maps into μ_s, then

$$M=\frac{\ln(1+u_s)}{\ln(1+\mu_s)},$$

which determines M in terms of the expansion. It follows from (9.19) that

$$F(\mu)=\frac{1}{\alpha^4}\frac{1+\mu}{(1+u)^2}\left(\frac{du}{d\mu}\right)^2 \tag{9.48}$$

$$=\frac{M^2}{\alpha^4(1+\mu)}. \tag{9.49}$$

Thus the form of the generating function written in terms of $\mu=U/\alpha^2$ is unaffected by radial or longitudinal expansion or contraction. The initial field is

$$b_z=\pm\frac{(8\pi)^{\frac{1}{2}}}{1+\varpi^2/L^2},\qquad b_\phi=\pm\frac{(8\pi)^{\frac{1}{2}}\varpi/L}{1+\varpi^2/L^2}. \tag{9.50}$$

The final field (B_z, B_ϕ) is the same function of μ, apart from the factor M/α^2 in magnitude and the change of scale from L to αL. They are shown in Fig. 9.2 as a function of radius $\mu^{\frac{1}{2}}$.

If initially the field is confined by the pressure differential Δp, it follows that

$$1+u_s=1/\Delta p. \tag{9.51}$$

Finally, when the confining pressure is ΔP, the tube is confined to $\mu \leqslant \mu_s$ where

$$1+\mu_s=M^2/\alpha^4\Delta P. \tag{9.52}$$

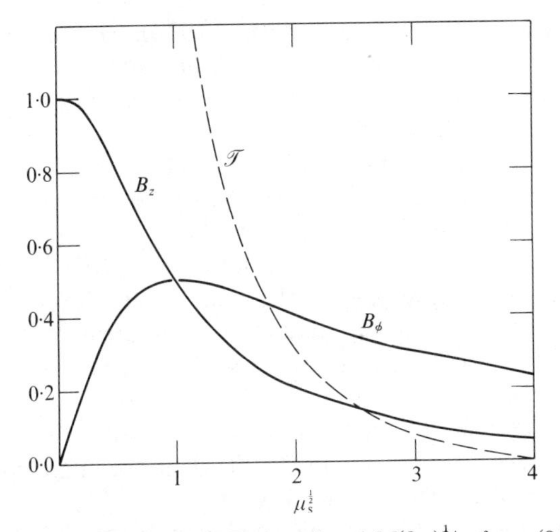

FIG. 9.2. A plot of the fields B_z and B_ϕ in units of $M(8\pi)^{\frac{1}{2}}/\alpha$ from (9.49) and the total tension (dashed line) in units of $(\pi L^2/\alpha^2)\ln^2(1+u_s)$ from (9.54), as a function of $\mu_s^{\frac{1}{2}}$. The total field pressure, from (9.48) in units of M^2/α^4, is identical with B_z in units of $M(8\pi)^{\frac{1}{2}}/\alpha^2$.

Noting that μ_s is related to u_s through (9.47), it follows from (9.51) and (9.52) that

$$\Delta p = (\alpha^4 \Delta P/M^2)^M, \tag{9.53}$$

which serves to determine M, given α, Δp, and ΔP.

As an example, suppose that $\alpha^4\Delta P = \epsilon\Delta p$ where $\epsilon \ll 1$, so that the tube expands. Then, it follows with the aid of (9.52) that

$$M = \frac{+\ln(1+\mu_s) - 2M\ln M}{\ln(1+\mu_s)+\ln(1/\epsilon)}$$

$$\cong \frac{\ln(1+\mu_s)}{\ln(1/\epsilon)}\left\{1 - \frac{2M\ln \mathrm{M}}{\ln(1+\mu_s)} - \frac{\ln(1+\mu_s)}{\ln(1/\epsilon)} + \dots\right\}$$

$$\cong \frac{\ln(1+\mu_s)}{\ln(1/\epsilon)}\left[1 + \frac{2}{\ln(1/\epsilon)}\ln\left\{\frac{\ln(1/\epsilon)}{\ln(1+\mu_s)}\right\} + \dots\right].$$

The total tension in the tube follows from (9.8),

$$\mathcal{T} = \frac{\pi L^2}{\alpha^2}\frac{\ln^2(1+u_s)}{\ln(1+\mu_s)}\left\{\frac{3\mu_s}{(1+\mu_s)\ln(1+\mu_s)} - 1\right\}. \tag{9.54}$$

The tension declines to zero as μ_s increases to 15·8, as shown in Fig. 9.2. Hence expansion of the tube to $\mu_s = 15{\cdot}8 (C/\alpha L = 3{\cdot}98)$ destroys stability.

9.5. Variation of radius along the tube

A slightly different problem is the equilibrium of a slender flux rope extending through an inhomogeneous atmosphere so that the confining pressure Δp varies along the tube (Parker 1974). The problem is superficially not unlike that of simple dilatation and extension treated above, although the equilibrium of the tube has some remarkable properties in the expanded portion of the tube.

Consider first a flux tube with uniform radius c for $z<-h$ and uniform radius C for $z>+h$. The radius may vary in any arbitrary manner throughout $-h<z<+h$, as sketched in Fig. 9.3. Then how is the field distribution $b_\phi(\varpi)$, $b_z(\varpi)$ at $z=-\infty$ related to $B_\phi(\Pi)$, $B_z(\Pi)$ at $z=+\infty$? There are three exact integrals of the equilibrium equations that relate the conditions at $z=\pm\infty$. The first is the uniform extension of fluid pressure along the magnetic lines of force, so that, if the fluid pressure is uniform across the tube at $z=-\infty$, it is also uniform at $z=+\infty$. It follows, then, that the field in the tube of uniform radius c at $z=-\infty$ can be described exactly by (9.11), while the field in the uniform radius C at $z=+\infty$ is described exactly by (9.13).

The second exact integral is the conservation of longitudinal flux in each annulus $(\varpi, \varpi+\mathrm{d}\varpi)$, $(\Pi, \Pi+\mathrm{d}\Pi)$ between $z=\pm\infty$. This condition is expressed by (9.15) and (9.17). The third exact integral is the conservation of torque along the annulus so that

$$\varpi\frac{b_\phi b_z}{4\pi}2\pi\varpi\,\mathrm{d}\varpi=\Pi\frac{B_\phi B_z}{4\pi}2\pi\Pi\,\mathrm{d}\Pi. \tag{9.55}$$

This condition replaces the conservation of azimuthal flux (9.16). Any unbalanced torsion shifts coils of field along the tube, from the regions of small radius to the regions of large radius, until (9.55) is satisfied.

Divide (9.15) into (9.55) and square the result so that the field can be written in terms of the generating functions. The result is

$$u^2\frac{\mathrm{d}f}{\mathrm{d}u}=U^2\frac{\mathrm{d}F}{\mathrm{d}U} \tag{9.56}$$

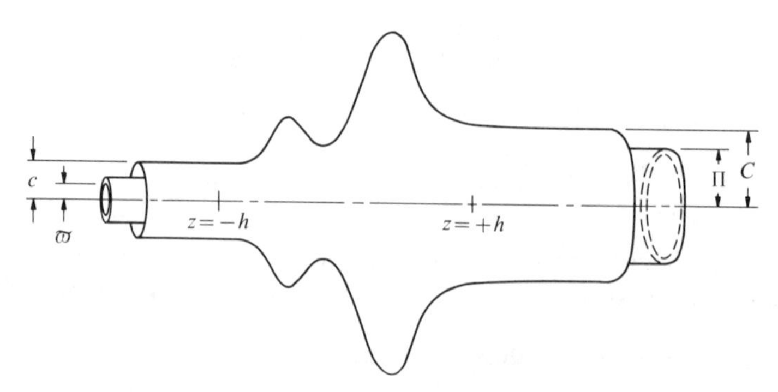

FIG. 9.3. A schematic drawing of the flux tubes of radius c and C joined through an arbitrary variation of radius in $-h<z<+h$.

to be solved simultaneously with (9.17). Divide (9.56) by U and subtract from (9.17), obtaining

$$F=\left(\frac{\mathrm{d}u}{\mathrm{d}U}\right)^2\frac{\mathrm{d}(uf)}{\mathrm{d}u}-\frac{u^2}{U}\frac{\mathrm{d}f}{\mathrm{d}u}. \tag{9.57}$$

Differentiate with respect to U and use the result to eliminate $\mathrm{d}F/\mathrm{d}U$ from (9.56), obtaining the equation

$$0=\frac{\mathrm{d}u}{\mathrm{d}U}\left\{2\frac{\mathrm{d}(uf)}{\mathrm{d}u}\frac{\mathrm{d}^2u}{\mathrm{d}U^2}+\frac{\mathrm{d}^2(uf)}{\mathrm{d}u^2}\left(\frac{\mathrm{d}u}{\mathrm{d}U}\right)^2-\frac{1}{U}\frac{\mathrm{d}}{\mathrm{d}u}\left(u^2\frac{\mathrm{d}f}{\mathrm{d}u}\right)\right\}. \tag{9.58}$$

The equation can be satisfied by $\mathrm{d}u/\mathrm{d}U=0$ for which $u=$ *constant* and $F\propto 1/U$, giving a purely azimuthal field, $B_z=0$, $B_\phi\propto 1/\varpi$. Alternatively the differential expression in braces must vanish giving $u=u(U)$. In that case the example $f=1-u$, $0\leq u\leq\frac{1}{2}$ treated in §9.3 reduces (9.58) to

$$0=\frac{\mathrm{d}u}{\mathrm{d}U}\left\{(1-2u)\frac{\mathrm{d}^2u}{\mathrm{d}U^2}-\left(\frac{\mathrm{d}u}{\mathrm{d}U}\right)^2+\frac{u}{U}\right\}. \tag{9.59}$$

This equation is not subject to exact integration by elementary methods. So consider the two cases of extreme compression $C\ll c$ $(u/U,\ \mathrm{d}u/\mathrm{d}U\gg 1)$ and extreme expansion $C\gg c$ $(u/U,\ \mathrm{d}u/\mathrm{d}U\ll 1)$.

9.5.1. *Extreme compression of the flux tube* $(C\ll c)$

If the tube at $z=+\infty$ is enormously compressed relative to $z=-\infty$, then $\mathrm{d}u/\mathrm{d}U$ and u/U are both large, presumably to the same order. Then drop the linear term u/U from (9.59) compared to $\mathrm{d}^2u/\mathrm{d}U^2$. The equation is integrable, yielding

$$2u=1-[1-\{1-(1-2u_s)^{\frac{3}{2}}\}U/U_s]^{\frac{2}{3}}$$

where the constants of integration are chosen so that $u(U_s)=u_s$ and $u(0)=0$. Consider the complete tube so that $u_s=\frac{1}{2}$ and

$$2u=1-(1-U/U_s)^{\frac{2}{3}}. \tag{9.60}$$

It is readily shown that

$$\frac{\mathrm{d}u}{\mathrm{d}U}=\frac{1}{3U_s(1-U/U_s)^{\frac{1}{3}}},$$

which is $O(1/U_s)$ for $U/U_s\ll 1$, and increases without bound as $U/U_s\to 1$. Hence the outer layer of the tube, of radius c, where b_z goes to zero, is compressed into a thin layer immediately inside the surface of the small radius C. The generating function follows from (9.57) as

$$F(U)=\frac{1}{9U_s^2}+\frac{1-(1-U/U_s)^{\frac{2}{3}}}{4U}$$

$$\cong\frac{1}{9U_s^2}\{1+O(U_s)\} \tag{9.61}$$

to the order considered. Then the longitudinal field is

$$B_z^2/8\pi\cong 1/9U_s^2, \tag{9.62}$$

The azimuthal field is much weaker

$$B_\phi^2/8\pi=(U/U_s)O(U_s^{-1})$$

and cannot be computed properly to the present order. The flux tube is essentially free of torsion where strongly compressed.

9.5.2. Extreme expansion of the flux tube $(C\gg c)$

If the tube at $z=+\infty$, is enormously expanded compared to $z=-\infty$, then du/dU and u/U are both very small, so that the second-order term $(du/dU)^2$ may be dropped from (9.59). The expansion enormously enhances the degree of twisting, so that it is sufficient to limit discussion to small values of u, for which the twisting at $z=-\infty$ is slight. Neglecting $2u$ compared to one, (9.59) reduces to

$$\frac{d^2u}{dU^2}+\frac{u}{U}=0. \tag{9.63}$$

The solution is

$$u(U)=C_1U^{\frac{1}{2}}J_1(2U^{\frac{1}{2}})+C_2U^{\frac{1}{2}}Y_1(2U^{\frac{1}{2}}),$$

where C_1 and C_2 are arbitrary constants and J_1 and Y_1 represent Bessel functions of the first and second kind. With the boundary conditions that $u(0)=0$ and $u(U_s)=u_s$, the solution is

$$u(U)=u_s\frac{U^{\frac{1}{2}}J_1(2U^{\frac{1}{2}})}{U_s^{\frac{1}{2}}J_1(2U_s^{\frac{1}{2}})}. \tag{9.64}$$

Then

$$\frac{du}{dU}=\frac{u_sJ_0(2U^{\frac{1}{2}})}{U_s^{\frac{1}{2}}J_1(2U_s^{\frac{1}{2}})} \tag{9.65}$$

and it follows from (9.57) that

$$F(U)=\frac{u_s^2\{J_0^2(2U^{\frac{1}{2}})+J_1^2(2U^{\frac{1}{2}})\}}{U_sJ_1^2(2U_s^{\frac{1}{2}})}. \tag{9.66}$$

Hence

$$\frac{B_z^2}{8\pi}=\frac{u_s^2 J_0^2(2U^{\frac{1}{2}})}{U_s J_1^2(2U_s^{\frac{1}{2}})},\qquad \frac{B_\phi^2}{8\pi}=\frac{u_s^2 J_1^2(2U^{\frac{1}{2}})}{U_s J_1^2(2U_s^{\frac{1}{2}})}. \tag{9.67}$$

The fields B_z and B_ϕ, and the total pressure F, are plotted in Fig. 9.4 out to the first zero of $J_0(2U^{\frac{1}{2}})$ at $U\equiv U_0=1{\cdot}4458$ where $\Pi/L=U^{\frac{1}{2}}=1{\cdot}2024$.

It is evident from inspection of (9.67) that the longitudinal field B_z falls to zero at $U=U_0$. Suitably choosing the sign to avoid negative B_z, the field rises to a maximum beyond and falls to zero again at $U=7{\cdot}618$ at the second zero of J_0, etc. This is odd, in view of the fact that the longitudinal field where the tube is compressed to $u_s \ll 1$ is essentially uniform, $b_z^2/8\pi \cong 1$. The unphysical behaviour of (9.67) becomes absurdity in (9.65), which yields $du/dU<0$ beyond U_0, whereas $u(U)$ and $U(u)$ represent a progressive mapping of one radius onto another. Both $u(U)$ and $U(u)$ must be non-decreasing functions of their arguments, and single-valued. Altogether eqns (9.64)–(9.67) are physically acceptable for

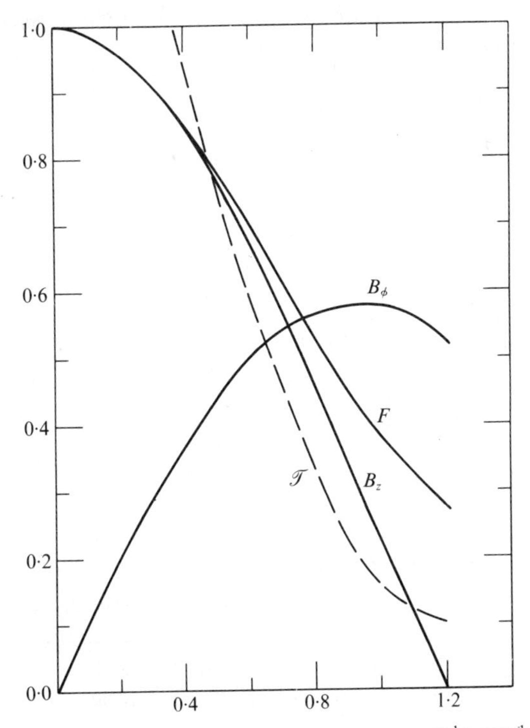

FIG. 9.4. A plot of the fields B_ϕ and B_z in units of $(8\pi U_s/u_s^2)^{\frac{1}{2}} J_1(2U_s^{\frac{1}{2}})$ as a function of radius $U^{\frac{1}{2}}$ from (9.67), together with the total pressure in units of $10(u_s^2/U_s)J_1^2(2U_s^{\frac{1}{2}})$ from (9.66). The total tension in units of $\pi L^2 u_s^2$ from (9.74) (dashed line) is given as a function of $U_s^{\frac{1}{2}}$.

expansion of the tube out to $U_s \leqslant U_0 = 1{\cdot}4458$. They are unacceptable beyond. But there is no physical obstacle to expanding the tube beyond $U_s = U_0$. The confining pressure increment is $\Delta P = F(U_s)$, with a value

$$\Delta P = u_s^2/U_0 \tag{9.68}$$

at $U = U_0$. If ΔP is made smaller than this, the tube expands beyond U_0.

Noting from (9.65) that $\mathrm{d}u/\mathrm{d}U = 0$ at U_0, it is evident from (9.59) that

$$\mathrm{d}u/\mathrm{d}U = 0 \tag{9.69}$$

is the solution for $U > U_0$, so that $u = u_s$ for all $U > U_0$. Then

$$F(U) = u_s^2/U \tag{9.70}$$

and the field is entirely azimuthal, with $B_z = 0$ and

$$B_\phi^2/8\pi = u_s^2/U. \tag{9.71}$$

There is no longitudinal flux beyond U_0. The lines of force of the azimuthal field where $U_s > U_0$ are drawn entirely from the surface $u = u_s$ of the tube. Altogether, then, where the expansion exceeds U_0, the equilibrium configuration is

$$\frac{B_z^2}{8\pi} = \frac{u_s^2 J_0^2(2U^{\frac{1}{2}})}{U_0 J_1^2(2U_0^{\frac{1}{2}})}, \qquad \frac{B_\phi^2}{8\pi} = \frac{u_s^2 J_1^2(2U^{\frac{1}{2}})}{U_0 J_1^2(2U_0^{\frac{1}{2}})} \tag{9.72}$$

with

$$F(U) = \frac{u_s^2\{J_0^2(2U^{\frac{1}{2}}) + J_1^2(2U^{\frac{1}{2}})\}}{U_0 J_1^2(2U_0^{\frac{1}{2}})} \tag{9.73}$$

out to U_0, irrespective of how far beyond U_0 the tube is expanded[6]. Note that numerically we have $U_0^{\frac{1}{2}} J_1(2U_0^{\frac{1}{2}}) = 0{\cdot}624$. The field beyond U_0 is an azimuthal sheath, given by (9.70) and (9.71), with the outer boundary location $\Pi = C$ determined by the confining pressure $\Delta P = u_s^2/U_s$.

The lines of force that make up the azimuthal sheath in the portion of the tube expanded beyond U_0 arrived there by slithering along the surface of the tube from the less expanded portions as indicated in Fig. 9.5. The topology of the lines is invariant, of course, because we assume infinite electrical conductivity. The helical coils are propelled onto the expanded portion, to achieve the equilibrium configuration required by (9.59), by the torque transmitted along the helical field. The surface of field forms a coil spring, which is weak in the expanded portion of the tube and strong in the compressed portion. Hence the spring in the

[6] This asymptotic form is, oddly, just the special force-free field $\nabla \times \mathbf{B} = 2\mathbf{B}/L$, wherein the electric current is everywhere parallel and proportional to the magnetic field.

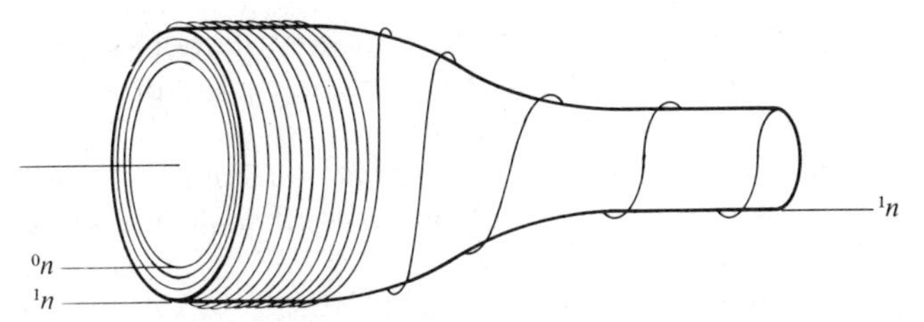

FIG. 9.5. A sketch of the magnetic lines of force from the surface of the flux rope sliding along the tube and winding themselves onto the expanded portion of the tube, where $U_s > U_0$, to form an azimuthal sheath.

compressed portion dominates and relieves its tension by *unwinding*, which then winds up the coil in the expanded portion where the field is too weak to resist. The transfer of coils proceeds at essentially the Alfven speed, the shifting coils of field being nothing more than torsional Alfven waves.

Now consider the tension and stability of the flux tube. The tension in the tube out to $U_s(\leq U_0)$ is readily shown from (9.7) and (9.66) to be

$$\mathcal{T} = 2\pi \int_0^C \mathrm{d}\Pi\Pi \left\{ \frac{B_z^2 - B_\phi^2}{8\pi} + F(U_s) \right\},$$

where the first term in the integrand represents the magnetic stress and $F(U_s)$ represents the contribution of the reduced gas pressure within the flux tube. It is readily shown from (9.66) that the magnetic contribution is

$$\mathcal{T}_M = \pi L^2 u_s^2 J_0(2U_s^{\frac{1}{2}})/U_s^{\frac{1}{2}} J_1(2U_s^{\frac{1}{2}}),$$

which falls to zero as U_s increases to $U_0 = 1{\cdot}4458$. The contribution from the pressure is

$$\mathcal{T}_p = \pi L^2 U_s F(U_s)$$

so that the total tension is

$$\begin{aligned} \mathcal{T} &= \pi L^2 \{ u_s^2 J_0(2U_s^{\frac{1}{2}})/U_s^{\frac{1}{2}} J_1(2U_s^{\frac{1}{2}}) + U_s F(U_s) \} \\ &= \frac{\pi L^2 u_s^2}{J_1^2(2U^{\frac{1}{2}})} \{ J_0^2(2U_s^{\frac{1}{2}}) + J_1^2(2U_s^{\frac{1}{2}}) + J_0(2U_s^{\frac{1}{2}}) J_1(2U_s^{\frac{1}{2}})/U_s^{\frac{1}{2}} \}, \end{aligned} \tag{9.74}$$

and is plotted in Fig. 9.4 as a function of U_s for $U_s \leq U_0$. As U_s increases to U_0, $\mathcal{T}_M$ declines to zero and the total tension takes on the value

$$\begin{aligned} \mathcal{T} &= \pi L^2 U_0 F(U_0) \\ &= \pi L^2 u_s^2. \end{aligned} \tag{9.75}$$

If the radius of the type is expanded beyond U_0, the field within the longitudinal core $U < U_0$ remains unchanged, contributing nothing, therefore, to the total tension. The confining pressure increment declines as

$F(U_s) = u_s^2/U_s$ with $U_s > U_0$, so that the total contribution of the longitudinal core is the tension $\pi L^2 U_0 F(U_s) = \pi L^2 u_s^2 U_0/U_s$ produced by the reduced pressure. The azimuthal magnetic field in the sheath $U_0 < U < U_s$ contributes only a compression (negative tension), in the amount $B_\phi^2(\Pi)/8\pi$. The total contribution of the azimuthal sheath, including the reduced gas pressure $F(U_s)$, is

$$\mathcal{T}_\phi = \pi L^2 \int_{U_0}^{U_s} \mathrm{d}U\{-B_\phi^2(\Pi)/8\pi + F(U_s)\}$$

$$= \pi L^2 u_s^2\{1 - U_0/U_s - \ln(U_s/U_0)\} \tag{9.76}$$

with $F(U) = u_s^2/U$. The total tension in the longitudinal core and azimuthal sheath is, then,

$$\mathcal{T} = \pi L^2 u_s^2\{1 - \ln(U_s/U_0)\}. \tag{9.77}$$

The total tension falls to zero when U_s reaches $eU_0 = 3{\cdot}93$, and is negative (representing a net compression) for larger U_s. Evidently the azimuthal sheath cannot be extended beyond $C/L = 1{\cdot}98$, if indeed it can develop as far as that.

As a matter of fact, we saw in §9.2 that at least in one special case instability occurs before the total magnetic stress

$$\mathcal{T}_M = 2\pi \int_0^C \mathrm{d}\Pi\,\Pi(B_z^2 - B_\phi^2)/8\pi \tag{9.78}$$

falls to zero. The magnetic tension $\mathcal{T}_M$ is less than the total tension by the confining pressure increment ΔP. For $U_s < U_0$ the magnetic tension is

$$\mathcal{T}_M = \frac{\pi L^2 u_s^2}{U_s J_1^2(2U_s^{\frac{1}{2}})} \int_0^{U_s} \mathrm{d}U\{J_0^2(2U^{\frac{1}{2}}) - J_1^2(2U^{\frac{1}{2}})\}$$

$$= \frac{\pi L^2 u_s^2 J_0(2U_s^{\frac{1}{2}})}{U_s^{\frac{1}{2}} J_1(2U_s^{\frac{1}{2}})}, \tag{9.79}$$

which goes to zero at $U_s = U_0$. If this is a better criterion for the onset of instability than $\mathcal{T} = 0$, it raises the question whether the azimuthal sheath ever appears in nature.

The azimuthal sheath is built up of coils of field that slither along the tube from the concentrated to the expanded regions. Since they come from an infinitesimal surface layer of the concentrated portion of the tube, it must take a very long time for them to accumulate on the expanded portion to build up the azimuthal sheath required for static equilibrium (Parker 1974). We would expect that, when a segment of a tube is first expanded, torsional equilibrium along the tube is not satisfied, but is established only after a period comparable to the Alfven transit time along the tube. For a slender tube this may be long compared to the time over which the expansion occurred. When the segment is first expanded, the field in the tube is approximately in the state (9.42) or (9.50). The expansion may be to arbitrarily large radii if the confining pressure ΔP is sufficiently small. Then with the passage of time, and the redistribution of the helical lines of force along the tube, the distribution tends toward (9.67) and (9.68). If the expansion exceeds U_0, the final redistribution of surface lines of force to build up the azimuthal sheath—compressing the longitudinal field back inside U_0—takes a very long time. Hence, the expansion of a portion of a flux rope is followed by a long period of active redistribution of field coils along the rope to the expanded section. If the initial expansion (without torsional equilibrium along the tube) approaches $U_s = 15{\cdot}8$, then the tension (9.54) goes to zero, so that instability, and then dynamical dissipation and line cutting between neighbouring loops of the corkscrew rope, are an evident result. For so large an expansion the re-establishment of equilibrium is a complicated procedure. If the line cutting proceeds fast enough, it may largely break the helical topology, reducing the twisting of the entire tube so that equilibrium is eventually achieved with only a modest torsion in the expanded portion and essentially none elsewhere. The result would be the loss of a large portion of the initial magnetic energy, the loss going into heat and motions of the fluid.

A more modest initial expansion, say to $U_s = 10$, might avoid the initial instability, but would become unstable as torsional equilibrium is established, with $\mathcal{T}_M$ vanishing for $U_s = 1{\cdot}44$ and $\mathcal{T}$ for $U_s = 3{\cdot}93$. It is an interesting question whether the expanded portion of the tube for $U_s > U_0 = 1{\cdot}44$ can survive its instability to approach hydrostatic equilibrium without having much of the total twisting destroyed. Or does dynamical line cutting between loops of the corkscrew always cut back the total twisting, reducing u_s by increasing L to where $U_s < U_0$? Unfortunately these important dynamical questions lie beyond the elementary problems of static equilibrium treated here. Our illustrations indicate the singular nature of the equilibria, from which we infer the inevitable appearance of contortions and instabilities, without, however, telling us very much about the details of those dissipative effects. An alternative approach is to observe the fields in action in the sun, treating the sun as the ultimate analogue computer. Unfortunately the sun, the active stars, and the galaxy 'compute out' problems of magnetic activity under conditions that are often so complicated as to obscure the basic principles that it is our purpose to discover.

The general conclusion from the calculations is that a flux rope is most strongly twisted where it is expanded, and twisted but little where it is compressed. If the expansion is extreme ($U_s > U_0$), then the trend toward torsional equilibrium along the tube builds up an azimuthal sheath on the expanded portions of the tube. Instabilities and dynamical line cutting may destroy the azimuthal sheath as rapidly as it is produced, with the result that no static equilibrium may be possible so long as there remains any significant twisting any where along the flux rope.

It follows from these general principles that the flux ropes that emerge through the surface of the sun and expand into the essentially empty space above the photosphere capture most of the twisting from the compressed portions beneath the surface. The expanded portions above the surface *sometimes* show torsion, up to pitch angles of $\vartheta \cong \pi/4$ (see, for instance, Nakagawa *et al.* 1971; Nakagawa *et al.* 1973). The fact that the torsion is often small above the surface suggests that there is very little ($\vartheta \ll 1$) in the concentrated flux ropes beneath the surface. Even $\vartheta = \pi/4$ where the field is expanded to 10^2 G above a sunspot extrapolates to $\vartheta \cong 0{\cdot}1$ back into the sun where the field is compressed to 4×10^3 G or more. The failure to observe any ϑ near to $\pi/2$ is probably a result of the dynamical dissipation that sets in when the tension $\mathcal{T}$ is reduced near to zero. Much of the observed activity of flux ropes emerging through the solar photosphere is to be understood as the direct manifestation of the dynamical dissipation (Parker 1974).

9.5.3. *Invariant field configuration*

The limiting form (9.66) for the generating function is an invariant, as may readily be demonstrated from (9.58) by putting (Parker 1976)

$$f(u) = J_0^2(2u^{\frac{1}{2}}) + J_1^2(2u^{\frac{1}{2}}). \tag{9.80}$$

Then if $\mathrm{d}u/\mathrm{d}U \neq 0$, (9.58) becomes

$$J_0(2u^{\frac{1}{2}})\frac{\mathrm{d}^2 u}{\mathrm{d}U^2} - \frac{J_1(2u^{\frac{1}{2}})}{u^{\frac{1}{2}}}\left\{\left(\frac{\mathrm{d}u}{\mathrm{d}U}\right)^2 - \frac{u}{U}\right\} = 0. \tag{9.81}$$

It is readily shown that

$$u^{\frac{1}{2}}J_1(2u^{\frac{1}{2}}) = QU^{\frac{1}{2}}J_1(2U^{\frac{1}{2}}) \tag{9.82}$$

is an exact solution of (9.81), where Q is an arbitrary constant[7]. It follows from (9.57) that

$$F(U) = Q^2\{J_0^2(2U^{\frac{1}{2}}) + J_1^2(2U^{\frac{1}{2}})\}, \tag{9.83}$$

which is the same form as $f(u)$, q.e.d.

[7] More generally write $Z_n(2u^{\frac{1}{2}}) \equiv c_1 J_n(2u^{\frac{1}{2}}) + c_2 Y_n(2u^{\frac{1}{2}})$ with $f = Z_0^2 + Z_1^2$. Then $u^{\frac{1}{2}}Z_1(2u^{\frac{1}{2}}) = QU^{\frac{1}{2}}Z_1(2U^{\frac{1}{2}})$ in place of (9.79), and $F(U) = Q^2\{Z_0^2(2U^{\frac{1}{2}}) + Z_1^2(2U^{\frac{1}{2}})\}$ in place of (9.80), etc.

The field components are

$$b_z^2/8\pi = J_0^2(2u^{\frac{1}{2}}), \qquad b_\phi^2/8\pi = J_1^2(2u^{\frac{1}{2}}) \tag{9.84}$$

$$B_z^2/8\pi = Q^2J_0^2(2U^{\frac{1}{2}}), \qquad B_\phi^2/8\pi = Q^2J_1^2(2U^{\frac{1}{2}}). \tag{9.85}$$

With the boundary condition $U = U_s$ for $u = u_s$, (9.82) yields

$$Q = u_s^{\frac{1}{2}}J_1(2u_s^{\frac{1}{2}})/U_s^{\frac{1}{2}}J_1(2U_s^{\frac{1}{2}}) \tag{9.86}$$

and

$$\frac{u^{\frac{1}{2}}J_1(2u^{\frac{1}{2}})}{u_s^{\frac{1}{2}}J_1(2u_s^{\frac{1}{2}})} = \frac{U^{\frac{1}{2}}J_1(2U^{\frac{1}{2}})}{U_s^{\frac{1}{2}}J_1(2U_s^{\frac{1}{2}})}. \tag{9.87}$$

The total tension is

$$\mathcal{T} = \pi L^2 u_s J_1^2(2u_s^{\frac{1}{2}})\left\{1 + \frac{J_0(2U_s^{\frac{1}{2}})}{U_s^{\frac{1}{2}}J_1(2U_s^{\frac{1}{2}})} + \frac{J_0^2(2U_s^{\frac{1}{2}})}{J_1^2(2U_s^{\frac{1}{2}})}\right\} \tag{9.88}$$

which is essentially (9.74) again. The total flux is

$$\Phi = \pi L^2(8\pi)^{\frac{1}{2}}u_s^{\frac{1}{2}}J_1(2u_s^{\frac{1}{2}}). \tag{9.89}$$

It is evident from (9.85) that the longitudinal field b_z falls to zero, and the field is entirely azimuthal, at the first zero of J_0, where $u = U_0$. To avoid the uninteresting complications of another layer of field beyond, the solution is limited to $u_s < U_0$, so that the longitudinal field has the same sign everywhere across $u(U)$. It follows from $\nabla \cdot \mathbf{B} = 0$ that the longitudinal field must have the same sign everywhere else along the tube.

Now the invariant solution leads again to the azimuthal sheath when the expansion exceeds $U_0 = 1{\cdot}4458$, of course. It follows from (9.82) that

$$\frac{\mathrm{d}u}{\mathrm{d}U} = \frac{QJ_0(2U^{\frac{1}{2}})}{J_0(2u^{\frac{1}{2}})},$$

from which it follows that expansion to $U_s = U_0 = 1{\cdot}4458$ causes $\mathrm{d}u/\mathrm{d}U$ to vanish, with the physical absurdity that $\mathrm{d}u/\mathrm{d}U < 0$ if U_s goes beyond U_0. Expansion to $U_s = U_0$ occurs if the confining pressure increment falls to

$$\Delta P = F(U_0),$$
$$= Q^2J_1^2(2U_0^{\frac{1}{2}}).$$

Using (9.82) to obtain Q in terms of u_s and U_0, it follows that

$$\Delta P = u_s J_1^2(2u_s^{\frac{1}{2}})/U_0. \tag{9.90}$$

Expansion beyond U_0 occurs if the confining pressure falls below this level, leading to the azimuthal sheath

$$F(U) = u_s J_1^2(2u_s^{\frac{1}{2}})/U = B_\phi^2/8\pi. \tag{9.91}$$

Thus (9.82) applies out to $U = U_0$, with $u = u_s$ for $U > U_0$. The position of the outer boundary U_s is determined by the confining pressure $\Delta P = F(U_s)$, so that[8]

$$U_s = u_s J_1^2(2u_s^{\frac{1}{2}})/\Delta P. \tag{9.92}$$

It should be noted again that the equilibrium field in the inner core, $U < U_0$, does not change as U_s increases beyond U_0. The only effect of declining ΔP, and increasing U_s beyond U_0, is the increasing thickness of the azimuthal sheath. Within $U < U_0$ the fields are

$$\frac{B_z^2}{8\pi} = \frac{u_s J_1^2(2u_s^{\frac{1}{2}})}{U_0 J_1^2(2U_0^{\frac{1}{2}})} J_0^2(2U^{\frac{1}{2}}),$$

$$\frac{B_\phi^2}{8\pi} = \frac{u_s J_1^2(2u_s^{\frac{1}{2}})}{U_0 J_1^2(2U_0^{\frac{1}{2}})} J_1^2(2U^{\frac{1}{2}}). \tag{9.93}$$

The inner core is determined only by conditions in the compressed portions of the tube where $u_s < U_0$.

The total tension including both the longitudinal core ($U < U_0$) and azimuthal sheath ($U_0 < U < U_s$), is essentially the value (9.77) already computed,

$$\mathcal{T} = \pi L^2 u_s J_1^2(2u_s^{\frac{1}{2}})\{1 - \ln(U_s/U_0)\}. \tag{9.94}$$

9.6. Vertical extension of a flux rope

In §§8.2, 8.3, and 8.6 we worked out the equilibrium configuration of an elemental flux tube (without torsion) in a stratified atmosphere. The calculations show that there is no equilibrium if the 'ends' of the tube are anchored more than a few scale heights apart in the *horizontal* direction. If the feet are more widely separated, then the magnetic buoyancy overpowers the tension and the tube floats up through the atmosphere at approximately the Alfven speed (see §8.7). In this section we take up the same general question for the twisted flux tube. Equilibrium proves to be even more elusive than with the untwisted tube. Magnetic buoyancy overpowers the tension if the flux tube extends *vertically* more than a few scale heights, irrespective of how close together the feet may be anchored. The apex of the tube floats upward, expanding, twisting, reducing the tension to zero, and taking the tube farther from equilibrium.

[8] This calculation assumes, as in previous examples (Parker 1975a), that there is some slight longitudinal component of field in the azimuthal sheath so that longitudinal hydrostatic equilibrium can be established and (9.95) is applicable. If (9.95) is not applicable, then an endless variety of circumstances can be imagined. One such case is treated in §12.4 where the curious result is shown that, if B_z is identically zero, there is no convective equilibrium in the azimuthal sheath in the presence of *any* temperature gradient.

Consider again a slender flux tube whose radius $R(z)$ is small compared to the local scale height $\Lambda = kT/mg$. It is clear that in hydrostatic equilibrium the radius of the flux tube increases continuously and monotonically with height in the atmosphere. For instance, in the limit of small twisting the equilibrium radius increases exponentially with a scale height equal to 4Λ (see §8.2); for strong twisting the radius increases with a scale height 2Λ. Hence $\partial R/\partial z = O(R/\Lambda)$, and, for a tube of sufficiently small radius R, the taper $\partial R/\partial z$ can be made arbitrarily small. Hence the representation (9.11) and (9.13) for a tube of uniform cross-section is applicable to the slender ($\partial R/\partial z \ll 1$) tube, neglecting terms $O(R/\Lambda)$ compared to one. The mathematical formalism developed in §9.5 for the exact solution of the tube with uniform, but different, radii at $z = -\infty$ and $z = +\infty$ is directly applicable to the problem of the slender tube of slowly varying radius. The three exact integrals relate conditions at two positions, say $z = 0$ and $z = z_1$, along the flux tube, and the forms (9.11) and (9.13) provide an approximate representation of the integrals at the two levels. The only difference is that now, with a gravitational field present, the first integral is not uniform pressure but rather the barometric law $\partial p/\partial z = -\rho g$ along each line of force. Thus, if the slender tube is assumed to be in local thermodynamic equilibrium with its surroundings, the pressure declines with height inside the tube in the same proportion as without. If the ambient pressure in the atmosphere is

$$P(z) = P(0) \exp\left\{-\int_0^z dz'/\Lambda(z')\right\},$$

then the confining differential pressure exerted on the surface of the tube is given exactly by

$$\Delta P(z)/\Delta P(0) = \exp\left\{-\int_0^z dz'/\Lambda(z')\right\} \equiv G(z), \tag{9.95}$$

as with the torsion-free tube (see §8.2).

We now use the approximate relations (9.11) and (9.13) to evaluate the three integrals at $z = 0$ and $z = z_1$. Neglecting terms $O(R/\Lambda)$ compared to one, the boundary condition at $z = 0$ is

$$f(u_s) = \Delta P(0)$$

while at any other position z,

$$\begin{aligned} F(U_s) &= \Delta P(z) \\ &= f(u_s)G(z). \end{aligned}$$

Conservation of longitudinal flux and torque yield (9.17) and (9.56), respectively. The invariant form (9.80) for $f(u)$ is as convenient as any (see other examples in Parker 1975a). Then $F(U)$ is given by (9.83) and (9.86), and it follows that

$$\frac{1}{U_s}\left\{1+\frac{J_0^2(2U_s^{\frac{1}{2}})}{J_1^2(2U_s^{\frac{1}{2}})}\right\}=\frac{G(z)}{u_s}\left\{1+\frac{J_0^2(2u_s^{\frac{1}{2}})}{J_1^2(2u_s^{\frac{1}{2}})}\right\}. \tag{9.96}$$

Solution of this transcendental equation yields U_s in terms of $G(z)$. We may as well suppose that the tube at $z=0$ is twisted but little ($u_s \ll 1$) because the torsion increases rapidly with height and soon reaches large values anyway. Taking $u_s \ll 1$, then, it follows that within a few scale heights U_s is also small and (9.96) reduces to

$$U_s^2 \cong u_s^2/G(z)$$

neglecting terms of the order of u_s compared to one. The tube radius increases upward with a scale height 4Λ, while the field declines with a scale height 2Λ, as already shown by (8.8).

The radius of the tube increases with height, with U_s reaching the value $U_0=1{\cdot}4458$, where the azimuthal sheath begins, at the level (call it $z=z_0$) where $G(z)$ falls to

$$\begin{aligned} G(z_0) &= \frac{u_s}{U_0}\{1+J_0^2(2u_s^{\frac{1}{2}})/J_1^2(2u_s^{\frac{1}{2}})\}^{-1} \\ &\cong (u_s^2/U_0)\{1+O(u_s)\}, \end{aligned} \tag{9.97}$$

Note that the buoyant force per unit length at the height z is

$$\mathcal{F}(z) = \pi C^2(z) g \Delta\rho(z), \tag{9.98}$$

$$= \pi L^2 U_s(z)\Delta P(z)/\Lambda(z), \tag{9.99}$$

$$= \pi L^2 U_s(z) F(U_s)/\Lambda(z), \tag{9.100}$$

$$= (A/\Lambda)B_s^2/8\pi,$$

where A is the cross-sectional area $\pi C^2(z)$. This last expression is identical with (8.58), of course, since the relevant field is B_s at the surface of the tube. Equilibrium requires that $\mathcal{F}$ be balanced by the upward decline of $\mathcal{T}$ given by (9.88),

$$\mathrm{d}\mathcal{T}/\mathrm{d}z = -\mathcal{F}. \tag{9.101}$$

It is an elementary but tedious exercise to differentiate (9.88) with respect to U_s, and (9.96) with respect to z to obtain $\mathrm{d}U_s/\mathrm{d}z$ to show that (9.101) is satisfied, i.e. to show the internal consistency of the Maxwell stresses. The point is that if $\mathcal{F}$ does not diminish sufficiently rapidly with height, then (9.101) will soon see $\mathcal{T}$ diminish to negative values. There is, then, no tension to oppose the buoyancy[9], and hence no equilibrium. Such is the case in the azimuthal sheath.

[9] The formal mathematical possibility of holding down the buoyancy by compression from above is not an available option in most physical circumstances.

The azimuthal sheath begins at the level $z=z_0$ where $U_s=U_0$ and extends upward from there. Above $z=z_0$, $F(U_s)$ is given by (9.91) so that the condition $F(U_s)=\Delta P(z)$ reduces to

$$U_s=u_sJ_1^2(2u_s^{\frac{1}{2}})/f(u_s)G(z). \tag{9.102}$$

The radius increases rapidly upward, with a scale height of only 2Λ.

Now, the magnetic buoyancy per unit length is given by (9.100), and can be written

$$\mathscr{F}(z)=\pi L^2u_sJ_1(2u_s^{\frac{1}{2}})/\Lambda. \tag{9.103}$$

This is a positive quantity and essentially independent of height (except insofar as $\Lambda=kT/Mg$ may depend upon z). The buoyancy of the azimuthal sheath alone is

$$\mathscr{F}_\phi(z)=(\pi L^2/\Lambda)u_sJ_1(2u_s^{\frac{1}{2}})(U_s-U_0)/U_s. \tag{9.104}$$

The azimuthal sheath is free to slide vertically on the inner core $U\leqslant U_0$, so we consider the stresses within the sheath independently. The tension in the azimuthal sheath (see (9.76) and (9.94)) is

$$\mathscr{T}_\phi(z)=-\pi L^2u_sJ_1^2(2u_s^{\frac{1}{2}})\{1-\ln(U_s/U_0)-U_0/U_s\} \tag{9.105}$$

Then with U_s given by (9.102), it follows that

$$\mathrm{d}U_s/\mathrm{d}z=U_s/\Lambda,$$

and it is readily shown that

$$\mathrm{d}\mathscr{T}_\phi/\mathrm{d}z=-\mathscr{F}_\phi,$$

There is, then, a self-consistent hydrostatic equilibrium up to the level where $\mathscr{T}$, given by (9.94), falls to zero. Denoting this level by $z=z_1$, it is evident from (9.94) that $U_s=2{\cdot}72\ U_0$ at z_1, and that $z_1=z_0+\Lambda$, so that the tension goes negative only one scale height above z_0. There is no real physical possibility for holding down the flux tube above $z=z_1$. Magnetic buoyancy takes over, running away with the tube to the top of the atmosphere. Any slender tube—even with a very small pitch angle at $z=0$—extending more than a few scale heights above $z=0$, is susceptible to this runaway buoyancy.

As an illustrative example suppose that the pitch angle is $\vartheta=10°$ at $z=0$. It follows from (9.84) that $u_s^{\frac{1}{2}}\cong 0{\cdot}17$ and from (9.80) that $\Delta P(0)\cong 1{\cdot}0$. It follows from (9.90) that the pitch angle increases to $\pi/2$ at z_0 where ΔP falls to $u_s^2/U_0=1/47{\cdot}6$ only four scale heights above $z=0$. The total tension vanishes in only one more scale height. Thus in five scale heights the twisted rope changes from a 10° pitch angle to catastrophe, with its upper portions ($z>z_1$) running away up the atmosphere.

9.7. An idealized example of a vertical flux rope

There is a particularly simple idealized example showing the runaway character of the non-equilibrium. Imagine a slender flux tube free of torsion, but which has a certain amount of azimuthal flux wrapped around it as a sheath. Imagine this magnetic cable in an atmosphere which buoys it upward, with the feet of the cable anchored in some suitably dense stratum at $z=0$. The azimuthal sheath is concentrated by the buoyant forces at the apex of the tube (discussed at length in Chapter 8 and sketched in Fig. 9.6) the relevant formulae being (8.8) and (8.9) for a field strength $B_z(0)$ and tube radius $R(0)$ at the reference level $z=0$. Suppose that the azimuthal sheath begins at z_0 where the radius is R_0. Above $z=z_0$, then, the longitudinal core is of uniform radius R_0 and uniform field strength $B_z(z_0)$, while the field in the azimuthal sheath is $B_\phi(\varpi)=B_z(z_0)(R_0/\varpi)$ out to $\varpi=R_s(z)$ where $B_\phi^2(\varpi)=8\pi\Delta P(z)$. Hence with (9.95) it follows that

$$R_s(z)=\mathrm{R}_0\exp\left\{\frac{1}{2}\int_{z_0}^{z}\frac{\mathrm{d}z'}{\Lambda(z')}\right\}. \tag{9.105}$$

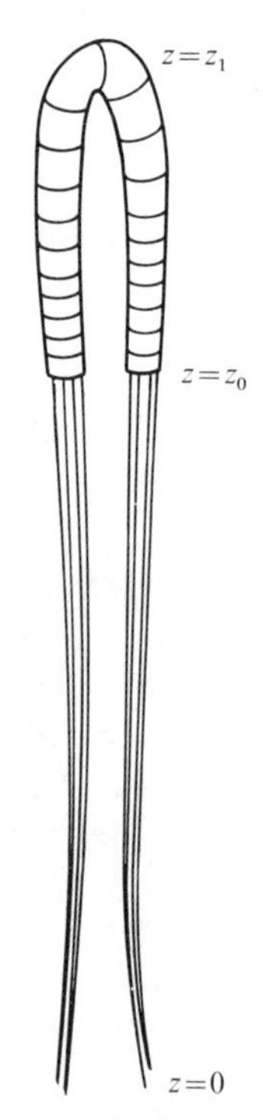

FIG. 9.6. A sketch of the idealized flux tube composed of a purely longitudinal core surrounded above the level $z=z_0$ by a purely azimuthal sheath.

The total tension in the cable above z_0 is readily shown to be

$$\mathcal{T}(z)=\frac{B_z^2(z_0)}{4\pi}\pi R_0^2\left\{1-\frac{1}{2}\int_{z_0}^{z}\frac{dz'}{\Lambda(z')}\right\},$$

which vanishes at $z=z_1$ just two scale lengths above z_0. The total azimuthal flux in the sheath between its lower edge at z_0 and the point of runaway at z_1 is

$$\begin{aligned}\Phi_\phi &= \int_{z_0}^{z_1} dz\int_{R_0}^{R(z)} d\varpi B_\phi,\\ &= B_z(z_0)R_0\int_{z_0}^{z_1} dz\int_{R_0}^{R(z)} d\varpi/\varpi,\\ &= B_z(z_0)R_0\int_{z_0}^{z_1} dz \ln\left(\frac{R(z)}{R_0}\right),\\ &= \tfrac{1}{2}B_z(z_0)R_0\int_{z_0}^{z_1} dz\int_{z_0}^{z} dz'/\Lambda(z').\end{aligned}$$

If Λ is independent of height, then, with $z_1=z_0+2\Lambda$, it is readily shown that

$$\begin{aligned}\Phi_\phi &= 2B_z(z_0)R_0\Lambda\\ &= 2\Phi\Lambda/\pi R_0\end{aligned} \tag{9.106}$$

where Φ is the total longitudinal flux $\pi R_0^2 B_z(z_0)$.

The runaway flux is relatively modest, amounting to the azimuthal flux produced by twisting the tube of radius R_0 over a distance of *only one scale height* to a pitch angle of the order of $\pi/4$. Note further that Φ_ϕ is a declining function of the height at which the azimuthal sheath begins, because Φ is constant, and Λ usually is constant or decreasing with height in most astrophysical bodies, while R_0 increases according to (8.9) with a scale height 4Λ. Thus the critical Φ_ϕ for runaway declines with height with a scale not larger than 4Λ. When runaway begins, then, conserving the total azimuthal flux, the system moves farther and farther beyond equilibrium.

We can see, then, that a cable with an azimuthal sheath may be in equilibrium provided that it does not raise its head above the level where the azimuthal sheath is sufficient to runaway with it to the top of the atmosphere. Once the tube extends above that critical level, it is doomed to be carried up out of the body.

Just how fast the non-equilibrium buoyancy will actually carry the rope upward is a difficult theoretical dynamical question, which our present equilibrium studies cannot answer. The Alfven speed is the characteristic velocity, but the precise answer depends upon the distance that helical coils of field are shifted along the flux rope, in competition with line cutting and dissipation in the expanded portion of the tube. The basic point is the qualitative assertion that twisted magnetic flux tubes are continually active, and are soon irretrievably lost from astrophysical bodies, even more rapidly than the elusive untwisted tube, treated in Chapter 8. The ephemeral bipolar regions (Harvey *et al.* 1975) appearing in the solar photosphere and giving rise to the x-ray bright spots (Vaiana *et al.* 1973) are presumably a direct consequence of this general principle. There is no possibility of holding them down because $\mathcal{T}<0$, and once $\mathcal{T}<0$, the apex buckles and kinks, becoming the seat of the dynamical dissipation that gives the enhanced x-ray emission (Parker, 1975a).

9.8. Flattening of a rising flux tube

The rate of rise of a slender tube of circular cross-section in an atmosphere unstable to convection was estimated in §8.7 to be of the order of the Alfven speed, as given by (8.60). It is this rapid rate of rise that leads to such severe non-equilibrium when the buoyancy overcomes the tension in the tube. Now the calculations in §8.7 were based on a flux tube without torsion. When a tube is twisted, but still force-free, the buoyancy is given by the expression (9.98). This expression is identical with (8.58) for the untwisted tube if it is remembered that the relevant field strength B is the field at the surface of the tube. Thus the characteristic Alfven speed is to be computed for the surface field and the external density.

It was pointed out in §8.7 that the dynamical pressure of the external fluid on the rising horizontal tube tends to flatten the tube so that it becomes a horizontal ribbon whose width extends across the direction of rise. In the twisted tube the flattening is resisted by the stresses in the azimuthal component of the field, limiting the eccentricity of the cross-section to finite values. To work out what that is, consider the irrotational flow of an ideal fluid around the elliptic cylinder

$$\frac{x^2}{a^2\cosh^2\xi_0}+\frac{y^2}{a^2\sinh^2\xi_0}=1$$

using the elliptic coordinates (see, for instance, Jeffreys and Jeffreys 1950; Morse and Feshbach 1953)

$$x=a\cosh\xi\cos\eta,\qquad y=a\sinh\xi\sin\eta.$$

The cylinder is given by $\xi=\xi_0$. Then if the fluid velocity $\mathbf{v}$ outside the cylinder is written as $-\nabla\Phi$, the velocity potential Φ satisfies (8.70) and, for incompressibility, $\nabla^2\Phi=0$. In the elliptical coordinates (ξ,η) this is

$$\partial^2\Phi/\partial\xi^2+\partial^2\Phi/\partial\eta^2=0.$$

Suppose that the external fluid moves with uniform velocity v_0 in the y-direction at large distance from the cylinder. Then the solution which reduces to $\Phi=-v_0 y$ at infinity and has $v_\xi=0$ on the boundary $\xi=\xi_0$ is

$$\Phi=-v_0 a\{\sinh \xi+\tfrac{1}{2}(1+\exp 2\xi_0)\exp(-\xi)\}\sin \eta.$$

The components of the fluid velocity are

$$v_\xi=-(1/h)\partial\Phi/\partial\xi, \qquad v_\eta=-(1/h)\partial\Phi/\partial\eta$$

where h is the scale factor $(a/2)^{\frac{1}{2}}(\cosh 2\xi-\cos 2\eta)^{\frac{1}{2}}$.

For a steady flow the fluid pressure follows from (8.70) as

$$p=p_0+\tfrac{1}{2}\rho(v_0^2-v^2),$$

where p_0 is the fluid pressure at infinity where the velocity is v_0. Hence, on the boundary of the cylinder, where $v_\xi=0$,

$$p-p_0=\tfrac{1}{2}\rho v_0^2\left\{1-\frac{2(\sinh \xi_0+\cosh \xi_0)^2 \cos^2\eta}{\cosh 2\xi_0-\cos 2\eta}\right\} \qquad (9.107)$$

The pressure at the two stagnation points ($x=0$, $y=\pm a \sinh \xi_0$; $\xi=\xi_0$, $\eta=\pi/2$) is higher than the pressure at the sides ($x=\pm a \cosh \xi_0$, $y=0$; $\xi=\xi_0$, $\eta=0$, π) by the amount

$$\Delta p=\tfrac{1}{2}\rho v_0^2(1+\coth \xi_0)^2. \qquad (9.108)$$

It is this pressure difference that tends to flatten the otherwise circular cross-section of the twisted flux rope. For a given v_0 the pressure difference increases monotonically with the degree of flattening, from $2\rho v_0^2$ for a circle ($\xi_0\to\infty$) to $\rho v_0^2/2\xi_0^2$ for a very eccentric ellipse ($\xi_0\to 0$). The large pressure difference for the eccentric ellipse is principally the consequence of the enormous flow velocity around the extreme edges, giving a large pressure reduction there.

Now consider the equilibrium cross-section of the slender twisted flux tube extending across the flow. The tube is flattened to some degree by the pressure differences exerted on the tube by the flow. Suppose, then, that in polar coordinates the cross-section of the tube is

$$R(\phi)=\sum_{n=0}^{\infty} R_n\epsilon^n \cos 2n\phi,$$

where ϵ is some small parameter. We suppose that the flow does not greatly flatten the tube, so that to a first approximation the first two terms of the series are an adequate representation,

$$R(\phi)=R_0+\epsilon R_1 \cos 2\phi. \qquad (9.109)$$

To this order the boundary is an ellipse, with semi-major axis $R_0+\epsilon R_1$ in the x-direction, and semi-minor axis $R_0 - \epsilon R_1$ in the y-direction, corresponding to

$$\coth \xi_0 = 1 + \epsilon R_1/R_0 + O(\epsilon^2)$$

or

$$\xi_0 \cong \tfrac{1}{2} \ln(2R_0/\epsilon R_1).$$

The pressure difference (9.108) is, then,

$$\Delta p \cong 2\rho v_0^2 \{1 + O(\epsilon)\}. \tag{9.110}$$

The field within the tube follows from solution of (6.9) after choosing the form for the total pressure

$$F(A) = p + B_z^2/8\pi$$

of the fluid and the longitudinal field B_z. The vector potential A yields the transverse components of the field

$$B_\varpi = (1/\varpi)\partial A/\partial\phi, \qquad B_\phi = -\partial A/\partial\varpi$$

in the cross-section as a result of the twisting and flattening of the tube. The total pressure declines monotonically with increasing radius ϖ because it is opposed by the inward force of the tension in B_ϕ (see (6.10)–(6.14)). To guarantee this, regardless of the sign of A, write

$$8\pi F(A) = \beta_0^2 - \sum_{n=1}^{\infty} \beta_n^2 A^{2n}.$$

To keep the mathematics as simple as possible, truncate the series after $n = 1$, so that (6.9) becomes the linear equation

$$\nabla_{xy}^2 A - \beta_1^2 A = 0. \tag{9.111}$$

The solution that is finite at the origin and tangential to the boundary (9.109) is[10]

$$A = B_\perp R_0 \{I_0(\beta_1\varpi) - \epsilon C I_2(\beta_1\varpi)\cos 2\phi\} \tag{9.112}$$

where

$$C \equiv \beta_1 R_1 I_1(\beta_1 R_0)/I_2(\beta_1 R_0).$$

Then

$$B_\varpi = 2\epsilon C B_\perp (R_0/\varpi) I_2(\beta_1\varpi)\sin 2\phi,$$
$$B_\phi = B_\perp \beta_1 R_0 \{-I_1(\beta_1\varpi) + \epsilon C I_2'(\beta_1\varpi)\cos 2\phi\}.$$

[10] The field eqn (9.111) can be solved exactly for the elliptical boundary of arbitrary eccentricity using elliptical coordinates. The solution involves Mathieu functions of imaginary argument—the modified Mathieu functions. The elegance of the exact solution seems not worth the enormously increased computational effort, since the boundary is elliptical only to first order in ϵ anyway.

Neglecting terms second order in ϵ the pressure of the magnetic field B_s at the boundary is just $B_\phi^2/8\pi$, which to first order is

$$\frac{B_s^2}{8\pi}=\frac{B_\perp^2}{8\pi}(\beta_1 R_0)^2\{I_1^2-2\epsilon C(I_1 I_2'-I_2 I_1')\cos 2\phi\}, \tag{9.113}$$

where the modified Bessel functions are understood to have the argument $\beta_1 R_0$. The boundary condition is, of course, that the total pressure be continuous across the boundary. Hence the magnetic pressure must have the same variation around the cylinder as the external fluid pressure (9.107). For small ϵ (large ξ_0), we have $\eta \cong \phi$, so that it is necessary only to write $2\cos^2\eta \cong 1+\cos 2\phi$ in (9.107) and equate coefficients of $\cos 2\phi$ in (9.107) and (9.113). The result is the equilibrium eccentricity

$$\begin{aligned}\epsilon\frac{R_1}{R_0}&=\frac{4\pi\rho v_0^2 I_2}{B_\perp^2(\beta_1 R_0)^3 I_1(I_1 I_2'-I_2 I_1')}\\&\cong\frac{4\pi\rho v_0^2 I_1 I_2}{B_s^2\beta_1 R_0(I_1 I_2'-I_2 I_1')}.\end{aligned} \tag{9.114}$$

The circumstance that $B_z = constant$ and β_1 is so small that $\beta_1 R_0 \ll 1$ represents a flux tube with a uniform twist[11],

$$B_\phi = -B_s\varpi/R_0$$

with

$$B_s=\tfrac{1}{2}(\beta_1 R_0)^2 B_\perp.$$

The lines of force all make one revolution around the tube in the distance $\lambda = 2\pi B_z/B_0$. The streaming of fluid past the tube, i.e. the buoyant rise of the tube through the fluid, produces the eccentricity

$$\epsilon R_1/R_0=(4\pi\rho v_0^2/B_s^2)\{1+O(\beta_1^2 R_0^2)\}. \tag{9.115}$$

The eccentricity is small if the rate of rise v_0 is small compared to the Alfven speed computed for the azimuthal component of the field in the tube and the external density ρ. When the rate of rise becomes comparable to the Alfven speed, the eccentricity becomes large, and, of course, our approximation $\epsilon R_1 \ll R_0$ becomes poor. In a stable atmosphere (see §8.8) the rate of rise is quite small, so that the tube is not strongly flattened and (9.115) is a good approximation. In an atmosphere that is

[11] In this case the pressure within the tube is $p(\varpi)=F(A)-B_z^2/8\pi \cong \{\beta_0^2-\beta_1^2 R_0^2 B_\perp^2(1+\tfrac{1}{2}\beta_1^2\varpi^2)-B_z^2\}/8\pi$ to lowest order in ϵ and $\beta_1 R_0$. The external pressure is $p(R_0)+(B_s^2+B_z^2)/8\pi$ so that the reduction in gas pressure within the tube is $\Delta p(\varpi)=\{B_z^2-B_s^2+2B_s^2\varpi^2/R_0^2\}/8\pi$. The magnetic buoyancy per unit length is then readily shown to be $\mathscr{F}=(\pi R_0^2/\Lambda)B_z^2/8\pi$, which is identical with (8.58) so that the rate of rise is given by (8.60).

convectively unstable (see §8.7) the rise may be a significant fraction of the Alfven speed, producing significant flattening.

We should go on to say that the convectively unstable atmosphere will undoubtedly be in a state of turbulent convection, as in the convective zone of the sun, so that the rapidly rising tube is buffeted by the irregular convective motions and the rate of rise retarded to some degree by the effective turbulent viscosity (see discussion in §8.8 and Goldstein 1938). The tube that is not strongly twisted will be subject to fragmentation as a result of both the local turbulence and the flattening caused by the buoyant rise.

We have noted that the flux ropes emerging through the surface of the sun are not always strongly twisted. Considering the enormous radial expansion they have undergone this would imply that either the twisting has been enormously reduced in the apex by rapid line cutting, or the ropes are not strongly twisted beneath the surface. Without strong twisting beneath the surface, it is not surprising that the emerging flux ropes appear fragmented into isolated tubes, down to the limit of resolution of the observing magnetograph.

9.9. Summary

In this chapter we have explored the behaviour of the twisted flux rope when its cross-section varies with time or with position along the rope. The outstanding characteristic is the increased pitch angle of the lines of force associated with an increase in radius of the tube. The expanding radius, then, quickly reduces the tension in the rope to zero and below, so that there is no tension available to maintain a taut configuration of the tube or to oppose the magnetic buoyancy. The buoyancy carries the tube upward through the atmosphere, further expanding the tube and aggravating the effect. At the same time, the reduction of the net tension in the tube makes it possible for the tube to relieve some of its internal torsion by taking up a corkscrew form, leading to rapid line cutting etc. if the corkscrew becomes so extreme that adjacent coils touch.

The calculations so far are limited to static conditions, from which we demonstrate the instability and non-equilibrium of the flux rope. The dynamical nature of the buoyant rise, once the tension $\mathcal{T}$ no longer overpowers the buoyant force $\mathcal{F}$, and once the tube takes up a corkscrew form, is beyond the scope of the present work. It depends upon the total torsion along the tube and on the rate at which the coils, propagating as Alfven waves, shift the torsion along the tube in response to the varying torque. We can only conjecture that the rise of a tube through a convecting atmosphere is a rapid affair, with complicated fragmentation of the tube. The conjecture is consistent with the observation (Harvey *et al.* 1975) of complex magnetic field configurations emerging through the solar photosphere.

Generally speaking, magnetic fields are generated deep within astrophysical bodies by the non-uniform rotation and cyclonic convection (Parker 1955b, 1971a,b). They escape by turbulent diffusion and by their own intrinsic magnetic buoyancy (see, for instance, Parker 1955a, 1975b). Hence, the fields observed at the surface of stars are transformed by their expansion before arrival at the surface. Their torsion has been enhanced by the radial expansion, and perhaps reduced by rapid line cutting. The modest torsion exhibited by the flux tubes emerging through the surface of the sun suggests that there is relatively little torsion in the flux tubes at depth. It is frustrating that we cannot see into the sun to check this theoretical conclusion.

It is amusing to note that the galaxy is an exception, it being the one transparent astrophysical body with its own intrinsic field. We see its internal fields, as they are generated by the turbulent interstellar medium. Indeed, in some respects the interstellar gas is *too* transparent. We see so far that the integrated signal along the line of sight contains little or no information on the local fluctuations of the field; we have little or no quantitative information on the extension of the fields outside the galaxy where they evidently form a galactic halo.

Altogether it is clear that an astrophysical body with internal turbulence or convection, and subject to non-uniform rotation, so that it generates a magnetic field within itself (see Chapter 19), can have no peace, for the magnetic organism that the body creates within itself has no equilibrium. The tendency for the field to escape its host is the cause of much, if not all, of the magnetic activity of the sun and other stars, and perhaps of some galaxies. The unstable, non-equilibrium state of any flux rope inevitably produces the line cutting and rapid dissipation that we call magnetic activity, with its production of superheated gases and fast particles (Parker 1975a, b; Glencross 1975).

References

Cheng, C. C. (1977). *Astrophys. J.* **213,** 558.

Furth, H. P., Rutherford, P. H. and Selberg, H. (1973). *Phys. Fluids* **16,** 1054.

Glencross, W. M. (1975). *Astrophys. J.* **199,** L53.

Goedbloed, J. P. and Sakanaka, P. H. (1974). *Phys. Fluids* **17,** 908.

Goldstein, S. (1938). *Modern developments in fluid dynamics*, vol. 1, pp. 15, 16. Clarendon Press, Oxford.

Harvey, K. L., Harvey, J. W., and Martin, S. F. (1975). *Solar Phys.* **40,** 87.

Jeffreys, H. and Jeffreys, B. S. (1950) *Methods of mathematical physics*, p. 420. Cambridge University Press.

Kadomtsev, B. B. (1966). *Rev. Plasma Phys.* **2,** 153.

KRUSKAL, M. and TUCK, J. L. (1958). *Proc. Roy. Soc. Ser. A* **245,** 222.
MOLODENSKY, M. M. (1974). *Solar Phys.* **39,** 393.
—— (1975). *ibid.* **43,** 311.
—— (1976). *ibid.* **49,** 279.
MORSE, P. M. and FESHBACH, H. (1953). *Methods of theoretical physics*, vol. II, pp. 1195–1206. McGraw-Hill, New York.
NAKAGAWA, Y., RAADU, M. A., BILLINGS, D. E., and MCNAMARA, D. (1971). *Solar Phys.* **19,** 72.
——, —— and HARVEY, J. W. (1973). *Solar Phys.* **30,** 421.
PARKER, E. N. (1955a). *Astrophys. J.* **121,** 491.
—— (1955b). *ibid.* **122,** 293.
—— (1971a). *ibid.* **163,** 255.
—— (1971b). *ibid.* **166,** 295.
—— (1974). *ibid.* **191,** 245.
—— (1975a). *ibid.* **201,** 494.
—— (1975b). *ibid.* **202,** 523.
—— (1976). *Astrophys. Space Sci.* **44,** 107.
RAADU, M. A. (1972). *Solar Phys.* **22,** 425.
ROBERTS, P. H. (1956). *Astrophys. J.* **124,** 430.
SAKANAKA, P. H. and GOEDBLOED, J. P. (1974). *Phys. Fluids* **17,** 919.
SHAFRANOV, V. D. (1957). *J. Nucl. Energy II* **5,** 86.
—— (1970). *Soviet Phys. Tech.* **15,** 175.
VAIANA, G. S., KRIEGER, A. S., and TIMOTHY, A. F. (1973). *Solar Phys.* **32,** 81.
WARE, A. A. and HAAS, F. A. (1966). *Phys. Fluids* **9,** 956.

10

THE ISOLATION AND CONCENTRATION OF MAGNETIC FLUX TUBES

10.1. The appearance of isolated flux tubes in nature

THE slender flux tubes treated in Chapters 7, 8, and 9 are a convenient device for discovering and studying the basic properties of magnetic fields. Any broad continuous field $B_i(r)$ may be decomposed into a family of close-packed, slender, elemental tubes of flux, so that the properties of the individual tube are part of the overall structure of any broad distribution of field. Many of the basic characteristics of the thick twisted tube are illustrated by the slender twisted tube. However, it has gradually come to light in the past decade that the isolated slender flux tube has an existence in its own right.

It was discovered over two decades ago that the sun has a general magnetic field, of some 1–10 G at the poles and, in a more irregular pattern, over much of the middle and low latitudes (Babcock and Babcock 1955; Altschuler *et al.* 1974). More recently it has been discovered that these broad field distributions are not what they seem. Rather than the broad continuous distribution recorded by the magnetometer, the general magnetic field of the sun is made up of many small, very intense, and widely separated flux tubes. The individual tubes in quiet regions are too small to be resolved, but the evidence is now convincing, based on comparative studies of the Zeeman broadening of the profiles of different lines, that most of the net magnetic flux through the solar photosphere is concentrated into small intense tubes of flux located in the boundaries of the granules and supergranules (Leighton 1963; Simon and Leighton 1964; Sheeley 1967, Beckers and Schröter 1968; Livingston and Harvey 1969, 1971; Sawyer 1971; Simon and Noyes 1971; Howard and Stenflo 1972; Frazier and Stenflo 1972). The observational questions are numerous, not the least of which is the direct determination of the magnetic field in the individual flux tubes. It is important to know whether the field is 2000 G, with a pressure of $1{\cdot}6\times10^5$ dyn cm^{-2}, as suggested by Livingston and Harvey (1969), Stenflo (1973), and Chapman (1973) or whether it is 'only' 1500 or 1000 G, with pressures of $0{\cdot}9\times10^5$ and $0{\cdot}4\times10^5$ dyn cm^{-2} respectively. The accumulation of evidence is 1500–2000 G, but firm confirmation is desirable.

The concentration of field into isolated flux tubes occurs spontaneously, in opposition to the considerable magnetic pressure of the concentrated field. The phenomenon challenges our understanding of the basic physics of flux tubes. The present chapter explores the thermal and dynamical effects in flux tubes extending vertically across a turbulent convective zone in order to establish an understanding of the effect.

In the sun there is very little of the net magnetic flux that escapes concentration into isolated tubes. The estimates are that at least 90 per cent of the flux in active regions is concentrated into the individual flux tubes forming the magnetic knots in the dark intergranule lanes (Beckers and Schröter 1968) and an even larger fraction in quiet regions in the supergranule boundaries (Howard and Stenflo 1972; Frazier and Stenflo 1972). In addition to the net flux in the isolated tubes, there is everywhere present a fluctuating background field of ± 1–3 G (Livingston and Harvey 1969; Frazier and Stenflo 1972). The ephemeral bipolar regions (Harvey *et al.* 1975) associated with the x-ray bright points (Vaiana *et al.* 1973; Golub *et al.* 1974) are among the more singular features of background field, each representing the emergence of a small isolated and concentrated flux tube through the surface of the sun (see Chapters 7 and 9 for discussion of the basic physical principles).

The total flux in the individual filament in the supergranule boundaries is estimated to be of the order of 3×10^{18} Mx, although there is a suspicion that this number, inferred from observations, may represent the total of two or more neighbouring flux tubes rather than a single tube. In that case 10^{18} Mx would appear to be closer to the correct value. The tubes evidently involve field strengths of 1500–2000 G (see the development of this question in Livingston and Harvey 1969, 1971; Sawyer 1971; Howard and Stenflo 1972; Stenflo 1973; Chapman 1973). Thus, a total flux of 1–3×10^{18} Mx implies a tube radius of 100–250 km. It is suggested that the individual tubes are remarkably alike in size and form. The field spreads out rapidly with height above the photosphere (Simon and Noyes 1971; Frazier and Stenflo 1972), and Frazier and Stenflo suggest that each tube may be surrounded by a local background field of opposite sign. However, from the fact that the tubes together carry the net flux through any general region of the surface of the sun, we must presume that the returning local background field of each tube represents only a fraction of the total flux in the tube.

Frazier (1970) points out that the individual magnetic flux tubes are concentrated in the corners where three or more supergranules meet. He observes downdrafts of the order of $0{\cdot}1$ km s^{-1} in the corners in coincidence with the fields and with the bright spots in the Ca^{+} network. The concentration of fields in the corners is not surprising, since a 'cork' floating on the surface of the sun would be swept into the downdrafts in the corners and trapped at the surface there.

If the fields in the magnetic filaments in a given region of the photosphere all have the same sign, then a typical mean field of 3 G over a broad quiet region requires one flux tube of 3×10^{18} Mx for each 10^{18} cm^2 of surface area. The characteristic separation of neighbouring tubes is then 10^4 km (to be compared with tube diameters of 2×10^2 km and supergranule diameters of 2–3×10^4 km). The surface area of the sun is 6×10^{22} cm^2, suggesting a total of nearly 10^5 isolated flux tubes through the surface of the sun at any one time.

The restriction of the flux tubes to the junctions of supergranule boundaries, and the total number of 10^5, suggest that the individual magnetic flux tubes are closely associated with the spicules and with the fine mottles observed in H_α. The facts are that the flux tubes and spicules are both concentrated in the Ca network in the supergranule boundaries, and particularly in the corners where three or more supergranules meet (Frazier 1970). The spicules and flux tubes appear in about equal numbers, there being about 10^5 spicules (Beckers 1963; Gibson 1973; Lynch *et al.* 1973) and 10^5 flux tubes extending outward through the surface of the sun at any given time. The concentration of the spicule jet cannot be explained in the absence of strong magnetic field (Parker 1964; Unno *et al.* 1974), so the spicule presumably occurs in a vertical flux tube. But 90 per cent or more of the magnetic flux is concentrated in the individual flux tubes, with only relatively weak fields in the broad regions between. There seems to be no choice, then, but to believe that the spicules are to be identified with the individual isolated flux tubes. Evidently there is a recurring eruption of gas upward along each filament every 10 min or so ($0{\cdot}5$–$1{\cdot}0\times10^3$ s) to produce the observed spicule phenomenon (Parker 1964, 1974b).

The final point is, then, that Beckers (1963) has shown that the fine mottles observed in H_α are, in fact, the spicules. If association of the spicules with the isolated flux tubes is correct, it follows that the H_α mottles are photographs of the individual flux tubes, as outlined by the gases trapped within them (Parker 1974b). Veeder and Zirin (1970) and Zirin (1972) have pointed out that high resolution H_α pictures generally show the detailed structure of the magnetic field in the solar photosphere and chromosphere. To this should be added the more specific statement that photographic pictures of mottles against the disc of the sun, or spicules at the limb, may, in fact, be pictures of the magnetic flux tubes that transport more than 90 per cent of the total flux through the photosphere.

In active regions on the sun, and particularly in association with sunspots, the individual flux tubes are of somewhat larger diameter than in quiet regions, and have been identified by Beckers and Schröter (1968) in association with a sharp decrease in the depth of some of the spectral lines. In this way they were able to pinpoint the magnetic flux tubes or knots on the visible surface of the sun. They showed that the knots have radii of about 500 km and field strengths of 1100–1400 G (10^{19} Mx). Most of the magnetic knots were observed in the dark intergranule lanes in association with downdrafts. They observed some 2×10^3 magnetic knots associated with a large unipolar spot, within a distance of $1{\cdot}4 \times 10^5$ km of the spot. The individual knots showed both positive and negative polarities, with an overall preference by about two to one for the polarity opposite to the sunspot. Thus the net flux carried by the 2×10^3 knots was 7×10^{21} Mx, about equal in magnitude, but opposite in sign, to the flux through the sunspot. It appears that knots and pores may be closely related, the dark pore being the appearance of a magnetic knot whose field has increased to 1500 G while its radius has grown to 750 km or more. Zwaan (1978) in a recent paper points out that the typical bipolar sunspot region grows through the continued emergence of individual tubes of force between the two regions of opposite polarity. Each small tube of force breaks through the surface in a concentrated form, forming two magnetic knots of opposite polarity. During the growth phase of the bipolar region the knots move toward other knots and pores of the same polarity, their continual coalescence building up patches of field to form the final large sunspots. The formative stage ends when the emergence of new flux ceases. After that time the individual magnetic knots cease to show the strong affinity for other knots of the same polarity, and generally move apart. Sunspots decay largely by the departure of small knots and pores from their perimeter. There is, then, a close association between magnetic knots, pores, and sunspots. The connection is evolutionary, with the spot forming from the coalescence of knots and eventually breaking up into knots when its life has run its course. All the stages of development from knots to spots involve the remarkable concentration of field to densities in excess of 10^3 G. The mutual attraction and coalescence of knots of like polarity is surprising in view of the net repulsion of their fields above the surface of the sun. That effect was treated in §8.9. The present chapter is concerned with the spontaneous concentration of the magnetic field within a single flux tube.

Now the observation that the magnetic field extending through the surface of the sun is concentrated into isolated and intense flux tubes is one of the most remarkable discoveries of new physics in the past decade. The energy state of the field at the surface of the sun is some 10^5 times higher than the expected broad distribution allowed by the overall mean; the field is concentrated to about 1500 G in regions where, typically, the mean field is only 5 G or so. The concentration is the rule rather than the exception and defies conventional notions on magnetic fields, whose behaviour in the laboratory is always to spread out under the influence of the isotropic pressure $B^2/8\pi$ to fill all the space allowed by the tension $B^2/4\pi$ along the lines of force. In a passive fluid the molecular diffusivity, and more important, the hydromagnetic exchange instability, disperse the field, permitting a trend toward degradation of the field energy and an approach to homogeneity. We might also expect turbulent diffusion to produce a similar trend toward reducing the extraordinary field energies of the concentrated state observed in the sun. But nature would not have it so, apparently. In the one layer of the sun that we can observe—the photosphere—the opposite prevails.

The problem presents itself in two stages. The first stage is the question of why the regions of the sun organize themselves into those with field and those without. The second stage then, is why are the fields so incredibly intense where they exist? To consider the first question, there are two lines of thought. The first is that the fields that appear at the surface of the sun represent the magnetic debris escaping from the broad fields generated deep in the convective zone by the solar dynamo (Parker 1975a; §8.7). On the basis of the general Rayleigh–Taylor instability of any magnetic field in a compressible medium (see Chapter 13) we would expect some gross fragmentation of the fields escaping from any astronomical body. The fragments appear briefly at the surface as they escape into space (Parker 1975b; §13.7). But it is not clear that the simple idea of escaping magnetic debris can explain why many regions are so completely devoid of magnetic field. A more general muddle would be expected.

Is there, then, some additional hydromagnetic effect that causes field-free fluid and field-impregnated fluid to separate, progressively concentrating the field into isolated bundles with field-free fluid between? For instance, a steady rotation ω of a finite volume of fluid eventually expels the magnetic field from the fluid, over a period less than the diffusion time (Parker 1963; Weiss 1966; see Chapter 16). The effect is easy to understand. The large-scale external magnetic field is, in fact, only an alternating magnetic field $\exp(i\omega t)$ when viewed from the rotating volume. So any initial internal field decays away and the external field penetrates only the characteristic skin depth $(\eta/\omega)^{\frac{1}{2}}$, where η is the effective diffusion coefficient. The same effect applies to a volume subject to random, rather than steady, rotation. From the frame of reference of the rotating volume, the external field fluctuates in a random manner and so does not penetrate far into the volume. Is it true, then, that there are extensive field-free volumes of fluid in the sun that remain field-free because they circulate in closed regions and so admit no fields, maintaining magnetic purity through hydrodynamic apartheid?

Clearly the interior of an element of fluid participating in turbulent mixing cannot remain uncontaminated by exterior fields. The element of volume is drawn into a ribbon of ever-diminishing thickness (Reid 1955; Batchelor 1955) so that its interior is increasingly exposed to the exterior. But might not there be relatively closed circulations on the sun, as there seems to be in some planetary atmospheres, as suggested by the bands and the red spot on Jupiter? There is no clear answer to this question at the present time. As we shall see, the second stage by which the fields are so extraordinarily concentrated, plays a role in the first stage. That concentration effect aids any natural hydrodynamic apartheid.

Altogether, then, it seems that there may be several contributing factors, starting with the fragmentation of the magnetic debris as it comes up from the broad fields generated in the lower convective zone. Then the granules and supergranules are localized and coordinated in some degree, so that the magnetic tubes of flux are swept into the downdrafts between the rising cells by the generally converging flows across the visible surface of the sun. Some further mechanism then concentrates the fields in the downdrafts to 1500 G so that the field resists being mixed freely about by the shifting patterns of circulation of the granules and supergranules. In this way the convective overturning, with the help of the concentrating mechanism, excludes the field from most of the fluid without necessarily having a circulation pattern that is entirely closed. Once the lines of force are delivered into the downdraft the field (for its own reasons elaborated in §10.11) shrinks from the passing fluid into a state of hydrodynamic isolation and magnetic concentration.

Consider then, the formation of the intense isolated flux tubes on a step-by-step basis. Concentration of field into the downdrafts at the boundaries of supergranules was discovered years ago, and is to be understood as a direct consequence of the sweep of the convection across the surface of the supergranule toward the boundaries (Leighton *et al.* 1962; Noyes and Leighton 1963; Simon and Leighton 1964; Parker 1963, 1973; Clark 1966, 1968; Wentzel and Solinger 1967). The supergranule velocities v of 0.3 km s^{-1} at the visible surface (where $\rho = 3\times 10^{-7}$ g cm^{-3}) yield a dynamical pressure $\frac{1}{2}\rho v^2 = 1{\cdot}3\times 10^2$ dyn cm^{-2}. They might be expected to compress the fields in the boundaries to a comparable pressure, $B^2/8\pi = \frac{1}{2}\rho v^2$, or about 55 G. The magnetic knots are found in the downdrafts in boundaries between granules, where the converging flows may be 3 km s^{-1}. The dynamical pressure is $1{\cdot}3\times 10^4$ dyn cm^{-2}, perhaps compressing the field to 550 G. But this is nowhere near the field strengths of 1500 G or more inferred from observation, whose pressures are ten times larger. Magnetic fields of 1500 G have pressures of $0{\cdot}9\times 10^5$ dyn cm^{-2}, while the gas pressure at the visible surface of the photosphere is about 2×10^5 dyn cm^{-2} rising to 5×10^5 dyn cm^{-2} at a depth of 200 km beneath the surface. Thus the magnetic pressure is a substantial fraction of the ambient gas pressure. The gas compresses the field only insofar as the total pressure outside the flux tube exceeds the gas pressure inside. The confinement may, therefore, be attributed to an enhanced dynamical pressure outside, or a reduced pressure inside, or a combination of the two.

It is not uncommon to encounter the suggestion that the flux tubes are strongly twisted, in such a way as to be force-free, $(\nabla \times \mathbf{B}) \times \mathbf{B} = 0$, and exert no force on the fluid, thereby 'explaining' their intense concentration. It was shown in the general development of twisted flux tubes in §9.1 that twisting does not concentrate the field in a suitable manner. For a given confining pressure increment Δp at the surface of the tube, the mean longitudinal field *declines* as the tube is twisted. The *peak* longitudinal field, on the axis of the tube, can be increased, but the *mean* longitudinal field *decreases* (see discussion at the end of §9.1) and the 1500–2000 G inferred from observations is believed to refer to the mean, rather than the peak, field.

To consider the real possibilities, there are the dynamical effects of the external and internal fluid motions that may contribute to the compression of the field. There is the possibility of a temperature reduction inside, so that the upper end of the flux tube, at the visible surface of the sun, is evacuated by gravity. The basic physics of the dynamical effects of both external turbulence and longitudinal flow in a flux tube is illustrated in §§10.3–10.9 with a number of idealized examples. All these effects occur in varying degrees in the flux tubes in the sun, and in other stars, and probably also in the gaseous disc of the galaxy. The degree to which they contribute to compressing the field depends quantitatively on the fluid velocities, because none of them produce fields significantly in excess of the equipartition value $B^2 = 4\pi\rho v^2$. Thus, for instance, if there are downdrafts in the flux tubes in the sun of 6–8 km s^{-1}, the concentration to 1500 G may be a direct dynamical consequence. However, until suitably strong downdrafts are observed, such ideas are only conjectural. It is an open observational question as to what effects of concentration arise in the magnetic field of the galaxy.

Consider, then, the possibility that a reduced gas temperature within the flux tube may be a prime factor in the concentration of flux tubes in the sun (see §8.3). In that case gravity evacuates the interior of the tube, so that the magnetic field is compressed by the pressure p of the surrounding atmosphere, as described by (8.10) and (8.11). The asymptotic field strength at great heights in the atmosphere is $B^2/8\pi \sim p$. It is clear, then, that cooling might be the cause of the observed concentration (although we are then confronted with the question as to the cause of the cooling). A 20 per cent reduction of temperature within the flux tube over five pressure scale heights (about 1500 km measured downward from the visible surface of the sun) would produce the desired concentration of the field at the surface. Incidentally, in this circumstance the strongest field concentration occurs unobserved at a depth of three or four scale heights below the surface. So slight a temperature reduction ($\lesssim 10^3$ K) over the 400-km diameter of the flux tube, partially obscured by the overlying photosphere, would be inconspicuous among the 500° temperature variations of the granules.

When we inquire as to the source of cooling, there are several possibilities. It has been suggested that the generation of Alfven waves through convective overstability in the first thousand or so km beneath the surface of the sun may remove a significant amount of heat from the flux tube (Parker 1976a; Roberts 1976), along the lines proposed earlier for the cooling and compression of the magnetic fields of sunspots and pores (see Parker 1974c; and references therein). There is a simpler effect, however, stemming directly from the observed downdraft and the superadiabatic temperature gradient beneath the photosphere, that produces strong cooling and is described briefly in §§10.10 and 10.11. The effect appears to be unavoidable and, consequently, would seem to be the principal cause of the spontaneous concentration of the flux tubes in the sun to such extreme values. By inference, the same effect is to be expected in any star cool enough to have a strong convective zone beneath its surface. Consider, then, the variety of the effects in a flux tube in a convecting fluid.

10.2. The concentration of magnetic field in converging flows

The surface concentration of the vertical magnetic field in the downdraft between two convective cells, sketched in Fig. 10.1, is expected on the basis of elementary hydromagnetic theory (Parker 1963, 1973). The lines of force move with the fluid, so that they are swept toward the downdraft. The field offers no sensible resistance to the concentration until the magnetic stress density $B^2/8\pi$ approaches the Reynolds stress density $\frac{1}{2}\rho v^2$ of fluid. Hence for weak fields ($B^2 \ll 4\pi\rho v^2$) the accumulation of field is only a kinematical problem, adequately described by the hydromagnetic equation

$$\partial \mathbf{B}/\partial t = \nabla \times (\mathbf{v} \times \mathbf{B}) \tag{10.1}$$

in which the velocity $\mathbf{v}$ is specified without regard for $\mathbf{B}$. It is sufficient for the present purposes of illustration to limit the discussion to two dimensions. Then let y be the vertical coordinate and x the horizontal coordinate, with the magnetic field expressed as

$$B_x = +\partial A/\partial y, \qquad B_y = -\partial A/\partial x.$$

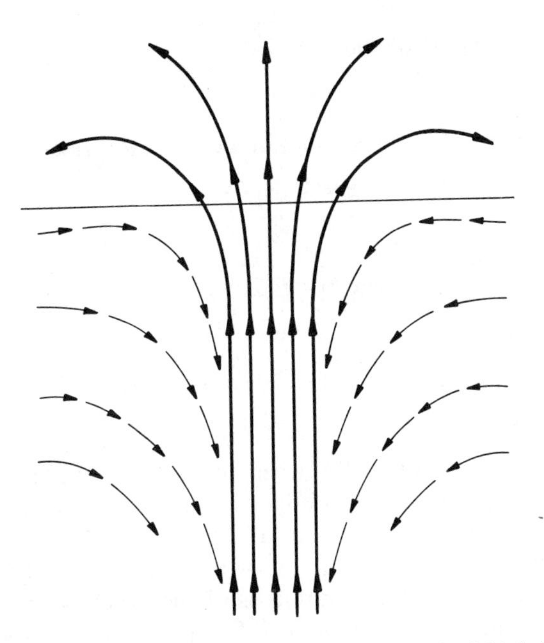

FIG. 10.1. A sketch of the concentration of vertical magnetic field (solid lines) in the downdraft (broken lines) between two supergranules.

Then (10.1) can be integrated once to give

$$\frac{dA}{dt} \equiv \frac{\partial A}{\partial t} + v_x \frac{\partial A}{\partial x} + v_y \frac{\partial A}{\partial y} = 0. \tag{10.2}$$

The general solution follows immediately in terms of the Lagrangian coordinate $x_i(X_k, t)$ at time t of the element of fluid initially at X_i,

$$A(x_i, t) = A(X_i, 0). \tag{10.3}$$

Consider, then, some illustrative examples. Suppose that the magnetic field is initially vertical, so that $B_x = 0$, $A = F(x)$, $B_y = -F'(x)$, in a steady converging flow of the form $v_x = v_x(x)$. Conservation of fluid requires that the y-component of the velocity satisfy the equation

$$\partial \rho v_x / \partial x + \partial \rho v_y / \partial y = 0.$$

But v_y does not enter into the solution (10.3) of (10.2) because A is initially independent of y. The x-coordinate of an element of fluid is given by

$$t = \int_X^x dx / v_x(x).$$

The converging flow is of the form

$$v_x = -x/\tau$$

in the neighbourhood of the stagnation point, from which it follows that

$$x = X\exp(-t/\tau),$$

and

$$A(x, t) = F(x\exp(t/\tau)).$$

Hence B_x remains zero and

$$\begin{aligned} B_y(x, t) &= -\exp(t/\tau)F'(x\exp(t/\tau)) \\ &= \exp(t/\tau)B_y(x\exp t/\tau, 0). \end{aligned}$$

The field grows exponentially with time while the horizontal scale $\exp(-t/\tau)$ is compressed exponentially with time. Thus a typical supergranule velocity of $0{\cdot}3\ \mathrm{km\,s^{-1}}$ at a distance of 5×10^3 km from the boundary yields a characteristic growth time of $\tau = 1{\cdot}6\times10^4$ s. The characteristic life of the supergranule is of the order of 10^5 s, so the field is quickly concentrated from a few G up to the 50–100 G that can oppose, and locally halt, the converging flow at the surface of the sun.

As a matter of fact, the supergranule pattern continually drifts with a small velocity $u \cong 0{\cdot}2\ \mathrm{km\,s^{-1}}$, so that it is of interest to consider the converging flow

$$v_x = (x - ut)/\tau$$

that moves steadily with a velocity u. It is readily shown that

$$X = \{x - u(t-\tau)\}\exp(t/\tau) - u\tau.$$

Then

$$B_y(x, t) = \exp(t/\tau)B_y[\{x - u(t - \tau(1 - \exp(-t/\tau)))\}\exp(t/\tau), 0] \tag{10.4}$$

Again the scale of the field diminishes as $\exp(-t/\tau)$ while the field strength increases as $\exp(t/\tau)$. The field also undergoes a net displacement $\{t - \tau(1 - \exp(-t/\tau))\}u$.

In view of the rapid rise of the field and field pressure with the passage of time, it is evident that the simple kinematical approach must very soon become inadequate. To take into account the reaction of the field against the converging flow in a very simple way, suppose that the fluid motion is unaffected until the vertical field reaches a strength B_1, whereupon it arrests the converging motion. Suppose, then, that fluid converges on $x = 0$ from both sides, accumulating a band of field B_1, of thickness $2wt$, where w is a characteristic velocity to be determined from the rate at which flux is carried into the band from either side. Then $v_x = 0$ within the band of field $-wt < x < +wt$ and

$$v_x = \begin{cases} -(x-wt)/\tau & \text{for} \quad x > +wt \\ -(x+wt)/\tau & \text{for} \quad x < -wt \end{cases}$$

on either side. The solution outside the growing band is just (10.4) with u replaced by $\pm w$. Thus, for instance, if the initial field were uniform, with $B_y(x, 0) = B_0 < B_1$, it would reach B_1 simultaneously everywhere after the passage of a time $t = \tau \ln(B_1/B_0)$. If, on the other hand, the initial field declines inversely with distance away from $x = 0$, then write

$$B_y(x, 0) = B_1 w\tau/(w\tau + |x|)$$

so that the band of field of strength B_1 has vanishing width at time $t = 0$. It follows that for subsequent times

$$B_y(x, t) = B_1 w\tau/\{x - w(t - \tau)\}$$

in $x > +wt$, with a similar expression for $B_y(x, t)$ in $x < -wt$. If the field at a distance L from the band is αB_1 where $\alpha < 1$, then

$$w\tau = L\alpha/(1-\alpha).$$

As a specific example, suppose that the limiting field is 60 G, and it is accumulated from a background field of 2 G at a distance of $L = 2 \times 10^3$ km, where the converging flow is $0{\cdot}3$ km s^{-1}. Then it is readily seen that $\alpha = \frac{1}{30}$ and $w\tau = 70$ km with $\tau = 7 \times 10^3$ s so that the rate of growth of the thickness of the band of field is $2w = 2 \times 10^3$ cm s^{-1} reaching a thickness of 4×10^2 km in a time of 2×10^4 s.

These few examples suffice to illustrate the sweeping of weak fields into the supergranule boundaries. Other examples are easily constructed or can be found in the references cited in §10.1. Resistive diffusion of the fluid across the vertical field has been included in the work of Clark (1966, 1968). The reduced electrical conductivity of the solar photosphere ($\sigma \cong 10^9$–10^{10} s^{-1}, Kopecky and Obridko 1968; Oster 1968 and references therein) leads to rapid diffusion in some cases. Indeed the characteristic diffusion velocity $c^2/\pi a\sigma$ across a distance a comparable to the 200-km radius of one of the concentrated flux tubes is $0{\cdot}16$ km s^{-1} in the thin photospheric layer where σ is as small as 10^9 s^{-1}. The crossing time is only 10^3 s. The thin photospheric layer of poorly conducting gas outside the tube pours into the partially evacuated interior like flood waters into a storm sewer. The effect is not without interest and merits further investigation.

There is a question that is the complement of the concentration of magnetic field in the downdraft at the supergranule boundaries, and that is the field brought to the surface by the upwelling at the centre of the supergranule cell. We must ask, then, down to what depth does the supergranule extend? Evidently the main azimuthal field of the sun lies below the bottom of the supergranule. Any part of the field that extends up into the supergranule would presumably be brought to the surface, and packed into the boundary of the supergranule cell. Thus, for instance, the ephemeral bipolar regions appear to be small strands of field whose magnetic buoyancy, together with the upwelling of the supergranule, has brought them to the surface. We have suggested (see §8.7; Parker 1975a) that magnetic buoyancy raises the general azimuthal field of the sun so rapidly to the surface that the field must be limited to 10^2 G or less and produced at depths of $1{\cdot}5 \times 10^5$ km or more. If that is indeed the case, then the supergranules probably do not dip into the azimuthal field (Parker 1973). Only in this way can we explain the generation of magnetic field in the convective core of the sun and the observed relative absence of upwelling field in the supergranule cell.

10.3. A magnetic flux tube in a turbulent fluid

The intense isolated flux tubes in the supergranule boundaries in the solar photosphere extend downward for $1\text{–}2 \times 10^3$ km through the vigorous turbulent convection of the ionization zone below the surface. Thus the first task is to explore the dynamical effects of external turbulence on an isolated tube of flux. We shall, therefore, treat a number of idealized circumstances to discover and illustrate the basic effects that might be expected. It is premature to construct models of the phenomenon as it actually occurs in the sun, for reasons that will become clear below, but it is advisable to have the basic facts and figures in mind so that we can judge the effectiveness of the various dynamical processes as we work them out in the succeeding sections.

The vertical and horizontal components of the gas velocities in the granules and supergranules are observed only at the visible surface of the sun. We are, then, free to guess how they vary with depth. Waddell (1958) inferred from his observations of the granules that the vertical component of the motion at the surface has r.m.s. values of $1{\cdot}8\ \mathrm{km\,s^{-1}}$ while in the horizontal direction the r.m.s. velocity is $3{\cdot}0\ \mathrm{km\,s^{-1}}$, giving a total r.m.s. velocity of $3{\cdot}5\ \mathrm{km\,s^{-1}}$. Beckers's (1968) observations indicate $3\ \mathrm{km\,s^{-1}}$ r.m.s., or $6\ \mathrm{km\,s^{-1}}$ velocity differences between granule centres and boundaries. The motions observed in supergranules are horizontal flow of the order of $0{\cdot}4\ \mathrm{km\,s^{-1}}$ (Leighton 1963). There are no observations that allow direct inference of convective or turbulent velocities below the visible surface. The mixing length theory of the convective zone (Böhm–Vitense 1958; Mihalas 1965; Spruit 1974) suggests numbers of the general order of magnitude of $1\ \mathrm{km\,s^{-1}}$. The important point is that the observed 3 km s^{-1} in the photosphere where $\rho \sim 3 \times 10^{-7}$ g cm^{-3} yields Reynolds stresses (or energy density) of the order of $\frac{1}{2}\rho v^2 = 1{\cdot}3 \times 10^4$ dyn cm^{-2}, comparable to the magnetic pressure $B^2/8\pi$ of about 600 G. Thus the surface velocities presently inferred from observation lead to dynamical pressures that are inadequate by a factor of ten to compress the fields to the 1500–2000 G inferred from observations. There are, however, the two possibilities, that (a) increased resolution may discover much higher gas velocities in concentrated regions (see, for instance, Beckers 1976; Deubner 1976) and/or (b) the velocities at the surface may also exist at depths of 10^3 km where the fluid density is $\rho \sim 10^{-5}$ g cm^{-3} and 2 km s^{-1} yields dynamical pressures of 2×10^5 dyn cm^{-2}, equivalent to 2200 G (Meyer *et al.* 1974). With this brief review of the observational facts and inferences of the dynamical situation in the solar photosphere, we proceed to the theoretical investigation of the effects of external turbulence on the internal state of a flux tube.

To formulate the essentials of the problem, and yet keep the problem tractable, we consider again the slender flux tube, with a width or diameter small compared to the scale of variation along the tube. Then the transverse components of both the magnetic field and the velocity within the field can be neglected to good approximation. The dynamical equations describing conditions within the tube reduce to those for a tube lying along a straight path, say the z-axis. The tube is massaged by the pressure fluctuations exerted on it by the exterior turbulence, with dynamical effects upon the gas and field in its interior. We shall specify the external pressure $P(z, t)$ exerted on the tube, to illustrate the effects. It makes little difference whether $P(z, t)$ is thought of as the pressure in the turbulent fluid along a straight or contorted path.

The calculations in this and later sections show the not surprising result that the equipartition condition $B^2/8\pi \cong \frac{1}{2}\rho v^2$ is an approximate upper limit on the enhancement of the field by the external turbulence, but there is no simple universal relation. In some circumstances the mean square field is increased by $4\pi\rho v^2$, while the square of the mean field is increased much less and may actually decrease. In other circumstances, both may in principle be increased by more than $4\pi\rho v^2$, as a consequence of accumulated effects along the flux tube. And finally, we shall see that infinitesimal convective forces in an ideal fluid may increase both the magnetic field and the fluid velocity to unlimited values. So we must proceed carefully, constructing the formal mathematical solutions to the idealized circumstances that best illustrate the various possible effects.

There are two easily identifiable effects of external turbulence on the internal field of a concentrated flux tube. The simplest is turbulent massage which alternately squeezes the flux tube here and there along its length, causing local expansion or compression of the tube with an associated motion of the fluid back and forth along the lines of force within the tube. The Bernoulli effect in the fluid motion within the tube reduces the internal gas pressure relative to the external gas pressure, leading to an increase in the equilibrium magnetic pressure. The effect is readily demonstrated by formal solution of the hydrodynamic equations.

The other effect occurs when the external turbulence is carried in a general flow along the flux tube, so that the massage becomes a systematic stroking of the tube in a single direction. It will be remembered that the flux tubes observed in the sun are confined, by the converging flows on the surface, to the convective *downdrafts* between granules and supergranules so that they are subject to downward stroking by the external turbulence. The effect is a downward pumping of the fluid within the tube, partially evacuating the interior and compressing the magnetic field, and is treated in idealized form in §10.5.

10.4. The Bernoulli effect within a flux tube

Consider the dynamical effects within a flux tube as a consequence of a periodic pressure applied to the surface of the tube (Parker 1974b). It is sufficient to treat the problem in two dimensions, considering a slender flux tube extending along the z-axis from $z=-L$ to $z=+L$. The surface of the tube is $y=Y(z,t)$ and the field strength $B(z,t)$ is assumed to be uniform across the small width $2Y(z,t)$ so that conservation of flux implies

$$B(z,t)Y(z,t)=B_0Y_0 \tag{10.5}$$

where B_0 and Y_0 are the field strength and width, respectively, at the distant ends of the tube $z=\pm L$. Local lateral pressure balance leads to the relation

$$P(z,t)=p(z,t)+B^2(z,t)/8\pi \tag{10.6}$$

between the variable fluid pressure $P(z,t)$ applied to $y=\pm Y(z,t)$ by the external turbulence and the internal fluid pressure $p(z,t)$. For simplicity we suppose the fluid to be inviscid and incompressible. The restriction to a slender tube means that $\partial Y/\partial z \ll 1$, so that the transverse components of the field and fluid velocity are small, $O(\partial Y/\partial z)$, and the tranverse forces per unit volume are small, $O\{(\partial Y/\partial z)^2\}$. The equations for conservation of momentum and fluid along the flux tube are then just

$$\partial u/\partial t+u\partial u/\partial z+(1/\rho)\partial p/\partial z=0 \tag{10.7}$$

$$\partial Y/\partial t+\partial uY/\partial z=0 \tag{10.8}$$

for the velocity $u(z,t)$ along the tube.

To keep out extraneous effects from the ends of the tube, we suppose that the pressure difference there between the exterior and the interior of the tube has the fixed value $P_0 - p_0$. Then at $z = \pm L$ transverse hydrostatic equilibrium requires that

$$P_0 - p_0 = B_0^2/8\pi. \tag{10.9}$$

Suppose further, that there is no expulsion of fluid from the tube, $u(\pm L, t) = 0$, and that there is no progressive displacement of the fluid within the tube, so that

$$\langle u(z, t)\rangle = 0, \qquad \langle \partial u/\partial t\rangle = 0, \tag{10.10}$$

where the angular brackets denote the time average at the position z.

Now (10.6) and (10.7) can be combined to eliminate p yielding

$$\rho \partial u/\partial t + (\partial/\partial z)(P + \tfrac{1}{2}\rho u^2 - B^2/8\pi) = 0 \tag{10.11}$$

Averaging over time eliminates $\partial u/\partial t$. Integration over z yields

$$\langle P + \tfrac{1}{2}\rho u^2 - B^2/8\pi\rangle = \mathit{constant}. \tag{10.12}$$

But if u vanishes at the distant ends of the tube, where the external pressure is P_0 and the field is B_0, it follows that the *constant* is $P_0 - B_0^2/8\pi = p_0$. Hence,

$$\langle B^2\rangle/8\pi = \langle P\rangle - p_0 + \tfrac{1}{2}\rho\langle u^2\rangle \tag{10.13}$$

at any point along the flux tube. The conclusion is that the massaging of the tube by the external turbulence enhances the mean magnetic field pressure above the mean ambient fluid pressure difference $\langle P\rangle - p_0$ by an amount equal to the mean kinetic energy density of the turbulence, as a consequence of the Bernoulli effect.

One is tempted to halt the inquiry at this point with the optimistic conclusion that we have demonstrated how the field pressure is enhanced by the external turbulence. But it must be remembered that the observer sees the field itself rather than the mean square, or root mean square, field, and the above result gives only the mean square. As we will soon see, the mean field $\langle B\rangle$ behaves rather differently from the square root of (10.13).

Equations (10.5)–(10.8) can be solved quite generally to give $u(z, t)$ and the associated $p(z, t)$ in terms of $Y(z, t)$. First of all note that $B(z, t)$ follows from $Y(z, t)$ through (10.5), so the problem is to compute the external pressure fluctuations $P(z, t)$ responsible for a given $B(z, t)$ or $Y(z, t)$. Integrating (10.8) from some arbitrary reference point $z = a$, it follows that $u(z, t)$ is related to $Y(z, t)$ by

$$u(z, t) = \{u(a, t)Y(a, t) - \int_a^z dx \partial Y(x, t)/\partial t\}/Y(z, t). \tag{10.14}$$

Integrating (10.11) from $z=a$ yields

$$P(z,t)+\tfrac{1}{2}\rho\{u^2(z,t)-V_A^2Y_0^2/Y^2(z,t)\}$$
$$=P(a,t)+\tfrac{1}{2}\rho\{u^2(a,t)-V_A^2Y_0^2/Y^2(a,t)\}$$
$$-\rho\int_a^z \mathrm{d}x\,\partial u(x,t)/\partial t \quad (10.15)$$

after using (10.5) to eliminate B. The velocity V_A is the basic Alfven speed $B_0/(4\pi\rho)^{\frac{1}{2}}$ computed at the undisturbed ends of the flux tube. The internal pressure $p(z,t)$ follows from (10.6) and (10.15) as

$$p(z,t)+\tfrac{1}{2}\rho u^2(z,t)=p(a,t)+\tfrac{1}{2}\rho u^2(a,t)-\rho\int_a^z \mathrm{d}x\,\partial u(x,t)/\partial t. \quad (10.16)$$

These relations serve to determine $u(z,t)$, $B(z,t)$, $P(z,t)$, and $p(z,t)$ in terms of the tube width $Y(z,t)$. Suppose, then, that the tube width is given by

$$Y(z,t)=Y_0\{1+n\cos\omega t\cos kz\exp(-\epsilon k|z|)\} \quad (10.17)$$

where $n<1$, so that $Y(z,t)$ is positive for all z and t, and ϵ is small but positive and non-vanishing so that in the limit of large L the width $Y(\pm L,t)$ at the ends is just Y_0. Thus, the tube width has a time average value of $2Y_0$ everywhere along its length, but in the central region, $z^2<1/\epsilon^2k^2$ there are standing waves of amplitude nY_0 on each surface. The waves die out toward the ends of the tube. The volume of the tube is

$$V(t)=2\int_0^L \mathrm{d}z\,Y(z,t)$$
$$=2Y_0\left[L+\frac{n\cos\omega t}{k(1+\epsilon^2)}\{\epsilon(1-\cos kL\exp(-\epsilon kL))\right.$$
$$\left.+\sin(kL)\exp(-\epsilon kL)\}\right]$$
$$=2Y_0L\{1+O(\epsilon,\exp-\epsilon kL)\}.$$

In the present limits of $\epsilon\to 0$, $\epsilon kL\to\infty$, the volume is constant and there is no flow at the ends if we choose the reference point $z=a$ to lie at the symmetric midpoint $z=0$, where $u(0,t)=0$. Then $u(z,t)=-u(-z,t)$. It is sufficient to treat only the half $z>0$, the results for the other half of the tube following from the symmetry. The requirement that the tube be slender is $kY_0\ll 1$.

It follows from (10.14) that

$$u(z,t)=\frac{n\omega\sin\omega t}{k(1+\epsilon)^2}\left\{\frac{\epsilon+(\sin kz-\epsilon\cos kz)\exp(-\epsilon kz)}{1+n\cos kz\cos\omega t\exp(-\epsilon kz)}\right\}. \quad (10.18)$$

For large positive z, then,

$$u(z, t) \sim (\epsilon n\omega/k)\sin \omega t \times \{1+O(\epsilon)\}$$

which satisfies $u(\pm L, t)=0$ in the present limit of small ϵ. For ϵ arbitrarily small there is a broad region $z^2 < 1/\epsilon^2 k^2$ in which the velocity has the simple periodic form

$$u(z, t)=\frac{n\omega \sin \omega t \sin kz}{k\{1+n \cos \omega t \cos kz\}} \tag{10.19}$$

representing standing waves. Note, finally that the time average of (10.18) is zero, so that the time average of $\partial u/\partial t$ is also zero and, hence,

$$\left\langle \int_0^z \mathrm{d}x \partial u(x, t)/\partial t \right\rangle = 0$$

which will be used in some of the manipulations that follow.

We need the mean square of the velocity in the central region $z^2 \ll 1/\epsilon^2 k^2$. It is readily shown that the time average there of $u^2(z, t)$ is

$$\langle u^2(z, t)\rangle = (\omega^2/k^2)\tan^2 kz[1/\{1-n^2\cos^2 kz\}^{\frac{1}{2}}-1],$$

upon noting that

$$\frac{1}{\pi}\int_0^\pi \frac{\mathrm{d}\tau \sin^2\tau}{(1+\alpha \cos \tau)^2} = \frac{1}{\alpha^2}\left\{\frac{1}{(1-\alpha^2)^{\frac{1}{2}}}-1\right\}.$$

The average over z yields

$$\langle\!\langle u^2 \rangle\!\rangle = \frac{2\omega^2}{\pi k^2}\int_0^{\pi/2} \mathrm{d}u \tan^2 u \left\{\frac{1}{(1-n^2\cos^2 u)^{\frac{1}{2}}}-1\right\}.$$

Thus for $n \ll 1$,

$$\langle\!\langle u \rangle\!\rangle = \frac{n^2\omega^2}{4k^2}\left\{1+3n^2/2^4+5n^4/2^6+175n^6/2^{12}+\dots \right.$$
$$\left. +\frac{(2m+1)!!}{(2m+2)!!}\frac{(m-\frac{1}{2})(m-\frac{3}{2})\dots\frac{3}{2}}{(m+1)!}n^{2m}+\dots\right\}, \tag{10.20}$$

while for $n \to 1$ we have

$$\begin{aligned}\langle\!\langle u^2 \rangle\!\rangle &= \frac{2\omega^2}{\pi k^2}\int_0^{\pi/2} \mathrm{d}u \tan^2 u(\csc u - 1)\\ &= \frac{2\omega^2}{\pi k^2}\int_0^{\pi/2}\frac{\mathrm{d}u \sin u}{1+\sin u}\\ &= \frac{\omega^2}{k^2}\left(1-\frac{2}{\pi}\right).\end{aligned} \tag{10.21}$$

The external pressure responsible for the variations of $Y(z, t)$ and the non-vanishing of $u(z, t)$ follows from (10.15) as

$$P(z,t) = P(0,t) + \tfrac{1}{2}\rho V_A^2[\{1+n\cos kz\cos\omega t\exp(-\epsilon kz)\}^{-2} - \{1+n\cos\omega t\}^{-2}] - \tfrac{1}{2}\rho u^2(z,t) - \rho\int_0^z \mathrm{d}x\,\partial u(x,t)/\partial t. \qquad (10.22)$$

Then noting that

$$\frac{1}{\pi}\int_0^\pi \frac{\mathrm{d}\tau}{(1+\alpha\cos\tau)^2} = \frac{1}{(1-\alpha^2)^{\frac{3}{2}}},$$

it follows from (10.22) that the time average pressure[1] is

$$\langle P(z,t)\rangle = \langle P(0,t)\rangle - \tfrac{1}{2}\rho\langle u^2(z,t)\rangle - \tfrac{1}{2}\rho V_A^2[(1-n^2)^{-\frac{3}{2}} - \{1-n^2\cos^2 kz\exp(-2\epsilon kz)\}^{-\frac{3}{2}}]. \qquad (10.23)$$

Then since u is negligible, $O(\epsilon V_A)$, at the ends of the tube, it follows that

$$\langle P(0,t)\rangle = \langle P(\pm L,t)\rangle + \tfrac{1}{2}\rho V_A^2\{(1-n^2)^{-\frac{3}{2}} - 1\} \qquad (10.24)$$

and

$$\langle P(z,t)\rangle = P_0 - \tfrac{1}{2}\rho u^2(z,t) + \tfrac{1}{2}\rho V_A^2[\{1-n^2\cos^2 kz\exp(-2\epsilon kz)\}^{-\frac{3}{2}} - 1]. \qquad (10.25)$$

In the central region $z^2 \ll 1/\epsilon^2k^2$, where $\exp(-\epsilon kz) \cong 1$, the average over z yields

$$\langle\langle P(z,t)\rangle\rangle = \langle P(\pm L,t)\rangle - \tfrac{1}{2}\rho\langle\langle u^2(z,t)\rangle\rangle - \tfrac{1}{2}\rho V_A^2\left\{1 - \frac{2}{\pi}\frac{\mathrm{d}}{\mathrm{d}n}nK(n)\right\}, \qquad (10.26)$$

where $K(n)$ is the complete elliptic integral of the first kind. The time average of the internal fluid pressure follows from (10.16) as

$$\langle p(z,t)\rangle = \langle p(0,t)\rangle - \tfrac{1}{2}\rho\langle u^2(z,t)\rangle. \qquad (10.27)$$

Since $u^2(\pm L, t)$ is small, $O(\epsilon^2 V_A^2)$, it follows that

$$\langle p(\pm L,t)\rangle = \langle p(0,t)\rangle$$

and

$$\langle p(z,t)\rangle = \langle p(\pm L,t)\rangle - \tfrac{1}{2}\rho\langle u^2(z,t)\rangle. \qquad (10.28)$$

[1] Note that the pressure $P(\pm L, t)$ at the ends oscillates with time but is uniform over space, so that it has no effect on the fluid velocity or the magnetic field there. The pressure difference $P - p$ maintains the constant value $P_0 - p_0$.

showing the reduction of the interior pressure by the Bernoulli effect. This same expression holds for the average over the central region. The time-averaged magnetic field follows directly from (10.5) and (10.17)

$$\langle B(z,t)\rangle = B_0\{1-n^2\cos^2 kz\,\exp(-2\epsilon kz)\}^{-\frac{1}{2}} \tag{10.29}$$

upon noting that

$$\frac{1}{\pi}\int_0^{\pi}\frac{\mathrm{d}\tau}{1+\alpha\cos\tau}=\frac{1}{(1-\alpha^2)^{\frac{1}{2}}}. \tag{10.30}$$

The mean value of the time average field over the central region $z^2 \ll 1/\epsilon^2 k^2$, wherein $\exp(-\epsilon k\,|z|)\cong 1$, is readily shown to be

$$\begin{aligned}\langle\langle B(x,t)\rangle\rangle &= \frac{2B_0}{\pi}\int_0^{\pi/2}\frac{\mathrm{d}u}{(1-n^2\cos^2 u)^{\frac{1}{2}}}\\ &= \frac{2B_0}{\pi}K(n).\end{aligned} \tag{10.31}$$

Incidentally the mean square field is

$$\langle B^2(z,t)\rangle = B_0^2\{1-n^2\cos^2 kz\,\exp(-2\epsilon kz)\}^{-\frac{3}{2}}$$

upon noting (10.25). The average of this over the central region is

$$\langle\langle B^2(z,t)\rangle\rangle = B_0^2\frac{2}{\pi}\frac{\mathrm{d}}{\mathrm{d}n}nK(n). \tag{10.32}$$

Combining (10.26) and the space average of (10.28), and using (10.9) and $\rho V_{\mathrm{A}}^2 = B_0^2/4\pi$, it follows that

$$\begin{aligned}\langle\langle P\rangle\rangle - \langle\langle p\rangle\rangle &= \frac{B_0^2}{8\pi}\frac{2}{\pi}\frac{\mathrm{d}}{\mathrm{d}n}nK(n),\\ &= \langle\langle B^2\rangle\rangle/8\pi,\end{aligned} \tag{10.33}$$

with the help of (10.32) in the last step. This is, of course, just the combined time and space average of (10.13).

Now, we know from (10.13) that the mean square field is enhanced above $P_0 - p_0$ by the amount of the Bernoulli effect $\langle\frac{1}{2}\rho u^2\rangle$. The important question is the enhancement of the mean field, for a given mean external pressure $\langle\langle P\rangle\rangle$ and applied internal pressure p_0. In the static case

$$\langle\langle B\rangle\rangle^2/8\pi = \langle\langle P\rangle\rangle - \langle p(\pm L,t)\rangle \tag{10.34}$$

(where $\langle\langle B\rangle\rangle = B_0$ and $\langle\langle P\rangle\rangle = P_0$, of course). In the dynamical case it follows from (10.26) and (10.31) that

$$\langle\langle B\rangle\rangle^2/8\pi - \{\langle\langle P\rangle\rangle - p_0\} = \frac{B_0^2}{8\pi}\left[\left\{\frac{2}{\pi}K(n)\right\}^2 - \frac{2}{\pi}\frac{\mathrm{d}}{\mathrm{d}n}nK(n)\right] + \tfrac{1}{2}\rho\langle\langle u^2\rangle\rangle \tag{10.35}$$

with $\frac{1}{2}\rho\langle\langle u^2\rangle\rangle$ given by (10.20). For small n

$$(2/\pi)K(n) \cong 1 + \tfrac{1}{4}n^2 + 9n^4/64 + \ldots$$

so that, neglecting terms fourth order in n, the result is

$$\langle\langle B\rangle\rangle^2/8\pi = \langle\langle P\rangle\rangle - p_0 + \tfrac{1}{2}\rho\frac{n^2}{4}(\omega^2/k^2 - V_A^2). \tag{10.36}$$

On the other hand, if $n \lesssim 1$, then

$$K(n) \cong \ln 4 - \tfrac{1}{2}\ln(1-n^2) + O\{(1-n^2)\ln(1-n^2)\},$$

and, with the aid of (10.21),

$$\langle\langle B\rangle\rangle^2/8\pi = \langle\langle P\rangle\rangle - p_0 + \tfrac{1}{2}\rho\left\{\frac{\omega^2}{k^2}\left(1-\frac{2}{\pi}\right) - V_A^2\frac{2}{1-n^2}\right\}. \tag{10.37}$$

To interpret these mathematical results, note that the phase velocity ω/k associated with the externally applied pressure fluctuations is comparable to the velocity v of the external turbulence. If we suppose, then, that the magnetic field B_0 at the ends of the tube is very weak, so that $V_A \ll v$, (10.36) reduces to

$$\begin{aligned}\langle\langle B\rangle\rangle^2/8\pi &\cong \langle\langle P\rangle\rangle - p_0 + \tfrac{1}{2}\rho v^2 n^2/4 \\ &\cong \tfrac{1}{2}\rho v^2 n^2/4\end{aligned} \tag{10.38}$$

with the replacement of ω/k by the characteristic turbulent velocity v. Under the same circumstances (10.37) becomes

$$\begin{aligned}\langle\langle B^2\rangle\rangle/8\pi &\cong \langle\langle P\rangle\rangle - p_0 + \tfrac{1}{2}\rho v^2(1-2/\pi) \\ &= \tfrac{1}{2}\rho v^2(1-2/\pi)\end{aligned} \tag{10.39}$$

so long as n does not approach too closely to one. Comparing these two expressions with the static situation, given by (10.34), it may be seen that the mean field is enhanced from a weak value B_0 up to some *fraction* of the equipartition value. The fraction is less than the upper limit $(1-2/\pi)^{\frac{1}{2}} \cong 0{\cdot}6$ imposed by (10.39).

On the other hand, if the initial field B_0 is not weak, but instead is near the equipartition value $V_A^2 \cong v^2 \cong \omega^2/k^2$, then n will be small compared to one and (10.36) tells us that $\langle\langle B\rangle\rangle^2 \cong B_0^2$. There is then no significant enhancement of the field beyond the initial value B_0 near equipartition.

Altogether, then, it appears that the fluid motion induced within the tube by the turbulent massage of the external turbulence increases the *mean* field only to a fraction of the value $(4\pi\rho v^2)^{\frac{1}{2}}$ that one might have hoped from conventional arguments on equipartition of energy. On the other hand, the *mean square* field is increased by the expected amount, as indicated by (10.13), because of the peaks in the time average where kz is equal to an integral number times π. We presume that it is the mean, rather than the r.m.s. field, that is inferred from observation, so we must look further.

10.5. Turbulent pumping of a flux tube

Consider the dynamical effects within a flux tube as a consequence of travelling pressure fluctuations applied to the surface of the tube (Parker 1974a) by the turbulence outside. We have in mind the effects that arise when the turbulence outside the tube, represented by standing waves in §10.4, is convected along the tube with a velocity w relative to the gas inside. Such circumstances would arise if there were a large-scale bulk motion of the turbulent fluid outside. In the sun, for instance, the flux tubes appear to stand vertically in the strong downdrafts at the junctions of three or more supergranules, so that the turbulent external fluid is generally streaming downward along the tube.

It is sufficient to treat the same slender flux tube as in §10.4. The fluctuations in the width $2Y(z, t)$ as a consequence of the variable pressure $P(z, t)$ exerted on the tube by the external turbulence, are assumed to have a uniform velocity w in the z-direction along the tube so that the tube boundary has the simple form $y = \pm Y(z - wt, t)$. The question is under what circumstances the travelling deformation of the boundary induces a net motion, and partial evacuation, of the fluid within. The problem is not unlike that of rowing a boat on the surface of a lake. The oars of the boat deform the surface of the water and when moved, the deformation transmits force (momentum) to the water of the lake. But we wish to treat the problem quantitatively with the simplification of an ideal fluid. It is a well known theoretical fact that the motions of an inviscid fluid initially at rest remain irrotational, so that there can be neither drag nor lift on any body moving steadily through the fluid. Drag and lift are a consequence of vorticity, needing some slight viscosity to bring them about. So we cannot row a boat on the surface of the inviscid lake[2]. We cannot row, that is, unless we are clever enough to use oars with blades of infinitesimal thickness. One introduces vorticity into real fluids with oars and paddles of finite thickness, and into ideal fluids with appendages of vanishing thickness. In view of the intractability of the real problem, with a small amount of viscosity and much turbulent vorticity, we study the drag on the ideal fluid through vortex sheets introduced by infinitely thin partitions. The idealizations allow us to proceed with a minimum of calculation to a rough estimate of the net pumping of the fluid along the tube.

[2] Although we can coast any distance with constant velocity by pushing off from the shore or hurling an oar from the boat.

Suppose, then, that the tube has uniform width Y_0 everywhere except at $z-wt=0$, where there is a thin partition extending from the wall $y=\pm Y_0$ inward to $y=\pm Y_0(1-n)$ where, as in §10.4, $0<n<1$. Thus, the width is $2Y_0$ everywhere except at $z=wt$, where the width is $2Y_0(1-n)$. The flow of fluid and field through the constriction is easily treated, using a conformal mapping in the frame of reference of the constriction at $z=wt$. Denote the fluid velocity by u_i in the frame $z=wt$ of the moving constriction. Then the momentum equation is

$$\rho u_j \frac{\partial u_i}{\partial x_j}+\frac{\partial}{\partial x_i}\left(p+\frac{B^2}{8\pi}\right)=\frac{B_j}{4\pi}\frac{\partial B_i}{\partial x_j} \tag{10.40}$$

for steady flow, while the hydromagnetic induction equation reduces to

$$u_j\frac{\partial B_i}{\partial x_j}=B_j\frac{\partial u_i}{\partial x_j} \tag{10.41}$$

with

$$\partial u_i/\partial x_i=0, \qquad \partial B_i/\partial x_i=0. \tag{10.42}$$

Write the magnetic field in the form

$$B_i=\alpha(4\pi\rho)^{\frac{1}{2}}u_i \tag{10.43}$$

where α is a numerical constant not equal to one (but see, §10.6). The hydromagnetic equation (10.41) is automatically satisfied, the curl of B_i is zero everywhere throughout the fluid (but not at the boundaries, of course), and the force $(\nabla\times\mathbf{B})\times\mathbf{B}/4\pi$ exerted on the fluid by the field vanishes so that (10.40) reduces to the Euler equation

$$\nabla(p+\tfrac{1}{2}\rho u^2)=0 \tag{10.44}$$

for irrotational flow. The fluid pressure is determined by the divergence of (10.44). Noting that

$$\nabla^2 u^2=2(\partial u_i/\partial x_j)(\partial u_i/\partial x_j)$$

for an irrotational flow, the result can be written in either of two ways,

$$\nabla^2(p+\tfrac{1}{2}\rho u^2)=0 \quad \text{or} \quad \nabla^2 p=-\rho(\partial u_i/\partial x_j)(\partial u_i/\partial x_j).$$

There is no viscous term in (10.44) of course, so if the fluid motion is taken to be irrotational, with uniform velocity $-w$ along the tube at $z=+\infty$, then the motion is irrotational everywhere along the tube and the velocity u_i can be written in terms of the velocity potential,

$$u_i=-\partial\phi/\partial x_i. \tag{10.45}$$

In view of (10.42) it follows that $\nabla^2\phi=0$ and u_i can be written in terms of the stream function ψ,

$$u_z = +\partial\psi/\partial y, \qquad u_y = -\partial\psi/\partial z \tag{10.46}$$

with $\nabla^2\psi = 0$. The magnetic field follows from either ϕ or ψ through (10.43).

With the boundary condition that the normal components of u_i and B_i vanish on $y = \pm Y(z, t)$, ψ and ϕ are readily obtained from a conformal mapping (see for instance, Smythe 1939) as the real and imaginary parts of

$$\psi + i\phi = \frac{2}{\pi} wY_0 \sin^{-1}\left\{\frac{1}{s}\sin(\pi\zeta/2)\right\} \tag{10.47}$$

where

$$\zeta = \eta + i\xi, \qquad \eta = y/Y_0, \qquad \xi = (z - wt)/Y_0,$$
$$s = \sin(\pi(1-n)/2). \tag{10.48}$$

The stream function and velocity potential are given by

$$\psi = \frac{2}{\pi} Y_0 w \sin^{-1}\left\{\frac{2\sin(\pi\eta/2)\cosh(\pi\xi/2)}{R+S}\right\} \tag{10.49}$$

$$\phi = \frac{2}{\pi} Y_0 w \cosh^{-1}\{\tfrac{1}{2}(R+S)\}, \tag{10.50}$$

where R and S are positive quantities with the values

$$R \equiv \left[\left\{1 + \frac{1}{s}\sin(\pi\eta/2)\cosh(\pi\xi/2)\right\}^2 + \frac{1}{s^2}\cos^2(\pi\eta/2)\sinh^2(\pi\xi/2)\right]^{\frac{1}{2}},$$

$$S \equiv \left[\left\{1 - \frac{1}{s}\sin(\pi\eta/2)\cosh(\pi\xi/2)\right\}^2 + \frac{1}{s^2}\cos^2(\pi\eta/2)\sinh^2(\pi\xi/2)\right]^{\frac{1}{2}}.$$

The upper boundary of the tube $\eta = 1$ corresponds to the streamline $\psi = wY_0$, while the centre line (the z-axis) corresponds to $\psi = 0$, with the lower boundary $\eta = -1$ as $\psi = -wY_0$. The velocity potential along either boundary is

$$\phi = \frac{2}{\pi} wY_0 \cosh^{-1}\left\{\frac{1}{s}\cosh(\pi\xi/2)\right\}.$$

On the centre line (the z-axis)

$$\phi = \frac{2}{\pi} wY_0 \cosh^{-1}\left\{1 + \frac{1}{s^2}\sinh^2(\pi\xi/2)\right\}$$

so that the velocity is

$$u_x = -\frac{w\cosh(\pi\xi/2)}{\{s^2+\sinh^2(\pi\xi/2)\}^{\frac{1}{2}}} \tag{10.51}$$

in the frame moving with the partition.

Now suppose that for $t<0$ there is no constriction of the channel so that $v_z = -w$ everywhere within the channel. At time $t=0$ the thin partition is pushed in at $z = wt$ so that n increases smoothly and continuously up to a value n_0 in a time τ, say

$$n(t) = \begin{cases} 0 & \text{for} \quad t<0 \\ n_0 t/\tau & \text{for} \quad 0<t<\tau \\ n_0 & \text{for} \quad t>\tau. \end{cases} \tag{10.52}$$

The streamlines are pinched in by the partition, as shown in Fig. 10.2 for $n = \frac{1}{2}$. The partition produces a variation of pressure within the fluid in the neighbourhood of the partition. The total pressure head Δp is readily computed by integrating from $\xi = -\infty$ to $\xi = +\infty$ along the z-axis (or any other convenient streamline) yielding

$$\begin{aligned}\Delta p &= Y_0 \int_{-\infty}^{+\infty} \mathrm{d}\xi(\partial p/\partial z)_{y=0} \\ &= -Y_0 \int_{-\infty}^{+\infty} \mathrm{d}\xi \left(\rho \frac{\partial u_z}{\partial t} + \frac{\partial}{\partial z} \tfrac{1}{2}\rho u_z^2\right)_{y=0} \\ &= -\rho Y_0 \int_{-\infty}^{+\infty} \mathrm{d}\xi \left(\frac{\partial u_z}{\partial t}\right)_{y=0}.\end{aligned}$$

The acceleration follows from (10.51) as

$$\frac{\partial u_z}{\partial t} = \frac{w\cosh(\pi\xi/2)}{\{s^2+\sinh^2(\pi\xi/2)\}^{\frac{3}{2}}} s \frac{\mathrm{d}s}{\mathrm{d}t}$$

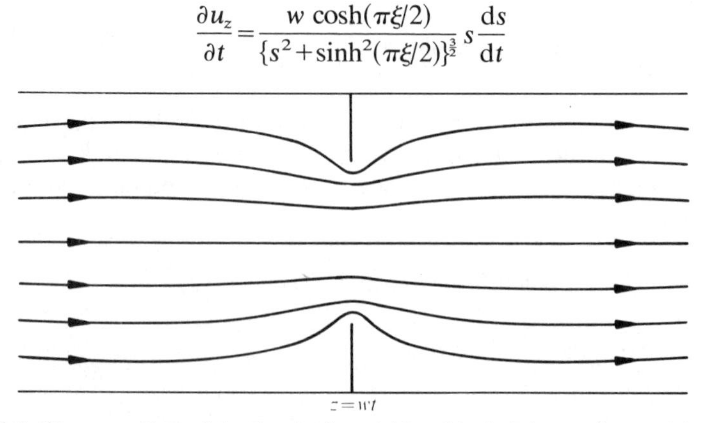

FIG. 10.2. The streamlines of the flow in the neighbourhood of the moving partition $z = wt$ for the case $n = 0{\cdot}5$.

where

$$\frac{\mathrm{d}s}{\mathrm{d}t} = -\frac{\pi}{2}\frac{\mathrm{d}n}{\mathrm{d}t}\cos\{\pi(1-n)/2\}.$$

Thus

$$\Delta p = -\rho w Y_0 s \frac{\mathrm{d}s}{\mathrm{d}t}\int_{-\infty}^{+\infty}\frac{\mathrm{d}\xi\cosh(\pi\xi/2)}{\{s^2+\sinh^2(\pi\xi/2)\}^{\frac{3}{2}}}$$

$$= -\frac{4}{\pi}\rho w Y_0 \frac{\mathrm{d}(\ln s)}{\mathrm{d}t}$$

$$= +2\rho w Y_0 \frac{\mathrm{d}n}{\mathrm{d}t}\tan(\pi n/2). \qquad (10.53)$$

Suppose, then, that there is one partition every l-cm along the tube, where l is rather greater than Y_0 so the concentration of flow through each constriction is not greatly affected by the constrictions at a distance l on either side. In this case the mean pressure gradient along the tube is $\Delta p/l$ during the insertion of the partitions, where Δp is given by (10.53).

Suppose, now, that the partitions are withdrawn. If they were of finite thickness, the flow of the ideal fluid would return to its initial uniform state $u_z = -w$ with the disappearance of the partitions. But the partitions are of vanishing thickness, so that, just as in real turbulent flows, the flow is not reversible. Withdrawal of the partitions leaves a vortex sheet where the partition had been (see Fig. 10.2) and the flow is unaffected by the withdrawal. Thus, insertion and withdrawal of the partitions introduces a net pumping or sweeping of the fluid along the tube that can be blocked only by an opposing pressure gradient $-\Delta p/l$.

Over a distance L the total pressure reduction is $\Delta pL/l$. In terms of external turbulence composed of eddies of diameter l and velocity v, all convected at velocity w along the tube, the characteristic rise time of the constrictions is $\tau = l/v$ with

$$\frac{\mathrm{d}}{\mathrm{d}t}nY_0 \cong \frac{nY_0}{\tau}.$$

With $n = \frac{1}{2}$ (10.53) gives

$$\Delta p \cong \rho w v Y_0/l. \qquad (10.54)$$

Thus, in principle, over a suitably long distance L, the total pressure difference $\rho w v Y_0 L/l^2$ can be very large. The accumulated sweeping effect of the moving constrictions greatly reduces the pressure inside the tube, and the magnetic field B is compressed to some fraction of the ambient external gas pressure.

In practice, however, the moving constrictions are relatively ineffective when $B^2/8\pi$ exceeds the kinetic energy density $\frac{1}{2}\rho v^2$ of the external turbulence. To give an idea of the accumulation of the effect over long distances note that the external turbulent pressure fluctuations, of order $\frac{1}{2}\rho v^2$ above the ambient pressure, compress the mean field B up to the value B' where (see (10.6))

$$\frac{B'^2}{8\pi} \cong \frac{B^2}{8\pi} + \tfrac{1}{2}\rho v^2.$$

Conservation of flux across the slender tube indicates that the width of the tube is reduced from Y_0 to $(1-n_0)Y_0$ where

$$B'(1-n_0)Y_0 = Y_0 B$$

or

$$n_0 = 1 - \{B^2/(B^2 + 4\pi\rho v^2)\}^{\frac{1}{2}}.$$

Then (10.53) yields

$$\Delta p = 2\rho w v (Y_0/l) n_0 \tan(\pi n_0/2).$$

The fluid is swept in the positive z-direction by the moving constrictions so that the tube is increasingly evacuated toward negative z, the mean rate of reduction of fluid pressure being $\Delta p/l$. The sum of the fluid and magnetic pressure is equal to the ambient external pressure, which is presumed to be independent of z. Hence the magnetic pressure increases toward negative z at the rate

$$-\frac{\mathrm{d}}{\mathrm{d}z}\frac{B^2}{8\pi} = (2\rho w v n_0 Y_0/l^2)\tan(\pi n_0/2).$$

Where the tube is so strongly evacuated that $B^2/8\pi > \frac{1}{2}\rho v^2$, it follows that

$$n_0 \cong 2\pi\rho v^2/B^2 \ll 1$$

so that

$$-\frac{\mathrm{d}}{\mathrm{d}z}\frac{B^2}{8\pi} \cong \frac{\pi\rho w v Y_0}{4l^2}\left(\frac{\frac{1}{2}\rho v^2}{B^2/8\pi}\right)^2.$$

Integrating over z yields

$$\frac{B^2/8\pi}{\frac{1}{2}\rho v^2} \cong \left\{\frac{3\pi w Y_0}{2lv}\frac{z_0 - z}{l}\right\}^{\frac{1}{3}}. \tag{10.55}$$

The magnetic pressure increases only as the cube root of the distance along the tube. In order to increase $B^2/8\pi$ by a factor of ten above the value of $\frac{1}{2}\rho v^2$ for the solar granules, the sweeping would have to extend along the tube for a distance of the order of $10^3 l$, or 10^6 km. We conclude that the turbulence carried in the downdraft along the flux tube produces field energy densities up to, but not significantly greater than, the turbulent energy density. Thus the observed photospheric granule velocities of 3 km s^{-1} where $\rho \cong 3\times10^{-7}$ g cm^{-3} may concentrate fields to densities of the order of 600 G. Since field pressures ten times greater are inferred from the magnetic observations, something remains yet to be discovered.

10.6. Fluid motion within a flux tube: incompressible case

Having failed to find a mechanism whereby the turbulence outside the flux tube can evacuate the tube significantly beyond the equipartition value $B^2 = 4\pi\rho v^2$, consider what effects might arise from the hydrodynamic forces within the tube. Again we find that there are no flows that push $B^2/8\pi$ beyond the kinetic energy density $\frac{1}{2}\rho v^2$ of the fluid. But there are some intriguing possibilities for the local increase of $\frac{1}{2}\rho v^2$ and $B^2/8\pi$ together to arbitrarily large values (in an ideal fluid) by the application of arbitrarily weak convective forces (Parker 1976b, c). This raises the question of whether the concentrated and unresolved magnetic fields of the small flux tubes in the supergranule boundaries are associated with concentrated, and so far unconfirmed, high-speed fluid motions (see, for instance, Deubner 1976). The question can, of course, only be decided by observation. The purpose here is to develop the basic physics of the effect. If observations should confirm the existence of the fluid motions in the flux tubes in the solar photosphere, we would then be in a position to develop quantitative models. The possibility of a direct and intimate connection between the concentration of field and the eruptive spicule phenomenon suggests that high fluid velocities within the tube should not be discarded out of hand.

In this section we consider the idealized case of the quasi-stationary flow v_i of an incompressible, inviscid, infinitely-conducting fluid along a magnetic field B_i (Parker 1976a). The dynamical equations are

$$\rho\frac{\partial v_i}{\partial t}+\rho v_j\frac{\partial v_i}{\partial x_j}+\frac{\partial}{\partial x_i}\left(p+\frac{B^2}{8\pi}\right)=\frac{B_j}{4\pi}\frac{\partial B_i}{\partial x_j}, \tag{10.56}$$

$$\frac{\partial B_i}{\partial t}+\frac{\partial}{\partial x_j}v_jB_i=\frac{\partial}{\partial x_j}v_iB_j, \tag{10.57}$$

$$\partial v_i/\partial x_i=0, \qquad \partial B_i/\partial x_i=0. \tag{10.58}$$

These equations possess the well-known (Lundquist 1950) stationary equipartition solution

$$v_i=\pm B_i/(4\pi\rho)^{\frac{1}{2}}, \qquad p+B^2/8\pi=\textit{constant}. \tag{10.59}$$

Chandrasekhar (1956b, 1961) has shown that an infinitesimal perturbation of the system leads to undamped oscillation about the equilibrium, so that the equipartition solutions may be said to be stable.

The physical basis for the solution is simple enough. The fluid streams along the field so that the Bernoulli formula $p+\frac{1}{2}\rho v^2 = constant$ is applicable along any magnetic line of force. The pressure reduction where v is large exactly compensates for the increased magnetic pressure at the same place, satisfying (10.59). The centrifugal force of the fluid streaming along the field lines with curvature K exactly compensates the centripetal force of the tension $B^2/4\pi$ in those same lines, $\rho v^2 K = KB^2/4\pi$.

Imagine, then, a slender flux tube with weak uniform magnetic field B_0 over the cross-section A_0 extending between the planes $z=\pm L$. There is a slow motion of the fluid with velocity $V_0 = B_0/(4\pi\rho)^{\frac{1}{2}}$ along B_0. Suppose that there is at our disposal the means to massage this tube with some small external pressure difference δP, applied along the surface of the tube, as well as at the ends. Consider what changes in the field and fluid velocity such gentle manipulation can accomplish.

If the cross-sectional area of the tube is A while the scale of variation of field and cross-section along the tube has a characteristic scale λ large compared to the radius $(A/\pi)^{\frac{1}{2}}$ of the tube, then the transverse components of field and velocity within the tube are small compared to the longitudinal components by the factor $A^{\frac{1}{2}}/\lambda$ and can be neglected. The equation of motion for the longitudinal flow velocity v along the tube is

$$\frac{\partial v}{\partial t}+\frac{\partial}{\partial z}\left(\frac{p}{\rho}+\tfrac{1}{2}v^2\right)=0 \tag{10.60}$$

while the hydromagnetic equation is

$$\frac{\partial B}{\partial t}+v\frac{\partial B}{\partial z}-B\frac{\partial v}{\partial z}=0. \tag{10.61}$$

Conservation of magnetic flux requires

$$AB = A_0 B_0 \tag{10.62}$$

and conservation of fluid,

$$\frac{\partial A}{\partial t}+\frac{\partial}{\partial z}(vA)=0. \tag{10.63}$$

Substitution of (10.62) into (10.63) to eliminate A in favour of B yields (10.61), of course.

The total pressure of the fluid and field outward against the sides of the tube is $p+B^2/8\pi$. This is equal to the total external pressure $p_0+B_0^2/8\pi+\delta P$, exerted inward on the tube, so that δP is

$$\delta P(z,t)=p(z,t)+B^2(z,t)/8\pi-(p_0+B_0^2/8\pi), \tag{10.64}$$

and is presumed to be at our disposal. It is this small pressure difference δP that is to manipulate the tube. It is a straightforward procedure to solve (10.60) and (10.61) for the pressure $\delta P(z, t)$ associated with any particular deformation of the tube. Keeping in mind that we are interested in producing a slow deformation of the tube while remaining in the neighbourhood of the equipartition solution (10.59), write

$$v = V + u, \qquad V = +B/(4\pi\rho)^{\frac{1}{2}} \tag{10.65}$$

with the supposition that u is small. Then (10.61) reduces exactly to

$$\frac{\partial}{\partial t}\frac{1}{V} + \frac{\partial}{\partial z}\frac{u}{V} = 0.$$

Integration over z yields

$$u(z, t) = -V(z, t)\int_{-L}^{z} \mathrm{d}s \frac{\partial}{\partial t}\frac{1}{V(s, t)} \tag{10.66}$$

with the boundary condition that the fluid is undisturbed ($u = 0$, $V = V_0$, $p = p_0$) at $z = -L$. Integrating (10.60) over z leads to

$$p(z, t) + \tfrac{1}{2}\rho v^2(z, t) = p_0 + \tfrac{1}{2}\rho V_0^2 - \int_{-L}^{z} \mathrm{d}s \frac{\partial v(s, t)}{\partial t}. \tag{10.67}$$

Hence, with the definition (10.64), the field and velocity changes in the tube are related to the externally applied pressure δP by

$$\begin{aligned}
\delta P(z, t)/\rho &= -uV - \tfrac{1}{2}u^2 - \int_{-L}^{z} \mathrm{d}s \frac{\partial v(s, t)}{\partial t} \\
&= -\int_{-L}^{z} \mathrm{d}s \frac{\partial V(s, t)}{\partial t}\left\{1 + \frac{V^2(z, t)}{V^2(s, t)}\right\} \\
&\quad -\tfrac{1}{2}V^2(z, t)\left\{\int_{-L}^{z} \frac{\mathrm{d}s}{V^2(s, t)}\frac{\partial V(s, t)}{\partial t}\right\}^2 \\
&\quad + \int_{-L}^{z} \mathrm{d}s \left\{\frac{\partial V(s, t)}{\partial t}\int_{-L}^{s} \mathrm{d}s' \frac{\partial}{\partial t}\frac{1}{V(s', t)}\right. \\
&\quad \left. + V(s, t)\int_{-L}^{s} \mathrm{d}s' \frac{\partial^2}{\partial t^2}\frac{1}{V(s', t)}\right\}.
\end{aligned}$$

This can be simplified in keeping with the spirit of the present calculation that the deformation is slow, so that $\partial V/\partial t \ll V\partial V/\partial z$. Then

$$\int_{-L}^{z} \mathrm{d}s \frac{\partial V(s, t)}{\partial t} \ll \tfrac{1}{2}V^2(z, t),$$

$$\int_{-L}^{z} \mathrm{d}s \frac{\partial^2 V(s, t)}{\partial t^2} \ll \tfrac{1}{2}V(z, t)\frac{\partial V(z, t)}{\partial t},$$

etc. and all terms second order in the time derivative of V can be neglected, yielding the first-order expression

$$\delta P(z, t) \cong -\rho \int_{-L}^{z} \mathrm{d}s \frac{\partial V(s, t)}{\partial t} \left\{1 + \frac{V^2(z, t)}{V^2(s, t)}\right\} \tag{10.68}$$

for the pressure $\delta P(z, t)$ to bring about the change $\partial V(z, t)/\partial t$. The pressure δP decreases with z along any part of the tube where the field is increasing, $\partial V/\partial t > 0$, and the magnitude of δP is proportional to the rate of increase of V. It follows that an applied pressure δP of arbitrarily small magnitude can manoeuvre the flux tube into any intense field configuration provided only that δP is applied for a sufficiently long period of time. In simple terms, the fluid motion V along the tube provides the mechanical advantage. This necessary motion is transformed, by the conservation of flux and fluid along the tube, so that the system remains in the neighbourhood of the equipartition solution, and the mechanical advantage continues throughout the deformation. The deviation $u(z, t)$ of the fluid velocity from the equipartition solution is small, being first order in $\partial V/\partial t$.

It is evident, then, that a weak field B_0 and small fluid velocity V_0 in a non-dissipative system can, over a period of time, be transformed into a field B of arbitrarily large intensity by an externally applied pressure $\delta P(z, t)$ of arbitrarily small magnitude. In view of the fact that the system remains near the stable equipartition solution throughout the transformation, there should be no strong instabilities. Clearly any final equipartition state is stable.

To illustrate the nature of the external pressure necessary to produce a simple compression of the tube, consider the simple example

$$V(z, t) = V_0\{1 + \alpha(t)\exp(-|z|/l)\} \tag{10.69}$$

wherein $V(-L, t) = V_0$ and the weak magnetic field B_0 is compressed over a region of characteristic length $2l$ in the vicinity of $z = 0$, to a maximum value $B_0[1 + \alpha(t)]$ at $z = 0$. Supposing that L is so large that $\exp(-L/l) \cong 0$, it follows from (10.68) that

$$\delta P(z, t) = -\rho l V_0 \frac{\mathrm{d}\alpha}{\mathrm{d}t}\{2 + \alpha \exp(z/l)\}\exp(z/l),$$

in the region $z < 0$, while for $z > 0$,

$$\delta P(z, t) = -\rho l V_0 \,\mathrm{d}\alpha/\mathrm{d}t$$
$$\times \left[2 + \alpha + \{1 - \exp(-z/l)\}\left\{1 + \frac{1 + \alpha \exp(-z/l)}{1 + \alpha}\right\}\right].$$

Then if the field is increasing with time ($\mathrm{d}\alpha/\mathrm{d}t > 0$), the pressure decreases

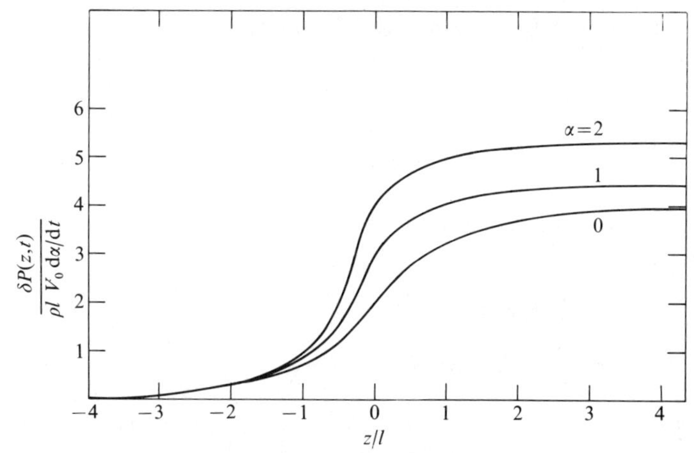

FIG. 10.3. A plot of the externally applied pressure $\delta P(z, t)$ in units of $-\rho l V_0\, \mathrm{d}\alpha/\mathrm{d}t$, causing growth of the compressed field (10.69) at the rate $V_0\, \mathrm{d}\alpha/\mathrm{d}t$ at $z=0$.

monotonically with z. The total pressure difference between the upper and lower ends of the tube is

$$\delta P(L, t) = -\rho l V_0 \frac{\mathrm{d}\alpha}{\mathrm{d}t} \frac{(2+\alpha)^2}{1+\alpha}. \tag{10.70}$$

The pressure $\delta P(z, t)$ is plotted in Fig. 10.3 in units of $-\rho l V_0\, \mathrm{d}\alpha/\mathrm{d}t$ for the initial departure of the tube from uniformity ($\alpha=0$) and for compression of the field at $z=0$ to $2B_0(\alpha=1)$, $3B_0(\alpha=2)$, etc. The reader may easily explore a variety of other cases. The basic principles are adequately illustrated by the present elementary example and show unlimited growth of field.

10.7. Field compression with convective forces: incompressible case

Suppose that the flux tube extends upward through a region of convective instability (as in the region immediately below the solar photosphere) so that the fluid motion along the flux tube is subject to a force proportional to the equipartition velocity V. In place of (10.60) write

$$\frac{\partial v}{\partial t} + \frac{\partial}{\partial z}\left(\frac{p}{\rho} + \tfrac{1}{2}v^2\right) = \frac{V}{\tau} \tag{10.71}$$

where τ is a constant, measuring the characteristic acceleration time of the fluid by the convective forces. With the boundary condition that the deviation $u(z, t)$ from the equipartition velocity $V(z, t)$ does not vanish at $z=-L$, the solution of (10.61) becomes

$$\frac{u(z,t)}{V(z,t)} = \frac{u(-L,t)}{V(-L,t)} - \int_{-L}^{z} \mathrm{d}s \frac{\partial}{\partial t} \frac{1}{V(s,t)}$$

in place of (10.66). Integrating (10.71) yields

$$p(z,t)+\tfrac{1}{2}\rho v^2(z,t)=p(-L,t)+\tfrac{1}{2}\rho v^2(-L,t)-\rho\int_{-L}^{+L}\mathrm{d}s\left\{\frac{\partial V(s,t)}{\partial t}-\frac{V(s,t)}{\tau}\right\}$$

in place of (10.67), upon neglecting terms second order in the time derivative.

The pressure difference necessary to carry out the deformation is (10.64)

$$\begin{aligned}\delta P(z,t)=&\,p(-L,t)+\tfrac{1}{2}\rho V^2(-L,t)-p_0-\tfrac{1}{2}\rho V_0^2\\&+\rho\left[V(-L,t)u(-L,t)-V^2(z,t)\frac{u(-L,t)}{V(-L,t)}\right.\\&\left.+V^2(z,t)\int_{-L}^{z}\mathrm{d}s\frac{\partial}{\partial t}\frac{1}{V(s,t)}-\int_{-L}^{z}\mathrm{d}s\left\{\frac{\partial V(s,t)}{\partial t}-\frac{V(s,t)}{\tau}\right\}\right].\end{aligned} \tag{10.72}$$

Suppose that the compression of the field is to arise spontaneously, with no externally applied manipulative forces, $\delta P=0$. Then (10.72) reduces to an equation for $V(s,t)$. The solution is particularly simple if it is assumed that the field at $z=-L$ is very weak and remains unperturbed, $V_0\cong 0$, $u(-L,t)\cong 0$, for in that case

$$0=V^2(z,t)\int_{-L}^{z}\mathrm{d}s\frac{\partial}{\partial t}\frac{1}{V(s,t)}-\int_{-L}^{z}\mathrm{d}s\left\{\frac{\partial V(s,t)}{\partial t}-\frac{V(s,t)}{\tau}\right\}. \tag{10.73}$$

There is a similarity solution of the form

$$V(z,t)=F(z)\exp(\sigma t),$$

for which

$$0=\sigma F(z)\int_{-L}^{z}\frac{\mathrm{d}s}{F(s)}+\left(\sigma-\frac{1}{\tau}\right)\int_{-L}^{z}\mathrm{d}sF(s).$$

Differentiate once with respect to z and divide through by $2\sigma F(z)F'(z)$. Then differentiate again with respect to z. The results can be written

$$F'/F=(1-1/2\sigma\tau)F''/F'.$$

The solution is

$$F(z)=(C_1z+C_2)^{1-2\sigma\tau},$$

where C_1 and C_2 are arbitrary constants. Since $V(-L,t)=V_0\cong 0$, write the solution as

$$V(z,t)=V_1(1+z/L)^{1-2\sigma\tau}\exp(\sigma t)$$

for an initial velocity V_1 at $z = 0$. Substituting this result back into (10.73) yields an identity, so that there is no dispersion relation relating σ to τ. The growth rate σ is determined by the initial form of $V(z, t)$ together with the strength $1/\tau$ of the convective forces.

The deviation from the equipartition state is

$$u(z, t) = (L/2\tau)(1 + z/L)$$

which is independent of time and is small for small convective forces (large τ). The fluid pressure is

$$\begin{aligned} p(z, t) &= p(-L, t) - \tfrac{1}{2}\rho V^2(z, t) \\ &= p(-L, t) - \tfrac{1}{2}\rho V_1^2(1 + z/L)^{2(1-2\sigma\tau)}\exp(2\sigma t), \end{aligned}$$

so that the fluid pressure within the tube declines with height and with time.

A particularly interesting special case is $2\sigma\tau = 1 - \epsilon$ where $\epsilon \ll 1$. Then the growth time of the field is $2\tau(1 + \epsilon + \ldots)$ and the field and velocity have the form

$$V(z, t) = V_1(1 + z/L)^{\epsilon}\exp\{(1-\epsilon)t/2\tau\}$$

with

$$p(z, t) = p_0 - \tfrac{1}{2}\rho V_1^2(1 - z/L)^{2\epsilon}\exp\{(1-\epsilon)t/2\tau\}.$$

The flux tube is relatively broad at $z = -L$ where the fluid enters, but has a smaller and nearly uniform cross-section elsewhere. The variation of field strength B, and radius $A^{\frac{1}{2}}$, along the tube is plotted in Fig. 10.4 for

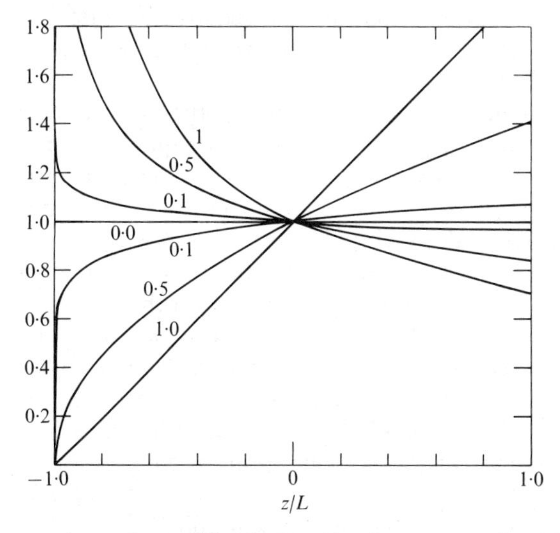

FIG. 10.4. The curves increasing monotonically with z/L represent the field strength, of Alfven speed $V(z, t)/V_1$, while the decreasing curves represent the relative radius of the tube. The values of $\epsilon = 1 - 2\sigma\tau$ are indicated on each curve.

$\epsilon = 0$, 0·1, 0·5, and 1·0. The important point is that the field grows exponentially with time given only that the convective force $1/\tau$ is non-vanishing. Any prescribed field strength is reached in the course of time.

We could go on from here to explore other solutions and cases, generally working toward the illustration of a complete self-consistent solution for a body of fluid in which the magnetic field is brought into a compressed state by the convective forces. In particular, it is desirable to demonstrate the limits for dynamical stability. However, there is a more elementary problem that needs attention, and that is the nature of the equilibrium solutions in a stratified atmosphere where the gas density varies along the flow.

10.8. Fluid motion within a flux tube: stratified atmosphere

The analysis of §§10.6 and 10.7 illustrates the dynamical evolution of equipartition flow from weak to strong fields under the influence of forces of arbitrarily small magnitude. The field within the flux tube grows slowly, but it grows without limit. Arbitrarily large field strengths—and associated flow velocities, of course—can be achieved by the most modest of forces. This remarkable conclusion is, of course, valid only in the circumstances of the ideal incompressible fluid for which it was calculated. When we turn to the question of the evolution of a flux tube extending vertically across a stratified atmosphere with pressure scale height $\Lambda = kT/mg$, there are limitations introduced by the fact that equipartition $\frac{1}{2}\rho v^2 = B^2/8\pi$ can ordinarily be achieved at only a single height in the atmosphere (Parker 1976c).

Consider, then, a slender, isolated flux tube of cross-section $A(s)$ and field strength $B(s)$ extending along some path in an atmosphere of temperature $T(z)$ and mean particle mass m in a gravitational acceleration g in the negative z-direction. Distance along the tube is denoted by s, making an angle $\phi(s)$ with the vertical z-direction.

It is clear that if the path of the flux tube were horizontal, then the fluid density would be uniform along the tube and the analysis of the previous sections would be applicable initially. But we do not expect to find horizontal convective forces, so it is not clear what might drive the fluid velocity in a horizontal tube toward the concentrated form. In any case the flux tubes in the sun are more or less vertical, as are the convective forces, so it is the vertical configuration on which we concentrate attention.

Then conservation of magnetic flux requires

$$A(s)B(s) = A_0B_0, \tag{10.74}$$

while conservation of matter leads to

$$A(s)\rho(s)v(s) = A_0\rho_0 v_0 \tag{10.75}$$

for the steady fluid velocity $v(s)$ along the tube in the gas density $\rho(s)$. Dividing (10.74) into (10.75) yields the basic conservation condition

$$\rho(s)v(s)/B(s) = \rho_0 v_0/B_0.$$

Hence the ratio of the kinetic energy density to the magnetic energy density varies inversely with the density

$$\frac{\frac{1}{2}\rho v^2}{B^2/8\pi} \propto \frac{1}{\rho},$$

The ratio of the kinetic energy to the magnetic energy increases with height. Hence the equipartition condition $B_i = (4\pi\rho)^{\frac{1}{2}} v_i$ can hold at only one height in the atmosphere. As we shall soon see, this restricts the growth of intense fields and limits them to the general vicinity of the height where $\rho v^2 = B^2/4\pi$. It will be shown that this can occur only in the neighbourhood of the level where $\mathrm{d}\Lambda/\mathrm{d}z = -\frac{1}{2}$, which condition is satisfied only in a narrow interval of height in the solar photosphere. But, curiously, this happens to be at just the level where the intense magnetic fields are observed. Altogether, then, we are interested in a flux tube which extends from great depth, where $\rho v^2 \ll B^2/4\pi$ upward across the height where $\mathrm{d}\Lambda/\mathrm{d}z = -\frac{1}{2}$, to somewhere above, where $\rho v^2 > B^2/8\pi$.

To illustrate the dynamical equilibrium along a flux tube in a stratified atmosphere, in which the ambient gas density varies with height z as a consequence of the gravitational acceleration g, note that the equation of motion is

$$v\frac{\mathrm{d}v}{\mathrm{d}s} + \frac{1}{\rho}\frac{\mathrm{d}p}{\mathrm{d}s} + g\cos\phi = 0$$

where s denotes distance measured along the tube and $\phi(s)$ is the angle between the axis of the tube and the vertical. Since $\mathrm{d}z = \mathrm{d}s\cos\phi$, the equation can be written

$$v\frac{\mathrm{d}v}{\mathrm{d}z} + \frac{1}{\rho}\frac{\mathrm{d}p}{\mathrm{d}z} + g = 0 \tag{10.76}$$

for any inclination of the tube. Suppose that the temperature of the gas flowing along inside the tube has the same value $T(z)$ as the gas outside, where the pressure is $P(z) = P(0)f(z)$ with

$$f(z) = \exp\left\{-\int_0^z \mathrm{d}z' \frac{mg}{kT(z')}\right\}.$$

It is convenient to denote height z in units of the scale height $\Lambda_0 = kT_0/mg$ at the reference level $z=0$, so that $z=\Lambda_0\zeta$, and to write the temperature in units of T_0, so that $T(z)=T_0\theta(z)$. Then

$$f(\zeta)=\exp\left\{-\int_0^\zeta d\xi/\theta(\xi)\right\}. \quad (10.77)$$

The field density $B(z)$ is determined by the difference between the external pressure $P(z)$ and the internal pressure $p(z)$,

$$B^2(z)/8\pi = P(z)-p(z).$$

If the field is measured in units of its strength B_0 at $z=0$, so that $B(z)=B_0\psi(z)$, then the condition for lateral pressure balance can be written

$$p(\zeta)=P(0)\{f(\zeta)-n\psi^2(\zeta)\} \quad (10.78)$$

where $n\equiv B_0^2/8\pi P(0)$. Write (10.75) as

$$\begin{aligned} v &= v_0\psi\rho_0/\rho \\ &= v_0\psi\theta p_0/p \\ &= v_0(1-n)\psi\theta/(f-n\psi^2). \end{aligned} \quad (10.79)$$

The equation of motion (10.76) can now be written

$$\frac{d}{d\zeta}\left\{\frac{\mu(1-n)^2\psi^2\theta^2}{(f-n\psi^2)^2}\right\}+\theta\frac{d}{d\zeta}\ln(f-n\psi^2)+1=0 \quad (10.80)$$

where $\mu\equiv mv_0^2/2kT_0$.

The solution of (10.80) is easily worked out for a number of special cases. Thus, for instance, for the isothermal case, for which $\theta=1$, it follows from (10.77) that $f=\exp(-\zeta)$, and from (10.80) that

$$\mu\left\{\frac{(1-n)^2\psi^2}{(f-n\psi^2)^2}-1\right\}+\ln\left\{\frac{f-n\psi^2}{f(1-n)}\right\}=0 \quad (10.81)$$

since f and ψ are both equal to one at $z=0$. As another example, the special case $\theta=1-\frac{1}{2}\zeta$ is immediately integrable, yielding $f=\theta^2=\psi^2$, so that the field pressure declines in direct proportion to the gas pressure.

In the weak field limit, with n, $\mu \ll 1$, (10.80) reduces to

$$\frac{1}{2}\frac{d(\ln\psi^2)}{d\zeta}=\frac{1+d\theta/d\zeta-nf/2\mu\theta}{\theta(nf/\mu\theta-1)}\{1+O(n\psi^2/f)\} \quad (10.82)$$

over the range of θ for which $n\psi^2 \ll f$. Thus, the weak-field problem is reduced to a quadrature. For the isothermal case, $\theta=1$, the weak-field solution is

$$\psi^2(\zeta)=\frac{(n-\mu)f^2(\zeta)}{nf(\zeta)-\mu}. \tag{10.83}$$

It is evident that (10.83) increases without bound as $f(\zeta)$ approaches μ/n, so that the exact solution (10.81) must be used in that neighbourhood.

We are interested as much in the general structure of the solutions of (10.80) as in the individual special solutions and examples. To consider the general structure note that $\theta(\zeta)$ can be expressed in terms of $f(\zeta)$, and note from (10.77) that $\mathrm{d}f/\mathrm{d}\zeta=-f/\theta$. Then (10.80) can be written

$$\frac{\mathrm{d}(\ln\psi^2)}{\mathrm{d}\zeta}=\frac{F(\psi^2,f)}{G(\psi^2,f)}, \tag{10.84}$$

where

$$F(\psi^2,f)\equiv\frac{2\mu(1-n)^2\theta f}{(f-n\psi^2)^2}-n+\frac{2\mu(1-n)^2\theta}{f-n\psi^2}\frac{\mathrm{d}\theta}{\mathrm{d}\zeta}, \tag{10.85}$$

$$G(\psi^2,f)\equiv\theta\left\{n-\frac{\mu(1-n)^2\theta(f+n\psi^2)}{(f-n\psi^2)^2}\right\}. \tag{10.86}$$

It is evident that the magnetic field pressure ψ^2 passes through an extremum at the zeros of F, while $\mathrm{d}\zeta/\mathrm{d}\psi^2$ vanishes and ζ passes through an extremum at the zeros of G. Thus, for instance, if G vanishes at $\zeta=\zeta_\mathrm{a}$, where $\psi(\zeta_\mathrm{a})\equiv\psi_\mathrm{a}$, $f(\zeta_\mathrm{a})\equiv f_\mathrm{a}$, the solution $\psi^2(\zeta)$ is confined to one side of ζ_a, where it may be double-valued. If $C\equiv(\partial G/\partial\psi^2)_\mathrm{a}$ and $D\equiv(\partial G/\partial f)_\mathrm{a}$, where again the subscript denotes the value at $\zeta=\zeta_\mathrm{a}$, then (10.84) becomes

$$\frac{\psi_\mathrm{a}^2\theta_\mathrm{a}}{f_\mathrm{a}}\frac{\mathrm{d}f}{\mathrm{d}\psi^2}=-\frac{C(\psi^2-\psi_\mathrm{a}^2)+D(f-f_\mathrm{a})}{F_\mathrm{a}}$$

in the neighbourhood of $(f_\mathrm{a},\psi_\mathrm{a}^2)$. The solution is

$$C(\psi^2-\psi_\mathrm{a}^2)+D(f-f_\mathrm{a})=\frac{CF_\mathrm{a}\psi_\mathrm{a}^2\theta_\mathrm{a}}{Df_\mathrm{a}}\left[1-\exp\left\{-\frac{Df_\mathrm{a}}{F_\mathrm{a}\psi_\mathrm{a}^2\theta_\mathrm{a}}(\psi^2-\psi_\mathrm{a}^2)\right\}\right].$$

Expand the exponential in ascending powers of $\psi^2-\psi_\mathrm{a}^2$. To lowest order,

$$f-f_\mathrm{a}=-C(f_\mathrm{a}/2F_\mathrm{a}\psi_\mathrm{a}^2\theta_\mathrm{a})(\psi^2-\psi_\mathrm{a}^2)^2,$$

so that f goes through a maximum or minimum (depending upon the sign of C/F_a) and returns from whence it came. Thus there is no dynamical solution extending through the atmosphere from deep down ($f\gg1$, $G>0$) up to the top ($f\ll1$, $G<0$). The only solution extending across the entire atmosphere is hydrostatic equilibrium, $\mu=0$.

The only way to avoid the turn-around at ζ_a is for F to have a zero at ζ_a as well. Then if *both* F and G have simple zeros at ζ_a, their ratio is finite and (10.84) provides a smooth continuous solution across the *critical point*, which we now denote by ζ_c (i.e. $\zeta_a \equiv \zeta_c$ if G and F both have simple zeros there). Expand F and G about f_c and ψ_c so that (10.84) can be written

$$\frac{d\psi^2}{df} = \frac{(\psi^2 - \psi_c^2)A + (f - f_c)B}{(\psi^2 - \psi_c^2)C + (f - f_c)D}$$

to lowest order, where $A \equiv (\theta_c\psi_c^2/f_c)(\partial F/\partial\psi_c^2)$, $B \equiv (\theta_c\psi_c^2/f_c)(\partial F/\partial f)_c$, and C and D are the corresponding derivatives of G, as already defined. Define the function $S(f - f_c)$ as

$$S(f - f_c) \equiv (\psi^2 - \psi_c^2)/(f - f_c).$$

The variables S and $f - f_c$ are separable and the solution of the equation is

$$(f - f_c)(S - a)^{\alpha}(S - b)^{\beta} = Q$$

where Q is an arbitrary constant, a and b are the roots of $CS^2 + (D - A)S - B$, so that

$$2Ca = A - D + \{(A - D)^2 + 4BC\}^{\frac{1}{2}},$$
$$2Cb = A - D - \{(A - D)^2 + 4BC\}^{\frac{1}{2}},$$

and the exponents α and β are

$$2\alpha = 1 + (A + D)/\{(A - D)^2 + 4BC\}^{\frac{1}{2}},$$
$$2\beta = 1 - (A + D)/\{(A - D)^2 + 4BC\}^{\frac{1}{2}},$$

The solutions $S = a$, $S = b$ for $Q = 0$ are of particular interest, representing smooth passage across the critical point $f = f_c$. In the near neighbourhood of f_c, there are the two solutions

$$2C(\psi^2 - \psi_c^2) = (a, b) \times (f - f_c). \tag{10.87}$$

Thus there is a solution extending continuously from the bottom to the top, across the critical point in the middle, provided only that F and G have common zeros. It is this requirement on the zeros of F and G that determines the ratio μ/n of the kinetic energy density to the field energy density $\frac{1}{2}\rho_0 v_0^2/B_0^2/8\pi$ at $z = 0$.

Consider, then, the conditions necessary for the simultaneous vanishing of F and G. Put $F = G = 0$ and solve (10.86) for n, using the result to eliminate the term n from (10.85). The result is

$$\frac{2\mu(1-n)^2}{(f - n\psi^2)^2}\left(\frac{d\theta}{d\zeta} + \frac{1}{2}\right) = 0$$

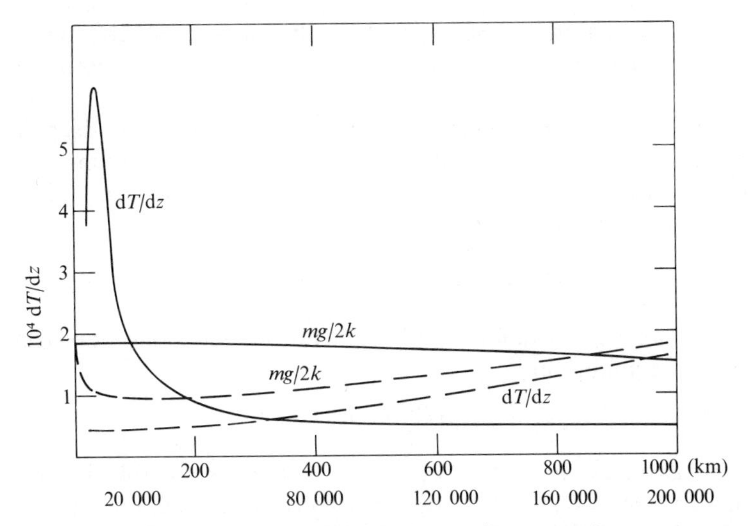

FIG. 10.5. The solid curve represents dT/dz in the sun (Spruit 1974) from the base of the photosphere to a depth of 1000 km, while the dashed curve represents dT/dz to the base of the convective zone at a depth of 2×10^5 km. The two curves labelled $mg/2k$ represent the critical values $d\theta/d\zeta=-\frac{1}{2}$ at the respective heights in the convective zone. The peak of dT/dz at a depth of about 40 km corresponds to $d\theta/d\zeta\cong-1\cdot5$.

requiring[3] that $d\theta/d\zeta=-\frac{1}{2}$, i.e. that

$$\frac{d}{dz}\frac{kT}{mg}=-\tfrac{1}{2}. \tag{10.88}$$

The scale height, $\Lambda=kT/mg$, must vary half as fast as the height z. The existence of a suitable dynamical solution, then, depends upon the temperature declining sufficiently rapidly with height. Before going on with the development of the mathematical properties of the solutions it is appropriate to inquire whether the necessary condition is ever satisfied in nature.

Consider the sun. The solid curve in Fig. 10.5 gives dT/dz in the outer 1000 km of the convective zone, based on the model given by Spruit (1974). The nearly horizontal solid line represents the critical value (10.38), $mg/2k=1\cdot86\times10^{-4}\,\mathrm{K\,cm^{-1}}$ for $g=2\cdot7\times10^4\,\mathrm{cm\,s^{-2}}$ and a gas composed of 95 per cent hydrogen and 5 per cent helium by number. The dashed curves represent dT/dz from the top of the convective zone down to its base at a depth of 2×10^5 km, and the critical value $mg/2k$ as a function of depth, allowing for the increase of gravity and the decrease of mean molecular weight from the increasing ionization. It is evident that

[3] The alternatives are $\mu=0$ (hydrostatic solution), or $n=1$ (vanishing gas density within the tube so that v is undefined and the result is a special case of $\mu=0$.)

the critical condition (10.88) is met in the uppermost 100 km of the convective zone. And this is precisely where the intense flux tubes are observed.

We are particularly interested in the circumstance that deep in the atmosphere the kinetic energy density of the fluid motions is small compared to the magnetic energy density, which is in turn small compared to the thermal energy density,

$$\tfrac{1}{2}\rho v^2 \ll B^2/8\pi \ll P.$$

In this region the solution to the dynamical equations approximates hydrostatic equilibrium

$$\psi^2(\zeta) \propto f(\zeta) \tag{10.89}$$

with $n\psi^2(\zeta) \ll f(\zeta)$. Assume that n and μ are comparable in order of magnitude. Suppose that, as in the solar convective zone, the temperature scale height is large compared to the pressure scale height so that $d\theta/d\zeta \ll 1$. Then it is readily seen by inspection of (10.85) and (10.86) that, deep in the atmosphere (where $f \gg 1$, $n\psi^2$), the function F is negative and G is positive. It is also evident from the facts that $f + n\psi^2 > f$ and $d\theta/d\zeta < 1$ that as f declines upward through the atmosphere, the function G reaches its zero before F. In this situation there is no dynamical solution extending up past the zero of G. The solution turns around at $\zeta = \zeta_0$. Only the completely hydrostatic solution $\mu = 0$ extends all the way, since in that limiting case ζ_0 moves up to $\zeta = +\infty$.

Suppose, then, that there is a thin stratum in the upper atmosphere, at some level $\zeta = \zeta_c$ in which the temperature declines sufficiently rapidly so that $d\theta/d\zeta = -\tfrac{1}{2}$. There is a value for the flow velocity, characterized by $\mu = mv_0^2/2kT_0$, such that the zero of G occurs at $\zeta = \zeta_c$, at which F also vanishes, so that the solution extends smoothly across the critical point ζ_c as described by (10.87). Thus the construction of the dynamical solution extending from the bottom to the top of the atmosphere is an eigenvalue problem which determines the value of μ/n.

The problem is particularly simple in the weak field case, for which (10.82) is the appropriate equation. It is clear that the denominator of the right-hand side vanishes where $nf = \mu\theta$. In order for the solution to extend across ζ_c, then, nf must have the value $\mu\theta$ at ζ_c, in which case

$$d(\ln \psi^2)/d\zeta = -1/2\theta$$

at the critical point. But presumably the temperature $\theta(\zeta)$ is specified, so that ζ_c is determined by the condition $(d\theta/d\zeta)_c = -\tfrac{1}{2}$. The function $f(\zeta)$ follows from $\theta(\zeta)$ through (10.77). Hence, for the zero of the denominator of (10.82) to coincide with $d\theta/d\zeta = -\tfrac{1}{2}$, n/μ must have the value

$$n/\mu = \theta(\zeta_c)/f(\zeta_c). \tag{10.90}$$

That is to say, the characteristic flow velocity v_0 is determined by

$$\begin{aligned} v_0 &= V_A(\mu/n)^{\frac{1}{2}} \\ &= V_A(f_c/\theta_c)^{\frac{1}{2}} \end{aligned} \tag{10.91}$$

where $V_A \equiv B_0/(4\pi\rho_0)^{\frac{1}{2}}$ is the Alfven speed at the reference level $\zeta = 0$ where the fluid velocity is v_0.

To illustrate the general nature of the dynamical solution in a stratified atmosphere, consider the simple situation in which the temperature declines upward according to

$$\theta(\zeta) = \frac{1 - \exp\{h(\zeta - \zeta_1)\}}{1 - \exp(-h\zeta_1)} \tag{10.92}$$

from $\zeta = -\infty$ to $\zeta = \zeta_2 < \zeta_1$, above which the temperature has the fixed value

$$\theta_2 = \frac{1 - \exp\{h(\zeta_2 - \zeta_1)\}}{1 - \exp(-h\zeta_1).} \tag{10.93}$$

The temperature is essentially uniform in the lower part of the atmosphere and would fall to zero at $\zeta = \zeta_1$ were it not for the intervention of the isothermal atmosphere at $\zeta_2 (< \zeta_1)$. The ambient pressure follows from (10.77) as

$$f(\zeta) = \theta(\zeta)^{\{1-\exp(-h\zeta_1)\}/h} \exp[-\{1 - \exp(-h\zeta_1)\}\zeta]. \tag{10.94}$$

In order that there be a critical point, it is necessary that $\mathrm{d}\theta/\mathrm{d}\zeta$ fall to $-\frac{1}{2}$ in $\zeta < \zeta_2$. Hence, it must be required that

$$2h \exp\{h(\zeta_2 - \zeta_1)\} > 1 - \exp(-h\zeta_1).$$

Since $\zeta_2 < \zeta_1$, this can be satisfied if and only if

$$2h > 1 - \exp(-h\zeta_1). \tag{10.95}$$

For the weak field case, $n\psi^2 \ll f$, the magnetic field follows directly from integration of (10.82). If

$$h = 1 - \exp(-h\zeta_1), \tag{10.96}$$

then the integration is elementary. The temperature is

$$h\theta(\zeta) = 1 - (1 - h)\exp(h\zeta)$$

and the pressure is

$$f(\zeta) = \theta(\zeta)\exp(-h\zeta).$$

Since ψ is normalized to unity at $\zeta=0$, the magnetic field is

$$\psi^2(\zeta)=\frac{h\exp(-h\zeta)}{1-\exp\{-h(\zeta_1-\zeta)\}}. \tag{10.97}$$

The critical point lies at

$$h\zeta_c=-\ln\{2(1-h)\}. \tag{10.98}$$

Evidently ζ_c can be positive or negative depending upon whether h is greater or less than one-half. It follows with the choice (10.96) that ζ_c is related to ζ_1 by

$$\zeta_c=\zeta_1-(1/h)\ln 2. \tag{10.99}$$

Finally note that the temperature and pressure at the critical point are $\theta_c=1/2h$ and $f_c=2(1-h)$ respectively, while the magnetic field is $\psi_c^2=4h(1-h)$. The field is a minimum at $\zeta=\zeta_c$. The eigenvalue follows from (10.90) or (10.91) as

$$\mu/n=2(1-h). \tag{10.100}$$

This ratio must be positive on physical grounds, requiring that $0<h<1$. It is obvious from (10.97) that for large negative ζ

$$\psi^2(\zeta)\sim hf(\zeta),$$

while for $\zeta\rightarrow\zeta_1$

$$\begin{aligned}\psi^2(\zeta)&\sim(1-h)/(\zeta_1-\zeta)\\&=\psi_c^2/4h(\zeta_1-\zeta).\end{aligned} \tag{10.101}$$

Altogether, then, the magnetic pressure at large depth is proportional to the ambient gas pressure, declining with height with a scale height Λ_0/h. The field passes through a minimum at ζ_c, and, in the weak field approximation, increases without bound toward the top $\zeta=\zeta_1$. Above ζ_2 the temperature is uniform, with value

$$h\theta_2=1-(1-h)\exp(\alpha\zeta_2),$$

and the pressure is

$$f(\zeta)=f_2\exp\{-(\zeta-\zeta_2)/\theta_2\}$$

where $f_2=\theta_2\exp(-\alpha\zeta_2)$. Note from (10.96), (10.98), and (10.100) that with $\zeta_c<\zeta_2<\zeta_1$ we have

$$2(1-h)>\exp(-h\zeta_2)>1-h$$

and

$$nf_2/\mu\theta_1=\{1/2(1-h)\}\exp(-\alpha\zeta_2)<1.$$

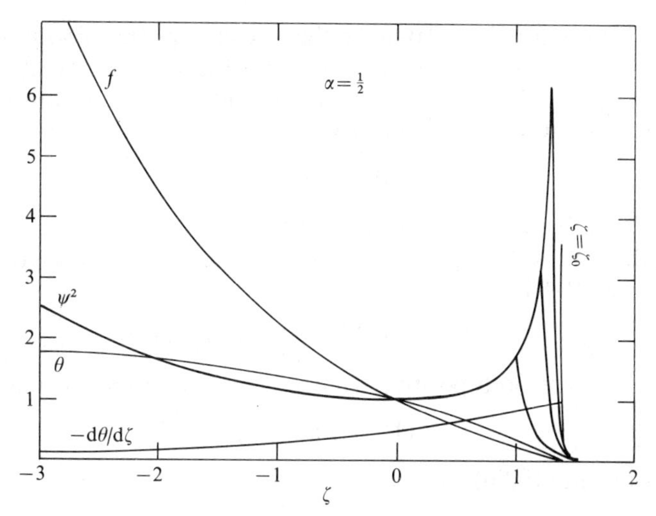

FIG. 10.6. The heavy curve, labelled ψ^2, represents the magnetic field pressure normalized to unity at $\zeta=0$, as given by (10.97) for $h=0\cdot5$. The pressure f and temperature θ, all normalized to unity at $\zeta=0$, are also shown. An isothermal atmosphere is appended at $\zeta=1\cdot0$, $1\cdot2$, and $1\cdot3$, with the corresponding drop in ψ^2. The light vertical line at the right-hand end corresponds to $\zeta=\zeta_0=1\cdot386$.

Hence $nf/\mu\theta_2<1$ for all $\zeta>\zeta_2$, and it follows from (10.82) that ψ^2 declines monotonically with increasing height above ζ_1. It is readily shown from integration of (10.82) that the magnetic field in the isothermal region above ζ_2 is

$$\psi^2(\zeta)=\psi_2^2\frac{\{2(1-h)-\exp(-h\zeta_2)\}\exp\{(\zeta_2-\zeta_1)/\theta_2\}}{2(1-h)\exp\{(\zeta-\zeta_2)/\theta_2\}-\exp(-\alpha\zeta_1)}$$

where ψ_2^2 is given by (10.97) evaluated at ζ_2.

The field is plotted in Fig. 10.6 for the special case $h=0\cdot5$, for which $\zeta_c=0$ and $\zeta_1=1\cdot386$. The field above ζ_2 is shown for $\zeta_2=1$, $1\cdot2$, $1\cdot3$. Figure 10.6 also shows $\theta(\zeta)$ and $f(\zeta)$ (without the isothermal plateau above ζ_2). The important feature is the sharp rise of the magnetic field above the critical point at $\zeta=0$, as indicated by (10.101). In the present weak field approximation, the increase becomes arbitrarily large as θ_2 declines below the value θ_c at the critical point. Thus the limiting increase of a weak field depends quantitatively upon the detailed temperature structure of the atmosphere. In the sun (see Fig. 10.5) the critical point, at $d\theta/d\zeta=-\frac{1}{2}$, lies very close under the photosphere and temperature minimum, admitting the possibility of strong concentration.

10.9. The limits on field compression in a stratified atmosphere

Consider now a simple example using the exact equations for strong fields in order to see what the real limits on field intensification in a stratified atmosphere are. To make the calculations tractable suppose that the temperature is uniform with a value $\theta = 1$ in $\zeta < 0$, and uniform with a value $\theta = \theta_1 \equiv 1 - \frac{1}{2}\zeta_1$ in $\zeta > \zeta_1 > 0$. In the intervening region, $0 < \zeta < \zeta_1$, let the temperature decline linearly

$$\theta(\zeta) = 1 - \tfrac{1}{2}\zeta,$$

so that $d\theta/d\zeta$ is just equal to the critical value $-\frac{1}{2}$.

Note that in $\zeta < 0$ the pressure is just $f(\zeta) = \exp(-\zeta)$. Hence deep in the atmosphere (10.81) reduces to

$$n\psi^2(\zeta) \sim f(\zeta)\{1 - (1-n)\exp\mu\}. \tag{10.102}$$

But ψ^2 and f are both positive. Hence it must be required that

$$(1-n)\exp\mu < 1. \tag{10.103}$$

It follows from (10.86) that

$$G \sim n\{1 + O(1/f)\}$$

and is positive. Therefore, G must remain positive all the way up to $\zeta = 0$, if there is to be a dynamical solution. Hence at $\zeta = 0$, where $\psi^2 = f = \theta = 1$,

$$G = n - \mu(1+n) > 0$$

or

$$\mu < n/(n+1). \tag{10.104}$$

This upper limit on μ automatically satisfies (10.103). Hence the dynamical solution will have an eigenvalue which satisfies (10.104).

In the transition region $(0, \zeta_1)$ the solution is $f = \theta^2 = \psi^2 = (1 - \frac{1}{2}\zeta)^2$. The solution must cross the critical point somewhere in $(0, \zeta_1)$, so that G must be negative at ζ_1. It follows from (10.86) that

$$G(\zeta_1) = n\theta_1 - \mu(1+n).$$

If this is to be negative, then

$$\theta_1 < \mu(1+n)/n < 1,$$

i.e.

$$\zeta_1 > 2\{1 - \mu(1+n)/n\}$$

if there is to be a dynamical solution extending across the critical point. The critical point lies at the zero of G, so that (10.86) yields

$$n\theta_c = \mu(1+n).$$

Hence

$$\zeta_c = 2\{1 - \mu(1+n)/n\} \geqslant 0$$

or

$$\frac{\mu}{n}=\frac{1-\frac{1}{2}\zeta_c}{1+n}\leqslant 1. \tag{10.105}$$

The location of the critical point is limited only to the interval $(0, \zeta_1)$ in this example, so that μ/n is not precisely fixed but is only limited, as indicated by (10.105). If we fix ζ_c by letting ζ_1 decrease to zero, then, since $0<\zeta_c<\zeta_1$, it follows that μ/n takes on the limiting value $1/(1+n)$ allowed by (10.104).

In the isothermal region $\zeta>\zeta_1$ the pressure is

$$f(\zeta)=f_1\exp\{-(\zeta-\zeta_1)/\theta_1\}$$

and the field is given by

$$\frac{\mu(1-n)^2\theta_1^2\psi^2}{(f-n\psi^2)^2}+\theta_1\ln\left(\frac{f-n\psi^2}{f/\theta_1^2}\right)=\mu+\theta_1\ln\{(1-n)\theta_1^2\}.$$

As $\zeta\to+\infty$ the pressure goes exponentially to zero and

$$\psi^2\sim\frac{\mu+\theta_1\ln(1-n)}{\mu\theta_1^2(1-n)^2}f^2.$$

The magnetic field pressure ψ^2 declines as the square of the gas pressure.

The field declines monotonically in $\zeta>0$ from its maximum $\psi=1$ at $\zeta=0$. The interesting physical question is the factor by which the field increases up to $\zeta=0$. Note the asymptotic solution (10.102) deep in the atmosphere, which can be rewritten as

$$B^2(z)/8\pi\sim P(z)\{1-(1-n)\exp\mu\}.$$

On the other hand, at $\zeta=0$

$$B^2(0)/8\pi=nP(0).$$

Thus the ratio of field pressure to gas pressure is enhanced as ζ increases to zero by the factor

$$R=n/\{1-(1-n)\exp\mu\}. \tag{10.106}$$

The enhancement factor is largest when μ assumes the maximum value allowed by (10.104),

$$R_{\max}=n/[1-(1-n)\exp\{n/(n+1)\}] \tag{10.107}$$

plotted in Fig. 10.7 as a function of n. Note that n is just the ratio of the magnetic pressure to the ambient gas pressure evaluated at the reference level $\zeta=0$. In the present example the ratio of field to gas pressure is a maximum at $\zeta=0$, so n represents the maximum value of that ratio. It is evident from (10.107) and Fig. 10.7 that the enhancement factor $R_{\max}$

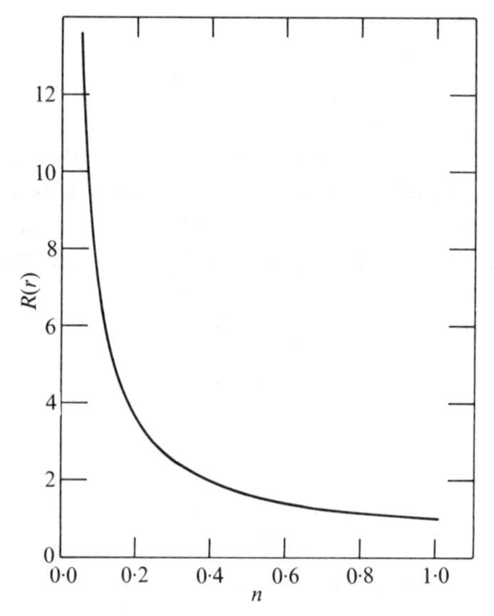

FIG. 10.7. A plot of the dynamical enhancement R_{max}(eqn (10.107)) of the magnetic pressure relative to the ambient gas pressure as a function of the ratio $n \equiv B_0^2/8\pi P(0)$ of field pressure to gas pressure at $\zeta = 0$.

increases without bound as n declines toward zero ($R_{max} \sim 2/3n$) (see §10.8) and decreases to one as n increases to 1. The intense flux tubes of 1500 G in the sun involve $n \cong 0{\cdot}4$ at the critical point where $d\theta/d\zeta = -\frac{1}{2}$ about 10^2 km below the visible surface. The enhancement factor is $R_{max} \cong 2$ in this case.

In Fig. 10.8 the gas pressure $f(\zeta)$ is plotted, together with the field pressure $\psi^2(\zeta)$, for $n = 0{\cdot}1$, $0{\cdot}2$, and $0{\cdot}4$ with μ equal to the maximum value $n/(n+1)$. The field above $\zeta = 0$ may vary in a variety of ways. If, for instance $\zeta_1 = 2$, then the temperature falls linearly to zero at ζ_1 and the decline of ψ^2 is the same for all n. The field is also shown above $\zeta = 0$ for $n = 0{\cdot}1$ and the arbitrary choice $\zeta_1 = 0{\cdot}2005$ so that $\theta_1 = 0{\cdot}90$ and $f_1 = \frac{9}{11}$, and for $n = 0{\cdot}2$ with $\zeta_1 = 0{\cdot}406$ so that $\theta_1 = 0{\cdot}797$ and $f_1 = \frac{2}{3}$. The decline of ψ^2 above the critical point at $\zeta = 0$ is of secondary interest at the moment.

The principal point of interest is the sudden upsurge of the magnetic pressure ψ^2 over the gas pressure toward the critical point at $\zeta = 0$, seen in Fig. 10.8. It is important to note, however, that the upsurge, by the enhancement factor R_{max}, is limited to a factor of about 2 for $n = 0{\cdot}4$, corresponding to about 1500 G at $\zeta = 0$. The dynamical compression of the field is not insignificant, then, nor is it adequate to accomplish the entire feat of compression from a few hundred G to 1500 G entirely on its own. The dynamical enhancement may function in partnership with other effects, such as reduced temperature, to bring about the extraordinary compression of field indicated by the observations.

It is obvious that if the dynamical effects are to be combined with cooling, then the analysis should be repeated without the restraining assumption that the temperature within the flux tube is equal to the temperature outside. This leads to mathematical difficulties in the form of movable critical points. Another obvious theoretical question is the stability of the equilibrium dynamical solution. Only the equilibrium solutions have been presented here; the very difficult question of their stability has been put off to some future time. Chandrasekhar (1961) has demonstrated stability of the equipartition solution in an incompressible fluid. In the stratified atmosphere stability is expected below the critical point, where the tension in the field dominates the Reynolds stresses ($\rho v^2 < B^2/4\pi$). At some height above the critical point, however, the Reynolds stress dominates; the Reynolds stress increases upward relative to the magnetic stress as ρ^{-1}. Consequently, the upper end of the flow is unstable (Geronicholas 1977b). The resulting turbulent instability above the critical point may be the cause of the spicule phenomenon higher up in the atmosphere.

If the dynamical enhancement plays a role in the concentration of the field, then the updrafts and/or downdrafts within the individual flux tubes must be comparable to the Alfven speed. A field of 1500 G at $\zeta = 0$

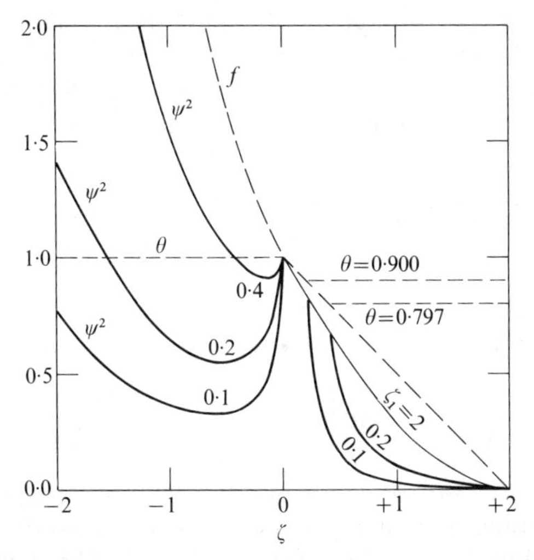

FIG. 10.8. A plot of magnetic field pressure ψ^2 (given by eqn (10.81)) normalized to unity at $\zeta = 0$ for $n = 0{\cdot}1$, $0{\cdot}2$, and $0{\cdot}4$. Above $\zeta = 0$ the temperature declines with height as $\theta = 1 - \frac{1}{2}\zeta$, becoming isothermal again at $\zeta_1 = 0{\cdot}2005$ in the case $n = 0{\cdot}1$ and at $\zeta_1 = 0{\cdot}405$ for $n = 0{\cdot}2$. For $\zeta_1 = 2$, ψ^2 follows the part of the curve labelled $\zeta_1 = 2$ for all n. The temperatures are shown by the dashed line and the pressure $f(\zeta)$ by the dotted line for $\zeta < 0$. For $\zeta > 0$, the pressure coincides with the curve for ψ^2 labelled $\zeta_1 = 2$.

where $\rho = 3{\cdot}5 \times 10^{-7}\ \mathrm{g\,cm^{-3}}$ (Spruit 1974) corresponds to a speed of $7\ \mathrm{km\,s^{-1}}$. Deubner (1976) has pointed out that some observations of the downdrafts in association with the intense but unresolved flux tubes suggest the possibility of extreme velocities. Confirmation or denial of high velocities is the critical observational test, obviously. If observations should confirm strong downdrafts, the next theoretical step would be to develop a quantitative model.

10.10. Reduced temperatures and field concentration

10.10.1. General ideas and possibilities

The sections above have examined the fluid motions within a slender flux tube, and the compression of the field, as a consequence of external pressure fluctuations and as a consequence of quasi-steady flow along the tube. The effects are many and varied, but none surpasses in any significant degree the equipartition relation $B^2 = 4\pi\rho v^2$. Hence, fluid velocities of $6\text{–}8\ \mathrm{km\,s^{-1}}$ would be required to produce the observed 1500 G fields. The observed gas velocities at the surface of the sun are not so large. Hence, it would seem that the fields are compressed by some mechanism other than a direct confrontation of the Reynolds and Maxwell stresses. The alternative is a temperature reduction ΔT of the gas within the flux tube, causing a partial evacuation of the tube by the forces of gravity. If $T(z)$ is the ambient temperature, while the temperature within the tube is $T(z) - \Delta T(z)$, then it follows from (8.7) (or (8.10)) that the ratio of the magnetic field pressure in a slender flux tube to the ambient gas pressure $p(z)$ outside varies with height according to

$$\frac{B^2(z)}{8\pi p(z)} = 1 - \left\{1 - \frac{B^2(z_0)}{8\pi p(z_0)}\right\}\exp\left\{-\int_{z_0}^{z} \frac{\mathrm{d}z\,\Delta T(z)}{z_0\Lambda(z)[T(z) - \Delta T(z)]}\right\} \tag{10.108}$$

above the base level z_0, where $\Lambda(z)$ is the pressure scale height kT/mg. If, for instance, the field is weak at z_0, so that $B^2(z_0) \ll 8\pi p(z_0)$, then

$$\frac{B^2(z)}{8\pi p(z)} \cong 1 - \exp\left\{-\int_{z_0}^{z} \frac{\mathrm{d}z\,\Delta T(z)}{\Lambda(z)[T(z) - \Delta T(z)]}\right\} \tag{10.109}$$

(see also (8.13)). As already noted, even a modest temperature decrease $\Delta T/T \ll 1$, extending over many scale heights can produce a significant pressure reduction within the tube, so that $B^2/8\pi$ may be a major fraction of the ambient gas pressure p.

What, then, might produce a temperature decrease within an intense magnetic flux tube extending more or less vertically upward to the surface of the sun? Sunspots are cool at the surface and the compression of the field to 3000 G may be understood on that basis (Parker 1955a; Schlüter and Temesvary 1958; Deinzer 1965). Perhaps the small flux tubes are to be understood as mini-sunspots, then. Unfortunately there has not been a definitive demonstration of a sunspot cooling mechanism of such a nature as to produce the observed field concentration. The *observed* cooling is adequate if it extends to a depth of $1{\cdot}2 \times 10^3$ km, but no mechanism for such cooling has been shown. It was pointed out by Biermann (1941) that the strong field of the sunspot suppresses the convective transport of heat, which is the principal energy transport throughout the convective core. Choking off the heat supply from below clearly produces a cool surface, in agreement with observations, but it also dams up the heat below the sunspot so that the gas is hotter, rather than cooler, than the ambient temperature $T(z)$, thereby dispersing the field, so far as we have been able to work out (Parker 1974c, 1976a; Boruta 1977). A simple model is constructed in §10.10.2 to illustrate the difficulty.

In view of the non-existence of any formal example illustrating the formation of a sunspot as a consequence of magnetic inhibition, we suggested that the convective generation of Alfven waves in the strong field immediately below the sunspot might consume so much energy as to produce the observed cooling and the attendant concentration of field (Parker 1974c, 1975c; Boruta 1977). The idea was originally proposed by Danielson (1962) to explain the missing energy flux of sunspots. If the outflow of heat is blocked by the magnetic field, one might expect the heat to flow around the edges of the spot, appearing as a bright ring immediately outside. There is no bright ring, so there was a serious question as to the fate of the energy that does not appear in the dark sunspot umbra.

Danielson suggested that the convective forces convert the heat into Alfven waves, which then propagate along the magnetic field up into the atmosphere above the spot and/or back down into the interior of the sun. Presumably, the waves propagate with but little dissipation. Their ultimate fate is not of direct interest to the sunspot question. The idea was developed by Savage (1969), Musman (1967) and Moore (1973) who showed that, with the partial reflection of Alfven waves in the steep density gradients near the visible surface, the convecting fluid is subject to overstability (Thompson 1951; Chandrasekhar 1952, 1953, 1956a, 1961), generating Alfven waves. The linearized theory cannot calculate the strength of the resulting Alfven waves, of course. We suggested that the idea should be turned around (Parker 1974c, 1975c) with the generation of Alfven waves the principal cause of the cooling of the sunspot, rather than merely a device for accounting for some missing energy. About 75 per cent of the normal photospheric brightness is missing in the umbra of a sunspot (3900 K as opposed to the normal 5600 K), so the convective heat engine must be approximately 75 per cent efficient. In this view the magnetic inhibition of convective heat transport becomes a secondary effect, helping to shape the temperature pattern of the sunspot, with its sharp edges, etc. but it is not the principal cause of the sunspot. Boruta (1977) has illustrated the formation of a sunspot by Alfven wave cooling in terms of the growth of a small perturbation in an initially uniform magnetic field.

The idea that sunspots are cooled principally by the convective generation of Alfven waves has not been confirmed by observation, although some strict limits have been placed. Unfortunately there is no feasible way to calculate how much wave energy should appear above a sunspot. Geronicholas (1977a) has shown that upward propagating waves are partially reflected back down into the sun in the steep density gradient at the photosphere, the amount of the reflection increasing rapidly with the period and wave length of the waves. The observations of Beckers (1976) and Beckers and Schneeberger (1977) put limits on the Alfven wave flux above a sunspot, but the limits are not incompatible with the expected wave flux if one takes account of the reflections in the low chromosphere (where the wavelength is large because of the declining gas density). Cowling (1976) asserts that 75 per cent efficiency is too much to expect of any heat engine, but offers no concrete alternative (Parker, 1977). It is interesting to note, too, that Spruit (1977) has shown that the absence of a bright ring around a sunspot does not necessarily imply that there is any heat flux missing. The convective heat transport at depths of 10^4 km in the sun is so effective that the heat blocked by the magnetic field of the sunspot spreads out over many times the radius of the sunspot in the relatively opaque surface layer of the sun, contributing a brightness enhancement so diffuse as to be well below the level of detectability.

Roberts (1976) has pointed out that convective overstability is to be expected in the slender flux tubes in granule and supergranule boundaries. Hence one expects at least some modest cooling. Even a small reduction of temperature, by some 10 per cent over 5 scale heights, is sufficient to give fields in excess of 10^3 G at the photosphere. The heat engine would need an efficiency of only 35 per cent. The 10 per cent reduction of temperature is comparable to the temperature reduction in the intergranule dark lanes and would be undetectable[4].

Altogether this leaves the theory of the concentration of magnetic flux

[4] The flux tubes in the supergranule boundaries are associated with bright dots, which are, in part, of chromosphere origin (as a consequence of the dissipation of Alfven waves perhaps). Spruit (1977) has pointed out, however, that a slender flux tube with an internal temperature below the ambient temperature of each level may appear bright because we see deeper into the sun along the tube, to where the ambient temperature is higher.

tubes in an untidy state. The dynamical effects treated above are real, but observational estimates of the fluid velocities imply that their dynamical consequences are not of major importance in achieving 1500 G. The theory of the generation of Alfven waves indicates that waves are produced in the small flux tubes and in sunspots, but neither theory nor observations can show yet that there are enough waves produced to provide the necessary cooling. It would seem advisable, then, to look further into the physics of the flux tubes in a convective environment.

§10.10.2 presents a simple linearized model of the magnetic inhibition of convective heat transport and its effect on the growth of magnetic concentrations from an initial uniform magnetic field. The result is a general stability of the field, with any initial inhomogeneities decaying rather than growing. The magnetic inhibition of convective heat transport by itself contributes nothing to initiating the concentration of magnetic fields. In §10.11, then, we return to the problem of the concentration of fields in the small flux tube and the magnetic knot, pointing out that their universal association with downdrafts, together with the magnetic inhibition of convection, suggests a new approach to the physics of field concentration, in which the superadiabatic gradient in the convective zone plays a direct role. The theory shows that field concentration in downdrafts at the top of a convective zone is a direct and automatic consequence of the downdraft and the superadiabatic gradient.

10.10.2. Magnetic inhibition of heat flow and the concentration of fields

Consider the temperature variations beneath the surface of the sun arising from the magnetic inhibition of convective heat transport. There are a number of examples that can be treated to illustrate the basic physical principles (cf. Parker 1976a). Suppose, for instance, that there is an initial vertical uniform magnetic field B_0 extending upward in the z-direction across a highly conducting atmosphere with a uniform temperature gradient, so that in barometric equilibrium,

$$T(z)=T_0(1-z/\lambda),\qquad p(z)=p_0(1-z/\lambda)^{\alpha},\qquad \rho(z)=\rho_0(1-z/\lambda)^{\alpha-1}.$$

The scale height of the atmosphere at $z=0$ is $\Lambda_0=kT_0/mg$, and it is readily shown from the barometric relation that $\lambda=\alpha\Lambda_0$. Suppose, then, that the system is perturbed slightly, so that the magnetic field, while remaining nearly vertical, has a magnitude $B_0+\delta B(x, y, z)$. As a consequence of the change in field strength, suppose that the normal uniform heat transport coefficient κ_0 becomes $\kappa_0(1-q)$ where the reduction q is proportional to the relative change in the magnetic pressure,

$$q(x, y, z)=\beta^2\frac{B^2(z)-B_0^2}{8\pi p(z)}\cong\beta^2\frac{B_0\delta B(x, y, z)}{4\pi p(z)}$$

where β is a numerical constant. The change in the magnetic pressure is a consequence of the change in the total temperature, written $T(z)-\delta T(x, y, z)$, through the barometric relation (10.109). Remembering that (10.109) gives the variation of the magnetic pressure from the ambient value, i.e. $(B^2-B_0^2)/8\pi p$, it follows for infinitesimal δT that

$$q(x, y, z)=\frac{\beta^2 T_0}{\Lambda_0}\int_{z_0}^{z}\frac{\mathrm{d}z'\delta T(x, y, z')}{T^2(z')}. \tag{10.110}$$

The heat flow equation is

$$\rho(z)C\left(\frac{\partial\delta T}{\partial t}+v_z\frac{\mathrm{d}T}{\mathrm{d}z}\right)=\nabla.\{\kappa_0(1-q)\nabla(T-\delta T)\}. \tag{10.111}$$

With $\nabla^2 T=0$ the linearized form of this equation is

$$\frac{\partial\delta T}{\partial t}-v_z\frac{T_0}{\lambda}=\frac{\kappa_0}{\rho(z)C}\left\{\nabla^2\delta T-\frac{\mathrm{d}q}{\mathrm{d}z}\frac{T_0}{\lambda}\right\}.$$

Choose the level $z=0$ so as to be the free radiative surface. Then at $z=0$ the boundary condition is

$$\kappa_0=aT_0^3\lambda \tag{10.112}$$

for the equilibrium, and

$$-\lambda\frac{\partial\delta T}{\partial z}=4\delta T-\frac{\alpha\beta^2}{\lambda}\int_{z_0}^{1-z_0/\lambda}\mathrm{d}z\delta T/(1-z/\lambda)^2$$

for the perturbation.

We are interested in unstable solutions in which the temperature perturbation δT grows exponentially with time so that the field begins to gather itself into bunches. It follows from (10.111) that a solution of the form

$$\delta T=T_0\theta(z)\exp(ik_x x+ik_y y+t/\tau)$$

yields

$$\left\{\frac{1}{\tau}+\frac{\kappa_0}{\rho(z)C}\left(k_x^2+k_y^2+\frac{\beta^2}{\Lambda_0\lambda(1-z/\lambda)^2}-\frac{\mathrm{d}^2}{\mathrm{d}z^2}\right)\right\}\theta=\frac{v_z}{\lambda}$$

It is evident that the term $1/\tau$ for the growing mode makes a positive contribution to the coefficient of θ, equivalent to the diffusion $k_x^2+k_y^2$. It is sufficient, then, to look for neutral stability, $1/\tau=0$, because if there are no neutral solutions, then it may be presumed that there are no growing solutions. Writing $\xi=1-z/\lambda$, so that the free surface is at $\xi=1$, the differential equation reduces to $v_z=0$ and

$$\mathrm{d}^2\theta/\mathrm{d}\xi^2-\{\lambda^2(k_x^2+k_y^2)+\alpha\beta^2/\xi^2\}\theta=0, \tag{10.113}$$

and the radiative boundary condition at $\xi = 1$ is

$$\frac{d\theta}{d\xi} = 4\theta - \alpha\beta^2 \int_1^{\xi_0} \frac{d\xi\theta}{\xi^2}. \tag{10.114}$$

Writing $k^2 = k_x^2 + k_y^2$, the solutions of (10.113) are expressible as

$$\theta = \xi^{\frac{1}{2}} K_p(k\lambda\xi), \qquad \xi^{\frac{1}{2}} I_p(k\lambda\xi)$$

in terms of the modified Bessel functions of order p, where $p^2 = \alpha\beta^2 + \frac{1}{4}$.

Suppose, then, that the base level $z = z_0$ is taken to lie very deep in the atmosphere ($z_0 \gg \lambda$) where the perturbation goes to zero. Then for $k\lambda > 0$, the function I_p is excluded and the solution is just

$$\theta = \xi^{\frac{1}{2}} K_p(k\lambda\xi).$$

To treat the case for strong inhibition of heat transport, let $\alpha\beta^2 = 6$ so that $p = \frac{5}{2}$. The integral in (10.114) is readily evaluated, yielding the boundary condition

$$K_{\frac{5}{2}}(k\lambda) + (\tfrac{1}{6}k\lambda - 1/k\lambda) K_{\frac{3}{2}}(k\lambda) = 0. \tag{10.115}$$

Given that

$$K_{\frac{3}{2}}(x) = (\pi/2x)^{\frac{1}{2}}(1 + 1/x)\exp(-x),$$

$$K_{\frac{5}{2}}(x) = (\pi/2x)^{\frac{1}{2}}(1 + 3/x + 3/x^2)\exp(-x),$$

this requirement reduces to

$$(k\lambda)^3 + 7(k\lambda)^2 + 12k\lambda + 12 = 0. \tag{10.116}$$

There are no positive roots to this equation and hence no solutions satisfying the boundary conditions with neutral stability. It is readily shown that there are stable solutions for which δT and δB relax back to zero. It can also be shown that, if the base level z_0 is placed at a finite depth and a fixed perturbation is introduced there, the perturbation extends through to the surface, but greatly reduced (see also Boruta 1977).

The absence of growing perturbations is no surprise, of course. It can be shown (Parker 1976a) that the blockage of the heat flow in any element of volume δV causes the element of volume to introduce a perturbation in the local temperature field, with the form of a dipole pointing in the direction opposite to ∇T. Hence the temperature is increased on the hot side of δV and reduced on the cool side. So in an object such as a star with an internal source of heat, it is the lower side of δV that is hotter. Hence integration of the barometric relation (10.109) upward through the atmosphere first encounters an increased temperature, resulting in a reduced field at the centre of the element of volume

δV. A reduced field means reduced inhibition of convective heat transport, contrary to the initial assumption that the heat flow is blocked by δV.

It has yet to be demonstrated whether the magnetic inhibition of heat flow plays a direct role in the formation of sunspots and pores. It does play a role in the concentration of field in the small flux tubes in supergranule boundaries, however, but in quite a different way from the conventional application to sunspots, as we shall see.

10.11. The superadiabatic effect

10.11.1. General conditions within a vertical flux tube

The small concentrated flux tubes in quiet regions, and the somewhat larger tubes producing the magnetic knots in active regions, are found exclusively in downdrafts, in the boundary between supergranules, and in the dark lanes between granules. Consider the effects expected as a consequence of a downdraft within the individual flux tube.

The first point to note is the well known fact that the temperature gradient in the top few thousand km of the convective zone of the sun (or other star of moderate surface temperature) is significantly steeper than the adiabatic temperature gradient. The strong superadiabatic gradient is a direct consequence of the high opacity and large outward energy flux from the interior. The opaque gas is forced by the high heat flux into turbulent convection. The consequent eddy viscosity and eddy heat transport strongly damp the convection, so that the temperature gradient must be strongly superadiabatic in order to force enough convection to handle the heat flux.

It is also a fact that the turbulent velocities in the first few thousand km beneath the surface of the sun are comparable to the Alfven speed in fields of only 200–300 G. Hence the observed fields of 1500 G are very much stronger than the turbulence ($B^2 \gg 4\pi\rho v^2$) so that the turbulent convection is very strongly suppressed and constrained by the field. The fluid in the field may oscillate up and down along the lines of force, but does not mix freely. Hence we return to Biermann's (1941) point that convective heat transport is strongly inhibited within the magnetic field. The idea has led to a number of investigations of the suppression of convective heat transport along the lines of force of a vertical magnetic field (Walen 1949; Cowling 1953). The researches of Thompson (1951) and Chandrasekhar (1952, 1953, 1956a, 1961) established the existence of overstability and the production of oscillatory motions in strong vertical fields, as well as convective rolls aligned along the horizontal fields. Cowling (1957) pointed out, however, that oscillatory motions involve no mixing of fluid and so are ineffective in transporting heat. It is easy to show (Parker 1978) that the effective heat transfer by oscillatory motions does not exceed the order of the radiative heat transfer. In the present problem we are concerned more with the heat flow across the lines of force, into the flux tube, than with the vertical transport along the lines, because we are working with a slender flux tube (radius $\lesssim 400$ km) of considerable length (~ 4000 km). The two-dimensional interchange of vertical lines of force would transport heat into the flux tube, then. Such long convective rolls, extending along the lines of force, are excited by convective forces in *horizontal* fields (Chandrasekhar 1961), but not in the *vertical* fields with which we are presently concerned. Hence there seems little possibility for significant convective heat transfer into a concentrated vertical flux tube. The convective transfer may be as small as the radiative transfer. The radiative transfer coefficient is smaller than the normal convective transfer coefficient by a factor of about 10^5 at depths of 10^3–10^4 km below the surface of the sun. Hence, the gas within the concentrated flux tube is an extraordinary thermal insulator compared to the gas outside the flux tube. Indeed, with downdrafts of the order of 10^2 m s^{-1} or more suggested by observation (Frazier 1970; Deubner 1976), a reduction of convective transport by the factor $\epsilon = 10^{-4}$ is sufficient for the gas in the downflow to behave as a thermal insulator so that its temperature varies approximately adiabatically.

Now the general magnetic field through any region of the solar photosphere is swept by the fluid into the downdraft at the junction of three or four supergranules. The gas outside the tube descends in the downflow, mixing with the surrounding fluid as it goes down, so that its temperature increases with depth more or less as computed from the familiar mixing length models of the convective zone (see, for instance, Spruit 1974). The gas *inside* the tube descends, but it is such an excellent thermal insulator that its temperature increases nearly adiabatically with depth. The superadiabatic temperature gradient in the normal convection outside the flux tube is so much steeper than the adiabatic gradient of the gas descending inside that the gas inside rapidly becomes cooler than the gas outside at the same depth. To a first approximation, neglecting heat transport inside the flux tube, the internal temperature $T_{\rm i}$ varies along the adiabatic gradient,

$$\mathrm{d}T_{\rm i}/\mathrm{d}z = (\mathrm{d}T/\mathrm{d}z)_{\rm ad}.$$

Descending from z_2 to z_1, the temperature rises by

$$T_{\rm i}(z_1) - T_{\rm i}(z_2) = -\int_{z_1}^{z_2} \mathrm{d}z(\mathrm{d}T/\mathrm{d}z)_{\rm ad},$$

remembering that with z measured positive in the upward direction,

$(\mathrm{d}T/\mathrm{d}z)_{\mathrm{ad}}$ is a negative quantity. Hence if the gas has the same temperature $T(z_2)$ inside and outside at the level z_2, the gas inside is cooler by the amount

$$\Delta T(z_1, z_2) = \int_{z_1}^{z_2} \mathrm{d}z \left\{ \left(\frac{\mathrm{d}T}{\mathrm{d}z}\right)_{\mathrm{ad}} - \frac{\mathrm{d}T}{\mathrm{d}z} \right\}$$

at the lower level z_1.

The rate of descent of the gas may be very slow. The downdraft inside the tube need only be large enough that the reduced heat transport does not warm it up to the ambient temperature $T(z)$ outside. As a matter of fact the reduced temperature of the gas within the tube causes it to sink independently of the downflow of gas outside. The larger is ΔT, the stronger is the downward force causing the internal gas to descend. We expect, therefore, that the downflow continues within the flux tube, maintaining a temperature reduction that is a substantial fraction of the idealized value above. Hence we suggest (Parker 1978) that: (a) *a strongly reduced temperature within a flux tube in a downdraft is inevitable*, (b) *the reduced temperature is responsible for the evacuation of the flux tube and the compression of the field to* 1500 G *or more*; and (c) *the reduced temperature may also be the cause of the very intense downdrafts in the fields inferred by Deubner* (1976). The field strength in the flux tube follows from the temperature reduction ΔT through (10.109). A small temperature reduction over many scale heights produces a magnetic field pressure that may be a substantial fraction of the ambient gas pressure (Parker 1955a, 1976a). Descending only a few thousand km from the surface of the sun reduces the temperature by 10^3 K or more below the ambient value, so that $B^2(z)/8\pi p(z)$ exceeds 0·5 and $B \sim 1500$ G at the surface.

10.11.2. Heat transport and adiabatic cooling

Consider the heat transport coefficients representative of radiative and convective effects in the first several thousand km beneath the surface of the sun. The heat transport coefficient K is defined such that the heat flux is $\mathbf{F} = -K\nabla T$ erg cm^{-2} s^{-1}. For the radiative heat transport coefficient the standard expression is

$$K_{\mathrm{r}} = 4acT^3/3\kappa\rho$$

where a is Stephan's constant for black-body radiation and κ is the Rosseland mean opacity for the medium (cf. Chandrasekhar 1939). Mixing length theory decrees that the convective heat transport is of the order of $\frac{1}{3}\rho C_{\mathrm{p}} v l \{(\mathrm{d}T/\mathrm{d}z)_{\mathrm{a}} - \mathrm{d}T/\mathrm{d}z\}$ where C_{p} is the specific heat at constant pressure, v is the rms turbulent velocity, and l is the characteristic eddy size (mixing length). Convention equates l to the scale height $\Lambda(z)$.

More elaborate expressions are sometimes employed. Values for an effective vertical heat transport coefficient K_c are readily obtained from Spruit's model of the convective zone (Spruit 1974) by noting that, at depths of 10^2 km or more below the surface of the sun, convective transport is the dominant effect and by defining the heat transport as $-K_c\{dT/dz-(dT/dz)_{ad}\}$. Since $F=6\times10^{10}$ erg cm^{-2} s^{-1}, the temperature gradient obtained from Spruit's tables yields K_c. The results are presented in Table 10.1 with depth $(-z)$ measured downward from unit optical depth in the continuum. The important point is the small value of the radiative transport below depths of 100 km compared to the normal convective transport.

Consider, then, the cooling of an insulated slab of fluid, $x^2<a^2$, moving in the z-direction up a temperature gradient; consider the idealized circumstance of an incompressible fluid with an initial temperature distribution $T_0(1+z/h)$. Suppose that the heat transport coefficient in the slab $x^2<a^2$ has the uniform value ϵK, with specific heat ρC. The heat transport outside the slab, $x^2>a^2$, is K and is extremely large compared to ϵK. Denote the thermometric conductivity by $Q=\epsilon K/\rho C$. At time $t=0$ the column is set in motion in the positive z-direction with a uniform velocity v. The transfer equation within the slab is

$$\frac{\partial T_i}{\partial t}+v\frac{\partial T_i}{\partial z}=Q\left(\frac{\partial^2 T_i}{\partial x^2}+\frac{\partial^2 T_i}{\partial z^2}\right). \tag{10.117}$$

The external temperature remains at the initial value $T_e=T_0(1+z/h)$ as a consequence of the high thermal conductivity. The temperature within the slab can be written

$$T_i(x, z, t)=T_0\{f(x, z)+\theta(x, z, t)\} \tag{10.118}$$

where $T_0 f(x, z)$ is the asymptotic steady state temperature (as $t\to\infty$)

TABLE 10.1

The specific heat per unit volume and the convective radiative heat transport coefficients as a function of depth below the visible surface of the sun, taken from Spruit (1974)

Depth (km)	K_r (erg cm s K)	K_c (erg cm s K)	*Specific heat* ρC_p (erg cm^3 K)
33	$2{\cdot}8\times10^{13}$	$1{\cdot}5\times10^{14}$	60
120	$1{\cdot}22\times10^{12}$	8×10^{14}	$2{\cdot}1\times10^{2}$
1×10^{3}	$1{\cdot}5\times10^{10}$	$1{\cdot}7\times10^{16}$	$3{\cdot}9\times10^{3}$
1×10^{4}	$3{\cdot}3\times10^{9}$	3×10^{18}	$3{\cdot}2\times10^{5}$
2×10^{4}	$1{\cdot}4\times10^{10}$	1×10^{19}	$1{\cdot}7\times10^{6}$

satisfying

$$v\frac{\partial f}{\partial z}=Q\left(\frac{\partial^2 f}{\partial x^2}+\frac{\partial^2 f}{\partial z^2}\right)$$

and θ is the initial transient caused by setting the slab in motion. It is readily shown that

$$f(x, z)=1+z/h-(v/2Qh)(a^2-x^2). \tag{10.119}$$

Hence, with the initial temperature distribution $T_0(1+z/h)$, it follows that the initial form of θ is

$$\theta(x, z, 0)=(v/2Qh)(a^2-x^2). \tag{10.120}$$

The desired solution of (10.117) is then (cf. Carslaw and Jaeger, 1959)

$$\theta(x, z, t)=\frac{v}{2Qh}\sum_{n=1}^{\infty} C_n \cos(n-\tfrac{1}{2})\frac{\pi x}{a}\exp\{-Q(n-\tfrac{1}{2})^2\pi^2 t/a^2\}.$$

It is readily seen that this form satisfies (10.117) and vanishes at $x^2=a^2$ so that the temperature is continuous across the faces of the slab. The coefficients C_n are readily evaluated using (10.120), with the final result

$$\theta(x, z, t)=\frac{2a^2 v}{Qh\pi^3}\sum_{n=1}^{\infty}\frac{(-1)^n\cos(n-\tfrac{1}{2})\pi x/a}{(n-\tfrac{1}{2})^3}\exp\{-Q(n-\tfrac{1}{2})^2\pi^2 t/a^2\}. \tag{10.121}$$

Note, then, that the temperature on the midplane ($x=0$) of the slab is lower than the temperature of the external medium at the same level z by the amount

$$\Delta T=T_0(va^2/2Qh)\left[1-\frac{4}{\pi^3}\sum_{n=1}^{\infty}\frac{(-1)^{n-1}}{(n-\tfrac{1}{2})^3}\exp\{-Q(n-\tfrac{1}{2})^2\pi^2 t/a^2\}\right].$$

The relaxation time to the final asymptotic state is $4a^2/\pi^2 Q$, and the final asymptotic temperature difference is

$$\Delta T=T_0 a^2 v/2Qh. \tag{10.122}$$

The adiabatic temperature gradient of the incompressible fluid is zero, so the difference between the ambient gradient $\mathrm{d}T/\mathrm{d}z$ and the adiabatic gradient is T_0/h. Hence ΔT is equal to $a^2 v/2Q$ times the difference in the temperature gradients.

To treat a specific example, note that Table 10.1 gives the ambient thermometric heat transport coefficient at a depth of 10^3 km as $K_c/\rho C_p=4\times10^{12}\ \mathrm{cm^2\,s^{-1}}$. The temperature gradient is $T_0/h=3{\cdot}4\times10^{-6}\ \mathrm{K\,cm^{-1}}$. For a thin slab, say $a=10^2$ km, then, the relaxation time is $10/\epsilon$ s and the final reduction is

$$\Delta T=0{\cdot}4\times10^{-4}v/\epsilon. \tag{10.123}$$

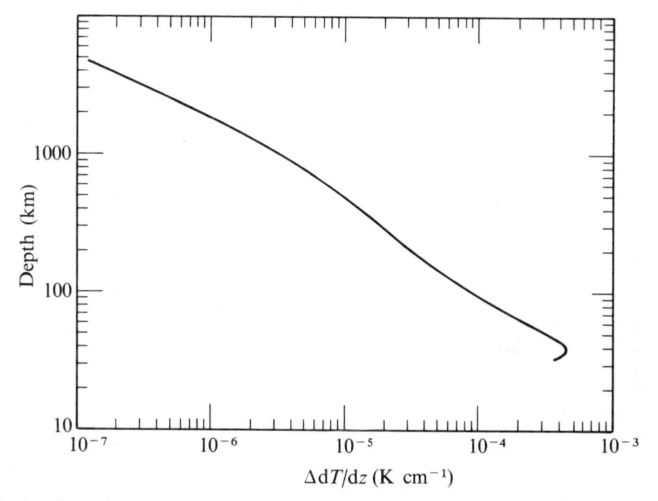

FIG. 10.9. A plot of the excess $\Delta\, dT/dz$ of the temperature gradient over the adiabatic gradient as a function of depth below the surface of the sun. It is this excess that causes the reduced temperature within flux tubes.

Thus with ϵ as small as 10^{-4} and a downdraft of $10^2\ \mathrm{m\,s^{-1}}$ (cf. Frazier 1970) there arises the enormous temperature difference of 4×10^3 K (the ambient temperature at a depth of 10^3 km is 15×10^3 K). The difference is so large because it is the asymptotic value for an infinitely long slab.

10.11.3. Conditions beneath the surface of the sun

Consider the difference ΔT between the temperature of gas flowing downward in a flux tube and the ambient gas temperature $T(z)$ of the convective zone of the sun. The ambient and the adiabatic temperature gradients d(ln T)/d(ln p) are available from Spruit (1974), so that

$$\Delta(dT/dz) = dT/dz - (dT/dz)_{ad}$$
$$= \frac{\mu m g}{k}\left\{\frac{d(\ln T)}{d(\ln p)} - \left(\frac{d(\ln T)}{d(\ln p)}\right)_{ad}\right\}, \qquad (10.124)$$

where g is the gravitational acceleration ($2{\cdot}7\times10^4\ \mathrm{cm\,s^{-2}}$), m is the mass of the hydrogen atom, and μ is the molecular weight (available from Spruit 1974). The difference $\Delta\, dT/dz$ is plotted in Fig. 10.9 as a function of depth. Integrating downward from a depth of 33 km (below unit optical depth) the resulting ΔT is shown in Fig. 10.10 by the solid curve in units of thousands of degrees. The integral

$$I \equiv \frac{\mu m g}{k}\int \frac{dz\,\Delta T}{T(T-\Delta T)} \qquad (10.125)$$

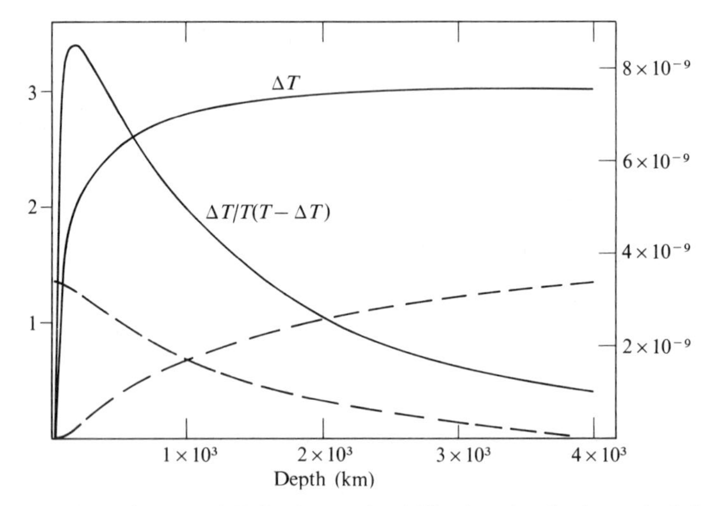

FIG. 10.10. The solid curve ΔT (in thousands of K), given by the integral of the excess $\Delta\, dT/dz$ downward from a depth of 33 km below the surface, represents the temperature reduction of an adiabatic volume of gas below the ambient temperature. The curve labelled $\Delta T/T(T-\Delta T)$ is the integrand of (10.125) including the factor $\mu mg/k$ with the ordinate measured on the right-hand side of the figure. The two dashed curves represent the integral downward from 33 km and upward from 4000 km.

in the exponential in (10.109) is given by the two dashed curves in Fig. 10.10, integrating downward from the depth of 33 km, and upward from a depth of 4000 km. The integrand $\Delta T/T(T-\Delta T)$ is also shown. Figure 10.11 shows the ratio $B^2/8\pi p$, and the field strength B in units of 10^4 G, as a function of depth under the simple assumption that the internal and external gas pressures are equal (so that $B=0$) at a depth of 4000 km. The depth of 4000 km was chosen quite arbitrarily for the present illustration. Since we do not know the velocity of the downdraft nor the precise value of the heat transport reduction factor ϵ, there is no way to determine how far down into the sun the downdraft extends. The depth of 4000 km yields a surface field of 1700 G, comparable to the observed values. A greater depth would produce a stronger field. The example is sufficient to illustrate the physical principles.

10.11.4. Discussion

It would appear from the foregoing that the concentration of the general magnetic field of the sun into isolated flux tubes is a consequence of the supergranules and of the strong superadiabatic temperature gradient in the first few thousand km beneath the surface. The supergranules sweep the lines of force into the junctions of their boundaries. The field is compressed to a few hundred G by the converging flows. The gas within the field is part of the general downdraft in the boundary. The suppression of convective heat transport by the field allows the temperature of the gas flowing downward within the field to fall below the temperature of the ambient gas outside the tube. The downdraft within the field is then enhanced, and the upper portion of the field is strongly evacuated by gravity, producing the observed concentration to 1500 G. For lack of a better term we call this process the *superadiabatic effect.*

Given that the breakup and compression of the general solar magnetic field into intense isolated bundles is the direct consequence of the supergranules and the superadiabatic temperature gradient, it follows that the same breakup and compression is to be expected in the fields of other G-stars. The existence of magnetic fields in other G-stars is an inference, of course, from the fact that the sun has a field. Indeed, since the magnetic field of the sun is a consequence of the convection and circulation (Parker 1955b, 1957, 1971; Steenbeck *et al.* 1966; Leighton 1969; Steenbeck and Krause 1969), we may infer that all main sequence stars with convective zones, i.e. with surface temperatures below about 8000 K, have magnetic fields. And, if we correctly understand the cause of the breakup and compression of the general field of the sun, we expect that the fields in such stars are also subject to breakup into isolated flux tubes in the boundaries of the dominant convective cells. We expect, too, that the tubes are compressed by the barometric consequences of the downdraft, along the lines illustrated in the sections above. It would be interesting to resolve the fields at the surface of an M-dwarf, for

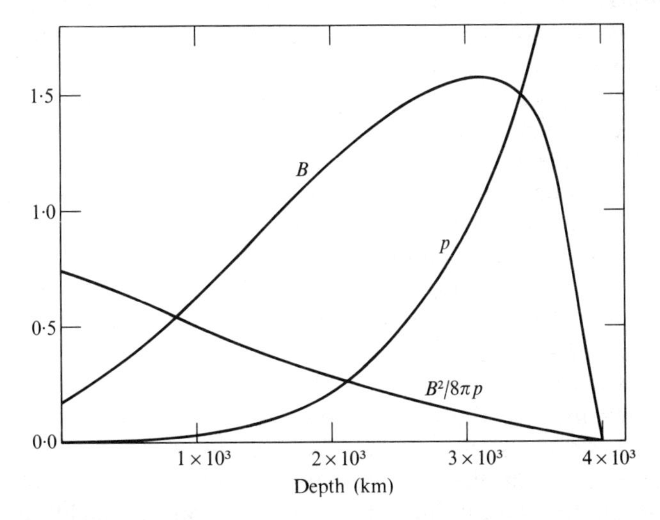

FIG. 10.11. The ratio of magnetic pressure to gas pressure, the gas pressure in units of 10^8 dyn cm^{-2}, and the magnetic field in units of 10^4 G as a function of depth below the surface of the sun.

instance,—say, a U.V. Ceti flare star—to see how extreme the effects might be there. Such stars convect all the way to their centres and are known, from their enormous flare-like outbursts, to possess strong magnetic fields. However, reality decrees that we pursue the basic physics with the aid of the nearest star, the sun, which, with ingenuity and hard work, can be studied in close detail.

The general occurrence of the superadiabatic effect in the magnetic field of the sun has extensive implications for the structure of the solar atmosphere. First of all, the superadiabatic effect produces strong downdrafts but not updrafts. In an updraft the magnetic field is dispersed at the surface by both the diverging flow and the enhanced temperature of the rising gas. Only in the downdraft is the sign of the effect such as to intensify the field and the downdraft. In this way, then, we can understand the observational fact that there are strong downdrafts of 0·1–3 $km\,s^{-1}$ (Frazier 1970; Deubner 1976) in the junctions of supergranule boundaries where the fields are concentrated, with no corresponding intense updrafts anywhere.

The preference for downdrafts in magnetic fields has implications for overstability and the emission of Alfven waves. The problem of overstability in a downdraft has not yet been worked out, but, as a guess, it would seem that the strong downdraft favors the downward emission over upward emission. Thus, unfortunately, the emission of Alfven waves and the additional cooling of the flux tube becomes an elusive observational question. The theoretical evidence (Roberts 1976) is that Alfven waves are produced. The problem is to determine to what extent. The generation of Alfven waves has two consequences of possible importance. One is the cooling of the region of emission. The other is the heating of the tenuous atmosphere in the field above the surface of the sun. The problem needs further investigation.

The superadiabatic effect in a slender flux tube (radius $\sim 10^2$ km) operates in a downdraft of 10 $m\,s^{-1}$ or more. The observations of Frazier (1970) and Deubner (1976) show values from $10^2\,m\,s^{-1}$ to more than $10^3\,m\,s^{-1}$, which appear entirely adequate for the superadiabatic effect. The general circulation pattern for the downdrafts within the flux tubes raises some interesting questions. Figure 10.12 is a schematic drawing of the lines of force of the general field at the surface of the sun. The lines spread out above the surface and fill essentially all of the volume. The solid arrows indicate the general downward flow of gas along the lines. The question is the source of that gas. Somewhere gas must flow upward from the surface of the sun, perhaps in sheets between the fields of neighbouring flux tubes. Such sheets of gas would be highly unstable to the hydromagnetic exchange instability. The breakup of the sheet, and the subsequent mixing of gas into the field, may be the source of gas

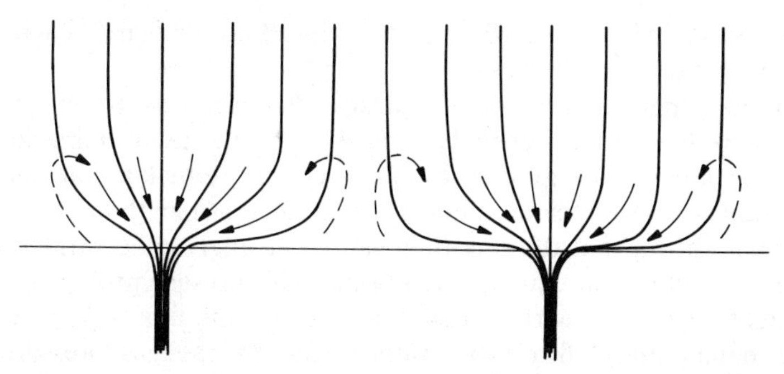

FIG. 10.12. A sketch of the magnetic field of isolated concentrated flux tubes where there is a net magnetic flux through the surface of the sun. The solid arrows represent the downdraft within the field. The dashed arrows are the conjectured upward flow supplying the downdraft.

supplying the downdraft (Giovanelli, 1977). The upward flow in this case would proceed as indicated by the dashed arrows.

The flow, driven by the downdraft in the concentrated flux tube, would contribute through instability and dissipation to the heating of the atmosphere above the surface. To estimate an upper limit to the heating of the atmosphere above, note that a downdraft of only 10^2 m s^{-1} working against the pressure differential $\Delta p = B^2/8\pi \cong 10^5$ dyn cm^{-2} does work at a rate of 10^9 erg cm^{-2} s^{-1}. If this occurs over the fraction 1/300 of the solar surface (in a region where the *mean* field is 5 G and the field in the flux tubes is 1500 G), the mean is 3×10^6 erg cm^{-2} s^{-1} over the surface of the sun (for a total of 2×10^{29} erg s^{-1}). This is enough energy to contribute significantly to the heating of the solar corona and chromosphere.

The flow of gas along and across the lines of force may involve dynamical effects. For suppose that the gas crosses over the lines of force only high up in the chromosphere or in the low corona. Then below that level the fluid streams along the lines of force. Conservation of matter and magnetic flux lead to the condition

$$\rho v/B = \rho_0 v_0/B_0, \qquad v/V_A = (v_0/V_{A0})(\rho_0/\rho)^{\frac{1}{2}}$$

along each elemental flux tube, where the subscript zero denotes the value at the surface of the sun where $\rho_0 = 3\times10^{-7}$ g cm^{-3}, $B_0 = 1500$ G, and $V_{A0} = 7{\cdot}5$ km s^{-1}. Typical observational values for downdrafts are $v_0 = 10^2$ m s^{-1} or more (Frazier 1970; Deubner 1976) in association with the flux tubes. Consider, then, the flow velocity along the same lines of force at some height z above the surface where the field has spread out to a typical mean value of, say, $B = 5$ G. The velocity increases upward, inversely proportional to the gas density ρ, becoming equal to the Alfven speed where the density has fallen to $0{\cdot}5\times10^{-10}$ g cm^{-3} in the low chromosphere. At that level $v = V_A = 2$ km s^{-1}. Higher up in the chromosphere where $\rho = 10^{-12}\,\mathrm{g\,cm^{-3}}$ the fluid velocity reaches $v = 10^2\,\mathrm{km\,s^{-1}}$ and $V_A = 14\,\mathrm{km\,s^{-1}}$.

It follows, then, that even so modest a downdraft as $10^2\,\mathrm{m\,s^{-1}}$ at the photosphere leads to powerful dynamical effects in the low chromosphere where the kinetic energy density $\frac{1}{2}\rho v^2$ becomes equal to the magnetic energy density. The dynamical effects are treated at length in §§10.6–10.9 where it is shown that the flux tube can be strongly concentrated in the neighbourhood of the equipartition point. The present problem is modified slightly from the earlier treatment because the flux tube is confined by the neighbouring flux tubes rather than by the gas pressure. The condition (10.88) for a critical point is not met anywhere above the photosphere, so there is no transition from strong to weak field in steady flows. Hence, the dynamical effects are not entirely clear except that they may be large. The dynamical interaction may be unstable leading to intermittent flow and violent oscillations of the field. It must not be forgotten that the spicules are associated with the intense flux tubes (Beckers 1968; Gibson 1973; Parker 1976a), representing variations in the gas flow in the strong magnetic field (for one very simple possibility see Parker 1964).

The final point concerns the flow velocity of $v = 10^2\,\mathrm{km\,s^{-1}}$ where $\rho = 10^{-12}\,\mathrm{g\,cm^{-3}}$ in the chromosphere. No such velocities are observed. Hence, we conclude that the hydromagnetic exchange instability, or some other disorganization effect, has caused the gas to cross over the lines of force at some *lower* level. It would seem, then, that the strong dynamical effects of the downflow in the low chromosphere may be complicated by instabilities higher in the chromosphere, so that the origin of the spicule is very complicated indeed. Both the updraft that supplies the downflow, and the downflow itself must be treated to get at the overall picture of the state of the field and the fluid motions. The picture is much more complicated than the conventional notions of the structure of the chromosphere.

As a final comment, it is not clear in what way the filigree structure in the photosphere (Dunn and Zirker 1973) fits into the picture. Altogether, it is evident that the superadiabatic effect is a complicated meteorological phenomenon, with consequences for the solar atmosphere and fields above, and for the magnetic fields far below, the photosphere.

The theoretical problems are sufficiently difficult that observational guidance is essential. Both vertical and horizontal motions as a function of both horizontal and vertical position are needed, ideally with a spatial resolution of a hundred km. The velocity cannot be directly observed to such small scales, but perhaps detailed studies of line profiles (cf. Beckers and Morrison 1970) would help in determining the statistical distribution of velocities. Particularly interesting and important for the theoretical development is the nature of the downdraft as it sinks several thousand km below the visible surface of the sun. The maximum field strengths evidently occur well below the surface, where both the field and the gas flow are out of reach of observation. Only the most extensive and precise modelling of the heat transfer and the hydrodynamics can hope to work out the quantitative details of the conditions beneath the surface.

This is perhaps the appropriate place to remark that the theoretical questions posed by the extreme concentration of the flux tubes in the supergranule boundaries should not divert us entirely from other questions such as the common appearance of isolated *bipolar* regions (emerging flux tubes) in the sun, and presumably in other stars. The ephemeral bipolar regions (Harvey *et al.* 1975) appear within the supergranules and are made up of fields concentrated to the not insignificant strength of 10^2 G. The ephemeral bipolar regions show no signs of the cooling that characterizes sunspots, but otherwise they behave like miniature bipolar sunspot groups, complete with miniature flares (Vaiana *et al.* 1973). They are a subject in themselves to be considered in parallel with the sunspot group. Their existence indicates the broad range of intensities and scales over which the bipolar active region phenomenon occurs, with flares ranging from bright points below the limit of resolution ($\lesssim 400$ km) and involving energies of perhaps 10^{27} erg, to the largest flares of 10^5 km extent and 10^{32} erg. The ephemeral bipolar region is to be understood as an emerging buoyant flux tube, that owes its appearance to the buoyant non-equilibrium of the general solar magnetic fields generated far beneath the surface, and its activity to their topological non-equilibrium.

References

ALTSCHULER, M. D., TROTTER, D. E., NEWKIRK, G., and HOWARD, R. (1974). *Solar Phys.* **39,** 3.

BABCOCK, H. D. and BABCOCK, H. W. (1955). *Astrophys. J.* **121,** 349.

BATCHELOR, G. K. (1955). *Proc. Cambridge Phil. Soc.* **51,** 361.

BECKERS, J. M. (1963). *Astrophys. J.* **138,** 648.

—— (1968). *Solar Phys.* **3,** 258, 367.

—— (1976). *Astrophys. J.* **203,** 739.

—— and MORRISON, R. A. (1970). *Solar Phys.* **14,** 280.

—— and SCHNEEBERGER, T. J. (1977). *Astrophys. J.* **215,** 356.

—— and SCHRÖTER, E. H. (1968). *Solar Phys.* **4,** 142, 165.

BIERMANN, L. (1941). *Vierteljahrsschr. Astron. Ges.* **76,** 194.

BÖHM-VITENSE, E. (1958). *Z. Astrophys.* **46,** 108.

BORUTA, N. (1977). *Astrophys. J.* **215,** 364.

CARSLAW, H. S. and JAEGER, J. C. (1959). *Conduction of heat in solids*, §§2.5, 2.9. Clarendon Press, Oxford.

CHANDRASEKHAR, S. (1939). *An introduction to the study of stellar structure. University of Chicago Press.*

—— (1952). *Phil. Mag.* **43,** 501.

—— (1953). *Proc. Roy. Soc. Ser. A* **216,** 293.

Chandrasekhar, S. (1956a). *Phil. Mag.* **45,** 1117.
—— (1956b). *Proc. Nat. Acad. Sci. U.S.A.* **42,** 273.
—— (1961). *Hydrodynamic and hydromagnetic stability.* Clarendon Press, Oxford.
Chapman, G. (1973). *Astrophys. J.* **191,** 255.
Clark, A. (1966). *Phys. Fluids* **9,** 485.
—— (1968). *Solar Phys.* **4,** 386.
Cowling, T. G. (1953). *The sun* (ed. G. P. Kuiper), pp. 569–572. University of Chicago Press.
—— (1957). *Magnetohydrodynamics.* John Wiley and Sons, New York.
—— (1976). *Mon. Not. Roy. Astron. Soc.* **177,** 409.
Danielson, R. E. (1962). *Astron. J.* **67,** 574.
Deinzer, W. (1965). *Astrophys. J.* **141,** 548.
Deubner, F. (1976). *Astron. Astrophys.* **47,** 475.
Dunn, R. B. and Zirker, J. B. (1973). *Solar Phys.* **33,** 281.
Frazier, E. N. (1970). *Solar Phys.* **14,** 89.
—— and Stenflo, J. O. (1972). *Solar Phys.* **27,** 330.
Geronicholas, E. (1977a). *Astrophys. J.* **211,** 966.
—— (1977b). ibid. **214,** 607.
Gibson, E. G. (1973). *The quiet sun, NASA SP*-303 p. 236. U.S. Gov't. Printing Office, Washington, D.C.
Giovanelli, R. G. (1977). *Solar Phys.* **52,** 315.
Golub, L., Krieger, A. S., Silk, J. K., Timothy, A. F., and Vaiana, G. S. (1974). *Astrophys. J. Lett.* **189,** L93.
Harvey, K. L., Harvey, J. W., and Martin, S. F. (1975). *Solar Phys.* **40,** 87.
Howard, R. and Stenflo, J. O. (1972). *Solar Phys.* **22,** 402.
Kopecky, M. and Obridko, V. (1968). *Solar Phys.* **5,** 354.
Leighton, R. B. (1963). *Ann. Rev. Astron. Astrophys.* **1,** 19.
—— (1969). *Astrophys. J.* **156,** 1.
—— Noyes, R. W., and Simon, G. W. (1962). *Astrophys. J.* **135,** 474.
Livingston, W. and Harvey, J. (1969). *Solar Phys.* **10,** 294.
—— ——(1971). *Solar magnetic fields, IAU Symposium No.* 43 (ed. R. Howard), p. 51. Reidel, Dordrecht.
Lundquist, S. (1950). *Ark. Fys.* **2,** 361.
Lynch, D. K., Beckers, J. M., and Dunn, R. B. (1973). *Solar Phys.* **30,** 63.
Meyer, F., Schmidt, H. U., Weiss, N. O., and Wilson, P. R. (1974). *Mon. Not. Roy. Astron. Soc.* **169,** 35.
Mihalas, D. (1965). *Astrophys. J.* **141,** 564.
Moore, R. L. (1973). *Solar Phys.* **30,** 403.
Musman, S. A. (1967). *Astrophys. J.* **149,** 201.
Noyes, R. W. and Leighton, R. B. (1963). *Astrophys. J.* **138,** 631.
Oster, L. (1968). *Solar Phys.* **3,** 543.
Parker, E. N. (1955a). *Astrophys. J.* **121,** 491.
—— (1955b). ibid. **122,** 293.
—— (1957). *Proc. Nat. Acad. Sci.* U.S.A. **43,** 8.
—— (1963). *Astrophys. J.* **138,** 552.
—— (1964). ibid. **140,** 1170.
—— (1971). ibid. **164,** 491.
—— (1973). ibid. **186,** 643, 665.
—— (1974a). ibid. **189,** 563.
—— (1974b). ibid. **190,** 429.

—— (1974c). *Solar Phys.* **36,** 249; **37,** 126.
—— (1975a). *Astrophys. J.* **198,** 205.
—— (1975b). ibid. **202,** 523.
—— (1975c). *Solar Phys.* **40,** 275. 291.
—— (1976a). *Astrophys. J.* **204,** 259.
—— (1976b). ibid. **210,** 810.
—— (1976c). ibid. **210,** 816.
—— (1977). *Mon. Not. Roy. Astron. Soc.* **179,** 93P.
—— (1978). *Astrophys. J.* **221,** 368.
REID, W. H. (1955). *Proc. Cambridge Phil. Soc.* **51,** 350.
ROBERTS, B. (1976). *Astrophys. J.* **204,** 268.
SAVAGE, B. D. (1969). *Astrophys. J.* **156,** 707.
SAWYER, C. (1971). *Solar magnetic fields, IAU Symposium No.* 43 (ed. R. HOWARD) p. 316. Reidel, Dordrecht.
SCHLÜTER, A. and TEMESVARY, ST. (1958). *Electromagnetic phenomena in cosmical physics, IAU Symposium No.* 6 (ed. B. LEHNERT). Cambridge University Press.
SHEELEY, N. R. (1967). *Solar Phys.* **1,** 171.
SIMON, G. W. and LEIGHTON, R. B. (1964). *Astrophys. J.* **140,** 1120.
—— and NOYES, R. W. (1971). *Solar magnetic fields, IAU Symposium No.* **43,** (ed. R. HOWARD), p. 663. Reidel, Dordrecht.
SMYTHE, W. R. (1939). *Static and dynamic electricity,* 1st. edn., §4.22. McGraw-Hill, New York.
SPRUIT, H. C. (1974). *Astrophys. J.* **34,** 277.
—— (1977). *Solar Phys.* 7 **55,** 3
STEENBECK, W. and KRAUSE, F. (1969). *Astron. Nach.* **291,** 49.
STEENBECK, W., KRAUSE, F., and RÄDLER, K. H. (1966). *Zeit f. Naturforsch,* **21a,** 369.
STENFLO, J. O. (1973). *Solar Phys:* **32,** 41.
THOMPSON, W. B. (1951). *Phil. Mag.* **42,** 1417.
UNNO, W., RIBES, E., and APPENZELLER, I. (1974). *Solar Phys.* **35,** 287.
VAIANA, G. S., KRIEGER, A. S. and TIMOTHY, A. F. (1973). *Solar Phys.* **32,** 81.
VEEDER, G. J. and ZIRIN, H. (1970). *Solar Phys.* **12,** 391.
WADDELL, J. H. (1958). *Astrophys. J.* **127,** 284.
WALEN, C. (1949). *On the vibrationary rotation of the sun.* Henrik Lindstahl's Bokhandel, Stockholm.
WEISS, N. O. (1966). *Proc. Roy. Soc. Ser.* A **293,** 310.
WENTZEL, D. G. and SOLINGER, A. B. (1967). *Astrophys. J.* **148,** 877.
ZIRIN, H. (1972). *Solar Phys.* **22,** 34.
ZWAAN, C. (1978). *Solar Phys.* (in press).

11

THE TOPOLOGY OF MAGNETIC LINES OF FORCE

11.1. General properties of lines of force

THE topological properties of the lines of force of a magnetic field $B_i(\mathbf{r})$ determine the activity of that field. For all but the most symmetrical topologies the field possesses no equilibrium, being compelled to a life of dynamical dissipation. For that reason it is essential to consider the topological properties of magnetic field lines before taking up the general question of dynamical non-equilibrium in Chapters 13 and 14. As a matter of fact, the topology of magnetic fields is a subject of some interest in its own right. The lines of force are generally stochastic (Jokipii 1966; Jokipii and Parker 1969a; Parker 1969) and, if confined to a finite volume of space, they are also ergodic (Jokipii and Parker 1969a). The random walk of neighbouring lines of force plays a fundamental role in the propagation of cosmic rays through the galaxy where the individual particles are closely constrained to motion along the lines. The diffusion of cosmic ray particles across the mean field direction relies upon the random walk of the individual lines of force relative to the mean field (Getmantsev 1963; Jokipii 1966; Parker 1968b; Jokipii and Parker 1969b; Lingenfelter *et al.* 1971). Thus the general inflation of the galactic field by the cosmic rays is an immediate consequence of the stochastic property of magnetic lines of force (Parker 1965, 1966, 1968a, 1969). The inflation of the field is the principal means of escape of the cosmic rays from the galaxy and is evidently the source of a halo of cosmic rays and magnetic field surrounding the galaxy. In most astrophysical circumstances the pattern in which the lines of force wind about their neighbours varies along the field and is the principal source of the dynamical activity of the magnetic fields of stars and galaxies. It is important, therefore, to understand the universal stochastic properties of the magnetic fields that appear in nature.

We recall that the lines of force of a field $B_i(\mathbf{r})$ are the instantaneous family of solutions, conveniently symbolized by

$$f_1(\mathbf{r}) = \Lambda_1, \qquad f_2(\mathbf{r}) = \Lambda_2, \tag{11.1}$$

of the two equations

$$\mathrm{d}x/B_x = \mathrm{d}y/B_y = \mathrm{d}z/B_z, \tag{11.2}$$

where Λ_1 and Λ_2 are constant on each individual line of force. The properties of the lines of force were discussed at some length in §4.2. In a fluid with infinite electrical conductivity, the topology of the lines is invariant, the lines of force being carried with the fluid. In the presence of resistivity the topology is not invariant (see illustrations in Parker and Krook 1956; Parker 1963b; Weiss 1966). It is possible to define the lines of force at each instant of time, but there is often no unique way by which the lines at successive instants can be related to each other. In particular, an unsymmetric topology may evolve rapidly as a consequence of the associated dynamical dissipation, even in very highly conducting fluids. But that is a topic for a later chapter. Here we confine attention to the topology at any instant of time.

The fundamental fact of the topology of $B_i(\mathbf{r})$ is that, in the absence of magnetic monopoles the field is solenoidal,

$$\partial B_i/\partial x_i = 0. \tag{11.3}$$

The magnetic lines of force can end only on a magnetic charge, whose apparent non-existence implies that magnetic lines of force do not end. It is not impossible that the magnetic lines of force close on themselves, but in any case the lines do not terminate. In the hypothetical circumstance that a magnetic field $B_i(\mathbf{r})$ without singularities is confined to a finite region of a two-dimensional space, the lines of force make up a family of closed curves, $f(x, y) = \Lambda$. However, magnetic fields in the real three-dimensional physical world are not constrained to so high a degree of symmetry. Almost all of the lines of force of a field in three dimensions fail to close. To put it in simple terms, the cross-sectional area of a mathematical line is zero[1]. Then, noting that a line may assume almost any configuration in space, we can see that it is highly unlikely that any

[1] In slightly distorted form this basic fact played an important role in scientific journalism some years ago, under circumstances sufficiently amusing to merit repeating. The National Aeronautics and Space Administration planned to release to the press statements of the scientific purpose of the various instruments to be carried into interplanetary space on a certain spacecraft. However, they experienced some difficulty in obtaining a statement from the team of scientists responsible for the magnetometer. With the deadline drawing near the project manager was forced to desperate measures. So he prepared an alternative statement which was to be published if the magnetometer team did not supply him with a suitable statement of their own. The manager's statement noted that the magnetic field carried in the solar wind is, after all, of no great importance, as may readily be seen from the fact that the field strength is only 5×10^{-5} G at the orbit of Earth, corresponding to one line of force through each $2\ \mathrm{m}^2$. That is to say, the lines of force are separated in space by distances of the order of $\sqrt{2}$ m. The lines have no intrinsic width of their own, so the chance of a cosmic ray proton hitting one of the lines arises solely from the finite radius of the proton. This being 10^{-13} cm, the mean free path for a proton colliding with a line of force is 2×10^{15} cm or about 10^2 a.u., i.e. far larger than the dimensions of the inner solar system where cosmic ray modulation was being studied. Faced with the prospect of this press release the magnetometer team found time to write a rational statement on magnetic fields in the solar wind.

line should run into its own tail to form a closed curve. In almost all cases, then, an individual line is of infinite length.

In the unlikely circumstance that the magnetic field is confined to a finite volume V in three-dimensional space, (i.e. the normal component vanishes everywhere over the surface S of the volume V) the lines of force within V are ergodic in almost all cases. That is to say, the line of force through any point P_1 in V also passes, somewhere along its length, within some arbitrarily small distance δ of any other point P_2 in the field within V. In this case the functions $f_1(\mathbf{r})$ and $f_2(\mathbf{r})$ are infinitely repetitive and multiple-valued. Given a value for Λ_1 and Λ_2, and a value of z for which there is one solution (x, y) of (11.1), there will in almost all cases be infinitely many pairs of numbers (x, y) that are solutions. Only if symmetries are imposed are f_1 and f_2 not infinitely repetitive and the lines of force not ergodic within V.

Ergodic fields have some interesting properties, such as a general absence of hydrostatic equilibrium. It is easily seen from (6.1) that, in the absence of gravity or other external forces, the fluid pressure is uniform along a line of force, $\mathbf{B} \cdot \nabla p = 0$. If the line of force fills the volume V occupied by the field, and if p is to be a continuous function of position, then p is uniform throughout V. Hence $\nabla p = 0$ and (6.1) reduces to $(\nabla \times \mathbf{B}) \times \mathbf{B} = 0$. The field has no forces exerted on it by the fluid anywhere throughout its entire volume. But it was shown in §5.3 that a magnetic field tends to expand so that inward forces must be exerted if the field is to be confined in equilibrium. Hence there is no equilibrium for an ergodic field. The various loops and coils of the field try to expand outward through the fluid, leading to $\nabla p \neq 0$. The pressure gradient causes a continual shifting of fluid along the magnetic lines of force so that gradually the loops and coils are able to free themselves from the fluid and expand outward through the surface S of the volume. With the passage of time the field slowly extricates itself from the fluid in V by the simple process of transferring the fluid along the ergodic lines (or line) of force to some single position, call it P_0, leaving all the rest of the field free of fluid.

The importance of ergodic fields in the astrophysical universe is not clear, because there is probably no volume of space that is entirely isolated magnetically. In no instance of which we are aware can the internal field be declared to be ergodic in the precise mathematical sense of the word. The magnetic field of Earth is partially isolated, with at least 90 per cent of the lines of force through the surface turning around and re-entering the planet, while some 5–10 per cent extend into the geomagnetic tail and connect in whole or in part to the magnetic field of the solar wind (Lanzerotti 1972; Paulikas 1974; Frank *et al.* 1976). Similarly, the magnetic field of the sun is only partially isolated. The strong fields of active regions are closed, re-entering the sun after extending above the surface, but the magnetic flux emerging in quiet regions is extended out into interstellar space by the solar wind. The external magnetic field of the galaxy is not observed (but there is evidence for an extended magnetic halo) so little can be said. The basic theoretical principle is that, of course, if left to itself, the magnetic flux extending out through the surface of an astrophysical body actively seeks the lowest external energy state available to it, and that is the closed form of a dipole. In most cases the field carries on dynamical line cutting to achieve this lower energy state, so that, if left to itself, the magnetic field of an isolated static astrophysical body is closed.

On the other hand, very few astrophysical bodies are so placid as to reach the ultimate closed equilibrium configuration. At least some small portion of the lines of force are forced open by escaping gases. The solar wind is the most familiar example (Parker 1963a). There may be a similar outflow of gas from the nucleus of the galaxy forcing open any lines of force of the nuclear field (Johnson and Axford 1971). Altogether, then, there appears to be no physical situation allowing clear application of the theoretical ergodic properties of the isolated region of magnetic field.

It is amusing to note, however, that if, contrary to the best values available from observation (Gott *et al.* 1974), one were to assume matter densities so high as to close the universe, then on the scale of the cosmos the field would be ergodic. But as with other aspects of the question of an open or closed universe, the result is without local physical consequences (ignoring the profound psychological impact that the ideas seem to have in some quarters).

On the whole it appears that, instead of fields occupying closed cells in space, the opposite situation obtains. In no circumstance are the lines of force of any magnetic field in the real world fully confined to a finite region of space. At least some small fraction of the magnetic flux wanders away into other parts of space, never to return.

To illustrate the expected universal wandering of magnetic lines of force in a three-dimensional space, suppose that we were to take a snapshot of the universe at some suitably defined instant of time t and then explore the topology of the lines of force in that 'instantaneous' picture. As an exercise consider the point P on the page in front of us.

$$\cdot P$$

The point defines a line of force of Earth's magnetic field. We contrive to follow the line to see where it leads. We expect that the line passes many times around through Earth and out into the magnetosphere, building up the preliminary appearance of an ergodic field. But if that is the case, then sooner or later the journey along the line of force must lead up through the surface of Earth at high latitude on a line that passes out into the solar wind. From there the line of force travels either to the sun or into interstellar space. If to the sun, then again we journey around many times through the body of the sun. If the line does not escape from the sun, then as our journey continues it begins to build up the appearance of an ergodic line. Hence, sooner or later, we will find ourselves coming out of the sun on a line of force extended by the solar wind through the solar system into interstellar space. We do not know that all of the lines of force extended by the solar wind into the outer solar system connect into the galactic field in interstellar space (see illustrative models in Parker (1963a), Yu (1974), and references therein). Some may return to the sun, in which case the story is repeated. In almost all cases, however, the line eventually escapes into interstellar space. The general idea then repeats itself in the magnetic field of the galaxy.

This author does not have the vision to fill in the details of the journey around, and in and out, of the gaseous disc of the galaxy (Parker 1969). But the general principles are clear. If intergalactic space is not entirely free of gas and field so that the galaxy is not a completely magnetically-isolated system, then the line eventually leaves the galaxy to extend out through the cosmos along a path that we cannot guess. The wandering of a line of force in the infinite three-dimensional space in which we evidently exist (Gott *et al.* 1974) raises some interesting conceptual questions that we are not prepared to treat.

On the other hand, if there are no fields in intergalactic space, then, of course, the galaxy is magnetically isolated and we expect the field to be ergodic throughout the galaxy. In that unlikely circumstance the journey would take us near, and ultimately through, each of the 10^{12} stars, large and small, within the galaxy. A Cook's tour *ad infinitum!*

Coming back to the remarks at the beginning of the chapter, the wandering of magnetic lines of force through space is of primary concern to the theory of propagation of individual charged particles, such as cosmic rays, through the solar system and through the galaxy. In the limit of small mass m the trajectories of a classical particle of charge q are precisely the lines of force of the static magnetic field in our instantaneous picture of the universe. We suppose that the particle is initially projected precisely along the local line of force with the velocity $w_{\parallel}$. Then the particle has no diamagnetic moment $\mu = \frac{1}{2}mw_{\perp}^2/B$ because the ratio of its perpendicular velocity component to its parallel velocity component $w_{\perp}/w_{\parallel}$ is zero. The cyclotron radius of the particle is zero, so that the particle is not repelled as it enters regions of strong field and the drift of the particle across the lines of force is zero. The trajectory reduces precisely to the line of force, the particle travelling the full length of the line. In the real world, of course, the field is not static, as in the snapshot under discussion. What is more, cosmic rays are so numerous that their collective pressure may be a dominant dynamical consideration (Parker 1969) in shaping the form and motion of the gas and field. But so long as the cyclotron radius of the particles is small compared to the scale of variation $B/|\nabla B|$ of the magnetic field, the lines of force are a useful approximation to the trajectories of the individual particles. In particular, if cosmic rays pass freely in and out of the solar system, then there must be a substantial connection of the lines of force from the solar wind into interstellar space, etc. (see discussion in Parker (1968b), Jokipii and Parker (1969b)). With this brief guide to the infinite wandering of an individual line of force, the next question is the relative wandering of two neighbouring lines.

11.2. Stochastic properties of lines of force

Consider how two neighbouring lines of force wander apart as a result of local random fluctuations in the magnetic field. To fix ideas consider the line of force through the point P above, and through a neighbouring point P' at a small distance δs on the page from P. The point P' serves as the starting point of another journey along a line of force through the cosmos. The journey along the line through P' exhibits the same general features as along the line through P, but eventually the two lines wander apart so that the quantitative aspects of the journey become quite different. We consider here how rapidly the paths of the two journeys diverge.

Jokipii (1966) pointed out that the power spectrum of the magnetic fluctuations evaluated at zero wave number represents the random walk of the magnetic lines of force. In the context of the lines of force through P and P', we note merely that the field at any point on the line through P' differs slightly from the field on the line through P because of local field gradients. The random variations of the local gradients along the line of force cause the lines to random walk relative to each other. Therefore, they walk away from each other. To illustrate the problem in quantitative terms consider the idealized circumstance that the line through P lies along the z-axis, on which the field density is $B(z)$. Then in the near neighbourhood of the z-axis, the field can be represented by

$$B_x = x\,\partial B_x/\partial x + y\,\partial B_x/\partial y + \dots$$

$$B_y = x\,\partial B_y/\partial x + y\,\partial B_y/\partial y + \dots$$

$$B_z = B(z) + x\,\partial B_z/\partial x + y\,\partial B_z/\partial y + \dots$$

to sufficient approximation. The coefficients are all functions of z, with (11.3) leading to

$$\partial B_x/\partial x + \partial B_y/\partial y + \mathrm{d}B/\mathrm{d}z = 0.$$

With one line of force along the z-axis, the equations for the neighbouring line, with coordinates $x(z)$, $y(z)$, are

$$\frac{dx}{dz}=\frac{B_x}{B_z}, \qquad \frac{dy}{dz}=\frac{B_y}{B_z},$$

so that to first order

$$dx/dz=\alpha_{xx}x+\alpha_{xy}y, \qquad dy/dz=\alpha_{yx}x+\alpha_{yy}y, \tag{11.4}$$

where the coefficients are

$$\alpha_{ij}(z)=B^{-1}\,\partial B_i/\partial x_j \qquad i,j=x,y. \tag{11.5}$$

Suppose that the field $\alpha_{ij}(z)$ varies at random with z, with a characteristic correlation length λ. Suppose further that the $\lambda\,|\alpha_{ij}(z)|\ll 1$ so that $x(z)$ and $y(z)$ change but little over one correlation length. It follows that the point $(x(z), y(z))$ random walks as z varies along the line of force. The probability distribution $\psi(x, y, z)$ of $x(z)$, $y(z)$ is described by a Fokker–Planck equation

$$\frac{\partial\psi}{\partial z}=\frac{\partial^2}{\partial x_i\,\partial x_j}\left\{\frac{\langle\Delta x_i\,\Delta x_j\rangle}{2\Delta z}\psi\right\} \tag{11.6}$$

where $\langle\Delta x_i\,\Delta x_j\rangle/\Delta z$ is the mean change of x_i and x_j in the distance Δz along the lines of force (see, for instance, Chandrasekhar (1943)). The $\langle\Delta x_i\,\Delta x_j\rangle/\Delta z$ can be related by standard computational methods to the two-point correlation function R_{ijkl} of the $\alpha_{ij}(z)$,

$$R_{ijkl}(\zeta)=\langle\alpha_{ij}(z)\alpha_{kl}(z+\zeta)\rangle=R_{klij}(\zeta) \tag{11.7}$$

We presume that the *statistical* properties of the $\alpha_{ij}(z)$ are locally independent of z so that the correlation is a function only of the separation ζ of the two points. The mean value is most conveniently taken over an ensemble of systems. The standard procedure to calculate $\langle\Delta x_i\,\Delta x_j\rangle/\Delta z$ is to integrate (11.4) over z for a distance Δz such that $\Delta z\gg\lambda$ but $\Delta z\,|\alpha_{ij}|\ll 1$. Hence over Δz there is the small change in x,

$$\Delta x=x\int_0^{\Delta z}dz'\alpha_{xx}(z')+y\int_0^{\Delta z}dz'\alpha_{xy}(z').$$

Hence

$$\begin{aligned}(\Delta x)^2=x^2&\int_0^{\Delta z}dz'\alpha_{xx}(z')\int_0^{\Delta z}dz''\alpha_{xx}(z'')\\&+2xy\int_0^{\Delta z}dz'\alpha_{xx}(z')\int_0^{\Delta z}dz''\alpha_{xy}(z'')\\&+y^2\int_0^{\Delta z}dz'\alpha_{xy}(z')\int_0^{\Delta z}dz''\alpha_{xy}(z'').\end{aligned}$$

Let $z'' = z' + \zeta$ so that

$$(\Delta x)^2 = x^2 \int_0^{\Delta z} dz' \alpha_{xx}(z') \int_{-z'}^{\Delta z - z'} d\zeta\, \alpha_{xx}(z' + \zeta) + \ldots.$$

Then with Δz large compared to the correlation length λ but still sufficiently small that $\Delta z\, |\alpha_{ij}| \ll 1$ (which is possible because we have assumed that $\lambda\, |\alpha_{ij}| \ll 1$) such quantities as $\alpha_{xx}(z')$ and $\alpha_{xx}(z' + \Delta z)$ are uncorrelated, and their ensemble average vanishes. Hence in computing $\langle(\Delta x)^2\rangle$ the lower limit on the second integral can be extended all the way to $-\infty$ because for all $\zeta < -(z' + \lambda)$, the quantities $\alpha_{xx}(z')$ and $\alpha_{xx}(z' + \zeta)$ are uncorrelated and so have zero mean. Similarly, the upper limit can be extended to $+\infty$. The only error occurs in the neighbourhood of $z' = 0$ for the lower limit or $z' = \Delta z$ for the upper limit. But that neighbourhood is of width λ and is, therefore, only a small fraction of the complete range of integration of Δz. Thus neglecting terms $O(\lambda/\Delta z)$ compared to one, we have

$$\langle(\Delta x)^2\rangle \cong x^2 \int_0^{\Delta z} dz' \int_{-\infty}^{+\infty} d\zeta \langle \alpha_{xx}(z')\alpha_{xx}(z' + \zeta)\rangle + \ldots$$

$$= x^2 \int_0^{\Delta z} dz' \int_{-\infty}^{+\infty} d\zeta R_{xxxx}(\zeta) + \ldots.$$

The integration over z' is now trivial, the final result being

$$\langle \Delta x_i\, \Delta x_j \rangle / \Delta z = x_k x_l \int_{-\infty}^{+\infty} d\zeta R_{ijkl}(\zeta), \qquad (11.8)$$

where repeated indices are summed over 1 and 2.

Suppose, as a simple example, that the field gradients $\alpha_{ij}(z)$ are statistically independent of each other. Then write

$$\langle(\Delta x)^2\rangle/\Delta z = a_1^2 x^2 + a_2^2 y^2$$

$$\langle \Delta x\, \Delta y \rangle / \Delta z = 0$$

$$\langle(\Delta y)^2\rangle/\Delta z = b_1^2 x^2 + b_2^2 y^2$$

where

$$a_1^2 = \int_{-\infty}^{+\infty} d\zeta R_{xxxx}(\zeta), \qquad a_2^2 = \int_{-\infty}^{+\infty} d\zeta R_{xyxy}(\zeta), \qquad (11.9)$$

$$b_1^2 = \int_{-\infty}^{+\infty} d\zeta R_{yxyx}(\zeta), \qquad b_2^2 = \int_{-\infty}^{+\infty} d\zeta R_{yyyy}(\zeta). \qquad (11.10)$$

Then for complete isotropy of fields and gradients we have $a_1^2 = a_2^2 = b_1^2 = b_2^2 \equiv a^2$. Equation (11.6) reduces to

$$\frac{\partial \psi}{\partial z} = \frac{a^2}{2} \left(\frac{\partial^2}{\partial x^2} + \frac{\partial^2}{\partial y^2} \right) (x^2 + y^2) \psi. \qquad (11.11)$$

The variables are separable in polar coordinates. Introduce the dimensionless coordinate $\xi = a^2 z/2$ and the polar coordinates (ϖ, ϕ), so that

$$\frac{\partial \psi}{\partial \xi} = \left(\frac{1}{\varpi}\frac{\partial}{\partial \varpi}\varpi\frac{\partial}{\partial \varpi} + \frac{1}{\varpi^2}\frac{\partial^2}{\partial \phi^2}\right)\varpi^2\psi. \tag{11.12}$$

Let

$$f = \varpi^2 \psi \tag{11.13}$$

and

$$\varpi = \exp(u). \tag{11.14}$$

Then

$$\partial f/\partial \xi = \partial^2 f/\partial u^2 + \partial^2 f/\partial \phi^2. \tag{11.15}$$

This equation has the same form as the heat flow equation in rectangular coordinates, whose solutions are well known (see, for instance, Carslaw and Jaeger (1959)).

Suppose, then, that at $\xi = 0$ the line of force has coordinates (ϖ_0, ϕ_0), so that

$$\psi(\varpi, \phi, 0) = \delta(\varpi - \varpi_0)\delta(\phi - \phi_0)/2\pi\varpi_0.$$

The appropriate form of the solution for $\xi > 0$ is

$$f = \frac{C}{|\xi|}\exp\left\{-\frac{(u-u_0)^2 + (\phi - \phi_0)^2}{4\,|\xi|}\right\},$$

where C is an arbitrary constant and $u_0 = \ln(\varpi_0)$. Hence

$$\psi(\varpi, \phi, \xi) = \frac{1}{4\pi\,|\xi|\,\varpi^2}\exp\left\{-\frac{\ln^2(\varpi/\varpi_0) + (\phi - \phi_0)^2}{4\,|\xi|}\right\} \tag{11.16}$$

where ϕ extends from $-\infty$ to $+\infty$ so that the probability density at any point ϕ is

$$\sum_{n=-\infty}^{+\infty} \psi(\varpi, \phi + 2\pi n, \xi).$$

Denote by $\Phi(\varpi, \xi)\,d\varpi$ the probability that the separation lies in the interval $(\varpi, \varpi + d\varpi)$. Then

$$\Phi(\varpi, \xi)\,d\varpi = \varpi\,d\varpi\int_{-\infty}^{+\infty} d\phi\,\psi(\varpi, \phi, \xi)$$

and

$$\Phi(\varpi, \xi) = \frac{1}{(4\pi\,|\xi|)^{\frac{1}{2}}\varpi}\exp\left\{-\frac{\ln^2(\varpi/\varpi_0)}{4\,|\xi|}\right\}.$$

The evolution of this probability distribution with increasing distance ξ is plotted in Fig. 11.1, showing the rapid flow of the probability into a long tail extending to large radius ϖ. The mean square displacement of the line from the z-axis is

$$\langle \varpi^2 \rangle = \int_0^\infty \mathrm{d}\varpi\varpi \int_{-\infty}^{+\infty} \mathrm{d}\phi\, \varpi^2 \psi(\varpi, \phi, \xi) \tag{11.17}$$

$$= \varpi_0^2 \exp 4\,|\xi|. \tag{11.18}$$

Thus the r.m.s. separation of the two lines of force increases exponentially, as $\exp(2a^2\,|z|)$, with distance z along the lines of force.

When, after a distance $z = z_1$, the two lines of force become separated by a distance comparable to the correlation length of the magnetic field, the simple expansion of the field about the position of one line can no longer be applied. In that case one writes the field in terms of the three independent fluctuating components $\delta B_x(z)$, $\delta B_y(z)$, $\delta B_z(z)$ at the position of each line, where the field $\delta B_i(z)$ is a random function of z. The two lines random walk independently of each other with increasing z. It is sufficient, then, to consider the position of either one relative to the z-axis. The change in the position of either of the two lines over a distance Δz is

$$\Delta x \cong \int_0^{\Delta z} \mathrm{d}z'\delta B_x(z')/B_z \tag{11.19}$$

$$\Delta y \cong \int_0^{\Delta z} \mathrm{d}z'\delta B_y(z')/B_z. \tag{11.20}$$

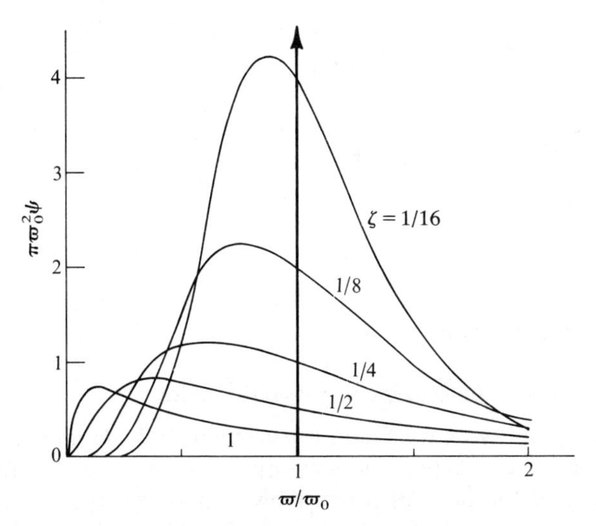

FIG. 11.1. A plot of the probability distribution of the radial separation of the two lines of force at successive distances ξ along the lines from their separation $\varpi = \varpi_0$ at $\xi = 0$.

Then if $\delta\beta_i \equiv \delta B_i/B_z$, it follows that

$$(\Delta x)^2 = \int_0^{\Delta z} dz' \delta\beta_x(z') \int_0^{\Delta z} dz'' \delta\beta_x(z'') \tag{11.21}$$

$$= \int_0^{\Delta z} dz' \delta\beta_x(z') \int_{-z'}^{\Delta z - z'} d\zeta \delta\beta_x(z' + \zeta). \tag{11.22}$$

Assume again that $|\delta\beta_i| \ll 1$, and extend the interval Δz to many times the correlation length. The result is again the standard form (11.8)

$$\langle \Delta x_i \, \Delta x_j \rangle / \Delta z = \int_{-\infty}^{+\infty} d\zeta R_{ij}(\zeta) \tag{11.23}$$

written now in terms of the correlation function for $\beta_i(z)$,

$$R_{ij}(\zeta) = \langle \beta_i(z) \beta_j(z + \zeta) \rangle. \tag{11.24}$$

The Fokker–Planck equation for the probability distribution $\psi(x, y, z)$ of the two lines (as a function of distance z along the lines) is (11.6) again.

In the absence of torsion we expect δB_x and δB_y to be uncorrelated. For simplicity suppose that the fluctuations are statistically homogeneous and isotropic, so that $\langle \Delta x^2 \rangle / \Delta z = \langle \Delta y^2 \rangle / \Delta z = \text{constant}$. Then if $\xi = \frac{1}{2}(z - z_1)\langle \Delta x^2 \rangle / \Delta z$, the Fokker–Planck equation is

$$\partial\psi/\partial\xi = \partial^2\psi/\partial x^2 + \partial^2\psi/\partial y^2. \tag{11.25}$$

If the position of the line is (x_1, y_1) at $\xi = 0$, then for $\xi > 0$ the distribution is

$$\psi(x, y, \xi) = \frac{1}{4\pi\xi} \exp\left\{ -\frac{(x - x_1)^2 + (y - y_1)^2}{4\xi} \right\}. \tag{11.26}$$

The characteristic width of the distribution is $2\xi^{\frac{1}{2}}$. A similar probability distribution applies to the position of the other line, initially at (x_2, y_2). The mean square separation s^2 of the lines for $\xi > 0$ is, then,

$$s^2 = (x_2 - x_1)^2 + (y_2 - y_1)^2 + 8\xi \tag{11.27}$$

growing linearly with increasing distance ξ along the field. The two lines of force go their separate ways.

11.3. The random component of magnetic fields

Up to this point we have taken for granted that the magnetic fields in the turbulent universe have random components, that lead to the stochastic wandering of the lines of force. Now that we have seen some of the consequences of the random component, it is desirable to go back for a closer look at the universal appearance of a random component of the magnetic field. To get to the heart of the matter, suppose that in some early 'Garden of Eden' the magnetic fields were not stochastic. The innocent field topologies were closed, and neighbouring lines of force remained neighbours in all cases. The question is whether there are effects, occurring universally, that would destroy the simple topology, converting it to the stochastic patterns of the present tumultous world.

11.3.1. Local random components

The simplest illustration of the origin of local random components is the supergranulation in the solar photosphere and its effect on the magnetic lines of force extending in the solar wind out through the solar system (Jokipii and Parker 1969a). The supergranules consist of horizontal motions at the surface of the sun caused by the upwelling of fluid from the convective zone. The horizontal velocities have an r.m.s. value of about $0{\cdot}4\ \mathrm{km\,s^{-1}}$, over the characteristic scale $L=2\times10^4$ km of the radius of the supergranule cell. The characteristic life of the motion is $\tau\cong10^5$ s. Leighton (1964, 1969) has pointed out that this horizontal motion of the fluid leads to a random walk of the magnetic lines of force of the solar magnetic field with an effective horizontal diffusion coefficient of the order of $0{\cdot}1L^2/\tau\cong4\times10^{12}\ \mathrm{cm^2\,s^{-1}}$. He points out that this random walk has direct consequences for the dispersal of the solar magnetic field over the period of the sunspot cycle. We would expect that the granules, for which $L\cong5\times10^7$ cm and $\tau\cong3\times10^2$ s, have a similar effect, with a comparable value $0{\cdot}1L^2/\tau\cong1\times10^{12}\ \mathrm{cm^2\,s^{-1}}$. The point with which we are concerned is that, ignoring resistive diffusion, the feet of the lines of force at the surface of the sun undergo a random walk while the lines are carried out through space with the speed v_w of the wind. It is sufficient for the present purposes of illustration to consider a uniform radial wind of velocity v_w extending the field radially from a non-rotating sun (for a more detailed discussion, see Jokipii and Parker (1969a)). Then in a time Δt the pattern of the lines is carried outward a distance $\Delta r=v_w\Delta t$ from the sun. In a radial wind the transverse dimensions of the pattern increase linearly with radial distance r. Thus, while the horizontal random walk has a mean square value

$$\langle(\Delta x)^2\rangle=L^2\Delta t/\tau \tag{11.28}$$

at the photosphere in a time Δt, the radial transport of this walk to a distance r in a time $\Delta t\cong r/v_w$ produces a mean square displacement

$$\langle(\Delta x)^2\rangle/\Delta z=(L^2/\tau v_w)(r/r_\odot)^2, \tag{11.29}$$

where $r_\odot$ is the radius of the sun. In terms of angular displacement $\Delta\phi=\Delta x/r$, it follows that

$$\langle(\Delta\phi)^2\rangle=L^2r/\tau v_w r_\odot^2. \tag{11.30}$$

With a uniform solar wind velocity of $v_w = 4\times10^2$ km s^{-1}, it follows that $L^2/\tau v_w \cong 10^6$ cm. Hence with $r_\odot = 0{\cdot}7\times10^{11}$ cm the r.m.s. transverse displacement at the orbit of Earth ($r = 1$ a.u. $= 1{\cdot}5\times10^{13}$ cm $= 220\, r_\odot$) is $\langle(\Delta x)^2\rangle/\Delta z = 5\times10^{10}$ cm. This corresponds to $\langle(\Delta x)^2\rangle^{\frac{1}{2}} = 0{\cdot}9\times10^{12}$ cm or $\langle(\Delta\phi)^2\rangle^{\frac{1}{2}} = 0{\cdot}06$.

Jokipii's point (Jokipii 1966, 1967, 1968) that the power at zero frequency in the fluctuations of the interplanetary field represents the random walk of the magnetic lines of force, allows a direct computation of the rate of random walk from direct spacecraft observations of the interplanetary field. Jokipii and Coleman (1968) deduced the value $\langle(\Delta\phi)^2\rangle^{\frac{1}{2}} = 0{\cdot}06$, in close agreement with that estimated from the supergranule motions at the sun[2].

Now the foregoing example was chosen for its simplicity, involving the direct manipulation of one end of the lines of force at the sun while the other end of the line is extended out into space by the solar wind. The results sketched in Fig. 11.2, are obvious. But, as a matter of fact, the example is incomplete, because it does not involve the reconnection of the lines of force. The global topology of the lines is unaffected, because with infinite conductivity the ultimate topological connections at either end of the lines in interplanetary space are preserved; the random walk of the lines outside the sun is precisely reversed beneath the surface of the sun.

11.3.2. Random component in turbulence

Consider, then, a more complete, if less vivid, example in which the lines of force are displaced by a random turbulent velocity $v_i(\mathbf{r}, t)$ in a field with finite resistivity. The overall topology is then altered by the resistive diffusion from the initial uniform field to a stochastic topology after a finite period of time.

Consider an infinite space filled with a homogeneous incompressible fluid with a high, but finite, electrical conductivity σ, through which there is a uniform magnetic field B_0 in the z-direction. For a short period of time t, the fluid is subject to a turbulent velocity $v_i(x, y, z, t)$, displacing the element of fluid initially at X_i to a final position $x_i(X_k)$ at time t. We suppose that the velocity is bounded and statistically homogeneous, so that $x_i(X_k)$ is a bounded random function of X_k with a correlation length L. In particular, if the turbulent velocity is subject to the upper bound

[2] Jokipii and Parker (1969a) deduced the somewhat larger value, $\langle(\Delta\phi)^2\rangle^{\frac{1}{2}} = 0{\cdot}3$ from the observed spreading of 1–10 MeV protons (Fan *et al.* 1968) streaming outward from the sun. But more recent observations show that particle streams beyond the orbit of earth involve local acceleration of particles in interplanetary space, so that their width may in fact be much larger than the $\langle(\Delta\phi)^2\rangle^{\frac{1}{2}}$ of the lines of force alone.

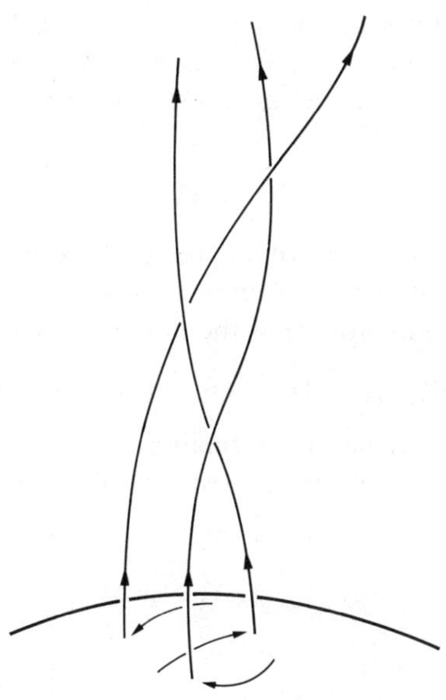

FIG. 11.2. A schematic drawing of the magnetic lines of force extended out from the supergranules at the surface of the sun by the solar wind, showing the interlacing of the lines as a result of the random walk of their feet at the sun.

$|v_i| < v_0$, then the net displacement in the allotted time t_1, is subject to the upper bound $|x_i(X_k) - X_i| < v_0 t_1$. To keep the calculations as simple as possible suppose that the fluid motions are confined to planes parallel to the xy-plane. Then $v_z(x, y, z, t) = 0$ and

$$\partial v_x/\partial x + \partial v_y/\partial y = 0. \tag{11.31}$$

The magnetic field at time t_1, is given by (4.32) as

$$B_x(x, y, z) = B_0 \partial x/\partial z, \tag{11.32}$$

$$B_y(x, y, z) = B_0 \partial y/\partial z, \tag{11.33}$$

$$B_z(x, y, z) = B_0, \tag{11.34}$$

upon noting that $z = Z$. The initial lines of force were all parallel to the z-axis, with X and Y constant. After the displacement they are given by the coordinate mapping

$$x = x(X, Y, z), \qquad y = y(X, Y, z) \tag{11.35}$$

represented by the Lagrangian coordinates of the elements of fluid on each initial line of force.

We suppose that t_1 is exceedingly small and/or the electrical conductivity extraordinarily high so that diffusion is negligible during the period of displacement, i.e. the diffusion length $(\eta t)^{\frac{1}{2}}$ characteristic of the time t_1 is small compared to the scale of variation of $B_i(x_k)$. However, following the period of displacement, the fluid is held motionless for a sufficiently long period of time that resistive diffusion occurs. The behaviour of the field is described by (4.15),

$$\partial \mathbf{B}/\partial t = \eta \nabla^2 \mathbf{B},$$

leading to a continuous relaxation of the field. To illustrate the qualitative effects of the diffusion, expand **B** in a Taylor series about the initial time $t = t_1$. Then, the initial evolution of the field away from (11.32)–(11.33) is

$$\delta \mathbf{B}(x, y, z, t) \cong \eta (t - t_1) \nabla^2 \mathbf{B}(x, y, z, t_1) \qquad (11.36)$$

after a small time $t - t_1$. The magnetic lines of force are altered slightly by the diffusion. The line through (x, y) at $z = 0$ changes from (11.35) to

$$x = x(X, Y, z) + \Delta x,$$

$$y = y(X, Y, z) + \Delta y,$$

which are the integrals of the equations

$$\frac{\mathrm{d}(x + \Delta x)}{\mathrm{d}z} = \frac{B_x(x, y, z) + \delta B_x(x, y, z)}{B_0},$$

$$\frac{\mathrm{d}(y + \Delta y)}{\mathrm{d}z} = \frac{B_y(x, y, z) + \delta B_y(x, y, z)}{B_0},$$

along the magnetic lines of force.

These exact equations for the line of force can be solved approximately when $|\delta \mathbf{B}| \ll |\mathbf{B}|$ by integrating along the *initial* line of force (11.35). The approximation remains valid until the integration proceeds so far that Δx_i is no longer small compared to the scale of variation of the initial field. But for small $\eta(t - t_1)$ there is very little deviation from the initial lines of force so that can be a very large distance. We proceed on the assumption that $t - t_1$ is chosen to be sufficiently small as to justify any values of z that we may care to use. Then

$$\frac{\mathrm{d}\Delta x}{\mathrm{d}z} \cong \frac{\delta B_x(x, y, z)}{B_0}, \qquad \frac{\mathrm{d}\Delta y}{\mathrm{d}z} \cong \frac{\delta B_y(x, y, z)}{B_0}$$

with x and y given by (11.35). With (11.36) and then (11.32) and (11.33), this becomes

$$\frac{\mathrm{d}\Delta x}{\mathrm{d}z} = \eta(t - t_1) \nabla^2 f_x(x, y, z), \qquad \frac{\mathrm{d}\Delta y}{\mathrm{d}z} = \eta(t - t_1) \nabla^2 f_y(x, y, z),$$

where f_i represents the derivative of the Langragian coordinate x_i with respect to z. It follows that the displacement of the line of force by diffusion grows with z according to

$$\Delta x = \eta(t - t_1) \int_0^z dz' \nabla^2 f_x(x, y, z'), \tag{11.37}$$

$$\Delta y = \eta(t - t_1) \int_0^z dz' \nabla^2 f_y(x, y, z'). \tag{11.38}$$

We suppose that the random displacement of field and fluid (11.26) was chosen in such a way that the self-correlation of $\partial x_i/\partial z$ is limited to some finite scale L, so that the correlation function

$$S_{ij}(\zeta) = \langle f_i(x_k) f_j(x_k + \zeta) \rangle \tag{11.39}$$

goes rapidly to zero beyond $|\zeta_k| = L$. Then we may proceed with the standard techniques for computing the progressive increase of the mean square (ensemble average) of Δx_i with z, beginning with the identity

$$\Delta x_i \Delta x_j = \eta^2 (t - t_1)^2 \int_0^z dz' \nabla^2 f_i[x(z'), y(z'), z'] \int_0^z dz'' \nabla^2 f_j\{x(z''), y(z''), z''\}.$$

As in the previous example, let $z'' = z' + \xi$ and suppose that $z \gg L$. Then upon taking the mean, the limits of the integration over ξ can be extended to $\pm\infty$,

$$\langle \Delta x_i \, \Delta x_j \rangle = \eta^2 (t - t_1)^2 \int_0^z dz' \int_{-\infty}^{+\infty} d\xi \langle \nabla^2 f_i\{x(z'), y(z'), z'\}$$
$$\times \nabla^2 f_j\{x(z' + \xi), y(z' + \xi), z' + \xi\}\rangle.$$

It is shown in (A2)–(A5) of Appendix A, Chapter 17 that if S_{ij} is defined by (11.39), then the auto-correlation function for $\nabla^2 f_i$ is

$$\langle \nabla^2 f_i(x_k) \nabla^2 f_j(x_k + \zeta_k) \rangle = \nabla^2 \nabla^2 S_{ij}(\zeta_k).$$

Hence

$$\langle \Delta x_i \, \Delta x_j \rangle = \eta^2 (t - t_1)^2 \int_0^z dz' \int_{-\infty}^{+\infty} d\xi \nabla^2 \nabla^2 S_{ij}(\zeta_k)$$
$$= \eta^2 (t - t_1)^2 z \int_{-\infty}^{+\infty} d\xi \nabla^2 \nabla^2 S_{ij}(\zeta_k), \tag{11.40}$$

where $\zeta_i = x_i(z' + \xi) - x_i(z')$. The integral is more complicated than in the previous case (11.22), because it is along the deformed line of force (11.35). It reduces to that simpler form if we suppose that the turbulent displacement of the fluid is small compared to the scale of the turbulent eddies. The point is that the integral is generally positive, representing the integral of the autocorrelation of the random function $\nabla^2 f_i$, so that $\langle \Delta x_i \, \Delta x_j \rangle$ progressively increases with z along the lines of force. The diffusion reconnects the initial uniform topology of the lines of force (11.35) into a stochastic topology in which the neighbouring lines of force wander apart and wind among each other with the random displacement Δx_i. Since turbulence is a universal companion of magnetic fields in the conducting gases in the cosmos, we expect that all lines of force have the stochastic property exhibited by this example.

11.3.3. Random component in inhomogeneous resistivity

The fact must not be overlooked that turbulence is not necessary for the existence of a random component of the magnetic field. We are accustomed to carrying out calculations with a uniform resistive diffusion coefficient η, for reasons of computational convenience. As a matter of fact there is no reason to expect η to be uniform. One glance at the 10 per cent temperature fluctuations across the granules in the sun (Schwarzschild 1959), or the spicules in the lower corona, or the general small-scale fluctuations in the solar wind, is sufficient to establish the general 'turbulent' inhomogeneity of η. It is readily shown that random fluctuations in η lead to stochastic lines of force.

To illustrate the effect consider a magnetic field $B(x)$ in the z-direction which varies in the transverse x-direction. To keep the algebra as simple as possible, suppose that the medium is fixed in position so that it does not respond to changing magnetic stresses. Then the field is described by (4.12)

$$\partial \mathbf{B}/\partial t + \nabla \times (\eta \nabla \times \mathbf{B}) = 0.$$

The random variations of the diffusion coefficient $\eta(\mathbf{r})$ are assumed to be statistically homogeneous and isotropic, with a single characteristic scale and correlation length l small compared to the scale of variation L of the mean field $\mathbf{e}_z B(x)$. Writing $\eta(\mathbf{r}) = \eta_0 + \delta\eta(\mathbf{r})$, let $\langle \delta\eta \rangle = 0$ so that the mean diffusion coefficient is just the constant η_0. Then if $\mathbf{B}$ is written as $\mathbf{e}_z B(x) + \delta\mathbf{B}(\mathbf{r}, t)$ the equations reduce to

$$\left(\frac{\partial}{\partial t} - \eta_0 \nabla^2\right) \delta B_x \cong -\frac{\partial \delta\eta}{\partial z} B'(x) + O\left(\frac{\eta_0 B}{L^2}, \frac{\delta\eta \delta B}{l^2}\right),$$

$$\left(\frac{\partial}{\partial t} - \eta_0 \nabla^2\right) \delta B_y = 0,$$

$$\left(\frac{\partial}{\partial t} - \eta_0 \nabla^2\right) \delta B_z = +\frac{\partial \delta\eta}{\partial x} B'(x) + O\left(\frac{\eta_0 B}{L^2}, \frac{\delta\eta \delta B}{l^2}\right).$$

These equations are valid so long as $\delta\eta \ll \eta_0$ and $|\delta\mathbf{B}| \ll B$. If $\delta\mathbf{B}(\mathbf{r}) = 0$ at time $t = 0$, then subsequently

$$\delta B_x(\mathbf{r}, t) = -\int \mathrm{d}^3\mathbf{r}' \int_0^t \mathrm{d}t' G(\mathbf{r}, \mathbf{r}'; t, t') B'(x') \frac{\partial \delta\eta(\mathbf{r}')}{\partial z'},$$

where G is the familiar Green's function

$$G(\mathbf{r}, \mathbf{r}'; t, t') = \frac{1}{\{4\pi\eta(t-t')\}^{\frac{3}{2}}} \exp\left\{-\frac{(\mathbf{r}-\mathbf{r}')^2}{4\eta(t-t')}\right\}$$

for the heat flow equation. There is a similar expression for δB_z, and $\delta B_y = 0$. Since $\langle\delta\eta\rangle = 0$, it follows that $\langle\delta B_x\rangle = 0$. For times sufficiently small that $4\eta t \ll L^2$, the width of G is small compared to the scale of variation of $B(x)$, so that $B(x)$ may be expanded as $B_0(1 + x/L + \ldots)$ and only the first non-vanishing term retained. For t so small that $4\eta t \ll l^2$, G is effectively a delta function and the field grows linearly with time

$$\delta B_x(\mathbf{r}, t) = -(B_0/L)t\, \partial\delta\eta(\mathbf{r})/\partial z.$$

Then δB_x has the same random properties as $\partial\delta\eta(\mathbf{r})/\partial z$. Ultimately for $L \gg l$ the diffusion reaches a quasi-equilibrium, for which $\partial/\partial t$ decreases from its initial magnitude, of the order of η/l^2, to the very much smaller η/L^2. Then in that limit

$$\nabla^2 \delta B_x \cong (B_0/\eta_0 L)\, \partial\delta\eta/\partial z$$

and

$$\delta B_x(\mathbf{r}) = -\frac{B_0}{4\pi\eta_0 L} \int \mathrm{d}^3\mathbf{r}' \frac{\partial\delta\eta(\mathbf{r}')/\partial z}{|\mathbf{r}-\mathbf{r}'|},$$

etc. Thus $\delta B_x(\mathbf{r})$ may be thought of as the Coulomb potential of the random charge distribution $\partial\delta\eta/\partial z$. As a matter of fact, the notorious non-convergence of the Coulomb potential restricts the choice of the random function $\partial\delta\eta/\partial z$ within the present approximation $|\delta\mathbf{B}| \ll B$. For if we do not require that the mean value of $\partial\delta\eta/\partial z$ over any volume V goes rapidly to zero as V exceeds some dimension b small compared to L—i.e. something equivalent to Debye shielding—then the integrand fails to converge, and δB_x may attain enormous values that are determined by the external dimensions L of the region. The approximation $|\delta\mathbf{B}| \ll B_0$ may be jeopardized and a more complete integration of the equation for the stationary field

$$\nabla \times (\eta \nabla \times \mathbf{B}) = 0$$

must be sought. Only if the auto-correlation of $\partial\delta\eta/\partial z$ becomes suitably negative between points separated by distances $O(l)$ so that the volume integral of $\partial\delta\eta/\partial z$ goes rapidly to zero beyond l, is the present formulation valid. The point important to the present discussion is that $\partial\delta\eta/\partial z$ is a random function of position. The large magnitude of $\delta\mathbf{B}$ in a fully random variation of $\partial\delta\eta/\partial z$ has interesting implications that are taken up elsewhere.

For present purposes it is sufficient to consider δB_x for $\eta t \ll l^2$. Then

$$\delta B_x(\mathbf{r}) = -\frac{B_0 t}{L}\left\{\delta\eta_z(\mathbf{r}) - \frac{\eta_0 t}{2}\nabla^2\delta\eta_z + \ldots + \frac{(-\eta_0 t)^n}{(n+1)!}(\nabla^2)^n\delta\eta_z + \ldots\right\},$$

where the subscript z denotes the derivative with respect to z. To illustrate the random walk of the lines of force, consider a field with precisely the spatial distribution of the first-order term. Say

$$\delta B_x(\mathbf{r}) = \epsilon B_0 \delta\eta_z(\mathbf{r}).$$

The higher order terms contribute further to the random walk, of course, but the lines of this field are already stochastic[3]. The magnetic line of force through the origin has coordinates $(\Delta x(z), 0, z)$ elsewhere, with

$$\frac{\mathrm{d}\Delta x}{\mathrm{d}z} = \frac{\delta B_x(\mathbf{r})}{B_0(1+\Delta x/L)}.$$

Expand $\delta\eta_z(\mathbf{r})$ in ascending powers of Δx, with the notation

$$\delta\eta_z(\mathbf{r}) = \delta\eta_z(z) + \delta\eta_{zx}(z)\Delta x + \ldots,$$

where the subscripts denote differentiation and the coefficients are evaluated at $(0, 0, z)$. Then, neglecting terms second order in Δx,

$$\mathrm{d}\Delta x/\mathrm{d}z = \epsilon\delta\eta_z(z) - \epsilon[\delta\eta_z(z)/L - \delta\eta_{zx}(z)]\Delta x + \ldots .$$

Integrating, this becomes

$$\Delta x(z) = \epsilon \exp\{\epsilon[\delta\eta_x(z) - \delta\eta(z)/L]\}\int_0^z \mathrm{d}z' f(z'),$$

where $f(z)$ is the random function

$$f(z) \equiv \delta\eta_z(z)\exp[-\epsilon\{\delta\eta_x(z) - \delta\eta(z)/L\}].$$

It is instructive to see what variations of $\delta\eta(z)$ are responsible for the

[3] The more fastidious reader may wish to carry through the formal expansion in powers of t, obtaining

$$\Delta x(z) = -\frac{t}{L}\left[\{\delta\eta(z) - \delta\eta(0)\} - \tfrac{1}{2}\eta_0 t\{\nabla^2\delta\eta(z) - \nabla^2\delta\eta(0)\}\right.$$
$$+\frac{t}{2L^2}\{\delta\eta(z) - \delta\eta(0)\}^2 + \frac{t\delta\eta(0)}{L}\{\delta\eta_x(z) - \delta\eta_x(0)\}$$
$$\left. -\frac{t}{L}\int_0^z \mathrm{d}z'\delta\eta(z')\delta\eta_{zx}(z')\right] + \mathrm{O}(t^3).$$

The integrand is readily transformed into $\delta\eta_x\delta\eta_z$ by integration by parts, so that the lowest order contribution to the accumulative random walk is the integral of $\delta\eta_x\delta\eta_z$.

accumulation of the displacement. Expanding the exponential in $f(z)$ in ascending powers of ϵ there are a number of terms, some of which are perfect differentials. Since $\delta\eta$ is a bounded variation, their integration leads to no accumulated displacement. Altogether

$$\begin{aligned}\Delta x(z) = \epsilon \exp[\epsilon\{\delta\eta_x(z) - \delta\eta(z)/L\}]\\ \times\left(\sum_{n=1}^{\infty}\left(\frac{\epsilon}{L}\right)^{n-1}\frac{1}{n!}\{\delta\eta_z^n(z) - \delta\eta_z^n(0)\}\right.\\ -\epsilon\int_0^z dz'\delta\eta_z(z')\delta\eta_x(z')\\ +\frac{\epsilon^2}{2!}\int_0^z dz'\delta\eta_z(z')\{\delta\eta_x^2(z') - 2\delta\eta_x(z')\delta\eta(z')/L\}\\ -\frac{\epsilon^3}{3!}\int_0^z dz'\delta\eta_z(z')\{\delta\eta_x^3(z') - 3\delta\eta_x^2(z')\delta\eta(z')/L\\ \left.+3\delta\eta_x(z')\delta\eta^2(z')/L^2\} + \ldots\right).\end{aligned}$$

We suppose that $\partial\delta\eta/\partial x$ and $\partial\delta\eta/\partial z$ are statistically independent, so the first term contributing to an accumulative displacement is the integral of $\delta\eta_x\delta\eta_z$. It follows that the accumulated mean square displacement over a large distance z is

$$\langle(\Delta x)^2\rangle \cong \epsilon^4\int_0^z dz'\int_0^z dz''\langle\delta\eta_x(z')\delta\eta_z(z')\delta\eta_x(z'')\delta\eta_z(z'')\rangle$$

where we neglect non-cumulative terms $O(\epsilon^2)$. Then if $C(\zeta)$ is the correlation function for $\delta\eta_x\delta\eta_z$,

$$C(\zeta) = \langle\delta\eta_x(z)\delta\eta_z(z)\delta\eta_x(z+\zeta)\delta\eta_z(z+\zeta)\rangle,$$

It is readily shown that

$$\langle(\Delta x)^2\rangle = 2\epsilon^4 z\int_0^{\infty} d\zeta C(\zeta).$$

Altogether, then, the random fluctuations $\delta\eta$ lead to random fluctuations in the magnetic lines of force. There is a component of the fluctuations which leads to a random walk of the lines with an accumulative mean square displacement.

For larger values of time the field goes over into the asymptotic Coulomb integral already mentioned. The random walk may then be very rapid, its magnitude depending sensitively on the statistical properties of $\delta\eta$.

In as much as temperature fluctuations are the rule, particularly in the presence of magnetic fields, we expect to find inhomogeneous resistivity.

That alone is sufficient to cause random walk of the lines, i.e. random variation of the winding pattern of lines of force about their neighbours.

As we shall see in Chapters 15 and 16, the turbulent and resistive conditions that are essential for the initial generation of the magnetic fields in Earth and other planets, in the sun and other stars, and presumably in the Milky Way and other galaxies, are also the conditions that produce the random variation of winding pattern of those fields. We conclude, therefore, that all the magnetic fields generated in the cosmos have stochastic topologies. There is no way to introduce the extreme uniformity and symmetry that would be required for them to be otherwise. The large-scale field, however regular its overall pattern may be, contains random components of significant strength over comparable and smaller scales. This fact leads to the conclusion in Chapter 14 that every astrophysical magnetic field is dynamically active. The random components of the field are subject to active dissipation that is orders of magnitude more rapid than simple resistive decay of the overall field. The high rate of the dynamical dissipation often produces suprathermal particles and their many observable effects.

11.4. An illustration of random walk

It will be useful for the later developments on dynamical non-equilibrium to present a simple illustrative example of the random walk of a magnetic line of force, showing how the line wanders about within the field as a whole. The line of force, or the elemental tube of flux, it will be remembered, is the path for the free passage of fluid. In magnetohydrostatic equilibrium the lines of force are paths of uniform pressure through the field ($\mathbf{B}\cdot\nabla p=0$ in the absence of external force). It is the random wandering of these paths of uniform pressure through the large-scale pattern of the field that in the real world prevents the establishment of the conventional ideal equilibria discussed at length in Chapter 6.

As an example suppose that the principal component of the magnetic field is the constant field B_0 in the z-direction. To this uniform field add the simple uniform twist $b_\phi=\epsilon B_0\varpi/a$ over a distance L_1 at the regular intervals $0<z-n(L_1+L_2)<L_1$ where $n=0, 1, 2, \ldots$. Over the intervening space, of length L_2, introduce the dipole field

$$b_\varpi=\epsilon B_0(a^2/\varpi^2)\cos\phi, \qquad b_\phi=\epsilon B_0(a^2/\varpi^2)\sin\phi$$

described by the vector potential

$$A=\epsilon B_0(a^2/\varpi)\sin\phi. \tag{11.49}$$

The pattern of the fields is sketched in Fig. 11.3. In the regions of uniform twist every line of force undergoes the angular rotation $\epsilon L_1/a$ about the z-axis while $\varpi=\text{constant}$. In the regions of the dipole field the

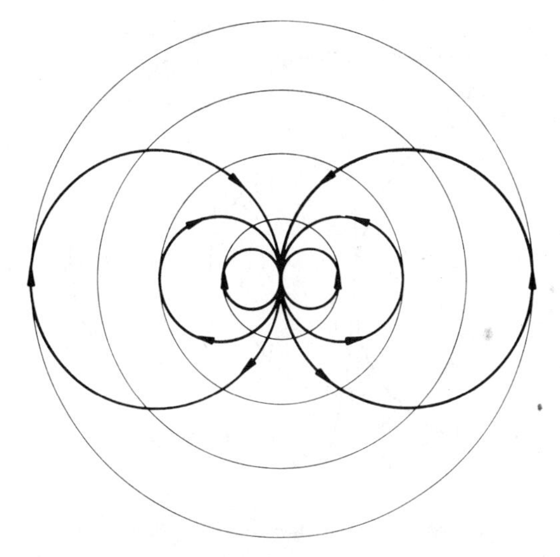

FIG. 11.3. A superposition of the winding pattern of the magnetic lines about the z-axis in the region of field subject to a uniform twist (light lines, concentric circles) and the region of field with the double twist (heavy lines).

displacement consists of a twist, in opposite sense on either side of $\phi=0$ along the lines $A=$ constant. A line of force entering the dipole field at $z=L_1$, with coordinates (ϖ_1, ϕ_1) leaves the region at $z=L_1+L_2$ with coordinates (ϖ_2, ϕ_2), where $A=$ *constant* leads to the relation

$$\sin\phi_1/\varpi_1=\sin\phi_2/\varpi_2$$

while integration of $\varpi\, d\phi = dz\, b_\phi/B_0$ yields

$$\phi_2-\phi_1+\tfrac{1}{2}\{\sin(2\phi_1)-\sin(2\phi_2)\}=2\epsilon(L_2/a)(a/\varpi_1)^3\sin^3(\phi_1).$$

The projection of the line of force through the point $\varpi/a=1$, $\phi=0$ at $z=0$ onto the (ϖ, ϕ) plane is shown in Fig. 11.4 for the special case where the rotation $\epsilon L_1/a$ is $\pi/3$ and the dipole field has the strength $\epsilon L_2/a=4$. The numbers on the graph represent the values of n at the positions of the projection at $z=n(L_1+L_2)$ beginning with $n=0$ where $\varpi=a$ and $\phi=0$. It is clear from the example that the line of force performs a complicated walk about the z-axis, the field pattern changing constantly along the field. The basic point to be established by the example is the penetration of the line of force into new regions when the field pattern changes along the field. With one fixed pattern the projection of the line executes a simple closed curve, of course, but with the combination of two simple, but different, patterns the path takes on a completely different character, undergoing erratic excursions over a broad region.

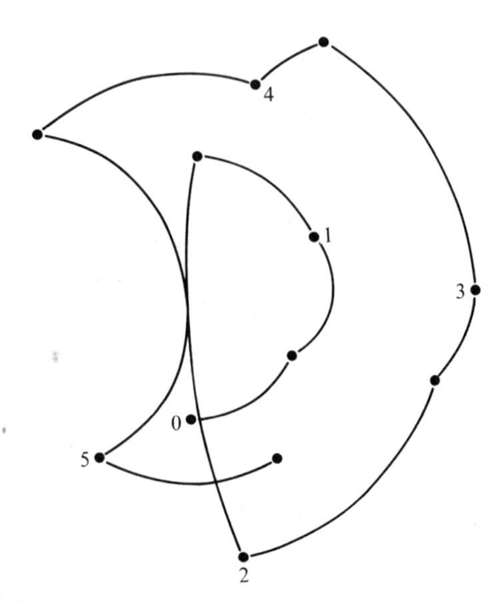

FIG. 11.4. The projection of a line of force onto the $z=0$ plane, beginning at $\varpi=a$, $\phi=0$ at $z=0$ and subject to the uniform twist $\epsilon L_1/a=\pi/3$ for a distance L_1 along the z-axis, followed by the double twist $\epsilon L_2/a=4$ for a distance L_2. The cycle then repeats n times, with the value of n indicated at the position of the line at the end of the nth cycle.

The example raises a number of interesting questions that we are not prepared to answer. Are the excursions bounded in the present example, and is the walk random in any sense? Or do the excursions lead to some systematic migration?

It is evident that the mathematical techniques developed to treat the trajectories of particles in phase space might be fruitfully applied to the behaviour of magnetic lines of force (cf. Dragt 1965, 1966; Arnold and Avez 1968; Dragt and Finn 1975). Thus, for a particle with momentum p_i subject to a force F_i we write

$$\mathrm{d}p_i/\mathrm{d}t=F_i.$$

Or in one dimension

$$\mathrm{d}p^2/\mathrm{d}x=2F.$$

For a magnetic line of force of the field (B_x, B_y, B_0) there is the similar form

$$\mathrm{d}x_i/\mathrm{d}z=B_i/B_0$$

with $i=1, 2$. Fortunately, we do not need the answers to these questions to go on to the proof of non-equilibrium.

References

ARNOLD, V. I. and AVEZ, A. (1968). *Ergodic problems of classical mechanics*, Benjamin, New York.

CARSLAW, H. S. and JAEGER, J. C. (1959). *Conduction of heat in solids*, Clarendon Press, Oxford.

CHANDRASEKHAR, S. (1943). *Rev. Mod. Phys.* **15,** 1.

DRAGT, A. J. (1965). *Rev. Geophys.* **3,** 255.

—— (1966). ibid. **4,** 112.

—— and FINN, J. M. (1975). *Tech. Rep. No. 75–098*, 1 July. University of Maryland.

FAN, C. Y., PICK, M., PYLE, K. R., SIMPSON, J. A., and SMITH, D. R. (1968). *J. Geophys. Res.* **73,** 1555.

FRANK, L. A., ACKERSON, K. L., and LEPPING, R. P. (1976). *J. Geophys. Res.*, **81,** 5859.

GETMANTSEV, G. G. (1963). *Soviet Astron.—A. J.* **6,** 477.

GOTT, J. R., GUNN, J. E., SCHRAMM, D. N., and TINSLEY, B. M. (1974). *Astrophys. J.* **194,** 543.

JOHNSON, H. E. and AXFORD, W. I. (1971). *Astrophys. J.* **165,** 381.

JOKIPII, J. R. (1966). *Astrophys. J.* **146,** 480.

—— (1967). ibid. **149,** 405.

—— (1968). ibid. **152,** 997.

—— and COLEMAN, P. J. (1968). *J. Geophys. Res.* **73,** 5495.

—— and PARKER, E. N. (1969a). *Astrophys. J.* **155,** 777.

—— —— (1969b) ibid. 799.

LANZEROTTI, L. J. (1972). *Rev. Geophys. Space Sci.* **10,** 379.

LEIGHTON, R. B. (1964). *Astrophys. J.* **140,** 1559.

—— (1969). ibid. **156,** 1.

LINGENFELTER, R. E., RAMATY, R., and FISK, L. A. (1971). *Astrophys. Lett.* **8,** 93.

PARKER, E. N. (1963a). *Interplanetary dynamical processes*, Wiley-Interscience, New York.

—— (1963b). Astrophys. J. **138,** 552.

—— (1965). ibid. **142,** 584.

—— (1966). ibid. **145,** 811.

—— (1968a). *Stars and stellar systems, Vol. VII. nebulae and interstellar matter* (ed. B. M. MIDDLEHURST and L. H. ALLER), vol. VII, chap. 14. University of Chicago Press.

—— (1968b). *J. Geophys. Res.* **21,** 6842.

—— (1969). *Space Science Rev.* **9,** 651.

—— and KROOK, M. (1956). *Astrophys. J.* **124,** 214.

PAULIKAS, G. A. (1974). *Rev. Geophys. Space Sci.* **12,** 117.

SCHWARZSCHILD, M. (1959). *Astrophys. J.* **130,** 345.

WEISS, N. O. (1966). *Proc. Roy. Soc. Ser. A* **293,** 310.

YU, G. (1974). *Astrophys. J.* **187,** 75.

12

NON-EQUILIBRIUM OF INVARIANT FIELDS

12.1. General remarks on non-equilibrium

THE non-equilibrium of magnetic fields appears in a variety of forms. We have already noted the runaway ability of flux tubes in a stratified atmosphere in §§8.6–8.8, 9.6, and 9.7. There are, in fact, some very simple and completely symmetric magnetic configurations that possess no equilibrium, and they are the subject of the present chapter. As we shall see, a temperature gradient in the absence of gravity, or a fixed boundary with vacuum beyond, is sufficient to disrupt the equilibrium, causing non-equilibrium of the field and fluid.

It can be stated quite generally that a suitable break in symmetry is the cause of non-equilibrium. Thus in a purely hydrodynamic fluid, involving no magnetic field, there is no available equilibrium state if the opposing forces do not line up properly. For instance, it is well known (see Chandrasekhar 1961) that there can be an equilibrium state in a fluid subject to thermal expansion and a gravitational potential ψ only if the temperature gradient $\partial T/\partial x_i$ and the potential gradient $\partial\psi/\partial x_i$ are parallel. We should add that the equilibrium, when they are parallel, is unstable unless suitably damped by viscosity. But instability is quite a different effect from non-equilibrium.

When there is a magnetic field present in a compressible fluid, there can be no equilibrium unless the gravitational force $-\partial\psi/\partial x_i$ is parallel to the magnetic force $\partial M_{ij}/\partial x_j$. When they are parallel, there may be an equilibrium, but the equilibrium is in many cases subject to the Rayleigh–Taylor instability as a result of the magnetic buoyancy (see Chapter 13). It will be shown below that a temperature gradient $\partial T/\partial x_i$ that is not parallel to $\partial M_{ij}/\partial x_j$ generally destroys the possibility for equilibrium, forcing the system into a state of motion for as long as the temperature gradient is maintained.

Altogether, then, it is not sufficient for equilibrium that the magnetic field possess some high degree of symmetry or invariance. It is necessary that the other effects present, such as the thermal expansion of the fluid, or the gravitational forces, also be properly symmetrical and aligned. Examples of non-equilibrium, where the temperature or the boundaries break the invariance, are given in the sections that follow, with particular application to fields invariant along the lines of force $B_j\,\partial B_i/\partial x_j = 0$. The uniform field and the azimuthal field are particularly convenient for illustrating the effects.

It should be pointed out that the topology of the magnetic field plays a critical role in the various types of non-equilibrium. The present chapter deals with symmetric closed topologies. In Chapter 14 we consider the general dynamical non-equilibrium that arises when the field topology lacks invariance, the topology varying along the general direction of the field (Parker 1972; Yu 1973).

It should be noted that there are non-equilibria that have nothing to do with topology, wherein the topology possesses equilibrium states but the fluid is initially distributed along the field in such a way as to cause non-equilibrium. Such problems often arise in astrophysical circumstances, and numerous examples have been treated in the literature. They are not our principal concern here, so we note briefly only a few well known examples to illustrate their nature and to distinguish them from the circumstances where no equilibrium exists for the particular topology.

In the most trivial case, in the absence of gravity, a magnetic field $B(x, y)$ in the z-direction in a gas whose pressure $p(x, y)$ does not satisfy $p+B^2/8\pi = constant$, is clearly out of equilibrium, but, equally clearly, it will pass into equilibrium in a characteristic time equal to the transit time of fast mode (compressional) hydromagnetic waves across the region, in which the pressure and the field readjust to equilibrium. We must then wait for the fast mode to damp out, of course, but the equilibrium is immediately available. Indeed, the hydromagnetic waves may be thought of as stable oscillations of finite amplitude about the final equilibrium.

To point out a well known but slightly less trivial example, it is obvious that in the presence of a gravitational field the condition $p+B^2/8\pi =$ *constant* is not enough for equilibrium. The equilibrium equation for an incompressible fluid of fixed but variable density ρ is

$$0=\nabla(p+B^2/8\pi)+\rho\nabla\psi \qquad (12.1)$$

and requires that $\rho=\rho(\psi)$. This may be shown by taking the curl of the equation, yielding

$$\nabla\rho\times\nabla\psi=\frac{\partial(\rho, \psi)}{\partial(x, y)}=0$$

so that the only solution is $\rho=\rho(\psi)$. Thus, as in any fluid, dense regions sink and tenuous regions rise. But once that gravitational separation of heavy and light fluid has taken place (in a finite space) the fluid is trapped in an equilibrium and can, in principle, settle down to an eternity of stasis.

Similarly a horizontal magnetic flux tube twisted more in one place along its length than another is not in equilibrium, but the transport of helical coils of field along its length equalizes the twisting in approximately one Alfven transit time, and the tube resides in static equilibrium thereafter.

These non-equilibria commonly occur in nature, and in view of their simplicity are taken for granted. We are after more subtle and exotic game. We are concerned with field configurations whose topology rules out equilibrium, so that in a highly conducting fluid there is no readjustment and redistribution that can bring the system into equilibrium. The present chapter takes up the non-equilibrium of a magnetic field with axial symmetry. It is a dynamical non-equilibrium that occurs in the azimuthal sheath of any expanded twisted flux rope extending upward through a star. The next chapter takes up a slightly different kind of non-equilibrium, involving a horizontal magnetic field confined in a slab of gas, representing the magnetic field in the gaseous disc of the galaxy, or beneath the surface of a star. There exists a single equilibrium state, but that equilbrium is unstable, taking the system farther and farther from equilibrium into a state of ever increasing chaos over ever diminishing dimensions. No ultimate stable equilibrium is available until resistive dissipation in the small-scale disorder destroys the large-scale field.

The examples prepare the way for the discussion of the dynamical non-equilibrium of non-symmetric (non-invariant) fields in Chapter 14.

12.2. Equilibrium of an axi-symmetric azimuthal field

The axi-symmetric azimuthal field affords the simplest illustration of non-equilibrium of a field and fluid in which the Lorentz force $\partial M_{ij}/\partial x_j$ exerted by the field on the fluid is not parallel to the temperature gradient $\partial T/\partial x_i$ across the fluid. It should be noted that in Chapter 9 we considered the hydrostatic equilibrium of an axi-symmetric azimuthal field $B_\phi(\varpi)$ with the assumption that there was some slight vestige of the longitudinal component $B_z(\varpi)$ remaining to provide barometric equilibrium of the gas in the z-direction along the field. The assumption was based on the evolution of the field into the limiting azimuthal form from an initial helical form $b_\phi(\varpi)$, $b_z(\varpi)$ so that in any finite time the longitudinal component may be small but non-vanishing. We now explore the alternative in which B_z is identically zero. The field is entirely azimuthal and the infinitely-conducting fluid is tied to the azimuthal lines of force. If the equation of state of the fluid contains the effects of thermal expansion, $(\partial\rho/\partial T)_p \neq 0$, then any variation of temperature along the z-axis causes a convective non-equilibrium (Parker 1975).

To treat the simplest possible case suppose that the field $B_\phi(\varpi, z)$ is embedded in a fluid of high electrical conductivity in the absence of gravitational forces. Then the equations for hydrostatic equilibrium are

$$0=\frac{\partial}{\partial \varpi}\left(p+\frac{B_\phi^2}{8\pi}\right)+\frac{B_\phi^2}{4\pi\varpi}, \tag{12.2}$$

$$0=\frac{\partial}{\partial z}\left(p+\frac{B_\phi^2}{8\pi}\right) \tag{12.3}$$

where the last term in (12.2) represents the inward force of the tension $B_\phi^2/4\pi$ of the azimuthal field extending around the radius ϖ. Take the curl of these equations, differentiating (12.2) with respect to z, (12.3) with respect to ϖ, and subtracting. The result is

$$\frac{\partial}{\partial z}\frac{B_\phi^2}{4\pi}=0. \tag{12.4}$$

It follows from (12.3) that

$$\partial p/\partial z=0. \tag{12.5}$$

In equilibrium, then, the fluid pressure and the field both depend only on ϖ, both being uniform in the z-direction. The form of the field is given by eqn (9.1).

12.3. Non-equilibrium in a finite volume

Consider an azimuthal field in a finite volume, such as a cylinder of finite length, or a sphere, centred on the origin. There is no azimuthal field in the vacuum outside, so B_ϕ vanishes on the surface. But if B_ϕ vanishes on the surface, then it follows from (12.4) that it vanishes everywhere throughout the interior. A non-vanishing azimuthal field of any form precludes equilibrium within the volume. This circumstance applies to the azimuthal field in the fluid metal core of a planet, for instance.

Consider, then, what happens if at time $t=0$ there is a non-vanishing field B_ϕ within the volume. To fix ideas, suppose that the fluid in which the field is embedded is confined to the cylinder $-h<z<+h$, $\varpi<a$. The cylinder is surrounded by an infinite empty space. The fluid within the cylinder has an electrical resistivity that may be arbitrarily small but non-vanishing. Therefore, the normal component of the electric current must fall *continuously* to zero as one approaches the boundary of the fluid from the inside. In view of the fact that the field is independent of both ϕ and z, the only non-vanishing component of the curl yields the current density

$$j_z=\frac{c}{4\pi}\frac{1}{\varpi}\frac{\partial}{\partial \varpi}\varpi B_\phi,$$

and in equilibrium this is independent of z, of course, because B_ϕ is independent of z. The exterior vacuum requires that j_z vanish on the ends $z = \pm h$. Hence it must vanish throughout, which requires $B_\phi \alpha 1/\varpi$. But in any real physical situation B_ϕ must be finite everywhere and vanish, rather than diverge, on the z-axis ($\varpi \to 0$). So the equilibrium cannot satisfy the boundary condition.

Suppose, then, that we go ahead to set up some simple initial field, say $B_\phi = B(\varpi)$ in the cylinder with the equilibrium pressure $p = p_0 - B^2(\varpi)/8\pi$. What happens? At time $t = 0$ all is in equilibrium, but after the elapse of any finite time t the magnetic field has diffused out the ends of the cylinder from a depth $\sim(4\eta t)^{\frac{1}{2}}$, where η is the familiar resistive diffusion coefficient $c^2/4\pi\sigma$. So long as $4\eta t \ll a^2$, the field in the neighbourhood of either end is given by

$$B_\phi(\varpi, z, t) = B(\varpi)\operatorname{erf}\left\{\frac{z \pm h}{(4\eta t)^{\frac{1}{2}}}\right\}. \tag{12.6}$$

Within the thin boundary layer of thickness $(4\eta t)^{\frac{1}{2}}$, B_ϕ is substantially less than the equilibrium $B(\varpi)$ elsewhere in the cylinder, and falls to zero at $z = \pm h$. The boundary layer is squeezed against $z = h$ with the pressure $p + B_\phi^2/8\pi$ exerted on it by the field and fluid farther in. Hence, at the boundary, where B_ϕ falls to zero, the fluid alone must sustain the full pressure p_h,

$$p_h = p_0 - B^2(\varpi)/8\pi.$$

Since the azimuthal field must vanish on the z-axis, p_h must be a strongly varying function of radial distance ϖ, decreasing outward from the z-axis. Hence within the boundary layer of thickness $(4\eta t)^{\frac{1}{2}}$ the fluid is accelerated radially outward along the ends of the cylinder, removing the field-free fluid from the boundary layer. The evacuation of fluid from the boundary layer brings more field up to the ends, through which the field escapes, and leads to further ejection of fluid from the boundary layer, etc. It is evident that the system is in a dynamical state with a continual ejection of fluid from the boundary layer at the ends.

The problem is an example of the general neutral sheet non-equilibrium discussed in Chapter 14. In that context, note that the boundary conditions $B_\phi = 0$ on $z = \pm h$ would be unchanged if the vacuum outside the ends of the cylinder were replaced by the ends of cylinders in which B_ϕ were of opposite sign. In that case the same ejection of fluid, from between the two opposite fields on either side of $z = \pm h$, would occur as in the present case with vacuum outside.

It should be noted further that the dynamical non-equilibrium at the ends of the cylinder is closely analogous to the Eckman layer that forms, for instance, in the bottom of a teacup when vigorously stirred round and round. The flow in the Eckman layer is radially inward, rather than outward, of course, because the Reynolds stress—the centrifugal force exerted by the circular motion of the tea—is radially outward, of opposite sign to the inward force $B_\phi^2/4\pi\varpi$ exerted by the magnetic field.

It is clear that the non-equilibrium applies to the azimuthal magnetic field (presumably of some 10^2 G) trapped in the molten metal core of Earth and surrounded by the relatively non-conducting silicate mantle. In the absence of any relative rotation of the core and mantle there would be a sheet of fluid flowing radially outward from either pole at the surface of the core. Recalling that in the Earth the whole system is rotating, it is evident that the Coriolis force on the radial outflow would lead to azimuthal acceleration, with the surface layer moving westward relative to the core.

Now relative rotation of the core and mantle would lead to an opposing centrifugal force to balance the non-equilibrium magnetic field. Benton and Loper (1969) have explored the general problem of the boundary layer at an insulating boundary and Loper (1972) has treated the complex non-linear dynamical problem posed by the non-equilibrium just described, working out the stationary solution to the dynamical equations in the presence of both an axial and an azimuthal field.

To illustrate in a simple way the fluid motions produced by the imbalance of forces consider the onset of motion from a state of rest in which the field satisfies the equilibrium equations (12.2) and (12.3) at $t=0$. We might think of the problem as $\eta=0$ for $t<0$, so that there is no diffusion whatever at the boundaries, switching on some small resistivity η when $t=0$ so that diffusion begins and non-equilibrium appears in the system. Then (apart from a small neighbourhood $\varpi<(4\eta t)^{\frac{1}{2}}$ of the z-axis) the field in a thin layer immediately inside the end $z=h$ begins to evolve away from the initial form $B(\varpi)$ as described by (12.6). Then B_ϕ is no longer independent of z, and the equilibrium equations are not satisfied, leading to fluid motion $(v_\varpi, 0, v_z)$. The motion is confined to the boundary layer of characteristic thickness $\epsilon=(4\eta t)^{\frac{1}{2}}$, and consists of a radial acceleration caused by the mean pressure gradient $B^2(a)/8\pi a$ over the radius a. Hence, in order of magnitude the radial acceleration is $B^2(a)/4\pi\rho a$, and after a time t,

$$v_\varpi=\mathrm{O}(V_\mathrm{A}^2 t/a)$$

where V_A is the characteristic Alfven speed $B(a)/(4\pi\rho)^{\frac{1}{2}}$. Conservation of fluid requires a longitudinal velocity v_z of the order of

$$v_z=\mathrm{O}(v_\varpi\epsilon/a).$$

In the initial departure from equilibrium ($t\ll a/V_\mathrm{A}$) the non-linear terms can be neglected and the equations of motion are

$$\rho \frac{\partial v_\varpi}{\partial t} = -\frac{\partial}{\partial \varpi}\left(p + \frac{B^2}{8\pi}\right) - \frac{B^2}{4\pi\varpi}, \tag{12.7}$$

$$\rho \frac{\partial v_z}{\partial t} = -\frac{\partial}{\partial z}\left(p + \frac{B^2}{8\pi}\right). \tag{12.8}$$

For the initial onset of field diffusion and fluid motion ($\epsilon \ll a$) the longitudinal flow v_z into the boundary layer is small compared to the radial motion, so that the acceleration $\partial v_z/\partial t$ is negligible and

$$\frac{\partial}{\partial z}\left(p + \frac{B^2}{8\pi}\right) = 0.$$

Hence

$$p + B^2/8\pi = f(\varpi),$$

where $f(\varpi)$ is an arbitrary function of ϖ. The equation for the radial acceleration becomes

$$\rho \frac{\partial v_\varpi}{\partial t} = f'(\varpi) - B^2(\varpi, t)/4\pi\varpi.$$

Integrating over time, the radial velocity produced by the azimuthal field (12.6) is

$$\begin{aligned}\rho v_\varpi(\varpi, z, t) &= f'(\varpi)t - \frac{1}{4\pi\varpi}\int_0^t \mathrm{d}t' B^2(\varpi, t') + \mathrm{O}(t^2)\\ &= f'(\varpi)t - \frac{B^2(\varpi)(h-z)^2}{8\pi\eta\varpi}\int_\xi^\infty \frac{\mathrm{d}s}{s^3}\operatorname{erf}^2(s)\end{aligned}$$

where $s = (h-z)/(4\eta t')^{\frac{1}{2}}$ and $\xi = (h-z)/(4\eta t)^{\frac{1}{2}}$. But far from the boundary ($\xi \gg 1$) the fluid is at rest, $v_\varpi = 0$. Noting the asymptotic form

$$\operatorname{erf} \xi \cong 1 - \frac{\exp(-\xi^2)}{\pi^{\frac{1}{2}}\xi}\left(1 - \frac{1}{2\xi^2} + \ldots\right),$$

there follows the result

$$\int_\xi^\infty \frac{\mathrm{d}s}{s^3}\operatorname{erf}^2 s \cong \frac{1}{2\xi^2}$$

for $\xi \gg 1$. Hence the condition that $v_\varpi = 0$ far from the boundary leads to the exact result

$$f'(\varpi) = B^2(\varpi)/4\pi\varpi$$

and

$$v_\varpi = \frac{B^2(\varpi)t}{4\pi\rho\varpi}\left\{1 - 2\xi^2\int_\xi^\infty \frac{\mathrm{d}s}{s^3}\,\mathrm{erf}^2 s\right\}. \tag{12.9}$$

The quantity in braces, giving the initial velocity profile across the boundary layer, is plotted in Fig. 12.1.

If instead of an azimuthal field $B(\varpi)$ there is a relative rotation of the fluid, so that at time $t=0$ there is present the azimuthal velocity $v(\varpi)$, then the radial velocity initiated by the non-equilibrium is inward rather than outward. The equation of motion for the azimuthal velocity is

$$\frac{\partial v_\phi}{\partial t} + \frac{v_\varpi}{\varpi}\frac{\partial}{\partial \varpi}\varpi v_\phi + v_z\frac{\partial v_\phi}{\partial z} = \nu\left(\frac{1}{\varpi}\frac{\partial}{\partial \varpi}\varpi\frac{\partial}{\partial \varpi} - \frac{1}{\varpi^2} + \frac{\partial^2}{\partial z^2}\right)v_\phi$$

with the uniform kinematic viscosity $\nu = \mu/\rho$. For the initial departure from $v_\phi = v(\varpi)$, the thickness of the boundary layer is $(4\nu t)^{\frac{1}{2}}$ and the derivative $\partial/\partial z = \mathrm{O}\{1/(4\nu t)^{\frac{1}{2}}\}$ dominates the others. The radial velocity is small, $\mathrm{O}(t)$, and v_z is small $\mathrm{O}\{(4\nu t)^{\frac{1}{2}}/a\}$ compared to the radial velocity. Altogether

$$\frac{\partial v_\phi}{\partial t} \cong \nu\frac{\partial^2 v_\phi}{\partial z^2},$$

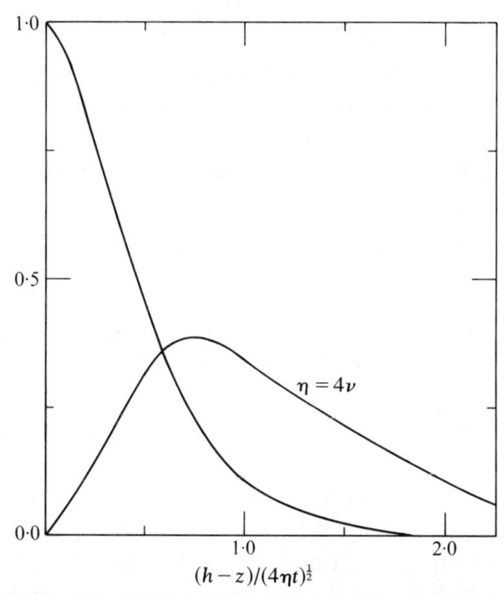

FIG. 12.1. The initial radial velocity profiles across the boundary layer at the ends of the cylinder. The unlabelled curve represents the velocity profile when there is an azimuthal magnetic field but no rotation of the fluid relative to the boundary. It represents the negative of the velocity profile when there is rotation but no field. The curve labelled $\eta = 4\nu$ corresponds to equal azimuthal field and rotation $B = (4\pi\rho)^{\frac{1}{2}}v$ with $\eta = 4\nu$.

with the solution

$$v_\phi(\varpi, z, t) = v(\varpi)\mathrm{erf}(\zeta)$$
$$\zeta = (h-z)/(4\nu t)^{\frac{1}{2}}$$

in exactly the same form as (12.6). The requirement $\partial v_\phi/\partial z = 0$ for dynamical equilibrium is violated for $t > 0$, leading to the radial flow that makes up the familiar Eckman layer. The radial component of the equation of motion is

$$\frac{\partial v_\varpi}{\partial t} = -\frac{1}{\rho}\frac{\partial p}{\partial \varpi} + \frac{v_\phi^2}{\varpi}$$

while the z-component is

$$\frac{\partial v_z}{\partial t} = -\frac{1}{\rho}\frac{\partial p}{\partial z}.$$

These two equations have exactly the same form as (12.7) and (12.8), except that the sign of the Reynolds stress term $+v_\phi^2/\varpi$ (the centrifugal force) is the reverse of the Maxwell stress $-B_\phi^2/4\pi\varpi$. The solution corresponding to (12.9), is the radial flow

$$v_\varpi = -\frac{v^2(\varpi)t}{\varpi}\left\{1 - 2\zeta^2\int_\zeta^\infty \frac{\mathrm{d}s}{s^3}\mathrm{erf}^2(s)\right\}. \tag{12.10}$$

The velocity profile across the growing Eckman boundary layer is given by the quantity in braces, which is of the same form as in (12.9), plotted in Fig. 12.1.

It is obvious that an initial magnetic field $B(\varpi)$ and differential rotation $v(\varpi)$ together yield the radial flow

$$v_\varpi = \frac{t}{\varpi}\left[\frac{B^2(\varpi)}{4\pi\rho}\left\{1 - 2\xi^2\int_\xi^\infty \frac{\mathrm{d}s}{s^3}\mathrm{erf}^2(s)\right\} - v^2(\varpi)\left\{1 - 2\zeta^2\int_\zeta^\infty \frac{\mathrm{d}s}{s^3}\mathrm{erf}^2(s)\right\}\right].$$

As a special example, suppose that the initial azimuthal velocity $v(\varpi)$ is just equal to the Alfven speed $B(\varpi)/(4\pi\rho)^{\frac{1}{2}}$. Then

$$v_\varpi = \frac{t}{\varpi}\frac{B^2(\varpi)}{2\pi\rho}\left\{\zeta^2\int_\zeta^\infty \frac{\mathrm{d}s}{s^3}\mathrm{erf}^2(s) - \xi^2\int_\xi^\infty \frac{\mathrm{d}s}{s^3}\mathrm{erf}^2(s)\right\}. \tag{12.11}$$

Thus, for the unlikely circumstance that the viscous and resistive diffusion coefficients are equal, $\nu = \eta$, the system remains in equilibrium $v_\varpi = 0$. Otherwise the centrifugal and magnetic forces do not balance, and there is an increasing radial flow in the boundary layer, with characteristic thickness $2(\eta^{\frac{1}{2}} + \nu^{\frac{1}{2}})t^{\frac{1}{2}}$ at the ends $z = \pm h$ of the cylinder. The radial velocity profile across the boundary layer is represented by the quantity in brackets, and is plotted in Fig. 12.1 for $\eta = 4\nu$. It is evident by inspection that interchanging the values of η and ν has the effect of changing the sign of v_ϖ. It is also evident that if $v(\varpi)$ and $B(\varpi)/(4\pi\rho)^{\frac{1}{2}}$ are not equal, the radial velocity may change sign across the boundary layer.

These examples suffice to illustrate the non-equilibrium. The more difficult problem of the dynamical steady state of the boundary layer resulting from the non-equilibrium, and the stability of that steady state, has been worked out by Loper (1972), in the context of the problem presented by the azimuthal field at the surface of the liquid metal core of Earth.

12.4. Non-equilibrium in the presence of thermal expansion

If the density of the conducting fluid in which B_ϕ is embedded is subject to thermal expansion $\rho = \rho(p, T)$ (an ideal gas being an example appropriate to astrophysical circumstances) then any variation of temperature along the z-axis causes non-equilibrium. To avoid the non-equilibrium discussed in §12.3, suppose that the conducting fluid either extends to infinity or is bounded by an infinitely conducting medium. Take the fluid to be an ideal gas, with the mean mass of its individual molecules equal to m. Then the pressure and density are related by $p = \rho kT/m$.

A variety of densities and temperatures can be employed for equilibrium, the only requirement being that the product ρT be independent of ϕ and z so as to satisfy (12.5). Any other choice results in non-equilibrium and convective motion. Indeed, the choices yielding non-equilibrium are so numerous as to be uninteresting as a general class. Any sort of motion can be produced by suitably loading the field with density $\rho(\varpi, z)$ or heating to a temperature $T(\varpi, z)$. But these are the trivial cases mentioned in §12.1. The interesting situation is the simplest circumstance in which all the lines of force are loaded equally with gas (the same mass per unit flux) to avoid any artificial prejudice toward non-equilibrium. Considering an azimuthal flux tube of cross-section δA and radius ϖ, the total magnetic flux is

$$\delta\Phi = B_\phi(\varpi, z)\delta A,$$

while the total mass is

$$\delta M = 2\pi\varpi\delta A\rho(\varpi, z).$$

Equal loading means that $\delta M/\delta\Phi$ is the same for all flux tubes, so that, for an infinitely conducting fluid

$$\varpi\rho(\varpi, z)/B(\varpi, z) = a\rho_0/B_0 = constant \tag{12.12}$$

throughout. Using this expression to write B_ϕ in terms of ρ, (12.4) becomes $\partial\rho/\partial z = 0$, so that (12.5) reduces to $\partial T/\partial z = 0$. Equilibrium exists only if the temperature is uniform along the axis of the field. If $\partial T/\partial z \neq 0$, there will of necessity be convective motions (produced by the magnetic stresses rather than gravitational forces for, it will be remembered, we have put g equal to zero). The convection arises because, according to (12.12), B_ϕ is larger, and the inward pressure gradient produced by the tension $B_\phi^2/4\pi$ in the field is larger, where the gas is cooler and denser. Hence the larger radial pressure difference produced by $B_\phi^2/4\pi\varpi$ in the cool region is not compensated by the smaller pressure difference in the hot region, where ρ, and hence B_ϕ, are smaller. The effect is analogous to the convection in a field-free fluid in the presence of gravity, where the pressure gradient $-\rho g$ in cool regions (where ρ is larger than normal) cannot be compensated by $-\rho g$ in hot regions (where ρ is smaller than normal).

Chandrasekhar (1975) has made the interesting point that the equilibrium requirement $\partial T/\partial z = 0$ in the azimuthal field is analogous to the Taylor–Proudman theorem in a rotating system. In a system rotating about the z-axis, the steady dynamical equilibrium fluid velocity v_i is restricted to $\partial v_i/\partial z = 0$. The centrifugal force and the Coriolis force play roles similar to the magnetic pressure gradient $(\partial/\partial\varpi)B_\phi^2/8\pi$ and the curvature stress $B_\phi^2/4\pi\varpi$, respectively. A similar restriction, $\partial v_i/\partial z = 0$, applies to the steady equilibrium motion in the presence of a strong uniform magnetic field parallel to the z-axis (see discussion of these effects in Chandrasekhar 1961).

Coming back to the non-equilibrium caused by the stresses $(1/4\pi)B_j\,\partial B_i/\partial x_j$ exerted on the fluid by the tension in the field, it is clear that our simple example of the azimuthal field is but a special case of a more general non-equilibrium in the presence of temperature gradients that are not parallel to the Lorentz force. It is clear that the small longitudinal component of field b_z in Chapter 9, providing barometric equilibrium along the z-direction, greatly simplified that illustrative example. Without it, the situation is much more complex.

Consider, then, the convective motions produced by $T(z)$ in the azimuthal field. The general form of the fluid motions produced by a temperature gradient $\partial T/\partial z$ in the azimuthal field of §12.2 is readily illustrated from the linearized equations of motion

$$\rho\frac{\partial v_\varpi}{\partial t} = -\frac{\partial}{\partial\varpi}\left(p+\frac{B_\phi^2}{8\pi}\right)-\frac{B_\phi^2}{4\pi\varpi},$$

$$\rho\frac{\partial v_z}{\partial t} = -\frac{\partial}{\partial z}\left(p+\frac{B_\phi^2}{8\pi}\right)$$

for the departure of the fluid from an initial position of rest. Then write

$v_i(\varpi, z, t) = u_i(\varpi, z)t/\tau$ so that

$$\frac{\rho u_\varpi}{\tau} = -\frac{\partial}{\partial \varpi}\left(p + \frac{B_\phi^2}{8\pi}\right) - \frac{B_\phi^2}{4\pi\varpi}, \tag{12.13}$$

$$\frac{\rho u_z}{\tau} = -\frac{\partial}{\partial z}\left(p + \frac{B_\phi^2}{8\pi}\right). \tag{12.14}$$

The non-linear terms $\rho v_j\, \partial v_i/\partial x_j$ are small, $O(t^2/\tau^2)$, for $v_i \ll V_A$ and so may be neglected for that period. These equations apply also to the limiting velocity field in the event that there is a frictional drag $-\rho v_i/\tau$ introduced on the right-hand side, and the time derivative put equal to zero on the left-hand side, of the equations of motion. If τ is sufficiently small, the limiting velocity remains small and the non-linear terms can be neglected for all t. The more difficult problem in which the convective velocity is limited by viscosity and/or modification of $T(\varpi, z)$ through the convective heat flow equation is not without interest. But it is a digression from our basic purpose, to illustrate the absence of equilibrium, so it will not be pursued here.

Conservation of fluid and azimuthal magnetic field, related to each other through (12.12), both require that $\partial \rho u_j/\partial x_j = 0$, or

$$\frac{1}{\varpi}\frac{\partial}{\partial \varpi}\varpi \rho u_\varpi + \frac{\partial}{\partial z}\rho u_z = 0. \tag{12.15}$$

To guarantee conservation of fluid, then, write ρu_i in terms of the stream function $\Psi(\varpi, z)$,

$$\varpi \rho u_\varpi = +a\rho_0\, \partial\Psi/\partial z, \qquad \varpi \rho u_z = -a\rho_0\, \partial\Psi/\partial\varpi.$$

Then (12.15) is automatically satisfied. The stream lines are given by $\Psi = constant$.

The curl of (12.13) and (12.14) yields

$$\varpi \frac{\partial}{\partial \varpi}\frac{1}{\varpi}\frac{\partial \Psi}{\partial \varpi} + \frac{\partial^2 \Psi}{\partial z^2} = -\frac{\tau}{a\rho_0}\frac{\partial}{\partial z}\frac{B_\phi^2}{4\pi}, \tag{12.16}$$

showing the z-variation of the magnetic stress $B_\phi^2/4\pi$ to be the direct cause of the fluid motion Ψ. To eliminate the velocity from the equations and obtain an expression for the configuration of ρ and B_ϕ, use (12.13) and (12.14) to eliminate u_i from (12.15). The result is

$$\left\{\frac{1}{\varpi}\frac{\partial}{\partial \varpi}\varpi\frac{\partial}{\partial \varpi} + \frac{\partial^2}{\partial z^2}\right\}\left(p + \frac{B_\phi^2}{8\pi}\right) + \frac{1}{\varpi}\frac{\partial}{\partial \varpi}\frac{B_\phi^2}{4\pi} = 0.$$

Then write p as $\rho kT/m$ and use (12.12) to express B_ϕ in terms of ρ. The result is the non-linear partial differential equation

$$\left\{\frac{1}{\varpi}\frac{\partial}{\partial \varpi}\varpi\frac{\partial}{\partial \varpi} + \frac{\partial^2}{\partial z^2}\right\}\left\{\frac{\rho kT}{m} + \frac{B_0^2}{8\pi}\frac{\varpi^2}{a^2}\frac{\rho^2}{\rho_0^2}\right\} + \frac{B_0^2}{4\pi\varpi}\frac{\partial}{\partial \varpi}\left(\frac{\varpi^2}{a^2}\frac{\rho^2}{\rho_0^2}\right) = 0, \tag{12.17}$$

giving $\rho(\varpi, z)$ in terms of $T(z)$.

The general solution of (12.17) is difficult, but the weak field case is easy and entirely adequate for illustrating the basic form of the non-equilibrium. We note merely that if $B_\phi^2/8\pi \ll p$ then B_ϕ alters p but little and, to a first approximation, $p \cong p_0$, where p_0 is a constant. Hence ρ varies inversely with the temperature, and (12.12) becomes a relation for $B_\phi^2/8\pi$ in terms of the temperature,

$$\frac{B_\phi^2}{8\pi} \cong \frac{B_0^2}{8\pi}\left\{\frac{\varpi}{a}\frac{T_0}{T(z)}\right\}^2 \{1+O(B_\phi^2/8\pi p)\}.$$

This is all we need to evaluate the source term on the right-hand side of (12.16), leading to

$$\varpi\frac{\partial}{\partial\varpi}\frac{1}{\varpi}\frac{\partial\Psi}{\partial\varpi}+\frac{\partial^2\Psi}{\partial z^2} = -\frac{\tau}{a\rho_0}\frac{B_0^2}{4\pi}\frac{\varpi^2}{a^2}\frac{\partial}{\partial z}\left[\left\{\frac{T_0}{T(z)}\right\}^2\right].$$

This equation is readily solved for Ψ. Only for a uniform temperature, $\partial T/\partial z = 0$, is there a static solution $\Psi = 0$. In general, the solution of the inhomogeneous equation can be written

$$\Psi_i = -\frac{\tau}{a\rho_0}\frac{B_0^2}{8\pi}\frac{\varpi^2}{a^2}\int_0^z \mathrm{d}\zeta \frac{T_0^2}{T^2(\zeta)}.$$

The solution of the homogeneous equation is made up of any sum of the functions

$$\Psi_0 = (A_1\varpi^2 + A_2)(z - z_0),$$
$$\Psi_k = \varpi\{B_1 I_1(k\varpi) + B_2 K_1(k\varpi)\}\{B_3 \sin(kz) + B_4 \cos(kz)\},$$
$$\Psi_q = \varpi\{C_1 J_1(q\varpi) + C_2 Y_1(q\varpi)\}\{C_3 \sinh(qz) + C_4 \cosh(qz)\},$$

over k and q. Note that if u_z is to be finite on the z-axis, then Ψ must vanish at least as fast as ϖ^2 as $\varpi \to 0$, requiring that A_2, B_2, and C_2 be set equal to zero.

As a specific example suppose that the field and the infinitely conducting fluid are confined within the circular cylinder $\varpi = R$ between the planes $z = 0, h$, so that v_ϖ vanishes on $\varpi = R$ and v_z vanishes on $z = 0, h$. Choose Ψ so that

$$\Psi = \Psi_i + \Phi_1 + \Phi_2 \qquad (12.17)$$

where

$$\Phi_1 = A\varpi^2 z \qquad (12.18)$$

$$\Phi_2 = \varpi \sum_{n=1}^{\infty} B_n I_1(n\pi\varpi/h)\sin(n\pi z/h). \qquad (12.19)$$

The boundary condition $v_z = 0$ on $z = 0$ is automatically satisfied by each term separately. The condition that $v_z = 0$ on $z = h$ requires that the constant A have the value

$$A = \frac{\tau h}{a^3 \rho_0} \frac{B_0^2}{4\pi} \left\langle \frac{T_0^2}{T^2} \right\rangle \tag{12.20}$$

where the angular brackets denote the mean over $(0, h)$. The boundary condition that v_ϖ vanish at $\varpi = R$ yields

$$\sum_{n=1}^{\infty} B_n \frac{n\pi R}{h} I_1\left(\frac{n\pi R}{h}\right) \cos\left(\frac{n\pi z}{h}\right) = \frac{\tau R^2}{a^3 \rho_0} \frac{B_0^2}{8\pi} \left\{ \frac{T_0^2}{T^2(z)} - \left\langle \frac{T_0^2}{T^2} \right\rangle \right\}.$$

The mean value of the right-hand side is zero and it is readily shown that

$$B_n = \frac{\tau B_0^2 R}{2\pi^2 a^3 \rho_0 n I_1(n\pi R/h)} \int_0^h \mathrm{d}\zeta \cos\left(\frac{n\pi\zeta}{h}\right) \times \left\{ \frac{T_0^2}{T^2(\zeta)} - \left\langle \frac{T_0^2}{T^2} \right\rangle \right\}. \tag{12.21}$$

Together (12.20) and (12.21) determine the coefficients in (12.17) once the temperature $T(z)$ is specified.

The example

$$T(z) = T_0/(1 + z/L)^{\frac{1}{2}} \tag{12.22}$$

is adequate for our purposes, yielding

$$\rho(z) = \rho_0(1 + z/L)^{\frac{1}{2}}$$

and

$$\langle T_0^2/T^2 \rangle = 1 + h/2L.$$

Then

$$\Psi = \frac{\tau B_0^2}{8\pi a^3 \rho_0 L} \left[\varpi^2 z(h - z) - \frac{8Rh^2\varpi}{\pi^3} \sum_{n=1}^{\infty} \frac{I_1\{(2n-1)\pi\varpi/h\} \sin\{(2n-1)\pi z/h\}}{(2n-1)^3 I_1\{(2n-1)\pi R/h\}} \right]$$

and

$$u_\varpi = + \frac{\tau B_0^2}{8\pi a^2 L \rho_0 (1 + z/L)^{\frac{1}{2}}} \left[\varpi(h - 2z) - \frac{8Rh}{\pi^2} \sum_{n=1}^{\infty} \frac{I_1\{(2n-1)\pi\varpi/h\} \cos\{(2n-1)\pi z/h\}}{(2n-1)^2 I_1[(2n-1)\pi R/h]} \right], \tag{12.23}$$

$$u_z = - \frac{\tau B_0^2}{8\pi a^2 L \rho_0 (1 + z/L)^{\frac{1}{2}}} \left[2z(h - z) - \frac{8Rh}{\pi^2} \sum_{n=1}^{\infty} \frac{I_0\{(2n-1)\pi\varpi/h\} \sin\{(2n-1)\pi z/h\}}{(2n-1)^2 I_1\{(2n-1)\pi z/h\}} \right]. \tag{12.24}$$

It is readily shown that these expressions satisfy the boundary conditions, noting that

$$h-2z=\frac{8h}{\pi^2}\sum_{n=1}^{\infty}\frac{\cos\{(2n-1)\pi z/h\}}{(2n-1)^2}.$$

The velocity profile across the ends $z=0$, h is shown in Fig. 12.2, along with the velocity v_z on the axis ($\varpi=0$) and the outer surface $\varpi=R$, all for the special case that $h=R$. For $\partial T/\partial z<0$, as in the present example, the fluid moves toward negative z (higher temperature) along the z-axis and toward positive z (lower temperature) near the outer walls $\varpi=R$. The direction of flow follows quite generally from the driving force

$$\mathbf{F}=-\mathbf{e}_\varpi B_\phi^2/4\pi\varpi$$

on the right-hand side of (12.13). The curl of this is

$$\begin{aligned}\nabla\times\mathbf{F}&=-\mathbf{e}_\phi(1/\varpi)(\partial/\partial z)B_\phi^2/4\pi\\&=-\mathbf{e}_\phi\frac{B_0^2}{4\pi}\frac{\varpi^2}{a^2}\frac{\mathrm{d}}{\mathrm{d}z}\frac{T_0^2}{T^2},\end{aligned}$$

which is negative in the present example, giving the non-equilibrium circulation just described.

The more general problem, including a gravitational acceleration in the z-direction has been discussed elsewhere (Parker 1975), but there are no qualitative differences of physical interest, so we do not pursue that more complicated problem here.

This illustrative example has been presented to show that azimuthal fields, such as the azimuthal sheath of the twisted flux tubes investigated

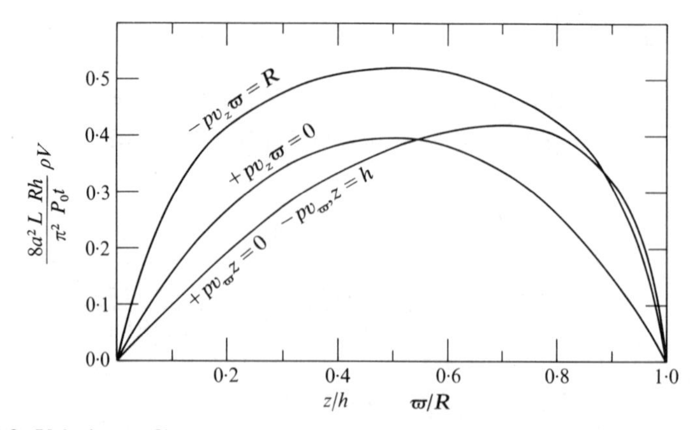

FIG. 12.2. Velocity profiles across the top and bottom of the cell $z=0$, h, up the centre $\varpi=0$, and down the side $\varpi=R$ for the special case $h=R$. The radial flow is an odd function of $z-\frac{1}{2}h$.

in Chapter 9, generally have no equilibrium. The hypothetical barometric equilibrium along the lines of force employed in Chapter 9 can establish itself only in ideally placid circumstances. Without it, the vertical temperature gradients that obtain in nearly all astrophysical bodies prevent equilibrium. We have suggested (Parker 1975) that the non-equilibrium of the azimuthal sheath may play a role in the activity of the intense flux tubes extending through the surface of the sun. In particular the ephemeral bipolar magnetic regions (Harvey *et al.* 1975) that give rise to the x-ray bright spots (Vaiana *et al.* 1973; Golub *et al.* 1974) may owe their activity in part to such non-equilibrium convection, as well as to the kinking instability that arises when the azimuthal sheath reduces the net tension $\mathcal{T}$ in the apex of the flux tube to zero (see §§9.6 and 9.7).

References

BENTON, E. R. and LOPER, D. E. (1969). *J. Fluid Mech.* **39,** 561.

CHANDRASEKHAR, S. (1961). *Hydrodynamic and hydromagnetic stability*, Clarendon Press, Oxford.

—— (1975). private communication.

GOLUB, L., KRIEGER, A. S., SILK, J. K., TIMOTHY, A. F., and VAIANA, G. S. (1974). *Astrophys. J. Lett.* **189,** L93.

HARVEY, K. L., HARVEY, J. W., and MARTIN, S. F. (1975). *Solar Phys.* **40,** 87.

LOPER, D. E. (1972). *Physics Earth Planet. Int.* **6,** 405.

PARKER, E. N. (1972). *Astrophys. J.* **174,** 499.

—— (1975). ibid. **201,** 502.

VAIANA, G. S., KRIEGER, A. S., and TIMOTHY, A. F. (1973). *Solar Phys.* **32,** 81.

YU, G. (1973). *Astrophys. J.* **181,** 1003.

13

THE BREAKUP AND ESCAPE OF SUBMERGED MAGNETIC FIELDS

13.1. Unstable effects of magnetic buoyancy

THUS far we have accepted the observed fact that magnetic fields in nature break up into separate flux tubes. This fact is sufficient motivation to work out the elementary properties of the individual flux tube in a stratified atmosphere. However the observed separation of solar fields into individual tubes, with field-free gaps between, is contrary to the elementary expectation for the magnetic field. It was emphasized in §§5.3, 5.4, 6.1, and 8.4 that the isotropic pressure $-\delta_{ij}B^2/8\pi$ causes the field to expand to fill all the available space. Thus the gaps between flux tubes are inexplicable unless some additional dynamical force comes into play to create them in opposition to the pressure of the field. The separation of the field continuum into tubes is universal in the environment of the solar envelope.

It was noted in §10.2 that converging flows separate and concentrate fields at the surface of the sun. But the fields in the sun are already separated into isolated tubes when they first emerge through the surface in the divergent upwelling of fluid near the centres of the supergranules. So it appears to be an intrinsic property of the field itself, rather than the general convection, that is the initial cause of the breakup.

As a matter of fact, it is again the magnetic pressure $B^2/8\pi$ that causes the continuum field to break up into separated flux tubes. Operating alone, the magnetic pressure tends to smooth out the field but in combination with a gravitational acceleration g it has the opposite effect. It is the magnetic buoyancy of $B^2/8\pi$ that causes the separation of B into individual flux tubes. Once separated into individual tubes there is no static equilibrium available (see §§8.6–8.8, 9.6 and 9.7). The field embarks on an irreversible path of activity, for which the only termination is the eventual exhaustion of the available magnetic and gravitational energy.

To illustrate the first breakup mechanism, imagine a horizontal magnetic field submerged below the visible surface of a star. Initially the magnetic field occupies a thick stratum with an upper boundary across which the field declines rapidly to zero. The local condition for hydros tic equilibrium is

$$\frac{d}{dz}\left(p+\frac{B^2}{8\pi}\right)=-\rho g$$

for a horizontal magnetic field B in a gas with pressure p and density ρ subject to a gravitational acceleration g. The abrupt upward decrease of field pressure at the upper boundary is compensated by an increase in gas pressure, and hence in gas density. If the field decreases to zero over a distance small compared to one scale height, then the gas pressure abruptly increases by the amount $\Delta p = B^2/8\pi$, and the density by

$$\begin{aligned}\Delta\rho &= (m/kT)B^2/8\pi,\\ &= (1/g\Lambda)B^2/8\pi,\end{aligned}$$

where Λ is the scale height mg/kT for a gas with mean molecular mass m. Thus the gas above the field is denser by the amount $\Delta\rho$ than the gas within the field, leading to overturning with tongues of the dense field-free fluid extending downward through the field. The field is broken into vertical sheets, which rise into the fluid above and become separated from the main body of the field below. It is the familiar Rayleigh–Taylor instability, whenever a denser fluid overlies a lighter.

There is a second effect, involving magnetic buoyancy, that is quite different in character and does not depend upon the cut-off of field with height. Basically the idea (Parker 1966, 1969) is that in hydrostatic equilibrium the field is confined by the weight of the gas in which it is embedded; the gas is supported in part by the field. The conducting gas 'hangs' on the level lines of the field. Consider, then, what happens if the lines of force are disturbed slightly, causing them to undulate. Then the gas tends to slide downward along the undulations out of the raised portions into the low places. This unburdens the raised portions of the lines of force, permitting them to expand upward further, at the same time increasing the burden at the low places and depressing them downward. It is evident that, unless the gas is very stable against convective overturning, the effect 'runs away' and the field becomes increasingly wavy. As we shall see, the undulations develop along thin tubes and vertical sheets, so that the field tends to break up into separate laminations.

These effects are explored and illustrated through idealized, quantitative examples in the sections below. The large-scale horizontal equilibrium field is broken into individual flux tubes for which there is then no equilibrium at all. The undulations of the individual tubes extend over dimensions greater than $2\pi\Lambda$, so that there is no possibility of preventing their raised portions from running away to the top of the atmosphere, as demonstrated in §§8.6–8.8. The instability removes any possibility for further equilibrium. The initial equilibrium is unstable, and the instability leads the system into increasing chaos, of which the separation into individual flux tubes is only the first step.

13.2. Breakup of the sharp upper boundary of a submerged field

Consider the local instability of the upper boundary of a horizontal magnetic field $\mathbf{e}_y B$ embedded in a conducting fluid in a gravitational field g in the negative z-direction. The field is confined to the half space $z<0$, the fluid density having the value ρ within the fluid and $\rho+\Delta\rho$ above. The density increment across the boundary of the field is just that amount necessary to compensate for the abrupt drop in magnetic pressure,

$$\Delta\rho/\rho = V_A^2/2u^2 \tag{13.1}$$

where V_A is the Alfven speed $B/(4\pi\rho)^{\frac{1}{2}}$ and u is the characteristic thermal velocity $(kT/m)^{\frac{1}{2}}$. As will be shown below, the most unstable modes are those of small wavelength, so consider the local instability over dimensions small compared to the scale height. Then ρ, p, and B may be taken to be uniform over the region of consideration except at the discontinuity $z=0$.

Introduce a small perturbation, involving a velocity $\mathbf{v}$ and field increment $\mathbf{b}$. Then if the locally uniform horizontal equilibrium field B is in the y-direction, the linearized hydromagnetic induction equations are

$$\partial\mathbf{b}/\partial t = B\,\partial\mathbf{v}/\partial y, \tag{13.2}$$

while the linearized equations of motion are

$$\nabla\cdot\mathbf{v}=0 \tag{13.3}$$

$$\rho\frac{\partial v_x}{\partial t} = -\frac{\partial \delta p_1}{\partial x} + \frac{B}{4\pi}\left(\frac{\partial b_x}{\partial y} - \frac{\partial b_y}{\partial x}\right), \tag{13.4}$$

$$\rho\frac{\partial v_y}{\partial t} = -\frac{\partial \delta p_1}{\partial y}, \tag{13.5}$$

$$\rho\frac{\partial v_z}{\partial t} = -\frac{\partial \delta p_1}{\partial z} + \frac{B}{4\pi}\left(\frac{\partial b_z}{\partial y} - \frac{\partial b_y}{\partial z}\right), \tag{13.6}$$

in the region $z<0$ where the field is non-vanishing. The pressure fluctuation associated with the perturbation is denoted by δp_1. The velocity field may be decomposed into a rotational and an irrotational part,

$$\mathbf{v} = \mathbf{w} - \nabla\Psi, \qquad \nabla^2\Psi = \nabla\cdot\mathbf{w} = 0.$$

Then the vorticity is $\boldsymbol{\omega} = \nabla\times\mathbf{w}$. The curl of (13.4)–(13.6) yields

$$\partial\boldsymbol{\omega}/\partial t = (B/4\pi\rho)(\partial/\partial y)\nabla\times\mathbf{b}.$$

With the aid of (13.2) this becomes

$$\partial^2\boldsymbol{\omega}/\partial t^2-(B^2/4\pi\rho)\partial^2\boldsymbol{\omega}/\partial y^2=0.$$

Thus the vorticity propagates along the basic field in the y-direction with the Alfven speed $V_A=B/(4\pi\rho)^{\frac{1}{2}}$. If $\boldsymbol{\omega}$ is of the form $\exp i(\omega t+\mathbf{k}.\mathbf{r})$, the dispersion relation is simply $\omega^2=k_y^2V_A^2$. The wave number k_y must be real (so that the solution is bounded as $y^2\to\infty$) so the only solutions are oscillatory (ω real). This Alfven mode is familiar (see §§7.3 and 8.9) and irrelevant for the present discussion.

Consider, then, the irrotational motion

$$\mathbf{v}=-\nabla\Psi,\qquad \nabla^2\Psi=0$$

with solutions of the form

$$\Psi=C\exp\{t/\tau+ik_xx+ik_yy+(k_x^2+k_y^2)z\},\tag{13.7}$$

which go to zero at $z=-\infty$. Then

$$\begin{aligned}b_x&=+B\tau k_xk_y\Psi,\\ b_y&=+B\tau k_y^2\Psi,\\ b_z&=+iB\tau k_y(k_x^2+k_y^2)^{\frac{1}{2}}\Psi,\\ \delta p_1&=+\rho\Psi/\tau.\end{aligned}\tag{13.8}$$

In the region $z>0$ where there is no field and the fluid density is $\rho_2=\rho+\Delta\rho$, the equations of motion are

$$\rho_2\partial\mathbf{v}/\partial t=-\nabla\delta p_2,\qquad \nabla.\mathbf{v}=0\tag{13.9}$$

where δp_2 denotes the pressure perturbation. The fluid is inviscid and initially without vorticity, so the flow is irrotational for all subsequent times,

$$\mathbf{v}=-\nabla\Phi,\qquad \nabla^2\Phi=0.\tag{13.10}$$

Then from the y-component of (13.9) it is readily shown that

$$\delta p_2=\rho_2\Phi/\tau.\tag{13.11}$$

Let

$$\Phi=S\exp\{t/\tau+ik_xx+ik_yy-(k_x^2+k_y^2)^{\frac{1}{2}}z\},\tag{13.12}$$

where S is a constant, so that the solution vanishes as $z\to+\infty$.

The boundary conditions are that the vertical component of the velocity and the total pressure $\delta p+Bb_y/4\pi$ are continuous across the undulating boundary $z=\xi(x,y,t)$ of the field. Since $v_z=\partial\xi/\partial t=\xi/\tau$, it follows that

$$\xi(x,y,t)=\tau(k_x^2+k_y^2)^{\frac{1}{2}}\Phi(x,y,0,t).$$

For small amplitudes, then, v_z is continuous across $z=0$, but in evaluating the mathematical forms of the pressure perturbations δp_1 and δp_2 we must take into account the weight of the fluid between $z=0$ and $z=\xi$. Thus if $(\delta p_1)_0+B(b_y)_0/4\pi$ denotes the total pressure at $z=0$, the pressure at $z=\xi$ is smaller by $g\rho_1\xi$, while if $(\delta p_2)_0$ is the pressure of the field-free fluid evaluated at $z=0$, the pressure at $z=\xi$ is $(\delta p_2)_0-g(\rho+\Delta\rho)\xi$. Equating these two expressions yields the boundary condition

$$(\delta p_2)_0=(\delta p_1)_0+B(b_y)_0/4\pi+g\Delta\rho\xi. \tag{13.13}$$

It follows, then, from (13.7), (13.9), (13.11), and (13.12) that the continuity of v_z requires that

$$C+S=0,$$

while (13.13) requires the condition

$$C(\rho+\tau^2B^2k_y^2/4\pi)+S\{\tau^2g\Delta\rho(k_x^2+k_y^2)^{\frac{1}{2}}-\rho-\Delta\rho\}=0.$$

Setting the determinant of the coefficients equal to zero yields

$$\frac{1}{\tau^2}=\frac{g\Delta\rho(k_x^2+k_y^2)^{\frac{1}{2}}-\rho V_A^2k_y^2}{2\rho+\Delta\rho}. \tag{13.14}$$

There are growing modes ($\tau>0$) for those wave numbers for which

$$k_y^2/(k_x^2+k_y^2)^{\frac{1}{2}}<(g/V_A^2)\Delta\rho/\rho.$$

The most unstable modes are those for which $k_y=0$. They do not deform the lines of force (which would require stretching the lines in opposition to the tension $B^2/4\pi$) but involve only a simple exchange of field and field-free fluid. The growth rate is

$$\begin{aligned}\frac{1}{\tau^2}&=\frac{g\Delta\rho k_x}{2\rho+\Delta\rho}\\&=\frac{gk_x}{1+4u^2/V_A^2}\end{aligned} \tag{13.15}$$

upon using (13.1). The magnetic field breaks up into separate flux tubes at its upper boundary.

This example serves to illustrate the general physical principles of the breakup. It does not predict the diameters of the flux tubes expected to emerge from the break, however. For such quantitative purposes a much more elaborate calculation is required, and it is not clear that a definitive theoretical solution of the problem is tractable. First of all, the present calculation is linearized, so that it is applicable only to the onset of the breakup. It certainly does not follow the evolution of the field all the way into individual flux tubes. But even if we overlook this difficulty, adopting instead the common view that the dimensions of the eventual separate flux tubes are determined by the scale most unstable at the outset, there are still many effects being ignored. The most obvious is the intrinsic thermal convective instability of the gases beneath the visible surface of most stars. Another is the turbulent environment produced by the thermal instability. The characteristic scale of the turbulence undoubtedly influences the scale of the breakup of the magnetic field. The standard ploy is to parametrize the turbulence in terms of the eddy velocity and scale, with the assertion that the principal effect of turbulence is one of diffusion—turbulent transport. The quantitative validity of such exercises is dubious. But in the present problem they are sufficient to provide some qualitative idea of the limitations of the calculation completed above.

Returning to (13.15), then, we observe that the breakup proceeds with increasing rapidity the smaller the transverse dimensions of the perturbation. Thus the dominant breakup occurs over arbitrarily small scales in the ideal fluid. Clearly, then viscosity and resistive diffusion must be included, providing a cut-off at small dimensions (large k_x).

It is a straightforward but tedious exercise to introduce viscosity and resistivity into the equations for the most unstable modes ($k_y = 0$). The calculation for a viscous fluid has already been done by Chandrasekhar (1961) in a slightly different context, wherein the fluid is incompressible and the field extends over all z rather than just $z < 0$. But for $k_y = 0$ and a given density discontinuity $\Delta\rho$, Chandrasekhar's equations describe exactly the present problem. In the present case, of course, $\Delta\rho$ is related to the magnetic field through (13.1), but that fact does not enter into the calculation of the growth rate $1/\tau$ in terms of $\Delta\rho$. The resulting dispersion relation is relatively complicated, so that the calculation of the transverse wave number k_x for the most unstable modes is tedious. It is sufficient for present purposes to note that the only characteristic length to be constructed from the basic physical parameters, viz. density ρ, viscosity μ, and gravitational acceleration g, is the combination $(\mu^2/\rho^2 g)^{\frac{1}{3}}$. The only dimensionless ratio is $\Delta\rho/\rho$. Thus in the presence of viscosity the characteristic wave number for maximum growth must be given by a functional form

$$k_x(\mu^2/\rho^2 g)^{\frac{1}{3}} = f(\Delta\rho/\rho).$$

Chandrasekhar finds that $f(\Delta\rho/\rho)$ varies but little with $\Delta\rho/\rho$ so that, within a factor of two,

$$k_x \cong 0{\cdot}2(\rho^2 g/\mu^2)^{\frac{1}{3}}. \tag{13.16}$$

The same argument applied to the resistive diffusion coefficient yields

$$k_x(\eta^2/g)^{\frac{1}{3}} = h(\Delta\rho/\rho)$$

or

$$k_x = \mathrm{O}\{(g/\eta^2)^{\frac{1}{3}}\}.$$

The larger of the two diffusion coefficients, μ/ρ or η, dominates and determines the wave number for the most rapid growth.

Recalling that eddy transport is the dominant diffusion effect beneath the surface of most stars[1], it appears that the effective values of the viscosity and the magnetic field diffusion coefficient η are both of the order of $\lambda V(\lambda)$ where $V(\lambda)$ is the turbulent velocity over the characteristic dimension $\lambda = \pi/k_x$ of the ordered motions. Then the effective eddy viscosity is

$$\mu_e \cong 0{\cdot}2\rho\lambda V(\lambda).$$

With this estimate of the effective viscosity (13.16) becomes

$$\lambda = \pi/k_x \cong 10^2 V^2(\lambda)/g.$$

Then, for instance, at a depth of 10^4 km beneath the surface of the sun, where the characteristic turbulent velocity is estimated (Spruit 1974) to be of the order of $0{\cdot}1\ \mathrm{km\,s^{-1}}$ over dimensions comparable to the scale height (3×10^8 cm, for $T = 7\times10^4$ K, $g = 2{\cdot}7\times10^4\ \mathrm{cm\,s^{-2}}$, $\rho = 0{\cdot}75\times10^{-3}\ \mathrm{g\,cm^{-3}}$), the result is $\lambda \cong 4$ km. This scale is so small that, in fact $V(\lambda)$ is much smaller than $0{\cdot}1\ \mathrm{km\,s^{-1}}$, so that λ must be even smaller. The point is that the dimensions are so small that although the idealization of the upper boundary of the field by a discontinuity illustrates the physical effect, it does not permit a quantitative estimate of the scale of the flux tubes produced by the instability. The flux tubes are so slender that the thickness of the field boundary enters into the calculation, and contributes to the determination of the scale of the breakup. A simple calculation in which the large-scale field cuts off exponentially with a scale h, is outlined in the next section.

13.3. Breakup of diffuse upper boundary of a submerged field

To illustrate restriction of the breakup of the upper boundary of a horizontal field over scales only a little smaller than the scale of the upward decline of the field strength, consider again a horizontal magnetic field $\mathbf{e}_y B(z)$ embedded in an isothermal gas of density $\rho(z)$ and pressure $p(z) = u^2\rho(z)$ where u is the characteristic thermal velocity $(kT/m)^{\frac{1}{2}}$. The

[1] The molecular viscosity for fully ionized hydrogen is $\mu = 10^{-16}T^{\frac{5}{2}}$ g cm s (Chapman 1954) while the resistive diffusion coefficient is $\eta = 10^{13}T^{\frac{3}{2}}\ \mathrm{cm^2\,s}$ (Cowling 1953).

gas is subject to the gravitational acceleration g in the negative z-direction. Over vertical scales small compared to the scale height $\Lambda = kT/mg$, the condition for hydrostatic equilibrium is uniform total pressure,

$$P_0 = u^2\rho(z) + B^2(z)/8\pi.$$

Consider the simple circumstance in which the magnetic field pressure has a uniform value $B_0^2/8\pi$ below $z=0$ and declines across the interval $0<z<h(\ll\Lambda)$ to the uniform value $B_1^2/8\pi$ throughout $z>h$. If the gas density has the uniform value ρ_0 in $z<0$, then the total pressure is

$$P_0 = \rho_0 u^2 + B_0^2/8\pi.$$

With the upward decline of field the gas density increases to the uniform value

$$\rho_1 = \rho_0 + (B_0^2 - B_1^2)/8\pi u^2$$

at $z=h$, and is uniform above. A convenient equilibrium form for the transition across the interval $(0, h)$ is

$$B^2(z)/8\pi = B_0^2/8\pi + \rho_0 u^2\{1-\exp(\alpha z/h)\}$$
$$\rho(z) = \rho_0 \exp(\alpha z/h)$$

where

$$\exp(\alpha) = \rho_1/\rho_0 = 1 + (B_0^2 - B_1^2)/8\pi\rho_0 u^2.$$

This equilibrium is unstable to convective overturning, the most unstable modes involving only motions (v_x, v_z) perpendicular to the magnetic field, as in the example for the sharp boundary treated in §13.2. For such motions the velocity field is essentially divergence-free, so that it is possible to write

$$v_x = +\partial\psi/\partial z, \qquad v_z = -\partial\psi/\partial x.$$

The condition of uniform total pressure is preserved, and the field strength and fluid density of an element of fluid do not vary as the element is displaced. Writing the total density as $\rho(z)+\delta\rho(x, z, t)$ and the total field as $\mathbf{e}_y\{B(z)+b(x, z, t)\}$ for the perturbed state, it follows that in the linear approximation

$$\frac{\partial\delta\rho}{\partial t} + v_z\frac{d\rho}{dz} = 0, \qquad \frac{\partial b}{\partial t} + v_z\frac{dB}{dz} = 0.$$

The equations of motion are

$$\rho(z)\frac{\partial v_x}{\partial t} = -\frac{\partial\delta P}{\partial x},$$

$$\rho(z)\frac{\partial v_z}{\partial t} = -\frac{\partial\delta P}{\partial z} - g\delta\rho,$$

where δP is the total pressure perturbation

$$\delta P = u^2\delta\rho + Bb/4\pi.$$

To obtain a solution for the equations of motion, differentiate the x-component with respect to z and the z-component with respect to x and form the difference, thereby eliminating δP. Then differentiate with respect to time and eliminate $\delta\rho$. The result can be written as

$$\frac{\partial^2}{\partial t^2}\left\{\frac{\partial^2\psi}{\partial z^2}+\frac{1}{\rho}\frac{d\rho}{dz}\frac{\partial\psi}{\partial z}+\frac{\partial^2\psi}{\partial x^2}\right\}=\frac{g}{\rho}\frac{d\rho}{dz}\frac{\partial^2\psi}{\partial x^2}. \tag{13.17}$$

In the uniform density ρ_0 in $z<0$, the equation reduces to $\nabla^2\psi=0$. The appropriate solution is

$$\psi = C_0\exp(t/\tau + ikx + kz)$$

so that the perturbation vanishes at $z=-\infty$. It is readily shown from the equations of motion that

$$\delta P = i\rho_0\psi/\tau.$$

In the uniform density ρ_1 in $z>h$ the solution is

$$\psi = C_3\exp(t/\tau + ikx - kz)$$

with

$$\delta P = -i\rho_1\psi/\tau.$$

In the transition region $0<z<h$ the logarithmic gradient of the density is α/h, so that if

$$\psi = C\exp(t/\tau + ikx + qz),$$

it follows that

$$q^2 + \alpha q/h + k^2(g\tau^2\alpha/h - 1)\tau^2 = 0.$$

In terms of the two quantities

$$q_{1,2} = (k^2 - g\tau^2\alpha k^2/h + \alpha^2/4h^2)^{\frac{1}{2}} \mp \alpha/2h$$

write

$$\psi = \exp(t/\tau + ikx)\{C_1\exp(q_1 z) + C_2\exp(-q_2 z)\}$$

for which

$$\delta P = (i\rho_0/k\tau)\exp(\alpha z/h)\exp(t/\tau + ikx)$$
$$\times\{C_1 q_1\exp(q_1 z) - C_2 q_2\exp(-q_2 z)\}.$$

The boundary conditions are that the normal component of the velocity $v_z = -ik\psi$ and the total pressure δP are continuous across the boundaries

$z=0$, h. It is readily shown, then, that

$$C_0=C_1+C_2, \qquad C_3=C_1\exp(q_1h)+C_2\exp(-q_2h)$$
$$kC_0=q_1C_1-q_2C_2$$
$$kC_3=-C_1q_1\exp(q_1h)+C_2q_2\exp(-q_2h).$$

The determinant of the coefficients is

$$(k+q_1)(k+q_2)\exp(q_1h)-(k-q_1)(k-q_2)\exp(-q_2h)=0,$$

more conveniently written as

$$\tanh\{\tfrac{1}{2}h(q_1+q_2)\}=-k(q_1+q_2)/(k^2+q_1q_2).$$

or

$$\tanh\{h^2k^2(1-\alpha g\tau^2/h)+\tfrac{1}{4}\alpha^2\}^{\frac{1}{2}}=-\frac{2\{h^2k^2(1-\alpha g\tau^2/h)+\frac{1}{4}\alpha^2\}^{\frac{1}{2}}}{hk(2-\alpha g\tau^2/h}.$$

This dispersion relation determines the growth rate $1/\tau$ in terms of k.

In the limit of large scales, $hk\rightarrow 0$, we expect that the thickness h of the transition region is unimportant, so that the growth rate reduces to (13.15). To show that this is indeed the case, consider the limit $hk\rightarrow 0$ with $1/\tau^2$ becoming small O(k). Write

$$1/\tau^2=\beta\alpha gk$$

where β is a number of the order of unity. Then to lowest order in hk the dispersion relation reduces to

$$\alpha\beta=\tanh(\tfrac{1}{2}\alpha).$$

For a small density increment α, then, $\beta=\frac{1}{2}$, and

$$1/\tau^2\cong\tfrac{1}{2}\alpha gk\cong gk(\rho_1-\rho_0)/(\rho_1+\rho_0) \qquad (13.18)$$

in agreement with (13.15). The growth rate increases with increasing wave number.

To explore the general case, consider $\alpha g\tau^2/h$ to be real and positive but limited to

$$\alpha g\tau^2/h<1+\alpha^2/4h^2k^2.$$

Then the radical on each side of the dispersion relation is real and positive. The left-hand side is positive. If $\alpha g\tau^2/h<2$, the right-hand side is negative, and there is no possibility for a root. If $\alpha^2/4h^2k^2>1$, so that there is the possibility that

$$2<\alpha g\tau^2/h<1+\alpha^2/4h^2k^2,$$

then the right-hand side is positive and there is a root. But our main point of concern is the behaviour at large wave numbers, $kh \gg 1$, to see the limitations on the increase of the growth rate $1/\tau$ with increasing wave number. So suppose that

$$\alpha g\tau^2/h > 1 + \alpha^2/4h^2k^2,$$

for which the dispersion relation becomes

$$\tan\{h^2k^2(\alpha g\tau^2/h - 1) - \tfrac{1}{4}\alpha^2\}^{\frac{1}{2}} = -\frac{2\{h^2k^2(\alpha g\tau^2/h-1)-\frac{1}{4}\alpha^2\}^{\frac{1}{2}}}{hk(2-\alpha g\tau^2/h)}.$$

For $hk \gg 1$ the tangent oscillates rapidly between $\pm\infty$ with increasing hk so that there are many roots in the neighbourhood of

$$\alpha g\tau^2/h = 1 + \alpha^2/4h^2k^2 \cong 1.$$

Hence

$$1/\tau^2 \sim \alpha g/h. \tag{13.19}$$

The growth rate is independent of wave number for $hk \gg 1$. The growth time τ can be represented in terms of the Alfven speed $V_A = B_0/(4\pi\rho)^{\frac{1}{2}}$. Put $B_1 = 0$ so that for $\alpha \ll 1$, $V_A^2 = \alpha u^2$. Then, noting that the scale height is $\Lambda = u^2/g$, it follows that

$$\tau^2 = \Lambda h/V_A^2,$$

so that the growth time is the Alfven transit time across the harmonic mean of h and Λ.

Were the calculation to include viscous and resistive dissipation[2], then, the growth rate would decline again with further increase of k. Very roughly a kinematic viscosity ν decreases the growth rate by $-\nu k^2$. The effective eddy viscosity in the convective zone of the sun, or other star, introduces a non-negligible dissipation, so that hk is significantly limited. But it must be recognized, too, that the convective instability of the outer envelope of the sun, and the turbulent eddies produced by the instability, enhance and influence the rate and scale of the breakup. Any quantitative statement is therefore difficult.

In conclusion, then, it is clear that a submerged magnetic field tends to

[2] A viscosity μ introduces a force term $(\partial/\partial x_j)\mu(\partial v_i/\partial x_j + \partial v_j/\partial x_i)$ on the right-hand side of the equations of motion so that in place of (13.17) the equation is

$$\frac{\partial^2}{\partial t^2}\left(\rho\frac{\partial^2\psi}{\partial z^2}+\frac{d\rho}{dz}\frac{\partial\psi}{\partial z}+\rho\frac{\partial^2\psi}{\partial x^2}\right)-\frac{\partial}{\partial t}\left\{\mu\nabla^2\nabla^2\psi\right.$$
$$\left.+\left(\frac{d^2\mu}{dz^2}+2\frac{d\mu}{dz}\frac{\partial}{\partial z}\right)\left(\frac{\partial^2\psi}{\partial z^2}-\frac{\partial^2\psi}{\partial x^2}\right)\right\}=g\frac{d\rho}{dz}\frac{\partial^2\psi}{\partial x^2}.$$

If, then, the kinematic viscosity $\nu = \mu(z)/\rho(z)$ is assumed to be uniform, the solution proceeds as before.

break up at its upper boundary over scales somewhat smaller than the boundary thickness. There seems to be no possibility for avoiding the breakup, which proceeds with a characteristic time of the order of the thickness divided by the Alfven speed.

The scale of the magnetic debris appearing at the surface of the sun, often greatly expanded from the dimensions in the dense regions of their origin far beneath the surface (Parker 1973, 1975a), appears to be accounted for satisfactorily by this qualitative conclusion. If the submerged azimuthal field has a characteristic thickness of 10^5 km, at a depth of $1–2\times10^5$ km, then the general scale of 10^4 km of flux emerging through the surface is not unexpected. The *very small* scales that develop later are probably a different effect, associated with the turbulence immediately beneath the surface and with the intensification of flux tubes by whatever mechanisms are responsible (see Chapter 10).

13.4. Breakup throughout a submerged field

Consider, then, the second manifestation of magnetic buoyancy, which breaks up the field throughout the entire volume. The effect depends in no way upon the existence of an upper boundary of the field. The vertical dimensions of the field are comparable to the scale height of the gas so that the barometric variations of gas density with height must be included in the calculation. To avoid the effects of an upper boundary on the field (treated in the section above) suppose that ratio of the field pressure $B^2/8\pi$ to gas pressure p has the uniform value α throughout the atmosphere. To keep the calculation as simple as possible consider an isothermal atmosphere in a uniform gravitational field g in the negative z-direction. If m is the mean mass per molecule and the temperature is T, then it is again convenient to define the characteristic thermal velocity $u=(kT/m)^{\frac{1}{2}}$ so that the equilibrium pressure p and density ρ are related by $p=\rho u^2$. Embedded in the atmosphere is a horizontal magnetic field $\mathbf{e}_y B(z)$ in the y-direction, with its pressure proportional to p,

$$B^2(z)/8\pi=\alpha\rho u^2$$

where α is a constant. It follows that the Alfven speed $V_A=B(z)/\{4\pi\rho(z)\}^{\frac{1}{2}}$ has the uniform value $(2\alpha)^{\frac{1}{2}}u$ throughout the atmosphere. It is shown below that the magnetic field does not cause a simple convective overturning, as in the examples of §§13.3 and 13.4 when the field pressure declines more rapidly with height than the gas pressure. Indeed the field now inhibits the simple transverse overturning. The effect is of quite a different nature.

Now hydrostatic equilibrium requires that

$$(\mathrm{d}/\mathrm{d}z)(p+B^2/8\pi)=-\rho g \tag{13.20}$$

so that

$$p(z)/p_0 = \rho(z)/\rho_0 = B^2(z)/B_0^2 = \exp(-z/\Lambda) \tag{13.21}$$

where the scale Λ is

$$\Lambda = (1+\alpha)u^2/g. \tag{13.22}$$

Suppose, then, that the equilibrium is perturbed slightly, so that the gas has a small velocity $\mathbf{v}$ associated with the density and pressure perturbations $\delta\rho$ and δp. Then conservation of matter requires that the time derivative of the density perturbation of a moving element of fluid be given by

$$\frac{d\delta\rho}{dt} \equiv \frac{\partial\delta\rho}{\partial t} + v_z\frac{d\rho}{dz} = -\rho\left(\frac{\partial v_z}{\partial x} + \frac{\partial v_y}{\partial y} + \frac{\partial v_z}{\partial z}\right). \tag{13.23}$$

Then if the pressure varies as the γ power of the density, $p + \delta p \propto (\rho + \Delta\rho)^\gamma$, it follows that δp, for a moving element of fluid, varies with $\delta\rho$ according to

$$d\delta p/dt = \gamma u^2\, d\delta\rho/dt \tag{13.24}$$

in the linear approximation. Then, noting the z-dependence of ρ and p from (13.21), it follows that

$$\partial\delta\rho/\partial t = -\rho\{\partial v_x/\partial x + \partial v_y/\partial y + (\partial/\partial z - 1/\Lambda)v_z\}, \tag{13.25}$$

$$\partial\delta p/\partial t = \gamma u^2 \partial\delta\rho/\partial t - (\gamma - 1)u^2\rho v_z/\Lambda. \tag{13.26}$$

Denote the perturbation of the magnetic field by $\mathbf{b}$. Then the hydromagnetic induction equations are

$$\partial b_x/\partial t = +B\partial v_x/\partial y, \tag{13.27}$$

$$\partial b_y/\partial t = -B\partial v_x/\partial x - B(\partial/\partial z - 1/2\Lambda)v_z, \tag{13.28}$$

$$\partial b_z/\partial t = +B\partial v_z/\partial y. \tag{13.29}$$

The equations of motion are

$$\rho\frac{\partial v_x}{\partial t} = -\frac{\partial\delta p}{\partial x} + \frac{B}{4\pi}\left(\frac{\partial b_x}{\partial y} - \frac{\partial b_y}{\partial x}\right), \tag{13.30}$$

$$\rho\frac{\partial v_y}{\partial t} = -\frac{\partial\delta p}{\partial y} - \frac{B}{4\pi}\frac{b_z}{2\Lambda}, \tag{13.31}$$

$$\rho\frac{\partial v_z}{\partial t} = -\frac{\partial\delta p}{\partial z} + \frac{B}{4\pi}\left\{\frac{\partial b_z}{\partial y} - \left(\frac{\partial}{\partial z} - \frac{1}{2\Lambda}\right)b_y\right\} - g\delta\rho. \tag{13.32}$$

We are interested in the internal instability in a broad stratified atmosphere. To exclude the possibility of external effects intruding from the dense atmosphere at $z = -\infty$, suppose that the fluid is bounded below by a fixed boundary $z=0$. Then consider solutions of the form

$$\delta\rho=\rho_0 D\exp\{t/\tau+ik_x x+ik_y y+ik_z z+(s-2)z/2\Lambda\},$$
$$v_i=uC_i\exp\{t/\tau+ik_x x+ik_y y+ik_z z+sz/2\Lambda\},$$
$$b_i=B_0 A_i\exp\{t/\tau+ik_x x+ik_y y+ik_z z+(s-1)z/2\Lambda\},$$

where k_x, k_y, k_z, and s are all real. The parameter s is to be chosen later so as to satisfy the requirement of a fixed boundary at $z=0$.

It is convenient to introduce the dimensionless growth rate $\Omega=\Lambda/u\tau$ and the dimensionless wave numbers $q_i=\Lambda k_i$ and $iQ=ik_z\Lambda+\frac{1}{2}(s-1)$. Then the induction equations are

$$\Omega A_1=+iq_2C_1,$$
$$\Omega A_2=-iq_1C_1-iQC_3,$$
$$\Omega A_3=+iq_2C_3.$$

Eliminating δp and b_i from the equations of motion leads to

$$\Omega_{12}^2C_1-iq_1(\gamma-1+2\alpha iQ)C_3=-iq_1\gamma\Omega D,$$
$$\Omega^2C_2+iq_2(1+\alpha-\gamma)C_3=-iq_2\gamma\Omega D,$$
$$-2\alpha iq_1(iQ-\tfrac{1}{2})C_1+\Omega_{23}^2C_3=-\{\gamma(iQ-\tfrac{1}{2})+1+\alpha\}\Omega D,$$

where

$$\Omega_{12}^2\equiv\Omega^2+2\alpha(q_1^2+q_2^2),$$
$$\Omega_{23}^2\equiv\Omega^2+2\alpha q_1^2-2\alpha iQ(iQ-\tfrac{1}{2})-(\gamma-1)(iQ-\tfrac{1}{2}).$$

Solving these three equations for C_i yields

$$C_1=-D(iq_1\Omega/R)[\gamma\Omega_{23}^2+\{\gamma(iQ-\tfrac{1}{2})+1+\alpha\}(2\alpha iQ+\gamma-1)],$$
$$C_2=-D(iq_2/\Omega R)[\gamma\Omega_{12}^2\Omega_{23}^2-\Omega_{12}^2(1+\alpha-\gamma)\{\gamma(iQ-\tfrac{1}{2})+1+\alpha\}$$
$$-4\alpha^2\gamma q_1^2(Q^2+\tfrac{1}{4})],$$
$$C_3=-D(\Omega/R)[\Omega_{12}^2\{\gamma(iQ-\tfrac{1}{2})+1+\alpha\}-2\alpha\gamma q_1^2(iQ-\tfrac{1}{2})],$$

where

$$R\equiv\Omega_{12}^2\Omega_{23}^2+2\alpha q_1^2(iQ-\tfrac{1}{2})(2\alpha iQ+\gamma-1).$$

Substituting these results for C_i into the equation of continuity yields (if $D\neq 0$) the dispersion relation

$$\Omega^4+\Omega^2(2\alpha+\gamma)(q_2^2+Q^2+\tfrac{1}{4})+q_2^2\{2\alpha\gamma(q_2^2+Q^2+\tfrac{1}{4})-(1+\alpha)(1+\alpha-\gamma)\}$$
$$+(q_1^2/\Omega_{12}^2)[\gamma\Omega^4+\Omega^2\{2\alpha\gamma q_2^2-2\alpha(2\alpha+\gamma)Q^2+\gamma-1+\tfrac{1}{2}\alpha\gamma\}$$
$$-4\alpha^2\gamma q_2^2(Q^2+\tfrac{1}{4})]=0. \tag{13.33}$$

We have written the dispersion relation in this way to separate the effect of q_1^2, taking advantage of the fact that Ω_{12}^2 is a positive quantity for the unstable ($\Omega^2>0$) solutions in which we are interested.

The dispersion relation is a cubic in Ω^2. In order that the solution can be fitted to a fixed boundary at $z=0$, it is necessary that the dispersion relation be an even function of k_z. Since (13.33) is an even function of Q, this can be accomplished if and only if $s=1$ so that $Q=k_z\Lambda=q_3$. Hence Q is real and the coefficients of the cubic equation for Ω^2 are real. A real positive root for Ω^2 yields an unstable mode; a real negative root yields an oscillatory (stable) mode. We expect two roots to correspond to fast and slow hydromagnetic waves modified by g (see §§7.3 and 7.4). Since the atmosphere itself is stable ($\gamma>1$), these modes may be considered alternatively as internal gravity waves modified by magnetic stresses. The discussion will be limited to positive roots, for which there is a growing mode.

Consider first whether the system is unstable to transverse convective overturning, as treated in §§13.2 and 13.3. The most unstable modes were those which overturn the lines of force without stretching them ($k_y=0$). In this case the dispersion relation reduces to

$$\Omega^2[\Omega^4+\Omega^2(2\alpha+\gamma)(q_1^2+Q^2+\tfrac{1}{4})+q_1^2\{\alpha(\alpha+\gamma)+\gamma-1\}]=0.$$

This equation for Ω^2 has a positive root if and only if the last term in the braces is negative,

$$1-\gamma>\alpha(\alpha+\gamma), \quad \text{i.e.} \quad \gamma<1-\alpha.$$

In the absence of a magnetic field ($\alpha=0$) we know that the atmosphere is unstable to convection if $\gamma<1$, in agreement with the present result. Hence adding the magnetic field ($\alpha>0$) has a stabilizing effect (because the magnetic pressure varies as the square of the density, producing an effective value of γ equal to two). The region is stabilized by the magnetic field to transverse overturning, contrary to the situation at the upper boundary discussed in §13.2.

So consider the modes with a non-vanishing longitudinal component, $q_2\neq 0$. Indeed, if we put $q_1=0$ to remove any vestige of the transverse overturning, (13.33) reduces to

$$\Omega^4+\Omega^2(2\alpha+\gamma)(q_2^2+Q^2+\tfrac{1}{4})$$
$$-q_2^2\{(1+\alpha)(1+\alpha-\gamma)-2\alpha\gamma(q_2^2+Q^2+\tfrac{1}{4})\}. \tag{13.34}$$

Assuming Q^2 to be real, it follows that there is a real positive root if and only if the curly bracket in the last term is positive. Thus instability extends over the range of wave numbers

$$0<q_2^2+Q^2<(1+\alpha)(1+\alpha-\gamma)/2\alpha\gamma-\tfrac{1}{4}.$$

This range is non-vanishing if and only if γ is less than

$$\gamma < (1+\alpha)^2/(1+\tfrac{3}{2}\alpha).$$

For weak fields, $\alpha \ll 1$, the criterion is

$$\gamma < 1 + \tfrac{1}{2}\alpha + \tfrac{1}{4}\alpha^2 + \ldots,$$

while for strong fields ($\alpha \gg 1$) the requirement is just

$$\gamma < \tfrac{2}{3}\alpha\left(1 + \frac{4}{3\alpha} + \frac{1}{9\alpha^2} + \ldots\right),$$

and no matter how stable the gas may be against convection, i.e. for any real gas ($\gamma < 3$), magnetic buoyancy of the field produces instability.

The last term in (13.34) contains the factor q_2^2, so that instability increases with increasing q_2^2 so long as the term in square brackets does not decline. But Q^2 appears only in the square brackets and diminishes the value of the brackets. Hence Q^2 has only a stabilizing effect, for the simple reason that if $Q = q_3 \neq 0$, then the sign of b_y alternates with increasing z. The field $B + b_y$ is alternately compressed and expanded in opposition to its pressure.

As a matter of fact, the most unstable modes are for q_1 large rather than small. Noting that for Ω^2 and q_2^2 of the order of one,

$$\lim_{q_1 \to \infty} (q_1^2/\Omega_{12}^2) = \frac{1}{2\alpha},$$

(13.33) reduces to

$$(2\alpha+\gamma)\Omega^4 + \{4\alpha(\alpha+\gamma)(q_2^2+\tfrac{1}{4}) + \gamma - 1\}\Omega^2 + 2\alpha q_2^2\{2\alpha\gamma q_2^2 - (1+\alpha)(1+\alpha-\gamma)\} = 0. \tag{13.35}$$

Note that Ω^2 is now independent of Q^2. Introduction of a rapid variation in the x-direction removes the inhibitory effect of Q. This quadratic equation for Ω^2 has a positive root if and only if

$$2\alpha\gamma q_2^2 < (1+\alpha)(1+\alpha-\gamma).$$

Thus there is a positive root for some suitably small value of q_2 provided only that $\gamma < 1+\alpha$ so that the right-hand side of the inequality is positive. That is to say, there is an unstable mode provided that γ is not so large as to stabilize the magnetic buoyancy. When $\gamma < 1+\alpha$ there is always instability at suitably small q_2. If the atmosphere is convectively neutral $\gamma = 1$, then all wave numbers in the range

$$q_2^2 < \tfrac{1}{2}(1+\alpha)$$

are unstable.

We are interested primarily in the instability caused by the magnetic field, so for the remainder of the discussion put $\gamma = 1$ to remove the convective forces of the gas from the problem. Consider first the modes which involve no transverse overturning ($q_1^2 = 0$). Then the dispersion relation is (13.34). To obtain the fastest growing modes let Q be very small. Then all modes for which

$$0 < q_2^2 < \tfrac{1}{4} + \tfrac{1}{2}\alpha$$

are unstable, and the growth rate is given by (13.34) as

$$2\Omega^2 = [(1+2\alpha)^2(q_2^2+\tfrac{1}{4})^2 + 4\alpha q_2^2\{1+\alpha-2(q_2^2+\tfrac{1}{4})\}]^{\frac{1}{2}} - (1+2\alpha)(q_2^2+\tfrac{1}{4}). \tag{13.36}$$

The growth rate Ω^2 is plotted in Fig. 13.1 (solid lines) as a function of q_2^2 for $\alpha = 0{\cdot}5$, $1{\cdot}0$, and $1{\cdot}5$. When $\alpha \ll 1$, the growth rate is

$$\Omega^2 = 2\alpha q_2^2(\tfrac{1}{4} - q_2^2)/(\tfrac{1}{4} + q_2^2),$$

neglecting terms $O(\alpha^2)$. The maximum growth rate arises for

$$q_2^2 = \tfrac{1}{4}(\sqrt{2}-1) = 0{\cdot}1035, \qquad q_2 = 0{\cdot}321,$$

and is

$$\Omega_{\max}^2 = \tfrac{1}{2}\alpha(3-2\sqrt{2}) = 0{\cdot}0858\alpha \tag{13.37}$$

Suppose, on the other hand, that the transverse wave number q_1, is very large. Then the dispersion relation reduces to (13.35) (with $\gamma = 1$),

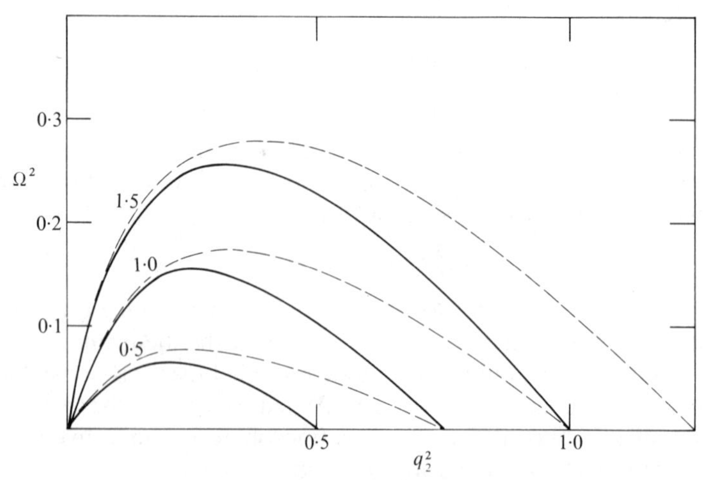

FIG. 13.1. A plot of Ω^2, where Ω is the growth rate $\Lambda/u\tau$, as a function of the square of the horizontal wave number $q_2 \equiv k_y\Lambda$ for $\alpha = 0{\cdot}5$, $1{\cdot}0$, and $1{\cdot}5$. The solid line represents Ω^2 for vanishing transverse wave number ($q_1 \equiv k_x\Lambda = 0$) while the broken line represents Ω^2 for large transverse wave number ($q_1 \gg 1$).

with the positive root

$$\Omega^2=[\alpha/(1+2\alpha)][\{4(1+\alpha)^2(q_2^2+\tfrac{1}{4})^2+2(1+2\alpha)(1+\alpha-2q_2^2)q_2^2\}^{\frac{1}{2}}$$
$$-2(1+\alpha)(q_2^2+\tfrac{1}{4})] \tag{13.38}$$

up to $q_2^2=\frac{1}{2}+\frac{1}{2}\alpha$. Thus, instability extends to larger values of q_2^2, by the amount $\frac{1}{4}$, for $q_1^2\gg 1$ (as compared to $q_1^2\ll 1$), and Q^2 has no inhibiting effects, so that modes with large Q^2 may appear. The growth rate Ω^2 is plotted in Fig. 13.1 (broken lines) as a function of q_2^2 for $\alpha=0{\cdot}5$, $1{\cdot}0$, and $1{\cdot}5$. When $\alpha\ll 1$,

$$\Omega^2=2\alpha\{(q_2^2+\tfrac{1}{16})^{\frac{1}{2}}-(q_2^2+\tfrac{1}{4})\}$$

neglecting terms $O(\alpha^2)$. The maximum growth rate arises for

$$q_2^2=\tfrac{3}{16},\qquad q_2=0{\cdot}433$$

and is

$$\Omega_{\max}^2=\tfrac{1}{8}\alpha, \tag{13.39}$$

to be compared with (13.37).

Altogether, then, the strongest instability occurs for $q_1\equiv k_x\Lambda\gg 1$. The instability is then indifferent to the vertical wave number $q_3\equiv k_z\Lambda$. For $0<\alpha\lesssim 2$, the maximum growth rate occurs for $q_2\cong 0{\cdot}5$ and is

$$\Omega_{\max}\cong 0{\cdot}4\alpha^{\frac{1}{2}} \tag{13.40}$$

i.e.

$$1/\tau\cong 0{\cdot}3V_A/\Lambda. \tag{13.41}$$

The growth time is of the order of three times the Alfven transit time across one scale height. The entire field breaks up rapidly into thin elements aligned along the field ($q_1^2\gg q_2^2$).

The unstable modes are easily described. Recalling that $s=1$ and $Q=k_z\Lambda$ if the solutions are to be cutoff from effects at $z=-\infty$, combine solutions for $\pm k_z$. The vertical component of the velocity can then be written

$$v_z=uC_3\exp(t/\tau)\sin(k_zz)\exp(ik_xx+ik_yy+z/2\Lambda)$$

so that $v_z=0$ at $z=0$. For finite k_x the most unstable modes occur for $k_z\Lambda\ll 1$, i.e. $Q\to 0$. The growing perturbation extends upward from the fixed boundary at $z=0$ to $z=+\infty$, or to another fixed boundary where $k_zz=n\pi$ $(n=1,2,3,\ldots)$. The amplitude of the kinetic energy $\frac{1}{2}\rho v^2$ is independent of height, with ρ declining exponentially as $\exp(-z/\Lambda)$ while the velocity increases as $\exp(+z/2\Lambda)$. (For examples wherein the kinetic energy decreases with height see Lerche and Parker 1967). The important point is that the instability is a result of the general magnetic buoyancy throughout the region and depends for its existence on the displacement of fluid along the field; the instability vanishes when the wave number k_y along the field is put equal to zero. The breakup occurs because the magnetic field supplies the fraction $\alpha/(1+\alpha)$ of the total support of the atmosphere against gravity. Hence, as already noted, if there are undulations in the magnetic lines of force, the gas tends to slide downward along the field from the peaks into the valleys. This relieves the weight on the peaks and increases the weight in the valleys, further enhancing the undulations. The gravitational energy of the matter declines because it moves downward, while the energy in the magnetic field declines because it expands upward. Illustrative examples of the rush of gas into the magnetic valleys, and the associated energy changes, may be found in Parker (1967a). The characteristic growth time τ is of the order of the Alfven transit time over three scale heights. The field breaks up most readily into thin undulating ropes and vertical sheets ($q_1^2 \gg 1$) so that the regions of compressed field in one striation are able to expand into the regions of reduced field in the striations on either side.

Note from (13.39) that the growth rate is bounded as q_1^2 increases. There is not the strong tendency to small scales that appears in the Rayleigh–Taylor instability at the upper boundary of a field, with the growth rate (13.15) increasing as $q_2^{1/2}$. In the present case the breakup occurs over a broad spectrum extending from q_1^2 only somewhat larger than one all the way to the viscous (or resistive) cutoff (which has not been included in the present calculations[3]). The scale of the growing undulations along the field involves wave lengths in the general neighbourhood of 15Λ.

Thus, in the initial linear departure of the system from equilibrium, the undulating tubes of flux are immediately subject to the buoyant catastrophe that arises when the wave length exceeds $2\pi\Lambda$ (see (8.52) and §§8.6–8.8, 9.6, and 9.7). It appears that a horizontal magnetic field in a quiescent atmosphere cannot be constrained for long from breaking up into independent flux tubes, subject to runaway magnetic buoyancy.

In an atmosphere that is strongly unstable to convection in the absence of a magnetic field (represented by $\gamma < 1$ in the present calculation) the

[3] Gilman (1970) and Roberts and Stewartson (1974, 1975) have inquired further into the problem of the transverse wave number, including the effects of rotation and diffusion and viscosity. Their work analyzes the problem in sufficient depth to show the complex dependence on the detailed circumstances. Lerche (1967a) treated the effect of rotation on the two-dimensional instability. Lerche and Parker (1968) treated the problem of the instability of a submerged field whose direction is a function of height in the atmosphere (a rotational shear). In neither case were there any significant stabilizing effects to be found. The instability proceeded essentially as outlined here (see comments in §13.8).

magnetically driven breakup sketched here is complicated by turbulent convection. Indeed, that appears to be the situation in the convective zone of the sun. In that case the convection adds to the breakup of the field and the mixing of magnetic flux to the surface. The point of the present calculation is to demonstrate that, even if nothing else is available to break up the field, the magnetic field alone is of such a nature as to break itself into separate flux tubes. Only strong constraints from the atmosphere itself ($\gamma > 1+\alpha$) can prevent the breakup. The outer envelopes of most stars, and the gaseous disc of the galaxy, where the known magnetic fields occur, have neutral convective stability, if they are not actively unstable. So we expect in most circumstances that the submerged fields, one way or another, break up into flux tubes as a consequence of their magnetic buoyancy. This theoretical fact is sufficient to account for the observed manifestation of the solar magnetic fields as many separate flux tubes rather than the simple continuum that one traditionally expected.

It would be interesting to include viscous and resistive diffusion in the equations to obtain an expression for the transverse wave number giving the maximum growth rate. But the calculation is extremely tedious, and its direct quantitative application to the turbulent astrophysical environment is dubious.

13.5. Equilibrium of a non-uniform atmosphere and field

The next question is whether there is some non-uniform equilibrium configuration into which the instability might somehow eventually convey the system, following departure from the initial uniform equilibrium (13.21). The answer would appear to be that (in the absence of extraordinarily heavy viscous damping) there is no haven in which the system may hope to find peace. An examination of the circumstances fails to find any without instability (Asseo *et al.* 1978). Considerations on the energy of the system indicate that the shifting of fluid back and forth along the lines of force goes on until the field is free of the fluid and the atmosphere is composed solely of field-free gas at the bottom of the available space. The characteristic time for reduction of the field energy is of the order of a few times the dynamical time (13.41).

To develop a clear picture of the problem we digress for a moment to examine the equilibrium of the field and gas system (13.21) for other than a uniform horizontal distribution of the field. In the linear approximation, any equilibrium that differs from the initial uniform equilibrium (13.21) must have $\Omega^2 = 0$, so that (13.33) reduces to

$$(q_1^2+q_2^2)\{2(q_2^2+Q^2+\tfrac{1}{4})-(1+\alpha)\}-2q_1^2(Q^2+\tfrac{1}{4})=0.$$

Thus for small transverse wave number $q_1^2 \ll 1$,

$$q_2^2 + Q^2 = \tfrac{1}{4}(1+2\alpha),$$

while for large transverse wave number

$$q_2^2 = \tfrac{1}{2}(1+\alpha) \tag{13.42}$$

and Q^2 is arbitrary. But the instability does not take the system into these wave numbers. The dominant instability (13.39) occurs for large q_1^2 with $q_2^2 = \frac{3}{16}$, whereas the wave numbers for equilibrium are the larger wave numbers for neutral stability. So it is not evident that any such equilibria are available to the system. Once the system begins to break up, it goes into concentrations of gas more massive and widely separated than the field can support.

What, then, is the nature of the non-linear regime, when the departures from the initial uniform equilibrium (13.21) are large? Using the methods outlined in §6.7.3 consider the two-dimensional equilibrium for which $\partial/\partial x = 0$. Then, for an isothermal atmosphere $p = \rho u^2$, (6.46) reduces to

$$\nabla_{yz}^2 A + 4\pi\rho_0 u^2 f'(A)\exp(-z/\lambda) = 0 \tag{13.43}$$

with

$$\rho = \rho_0 f(A)\exp(-z/\lambda), \qquad B_y = +\partial A/\partial z, \qquad B_z = -\partial A/\partial y,$$

where λ is the scale height u^2/g for the gas alone. The choice

$$f(A) = (A_0/A)^\nu \tag{13.44}$$

where ν is a positive number, is the form of $f(A)$ that preserves the ratio of fluid mass to flux along the individual lines of force from the initial equilibrium (13.21), for which $\rho/B \propto A$ (Mouschovias 1974). Thus the form represents a possible state of the system (13.21). Equation (13.43) becomes

$$\left(\frac{A}{A_0}\right)^{\nu+1} \nabla_{yz}^2 \frac{A}{A_0} = \frac{4\pi\rho_0 u^2 \nu}{A_0^2}\exp\left(-\frac{z}{\lambda}\right).$$

To avoid solutions which diverge as $y \to \pm\infty$ consider the form

$$A(y, z) = A_0 Y(y)\exp(-z/\lambda(\nu+2)\}. \tag{13.45}$$

With this form the equation can be written as

$$\mathrm{d}^2 Y/\mathrm{d}\zeta^2 = -Y + \nu H^{\nu+2}/2Y^{\nu+1} \equiv -\mathrm{d}V/\mathrm{d}Y \tag{13.46}$$

where

$$V(Y) \equiv \tfrac{1}{2}Y^2 + H^{\nu+2}/2Y^\nu, \tag{13.47}$$

$$H^{\nu+2} \equiv (8\pi\rho_0 u^2\lambda^2/A_0^2)(\nu+2)^2, \tag{13.48}$$

$$\zeta \equiv y/\lambda(\nu+2). \tag{13.49}$$

Equation (13.46) is essentially an equation of motion with ζ the time variable and V the potential energy. The solution reduces to the quadrature

$$2^{\frac{1}{2}}\zeta = \pm \int \mathrm{d}Y/\{E - V(Y)\}^{\frac{1}{2}} \tag{13.50}$$

where E is a constant, analogous to total energy. The minimum value of $V(Y)$ occurs at $Y \equiv Y_0 = H(\nu+2)^{1/(\nu+2)}$ and is

$$V(Y_0) = \tfrac{1}{2}H(\nu/2)^{2/(\nu+2)}(1+2/\nu). \tag{13.51}$$

The solutions are real, and oscillate about Y_0, provided that $E > V(Y_0)$. If E is only a little larger than the minimum value $V(Y_0)$ write

$$Y = Y_0(1+\epsilon\xi) \tag{13.52}$$

with $\epsilon \ll 1$. Then (13.46) reduces to

$$\mathrm{d}^2\xi/\mathrm{d}\zeta^2 + (2+\nu)\xi = 0, \tag{13.53}$$

and a convenient solution is

$$\xi = \sin(2+\nu)^{\frac{1}{2}}\zeta \tag{13.54}$$

$$= \sin\{y/\lambda(\nu+2)^{\frac{1}{2}}\}. \tag{13.55}$$

Then

$$A = A_0 Y_0[1+\epsilon \sin\{y/\lambda(\nu+2)^{\frac{1}{2}}\}], \tag{13.56}$$

$$B_y = -A/\lambda(\nu+2), \tag{13.57}$$

$$B_z = -A_0 Y_0\{\epsilon/\lambda(\nu+2)^{\frac{1}{2}}\}\cos\{y/\lambda(\nu+2)^{\frac{1}{2}}\} \times \exp\{-z/\lambda(\nu+2)\}, \tag{13.58}$$

$$\rho = (\rho_0/Y_0^{\nu})[1-\nu\epsilon \sin\{y/\lambda(\nu+2)^{\frac{1}{2}}\}] \times \exp\{-z/\lambda(\nu+2)\}. \tag{13.59}$$

If ρ_0 is defined as the mean density at $z=0$, then $Y_0=1$ and $H=(2/\nu)^{1/(\nu+2)}$. It follows from (13.48) that

$$A_0^2 = 4\pi\rho_0 u^2\lambda^2\nu(\nu+2)^2. \tag{13.60}$$

The ratio of magnetic field pressure to gas pressure at any height z is designated by the constant α. It follows from (13.57) and (13.59) that $\alpha = \frac{1}{2}\nu$. It is evident from (13.22) and $\lambda = u^2/g$ that

$$\Lambda = \tfrac{1}{2}(\nu+2)\lambda$$

and from (13.60) that

$$A_0 = 2B_0\Lambda.$$

The horizontal wave number is $k_y = 1/\lambda(\nu+2)^{\frac{1}{2}}$ so that $q_2 \equiv k_y\Lambda = [(\alpha+1)/2]^{\frac{1}{2}}$ and is precisely the value (13.42). The horizontal wave length is $2\pi\lambda(\nu+2)^{\frac{1}{2}} = 4\pi\Lambda/(\nu+2)^{\frac{1}{2}}$ and is, therefore, rather larger than the scale height Λ for all $\nu \lesssim 16\pi^2 - 2$. It is evident that for $\epsilon \ll 1$, we have recovered the linearized static equilibrium treated at the beginning of this section by putting $\Omega = 0$ in the dynamical equations. We have shown, through the quadrature (13.50), how the static equilibrium with small amplitude fits into the general family of equilibria.

It is now a simple matter to show for the general case (13.50) that the loading of the lines of force with matter is preserved from (13.21), for which the mean mass (per unit length in the y-direction) per unit flux is proportional to the number of lines of force from the top of the atmosphere at $z = +\infty$. The vertical separation of the lines of force A and $A + \delta A$ is $\delta z = \delta A/|\partial A/\partial z| = \delta A/|B_y|$ so that the mean mass between the lines over one wave length L is

$$\frac{\delta M}{\delta A} = \frac{1}{L|B_y|}\int_0^L \mathrm{d}y\,\rho(y, z)$$

the integration being along $A = constant$. With the aid of (13.44) and (13.45) this becomes

$$\frac{\delta M}{\delta A} = \lambda(\nu+2)\frac{\rho_0 A}{LA_0^2}\int_0^L \frac{\mathrm{d}Y}{Y(y)^{\nu+2}}, \qquad (13.61)$$

which is directly proportional to A, the number of lines of force to the top of the atmosphere. Q.E.D.

Going back to the general solution (13.50), a particularly simple example arises for $\nu = 2$, for which

$$Y = \pm\{E + (E^2 - H^4)^{\frac{1}{2}} \sin 2(\zeta - \zeta_0)\}^{\frac{1}{2}}.$$

The solution is real if and only if $E^2 > H^4$. This solution corresponds to equal magnetic and gas pressures, $\alpha = 1$. The horizontal coordinate reduces to $\zeta = y/4\lambda = y/2\Lambda$, while $H = 1$ and $A_0 = (32\pi\rho_0 u^2\Lambda^2)^{\frac{1}{2}}$. The horizontal wave length is $2\pi\Lambda$, and is large compared to the scale height. The unstable modes, it will be recalled, occur for somewhat larger wave lengths. The solution is

$$A = A_0[E + (E^2 - 1)^{\frac{1}{2}} \sin\{(y - y_0)\Lambda\}]^{\frac{1}{2}} \exp(-z/2\Lambda), \qquad (13.62)$$

$$B_y = -A/2\Lambda, \qquad (13.63)$$

$$B_z = -\frac{(A_0/\Lambda)(E^2-1)^{\frac{1}{2}} \cos[(y - y_0)/\Lambda]\exp(-z/2\Lambda)}{[E + (E^2-1)^{\frac{1}{2}} \sin\{(y - y_0)/\Lambda\}]^{\frac{1}{2}}} \qquad (13.64)$$

$$\rho = \rho_0 \frac{\exp(-z/\Lambda)}{[E + (E^2-1)^{\frac{1}{2}} \sin\{(y - y_0)/\Lambda\}]^{\frac{1}{2}}} \qquad (13.65)$$

In the simple separable solution (13.45) employed here, the magnetic lines of force $A = constant$ follow the same pattern at all heights (other forms may be found in the literature, Parker 1968a; Mouschovias 1974; Low 1975a, b). The lines of force are plotted in Fig. 13.2 for $E = 2^{\frac{1}{2}}$ and $E = 4$. The horizontal variation of the gas density is also shown in Fig. 13.2. Note the extreme concentration of gas, and the extreme upward expansion of the field between the concentrations, as E increases. Both the magnetic energy and the gravitational potential energy have decreased significantly.

An extreme example follows for a cold gas ($\alpha \gg 1$), so that $\lambda \ll \Lambda$ and $\nu \gg 1$. Then the gas sinks down along the lines of force to concentrate in vertical sheets. The magnetic field is, then, a potential field between the sheets ($\nabla \times \mathbf{B} = 0$). It follows that

$$A = -2\Lambda B_0 \cos(y/2\Lambda)\exp(-z/2\Lambda)$$

so that

$$B_y = +B_0 \cos(y/2\Lambda)\exp(-z/2\Lambda),$$
$$B_z = -B_0 \sin(y/2\Lambda)\exp(-z/2\Lambda).$$

If the surface density of each sheet of gas is $s(z)\,\mathrm{g\,cm^{-2}}$, located at $y = \pm(2n+1)a \quad (n = 0, 1, 2, \ldots)$, then the condition for hydrostatic equilibrium is that the upward stress (dyn $\mathrm{cm^{-2}}$) of the field on each side of the sheet must balance the weight sg of the sheet. Hence at $y = a$,

$$2B_y(a, z)B_z(a, z)/4\pi = s(z)g,$$

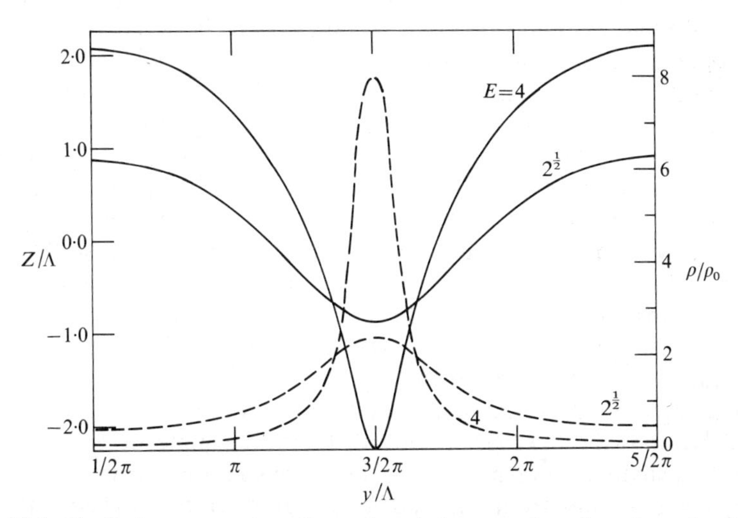

FIG. 13.2. The broken curves show the weak and strong condensation of gas density ρ/ρ_0 for $E = 2^{\frac{1}{2}}$ and $E = 4$, respectively, based on (13.65). The solid curves are the associated magnetic lines of force $\Lambda = constant$, from (13.62).

or

$$s(z) = (B_0^2/2\pi g)\sin(a/2\Lambda)\cos(a/2\Lambda)\exp(-z/\Lambda).$$

Note that the maximum weight per unit flux B_0 occurs for the separation $2a = \pi\Lambda$ and declines to zero as $2a$ approaches the maximum possible separation $2\pi\Lambda$. This maximum separation is half the maximum $4\pi\Lambda$ for $\alpha = 1(\nu = 2)$ given by (13.62)–(13.65) because, in the present case, the gas is wholly dependent upon the magnetic field for support, whereas for $\alpha = 1$ the thermal motions of the gas supported half the weight. For the cold gas ($\alpha \gg 1$) note that, as a increases toward the maximum value $\pi\Lambda$, the matter sinks lower and lower relative to the field. This is readily shown by noting that the lines of force are given by $A = constant$, so that the vertical coordinate of the line of force at the height $z = z_0$ at the apex of the line of force (at $y = \pm 2na$ midway between the sheets of matter) is depressed a distance

$$h = 2\Lambda \ln\{\sec(a/2\Lambda)\}$$

to $z = z_0 - h$ at the sheets of matter. The depression becomes large without bound as a increases to $\pi\Lambda$. It is obvious, then, that inhomogeneities in the gas developing from the initial equilibrium (13.21) over separations greater than $2\pi\Lambda$ are unstable, and form concentrations that cannot be supported by the field in any way. There is no equilibrium ahead, short of reaching the bottom of the available space.

In §13.7 we treat an example with a well defined lower boundary so as to allow a unique definition of the energy reduction as the gas concentrates into packets and the field expands upward between.

Altogether, then, the general equilibrium (13.50) does not lie in the direct path of the dynamical evolution of the system away from (13.21), the horizontal wave number q for equilibrium being larger than the wave numbers of the developing instability. There is simply a limit to how much weight the field can support. The next point is that the non-uniform equilibrium, whether accessible or not, is unstable, so that the system could not remain long in it, even if it could get there in the first place. It is simply a fact that every non-uniform equilibrium configuration of the system is unstable to the same dynamical effects that plagued the initial uniform equilibrium (13.21) and drove it into non-uniformity. To see this note that (except when completely cold, $\alpha = \infty$) the gas extends continuously along the lines of force as they arch up from one gas concentration and down into the next. Therefore, if the ideal symmetric equilibrium is perturbed slightly, lowering one pocket of gas relative to its neighbours, the gas will stream from the neighbours up over the arched lines and down into the depressed pocket. This depresses the pocket further, while unburdening the neighbouring concentrations. The familiar instability carries on just as when starting from the uniform field (13.21). It would appear, then, that the gas must eventually accumulate in pockets of increasing mass and separation[4]. The rate of transfer may decline with increasing separation and decreasing density at the apex of the lines of force arching upward between. But the transfer never stops. With the increasing concentration of the gas into fewer and fewer pockets at wider separations, the field between is increasingly free to expand upward and decrease its energy, while the gas settles lower in the gravitational field. It is this declining magnetic and gravitational energy that is available to transport the gas to form the growing concentrations.

It is a simple exercise to compute the total energy available. To define the problem clearly, suppose that there is a rigid boundary, at $z=0$. Then the lowest distribution to which the gas can condense is $\exp(-gz/u^2)$. The initial potential energy of the density distribution (13.21) above $z=0$ is

$$\Phi_0 = \rho_0(1+\alpha)^2 u^4/g$$

per unit area, while the final energy of the same amount of gas for $\exp(-u^2 z/g)$ is

$$\Phi_1 = \rho_0(1+\alpha)u^4/g.$$

The total available gravitational energy is, therefore

$$\begin{aligned} \Delta\Phi &= \Phi_0 - \Phi_1, \\ &= \rho_0\alpha(1+\alpha)u^4/g, \\ &= \Phi_0\alpha/(1+\alpha). \end{aligned} \tag{13.62}$$

If the gas congregates in widely separated locations (which then rest on $z=0$), the field can expand without limit in the broad regions between, so that the initial magnetic energy of the field (13.21)

$$\begin{aligned} \mathcal{M} &= \Lambda B_0^2/8\pi, \\ &= \rho_0\alpha(1+\alpha)u^4/g, \\ &= \Delta\Phi. \end{aligned} \tag{13.63}$$

falls toward a lower limit of zero. Thus the total available energy is $2\Delta\Phi$. (For detailed calculations for specific configurations see Parker 1967a, Appendix I.)

The energy transfers gas along the lines of force into increasingly widely spaced and massive pockets. The system comes to rest only in the limit of large time, when the magnetic field has expanded almost

[4] Separations wider than about $4\pi\Lambda$ can be near equilibrium only if there is some underlying structure to support them. As already noted, the similarity condition (13.45) admits of no supporting plane boundary and no separations greater than about $4\pi\Lambda$ for α, ν of the order of unity.

everywhere along the field to negligible energies and the gas has settled down to the thermal distribution $\exp(-gz/u^2)$ in the concentrated pockets. The magnetic field energy decays in a period characterized by the dynamical instability time (13.41) and is not restricted to the (usually) long resistive decay time (§§4.2, 4.5, and 4.6)$t \cong \Lambda^2\sigma/c^2$.

The effect is particularly striking in the gaseous disc of the galaxy, for which the resistive decay time is of the order of 10^{20} years (ambipolar diffusion (see §4.6) may decrease this to 10^{10}–10^{11} years) while the dynamical time Λ/V_A is of the order of only 10^7 years. The buoyant escape from a depth of 2×10^5 km beneath the surface of the sun is estimated (§§8.6 and 8.7) to be only about ten years. Thus the breakup of a submerged field in a star or galaxy, and the subsequent buoyant escape of that flux tube, leads to rapid decay of the magnetic field. The dynamical effect guarantees the destruction of any initial (primordial) strong fields, irrespective of the large electrical conductivity and geometrical dimensions of the system. Turbulent transport may also contribute to the rapid decay of the field, and is taken up in Chapter 16. It is particularly important in the decay of weak fields, wherein the buoyancy effects are dominated by the convection and turbulence of the medium. For strong fields ($\alpha \cong 1$) the buoyancy effect is the principal means of escape.

In view of the major role played by magnetic buoyancy in the behaviour of the magnetic fields presently found in stars and galaxies, the buoyant escape of field merits further study. In particular, we should develop a clear understanding of the decay of the field energy trapped in an atmosphere, as distinct from the decay of the net flux trapped in the atmosphere. The energy decays, as described above, when the field expands upward through the atmosphere everywhere except at a few isolated locations in the atmosphere where the total flux is trapped in the gas. With the aid of some idealized examples, we shall illustrate how the net flux is then able to escape.

As a matter of fact there are *several* dissipative effects introduced by the dynamical breakup of a large-scale submerged field into individual flux tubes. In addition to the dynamical escape of the flux, there is enhanced diffusion of the gaseous atmosphere downward through the field as a consequence of the reduced internal scales of the breakup. The effect is not negligible in the cool interstellar clouds and is taken up briefly in the next section.

13.6. Enhanced diffusion

The characteristic escape time τ for the equilibrium field (13.21) in an isothermal atmosphere as a consequence of a diffusion coefficient η is

$$\tau = O(\Lambda^2/\eta).$$

For resistive diffusion over a scale height 5×10^3 km in a stellar envelope at 10^5 K, it follows that $\eta \cong 2\times10^5\ \text{cm}^2\,\text{s}^{-1}$ and $\tau = 10^{12}\ \text{s} = 3\times10^4$ years. In the gaseous disc of the galaxy where the scale height is $\Lambda \cong 150$ pc, and the ambipolar diffusion coefficient might be $10^{23}\ \text{cm}^2\,\text{s}^{-1}$, the characteristic escape time is of the order of $\tau \cong 3\times10^{18}\ \text{s} = 10^{11}$ years.

But the calculations of §13.4 show that the field breaks up into thin filaments in a characteristic time given by (13.41) as $3\Lambda/V_A$. A field of 10^2 G at a depth of 10^4 km in a stellar envelope where $\rho \cong 10^{-3}\ \text{g cm}^{-3}$ yields $V_A = 10^3\ \text{cm s}^{-1}$ and a characteristic instability time of $3\Lambda/V_A \cong 10^6$ s. Thus, after a period long compared to 10^6 s—say after 10^7 s—we would expect the field to be broken into a fabric of close packed, undulating flux tubes, with characteristic transverse dimensions small compared to one scale height, $q_1 = k_x\Lambda \gg 1$. The growth of the transverse wave number is limited only by diffusion, presumably with the characteristic diffusion rate limited to some fraction of the dynamical growth rate (13.41), $\eta q_1^2 \lesssim 1/\tau$ or

$$q_1 \lesssim (0{\cdot}3 V_A \Lambda/\eta)^{\frac{1}{2}}.$$

The characteristic diffusion time across a single filament is, then,

$$\begin{aligned}\tau_f &= O(1/\eta k_x^2),\\ &= O(3\Lambda/V_A).\end{aligned} \tag{13.64}$$

It follows that the gas caught in the depressed portion of a flux tube can escape downward as soon as it diffuses the small transverse distance π/k_x to the apex of an undulation of a neighbouring tube. The gas is then free to slide down along the lines of force of the neighbouring tube, etc. If the undulations have developed to an amplitude of the order of Λ, the gas progresses downward a distance Λ before reaching the bottom of that filament. The whole process takes a time of the order of τ_f plus the free-fall time. The mean rate of downward motion is then of the order of $\Lambda/\tau_f \cong 10^{-1} V_A$. The field is able to escape from the gas, and vice versa, in a period determined more by the dynamical time Λ/V_A than by the diffusion time Λ^2/η. In the stellar envelope Λ/V_A is 10^6 s instead of the 10^{12} s for diffusion over Λ, while in the gaseous disc of the galaxy Λ/V_A is of the order of 5×10^{14} s instead of 3×10^{18} s. The effect may, therefore, be of some importance in those circumstances where more rapid means of escape are not available (Parker 1967b).

13.7. Dynamical escape of magnetic flux

The dynamical buoyancy of the horizontal magnetic field submerged beneath the surface of a gravitating body is one of the principal means of escape of the field from the body. It has been shown in the foregoing sections how in times of the order of Λ/V_A the field breaks up into flux tubes and those flux tubes arch upward to form re-entrant loops extending above the 'surface' of the body. The magnetic energy is greatly diminished, by the amount (13.63). But of course, the lines of force remain captive where they dip through the pockets of concentrated gas, as sketched in Fig. 13.3. So there is, as yet, no magnetic flux lost from the system. A simple mechanical restoration of the gas to its initial distribution (13.21) would restore the field as well. But, as a matter of fact, when the flux tubes arch upward, magnetic fields of opposite sign come in contact with one another. Rapid reconnection of the lines of force across the neutral sheet between the opposite fields alters the topology of the lines and permits their escape from the system (Parker 1975b). So although no magnetic flux has escaped when the system reaches the state illustrated in Figs 13.3 and 13.4, there is a net loss of flux in the next stage of the development. If the reader will accept some additional idealization of the system, we can proceed by elementary methods to illustrate the next stage in the evolution of the system, wherein there is a net loss of magnetic flux. In particular, we consider the circumstance in which the system is constrained to two dimensions, as sketched in Fig. 13.3, and the magnetic field is confined by the weight of a layer of field-free fluid resting on top of the field instead of being distributed throughout the field. This idealization greatly facilitates the calculations without compromising the basic problem of the escape of magnetic flux from beneath an overburden of conducting gas. The continuity of the confining gas is maintained so that no lines of force can escape through gaps.

The two-dimensional problem is sufficient. We beg the reader's indulgence at this point in switching the assignment of coordinates so that

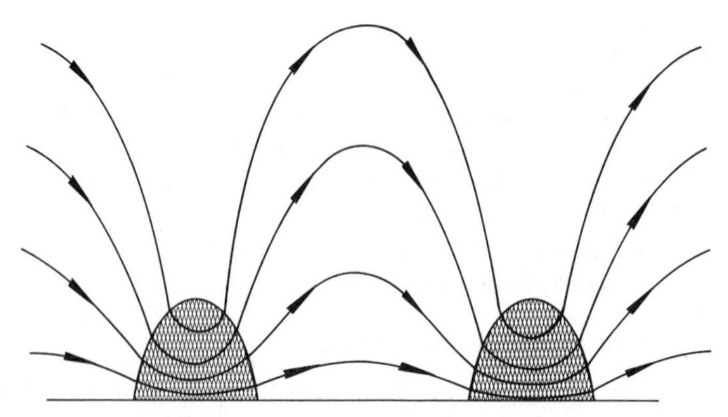

FIG. 13.3. A sketch of the magnetic lines of force of an initial horizontal field held captive in the gas concentrations represented by the cross-hatching.

the x-axis lies along the initial horizontal lines of force, while the y-axis is vertically upward. Thus the y-and z-coordinates of the previous sections become the x-and y-coordinates, respectively, of the present section. The motivation for the change is the conventional notation $z=x+iy$ in the conformal mappings by which the magnetic field configurations are computed. Thus z becomes the familiar complex variable.

In these terms, then, suppose that initially the uniform horizontal magnetic field $\mathbf{B}=\mathbf{e}_x B_0$ extends from $y=-\infty$ up to $y=+h$, where it is confined by the weight of a thin sheet of dense, infinitely conducting fluid (liquid or cold gas) with a surface density s such that

$$sg = B_0^2/8\pi. \tag{13.65}$$

The field in $y<h$ is pervaded by a tenuous hot gas of negligible density, weight, and pressure but with enormous electrical conductivity so as to preserve the topology of the field. The magnetic field is curl-free, of course, except where it presses against the dense overlying fluid.

13.7.1. *Breakup of confining fluid sheet*

Now it is clear from the development of §§13.2–13.5 that the overlying sheet of fluid is unstable. The fluid tends to run into any depressions in the upper boundary of the perturbed field, uncovering the field elsewhere and allowing it to bulge upward. A slight variation of the mathematics of §13.2 gives the growth rate of the instability. Denoting the upper boundary of the field by $y=h+\xi(x,t)$ and the perturbed field by (B_0+b_x, b_y), the coincidence of the upper boundary with the magnetic lines of force yields

$$\partial\xi/\partial x = b_y/B_0 \tag{13.66}$$

in the usual linear approximation. The horizontal motion of the fluid is a consequence of the slope of the upper boundary, so that

$$\partial v_x/\partial t = -g\partial\xi/\partial x, \tag{13.67}$$

while the surface density $s+\delta s$ of the layer varies as

$$\partial\delta s/\partial t = -s\,\partial v_x/\partial x. \tag{13.68}$$

The vertical acceleration of the layer is

$$s\,\partial^2\xi/\partial t^2 = B_0 b_x/4\pi - g\delta s. \tag{13.69}$$

Write the perturbation field (b_x, b_y) in terms of a vector potential A, with

$$b_x = +\partial A/\partial y, \qquad b_y = -\partial A/\partial x. \tag{13.70}$$

Then, since the field is curl-free,

$$\partial^2 A/\partial x^2 + \partial^2 A/\partial y^2 = 0.$$

Write the vector potential as

$$A=(CB_0/k)\exp(t/\tau)\exp\{ikx+k(y-h)\} \tag{13.71}$$

so that A vanishes at $y=-\infty$. The amplitude factor C is a dimensionless constant. It follows from (13.66) that $A(x, 0, t)=-B_0\xi(x, t)$. The dispersion relation follows from (13.69), upon using (13.67) and (13.68) to eliminate v_x and δs in favour of ξ. The result can be written

$$1/\tau^4+2gk/\tau^2-g^2k^2=0$$

upon using (13.65), so that the growth rate of the instability is

$$1/\tau=\{gk(2^{\frac{1}{2}}-1)\}^{\frac{1}{2}}. \tag{13.72}$$

The breakup occurs with a characteristic time of the order of the free-fall time over the horizontal scale of the breakup.

13.7.2. Equilibrium configuration

Suppose that the breakup of the confining fluid occurs for some single wave number $k=\pi/L$ so that the surface layer of fluid separates into equally spaced clumps separated by a distance $2L$, each clump of mass $m=2Ls=B_0^2L/4\pi g$. To keep the mathematics as simple as possible, suppose that the separation L and the volume density of the fluid are together sufficiently large that the width and thickness of each clump (portrayed in Fig. 13.3) is small compared to L. Each clump may then be approximated by a slender rod[5] lying across the top of the field. The magnetic field confined beneath the regularly spaced clumps (rods) is then readily obtained by a conformal mapping of an initial complex z_0-plane ($z=x+iy$), in which the potential function is

$$W(z_0)=2LB_0z_0/\pi,$$

into the final z-plane[6]. If, as a consequence of the high electrical conductivity of the fluid, the topology of the lines of force is preserved, then the appropriate mapping is (Smythe 1939)

$$\exp(i\pi z/2L)=\cos z_0.$$

The potential function in the final z-plane is then

$$W(z)=(2LB_0/\pi)\arccos\{\exp(i\pi z/2L)\}. \tag{13.73}$$

[5] Rods of finite radius can be treated (Richmond 1923; Smythe 1939), with considerable increase in computational effort and with little effect on the conclusions.

[6] There is no connection between this mathematical potential function, which is about to be mapped onto the z-plane, and the initial uniform field in $y<h$.

Writing

$$W(z) = U(x, y) + iV(x, y)$$

where U and V are real, it can be shown that

$$2\sin^2\left(\frac{\pi U}{LB_0}\right) = [\{1-\exp(-Y)\}^2 + 4\sin^2 X \exp(-Y)]^{\frac{1}{2}} + 1 - \exp(-Y), \quad (13.74)$$

$$2\sinh^2\left(\frac{\pi V}{LB_0}\right) = [\{1-\exp(-Y)\}^2 + 4\sin^2 X \exp(-Y)]^{\frac{1}{2}} - 1 + \exp(-Y), \quad (13.75)$$

where, for convenience,

$$X \equiv \pi x/2L, \qquad Y \equiv \pi y/L. \quad (13.76)$$

The function V is the vector potential, and U is the scalar potential, of the magnetic field. The lines of force are given by $V = constant$, and are plotted in Fig. 13.4. For $Y \to +\infty$ the asymptotic form of the vector potential is

$$V \sim (2LB_0/\pi)\sin X \exp(-\tfrac{1}{2}Y) + O\{\exp(-Y)\}, \quad (13.77)$$

while for $Y \to -\infty$

$$V \sim B_0 y - (2LB_0/\pi)\{\ln 2 - (\tfrac{1}{4} - \tfrac{1}{2}\cos^2 X \exp(Y)\} + O\{\exp(\tfrac{3}{2}Y)\}. \quad (13.78)$$

Thus the field has the uniform value B_0 at $y = -\infty$ and declines exponentially to zero at $y = +\infty$. The weight of the rods, located at $y = 0$ and $x = \pm(2n+1)L$ $(n = 0, 1, 2, \ldots)$ is just the weight of the initial sheet (13.65). Hence it is sufficient to confine the field pressure $B_0^2/8\pi$ exerted from $y = -\infty$. The line of force in contact with the lower surface of each rod is $V = 0$. If the rods were expanded back to a uniform sheet again, confining the field as they did before the breakup, the flux density would have the uniform value B_0 all the way to the sheet. The field at large negative y would be unaffected, of course, with the vector potential given by (13.77) as

$$V \sim B_0 y - (2LB_0/\pi)\ln 2$$

for the uniform field all the way to the confining sheet. The uppermost line of force $V = 0$ would then lie at $y = h$ where

$$h = 2\{\ln 2\}L/\pi. \quad (13.79)$$

This would be the position of the fluid, then, were it restored to a uniform sheet.

13.7.3. *Energy decrease*

In passing from the initial uniform field to the equilibrium (13.75) the fluid has descended a distance h in the gravitational field g. The change in the gravitational potential energy of the matter is

$$\Delta\Phi = -gsh \text{ erg cm}^{-2}$$
$$= -1{\cdot}386(B_0^2/8\pi)L/\pi. \tag{13.80}$$

It is a straightforward procedure to calculate the decrease in magnetic energy. The magnetic energy of the initial uniform field above some base level $y = y_0 < 0$ is just

$$\mathcal{M}_0 = (h - y_0)B_0^2/8\pi. \tag{13.81}$$

The subsequent mean magnetic energy above the level y_0 is

$$\mathcal{M} = \frac{1}{L}\int_0^L \mathrm{d}x \int_{y_0}^{\infty} \mathrm{d}y \frac{B^2}{8\pi}, \tag{13.82}$$

with

$$B^2 = (\nabla V)^2$$
$$= |\mathrm{d}W/\mathrm{d}z|^2$$

where W is given by (13.73). It is readily shown that

$$\mathcal{M} = \frac{B_0^2}{8\pi}\frac{L}{\pi}\int_0^{\pi} \mathrm{d}X \int_0^{\psi_0} \frac{\mathrm{d}\psi}{(1 - 2\psi\cos X + \psi^2)^{\frac{1}{2}}},$$

where now $X = \pi x/L$ and $\psi = \exp(-Y)$ with $Y = \pi y/L$, $Y_0 = \pi y_0/L$. Expand the integrand in Legendre polynomials, noting that for $Y_0 < 0$ we have $\psi_0 > 1$ so that two separate expansions must be used, one for $\psi < 1$ and the other for $\psi > 1$. Then the integration over ψ can be performed immediately, with the result

$$\mathcal{M} = \frac{B_0^2}{8\pi}\frac{L}{\pi}\left[1 + \ln(\psi_0) + \frac{1}{\pi}\sum_{n=1}^{\infty}\frac{1}{n}\left\{\frac{2n+1}{n+1} - \frac{1}{\psi_0^n}\right\}\int_0^{\pi} \mathrm{d}X P_n(\cos X)\right],$$

Noting that

$$\int_0^{\pi} \mathrm{d}X P_n(\cos X) = 0 \qquad n \text{ odd}$$
$$= \pi(2n)!/2^{2n}(n!)^2 \qquad n \text{ even},$$

replace n by $2n$ in the sum, so that the sum runs over all integral n. It follows that

$$\mathcal{M} = \frac{B_0^2}{8\pi}\frac{L}{\pi}\left[1 + \ln(\psi_0) + \sum_{n=1}^{\infty}\frac{(4n)!}{2^{4n}2n(2n!)^2}\left\{\frac{4n+1}{2n+1} - \frac{1}{\psi_0^{2n}}\right\}\right]. \tag{13.83}$$

Let Y_0 become large and negative, so that below Y_0 the field is unaffected by the perturbations at its upper surface. Then $\psi_0 \gg 1$ and

$$\mathcal{M} \sim \frac{B_0^2}{8\pi}\frac{L}{\pi}\left\{1 - Y_0 + \sum_{n=1}^{\infty}\frac{(4n)!(4n+1)}{2^{4n}2n(2n+1)(2n!)^2}\right\}.$$

The energy change in the magnetic field is

$$\Delta\mathcal{M} \equiv \mathcal{M} - \mathcal{M}_0$$

for $y_0 \to -\infty$, or

$$\Delta\mathcal{M} = \frac{B_0^2}{8\pi}\frac{L}{\pi}\left\{\frac{\pi h}{L} - 1 - \sum_{n=1}^{\infty}\frac{(4n)!(4n+1)}{2^{4n}2n(2n+1)(2n!)^2}\right\}$$

The sum converges slowly, to a value 1·0. With h given by (13.79) the energy change is

$$\Delta\mathcal{M} = -\frac{B_0^2}{8\pi}\frac{L}{\pi}2(1 - \ln(2))$$

$$= -0{\cdot}6(B_0^2/8\pi)L/\pi. \tag{13.84}$$

This is about half the change in the gravitational potential energy of the confining matter, given by (13.80). The total energy change is the decrease

$$\Delta\Phi + \Delta\mathcal{M} = -2{\cdot}0(B_0^2/8\pi)L/\pi. \tag{13.85}$$

13.7.4. Further evolution of the field

As a matter of fact the evolution of the system does not stop with the state shown in Fig. 13.4. There are several additional manoeuvres by which the field progresses toward still lower energies. For instance, were it not for the constraint to two dimensions, the field would break up into tubes and sheets as described in §13.4. Within the constraint to two dimensions, it is evident that the more widely spaced (larger L) are the pockets of matter confining the field, the deeper they sink into the field and the more energy is released. Hence, as already noted, the periodic equilibrium (13.75) for equally spaced rods is unstable (Parker 1967a), the rods tending to join neighbours to form clumps of increasing weight and horizontal spacing, with the growing clumps sinking progressively deeper into the field and the field bulging further upward between. The qualitative picture, illustrated by Fig. 13.4, remains the same, but with the characteristic dimensions increasing with time. None of these effects lead directly to freeing the lines of force from under the gas, of course.

Consider, then, the effect that permits magnetic lines of force to escape from underneath the confining rods of matter. For simplicity suppose that the rods are constrained to their positions $x = \pm(2n+1)L$, so that the

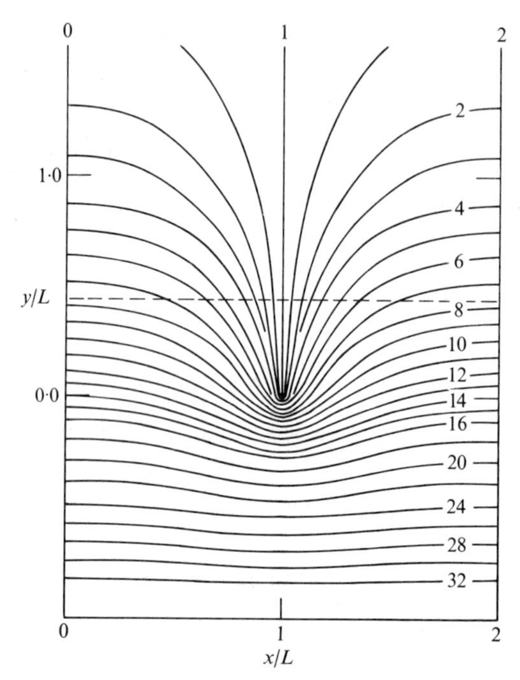

FIG. 13.4. The magnetic lines of force for the initial configuration of a single cell ($0<x<2L$) of the periodic field (13.74) and (13.75) confined by a grid of slender rods (line currents) at $x=\pm(2n+1)L$, $y=0$. The numbers on the lines indicate the vector potential V in units of $LB_0/8\pi$. The broken line represents the position of the field surface if the system were restored to its initial horizontal uniformity.

static equilibrium of Fig. 13.4 is preserved. Then note that the field configuration shown in Fig. 13.4 is in equilibrium everywhere except on the vertical plane above each rod, across which the magnitude of the field is uniform and non-zero, while the direction reverses discontinuously. There is nothing to keep the two oppositely directed fields apart, with the result that, no matter how high the electrical conductivity of the fluid, the fields diffuse across the discontinuity and rapidly dissipate each other. It is the familiar neutral sheet dissipation, with the fluid continually squeezing out from between the two opposite fields, allowing them to come together and annihilate. The lines of force reconnect rapidly, because the characteristic scale of the transition is so small. The effect is treated at length in Chapter 15. Theoretical considerations indicate that the reconnection proceeds at some fraction $O(10^{-1})$ of the Alfven speed (Petscheck 1964; Axford 1967; Sonnerup 1970; Fukao and Tsuda 1973a, b; Vasyliunas 1975; Roberts and Priest 1975).

In this connection we should not overlook the observational fact (Sheeley *et al.* 1975) that, no matter how complex the magnetic field topologies that emerge through the visible surface of the sun, rapid reconnection of the magnetic lines of force quickly reduces the field above the surface toward the vacuum configuration, $\mathbf{B} = -\nabla U$, having the lowest energy for the given flux pattern coming through the surface. Evidently the continual activity of complex magnetic regions on the sun—the eruptions, flares, plages, x-ray bright points—is the direct result of the rapid reconnection. The close association with the topology of the field is treated in Chapter 14.

The important point is the simple fact that magnetic fields are not long in approaching an overall equilibrium, by reconnection across neutral sheets, no matter how high the electrical conductivity. The reconnection of magnetic lines of force across the plane above each rod alters the topology of the lines so that the reconnected lines are free to escape, as sketched in Fig. 13.5.

The dynamical reconnection continues until the discontinuity is removed from the field. The equilibrium configuration, in which the discontinuity has completely disappeared, is shown in Fig. 13.6. The equilibrium field is easily calculated (Richmond 1923; Smythe 1939) for rods of circular cross-section. The magnetic lines of force are the same as the equipotential lines in an electrostatic field bounded by a grid of rods. To compute one cell of the electrostatic field, place charges in the z_0-plane at

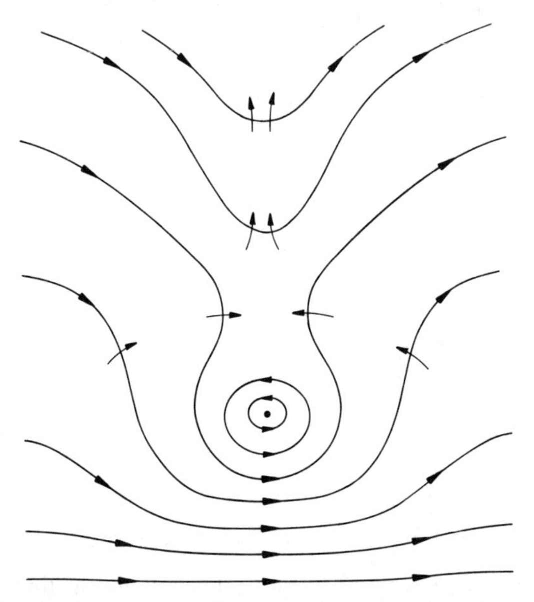

FIG. 13.5. A sketch of the intermediate state (between Figs 13.4 and 13.6) of field annihilation and rapid reconnection in the region above the rod. The arrows indicate the direction of motion of the lines of force. The lines of force above the rod, but reconnected so that they no longer pass under the rod, are free to escape from the system.

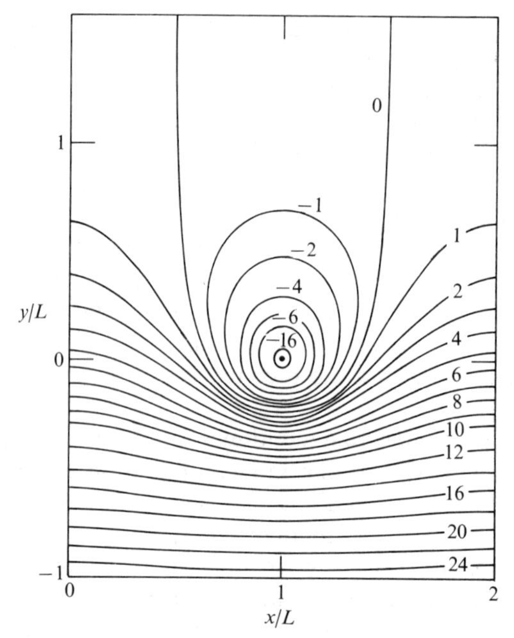

FIG. 13.6. The magnetic lines for the final configuration of a single cell ($0<x<2L$) of the periodic field (13.87) confined by a grid of slender rods (line currents) at $x=(2n+1)L$, $y=0$. The number on the lines indicate the vector potential U in units of $LB_0/8\pi$.

$z_0 = \pm i2L$, for which the potential function is

$$W(z_0) = \ln(1 + z_0^2/4L^2).$$

Then introduce the mapping

$$\pi z = 2L \ln(z_0/2L)$$

to the z-plane, taking the positive real z_0-axis into the real z-axis and the negative real z-axis into $y = +L$, with the field confined between. The potential function becomes

$$W(z) = \ln\{1 + \exp(\pi z/L)\}, \qquad (13.86)$$

so that, with $W = U + iV$, we have

$$\exp(2U) = 1 + 2\exp X \cos Y + \exp(2X),$$

$$\cot V = (1 + \exp X \cos Y)/\exp X \sin Y,$$

where again $X = \pi x/L$ and $Y = \pi y/L$. The magnetic lines of force are given by $U = constant$. The entire picture must now be rotated by 90° to conform to the previous example (13.75) shown in Fig. 13.4 ($X \to -Y$, $Y \to +X$). Replace U and V by $\pi U/LB_0$ and $\pi V/LB_0$, respectively, so that U represents flux per unit distance and the field has the uniform value $B_0 = -\partial U/\partial Y$ far below the rods ($Y \to -\infty$) which are spaced along the x-axis at $x = \pm(2n+1)L$. Figure 13.6 is a plot of the field in the region $0 < x < 2L$, corresponding to the cell shown in Fig. 13.4. The lines of force are given by $U = \textit{constant}$ with

$$\exp(2\pi U/LB_0) = 1 + 2\exp(-Y)\cos X + \exp(-2Y). \qquad (13.87)$$

The lines of force that have undergone reconnection appear as $U < 0$, and circle the rod at $x = L$, $y = 0$. The dividing line $U = 0$ follows the path $\exp(-Y) + 2\cos X = 0$ for finite X and Y. The total flux reconnected depends upon the radius R of the rods. The calculation leading to (13.87) for $R \to 0$ is valid so long as $R \ll L$ with the rod surface represented by one of the nearly circular lines of force $U = \textit{constant}$ enclosing the point $x = L$, $y = 0$. (For a treatment of the problem with R comparable to L, see Richmond 1923). The total reconnected flux ϕ circling the rod is just equal to $-U$ at the under-surface of the rod ($X = \pi$, $Y = -\pi R/L$) so that

$$\begin{aligned} \phi &= -U(L, -R) \\ &= B_0(L/\pi)[\ln(L/\pi R) - \pi R/2L + O\{(\pi R/2L)^2\}]. \qquad (13.88) \end{aligned}$$

Thus ϕ becomes large without limit as $R/L \to 0$. If, for instance, $R = L/4\pi$, then $\phi = 0{\cdot}44LB_0$. The reconnection of the lines frees a substantial amount of flux from beneath the material rods.

It should be noted, then, that no lines of force have passed across the rods. The original number of lines is still pinned beneath them. The difference is that a number ϕ have closed themselves around the rod, so that they no longer extend in the x-direction from one rod to the next. They close locally and make no contribution to the net flux. An observer stationed far out on the positive y-axis would record the outward passage of ϕ lines of force, freed from under the rods at the time their reconnection occurred.

The reconnection and escape of the magnetic flux ϕ is accompanied by a further decrease of the energy of the system, of course. It is obvious from Figs 13.4 and 13.6 that the magnetic energy has decreased. It is also easy to show that the matter has settled downward in the process, decreasing the gravitational energy. To compute the downward displacement of the rods note that the total flux between the under-surface of the rod and some basic level $y = y_0$ of large negative y is

$$U(L, y_0) - U(L, -R) = \phi - B_0 y_0$$

in the limit of large negative y_0. Were this field uniform, with strength B_0, all the way from y_0 to the level $y = l$, then l would have a value such that

$$(l - y_0)B_0 = \phi - B_0 y_0$$

or

$$l=(L/\pi)\ln(L/\pi R).$$

That is to say, the rods are now a distance l below the initial top of the field. Comparison with (13.79) shows that the rods are a distance

$$l-h=\{\ln(L/\pi R)-2\ln(2)\}L/\pi$$

farther down than before the reconnection.

To go on, then, the magnetic field circling each material rod involves no net flux. The lines of force around each rod form a closed system, subject to dissipation by turbulent diffusion (see Chapter 16) and their own internal buoyant instability. The relaxation of the artificial constraint to two dimensions is alone sufficient to produce instability merely as a result of the net compressive stresses transmitted along the rods by the encircling field. So in one way or another the encircling field will free itself. But the equilibrium depicted in Fig. 13.6 requires that the total flux circling the rod outside its radius R must be ϕ, given by (13.88). If there is less, then the horizontal flux trapped below the rod is not prevented from bulging up through the gaps between the rods and reconnecting above, as sketched in Fig. 13.5. Hence the continued rate of escape of horizontal field from beneath the grid of rods is equal to the rate of which the lines of force circling around each rod are able to dissipate and/or escape. That rate of dissipation depends entirely on the special conditions of each circumstance. About all we can say here is that we can think of no circumstances wherein the instabilities discussed in these chapters can be avoided sufficiently to give a characteristic escape time greater than $O(L/V_A)$ where $V_A=B_0/(4\pi\rho)^{\frac{1}{2}}$.

13.8. Effects of cosmic rays

The magnetic field of the galaxy is inflated by the cosmic ray gas, which adds significantly to the thickness of the gas and field distribution in the disc of the galaxy, and which contributes substantially to the dynamical instability and escape of the magnetic field from the disc (Parker 1958, 1966, 1969). Thus the escape rate of the magnetic field in the gaseous disc is somewhat higher than we would estimate from the instability of the magnetic field alone (§13.6) and the turbulent diffusion alone. How much higher is hard to estimate, particularly since so little is known of the rate of production of cosmic rays in the disc. The general physical effect is worthy of our attention, however, and the day may come when improved knowledge of the cosmic ray sources (from studies of their isotopic abundances and from their γ-ray emission) will permit a more quantitative treatment of the problem.

The pressure P of the cosmic ray gas is comparable to the pressure

$B^2/8\pi$ of the galactic field, and hence is comparable to the dynamical pressure of the turbulent interstellar gas. The cosmic ray pressure has been maintained at the position of the sun at least for the past 10^9 years, so it is assumed that the inflation of the galactic magnetic field is near an equilibrium state, with cosmic rays escaping from the disc of the galaxy at about the same rate that they are produced within the disc (see review in Parker 1968c). The cosmic ray gas has so little mass that it is not significantly affected by the gravitational acceleration g perpendicular to the disc. The cosmic ray gas, then, plays a role much like the buoyancy of the magnetic field in producing and inflating the upward bulges of the perturbed horizontal equilibrium field.

To illustrate the effects of the cosmic ray gas on the equilibrium and instability of a horizontal magnetic field in an isothermal atmosphere, confined by a gravitational acceleration g, suppose that the cosmic ray gas pressure P is proportional to the gas pressure $p \equiv \rho u^2$

$$P = \beta p$$

where β is a constant, in the same way that magnetic pressure was taken in §13.4 to be proportional to p. Then the total pressure is $p + B^2/8\pi + P = \rho u^2(1+\alpha+\beta)$ and the barometric equilibrium leads to the relation

$$\Lambda = (1+\alpha+\beta)u^2/g$$

for the pressure scale height Λ in an atmosphere in which u^2 and g are constant throughout. This fundamental relation must be kept in mind when inferring the strength of the galactic magnetic field from the observed Faraday rotation measures, etc. (Parker 1965, 1966). Strong fields imply large Λ (see discussion in Chapter 22).

When the equilibrium is perturbed, the cosmic rays move freely along the lines of force so that P remains uniform along each line. Neglecting second order terms, the volume of each elemental tube of flux is unchanged by the perturbation, so that the equation for the cosmic ray pressure $(P+\delta P)$ is (Parker 1966, 1969)

$$(\mathrm{d}/\mathrm{d}t)(P+\delta P) = 0$$

or, to first order in the perturbations,

$$\partial \delta P/\partial t + v_z \, \mathrm{d}P/\mathrm{d}z = 0.$$

Note that

$$\frac{1}{P}\frac{\mathrm{d}P}{\mathrm{d}z} = \frac{1}{\rho}\frac{\mathrm{d}\rho}{\mathrm{d}z} = \frac{1}{p}\frac{\mathrm{d}p}{\mathrm{d}z} = \frac{2}{B}\frac{\mathrm{d}B}{\mathrm{d}z} = -\frac{g}{(1+\alpha+\beta)u^2} \equiv -\frac{1}{\Lambda}.$$

The variation of the pressure of the thermal gas is given by (13.24)–(13.26) and the variation of the magnetic field by (13.27)–(13.29). The equations of motion (13.30)–(13.32) are modified by the addition of the cosmic ray pressure variation δP, so that δp is replaced by $\delta p + \delta P$. Proceeding exactly as in §13.4, with $s = 1$ and $k_x = 0$, it is readily shown that the dispersion relation for the two-dimensional normal modes is

$$\Omega^4 + b\Omega^2 - c = 0,$$

where

$$b = (2\alpha + \gamma)(q_2^2 + q_3^2 + \tfrac{1}{4})$$
$$c = q_2^2\{(1 + \alpha + \beta - \gamma)(1 + \alpha + \beta) - 2\alpha\gamma(q_2^2 + q_3^2 + \tfrac{1}{4})\}.$$

Instability arises whenever Ω^2 is positive so that there is at least one real positive root $1/\tau$. A necessary and sufficient condition for instability $(\tau > 0)$ is $c > 0$, so that the root

$$\Omega^2 = \tfrac{1}{2}\{(b^2 + 4c)^{\frac{1}{2}} - b\}$$

is positive. Hence when

$$(1 + \alpha + \beta - \gamma)(1 + \alpha + \beta) > 2\alpha\gamma(q_2^2 + q_3^2 + \tfrac{1}{4}),$$

the system is unstable. Noting that the cosmic ray pressure β appears only on the left-hand side of the inequality, it is evident that its effect is to enhance the instability. The larger is β, the larger may be the index γ and the wave numbers q_2 and q_3 for instability. Since b is independent of β, and c increases with β, it is evident that the growth rate of the instability increases with increasing β, all other parameters being the same. In contrast note that the magnetic field strength α appears with both signs, because the magnetic field has tension as well as pressure. The tension inhibits the instability at large wave numbers. In the limit of small α, with only the cosmic rays to inflate the thermal gas beyond its own intrinsic scale height of u^2/g, the system is unstable for all wavelengths provided only that $\gamma < 1 + \beta$. Since γ is generally comparable to, or less than, one in the interstellar gas, because the equilibrium temperature is reduced with increasing temperature (Savedoff and Spitzer 1950; Parker 1953), there is always an instability, the maximum growth rate

$$\Omega_{\text{max}}^2 \sim \frac{q_2^2}{q_2^2 + q_3^2} \frac{(1 + \alpha + \beta - \gamma)(1 + \alpha + \beta)}{\gamma}$$

arising in the limit of large wave number q_2.

For α and β of the order of unity (as in the galaxy), the growth time of the instability is of the general order of $\Omega = 1$, so that $1/\tau = O(u/\Lambda) = O(V_A/\Lambda) = O(g^{\frac{1}{2}}/\Lambda^{\frac{1}{2}})$, where V_A is the Alfven speed in the equilibrium field. The time of growth is characterized by the sound or Alfven transit time, or the free-fall time, across one scale height. To treat a specific case, note that $\alpha \cong \beta \cong 0{\cdot}5$ in the gaseous disc of the galaxy (for a magnetic field of 3×10^{-6} G, a cosmic ray energy density of $1\,\text{eV cm}^{-3}$, and an r.m.s. three-dimensional turbulent velocity $v = 10\,\text{km s}^{-1}$, $u = v/\sqrt{3}$ in a mean interstellar gas density of 2 hydrogen atoms cm^3). Then for isothermal perturbations, $\gamma = 1$, all wave numbers less than

$$q_2^2 + q_3^2 < \tfrac{7}{4}$$

are unstable. For $q_3 = 0$, all wavelengths along the field in excess of $4{\cdot}75\Lambda$ are unstable ($q_2 = 7^{\frac{1}{2}}/2$). Thus $q_2^2 = \frac{7}{4}$, $q_3^2 = 0$ yields marginal stability ($\Omega = 0$). The maximum growth rate is $\Omega = 0{\cdot}58$ in the neighbourhood of the wavelength $9\Lambda(q_2^2 = 0{\cdot}5,\ q_3^2 = 0)$. The growth rate is, then, $1/\tau \cong 0{\cdot}6u/\Lambda$. With $u = 10\,\text{km s}^{-1}$ and $\Lambda = 150$ pc the characteristic growth time is about 10^{15} s or 3×10^7 years. The cosmic rays and the magnetic field contribute about equally to the instability.

The effect of the transverse wave number k_x, put equal to zero so far in the calculations of the cosmic ray instability, is readily demonstrated in the weak field limit ($\alpha \to 0$). It can be shown (Parker 1967b) that the dispersion relation is, then,

$$\Omega^4 + B\Omega^2 - C = 0$$

where

$$B = \gamma(q_1^2 + q_2^2 + q_3^2 + \tfrac{1}{4}),$$
$$C = (1 + \beta - \gamma)(1 + \beta)(q_1^2 + q_2^2).$$

The transverse wave number q_1 appears inseparably from q_2, indicating that it enhances the growth rate in the same way as q_2. In the limit of large wave number (k_x, k_y) the growth rate Ω increases to the asymptotic value $(C/B)^{\frac{1}{2}}$. In contrast, when $\alpha > 0$, $\beta = 0$ the instability increases with increasing k_x at a rate proportional to k_z, and declines to zero with increasing k_y because of the tension in the lines of force.

A variety of examples have been worked out in the literature to illustrate the many facets of the combined magnetic field, cosmic ray instability of the gaseous disc of the galaxy. The instability is stronger by the factor g^2/GB_0^2 than the self-gravitation of the interstellar gas (Jean's gravitational instability). With the galactic field B_0 of the order of 3×10^{-6} G and $g = 3\times10^{-9}\,\text{cm s}^{-2}$, and the value $G = 6{\cdot}66\times10^{-8}\,\text{cm}^3\,\text{g}^{-1}\,\text{s}^{-2}$ for the gravitational constant, the factor is equal to 15. The essential nature of the cosmic ray, magnetic field instability is unaffected by the rotation of the galaxy (Shu 1974), by the variation of the gravitational acceleration with height (Parker 1966), by slabs of horizontal field lying at right angles upon each other (Parker 1967b), by a continuous rotation of the field direction with height z in the gravitational field (Lerche and Parker 1968), by deformation of the fields into the azimuthal direction around a cylindrical body with rotational symmetry (Parker 1966) or by resistivity and viscosity (Parker 1967b). The problem can be treated by the simple normal mode analysis given here, or as an initial value problem with $v_i = 0$ at $z = +\infty$ (Lerche and Parker 1967) with comparable results. The enhanced diffusion of the magnetic field as a result of the perturbed magnetic configuration has been treated (Parker 1967b), showing that it contributes to the escape of the lines of force from the gaseous disc. The decline of the energy of the system with the increased clumping of the gas has been studied and illustrated with a number of idealized examples (Parker 1967a; Lerche 1967b; Mouschovias 1974). There is no final equilibrium state in which the gas field, cosmic ray system can come to rest. Formally the equations indicate that the general turbulence caused by the instability goes on indefinitely, the final asymptotic state being simply the disengagement of the field and cosmic rays from the thermal gas. Evidently the clumping of the gas in the low places along the field increases until hot stars are formed, whose radiation then disperses the gas and the instability starts all over again (Parker 1968a, b). Mouschovias *et al.* (1974) point out that the formation of large cloud complexes at intervals of a kpc along the galactic magnetic field are to be understood as the direct consequence of the instability.

The dynamical instability produced by the cosmic rays and magnetic field is the principal means of escape of the cosmic rays from the magnetic field of the galaxy (Parker 1965, 1966, 1969). The cosmic ray pressure P builds up in the disc until it approaches the magnetic stress density $B^2/8\pi$, whereupon the cosmic rays inflate the upward bulges of the field into extended bubbles, which eventually escape from the galaxy by rapid reconnection, as outlined in §13.7. The escape of cosmic rays from deep in the disc is facilitated by the random walk of the magnetic lines of force (chapter 10), which allows the cosmic rays to pass near the surface of the disc at random intervals of the order of a kpc along the galactic field (Jokipii and Parker 1969). The topic is taken up again in chapter 22.

13.9. Summary

The illustrative examples computed at such length in the foregoing sections of this chapter illustrate the irrepressible buoyancy of magnetic fields (see also §§8.6–8.8). It was emphasized in §§5.3 and 5.4 that a magnetic field can be contained only when suitably held down, by submergence in a gravitating body. Now we find that the field is a very 'slippery' thing to hang on to. Given a time very long compared to Λ/V_A the field generally escapes the grasp of the gravitating fluid object. Special circumstances involving strongly stable, placid, conducting fluids can be invented to retain fields for long periods, but in the real universe of convective stellar envelopes and turbulent gaseous discs of galaxies, such ideal circumstances are rare. Not even the stable, non-convecting core of a star can hang on to a strong field for long (see Parker 1974; and §8.8), although weak fields could be hidden there if that is of any interest.

Altogether, then, even the best magnetohydrostatic equilibrium is unstable, becoming highly dissipative after a short period of time. Magnetic fields upset whatever creates and holds them. They are the prodigal children of rotating convective astrophysical bodies. They are the cosmic 'escape artists' with a remarkably versatile repertoire of tricks. The tricks are the magnetic activity that we find on all magnetic astrophysical bodies. The universal existence of magnetic fields is, then, testimony to their copious generation within astrophysical bodies. Their existence, like that of so many biological creatures, is an active struggle leading to oblivion. Were nature to fail to provide rapid reproduction[7] of the fields from the environment, the fields would quickly die out and cease to complicate our universe.

The chapters which follow are directed to further means of dissipation of the wily magnetic field, and then to the circumstances that cause its abundance in spite of its profligate habits. Once we have outlined the various modes of internal dissipation and eventual escape, the question will be whether the theory for replenishing the fields is adequate to explain their general existence.

References

Asseo, E., Cezarsky, C. J., Lachieze-Rey, M. and Pellat, R. (1978), *Astrophys. J. Lett.* **225,** L21.
Axford, W. I. (1967). *Space Sci. Rev.* **7,** 149.
Chandrasekhar, S. (1961). *Hydrodynamic and hydromagnetic stability*, pp. 94, 97. Clarendon Press, Oxord.
Chapman, S. (1954). *Astrophys. J.* **120,** 151.
Cowling, T. G. (1953). *The sun* (ed. G. P. Kuiper). University of Chicago Press.
Fukao, S. and Tsuda, T. (1973a). *Planet. Space Sci.* **21,** 1151.
—— —— (1973b). *J. Plasma Phys.* **9,** 409.
Gilman, P. (1970). *Astrophys. J.* **162,** 1019.
Jokipii, J. R. and Parker, E. N. (1969). *Astrophys. J.* **155,** 777, 799.
Lerche, I. (1967a). *Astrophys. J.* **148,** 415.
—— (1967b). ibid. **149,** 395, 553.
—— and Parker, E. N. (1967). *Astrophys. J.* **149,** 559.
—— ——(1968). ibid. **154,** 515.
Low, B. C. (1975a). *Astrophys. J.* **197,** 215.
—— (1975b). ibid. **198,** 211.
Mouschovias, T. Ch. (1974). *Astrophys. J.* **192,** 37.
——, Shu, F. H. and Woodward, P. R. (1974). *Astron. Astrophys.* **33,** 73.
Parker, E. N. (1953). *Astrophys J.* **117,** 169, 431.
—— (1958). *Phys. Rev.* **109,** 1329.

[7] The proper word is 'reproduction' rather than 'production' because the hydromagnetic induction equation is linear in the magnetic field.

—— (1965). *Astrophys. J.* **142,** 584.
—— (1966) ibid. **145,** 811.
PARKER, E. N. (1967a). ibid. **149,** 517.
—— (1967b). ibid. **149,** 535.
—— (1968a). ibid. **154,** 57.
—— (1968b). ibid. 875.
—— (1968c). Nebulae and interstellar matter, *Stars and stellar systems* (ed. B. M. MIDDLEHURST and L. H. ALLER), vol. VII, chap. XIV. University of Chicago Press.
—— (1969). *Space Sci. Rev.* **9,** 651.
—— (1973). *Astrophys. J.* **186,** 665.
—— (1974). *Astrophys. Space Sci.* **31,** 261.
—— (1975a). *Astrophys J.* **198,** 205.
—— (1975b). ibid. **202,** 523.
PETSCHEK, H. (1964). *AAS–NASA symposium on the physics of solar flares* (ed. W. HESS), p. 245. U.S. Government Printing Office, Washington.
RICHMOND, H. W. (1923) *Proc. Lond. Math. Soc., Ser. 2,* **22,** 389.
ROBERTS, B. and PRIEST, E. R. (1975). *J. Plasma Physics* **14,** 417.
ROBERTS, P. H. and STEWARTSON, K. (1974). *Phil. Trans Roy. Soc. Ser. A* **277,** 297.
—— —— (1975). *J. Fluid Mech.* **68,** 447.
SAVEDOFF, M. and SPITZER, L. (1950). *Astrophys. J.* **111,** 593.
SHEELEY, JR., N. R., BOHLIN, J. D., BRUECKNER, G. E., PURCELL, J. D., SCHERRER, V., and TOUSEY, R. (1975). *Solar Phys.* **40,** 103.
SHU, F. H. (1974). *Astron. Astrophys.* **33,** 55.
SMYTHE, W. R. (1939). *Static and dynamic electricity*, 1st edn., pp. 84, 99. McGraw-Hill, New York.
SONNERUP, B. U. A. (1970). *J. Plasma Phys.* **4,** 161.
SPRUIT, H. C. (1974). *Solar Phys.* **34,** 277.
VASYLIUNAS, V. M. (1975). *Rev. Geophys. Space Phys.* **13,** 303.

14

NON-EQUILIBRIUM IN FIELD TOPOLOGIES LACKING INVARIANCE

14.1. General remarks on field topology and non-equilibrium

HAVING dealt thus far with symmetric (invariant) field topologies, consider now the consequences of a departure from symmetry. Whereas magnetostatic equilibrium is possible under special circumstances in fields with suitable invariance, it turns out that there can be no equilibrium when topological invariance is lacking (Parker 1972; Yu 1973). The field without topological invariance exists in a dynamical state of activity and dissipation, which continues until the variant components of the field are destroyed and the topology becomes suitably invariant.

It was pointed out in Chapter 11 that the magnetic fields in nature are generally stochastic, lacking the symmetric topologies necessary for equilibrium. Hence astrophysical fields are condemned, like Sisyphus, to an 'eternity' of activity. In an isolated system with no energy sources the dissipation of electrical resistivity provides eventual equilibrium of course. There is only a finite amount of energy in the non-symmetric component of the topology so that, if nothing else interferes, the active non-equilibrium must run down. But in nature the dynamical reduction of the topology goes on in competition with the many turbulent regenerative and diffusive effects that complicate the topology, as indicated in Chapter 11, and remove the field from the invariant form required for equilibrium. Altogether the topological non-equilibrium plays a basic role in turbulent convecting fluids, generally cutting the mutually entangling lines of force, and preventing the enormous stretching and intensification of fields that would otherwise occur. The topological non-equilibrium is the cause of the concentrated transient dissipation—magnetic activity—observed in association with fields in the astrophysical universe. Together with the non-equilibrium and instability of most symmetric equilibrium field patterns, the topological non-equilibrium guarantees that few, if any, magnetic fields can ever be free of dynamical activity.

Consider, then, the magnetohydrostatic equilibrium of a magnetic field with a well defined topology. To be clear as to the form of the topology, imagine that there is initially simply a uniform magnetic field B_0 in the z-direction filling an infinite volume. The volume extends to infinity in the x- and y-directions perpendicular to the z-axis. We are concerned with the equilibrium deep in the interior of the volume, far from boundaries, where the field and fluid stresses are in mutual balance among themselves, without disturbance from externally imposed boundary forces. Suppose, then, that the volume is terminated in the z-direction by the planes $z = \pm L$ of infinitely conducting material in which the lines of force of the field are permanently affixed. The region of uniform field B_0 between the planes $z = \pm L$ is filled with a fluid of infinite electrical conductivity and uniform density ρ.

Now hold the plane $z = -L$ fixed while the plane $z = +L$ is continuously mapped onto itself,

$$dx/dt = f(x, y, t), \qquad dy/dt = g(x, y, t),$$

where f and g are continuous single-valued functions of x, y, and t. After a time each point initially at (x_0, y_0) lies at (x_t, y_t) having made the journey by some tortuous path. We may suppose that the mapping of the plane $z = +L$ into itself is incompressible, so that

$$\frac{\partial f}{\partial x} + \frac{\partial g}{\partial y} = 0, \qquad \frac{\partial(x_t, y_t)}{\partial(x_0, y_0)} = 1.$$

The ends of the magnetic lines of force remain fixed in $z = -L$ but follow the convolutions of $z = +L$ so that the winding pattern (the topology) of the convolution of the points on $z = +L$ is imposed on the magnetic lines of force between $z = -L$ and $z = +L$ (after passage of a suitable time in excess of the Alfvén transit time $2L(4\pi\rho)^{\frac{1}{2}}/B_0$, of course). Figure 14.1 is a sketch of the lines of force.

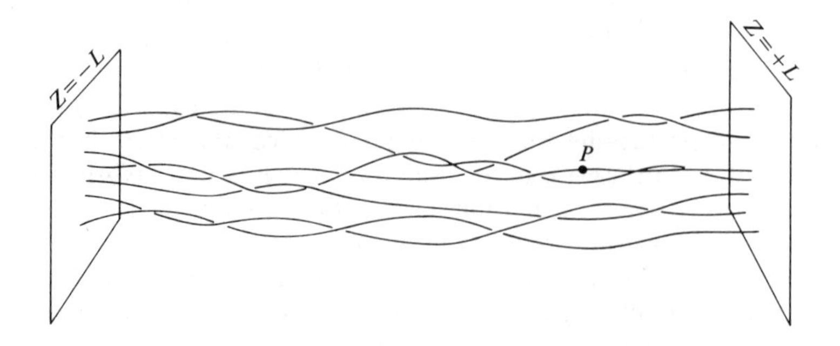

FIG. 14.1. A sketch of the magnetic lines of force extending from the fixed plane $z = -L$ to the plane $z = +L$ across the intervening volume of conducting fluid. The positions at which the lines cross the plane $z = +L$ have been manipulated in some complicated and arbitrary way so that the lines of force wind round each other throughout the volume between the two planes.

We suppose that the functions f and g are bounded in such a way that the net excursion of any point on $z=+L$ is bounded by a length l which represents the characteristic transverse scale of the variations introduced into the field,

$$(x_t-x_0)^2+(y_t-y_0)^2<l^2.$$

The length $2L$ of the lines of force is taken to be very large compared to the internal scale l. Indeed, we make L arbitrarily large compared to the total path length $\mathcal{L}$,

$$\mathcal{L}\equiv\int_0^t dt'[f^2\{x(t'),y(t'),t'\}+g^2\{x(t'),y(t'),t'\}]^{\frac{1}{2}}$$

transversed by any given point on $z=+L$, although $\mathcal{L}$ may be large compared to the net displacement of any point. Generally

$$l\ll\mathcal{L}\ll L.$$

The lines of force may be wound and braided many times around neighbouring lines over the length $2L$, but the lines deviate by only a small angle from the z-direction because the windings are spread over the long length $2L$. This is no more than the restriction that the local perturbation to the field is small compared to the total field. Generally speaking, n revolutions about a radius l yields an angular deviation of the order of $n\pi l/L\cong\mathcal{L}/L\ll 1$.

The restriction to small perturbations ($\mathcal{L}/L\ll 1$) is necessary in order that the z-direction not lose its identity. We must know relative to what direction we are to consider invariance of the field.

The question, then, is the circumstances under which the mapping, and the wrapping of the lines of force about each other, leaves the field with a topology for which there is a formal solution of the equilibrium equations

$$\frac{\partial}{\partial x_i}\left(p+\frac{B_jB_j}{8\pi}\right)=\frac{B_j}{4\pi}\frac{\partial B_i}{\partial x_j},\qquad \frac{\partial B_j}{\partial x_j}=0. \tag{14.1}$$

Of course, when the topology of the lines of force prohibits an equilibrium solution, the question arises as to the physical reasons for the absence of equilibrium and, finally, the dynamical state implied by the failure of equilibrium. But for the moment, consider only the question of the existence of solutions of (14.1).

We begin by noting that the only known solutions of (14.1) are invariant with respect to at least one coordinate, $\partial B_i/\partial\xi=0$ with $\xi=z$ for translational invariance and $\xi=\phi$ for rotational invariance (axial symmetry), as described in Chapter 6. The procedure adopted here will be to examine fields with topologies in the near neighbourhood of known equilibrium solutions, looking for solutions lacking invariance. Consider, for instance, the field $B_i(x, y)+\epsilon b_i(x, y, z)$ (where $\epsilon \ll 1$) in the neighbourhood of the invariant equilibrium $B_i(x, y)$. As we shall see, solutions to (14.1) exist only when $\partial b_i/\partial z = 0$, so that the total field $B_i + \epsilon b_i$ has the same invariance as B_i for equilibrium. A similar result obtains for axial symmetry ($\partial/\partial\phi = 0$). Close scrutiny of the internal conditions within the fields (Parker 1972; Yu 1973) shows that non-equilibrium, when the invariance is lacking, is of the form of neutral point reconnection of the lines of force.

To put the result in the context of the present geometry, equilibrium of the magnetic field extending across a volume of space with dimensions $2L$ large compared to the characteristic transverse scale of variation l of the field, exists only when the local winding pattern (the topology) of the field is invariant along the large-scale direction of the field.

To begin, then, we note that in a three-dimensional space the equilibrium equations (14.1) make up a set of four, sufficient to determine the allowed forms of the four unknown quantities p and B_i. We observe that the equations are quasi-linear (i.e. linear in the derivatives of p and B_i).

Consider, then, the characteristics of (14.1). Generally speaking, the specification of the four unknown quantities p and B_i along some curve in space together with the four equations (14.1) is sufficient to compute eight of the derivatives of p and B_i at any point on the curve. The characteristics are those special curves on which it is not possible to compute the derivatives from the information given. To carry out the calculation for the general case, consider the curve with direction cosines γ_i through some general point x_k. Denote by ds the element of arc length along the curve so that

$$dx_k = \gamma_k \, ds.$$

Then at the neighbouring point $x_k + dx_k$ on the curve the quantities p and B_i have changed by the specified amounts dp and dB_i, related to the derivatives by

$$dp = \partial p/\partial x_k \gamma_k \, ds,$$

$$dB_i = \partial B_i/\partial x_k \gamma_k \, ds.$$

Together with the equilibrium equations (14.1) these relations constitute eight linear equations in the derivatives $\partial p/\partial x_k$ and $\partial B_i/\partial x_k$, allowing computation of any eight of the twelve derivatives. Suppose, then, that in addition to specification of p and B_i on the curve, we also specify $\partial p/\partial z$, $\partial B_i/\partial z$, so that the eight derivatives of p and B_i with respect to x and y can be calculated. The characteristics are those curves along which the calculation is not possible, i.e. they are the curves for which the determinant of the coefficients vanishes. It may be seen by inspection of (14.1) that the determinant D of the coefficients of the derivatives is

$$D=\begin{vmatrix} 1 & 0 & 0 & -B_y & +B_y & 0 & B_z & 0 \\ 0 & 1 & 0 & +B_x & -B_x & 0 & 0 & B_z \\ 0 & 0 & 0 & 0 & 0 & 0 & -B_x & -B_y \\ 0 & 0 & 1 & 0 & 0 & 1 & 0 & 0 \\ \gamma_x & \gamma_y & 0 & 0 & 0 & 0 & 0 & 0 \\ 0 & 0 & \gamma_x & \gamma_y & 0 & 0 & 0 & 0 \\ 0 & 0 & 0 & 0 & \gamma_x & \gamma_y & 0 & 0 \\ 0 & 0 & 0 & 0 & 0 & 0 & \gamma_x & \gamma_y \end{vmatrix} \tag{14.2}$$

after removing superfluous factors of 4π. The first column represents the coefficients of $\partial p/\partial x$, the second column $\partial p/\partial y$, the third $\partial B_x/\partial x$, etc. Equating D to zero yields

$$(\gamma_x^2+\gamma_y^2)(B_y\gamma_x-B_x\gamma_y)^2=0. \tag{14.3}$$

The first factor is $(\mathrm{d}x/\mathrm{d}s)^2+(\mathrm{d}y/\mathrm{d}s)^2$ and represents the elliptic characteristics, $\mathrm{d}y=\pm i\,\mathrm{d}x$. The second factor represents the hyperbolic characteristics,

$$\mathrm{d}x/B_x=\mathrm{d}y/B_y$$

representing, obviously, the projection of the magnetic lines of force on the xy-plane.

The next step is to work out necessary conditions for equilibrium. We begin with the simple case of fields in the neighbourhood of a uniform field and progress from there to more complicated circumstances.

14.2. Fields in the neighbourhood of invariant topologies

Consider the magnetic field $B_i(x_k)$ that is a solution of the equilibrium equations (14.1) throughout some extended volume of dimensions L. All known solutions of the equilibrium equations have some invariance, generally translational or rotational. So consider the case that $\partial B_i/\partial z=0$. Then B_i is given by (6.4) and (6.9) in terms of the vector potential A with

$$B_x=+\partial A/\partial y, \qquad B_y=-\partial A/\partial x, \qquad B_z=B_z(A), \qquad P=P(A), \tag{14.4}$$

$$\nabla_{xy}^2A+F'(A)=0, \qquad F(A)=4\pi P(A)+\tfrac{1}{2}B_z^2(A). \tag{14.5}$$

The projection of the lines of force onto the xy-plane is just $A(x, y)=$ *constant.* We are curious to know whether there is an equilibrium magnetic field configuration, in the near neighbourhood of $B_i(x_k)$, which does not have the invariance $\partial/\partial z=0$. Consider, then, the total field

$$B_i(x, y)+\sum_{n=1}^{\infty}\epsilon^n{}_n b_i(x, y, z) \tag{14.6}$$

and the associated fluid pressure

$$P(A)+\sum_{n=1}^{\infty}\epsilon^n{}_np(x, y, z) \tag{14.7}$$

where ϵ is some suitably small dimensionless parameter. Substituting into (14.1) the terms first order in ϵ become

$$\frac{\partial}{\partial x_i}(4\pi{}_1p+B_j{}_1b_j)=B_j\frac{\partial{}_1b_i}{\partial x_j}+{}_1b_j\frac{\partial B_i}{\partial x_j}, \tag{14.8}$$

$$\partial{}_1b_j/\partial x_j=0. \tag{14.9}$$

The characteristics of these four equations are again just (14.3), of course. The equations second and higher order in ϵ are easily generated. It is readily shown that the equations for ${}_nb_i$ and ${}_np$ have the same characteristics (14.3) in terms of the fields of order $n-1$. The equations (14.8) and (14.9) are homogeneous in ${}_1b_i$ and ${}_1p$, and the coefficients of ${}_1b_i$ and ${}_1p$ are functions only of x and y. It follows at once that z is a separable coordinate and the z-dependence is exponential,

$${}_1b_i=\beta_i(x, y)\exp(ikz), \tag{14.10}$$

$${}_1p=q(x, y)\exp(ikz). \tag{14.11}$$

Indeed, this form applies to ${}_nb_i$ and ${}_np$ to all orders. We are interested in bounded solutions throughout the broad interior of a large volume. Hence those solutions for which k has an imaginary part are excluded because, if their amplitude is not zero at any point deep in the interior of the volume, then they increase without bound as $z^2<L^2$ becomes large without limit. The wave number k, then, must be real or zero.

To take up the simplest case first, suppose that $B_x=B_y=0$ while $B_z=B_0$ is a constant, so that ${}_1b_i$ represents perturbations on a uniform field. This is the basic problem posed by the scenario outlined in §14.1 and sketched in Fig. 14.1, in which the magnetic lines of a uniform field are manipulated by displacements at $z=+L$. From the mathematical point of view the example is particularly interesting because it can be worked out easily to all orders in ϵ.

With $B_x=B_y=0$, the hyperbolic characteristics (the magnetic lines of force) drop out, and the only factor in (14.3) that does not vanish automatically is $\gamma_x^2+\gamma_y^2$. Thus the equations are fully elliptic. This fact appears immediately in the solution.

The first order equations (14.8) and (14.9) reduce to

$$\frac{\partial}{\partial x_i}(4\pi{}_1p+B_0{}_1b_z)=B_0\frac{\partial{}_1b_i}{\partial z}, \qquad \frac{\partial b_{1j}}{\partial x_j}=0 \tag{14.12}$$

The nth order equations are readily shown to be

$$\frac{\partial}{\partial x_i}(4\pi\,{}_np+B_0\,{}_nb_z+{}_nf)=B_0\frac{\partial\,{}_nb_i}{\partial z}+{}_ng_i,\qquad \frac{\partial\,{}_nb_j}{\partial x_j}=0, \tag{14.13}$$

where

$$2\,{}_{2n}f={}_nb_j\,{}_nb_j+2\,{}_{n-1}b_j\,{}_{n+1}b_j+\ldots+2\,{}_1b_j\,{}_{2n-1}b_j, \tag{14.14}$$

$${}_{2n+1}f={}_nb_j\,{}_{n+1}b_j+{}_{n-1}b_j\,{}_{n+2}b_j+\ldots+{}_1b_j\,{}_{2n}b_j, \tag{14.15}$$

$${}_ng_i={}_1b_j\frac{\partial\,{}_{n-1}b_i}{\partial x_j}+{}_2b_j\frac{\partial\,{}_{n-2}b_i}{\partial x_j}+\ldots+{}_{n-1}b_j\frac{\partial\,{}_1b_i}{\partial x_j}, \tag{14.16}$$

$${}_1g_i=0.$$

There are a variety of ways to establish that the only solutions for ${}_np$ and ${}_nb_i$ involve $\partial/\partial z=0$. The divergence of (14.12), together with the vanishing of the divergence of ${}_1b_i$, yields

$$\nabla^2(4\pi\,{}_1p+B_0\,{}_1b_z)=0.$$

The only bounded solution in an infinite space is

$$4\pi\,{}_1p+B_0\,{}_1b_z=\textit{constant}.$$

Substitution of this into (14.12) yields $\partial\,{}_1b_i/\partial z=0$.

As an alternative procedure, note that the z-component of (14.12) is just the statement that $\partial p/\partial z=0$. Hence if $k\neq 0$, (14.11) requires that $q(x,y)=0$, and hence ${}_1p=0$. Then the first two components of (14.12) become

$$\partial\,{}_1b_z/\partial x-\partial\,{}_1b_x/\partial z=0,$$

$$\partial\,{}_1b_z/\partial y-\partial\,{}_1b_y/\partial z=0,$$

and all three components of the curl of ${}_1b_i$ vanish. Hence ${}_1b_i=-\partial\Phi/\partial x_i$ with $\nabla^2\Phi=0$. The only bounded solution is $\Phi=\textit{constant}$, for which case ${}_1b_i=0$. The only way to avoid this trivial solution is to put $k=0$, so that $\partial\,{}_1b_i/\partial z=0$.

Note that if $\partial\,{}_1b_z/\partial z=0$, then (14.9) becomes

$$\partial\,{}_1b_x/\partial x+\partial\,{}_1b_y/\partial y=0$$

so that ${}_1b_x$ and ${}_1b_y$ are expressible in terms of a vector potential ${}_1a(x,y)$,

$${}_1b_x=+\partial a/\partial y,\qquad {}_1b_y=-\partial a/\partial x. \tag{14.17}$$

Consider, then, the higher order equations. The divergence of (14.13) for $n=2$ yields

$$\nabla^2(4\pi\,{}_2p+B_0\,{}_2b_z+{}_2f)=\frac{\partial}{\partial x_j}\,{}_2g_j.$$

Note that ${}_2g_i$ depends only upon ${}_1b_i$ and so is independent of z. Then the derivative with respect to z yields

$$\nabla^2 \frac{\partial}{\partial z}(4\pi\,{}_2p + B_0\,{}_2b_z + {}_2f) = 0.$$

The only physically acceptable solution of Laplace's equation is a constant,

$$\frac{\partial}{\partial z}(4\pi\,{}_2p + B_0\,{}_2b_z + {}_2f) = \textit{constant}.$$

But if this is integrated with respect to z, it diverges as $z \to \pm\infty$ unless the constant is zero. Setting the constant equal to zero and integrating yields

$$4\pi\,{}_2p + B_0\,{}_2b_z + {}_2f = c(x, y) \tag{14.18}$$

where c is an arbitrary function of x and y. But with solutions of the form (14.10) and (14.11), (14.18) can be satisfied only if $k = 0$ (i.e. $\partial/\partial z = 0$) or if $c(x, y) = 0$. To avoid, if we can, the conclusion that $\partial/\partial z = 0$, suppose that $c(x, y) = 0$. Then (14.18) is satisfied for the moment. But note that the z-component of (14.13) is

$$\frac{\partial}{\partial z}(4\pi\,{}_2p + {}_2f) = {}_2g_z. \tag{14.19}$$

Note from (14.16) that ${}_2g_i$ depends only on ${}_1b_i$ and hence is a function only of x and y. Hence integrating (14.19) over z yields

$$4\pi\,{}_2p + {}_2f = {}_2g_z(x, y)z + g(x, y) \tag{14.20}$$

where $g(x, y)$ is an arbitrary function of x and y. Hence ${}_2g_z = 0$, if divergence as $z \to \pm\infty$ is to be avoided. With solutions of the form (14.10) and (14.11), the relation (14.20) requires either $k = 0$ or $g(x, y) = 0$. Again we avoid $\partial/\partial z = 0$ and assume the latter, leading to $4\pi\,{}_2p + {}_2f = 0$. Then (14.18) reduces to ${}_2b_z = 0$, and the end of the argument is near. The x- and y-components of (14.13) reduce to

$$B_0 \frac{\partial\,{}_2b_x}{\partial z} + {}_2g_x = 0, \qquad B_0 \frac{\partial\,{}_2b_y}{\partial z} + {}_2g_y = 0. \tag{14.21}$$

Note again that ${}_2g_i$ is independent of z because it depends only upon ${}_1b_i$. Then integration over z forces the conclusion that ${}_2g_x = {}_2g_y = 0$ if ${}_2b_x$ and ${}_2b_y$ remain bounded as $z \to \pm\infty$. Then it follows from (14.21) that $\partial\,{}_2b_x/\partial z = \partial\,{}_2b_y/\partial z = 0$. Altogether, then, we cannot avoid the conclusion that

$$\partial\,{}_2b_i/\partial z = 0, \qquad {}_2g_z = 0. \tag{14.22}$$

Having shown that ${}_2b_i$ is independent of z, it follows that ${}_3g_i$, which depends only upon ${}_1b_i$ and ${}_2b_i$, is independent of z. The entire sequence of arguments that lead to (14.22) for $n=2$ may now be applied to (14.13) for $n=3$, with the same result, etc. It follows, then, that

$$\partial\,{}_nb_i/\partial z=0, \qquad {}_ng_z=0 \tag{14.23}$$

for all n. Since $\partial\,{}_nb_j/\partial x_j=0$, it follows that ${}_nb_x$ and ${}_nb_y$ can be expressed in terms of a vector potential as $+\partial\,{}_na/\partial y$ and $-\partial\,{}_na/\partial x$, respectively.

It follows to all orders in n that the only equilibrium fields in the neighbourhood of a uniform field are invariant along the uniform field. The winding pattern of ${}_nb_i$ about the uniform field must not vary along B_0.

To explore the equilibrium solutions a little further note that if $k=0$ and $\partial\,{}_2b_i/\partial z=0$, then $c(x, y)$ and $g(x, y)$ in (14.18) and (14.20) need not vanish. Nor need we assume that ${}_ng_x={}_ng_y=0$. It was only in an attempt to avoid $k=0$ that they were put equal to zero. Equation (14.21) is not applicable because it was based on the unnecessary assumption that $c(x, y)=0$. We must go back to (14.13) and begin anew. For $n=2$, (14.13) reduces to

$$\frac{\partial}{\partial x}(4\pi\,{}_2p+B_0\,{}_2b_z+{}_1b_j\,{}_1b_j/2)=\left({}_1b_x\frac{\partial}{\partial x}+{}_1b_y\frac{\partial}{\partial y}\right){}_1b_x,$$

$$\frac{\partial}{\partial y}(4\pi\,{}_2p+B_0\,{}_2b_z+{}_1b_j\,{}_1b_j/2)=\left({}_1b_x\frac{\partial}{\partial x}+{}_1b_y\frac{\partial}{\partial y}\right){}_1b_y,$$

$$0={}_2g_z=\left({}_1b_x\frac{\partial}{\partial x}+{}_1b_y\frac{\partial}{\partial y}\right){}_1b_z.$$

Consider the last of these equations. In view of (14.17) the characteristics are $a(x, y)=$ *constant*. Hence

$${}_1b_z={}_1b_z(a). \tag{14.24}$$

The curl of the first two components yields

$$\left({}_1b_x\frac{\partial}{\partial x}+{}_1b_y\frac{\partial}{\partial y}\right)\nabla^2_{xy}a=0,$$

which has as its solution

$$\nabla^2_{xy}(a)=-h'(a) \tag{14.25}$$

where $h(a)$ is an arbitrary function of a and $h'(a)$ denotes the derivative with respect to the argument. The first two components can now be rewritten as

$$\frac{\partial}{\partial x}\{4\pi\,{}_2p+B_0\,{}_2b_z+\tfrac{1}{2}\,{}_1b_z^2(a)-h(a)\}=0,$$

$$\frac{\partial}{\partial y}\{4\pi\,{}_2p+B_0\,{}_2b_z+\tfrac{1}{2}\,{}_1b_z^2(a)-h(a)\}=0,$$

from which it follows that

$$h(a) = 4\pi\,{}_2p + B_0\,{}_2b_z + \tfrac{1}{2}\,{}_1b_z^2(a) + constant. \tag{14.26}$$

The equivalence of (14.24)–(14.26) to (14.4) and (14.5) is obvious. They represent the condition that the stresses introduced by ${}_1b_x$, ${}_1b_y$, and ${}_2b_z$ can be balanced by the pressures ${}_1p$ and ${}_2p$ that extend uniformly along the lines of force of the next lower order field.

Now it has been shown that $\partial\,{}_nb_i/\partial z = 0$ to all orders in n. But there are several obvious questions that confront us. First of all we wonder whether equilibrium requires that b_i be invariant in the z-direction or in the direction of the local lines of force of the equilibrium field B_i. The calculation for B_i in the z-direction does not distinguish the two possibilities. The question is non-trivial, as may be seen by considering a rope made in the conventional manner with three twisted strands of hemp wrapped around each other. Such a rope of magnetic field may be constructed by starting with an equilibrium field B_i in the form of a single straight twisted flux tube. Then introduce the small perturbation ϵb_i so that the original tube, imagined to be composed of three spiral sectors extending along the lines of force, has each sector twisted slightly to form a twisted strand. The final rope has a topological pattern that is invariant along the spiral lines of B_i, but the field is clearly not invariant in the z-direction. The question whether equilibrium exists is not answered by the calculations thus far.

We may go on to ask what is the necessary invariance for equilibrium with axial symmetry ($\partial/\partial\phi = 0$) etc. instead of translational symmetry ($\partial/\partial z = 0$). With axial symmetry the problem is essentially one of equilibrium confinement of plasma in a toroidal magnetic configuration in the laboratory. It has been studied at length and it can be shown that there is no equilibrium unless the lines of force are confined to closed toroidal surfaces (see, for instance, Artsimovich 1964; Yu 1973).

In an entirely different direction it must be noted that we are searching for general equilibrium. Hence we can expect to find only those more general equilibria that are analytic functions of ϵ. The expansion cannot discover equilibria which do not converge uniformly to the zero order equilibrium. Thus there may be special cases, or even whole classes, of solutions that slip through our mathematical net. It would be interesting if examples could be found. To explore the more general questions, the sections that follow are directed toward exploration of equilibrium requirements in the neighbourhood of non-uniform equilibrium fields of the general form (14.4) and (14.5).

14.3. Equilibrium in the neighbourhood of $B_z(x, y)$

Having determined $\partial b_i/\partial z = 0$ to be a necessary condition for equilibrium of a field in the neighbourhood of a uniform field, as described in

§14.1, let us go on to consider fields in the near neighbourhood of $B_x = B_y = 0$, $B_z = B_z(x, y)$. The fluid pressure is given by the zero order equilibrium

$$P(x, y) + B_z^2(x, y)/8\pi = constant. \tag{14.27}$$

We suppose for convenience that, although $B_z(x, y)$ may vary widely, it does not vanish and change sign. Write the field as $(\epsilon b_x, \epsilon b_y, B_z + \epsilon b_z)$ with $b_i = b_i(x, y, z)$ and the total pressure as $P + \epsilon p(x, y, z)$. Then to first order in ϵ, the equilibrium equations are

$$\frac{\partial}{\partial x}(4\pi p + B_z b_z) = B_z \frac{\partial b_x}{\partial z} \tag{14.28}$$

$$\frac{\partial}{\partial y}(4\pi p + B_z b_z) = B_z \frac{\partial b_y}{\partial z} \tag{14.29}$$

$$\frac{\partial}{\partial z} 4\pi p = \left(b_x \frac{\partial}{\partial x} + b_y \frac{\partial}{\partial y}\right) B_z \tag{14.30}$$

$$\partial b_j / \partial x_j = 0. \tag{14.31}$$

These four equations have only the elliptic characteristics $\gamma_x^2 + \gamma_y^2 = 0$. It follows from (14.30) and (14.31) that

$$\frac{\partial}{\partial z}(4\pi p + B_z b_z) = -\Psi \tag{14.32}$$

where

$$\Psi = B_z^2 \left(\frac{\partial}{\partial x}\frac{b_x}{B_z} + \frac{\partial}{\partial y}\frac{b_y}{B_z}\right). \tag{14.33}$$

Then differentiate (14.28) and (14.29) with respect to z and use (14.32) to eliminate $4\pi p + B_z b_z$. The result is

$$\frac{1}{B_z^2}\frac{\partial \Psi}{\partial x} + \frac{\partial^2}{\partial z^2}\frac{b_x}{B_z} = 0, \qquad \frac{1}{B_z^2}\frac{\partial \Psi}{\partial y} + \frac{\partial^2}{\partial z^2}\frac{b_y}{B_z} = 0.$$

Differentiate the first of these with respect to x and the second with respect to y and add. The result can be written

$$\frac{\partial}{\partial x}\frac{1}{B_z^2}\frac{\partial \Psi}{\partial x} + \frac{\partial}{\partial y}\frac{1}{B_z^2}\frac{\partial \Psi}{\partial y} + \frac{\partial}{\partial z}\frac{1}{B_z^2}\frac{\partial \Psi}{\partial z} = 0.$$

This form is totally elliptic. In an infinite space its only bounded solutions are constants,

$$\Psi = C$$

Hence from (14.33) it follows that

$$\frac{\partial}{\partial x}\frac{b_x}{B_z} + \frac{\partial}{\partial y}\frac{b_y}{B_z} = \frac{C}{B_z^2}.$$

This relation states that the divergence of the vector b_x/B_z, b_y/B_z has the same sign for all x and y. But b_x/B_z and b_y/B_z must be bounded in the x- and y-directions, requiring that $C=0$. Hence b_x and b_y are expressible in terms of a stream function, or vector potential, a,

$$b_x = +B_z\ \partial a/\partial y, \qquad b_y = -B_z\ \partial a/\partial x.$$

Now differentiate (14.28) with respect to y and (14.29) with respect to x and subtract, so as to eliminate $4\pi p + B_z b_z$. The result is

$$ik\left(\frac{\partial}{\partial y} B_z b_x - \frac{\partial}{\partial x} B_z b_y\right) = 0$$

which becomes

$$ik\left(\frac{\partial}{\partial x} B_z^2 \frac{\partial a}{\partial x} + \frac{\partial}{\partial y} B_z^2 \frac{\partial a}{\partial y}\right) = 0.$$

If ik (in (14.10) and (14.11)) does not vanish, then the only bounded solution is $a = constant$, yielding $b_i = 0$. It is evident then that the only non-trivial solutions are of the form

$$\partial b_i/\partial z = 0, \qquad \text{i.e. } k = 0.$$

It follows from (14.28) and (14.29) that

$$4\pi p(x, y) + B_z(x, y) b_z(x, y) = constant.$$

It follows from (14.31) that b_x and b_y can be written

$$b_x = +\partial\theta/\partial y, \qquad b_y = -\partial\theta/\partial x.$$

Then (14.30) becomes

$$\frac{\partial\theta}{\partial y}\frac{\partial B_z}{\partial x} - \frac{\partial\theta}{\partial x}\frac{\partial B_z}{\partial y} = 0,$$

so that

$$\theta = \theta(B_z) \quad \text{or} \quad B_z = B_z(\theta),$$

etc.

14.4. Coordinates and equations for the general field invariant to translation

Consider the equilibrium of magnetic fields in the neighbourhood of the general equilibrium field described by (14.4) and (14.5) representing the most general equilibrium with the translational invariance $\partial/\partial z = 0$. To set up the problem in unambiguous form, restrict attention to regions within the equilibrium field where there are no special directions besides the z-direction. Thus, for instance, there is no net flux extending in the x- or y-direction across the region. For if there were, the flux would define one or more directions besides the z-direction and the problem becomes much more complex. Thus, within the region essentially all of the lines of force of $B_x = +\partial A/\partial y$, $B_y = -\partial A/\partial x$ form closed curves, $A = constant$. No more than an occasional singular line, representing no net magnetic flux, penetrates across the region. The net flux in the x- (or y-) direction must average to zero over dimensions of the order of the finite scale of variation l of B_i. The closed curves themselves have dimensions of the order of l. Figure 14.2 is a sketch of the topology of the lines of force of B_i projected onto the xy-plane. The lines of force $A(x, y) = constant$ may be thought of as level contours drawn on terrain with elevation $A(x, y)$. All contours, except for an occasional special singular path, form closed curves and do not lead across the 'hills'.

The equations for the first-order deviation b_i from B_i are (14.8) and (14.9) whose solutions are of the form (14.10) and (14.11). Note from (14.3) that the lines of force, given by $A(x, y) = constant$, in the xy-plane are characteristics of (14.8) and (14.9). Indeed they are degenerate characteristics. This suggests that $A(x, y)$ would be a useful coordinate. Introduce the orthogonal coordinate $S(x, y)$, so that $\nabla S \,.\, \nabla A = 0$ and

$$\frac{\partial S}{\partial x} = +f\frac{\partial A}{\partial y}, \qquad \frac{\partial S}{\partial y} = -f\frac{\partial A}{\partial x} \tag{14.34}$$

FIG. 14.2. A sketch of the projection of the lines of force of B_i onto the xy-plane. The solid curves represent the contours $A = constant$, while the dashed curves represent the orthogonal coordinate lines $S = constant$.

where f is a function of x and y. It is always possible to find an $f(x, y)$ such that the orthogonal coordinate $S(x, y)$ can be constructed. The mathematical problem is equivalent to finding the integrating factor for the ordinary differential expression of the form $P\,dx + Q\,dy = 0$. Differentiating $\partial S/\partial x$ with respect to y and $\partial S/\partial y$ with respect to x, it follows from (14.34) that

$$\frac{\partial\{\ln(f)\}}{\partial x}\frac{\partial A}{\partial x} + \frac{\partial\{\ln(f)\}}{\partial y}\frac{\partial A}{\partial y} = F'(A)$$

upon using (14.5) to replace $\nabla_{xy}^2 A$. The solution of this equation for f can always be written in terms of the integral of the inhomogeneous term along the characteristics $S = constant$. The curves $S = constant$ represent the paths that run straight up and down the hills $A(x, y)$, as indicated by the dashed curves in Fig. 14.2. For use in later manipulations note that

$$B_x\frac{\partial}{\partial x} + B_y\frac{\partial}{\partial y} = f\,|\nabla A|^2\frac{\partial}{\partial S}. \tag{14.35}$$

Now the pressure P is constant along the magnetic lines of force of the field B_i,

$$B_j\,\partial P/\partial x_j = 0.$$

The pressure $P + \epsilon p$ is uniform along the lines of force of the neighbouring configuration $B_i + \epsilon b_i$ so that to first order

$$B_j\,\partial p/\partial x_j + b_j\,\partial P/\partial x_j = 0. \tag{14.36}$$

This same expression can be deduced directly from (14.8) by forming the scalar product with B_i.

14.5. Equilibrium in the neighbourhood of a force-free field

As a preliminary exercise to the assault on the general case, consider the circumstance that B_i is a force-free field, so that $\partial P/\partial x_i = 0$. The perturbation ϵb_i is not restricted to the force-free form, of course, so the determinant for the characteristics of (14.8) and (14.9) is again (14.2), giving both elliptic and hyperbolic characteristics. As we shall soon see, the hyperbolic characteristics (the lines of force $A = constant$) play the deciding role in determining equilibrium.

Given that $\partial P/\partial x_i = 0$, (14.36) reduces to

$$B_j \partial p/\partial x_j = 0 \tag{14.37}$$

Using (14.11) and (14.35) this can be written as

$$f\,|\nabla A|^2\,\partial p/\partial S + ikB_z(A)p = 0. \tag{14.38}$$

Integrating over S along the contour of constant A, starting at some suitable position $S = S_0$, leads to

$$p(A, S) = p(A, S_0)\exp\left\{-ikB_z(A)\int_{S_0}^{S} dS/f\,|\nabla A|^2\right\} \qquad (14.39)$$

But, as noted above, the contours $A = constant$ are all closed, so that the integration proceeds all the way around and back to the starting place at $S = S_0$. The pressure p is a physical quantity, finite, continuous, and single valued. Hence the only acceptable solutions (14.39) are those which are periodic around the contour, returning $p(A, S)$ to its initial value $p(A, S_0)$ upon arriving back at the starting point. This requires

$$kB_z(A)\oint dS/f\,|\nabla A|^2 = \pm 2n\pi \qquad (14.40)$$

where $n = 0, 1, 2, \ldots$ To illustrate the consequences of (14.40) note that the integral can be transformed into a conventional line integral by writing

$$\begin{aligned} dS &= \frac{\partial S}{\partial x}dx + \frac{\partial S}{\partial y}dy \\ &= \nabla S \cdot \mathbf{ds} \\ &= |\nabla S|\,ds \\ &= f\,|\nabla A|\,ds \end{aligned}$$

where ds is the element of arc length $\{(dx)^2 + (dy)^2\}^{\frac{1}{2}}$ along the contour, and the vector arc length $\mathbf{ds}$ is in the direction of ∇S. Then (14.40) becomes

$$kB_z(A)\oint ds/|\nabla A| = \pm 2n\pi. \qquad (14.41)$$

The integrand is positive and the integral non-zero. The combination $B_z(A)\oint ds/|\nabla A|$ is not an integral of (14.4) and (14.5); it is a continuous function of A and varies continuously from one contour to the next. Therefore, if (14.40) or (14.41) is satisfied on some special contour $A = A_0$, it is generally not satisfied on any neighbouring contour $A = A_0 + \delta A_0$. The only way that (14.41) can be satisfied on all contours is to put $k = 0$. Then $n = 0$ and $p(A, S)$ is a function only of A. We arrive at the conclusion that the only physically acceptable (single-valued) solutions of the equilibrium equations (14.4) and (14.5) are those for which $k = 0$, i.e.

$$\partial b_i/\partial z = 0 \qquad (14.42)$$

The equilibrium field is invariant with respect to the principal direction of the field. In that case the total field $B_i + \epsilon b_i$ is just another case of (14.4) and (14.5).

14.6. Equilibrium in the neighbourhood of the general invariant field

Now we are ready to turn to the general problem posed in §14.2. Consider a magnetic field $B_i(x, y)+\epsilon b_i(x, y, z)$ in the neighbourhood of the general equilibrium field $B_i(x, y)$ satisfying (14.4), (14.5), and (14.36). Then in place of (14.37) and (14.38), (14.36) leads to

$$f\,|\nabla A|^2 \frac{\partial p}{\partial S}+ikB_z(A)p+\left(b_x\frac{\partial A}{\partial x}+b_y\frac{\partial A}{\partial y}\right)P'(A)=0 \qquad (14.43)$$

upon noting (14.35) and (14.4). This equation, like (14.38), can be integrated to give $p(A, S)$ in terms of an integral along $A=constant$.

Note that

$$b_x\frac{\partial A}{\partial x}+b_y\frac{\partial A}{\partial y}=b_\perp\,|\nabla A|$$

where $b_\perp$ is the component of (b_x, b_y) perpendicular to the contour $A=constant$. Define the function

$$H(A, S)=B_z(A)/f\,|\nabla A|^2 \qquad (14.44)$$

which is specified entirely by the choice of B_z. Then define the function

$$r(A, S)=-b_\perp P'(A)/f\,|\nabla A|.$$

Since r is made up of the magnetic fields b_x and b_y, it must be single-valued on each contour $A=constant$. That is to say, $r(A, S)$ is a periodic function around each closed contour.

Altogether, then, the differential equation (14.44) can be written

$$\partial p/\partial S+ikH(A, S)p=r(A, S). \qquad (14.45)$$

In terms of the pressure $p(A, S_0)$ at one point S_0 on the contour $A=constant$, the pressure elsewhere on the contour is

$$p(A, S)=\left[p(A, S_0)+\int_{S_0}^{S} dS'r(A, S')\exp\left\{ik\int_{S_0}^{S'} dS''H(A, S'')\right\}\right] \times\exp\left\{-ik\int_{S_0}^{S} dS'H(A, S')\right\}, \qquad (14.46)$$

To be compared with the simpler expression (14.39) arising when B_i is a force-free field.

Now, as in §14.5, the only physically acceptable solutions for $p(A, S)$ are those that are continuous, single-valued functions of S around each contour $A=constant$, i.e. periodic around each contour. The solutions (14.46) of (14.45) generally do not have this property, because the inhomogeneous term $r(A, S)$ in (14.45) has just the required periodicity, so the *solution* of the inhomogeneous equation does not have the required periodicity for almost all values of A. It is a special case of Floquet's theorem (see, for instance, Ince 1926). The only way to avoid the problem is to put $k = 0$ so that

$$p(A, S) = p(A, S_0) + \int_{S_0}^{S} dS' r(A, S').$$

Suitable choice of $b_\perp$, so that $\oint dS' r(A, S') = 0$, leads to a single-valued solution. The total field $B_i + \epsilon b_i$ is then independent of z and is just another example of the family of two-dimensional equilibria, from which B_i was chosen in the first place.

It is instructive in the present circumstance to explore the breakdown of the solution in some detail, rather than appealing to general mathematic theorems. The difficulty with the solution for $p(A, S)$ when $k \neq 0$ arises from the fact that there are two ways in which the pressure $p(A, S)$ at $kz = 2\pi$ can be calculated in terms of the pressure distribution at $kz = 0$. One method is to note from (14.10) and (14.11) the general periodicity of the pressure perturbation. The other is to note that the pressure is uniform along the magnetic lines of force. The two methods must give the same result.

Consider, then, the contour

$$A(x, y) = A_0$$

in the $z = 0$ plane. Given that the pressure is $p(A_0, S)$ around this contour at $z = 0$, it follows from (14.11) that the pressure is also $p(A_0, S)$ at $kz = 2\pi$. Thus at both $z = 0$ and $z = 2\pi/k$ the total pressure on $A = A_0$ is

$$P(A_0) + \epsilon p(A_0, S). \tag{14.47}$$

But we also can calculate the pressure at $kz = 2\pi$ from the fact (14.36) that the pressure $P + \epsilon p$ is constant along the magnetic lines of force. Consider, then, the magnetic lines of force of $B_i + \epsilon b_i$ that pass through $A = A_0$ at $z = 0$. To a first approximation, neglecting terms $O(\epsilon)$, they are the lines of B_i that spiral around the cylindrical surface (14.46) as they extend in the z-direction (see Fig. 14.3). The line through (A_0, S) at $z = 0$ rotates around to $[A_0, S + \Delta S(A_0, S)]$ at $k_z = 2\pi$. Of course, with the inclusion of the field ϵb_i the lines deviate slightly from this, by the amount

$$d\Delta x_i/dz = \epsilon b_i/B_z(A) + O(\epsilon^2),$$

integrated along the zero order line of force,

$$dx/dz = B_x/B_z(A), \qquad dy/dz = B_y/B_z(A),$$
$$A(x, y) = A_0.$$

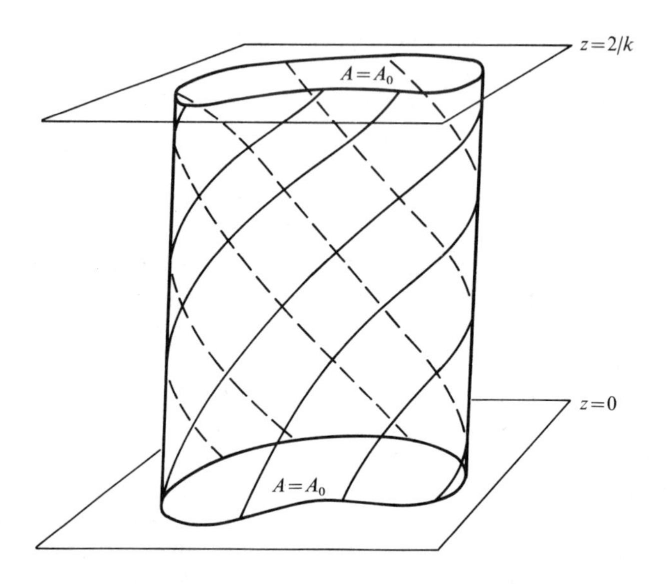

FIG. 14.3. A sketch of the cylinder $A(x, y) = A_0$ between the planes $z = 0, 2\pi/k$. The spiral lines represent the magnetic lines of force of the basic equilibrium field B_i.

From (14.10) it follows that

$$d\Delta x_i/dz = \{\epsilon q_i(x, y)/B_z(A_0)\}\exp(ikz).$$

The integration of $\exp(ikz)$ alone would give $\Delta x_i = 0$ at $k_2 = 2\pi$, but x and y are functions of z as well, so that in general Δx_i does not vanish at $k_z = 2\pi$. Hence the curve (14.46) at $z = 0$ maps into some slightly different contour,

$$A(x, y) + \epsilon a(x, y) = A_0 \tag{14.48}$$

at $k_z = 2\pi$. The pressure (14.47) at $z = 0$ maps along the lines of force onto the contour (14.48) so that it is

$$P(A_0 - \epsilon a) + \epsilon p\{A_0, S + \Delta S(A_0, S)\}$$
$$= P(A_o) + \epsilon p\{A_0, S + \Delta S(A_0, S)\} - P'(A_0)\epsilon a \tag{14.49}$$

at $kz = 2\pi$, neglecting terms $O(\epsilon^2)$.

Now the total magnetic flux through (14.48) at $kz = 2\pi$ is the same as through (14.46) at $kz = 0$. Hence for most choices of A_0, the contour (14.48) does not lie wholly outside, or wholly inside, (14.46). For if it lay outside it would encompass more flux, and if inside, less flux.

If, for instance, the contour (14.48) lies a small distance $\epsilon h(A_0, S)$

outside (14.46), then the extra flux is

$$\delta\Phi = \epsilon \oint \mathrm{d}s\, h(A_0, S) B_z(A_0) + O(\epsilon^2)$$
$$= \epsilon B_z(A_0) \oint \frac{\mathrm{d}S\, h(A_0, S)}{f|\nabla A|} + O(\epsilon^2).$$

The flux of ϵb_z contributes only to second order in ϵ. So (14.48) must either coincide exactly with (14.46), or lie partially inside and partially outside (14.46), crossing (14.46) in at least two places.

If (14.48) coincides with (14.46), then $a(x, y) = 0$ and (14.49) reduces to

$$p(A_0) + \epsilon p\{A_0, S + \Delta S(A_0, S)\}, \tag{14.50}$$

which must be identical with (14.47)[1]. Remembering that $p(A_0, S)$ is periodic around the contour (14.46), it is evident that this expression does not agree with (14.46) unless ΔS represents an integral number of revolutions around the cylinder (14.46) as the lines of B_i pass from $kz = 0$ to $kz = 2\pi$. Generally $\Delta S(A_0, S)$ is a continuous function of A_0, varying continuously within some finite range of values. Thus for a given choice of k there may well be one or more values of A_0 for which ΔS represents an integral number of rotations. But for most values of the continuum A_0, ΔS does not. It follows, then, that the projection of the pressure along the lines of force gives a pressure variation (14.50) at $kz = 2\pi$ that does not agree with (14.47) obtained from the basic form (14.10) and (14.11) of the solutions of the equilibrium equations (14.8) and (14.9). The only way to avoid the contradiction is to move $z = 2\pi/k$ out of the region, all the way to infinity, by putting $k = 0$.

On the other hand, suppose that the contour (14.48) does not coincide with (14.46). Then the contours must cross in at least two places, call them (A_0, S_1) and (A_0, S_2). Since the two points lie on (14.46), it follows that $a = 0$ at the two points. Hence (14.49) reduces to (14.50) at (A_0, S_1) and (A_0, S_2), which must correspond to (14.47) at the same points. But this can be if and only if $\Delta S(A_0, S_1)$ and $\Delta S(A_0, S_2)$ both represent an integral number of rotations around the contour. Again, there may be special values of A_0 on which ΔS somewhere assumes integral values, but for most A_0, $\Delta S(A_0, S_1)$ and $\Delta S(A_0, S_2)$ are not integers, and the pressure projected along the lines of force does not agree with the pressure (14.47) projected along the coordinates. The only resolution of the difficulty is to push $z = 2\pi/k$ off to infinity, putting $k = 0$. Thus, in the general case, we are led to the conclusion that the invariance

$$\partial b_i / \partial z = 0 \tag{14.51}$$

[1] Note that in the force-free field treated in §14.5, for which $P'(A) = 0$, (14.49) reduces to (14.50) regardless of whether a is zero or not.

is a necessary condition for equilibrium. Any field in which the winding pattern changes along the field, so that (14.51) is excluded by the topology, cannot be in equilibrium.

14.7. Discussion of equilibrium and non-equilibrium

The development of the invariance requirement for magnetostatic equilibrium has proceeded through a sequence of special cases, culminating in the statement (14.51) for the general field (14.6). The analysis is general, assuming only that the magnetic field is analytic in its deviation ϵ from the invariant field $B_i(x, y)$.

The next question would be, of course, whether there exist special equilibrium solutions of (14.1) that are not analytic functions of ϵ in the neighbourhood of the invariant equilibrium field $B_i(x, y)$ as assumed in (14.6) and (14.7). This would be equivalent to the solution with $\epsilon = O(1)$. We conjecture that there may be no clear answer because if ϵ is not small, the whole question of an invariant direction becomes ambiguous. But it would be interesting to see if one or more counter examples could be constructed.

The calculations show that a magnetic field extending in the z-direction across a broad volume of space can be in equilibrium only if the pattern of winding of the lines of force about each other does not vary in the z-direction along the field. This invariance is a very special requirement and, in view of the discussion in Chapter 11, it would appear that there are no circumstances in nature where one could expect magnetic fields with the necessary invariant topology. Chapter 11 presented illustrations of the universal destruction of invariance by the common inhomogeneous velocity field and diffusion coefficient. The conclusion is thrust upon us that in most astrophysical circumstances the magnetic forces cannot be put into balance with the gas pressure: Without an invariant topology, $(\nabla \times \mathbf{B}) \times \mathbf{B}$ cannot be expressed as the gradient of a scalar. The fluid is accelerated.

Now in a highly conducting ideal fluid the general topology over large dimensions is frozen into the fluid. Fluid motions do not alter the topology. So the non-equilibrium condition continues, and the fluid acceleration continues, without eliminating the topological basis for the non-equilibrium.

It appears that the topological non-equilibrium is the basis for the continued activity and dissipation of the variable magnetic fields in the universe. It is essential, therefore, to understand the nature of the non-equilibrium if we are to understand the universal activity of the ubiquitous magnetic field. First of all, as is clear from the formal arguments presented in §14.5, the non-equilibrium arises because the fluid pressure is invariant along the lines of force of the magnetic field. The lines of force generally rotate, or spiral, about each other as they progress across the region, with the result that the pressure pattern spirals too. It is, then, incompatible for the rotation pattern to change along the field. For if there are two or more patterns, then the pressure configuration of one is projected along the lines into the other, with which it is generally not compatible. Figure 11.4 (p. 296) is an example of the wandering of a magnetic line as a result of a simple repetitive change of winding pattern. It is readily seen that pressure pattern is soon scrambled as it is projected along the lines of force.

There is another way to demonstrate the impossibility of equilibrium in changing winding patterns, based on a theorem worked out by Yu (1973). We begin by remarking that in any winding pattern the separation of neighbouring lines on any given cylindrical surface (14.46) undergoes no progressive increase along the field. After each revolution their separation returns to the same value. But in progressing from one winding pattern to another, two neighbouring lines travelling together with fixed separation in one pattern generally are not so fortunate as to remain close travelling companions in the other pattern. In most cases their separation increases without bound so long as the second pattern continues.

To show the difficulty that arises from this, consider the neighbouring isobaric surfaces p and $p+\delta p$, separated by an infinitesimal distance δh. Consider the magnetic flux tube (confined between the two isobaric surfaces) whose cross-section is a parallelogram, with infinitesimal altitude δh and width δw, as sketched in Fig. 14.4. The vertices of the parallelogram are defined by two lines of force separated by δw in each isobaric surface. The width and altitude are presumed to vary along the tube. The cross-sectional area is $\delta h\,\delta w$, and conservation of magnetic flux

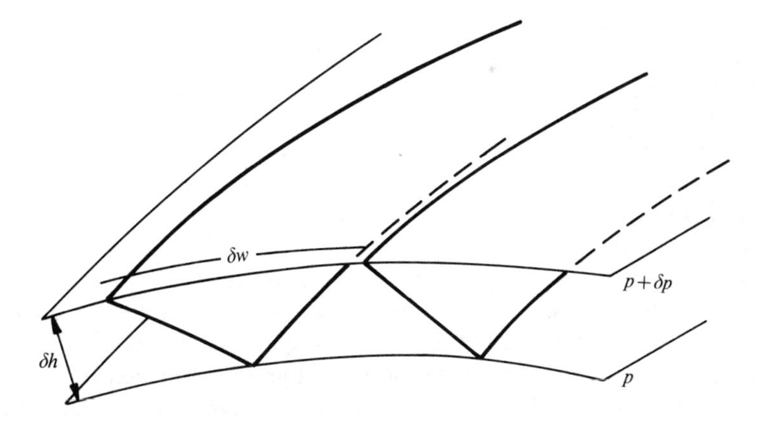

FIG. 14.4. A sketch of the lines of force in the two isobaric surfaces p and $p+\delta p$.

leads to the relation

$$B\,\delta h\,\delta w = \delta\Phi_0$$

with the field density B, where $\delta\Phi_0$ is a constant. The pressure gradient $\delta p/\delta h$ varies inversely with δh. For hydrostatic equilibrium the pressure gradient is balanced by the Lorentz force $\mathbf{j}\times\mathbf{B}/c$. If we denote by $j_\perp$ the component of the current density perpendicular to $\mathbf{B}$, then equating the pressure gradient and the Lorentz force yields

$$\delta p/\delta h = j_\perp B/c$$

$$= j_\perp\,\delta\Phi_0/c\,\delta h\,\delta w.$$

Hence

$$j_\perp = c\,\delta p\,\delta w/\delta\Phi_0.$$

Since δp is constant along the tube, it follows that $j_\perp$ is proportional to the separation δw of the two lines of force in each isobaric surface that bound the tube.

Now we expect that the current density $j_\perp$ is bounded. Indeed, from the fact that $4\pi\mathbf{j} = c\nabla\times\mathbf{B}$ it follows that $j_\perp$ is bounded by something of the order of $c\,|\mathbf{B}|/4\pi l$ where l is the scale of variation of $\mathbf{B}$. But if $j_\perp = O(cB/l)$ in the pattern where the separation is δw, then in the other pattern where the separation increases without bound, hydrostatic equilibrium leads to the conclusion that $j_\perp$ increases without bound. The unbounded $j_\perp$ arises from the unbounded increase of $\delta p/\delta h$ as δw increases. The isobaric surface $p+\delta p$ approaches the surface p.

This, then, is another way to illustrate the catastrophe that occurs when the topological pattern of the field varies along the lines of force. The unbounded increase in $j_\perp$ implies an unbounded increase in field gradients—more precisely, an unbounded increase in the curl of the field.

To show a specific example of the non-equilibrium in its simplest form, consider a long straight flux tube of radius $\varpi = a$ and uniform longitudinal field B_0 extending along the z-azis. The magnetic field is embedded in a fluid in which the pressure P is adjusted (across the distant ends of the tube) to be uniform with value P_0. The electrical conductivity is so high that the resistive decay across the diameter of the tube—with a characteristic time $a^2\sigma/c^2$—is too slow to have sensible effects. Now suppose that the tube is twisted slightly and uniformly so that the line of force through the point (ϖ_0, ϕ_0) at $z=0$ has coordinates $(\varpi_0, \phi - kz)$ elsewhere along the tube. Then imagine that the infinitely conducting spiral partition $\phi = kz$ is inserted between the lines of force along the tube, dividing the tube into two equal spiral halves. The field is continuous across the partition, of course. The tube and the spiral partition are sketched in Fig. 14.5. We suppose that the spiral is gentle ($ka \ll 1$), making only the small angle ka with the z-direction. Finally, then, suppose that the field in each side of the spiral partition is twisted

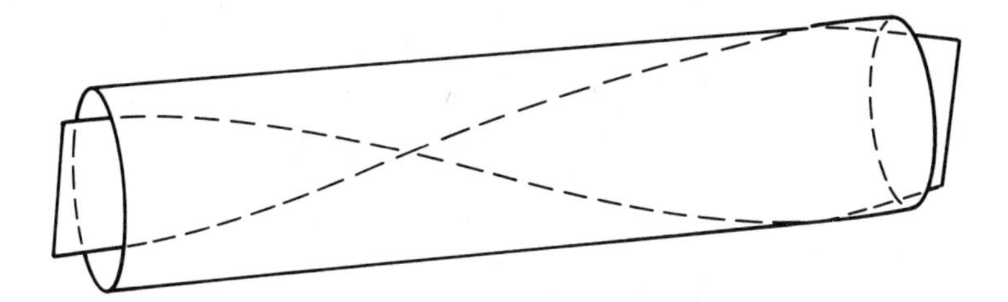

FIG. 14.5. A sketch of the flux tube confined within the cylinder $\varpi \leq a$ and divided by the spiral partition $\phi = kz$.

separately but equally. Then neither side has the twist $\phi = kz$ anymore, with the result that the field jumps discontinuously across the partition $\phi = kz$. It follows that the field pattern is no longer invariant with respect to the coordinate z: If nothing else, the field jumps discontinuously along any line $\varpi = \varpi_0$, $\phi = \phi_0$ where the line intersects the partition at $kz = \phi_0 \pm n\pi$ where $n = 0, 1, 2, 3, \ldots$.

The local equilibrium of the strongly twisted field on either side of the partition is affected but little by the gentle spiral of the partition ($ka \ll 1$), so that the equilibrium is described locally by (14.4) and (14.5), neglecting terms $O(ka)$. For the special choice

$$F(A) = \tfrac{1}{2}K^2A^2 + 4\pi P_0 + \tfrac{1}{2}B_0^2 \tag{14.52}$$

it follows that

$$A = \mp B_1 J_1(K\varpi)\sin(\phi - kz) \tag{14.53}$$

$$B_\varpi = \mp B_1 \frac{J_1(K\varpi)}{K\varpi}\cos(\phi - kz) \tag{14.54}$$

$$B_\phi = \pm B_1 J_1'(K\varpi)\sin(\phi - kz) \tag{14.55}$$

$$B_z = B_0, \qquad P = P_0 + K^2A^2/8\pi.$$

The upper sign is to be used on one side of the partition and the lower on the other, depending upon which way the two bundles of field are twisted. The wave number K is chosen so that Ka is the first zero of J_1 to satisfy the boundary condition $B_\varpi = 0$ on $\varpi = a$. The projection of the lines of force onto the $(\varpi, \phi - kz)$-plane is shown in Fig. 14.6. There are two important points to note. Both are a consequence of the general topology of the field and depend in no way upon the special form (14.52) employed in the present illustration. The first is that B_ϖ reverses sign discontinuously across the partition $\phi = kz$. The second is that there is a line of force ($A = 0$) extending around the boundary $\varpi = a$, $\phi = kz$ of each bundle of field. It follows that the fluid pressure $P(A)$ is uniform around the boundary.

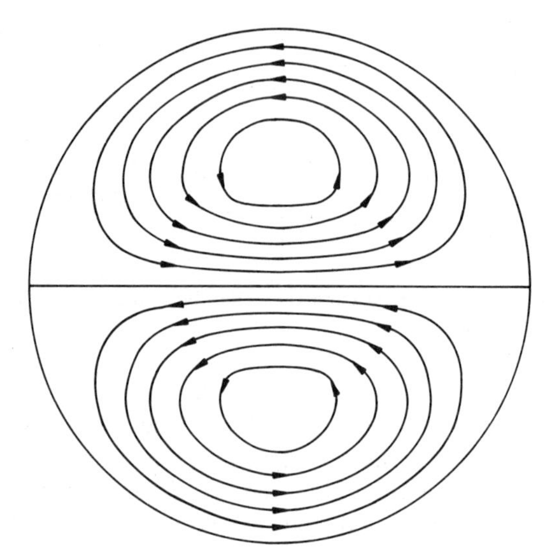

FIG. 14.6. A map of the lines of force, $A(x, y) = constant$, of the twisted fields on either side of the spiral partition.

Now remember that the two twisted bundles of flux on either side of the partition spiral around each other. It is clear that the tension $(O(B_0^2/8\pi)$ assuming that $B_1 \ll B_0)$ in each bundle presses the two together as they spiral about each other. The tension also pulls each bundle away from the outer wall $\varpi = a$ slightly. Hence, as a result of the general spiral ka, the pressure of the field exerted on $\phi = kz$ is slightly higher than the pressure on the outer surface $\varpi = a$. There is, then, a minor readjustment of the equilibrium, compressing and increasing B_ϖ slightly at the partition and reducing B_ϕ slightly at the outer boundary.

Now remove the partition so that no material boundary interferes with the equilibrium of the fields. The field B_ϖ changes sign discontinuously across $\phi = kz$. But, of course, for any *finite* electrical conductivity, no matter how large, B_ϖ goes rapidly but continuously through zero across $\phi = kz$. No matter how small the resistivity η, after some very long time t the field falls to zero across a thin layer of thickness $(\eta/t)^{\frac{1}{2}}$ at $\phi = kz$. With the decay of the field across this thin transition layer, the full force of the field pressure on either side is exerted on the fluid there. This is, then, similar to the non-equilibrium at the ends of the cylinder discussed in §12.2. The fluid on $\phi = kz$ is squeezed harder, by an amount of the order of $kaB_0^2/8\pi$, and on $\varpi = a$ the fluid is squeezed less. But the lines of force extend around $\phi = kz$ and $\varpi = a$, so if the fluid is squeezed harder in the thin spiral ribbon $\phi = kz$ than at the outer wall of the cylinder, the fluid flows out of the thin layer between the two fields, along the lines of force into the region of reduced pressure at the outer wall. This readjustment of fluid in no way relieves the situation, of course. The tension continues to press the two bundles together, so that the fluid continues to squeeze out from between the two opposite B_{ϖ} on either side of $\phi = kz$. The gradient $(1/\varpi)\partial B_{\varpi}/\partial\phi$ becomes steeper without limit. Eventually, then, no matter how high the electrical conductivity, the gradient becomes so steep that dissipation proceeds at a pace limited only by the rate at which the fluid squeezes out from between the two opposite B_{ϖ}. The dissipation continues so long as the opposing radial components B_{ϖ} face each other across $\phi = kz$, i.e. so long as the two bundles have relative torsion. The dissipation is at the expense of B_{ϖ} (and the associated B_{ϕ}), and ceases only when the torsion in the two bundles is reduced to the same value $\phi = kz$ as the winding of the bundles about each other. In that case $\partial B_i/\partial z = 0$ and equilibrium is achieved; the dissipation, depending upon the enormous gradient $\partial B_{\varpi}/\partial\phi$ at $\phi = kz$ between the two opposite B_{ϖ} on either side, ceases.

One finds from the study of specific cases that the non-equilibrium caused by changes in the winding pattern along the general direction of the field is generally of the form illustrated by this simple example. Because of the change in topology of the field, there are some elemental flux tubes within the field extending between regions requiring different total pressures for equilibrium. The pressure extends uniformly along the tubes and somewhere the pressure is not adequate to keep two opposite field components from approaching each other. Then, no matter how large the electrical conductivity, the fluid is squeezed out from between, the field gradient increases without limit, and dissipation and line cutting proceed as rapidly as the expulsion of fluid.

14.8. Dynamical non-equilibrium in field topologies without invariance

Consider the general conditions on the fluid pressure necessary for equilibrium of the magnetic field. Pick any point Q within a magnetic field $B_i(x_k)$. Introduce a local cartesian coordinate system (ξ, η, ζ) with its origin at Q, and rotate the axes so that the ζ-axis lies along the field at Q, as illustrated in Fig. 14.7. Denote the magnitude of the field at Q by B. Then the ξ- and η-components of the field vanish at Q, while $B_{\zeta} = B$. If the scale of variation of B_i is l, then within some suitably small neighbourhood of $Q(\xi^2, \eta^2, \zeta^2 \ll l^2)$ the field can be adequately represented by the expansions

$$B_{\xi} = B_{\xi\xi}\xi + B_{\xi\eta}\eta + B_{\xi\zeta}\zeta + \ldots$$
$$B_{\eta} = B_{\eta\xi}\xi + B_{\eta\eta}\eta + B_{\eta\zeta}\zeta + \ldots$$
$$B_{\zeta} = B + B_{\zeta\xi}\xi + B_{\zeta\eta}\eta + B_{\zeta\zeta}\zeta + \ldots$$

where B_{ij} denotes $\partial B_i/\partial x_j$ evaluated at Q.

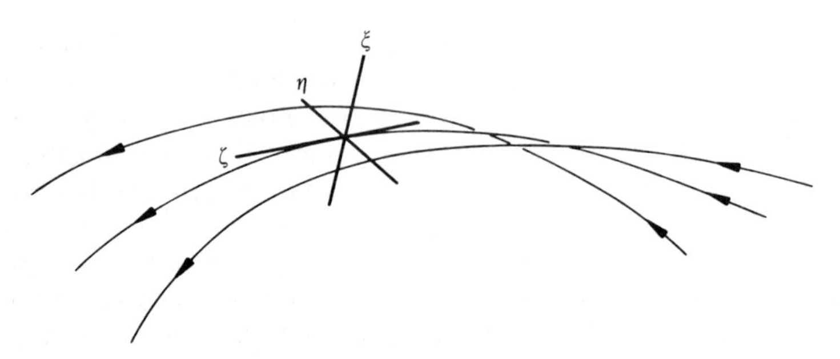

FIG. 14.7. The local Cartesian coordinate system (ξ, η, ζ) with the ζ-axis tangent to the field $B_i(x_k)$ at the location of the origin.

To keep the problem as simple as possible, drop the non-essential features such as the transverse variation of $B_\zeta(B_{\zeta\xi} = B_{\zeta\eta} = 0)$ and the longitudinal variation of $B_i(B_{i\zeta} = 0)$. These reductions greatly simplify the discussion of the open or closed local topology of the lines of force, without omitting any essential physics. Then

$$B_{\xi\xi} + B_{\eta\eta} = 0$$

so that B_ξ and B_η can be expressed as

$$B_\xi = +\partial A/\partial\eta, \qquad B_\eta = -\partial A/\partial\xi \tag{14.56}$$

in terms of the vector potential

$$A = \tfrac{1}{2}\{(B_{\xi\xi} - B_{\eta\eta})\xi\eta + B_{\xi\eta}\eta^2 - B_{\eta\xi}\xi^2\}. \tag{14.57}$$

The projection of the lines of force on the plane $\zeta = 0$ (perpendicular to the main field direction) is $A(\xi, \eta) = constant$.

Now rotate the coordinate system about the ζ-axis by an angle θ, so that the new transverse coordinates x and y are related to ξ and η by

$$\xi = x \cos\theta + y \sin\theta, \qquad \eta = -x \sin\theta + y \cos\theta.$$

Choosing θ such that

$$\tan(2\theta) = (B_{\xi\xi} - B_{\eta\eta})/(B_{\eta\xi} + B_{\xi\eta}),$$

the vector potential reduces to

$$A = \tfrac{1}{4}x^2(B_{\xi\eta} - B_{\eta\xi} - C) + \tfrac{1}{4}y^2(B_{\xi\eta} - B_{\eta\xi} + C) \tag{14.58}$$

where C is the positive square root

$$C = +\{(B_{\xi\xi} - B_{\eta\eta})^2 + (B_{\xi\eta} + B_{\eta\xi})^2\}^{\frac{1}{2}}. \tag{14.59}$$

The coefficients of x^2 and y^2 have the same sign in (14.58), and the lines of force are a family of ellipses, if

$$C^2 < (B_{\xi\eta} - B_{\eta\xi})^2.$$

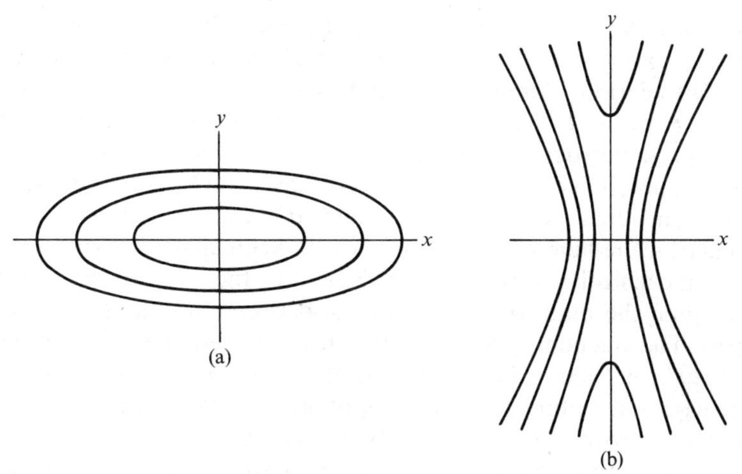

FIG. 14.8. A sketch of the magnetic lines of force (a) when $B_{yx} > 0$, $B_{xy} < 0$ and (b) when $B_{yx} > B_{xy} > 0$.

It can be shown that this condition is equivalent to

$$B_{\eta\xi} B_{\xi\eta} < -B_{\xi\xi}^2. \tag{14.60}$$

The field configuration is illustrated in Fig. 14.8a. The important point is that, if the lines of force are ellipses, then they are closed in the neighbourhood of the ζ-axis. The fluid is locally confined and, together with the longitudinal field B, may be compressed to any pressure necessary for equilibrium. Thus, with $\partial/\partial\zeta = 0$, (14.60) is a sufficient condition for local equilibrium. The inequality (14.60) is satisfied only if $B_{\eta\xi}$ and $B_{\xi\eta}$ are of opposite sign and of sufficiently large magnitude, implying that there is torsion in the field as it extends along the ζ-axis.

To show the limitation more directly note that the torsion $(\nabla \times \mathbf{B})_\zeta$ is

$$(\nabla \times \mathbf{B})_\zeta = B_{\eta\xi} - B_{\xi\eta}.$$

Suppose that $(\nabla \times \mathbf{B})_\zeta$ is positive as a consequence of $B_{\eta\xi}$ being positive and $B_{\xi\eta}$ being negative. Then (14.60) provides the lower limit

$$(\nabla \times \mathbf{B})_\zeta > B_{\eta\xi} + B_{\xi\xi}^2 / B_{\eta\xi}. \tag{14.61}$$

The right-hand side of this equality has a minimum value $2|B_{\xi\xi}|$ for $B_{\eta\xi} = -B_{\xi\eta} = |B_{\xi\xi}|$. Thus the minimum torsion $(\nabla \times \mathbf{B})_\zeta$ to close the field is $2|B_{\xi\xi}|$. It occurs when $B_{\eta\xi}$ and $B_{\xi\eta}$ are equal in magnitude. If $B_{\eta\xi}$ and $B_{\xi\eta}$ are not equal in magnitude, then more torsion is required to close the field lines into ellipses[2] about the ζ-axis.

[2] Note that the lines form circles only in the special case when $B_{\xi\eta} = -B_{\eta\xi}$ and $B_{\xi\xi} = B_{\eta\eta} = 0$.

Now consider the situation when the torsion is below the limit (14.61). Then the coefficients of x^2 and y^2 in (14.58) have opposite sign and the lines of force form a family of hyperbolas, as illustrated in Fig. 14.8b. The lines are not closed. They extend out of the region and so may be subject to fluid pressures determined by conditions elsewhere in the field. If the pressures, determined elsewhere, do not satisfy the local requirements for equilibrium, then non-equilibrium appears and the fields evolve. In particular, if the fluid pressure is too low to keep apart the opposite fields B_x facing each other across the x-axis, and/or the opposite B_y across the y-axis, then the fluid squeezes out from between and the field gradients steepen. The steepening goes on without limit, leading to dissipation and line cutting, no matter how large the electrical conductivity σ, etc.

The essential feature of the non-equilibrium is the absence of sufficient torsion to overcome $B_{\xi\xi}$ and close the local topology. In the simplest case, then, put the non-essential $B_{\xi\xi}$ equal to zero, so that (14.58) reduces to

$$A(x, y) = \tfrac{1}{2}B_{xy}y^2 - \tfrac{1}{2}B_{yx}x^2. \tag{14.62}$$

In this case the coordinate rotation θ is zero so that $\xi = x$, $\eta = y$. Then, with the present condition $\partial/\partial\zeta = 0$ the equilibrium conditions (14.4) and (14.5) are appropriate. They reduce to[3],

$$\begin{aligned} F'(A) &= -\nabla^2_{xy}A \\ &= B_{yx} - B_{xy}, \end{aligned}$$

so that the total pressure is

$$F(A)/4\pi = P_0 + (B_{yx} - B_{xy})A/4\pi \tag{14.63}$$

where P_0 is a constant, of no particular physical interest. Equation (14.63) gives the pressure for equilibrium. Thus, if the pressure (piped into the region along the lines of force from elsewhere) has the form (14.63), the local field is in equilibrium. If the pressure is different, the field is not in equilibrium. In particular, if B_{yx} and B_{xy} have the same sign and the pressure is below that required for equilibrium, then no equilibrium is available. The field gradients increase without bound.

To illustrate the dynamical growth of field gradients, suppose that the pressure P does not satisfy (14.63), but is instead the fraction κ of the equilibrium value (14.63). Then the local field configuration begins to evolve according to the familiar dynamical equations

$$\rho\left(\frac{\partial v_i}{\partial t} + v_j\frac{\partial v_i}{\partial x_j}\right) = -\frac{\partial}{\partial x_i}\left(p + \frac{B^2}{8\pi}\right) + \frac{B_j}{4\pi}\frac{\partial B_i}{\partial x_j}.$$

Consider a uniform incompressible, inviscid, infinitely conducting fluid.

[3] It will be remembered that we are assuming that $B_{\zeta\xi} = B_{\zeta\eta} = 0$.

To avoid the non-linear terms $v_j\,\partial v_i/\partial x_j$, suppose that there is a frictional drag $-\rho v_i/\tau$ which enforces low velocities upon the system, so that $v_j\,\partial v_i/\partial x_j$ is small compared to $\rho\,\partial v_i/\partial t$ and/or $\rho v_i/\tau$. Suppose that the externally imposed pressure (extending into the region along the lines of force) is applied to the region at the ellipse

$$x^2/a^2+y^2/b^2=1. \tag{14.64}$$

Then consider the fluid motions and field evolution only inside this ellipse. Write the local pressure as

$$P=\kappa\Omega A/4\pi+P_1(t)(x^2/a^2+y^2/b^2-1). \tag{14.65}$$

Thus on the ellipse (14.64) the pressure is just the number κ times the externally applied pressure (14.63), with $P_0=0$ and Ω equal to the initial value of $B_{yx}-B_{xy}$. The second term represents local pressure variations within the boundary ellipse, with P_1 a function of time. The equations of motion become, then

$$\rho\left(\frac{\partial v_x}{\partial t}+\frac{v_x}{\tau}\right)=-\frac{\partial P}{\partial x}-\frac{B_y}{4\pi}\left(\frac{\partial B_y}{\partial x}-\frac{\partial B_x}{\partial y}\right),$$

$$\rho\left(\frac{\partial v_y}{\partial t}+\frac{v_y}{\tau}\right)=-\frac{\partial P}{\partial y}+\frac{B_x}{4\pi}\left(\frac{\partial B_y}{\partial x}-\frac{\partial B_x}{\partial y}\right).$$

Since the fluid is incompressible, write

$$v_x=+\partial\psi/\partial y,\qquad v_y=-\partial\psi/\partial x, \tag{14.66}$$

and let

$$\psi=-D(t)xy, \tag{14.67}$$

where D is a function of time. The equations of motion now reduce to

$$4\pi\rho\left(\frac{\mathrm{d}}{\mathrm{d}t}+\frac{1}{\tau}\right)D=-B_{yx}\{\kappa\Omega+B_{xy}-B_{yx}\}+\frac{8\pi P_1(t)}{a^2} \tag{14.68}$$

$$4\pi\rho\left(\frac{\mathrm{d}}{\mathrm{d}t}+\frac{1}{\tau}\right)D=-B_{xy}\{\kappa\Omega+B_{xy}-B_{yx}\}-\frac{8\pi P_1(t)}{b^2}. \tag{14.69}$$

Subtracting the first from the second yields

$$8\pi P_1(1/a^2+1/b^2)=(B_{yx}-B_{xy})(\kappa\Omega+B_{xy}-B_{yx}) \tag{14.70}$$

relating $P_1(t)$ to the time-varying spatial derivatives B_{xy} and B_{yx}. Eliminating P_1 from (14.68) yields

$$4\pi\rho\left(\frac{\mathrm{d}}{\mathrm{d}t}+\frac{1}{\tau}\right)D=\frac{a^2B_{yx}+b^2B_{xy}}{a^2+b^2}(B_{yx}-B_{xy}-\kappa\Omega). \tag{14.71}$$

The field varies according to the hydromagnetic equation

$$\partial A/\partial t + v_x\, \partial A/\partial x + v_y\, \partial A/\partial y = 0$$

in a highly conducting fluid. With A given by (14.62) and v_x and v_y by (14.66) and (14.67) it follows that

$$\mathrm{d}B_{xy}/\mathrm{d}t = -2DB_{xy}, \qquad \mathrm{d}B_{yx}/\mathrm{d}t = +2DB_{yx}. \qquad (14.72)$$

Eqns (14.72) determine the time evolution of the field.

Consider the simple case that the external pressure is applied on the circle of radius a, so that $b = a$. If the frictional drag is large, the fluid motion is small and always near its terminal velocity so that $\mathrm{d}/\mathrm{d}t \ll 1/\tau$. Then (14.71) yields

$$D = (\tau/4\pi\rho)(B_{yx} + B_{xy})(B_{yx} - B_{xy} - \kappa\Omega).$$

Let

$$U = B_{yx} + B_{xy}, \qquad V = B_{yx} - B_{xy} \qquad (14.73)$$

and introduce the time coordinate $\mu = \tau t/4\pi\rho$. Then the initial value of V is just Ω, which we presume, for convenience, to be positive. Then

$$D = (\tau/4\pi\rho)U(V - \kappa\Omega).$$

Adding and subtracting the two eqns (14.72) yields

$$\mathrm{d}U/\mathrm{d}\mu = -UV(\kappa\Omega - V), \qquad \mathrm{d}V/\mathrm{d}\mu = -U^2(\kappa\Omega - V). \qquad (14.74)$$

Hence

$$\mathrm{d}U/\mathrm{d}V = V/U,$$

so that

$$U^2 = V^2 + G \qquad (14.75)$$

where G is a *constant*. The second of the two equations in (14.74) becomes, then,

$$\mathrm{d}V/\mathrm{d}\mu = (V - \kappa\Omega)(V^2 + G). \qquad (14.76)$$

Integration of this equation depends upon the sign and magnitude of G. We have already pointed out, on physical grounds, that the field gradients run away, steepening without limit with the passage of time μ, in the case that the fluid pressure applied to the lines of force between two opposite fields (B_{yx}, B_{xy} with the same sign in (14.62)) is below the equilibrium value (14.63), i.e. $\kappa < 1$. The other circumstances ($\kappa > 1$, or B_{yx}, B_{xy} of opposite sign) is stable.

Suppose first that B_{yx} and B_{xy} have opposite sign, so that the lines of force, given by A in (14.62), are ellipses. The field configuration is sketched in Fig. 14.8a. For convenience make B_{yx} positive while B_{xy} is negative. Then $V > U$ so let $G = -R^2$ with $0 \leq R \leq \Omega$, so that U is real. Equation (14.76) is

$$\mathrm{d}V/\mathrm{d}\mu = (V - \kappa\Omega)(V^2 - R^2)$$

and its integral can be written

$$\left\{\frac{V-\kappa\Omega}{\Omega(1-\kappa)}\right\}^{2R}\left(\frac{\Omega-R}{V-R}\right)^{\kappa\Omega+R}\left(\frac{V+R}{\Omega+R}\right)^{\kappa\Omega-R} = \exp\{2R(\kappa^2\Omega^2 - R^2)\mu\}.$$

For $\kappa > 1$, it is evident that initially, when $V = \Omega$, we have $R < V < \kappa\Omega$, so that $\mathrm{d}V/\mathrm{d}\mu < 0$. Hence V decreases, asymptotically approaching R in the limit of long time, $\mu \to +\infty$. For $\kappa < 1$, it follows that $V - \kappa\Omega > 0$ so that $\mathrm{d}V/\mathrm{d}\mu$ is positive. Then V decreases asymptotically to $\kappa\Omega$, in the limit $\mu \to +\infty$. The system evolves into a final equilibrium, as expected.

Now suppose that B_{xy} and B_{yx} in (14.62) have the same sign (say both are positive). Then the lines of force are hyperbolas, and the field is separated into distinct regions, as sketched in Fig. 14.8b. Since $U > V$, write $G = +S^2$. The differential equation (14.76) becomes

$$\mathrm{d}V/\mathrm{d}\mu = (V - \kappa\Omega)(V^2 + S^2)$$

and its integral can be written

$$(\kappa^2\Omega^2 + S^2)\mu = -(\kappa\Omega/S)\{\arctan(V/S) - \arctan(\Omega/S)\}$$
$$-\tfrac{1}{2}\ln\left(\frac{V^2+S^2}{\Omega^2+S^2}\right) + \ln\left(\frac{V-\kappa\Omega}{\Omega(1-\kappa)}\right)$$

If $\kappa > 1$, then initially $\mathrm{d}V/\mathrm{d}\mu < 0$. The quantity V decreases asymptotically to zero in the limit of large μ. As we expect, an excess fluid pressure in the region merely inflates the field until equilibrium is achieved.

On the other hand, when $\kappa < 1$, it follows that $\mathrm{d}V/\mathrm{d}\mu > 0$ and V decreases without bound. In the limit of large V, the time μ approaches the finite limiting value μ_1, given by

$$(\kappa\Omega^2 + S^2)\mu_1 = -(\kappa\Omega/S)\{\pi/2 - \arctan(\Omega/S)\}$$
$$+\tfrac{1}{2}\ln(\Omega^2 + S^2) - \ln\{\Omega(1-\kappa)\},$$

with

$$V \sim \{2(\mu_1 - \mu)\}^{-\frac{1}{2}}.$$

The quantity S is determined by the initial values of U and V, with

$$S^2 = U^2 - V^2 = 4B_{xy}B_{yx}$$

so that $0 < S^2 < \Omega^2$, with S^2 near zero if B_{xy}/B_{yx} is either large or small compared to one, and near Ω^2 if B_{xy} and B_{yx} are nearly equal. If

$B_{xy} = B_{yx}$, then the field is a potential field and its equilibrium does not depend upon fluid pressure. In that case $S^2 = \Omega^2 \to 0$ and

$$\mu_1 = [\ln\{2^{\frac{1}{2}}/(1-\kappa)\} - \pi\kappa/4]/\Omega^2(\kappa^2+1).$$
$$\to \infty$$

The growth rate goes to zero and the field is in a stable equilibrium as we would expect.

On the other hand, if, say, $B_{yx} \gg B_{xy} > 0$, then $S \to 0$ and the field configuration becomes

$$B_x \cong 0, \qquad B_y = B_{yx}x,$$

consisting of two opposite fields facing each other across the y-axis. The characteristic time to catastrophe is

$$\mu_1[\ln\{1/(1-\kappa)\} - \kappa]/\kappa^2\Omega^2.$$

If the applied pressure is only a little less than the equilibrium value $\kappa = 1$, say $\kappa = 1 - \epsilon$ where $\epsilon \ll 1$, then

$$\mu_1 \cong \Omega^{-2}\ln(1/\epsilon).$$

The runaway slows to zero as $\epsilon \to 0$. On the other hand, if the applied pressure is only a small fraction of that needed for equilibrium ($\kappa \ll 1$), then

$$\mu_1 \cong 1/2\Omega^2.$$

In order of magnitude Ω, as the initial value of V, is comparable to the characteristic field gradient B/l between the two opposite fields. Hence with $\mu = \tau t/4\pi\rho$, it follows that the characteristic time for runaway field gradients is

$$\begin{aligned} t_1 &= 4\pi\rho\mu_1/\tau \\ &= l^2 4\pi\rho/B^2\tau \\ &= l^2/V_A^2\tau \end{aligned}$$

where V_A is the characteristic Alfven speed $B/(4\pi\rho)^{\frac{1}{2}}$. Then l/V_A is the characteristic Alfven transit time t_A across the region and

$$t_1 = t_A^2/\tau.$$

Thus, were it not for the heavy drag $1/\tau$ introduced to linearize the equations, the time to catastrophe would be of the general order of magnitude of the Alfven transit time.

The catastrophe occurs whenever the fluid pressure extending into the region along the lines of force is insufficient to keep the opposite fields separated. The opposite fields approach each other, steepening the field gradient and increasing the electric current density $\mathbf{j}=(c/4\pi)\nabla\times\mathbf{B}$ without limit. Some time before μ_1 is reached and $\mathbf{j}$ becomes infinite, dissipation steps into the picture and field annihilation and line cutting occurs. The catastrophe occurs in every magnetic field that does not have the high degree of invariance necessary for magnetohydrostatic equilibrium. In the next chapter the dynamical non-equilibrium is explored further, together with the rapid dissipation to which it leads.

Coming back, then, to the point made at the beginning of the chapter, a magnetic field caught in the continual mixing and overturning of the gases in the sun, or in other stars, or in the galaxy, is subjected to an eternity of active reconnection. The general inhomogeneity of the gas, and the motion of the gas, continually complicates the topology of the lines of force, causing them to be increasingly stochastic and driving the topology into the varying forms subject to rapid reconnection.

References

ARTSIMOVICH, L. A. (1964). *Controlled thermonuclear reactions*, pp. 286–90. Gordon and Breach, New York.

INCE, E. L. (1926). *Ordinary Differential equations*, Longsmans, Green, and Co., New York. Section 15.7, p. 499. Reprinted by Dover Publications, New York.

PARKER, E. N. (1972). *Astrophys. J.* **174,** 499.

YU, G. (1973). *Astrophys. J.* **181,** 1003.

15

RAPID RECONNECTION OF MAGNETIC LINES OF FORCE

15.1. The general nature of rapid reconnection

WHEN the fluid pressure p between two oppositely directed fields $\pm B$ (Fig. 14.8b) is insufficient to keep the fields apart, the fluid squeezes from between and the two fields approach each other. The field gradient steepens and eventually the electric current density $(c/4\pi)\nabla\times\mathbf{B}$ becomes so large that there is strong resistive dissipation no matter how large the electrical conductivity σ. Local hydromagnetic instabilities may arise in the steep field gradient, breaking up and mixing the opposite fields, and further increasing the dissipation. Indeed, if the current sheet becomes as thin as the characteristic thermal ion cyclotron radius, the conduction electrons excite microturbulence and the effective resistivity becomes anomalously high, greatly enhancing the dissipation and reconnection of the lines of force of the two opposite fields across the layer. The topology of the magnetic lines of force, that is so permanently fixed in the highly conducting fluid elsewhere, is broken by the dissipation in the thin layer between the opposite fields.

The concept of neutral point annihilation and rapid reconnection of magnetic field came to light in connection with the questions posed by the observations of solar flares (Sweet 1969). Giovanelli (1947, 1948) discussed electrical discharges along neutral lines in a magnetic field, while Dungey (1953, 1958) was the first to urge that the x-type neutral point (Fig. 14.8b) has exotic electromagnetic properties, involving the rapid dissipation of magnetic field. Sweet (1958a, b) was the first to describe the phenomenon in terms of two opposite fields being pressed together, with the subsequent reconnection of the lines of force and the alteration of the topology of the two fields. Parker (1957) worked out the rate of reconnection from the basic conservation laws in terms of the strength B of the two opposite fields and the width $2L$ over which they are pressed together. The problem is a simple one. Under more or less steady conditions, what is the velocity w with which two opposite fields (sketched in Fig. 15.1) of characteristic dimension L, move steadily toward each other (from $y=\pm\infty$) and dissipate in the thin current sheet (of thickness $2l$ along the x-axis) between the two fields?

The total pressure $p+B^2/8\pi$ is uniform across the current sheet, so the fluid pressure p is highest on the central plane where B goes through

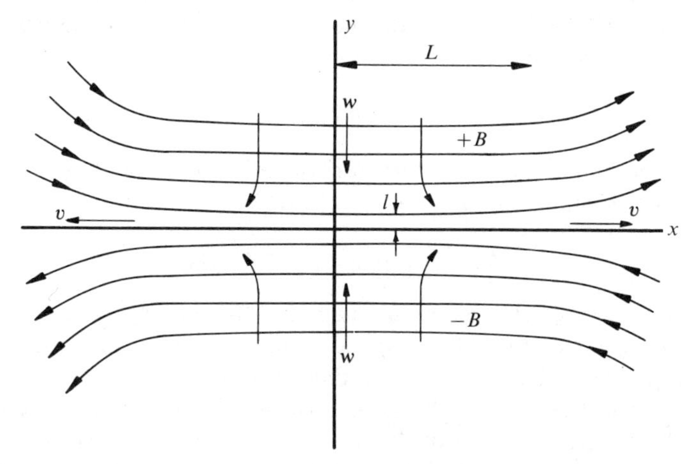

FIG. 15.1. A schematic drawing of two opposite magnetic fields $\pm B$ with characteristic scale L pressed together by their pressure $B^2/8\pi$ so as to squeeze the fluid from between, the fluid escaping along the $\pm x$-axis with a velocity v comparable to the Alfven speed. The fluid is ejected from the thin layer of thickness $2l$ across which the field changes from $+B$ to $-B$. In the steady state the two fields approach each other with a velocity w, the short arrows indicating the general direction of motion of the fluid.

zero. It is higher by the amount $\Delta p = B^2/8\pi$ equal to the magnetic pressure exerted against each side of the current sheet. This pressure excess ejects the fluid from between the two opposite fields. The ejection is along the lines of force in the $\pm x$-direction, along the midplane between the fields, with a velocity v in the x-direction that varies as

$$\rho v\, \partial v/\partial x + \partial p/\partial x = 0$$

under steady conditions. Considering the fluid to be incompressible, so that $\rho = constant$, and integrating along the x-axis from the origin at the centre, where $v = 0$, to a point outside the fields, the result is $\rho v^2 = 2\Delta p$, where Δp is the pressure excess $B^2/8\pi$. Hence the velocity of expulsion is just equal to the Alfven speed, $v = V_A = B/(4\pi\rho)^{\frac{1}{2}}$.

The net expulsion of fluid out both ends of the current sheet, of width $2l$, is at the characteristic rate $4lv$ cm^3/s^{-1}. Conservation of fluid requires that the net inflow from above and below balance the outflow. Hence

$$wL = V_A l. \tag{15.1}$$

The field changes by $2B$ from $+B$ to $-B$, across the thickness $2l$, so that the current density (perpendicular to the xy-plane) is essentially

$$j = cB/4\pi l.$$

The Ohmic dissipation across the width l is lj^2/σ and, in the steady state, is just sufficient to devour the influx of magnetic energy $wB^2/8\pi$ from either side. Hence, in terms of the resistive diffusion coefficient $\eta = c^2/4\pi\sigma$, it follows that

$$w = 2\eta/l, \tag{15.2}$$

which is, of course, just the characteristic diffusion velocity over a scale l. Solving for w and l, (15.1) and (15.2) give

$$w = 2V_A/R_m^{\frac{1}{2}}, \qquad l = 2L/R_m^{\frac{1}{2}} \tag{15.3}$$

where R_m is the characteristic magnetic Reynolds number $2LV_A/\eta$ defined in terms of the Alfven speed in the field B of dimension $2L$.

To appreciate the result, note that if the fields were not pressed together, or if they were in the same, instead of opposite, directions, then the smallest scale would be of the order of L instead of l. In that case the characteristic diffusion velocity would be $2\eta/L = 2V_A/R_m$. Thus the rate (15.3) at which the opposite fields merge is larger by $R_m^{\frac{1}{2}}$ than for passive diffusion alone. The magnetic Reynolds number is large compared to one in most astrophysical fields (for instance $R_m \cong 10^5$ for a pore in the sun, and 10^9 for a supergranule). The resistive dissipation and reconnection is enhanced by a large factor. But the rate of reconnection is still small compared to the Alfven speed, by the same large factor $R_m^{\frac{1}{2}}$.

If the fluid is a tenuous gas, rather than incompressible, then the annihiliation proceeds somewhat more rapidly (Parker 1963) because the gas is compressed between the fields and less volume of fluid needs to be squeezed from between the field. But, except in the most extreme conditions of very low density, compressibility is only a minor effect. Hence the discussion here is carried out for an incompressible fluid. Some of the many dynamical effects that occur in a low density gas are noted in §15.8.

Petschek (1964) made the fundamental point that there is no compelling reason to identify the width of the dissipation region with the overall dimensions L of the opposite fields. He suggested instead that when two opposite fields were pressed together, the field is just as likely to take the form shown in Fig. 15.2 as that shown in Fig. 15.1. The fields presumably meet only across a narrow apex, with a width λ small compared to the full width of the field. Thus the diffusion region shown in Fig. 15.1 occupies only the tiny area of width λ indicated by the dotted rectangle at the centre of Fig. 15.2. The effective magnetic Reynolds number of the small diffusion region need not be a large number, even though the Reynolds number is large for the field as a whole. The rate of merging might then be a significant, rather than an insignificant, fraction of the Alfven speed. Petschek suggested that the two opposite fields might be expected to form the narrow apexes (shown in Fig. 15.2) where they meet because, once

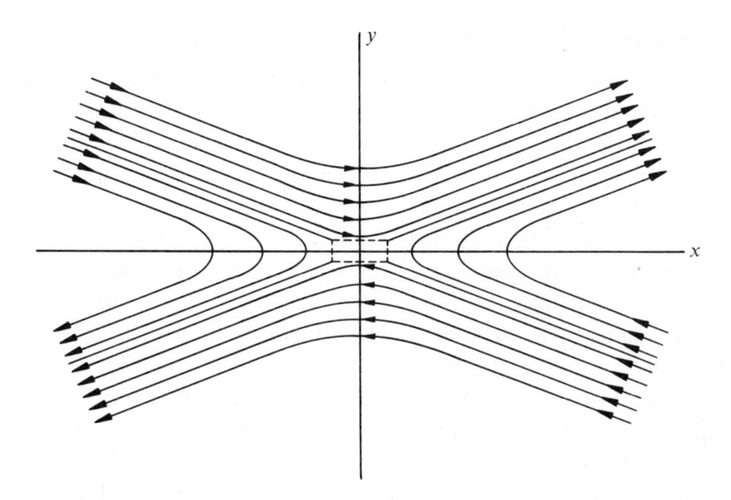

FIG. 15.2. The alternative configuration to Fig. 15.1 suggested by Petschek in which the opposite fields meet only across the narrow diffusion region of length λ ($\lambda \ll L$) and thickness l indicated by the dotted rectangle.

the lines of force are reconnected, their tension ejects the fluid vigorously from the central region, squirting the fluid out along the x-axis. The opposite fields are 'sucked' together at the origin by the local pressure reduction, as the fluid is vigorously pulled away in the x-direction by the reconnected lines of force. Petschek estimated that, depending upon external conditions, the merging rate could be anything from (15.3) up to a maximum of the order of $V_A/\ln(R_m)$ where R_m is the magnetic Reynolds number computed for the large-scale field. Since $\ln(R_m)$ is generally only 10–30 at the surface of stars, and no more than 40 or 50 in galaxies, it would follow that the rate of merging and reconnection of opposite magnetic fields may proceed as fast as 10^{-2}–10^{-1} times the Alfven speed.

The detailed arguments put forth by Petschek have been the subject of considerable discussion (Petschek and Thorne 1967), but it appears that the basic idea, and Petschek's arguments that lead to it, are sound. Rapid reconnection at some fraction of the Alfven speed is a general occurrence and is the basis for the most violent magnetic activity, such as the solar flare. Rapid reconnection occurs wherever any field component changes sign and is not held apart by suitable fluid pressure (see §14.8). The reconnection may proceed as rapidly as 10^{-2}–10^{-1} V_A. Rapid reconnection in the general topological nonequilibrium of magnetic fields lacking symmetry (Chapter 14) causes vigorous dissipation of the magnetic field. The generally non-symmetric field topologies produced in nature are continually active as a consequence of rapid reconnection.

15.2. Kinematical considerations

The basic physical principles of the merging and reconnection of two opposite magnetic fields are most easily developed beginning with some kinematical illustrations of the transport of magnetic field in an infinitely conducting fluid. Following this we will be in a position to appreciate the role of resistive diffusion, which is then taken up at length before going on to the dynamical considerations in §15.6.

Consider, then, two opposite fields carried against each other by the converging flow of fluid, as indicated in Fig. 15.1. To fix ideas, suppose that the magnetic field at $y = +h$ is uniform, with the value $B_x = +B_0$, $B_y = 0$, while at $y = -h$ the field is the opposite, $B_x = -B_0$, $B_y = 0$. The inward fluid motion is $v_x = 0$, $v_y = \mp v_0$ at $y = \pm h$, respectively, and we suppose that the fluid motion is mirror-symmetric about the x-axis, with $v_y = 0$ on $y = 0$. Nothing essential is omitted if we suppose that the fluid is incompressible and, at least for the moment, infinitely conducting. Then v_x and v_y can be expressed in terms of the stream function $\psi(x, y)$,

$$v_x = +\partial\psi/\partial y, \qquad v_y = -\partial\psi/\partial x. \tag{15.4}$$

The streamlines are given by $\psi(x, y) = \Lambda$. Denote by the infinitesimal vector $[\delta x(x, y), \delta y(x, y)]$ the separation of two points moving with the fluid at (x, y). The points are separated by $(\delta X, \delta Y)$ at $y = +h$. Then it follows from (4.32) that the magnetic field at (x, y) is

$$B_x(x, y) = B_0\delta x/\delta X, \qquad B_y(x, y) = B_0\delta y/\delta X \tag{15.5}$$

in $y > 0$, with similar expressions, but with opposite sign, in $y < 0$.

The essential point is that somewhere on the x-axis, there is a stagnation point where the incoming stream of fluid divides, flowing off to both the right and left along the x-axis. It is convenient to move the coordinate system so as to place the origin at the stagnation point. The stagnation point has the property that two neighbouring infinitesimal elements of fluid are separated there to go their opposite ways toward $x = \pm\infty$, so that $\delta x/\delta X$ increases without bound as the flow continues past the stagnation point. It follows from (15.5) that B_x increases without bound as the flow passes the stagnation point. The field increases because magnetic lines of force are transported into the region, by the converging flow, but cannot escape because they are permanently connected in the infinitely conducting fluid. So they pile up against the x-axis and the field strength increases without bound. The stagnation point is the key to the whole problem, therefore. A small region of resistivity at the stagnation point breaks the lines of force and removes the singular behaviour of the magnetic field. The region of resistivity may be of arbitrarily small width $2a$, so long as it encloses the critical point. The unbounded growth of field is then limited to something of the order of B_0h/a (see §15.3). Then, within the region of resistivity, the field breaks the original topological connection in the x-direction and assumes a new topology, connecting with the lines of force of the opposite field on the other side of the x-axis, and releasing the field from the catastrophe predicted by (15.5) for the infinitely conducting medium.

If η is the finite resistive diffusion coefficient in the region of dimension a, through which the fluid velocity is v, it is necessary only that the magnetic Reynolds number for the region, $r_m \equiv av/\eta$ be less than one, $\eta > av$, for the reconnection to be effective. For small a, v will be small too, and η need differ but little from zero.

This is the physical basis for understanding Petschek's basic point, that the essential question is the dissipation in the near neighbourhood of the stagnation point. The dissipation elsewhere is not essential, and usually negligible, because the overall magnetic Reynolds number $R_m = hV_A/\eta$ is large compared to one.

To be more quantitative, the separation δy of two points moving with the fluid velocity v_y along the y-axis varies with time according to

$$d\delta y/dt = \delta y \; \partial v_y/\partial y$$

for the symmetric conditions ($\partial v_y/\partial x = 0$ on the y-axis) treated here. But $dt = dy/v_y$, so

$$d\delta y/\delta y = (dy/v_y) \; \partial v_y/\partial y.$$

Integration yields

$$\ln(\delta y) = \ln(v_y) + \textit{constant},$$

so that if the initial separation was δY at $y = h$ where $v_y = -v_0$, it follows that

$$\delta y = -\delta Y v_y/v_0$$

subsequently. Conservation of fluid and symmetry about the y-axis ($\partial v_y/\partial x = 0$, etc) requires that the area $\delta x \delta y$ of any initial rectangle (δX, δY) carried in along the y-axis remains constant. Hence

$$\delta x = -\delta X v_0/v_y$$

so that, from (15.5)

$$B_x = -B_0 v_0/v_y. \tag{15.6}$$

Thus B_x grows with decreasing v_y as the fluid approaches the stagnation point. The hydrodynamic flow in the neighbourhood of a simple stagnation point is $\psi = (v_0/h)xy$. It follows that $v_y = -v_0 y/h$ so that

$$B_x = +B_0 h/y. \tag{15.7}$$

Hence, in the present example, B_x reaches the enormous value $B_0 h/a$ by the time it is carried to a small region of resistive diffusion in $y^2 < a^2$.

But in fact in the real dynamical flow B_x cannot increase significantly from its initial value B_0, because it is the pressure $B_0^2/8\pi$ that drives the fluid velocity. The field cannot compress itself to a strength in excess of B_0 without violating conservation of energy. So in real flows B_x must remain equal to B_0, or decline, as the field is carried in along the y-axis. Hence v_y must remain constant at $-v_0$, or increase. The classical hydrodynamic stagnation flow $\psi=(v_0/h)xy$ clearly does not have this enviable property. But if the flow is discontinuous as a consequence of a hydromagnetic shock, then $|v_y|$ does not decline as it approaches the x-axis, but instead makes an abrupt turn at the shock front and heads off to $x=\pm\infty$. Only within the small resistive region does v_y go continuously to zero, and there, with the strong diffusion, it is (15.1)–(15.3) that is applicable, rather than (15.5). So an essential part of Petschek's idea is the introduction of transverse hydromagnetic shocks (standing Alfven waves) so that the fluid velocity v_y remains constant all the way into the region of resistive dissipation. Without the standing Alfven waves there can be no self-consistent solution with a narrow apex because of the attendant growth of B_x with continuously declining v_y.

If we ask ourselves how small might the necessary region of resistive diffusion be, the answer is that it must encompass the region of decline of v_y. The shock thickness in a real fluid is limited by the resistivity to a few times η/v_0, and (see (15.2)) this is just the thickness of the diffusion region for merging opposite fields. Altogether, then, the fluid flow must be something like Fig. 15.3. If the merging is to proceed rapidly compared to (15.3), there is no other choice. With the necessary kinematical

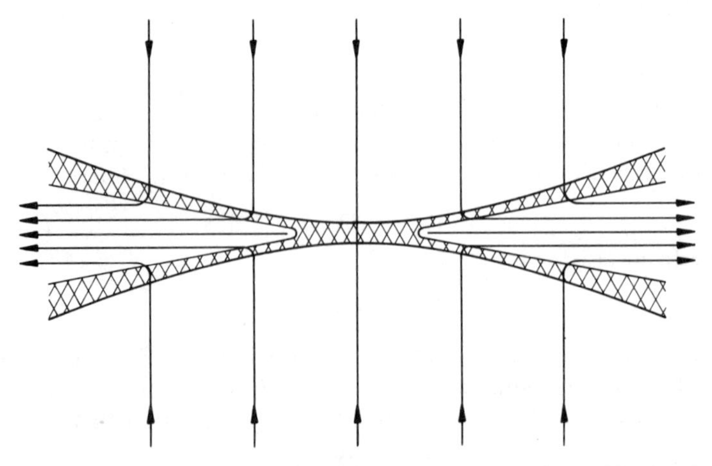

FIG. 15.3. The essential features of the fluid motion necessary for rapid reconnection of opposite fields. The cross-hatched regions indicate resistive diffusion, where the opposite fields meet in the neighbourhood of the origin, and at the sharp bend in the standing Alfven waves (shocks) where the fluid motion is deflected through a right angle.

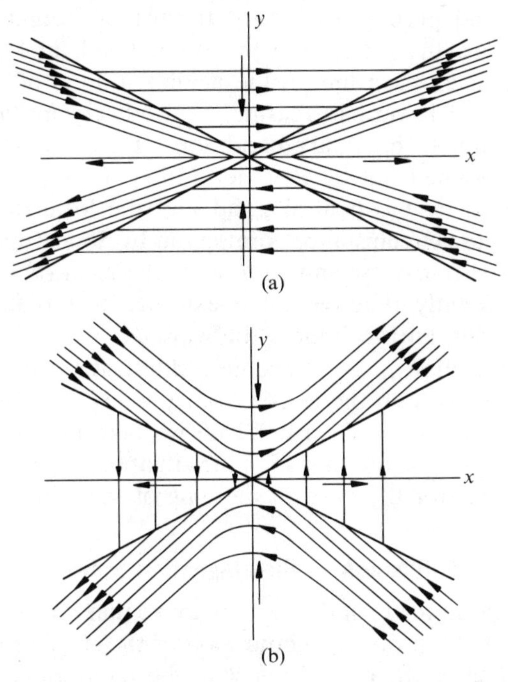

FIG. 15.4. Sketches of the two topological possibilities for a magnetic field carried in toward the x-axis, with the converging fluid flow (short straight arrows) deflected abruptly through oblique standing Alfven waves (heavy diagonal lines). In (a) the initial lines of force are parallel to the x-axis, i.e. inclined less steeply than the standing wave fronts. In (b) the initial lines are folded, forming a wedge that is more steeply inclined than the standing wave fronts.

configuration clearly in mind, then, the next question is whether the dynamics would have it so.

A sharp corner in the flow of an incompressible fluid in a transverse magnetic field is an Alfven wave, there being no other wave motion than the Alfven wave, produced by the tension in the lines of force. Such sharp crested waves are called transverse shocks. As soon as one tries to sketch the form of the corner Alfven wave, it becomes apparent that there are two possibilities, illustrated in Fig. 15.4. The magnetic lines of force being carried toward the origin may be inclined either more or less than the shocks, where the fluid turns a sharp corner. In Fig. 15.4(a) the lines of force are drawn horizontally, as if the incoming velocity v_y were uniform all the way in to the corner. In Fig. 15.4(b) the approaching lines are bent toward the origin, so that they are inclined more steeply than the Alfven wave corner.

The physical distinction between the two circumstances is evident from the fact that the Alfven wave appears first on each line of force at large x^2

in Fig. 15.4(a) and propagates inward toward the origin as the line is transported toward the origin. On the other hand, in Fig. 15.4(b) the wave appears first on each line in the neighbourhood of the y-axis and propagates outward toward increasing x^2. Both circumstances are self-consistent for suitable boundary conditions. Figure 15.4(b) is the circumstance first pointed out by Petschek (1964) in which the transverse shocks are caused by the neutral point and the local diffusion region. Figure 15.4(a) is the circumstance pointed out by Sonnerup (1970) which, with an additional transverse shock, permits the most rapid reconnection of field, but evidently requires some external perturbation (an inside corner in each quadrant) to initiate the waves.

The next two sections present examples illustrating the pile up of the magnetic field carried into the x-axis from $y = \pm\infty$ and the relief of that accumulation by diffusion. Then in §15.6 we take up the circumstances depicted in Fig. 15.4(b), and in § 15.7 the situation of Fig. 15.4(a), giving approximate values for the rates of merging of the magnetic fields.

15.3. The effects of diffusion at the stagnation point

Consider the point made in the section above that the resistive diffusion of the magnetic field in the neighbourhood of the stagnation point of the hydrodynamic flow is the crucial effect in the rapid merging and reconnection of two opposite fields $\pm B$ facing each other across the x-axis. Suppose that the fields are carried in an ideal, inviscid, incompressible fluid with steady velocity components (v_x, v_y) expressed in (15.4) in terms of the stream function $\psi(x, y)$. The flow lines are given by $\psi(x, y) = \Lambda$, of course. Embedded in this highly conducting fluid is the magnetic field

$$B_x = +\partial A/\partial y, \qquad B_y = -\partial A/\partial x. \tag{15.8}$$

There may also be a component of field B_z perpendicular to the xy-plane, satisfying

$$dB_z/dt \equiv \partial B_z/\partial t + v_x\, \partial B_z/\partial x + v_y\, \partial B_z/\partial y = 0. \tag{15.9}$$

But for the two-dimensional motion (v_x, v_y) of an incompressible fluid B_z appears only in the total pressure $p + B_z^2/8\pi$ in the dynamical equations, so that B_z has no effect on v_x and v_y for a given specification of the total pressure at the boundaries. Thus B_z will be ignored.

The induction equation (4.13) for B_x and B_y reduces to

$$\nabla\{v_x B_y - v_y B_x + \eta(\partial B_x/\partial y - \partial B_y/\partial x)\} = 0,$$

where η may be a function of position. Hence the quantity in brackets is a constant, which we denote by $-cE$. In terms of the vector potential A, then,

$$v_x \frac{\partial A}{\partial x} + v_y \frac{\partial A}{\partial y} - \eta\left(\frac{\partial^2 A}{\partial x^2} + \frac{\partial^2 A}{\partial y^2}\right) = cE. \tag{15.10}$$

As it stands, this equation is fully elliptic, with the characteristics $x \pm iy = constant$. But in regions where resistive diffusion can be neglected ($\eta = 0$) the characteristics are the stream lines

$$\psi(x, y) = \Lambda. \tag{15.11}$$

If $ds = (dx^2 + dy^2)^{\frac{1}{2}}$ denotes an element of distance measured along a characteristic curve $dx/v_x = dy/v_y$, then

$$dx/ds = v_x/v, \qquad dy/ds = v_y/v, \tag{15.12}$$

where $v = (v_x^2 + v_y^2)^{\frac{1}{2}}$. With $\eta = 0$, (15.10) reduces to

$$\begin{aligned} dA/ds &= (\partial A/\partial x)\, dx/ds + (\partial A/\partial y)\, dy/ds, \\ &= cE/v(s). \end{aligned}$$

Hence, integrating along the characteristic (15.11), it follows that

$$A(x, y) = cE \int_{\substack{s_0 \\ \psi=\Lambda}}^{s} \frac{ds}{v(s)} + F(\Lambda) \tag{15.13}$$

where s_0 is some convenient starting point at which $A = F(\Lambda)$. It is convenient in the present problem to use (15.12) to rewrite (15.13) in terms of an integral over x or y, giving the choices

$$A = cE \int_{\psi=\Lambda} \frac{dx}{v_x} + F(\Lambda), \tag{15.14}$$

$$A = cE \int_{\psi=\Lambda} \frac{dy}{v_y} + F(\Lambda). \tag{15.15}$$

Note that the integrand diverges in the neighbourhood of the stagnation point where v_x and v_y decline to zero.

Consider, then, the circumstance where the fluid flows in toward the x-axis from $y = \pm h$ where the magnetic field is $B_x = \pm B_0$, $B_y = 0$ and the fluid velocity is $v_x = 0$, $v_y = \mp v_0$, respectively. Then A is independent of x on $y = \pm h$. Hence $F(\Lambda)$ is independent of ψ and may as well be put equal to zero. Restricting attention to the upper half plane $y \geqslant 0$ (the lower half plane following from symmetry) consider again the simple hydrodynamic flow

$$\psi = v_0 xy/h, \qquad cE = -v_0 B_0 \tag{15.16}$$

appropriate for the neighbourhood of a stagnation point so that

$$v_x = +v_0 x/h, \qquad v_y = -v_0 y/h.$$

The stream lines are rectangular hyperbolas. It follows from (15.14) that

$$A = +B_0 h \ln(y/h), \qquad B_x = B_0 h/y, \qquad B_y = 0. \tag{15.17}$$

This is, of course, just (15.7) again. The field remains in the x-direction and increases like $1/y$ toward the x-axis. The configuration in the first quadrant, to which we devote our attention, is shown in Fig. 15.5.

The simplest illustration arises when the resistivity in the slab $y^2 < b^2$ is increased from zero to some very large value η (so that $v_0 b^2/h\eta \ll 1$). Then if R denotes the vector potential within the resistive region, (15.10) reduces to

$$(\partial^2/\partial x^2 + \partial^2/\partial y^2)R = 0. \tag{15.18}$$

In the present case where B_x is an odd function of y, it follows that R is an even function of y (because $B_x = \partial R/\partial y$), and must match smoothly to A (given by (15.17)) at the boundary $y = b$ of the resistivity. Hence R is independent of x, so that it must be of the form $C_1 + C_2 y$. But if R is to be an even function of y, then $C_2 = 0$ and $R = C_1$. The magnetic field vanishes within the resistive region. The magnetic field is given by (15.17) for $y > b$, the field vanishes in $y^2 < b^2$, and the field is equal to the negative of (15.17) in $y < -b$. Thus the field increases to $B_0 h/b$, just before it drops to zero at $y = b$. This maximum field can be very large if b is small, of course, because v_y becomes so small, but for any non-vanishing b, no matter how small, the field is finite.

Now consider what happens if resistive diffusion is restricted to the rectangle $-a < x < +a$, $-b < y < +b$ around the stagnation point at the

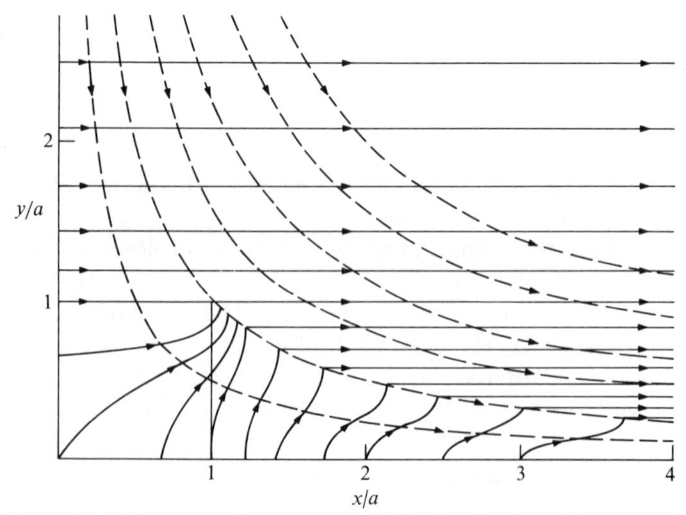

FIG. 15.5. A plot of the stream lines $\psi = v_0 xy/h$ (broken lines) and the magnetic lines of force based on (15.34) and (15.35) (solid lines) show the effect of the resistive rectangle $x^2, y^2 < a^2$ on the field transported in the otherwise infinitely conducting fluid.

origin, the rest of the fluid being infinitely conducting (Parker 1973b). With $\eta = 0$ elsewhere, it is clear that the magnetic field is unaffected on any stream lines that do not intersect the resistive rectangle (for which $\psi^2 > (v_0 ab/h)^2$). Nor is the field affected *upstream* from the rectangle on those stream lines that do intersect it. But the field carried into the rectangle is altered there and is not described by (15.14) either within the rectangle or downstream from it. The effects are shown in Fig. 15.5, illustrating the reconnection of the magnetic lines of force and the resulting downstream wake. To calculate the effects, denote by R the vector potential within the rectangle, and by W the vector potential in the downstream wake $\psi^2 < (v_0 ab/h)^2$. Then for the simple case that $\eta = \infty$ within the rectangle, (15.10) reduces to (15.18). Since $\eta = 0$ again in the downstream wake, W satisfies (15.13) and hence either (15.14) or (15.15), as does A. It is convenient to write the solution for the vector potential W in the downstream wake ($x > a$, $y < ab/x$) as

$$W = -B_0 h \ln(hx/ab) + G(\psi). \tag{15.19}$$

The boundary conditions are easily specified. The normal component of the magnetic field is continuous across the boundaries, which is most conveniently accomplished by making the vector potential continuous across the boundaries. Thus R is continuous with A along the upper and lower sides of the resistive rectangle, and with W along the right and left sides of the rectangle. Along the boundary $xy = \pm ab$ of the downstream wake, W and A must join continuously. Hence along the wake boundary, comparison of (15.17) and (15.19) leads to

$$G(\pm v_0 ab/h) = 0. \tag{15.20}$$

It should be noted, too, that ψ is an odd function of x and y, while the vector potential is an even function. Hence G must be an even function of ψ.

Now R satisfies (15.18), and is an even function of both x and y. The vector potential is given by (15.17) as $B_0 h \ln(b/h)$ on the upper side of the rectangle for $-a < x < +a$. Hence R must be independent of x there, so that the most general form for R that fits continuously to A is

$$R = B_0 h \ln(b/h) + \sum_{n=1}^{\infty} A_n \cosh\left\{(n - \tfrac{1}{2})\frac{\pi x}{b}\right\} \cos\left\{(n - \tfrac{1}{2})\frac{\pi y}{b}\right\}. \tag{15.21}$$

Matching R to W at $x = a$, where $\psi = v_0 ay/h$, leads to

$$G(v_0 ay/h) = \sum_{n=1}^{\infty} A_n \cosh\{(n - \tfrac{1}{2})\pi a/b\} \cos\{(n - \tfrac{1}{2})\pi y/b\}. \tag{15.22}$$

Note that this condition leads automatically to (15.20).

Altogether, then, the vector potentials (15.17), (15.19), and (15.21) with the condition (15.22) satisfy the requirement that the normal component of the field is continuous at all the boundaries. What is still lacking is a specification of the tangential component of the field that is trapped in the fluid as it flows out through the side of the rectangle $x=a$ and the electrical conductivity is switched on. The normal component of the Poynting vector, given by (5.6), is continuous across $x=a$ (see (5.8))[1]. The x-component of the Poynting vector is $Q_x = -cEB_y/4\pi$. With E constant it follows that $\partial R/\partial x = \partial W/\partial x$ at $x=a$, leading to

$$\begin{aligned}&-B_0h/a+(v_0y/h)G'(v_0ay/h)\\&=(\pi/b)\sum_{n=1}^{\infty}A_n(n-\tfrac{1}{2})\sinh\{(n-\tfrac{1}{2})\pi a/b\}\cos\{(n-\tfrac{1}{2})\pi y/b\}.\end{aligned} \quad (15.23)$$

To solve (15.22) and (15.23) for A_n and for G, differentiate (15.22) with respect to y and then eliminate G' from (15.23). The result is

$$\begin{aligned}B_0h/a=-(\pi/b)\sum_{n=1}^{\infty}A_n(n-\tfrac{1}{2})[&\sinh\{(n-\tfrac{1}{2})\pi a/b\}\cos\{(n-\tfrac{1}{2})\pi y/b\}\\&+(y/a)\cosh\{(n-\tfrac{1}{2})\pi a/b\}\sin\{(n-\tfrac{1}{2})\pi y/b\}].\end{aligned} \quad (15.24)$$

The slow convergence of this sum discourages an attempt at solution for A_n in terms of B_0. Instead, multiply (15.22) by $\cos\{(m-\tfrac{1}{2})\pi y/b\}$ and integrate from $-b$ to $+b$. It follows that

$$A_m\cosh\{(m-\tfrac{1}{2})\pi a/b\}=\frac{1}{b}\int_{-b}^{+b}dy\cos\{(m-\tfrac{1}{2})\pi y/b\}G(v_0ay/h). \quad (15.25)$$

Substituting this into (15.24) yields

$$\begin{aligned}B_0h/a=&-(\pi/b^2)\sum_{n=1}^{\infty}\int_{-b}^{+b}d\mu\cos\{(n-\tfrac{1}{2})\pi\mu/b\}G(v_0a\mu/h)\\&\times(n-\tfrac{1}{2})[\tanh\{(n-\tfrac{1}{2})\pi a/b\}\cos\{(n-\tfrac{1}{2})\pi y/b\}+(y/a)\sin\{(n-\tfrac{1}{2})\pi y/b\}].\end{aligned}$$

Note that

$$(n-\tfrac{1}{2})\cos\{(n-\tfrac{1}{2})\pi\mu/b\}=(b/\pi)(\partial/\partial\mu)\sin\{(n-\tfrac{1}{2})\pi\mu/b\}.$$

Then integrate by parts, taking advantage of (15.20). The result can be

[1] Note that the same condition does not apply at the upper boundary $y=b$ where magnetic field is continually released from the fluid as the fluid passes across $y=b$ and the conductivity is switched off. There the field is suddenly released and quickly dissipated (if the conductivity within the rectangle is small but not identically zero).

written

$$\frac{2bh^2B_0}{a^2v_0}=\int_{-b}^{+b}\mathrm{d}\mu G'(v_0a\mu/h)$$
$$\times\sum_{n=1}^{\infty}\left[\tanh\left\{(n-\tfrac{1}{2})\frac{\pi a}{b}\right\}\right.$$
$$\times\left\{\sin(n-\tfrac{1}{2})\frac{\pi}{b}(y+\mu)-\sin(n-\tfrac{1}{2})\frac{\pi}{b}(y-\mu)\right\}$$
$$\left.+\frac{y}{a}\left\{\cos\left\{(n-\tfrac{1}{2})\frac{\pi}{b}(y-\mu\right\}-\cos\left\{(n-\tfrac{1}{2})\frac{\pi}{b}(y+\mu)\right\}\right\}\right]. \qquad (15.26)$$

To treat the second term in braces, note the identity

$$\sum_{n=1}^{\infty}\cos\{(n-\tfrac{1}{2})s\}=\tfrac{1}{2}\sum_{n=-\infty}^{+\infty}\exp\{\mathrm{i}(n-\tfrac{1}{2})s\},$$
$$=\tfrac{1}{2}\exp(-\tfrac{1}{2}\mathrm{i}s)\sum_{n=-\infty}^{+\infty}\exp(\mathrm{i}ns),$$
$$=\pi\exp(-\tfrac{1}{2}\mathrm{i}s)\sum_{n=-\infty}^{+\infty}\delta(s-2\pi n).$$

Then, if $S(\pm)$ is defined as

$$S(\pm)\equiv\int_{-b}^{+b}\mathrm{d}\mu G'(v_0a\mu/h)\sum_{n=1}^{\infty}\cos\left\{(n-\tfrac{1}{2})\frac{\pi}{b}(y\pm\mu)\right\},$$

let $\phi=\pi\mu/b$ so that

$$S(\pm)=b\int_{-\pi}^{+\pi}\mathrm{d}\phi G'(v_0ab\phi/\pi h)\exp\{-\tfrac{1}{2}\mathrm{i}(\pi y/b\pm\phi)\}$$
$$\times\sum_{n=-\infty}^{+\infty}\delta(\pi y/b\pm\phi-2\pi n).$$

For $y^2<b^2$ the integration picks up only the δ-function for $n=0$, so that

$$S(\pm)=bG'(\pm v_0ay/h).$$

Now G' is an odd function of its argument. Hence

$$S(+)-S(-)=2bG'(v_0ay/h). \qquad (15.27)$$

But, from the definition of $S(\pm)$, this is obviously just the integral and sum of the second expression in braces on the right-hand side of (15.26)[2].

[2] For $y=\pm b$ the integration picks up half of the δ-function for $n=\pm 1$, giving a Stokes phenomenon at the end-points.

To treat the first term in curly brackets consider the sum

$$U(s) \equiv \sum_{n=-\infty}^{+\infty} \tanh\{(n-\tfrac{1}{2})\pi a/b\}\sin\{(n-\tfrac{1}{2})s\}. \tag{15.28}$$

The sum is not convergent when taken outside the integral, but it can be converted to a rapidly converging sum using Lighthill's theorem, that if

$$H(n) = \int_{-\infty}^{+\infty} \mathrm{d}m \exp(\mathrm{i}2\pi mn) G(m),$$

$$G(m) = \int_{-\infty}^{+\infty} \mathrm{d}n \exp(-\mathrm{i}2\pi mn) H(n),$$

then

$$\sum_{n=-\infty}^{+\infty} H(n) = \sum_{m=-\infty}^{+\infty} G(m).$$

Then let

$$H(n) \equiv \tanh\{(n-\tfrac{1}{2})\pi a/b\}\sin\{(n-\tfrac{1}{2})s\}.$$

It follows that

$$\begin{aligned}
G(m) &= \int_{-\infty}^{+\infty} \mathrm{d}n \exp(-\mathrm{i}2\pi mn)\tanh\{(n-\tfrac{1}{2})\pi a/b\}\sin(n-\tfrac{1}{2})s \\
&= \exp(-\mathrm{i}\pi m)\int_{-\infty}^{+\infty} \mathrm{d}\nu \exp(-\mathrm{i}2\pi m\nu)\tanh(\nu\pi a/b)\sin(\nu s) \\
&= \tfrac{1}{2}\exp(-\mathrm{i}\pi m)\int_{-\infty}^{+\infty} \mathrm{d}\nu \tanh(\nu\pi a/b) \\
&\qquad\qquad \times\{\sin\nu(s+2\pi m)+\sin\nu(s-2\pi m)\} \\
&= (b/4a)\exp(-\mathrm{i}\pi m)[\operatorname{csch}\{(s+2\pi m)b/2a\}+\operatorname{csch}\{(s-2\pi m)b/2a\}].
\end{aligned}$$

(see, for instance, Gradshteyn and Ryzhik 1965). Hence, with Lighthill's theorem,

$$\begin{aligned}
U(s) &= \frac{b}{4a}\sum_{m=-\infty}^{+\infty}(-1)^m[\operatorname{csch}\{(s+2\pi m)b/2a\}+\operatorname{csch}\{(s-2\pi m)b/2a\}] \\
&= \frac{b}{2a}\sum_{m=-\infty}^{+\infty}(-1)^m \operatorname{csch}\{(s+2\pi m)b/2a\}. \qquad (15.29)
\end{aligned}$$

Now it is readily seen from (15.28) that U is just twice the sum from $+1$ to ∞. Hence the integral and sum of the first brace on the right-hand side of (15.26) can be written in terms of the integral over G' multiplied by the sum (15.29). The result is

$$\frac{b}{2a}\int_{-b}^{+b} \mathrm{d}\mu G'\left(\frac{v_0 a\mu}{h}\right) \sum_{m=-\infty}^{+\infty} (-1)^m$$
$$\times\left[\operatorname{csch}\left\{\frac{\pi}{2a}(y+2mb+\mu)\right\}-\operatorname{csch}\left\{\frac{\pi}{2a}(y+2mb-\mu)\right\}\right].$$

Then noting that G' is an odd function of its argument, this sum can be reduced to two times the sum over the first csch.

Altogether, then, (15.26) can be written as

$$2h^2 B_0/av_0 = 2yG'(v_0 ay/h)$$
$$+\int_{-b}^{+b} \mathrm{d}\mu G'(v_0 a\mu/h) \sum_{n=-\infty}^{+\infty} (-1)^n \operatorname{csch}\left\{\frac{\pi}{2a}(y+2nb+\mu)\right\}. \tag{15.30}$$

This integral equation serves to determine G', and, together with (15.20), gives G. The coefficients A_n follow through (15.25).

Note that the terms $n=0$, ± 1 in the sum have simple poles. The pole in the $n=0$ term is at $\mu=-y$. Provided that $y \neq b$ we may use the principal value. For $y=b$ the integration does not cross the pole but stops on it, giving a divergent integral if $G'(v_0 ay/h)$ does not vanish at $y=b$. The terms $n=\pm 1$ have a pole only if $y=\pm b$, and again the integral diverges. Hence we must have

$$G'(\pm v_0 ab/h)=0 \tag{15.31}$$

in addition to (15.20) if the fields are to be finite. Hence both G and G' vanish on the boundaries of the downstream wake.

If $a \to \infty$, so that the rectangle becomes an infinite slab of zero conductivity and thickness $2b$ interposed between opposite fields, it is evident that G' becomes small $O(1/a^2)$ so that the left-hand side of (15.30) and the integral on the right-hand side are both small $O(1/a)$. It follows from (15.22) or (15.23) that the coefficients A_n vanish, and the magnetic field vanishes within the slab, as pointed out earlier.

The sum (15.30) converges rapidly for $b/a \geq O(1)$. To obtain a solution for general a and b, assume that $G(\psi)$ can be expanded in the form

$$G(\psi)=\{1-(h\psi/v_0 ab)^2\}^2 \sum_{n=0}^{\infty} a_n \psi^{2n}. \tag{15.32}$$

This is the most general expansion that is an even function of ψ and automatically satisfies (15.20) and (15.31) to terms fourth order in ψ. Then substitute into (15.30), keeping only the first term a_0 in the expansion in ψ. Evaluate the result at $y=0$, so that

$$-\tfrac{1}{2}b^2hB_0 = a_0\left\{\int_{-b}^{+b}\frac{\mathrm{d}\mu\,\mu(1-\mu^2/b^2)}{\sinh(\pi\mu/2a)}\right.$$

$$\left.+2\sum_{n=1}^{\infty}\int_{-b}^{+b}\frac{\mathrm{d}\mu\,\mu(1-\mu^2/b^2)}{\sinh(\pi\mu/2a+n\pi b/a)}\right\}. \tag{15.33}$$

The integrals are easily evaluated by numerical methods. For the special case $a=b$, it is readily shown that

$$a_0 = -0{\cdot}194B_0h.$$

It follows from (15.19) that

$$W = -B_0h\{\ln(hx/a^2)+0{\cdot}194(1-x^2y^2/a^4)\}. \tag{15.34}$$

Then from (15.25) and (15.21) it follows that

$$R(x,y) = B_0h\ln(b/h)$$

$$+3.1B_0h\sum_{n=1}^{\infty}\frac{(-1)^{n+1}\{1-3/(n-\frac{1}{2})^2\pi^2\}}{(n-\frac{1}{2})^3\pi^3\cosh(n-\frac{1}{2})\pi}$$

$$\times\cosh\{(n-\tfrac{1}{2})\pi x/a\}\cos\{(n-\tfrac{1}{2})\pi y/a\}. \tag{15.35}$$

The streamlines and the magnetic lines of force are plotted in Fig. 15.5. Figure 15.5 illustrates the accumulation of lines of force as they are swept against the x-axis by the inward flow from $y=\pm h$. The accumulation is broken by the square x^2, $y^2<a^2$ of high resistivity, reconnecting the lines of force across the x-axis and releasing them from the trap, to flow outward toward $x=\pm\infty$ in the downstream wake of the resistive square. The half width of the wake is $w(x)=a^2/x$. The field within the wake is described by the vector potential W, given by (15.34). It is readily shown that within the wake

$$B_x = 0.39B_0(1-y^2/w^2)hxy/a^2w,$$

$$B_y = B_0\{1-0{\cdot}388y^2/w^2+0{\cdot}388y^4/w^4\}h/x.$$

Thus B_y decreases as $1/x$ as the field is carried out along the x-axis. On the other hand, B_x vanishes at the centre line and at the boundaries of the wake, but has its maximum value at $y=w(x)/3^{\frac{1}{2}}$, where

$$B_x = 0{\cdot}15B_0hx/a^2.$$

This is to be compared to the field (15.17) in the absence of the resistive square, which would be $B_x = 3^{\frac{1}{2}}B_0hx/a^2$. The maximum field in the wake is the fraction 0.087 of the field at the same position were the resistive square not present.

Even more important than the quantitative reduction of the field is the alteration of the topology, so that the lines of force can be swept away from the origin toward $x = \pm\infty$. In the present illustrative example the reduced B_x still increases linearly with x along the wake, just as it does along any stream line in the absence of the resistive square and wake. But the increase is not essential. Rather it is an artifact of the arbitrary and artificial choice of the velocity field. The increase would be avoided if, for instance, the flow along the centre line of the wake (the x-axis) were accelerated slightly so that the point where a line of force crosses the x-axis catches up to the point where the line crosses the boundary of the wake (see Fig. 15.5). That slight modification of the flow in the wake would produce the field

$$B_x = 0, \qquad B_y = B_0 h/x$$

throughout the wake. This field declines to zero, instead of undergoing the unbounded increase (15.17) in the absence of the resistive square. Note that the fact of the reduction of field to zero with increasing distance along the wake is independent of the dimensions of the resistive region. The reduction of the field within the wake requires only that the resistive region exist at the stagnation point.

Altogether, then, the unbounded increase of field in the converging flow (15.16) of a fluid with high electrical conductivity can be avoided only if there is a region of resistive diffusion at the stagnation point of the converging flow. Without a resistive region the topology of the field is invariant and the lines of force convected into the region cannot escape but must accumulate. Only by resistive diffusion at the stagnation point can the lines be reconnected across the x-axis, so as to escape with the outflow of fluid.

15.4. The general effect of diffusion

Now consider some illustrative examples of the variation of the magnetic field in the converging flow (15.16), but with a finite and uniform diffusion coefficient η throughout the entire region (Parker 1973b). Then with the dimensionless variables

$$\xi = x(v_0/\eta h)^{\frac{1}{2}}, \qquad \zeta = y(v_0/\eta h)^{\frac{1}{2}}$$

the induction equation (15.10) reduces to

$$\partial^2 A/\partial\xi^2 - \xi\, \partial A/\partial\xi + \partial^2 A/\partial\zeta^2 + \zeta\, \partial A/\partial\zeta = -cEh/v_0. \qquad (15.36)$$

The variables are separable and the general solution of the homogeneous equation can be written variously as

$$A = \exp(-\tfrac{1}{2}\zeta^2) \sum_{n=-\infty}^{+\infty} \{A_n\,{}_1F_1(-\tfrac{1}{2}n; \tfrac{1}{2}; \tfrac{1}{2}\xi^2) + B_n\xi\,{}_1F_1(-\tfrac{1}{2}n + \tfrac{1}{2}; \tfrac{3}{2}; \tfrac{1}{2}\xi^2)\}$$
$$\times \{C_n\,{}_1F_1(\tfrac{1}{2}n + \tfrac{1}{2}; \tfrac{1}{2}; \tfrac{1}{2}\zeta^2) + D_n\zeta\,{}_1F_1(\tfrac{1}{2}n + \tfrac{1}{2}; \tfrac{3}{2}; \tfrac{1}{2}\zeta^2)\}$$
$$= \sum_{n=-\infty}^{+\infty} \{A_n\,{}_1F_1(-\tfrac{1}{2}n; \tfrac{1}{2}; \tfrac{1}{2}\xi^2) + B_n\xi\,{}_1F_1(-\tfrac{1}{2}n + \tfrac{1}{2}; \tfrac{3}{2}; \tfrac{1}{2}\xi^2)\}$$
$$\times \left\{G_n\,{}_1F_1\left(-\frac{n}{2}; \tfrac{1}{2}; -\tfrac{1}{2}\zeta^2\right\} + H_n\zeta\,{}_1F_1(-\tfrac{1}{2}n + \tfrac{1}{2}; \tfrac{3}{2}; -\tfrac{1}{2}\zeta^2)\right\} \qquad (15.37)$$

where the sum may be over discrete values of n, or an integration over a continuum. The second form is particularly convenient because when n is a positive even integer, the confluent hypergeometric functions reduce to Hermite polynomials in the terms even in ξ and ζ, and when n is a positive odd integer, they reduce to polynomials in the odd terms. For $n = 0$, the terms reduce to

$$A = \{A_0 + B_0 \int_0^{\xi} d\xi' \exp(\tfrac{1}{2}\xi'^2)\}\{G_0 + H_0 \operatorname{erf}(\zeta/2^{\frac{1}{2}})\}. \qquad (15.38)$$

The solution to the inhomogeneous equation is easily constructed. Let

$$A(\xi, \zeta) = F(\xi) + G(\zeta).$$

The variables are again separable and it is easily shown that

$$A(\xi, \zeta) = -(cEh/v_0)\{\nu I_1(\xi) + (1-\nu)I_2(\zeta)\} \qquad (15.39)$$

where ν is an arbitrary number and

$$I_1(\xi) \equiv \int_0^{\xi} du \exp(\tfrac{1}{2}u^2) \int_0^{u} dv \exp(-\tfrac{1}{2}v^2), \qquad (15.40)$$

$$I_2(\zeta) \equiv \int_0^{\zeta} du \exp(-\tfrac{1}{2}u^2) \int_0^{u} dv \exp(+\tfrac{1}{2}v^2). \qquad (15.41)$$

Numerical values for I_1 and I_2 are given in Table 15.1. The general solution is the sum of (15.37) and (15.39).

Now, the inhomogeneous term cE on the right-hand side of (15.10) and (15.36) represents the net rate of convection $\mathbf{v} \times \mathbf{B}$ of magnetic flux. We are interested in the circumstance that opposite fields $\pm B_0$ are convected together at the x-axis from some large distance $y = \pm h$ by the converging fluid velocity (15.16). In the simplest case, the field is in the $\pm x$-direction, with $B_x = \pm B_0$ at $y = \pm h$, respectively. To represent this put $\nu = 0$ so that (15.39) reduces to

$$A(\xi, \zeta) = -(cEh/v_0)I_2(\zeta), \qquad (15.42)$$

TABLE 15.1

X	$I_1(x)$	$I_2(x)$	$I_1'(x)$	$I_2'(x)$
0	0	0	0	0
0·25	0·03	0·03		
0·50	0·13	0·12	0·54	0·47
1·00	0·60	0·44	1·41	0·73
1·50	1·74	0·83	3·35	0·76
2·00	4·45	1·19	8·84	0·64
2·50	12·6	1·48	27·8	0·51
3·00	43·4	1·70	112·5	0·39
3·50	178	1·87	571	0·31
4·00	1 000	2·01	3 740	0·27
4·50	7 300	2·14	31 100	0·23
5·00	70 000	2·25	336 000	0·21

from which it follows that

$$B_x = -(cEh/v_0)(v_0/\eta h)^{\frac{1}{2}} \exp(-\tfrac{1}{2}\zeta^2) \int_0^\zeta dv \exp(\tfrac{1}{2}v^2),$$

$$B_y = 0$$

throughout the entire region $x^2 < h^2$. For $\zeta \gg 1$, then, the asymptotic form is

$$B_x \sim -(cE/v_0)(v_0 h/\eta)^{\frac{1}{2}}(1/\zeta + 1/\zeta^3 + \ldots)$$

This is, of course, just (15.7) and (15.17) again because diffusion is negligible over large dimensions ($\zeta \gg 1$). On the other hand, as the field is convected in close to the x-axis ($\zeta \ll 1$), it varies as

$$B_x \cong -(cE/v_0)(v_0 h/\eta)^{\frac{1}{2}}(\zeta - \tfrac{1}{3}\zeta^3 + \ldots)$$

and passes linearly through zero across the x-axis. The field is dominated by diffusion in the neighbourhood of the x-axis. With the field strength $B_x = B_0$ at $y = h$ far from the x-axis ($hv_0 \gg \eta$) where the inward velocity is $v_y = -v_0$, it follows that $cE = -v_0 B_0$ and

$$B_x = +B_0(hv_0/\eta)^{\frac{1}{2}} \exp(-\tfrac{1}{2}\zeta^2) \int_0^\zeta dv \exp(\tfrac{1}{2}v^2).$$

Altogether, then, the field strength increases as $1/y$ toward the x-axis until y becomes so small that diffusion comes into play. The field then passes through a maximum of 0.8 $(v_0 h/\eta)^{\frac{1}{2}} B_0$ at $\zeta \cong 1.3$ and declines to zero at the x-axis. The field is plotted in Fig. 15.6, together with the variation of the fluid pressure for small velocities, so that $p + B^2/8\pi \cong$ *constant.* The opposite fields on either side of the x-axis are carried together by the convection and obliterated by diffusion. They are not

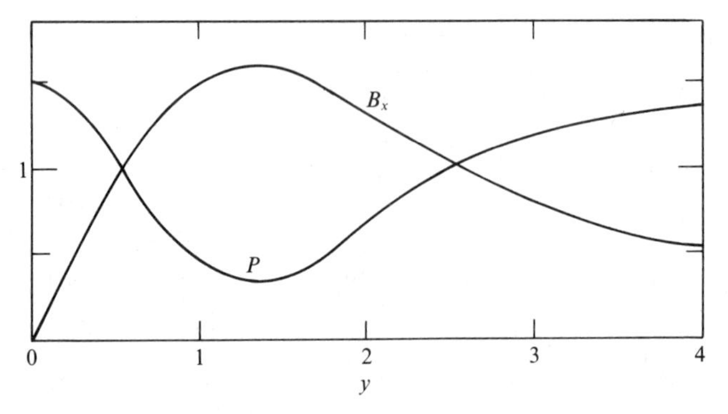

FIG. 15.6. A plot of the magnetic field B_x and fluid pressure variation p as a function of distance y from the neutral plane based on (15.42). Distance y is in units of the diffusion length $(\eta h/v_0)^{\frac{1}{2}}$.

convected out again along the x-axis, in the present example, but cancel each other through resistive diffusion upon approaching the x-axis.

Fields are never precisely antiparallel, as assumed in this idealized example, of course. Introducing a small y-component, so that the field configuration resembles that shown in Figs. 15.1 or 15.2 alters the situation entirely. Magnetic flux is destroyed only in the thin diffusion region $-1<\zeta<+1$ so the flux is conserved in the flow which does not pass through the diffusion. Then every line of force carried in along the y-axis is reconnected by the diffusion to its mirror image on the other side of the x-axis and is then convected out along the x-axis.

As as example, introduce the even term from (15.37) for $n=1$. Then,

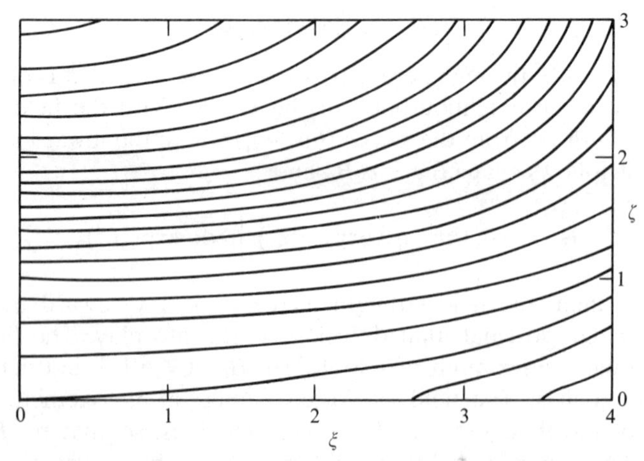

FIG. 15.7. A plot of the magnetic lines of force of two opposite fields brought together by the converging flow (15.16). The field is described by (15.43).

altogether,

$$A(\xi, \zeta) = B_0 h\{I_2(\zeta) + \epsilon(1-\xi^2)(1-\zeta^2)\}. \tag{15.43}$$

The magnetic lines of force are plotted in Fig. 15.7 for $\epsilon = 0.01$.

The illustrations of this and the preceding section (Figs 15.5–15.7) (see also Uberoi 1963) give some idea of the configuration of the magnetic field in the neighbourhood of the neutral stagnation point. The essential nature of the resistive diffusion in breaking the topology of the magnetic lines of force and permitting their escape in the outflow should be apparent. In the next section we move on to diffusion in kinematical velocity fields that resemble the large-scale flows shown in Fig. 15.4 outside the region of diffusion surrounding the neutral point.

15.5. Discontinuous velocity fields and diffusion

Consider the symmetric velocity field of which the first quadrant is sketched in Fig. 15.8, wherein the inflow of fluid from $y = \pm\infty$ is uniform ($v_x = 0$, $v_y = \mp v_0$) into $y = \pm x \tan \alpha$, whereupon the velocity turns through a right angle and flows out toward $x = \pm\infty$, with uniform velocity $v_x = \pm v_0 \cot \alpha$, $v_y = 0$. The fluid flow is incompressible and is conserved across $y = \pm x \tan(\alpha)$. Suppose that the fluid in the first quadrant carries a uniform field B_0 making an angle β with the x-direction so that $B_x = B_0 \cos \beta$, $B_y = B_0 \sin \beta$. An equal and opposite field is presumed carried in from $y = -\infty$, etc. To show what happens to the field (presumed to be frozen into the fluid by the high electrical conductivity) as it is carried across the discontinuity at $y = \pm x \tan \theta$, note from Fig. 15.8 that if an element of fluid on the stream line $x = a$ reaches $y = +x \tan \theta$ at time t,

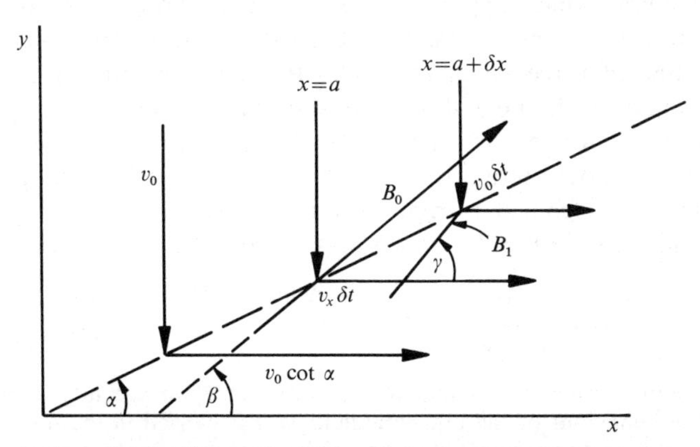

FIG. 15.8. A schematic drawing of the uniform fluid velocity $v_y = -v_0$ in the first quadrant, deflected through $\pi/2$ at the line $y = x \tan(\alpha)$, and flowing toward $x = \pm\infty$ with the uniform velocity $v_0 \cot(\alpha)$. The two lines labelled $x = a$ and $x = a + \delta a$ indicate the paths of two fluid elements on the lines of force of the field B_0. initially inclined at an angle β. Following deflection the inclination is changed to γ.

then the element of fluid on the same line of force at the stream line $x = a + \delta x$ reaches $y = x \tan \theta$ at time $t + \delta t$ where

$$\delta t = (\tan \beta - \tan \alpha)\delta x / v_0.$$

At this time, the first element of fluid has moved from $x = a$, $y = a \tan \alpha$ to the position $x = a + v_x\, \delta t = a + v_0 \cot \alpha\, \delta t$. Thus the line of force connecting the two elements of fluid, now at $(a + v_0 \cot \alpha\, \delta t,\ a \tan \alpha\}$ and $\{a + \delta x, (a + \delta x)\tan \alpha\}$, makes an angle γ with the x-direction, where

$$\begin{aligned} \tan \gamma &= \delta x \tan \alpha / \{\delta x - v_0 \cot \alpha \delta t\} \\ &= \tan^2 \alpha / \{2 \tan \alpha - \tan \beta\}. \end{aligned} \tag{15.44}$$

This relation determines the direction of the field after being carried across the discontinuity at $y = x \tan \alpha$. The magnitude B_1 of the field can be computed from the fact that the normal component is continuous across $y = x \tan \alpha$. The normal component of the field B_0 prior to arrival at the discontinuity is $B_0 \sin(\beta - \alpha)$. After crossing the continuity the normal component is $B_1 \sin(\gamma - \alpha)$. Equating these two components, and writing γ in terms of α and β with the aid of (15.44), it follows that

$$B_1 = B_0 \cos \beta \cot \alpha [\tan^4 \alpha + \{2 \tan \alpha - \tan \beta\}^2]^{\frac{1}{2}}. \tag{15.45}$$

Hence

$$B_x = B_0 \cos \beta \{2 - \tan \beta \cot \alpha\}. \tag{15.46}$$

$$B_y = B_0 \cos \beta \tan \alpha \tag{15.47}$$

after crossing the discontinuity[3].

Consider now what happens to the field as a consequence of resistive diffusion into the opposite field on the other side of the x-axis. Every magnetic line of force is carried up to the x-axis at the origin, so that reconnection of each line with its opposite number occurs at the origin no matter how large the electrical conductivity of the fluid. The development of the diffusion downstream from the origin is readily illustrated once the field is well away from the origin ($x \tan \alpha \gg \eta / v_0$). For then the diffusion layer is relatively thin and $\partial/\partial x \ll \partial/\partial y$ so that (15.10) reduces to

$$v_x\, \partial A/\partial x - \eta\, \partial^2 A/\partial y^2 \cong cE \tag{15.48}$$

[3] The problem is treated kinematically, so far, since we are interested only in studying the behaviour of the magnetic field as it is carried in the flow. But note that the discontinuity $y = x \tan(\alpha)$ is essentially an Alfven wave, whose velocity of propagation perpendicular to the front is just the Alfven speed computed in the normal component of the magnetic field. Hence the flow is dynamically self-consistent at the wave front $y = x \tan \alpha$, rather than merely kinematical, if $B_0 \sin |\alpha - \beta| = v_0 (4\pi\rho)^{\frac{1}{2}} \cos \alpha$.

with $cE = v_y B_x = -v_0 B_0 \cos\beta$ determined by the inflow of magnetic flux from $y = \pm\infty$. The solution of the inhomogeneous equation can be written

$$A(x, y) = v_0 B_0 \cos\beta\{(1-\nu)y^2/2\eta - (\nu x/v_0)\tan\alpha\}$$

where ν is a numerical constant, to be determined from the boundary conditions. The homogeneous equation is just the familiar parabolic heat flow equation, of course. The solution needed to treat the diffusion in the neighbourhood of the x-axis is

$$A(x, y) = v_0 B_0 \cos\beta\, C \int_0^y dy' \operatorname{erf}[y'\{v_0/4\eta x \tan\alpha\}^{\frac{1}{2}}]$$

where C is an arbitrary constant. Together, then, the total field is

$$B_x(x, y) = v_0 B_0 \cos\beta[(1-\nu)y/2\eta + C \operatorname{erf}\{y(v_0/4\eta x \tan\alpha)^{\frac{1}{2}}\}],$$
$$B_y(x, y) = v_0 B_0 \cos\beta[(\nu/v_0)\tan(\alpha) + C(\eta \tan\alpha/\pi x v_0)^{\frac{1}{2}}\{1 - \exp(-y^2 v_0/4\eta x \tan\alpha)\}].$$

Well back from the x-axis, where diffusion has not yet had a sensible effect[4], i.e. where

$$x \tan\alpha > y \gg (4\eta x \tan\alpha/v_0)^{\frac{1}{2}}, \qquad (15.49)$$

the above expressions reduce asymptotically to

$$B_x(x, y) \sim v_0 B_0 \cos\beta\{(1-\nu)y/2\eta + C\},$$
$$B_y(x, y) \sim B_0 \nu \cos\beta \tan\alpha.$$

These components must be identical with (15.46) and (15.47), from which it follows that $\nu = 1$ and

$$C = (2 - \tan\beta \cot\alpha)/v_0.$$

Hence

$$A(x, y) = B_0 \cos\beta[-x \tan\alpha + (2 - \tan\beta \cot\alpha) \int_0^y dy' \operatorname{erf}\{y'(v_0/4\eta x \tan\alpha)^{\frac{1}{2}}\}], \qquad (15.50)$$

and

$$B_x = B_0 \cos\beta(2 - \tan\beta \cot\alpha)\operatorname{erf}\{y(v_0/4\eta x \tan\alpha)^{\frac{1}{2}}\}, \qquad (15.51)$$
$$B_y = B_0 \cos\beta[\tan\alpha + (2 - \tan\beta \tan\alpha)(\eta \tan\alpha/\pi v_0 x)^{\frac{1}{2}} \times\{1 - \exp(-y^2 v_0/4\eta x \tan\alpha)\}]. \qquad (15.52)$$

[4] We neglect the slight rounding of the sharp corner in each line of force at $y = x \tan\alpha$.

Now the approximate equation (15.48), from which (15.51) and (15.52) are deduced, is valid for x sufficiently large that the characteristic thickness of the diffusion region $D(x)=(4\eta x \tan\alpha/v_0)^{\frac{1}{2}}$ is small compared to x, i.e. for

$$xv_0/4\eta \tan\alpha \gg 1.$$

Hence the results are not correct in the neighbourhood of the origin. Away from the origin, then, where the equations are valid, (15.52) reduces to

$$B_y = B_0 \cos\beta \tan\alpha[1+\mathrm{O}\{(\eta \tan\alpha/v_0 x)^{\frac{1}{2}}\}],$$

which is just (15.47). Thus the y-component of the field is essentially unaffected by diffusion, because B_y is uniform across the x-axis. As we would expect, it is B_x that is dissipated, because of its steep gradient across the x-axis. The x-component varies as $\mathrm{erf}\{y/D(x)\}$ across the x-axis. An example is shown in Fig. 15.9 where the lines of force for the special case $\beta=0$ are plotted in the xy-plane using the dimensionless coordinates $r=4v_0 x\tan(\alpha)/\eta$, $s=v_0 y/\eta$ so that

$$B_x = 2B_0\,\mathrm{erf}(s/r^{\frac{1}{2}}),\qquad B_y \cong B_0 \tan\alpha, \tag{15.53}$$

$$A = \frac{B_0\eta}{4v_0}\left\{-r+8r^{\frac{1}{2}}\int_0^{s/r^{\frac{1}{2}}} du\,\mathrm{erf}(u)\right\} \tag{15.54}$$

and α does not appear explicitly in the functional form of the vector potential. The calculations are valid for the field between the x-axis and the discontinuity $y=x\tan\alpha$ $(r=4s\tan^2(\alpha))$ for $r\gg 4\tan^2(\alpha)$.

Another example of particular interest is $\gamma=\pi/2$. It will arise later in Petschek's model of the neutral point annihilation in §15.6. With $\gamma=\pi/2$, $B_x=0$ it follows from (15.46) that

$$\tan\alpha = \tfrac{1}{2}\tan\beta. \tag{15.55}$$

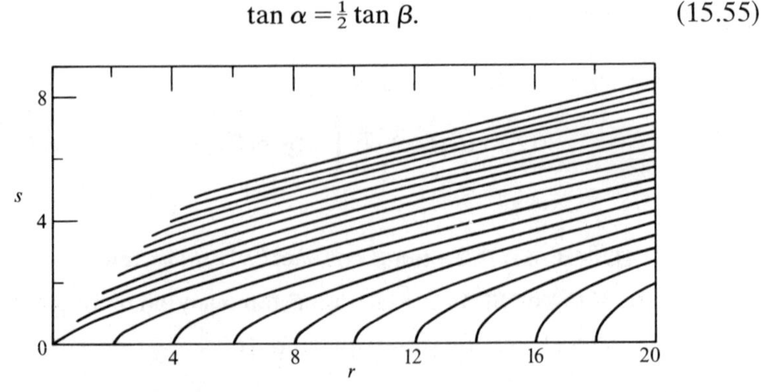

FIG. 15.9. The magnetic field lines of (15.53) and (15.54) in the first quadrant. The results are valid for $0<r<4s^2\tan(\alpha)$ and $r\gg\tan^2(\alpha)$, with the angle α a free parameter in the calculations. The lines are plotted at equal flux intervals.

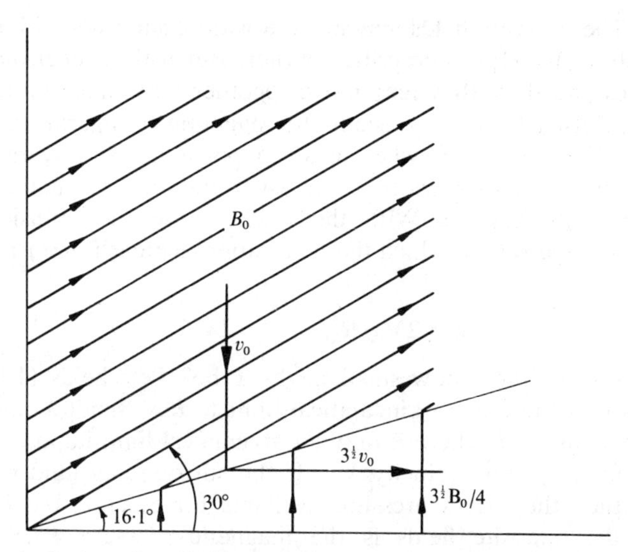

FIG. 15.10. A plot from (15.56) of the magnetic lines of force (initially at an inclination of 30°) as they are carried toward the x-axis in a flow that is abruptly deflected through $\pi/2$ on the line $y = \frac{1}{2}x/3^{\frac{1}{2}}$. The lines reconnect at the origin where the field gradient is infinitely steep.

Hence

$$B_y = \tfrac{1}{2} B_0 \cos \beta. \tag{15.56}$$

There is, then, no effect of resistive diffusion, except at the origin where the lines of force reconnect across the x-axis[5]. The y-component of the field is unaffected, and the x-component is zero. The field configuration in the first quadrant is shown in Fig. 15.10 for $\beta = 30°(\alpha = 16{\cdot}1°)$, for which the outflow velocity is $v_0 \cot \alpha = 3^{\frac{1}{2}} v_0$ and the field in the outflow is $B_y = B_0 3^{\frac{1}{2}}/4$.

Altogether, these examples—perhaps too numerous—should establish some idea of the behaviour of magnetic fields of opposite sign carried against each other by the converging motion of the fluid. With the physical picture firmly established in mind, consider the dynamics of such fields and flows.

15.6. Petschek's dynamical considerations

We treat the dynamics of the field configuration wherein two opposite fields $\pm B$ face each other across the x-axis without the necessary fluid pressure enhancement to keep them apart. The fields merge as described

[5] Anticipating the dynamical considerations of the next section note[3] that the flow is dynamically self-consistent at the discontinuity if $v_0 = \frac{1}{2} \sin \beta \, B_0/(4\pi\rho)^{\frac{1}{2}}$.

in §15.1. The opposite fields may move toward each other with a velocity w as small as $V_A/R_m^{\frac{1}{2}}$ as described earlier. But if they meet only over a very narrow width λ, they may move together very much more rapidly. Following Petschek, then, consider the magnetic configuration shown in Figs 15.2, 15.3, and 15.10, wherein two opposite fields of overall width L meet only over a narrow width λ, for which the local magnetic Reynolds number is $R_\lambda = 2\lambda V_A/\eta$. With the thickness of the diffusion region denoted by l, the rate at which the fields approach each other is given by (15.3) as

$$w = 2V_A/R_\lambda^{\frac{1}{2}}, \qquad l = 2\lambda/R_\lambda. \tag{15.57}$$

The question is, then, how small might λ be? Petschek's (1964) basic point was that there is no kinematical limit to how small λ may be and how large w may be. There is only the dynamical limit imposed on w by the maximum ejection velocity V_A. If the surrounding fluid pressure is uniform, then the excess pressure available for ejecting the fluid from between the opposite fields is the magnetic pressure $B^2/8\pi$ of the opposite fields. Hence the ejection velocity v is approximately equal to the Alfven speed $V_A = B/(4\pi\rho)^{\frac{1}{2}}$ and cannot be larger. Conservation of fluid requires (15.1), so that the merging velocity w is

$$w = V_A l/\lambda.$$

Thus if λ is as small as l, we might expect that w could be as large as V_A. That would be rapid annihilation and reconnection indeed! The question is whether nature would have it so. For suppose, as in Figs. 15.2 and 15.10, that two fields of width L approach each other with a velocity w. Denote by α the half-angle of the exit flow, outward along the x-axis, in which the outflow velocity is V_A. Then conservation of fluid over the whole region requires that

$$w = V_A \tan\alpha.$$

Hence if w were comparable to V_A, the half-angle would be $\alpha = \pi/4$. The fields, then, would meet only in the apex of a sharp corner. Indeed, it follows from (15.55) that the lines of force are folded back by an angle $\beta = 63{\cdot}5°$ from the x-axis. A magnetic field extends only very weakly into so sharp a corner. The tension in the lines of force around a sharp curve pulls the field strongly away from the corner. The field pressure in the corner, which is responsible for squeezing the fluid out from between the fields where they meet in the narrow diffusion region, is enormously reduced and the effective Alfven speed V_A to be used in (15.56) is greatly reduced from the value $B/(4\pi\rho)^{\frac{1}{2}}$ for the field as a whole. Evidently, then, there is some optimum value of λ, associated with some optimum angle α, for which the merging velocity w is a maximum, and evidently less than

V_A. The key to the question is how effectively a magnetic field extends into a sharp corner. The geometry is set forth in detail in Fig. 15.11.

To estimate the field in a wedge of half-angle $\pi/2-\alpha$ note that the two opposite fields of strength B, subject to uniform external fluid pressure, and slowly pushing the fluid in from each side with the small velocity w, will be in nearly potential form,

$$B_x = -\partial\Phi/\partial x, \qquad B_y = -\partial\Phi/\partial y,$$
$$\partial^2\Phi/\partial x^2 + \partial^2\Phi/\partial y^2 = 0.$$

Suppose, then, that, as sketched in Fig. 15.11 the lines of force make an angle β with the x-axis on either side, so that the field occupies a wedge of half-angle $\pi/2-\beta$. In polar coordinates $\varpi=(x^2+y^2)^{\frac{1}{2}}$, $\tan\phi = y/x$, the appropriate solution of Laplace's equation is

$$\Phi = -B_0 h(1-2\beta/\pi)(\varpi/h)^{1/(1-2\beta/\pi)}\cos\{(\phi-\beta)/(1-2\beta/\pi)\}$$

so that

$$B_\varpi = B(\varpi/L)^{2\beta/(\pi-2\beta)}\cos\{(\phi-\beta)/(1-2\beta/\pi)\}, \tag{15.58}$$
$$B_\phi = B(\varpi/L)^{2\beta/(\pi-2\beta)}\sin\{(\phi-\beta)/(1-2\beta/\pi)\}. \tag{15.59}$$

wherein the field has magnitude B at the distance $\varpi = L$ from the origin. The lines of force are confined between $\phi=\beta$ and $\phi=\pi-\beta$ with B_ϕ vanishing at $\phi=\beta,\ \pi-\beta$.

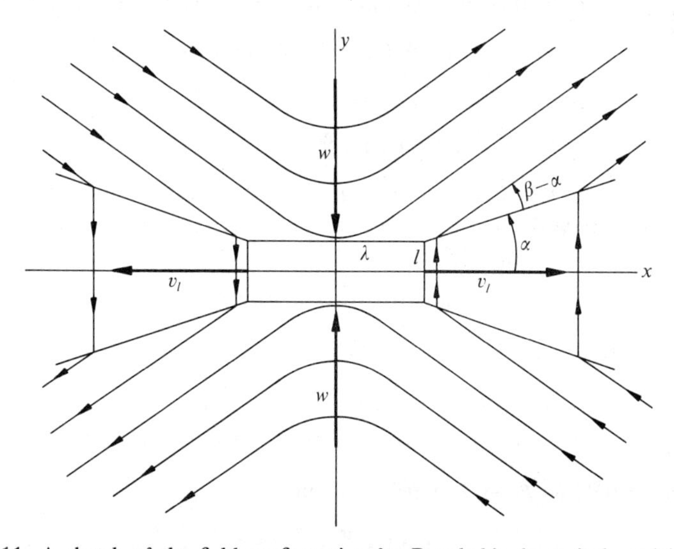

FIG. 15.11. A sketch of the field configuration for Petschek's dynamical model of rapid reconnection. The standing Alfven wave is inclined at an angle α relative to the x-axis, while the incoming field is inclined at an angle β. The rectangle with sides 2λ and $2l$ enclosing the origin represents the diffusion region, wherein resistive diffusion is important.

The field at the surface of the diffusion region $x=0,\ y=l$ is

$$B_l \equiv B_\phi(l, \pi/2) = B(l/L)^{2\beta/(\pi-2\beta)}.$$

It is the pressure of B_l that expels the fluid from the region, so that the velocity of expulsion is approximately

$$\begin{aligned} V_l &= B_l/(4\pi\rho)^{\frac{1}{2}}, \\ &= V_A(l/L)^{2\beta/(\pi-2\beta)}. \end{aligned} \tag{15.60}$$

It is the limitation on the velocity V_l that restricts β to relatively small values and makes $w < V_A$.

To proceed, then, conservation of fluid requires that $w\lambda = V_l l$ within the diffusion region and $w = V_l \tan\alpha$ across the shock outside. Hence, if the velocity fields are uniform and continuous, it follows that

$$w/V_l = l/\lambda = \tan\alpha. \tag{15.61}$$

The deflection of the fluid flow at $y = \pm x \tan\alpha$ represents a standing Alfven wave. The propagation speed of the wave perpendicular to the plane of the wave is $B_n/(4\pi\rho)^{\frac{1}{2}}$ where B_n is the normal component of the magnetic field. For steady conditions the propagation velocity is just equal to the normal component of the fluid velocity $w\cos\alpha$. At the corner $(x=\lambda,\ y=l)$ of the diffusion region, the field is approximately $B(\lambda/L)^{2\beta/(\pi-2\beta)}$ in the radial direction for the idealized field (15.58) and (15.59) and the further approximation that $l \ll \lambda$, $\alpha \ll 1$. Hence

$$w \cong V_A(\lambda/L)^{2\beta/(\pi-2\beta)}\sin(\beta-\alpha).$$

For $\alpha, \beta \ll 1$, then

$$w \cong V_A(\lambda/L)^{2\beta/\pi-2\beta)}(\beta-\alpha), \tag{15.62}$$

where we neglect terms $O(\alpha)$ and $O(\beta)$ compared to one in the various factors but carry the exponents of those factors without approximation. Equations (15.60) and (15.61) yield

$$w = \alpha V_A(l/L)^{2\beta/(\pi-2\beta)} \tag{15.63}$$

so that (15.62) and (15.63) together reduce to

$$\begin{aligned} \beta &= \alpha\{1+(l/\lambda)^{2\beta/(\pi-2\beta)}\} \\ &= \alpha(1+\alpha^{2\beta/(\pi-2\beta)}). \end{aligned}$$

Since $\alpha = O(\beta)$ as $\beta \to 0$, it follows that

$$\lim_{\beta\to 0} \alpha^{2\beta/(\pi-2\beta)} = 1$$

and

$$\beta = 2\alpha. \tag{15.64}$$

This was just the condition (15.55) deduced from flux conservation for this same configuration.

With the condition (15.2) connecting w and l, the basic relations are now all in hand. From (15.61), and then (15.2), it follows that

$$\begin{aligned}\alpha &= l/\lambda \\ &= 2\eta/w\lambda \\ &= 4(\eta/2LV_A)(L/\lambda)(V_A/w).\end{aligned}$$

Then in terms of the overall magnetic Reynolds number $R_m = 2LV_A/\eta$ and the result from (15.62) and (15.64) that

$$w = \alpha V_A(\lambda/L)^{2\beta/(\pi-2\beta)}, \tag{15.65}$$

we obtain

$$L/\lambda = (\tfrac{1}{4}\alpha^2 R_m)^{1-2\beta/\pi}, \tag{15.66}$$

Hence L/λ can be eliminated from (15.62), so that

$$w/V_A = \alpha^{1-8\alpha/\pi}(\tfrac{1}{4}R_m)^{-4\alpha/\pi} \tag{15.67}$$

upon writing $\beta = 2\alpha$ in the exponents. To find the optimum value of α, for maximum w/V_A, differentiate with respect to α and set the result equal to zero. It follows that

$$4\alpha = \pi\{\ln(\tfrac{1}{4}R_m) + 2 + 2\ln(\alpha)\}^{-1},$$

which can be iterated and written in terms of the continued logarithmic fraction

$$\begin{aligned}\alpha &= \cfrac{\pi/4}{\ln(\tfrac{1}{4}R_m) + 2 + 2\ln\left[\cfrac{\pi/4}{\ln(\tfrac{1}{4}R_m) + 2 + \ln\left\{\cfrac{\pi/4}{\ln(\tfrac{1}{4}R_m) + 2 + \ldots}\right\}}\right]} \\ &\cong \frac{\pi}{4\ln(\tfrac{1}{4}R_m)}\left\{1 - \frac{2}{\ln(\tfrac{1}{4}R_m)} - \frac{2}{\ln(\tfrac{1}{4}R_m)}\ln\left(\frac{\pi/4}{\ln(\tfrac{1}{4}R_m)}\right) + \ldots\right\}\end{aligned}$$

It follows that the maximum merging velocity w is

$$w/V_A \cong \pi/4e\ln(\tfrac{1}{4}R_m) \cong \pi/4e\ln(R_m), \tag{15.68}$$

neglecting terms of the order of $\ln(\ln(R_m))$ compared to $\ln(R_m)$. We write $e \equiv 2{\cdot}718$ for the base of the natural logarithm. To the same approximation it follows from (15.66) that

$$L/\lambda = \tfrac{1}{4}eR_m(\pi/4e\ln(R_m))^2. \tag{15.69}$$

From (15.2) it follows that

$$l/\lambda = (4/R_m)(V_A/w)L/\lambda.$$

With (15.68) and (15.69) this becomes

$$\tan \alpha = l/\lambda \cong \pi/4 \ln(R_{m}). \tag{15.70}$$

These results are based on the approximate potential form for the field, and on the approximate conservation relations (15.61). Hence they are an estimate of the merging velocity, rather than a precise result. For comparison, Petschek's estimate of the merging rate is larger than (15.68) by a factor e (see discussion in Vasyliunas 1975), and the result of Roberts and Priest (1975) is smaller by the factor $0{\cdot}21e = 0{\cdot}57$. Roberts and Priest carried out a much more sophisticated calculation of the magnetic field at the boundary of the diffusion region. But all the analyses are based on the approximate, order of magnitude, relations (15.2) and (15.61). So the factor $\pi/4e \ln(R_{m})$ must be considered uncertain to at least a factor of two. Vasyliunas (1975) gives an extended discussion of the problem and the reader is referred to that paper for a more detailed analysis.

A recent paper by Yeh (1976a) discusses the formal expansion of the fluid velocity and the magnetic field in the neighbourhood of the origin[6]. He notes in particular the poor convergence of any expansion schemes used so far, and points out that there is no compelling reason to require $\lambda \gg l$. Fukao and Tsuda (1973a, b) present several numerical experiments. They encounter severe numerical convergence problems as soon as the magnetic Reynolds number R_{m} is as large as the number of grid points with which the computation is carried out. They had available 200 grid points in the y-direction, so it is difficult to compare their results with the limiting case of $R_{m} \to \infty$ employed here. They obtain numerical solutions for w/V_{A} of 0·5 and 1·0 but whether this was the result of resistive or numerical diffusion is hard to say.

It appears from Petschek's analysis that there is rapid merging wherever there is a reversal of one of the components of the magnetic field along the lines of force on which the fluid pressure is inadequate to keep the reversed portions apart. The merging may proceed at any speed from $V_{A}/R_{m}^{\frac{1}{2}}$ up to something of the order of $\frac{1}{3}V_{A}/\ln(R_{m})$. So far as we are aware, no one has yet succeeded in carrying out the formal dynamical analysis to such a degree of precision that the merging velocity w can be directly related to the controlling physical conditions at the external boundaries of the field. So we can do no more than point out the possibility, and the likely occurrence, of the higher merging rates.

In this connection it should be noted that Sonnerup has worked out an interesting dynamical situation in which the merging velocity may be as large as V_{A}, in the limit of large magnetic Reynolds number R_{m}. Again

[6] See also the exact solutions of Uberoi (1963) and Yeh and Axford (1970) and the formal illustrations of Sonnerup and Priest (1975) and Priest and Cowley (1975).

the physical setting for such high merging rates is not entirely clear, but the possibility for $w \cong V_A$ is an important concept, and is the subject of the next section.

15.7. Sonnerup's dynamical solution

Petschek's analysis of the merging rate of opposite fields is based on the deflection of the fluid velocity through $\pi/2$ at a standing Alfven wave. The approaching field is folded into a wedge of half-angle $\pi/2-\beta$, as shown in Fig. 15.11, and it is the decrease of the field in the point of the wedge with increasing β that limits the rate of merging. The Alfven wave, at which the deflection occurs, is an oblique wave propagating outward, away from the diffusion region at the origin, as is readily seen from Fig. 15.11 if one pictures himself approaching the origin with the incoming fluid. The wave originates at the diffusion region, presumably as a consequence of the merging of field and the associated hydrodynamic stagnation point. One can imagine that the merging progresses initially at some very slow rate (15.3). Then, as the Alfven waves begin to form, the rate increases up to a value in the neighbourhood of (15.68).

There are other possibilities, of course, and Sonnerup (1970) pointed out that with two standing Alfven waves in each quadrant, instead of only one, a rather different situation may arise. He was able to work out a complete dynamical picture, with an entirely plausible diffusion region at the origin. The outstanding feature is that the approaching field may be uniform, rather than folded, so that no rate limit short of V_A is introduced into the problem. The field configuration is shown in Fig. 15.12,

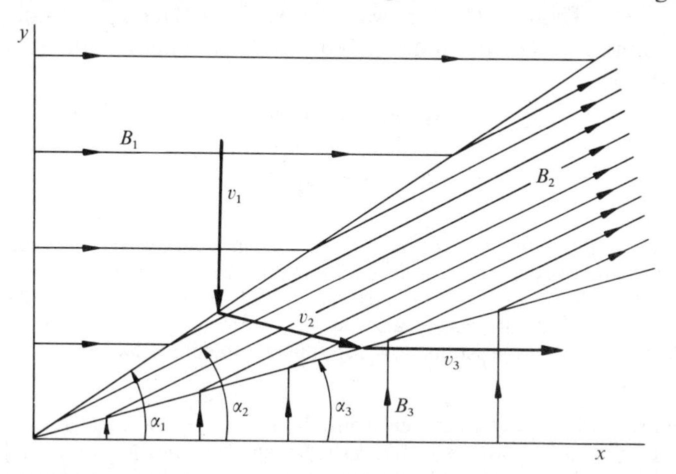

FIG. 15.12. The magnetic lines of force and the fluid velocity in the first quadrant in the form proposed by Sonnerup for rapid reconnection. The magnetic and velocity fields are uniform except where they cross the two standing Alfven waves at $y = x \tan \alpha_1$ and $y = x \tan \alpha_3$.

wherein a uniform field B_1 in the x-direction is carried in with the uniform velocity $w = v_1$. This field cannot be converted into a uniform outflow along the x-axis by a *single* standing Alfven wave, but it can be so converted by *two* standing waves. The impossibility of passing from the uniform field in the x-direction ($\beta = 0$) to a field in the y-direction ($\gamma = \pi/2$)[7] follows directly from (15.55). Putting $\beta = 0$ yields $\alpha = 0$, leading[5,8] to vanishing inflow velocity because $w = V_A \tan\alpha$. With two standing waves, on the other hand, the transformation of the uniform field in the x-direction to a uniform field in the y-direction is readily effected.

Consider, then, the velocity and field configuration shown in Fig. 15.12. It is sufficient to treat only the first quadrant, wherein a uniform field B_1 in the x-direction is carried in the fluid with the uniform velocity v_1 in the negative y-direction toward the x-axis. The flow encounters the plane Alfven wave front $y = x \tan\alpha_1$ and is deflected into $\mathbf{v}_2$, which is most conveniently denoted by its components v_x and $-v_y$, so that v_x and v_y are both positive quantities. The magnetic field is compressed to the uniform field B_z making an angle α_2 with the x-axis. The flow encounters a second Alfven wave front $y = x \tan\alpha_3$ and is deflected into the x-direction with velocity v_3. The magnetic field is rotated and expanded into the uniform field B_3 in the y-direction and is carried out toward $x = +\infty$ along the x-axis.

To show that the dynamics is self-consistent and to work out the relations between the various fields and velocities, suppose that the incoming field B_1 and velocity v_1 are specified. It will be convenient to represent the fields in terms of their equivalent Alfven velocities $V_1 \equiv B_1/(4\pi\rho)^{\frac{1}{2}}$, etc. Then, at the first wave front, $y = x \tan\alpha_1$, continuity of the normal components of the fluid velocity and the magnetic field yields the two conditions

$$v_1 \cos\alpha_1 = v_x \sin\alpha_1 + v_y \cos\alpha_1 \tag{15.71}$$

and

$$V_1 \sin\alpha_1 = V_2 \sin(\alpha_1 - \alpha_2), \tag{15.72}$$

[7] At first sight it might appear that $\gamma = \pi/2$ is an unnecessary restriction, with some other value, say $\gamma = \pi/3$ as a viable alternative. But that leads to a kink in the field where it crosses the x-axis and joins to the field from the fourth quadrant. The kink would accelerate the fluid from the uniform flow treated by Petschek, and no self-consistent flow for this circumstance has yet been demonstrated.

[8] As a matter of fact, it can be seen from (15.44) that there is no unique limiting angle α for $\gamma \to \pi/2$ and $\beta \to 0$, because, for small α and β, (15.44) reduces to $\tan\gamma = \alpha^2/(2\alpha - \beta)$. As $\gamma - \pi/2$ and $\tan\gamma$ become large, $2\alpha - \beta$ must become small compared to α^2. Approaching the limit in this way, the result is $2\alpha = \beta + O(\beta^{2+\epsilon})$ for $\beta \to 0$, whereas, if we set $\beta = 0$ at the outset, then $\tan\gamma = \frac{1}{2}\alpha$ and there is no solution for α as $\gamma \to \pi/2$.

respectively. The perpendicular velocity of the wavefront relative to the fluid is equal to the Alfven speed of the normal component of the magnetic field, so that

$$v_1 \cos \alpha_1 = V_1 \sin \alpha_1 \tag{15.73}$$

Finally, considering two elements of fluid on the same line of force separated by a distance δx, we calculate their relative position after passage through $y = x \tan \alpha_1$, as in deriving (15.44). The line of force connects the two elements after passage through the front, with the result that the inclination of the line of force is then α_2, where

$$\tan \alpha_2 = (v_1 - v_y)/(v_1 \cot \alpha_1 + v_x). \tag{15.74}$$

Upon arrival at the second front $y = x \tan \alpha_2$, continuity of the normal components of velocity and magnetic field yields

$$v_x \sin \alpha_3 + v_y \cos \alpha_3 = v_3 \sin \alpha_3 \tag{15.75}$$

and

$$V_2 \sin(\alpha_2 - \alpha_3) = V_3 \cos \alpha_3. \tag{15.76}$$

The wave velocity perpendicular to the front is

$$v_3 \sin \alpha_3 = V_3 \cos \alpha_3. \tag{15.77}$$

Finally, we consider again two points initially on the same line of force and moving with the fluid. Following passage through $y = x \tan \alpha_3$ one point must lie directly above the other, so that the line of force connecting them is in the y-direction. This leads to the condition that

$$v_3(\tan \alpha_2 - \tan \alpha_3) = v_y + v_x \tan \alpha_2. \tag{15.78}$$

These eight equations, then, determine v_x, v_y, v_3, V_2, V_3 and α_1, α_2, α_3 in terms of the incoming fluid velocity v_1 and field V_1. After some considerable algebraic manipulation the solution of (15.71)–(15.78) proves to be

$$v_x = V_1/2^{\frac{1}{2}}, \qquad v_y = v_1/2^{\frac{1}{2}}(1+2^{\frac{1}{2}}), \tag{15.79}$$

$$v_3 = V_1(1+2^{\frac{1}{2}}), \tag{15.80}$$

$$V_2 = \{v_1^2 + (1+2^{\frac{1}{2}})^2 V_1^2\}^{\frac{1}{2}}/2^{\frac{1}{2}}, \tag{15.81}$$

$$V_3 = v_1/(1+2^{\frac{1}{2}}), \tag{15.82}$$

$$\tan \alpha_1 = v_1/V_1 \tag{15.83}$$

$$\tan \alpha_2 = v_1/(1+2^{\frac{1}{2}})V_1, \tag{15.84}$$

$$\tan \alpha_3 = v_1/(1+2^{\frac{1}{2}})^2 V_1, \tag{15.85}$$

as can be verified by substitution back into the eight equations.

The total pressure $p + B^2/8\pi$ is constant across an Alfven wave, and therefore is constant throughout the entire quadrant. It follows that the uniform fluid pressure p_1 in the incoming fluid becomes

$$p_2 = p_1 + (B_1^2 - B_2^2)/8\pi$$
$$= p_1 - \tfrac{1}{4}\rho\{V_1^2(1+2^{\frac{3}{2}}) + v_1^2\} \tag{15.86}$$

after crossing the wave front at $y = x \tan \alpha_1$ and

$$p_3 = p_1 + \tfrac{1}{2}\rho\{V_1^2 - v_1^2/(1+2^{\frac{1}{2}})^2\} \tag{15.87}$$

after crossing $y = x \tan \alpha_3$ into the region of uniform outflow.

In summary, then, the dynamical requirements (15.71)–(15.78) are self-consistent, so that a complete solution exists. The solution exists for all merging velocities v_1 no matter how large compared to the Alfven speed V_A in the merging fields. The system can be driven by suitable external pressures at any rate that we wish. The total pressure $p + B^2/8\pi = p_1 + \frac{1}{2}\rho V_1^2$ is constant throughout the system, and the velocity of ejection of fluid from the system is given by (15.80) as $V_1(1+2^{\frac{1}{2}})$ for all inflow velocities $w = v_1$. Thus the fluid velocity increases through the two Alfven waves if the inflow v_1 is less than $V_1(1+2^{\frac{1}{2}})$ and decreases if greater, while the magnetic field decreases if $v_1 < V_1(1+2^{\frac{1}{2}})$ and increases if v_1 is greater. Note that the exit angle α_3 increases as the system is driven faster.

We are interested in the circumstance that the magnetic field drives the fluid, for which a necessary condition would be that the field energy decline through the passage across the two waves, i.e. $V_3 < V_1$. Hence the merging velocity is limited to

$$v_1 < V_1(1+2^{\frac{1}{2}}).$$

Whether v_1 is as large as $V_1(1+2^{\frac{1}{2}})$ depends upon external conditions.

The question of external conditions brings us to the basic dilemma posed by Sonnerup's dynamical model. It was noted above that Petschek's model involves Alfven waves moving outward from the diffusion region at the origin (as seen by the fluid). It is evident from inspection of Fig. 15.12 that the first wave in Sonnerup's model is moving *inward* toward the diffusion region. The wave must be initiated by some external disturbance (Sonnerup noted that a suitable corner in the external boundary would be sufficient), and it is not clear whether such a wave can be expected under ordinary circumstances in nature. What is more, the wave, initiated at some distance L from the origin, will not be perfectly sharp by the time $O(L/V_A)$ that it arrives at the origin, even if it were sharp to begin. In a time L/V_A resistive diffusion will have rounded the sharp angle in the field to a characteristic width of the order of $(\eta L/V_A)^{\frac{1}{2}}$. The thickness l of the diffusion region will not be less than the thickness $(\eta L/V_A)^{\frac{1}{2}}$ of the wave (see Fig. 15.3) so that from (15.2) we have a merging rate that is not more than something of the order of (15.3). A merging rate $v_1 = V_1$ would require that l be as small as L/R_m. Evidently the ideal circumstance treated by Sonnerup (1970) is possible only if the resistive diffusion is much smaller throughout the space than it is in the diffusion region around the origin, so that the first Alfven wave remains sharp and l can be small as L/R_m. The resistive diffusion coefficient in the diffusion region would have to be larger than the general η throughout the field by at least the factor $R_m^{\frac{1}{2}}$, where R_m is the general magnetic Reynolds number LV_A/η. The possibility for enhanced resistivity in the diffusion region exists in very intense field gradients wherein the electron conduction velocities become comparable to the ion sound speed (see, for instance, Linhart 1960; Alfven and Carlquist 1967; Hamberger and Jancarik 1970; Coppi and Mazzucato 1971). Thus, one should think of Sonnerup's model in connection with the powerful currents of the solar flare and perhaps the geomagnetic tail.

Sonnerup works out a plausible semi-quantitative model for the diffusion region in the neighbourhood of the origin on the basis of the model with infinitely sharp Alfven waves. He demonstrates again the principle that the diffusion region can shrink sufficiently to accommodate any merging velocity v_1 up to $V_1(1+2^{\frac{1}{2}})$.

It is evident from Fig. 15.12 that the first standing wave in Sonnerup's model converts the field into the wedge form needed for Petschek's model, shown in Fig. 15.11 (Sonnerup 1973). The conversion to the wedge form is accomplished by the dynamical stresses of the standing wave so that the field strength is not reduced in the apex of the wedge, as when the potential form of the field is employed. This is why Sonnerup's model gives merging velocities as large as the Alfven speed $V_1 \equiv V_A$. It is clear, therefore, that any dynamical circumstance that can thrust the lines of force into the point of a wedge and press that point against an opposite field, leads to rapid merging, at rates of the order of the Alfven speed V_A. The first Alfven wave in Sonnerup's model is such a dynamical device, if it can be initiated by some sufficient external effect.

Altogether, then, Sonnerup's model, whether it exists in nature or not, serves to illustrate the theoretical possibility for achieving merging rates comparable to the Alfven speed. From it we can see that such rapid merging, requires transporting the field undiminished to the doorstep of the diffusion region. Petschek's model illustrates another range of possibilities. It remains to be shown what external circumstances can push the field into the blunt wedge $\alpha = \pi/4 \ln(R_m)$ for the maximum merging rate (15.68) obtainable from Petschek's model, not to mention the $\alpha = O(1)$ provided by Sonnerup's model with merging rates as large as the Alfven speed. Yang and Sonnerup have recently extended the work to a compressible fluid (Yang and Sonnerup 1976, 1977).

Yeh and Axford (1970) present an elegant similarity solution of the complete dynamical hydromagnetic equations in the neighbourhood of the neutral point at the origin (Yeh 1976b). However, the solutions are well-behaved only over finite angular sectors about the origin, so that without external physical boundaries to cut off the solutions short of the singular lines, the solutions do not apply to real circumstances. Yeh and Dryer (1973) have extended the solutions to compressible fluids.

15.8. Instability and enhanced reconnection of fields

Thus far the discussion of rapid reconnection of opposite field components has treated steady laminar flows. The question naturally arises as to whether the flows are dynamically stable or unstable. If they are unstable, what effect might the instability have on the reconnection rates computed in the sections above?

It is a well known fact from laboratory plasma research that an ionized gas—a plasma—initially embedded in a magnetic field quickly escapes the clutches of the field. Careful examination has shown that there is always a dynamical instability that permits the escape. With careful design and construction it is generally possible in the laboratory to suppress the most active instabilities through introduction of suitable rigid conductors and currents from external sources. But in the astrophysical world there is no evident possibility for stabilization, suggesting that whenever a plasma is caught and compressed between opposite magnetic fields, the plasma is unstable and quickly wiggles out of the trap. This has the effect of bringing the opposite fields closer together and increasing the average merging rate. Hence considerable thought has been given to the problem, particularly in connection with the violent merging thought to be the cause of solar flares, so that a number of instability effects, on both large and small scales, are expected to enhance the reconnection rate. The dominant instability depends, of course, on the details of the particular circumstance, so no general quantitative statements are possible. But the enhancement of the merging rate can be large, as well as exhibiting the explosive onset observed in flares, so that the topic is important in understanding the universal rapid reconnection in the complex field topologies that fill the astrophysical universe (see discussion in §21.4).

In the most general terms laboratory plasma experiments have led to the empirical concept of Bohm diffusion (Bohm 1949), applicable to the escape of a plasma trapped in a magnetic field. Bohm diffusion is a consequence of several different small-scale instabilities within the plasma. The various instabilities may occur individually or in combination under various physical circumstances of temperature and pressure gradients, and current densities. The concept is simply that, for one reason or another, the ions (and/or the electrons) are scattered through a large angle once every cyclotron period. The effective diffusion coefficient is expressible in terms of the thermal velocity $u=(kT/m)^{\frac{1}{2}}$ and the charge q of the ions of mass m. The characteristic cyclotron radius of the ions is $r_c = muc/qB$ so that the diffusion coefficient is

$$\begin{aligned}\eta_B &= \nu u r_c \\ &= \nu(c/qB)kT\end{aligned}$$

wherever the plasma density and temperature gradients are steep. The quantity ν is a number, determined empirically to be somewhere in the neighbourhood of 0·1–1·0 (Bohm 1949; Artsimovich 1964). Note that the ion mass has cancelled out, so that the result is the same for electrons and singly-charged ions. It is not necessary, nor is it possible, to make a general assertion as to whether the ions or the electrons are more directly responsible for the anomalous Bohm diffusion.

With the Bohm diffusion coefficient η_B (15.1) and (15.2) yield a reconnection rate

$$\begin{aligned}w &= (2\eta_B V_A/L)^{\frac{1}{2}} \\ &= u(2\nu c/L\Omega_p)^{\frac{1}{2}} \qquad (15.88)\end{aligned}$$

where Ω_p is the ion plasma frequency $(4\pi\rho)^{\frac{1}{2}}q/m$. The merging rate is independent of the magnetic field strength B and is characterized by the effective sound velocity u, i.e. the ion thermal velocity.

To compare this result with the reconnection rate (15.3) for resistive diffusion alone, suppose that u and V_A are comparable. Then the effective magnetic Reynolds number in (15.88) is $R_B = 2L\Omega_p/\nu c$ and

$$w = 2V_A/R_B^{\frac{1}{2}},$$

to be compared with $R_m = 2LV_A/\eta$ in (15.3). Numerically

$$\Omega_p \cong 1{\cdot}3\times10^3 N^{\frac{1}{2}}, \qquad \eta \cong 10^7 T^{-\frac{3}{2}}, \qquad u \cong 0{\cdot}9\times10^4 T^{\frac{1}{2}}$$

for ionized hydrogen with a temperature T and N atom cm^{-3}. Hence

$$R_B/R_m = 0{\cdot}5\times10^{-4} N^{\frac{1}{2}}/T^2.$$

Bohm diffusion becomes particularly important ($R_B \ll R_m$) at low densities and high temperatures. Thus, for instance, in the chromosphere of the sun, where $N \cong 10^{12}\ \text{cm}^{-3}$ and $T \cong 2\times10^4$ K we have $R_B = 10^{-7} R_m$. The effective Reynolds number is reduced by a large factor. In the convective zone 10^4 km beneath the solar photosphere, where $N \cong 6\times10^{20}\ \text{cm}^{-3}$ and $T \cong 7\times10^4$ K, then $R_B \cong 2\times10^{-3} R_m$, and there is still a substantial reduction in the effective magnetic Reynolds number.

On the other hand, if we consider Petschek's model, where the effective diffusion rate appears only as $\ln(R_B)$ (see eqn (15.68)) then Bohm diffusion reduces the logarithm by 6–15 and the net effect on the merging rate is a factor of two, comparable to the present theoretical uncertainty in the numerical factor in (15.68).

Passing on from the general, and rather vague, concept of Bohm diffusion, a variety of specific instabilities have been suggested which serve to enhance the diffusion, and hence the merging rate. The resistive tearing modes (Furth *et al.* 1963; Jaggi 1963; Coppi and Friedland 1971; Biskamp and Schindler 1971) contribute substantially to the reduction of the effective magnetic Reynolds number. The resistive tearing mode involves the breakup of the field in the neighbourhood of the neutral line between oppositely directed field components. The original straight parallel lines of the opposite field reconnect into a number of neutral points and magnetic islands, in the manner illustrated in Fig. 15.13. The instability arises from the fact that the dissipation is very intense in the thin transition layer, of characteristic thickness h between the opposite fields $\pm B_0$ a distance h on either side in the initial configuration shown in Fig. 15.1. The initial merging velocity is given by (15.2). Suppose, then, that the system is perturbed slightly, so that the fields are pinched together a little more closely at intervals a along the neutral line. The diffusion and reconnection proceed more rapidly where the fields are pinched together, cutting the lines of force and rejoining them into the localized elongated loops sketched in Fig. 15.13. Once an elongated loop is formed, the tension in the lines of force tends to pull the loop into a more nearly circular form. The effect, then, is to pull the field and plasma away from the vicinity of each neutral point, sucking more field in from either side and further enhancing the reconnection there. Theoretical treatment of the effect is not an entirely elementary procedure (see the original paper by Furth *et al.* 1963) but the basic result for a classical homogeneous incompressible fluid can be reduced to a simple formula for the characteristic growth time τ,

$$\tau = O[(kh)^{\frac{2}{5}}(h^2/\eta)^{\frac{3}{5}}(h/V_A)^{\frac{2}{5}}]$$

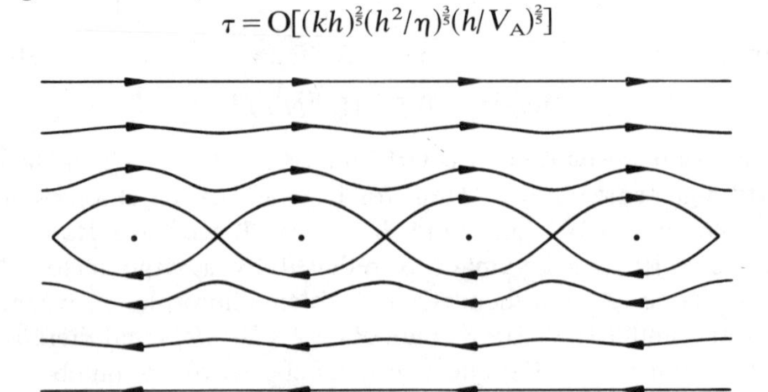

FIG. 15.13. A schematic drawing of the magnetic islands that form along the neutral sheet, between opposite components of the field, as a consequence of the resistive tearing instability.

in order of magnitude in the limit of long wavelength $a = 2\pi/k$ along the neutral line. Note that h^2/η is the characteristic diffusion time, while h/V_A is the characteristic Alfven transit time, across the transition layer thickness h. The Alfven speed is computed from the strength of the field B_0 at the distance h on either side of the neutral line. Thus the growth time is nearly the harmonic mean of the diffusion time and the very small Alfven transit time across the scale h, diminished further by the factor $(2\pi h/a)^{\frac{2}{5}} \ll 1$.

It may be expected, therefore, that any transition layer of thickness l between opposite fields is subject to its own internal tearing mode instability at the high rate $1/\tau$ computed for $l = a$. It is difficult to say exactly how the onset of the instability affects the overall rate of reconnection of field lines across the transition layer l, except to remark that it enhances the dissipation by further breakup of the field. One expects that the rate of merging of opposite fields is significantly enhanced by the instability, which has the effect of increasing the effective diffusion coefficient η_T in the thin dissipation layer. Perhaps the greatest uncertainty lies in the question of what happens after the initial growth of the magnetic islands, shown in Fig. 15.13. It would seem that the next step is to sweep the islands out along the neutral line from between the opposite fields, so that the two opposite fields can again confront each other. The general squeezing of the fluid from between the fields should take care of this satisfactorily, but the time of appearance of the next generation of islands, and their rate of growth is not readily calculated. Various authors have followed the evolution of the magnetic islands remaining in place after their initial formation (Dickman *et al.* 1969; Finn and Kaw 1977; Drake and Lee 1977; White *et al.* 1977). The islands are themselves unstable to merging, with the result that they coalesce, forming eventually into one large island. It may be some such effect that carries a broad region of confrontation between opposite fields, illustrated in Fig. 15.1, into the configuration shown in Fig. 15.2, wherein the merging rate is much accelerated. In any case, the development of the resistive tearing mode instability and the associated dynamical flow and field reconnection is a complicated problem. Indeed, it was pointed out early in the development of the theory of the tearing instability (Coppi 1964; Rosenbluth and Chang 1967; Hazeltine *et al.* 1975) that the small scale and increasing sharpness of the field perturbations may require inclusion of the finite cyclotron radius of the ions in the circumstances encountered in the geomagnetic tail and solar flares. But that lies beyond the scope of the present work.

Turning, then, to other effects that arise in the extreme conditions in the current sheet between opposite fields, we come next to the anomalous resistivity that arises when the field gradient becomes so steep that the electron conduction velocities exceed the ion thermal velocity $u=(kT/m)^{\frac{1}{2}}$. When the diffusion region becomes so thin that the conduction velocity exceeds u, the mean flow of the electrons excites the Alfven–Carlquist instability and/or the ion-sound wave instability (Linhart 1960; Alfven and Carlquist 1967; Hamberger and Friedman 1968; Sagdeev and Galeev 1969; Hamberger and Jancarik 1970; Burchenko *et al.* 1971; Coppi and Mazzucato 1971; Papadopoulos and Coffey 1974; Tange and Ichimaru 1974; Papadopoulos 1976). When the waves thus excited develop to large amplitudes, they scatter the electrons and may enormously increase the effective resistivity. Anomalous resistivity is believed to play an important role in the acceleration of electrons to many kV of energy in both the solar flare and in the geomagnetic field. In the solar flare the electrons produce the x-ray emission, while in the geomagnetic field they form the intermittent streams of electrons associated with the aurora and contribute to the general background of energetic trapped particles (see, for instance, Papadopoulos and Coffey 1974; Chang *et al.* 1976). Anomalous resistivity probably plays a role in the active magnetosphere of Jupiter too (Fillius and McIlwain 1974; Simpson *et al.* 1974; Trainor *et al.* 1974; Van Allen *et al.* 1974; Mihalov and Wolfe 1976). The region in and outside the pulsar magnetosphere, and such active supernova remnants as the crab nebula, probably also involve anomalous resistivity in some of their regions of concentrated currents, but one immediately runs into the lack of direct detailed observational information.

The conditions for the threshold and onset of plasma turbulence, above which anomalous resistivity enters the picture, are easily expressed. In the context of the present problem of the rapid reconnection of opposite fields of strength $\pm B$ across a diffusion region of thickness $2l$, note that the current density j in the diffusion region is given by

$$4\pi j \cong cB/l.$$

In terms of the number N of electrons per unit volume, each with a mean conduction velocity equal to the ion thermal velocity $u=(kT/m)^{\frac{1}{2}}$, it follows that $j=Neu$. In this circumstance, then,

$$l = cB/4\pi Neu,$$

when the conduction velocity is equal to the ion thermal velocity. The thickness of the diffusion region must be as small or smaller than this for the appearance of anomalous resistivity. Now, in terms of the ion thermal velocity $u=(kT/m)^{\frac{1}{2}}$, the characteristic ion cyclotron radius in the field B is $r_c = muc/eB$ for singly-ionized atoms. Hence the threshold thickness l can be written in terms of r_c as

$$\begin{aligned} l &= r_c(B^2/4\pi)/Nmu^2 \\ &= r_c V_A^2/u^2 = 4r_c(B^2/8\pi p) \end{aligned}$$

where V_A is the Alfven speed $B/(4\pi\rho)^{\frac{1}{2}}$ and the gas pressure p is $2Nmu^2$. The gas pressure in the diffusion region is at least as large as the magnetic pressure, so the thickness of the diffusion region must be reduced to dimensions comparable to the thermal ion cyclotron radius.

We can see at once that when this occurs, the gas is no longer characterized accurately as a classical fluid. There is no reason to think that the fluid approximation produces qualitative errors in the merging rate, but certainly a careful look at the problem from the point of view of a plasma made up of individual ions and electrons with finite cyclotron radii is desirable. A number of authors (Speiser 1965, 1967, 1970; Syrovatski 1966a, b, 1969, 1971; Sonnerup 1971; Anzer 1973) have explored the problem, motivated by both the question of the rate of reconnection of the magnetic field, and the question of the observed efficient acceleration of particles to high energy in the reconnection in the solar flare and in the magnetospheres of Earth and of Jupiter (Dungey 1961, 1963). But the very important and difficult theoretical problem of particle acceleration in reconnection in tenuous plasmas takes us away from the simple question of the reconnection rate in dense hydrodynamic fluids that is the concern of this chapter.

To go on to other well known laboratory plasma instabilities, then, consider the large-scale magnetohydrodynamic instabilities that plague the laboratory confinement of plasma. The fluting or interchange instability stands out as the major contributor, in which a sheet of plasma trapped between two (parallel or antiparallel) fields breaks up, the fluid exchanging places with tubes of magnetic flux on either side. Figure 15.1 is a sketch of the field and plasma configuration with a layer of nearly field-free fluid trapped along the x-axis between the opposite fields on either side. The interchange proceeds at some fraction of the Alfven speed V_A. In the context of rapid reconnection, the interchange instabilities enhance the merging rate by allowing the fluid between the opposite fields to escape out the sides of the diffusion region, so that the fluid need not be all ejected out the narrow ends. The fluid escaping out the sides is then free to flow away along the lines of force over a region much broader than the small thickness l of the diffusion region.

The reader is referred to the literature (Bernstein *et al.* 1958; Rosenbluth 1960; Artsimovich 1964; Boyd and Sanderson 1969) for a detailed presentation of the interchange instability. The review by Rosenbluth is particularly useful in the present context. It is well known that the simplest interchanges of field and fluid can be blocked by the introduction of a magnetic field perpendicular to the two opposite fields, i.e. perpendicular to the plane of Fig. 15.1. This is, of course, just the circumstance that we are treating in the present chapter, where one component of the field reverses, but the perpendicular component does not. The total field, then, undergoes a continuous rotation across the neutral sheet, so that flux tubes at different distances from the plasma sheet on the x-axis are not parallel and cannot interchange places. With suitably tailored fields the suppression of the interchange is successful. But the system is then unstable to long wavelength kink instabilities, for the general reason that the magnetic pressure $B^2/8\pi$ in all three dimensions dominates the tension $B^2/4\pi$ in the one direction along the lines of force. Hence in one dimension or another, there is necessarily compression and hence buckling (see discussion in §§5.3 and 5.4). In the *laboratory* the designer then stabilizes the system with the introduction of external rigid conductors. Without rigid conductors to confine the field, there is no known field configuration giving stable confinement of fluid.

Applying these principles, learned in the plasma laboratory, to the fluid making up the diffusion region between two opposite fields, it follows that the fluid escapes from the diffusion region principally by the interchange instability, or by the kinking instability if there is sufficient rotation of the field to stabilize the interchange instability. As a matter of fact, the thickness l of the diffusion region is so small in most cases that there is but little rotation of the field across it, so we expect the interchange instability, rather than the kink instability, to dominate the escape of fluid. It would appear, then, that the fluid caught between the opposite fields escapes from the trap at some fraction of the Alfven speed, causing the reconnection generally to proceed at a speed

$$w = \epsilon V_A$$

where ϵ is a number, presumably less than one (Parker 1973a; Uchida and Sakurai 1977). The general picture is shown schematically in Fig. 15.14, with the fluid escaping into the opposite fields on either side, and from there out along the lines of force. If the fluid penetrates a distance h into the field, then it escapes over a total width $h+l$ and (15.1) becomes $wL = V_A(h+l)$ while the diffusion is still limited to the dimensions l given in (15.2). Eliminating l leads to

$$w = \tfrac{1}{2}V_A\{h/L + (h^2/L^2 + 1/R_m)^{\frac{1}{2}}\}.$$

This reduces to (15.3) in the absence of the exchange instability ($h=0$). Presumably $h \gg l$ (i.e. $h \gg L/R_m^{\frac{1}{2}}$) and

$$w \cong V_A h/L.$$

There is no universal value for ϵ, for the rate depends completely on the details of the local conditions. Chao, *et al.* (1977) have worked out the rate for the plasma sheet in the magnetosphere of Earth, and find that the growth rate of the interchange instability may be as large as $10^{-1}\,s^{-1}$ when the plasma sheet is decoupled from the ionosphere by electric fields

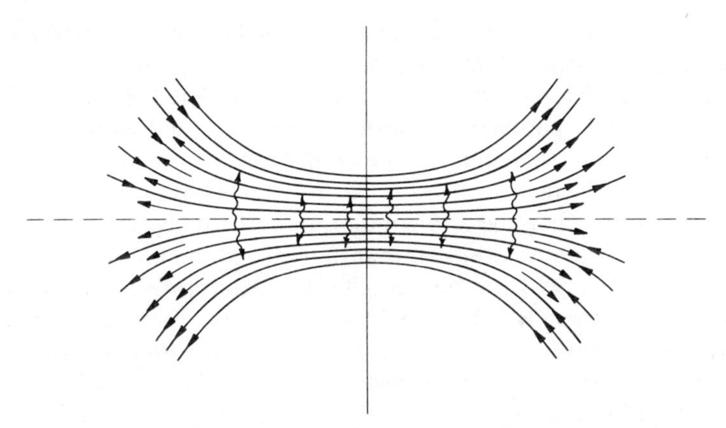

FIG. 15.14. A schematic drawing of the magnetic lines of force of two opposite fields and of the escape of the fluid trapped between via the hydromagnetic exchange instability. The short arrows indicate the escape of the fluid, first moving across the field as fluid and flux tubes are interchanged, and then streaming away along the magnetic lines of force.

on the auroral field lines during a magnetic substorm. In that case the reconnection would be enormously enhanced. The diffusion region is generally not a plane, as treated in the idealized sketches (see Fig. 15.1) but is curved. The fluid then escapes principally from the convex side, because for quasi-equilibrium of the field, the tension in the lines of force extending around the curve is balanced by a decline in the field pressure away from the diffusion region. Hence any blobs of fluid moving out the convex side, find themselves propelled away by the declining magnetic pressure. By the same token, the concave boundary on the other side of the diffusion region is relatively stable, so that little or no fluid escapes from that side. For the flat diffusion region treated in the formal discussion above, both sides have neutral stability, so that the escape is controlled by whatever large-scale fluctuations of the field and fluid may be present.

It has become apparent in the last few years that the disruptive instability in the laboratory tokamak plasma confinement device is of the nature of neutral point reconnection, involving the interaction of two or more helical hydromagnetic modes. The reconnection introduces random connections of the lines of force between neighbouring magnetic surfaces, as a consequence of the overlap of resonances and/or magnetic islands. The effect is particularly interesting here after studying the non-equilibrium of fields with topological variations, because the disruptive instability appears to start with a well-ordered field—a completely invariant field for which there is a definite equilibrium—and introduces a stochastic component for which there is, then, no equilibrium. An alternative possibility is that the topology of the initial field is not quite invariant, i.e. that there is some small stochastic component which causes non-equilibrium in which the neutral point reconnection grows slowly, as fluid is displaced along the lines of force by a small imbalance in the pressure. The first possibility is more complicated, and more interesting, however, for it would imply that the equilibrium of an invariant field can be unstable in such a way as to destroy the invariance and produce *non*-equilibrium. Then the non-equilibrium is driven by the energy of the maverick components, while the instability continues to destroy the invariance and to feed the non-equilbrium. The process continues until eventually the torsion in the field is removed. The situation in the tokamak is complicated, of course, by the closed toroidal geometry, in which lines of force re-enter the instability at a different place upon each circuit around the torus.

The astrophysical implications are fundamental to the activity of force-free magnetic fields. Thus, for instance, the fields of some sunspots and prominences show torsion, while others do not, and some torsional fields are active while others are not. Some helical fields in the laboratory escape the disruptive instability, while others do not, with no clear picture yet as to what makes the difference. There is the possibility, then, that the combined instability of invariant fields, and the non-equilibrium of fields lacking invariance, team up to produce dissipation of any deviation of the field from the simple potential field to a degree beyond anything described above. Fortunately, the disruptive instability in the tokamak is under intensive study (Rutherford 1973; Furth *et al.* 1973; Rosenbluth *et al.* 1973; von Goeler *et al.* 1974; Finn 1975; Jacobsen 1975; Hutchinson 1976).

In the absence of a unique value for ϵ in the turbulent environment experienced by most magnetic fields in the astrophysical world, we suggest that ϵ may generally rise to values of the order of 10^{-1}, and even larger in a variety of active circumstances (see, for instance, the observational evidence in Axford *et al.* 1965; Axford 1967; Dessler 1968, 1971).

The important point is, then, that the rapid reconnection often proceeds at a significant fraction of the Alfven speed whether the system evolves from Fig. 15.1 into Petschek's configuration (Fig. 15.11) or not. Presumably the most rapid merging would occur with the interchange instability operating in Petschek's configuration.

Altogether, then, the rapid reconnection of magnetic lines of force proceeds rapidly in any field lacking topological invariance. The lines of force may be cut and reconnected at a significant fraction of the Alfven speed, continually reducing the field toward symmetric topologies. The effect is particularly important in controlling the otherwise enormous growth of the mean square field in a turbulent fluid with high electrical conductivity. As we shall see in the next chapters, the turbulent diffusion of magnetic field is possible only because of such rapid reconnection. The observed rapid evolution of the complicated magnetic fields of active regions on the sun into relatively simple potential forms is another direct consequence of the universal rapid reconnection. And, of course, such explosive events as the geomagnetic substorm, solar brightenings and eruptions, solar flares, and perhaps even some of the explosive phenomena observed in active galaxies, are the direct result of rapid reconnection.

References

ALFVEN, H. and CARLQUIST, P. (1967). *Solar Phys.* **1**, 220.
ANZER, U. (1973). *Solar Phys.* **30,** 459.
ARTSIMOVICH, L. A. (1964). *Controlled thermonuclear reactions*, pp. 75–6. Gordon and Breach, New York.
AXFORD, W. I. (1967). *Space Sci. Rev.* **7,** 149.
—— PETSCHEK, H. E., and SISCOE, G. L. (1965). *J. Geophys. Res.* **70,** 1231.
BERNSTEIN, I. B., FRIEDMAN, E. A., KRUSKAL, M. D., and KULSRUD, R. M. (1958). *Proc. Roy. Soc. Ser. A* **244,** 17.
BISKAMP, D. and SCHINDLER, K. (1971). *Plasma Phys.* **13,** 1013.
BOHM, D. (1949). *The characteristics of electrical discharges in magnetic fields* (ed. A. GUTHERIE, and R. WAKERLING), chap. 2, 5. McGraw-Hill, New York.
BOYD, J. T. M. and SANDERSON, J. J. (1969). *Plasma dynamics*, p. 71. Barnes and Noble, New York.
BURCHENKO, P. Y., VOLKOV, E. D., RUDAKOV, V. A., SIZONENKO, V. L., and STEPANOV, K. N. (1971). *Intern. Conf. Plasma Physics and Controlled Thermonuclear Research*, paper CN 28/H-9. Madison, Wisconsin, I.A.E.A., Vienna.
CHAO, J. K., KAN, J. R., LUI, A. T. T., and AKASOFU, S.-I. (1977). *Planet. Space Sci.* **25,** 703.
COPPI, B. (1964). *Phys. Fuids* **7,** 1501.
—— and MAZZUCATO, E. (1971). *Phys. Fluids* **64,** 134.
—— and FRIEDLAND, A. B. (1971). *Astrophys. J.* **169,** 379.
DESSLER, A. J. (1968). *J. Geophys. Res.* **73,** 209, 1861.
—— (1971). ibid. **76,** 3174.
DICKMAN, D. O., MORSE, R. L., and NIELSON, C. W. (1969). *Phys. Fluids* **12,** 1708.
DRAKE, J. F. and LEE, Y. C. (1977). *Phys. Rev. Lett.* **39,** 453.
DUNGEY, J. W. (1953). *Phil Mag.* **44,** 725.
—— (1958). *Cosmic electrodynamics*, pp. 98, 125. Cambridge University Press.
—— (1961). *Phys. Rev. Lett.* **6,** 47.
—— (1963). *Planet. Space Sci.* **10,** 233.
FILLIUS, R. W. and MCILWAIN, C. E. (1974). *J. Geophys. Res.* **79,** 3587.
FINN, J. M. (1975). *Nucl. Fusion* **15,** 845.
—— and KAW, P. K. (1977). *Phys. Fluids* **20,** 72.
FUKAO, S. and TSUDA, T. (1973a). *Planet Space Sci.* **21,** 1151.
—— —— (1973b). *J. Plasma Phys.* **9,** 409.
FURTH, H. P., KILLEEN, J., and ROSENBLUTH, M. N. (1963). *Phys. Fluids* **6,** 459.
—— RUTHERFORD, P. H., and SELBERG, H. (1973). *Phys. Fluids* **16,** 1054.
GIOVANELLI, R. G. (1947). *Mon. Notic. Roy. Astron. Soc.* **107,** 338
—— (1948). ibid. **108,** 163.

GRADSHTEYN, J. S. and RYZHIK, I. M. (1965). *Tables of integrals, series, and products* 3.981 (6). Academic Press, New York.
HAMBERGER, S. M. and FRIEDMAN, M. (1968). *Phys. Rev. Lett.* **21,** 674.
—— and JANCARIK, J. (1970). *Phys. Rev. Lett.* **25,** 999.
HAZELTINE, R. D., DOBROTT, D., and WANG, T. S. (1975). *Phys. Fluids* **18,** 1778.
HUTCHINSON, I. H. (1976). *Phys. Rev. Lett.* **37,** 338.
JACOBSEN, R. A. (1975). *Plasma Phys.* **17,** 547.
JAGGI, R. K. (1963). *J. Geophys. Res.* **68,** 4429.
LINHART, J. G. (1960). *Plasma physics*, p. 173. North-Holland, Amsterdam.
MIHALOV, J. D. and WOLFE, J. H. (1976). *J. Geophys. Res.* **81,** 3412.
PAPADOPOULOS, K. (1976). *NRL Memo, Report 3275*, Naval Research Laboratory, Washington, D.C.
—— and COFFEY, T. (1974). *J. Geophys. Res.* **79,** 1558.
PARKER, E. N. (1957). *J. Geophys. Res.* **62,** 509.
—— (1963). *Astrophys. J. Suppl.* **8,** 177.
—— (1973a). *Astrophys. J.* **180,** 247.
—— (1973b). *J. Plasma Phys.* **9,** 49.
PETSCHEK, H. E. (1964). *The physics of solar flares, AAS-NASA Symposium*, NASA SP-50 (ed. W. N. HESS), p. 425. Greenbelt, Maryland 1963.
—— and THORNE, R. M. (1967). *Astrophys. J.* **147,** 1157.
PRIEST, E. R. and COWLEY, S. W. H. (1975). *J. Plasma Phys.* **14,** 271.
ROBERTS, B. and PRIEST, E. R. (1975). *J. Plasma Phys.* **14,** 417.
ROSENBLUTH, M. N. (1960). *Symposium on plasma physics*, (ed. F. CLAUSER), pp. 19–25. Addison-Wesley, Reading, Massachusetts.
—— and CHANG, D. B. (1967). *J. Geophys. Res.* **72,** 143.
—— DAGAZIAN, R. Y., and RUTHERFORD, P. H. (1973). *Phys. Fluids* **16,** 1894.
RUTHERFORD, P. H. (1973). *Phys. Fluids* **16,** 1903.
SAGDEEV, R. Z. and GALEEV, A. A. (1969). *Nonlinear plasma theory*, pp. 54, 103. Benjamin, New York.
SIMPSON, J. A., HAMILTON, D., LENTZ, G., MCKIBBEN, R. B., MOGRO-CAMPERO, A., PERKINS, M., PYLE, K. R., and TUZZOLINO, A. J. (1974). *J. Geophys. Res.* **79,** 3522.
SONNERUP, B. U. O. (1970). *J. Plasma Phys.* **4,** 161.
—— (1971). *J. Geophys. Res.* **76,** 8211.
—— (1973). *High energy phenomena on the sun–symposium proceedings*, NASA-GSFC Doc. X-693-73-193, p. 357.
—— and PRIEST, E. R. (1975). *J. Plasma Phys.* **14,** 283.
SPEISER, T. W. (1965). *J. Geophys. Res.* **70,** 1717, 4219.
—— (1967). ibid. **72,** 3919.
—— (1970). *Planet. Space Sci.* **18,** 613.
SWEET, P. A. (1958a). *Proc. IAU Symposium No. 6*, p. 123. Stockholm, 1956.
—— (1958b) *Nuovo Cim. Suppl.*, **8,** (10) 188.
—— (1969). *Ann. Rev. Astron. Astrophys.* **7,** 149.
SYROVATSKI, S. I. (1966a). *Sov. Astron. A. J.* **10,** 270.
—— (1966b). *Sov. Phys. JETP* **23,** 754.
—— (1969), *Solar flares and space research* (ed. C. DE JAGER and Z. SVESTKA), p. 346. North-Holland, Amsterdam.
—— (1971). *Sov. Phys. JETP* **33,** 933.
TANGE, T. and ICHIMARU, S. (1974). *Pub. phys. Soc. Jpn*, **36,** 1437.
TRAINOR, J. H., MCDONALD, F. B., TEEGARDEN, B. J., WEBBER, W. R., and ROELOF, E. C. (1974). *J. Geophys. Res.* **79,** 3600.

UBEROI, M. S. (1963). *Phys. Fluids* **6,** 1379.
UCHIDA, Y. and T. SAKURAI (1977). *Solar Phys.* **51,** 413.
VAN ALLEN, J. A., BAKER, D. N., RANDALL, B. A., and SENTMAN, D. D. (1974). *J. Geophys. Res.* **79,** 3559.
VASYLIUNAS, V. M. (1975). *Rev. Geophys. Space Phys.* **13,** 303.
VON GOELER, S., STODIEK, W., and SAUTHOFF, N. (1974). *Phys. Rev. Lett.* **33,** 1201.
YANG, C. K. and SONNERUP, B. U. O. (1976). *Astrophys. J.* **206,** 570.
—— —— (1977). *J. Geophys. Res.* **82,** 699.
YEH, T. (1976a). *Astrophys. J.* **207,** 837.
—— (1976b). *J. Geophys. Res.* **81,** 4524.
—— and AXFORD, W. I. (1970). *J. Plasma Phys.* **4,** 207.
—— and DRYER, M. (1973). *Astrophys. J.* **182,** 301.
WHITE, R. B., MONTICELLO, D. A., ROSENBLUTH, M. N., and WADDELL, B. V. (1977). *Phys. Fluids* **20,** 800.

16

EXCLUSION OF MAGNETIC FIELDS FROM CLOSED CIRCULATION PATTERNS

16.1. Magnetic fields in convecting fluid

THE preceding chapters have aimed at the dynamical non-equilibrium of magnetic fields, with particular attention to magnetic buoyancy and rapid reconnection as the basis for the continual escape and activity of the magnetic fields in astrophysical bodies. The exposition concentrated on the dynamical effects of the field on the fluid and the consequences for the field. We turn now to the dynamical effects of the fluid on the magnetic field, with particular attention to the effects of convection and turbulence. When a weak magnetic field finds itself in a fluid that is hydrodynamically active, so that the Reynolds stresses are able to compete with, or overpower, the Maxwell stresses ($\rho v^2 \gtrsim B^2/4\pi$), then there are a number of important consequences. The convection zone of a star is such a case, where the fluid is active irrespective of the presence and activity of magnetic fields. The stellar wind and the gaseous disc of the galaxy, not to mention the convective cores of planets, are all hydrodynamically active fluids in their own right.

The effect of the fluid motion on the magnetic field depends very much on whether the stream-lines have a closed and fixed topological pattern, in which there is no mixing of fluid perpendicular to the stream-lines, or whether the stream-lines are open or their connectivity variable so that the fluid mixes freely throughout the volume. In the absence of mixing, the closed circulation tends to exclude the field, whereas if there is a general mixing of fluid through the region, the effect on the large-scale field distribution is essentially diffusion. If the dominant eddies have a characteristic dimension L, velocity v, and life τ, presumably with $\tau \cong L/v$, then the diffusion is characterized by the eddy diffusivity $v^2\tau$. The treatment of the problem begins in the present chapter with the exclusion of field from the steady circulation of fluid in a closed pattern.

The treatment of a magnetic field in a given hydrodynamic flow is a kinematical problem, and in many respects, therefore, is simpler than the dynamical problems treated above. We are confronted with the solution of the linear eqns (4.12) or (4.13) for B_i given the velocity field v_i. But of course we deal with complex, and sometimes turbulent, velocity fields, so that v_i may be a complicated function of time and space, and the solution is anything but elementary. It is necessary, therefore, to give some thought to the physics before plunging into the mathematics.

To begin with the simplest effect, consider a uniform magnetic field B_0 embedded in a highly conducting fluid. At time $t=0$ the fluid is set in motion with a fixed continuous velocity pattern $v_i(\mathbf{r})$ with a characteristic scale L. To illustrate the effects, suppose that the magnitude of the velocity v is so large and/or the resistive diffusion coefficient η is so small that the magnetic Reynolds number $R_m = Lv/\eta$ is large compared to one, so that to a first approximation the lines of force are carried bodily with the fluid. Hence, any locally closed circulation pattern winds the lines of force about the region, with the subsequent field given by (4.32) or (4.35). In the case that the initial field is perpendicular to the axis of rotation of the region, so that the lines of force are wrapped around the outside of the rotating region, like rope around a rotating drum, then the important feature is that the successive layers of field are of alternate sign, representing the lines of force entering and leaving at diametrically opposite sides of the regions, as sketched in Fig. 16.1(a). On the other hand, if the initial lines of force extend lengthwise through the region, in the direction parallel to the axis of rotation, the lines are wrapped into coils of the *same* sign at each end, as sketched in Fig. 16.1(b).

In the former case the thickness l of each strand of field of a given sign is of the order of L/n after n rotations, occurring in a characteristic time

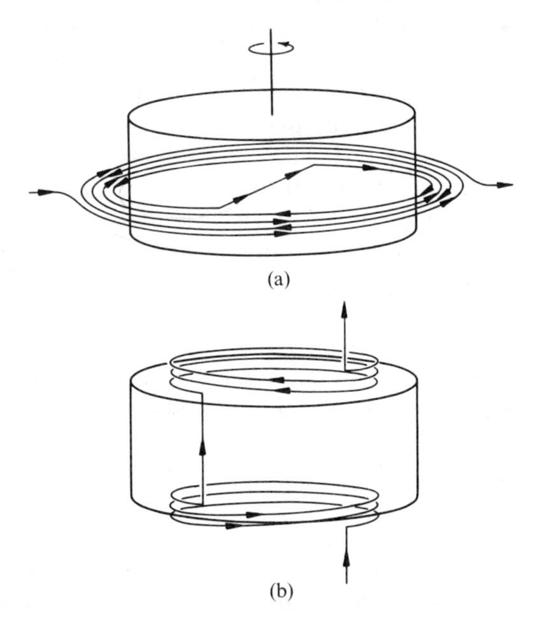

FIG. 16.1. A sketch of a line of force passing through a body rotating about a vertical axis and represented by the right circular cylinder. The line of force is shown after approximately two revolutions when the field extends across the body (a) perpendicular and (b) parallel to the axis of rotation.

$2\pi Ln/v$. Conservation of flux leads to the field strength B in each strand increasing according to $B \cong B_0 L/l = nB_0$, with the sign of the field reversing from one filament to the next. At the same time the effective magnetic Reynolds number of each filament declines from the initial $R_{\mathrm{m}} = Lv/\eta$ to $lv/\eta = R_{\mathrm{m}}/n$. The point is eventually reached when the diffusion term $\eta \nabla^2 B_i$, of the order of $\eta B/l^2 = (\eta B_0/L^2)n^3$, becomes comparable to the convection term $B_j\, \partial v_i/\partial x_j - v_j\, \partial B_i/\partial x_j$, of order vB_0/L. (This occurs after a time t when the characteristic diffusion time l^2/η across a filament becomes comparable to t.) Then, neglecting numerical factors of the order of unity, it follows that

$$n = R_{\mathrm{m}}^{\frac{1}{3}}, \qquad t = R_{\mathrm{m}}^{\frac{1}{3}} L/v, \tag{16.1}$$

at which time the r.m.s. field intensity in the region is

$$\langle B^2 \rangle^{\frac{1}{2}} = R_{\mathrm{m}}^{\frac{1}{3}} B_0. \tag{16.2}$$

With the continual decrease of l the filaments are subsequently destroyed by diffusion, leaving the interior of the region relatively free of field. The interior may be free of field in a time rather less than the characteristic diffusion time $L^2/\eta = R_{\mathrm{m}} L/v$ for the region, because of the application of reverse fields at the boundary, and because the internal non-uniform rotation reorients the initial field over the declining scale l. Altogether, we would expect the final asymptotic state to be approached after a period of a few times $t = R_{\mathrm{m}}^{\frac{1}{3}} L/v$. Since $R_{\mathrm{m}} \gg 1$ in most astrophysical circumstances, this is a time large compared to the rotation period L/v, of course, but exceedingly small compared to the overall diffusion time L^2/η for the same size region at rest. The exclusion effect depends upon there being little or no mixing in and out of the region during the time t.

The remarkable exclusion effect (Parker 1963, 1975; Weiss 1966; Parker 1966) becomes understandable if we view the field within the rotating region from the rotating frame of reference. From that frame the external field rotates around the region with an angular velocity ω of the order of v/L. The components of the external field applied to the boundary of the rotating region alternate with time, as $\exp(\mathrm{i}\omega t)$. But it is well known (Thomson 1893) that an externally applied alternating field penetrates only the characteristic skin depth $(\eta/\omega)^{\frac{1}{2}} \cong (\eta L/v)^{\frac{1}{2}} \cong L/R_{\mathrm{m}}^{\frac{1}{2}}$ into the region. Since $R_{\mathrm{m}} \gg 1$, the penetration is slight and the interior is free of magnetic field.

The exclusion of field from the interior does not occur for the component of the external field parallel to the axis of rotation (Sweet 1949; Cowling and Hare 1957) shown in Fig. 16.1(b). For in that case the external field is static in the rotating frame of reference and its penetration is maintained for all time. For the same reason, the field is retained in the moving fluid if the fluid motion is not closed on itself. If, for instance, the motion is in the form of a steady longitudinal stream, either parallel or perpendicular to the large-scale field, the field is not expelled from the moving fluid. For in that case, a simple transformation to a moving frame puts the stream at rest and the entire external universe in motion, so that there is no absolute statement as to which region is at rest and which is moving. Exclusion occurs only when the circulation forms a closed pattern and the fluid is continually reoriented (either by steady or random rotation) relative to the large-scale external field. Thus, for instance a spherical volume rotated at random, but without mixing, excludes the field from its interior.

These effects are illustrated with simple formal examples in the next section. The problem is particularly interesting when we come to address the question of the general absence of net magnetic flux extending through the interior of the supergranule cells observed in the solar photosphere. The net flux is observed to be concentrated into the boundaries of the supergranules. A number of effects contribute to the absence of flux from the interior. To decide whether the present effect contributes, it is necessary to decide whether the fluid involved in the convective cell, identified as an individual supergranule, is so self-contained a weather system that there is little or no mixing of external field into the cell by diffusion[1] over the characteristic period of exclusion $R_{\mathrm{m}}^{\frac{1}{3}}\tau$. The red spot on Jupiter appears to be a strongly isolated region of circulation, but it is difficult to know what to think of the supergranules. The form and location of a supergranule changes over periods of only 10^5 s. Is it possible, however, for the fluid within the cell to remain isolated for much longer periods? And in how many astrophysical objects that lie beyond the resolution of our observing instruments is the field excluded from a major portion of the fluid? Our efforts to understand their active magnetic behaviour are conventionally based on the quaint idea of a continuous distribution of field, as developed for the sun prior to the realization of the highly discontinuous distribution of the solar fields described in Chapter 10.

16.2. Magnetic effects in steady fluid motions

To illustrate the effects of steady local fluid motions on a large-scale magnetic field, described in §16.1, consider first the motion $v(x, y)$ in the z-direction of a highly conducting fluid. Such motion has no effect whatsoever on a magnetic field parallel to itself. Suppose, then, that at time $t=0$ the fluid is pervaded by a uniform magnetic field B_0 in the

[1] The turbulent diffusion caused by the small-scale turbulence is the dominant effect, so that the magnetic Reynolds number should be computed from the eddy diffusivity. But this takes us into the next chapter.

x-direction, perpendicular to the direction of fluid motion. The induction equation is

$$\partial B_x/\partial t = \partial B_y/\partial t = 0$$

$$\{\partial/\partial t - \eta(\partial^2/\partial x^2 + \partial^2/\partial y^2)\}B_z = B_0\, \partial v/\partial x.$$

Thus initially, before diffusion becomes significant ($t \ll L^2/\eta$), the field grows linearly with time $B_z(x, y, t) = B_0 t\, \partial v/\partial x$. The final asymptotic state ($t \gg L^2/\eta$) is described by

$$\partial^2 B_z/\partial x^2 + \partial^2 B_z/\partial y^2 = -(B_0/\eta)\, \partial v/\partial x$$

in which B_z has the same distribution as the electrostatic potential of a charge distribution $-(B_0/4\pi\eta)\, \partial v/\partial x$. If, for instance, the fluid motion consists of a circular column of fluid of radius a centred on the z-axis and moving with uniform velocity v_0 along the axis, then it is readily shown that

$$B_z(x, y) = (B_0 v_0 a^2 \cos\phi/2\eta) \times \begin{cases} \varpi & \text{for} \quad \varpi \leqslant a \\ a^2/\varpi & \text{for} \quad \varpi \geqslant a \end{cases}$$

where ϖ represents distance $(x^2+y^2)^{\frac{1}{2}}$ from the z-axis and ϕ is the azimuthal angle measured from the x-axis. The field strength B_z varies linearly as x across the moving column of field, and declines in the two-lobed form x/ϖ^2 outside.

If, on the other hand, the fluid motion consists of a slab of fluid $-a < x < +a$ with uniform velocity v_0 along the z-axis, then

$$B_z(x, y) = B_0 v_0 a\eta \begin{cases} 1 & \text{for} \quad x \geqslant a \\ x/a & \text{for} \quad -a \leqslant x \leqslant +a \\ -1 & \text{for} \quad x \leqslant -a. \end{cases} \tag{16.3}$$

The strength B_z varies linearly across the slab. It does not decline outside, because of the infinite width of the slab.

The important point is that the large-scale external field in the x-direction is unaffected by the fluid motion. It extends across the column of moving fluid, unaffected by the motion of the fluid.

Consider, then, the rotational motion $v_\phi(\varpi, z)$ about the z-axis. In the first instance, suppose that there is a uniform field B_0 in the z-direction, parallel to the axis of rotation, as sketched in Fig. 16.1(b). Then the only non-vanishing component of (4.13) is

$$\left\{\frac{\partial}{\partial t} - \eta\left(\frac{\partial^2}{\partial \varpi^2} + \frac{1}{\varpi}\frac{\partial}{\partial \varpi} - \frac{1}{\varpi^2} + \frac{\partial^2}{\partial z^2}\right)\right\} B_\phi = B_0 \frac{\partial v_\phi}{\partial z}.$$

The field in the z-direction is unaffected by the fluid motion, while the azimuthal component grows linearly with time and has the same spatial form as $\partial v_\phi/\partial z$ in the initial period of growth before diffusion becomes significant

$$(t \ll L^2/\eta) \qquad B_\phi(\varpi, z, t) = B_0 t\, \partial v_\phi/\partial z.$$

The final asymptotic steady state (in which $\partial/\partial t = 0$) of the azimuthal field is readily computed in simple cases. If, for instance, the rotation is confined to the slab $-a < z < +a$, in which the fluid velocity is a function only of radius, so that

$$v_\phi(\varpi, z) = \begin{cases} 0 & \text{if} \quad z^2 > a^2 \\ v(\varpi) & \text{if} \quad z^2 < a^2, \end{cases}$$

then

$$\partial v_\phi/\partial z = v(\varpi)\{\delta(z+a) - \delta(z-a)\}.$$

The field satisfies the equation

$$\{\partial^2/\partial\varpi^2 + (1/\varpi)\,\partial/\partial\varpi - 1/\varpi^2 + \partial^2/\partial z^2\}B_\phi = 0 \tag{16.4}$$

except on $z = \pm a$, across which B_ϕ is continuous and $\partial B_\phi/\partial z$ undergoes the discontinuous jump

$$\begin{aligned} \Delta(\partial B_\phi/\partial z) &\equiv \partial B_\phi/\partial z\big|_{\pm a-\epsilon}^{\pm a+\epsilon} \\ &= \pm B_0 v(\varpi)/\eta. \end{aligned} \tag{16.5}$$

This result follows immediately from the integration of (16.3) from $\pm a - \epsilon$ to $\pm a + \epsilon$, where ϵ is an infinitesimal. Integrating again leads to the continuity of B_ϕ.

In order that B_ϕ be finite at $\varpi = 0$ and vanish at $\varpi, z = \pm\infty$ for a bounded velocity $v(\varpi)$, consider solutions of (16.4) of the form $J_1(k\varpi)\exp(\pm kz)$. It is evident from the fact that the source (16.3) is an odd function of z that B_ϕ will also be an odd function, so write

$$B_\phi(\varpi, z) = \int_0^\infty dk\,k J_1(k\varpi) \begin{cases} f_1(k)\exp\{-k(z-a)\} & \text{for} \quad z \geqslant a \\ f_2(k)\sinh(kz) & \text{for} \quad -a \leqslant z \leqslant +a \\ -f_1(k)\exp\{k(z+a)\} & \text{for} \quad z \leqslant -a. \end{cases}$$

Matching the solutions at $z = \pm a$ leads to

$$f_1(k) = f_2(k)\sinh(ka). \tag{16.6}$$

The jump condition (16.5) requires that

$$\int_0^\infty dk\,k^2 J_1(k\varpi)\{f_1(k) + f_2(k)\cosh(ka)\} = -B_0 v(\varpi)/\eta$$

so that if

$$v(\varpi)=\int_0^\infty \mathrm{d}kkV(k)J_1(k\varpi), \tag{16.7}$$

then

$$V(k)=\int_0^\infty \mathrm{d}\varpi\varpi v(\varpi)J_1(k\varpi)$$

and it follows that

$$k\{f_1(k)+f_2(k)\cosh(ka)\}=-B_0V(k)/\eta. \tag{16.8}$$

Simultaneous solution of (16.6) and (16.8) yields

$$f_2(k)=-B_0V(k)/\eta k(\sinh ka+\cosh ka) \tag{16.9}$$

with $f_1(k)$ given by (16.6). It follows that for $z>a$

$$B_\phi=-\frac{B_0}{\eta}\int_0^\infty \mathrm{d}k\frac{V(k)\sinh(ka)J_1(k\varpi)\exp\{-k(z-a)\}}{\sinh(ka)+\cosh(ka)}, \tag{16.10}$$

while in $-a<z<+a$,

$$B_\phi=-\frac{B_0}{\eta}\int_0^\infty\frac{\mathrm{d}kV(k)\sinh(kz)J_1(k\varpi)}{\sinh(ka)+\cosh(ka)}. \tag{16.11}$$

One illustrative example will suffice for the present purposes. Suppose that the rotation is essentially uniform out to some radial distance $b(\gg a)$, beyond which it cuts off exponentially,

$$v(\varpi)=v_0(\varpi/b)\exp(-\varpi/b). \tag{16.12}$$

Then the rotating region is effectively a broad flat disc of thickness $2a$ and radius b. It follows (Watson 1958; §13.2 (6)) that

$$V(k)=3v_0kb^3/(1+k^2b^2)^{\frac{5}{2}}.$$

Then at intermediate radial distances $a\ll\varpi\ll b$, where the rotation is uniform, the rotational velocity being $v_0\varpi/b$, the Bessel function oscillates exceedingly rapidly with k except for sufficiently small values of k such that $k=\mathrm{O}(1/\varpi)\gg 1/a$. Hence the integral over k is determined principally by the integrand where $ka\ll 1$. Then (16.10) and (16.11) reduce to

$$B_\phi=-\frac{3B_0v_0b^3a}{\eta}\int_0^\infty\frac{\mathrm{d}kk^2J_1(k\varpi)\exp\{-k(z-a)\}}{(1+k^2b^2)^{\frac{5}{2}}},$$

$$B_\phi=-\frac{3B_0v_0b^3z}{\eta}\int_0^\infty\frac{\mathrm{d}kk^2J_1(k\varpi)}{(1+k^2b^2)^{\frac{5}{2}}}$$

respectively, upon neglecting terms $\mathrm{O}(ka)$ compared to one. Then since

(Watson 1958; §13.6 (2))

$$\int_0^\infty \frac{\mathrm{d}x\,x^2 J_1(\alpha x)}{(x^2+q^2)^{\frac{5}{2}}} = \frac{\alpha}{3q}\exp(-\alpha q),$$

it follows that in $-a<z<+a$,

$$B_\phi = -B_0 z v(\varpi)/\eta. \tag{16.13}$$

In the motionless fluid outside the rotating slab, $z>a$, but with $z \ll \varpi$, the exponential factor in the integrand can be set equal to one, and

$$B_\phi = -B_0 a v(\varpi)/\eta. \tag{16.14}$$

Comparing (16.13) and (16.14) with the result (16.3) for a slab of fluid moving with the uniform velocity v_0, it is evident that (16.1) applies locally to any part of a thin disc undergoing uniform rotation. The field varies linearly across the thickness of the disc, and is independent of z immediately outside the disc. At large z ($z \gg b$), the exponential factor $\exp\{-k(z-a)\}$ cuts off so rapidly with increasing k that $kb \ll 1$ over the range $k \lesssim 1/z$ in which the integrand is non-negligible, with the result that

$$B_\phi = \frac{-9B_0 v_0 b^3 a\varpi(z-a)}{\eta\{\varpi^2+(z-a)^2\}^{\frac{5}{2}}}. \tag{16.15}$$

At large distances the field declines as $(z-a)^{-4}$. This example represents, for instance, the rotation of the thin gaseous disc of a galaxy, or of a protostar system, surrounded by a fixed conducting medium (Parker 1975).

The topology of the lines of force is sketched in Fig. 16.1(b). The diffusion does not alter the basic topology of the initial winding in this case, although it causes the field to spread out somewhat and limits the accumulation of total azimuthal flux with the passage of time. The large-scale field parallel to the axis of rotation is unaffected by the rotation. The azimuthal field produced by the shear of the rotation is coiled at the ends of the rotating cylinder. The adjacent coils have the same sign so that they blend together into a continuous field.

Consider, then, what happens when the external field has a component perpendicular to the axis of rotation.

16.3. The exclusion of field by steady circulation

To illustrate the exclusion of the field component perpendicular to the axis of rotation, suppose that initially there is a large-scale uniform field B_0 in the x-direction throughout the highly conducting fluid. A circular column of fluid concentric about the z-axis is set in motion at time $t=0$, the fluid having the azimuthal velocity $v_\phi(\varpi)$. The magnetic lines of force extending through the rotating column are carried around with it, wrapping around the outside of the column in layers of alternating sign, as sketched in Fig. 16.1(a). It is evident that the rotation alters the x-component and introduces a y-component as well. It is convenient, therefore, to work with the vector potential A,

$$B_x = +\partial A/\partial y, \qquad B_y = -\partial A/\partial x.$$

Then integrating the induction equation (4.13), the result is

$$\partial A/\partial t + v_j\, \partial A/\partial x_j = \eta \nabla^2 A,$$

the integration constant being put equal to zero because there is no large-scale electric field of external origin impressed across the region. The final asymptotic steady state is described by

$$\left(\frac{\partial^2}{\partial \varpi^2} + \frac{1}{\varpi}\frac{\partial}{\partial \varpi} + \frac{1}{\varpi^2}\frac{\partial^2}{\partial \phi^2}\right) A = \frac{v(\varpi)}{\eta \varpi}\frac{\partial A}{\partial \phi}. \tag{16.16}$$

A variety of simple forms for $v(\varpi)$ permit an analytical solution of this equation (Parker 1963). Suppose, for instance, that the fluid rotates rigidly with angular velocity ω out to $\varpi = a$ and is at rest beyond. Then

$$v(\varpi) = v_0 \varpi / a$$

for $\varpi < a$, where $\omega = v_0/a$. Consider solutions of the form $A = F(\varpi)\exp(in\phi)$. It follows that

$$\mathrm{d}^2F/\mathrm{d}\xi^2 + (1/\xi)\,\mathrm{d}F/\mathrm{d}\xi + (1 - n^2/\xi^2)F = 0$$

where

$$\xi \equiv \mathrm{i}^{\frac{3}{2}} (nv/an)^{\frac{1}{2}} \varpi.$$

The solution is made up of a linear combination of $J_n(\xi)$ and $Y_n(\xi)$. We require solutions that are finite at the origin, so the choice is $J_n(\xi)$. In view of the complex argument of the Bessel functions, it is convenient to use the ber_n and bei_n functions, so that the vector potential is given by the real part of

$$A(\varpi, \phi) = \sum_{n=0}^{\infty} A_n[\mathrm{ber}_n\{(nv/an)^{\frac{1}{2}}\varpi\} + \mathrm{i}\,\mathrm{bei}_n\{(nv/a\eta)^{\frac{1}{2}}\varpi\}].$$
$$\times \exp(\pm \mathrm{i}n\phi) \tag{16.17}$$

In the region $\varpi > a$, where the fluid is motionless, the solution is $\varpi^{\pm n}\exp(\pm \mathrm{i}n\phi)$. We need a solution that reduces at infinity to the uniform value B_0 in the x-direction, so that $A \sim B_0 y = B_0 \varpi \sin\phi$. Therefore write

$$A = -\mathrm{i}B_0(\varpi + a^2C_1/\varpi)\exp(\mathrm{i}\phi). \tag{16.18}$$

It is evident, then, that with the impressed external field $A = B_0\varpi \sin\phi$, the only terms needed in (16.17) are those for $n = 1$.

Now define the wave number $q = (v/a\eta)^{\frac{1}{2}}$. Then $qa = (av/\eta)^{\frac{1}{2}}$, which is just the magnetic Reynolds number R_m of the rotating fluid. Since $R_m \gg 1$, use the asymptotic form

$$\mathrm{ber}_1(q\varpi) + \mathrm{i}\,\mathrm{bei}_1(q\varpi) \sim (2\pi q\varpi)^{-\frac{1}{2}} \exp\{(1+\mathrm{i})q\varpi/2^{\frac{1}{2}} + \mathrm{i}3\pi/8\}\{1 + \mathrm{O}(1/q\varpi\}.$$

except for ϖ so small that $q\varpi$ is not large. But, as we will soon see, the vector potential is exponentially small there, of the order of $\exp[-\mathrm{O}(qa)]$, so it matters not whether the asymptotic form is valid. Hence, within the rotating fluid write

$$A(\varpi, \phi) = A_1(2\pi q\varpi)^{-\frac{1}{2}} \exp\{(1+\mathrm{i})q\varpi/2^{\frac{1}{2}} + \mathrm{i}3\pi/8 + \mathrm{i}\phi\}.$$

Now the normal (radial) component of the magnetic field is continuous across $\varpi = a$, which is satisfied by making the vector potential continuous. The azimuthal component of the field is also continuous, but its radial gradient is not, because of the discontinuous jump in the azimuthal velocity. Integrate the azimuthal component of the induction equation

$$\eta\left(\frac{\partial^2}{\partial\varpi^2} + \frac{1}{\varpi}\frac{\partial}{\partial\varpi} - \frac{1}{\varpi^2} + \frac{1}{\varpi^2}\frac{\partial^2}{\partial\phi^2}\right)B_\phi + \frac{\partial}{\partial\varpi} v_\phi B_\varpi = 0$$

from $a - \epsilon$ to $a + \epsilon$. Since B_ϖ is continuous and v_ϕ jumps from v to zero, it follows that

$$\eta \left.\frac{\partial B_\phi}{\partial\varpi}\right|_{a-\epsilon}^{a+\epsilon} = vB_\varpi(a)$$

(as in (16.5)). Since $B_\phi = -\partial A/\partial\varpi$, we have

$$\partial^2 A/\partial\varpi^2 \,|_{a-\epsilon}^{a+\epsilon} = (v/\eta a)\, \partial A/\partial\phi. \qquad (16.19)$$

Then continuity of the vector potential across $\varpi = a$, with the vector potential given by (16.18) outside, yields

$$A_1(2\pi qa)^{-\frac{1}{2}} \exp\{(1+\mathrm{i})qa/2^{\frac{1}{2}} + \mathrm{i}3\pi/8\}$$
$$= -\mathrm{i}B_0 a(1 + C_1). \qquad (16.20)$$

The jump condition (16.19) leads to

$$A_1(2\pi qa)^{-\frac{1}{2}} \exp\{(1+\mathrm{i})qa/2^{\frac{1}{2}} + \mathrm{i}3\pi/8\}$$
$$+ 2B_0 C_1/aq^2 = \mathrm{i}B_0 a(1 + C_1).$$

Simultaneous solution of these two equations yields

$$C_1 = -(1 + \mathrm{i}/a^2q^2)^{-1} \cong -(1 - \mathrm{i}/a^2q^2)$$
$$A_1 = -B_0(C_1/aq^2)(2\pi qa)^{\frac{1}{2}} \exp[\{(1+\mathrm{i})aq/2^{\frac{1}{2}} + \mathrm{i}3\pi/8\}]$$

so that within the rotating column the vector potential is given by the real part of

$$A = \frac{B_0}{aq^2}\left(\frac{a}{\varpi}\right)^{\frac{1}{2}} \exp\left\{-\frac{q(a-\varpi)}{2^{\frac{1}{2}}}\right\} \exp \mathrm{i}\left\{\phi - \frac{q(a-\varpi)}{2^{\frac{1}{2}}}\right\}. \qquad (16.21)$$

Outside the rotating column the vector potential is given by the real part of

$$A = -\mathrm{i}B_0\{\varpi - a^2(1-\mathrm{i}/a^2q^2)/\varpi\}\exp(\mathrm{i}\phi),$$

so that

$$A = B_0\{(\varpi - a^2/\varpi)\sin\phi + (1/q^2\varpi)\cos\phi\}. \qquad (16.22)$$

The important point to note is the exponential decline

$$\exp\{-q(a-\varpi)/2^{\frac{1}{2}}\}$$

of the magnetic field inward from the boundary of the rotating column of fluid. This is the familiar decline of an alternating field of angular frequency $\omega = v/a$ applied to the boundary $\varpi = a$. For with

$$\partial \mathbf{B}_\phi/\partial t = \eta\, \nabla^2 \mathbf{B}_\phi$$

in the frame of reference of the rotating fluid, the local field, very near the boundary, is described by

$$\partial B_\phi/\partial t = \eta\, \partial^2 B_\phi/\partial \zeta^2$$

where $\zeta = a - \varpi$ is distance from the boundary. The solution for $B_\phi \propto \sin(\omega t)$ at the boundary is the real part of

$$\exp\{-(\omega/2\eta)^{\frac{1}{2}}\zeta\}\exp \mathrm{i}\{\omega t - (\omega/2\eta)^{\frac{1}{2}}\zeta\}.$$

Transforming to the fixed frame so that $\phi = \omega t$, and writing $\omega = v/a$, $q^2 = v/a\eta$, this becomes

$$\exp\{-q(a-\varpi)/2^{\frac{1}{2}}\}\exp \mathrm{i}\{\phi - q(a-\varpi)/2^{\frac{1}{2}}\} \qquad (16.23)$$

which is identical with (16.21) in the near neighbourhood of $\varpi = a$. This illustrates the physical explanation for the exclusion, that the fluid within the rotating region sees the external field as an alternating field, which is limited to the conventional skin depth $(\eta/\omega)^{\frac{1}{2}}$.

Note that if the inner portions of the cylinder or sphere rotate relative to the outer parts, then further exclusion of field by the inner rotation occurs. A simple example (Parker 1963) is given by the cylindrical velocity field $v_\phi = v(a/\varpi)^2$ where v is a constant. The azimuthal velocity increases monotonically with decreasing distance ϖ from the z-axis. If the field at $\varpi = +\infty$ is in the x-direction ($\phi = 0$) with magnitude B_0, then it is readily shown from (16.16) that

$$A(\varpi, \phi) = 2(a^2v/\eta)B_0[\mathrm{kei}_2\{(4a^2v/\eta\varpi)^{\frac{1}{2}}\}\sin\phi - \mathrm{ker}_2\{(4a^2v/\eta\varpi)^{\frac{1}{2}}\}\cos\phi]. \quad (16.24)$$

The lines of force are given by $A = constant$ and are plotted in Fig. 16.2.

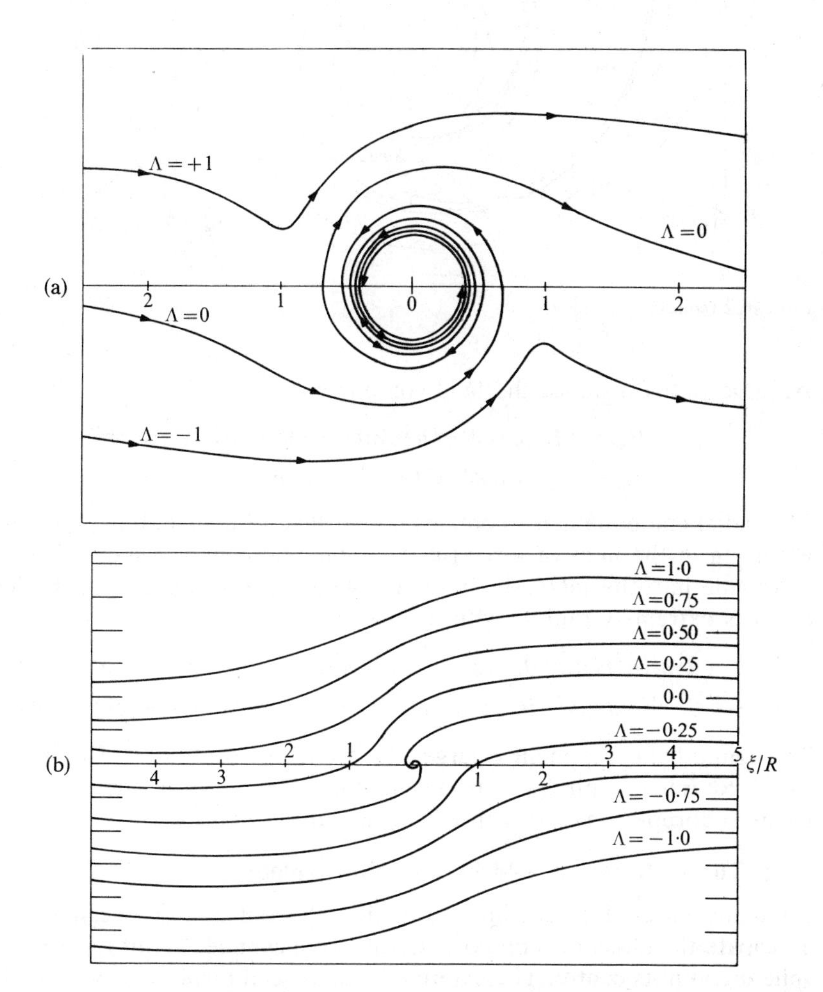

FIG. 16.2. The magnetic lines of force of the initially uniform field in the circular fluid motion $v_\phi = v(a/\varpi)^2$ (a) after a time $t = a/v$, showing how the lines of force are drawn into a spiral and intensified by the shear, with the field strength varying as $(a/\varpi)^3\cos(a/\varpi)^3$ toward the origin, and (b) the final asymptotic state as $t \to \infty$, with distance in units of a^2v/η, from (16.24). In (c) the asymptotic state of the lines of force in the neighbourhood of $A = 0$ is shown greatly magnified to illustrate the deflection of the lines out of the rotating region. The indicated values of $\Lambda = A$ are proportional to the total magnetic flux measured from the singular line that passes through the origin.

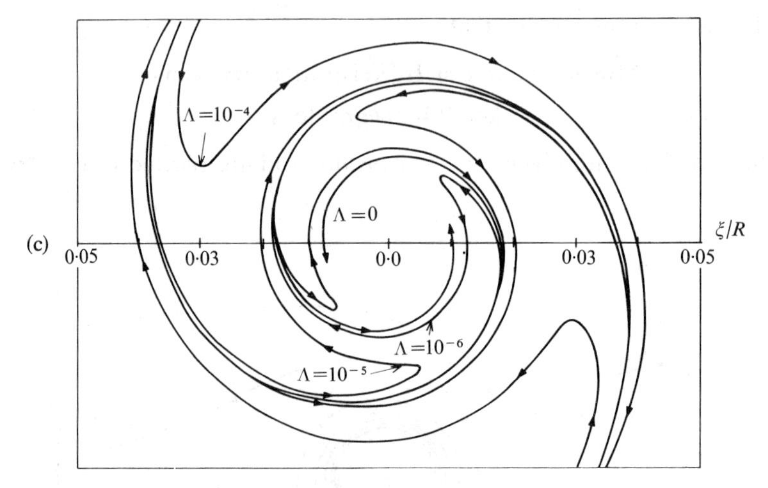

FIG. 16.2 (*cont'd.*)

At large radial distance the field components are

$$B_\varpi \sim +B_0\{\cos\phi + (a^2 v/\eta\varpi)\sin\phi + \mathrm{O}(a^4 v^2/\eta^2\varpi^2)\},$$
$$B_\phi \sim -B_0\{\sin\phi + \mathrm{O}(a^4 v^2/\eta^2\varpi^2)\}.$$

The radial component B_ϖ contains no net flux, of course, but it produces a net jog in the lines of force (in the x-direction) by a distance $2va^2/\eta$.

Moving into the increasingly rapid rotation about the z-axis, the field declines extremely rapidly with decreasing ϖ,

$$B_\varpi \sim +2^{\frac{1}{4}}\pi^{\frac{1}{2}}(a^2 v/\eta\varpi^2)B_0 \exp\{-(2a^2 v/\eta\varpi)^{\frac{1}{2}}\}\sin\{\phi - \pi/8 - (2a^2 v/\eta\varpi)^{\frac{1}{2}}\}$$
$$B_\phi \sim -2^{\frac{1}{4}}\pi^{\frac{1}{2}}(av/\eta\varpi^2)^{\frac{3}{2}}B_0 \exp\{-(2a^2 v/\eta\varpi)^{\frac{1}{2}}\}\cos\{\phi - \pi/8 - (2a^2 v/\eta\varpi)^{\frac{1}{2}}\}.$$

The azimuthal component is larger, by the factor $(av/\eta\varpi^2)^{\frac{1}{2}}$. The lines of force go into a spiral of increasing tightness, and the field strength declines abruptly inward with the form $\exp\{-(2a^2 v/\eta\varpi)^{\frac{1}{2}}\}$.

16.4. The exclusion of field by variable circulation

On the basis of these three examples, then, it is a simple matter to anticipate the effect of a random rotation of a cylinder about its axis, or a sphere about its centre. The external field, as seen by an observer moving with the cylinder or sphere, rotates randomly. If there is no preferred orientation of the cylinder, then the mean external field has no mean, as seen from the cylinder. So there can be no mean field extending across the cylinder (or sphere) beyond the skin depth of the lowest frequency Fourier component of the rotation. If that frequency is not zero, then the field is excluded from the deep interior.

A single example illustrates the effect. Suppose that we construct a local plane Cartesian coordinate system xy in some small region on the surface of a body, the coordinate system being fixed on the surface. In that frame of reference, suppose that the tangential components of the external field (B_x, B_y) impressed upon the surface are $B_x = B_0 \cos \phi$, $B_y = B_0 \sin \phi$ where ϕ is the fluctuating angle

$$\phi(t) = \int_0^t \mathrm{d}t' \Omega(t').$$

Suppose that the angular velocity $\Omega(t)$ varies as

$$\Omega(t) = \Omega_0 + \Omega_1 \cos(\alpha t)$$

so that

$$\phi(t) = \Omega_0 t + (\Omega_1/\alpha)\sin(\alpha t).$$

The orientation ϕ of the body oscillates with amplitude Ω_1/α about the continually varying position $\Omega_0 t$. This is not a random variation, of course, but it serves to illustrate the general consideration. Write

$$B(t) \equiv B_x(t) + \mathrm{i}B_y(t) = B_0 \exp(\mathrm{i}\phi).$$

The Fourier transformation of $B(t)$ is

$$\beta(\omega) = \int_{-\infty}^{+\infty} \mathrm{d}t \exp(-\mathrm{i}\omega t) B(t)$$

$$= \frac{B_0}{\alpha} \int_{-\pi}^{+\pi} \mathrm{d}\theta \exp\left\{-\mathrm{i}\frac{\omega - \Omega_0}{\alpha}\theta + \frac{\mathrm{i}\Omega_0}{\alpha}\sin\theta\right\}$$

$$\times \sum_{n=-\infty}^{+\infty} \exp \mathrm{i}\{2\pi(\omega - \Omega_0)/n\}$$

where $\theta = \alpha t$ and the integral over αt from $-\infty$ to $+\infty$ has been broken into intervals $(-\pi, +\pi)$, $(\pi, 3\pi)$, etc. The integral from $-\pi$ to $+\pi$ is just the Bessel function $2\pi J_\nu(\Omega_1/\alpha)$ of order $\nu = (\omega - \Omega_0)/\alpha$. The sum can be written as the sum of the Fourier transforms of the individual terms, using Lighthill's theorem, with the result that

$$\beta(\omega) = 2\pi B_0 J_\nu(\Omega_1/\alpha) \sum_{n=-\infty}^{+\infty} \delta(\omega - \Omega_0 - n\alpha).$$

The frequency spectrum consists of a series of spikes at $\omega = \Omega_0 + n\alpha$. Thus the field

$$B(t) = \frac{1}{2\pi} \int_{-\infty}^{+\infty} \mathrm{d}\omega \exp(\mathrm{i}\omega t)\beta(\omega)$$

can be written as a series

$$B(t) = B_0 \exp(\mathrm{i}\Omega_0 t) \sum_{n=-\infty}^{+\infty} J_n(\Omega_1/\alpha)\exp(\mathrm{i}n\alpha t).$$

Then since $J_n(x) = (-1)^n J_{-n}(x)$, it follows that

$$B_x(t) = B_0 \Bigg[J_0(\Omega_1/\alpha)\cos(\Omega_0 t)$$

$$+ \sum_{n=1}^{\infty} \{J_{2n}(\Omega_1/\alpha)\{\cos((2n\alpha+\Omega_0)t) + \cos((2n\alpha-\Omega_0)t)\}$$

$$+ J_{2n-1}(\Omega_1/\alpha)[\cos\{((2n-1)\alpha+\Omega_0)t\} - \cos\{((2n-1)\alpha-\Omega_0)t\}]\}\Bigg],$$

$$B_y(t) = B_0 \Bigg[J_0(\Omega_1/\alpha)\sin(\Omega_0 t)$$

$$+ \sum_{n=1}^{\infty} \{J_{2n}(\Omega_1/\alpha)\{\sin((2n\alpha+\Omega_0)t) - \sin((2n\alpha-\Omega_0)t)\}$$

$$+ J_{2n-1}(\Omega_1/\alpha)[\sin\{((2n-1)\alpha+\Omega_0)t\} + \sin\{((2n-1)\alpha-\Omega_0)t\}]\}\Bigg].$$

It is evident that the time average of both components B_x, B_y of the external field are zero, unless Ω_0/α is equal to an integer or zero, so that there is no penetration of field into the rotating body beyond the characteristic skin depth computed for the lowest frequency Ω_0 or $\Omega_0 \pm 2n\alpha$. If $\Omega_0 = 0$, however, then

$$B_x = B_0 \left\{ J_0(\Omega_1/\alpha) + 2\sum_{n=1}^{\infty} J_{2n}(\Omega_1/\alpha)\cos(2n\alpha t) \right\}$$

$$B_y = 2B_0 \sum_{n=1}^{\infty} J_{2n-1}(\Omega_1/\alpha)\sin((2n-1)\alpha t).$$

In this case B_x has a mean value $B_0 J_0(\Omega_1/\alpha)$, which penetrates, after a time, throughout the entire body. The example is not a random rotation, so that, as it turns out, the x-component of the external field has a d.c. component. The d.c. component penetrates through the body, the effective skin depth being infinite. If the amplitude Ω_1 of the change in the angular velocity is very large ($\Omega_1 \gg \alpha$), then the d.c. component is very small, of course, with $J_0(\Omega_1/\alpha) = \mathrm{O}\{(\alpha/\Omega_1)^{\frac{1}{2}}\}$, but there is no exclusion effect for $\Omega_0 = 0$. Only if the external field oscillates about a zero point ($\Omega_0 \neq 0$) is the field fully excluded. A random rotation of the body does just that, of course.

16.5. The distribution of field in steady convection

To illustrate the exclusion of field from regions of steady fluid motions, consider the effect of the periodic two-dimensional convective motions

$$v_x = v\sin(kx)\exp(-ky), \qquad v_y = v\cos(kx)\exp(-ky). \tag{16.25}$$

The motion is both irrotational and incompressible and can be described in terms of the velocity potential

$$\begin{gathered}\phi = -(v/k)\cos(kx)\exp(-ky)\\ v_x = -\partial\phi/\partial x, \qquad v_y = -\partial\phi/\partial y,\end{gathered} \tag{16.26}$$

and in terms of the stream function

$$\begin{gathered}\psi = -(v/k)\sin(kx)\exp(-ky)\\ v_x = +\partial\psi/\partial y, \qquad v_y = -\partial\psi/\partial x\end{gathered} \tag{16.27}$$

with

$$\nabla^2\phi = \nabla^2\psi = 0. \tag{16.28}$$

The fluid motion grows rapidly weaker with increasing y, so that for $ky \gg 1$ the fluid is essentially at rest. The motion grows rapidly stronger as y becomes large and negative. The stream lines are illustrated in Fig. 16.3.

We present a number of examples (Parker 1963) to illustrate the exclusion of a variety of field configurations. One is a vertical magnetic field, extending upward through the convective motions. In this case the field is confined to thin intense sheets in the downdrafts, and excluded from the rest of the region below the x-axis. Another is a horizontal field

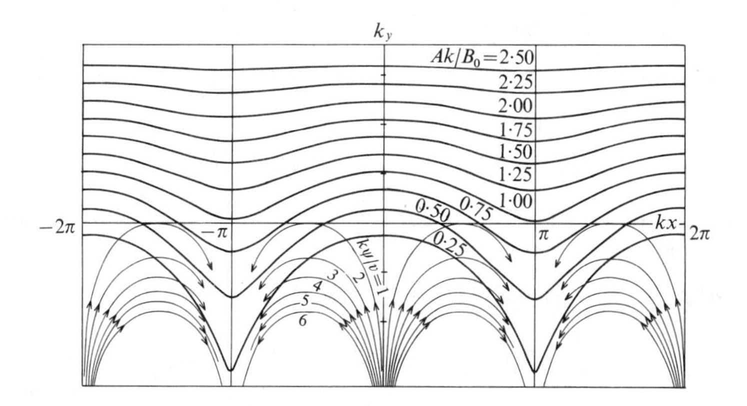

FIG. 16.3. The light lines represent the stream lines of the fluid motion (16.27). The heavy lines represent the lines of force of a horizontal field pressed down upon the top of the fluid motion, computed from (16.32).

extending across the top of the convection penetrating downward into the convection, particularly in the downdrafts. The final example is a strand of magnetic field uplifted from below by an updraft.

As a matter of computational convenience, slide the coordinate system in the y-direction so that the velocity amplitude v at $y=0$ is $v=\eta k$. Then the effective magnetic Reynolds number $(v_x^2+v_y^2)^{\frac{1}{2}}/\eta k$ is $\exp(-ky)$, being less than one in $y>0$ and greater than one in $y<0$.

Describing the magnetic field in terms of the vector potential A, as in §16.15, with no e.m.f. except that induced by v_i, the equation for the final asymptotic steady state of the field is

$$v_j\,\partial A/\partial x_j=\eta\nabla^2 A. \tag{16.29}$$

16.5.1. *Horizontal magnetic field*

As the first illustrative example, suppose that there is a uniform horizontal field B_0 in the x-direction in the region above the steady convection. This field presses downward on the tops of the convective cells. Consider in what way it extends downward into the region $y<0$ where the fluid motions are strong. It is obvious that the upwelling fluid at $kx=2n\pi$ $(n=0,\pm1,\pm2,\ldots)$ excludes the field, pushing it up to the top $(y\cong 0)$ of the convective region. The question is to what degree the field is carried down into the convection in the downdrafts at $kx=(2n+1)\pi$.

For the present example, the field equation (16.29) is most conveniently treated in cartesian coordinates. Let $\xi=kx$ and $\zeta=ky$. Then with $v=\eta k$, (16.29) becomes

$$\frac{\partial^2 A}{\partial\xi^2}+\frac{\partial^2 A}{\partial\zeta^2}-\left\{\sin\xi\frac{\partial A}{\partial\xi}+\cos\xi\frac{\partial A}{\partial\zeta}\right\}\exp(-\zeta)=0. \tag{16.30}$$

The variables are not separable in this form so let

$$A(\xi,\zeta)=f(\xi,u)\exp(-u\cos\xi)$$

where $u=\frac{1}{2}\exp(-\zeta)$. Then

$$\partial^2 f/\partial\xi^2+u^2\{\partial^2 f/\partial u^2+(1/u)\,\partial f/\partial u-f\}=0.$$

Separating the variables, it is readily shown that the solution can be expressed as

$$A(\xi,\zeta)=\exp(u\cos\xi)\Big[(C_1+C_3\xi)\{C_3I_0(u)+C_4K_0(u)\}$$
$$+\sum_{n=1}^{\infty}\{D_{1n}\sin(n\xi)+D_{2n}\cos(n\xi)\}\{D_{3n}I_n(u)+D_{4n}K(u)\}\Big] \tag{16.31}$$

where the C_i and the D_{ij} are arbitrary constants. For the problem at hand, the uniform field above the convection requires that $A \sim B_0 y$ as $y \to +\infty$ $(u \to 0)$. Since the uniform field is independent of x, it is represented by the terms for $n = 0$. Noting the asymptotic forms

$$I_0\{\tfrac{1}{2}\exp(-\zeta)\} \sim 1 + \tfrac{1}{4}\exp(-2\zeta) + \dots$$
$$K_0\{\tfrac{1}{2}\exp(-\zeta)\} \sim \zeta + \ln(4/\gamma) + \tfrac{1}{4}\{\zeta + 1 + \ln(4/\gamma)\}\exp(-2\zeta) + \dots$$

as $\zeta \to +\infty$, it is readily seen that the desired solution is (Parker 1963)

$$A(x, y) = B_0 k^{-1} \exp\{-\tfrac{1}{2}\cos(kx)\exp(-ky)\} K_0\{\tfrac{1}{2}\exp(-ky)\} \tag{16.32}$$

As y becomes large and negative, the asymptotic form of $K_0(u)$ is $(\pi/2u)^{\frac{1}{2}}\exp(-u)$, so that

$$A(x, y) \sim B_0(\pi^{\frac{1}{2}}/k)\exp[-\tfrac{1}{2}\{1 + \cos(kx)\}\exp(-ky) + \tfrac{1}{2}ky],$$

so that the field declines downward as

$$\exp\{\tfrac{1}{2}ky - \exp(-ky)\}$$

in the upwelling fluid and as $\exp(\frac{1}{2}ky)$ in the downdrafts. Thus, even in the downdrafts, where the field is pulled into the convection by the fluid motion, the field declines exponentially downward into the convection. The reason for the downward decline is transverse (horizontal) diffusion. The lines of force are plotted, together with the streamlines of the fluid motion, in Fig. 16.3 in flux units of $\Delta A = B_0 \pi^{\frac{1}{2}}/k$.

16.5.2. Vertical magnetic field

If there is a vertical magnetic field (y-direction) of uniform strength B_0 filling the region from $x = -\infty$ to $+\infty$ above the convection y, then the lines of force of the field must somehow pass downward through the convection to $y = -\infty$, presumably in the downdrafts. If there are no perturbing effects extending into the region from $x = \pm\infty$, then the lines of force will be separated into bundles of equal total flux $2\pi B_0/k$ in each downdraft[2]. As a solution let

$$A(\xi, \zeta) = -B_0 k^{-1}\left\{\xi + \sum_{n=1}^{\infty} A_n(\xi, \zeta)\right\}, \tag{16.33}$$

where the first term represents the uniform vertical field $B_y = B_0$ at $ky \gg 1$. The term A_n represents the effect of the interaction of the

[2] The obvious solution from (16.31) is $A(\xi, \zeta) = -B_0 k^{-1} \xi \exp\{\frac{1}{2}\cos\xi \exp(-\zeta)\} I_0\{\frac{1}{2}\exp(-\zeta)\}$ from which it follows that the field is B_0 in the vertical direction at $y \to +\infty$, and $\pi B_0\{1 - \frac{1}{2}\xi \sin\xi \exp(-\zeta)\}\exp(-\frac{1}{2}\zeta + \frac{1}{2}\exp(-\zeta)\}$ as y becomes large and negative. But note the linear increase with ξ of the term $\xi \sin\xi \exp(-\zeta)$.

velocity field with A_{n-1}, beginning with

$$\partial^2 A_1/\partial\xi^2 + \partial^2 A_1/\partial\zeta^2 = \sin\xi \exp(-\zeta). \tag{16.34}$$

Then for $n > 1$, (16.30) yields

$$\partial^2 A_n/\partial\xi^2 + \partial^2 A_n/\partial\zeta^2 = \sin\xi \exp(-\zeta)\,\partial A_{n-1}/\partial\xi + \cos\xi \exp(-\zeta)\,\partial A_{n-1}/\partial\zeta.$$

It is readily shown that

$$A_1 = -\tfrac{1}{2}(\zeta + \tfrac{1}{2})\sin\xi \exp(-\zeta), \tag{16.35}$$

$$A_2 = +\tfrac{1}{32}(2\zeta + \tfrac{1}{2})\sin 2\xi \exp(-2\zeta) \tag{16.36}$$

$$A_3 = -\tfrac{1}{64}(\eta + \tfrac{3}{4})\sin\xi \exp(-3\zeta) - \tfrac{1}{576}(3\zeta + \tfrac{1}{2})\sin 3\xi \exp(-3\zeta), \tag{16.37}$$

etc. The term A_n vanishes as $\exp(-n\zeta)$ as ζ becomes large and positive. As ζ becomes large and negative the field is concentrated strongly into the downdrafts at $\xi = (2n+1)\pi$, so that the solution is non-vanishing only in the near neighbourhood of $\xi = (2n+1)\pi$. It is readily shown by direct substitution into (16.30) that, neglecting terms of the order of $(\xi - \pi)^2$ and $\exp(\zeta)$ compared to one, the vector potential for large negative ζ in the neighbourhood of $\xi = \pi$ is given by (Parker 1963)

$$A(\xi, \zeta) \sim -\pi B_0 k^{-1} \operatorname{erf}\{2^{-\frac{1}{2}}(\xi - \pi)\exp(-\tfrac{1}{2}\zeta)\} \tag{16.38}$$

so that

$$B_x \sim B_0(\pi/2)^{\frac{1}{2}}(\xi - \pi)\exp\{-\tfrac{1}{2}\zeta - \tfrac{1}{2}(\xi - \pi)^2 \exp(-\zeta)\}. \tag{16.39}$$

$$B_y \sim B_0(2\pi)^{\frac{1}{2}} \exp\{-\tfrac{1}{2}\zeta - \tfrac{1}{2}(\xi - \pi)^2 \exp(-\zeta)\}. \tag{16.40}$$

Thus at a depth ζ the field is confined into columns in the downdrafts. The profile of the field strength in each column is Gaussian, and the characteristic width of the Gaussian is the very small quantity $2^{\frac{1}{2}} k^{-1} \exp(\frac{1}{2}\zeta)$ equal to $(2^{\frac{1}{2}}/k)R_m$ where R_m is the effective magnetic Reynolds number $\{(v_x^2 + v_y^2)^{\frac{1}{2}}/\eta k\} = \exp(-ky)$ at the depth y. The lines of force are plotted, together with the stream lines of the fluid motion, in Fig. 16.4.

16.5.3. Magnetic filaments

Consider finally a magnetic field of the form of the bipolar region observed on the sun, wherein a flux tube is carried to the surface by magnetic buoyancy and/or a convective upwelling of fluid. In the final steady state in the periodic convective pattern (16.25) the lines of force are confined to the downdrafts, of course, being swept into them by the converging flow of the fluid. So the lines of force extending up the downdraft at $\xi = \pi$, find their way back down at $\xi = -\pi$ and $\xi = +3\pi$, etc. Suppose, then, that at large negative ζ the field is vertically upward at

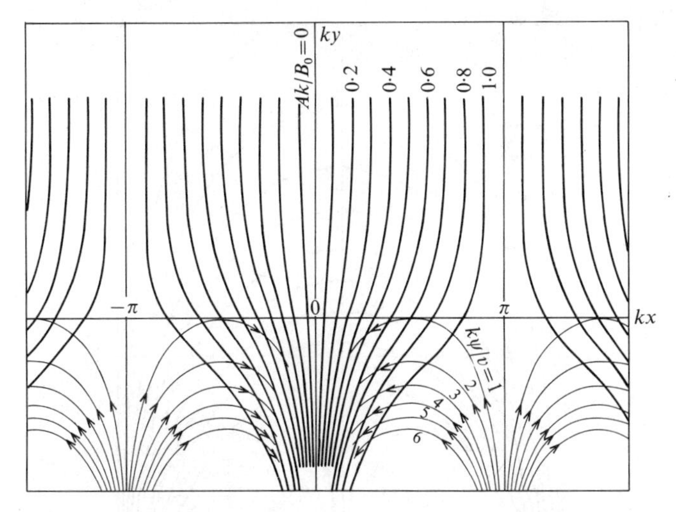

FIG. 16.4. The light lines represent the stream lines (16.27) and the heavy lines represent the lines of force of a vertical field extending upward through the fluid motion, computed from (16.35)–(16.38).

$\xi=(4n+1)\pi$ and downward at $(4n-1)\pi$. The solution of (16.30) which illustrates this case, with the field arching over the top of the convective upwellings at $\xi=2n\pi$, is most readily obtained using the velocity potential ϕ and stream function ψ as coordinates. For then $\partial\phi/\partial x=-\partial\psi/\partial y$, $\partial\phi/\partial y=+\partial\psi/\partial x$ and

$$v_j\,\partial A/\partial x_j=(\nabla\phi)^2\,\partial A/\partial\phi,$$
$$\nabla^2 A=(\nabla\phi)^2(\partial^2 A/\partial\phi^2+\partial^2 A/\partial\psi^2),$$

so that (16.30) reduces to

$$\partial^2 A/\partial\psi^2+\partial^2 A/\partial\phi^2+\partial A/\partial\phi=0. \tag{16.41}$$

The variables are separable and the general solution can be written as a sum of exponential functions of ψ and ϕ. It turns out, however, that the solution to the problem at hand can be written in the non-separable form (Parker 1963)

$$\begin{aligned}A&=B_0(\pi/k)\operatorname{erf}[2^{-\frac{1}{2}}\{\phi+(\psi^2+\phi^2)^{\frac{1}{2}}\}^{\frac{1}{2}}]\\&=B_0(\pi/k)\operatorname{erf}\{2^{-\frac{1}{2}}(1+\cos\xi)^{\frac{1}{2}}\exp(-\tfrac{1}{2}\zeta)\}\end{aligned} \tag{16.42}$$

It is readily shown by differentiation of $A(\psi,\phi)$ that this expression satisfies (16.39), and hence (16.30). The field components are

$$B_x=-B_0(\tfrac{1}{2}\pi)^{\frac{1}{2}}(1+\cos\xi)^{\frac{1}{2}}\exp\{-\tfrac{1}{2}\zeta-\tfrac{1}{2}(1+\cos\xi)\exp(-\zeta)\}, \tag{16.43}$$

$$B_y=+B_0\frac{(\tfrac{1}{2}\pi)^{\frac{1}{2}}\sin\xi}{(1+\cos\xi)^{\frac{1}{2}}}\exp\{-\tfrac{1}{2}\zeta-\tfrac{1}{2}(1+\cos\xi)\exp(-\zeta)\}. \tag{16.44}$$

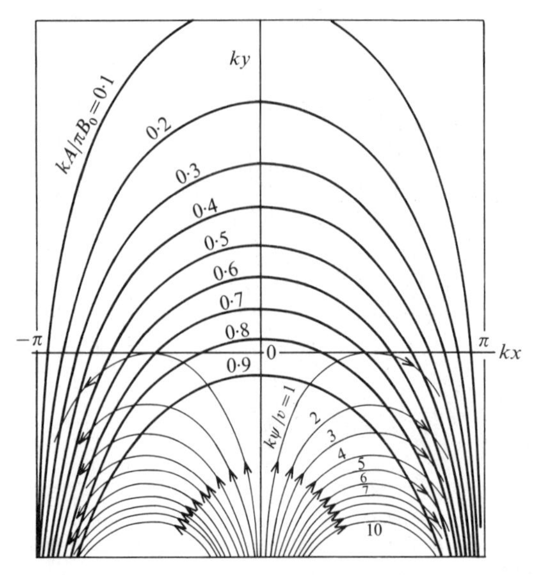

FIG. 16.5. The light lines represent the stream lines (16.27) and the heavy lines represent the lines of force of a strand of magnetic field stretched up over the top of an upwelling, computed from (16.42).

At large negative ζ, deep in the convective region, the field is confined to thin columns of characteristic thickness $2\exp(\frac{1}{2}\zeta)$ extending down the centre of each downdraft $\xi=(2n-1)\pi$. Thus, for instance, in the neighbourhood of $\xi=\pi$,

$$B_y \sim B_0\pi^{\frac{1}{2}}\exp\{-\tfrac{1}{2}\zeta-\tfrac{1}{4}(\xi-\pi)^2\exp(-\zeta)\}, \tag{16.45}$$

$$A \sim B_0(\pi/k)\operatorname{erf}\{\tfrac{1}{2}(\xi-\pi)\exp(-\tfrac{1}{2}\zeta)\}, \tag{16.46}$$

and $B_x \sim \frac{1}{2}(\xi-\pi)B_y$. The lines of force are plotted, together with the stream lines of the fluid motion, in Fig. 16.5. Note that the form of (16.45) is identical with (16.40), as we would expect (see also the illustrations in Clark 1965). The vertical component of the field is confined by any converging horizontal flow $v_x=-v_0kx$ to a horizontal profile $B_y \propto \exp\{-(v_0k/2\eta)x^2\}$ regardless of where the lines of force extend elsewhere in the space. In the convective flow (16.25), the effective value of v_0 is $v\exp(-ky)$, so that the characteristic width is

$$(2\eta/v_0k)^{\frac{1}{2}}=(2\eta/kv)^{\frac{1}{2}}\exp(\tfrac{1}{2}ky)=2^{\frac{1}{2}}\exp(\tfrac{1}{2}ky)$$

at any level y.

16.6. The onset of field exclusion

The manner in which the field goes about its exodus from the interior of a rotating region was described briefly in §16.1. The lines of force initially extending across the diameter of the region are wrapped about the region by the rotation, forming a roll of magnetic sheets of alternating sign. The thickness of each sheet declines inversely with time until diffusion becomes important and the field is obliterated, surviving only in the fluid outside the region of rapid rotation. This is the means by which the exclusion of field comes about, and the effect is sufficiently interesting in its own right to merit detailed study. It is particularly important to understand the effect if we consider to what degree the field is excluded from the supergranules by their continual overturning.

Some years ago Parker (1966) worked out the fields as analytical functions of space and time in rigidly rotating cylinders and spheres surrounded by a conducting medium (see also Herzenberg and Lowes 1957). He followed in detail the development and decay of the individual filaments of field as they wind about the rotating region. Weiss (1966) explored the development and eventual exclusion of field from regions of continuous fluid motion (thereby avoiding the discontinuity in the fluid velocity that occurs at the surface of the rigidly rotating cylinder or sphere) and produced computer print-outs of the evolution of the field at successive stages of development from an initial uniform state into the final excluded form. He considered the simple convective cell described by the stream function

$$\psi = \cos(kx)\cos(ky) \tag{16.47}$$

for k^2x^2, $k^2y^2 < \pi^2$ and employed numerical methods for the solution of (16.29)[3]. He pointed out the initial linear rise of field to the maximum (16.2) after a time of the order of (16.1) and showed the subsequent decline to the final asymptotic form with the field excluded from the central regions of the cell. The time evolution of a field, initially uniform and vertical, is shown, reproduced from his original paper, in Fig. 16.6. The magnetic Reynolds number is $R_m = 10^3$. The breakup of the filaments is particularly noteworthy, forming closed loops as the lines of force reconnect at various points along the windings. The example provides an excellent illustration of the reconnection of magnetic lines of force (Bullard 1949; Parker and Krook 1956) as a consequence of resistive diffusion. It may be seen from Fig. 16.6 that the final asymptotic form is reached, after a few small oscillations in the disintegrating loops in the interior, in a period of about three times the characteristic period (16.1). Figure 16.7 is a plot of the mean square field as a function of time for various magnetic Reynolds numbers up to 2000. Weiss goes on to treat a row of convective cells, etc. But the results for the single cell illustrate the essential character of the phenomenon.

The final asymptotic strength of the field compressed into the boundary

[3] It is a simple matter to show that there are no non-trivial elementary analytical solutions available.

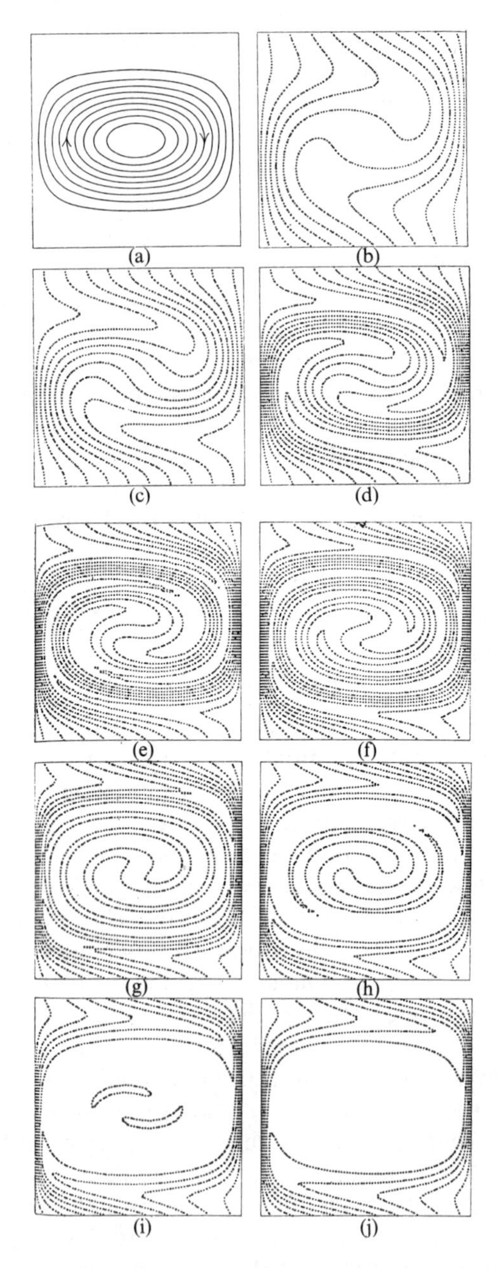

FIG. 16.6. The stream lines of the fluid velocity (16.47) are plotted in (a). The lines of force of the final asymptotic state for the modest magnetic Reynolds number $R_m = 40$ are shown in (b). For a magnetic Reynolds number $R_m = 10^3$ the development of the lines of force with time is shown in (c) through (j) at intervals of $\Delta t = 0{\cdot}5$, beginning with $t = 0{\cdot}5$ and ending near the asymptotic state with $t = 4{\cdot}0$, reproduced from Weiss (1966).

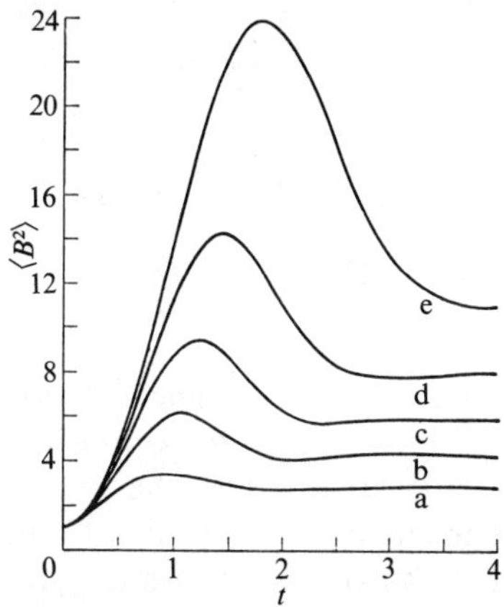

FIG. 16.7. The development of the mean square magnetic field with time in the velocity field (16.47) for magnetic Reynolds numbers (a) $R_m = 40$, (b) $R_m = 100$, (c) $R_m = 200$, (d) $R_m = 400$, (e) $R_m = 1000$, reproduced from Weiss (1966).

of the cell follows from the considerations at the end of §16.4, wherein it was pointed out that the lines of force extending upward through a horizontal converging flow $v_x = -vkx$ are compressed into a thin filament of characteristic width $1/R_m^{\frac{1}{2}}k$ where $R_m = v/k\eta$. Thus, if the field was initially B_0 over the entire region, then the peak field in the compressed ribbon is $B = B_0 R_m^{\frac{1}{2}}$. The field is essentially zero outside the ribbon, so that the mean square field, averaged over the entire region is, then, $B_0^2 R_m^{\frac{1}{2}}$, so that the r.m.s. field is $B_0 R_m^{\frac{1}{4}}$. It is smaller than the maximum field, at the intermediate time (16.1), by $R_m^{\frac{1}{12}}$. The increase of the mean square field through the maximum (16.2) and the subsequent decline to the final value $B_0^2 R_m$ is vividly displayed in Weiss's numerical calculation, shown in Fig. 16.7.

References

BULLARD, E. C. (1949). *Proc. Roy. Soc.* A **199,** 413.
CLARK, A. (1965). *Phys. Fluids* **7,** 1455.
COWLING, T. G. and HARE, A. (1957). *Quart. J. Mech. appl. Math.* **10,** 385.
HERZENBERG, A. and LOWES, F. J. (1957). *Phil. Trans. Roy. Soc. Ser. A* **249,** 507
PARKER, E. N. (1963). *Astrophys. J.* **138,** 552.
—— (1975). *Astrophys. Space Sci.* **24,** 279.
—— and KROOK, M. (1956). *Astrophys. J.* **124,** 214.
PARKER, R. L. (1966). *Proc. Roy. Soc, Ser. A* **291,** 60.
SWEET, P. A. (1949). *Mon. Notic. Roy. Astron. Soc.* **109,** 507.
—— (1950). ibid. **110,** 69.
THOMSON, J. J. (1893). *Recent researches in electricity and magnetism.* Clarendon Press. Oxford.
WATSON, G. N. (1958). *Theory of Bessel Functions,* Cambridge University Press, Cambridge.
WEISS, N. O. (1966). *Proc. Roy. Soc, Ser. A* **293,** 310.

17

MAGNETIC FIELDS IN TURBULENT FLUIDS

17.1. Introduction

HAVING described the exclusion of magnetic fields from closed circulation patterns of fluid, we turn now to the opposite circumstance wherein the fluid mixes at random throughout the space. The high electrical conductivity causes the magnetic field to mix at random with the fluid, so that a variety of phenomena occur, including turbulent diffusion of the mean field and production of new magnetic flux on both a small and a large scale. Indeed, the complexity is so great that there is not yet a general theory for the behaviour of magnetic fields in a turbulent fluid. On the other hand, the turbulent dissipation and the turbulent generation of the mean large-scale magnetic field play a central role in understanding the origin and the behaviour of the magnetic fields in astronomical bodies. So the subject of turbulence, however difficult, cannot be avoided if we wish to understand the 'life cycle' of the fields treated in the foregoing chapters.

In the absence of a general theory one is led to treat a variety of special cases, each sufficiently idealized to be tractable. With a suitably diverse assortment of special cases it is possible to establish that turbulent diffusion of magnetic fields is a real and powerful effect. In rapidly rotating bodies there is the additional effect that turbulent diffusion may be reduced, or even reversed, together with the generation of new magnetic flux, all as a consequence of the cyclonic eddies of which turbulence is composed in a rapidly rotating body. The topic is doubly fascinating because it deals directly with some of the most fundamental unsolved problems of classical hydrodynamics, such as the accumulation of strain in an element of fluid in a turbulent flow, and the correlation of the strain with the net displacement. These problems are not particularly important in the turbulent diffusion of scalar fields, such as smoke, but they are crucial for the diffusion of magnetic fields. The problem is difficult, and Kraichnan (1976a, b) has emphasized that its *general* treatment lies outside any presently available formal statistical approximation. It is a simple fact that what we know today of the turbulent diffusion and turbulent generation of magnetic fields rests on special limiting cases, such as the short–sudden approximation, the low magnetic Reynolds limit, and the analytical and numerical study of a sea of artificially prescribed turbulent eddies. This foundation is firm, if inelegant, because it is quantitative and because a diverse assortment of cases have been treated. A number of exotic effects have been discovered and there may well be additional effects still unknown. The subject has moved out of the speculative state, where it started with the recognition of hydromagnetics thirty years ago.

The understanding of magnetic fields in turbulent conducting fluids is of paramount importance to the astrophysical question of the origin of cosmic magnetic fields. The first step toward answering the question is to decide whether the observed magnetic fields are of contemporary or primordial origin. The planets, stars, and galaxies that populate the universe are believed to have been formed in the past from the collapse of diffuse clouds of gas into the condensed active forms that presently catch our attention. It is not unreasonable to suppose that the original diffuse gas clouds were pervaded by large-scale magnetic fields, much as clouds are today (Thorne 1967). Hence, it is not unreasonable to suppose that in the contraction of the gas to form a galaxy, the primordial field was carried with the gas into the final compressed state, so that each galaxy was born with a significant magnetic field of its own. Thus, for instance, isotropic contraction of a gas of density ρ compresses its internal magnetic field in proportion to $\rho^{\frac{2}{3}}$. If the gas contracted isotropically from an intergalactic density of 10^{-6} H-atoms cm^{-3} to a typical galactic density of 1 atom cm^{-3}, any magnetic field in the gas is compressed by a factor of 10^4, say from a primordial value of 10^{-9} G to 10^{-5} G within the galaxy. Similar compression might reasonably be expected in the formation of stars and planets from the diffuse interstellar gas where the magnetic fields are presently observed to be a few μG. If the primordial field captured by the galaxy is more or less permanently fixed in the galaxy thereafter, then the theory of the origin of the existing galactic field is simply treated. The field is 'primordial', originating in an early epoch of the universe and preserved by the high electrical conductivity of the gaseous disc of the galaxy. Unable to obtain direct information on conditions in the early pre-galactic epoch of the universe, there would be little more to be said. Of course the observed non-uniform rotation of the gaseous disc of the galaxy would shear the lines of force into an azimuthal field composed of many filaments of alternating signs, contrary to contemporary observation, but an agile mind can readily circumvent such difficulties (cf. Piddington 1970, 1972a, b, 1975b). The idea that the galactic magnetic field is of primordial origin has been widely accepted for many years (cf. Parker 1970a).

To examine the question of primordial fields more closely we begin with the planetary magnetic fields. In that case, the answer to the question of primordial origin is easy. The small size of the four inner planets, with liquid iron cores ($\sigma \cong 10^{16}\ s^{-1}$), leads to a resistive decay time of less than 10^5 years. The present age of the planets is in excess of 10^9 years, so any primordial fields disappeared long ago. Even Jupiter, with its large dimensions, ten times the size of Earth, and its interior of metallic hydrogen ($\sigma \cong 10^{17}\,\text{s}^{-1}$) cannot delay the resistive decay of magnetic fields beyond a characteristic time of about 10^8 years. So the present fields of the planets require a contemporary cause.

The resistive decay time for magnetic fields in stars is much longer, of the order of 10^{10} years, comparable to the age of the galaxy and most of its stars (Cowling 1953). The resistive decay time of a magnetic field in the gaseous disc of the galaxy is of the general order of 10^{24} years, and even ambipolar diffusion requires 10^{11} years to release the field from the clutches of the disc. It might seem, then, that a primordial explanation of stellar and galactic fields is possible or even unavoidable. However, it must be remembered that strong fields escape by their own devices, largely through magnetic buoyancy and the associated Rayleigh–Taylor instability (cf. Chapters 8, 9, and 13). Thus, the present-day galactic field of some 3×10^{-6} G escapes through its dynamical effects in a characteristic period of only 10^8 years, while fields of 10^2 G escape from the convective zone of the sun in only 10 years. A fragment of a primordial field (say, $B<10^7$ G) could be trapped in the stable core of a star for 10^{10} years, and perhaps some of the rigidly rotating fields of the magnetic stars are to be understood in this way. But the fields observed at the surface of the sun oscillate with a period of about 22 years. Evidently they are not of so fixed a character. Altogether, then, it would seem that the active fields of the sun and other stars, and the field of the galaxy all require a contemporary cause. There are, evidently, processes in planets, stars, and the gaseous disc of the galaxy that generate magnetic flux, because these objects continually lose magnetic flux and yet their fields are not exhausted at this late epoch in time.

The next question, then, is to consider what might create or amplify magnetic fields. To understand this question note that the large-scale behaviour of magnetic fields is described by the hydromagnetic equation (4.14). The equation is linear and homogenous in $\mathbf{B}$. There is no source term. That is to say, there is no outright creation of field in the hydromagnetic description of magnetic fields. The contemporary magnetic fields can be sustained against resistive and turbulent decay only by *amplification* of existing magnetic flux, rather than by the fresh creation of fields. Hence, if at any point in time the universe were devoid of magnetic fields, then, so far as hydromagnetic effects are concerned, there would be no magnetic field at any other time. The general existence of magnetic fields today implies, then, that there were magnetic fields of some sort in the primordial universe.

In this connection it is interesting to note the point made by Biermann

(1950), that there are microscopic thermal and inertial 'battery' effects in a moving ionized gas. The battery effects are overlooked by the hydromagnetic approximation of the gas as a classical resistive fluid. In particular, the isotherms in a rotating star do not coincide with the equipotential surfaces of the combined gravitational and centrifugal potentials. The unbalanced forces on the free electrons cause a small meridional (poloidal) current to flow, of the general order of $j = m\Omega^2 R\sigma/e$ where Ω is the angular velocity of the star of radius R and electrical conductivity σ. If a star is initially free of all magnetic fields, then over an extended period of time the electric current builds up an azimuthal (toroidal) magnetic field. For the sun an azimuthal field of 50 G might be built up in 4×10^9 years (Roxburgh 1966). However, any meridional (poloidal) magnetic fields, or any convection in the star, confound the effect. So it is doubtful that the contemporary fields of stars are produced by the battery effect. The importance of the battery effect is that it guarantees that, if all else fails, the stars and galaxies in the universe are seeded with magnetic fields. The battery effect guarantees that the powerful hydromagnetic process, described by (4.19), has an initial field with which to operate. It is not possible that **B** remain identically zero. Battery effects alone might produce as much as 10^{-10}–10^{-9} G in the gaseous disc of a galaxy and up to 50 G in stars (see Mestel and Roxburgh 1961; Cattani and Sacchi 1966; Roxburgh 1966; Cattani 1967; Thorne 1967; and references therein).

To continue, then, with the discussion of hydromagnetic effects, it should be pointed out that the diffusion and dissipation of magnetic fields play an essential role in the hydromagnetic amplification of magnetic fields (Parker 1955, 1970a; Leighton 1964, 1969; Steenbeck and Krause 1966). Since diffusion and dissipation are also the graveyard of all magnetic fields, this statement seems paradoxical at first sight; but there are several vital tasks performed by diffusion and dissipation, not unlike the essential role of individual death in the survival of biological species. Consider, for instance, the catastrophic long-term consequences to a vertebrate species that, upon reaching maturity, ceased to age, thereby being immortal. The species would soon overwhelm the environment, and its ability to evolve would be destroyed. In a similar vein the resistive dissipation of magnetic fields is necessary to control the otherwise unlimited growth of small-scale fields in the turbulent fluids. Without rapid destruction of the small-scale fields, they would soon overwhelm the turbulent motion, reducing the mixing of fluid and field to zero. Resistive dissipation avoids this catastrophe. The fields are drawn out by eddies of dimension L into ribbons that soon become very thin compared to L. The ribbons are then cut by the non-equilibrium rapid reconnection (neutral point annihilation, discussed in Chapter 15).

The second point is that the amplification of magnetic flux requires reconnection of lines of force over dimensions L of the individual eddies, and over dimensions as large as the scale λ of the entire system (in the lowest mode of field generation). Speaking broadly, the amplification of magnetic flux over a scale λ in the ordinary circumstance of the planets, stars, and galaxies, requires the formation of a field, and hence, the reconnection of the lines of force, over dimensions that are a significant fraction of λ (Parker 1955). Reconnection over a dimension λ as a consequence of resistive diffusion η requires a time of the general order of λ^2/η. Yet, to use the sun as an example, the fields are built up and destroyed in periods of 10 years, over dimensions of $R_\odot = 7\times10^5$ km across the convective zone in which the electrical conductivity exceeds $10^{13}\,\mathrm{s}^{-1}$ (appropriate for 10^4 K). With $\eta = 10^7\,\mathrm{cm}^2\,\mathrm{s}^{-1}$, then, and $\lambda = R_\odot$ it follows that $\lambda^2/\eta = 5\times10^{14}\,\mathrm{s} \cong 10^7$ years. A ten-year period would be impossible.

It is here, then, that the turbulent diffusion of magnetic fields enters the picture (Steenbeck and Krause 1966; Leighton 1969) as an essential part of the formation of magnetic fields by fluid motions. If there is a turbulent diffusion effect with a diffusion coefficient η_T of the same general order as the turbulent diffusion coefficient κ for scalar fields (such as smoke), then the amplification of weak magnetic fields to form the strong fields currently found in the sun and the galaxy follows automatically as a consequence of the convective motions and non-uniform rotation of those bodies. If, on the other hand, there is no turbulent diffusion effect, then we are presently without a viable explanation for the origin of the observed fields. Hence, the physics of turbulent diffusion of magnetic fields is crucial for understanding the astrophysical magnetic field. It is the subject of this chapter and of the next where we also look into the generation of magnetic fields.

Now the concept of turbulent diffusion of magnetic fields has been around for a long time (Sweet 1950). The diffusion of magnetic fields in a turbulent fluid is suggested by simple analogy to the turbulent diffusion of scalar fields (Taylor 1921; Batchelor 1959). On the other hand, Alfven (1947) and Biermann and Schlüter (1951) postulated, on an equally plausible basis, that the turbulence causes the small-scale magnetic fields to grow until they reach equipartition of energy with the turbulent eddies, $\langle B^2\rangle/8\pi = \frac{1}{2}\rho v^2$. The suggestion was based on the strong coupling between the Maxwell stresses M_{ij} and the Reynolds stresses R_{ij} and the similar form of two isotropic stress tensors

$$M_{ij} = -\delta_{ij}B^2/8\pi + B_iB_j/4\pi, \qquad R_{ij} = -\delta_{ij}(p + \tfrac{1}{2}\rho v^2) - \rho v_i v_j$$

in an incompressible fluid. But, on the other hand, the similarity of the vorticity equation (4.30) and the hydromagnetic equation (4.16) led

Batchelor (1950) and Chandrasekhar (1950, 1951) to suggest that the Fourier spectrum of the magnetic fields is similar to the vorticity spectrum, both spectra rising with increasing wave number to their peaks just before the dissipative cutoff. Hence, if there is to be equipartition of energy, it occurs only at the peak of the vorticity spectrum if the magnetic spectrum extends that far, i.e. if $\eta < \nu$. Various authors explored the analogy and its consequences (Moffatt 1961; Parker 1963; Pao 1963; Saffman 1963, 1964; Kazantsev 1967, 1968; see also the ideas suggested by Parker 1969; Vainshtein 1970; Nagarajan 1971). Altogether it would appear that the analogy provides a dilemma, rather than a solution, to the astronomical problem of the origins of fields.

Kraichnan and Nagarajan (1967) show from a detailed analysis of the interaction of the Fourier components of the magnetic and velocity fields that the general effect of a turbulent fluid on a magnetic field depends upon the sum over wave numbers of many positive and negative contributions, so that elementary statistical conjectures cannot provide a reliable solution to the problem. They make the basic point that no methods presently available are able to sum the effects with sufficient accuracy to determine the sign of the net effect in the general case.

At the present time there are available the results of calculations in the quasi-linear approximation and in the direct interaction approximation. Both lead to the result that the turbulent diffusion coefficient η_T for the mean magnetic field is identical to the coefficient for the mean scalar field. The quasi-linear approximation (Steenbeck and Krause 1969; Rädler 1969; Roberts 1971) is valid for small magnetic Reynolds number for the individual eddies, so that the turbulent velocity field produces only small local perturbations in the large-scale field. The direct interaction approximation (Kraichnan 1976a) gives an improved value for the diffusion coefficient for scalar fields, and the same value for magnetic fields, and one would expect the direct interaction approximation to be more generally applicable, i.e. not restricted to small magnetic Reynolds numbers. But Kraichnan goes on to show that the result of the direct interaction approximation for magnetic fields is not generally internally consistent. Local persistent helicity fluctuations make a large negative contribution to the effective diffusivity of the mean magnetic field, at the same time that they increase the turbulent diffusivity of the scalar field.

Some years ago Parker (1971) introduced a statistical disorder hypothesis that the accumulating strain of an element of fluid loses all memory of the initial position of the element of fluid. It follows immediately that the turbulent diffusion of the mean magnetic field, carried in strong turbulence with high magnetic Reynolds number, is identical with the diffusion of a scalar field, $\eta_T = \kappa$. Moffatt (1974) argued that the principle cannot be generally valid. Working in the limit of large magnetic

Reynolds number too, he pointed out that it is a serious question whether the effect of turbulence on the mean magnetic field can generally be characterized by diffusion and a simple scalar or tensor diffusion coefficient.

Kraichnan (1976a, b) has considerably illuminated the problem with quantitative studies of special cases of disordered velocity fields. His calculations show that, if the local helicity fluctuations are strong and persist for a period in excess of the usual characteristic time $1/\tau_0 = v_0 k_0$ for turbulence with velocity v_0 and wave number k_0, then the turbulent diffusion coefficient for magnetic fields is strongly reduced. If the correlation of the strong helicity persists for periods of $2\tau_0$, the effective diffusion coefficient may be reduced to negative values, causing the mean field to bunch together so that the local field grows unstably. As a matter of fact, both Moffatt and Kraichnan show specific examples which raise the question of whether either η_T or κ approach fixed limits with the passage of time. If we add to this some of the higher order dynamo coefficients (Parker 1955, 1970c) and the higher order effects caused by gradients in the statistical properties of the turbulence (Rädler 1968), the general theory becomes very complicated indeed.

Fortunately most circumstances in nature are approximated adequately, as it turns out, by the basic effects of turbulent diffusion, with $\eta_T = \kappa$, and a single dynamo coefficient. The more exotic effects may occur in special circumstances, such as bodies rotating rapidly compared to the turnover time of the individual convective cells. But the physicist is under no obligation to tackle the most complicated situations first. We shall content ourselves with learning to walk before we try to run. The main point is that most of the complicating effects arise when the cyclonic rotation (helicity) of the individual eddies and convective cells is persistent beyond the usual eddy life $\tau_0 = L/v = 1/v_0 k_0$. To be specific, the calculations show that the effects of persistent helicity within individual eddies, contributing to additional dynamo coefficients and to the reduction of the effective turbulent diffusion coefficient, are quadratic in the accumulated rotation Φ of the fluid. It is for this reason that they do not cancel out in turbulence with vanishing overall helicity (i.e. statistically mirror-symmetric). It is also the reason that they contribute but little in the ordinary case where $\Phi < 1$. In this case Kraichnan's calculations show that η_T approximates closely to the turbulent diffusion coefficient κ, even in the case of a fixed flow pattern—'frozen' turbulence. Hence, it appears that unless the helicity is extremely persistent as a consequence of some special circumstance, the disorder hypothesis is applicable; the accumulated strain is uncorrelated with the net displacement of the element of fluid, and $\eta_T \cong \kappa$.

The astrophysical implication is that in slowly rotating bodies such as the sun and the galaxy, the net effect of turbulence is to cause magnetic fields to diffuse with an effective coefficient η_T closely equal to the turbulent diffusion coefficient κ for scalar fields. The dynamo effects in such bodies can also be described well enough by a few simple coefficients. It also follows, of course, that in very rapidly rotating bodies such as Earth, and early-type stars, where there may be persistent helicity in the turbulent eddies, the higher order contributions must be included, perhaps seriously altering the simple quantitative picture that we build up for the slow rotaters.

In view of the inadequacy of any formal statistical techniques to deal with these effects in a general way, we make extensive use of specific models of random turbulent flows (Parker 1955, 1970a, c; Kraichnan 1970, 1976a, b) to give quantitative demonstration of the statistical effects. The quasi-linear approximation is the lowest non-trivial level of approximation. It is applicable to turbulence in which the magnetic Reynolds number Lv/η of the eddies is small so that the local perturbations δB_i of the large-scale field B_i remain small. The method is extremely useful for demonstrating the existence of turbulent diffusion and of the dynamo effect. Rädler (1968) has used it to show the many higher order effects that arise when the statistical properties of the turbulence vary with position. Hence much of the present chapter is devoted to the quasi-linear approximation in its many forms, to acquaint the reader with the general conclusions deduced from it and to prepare the way for reading the extensive literature on the subject.

Working in the limit of small magnetic Reynolds number so that $\delta B_i \ll B_i$ the quasi-linear approximation does not pick up the quadratic and higher order effects. To explore these higher order effects, we employ examples of two-dimensional turbulence, and the higher order 'short–sudden' approximation, both of which can be treated exactly. The short–sudden approximation was used in the first demonstration of the dynamo effect of cyclonic turbulence, i.e. turbulence with net helicity (Parker 1955). Starting with a large-scale magnetic field, the strong velocity field is switched on for a period of time τ_1 so brief ($\tau_1 \ll L^2/\eta$) that diffusion can be ignored. Large distortions may be produced in the field in this time because there is no restriction on the magnitude of the turbulent velocity. The characteristic displacement $v\tau_1$ may be as large or larger than the eddy size L. Following the brief period of displacement, the fluid is held motionless for a suitable time τ_2 ($\gg L^2/\eta$) during which the smaller-scale fields are obliterated by diffusion and there remains only a large-scale field again. The final large-scale field differs from the initial large-scale field by whatever average fields are produced by the turbulence. The process is repeated indefinitely to give the continuous evolution of the large-scale field with time. The effect has, in common with the small magnetic Reynolds number approximation, the property that the cycling time $\tau_1+\tau_2$ is large compared to the diffusion time L^2/η. But the short–sudden approximation extends to all orders because the δB_i are not limited to being small compared to the large-scale field B_i. Hence the short–sudden approximation obtains the higher order contributions that are missed by the quasi-linear approximation. The quasi-linear approximation is a special case of the short–sudden approximation, in the limit that $v\tau_1 \ll L$. The more extensive results of the short–sudden approximation reduce exactly to those of the quasi-linear approximation as $v\tau_1/L \rightarrow 0$. In the quasi-linear approximation the hydromagnetic equations can be solved because the interaction of v_i with δB_i is neglected (because of the small magnetic Reynolds number), leading to a complete and very convenient solution in terms of the general statistical properties of v_i. The artificial switching of velocity fields employed by the short–sudden approximation is avoided at the cost of the higher order effects.

The fact should not be overlooked, of course, that the short–sudden approximation misses the effect of successive eddies interacting with the δB_i of earlier eddies. To investigate the complete effect of a magnetic field carried in a turbulent flow, one resorts to the infinite conductivity limit, switching on the fluid velocity and letting it run steadily, ignoring all diffusion effects. Kraichnan (1976b) has employed this technique extensively in his numerical experiments. Of course the short–sudden approximation is readily adapted to the infinite conductivity limit too, because in the initial stage τ_1 the magnetic field is calculated ignoring resistive diffusion. We sometimes employ the results of the short–sudden approximation in this way, extending τ_1 to large values in the limit as $\eta \rightarrow 0$. But the limit $\eta = 0$, $t \rightarrow \infty$ raises some interesting mathematical questions.

The short–sudden approximation is developed in §18.3 where it is used to show all of the effects through second order in derivatives of the large-scale field. Indeed, some readers may wish to consult §18.3 directly for a concise summary of the final results. However, the extensive calculations and discussion that intervene are important for developing an understanding of the physics of the problem. In particular, the intervening discussion is essential to obtain an idea of the basic theoretical questions that are still outstanding, and an idea of the methods already used to attack the questions. An obvious division of the problem into categories can be made along the lines of the helicity of the eddies singly or collectively. The individual turbulent eddy, insofar as it can be defined, generally has a net helicity. In a non-rotating body we would expect that the mean helicity over many eddies is zero. On the other hand, most astrophysical bodies rotate, so that their internal turbulence and convection is cyclonic, i.e., has a net helicity such as the atmospheric circulation in either hemisphere of Earth.

The mean helicity of the eddies produces the basic dynamo effect, characterized by the zero-order dynamo coefficient, generating new magnetic flux from the old and amplifying the field as a whole. Whether there is an overall helicity or not, the helicity of the individual eddy is the basis for the higher order dynamo coefficients and for the reduction of turbulent diffusion. The present chapter begins the development with the lowest order terms in the absence of a mean helicity. The helicity fluctuations of the individual eddies contribute only in second order and so do not appear in the first examples, restricted to low magnetic Reynolds numbers or to two-dimensional motion. Large Reynolds numbers in three dimensions are taken up in Chapter 18 in the short–sudden approximation. Throughout the entire discussion the problem is treated kinematically, with the fluid velocity $v_i(x_k, t)$ specified and unaffected by the magnetic fields. The complete dynamical question has not yet been developed to a degree where it can be treated with any generality, although reference to the suppression of cyclonic motion by magnetic field will be made at appropriate places in the text.

A brief exposition of the problem of the turbulent mixing of magnetic field is appropriate here to illustrate the physical problem before us. It will serve as a useful guide in the navigation of the mathematical considerations of the succeeding sections. To fix ideas, then, consider a turbulent velocity field $v_i(x_k, t)$ with characteristic velocity v_0, scale $L = \pi/k_0$ and magnetic Reynolds number $R_m = Lv_0/\eta$. A large-scale magnetic field B_i extends across the fluid and is perturbed weakly or strongly by the turbulence, depending upon the magnitude of the magnetic Reynolds number.

When the magnetic Reynolds number is very large, $Lv_0 \gg \eta$, the magnetic field is carried in the fluid and mixed throughout the space along with the fluid, just as ink or smoke would be carried with the fluid. Indeed, if a thin line of ink were placed along a line of force in the fluid, the later configuration of the ink would map the line of force (see the discussion in §§4.2–4.4). It follows, therefore, that the lines of force are mixed throughout a turbulent conducting astrophysical body in the same manner and to the same degree as the chemical constituents. The lines of force are brought to the surface, or carried deep into the interior, to the same degree that the chemical constituents are mixed to the surface or carried into the interior.

There is more to the problem of the magnetic field, however, because the lines of force are stretched and rotated while being mixed so that the field strength in each element of fluid varies in a complicated manner while the element is transported at random through the volume. The equation of continuity for the density of ink mixed through an incompressible fluid is just

$$d\rho/dt = \partial\rho/\partial t + v_j\, \partial\rho/\partial x_j = 0. \tag{17.1}$$

The density of the ink is not affected by the transport, so that if the element of fluid initially at X_i is transported to $x_i(X_k, t)$ after a time t, the density of the ink upon arrival at x_i is equal to the initial value

$$\rho(x_k, t) = \rho\{X_k(x_n), 0\}. \tag{17.2}$$

For the magnetic field, on the other hand, (4.16) yields

$$\mathrm{d}B_i/\mathrm{d}t = B_j\, \partial v_i/\partial x_j \tag{17.3}$$

with $B_j\, \partial v_i/\partial x_j$ on the right-hand side instead of a zero. The formal solution is then (4.32) instead of (17.2),

$$B_i(x_k, t) = B_j(X_k, 0)\, \partial x_i/\partial X_j \tag{17.4}$$

with the additional factor $\partial x_i/\partial X_j$ representing the accumulated strain of the element of fluid. It is the strain factor $\partial x_i/\partial X_j$ that makes the problem of turbulent diffusion of magnetic field so complicated. We collect together in the sections that follow examples to illustrate the effects of the strain in a variety of circumstances. In the simplest case, in the presence of complete disorder, the factor $\partial x_i/\partial X_j$ in (17.4) averages to δ_{ij} so that (17.4) reduces to (17.2). Turbulent diffusion of magnetic field then proceeds in a fashion identical to the diffusion of scalar fields, with equal diffusion coefficients $\eta_{\mathrm{T}} = \kappa$. This is evidently the state of affairs for small magnetic Reynolds numbers, treated with the quasi-linear approximation, and for turbulence of large magnetic Reynolds numbers when the local helicity fluctuation is neither particularly large nor persistent.

The theory of turbulent diffusion of magnetic field is sufficiently complex that the exposition is limited initially to the simplest possible circumstances, treating turbulence that is statistically stationary in time, homogeneous over an infinite space, and isotropic. We avoid the non-essential effects of boundaries, and steep gradients in the strength of the turbulence. The treatment is limited to turbulence with mirror symmetry. There is no net helicity, $\langle \mathbf{v} \cdot \nabla \times \mathbf{v} \rangle = 0$. Thus, if one eddy possesses a modest helicity during its lifetime, it is assumed that at about the same time there is a nearby eddy with opposite helicity, so that at no time is there a net helicity over dimensions much larger than the characteristic length L of the individual eddy. The characteristic dimension of the large-scale magnetic field is denoted by λ and is taken to be large compared to L.

To get on with the problem, then, the important feature of mirror-symmetric, random turbulence is that it does not generate new magnetic flux. The turbulence scrambles the existing lines of force, cutting them at random perhaps, but producing no net flux. Thus, for instance, if there is initially a straight field $B_z(x, y)$ extending from $z = -\infty$ to $+\infty$ and confined within some distance $\varpi \equiv (x^2 + y^2)^{\frac{1}{2}} < b$ of the z-axis, then no matter how entangled the field may become at a later time, it is readily shown from (4.16) that the total flux across any plane $z = constant$,

$$F(t) = \int_{-\infty}^{+\infty} \mathrm{d}x \int_{-\infty}^{+\infty} \mathrm{d}y B_z(x, y, z, t), \tag{17.5}$$

is equal to the initial total flux

$$F(0) = \int_{-\infty}^{+\infty} \mathrm{d}x \int_{-\infty}^{+\infty} \mathrm{d}y B_z(x, y, z, 0).$$

Needless to say $\partial F/\partial z = 0$ as well as $\partial F/\partial t = 0$. This basic condition simplifies some of the later development of various special examples.

To begin the exploration of the behaviour of a magnetic field carried in a specified turbulent velocity field $v_i(x_k, t)$ consider the basic similarity of the magnetic field and the vorticity (Batchelor 1952). The magnetic field satisfies (4.16) and the vorticity $\omega = \nabla \times \mathbf{v}$ satisfies the identical form (4.30). In terms of the Lagrangian time derivative, the magnetic field satisfies (17.3) while the vorticity satisfies

$$\mathrm{d}\omega_i/\mathrm{d}t = \omega_j \, \partial v_i/\partial x_j. \tag{17.6}$$

Consider, then a free-running field of turbulence in which the inviscid fluid is coasting under its own inertia, so that ω_i varies as described by (17.6). It follows that an initial weak field B_i with the same initial configuration as ω_i is transported by v_i in exactly the same way as the vorticity. That is to say, if initially

$$B_i(X_k) = \alpha(X_n)\omega_i(X_k) \tag{17.7}$$

where $\alpha(X_k)$ is a function of position but constant along each vortex line,

$$B_j \, \partial\alpha/\partial X_j = 0,$$

then

$$B_i(x_k, t) = \alpha\{X_n(x_k)\}\omega_i(x_k, t) \tag{17.8}$$

at all subsequent times. The magnetic field is transported in exactly the same way as the vorticity. Hence, the mean large-scale magnetic field undergoes the same turbulent diffusion as the mean vorticity.

To provide a physical example, suppose that initially the velocity field is non-vanishing only within the distance $\varpi = b$ of the z-axis. Within this cylindrical volume there is a random component of the fluid velocity (for which $\langle \mathbf{v} \rangle = 0$, $\langle \nabla \times \mathbf{v} \rangle = 0$) composed of eddies of scale L somewhat smaller than the radius b. There is also a net mean angular velocity Ω about the z-axis (for which the mean curl is not zero, of course, although the mean helicity remains zero because of the random character of v_z, with $\langle v_z \rangle = 0$). Then it is possible to set up a magnetic field that is everywhere parallel to the vorticity, as described by (17.7), with a net magnetic flux in the z-direction. The configuration is sketched in Fig. 17.1.

Now colour the fluid initially within $\varpi \leq b$ with a suitable dye so that the fluid can be identified at any subsequent time t. The fluid is set in motion at time $t=0$ and is allowed to run free thereafter. The turbulence mixes the dyed fluid outward into the surrounding space ($\varpi > b$), and the initial rotation of the fluid is soon spread out through the surrounding medium. The total angular momentum is conserved but the mean angular velocity about the z-axis decreases with the increasing distance of each element of fluid from the z-axis and with the mixing of rotating fluid into the surrounding non-rotating fluid. The mean magnetic field declines along with the mean vorticity density as indicated by (17.8). The dispersal of the vorticity with the passage of time is readily demonstrated in the laboratory with a purely hydrodynamic experiment with Reynolds numbers up to at least 10^5. The dispersal of a vector field, such as the magnetic field, is then demonstrated through the relation (17.8).

As another example consider the simple circumstance in which an initial large-scale magnetic field $B_z(x, y)$ in the z-direction is subject to the three-dimensional incompressible velocity field $v_i(x, y, t)$ that does not vary in the z-direction. Then $\partial v_z/\partial z = 0$ so that v_x and v_y together

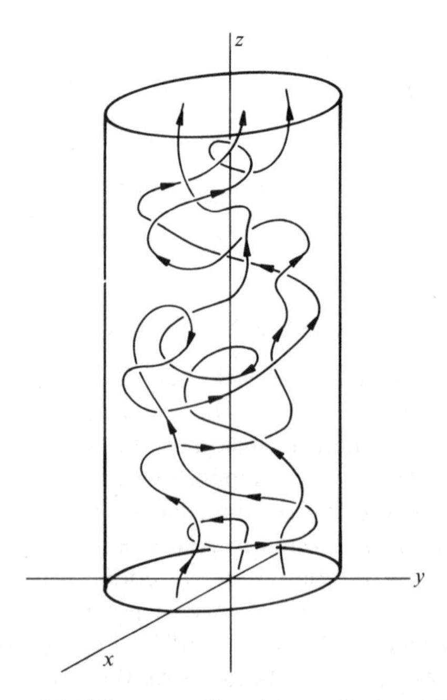

FIG. 17.1. The vortex and field lines extending along a cylinder of fluid with a net rotation.

represent an incompressible flow. Such circumstances might arise in nature where a strong field B_z constrains weak fluid motions. The velocity field $v_i(x, y, t)$ is the only fluid motion that does not deform the strong field. In this simple case $B_j\,\partial v_i/\partial x_j = 0$. Then (17.3) reduces to

$$dB_i/dt = 0. \tag{17.9}$$

Since B_x and B_y are initially zero, they remain so for all subsequent time, while B_z is non-vanishing and is transported in the same way as a scalar field (17.1), so that

$$B_z(x, y, t) = B_z(X, Y, 0) \tag{17.10}$$

where again (x, y) is the position of an element of fluid initially at (X, Y). Thus, as a consequence of $v_i(x, y, t)$ the mean field distribution spreads out with time in the same way as a scalar field, e.g. dye or ink or smoke. It is well known from the study of the turbulent diffusion of smoke that the time evolution of the local mean field density $\langle\rho\rangle$ can be described quite satisfactorily by a transport equation of the familiar form (Taylor 1921).

$$\partial\langle\rho\rangle/\partial t = \kappa\nabla^2\langle\rho\rangle \tag{17.11}$$

with the turbulent diffusion coefficient related to the scale (mixing length) L and velocity v_0 of the dominant eddies by

$$\kappa \cong 0{\cdot}4 v_0 L \tag{17.12}$$

(Kraichnan 1970, 1976a, b). It follows that to the same degree of accuracy

$$\partial\langle B_z\rangle/\partial t = \kappa\nabla^2\langle B_z\rangle. \tag{17.13}$$

These two examples illustrate some aspects of the turbulent diffusion of magnetic field. They represent a very restricted class of field and flow, however. As soon as B_i is not parallel to ω_i, or v_i depends upon the third coordinate z, the situation is much more complicated. For instance, a weak dependence of v_i on z (perhaps as a consequence of strong B_i) produces a slight deformation of the individual flux tubes from their initial straight lines as sketched in Fig. 17.2. The crossing of flux tubes leads to the dynamical rapid reconnection of the magnetic lines of force (Chapter 15). If B_z is strong and the dynamical reconnection keeps B_x and B_y weak, then it may still be argued that (17.13) is approximately correct. But the complication is obvious and, in any case, we do not include dynamical effects in the present kinematical treatment of the problem. Evidently some very general statistical hypotheses and approximations must be used to treat the behaviour of the large-scale field when v_i depends strongly on all three coordinates x_i. The hypotheses will be

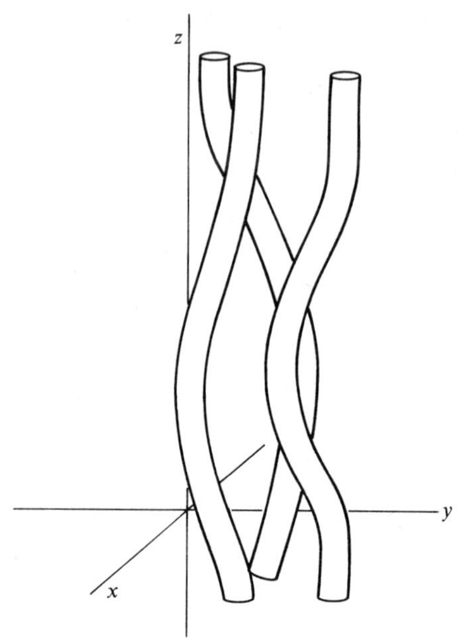

FIG. 17.2. The individual flux tubes in a magnetic field carried in a fluid in which the motions are parallel to the xy-plane and vary only slightly with z, showing the cross-over of neighbouring tubes.

justified and delimited by comparison of their consequences with the quantitative calculations of a number of special cases.

17.2. The quasi-linear approximation

The special case most widely treated in the literature is the quasi-linear approximation (Yaglom 1962; Meecham 1964; Frisch 1968). The method has its limitations because it neglects mode–mode coupling (Kadomtsev 1965; Bernstein and Engelmann 1966; Dolinsky and Goldman 1967; Fu 1973) which seems to be a valid approach in some cases (Kennel and Engelmann 1966; Kennel and Petschek 1966) but certainly not in others (Kraichnan 1961; Herring 1969). It seems that the quasi-linear approximation is valid for small magnetic Reynolds numbers Lv_0/η, because the linear diffusion term $\eta\,\nabla^2 B_i$ dominates the behaviour of the small-scale fluctuations δB_i of the field, causing δB_i to remain small compared to the mean field (Rädler 1968). The form into which the equations are cast by the quasi-linear approximation is particularly convenient to use because the description of the large-scale field is in terms of the two-point correlation of the turbulent velocity $\langle v_i(x_k, t)v_j(x_k+\zeta_k, t+s)\rangle$ whose properties are well known from the theory of isotropic homogeneous turbulence.

The standard procedure is to write the magnetic field $B_i(x_k, t)$ as the sum of the ensemble average field $\langle B_i(x_k, t)\rangle$ and the instantaneous local deviation $\delta B_i(x_k, t)$ of B_i from the mean $\langle B_i\rangle$,

$$B_i(x_k, t) = \langle B_i(x_k, t)\rangle + \delta B_i(x_k, t),$$
$$\langle \delta B_i(x_k, t)\rangle = 0.$$

Then the quasi-linear approximation applies to those cases where $v_i \delta B_j$ and its derivatives $\partial v_i \delta B_j/\partial x_i$, $\partial v_i \delta B_j/\partial x_j$ differ but little from, or oscillate rapidly about, their mean values, so that the difference can be neglected in the equations.

To develop the approximate equations, begin with the fact that the ensemble average of the hydromagnetic induction equation

$$\left(\frac{\partial}{\partial t} - \eta\nabla^2\right) B_i = B_j \frac{\partial v_i}{\partial x_j} - v_j \frac{\partial B_i}{\partial x_j} \tag{17.14}$$

yields

$$\left(\frac{\partial}{\partial t} - \eta\nabla^2\right) \langle B_i\rangle = \left\langle \delta B_j \frac{\partial v_i}{\partial x_j}\right\rangle - \left\langle v_j \frac{\partial \delta B_i}{\partial x_j}\right\rangle \tag{17.15}$$

in the turbulent velocity field v_i, for which $\langle v_i\rangle = 0$. Subtracting this from (17.14), the result is the exact equation

$$\left(\frac{\partial}{\partial t} - \eta\nabla^2\right) \delta B_i = \langle B_j\rangle \frac{\partial v_i}{\partial x_j} - v_j \frac{\partial \langle B_i\rangle}{\partial x_j}$$
$$+ \delta B_j \frac{\partial v_i}{\partial x_j} - v_j \frac{\partial \delta B_i}{\partial x_j} - \left\langle \delta B_j \frac{\partial v_i}{\partial x_j}\right\rangle + \left\langle v_j \frac{\partial \delta B_i}{\partial x_j}\right\rangle \tag{17.16}$$

for the magnetic fluctuations δB_i. Now suppose that for some reason $|\delta B_i| \ll |\langle B_i\rangle|$. Then the terms $\delta B_j\, \partial v_i/\partial x_j$ are generally small compared to $\langle B_j\rangle\, \partial v_i/\partial x_j$ and may be neglected in the equation. Such a situation obtains when the resistive diffusion coefficient η is large, so that the small-scale fluctuations δB_i decay rapidly and remain small within the individual turbulent eddies. The same *might* be expected to hold in an idealized turbulence with *small* η in which the correlation time τ_0 is small compared to the correlation length L divided by the r.m.s. turbulent velocity v_0. Then the fluid moves only the small distance $v_0\tau_0$ during the short life τ_0, and the deflection of the lines of force of $\langle B_i\rangle$ during one correlation time τ is small, $O(v_0\tau_0/L)$. (See discussion in Frisch 1968; Roberts 1971). For one reason or another, then, it may be conjectured that the second order terms can be neglected, producing the quasi-linear 'approximation'

$$\left(\frac{\partial}{\partial t} - \eta\nabla^2\right) \delta B_i = \langle B_j\rangle \frac{\partial v_i}{\partial x_j} - v_j \frac{\partial \langle B_i\rangle}{\partial x_j} \tag{17.17}$$

for the small-scale magnetic field δB_i. With this simplified equation it is possible to go forward with the computation, although it remains to be shown in each case that the field deduced from (17.17) is consistent with the assumption on which (17.17) is based. Unfortunately, the quasi-linear approximation is so facile and convenient as to be seductive, and not all its users have adequately tested their final results.

To follow, through, then with the exact solution (in the limit of large λ/L) of the approximate eqn (17.17) note that the solution of (17.17) is readily expressed in terms of the familiar Green's function G for the heat flow equation

$$G(\mathbf{r}, \mathbf{r}'; t, t') = \{4\pi\eta(t-t')\}^{-\frac{3}{2}} \exp\{-(\mathbf{r}-\mathbf{r}')^2/4\eta(t-t')\}. \qquad (17.18)$$

Assuming that the random component of the field is composed only of the small-scale fluctuations produced by the turbulence (i.e. for the case of no initial large-scale random field, or alternatively for large values of $t \gg L^2/\eta$ such that any initial fields have decayed) it follows that the formal solution of (17.17) is

$$\delta B_i(\mathbf{r}, t) = \int_{-\infty}^{t} \mathrm{d}t' \int \mathrm{d}^3\mathbf{r}' G(\mathbf{r}, \mathbf{r}'; t, t')$$

$$\left\{\langle B_k(\mathbf{r}', t')\rangle \frac{\partial v_i(\mathbf{r}', t')}{\partial x_k'} - v_k(\mathbf{r}', t') \frac{\partial \langle B_i(\mathbf{r}', t')\rangle}{\partial x_k'}\right\}. \qquad (17.19)$$

Substituting this expression for δB_i into the right-hand side of (17.15) the resulting equation for the mean field $\langle B_i \rangle$ is

$$\left(\frac{\partial}{\partial t} - \eta\nabla^2\right)\langle B_i\rangle = \int_{-\infty}^{t} \mathrm{d}t' \int \mathrm{d}^3\mathbf{r}' G\left\{\langle B_k(\mathbf{r}', t)\rangle \left\langle \frac{\partial v_i(\mathbf{r}, t)}{\partial x_j} \frac{\partial v_j(\mathbf{r}', t')}{\partial x_k'} \right\rangle\right.$$

$$\left. - \frac{\partial \langle B_j(\mathbf{r}', t')\rangle}{\partial x_k'} \left\langle \frac{\partial v_i(\mathbf{r}, t)}{\partial x_j} v_k(\mathbf{r}', t') \right\rangle\right\}$$

$$- \frac{\partial}{\partial x_j} \int_{-\infty}^{t} \mathrm{d}t' \int \mathrm{d}^3\mathbf{r}' G \left\{ B_k(\mathbf{r}', t') \right.$$

$$\times \left\langle v_j(\mathbf{r}, t) \frac{\partial v_i(\mathbf{r}', t')}{\partial x_k'} \right\rangle$$

$$\left. - \frac{\partial \langle B_i(\mathbf{r}', t')\rangle}{\partial x_k'} \langle v_j(\mathbf{r}, t) v_k(\mathbf{r}', t')\rangle \right\}. \qquad (17.20)$$

Hence the evolution of the mean field $\langle B_i \rangle$ follows from the two-point correlation tensor $\langle v_i(x_k, t) v_j(x_k', t')\rangle$ of the turbulent velocity. The basic properties of the correlation are well known (see, for instance, Batchelor 1953) and are reviewed briefly in Appendix A at the end of this chapter. It follows from (A7) that the first term in the integrand of (17.20) vanishes because of the incompressibility of the fluid.

The second and third terms are readily evaluated from their symmetry. With $x_i' = x_i + \zeta_i$, expand $B_j(x_m', t')$ about x_m so that

$$\langle B_j(x_m', t)\rangle = \langle B_j(x_m, t)\rangle + \zeta_n\, \partial\langle B_j(x_m, t)\rangle/\partial x_n + \ldots,$$

neglecting terms second order in L/λ, where, it will be recalled, λ is the large characteristic scale of the mean $\langle B_i\rangle$. Note that with mirror-symmetric turbulence the quantities $\langle(\partial v_i/\partial x_j)v_k'\rangle$ and $\langle v_j\, \partial v_i'/\partial x_k'\rangle$, given by (A3) and (A4) in terms of (A2), are odd functions of ζ_n, so that the integral over ζ_n vanishes. If multiplied by another odd function, such as the first-order term in the expansion of $\langle B_j\rangle$, then the integral over ζ_n does not vanish, of course. It follows that the second and third terms on the right-hand side of (17.20) reduce to

$$\int_0^\infty \mathrm{d}s \int \mathrm{d}^3\zeta_m G(\zeta, s)\left(\frac{\partial^2\langle B_j\rangle}{\partial x_k\, \partial x_n}\zeta_n\frac{\partial V_{ik}}{\partial \zeta_j} - \frac{\partial^2\langle B_k\rangle}{\partial x_j\, \partial x_n}\zeta_n\frac{\partial V_{ji}}{\partial \zeta_k}\right)$$

upon using (A3) and (A4) and writing $t' = t - s$, $x_n' = x_n + \zeta_n$. Noting that $V_{ij} = V_{ji}$, it follows that the two terms are identical but of opposite sign, so that they cancel exactly. Indeed, it is readily seen that the expansion of B_j in powers of ζ_n cancels to all orders in ζ_n. Hence (17.20) reduces to

$$\left(\frac{\partial}{\partial t} - \eta\nabla^2\right)\langle B_i\rangle = u^2\frac{\partial^2\langle B_i\rangle}{\partial x_j\, \partial x_k}\int_0^\infty \mathrm{d}s \int \mathrm{d}^3\zeta_m G(\zeta, s)\, V_{jk}(\zeta, s)$$

neglecting terms $O(L^3/\lambda^3)$ compared to one. With V_{ij} given by (A2) the equation reduces to

$$\left(\frac{\partial}{\partial t} - \eta\nabla^2\right)\langle B_i\rangle = \eta_T\nabla^2\langle B_i\rangle \tag{17.21}$$

where

$$\eta_T \equiv 4\pi u^2\int_0^\infty \mathrm{d}s \int \mathrm{d}^3\zeta_m G(\zeta, s)\zeta^2\{B(\zeta, s) + \tfrac{1}{3}\zeta^2 A(\zeta, s)\}, \tag{17.22}$$

in which we have made use of the fact that

$$\begin{aligned}\int \mathrm{d}^3\zeta_m G(\zeta, s)A(\zeta, s)\zeta_1^2 &= \tfrac{1}{3}\int \mathrm{d}^3\zeta_m G(\zeta, s)A(\zeta, s)\zeta^2\\ &= \frac{4\pi}{3}\int_0^\infty \mathrm{d}\zeta G(\zeta, s)A(\zeta, s)\zeta^4.\end{aligned}$$

Thus η_T represents the turbulent diffusion coefficient. The total effective diffusion coefficient is $\eta + \eta_T$. The quasi-linear approximation, then, predicts that the turbulent mixing of a vector field leads to a diffusion equation, just as with a scalar field. To illustrate the magnitude of the diffusion coefficient, consider the example for which the spatial dependence of A and B is

$$A(\zeta)=T(s)\frac{3L^3}{\zeta^5}\int_0^{\zeta/L} d\xi\, \xi^4 \exp(-\xi^2),$$

$$B(\zeta)=T(s)\left\{\exp(-\zeta^2/L^2)-\frac{L^3}{\zeta^3}\int_0^{\zeta/L} d\xi\, \xi^4 \exp(-\xi^2)\right\}$$

say, satisfying (A8). Then (17.22) reduces to

$$\eta_T=u^2\int_0^\infty \frac{ds\,T(s)}{(1+4\eta s/L^2)^{\frac{3}{2}}}. \tag{17.23}$$

For the simple case that $T(s)=1$ for $0<s<\tau$, and zero for $s>\tau$,

$$\eta_T=\frac{u^2L^2}{2\eta}\left\{1-\frac{L}{(L^2+4\eta\tau)^{\frac{1}{2}}}\right\}. \tag{17.24}$$

The general result of the quasi-linear approximation, that the turbulent mixing of the mean magnetic field is described exactly in the limit of large λ/L by a diffusion equation, is applicable, of course, only to those circumstances to which the quasi-linear approximation is valid. We expect the quasi-linear approximation to be valid for small magnetic Reynolds number, $v_0L/\eta\cong L^2/\eta\tau_0\ll 1$, because the fluctuations δB_i are then limited to small values by the resistivity. In this limit (17.24) reduces to

$$\eta_T=u^2L^2/2\eta=v_0^2L^2/6\eta. \tag{17.25}$$

In terms of the magnetic Reynolds number $R_m=v_0L/\eta$, this is

$$\eta_T=\tfrac{1}{6}R_m^2\eta. \tag{17.26}$$

This expression for the turbulent contribution to the resistivity (and valid for small R_m) was first obtained by Rädler (1968) and Moffatt (1970). In the limit of small R_m the turbulent mixing can have only a small effect compared to the resistive diffusion.

Now some authors have suggested (see, for instance, Rädler 1968; Roberts 1971; Lerche 1971) that the quasi-linear approximation may be valid with the sole restriction that the correlation time τ is small compared to the characteristic time L/v_0, because the δB_i produced by the displacement $v_0\tau$ is small, of the order of $B_iv_0\tau/L$. In this instance, then, the correlation functions A and B in (17.22) are non-vanishing only for $s\leqslant\tau$. Then for any η, we have, in the limit of small τ, that $\eta s/L^2\leqslant \eta\tau/L^2=1/R_m\ll L$. Therefore, during the short time τ the Green's function $G(\zeta, s)$ is essentially a Dirac delta function $\delta(\zeta_k)=\delta(\zeta)/4\pi\zeta^2$ and

(17.22) reduces to the general form[1]

$$\eta_T = u^2 \int_0^{\infty} \mathrm{d}s T(s) = \tfrac{1}{3} v_0^2 \tau. \tag{17.27}$$

In this circumstance the diffusion coefficient is identical with the general diffusion coefficient (B3) for a scalar field. Indeed, if one declares simply that the resistivity η is so small that the magnetic Reynolds number $R_m = Lv_0/\eta$ is large, then with $\tau \cong L/v$ the Green's function is a delta function, and (17.27) is the general result. But there is no reason whatever to think that the quasi-linear approximation has any validity at all in this case. Indeed, we would expect δB_i to become comparable to $\langle B_i \rangle$ within a single eddy life $\tau \cong L/v$.

As a matter of fact, when $R_m \gg 1$ a small correlation time is insufficient to justify the quasi-linear approximation. The difficulty is readily shown by integrating (17.17) in the limit of small $\eta\tau$ so that $G(\zeta, s)$ reduces to $\delta(\zeta)$. Then

$$\begin{aligned}\delta B_i(\mathbf{r}, t) &= \int_0^t \mathrm{d}t' \int \mathrm{d}^3\mathbf{r}' G(\mathbf{r},\mathbf{r}'; t, t') \left\{ \langle B_j(\mathbf{r}', t') \rangle \frac{\partial v_i(\mathbf{r}', t')}{\partial x_j'} \right. \\ &\qquad \left. - v_j(\mathbf{r}', t') \frac{\partial \langle B_i(\mathbf{r}', t') \rangle}{\partial x_j'} \right\} \\ &= \int_0^t \mathrm{d}t' \left\{ \langle B_j(\mathbf{r}, t') \rangle \frac{\partial v_i(\mathbf{r}, t')}{\partial x_j} - v_j(\mathbf{r}, t') \frac{\partial \langle B_i(\mathbf{r}, t') \rangle}{\partial x_j} \right\}\end{aligned}$$

Then, since $\langle B_i(\mathbf{r}, t) \rangle$ varies slowly with time, it can be taken outside the integral and

$$\delta B_i(\mathbf{r}, t) \cong \langle B_k(\mathbf{r}, t) \rangle \frac{\partial \xi_i(\mathbf{r}, t)}{\partial x_k} - \xi_k(\mathbf{r}, t) \frac{\partial \langle B_i(\mathbf{r}, t) \rangle}{\partial x_k} + \delta B_i(\mathbf{r}, 0) \tag{17.28}$$

where

$$\xi_i(\mathbf{r}, t) \equiv \int_0^t \mathrm{d}t v_i(\mathbf{r}, t)$$

is the accumulated displacement produced by $v_i(\mathbf{r}, t)$ at a fixed point. It would appear that ξ_i becomes large with the passage of time. After n correlation periods τ, in which the fluid undergoes a characteristic random displacement $v\tau$, the accumulated random walk $|\int \mathrm{d}t v_i|$ is of the order of $n^{\frac{1}{2}} v\tau$. Hence in a time t, $n = t/\tau$ and the mean displacement is of the order of

$$\langle \xi^2 \rangle^{\frac{1}{2}} = v_0 \mathrm{O}\{(t\tau)^{\frac{1}{2}}\}.$$

[1] The same result is obtained from the special case (17.23), of course, for $\eta\tau \ll L^2$.

It follows from (17.28) that[2]

$$|\delta B_i| = \langle B_i \rangle (v_0 \tau / L) \mathrm{O}\{(t/\tau)^{\frac{1}{2}}\}. \tag{17.29}$$

No matter how small τ is, δB_i does not remain small in the limit of large t. Thus there is no reason to think that $v_j \delta B_i - \langle v_j \delta B_i \rangle$ remains negligible on the right-hand side of (17.16). The only case that seems self-consistent is the limit of small magnetic Reynolds number, $v_0 L/\eta \ll 1$.

17.3. Fourier transformed equations

If the hydromagnetic equation (17.14) is expressed in terms of the Fourier transforms of v_i and B_i, then the final result is an integro-differential equation for which an obvious iteration procedure is available for solution. The method makes clear the nature of the quasi-linear approximation when the magnetic Reynolds number is small. As a matter of fact, several options are open for carrying out various approximate solutions. We could, for instance, work directly with the Fourier transform of the truncated equation (17.17) (see, for instance, Lerche 1971; Roberts and Soward 1975). We shall instead transform the exact equation (17.14) (see Vainshtein and Zeldovich 1972) and then introduce the various approximations for solution from the exact transformed equation.

The fluid velocity $v_i(\mathbf{r}, t)$ is again to be specified, and expressed in terms of its spatial Fourier transform $u_i(\mathbf{k}, t)$. The magnetic field, to be computed, is expressed in terms of its spatial Fourier transform $b_i(\mathbf{k}, t)$, so that

$$B_i(\mathbf{r}, t) = \int d^3\mathbf{k} \exp(i\mathbf{k} \cdot \mathbf{r}) b_i(\mathbf{k}, t),$$

$$b_i(\mathbf{k}, t) = \frac{1}{(2\pi)^3} \int d^3\mathbf{r} \exp(-i\mathbf{k} \cdot \mathbf{r}) B_i(\mathbf{r}, t),$$

$$v_i(\mathbf{r}, t) = \int d^3\mathbf{k} \exp(i\mathbf{k} \cdot \mathbf{r}) u_i(\mathbf{k}, t),$$

$$u_i(\mathbf{k}, t) = \frac{1}{(2\pi)^3} \int d^3\mathbf{r} \exp(-i\mathbf{k} \cdot \mathbf{r}) v_i(\mathbf{r}, t).$$

Then since B_i and v_i are real quantities, it follows that

$$b_i(\mathbf{k}, t) = b_i^*(-\mathbf{k}, t); \qquad u_i(\mathbf{k}, t) = u_i^*(-\mathbf{k}, t), \tag{17.30}$$

where the asterisk denotes the complex conjugate. Note that

$$\langle v_i \rangle = 0, \qquad \langle u_i \rangle = 0. \tag{17.31}$$

To proceed with the calculation, we begin with the complete hydromagnetic induction equation (17.14). The Fourier transform of the

[2] This result is worked out in more detail in §17.5.

equation converts the left-hand side to $(\partial/\partial t+\eta k^2)b_i$. On the right-hand side express B_k and v_k in terms of their Fourier transforms. The result can be written

$$\left(\frac{\partial}{\partial t}+\eta k^2\right)b_i(\mathbf{k}, t)=\frac{i}{(2\pi)^3}\int d^3\mathbf{r}\int d^3\mathbf{k}'\int d^3\mathbf{k}''\exp\{i(\mathbf{k}'+\mathbf{k}''-\mathbf{k})\,.\,\mathbf{r}\}$$
$$\times\{k_j'b_j(\mathbf{k}'', t)u_i(\mathbf{k}', t)-k_j''b_i(\mathbf{k}'', t)u_j(\mathbf{k}', t)\}.$$

The standard procedure is to integrate first over $\mathbf{r}$, yielding $(2\pi)^3\delta(\mathbf{k}'+\mathbf{k}''-\mathbf{k})$ and then over $\mathbf{k}'$, so that

$$\left(\frac{\partial}{\partial t}+\eta k^2\right)b_i(\mathbf{k}, t)=ik_j\int d^3\mathbf{k}'\{b_j(\mathbf{k}', t)u_i(\mathbf{k}-\mathbf{k}'', t)-b_i(\mathbf{k}'', t)u_j(\mathbf{k}-\mathbf{k}', t)\},\tag{17.32}$$

upon taking advantage of the fact that

$$k_jb_j(\mathbf{k}, t)=k_ju_j(\mathbf{k}, t)=0$$

as a consequence of the vanishing divergence of B_i and v_i. Denoting the right-hand side of (17.32) by $F_i(\mathbf{k}, t)$ it follows that

$$\left(\frac{\partial}{\partial t}+\eta k^2\right)b_i(\mathbf{k}, t)=F_i(\mathbf{k}, t),\tag{17.33}$$

and, hence, that

$$b_i(\mathbf{k}, t)=b_i(\mathbf{k}, 0)\exp(-\eta k^2t)+\int_0^t dt'F_i(\mathbf{k}, t')\exp\{-\eta k^2(t-t')\}.\tag{17.34}$$

With F_i a linear functional of b_i it is apparent that the exact equation (17.34) is a linear integral equation for b_i. A variety of techniques are available for solution, the most obvious being an iterative procedure with

$$b_i(\mathbf{k}, t)=\sum_{n=0}^{\infty}b_i^{(n)}(\mathbf{k}, t)\tag{17.35}$$

and

$$b_i^{(0)}(\mathbf{k}, t)=b_i(\mathbf{k}, 0)\exp(-\eta k^2t).\tag{17.36}$$

Then

$$b_i^{(1)}(\mathbf{k}, t)=ik_j\int_0^t dt'\exp\{-\eta k^2(t-t')\}$$
$$\times\int d^3\mathbf{k}\{b_j^{(0)}(\mathbf{k}', t')u_i(\mathbf{k}-\mathbf{k}', t')$$
$$-b_i^{(0)}(\mathbf{k}', t')u_j(\mathbf{k}-\mathbf{k}', t')\}\tag{17.37}$$

and

$$b_i^{(2)}(\mathbf{k}, t) = \mathrm{i}k_j \int_0^t \mathrm{d}t' \exp\{-\eta k^2(t-t')\} \int \mathrm{d}^3\mathbf{k}' \{b_j^{(1)}(\mathbf{k}', t')u_i(\mathbf{k}-\mathbf{k}', t') - b_i^{(1)}(\mathbf{k}', t')u_j(\mathbf{k}-\mathbf{k}', t')\}, \tag{17.38}$$

etc. The question is under what circumstances the series (17.35) converges.

It is evident that, if $\eta\tau$ is large compared to L^2, where L is the characteristic eddy diameter (correlation length), then the factor $\exp(-\eta k^2 t)$ should provide convergence. It is then sufficient to drop all terms for $n=3$ or greater. It is evident from (17.31) and (17.37) that the ensemble average of $b_i^{(1)}$ vanishes, leaving only $b_i^{(2)}$ beyond $b_i^{(0)}$. Use (17.36) to express $b_i^{(1)}$ in terms of $b_i(\mathbf{k}, 0)$ in (17.38). This truncation of the series is essentially the quasi-linear approximation. In the present context it appears as the direct consequence of large $\eta\tau/L^2$ in a systematic expansion procedure.

To the order considered, then, the second order term represents the effect of the turbulent fluid on the zero order field. The entire magnetic effect of the turbulent velocity v_i is, then,

$$\begin{aligned}\langle b_i^{(2)}(\mathbf{k}, t)\rangle = -k_j \int_0^t \mathrm{d}t' \int_0^{t'} \mathrm{d}t'' \int \mathrm{d}^3\mathbf{k}' \int \mathrm{d}^3\mathbf{k}'' \\ \times \exp\{-\eta(k^2t - k^2t' + k'^2t' - k'^2t'' + k''^2t'')\} \\ \times k_k'\{b_i(\mathbf{k}'', 0)\langle u_j u_k\rangle - b_j(\mathbf{k}'', 0)\langle u_i u_k\rangle\},\end{aligned} \tag{17.39}$$

where

$$\langle u_i u_j\rangle \equiv \langle u_i(\mathbf{k}-\mathbf{k}', t')u_j(\mathbf{k}'-\mathbf{k}'', t'')\rangle. \tag{17.40}$$

The first and the third terms in the integrand have cancelled because $\langle u_i u_j\rangle = \langle u_j u_i\rangle$.

The general properties of $u_i(\mathbf{k}, t)$ are well known and are summarized briefly in Appendix A of this chapter for the convenience of the reader. It follows directly from (A23), A(24),. (17.30), and (17.40) that in the absence of helicity ($\gamma = 0$) we have

$$\begin{aligned}\langle u_i u_j\rangle &= \langle u_i(\mathbf{k}-\mathbf{k}', t')u_j^*(\mathbf{k}''-\mathbf{k}', t'')\rangle \\ &= u^2\delta(\mathbf{k}+\mathbf{k}'')\beta(q, s)(\delta_{ij} - q_iq_j/q^2)\end{aligned}$$

where $s = t'-t''$ and $q_i = k_i - k_i'$. The integration over $\mathbf{k}''$ in (17.39) can now be carried out, leading to

$$\begin{aligned}\langle b_i^{(2)}(\mathbf{k}, t)\rangle = -u^2 b_i(\mathbf{k}, 0)k_jk_k \exp(-\eta k^2 t) \int_0^t \mathrm{d}t' \int_0^{t'} \mathrm{d}s \int \mathrm{d}^3\mathbf{q} \\ \times \beta(q, s)(\delta_{jk} - q_jq_k/q^2)\exp\{-\eta(k'^2 - k^2)s\}\end{aligned} \tag{17.41}$$

where $\mathrm{d}^3\mathbf{q} = \mathrm{d}^3\mathbf{k}'$ and we have written $k_k' = q_k + k_k$ with $q_k(\delta_{jk} - q_jq_k/q^2) = 0$.

To simplify the exposition, write $\beta(q, s)$ as $T(s)\beta(q)$. Now the truncation of (17.35) to $n \leqslant 2$ is a good approximation if $\eta k^2 \tau \gg 1$ so that the field $b_i^{(n)}(\mathbf{k}, t)$ depends upon $b_i^{(n-1)}(\mathbf{k}, t')$ only for t' very near to t. Note that the turbulent spectrum function $\beta(q)$ is a maximum in the general neighbourhood of $q = \pi/L$ and goes rapidly to zero as $q \to 0, \infty$. Hence, for all practical purposes, the convergence requirement is $\eta\tau/L^2 \cong \eta/vL \gg 1$, i.e. small magnetic Reynolds number. The integrations over s and t in (17.41) can be carried out immediately by noting that $\beta(q)$ is negligible for wave numbers k_i' close to k_i, for which $q_i = k_i' - k_i \ll \pi/L$. So, for values of q for which $\beta(q)$ is not negligible, the exponential factor $\exp\{-\eta(k'^2 - k^2)s\}$ declines rapidly with increasing s. The characteristic time of the decline is $1/\eta(k'^2 - k^2) \cong L^2/\eta \ll \tau$. Over this narrow region the time correlation $T(s)$ is essentially equal to one. Hence, for those wave numbers for which $\beta(q)$ is non-negligible we may put $T(s) = 1$ in the integrand and

$$\int_0^t \mathrm{d}t' \int_0^{t'} \mathrm{d}s T(s) \exp\{-\eta(k'^2 - k^2)s\}$$
$$= \frac{1}{\eta(k'^2 - k^2)} \left[t - \frac{1 - \exp\{-\eta(k'^2 - k^2)t\}}{\eta(k'^2 - k^2)} \right]$$
$$\cong \frac{t}{\eta(k'^2 - k^2)} \{1 + \mathrm{O}(L^2/\eta t)\} \tag{17.42}$$

for $t \gg \tau$ and $\eta t \gg L^2$. This result may now be substituted into (17.41). We are interested in the large-scale field ($k \ll \pi/L$), it will be remembered, so the integral over q can be evaluated taking advantage of the fact that the turbulent velocity transform $\beta(k)$ is negligible for the small wave numbers for which we wish to compute $b_i^{(2)}(\mathbf{k}, t)$. Thus, for $kL \ll 1$ and $qL = \mathrm{O}(1)$, it follows that the $k'^2 - k^2$ in the denominator of (17.40) can be written

$$k'^2 - k^2 = q^2 + 2\mathbf{k} \cdot \mathbf{q} \cong q^2.$$

It follows from (17.41) that

$$\langle b_i^{(2)}(\mathbf{k}, t)\rangle = -\frac{u^2 t}{\eta} b_i(\mathbf{k}, 0) \exp(-\eta k^2 t) \int \frac{\mathrm{d}^3\mathbf{q}}{q^2} \beta(q) \{k^2 - (\mathbf{k} \cdot \mathbf{q})^2/q^2\}$$
$$= -\frac{8\pi}{3} \frac{k^2 u^2 t}{\eta} b_i(\mathbf{k}, 0) \exp(-\eta k^2 t) \int_0^\infty \mathrm{d}q \beta(q).$$

Altogether, then

$$b_i(\mathbf{k}, t) = b_i(\mathbf{k}, 0) \exp(-\eta k^2 t)$$
$$\times \left\{ 1 - \frac{8\pi k^2 u^2 t}{3\eta} \int_0^\infty \mathrm{d}q \beta(q) + \ldots \right\} \tag{17.43}$$
$$\cong b_i(\mathbf{k}, 0) \exp\{-(\eta + \eta_{\mathrm{T}})k^2 t\}, \tag{17.44}$$

where

$$\eta_T = \frac{8\pi u^2}{3\eta}\int_0^\infty dq\beta(q). \tag{17.45}$$

The expansion (17.43) converges for $v^2t/\eta \ll 1$. The result (17.45) is to be compared with (17.25). The turbulent diffusion coefficient η_T given by (17.45) for large $\eta\tau/L^2$ is not proportional to v^2 i.e. to $\int dqq^2\beta(q)$, but rather to $\int dq\beta(q)$ which weights the larger Fourier components more heavily because the ability to deform and transport the lines of force declines in proportion to the magnetic Reynolds number $v/\eta q$. The total effect is small $O(L^2/\eta\tau)$ as η becomes large.

It was pointed out in §17.2 that the quasi-linear truncation is applied by some authors to the case where the correlation time τ is small ($\tau \ll L/v$), without any restriction on η. The argument is that the field perturbation produced by each eddy is small, $O(v\tau/L)$, and hence the higher order terms can be neglected. Without considering whether the truncation is justified, the consequences are readily computed from (17.41). Let

$$T(s) = \exp(-\pi s^2/4\tau^2) \tag{17.46}$$

with $\tau \ll L/v$. Then for those values of q for which $\beta(q)$ is non-negligible, $k'^2 - k^2 = O(\pi^2/L^2)$ and

$$\int_0^t dt' \int_0^{t'} dsT(s)\exp\{-\eta(k'^2-k^2)s\} \cong \tau t \tag{17.47}$$

for all $t \gg \tau$. Then (17.41) reduces to

$$\begin{aligned}\langle b_i^{(2)}(\mathbf{k}, t)\rangle &= -u^2\tau b_i(\mathbf{k}, 0)t\exp(-\eta k^2 t)\\ &\quad\times k_j k_k \int d^3\mathbf{q}\beta(q)(\delta_{jk} - q_j q_k/q^2)\\ &= -\frac{8\pi}{3}u^2\tau b_i(\mathbf{k}, 0)t\exp(-\eta k^2 t)k^2\int_0^\infty dqq^2\beta(q).\end{aligned}$$

With $u^2 = \frac{1}{3}v^2$ and the normalization (A25) the field fluctuation is

$$\langle b_i^{(2)}(\mathbf{k}, t)\rangle = -\tfrac{1}{3}v^2\tau b_i(\mathbf{k}, 0)k^2 t\exp(-\eta k^2 t).$$

It follows from (17.35) that, to second order,

$$\langle b_i(\mathbf{k}, t)\rangle = b_i(\mathbf{k}, 0)\exp(-\eta k^2 t)\{1 - \tfrac{1}{3}v^2k^2\tau t + \ldots\} \tag{17.48}$$

$$\cong b_i(\mathbf{k}, 0)\exp\{-(\eta + \eta_T)k^2 t\} \tag{17.49}$$

where

$$\eta_T \equiv \tfrac{1}{3}v^2\tau \tag{17.50}$$

represents the eddy diffusivity. The result is identical with (17.27), of course, deduced in the same limit from (17.22). Note that the expansion (17.48) converges for $v^2k^2\tau t<1$, but not for larger values of t.

Now the limit $\eta\tau\to 0$ implies either $\tau\to 0$, as employed above, or $\eta\to 0$ for large magnetic Reynolds numbers, as treated in the third example in §17.2. It was shown at the end of §17.2 that δB_i does not remain small unless the magnetic Reynolds number is small, whether τ is small or not. The implications and consequences of ordering the limits $\tau\to 0$ and $\eta\to 0$ have been examined by Moffatt (1976, 1978) through a number of illustrative examples, and the interested reader is referred to that article for further elucidation. Altogether the quasi-linear approximation, equivalent to the truncation of (17.35) beyond $n=2$, is able to demonstrate only that the turbulent mixing of magnetic field resembles the turbulent diffusion of a scalar field when the Reynolds number, and the effective mixing of field, are small. The effective turbulent diffusion coefficient is given by (17.45).

Suppose, then, that (17.35) is not truncated and the magnetic Reynolds number Lv/η of the individual eddies is not small. The exponential factor $\exp(-\eta k^2 t)$ does not cause the terms in (17.37) to diminsih rapidly with increasing η and more powerful techniques must be employed. Vainshtein (1970) has shown that there are diagram techniques available for the special case $\tau\ll L/v$.

For small τ, it follows that $T(t'-t'')=\tau\delta(t'-t'')$. Assuming Gaussian statistics, he uses diagram techniques for obtaining partial sums (see also Vainshtein and Zeldovich, 1972). Suitable rearrangement of the ordering of the terms in the sum leads eventually to the result that the turbulent mixing field produces a net diffusion of the mean field with the turbulent diffusion coefficient

$$\eta_T=\tfrac{1}{3}v^2\int_0^\infty dk\int_0^\infty ds\, k^2\beta(k,s)$$

$$=\tfrac{1}{3}v^2\tau\int_0^\infty dk\, k^2\beta(k) \tag{17.51}$$

where again $\beta(k,s)=T(s)\beta(k)$ with τ defined by $\int_0^\infty ds\, T(s)$. The validity of the result depends upon the ultimate *uniform* convergence of (17.37), rather than the rapid convergence necessary to justify the quasi-linear truncaton. The result, therefore, may be valid for $\eta_T>\eta$.

The deepest exploration of the ramifications of the quasi-linear truncation has been given by Roberts and Soward (1975) in their excellent review of the mathematical treatment in terms of Fourier transforms over both space and time. The several limiting forms that they work out for the turbulent diffusion coefficient, and the several contributions of the turbulent velocity (without mirror symmetry) to a large-scale electromotive force, show clearly the pitfalls once we depart from small magnetic Reynolds numbers ($\eta_T \ll \eta$). They point out that their eqns (3.29) and (3.35) yield different ratios of the three terms contributing to the large-scale e.m.f., compared to the results of Steenbeck *et al.* (1966). When Roberts and Soward go on to apply the quasi-linear truncation to the turbulence with a net mean helicity, they introduce a 'small kinematic viscosity ν' to obtain convergence. Working in the limit of small magnetic Reynolds numbers, where they expect the quasi-linear approximation to be valid, they show that their eqn (4.26) for the contribution to the general e.m.f. is proportional to $1/\nu$. The contribution depends entirely on the choice of the convergence parameter ν and increases without bound as $\nu \rightarrow 0$. Their eqn (4.26) is to be compared with their earlier (3.30), where the Lorentz force of the field on the fluid motion is neglected.

These examples suffice to illustrate the necessity for a careful consideration of the truncation involved in the quasi-linear approximation. One must be very sure that the result is reasonable, and consistent with the original approximation. It is not sufficient merely that the correlation time τ be small compared to L/v. The difficulties and limitations of the quasi-linear approximation have been illustrated and discussed at length by a number of authors (see, for instance, Kraichnan 1961; Frisch 1968; Herring 1969).

17.4. The infinite conductivity limit

The turbulent mixing of magnetic field has been treated, so far, in the low conductivity limit by the quasi-linear approximation. The treatment is complete, in that the transport properties are expressible in terms of the familiar two-point velocity correlation function. For intermediate and large conductivities, a summing technique gives (17.51). It is interesting to note that the high conductivity limit is also completely reducible, with the transport properties expressed in terms of Lagrangian correlations, i.e. the correlation of velocities and displacements at different times along the path of the individual element of fluid. Unfortunately, such correlations are not as well studied as the familiar Eulerian two-point correlation. The development is easily written down (see Parker 1971, 1973; Moffatt 1974) using the Cauchy integral (4.32) to compute the field $B_i(x_k, t)$ in the element of fluid at x_k at time t from the initial field $B_i(X_k, 0)$ in the same element of fluid at time $t = 0$ at the initial position X_k. Then the position of the fluid element x_k at time t is

$$x_k = x_k(X_n, t). \tag{17.52}$$

It is convenient to denote the net displacement in the time t by

$$\xi_k(x_n, t) = x_k(X_n, t) - X_k. \tag{17.53}$$

The accumulated strain of the fluid is described by $\partial\xi_k/\partial X_j = \partial x_k/\partial X_j - \delta_{kj}$. Note that the r.m.s. values of both ξ_k and $\partial\xi_k/\partial X_j$ increase without bound (as $t^{\frac{1}{2}}$ and $\exp(t/\tau)$, respectively).

Consider an ensemble of systems, each with the same initial field $B_i(X_k, 0)$ with characteristic scale λ large compared to the eddy diameter L. At time $t=0$ the turbulence is switched on and a small-scale random field is produced by the turbulence, in addition to the slow evolution of the mean field. The mean field follows from (4.16) as

$$\partial\langle B_i\rangle/\partial t = \epsilon_{ijk}(\partial/\partial x_j)\epsilon_{klm}\langle v_l B_m\rangle. \tag{17.54}$$

With the aid of the general solution (4.32) it follows that

$$\begin{aligned}\langle v_i(x_k, t)B_j(x_k, t)\rangle &= \langle v_i(x_n, t)B_k(X_n, 0)\partial x_j/\partial X_k\rangle && (17.55)\\ &= \langle v_i(x_n, t)(\delta_{jk} + \partial\xi_j/\partial X_k)B_k(X_n, 0)\rangle \\ &= \langle v_i(x_n, t)(\partial\xi_j/\partial X_k)B_k(X_n, 0)\rangle. && (17.56)\end{aligned}$$

The term in δ_{jk} reduces to $\langle v_i(x_n, t)B_j(X_n, 0)\rangle$, which vanishes because the present velocity at (x_n, t) is uncorrelated with the field at the initial position X_n.

Now there are several ways to proceed beyond this point. One scheme is to take advantage of the large-scale λ and slow rate of change with respect to time of the mean field (Parker 1970a, b; Moffatt 1974; Kraichnan 1976a). Another is to introduce some statistical hypothesis, such as complete disorder, and deduce the asymptotic effects on the mean field in the limit of large time. We explore the latter possibility first.

Following Parker (1973) consider an ensemble of systems of turbulent conducting fluid, all with the same initial field $B_i(X_k, 0)$. Denote by

$$\Psi(x_k - X_k, t)\, d^3X_k = \Psi(\xi_k, t)\, d^3X_k \tag{17.57}$$

the probability that the element of fluid at x_k at time t was initially in the element of volume d^3X_k at $t=0$. Then with the formal solution (4.32) it follows that the mean field at x_k at time t is

$$\langle B_i(x_k, t)\rangle = \int d^3X_n \Psi(\xi_k, t)B_j(X_k, 0)\langle\langle\partial x_i/\partial X_j\rangle\rangle \tag{17.58}$$

in terms of the initial field $B_i(X_k, 0)$. The quantity $\langle\langle\partial x_i/\partial X_j\rangle\rangle$ represents the mean strain for all those elements of fluid starting at X_k and arriving by various paths at x_k at time t. For comparison the mean density of an initial scalar field $\rho(X_k, 0)$ is

$$\langle\rho(x_k, t)\rangle = \int d^3X_n \Psi(x_k - X_k, t)\rho(X_j, 0). \tag{17.59}$$

The difference between (17.58) and (17.59) is the factor $\langle\langle\partial x_i/\partial X_j\rangle\rangle$.

To begin, consider the ensemble average of the accumulated strain of all fluid elements after a time t. Two points initially at the position X_i and $X_i+\delta_{1i}\delta X_1$, so that they are separated by an infinitesimal distance δX_1, are subsequently located at the points $x_i(X_k, t)$ and $x_i(X_k+\delta_{1k}\delta X_1, t) = x_i(X_k, t)+(\partial x_i/\partial X_1)\delta X_1$. Their separation is then

$$\delta x_i = (\partial x_i/\partial X_1)\delta X_1.$$

The separation δx_i is different in each member of the ensemble, of course, so that with the same δX_1 the ensemble average is

$$\langle \partial x_i/\partial X_1\rangle = \langle \delta x_i\rangle/\delta X_1.$$

But from the symmetry of isotropic homogeneous turbulence, it is evident that there is no preference for the point x_i to move in a specific direction from its initial position X_i, i.e. $\langle \xi_i\rangle = 0$. The same is true for the point $x_i+\delta x_i$. Hence the mean separation does not change as a consequence of the turbulent motion and

$$\langle \delta x_i\rangle = \delta X_i = \delta_{i1}\delta X_1. \tag{17.60}$$

Hence

$$\langle \partial x_i/\partial X_1\rangle = \delta_{i1}$$

and in general

$$\langle \partial x_i/\partial X_j\rangle = \delta_{ij}. \tag{17.61}$$

Now in (17.59) we need the mean value of $\partial x_i/\partial X_j$ for those elements of fluid starting at X_k and, by various routes, finding themselves at x_k at time t. This is only a small sub-ensemble of the whole. Consider, then, the experimental determination of $\partial x_i/\partial X_j$ for the sub-ensemble. The elements of fluid are made visible by replacing them with ink at the starting position X_i. Then the turbulence is switched on and, after a long time t, those elements of fluid instantaneously at x_i are examined. Their form, greatly distorted from the original, permits the observer at x_i to determine their accumulated strain $\partial x_i/\partial X_j$ during the tortuous transit from X_i to x_i. In particular, if the element was initially an infinitesimal rectangular parallelepiped with sides $(\delta X_1, \delta X_2, \delta X_3)$ aligned along the coordinate axes; then subsequently the element is an oblique parallelepiped with sides $\delta x_i^{(1)} = \delta X_1\, \partial x_i/\partial X_1, \ldots$. The mean value of $\partial x_i/\partial X_j$ for the sub-ensemble is denoted by $\langle\langle \partial x_i/\partial X_j\rangle\rangle$ and can be computed from the physical measurements of the sides $\delta x_i^{(1)}$, $\delta x_i^{(2)}$, $\delta x_i^{(3)}$ of the element of volume. If the measurements are performed with a small but fixed upper bound ϵ on the error, then the mean value $\langle\langle \partial x_i/\partial X_j\rangle\rangle$ can be determined to $O(\epsilon/\delta X_i) \ll 1$. Is it possible, then, in the limit of large t, for the observer at x_i to determine something about the relative location $X_i - x_i$ of the initial position of the elements of fluid from measurements of the accumulated strain at x_k at time t? More generally, considering the whole ensemble, could an observer, by making measurements of fixed uncertainty ϵ on the distorted elements of fluid at x_i after a very long time t sort out the elements on a statistical basis as to which were likely to have started from nearby or far away, or in the positive or negative j direction from x_i? If the answer to the question is affirmative, then the elements of fluid retain a non-decreasing memory of their initial position for all time t. On the other hand, if the memory declines with increasing t, so that measurements of increasing accuracy have to be made as time goes on to deduce the initial position, then eventually $\langle\langle \partial x_i/\partial X_i \rangle\rangle$ becomes independent of $x_i - X_i$ and is equal to the ensemble average,

$$\langle\langle \partial x_i/\partial X_j \rangle\rangle = \langle \partial x_i/\partial X_j \rangle = \delta_{ij} \tag{17.62}$$

for essentially all $x_i - X_i$. For this case, (17.58) reduces to

$$\langle B_i(x_k, t)\rangle = \int d^3X_k \Psi(x_k - X_k, t) B_i(X_k, 0). \tag{17.63}$$

But this is exactly the expression (17.59) for the mixing of a scalar field $\rho(x_k, t)$, satisfying (17.1). Hence with the statistical assumption that the deformation $\partial x_i/\partial X_j$ loses all memory of its relative initial position $X_i - x_i$ in the limit of large t, the turbulent mixing of the magnetic field leads to an evolution of the mean field that is identical with a scalar field. We write

$$\partial\langle B_i\rangle/\partial t = \eta_T \nabla^2 \langle B_i \rangle. \tag{17.64}$$

again for the evolution of the mean field $\langle B_i \rangle$ in the turbulent velocity field, where the diffusion coefficient η_T is identical with the eddy diffusion coefficient κ for a scalar field mixed through the same turbulence (Parker 1971). The simple disorder (amnesia) hypothesis leads to (17.62) and to the scalar diffusion of the mean magnetic field.

Moffatt (1974) has expressed doubts about the validity of the principle. He wrote the time rate of change of the magnetic field in terms of the Lagrangian displacement and strain and showed that the resulting statistical forms do not admit of a ready solution to the problem. Indeed, Moffatt raised the question of whether the long-term effect on the mean field can be described by a diffusion coefficient at all. Kraichnan (1976a) turned to an Eulerian formulation of the problem to avoid the non-convergent integrals encountered by Moffatt and obtained some very interesting results, which he then demonstrated by Monte Carlo calculations based on the Lagrangian formulation. The results merit careful examination[3].

[3] The problem is taken up again in §18.4.

First of all, Kraichnan considers fluid motions composed of eddies of characteristic velocity v_0 and wave number k_0 with which is associated the characteristic time $\tau_0 = 1/k_0 v_0$. Then he notes that for times τ_1 small compared to τ_0 the quasi-linear approximation provides a valid solution to (4.16) (because δB_i becomes comparable to B_i only after a time of the order of τ_0). As will be shown in §18.5, the result is equivalent to the time variation

$$\partial\langle\mathbf{B}\rangle/\partial t = \tau_1 v_0^2 \nabla^2 \langle\mathbf{B}\rangle - 2\tau_1 \nabla \times (\mu\langle\mathbf{B}\rangle) \tag{17.65}$$

at every point in space, where

$$\langle v_i v_j \rangle = v_0^2 \delta_{ij}, \qquad \langle v_i\, \partial v_j/\partial x_k \rangle = \mu \epsilon_{ijk}$$

so that μ represents the mean helicity of the fluid notions, with a characteristic time variation τ_2.

Consider, then, a sequence of motions for which the velocity is correlated only over a time interval τ_1 and the vorticity over an interval τ_2, both small compared to τ_0. It follows that (17.65) is a stochastic equation with random coefficients $\eta \equiv \tau_1 v_0^2$ and $\alpha = 2\tau_1 \mu$. Equation (17.65), as it stands, applies to an ensemble of systems to which a single value of η and α are to be applied. Consider, then, an ensemble of ensembles, over which α is a stationary, homogeneous, isotropic, random function of $\mathbf{r}$ and t with a mean value of zero. Applying the quasi-linear approximation to the solution of (17.65), the result is (Kraichnan 1976a)

$$\partial\langle\mathbf{B}\rangle/\partial t = \eta \nabla^2 \langle\mathbf{B}\rangle + \tau_2 \nabla \times (\langle \alpha \nabla \times \alpha \rangle \langle\mathbf{B}\rangle)$$

for the case where $\tau_1 \leqslant \tau_2 \ll (\langle\alpha^2\rangle^{\frac{1}{2}} k_2)^{-1}$, $(\eta k_2^2)^{-1}$. The wave number k_2 denotes the characteristic wave number of the random variable α. The operator ∇ applies to everything to the right of it, so that upon using the usual vector identities for $\nabla \times \alpha\,|\mathbf{B}|$ and $\nabla \times \nabla \times \mathbf{B}$, and the fact that $\nabla\langle\alpha\rangle = 0$, $\langle \alpha \nabla \alpha \rangle = 0$, it follows that

$$\partial\langle\mathbf{B}\rangle/\partial t = (\eta - \tau_2 \langle\alpha^2\rangle) \nabla^2 \langle\mathbf{B}\rangle. \tag{17.66}$$

Thus the effect of the mean square helicity $\langle\alpha^2\rangle$ is to reduce the effective diffusion coefficient below the value η deduced by conventional single application of the quasi-linear approximation. The double application shows that the helicity has the effect of reducing the turbulent diffusion coefficient for the vector magnetic field below the value for the scalar field.

We may worry about the long-term validity of the result (17.66) obtained by the quasi-linear approximation for small η, as noted following (17.29), but the result (17.65) may be presumed valid for periods of time large compared to τ_1 and approaching τ_0. That is sufficient to make the point. Thus the amnesia hypothesis and the equality of the scalar and vector diffusion coefficients seems to be valid in the limit of small local helicity fluctuations α, but when $\alpha \geqslant O(1)$ the hypothesis is in error and the vector diffusion coefficient is smaller.

To obtain quantitative results Kraichnan (1976b) went on with the Lagrangian formulation of the problem, based on (4.32), to demonstrate the effective diffusion coefficient for turbulence with mirror symmetry and for turbulence with maximum helicity α/v_0^2 with a prolonged life $\tau_2 > \tau_0$. The two extremes show the theoretical possibilities, ranging from equality of scalar and vector diffusion coefficients, $\eta = \kappa$, to a vector diffusion coefficient η that is negative so that the fluctuations in the mean field grow unstably.

The Lagrangian formulation of the problem begins with the fact that the mean field is of large-scale λ so that the field at X_k and x_k can be related by the expansion

$$B_i(X_k, t) = B_i(x_k, t) - \xi_n\, \partial B_i(x_k, t)/\partial x_n + \tfrac{1}{2}\xi_n\xi_m\, \partial^2 B_i(x_k, t)/\partial x_n\, \partial x_m + \ldots, \tag{17.67}$$

where $\xi_n = x_n - X_n$. It is clear that the expansion is valid for $|\xi_n| \ll \lambda$, but care must be exercised in passing to the limit $t \to \infty$ wherein ξ_n grows without bound.

Now, suppose that ξ_n denotes the net displacement $x_n - X_n$ of the element of fluid from $t = 0$ to time t. The general result (4.32) gives $B_i(x_k, t)$ in terms of $B_i(X_k, 0)$. Following Kraichnan (1976a), use (17.67) to write $B_j(X_k, 0)$ in terms of $B_j(x_k, 0)$. Then take the ensemble average, with the assumption that the initial field $B_j(X_k, 0)$, produced by circumstances in $t < 0$, is uncorrelated with the fluid motions following $t = 0$. The result is readily shown to be

$$\langle B_i(x_k, t)\rangle = \langle B_j(x_k, 0)\rangle\langle \partial x_i/\partial X_j\rangle - \gamma_{inj}\, \partial\langle B_j(x_k, 0)\rangle/\partial x_n + \xi_{inmj}\, \partial^2\langle B_j(x_k, 0)\rangle/\partial x_n\, \partial x_m, \tag{17.68}$$

$$\gamma_{ijk} \equiv \langle \xi_j\, \partial x_i/\partial X_k\rangle, \qquad \xi_{ijkm} \equiv \tfrac{1}{2}\langle \xi_j \xi_k\, \partial x_i/\partial X_m\rangle.$$

For turbulence that is statistically isotropic and homogeneous, note that

$$\langle \partial x_i/\partial X_j\rangle = \delta_{ij}, \qquad \gamma_{inj} = \epsilon_{inj}\gamma(t)$$

where $\gamma(t)$ is a scalar function of time, because of the generally increasing displacement ξ_i and strain $\partial x_i/\partial X_k$. The symmetry of ξ_{ijkm} on j and k leads to

$$\xi_{inmj} = \delta_{ij}\delta_{nm}\zeta(t) + (\delta_{in}\delta_{jm} + \delta_{im}\delta_{jn})\psi(t).$$

Contracting on the indices of γ_{inj} and ξ_{inmj} it is readily shown that

$$6\gamma(t) = \epsilon_{inj}\gamma_{inj}(t), \qquad 15\zeta(t) = 2\xi_{inni}(t) - \xi_{iinn}(t).$$

In view of the isotropy of the turbulence, γ and ζ can be written in terms of the individual components ξ_1 and ξ_2 as

$$\gamma(t)=\langle\xi_2\,\partial\xi_1/\partial X_3\rangle,\qquad \zeta(t)=\tfrac{1}{2}\langle\xi_2^2\rangle+\tfrac{1}{2}\langle\xi_2^2\,\partial\xi_1/\partial X_1\rangle.$$

Substituting these forms into (17.68) we find

$$\langle B_i(x_k,t)\rangle=\langle B_i(x_k,0)\rangle-\gamma(t)\epsilon_{ijn}\,\partial\langle B_n(x_k,0)\rangle/\partial x_j$$
$$+\zeta(t)\nabla^2 B_i(x_k,0).$$

Hence, differentiating with respect to t, it follows that

$$\frac{\partial}{\partial t}\langle B_i(x_k,t)\rangle=-\frac{\mathrm{d}\gamma}{\mathrm{d}t}\epsilon_{ijn}\frac{\partial\langle B_n(x_k,0)\rangle}{\partial x_j}+\frac{\mathrm{d}\zeta}{\mathrm{d}t}\nabla^2 B_i(x_k,0).$$

We need now to express $\langle B_i(x_k,0)\rangle$ in terms of $\langle B_i(x_k,t)\rangle$. Write

$$\langle B_i(x_k,0)\rangle=\langle B_i(x_k,t)\rangle+s\epsilon_{ijn}\,\partial\langle B_n(x_k,t)\rangle/\partial x_j$$

and substitute into the right-hand side. Equating terms in B_i and its derivatives yields $s=\gamma$. Hence

$$\partial\langle B_i\rangle/\partial t=-\alpha(t)\epsilon_{inj}\langle\partial B_j/\partial x_n\rangle+\eta(t)\nabla^2\langle B_i\rangle \tag{17.69}$$

where

$$\alpha(t)\equiv\mathrm{d}\gamma/\mathrm{d}t,\qquad \eta(t)\equiv\mathrm{d}\zeta/\mathrm{d}t+\tfrac{1}{2}\mathrm{d}\gamma^2/\mathrm{d}t.$$

The coefficient $\alpha(t)$ represents the dynamo effect of the helicity and is the subject of the next chapter.

Now consider two ensembles of systems, each with net helicity α, but with opposite signs of α. Thus if positive α refers to one ensemble, the other has $\mathrm{d}\gamma/\mathrm{d}t=-\alpha$, so that when we combine the two ensembles, the resulting ensemble has zero net helicity. Note that α depends upon the off-diagonal components of $\partial x_i/\partial X_j$. The quantity ζ depends upon the diagonal components of $\partial x_i/\partial X_j$, and is taken to be the same in both systems. Thus $\eta(t)$, depending only upon ζ and the square of γ, has the same value in both systems. Suppose now that the two ensembles are combined into one, so that for the combined system the value of γ, denoted by γ_0, would be zero, in spite of the fact that the combined system is made up of two halves with equal (but opposite) helicity. If the above analysis were repeated for the combined ensemble, for which $\gamma_0=0$, we would obtain a value for the coefficient

$$\eta_0=\mathrm{d}\zeta/\mathrm{d}t$$

where ζ is the same as in either of the two ensembles taken separately. Hence, in terms of γ^2 for the separate systems,

$$\eta_0=\eta-\tfrac{1}{2}\mathrm{d}\gamma^2/\mathrm{d}t$$
$$=\eta-\int_0^t \mathrm{d}s\,\alpha(s)\alpha(t).$$

The reduction of the diffusion coefficient by the square of the helicity is evident.

It should be noted, however, that the reduction of the diffusion coefficient by the helicity in an ensemble of systems in which each member has a net helicity α is not necessarily the same thing as the reduction in an ensemble of systems in which each member itself has zero helicity. Thus, as with the Eulerian formulation, consider an ensemble of ensembles, each member ensemble having zero net helicity, as in the paragraph above, and reduced diffusion coefficient η_0. Then (17.69) represents a stochastic equation with the random coefficient $\alpha(t)$. Using the quasi-linear approximation for times $0 \leqslant t < \tau_0$, the solution can be developed in the same manner as the solution of (17.65) leading to (17.66), with the result

$$\eta_0 = \eta - \int_0^t ds\langle\alpha(s)\alpha(t)\rangle. \tag{17.70}$$

The final quantitative demonstration of the dependence of the effective turbulent diffusion coefficient on the local helicity fluctuations was carried out with Monte Carlo methods based on the Lagrangian formulation developed above. Kraichnan (1976b) specified various velocity forms for the individual eddies and carried through an extensive array of numerical experiments with the aid of a computer to determine the precise values of α, η, and κ, as well as other quantities, for mirror-symmetric turbulence, as well as turbulence made up of individual eddies with strong helicity (the maximum possible helicity for a given characteristic velocity v_0 and wave number k_0).

Consider an eddy as a moving mass of fluid—with dimensions L and characteristic velocity v_0—with a motion w relative to the background fluid. Suppose that the duration of this coordinated eddy is τ_2, which may be longer or shorter than the characteristic time $\tau_0 = 1/\kappa_0 v_0$. The numerical experiments show that the scalar and vector turbulent diffusion coefficients κ and η_T are essentially equal until the helicity becomes a significant fraction of the maximum. For mirror-symmetric turbulence the two coefficients are equal $\kappa = \eta_T$ to within the statistical uncertainty of the calculation, even in the extreme case where the turbulent velocity pattern is fixed in time ($\tau_2 = \infty$, Kraichnan's $\omega_0 = 0$). On the other hand, when the individual eddies are given the maximum possible helicity, the effective diffusion coefficient for vector fields responds by declining with time as the calculation progresses. For times small compared to τ_0, η_T is equal to κ, but by the time τ_2 is as large as τ_0, the diffusion coefficient η_T for vector fields is noticeably reduced from κ for scalar fields. The decline of η_T continues as τ_2 is increased, so that η_T reaches zero and goes negative after a time τ_2 of two or three times τ_0. A negative diffusion coefficient implies unstable concentration of vector field.

Figure 17.3 is reproduced from Kraichnan's paper (Kraichnan 1976b) showing the time evolution of the scalar and vector diffusion coefficients, starting from $t=0$ when the turbulence is switched on. The curves were calculated for turbulence with maximum helicity within each eddy, and a time correlation of the form $\exp\{-\frac{1}{2}\omega_0^2(t-t')^2\}$. The vector diffusion coefficient $\eta_T(t)$ is represented by the curves 1, 3, and 5 for $\omega_0/v_0k_0=0$, 1, 2, respectively. The scalar coefficient $\kappa(t)$ is represented by curves 2, 4, and 6 for $\omega_0/v_0k_0=0$, 1, 2, respectively. Thus, curves 1 and 2 apply to frozen turbulence ($\omega_0=0$) with maximum helicity in each individual eddy. It is evident from 1 and 2 that $\eta_T(t)=\kappa(t)$ for t up to about $1/v_0k_0$, after which the vector diffusion coefficient drops away, becoming negative for $t \gtrsim 3/v_0k_0$. It is equally apparent from a comparison of curves 3 and 4 that the introduction of a finite correlation time of the order of $1/v_0k_0$ leads to comparable $\eta_T(t)$ and $\kappa(t)$. Curves 5 and 6 for $\omega_0=2k_0v_0$ show $\eta_T(t)=\kappa(t)$ to within the statistical uncertainties, for all t.

Kraichnan notes the continuing correlation of displacement and strain in the frozen ($\omega_0=0$) turbulence with maximally helical eddies, showing directly that the disorder hypothesis does not apply to that extreme case. On the other hand, the calculations show that in any but the extreme circumstance of maximum helicity and long correlation times the long-term memory of an element of fluid (contained in the accumulated strain $\partial x_i/\partial X_j$) for its initial position is poor, if not identically zero. Thus $\eta_T \cong \kappa$ is a useful working approximation except perhaps in rapidly rotating bodies where cyclonic effects may become strong and persistent. We shall have more to say on this problem in §18.4 after developing the effect of helicity in its own right.

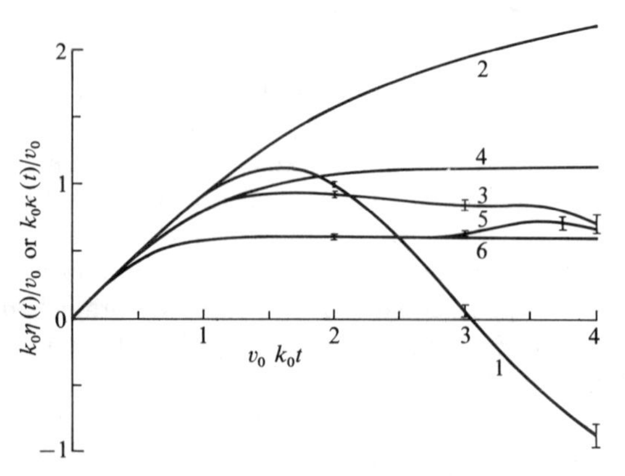

FIG. 17.3. The time evolution of the vector and scalar diffusion coefficients $\eta_T(t)$ and $\kappa(t)$, respectively, for maximally helical turbulence (reproduced from Kraichnan 1976b). The curves labelled 1, 3, and 5 represent $\eta_T(t)$ in units of v_0/k_0 for correlation times ∞, $1/v_0k_0$, and $1/2v_0k_0$, respectively ($\omega_0/v_0k_0=0$, 1, 2), while the curves labelled 2, 4, and 6 represent $\kappa(t)$ in units of v_0/k_0 for the same correlation times, respectively.

In this connection it should be noted that Roberts and Soward (1975), working from the quasi-linear approximation, present a general expression for the eddy diffusion coefficient in terms of the Fourier transform of the velocity correlation. Their result (their eqn (4.13)) leads to a negative turbulent diffusion coefficient when the peak of the Fourier transform occurs at frequencies ω different from zero[4].

The theory of turbulent diffusion of a scalar field is well known (Taylor, 1921; Batchelor, 1959), and Kraichnan's (1970, 1976b) Monte Carlo calculations show that the resulting coefficient κ is relatively insensitive to the shape of the turbulent spectrum, with

$$\kappa \cong (1{\cdot}2 \pm 0{\cdot}2) v_0/k_0 = (1{\cdot}2 \pm 0{\cdot}2) v_0^2 \tau_0 \tag{17.71}$$

where k_0 is the wave number of the largest eddies, with the r.m.s. velocity v_0 in any one direction. In terms of the r.m.s. total velocity, $V_0^2 = 3v_0^2$, this is $0{\cdot}4 V_0^2 \tau_0$. The only significant deviation from this is in the long-lived turbulence with maximal helicity where the value of κ grows with the progress of time, reaching about twice the above value after a time $t = 4\tau_0$, presumably as a consequence of the larger correlation time τ_2. One should perhaps write $v_0^2 \tau_2$ in such cases.

The result (17.51) obtained by Vainshtein (1970) is *essentially* the same as (17.70) but is restricted to small correlation times $\tau_1 \ll \tau_0$, while (17.70) and (17.71) are valid for the more realistic situation $\tau_1 \cong \tau_0$.

17.5. A physical example of turbulent mixing of magnetic field

The calculations cited in the section above establish that in many, if not all cases, turbulent mixing causes the mean large-scale magnetic field to diffuse in essentially the same manner as a scalar field, with an effective diffusion coefficient (17.70) approximately equal to (17.71). The mean field is mixed to the surface, or into the interior, of an astrophysical body to about the same degree as the chemical constituents. But before going on to the application of this basic effect there are a number of questions that come to mind. How is it, for instance, that the turbulent mixing of a magnetic field carried in a fluid with infinite electrical conductivity, as treated in §17.4, can produce diffusion without reconnection of the magnetic lines of force. The diffusive merging of different field configurations necessarily implies changing field topology (i.e. reconnection). And while we are asking questions, what happens to the small-scale fields

[4] Roberts and Soward *postulate* (their eqn (2.25)) that the maximum of the Fourier transform occurs at $\omega = 0$. On that basis they state that their diffusion coefficient is positive. But we see no fundamental objection to the possibility that the Fourier transform may be larger at some $\omega \neq 0$, allowing the intriguing possibility that the diffusion is negative.

generated by the individual turbulent eddies? (See special examples in Moffatt (1963) for isolated eddies).

The formal answer to the first question is that in an infinitely conducting fluid, it is the lines of force of the total field $B_i = \langle B_i \rangle + \delta B_i$ that do not reconnect, whereas it is the mean field $\langle B_i \rangle$ that is subject to turbulent diffusion. The lines of $\langle B_i \rangle$ reconnect. This answers the question in a formal sense, but does not illustrate the detailed behaviour of the complete field B_i. We have to see how δB_i behaves to go further, or to answer the second question. It is instructive, then, to go back to some idealized illustrative examples for which the calculations can be carried out in detail to show how the mean field $\langle B_i \rangle$ is displaced by the circulation and mixing of the fluid. What property of the velocity field is responsible for the turbulent displacement (diffusion) of the mean field? To answer this question consider a large-scale magnetic field with characteristic dimension λ, in which there is a gradient in the field density in a direction perpendicular to the local direction of the field. For convenience rotate the coordinate system so that the z-axis lies along the local mean field direction and the y-axis points in the direction of increasing field density. Locally, then, the large-scale field is of strength B_0 with

$$B_x, B_y = B_0 O\{(x^2 + y^2 + z^2)/\lambda^2\},$$
$$B_z = B_0(1 + y/\lambda + \ldots) \qquad (17.72)$$

over the dimension L of the local eddies, neglecting terms second order in L/λ compared to unity.

To examine the mixing and diffusion of field in its most elementary form, divide the space into identical contiguous cubical cells with side $2L$. Then with the same fluid motions in each cell, consider the cell x^2, y^2 $z^2 \leq L^2$. The fluid is given a velocity $v_i(x, y, z, t)$, with the velocity falling to zero at the boundaries of the cell, so that all the magnetic effects are confined to the interior of the cell. The cell, with its closed internal pattern of circulation, is a simple model of an eddy. The fluid motion is switched on for a time τ and switched off. Subsequently the pattern of cubical cells is relocated and the velocity is switched on again for a time τ, etc. Thus, even though the fluid remains within each cell during the time τ, there is a net mixing of fluid throughout the large region as one cell pattern is replaced by another at time intervals of τ.

We continue with the condition that the electrical conductivity of the fluid is high ($\eta \ll Lv \cong L^2/\tau$) so that the lines of force are carried bodily with the fluid and do not reconnect significantly over the time τ. The question is, what aspect of the circulation of the fluid is responsible for displacing the mean field in the direction $-\nabla B_z$. To be precise, the centre of gravity of B_z is the measure of the mean position of the field. How is the centre of gravity moved by the fluid?

To take the problem one step at a time, consider the two-dimensional motion $\{v_x(x, y, z), 0, v_z(x, y, z)\}$. In this case there is no motion of the fluid in the direction of the field gradient (the y-direction), so there can be no net displacement of field in the y-direction. The equation for the time variation of B_y is

$$dB_y/dt = 0.$$

With $B_y = 0$ initially, it follows that $B_y = 0$ for all t. The only non-vanishing components are B_x and B_z, most conveniently written

$$B_x = \partial A_y(x, y, z, t)/\partial z, \qquad B_z = -\partial A_y(x, y, z, t)/\partial x$$

so that their variation is described by

$$dA_y/dt = 0. \tag{17.73}$$

Since the fluid velocity falls to zero at the walls $x = \pm L$, it is evident that A_y does not change at the walls. Hence the magnetic flux across the cell does not change,

$$\frac{\partial}{\partial t}\int_{-L}^{+L} dx B_z(x, y, z, t) = \frac{\partial}{\partial t}\{A_y(-L, y, z, t) - A_y(+L, y, z, t)\}$$
$$= 0$$

$$\frac{\partial}{\partial t}\int_{-L}^{+L} dz B_x(x, y, z, t) = \frac{\partial}{\partial t}\{A_y(x, y, +L, t) - A_y(x, y, -L, t)\}$$
$$= 0$$

as is obvious from the fact that the lines of force move with the fluid in planes $y = constant$.

The first moments of the field do not vary. For instance, define the x-moment of B_z as

$$M_{xz} = \int_{-L}^{+L} dx \int_{-L}^{+L} dy \int_{-L}^{+L} dz x B_z$$
$$= -2L \int_{-L}^{+L} dx \int_{-L}^{+L} dz\{\partial(xA_y)/\partial x - A_y\}$$
$$= -2L \int_{-L}^{+L} dz x A_y \Big|_{-L}^{+L} + 2L \int_{-L}^{+L} dx \int_{-L}^{+L} dz A_y.$$

Since A_y at $x = \pm L$ has not changed from its initial value, it is evident that the first term on the right-hand side does not vary with time. From the fact (17.73) that A_y in each element of fluid does not vary, and the fluid remains in the plane $y = constant$ as it moves, it is evident that the second term on the right-hand side does not vary either. Initially $M_{xz} = 0$ so that $M_{xz} = 0$ for all subsequent times. It is readily shown that the first moment M_{yz}, given by the integral of yB_z over the cell, is also independent of time and maintains its initial value $8B_0L^5/3\lambda$. Altogether, the fluid velocity $(v_x, 0, v_z)$ causes no shift in the centre of gravity of the field distribution within the cell, and hence no net transport of field.

Consider, then, the velocity field $\{0, v_y(x, y, z), v_z(x, y, z)\}$, involving fluid motions constrained to surfaces $x = constant$. In that case $dB_x/dt = 0$ and B_x is zero for all time. Write $B_y = +\partial A_x/\partial z$ and $B_z = -\partial A_x/\partial y$, so that the evolution of the field is described by $dA_x/dt = 0$. It is evident that the mathematical calculations performed in the paragraph above for $(v_x, 0, v_z)$ can now be repeated for $(0, v_y, v_z)$, leading to the conclusion that there is no change in the total flux across the cell, and no shift in the centre of gravity. In neither case, then, do the fluid motions contribute a net displacement of the field. Both motions produce local small-scale fields, $(B_x, 0, B_z)$ and $(0, B_y, B_z)$, respectively, within the cell, of course, but such fields have vanishing mean over the cell.

Consider, finally, the remaining possibility $\{v_x(x, y, z), v_y(x, y, z), 0\}$. Then the magnetic field components vary as

$$dB_x/dt = B_x\, \partial v_x/\partial x + B_y\, \partial v_x/\partial y + B_z\, \partial v_x/\partial z, \tag{17.74}$$

$$dB_y/dt = B_x\, \partial v_y/\partial x + B_y\, \partial v_y/\partial y + B_z\, \partial v_y/\partial z, \tag{17.75}$$

$$dB_z/dt = 0. \tag{17.76}$$

It follows at once from a comparison of (17.76) with (17.1) that the z-component of the magnetic field diffuses in exactly the same manner as a scalar field. The question is how does the diffusion come about?

To take the investigation one step at a time, rewrite (17.74) as

$$\partial B_x/\partial t = (\partial/\partial y)(v_xB_y - v_yB_x) + (\partial/\partial z)v_xB_z.$$

Since v_i vanishes on the boundaries of the cell, the net flux of B_x across any plane $x = constant$ is constant in time,

$$\begin{aligned}\frac{\partial}{\partial t}\int_{-L}^{+L} dy \int_{-L}^{+L} dz\, B_x(x, y, z, t) &= \int_{-L}^{+L} dz (v_xB_y - v_yB_x)\Big|_{y=-L}^{y=+L} \\ &\quad + \int_{-L}^{+L} dy\, v_xB_z \Big|_{z=-L}^{z=+L} \\ &= 0.\end{aligned}$$

Initially there is no net flux of B_x, so that the mean remains zero. The same follows for B_y.

The integral of B_z across any plane $z = constant$ is also fixed in time, at the initial value $4L^2B_0$. But the first moment of B_z, (the integral of yB_z over the cell) varies. With the intial field (17.72) the initial value of the moment is

$$M_{yz}(0)=\int_{-L}^{+L}\mathrm{d}x\int_{-L}^{+L}\mathrm{d}y\int_{-L}^{+L}\mathrm{d}z\,yB_z$$
$$=8B_0L^5/3\lambda$$

as already noted, because of the asymmetric distribution of $B_z(y)$. Thus, for instance, if the fluid motion $(v_x, v_y, 0)$ consisted of each element of fluid being displaced in azimuth by $\pi/2$ around the cell $(x\rightarrow y, y\rightarrow -x)$, then B_z would rotate with it, as described by (17.76). Then $B_z = B_0(1-x/\lambda)$ and the moment M_{yz} falls to zero. The centre of gravity, initially at

$$x_0=0, \qquad y_0=M_{yz}/8L^3B_0=L^2/3\lambda, \qquad z_0=0$$

moves to $y=0$ (and $x=-L^2/3\lambda$). Rotation through another $\pi/2$ (so that $x\rightarrow -x$, $y\rightarrow -y$ altogether) brings the field around to $B_z=B_0(1-y/\lambda)$ and moves the centre of gravity to $x=0$, $y=-L^2/3\lambda$. There is then a final mean displacement of the magnetic lines of force by $2L^2/3\lambda$ in the direction of $-\nabla B_z$. As another example, suppose that the fluid within the cell is subject to the angular velocity $\omega=\Omega f(\varpi)$ about the z-axis out to the radial distance $\varpi=L$. The function $f(\varpi)$ has a value of unity at $\varpi=0$, where the fluid rotates with an angular velocity Ω, falling to zero at $\varpi=L$. Between $\varpi=L$ and the boundaries x, $y=\pm L$ the fluid is at rest. Then an element of fluid initially at $\varpi=\varpi_0(<L)$, $\phi=\phi_0$ arrives at $\varpi=\varpi_0$, $\phi=\phi_0+\Omega\tau f(\varpi)$ after a time τ, transporting the field strength $B_0\{1+(\varpi_0/\lambda)\sin\phi_0\}$ unchanged from the initial position. Hence at time τ, the field is

$$B_z(\varpi,\phi,\tau)=B_0[1+(\varpi/\lambda)\sin\{\phi-\Omega\tau f(\varpi)\}]. \tag{17.77}$$

The decline of $f(\varpi)$ with increasing ϖ means that the inner portions of the fluid rotate more rapidly than the outer portions, so the field becomes increasingly mixed around the cell as time increases. In the asymptotic limit, $\tau\rightarrow\infty$, it is obvious that B_z is distributed around the cell in threads of decreasing thickness, so that any form of local averaging or diffusion leads to the uniform value $\langle B_z\rangle\sim B_0$. This is reflected in the moment

$$M_{yz}(\tau)=\int_{-L}^{+L}\mathrm{d}x\int_{-L}^{+L}\mathrm{d}y\int_{-L}^{+L}\mathrm{d}z\,yB_z$$
$$=M_{yz}(0)\left[1-\frac{3\pi}{16}+\frac{3\pi}{4L^4}\int_0^L\mathrm{d}\varpi\varpi^3\cos\{\Omega\tau f(\varpi)\}\right]$$

which has the initial value $M_{yz}(0)=8L^5B_0/3\lambda$. In the limit of large $\Omega\tau$ the factor $\cos(\Omega\tau f(\varpi))$ oscillates over ϖ with increasing rapidity, so that the integrand averages to zero and the change in M_{yz} is

$$\Delta M_{yz}(\tau)\sim -M_{yz}(0)3\pi/16.$$

Now for $f(\varpi)=1-\varpi/L$, $1-\varpi^2/L^2$, $1-\varpi^4/L^4$ and 1 out to $\varpi=L$, with $f(\varpi)=0$ for $\varpi>L$, it is readily shown that

$$M_{yz}(\tau)=M_{yz}(0)\left[1-\frac{3\pi}{16}\left\{1-12\frac{\Omega^2\tau^2-2(1-\cos(\Omega\tau))}{\Omega^4\tau^4}\right\}\right], \quad (17.78)$$

$$M_{yz}(\tau)=M_{yz}(0)\left[1-\frac{3\pi}{16}\left\{1-2\frac{1-\cos(\Omega\tau)}{\Omega^2\tau^2}\right\}\right], \quad (17.79)$$

$$M_{yz}(\tau)=M_{yz}(0)\left[1-\frac{3\pi}{16}\left\{1-\frac{\sin(\Omega\tau)}{\Omega\tau}\right\}\right], \quad (17.80)$$

$$M_{yz}(\tau)=M_{yz}(0)\left\{1-\frac{3\pi}{16}(1-\cos(\Omega\tau))\right\}, \quad (17.81)$$

respectively. The variation of $M_{yz}(\tau)$ with time is shown in Fig. 17.4 for the four cases. Note that as $\tau\to\infty$

$$M_{yz}(\tau)\sim M_{yz}(0)(1-3\pi/16)=0{\cdot}411M_{yz}(0)$$

in the first three cases, as B_z is stirred indefinitely around the cell. The smearing of B_z around the cell takes place most quickly for $f(\varpi)=1-\varpi/L$ because the angular velocity changes at a uniform rate across the cell, whereas for $f(\varpi)=1-\varpi^2/L^2$ and $1-\varpi^4/L^4$ there is a progressively larger inner core subject to uniform angular velocity, in which the field is able to preserve its identity for a longer time. For $f(\varpi)=1$ the fluid in $\varpi<L$ rotates rigidly and the field remains coherent over the entire cell as

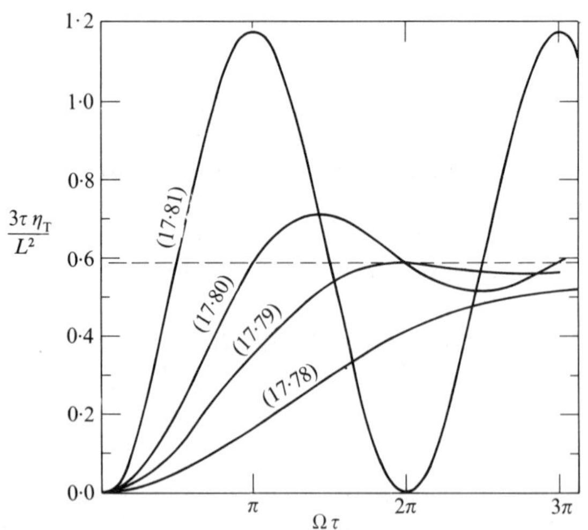

FIG. 17.4. A plot of η_T in units of $L^2/3\tau$ as a function of $\Omega\tau$. This is also a plot of $\Delta M_{yz}(\tau)/M_{yz}(0)$. The numbers on each curve refer to the appropriate equation number for $M_{yz}(\tau)$. The horizontal broken line represents the asymptotic limit for $t\to\infty$ (except for (17.81)).

it circulates round and round. In that case, $M_{yz}(\tau)$ oscillates forever between $M_{yz}(0)$ and $M_{yz}(0)(1-3\pi/8)$.

The equivalent diffusion coefficient for B_z is readily calculated from the change in the first moment M_{yz}. As already noted, it follows from (17.76) that B_z is transported by the turbulence in the same way as a scalar field, satisfying (17.1). It is well known (Taylor 1921), then, that the mean field satisfies the familiar diffusion equation

$$\partial\langle B_z\rangle/\partial t = \eta_T \nabla^2 \langle B_z\rangle$$

with the effective turbulent diffusion coefficient η_T equal to the eddy diffusivity for scalar fields. The flux of field is $-\eta_T \nabla B_z$. Hence, if in a time τ the moment M_{yz} changes by $\Delta M_{yz}(\tau) = M_{yz}(\tau) - M_{yz}(0)$, the mean displacement of field in the cell is $\Delta y = \Delta M_{yz}/8L^3 B_0$; the mean transport rate is $\Delta y/\tau$ multiplied by the mean field B_0. This rate of transport must equal $-\eta_T |\nabla B_z| = -\eta_T B_0/\lambda$, so that

$$\eta_T = -\lambda \Delta y/\tau = -\lambda \Delta M_{yz}(\tau)/8L^3 \tau B_0. \tag{17.82}$$

Thus we obtain an exact relation for η_T in terms of the shift in the first moments (17.78)–(17.81).

The whole problem can be treated formally, of course. According to (17.76) the field B_z is conserved. The electrical conductivity of the fluid is so high ($\eta\tau \ll L^2$) that B_z moves with the fluid, and the local rate of field transport is $F_i = v_i B_z$. Now, suppose that the three-dimensional grid of cells of fluid circulation is applied for a time τ, with the results calculated above. At the end of that time the field has changed from (17.72) to (17.77). Write the magnetic field as

$$B_z = \langle B_z\rangle + \delta B_z$$

where $\langle B_z\rangle$ denotes the large-scale field (17.72) and δB_z is defined as $B_z - \langle B_z\rangle$. Note that δB_z is generally not small. Then, during the circulation, the flux of field is

$$F_i = v_i(\langle B_z\rangle + \delta B_z).$$

The mean flux over the whole cell is

$$\langle F_i\rangle = \langle v_i \delta B_z\rangle.$$

Now the hydromagnetic equation (17.76) can be written

$$\begin{aligned}\partial B_z/\partial t &= -\{\partial v_x B_z/\partial x + \partial v_y B_z/\partial x\}\\ &= -\partial F_i/\partial x_i.\end{aligned}$$

Hence

$$\partial\langle B_z\rangle/\partial t = -\partial\langle F_i\rangle/\partial x_i.$$

Note, then, that the solution (17.4) based on the initial field (17.72), is

$$B_z(x_k, t) = B_0\{1 + Y(x_k, t)/\lambda\}$$

after a time t, where $Y(x_k, t)$ is the initial y-coordinate of the element of fluid at x_k at time t. It follows that

$$\delta B_z(x_k, t) = B_0\{Y(x_k, t) - y\}/\lambda.$$

Therefore, the average over the cell is

$$\langle F_y \rangle = \frac{B_0}{8\lambda L^3} \int_{-L}^{+L} dx \int_{-L}^{+L} dy \int_{-L}^{+L} dz\, v_y\{Y(x_k, t) - y\}.$$

Since $\int dx\, v_y = 0$, this reduces immediately to

$$\langle F_y \rangle = B_0 I / 8\lambda L^3, \tag{17.83}$$

where I is the integral

$$I = \int_{-L}^{+L} dx \int_{-L}^{+L} dy \int_{-L}^{+L} dz\, v_y Y(x_k, t). \tag{17.84}$$

To exhibit the physical significance of I, consider the first moment of B_z,

$$\begin{aligned} M_{yz} &\equiv \int_{-L}^{+L} dx \int_{-L}^{+L} dy \int_{-L}^{+L} dz\, y B_z(x_k, t) \\ &= B_0 \int_{-L}^{+L} dx \int_{-L}^{+L} dy \int_{-L}^{+L} dz\, y\{1 + Y(x_k, t)/\lambda\}. \end{aligned}$$

Then the time derivative is

$$\begin{aligned} \frac{dM_{yz}}{dt} &= \frac{B_0}{\lambda} \int_{-L}^{+L} dx \int_{-L}^{+L} dy \int_{-L}^{+L} dz\, y\left(v_x \frac{\partial Y}{\partial x} + v_y \frac{\partial Y}{\partial y}\right) \\ &= \frac{B_0}{\lambda} \int_{-L}^{+L} dx \int_{-L}^{+L} dy \int_{-L}^{+L} dz \left\{\frac{\partial}{\partial x}(y v_x Y) + \frac{\partial}{\partial y}(y v_y Y) - v_y Y\right\} \end{aligned}$$

for $\partial v_x/\partial x + \partial v_y/\partial y = 0$ in an incompressible flow. Since v_x vanishes at $x = \pm L$ and v_y vanishes at $y = \pm L$, it follows that

$$\partial M_{yz}/\partial t = -B_0 I/\lambda. \tag{17.85}$$

Thus

$$\langle F_y \rangle = -(1/8L^3)\, \partial M_{yz}/\partial t.$$

With $\partial\langle B_z \rangle/\partial y = \langle B_z \rangle/\lambda$ it follows that

$$\begin{aligned} \partial\langle B_z \rangle/\partial t &= -(\partial/\partial y)\langle F_y \rangle \\ &= +\frac{\partial}{\partial y} \eta_T \frac{\partial\langle B_z \rangle}{\partial y} \end{aligned} \tag{17.86}$$

with

$$\eta_T = I/8L^3 \tag{17.87}$$

which is the result (17.82), of course. This describes the transport of field during the first cycle of duration τ. The circulation in each cell may vary with time over the period τ (i.e. Ω may be a function of time) or the velocity may be uniform in time, being switched on abruptly at $t=0$ and off at $t=\tau$. If the former, then, replace v_y by its time average over τ.

Let the cycle of circulation of duration τ be repeated over and over. To outline the calculation for successive cycles (with $L \ll \lambda$) suppose that the circulation is switched off at the end of the $(n-1)$th cycle. The grid of circulation cells is then moved a random distance $O(L)$ in a random direction and switched on again for a time τ, with B_0 again denoting the local value of $\langle B_z \rangle$, and B_0/λ the local gradient of $\langle B_z \rangle$. With this scheme of things it follows that the small scale fields at the beginning of each cycle are uncorrelated with the velocity of the new cycle, but the velocity of each cycle is systematic and not at all random. We consider an ensemble of systems subject to the same statistical treatment in the previous cycles, with the location of the grid used in each system *unrelated* to the location in the others. But the *same* velocity grid is applied to all members of the ensemble for the new (the nth) cycle. An ensemble of such ensembles is then employed to go on to the $(n+1)$th cycle, etc. The method of summing is extravagant in its use of systems, but is a legitimate mathematical operation of course.

To carry through the formal calculation for each step, in which a single grid of fluid circulation is applied to all members of the ensemble, denote the field of the beginning of the nth cycle (at time $(n-1)\tau$) by $\langle B_z \rangle + \delta B_z(x_k, t_n)$ where $\langle B_z \rangle$ is, of course, just (17.72) and δB_z averages to zero over any square $2L \times 2L$ in the xy-plane. In the time s after $t_n (t = t_n + s)$ the field evolves to

$$B_z(x_k, s) = B_0\{1 + Y(x_k, s)/\lambda\} + \delta B_z\{X_k(x_l, s), t_n\}$$

so that the change of B_z from its configuration at time t_n is

$$\begin{aligned}\delta B_z(x_k, s) = (B_0/\lambda)\{Y(x_k, s) - y\}\\ + \delta B_z\{X_k(x_l, s), t_n\} - \delta B_z(x_k, t_n).\end{aligned}$$

It follows that

$$\begin{aligned}\langle v_y(x_k, s)\delta B_z(x_k, s)\rangle = (B_0/\lambda)\langle v_y(x_k, s)\{Y(x_k, s) - y\}\rangle\\ + \langle v_y(x_k, s)\delta B_z\{X_k(x_l, s), t_n\}\rangle - \langle v_y(x_k, s)\delta B_z(x_k, t_n)\rangle.\end{aligned}$$

Now v_y is the same for all members of the ensemble. Hence

$$\langle v_y(x_k, s)\delta B_z(x_k, t_n)\rangle = 0.$$

The average quantity $v_y(x_k, s)\delta B_z\{X_k(x_l, s), t_n\}$ also vanishes. It correlates v_y at the position x_k with the initial random field $\delta B_z(X_k, t_n)$ at the position $X_k(x_k, s)$ determined by v_i in the present cycle. But a selection of fields from the initial unrelated fields $\delta B_i(x_k, t_n)$ by any systematic process yields unrelated fields. Altogether, then

$$\langle v_y(x_k, s)\delta B_z(x_k, s)\rangle = (B_0/\lambda)\langle v_y(x_k, s)\{Y(x_k, s) - y\}\rangle.$$

The mean transport rate is then

$$\langle F_y\rangle = (B_0/\lambda)\langle v_y(x_k, s)\{Y(x_k, s) - y\}\rangle.$$

Since v_y and $Y(x_k, s)$ are the same in each member of the ensemble, the angular brackets imply only the average over the small-scale cell $2L\times 2L$. This brings us to (17.83) and the ensuing diffusion eqn (17.86) and diffusion coefficient (17.87). Thus the calculation for the first cycle is repeated for the nth, and the resulting diffusion equation for the large-scale mean field (17.86) is applicable for all t with the diffusion coefficient (17.82) or (17.87)[5]. In terms of the change in M_{yz} over the period τ of a given cycle, it follows that

$$\eta_T = -(L^2/3\tau)\Delta M_{yz}(\tau)/M_{yz}(0). \tag{17.88}$$

The effective diffusion coefficient is plotted in Fig. 17.4 in units of $L^3/3\tau$ as a function of $\Omega\tau$ for the four examples (17.78)–(17.81). If we take $\Omega\tau = \pi/2$ as characteristic of a real turbulent eddy of scale L during its life $\tau = L/v$, then η_T is equal to 0·048, 0·108, 0·21, and 0·59 times $L^2/3\tau$ for the four cases, respectively. For comparison with later results it is convenient to write η_T in terms of the mean square velocity v^2. Then with the average

$$v^2 = 2\pi\int_0^L d\varpi\varpi(\varpi\dot{\phi})/4L^2$$
$$= \frac{\pi\Omega^2}{2L^2}\int_0^L d\varpi\varpi^3 f^2(\varpi)$$

over the $2L\times 2L$ area of the cell, it is readily shown that $L^2\Omega^2 = \alpha v^2$

[5] An alternative scenario would be to suppose that after the brief period τ of circulation of the fluid, the system is held at rest for a time τ_1 that is large compared to the characteristic diffusion time L^3/η over a single cell (but suitably short compared to the large-scale diffusion time λ^2/η). The field fluctuations δB_i produced by the circulation have no scale associated with them that is larger than L, so they are lost during the period τ_1. The next cycle begins anew with $\delta B_i = 0$ and the large-scale field given by (17.72). The effective diffusion coefficient is then (17.88) with τ in the dominator replaced by $\tau + \tau_1 \cong \tau_1$. The algebraic result for η is the same as the quasi-linear result of §§17.2 and 17.3 for small magnetic Reynolds number, with η_T limited to values less than η. See (17.26).

where $\alpha = 120/\pi$, $48/\pi$, $24/\pi$, and $8/\pi$ for (17.78)–(17.81), respectively, leading to

$$\eta_T = \beta v^2 \tau \qquad (17.89)$$

where $\beta = 0{\cdot}25$, $0{\cdot}23$, $0{\cdot}21$, and $0{\cdot}20$, respectively. Thus in terms of v^2 the mean turbulent transport is relatively independent of the form $f(\varpi)$, the coefficient β varying by only 20 per cent over the entire range from $f = 1 - \varpi/L$ to $f = 1$. Indeed in the limit of small $\Omega\tau$, $\Delta M_{yz}(\tau)/\Delta M_{yz}(0)$ follows from (17.78)–(17.81) as $\Omega^2\tau^2$ multiplied by $\pi/160$, $\pi/64$, $\pi/32$, and $3\pi/32$, respectively so that (17.82), with the above values of $L^2\Omega^2$, yields

$$\eta_T = \tfrac{1}{4} v^2 \tau$$

in all four cases. In the opposite extreme $\Omega\tau \to \infty$ it follows that

$$\eta_T = 0{\cdot}196 L^2/\tau \to 0$$

for the three cases (17.78)–(17.80).

Coming back to the fact that in the present special case the large-scale magnetic field is dispersed by the turbulence in exactly the same way as a scalar field, it follows that the diffusion coefficient is calculable by the conventional methods for scalar fields. In that case we would have noted that the cells of fluid random walk over space, so that the large-scale distribution can be computed from a Fokker–Planck equation. Standard procedure would then be to note that the diffusion coefficient in the Fokker–Planck equation is $\frac{1}{2} l^2/\tau$ where l^2 is the mean square displacement in the time τ in the y-direction. Thus, for instance, the mean square displacement for $f(\varpi) = 1$ is

$$\begin{aligned} l^2 &= \frac{1}{4L^2} \int_0^{2\pi} \mathrm{d}\phi \int_0^L \mathrm{d}\varpi\varpi^3 \{\sin^2\phi(1 - \cos\Omega\tau)^2 + \cos^2\phi \sin^2(\Omega\tau)\} \\ &= (\pi L^2/8)(1 - \cos(\Omega\tau)), \end{aligned}$$

so that the effective diffusion coefficient is $(\pi L^2/16\tau)(1 - \cos(\Omega\tau)$, in agreement with the result for η_T obtained from (17.81) and (17.82).

It should be noted that the calculated values of η_T plotted in Fig. 17.4 are exact for the idealized eddies $\{v_x(x, y, z), v_y(x, y, z), 0\}$ employed. The large-scale magnetic field, and any scalar field, embedded in the fluid are mixed through the fluid with the effective diffusion coefficient $\eta_T = \kappa$.

It should not escape our attention, of course, that the sequence of circulating cells of fluid generates intense sheets of B_x and B_y of opposite sign at the top and bottom of each cell. These fields have no mean value over dimensions equal to the cell size $2L$, and no mean over the ensemble. After a time large compared to τ the characteristic scale of B_x and B_y becomes small compared to L as a consequence of the random sign of neighbouring sheets of B_x and B_y. Thus even a very small resistive diffusion η destroys them and halts their accumulation without effecting the large-scale field $\langle B_z \rangle$. This question is taken up in §17.6 in more general terms, where it is shown that rapid reconnection, as well as resistive diffusion, play an important role in limiting the small-scale fields.

The idealized example of turbulent diffusion considered in this section demonstrates that the diffusion is basically the result of the component of the fluid velocity in the y-direction (the direction of $\nabla\langle B_z \rangle$) together with the component in the x-direction (the direction $\langle \mathbf{B} \rangle \times \nabla\langle B_z \rangle$ perpendicular to both $\langle \mathbf{B} \rangle$ and $\nabla\langle \mathbf{B} \rangle$). The diffusion coefficient was calculated exactly for a variety of velocity profiles, leading to (17.78)–(17.81). The general result is $\eta_T = \beta\langle v^2 \rangle\tau$ with $\beta \cong 0{\cdot}20$–$0{\cdot}25$ for $\Omega\tau$ of the order of unity. The example illustrates in detail how the fields within each cell are distorted and displaced to produce the mean diffusion coefficient η_T.

It is not without interest to consider for a moment that the idealized example can be regarded as the lowest order term in an expansion of the general problem of turbulent diffusion of field in three-dimensional turbulence. Consider, for instance, the velocity field $v_i(x_k)$ for which $v_i B_i \neq 0$. Then, as has been demonstrated, the primary convective transport is produced by the two components v_x, v_y perpendicular to the large-scale field, B_z. The component v_z (parallel to the large-scale field) gives no net transport when combined with one or the other of v_x or v_y, but if v_x and v_y are both present, the addition of v_z as a third component has an effect through the higher order interaction of v_x and v_y with the small-scale fields produced by v_z and the interaction of v_z with the small-scale fields produced by v_x and v_y. It is the third component v_z that makes the general problem of the turbulent mixing of magnetic field so complex.

To develop a logical scheme for treating the general three-dimensional velocity field $v_i(x_k, t)$ consider the illustrative example for v_x and v_y as merely the lowest order interaction in the convective equation (17.3). Write

$$B_x = B_x^{(1)} + B_x^{(2)} + \ldots, \qquad B_y = B_y^{(1)} + B_y^{(2)} + \ldots,$$
$$B_z = B_z^{(0)} + B_z^{(1)} + B_z^{(2)} + \ldots,$$

where $B_z^{(0)}$ is a consequence of the basic mixing property of v_x and v_y so that

$$\mathrm{d}B_z^{(0)}/\mathrm{d}t = 0. \tag{17.90}$$

Then $B_i^{(1)}$ represents the field produced by the interaction of the three components of the fluid velocity with $B_z^{(0)}$; $B_i^{(2)}$ is the result of the interaction with $B_i^{(1)}$; etc. The hydromagnetic equation (17.3) then de-

decomposes into the equations

$$\mathrm{d}B_i^{(1)}/\mathrm{d}t = B_z^{(0)}\, \partial v_i/\partial z, \tag{17.91}$$

$$\mathrm{d}B_i^{(2)}/\mathrm{d}t = B_j^{(1)}\, \partial v_i/\partial x_j, \tag{17.92}$$

etc. This hierarchy of equations is useful, of course, only when it can be summed, or when it converges very rapidly so that it can be truncated e.g. when the first term $B_z^{(0)}$ is a useful estimate of B_z. We are interested principally in the ensemble average large-scale field $\langle B_z \rangle$. The zero order field $\langle B_z^{(0)} \rangle$ readily follows from (17.90). To compute the next order term $\langle B_z^{(1)} \rangle$ from (17.91), the ensemble average yields the term $B_z^{(0)}\, \partial v_i/\partial z$ on the right-hand side. We know $\langle B_z^{(0)} \rangle$, but we cannot compute $\langle B_z^{(0)}\, \partial v_i/\partial z \rangle$ unless the correlation between $B_z^{(0)}$ and $\partial v_i/\partial z$ is known explicitly.

The quasi-linear approximation (see, for instance, Frisch 1968) gets over the problem by supposing that B_i changes but little during the life τ of a single eddy, so that for a random $\partial v_i/\partial x_j$ the average effect is just the interaction of $\partial v_i/\partial x_j$ with the average field $\langle B_i \rangle$. This idealization is applicable when the magnetic Reynolds number of the individual eddies is small, so that δB_i relaxes quickly back to zero.

In the next chapter we carry through the calculation for the effect of small-scale turbulence on a large-scale field. The appropriate expression for the vector potential a_i added by the turbulence to the initial vector potential A_i is given by (18.36), where ξ_i is the Lagrangian displacement of the element of fluid at x_i at time t from its initial position at X_i. The diffusion term is the second order term. In the absence of helicity, $\langle \xi_k\, \partial \xi_j/\partial x_i \rangle = 0$ and only the diagonal terms of $\langle \xi_k \xi_l \rangle$ are non-vanishing. The diffusion term is the second order term, $\partial^2 A_j/\partial x_k\, \partial x_l$ whose coefficient in (18.36) is the turbulent diffusion coefficient η_T multiplied by the elapsed time τ_1. Hence (18.36) gives a turbulent diffusion coefficient as one-half the mean square displacement per unit time in the direction of diffusion. This is, of course, the same result as for the diffusion of scalar fields (Taylor 1921; Batchelor 1959; Kraichnan 1976a).

17.6. Growth of small-scale fields

Thus far in the development we have paid but little attention to the small-scale or random component δB_i. It was noted in eqns (17.28) and (17.29) that δB_i initially grows rapidly with time for small $\eta\tau$, so that application of the quasi-linear approximation is soon dubious. It was noted, too, that for small magnetic Reynolds number $L^2/\eta\tau$, the small-scale fields are limited to small values (compared to $\langle B_i \rangle$) by the resistivity. In contrast, the magnetic Reynolds number of the dominant eddies in most astrophysical circumstances is large, and we expect δB_i to become large, perhaps large compared to the mean field $\langle B_i \rangle$. Indeed, Piddington (1970, 1972a, b; 1975a, b) has argued that δB_i grows so large that the tension $(\delta B_i)^2/4\pi$ in the lines of force surpasses the Reynolds stress ρv^2, with the result that the field halts the turbulent mixing of fluid throughout the region and converts the fluid motion into local oscillations. He argues, therefore, that there is generally no turbulent diffusion of magnetic field, and by implication, no mixing of the chemical constituents. His ultimate conclusion is that turbulent diffusion is negligible in the sun and in the gaseous disc of the galaxy, so that the observed magnetic fields are to be regarded as primordial. This conclusion overlooks the rapid buoyant escape of fields from the convective zone of the sun and the gaseous disc of the galaxy (see Chapters 13 and 14). But the question raised, as to the increase of δB_i and the suppression of turbulent mixing, deserves attention, being of interest in its own right. The turbulent mixing of both magnetic fields and chemical constituents throughout the interiors of stars, and throughout the interstellar gas, is a basic effect that should be understood by the physicist. The mixing of chemical elements is a fundamental part of the theory of the origin of heavy elements through successive generations of stars.

The growth of small-scale fields in a turbulent fluid with high magnetic Reynolds number has been a topic of interest and discussion since hydromagnetic theory was established twenty-five years ago. The variety of conjecture on the net effect of random turbulence on a magnetic field, noted in the introductory remarks of this chapter, is a consequence of the complexity and uncertainty of the topic. The ideas range from rapid dissipation of the large-scale field, to the growth of small-scale elements to equipartition with the velocity, to the negative turbulent diffusion and exponential instability of the large-scale field in turbulence with strong local fluctuations of helicity.

Consider again the calculation of δB_i from (17.17). Restricting ourselves to periods of time over which the mean large-scale magnetic field $\langle B_i\rangle$ changes but little, and resistive diffusion can be neglected, integration of (17.17) yields

$$\delta B_i(t)=\delta B_i(0)+\langle B_j\rangle\int_0^t \mathrm{d}t'\,\frac{\partial v_i(t')}{\partial x_j}-\frac{\partial\langle B_i\rangle}{\partial x_j}\int_0^t \mathrm{d}t'\,v_j(t')$$

at any fixed point x_n. The mean square δB_i is, then,

$$\begin{aligned}\langle\delta B_i(t)\delta B_i(t)\rangle&=\langle\delta B_i(0)\delta B_i(0)\rangle\\&\quad+\langle B_j\rangle\langle B_k\rangle\int_0^t \mathrm{d}t'\int_0^t \mathrm{d}t''\left\langle\frac{\partial v_i(t')}{\partial x_j}\frac{\partial v_i(t'')}{\partial x_k}\right\rangle\\&\quad-2\langle B_j\rangle\frac{\partial\langle B_i\rangle}{\partial x_k}\int_0^t \mathrm{d}t'\int_0^t \mathrm{d}t''\left\langle\frac{\partial v_i(t')}{\partial x_j}v_k(t'')\right\rangle\\&\quad+\frac{\partial\langle B_i\rangle}{\partial x_j}\frac{\partial\langle B_i\rangle}{\partial x_k}\int_0^t \mathrm{d}t'\int_0^t \mathrm{d}t''\langle v_j(t')v_k(t'')\rangle\end{aligned}\qquad(17.93)$$

since $\langle v_j \rangle = 0$, $\langle \partial v_i/\partial x_j \rangle = 0$. Consider the last term first. It follows from (A1) and (A2) that, with $t'' = t - s$ and $V_{ij}(0, s) = \delta_{ij} B(0, s)$ the term can be written as

$$\int_0^t \mathrm{d}t' \int_0^t \mathrm{d}t'' \langle v_j(t') v_k(t'') \rangle = u^2 \delta_{jk} \int_0^t \mathrm{d}t' \int_{-(t-t')}^{t'} \mathrm{d}s B(0, s).$$

Recall that $B(0, s)$ is an even function of s with a maximum at $s = 0$ and a finite width of the order of the eddy life τ, beyond which $B(0, s)$ is essentially zero. Indeed, the characteristic eddy life is defined by (A22) in terms of the integral of $B(0, s)$ from $s = 0$ to $s = \infty$. In the limit of $t \gg \tau$, the limits t' and $-(t - t')$ on the integral over s can be extended to $+\infty$ and $-\infty$, respectively. The only error is in a neighbourhood (of width τ) at $t' = 0$ and at $t' = t$. The contribution from this neighbourhood is small $O(\tau/t)$ compared to the integral over the whole range. Hence, in the limit of large t/τ,

$$\int_0^t \mathrm{d}t' \int_{-(t-t')}^{t} \mathrm{d}s B(0, s) = 2 \int_0^t \mathrm{d}\tau' \tau = 2\tau t.$$

Note from (A.17) that $\langle v_k(t'')\, \partial v_i(t')/\partial x_j \rangle = 0$. The remaining integral in (17.93) is over $\langle \partial v_i/\partial x_j\, \partial v_i/\partial x_k \rangle$, which is not zero. It follows from (A2), (A5), (A14), and (A15), and the approximation $\alpha(s) \cong \alpha(0)$, that

$$\langle \partial v_i(t')/\partial x_j\, \partial v_i(t'')/\partial x_k \rangle = +5\alpha \delta_{jk} u^2 T(s).$$

Altogether, then

$$\langle \delta B_i(t) \delta B_i(t) \rangle = \langle \delta B_i(0) \delta B_i(0) \rangle + \\ + 2u^2 \tau t \{5\alpha \langle B_i \rangle \langle B_i \rangle + \partial \langle B_i \rangle/\partial x_j\, \partial \langle B_i \rangle/\partial x_j\}.$$

Since $\alpha = O(1/L^2)$, while $\langle B_i \rangle$ presumably varies on a much larger scale, the terms $\partial \langle B_i \rangle/\partial x_j$ can be dropped in the present approximation and

$$\langle \delta B_i(t) \delta B_i(t) \rangle = \langle \delta B_i(0) \delta B_i(0) \rangle + 10\alpha u^2 \tau t \langle B_i \rangle \langle B_i \rangle.$$

The mean square small-scale field δB_i grows linearly with time. Since $10\alpha u^2 \tau$ is of the order of $v^2 \tau/L^2$, it follows that $\langle \delta B_i \delta B_i \rangle$ increases by an amount comparable to the square of the mean field in a period of the order of $L^2/\tau v^2 = \tau_0^2/\tau$. This is just the result (17.29), of course. If τ is taken to be very small compared to τ_0, this may require a period much longer than the characteristic time $\tau_0 = L/v$. But it is clear again that the quasi-linear approximation must break down in the presence of large magnetic Reynolds numbers.

To carry the calculation farther, it is necessary to work with the exact equations. Note, then, that the exact equation (17.3) for the magnetic field is identical with the equation

$$\mathrm{d}\delta x_i/\mathrm{d}t = \delta x_j\, \partial v_i/\partial x_j \qquad (17.94)$$

for the stretching of an infinitesimal displacement vector δx_i carried with the fluid. Thus, if initially we choose a vector $\delta x_i(0)$ parallel to the magnetic field $B_i(0)$, so that

$$\delta x_i(0) = \gamma B_i(0),$$

then the two vectors move together with the fluid, with the Lagrangian coordinate $x_k(t)$, so that

$$\delta x_i(t) = \gamma B_i(t)$$

at all subsequent times. It follows that the computation of the total field in a turbulent fluid with high electrical conductivity is equivalent to the computation of the stretching of infinitesimal line segments carried with the fluid, the latter problem having received substantial attention (Batchelor 1952, 1955; Reid 1955; Cocke 1969; Kraichnan 1974) in the literature.

Consider, then, the time rate of change of the mean length l of an infinitesimal segment carried with the fluid. Whether $\partial v_i/\partial x_j$ causes l to increase or decrease, it is obvious from the linearity of (17.94) that $\mathrm{d}l/\mathrm{d}t$ must be directly proportional to l; the length of a segment n times longer varies n times faster. Hence

$$\mathrm{d}l(t)/\mathrm{d}t = f(t)l(t) \tag{17.95}$$

where $f(t)$ may be a function of time, but not of l. Since $f(t)$ is derived from $\partial v_i/\partial x_j$ in (17.93), it is evident that $f(t)$ is a random function of time, presumably with zero mean because $\langle \partial v_i/\partial x_j \rangle = 0$. An infinitesimal segment placed at random in a turbulent velocity field is equally likely to increase or decrease. Then

$$\left\langle \frac{1}{l}\frac{\mathrm{d}l}{\mathrm{d}t} \right\rangle = \langle f \rangle = 0,$$

a result deduced by Cocke (1969, 1971) on formal grounds, taking advantage of some new mathematical techniques for summing stochastic variables.

Integration of (17.95) yields

$$l(t) = l(0)\exp(F(t)) \tag{17.96}$$

where

$$F(t) \equiv \int_0^t \mathrm{d}t' f(t'), \qquad \langle F \rangle = \int_0^t \mathrm{d}t' \langle f(t') \rangle = 0.$$

It follows that

$$\langle l(t) \rangle = l(0)\langle \exp(F(t)) \rangle. \tag{17.97}$$

Now we shall soon see that the mean square value of F grows with time and eventually becomes large compared to one. The change in F in one correlation time is of the order of $\langle f^2\rangle^{\frac{1}{2}}\tau = O(v\tau/L)$. This is presumably of the order of unity or less, so that the change in F in one coherence time τ is small compared to F. It follows that, for suitably large values of t, the probability distribution $\Psi(F, t)$ of F is describable by the Fokker–Planck equation

$$\partial\Psi/\partial t = \tfrac{1}{2}\beta\, \partial^2\Psi/\partial F^2, \tag{17.98}$$

where $\tau\beta$ is the mean square variation of F in a period τ. If the initial distribution $\Psi(F, 0)$ is confined to some finite region $F^2 < F_0^2$, then for large values of $t(\beta t \gg F_0^2)$ the distribution takes up the asymptotic form

$$\Psi(F, t) = \frac{1}{(2\pi\beta t)^{\frac{1}{2}}}\exp\left(-\frac{F^2}{2\beta t}\right). \tag{17.99}$$

It follows that

$$\langle F^2\rangle \equiv \int_{-\infty}^{+\infty} \mathrm{d}F\Psi(F, t)F^2 = \beta t$$

and

$$\langle \exp(F)\rangle \equiv \int_{-\infty}^{+\infty} \mathrm{d}F\Psi(F, t)\exp(F) = \exp(\tfrac{1}{2}\beta t).$$

Hence the mean length (17.97)

$$\langle l(t)\rangle = l(0)\exp(\tfrac{1}{2}\beta t) \tag{17.100}$$

grows exponentially with time. The characteristic growth time is $2/\beta = O(L^2/v^2\tau) = O(\tau)$.

A more direct approach has been taken by Kraichnan who points out that an initial line segment, or magnetic field, in the $i = 1$ direction has a subsequent length, or strength, whose mean square is just $\langle(\partial x_i/\partial X_1)^2\rangle$, which he calculates by Monte Carlo techniques. The numerical results (Kraichnan 1976b, Fig. 5) are that the mean length grows exponentially with an effective value $\beta\tau_0$ in (17.100) very close to one. In three cases, ranging from the extreme of mirror symmetry to the extreme of maximum helicity, $\beta\tau_0$ varies only from 1·17 to 0·92, respectively. The mean is 1·04±0·12.

So the mean length of a line segment, and hence the r.m.s. small-scale magnetic field, grow exponentially with time, as $\exp(at/\tau)$ where a is a number of the order of unity. This raises a serious problem at once, because no matter how weak is the initial large-scale mean magnetic field, the continuing turbulent mixing and stretching soon increase the local magnetic field strength to where the Maxwell stress density $\langle(\delta B)^2\rangle/4\pi$

becomes comparable to the Reynolds stress density ρv^2 of the turbulent fluid. The magnetic field then resists any further stretching, and one might expect that the turbulent mixing would be significantly reduced from the diffusion coefficient obtained from the kinematical calculations of the previous section. This is the point of departure for Piddington's view, that there can be no turbulent diffusion and, hence, all stellar and galactic fields are primordial[6].

The first question is whether observations of turbulent convection in the presence of large-scale magnetic fields show any sign of the enormous small-scale fields predicted by (17.100), or any sign of inhibition by unseen fields. We can see the convection and turbulence in the solar photosphere, and we can see the mean magnetic fields, which are sometimes as strong as 10–10^2 G. But there is no indication that the turbulence is generally affected, and the turbulent mixing reduced, by the magnetic field[7]. In sunspots, to be sure, where $B > 2\times 10^3$ G, the normal photospheric convective motions—the granules—seem to be altered in character (see, for instance Beckers and Schultz 1972; Beckers 1976) but not as much as one might have expected with $B^2/8\pi$ some ten times larger than the observed $\frac{1}{2}\rho v^2$.

The important point is that normal granules with magnetic Reynolds numbers of the order of 10^4 ($L = 10^8$ cm, $v = 2\times 10^5$ cm s^{-1}, $\sigma \cong 10^{11}$ s^{-1}, $\eta = 10^9$ cm^2 s^{-1}), appear to be unaffected by large-scale fields of 10^2 G ($B^2/4\pi \cong 0{\cdot}1\rho v^2$).

In interstellar space (where $B^2/8\pi \cong \frac{1}{2}\rho v^2$) there is no evidence from spectroscopic studies of the heavy elements that the turbulent mixing is significantly inhibited by the magnetic field. The heavy elements are produced in point sources (supernovae) and yet seem to be widely distributed throughout the interstellar gas, indicating mixing times of no more than 10^9 years over hundreds of pc.

On the surface of it, there seems to be a contradiction, then. The exponential growth of small-scale magnetic fields would suggest that they quickly overpower the turbulence and halt the mixing. Yet there is no observational evidence that this happens. Indeed, the evidence suggests that the convective overturning and turbulent mixing are not greatly affected.

[6] As a matter of fact, the present galactic field would decay anyway in a time of the order of only 10^8 years as a consequence of its magnetic buoyancy (see §§13.1–13.4) so there is no evident prospect that the present galactic magnetic field is a surviving primordial field.

[7] Recent work of Busse (1975) illustrates how little the magnetic field affects the form of the convection in periodic two-dimensional convective cells at small Chandrasekhar numbers and modest magnetic Reynolds numbers.

The resolution of the dilemma posed by (17.100), and the resulting conclusion that

$$\langle(\delta B)^2\rangle \cong \langle B^2\rangle \exp(2at/\tau), \tag{17.101}$$

lies in the effects omitted in the idealized kinematical calculation leading to (17.100). There are three important effects that spring to mind.

The first is that the magnetic field is stretched into many thin, closely packed filaments as δB_i is increased in the successive eddies that shear and whirl the lines of force about. Roughly speaking, the thickness $h(t)$ of a ribbon of fluid decreases inversely with the increasing length and field strength (Reid 1955; Batchelor 1955). Eventually that thickness must be so small that diffusion and reconnection with neighbouring uncorrelated filaments becomes important. Beginning with a large-scale field, the neighbouring filaments, each initially with a cross dimension L (as determined by the first eddy), become uncorrelated after the first eddy life τ and remain uncorrelated, then, as their thickness $h(t)$ decreases with time. The neighbouring filaments cannot avoid the non-equilibrium line cutting (Chapters 15 and 16) which proceeds at a rate ϵV_A across the filament. Hence the filament is dispersed when the time $h(t)/\epsilon V_A$ needed to cut across the filament thickness $h(t)$ becomes as small as the characteristic stretching time. Now V_A is the Alfven speed in the local field $\langle(\delta B)^2\rangle^{\frac{1}{2}}$, and $\langle(\delta B)^2\rangle^{\frac{1}{2}}$ grows from the large-scale mean field $B(0)$ at a rate of the order of

$$\delta B(t) = B(0)\exp(at/\tau)$$

with

$$h(t) = L\exp(-at/\tau),$$

$$V_A(t) = V(0)\exp(+at/\tau)$$

where $V(0)$ denotes the Alfven speed in the large-scale field $B(0)$. Line cutting breaks up the filaments at least by the time that $\epsilon V_A \tau$ becomes equal to h, if not sooner. This occurs when, in order of magnitude,

$$\exp(2at/\tau) = L/\epsilon V(0)\tau, \tag{17.102}$$

at which time $\delta B(t)/B(0)$ is of the order of $\{L/\epsilon V(0)\tau\}^{\frac{1}{2}}$. With $\tau = L/v$ this is $\{v/\epsilon V(0)\}^{\frac{1}{2}}$, so that the characteristic Alfven speed reaches a maximum of the order of

$$V_A = \{vV(0)/\epsilon\}^{\frac{1}{2}} \tag{17.103}$$

before line cutting reduces the filament. It is evident from (17.103) that apart from the factor $\epsilon^{\frac{1}{2}}$, the maximum Alfven speed expected in the small-scale field is just the harmonic mean of the turbulent velocity v and the Alfven speed in the large-scale field $B(0)$. Thus, with ϵ of the order of

10^{-1}, the small-scale field produced by the turbulence from a weak large-scale field does not exceed the Reynolds stresses, even at its maximum. The average value of the mean square, small-scale fields is even less[8].

This theoretical result appears to be compatible with the observed behaviour of the solar magnetic fields and turbulence at the surface of the sun, already mentioned. The granules and the supergranules are generally unaffected by the magnetic fields, except in sunspots where the mean large-scale field is so strong ($V(0) \cong 10\ \text{km s}^{-1}$) as to block the normal photospheric turbulence (granules with $v \cong 2\ \text{km s}^{-1}$).

In interstellar space the galactic magnetic field provides an example where the mean large-scale field stresses appear to be equal to the Reynolds stress of the turbulence, with $V(0) \cong v \cong 10\ \text{km s}^{-1}$. Hence turbulent mixing may be inhibited there in some degree, although in view of the approximate uniformity of the chemical constituents (allowing for the depletion of the more refractory elements to form dust grains) the inhibition of mixing is by no means complete.

This brings us to the second point, that the calculations carried out in this chapter are kinematical, ignoring the Maxwell stresses, while the calculations lead to the conclusion that the local small-scale fields do not necessarily remain weak. In such cases we must turn our thoughts once more to the dynamical effects of strong fields worked out in the preceding chapters. A *weak* field, whether of large or small scale, escapes from a volume of turbulent convecting fluid only when mixed to the surface. But a *strong* field also escapes because of its intrinsic buoyancy, as described in Chapters 8 and 13. Taking into account the magnetic buoyancy, it follows that the local intense filaments of field δB_i rise up and out of the region of turbulence at a rate comparable to their internal Alfven speed $V_A = \delta B/(4\pi\rho)^{\frac{1}{2}}$. The buoyant rise is a systematic motion, so that even when $V_A < v$, the stronger field elements quickly disappear from the convecting region in which they are produced. We expect the buoyant loss in most circumstances to limit the growth of the local field δB_i to

[8] Formal theories of hydrodynamic turbulence have been constructed on the assumption that the dynamics can be formulated in terms of the spectrum function depending only on the magnitude of the wavenumber. The idealization has been extended to hydromagnetic turbulence, by including the spectrum function of the magnetic field in the equations. The results of such calculations (see, for instance, Nagarajan 1971) suggest that an initial weak magnetic field grows at all wave numbers smaller than the dissipative cutoff to reach equipartition with the velocity field. It is particularly interesting that the magnetic field grows at wave numbers smaller than those of the initial weak field, so that the turbulence has a net dynamo effect, producing large-scale field where initially there was none. We are not aware of any astrophysical circumstance where such effects are observed.

values such that V_A remains well below the turbulent velocity v. Hence the Maxwell stresses remain below the Reynolds stresses, and turbulent mixing, if hampered, is not altogether halted.

The final point is that in some way (see Chapter 10) the magnetic fields in the solar photosphere exhibit a remarkable ability to bunch into intense and widely separated flux tubes. The tubes, evidently of some 10^3 G, are so strong as to avoid being overturned and stretched by the individual convective cells (the granules and supergranules), so that they are immune to the intensification (17.101) that dominates a continuous field. The individual flux tubes are buffeted about by the supergranules, to be sure, being confined by the converging flow into the downdrafts between supergranule cells. The point of intersection of the individual tube with the surface of the sun is subject to random walk across the disc of the sun (with an effective diffusion coefficient $\eta_T \cong \frac{1}{3}L^2/\tau \cong 10^{13}$ cm^2 s^{-1} appropriate to the supergranules; Leighton 1964, 1969) so that the field is indeed subject to turbulent mixing and diffusion even if the individual tubes are too strong to be caught up in the local mixing and overturned by the local eddies. But the widely separated, intense flux tubes do not hamper the individual granules[9] and supergranules. In the circumstance of the solar photosphere, then, we are dealing with strong, rather than weak, fields even though the overall average field may be only a few G. The magnetic fields at the surface of the sun simply have a life-style that is different from the conventional concept of continuous fields treated in this chapter. One cannot help wondering in what other astrophysical circumstances, all removed so far in space that their small-scale structure cannot be resolved, the magnetic fields have a similar behaviour.

Altogether, it appears that nature evades the production of small-scale fields to the degree required to suppress turbulent mixing. The magnetic catastrophe suggested by the purely kinematical calculations is never approached, so that the universe is turbulent in spite of its magnetic fields.

17.7. Turbulent diffusion and primordial fields

The length of the present chapter reflects the incomplete nature of the theory of the turbulent diffusion of a large-scale magnetic field in a highly conducting fluid. The discourse illustrates the turbulent effects for a number of special cases, while pointing out the obstacles to theoretical progress on a more general front. Kraichnan (1976a, b) discusses the problem of a formal deductive theory, pointing out the inadequacy of existing statistical approximations to attack the general case at high

[9] The filigree (Dunn and Zirker 1973) is the only evidence of an effect as localized as the intense flux tubes.

magnetic Reynolds number. Turbulent diffusion of a large-scale magnetic field seems to depend upon subtle internal details of the fluid motion that are not provided by the conventional treatments of the hydrodynamics. Only the quasi-linear approximation is tractable in that sense, with the mean field satisfying a diffusion equation with a diffusion coefficient depending only upon the two-point velocity correlation tensor, familiar from the conventional theory of hydrodynamic turbulence. But the method is valid only for small magnetic Reynolds numbers, wherein the turbulent contribution to the diffusion is small compared to the resistive contribution. Altogether, then, the problem of turbulent diffusion at large magnetic Reynolds numbers, pertinent to most astrophysical circumstances, forces one to proceed from *a priori* statistical hypotheses on disorder and loss of memory, or from deductive calculations based on special idealized circumstances, such as the two-dimensional velocity field $v_x(x, y, z, t)$, $v_y(x, y, z, t)$ (§17.5) or the Monte Carlo calculations of Kraichnan for specific three-dimensional velocity fields (§17.4), or the short–sudden approximation (§18.3).

Interest in the problem of turbulent diffusion of vector fields is maintained by the intellectual challenge of so basic a problem in classical physics. The interest is sharpened by the general occurrence of magnetic fields in the convecting and turbulent astrophysical bodies in the universe about us. The electrical conductivity σ and physical dimensions λ of those bodies—the sun, the distant stars, and the gaseous disc of the galaxy—are so large that the resistive decay time $\lambda^2/\eta = 4\pi\lambda^2\sigma/c^2$ of the internal magnetic field in each body is, in most cases, as large or larger than the age of the body, leading to the natural suggestion that the magnetic fields are primordial. The idea would be (Cowling 1945, 1953) that the lines of force, presently observed where they extend through the surface of the sun, also pass through the core of the sun where they are firmly and 'eternally' held fast by the high electrical conductivity of the dense and motionless gas of the core (the resistive decay time is comparable to the present 5×10^9-year age of the sun). The lines of force were carried there from interstellar space by the high electrical conductivity of the interstellar gas that collapsed to form the sun, and have been held prisoner by the core ever since. A similar view can be entertained for the galaxy. Piddington (1975a, b) has described at length how one could imagine the observed oscillatory behaviour of the magnetic fields at the surface of the sun to be produced by suitable, large, torsional oscillations of the surface layers.

The turbulent diffusion and buoyant escape of magnetic fields compels a different interpretation of the magnetic fields existing today. Insofar as the principal effect of turbulence is to mix the magnetic field throughout the parent body to the same degree as the chemical constituents, there seems to be no possibility for retaining primordial fields to the present day. Effective mixing is implied by the uniform distribution of the basic stable elements, from which we conclude that the magnetic field is similarly mixed, and hence reduced. Certainly there is no way a primordial field can be retained for long in the gaseous disc of the galaxy, nor in the convective zone of the sun or other star. In fact it is not clear to what degree the lines of force are caught up in the initial formation of a star in the first place. The buoyancy and the attendant Rayleigh–Taylor instability, not to mention the fluting instability, must enormously reduce the flux through the gas in the process of star formation (for a discussion of current views on star formation see the review by Mestel 1965). Any flux which fails to escape, as a consequence of its buoyancy and the tension along the lines of force, and as a consequence of turbulent mixing in the collapsing gas cloud, may be presumed to be lost during the early convective state of the individual stars.

For those who prefer to begin with the *hypothesis* that there was a strong field embedded in the core of the sun at the time the sun reached the main sequence some 5×10^9 years ago, there are a number of considerations that constrain the behaviour of the field. First of all, a field of 10^8 G or more ($B^2/8\pi \gtrsim 2\times10^{-3}\,NkT$ at the centre of the sun where $N \cong 10^{26}$ particle cm^{-3}, $T = 1{\cdot}5\times10^7$ K) rises up through the stable core and is lost, as a consequence of its buoyancy (Parker 1974, see §8.8). A field of 10^7 G or less might be retained for 5×10^9 years, and we know of no objection to the postulate that there is at the present time a closed field configuration of 10^7 G or less hidden away in the deep interior of the sun. The postulate serves no useful purpose, however. Only fields in excess of 10^8 G can have any significant effect on the thermonuclear reactions in the core of the sun (Chitre *et al.* 1973; Snell *et al.* 1976) and such fields would have been lost long ago. The total flux in a field B_ϕ of 10^7 G in the core, say out to a radius $r = 3\times10^5$ km, can be no more than $\Phi = \frac{1}{2}\pi r^2 B_\phi = 1{\cdot}5\times10^{28}$ Mx. This is insufficient to make any appreciable contribution to the flux appearing at the surface of the sun, at a rate of the order of 10^{15} Mx s^{-1}, because at this rate of loss B_ϕ would be exhausted in only 10^6 years.

Piddington has urged that a weaker field of, say 10^2 G extends all the way through the core, appearing as a field of about 10 G at the surface of the sun. By assuming suitable torsional oscillations of the outer layers of the sun, the lines of force of this 'primordial' field through the core can be sheared to produce any desired azimuthal field at the surface. Horizontal shear does not affect the radial (vertical) component however, and it is observed that polar fields reverse in step with the familiar 11-year magnetic cycle of the sun. The idea can survive, then, only by arguing that no one has yet observed the vertical component of field at the poles.

Observations use the longitudinal Zeeman effect, measuring the component of the field in the line of sight. When viewing high latitudes at the sun the observer sees the horizontal, rather than the vertical component of the field from Earth. The game can continue until a spacecraft passing over the poles of the sun measures the radial component of the polar fields.

Altogether, the theory of magnetic buoyancy and dynamical instability of strong fields, and the turbulent diffusion of weak fields, suggests that there are no primordial fields surviving in the universe from the time when the stars or the galaxy were formed. The planetary magnetic fields have a resistive decay time (see Chapter 20) of the order of 10^4 years, for iron cores (in which $\sigma \cong 10^{16}$ e.s.u., $\eta = 10^4$ cm s^{-1}) with radii λ of the order of $1\text{–}3 \times 10^3$ km. Jupiter, whose interior is made up of metallic hydrogen and helium, may have a resistive decay time as long as 3×10^8 years (Stevenson 1974). But even this is short compared to the 5×10^9-year age of Jupiter.

For the sun, a turbulent diffusion coefficient of the order of $0{\cdot}2v^2\tau$ (see (17.66) and (17.89)) provides rapid dispersal of the magnetic field across the surface and rapid loss of weak field (with negligible buoyancy of its own) from the convective zone. To use observed values for the turbulent eddy life τ and velocity v, the granules (for which $v \cong 3$ km s^{-1} and $\tau \cong 10^3$ s, at the visible surface) suggest $\eta_T = 10^{13}$ cm^2 s^{-1}, with the supergranules (for which $v \cong 0{\cdot}3$ km s^{-1} and $\tau \cong 10^5$ s) leading to the same number. Hence the characteristic diffusion time across a distance of 2×10^5 km, equal to present estimates of the depth of the convective zone, is 2×10^7 s or about 8 months (ten solar rotations). Across one radian on the surface ($R_\odot = 7 \times 10^5$ km) the diffusion time is 10 years, playing an essential role in the redistribution of the flux that makes up the 11-year cycle of magnetic activity (Leighton 1964, 1969). It may be presumed that the magnetic fields of other stars are subject to similar turbulent mixing effects.

The mixing and diffusion of a weak magnetic field in the gaseous disc of the galaxy arises from the turbulence in the interstellar gas, for which $v \cong 10$ km s^{-1} and $L \cong 10^2$ pc. Then if $\tau \cong L/v$, it follows that $\tau = 3 \times 10^{14}$ s $= 10^7$ years and $\eta_T = 0{\cdot}6 \times 10^{26}$ cm^2 s^{-1}. The characteristic diffusion time across the scale height of the gaseous disc, estimated to be 200 pc in the neighbourhood of the sun, is 6×10^{15} s or 2×10^8 years. Thus a *weak* field trapped in the gaseous disc must soon decay and cannot survive the turbulent mixing for the estimated 10^{10}-year age of the galaxy. As a matter of fact, the galactic magnetic field inferred from observations (see, for instance, Manchester 1972) is relatively *strong*, of the order of 3×10^{-6} G, so that $B^2/4\pi = 0{\cdot}6 \times 10^{-12}$ dyn cm^{-2}. The Reynolds stresses of the interstellar gas motions are $\rho v^2 = 1{\cdot}6 \times 10^{-12}$ dyn cm^{-2} for a mean gas density of 1 H atom cm^{-3}. The field is nearly as strong as the turbulent motions, so that magnetic buoyancy and the magnetic Rayleigh–Taylor instability (see §§13.1–13.4) provide for rapid escape, at a rate characterized by the Alfven speed of 7 km s^{-1}. The escape time over the scale height of 200 pc can not be more than a few times the Alfven transit time (3×10^7 years) and must be presumed to be only 10^8 years or a little more.

Altogether, then, the original 'primordial' fields of ordinary stars and galaxies must have disappeared long ago. The present fields must be the product of present effects. There must be some process, in the core of Earth, in the core of Mercury, Jupiter, etc., in the convective zone of the sun and other stars, and in the gaseous disc of the galaxy, and in many other galaxies, that generates magnetic fields at a rate sufficient to balance the considerable loss rate. Thus, in the core of Earth, the amplification rate of the magnetic field must equal or exceed the resistive decay rate, with a characteristic time of about 3×10^4 years. In the sun the growth must be sufficient to replenish the fields faster than they can be mixed by the turbulence and faster than they can rise out of the turbulence as a consequence of their buoyancy, in periods of the order of a few years. It is no surprise, then, that the solar magnetic cycle is not longer than 11 years. Fields are lost in this time anyway, whether they go through a reversal or not. In the gaseous disc of the galaxy, where the effects progress on a grander scale, the sources of field must be sufficient to replenish the observed fields in a time of the order of 10^8 years.

The creation of magnetic flux by fluid motions is the subject of the next chapter, where it is shown that the expected fluid motions within any rotating convecting body are just such as to produce efficient generation of magnetic field. Indeed, the generation is so effective in most cases that the field grows until $B^2/8\pi$ becomes comparable to the Reynolds stresses. When that stage is reached the rate of loss is enhanced by magnetic buoyancy and the rate of generation is reduced by the reaction of the field back onto the non-uniform rotation and cyclonic convective motions that are producing the field. In the core of Earth, where the low compressibility of the molten metal rules out magnetic buoyancy, the balance appears to be between the Maxwell stresses $B^2/8\pi$, and that part of the coriolis force not balanced by the pressure. The occasional sudden reversals of the field of Earth suggest that the balance may sometimes be upset, but in any case the result is a surface dipole field of about 0·6 G for most (about 99 per cent) of the time.

In the sun the peculiar tendency for the fields to clump themselves into isolated intense filaments where they pass through the surface (see §10.1) confuses the issue to some degree. We may be certain that their buoyancy is an important factor in their loss, limiting fields to something of the order of 10^2 G deep in the convective zone. But the structure of the magnetic fields and the fluid motions, including the non-uniform rotation, below the surface are both so unclear at the present time that little more can be said.

In the gaseous disc of the galaxy, with $B^2/8\pi$ comparable to $\frac{1}{2}\rho v^2$, it appears that both magnetic buoyancy and the suppression of the interstellar turbulence by the magnetic stresses play a role in limiting the field to its present level of about 3×10^{-6} G in the general vicinity of the sun.

It is interesting to note that the same turbulent diffusion and internal rapid reconnection associated with the loss of magnetic field from astrophysical bodies (see Chapter 13) is an essential part of the efficient generation of magnetic field. The essential role of turbulent diffusion in the generation of magnetic field was first pointed out by Steenbeck and Krause (1966) (see also Leighton 1969; Parker 1971) as an essential ingredient in the generation of magnetic field in stellar objects with such enormous magnetic Reynolds numbers. Diffusion, so as to permit reconnection, is an essential part of the dynamo process (Parker 1955, 1957, 1970a, b; Parker and Krook 1956). Resistive diffusion is sufficient in planetary interiors, but woefully inadequate in the larger dimensions of stars and galaxies. Thus turbulent diffusion produces the need for contemporary generation of fields and is itself an essential ingredient in that generation. Without turbulence the origin and nature of the magnetic fields in the universe would be quite different.

Appendix A. Correlations in isotropic turbulence

The general properties of the two-point velocity correlation $\langle v_i(x_n, t)v_j(x_n+\zeta_n, t+s)\rangle$ are readily demonstrated (Batchelor 1953). It follows at once from the statistical homogeneity and stationary quality of the turbulence that the correlation must be independent of the position x_n and the time t, and can depend only upon the separation ζ_n and s in space and time. What is more, the correlation extends symmetrically into the past and the future,

$$\begin{aligned}\langle v_i(x_n, t)v_j(x_n\pm\zeta_n, t\pm s)\rangle &= \langle v_i(x_n\pm\zeta_n, t)v_j(x_n, t\pm s)\rangle\\ &= \langle v_i(x_n\pm\zeta_n, t\pm s)v_j(x_n, t)\rangle\\ &= u^2V_{ij}(\zeta, s) = u^2V_{ji}(\zeta, s)\end{aligned}\tag{A1}$$

where the $\pm$ signs are all independent of each other. The factor u^2 is defined to be the mean square velocity in any one dimension, and is introduced as a normalization factor so that $V_{ij}(0, 0)=\delta_{ij}$. If v^2 denotes the mean square velocity $\langle v_jv_j\rangle$, then clearly $u^2=\frac{1}{3}v^2$, in the special case of statistical isotropy. The function $V_{ij}(\zeta, s)$, being the product of two vectors v_i and v_j, is obviously a tensor. The statistical isotropy of the turbulence implies that V_{ij} is an isotropic tensor. If we assume for the moment that the turbulence is mirror-symmetric as well as isotropic (i.e. no mean helicity), then V_{ij} must be of the form

$$V_{ij}(\zeta, s) = A(\zeta, s)\zeta_i\zeta_j + B(\zeta, s)\delta_{ij}, \tag{A2}$$

where the scalar functions A and B depend only upon the magnitude of the separation, ζ_k and s.

To develop the properties of $V_{ij}(\zeta, s)$, note that

$$\begin{aligned}\langle v_i(x_n, t)\, \partial v_j(x'_n, t')/\partial x'_k\rangle &= \langle v_i(x_n, t)\, \partial v_j(x_n + \zeta_n, t+s)/\partial \zeta_k\rangle \\ &= u^2\, \partial V_{ij}/\partial \zeta_k.\end{aligned} \tag{A3}$$

On the other hand,

$$\begin{aligned}\langle \partial v_i(x_n, t)/\partial x_k v_j(x'_n, t')\rangle &= \langle \partial v_i(x'_n - \zeta_n, t)/\partial \zeta_k v_j(x'_n, t')\rangle \\ &= -u^2\, \partial V_{ij}/\partial \zeta_k.\end{aligned} \tag{A4}$$

It is evident, then, that

$$\left\langle \frac{\partial v_i(x_n, t)}{\partial x_l} \frac{\partial v_j(x'_n, t')}{\partial x'_k} \right\rangle = -u^2 \frac{\partial^2 V_{ij}}{\partial \zeta_l\, \partial \zeta_k}. \tag{A5}$$

Thus, for instance, in statistically homogeneous turbulence

$$\left\langle \frac{\partial v_i(x_n, t)}{\partial x_l} \frac{\partial v_j(x'_n, t')}{\partial x'_k} \right\rangle = -\left\langle v_i(x_n, t) \frac{\partial^2 v_j(x'_n, t')}{\partial x'_l\, \partial x'_k} \right\rangle$$

etc. For an incompressible fluid, $\partial v_j/\partial x_j = 0$ so that

$$\partial V_{ij}/\partial \zeta_j = \partial V_{ji}/\partial \zeta_j = 0 \tag{A6}$$

and

$$\left\langle \frac{\partial v_i(x_n, t)}{\partial x_j} \frac{\partial v_j(x'_n, t')}{\partial x'_k} \right\rangle = -u^2 \frac{\partial^2 V_{ij}}{\partial \zeta_k\, \partial \zeta_j} = 0. \tag{A7}$$

Applying (A6) to (A2), it follows that

$$0 = \zeta_i(4A + A'\zeta + B'/\zeta)$$

where the prime denotes differentiation with respect to ζ. For non-vanishing ζ_i, this condition reduces to

$$4A + A'\zeta + B'/\zeta = 0. \tag{A8}$$

To demonstrate the physical significance of the mathematical form (A2), consider the correlation $f(\zeta, s)$ of the components of the velocity v_ζ in the direction of the vector separation ζ_k. Then

$$\langle v_\zeta(x_k, t) v_\zeta(x_k + \zeta_k, t+s)\rangle \equiv u^2 f(\zeta, s). \tag{A9}$$

The correlation $g(\zeta, s)$ of the components $v_t(x_k)$ in any direction perpendicular to ζ_k is

$$\langle v_t(x_k, t)v_t(x_k+\zeta_k, t+s)\rangle \equiv u^2 g(\zeta, s)$$

with $f(0,0)=g(0,0)=1$. To relate f and g to A and B, rotate the coordinate axes so that the $i=1$ axis lies along ζ_k. Then $\zeta_k=\delta_{1k}\zeta$ and it follows from (A2) that

$$f(\zeta, s)=A(\zeta, s)\zeta^2+B(\zeta, s), \tag{A10}$$

$$g(\zeta, s)=B(\zeta, s). \tag{A11}$$

Hence

$$A=(f-g)/\zeta^2 \tag{A12}$$

and the function A is essentially the difference between the correlation functions of the components parallel and perpendicular to the separation ζ_k, while B is the correlation between the components perpendicular to ζ_k. It follows from (A8) that

$$g=f+\tfrac{1}{2}\zeta f'. \tag{A13}$$

Now a function is on the average less than its r.m.s. value, leading to Schwarz's inequality and the statement that

$$\begin{aligned} V_{\alpha\alpha}(\zeta, s) &= \langle v_\alpha(x_k, t)v_\alpha(x_k+\zeta_k, t+s)\rangle \\ &< \langle v_\alpha(x_k, t)^2\rangle^{\frac{1}{2}}\langle v_\alpha(x_k+\zeta_k, t+s)^2\rangle^{\frac{1}{2}} \\ &= V_{\alpha\alpha}(0, s)u^2 \end{aligned}$$

where the summation convention is not used with Greek indices. It follows that $V_{\alpha\alpha}(\zeta, s)$ has a maximum at $\zeta=0$, implying that $\partial V_{\alpha\alpha}/\partial\zeta_i$ vanishes at $\zeta=0$. The same conclusion follows for $f(\zeta)$ and $g(\zeta)$. Hence in the neighbourhood of $\zeta=0$, the functions f and g, related by (A13), can be expanded as

$$\begin{aligned} f(\zeta, s) &= T(s)\{1-\tfrac{1}{2}\alpha(s)\zeta^2+\ldots\}, \\ g(\zeta, s) &= T(s)\{1-\alpha(s)\zeta^2+\ldots\}, \end{aligned}$$

where $T(0)=1$ and $\alpha(0)=O(L^{-2})$. It follows from (A13) that (A10) and (A11) yield

$$A(\zeta, s)=\tfrac{1}{2}T(s)\alpha(s)+\ldots, \tag{A14}$$

$$B(\zeta, s)=T(s)\{1-\alpha(s)\zeta^2+\ldots\}. \tag{A15}$$

Thus, it follows from (A2) that

$$\langle v_i(x_k, s)v_j(x_k, s)\rangle = u^2 V_{ij}(0,0)=u^2\delta_{ij}, \tag{A16}$$

$$\partial V_{ij}/\partial\zeta_k=0, \tag{A17}$$

and

$$\partial^2 V_{ij}/\partial\zeta_n\,\partial\zeta_m=\alpha(0)\{\tfrac{1}{2}(\delta_{in}\delta_{jm}+\delta_{im}\delta_{jn})-2\delta_{ij}\delta_{nm}\} \tag{A18}$$

at $\zeta=0, s=0$. It is readily shown, then, that the mean square vorticity ($\omega_i=\epsilon_{ijk}\,\partial v_k/\partial x_j$) is related to α by

$$\begin{aligned}\langle\omega_i\omega_i\rangle &= -u^2\epsilon_{ijk}\epsilon_{inm}(\partial^2 V_{km}/\partial\zeta_j\,\partial\zeta_n)_{\zeta=0}\\ &= 15\alpha u^2\end{aligned} \tag{A19}$$

so that $\alpha(0)$ is a direct measure of the mean square vorticity.

Note that if the condition of mirror symmetry were to be dropped, admitting a net helicity for the turbulence, then, the most general isotropic correlation tensor is

$$V_{ij}(\zeta, s)=A(\zeta, s)\zeta_i\zeta_j+B(\zeta, s)\delta_{ij}+C(\zeta, s)\epsilon_{ijk}\zeta_k. \tag{A20}$$

It follows that $\partial V_{ij}/\partial x_n$, given by (A17) when $C(\zeta, s)=0$, has the additional terms $C(\zeta)\epsilon_{ijn}+C'(\zeta)\epsilon_{ijk}\zeta_k\zeta_n/\zeta$. The mean helicity is, then,

$$\begin{aligned}\epsilon_{ijk}\langle v_i\,\partial v_k/\partial x_j\rangle &= u^2\epsilon_{ijk}\,\partial V_{ik}/\partial\zeta_i\,|_{\zeta=0}\\ &= u^2C(0,0)\epsilon_{ijk}\epsilon_{ikj}\\ &= -6u^2C(0,0).\end{aligned} \tag{A21}$$

Finally, note that a convenient definition of the correlation time τ (the characteristic eddy life) is

$$\tau=\int_0^\infty \mathrm{d}s V_{11}(0, s)=\int_0^\infty \mathrm{d}sB(0, s). \tag{A22}$$

Now consider the correlation function of the fourier transform $u_i(\mathbf{k}, t)$ of the velocity $v_i(\mathbf{r}, t)$. The basic properties of the correlations are readily demonstrated, working directly with the transform relations and the correlation tensor (A20). It follows from the definition of u_i that

$$\begin{aligned}\langle u_i(\mathbf{k}, t)u_j^*(\mathbf{k}', t')\rangle &= \frac{1}{(2\pi)^6}\int \mathrm{d}^3\mathbf{r}\int \mathrm{d}^3\mathbf{r}' \exp \mathrm{i}(\mathbf{k}'\cdot\mathbf{r}'-\mathbf{k}\cdot\mathbf{r})\\ &\quad\times\langle v_i(\mathbf{r}, t)v_j(\mathbf{r}', t')\rangle\\ &= \frac{u^2}{(2\pi)^6}\int \mathrm{d}^3\mathbf{r}\int \mathrm{d}^3\mathbf{r}' \exp \mathrm{i}(\mathbf{k}'\cdot\mathbf{r}'-\mathbf{k}\cdot\mathbf{r})\\ &\quad\times(A\zeta_i\zeta_j+B\delta_{ij}+C\epsilon_{ijk}\zeta_k),\end{aligned}$$

where $\mathbf{r}'=\mathbf{r}+\boldsymbol{\zeta}$ so that the exponential factor in the integrand becomes $\exp\{\mathrm{i}(\mathbf{k}-\mathbf{k}')\cdot\mathbf{r}'\}\exp \mathrm{i}\mathbf{k}'\cdot\boldsymbol{\zeta}$. The integration over $\mathbf{r}'$ can then be carried out, leading to

$$\begin{aligned}\langle u_i(\mathbf{k}, t)u_j^*(\mathbf{k}', t')\rangle &= \frac{u^2}{(2\pi)^3}\delta(\mathbf{k}'-\mathbf{k})\int \mathrm{d}^3\boldsymbol{\zeta}\exp(-\mathrm{i}\mathbf{k}\cdot\boldsymbol{\zeta})\\ &\quad\{A(\zeta, s)\zeta_i\zeta_j+B(\zeta, s)\delta_{ij}+C(\zeta, s)\epsilon_{ijk}\zeta_k\}.\end{aligned}$$

Denote the Fourier transform of $A(\zeta, s)$ over space by $a(k, s)$, the transform of $B(\zeta, s)$ by $b(k, s)$, and the transform of $C(\zeta, s)$ by $c(k, s)$. It follows from the definition of $a(k\,s)$ that $\partial^2 a/\partial k_i\,\partial k_j$ is the Fourier transform of $-A\zeta_i\zeta_j$. It also follows that $\mathrm{i}\,\partial c/\partial k_n$ is the fourier transform of $C(\zeta)\zeta_n$. Altogether, then,

$$\langle u_i(\mathbf{k}, t)u_j^*(\mathbf{k}', t')\rangle = u^2\delta(\mathbf{k}-\mathbf{k}')\{-\partial^2 a/\partial k_i\,\partial k_j + b(k, s)\delta_{ij} + i\epsilon_{ijh}\,\partial c/\partial k_h\}.$$

Since a, b, and c are functions only of the magnitude k of the wave number, it follows that

$$\partial a/\partial k_i = a'k_i/k$$

where the prime denotes differentiation with respect to k. Hence

$$\partial^2 a/\partial k_i\,\partial k_j = a''k_ik_j/k^2 + (a'/k)(\delta_{ij} - k_ik_j/k^2),$$

$$\epsilon_{ijn}\,\partial c/\partial k_n = \epsilon_{ijn}k_nc'/k,$$

so that

$$\langle u_i(\mathbf{k}, t)u_j^*(\mathbf{k}', t')\rangle = u^2\delta(\mathbf{k}-\mathbf{k}')\{\delta_{ij}b - a''k_ik_j/k^2 + (a'/k)(\delta_{ij} - k_ik_j/k^2) + \mathrm{i}\epsilon_{ijn}k_nc'/k\}.$$

The incompressibility of the fluid, $\partial v_j/\partial x_j = 0$, implies that $u_jk_j = 0$. Hence

$$0 = k_j\langle u_iu_j\rangle,$$

requiring that $a'' = b$. Let $\beta(k, s) = b + a'/k$ and $\gamma = c'/k$ so that

$$\langle u_i(\mathbf{k}, t)u_j^*(\mathbf{k}', t')\rangle = u^2\delta(\mathbf{k}-\mathbf{k}') \times\{\beta(k, s)(\delta_{ij} - k_ik_j/k^2) + \mathrm{i}\gamma(k, s)\epsilon_{ijn}k_n\}. \quad \text{(A23)}$$

Noting that

$$\langle u_i(\mathbf{k}, t)u_j(\mathbf{k}', t')\rangle = \langle u_i(\mathbf{k}, t)u_j^*(-\mathbf{k}', t')\rangle, \quad \text{(A24)}$$

it follows that the mean square velocity is

$$\begin{aligned} v^2 \equiv \langle v_iv_i\rangle &= \int \mathrm{d}^3\mathbf{k} \int \mathrm{d}^3\mathbf{k}' \exp\{\mathrm{i}(\mathbf{k}+\mathbf{k}')\cdot\mathbf{r}\}\langle u_i(\mathbf{k}, t)u_i(\mathbf{k}', t)\rangle \\ &= 2u^2 \int \mathrm{d}^3\mathbf{k} \int \mathrm{d}^3\mathbf{k}' \exp\{\mathrm{i}(\mathbf{k}+\mathbf{k}')\cdot\mathbf{r}\}\beta(k, 0)\delta(\mathbf{k}+\mathbf{k}') \\ &= 2u^2 \int \mathrm{d}^3\mathbf{k}\beta(k, 0) \\ &= 8\pi u^2 \int_0^\infty \mathrm{d}kk^2\beta(k, 0). \end{aligned}$$

Then since $u^2 = \frac{1}{3}v^2$ it follows that

$$\int_0^\infty dkk^2\beta(k,0) = 3/8\pi. \tag{A25}$$

Appendix B. Turbulent diffusion of a scalar field

To compare the diffusion coefficients for scalar and vector fields, note that, had we applied the quasi-linear approximation to the turbulent diffusion of a scalar field $\rho(x_k, t) = \langle\rho\rangle + \delta\rho$, then the calculation would have begun with (17.1), the mean value of which is

$$\partial\langle\rho\rangle/\partial t = -\langle v_j\, \partial\delta\rho/\partial x_j\rangle. \tag{B1}$$

Subtracting this from (17.1) yields

$$\begin{aligned}\partial\delta\rho/\partial t &= -v_j\, \partial\langle\rho\rangle/\partial x_j + \langle v_j\, \partial\delta\rho/\partial x_j\rangle - v_j\, \partial\delta\rho/\partial x_j \\ &\cong -v_j\, \partial\langle\rho\rangle/\partial x_j.\end{aligned}$$

Hence

$$\delta\rho(x_k, t) = -\int_{-\infty}^{t} dt' v_j(x_k, t')\, \partial\langle\rho(x_k, t')\rangle/\partial x_j,$$

and with $t = t' + s$, (B1) becomes

$$\begin{aligned}\partial\langle\rho\rangle/\partial t = \int_0^\infty ds \Bigg\{\Bigg\langle v_j(x_k, t)\frac{\partial v_n(x_k, t-s)}{\partial x_j}\Bigg\rangle\frac{\partial\langle\rho\rangle}{\partial x_n} \\ + \langle v_j(x_k, t) v_n(x_k, t-s)\rangle\frac{\partial^2\langle\rho\rangle}{\partial x_j\,\partial x_n}\Bigg\}\end{aligned}$$

with $\langle\rho\rangle$ evaluated at $(x_n, t-s)$ in the integrand. It follows from (A17) that the first term in the integrand is identically zero. It follows from (A2), (A19), (A15), and (A22) that the second term yields

$$\partial\langle\rho\rangle/\partial t = D\nabla^2\langle\rho\rangle \tag{B2}$$

where the diffusion coefficient is

$$D = u^2\int_0^\infty dsT(s) = \tfrac{1}{3}v^2\tau. \tag{B3}$$

References

Alfven, H. (1947). *Mon. Notic. Roy. Astron. Soc.* **107,** 211.
Batchelor, G. K. (1950). *Proc. Roy. Soc., Ser. A* **201,** 405.
—— (1952). ibid. **213,** 349.
—— (1953). *The theory of homogeneous turbulence* Chapters II and III. Cambridge University Press.

—— (1955). *Proc. Cambridge Phil. Soc.* **51,** 361.
—— (1959). *J. Fluid Mech.* **5,** 113.
BECKERS, J. M. (1976). *Astrophys. J.* **203,** 739.
—— and SCHULTZ, R. (1972). *Solar Phys.* **27,** 61.
BERNSTEIN, I. B. and ENGELMANN, F. (1966). *Phys. fluids* **9,** 937.
BIERMANN, L. (1950). *Z. Naturforsch.* **52,** 65.
—— and SCHLÜTER, A. (1951). *Phys. Rev.* **82,** 863.
BUSSE, F. H. (1975). *J. Fluid Mech.* **71,** 193.
CATTANI, D. (1967). *Nuovo Cim. B* **152B,** 574.
—— and SACCHI, C. (1966). *Nuovo Cim.* **46,** 258.
CHANDRASEKHAR, S. (1950). *Proc. Roy. Soc., Ser. A* **204,** 435.
—— (1951). ibid. **207,** 301.
CHITRE, S. M., EZER, D., and STOTHERS, R. (1973). *Astrophys. Lett.* **14,** 37.
COCKE, W. J. (1969). *Phys. Fluids* **12,** 2488.
—— (1971). ibid. **14,** 1624.
COWLING, T. G. (1945). *Mon. Notic. Roy. Astron. Soc.* **105,** 166.
—— (1953). *Solar electrodynamics in the sun* (ed. G. P. KUIPER), University of Chicago Press.
DOLINSKY, A. and GOLDMAN, R. (1967). *Phys. Fluids* **20,** 1251.
DUNN, R. B. and ZIRKER, J. B. (1973). *Solar Phys.* **33,** 281.
FRISCH, U. (1968). *Wave propagation in random media* in *Probabilistic methods in applied mathematics* (ed. A. T. BARUCHA-REID), Vol. 1. Academic Press, New York.
FU, K. Y. (1973). *Plasma Phys.* **15,** 57.
HERRING, J. R. (1969). *Phys. Fluids* **12,** 39.
KADOMTSEV, B. B. (1965). *Plasma turbulence.* Academic Press, New York.
KAZANTSEV, A. P. (1967). *Zh. Eksp. Teor. Fiz.* **53,** 1806.
—— (1968). *Sov. Phys. JETP* **26,** 1031.
KENNEL, C. F. and ENGELMANN, F. (1966). *Phys. Fluids* **9,** 2377.
—— and PETSCHEK, H. E. (1966). *J. Geophys. Res.* **71,** 1.
KRAICHNAN, R. H. (1961). *J. Math. Phys.* **2,** 124.
—— (1970). *Phys. Fluids* **13,** 22.
—— (1974). *J. Fluid Mech.* **64,** 737.
—— (1976a). ibid. **75,** 657.
—— (1976b). ibid. **77,** 753.
—— and NAGARAJAN, S. (1967). *Phys. Fluids* **10,** 859.
LEIGHTON, R. B. (1964). *Astrophys. J.* **140,** 1559.
—— (1969). ibid. **156,** 1.
LERCHE, I. (1971). *Astrophys. J.* **166,** 627, 639.
MANCHESTER, R. N. (1972). *Astrophys. J.* **172,** 43.
MEECHAM, W. C. (1964). *J. Geophys. Res.* **69,** 3175.
MESTEL, L. (1965). *Quart. J. Roy. Astron. Soc.* **6,** 151, 265.
—— and ROXBURGH, I. W. (1961). *Astrophys. J.* **136,** 615.
MOFFATT, H. K. (1961). *J. Fluid Mech.* **11,** 625.
—— (1963). ibid. **17,** 225.
—— (1970). ibid. **41,** 435.
—— (1974). ibid **64,** 1.
—— 1976, *Advances in Applied Mechanics* **16,** 119
—— 1978. *Magnetic field generation in electrically conducting fluids*, Cambridge University Press, Cambridge.
NAGARAJAN, S. (1971). *Solar magnetic fields*, IAU Symposium No. 43, (ed. R. HOWARD), p. 487. Reidel, Dordrecht.

PAO, Y. H. (1963). *Phys. Fluids* **6,** 632.
PARKER, E. N. (1955). *Astrophys. J.* **122,** 293.
—— (1957). *Proc. Nat. Acad. Sci.* **43,** 8.
—— (1963). *Astrophys. J.* **138,** 826.
—— (1969). ibid. **157,** 1119, 1129.
—— (1970a). *Astrophys. J.* **160,** 383.
—— (1970b). *Ann. Rev. Astron. Astrophys.* **8,** 1.
—— (1970c). *Astrophys. J.* **162,** 665.
—— (1971). ibid. **163,** 279.
—— (1973). *Astrophys. Space Sci.* **22,** 279.
—— (1974). ibid. **31,** 261.
—— and KROOK, M. (1956). *Astrophys. J.* **124,** 214.
PIDDINGTON, J. H. (1970). *Aust. J. Phys.* **23,** 731.
—— (1972a). *Solar Phys.* **22,** 3.
—— (1972b). *Cosmic Electrodyn.* **3,** 50; **5,** 129.
—— (1975a). *Astrophys. Space Sci.* **35,** 269.
—— (1975b). ibid. **39,** 157.
RÄDLER, K. H. (1968). *Z. Naturforsch. A* **23,** 1841, 1851.
—— (1969). *Monats. Deutsch Akad. Wiss.* **11,** 194, 272.
REID, W. H. (1955). *Proc. Cambridge Phil. Soc.* **51,** 350.
ROBERTS, P. H. (1971). Dynamo theory In *Mathematical problems in the geophysical sciences* (ed. W. H. REID), pp. 129–206. American Mathematical Society, Providence, Rhode Island.
—— and SOWARD, A. M. (1975). *Astron. Nachr.* **296,** 49.
ROXBURGH, I. W. (1966). *Mon. Notic. Roy. Astron. Soc.* **132,** 201.
SAFFMAN, P. G. (1963). *J. Fluid Mech.* **16,** 545.
—— (1964). ibid. **18,** 449.
SNELL, R. L., WHEELER, J. C., and WILSON, J. R. (1976). *Astrophys. Lett.* **17,** 157.
STEENBECK, M. and KRAUSE, F. (1966). *Z. Naturforsch. A* **21,** 1285.
—— —— (1969). *Astron. Nachr.* **291,** 49, 271.
—— —— and RÄDLER, K. H. (1966). *Z. Naturforsch. A* **21,** 369.
STEVENSON, D. (1974). *Icarus* **22,** 403.
SWEET, P. A. (1950). *Mon. Notic. Roy. Astron. Soc.* **110,** 69.
TAYLOR, G. I. (1921). *Proc. Lond. Math. Soc.* **20,** 196.
THORNE, S. K. (1967). *Astrophys. J.* **148,** 51.
YAGLOM, A. M. (1962). *An introduction to the theory of stationary random functions,* Prentice-Hall, Englewood Cliffs, New Jersey.
VAINSHTEIN, S. I. (1970). *Zh. Eksp. Teor. fiz.* **58,** 153; *Sov. Phys. JETP* **31,** 87.
—— and ZELDOVICH, YA. B. (1972). *Sov. Phys. Uspelchi* **15,** 159.

18

THE GENERATION OF LARGE-SCALE FIELDS IN TURBULENT FLUIDS

18.1. Development of ideas on the origin of fields

MAGNETIC fields are found in nearly all astrophysical bodies where the means are available to look for them, from planets and stars, to galaxies. In few, if any, instances can the present fields be considered remnants of a primordial field, trapped in the collapsing gas from which the body was formed. Resistive diffusion destroys planetary fields, while magnetic buoyancy and turbulent diffusion dispose of both strong and weak fields in stars and galaxies. Only deep in the stable core of a star is it possible to preserve a fragment ($B < 10^7$ G) of a primordial field. In view of the general occurrence of active magnetic fields at the surface it is evident that there are one or more mechanisms to be found in most astrophysical objects producing magnetic flux at the present time. The challenge, then, is to discover what the mechanism, or mechanisms, may be.

The development of ideas on the origin of fields—originally only the geomagnetic field—runs an interesting course through the last several centuries. A study of the development reveals the role of false steps, critical appraisal, and new ideas. It is unfortunate that so much of the review literature on the origin of astrophysical magnetic fields is concerned solely with current mathematical formalism and not at all with the development of the basic physics, as though the mathematical equations arose through spontaneous creation, and their sole purpose was to produce a convergent series. The purpose of the present writing is to explore and expound the physical principles, with mathematics as a tool rather than a goal. Hence we present in concise form an account of the development of present ideas, and ultimately the development of the mathematical dynamo equations that arose from those ideas. The exposition goes on to point out hydromagnetic effects of higher order, that have yet to be incorporated into the dynamo equations, including Kraichnan's negative diffusion, and some of the higher order dynamo coefficients. All of these effects arise from basic physical causes that must be illustrated and understood in both physical and formal terms.

The question of the origin of magnetic field first arose in Western history when Gilbert (physician to Queen Elizabeth I of England) introduced the concept of the magnetic field of Earth over four hundred years ago. He showed that the field, as defined by the pointing of a freely suspended magnetic needle, was associated with the planet in precisely the same form as the field of a lodestone is associated with the stone. It was only natural, then, that he should reduce the two phenomena to one, with the declaration that Earth is a huge lodestone. The explanation accounted in a simple way for all the known facts for the succeeding three centuries, until it became clear in the late nineteenth century that ferromagnetic substances lose their magnetic properties at relatively modest temperatures (the Curie point) of several hundred degrees centigrade. The interior of Earth was known to be extremely hot. Hence it could not be ferromagnetic whether composed of iron, or iron oxide, or any other substance.

By the time the Curie point was known to physicists, there was a variety of atomic and electromagnetic effects, many of them suggesting possibilities for producing the geomagnetic field. The piezo-electric effect, the magnetostrictive effect, and the thermoelectric effect were all suggested at one time or another. Physicists did not fail to note the possibility for producing the magnetic field (observed until recently only at the surface of Earth) by imagining a sufficient separation of positive and negative electric charge across the non-conducting atmosphere. The rotation of Earth would carry the two shells of charge around each day to produce electric currents, and hence a magnetic field between and around the shells. In a more fundamental direction some authors imagined modifications to Maxwell's equations, with additional terms to produce magnetic field from the motion of matter or from gravitational fields. All these ideas, and more, were considered by the agile minds of physicists. A concise history of the various schemes that flared up, flourished, and died in the last hundred years can be found in an article by Blackett (1947).

Blackett (1947, 1952) made the most recent proposal to explain astrophysical magnetic fields as a fundamental property of rotating bodies, outside the conventional Maxwell theory of electromagnetism. The idea began with the fact that the magnetic dipole moment of Earth is $\mathcal{M}_E = 7\times10^{26}$ G cm^3 while the angular momentum is $\mathcal{L}_E = 7\times10^{40}$ g cm^2 s^{-1}, giving the ratio

$$\mathcal{M}_E/\mathcal{L}_E \cong 10^{-14}\ \mathrm{cm}^{\frac{1}{2}}\ \mathrm{g}^{-\frac{1}{2}}.$$

The angular momentum of the sun is of the order of 5×10^{48} g cm^2 s^{-1} while the magnetic moment, based on Hale's (1908a, b, 1913) (Hale *et al.* 1918) erroneous assertion that the polar fields are 50 G, was computed to be 7×10^{34} G cm^3, giving the ratio

$$\mathcal{M}_s/\mathcal{L}_s \cong 10^{-14}\ \mathrm{cm}^{\frac{1}{2}}\ \mathrm{g}^{-\frac{1}{2}}$$

similar to Earth. The first stellar magnetic field, other than the sun, to be determined was for the A-type star 78-Virginis, for which Babcock

(1947) measured a longitudinal Zeeman effect corresponding to a mean field of about 1500 G. From the estimated mass and radius of an A-star, and an assumed rotation period of one day, Blackett estimated the ratio of the magnetic moment to angular momentum to be $10^{-14}\ \mathrm{cm}^{\frac{1}{2}}\ \mathrm{g}^{-\frac{1}{2}}$, as for Earth and for the sun. This suggested a universal relationship between the magnetic moment and spin angular momentum of electrically neutral, macroscopic bodies.

There is, of course, a fundamental relation between the magnetic moment and spin angular momentum of microscopic charged particles. For the electron it is the Bohr magneton,

$$\mathcal{M}_e/\mathcal{L}_e = e/2mc = 0{\cdot}88 \times 10^7\ \mathrm{cm}^{\frac{1}{2}}\ \mathrm{g}^{-\frac{1}{2}} \tag{18.1}$$

which is enormously larger than the $10^{-14}\ \mathrm{cm}^{\frac{1}{2}}\ \mathrm{g}^{-\frac{1}{2}}$ for the macroscopic body. The large magnitude of the ratio for the electron is not surprising, of course, because the electron possesses an enormous electric charge density. Indeed, the ratio (18.1) follows directly from the classical equations of electricity and magnetism if we imagine that the electron is a sphere with suitable charge and mass distribution. On the other hand, the macroscopic body has no net charge (the earlier ideas of charge separation had been dispelled) but it has a gravitational field. Blackett noted that if the gravitational mass $G^{\frac{1}{2}}m$ were responsible for producing the magnetic field of a rotating body instead of the electric charge e, then we would expect the ratio of the magnetic moment to angular momentum to be smaller than the Bohr magneton by the ratio $G^{\frac{1}{2}}m/e = 4{\cdot}9 \times 10^{-22}$, yielding $0{\cdot}4 \times 10^{-14}\ \mathrm{cm}^{\frac{1}{2}}\ \mathrm{g}^{-\frac{1}{2}}$. This number is within a factor of four of the values for Earth, the sun, and 78-Virginis available to Blackett in 1947. Hence Blackett suggested that there exists the fundamental relation

$$\mathcal{M}/\mathcal{L} = \beta G^{\frac{1}{2}}/2c = 4 \times 10^{-15}\beta \tag{18.2}$$

between the magnetic moment and angular momentum of all electrically neutral gravitating bodies, where β is a number of the order of unity. He suggested that this ratio is of the same fundamental nature as the Bohr magneton (18.1).

The idea runs counter to the structure of modern gravitational and electromagnetic field theory, so it was not viewed with favour in many quarters. If correct, it would require serious, and difficult, revision of a general field theory that fitted a wide variety of known facts in a natural way. Blackett's idea collapsed when continuing observations showed that the field of 78-Virginis varies irregularly with time (see, for instance, Babcock 1960a).

Since then, it has been found that the polar fields of the sun are only 5–10 G, rather than 50 G (Babcock and Babcock 1955; Babcock 1963; see Stenflo 1970 for a critique of Hale's result). What is more, the polar fields of the sun reversed in 1958–1959 (Babcock 1959) and again in 1974, apparently in step with the sunspot cycle. The magnetic field of Earth is known now to reverse, at random intervals of 10^5–10^7 years (Runcorn 1955; Doell and Cox 1965; Wilson 1966, 1967; Cox and Dalrymple 1967). Altogether, then, astrophysical fields appear to be variable and often, if not always, reversible, ruling out the possibility of any fundamental relation with spin. An effect more along 'meteorological' lines would seem of the right character.

The next observational advances in astrophysical magnetism were made by Babcock (1947, 1958), and then others, showing several hundred stars with fields so strong (>100 G) as to be detectable in spite of the inability to resolve their discs. The observations (Babcock 1960a; Preston 1967, 1971) are based on measurement of the longitudinal Zeeman effect, yielding essentially the algebraic mean of the component of the field in the line of sight over the visible hemisphere of the star. In view of the problem of rotational line broadening, the detection of magnetic fields favours stars with low rotation and/or those viewed from a direction along the spin axis. The star HD 215441 presents to the observer at Earth a patch of field of about 3×10^4 G (Babcock 1960b) giving a clear separation of the Zeeman components of the line.

It is interesting to ask, then, what an observer might expect to see of the dipole field of an unresolved star. The external field of a uniformly magnetized sphere has the same form as the field of a dipole located at the centre of the sphere. Consider first the observed contribution of the component of the dipole parallel to the line of sight. If that component has a field strength $B_\parallel$ at the pole, then at an angle θ around the sphere from the pole the radial component is $B_r = B_\parallel \cos\theta$, and the meridional component is $B_\theta = \frac{1}{2}B_\parallel \sin\theta$. The component in the line of sight is $B_z = \frac{1}{2}B_\parallel(3\cos^2\theta - 1)$. If the surface of the star were uniformly bright at all view angles, i.e. ignoring limb darkening, the contribution of each annulus $(\theta,\ \theta + d\theta)$ to the total signal is just the projected area, proportional to $\sin\theta\cos\theta\, d\theta$. The average B_z is then readily calculated to be $\frac{1}{4}B_\parallel$. The mean field in the line of sight is one-fourth the polar field.

Consider the contribution of the dipole component perpendicular to the line of sight. There are as many lines of force leaving the surface on one side as entering at the other. Hence there is no contribution to the total signal. It follows that a dipole with a polar field B_0 at the surface of the star, inclined at an angle ϑ to the line of sight, contributes a mean field in the line of sight of $\frac{1}{4}B_0 \cos\vartheta$. Thus the peak at the poles may be very much stronger, by the factor $4\sec\vartheta$ than the observed mean field. If the field is quadrupole or higher order, then the factor $4\sec\vartheta$ is to be replaced by a larger number.

More sophisticated analyses of the observed mean field in the line of sight and its variation with time as the star spins on its axis, can be found in the literature (see Babcock 1958 and references therein). The inclusion of limb darkening can have significant effects, the area around the edge of the disc contributing correspondingly less.

The observed variability of most of the known magnetic stars indicates that their dipole moment, and higher moments, are inclined and offset from the spin axis, while we view the star from some other angle. The rotation of the entire star, with its eccentric magnetic structure accounts for much, if not all, of the variability.

The dipole fields of many stars are so strongly inclined to the spin axis as to suggest some systematic effect causing the nearly perpendicular orientation (see, for instance, the recent work of Mestel and Moss 1977; Moss 1977).

It should be noted, then, that the magnetic field of the sun, although highly erratic, shows a statistical alignment with the spin axis of the sun. It should be noted, too, that the field of the sun would be completely undetectable were it not for the observer's ability to resolve the disc. The longitudinal Zeeman broadening for the unresolved disc of the sun would yield no more than a fraction of a gauss, and so small a signal would be lost in the noise. Hence we cannot see the magnetic fields of other slowly rotating, main-sequence 'insignificant' stars like the sun. We presume that the sun is typical of its race, concluding, therefore, that most other slowly rotating main-sequence stars have comparable fields with comparable behaviour. Presumably the general fields of such stars are a few G, roughly aligned along the rotation axis, and varying periodically, like the 22-year magnetic cycle of the sun. Wilson's observations of the calcium emission, of stars young enough yet to show strong chromospheric lines, furnish important supporting evidence (Wilson 1966, 1968, 1967).

Certainly there are strong fields in some of the 'very insignificant' stars. The UV-Ceti flare stars (Joy and Humason 1949; Luyten 1949; Lovell 1971; Moffett 1975) are dM subdwarfs of small mass and miniscule luminosity, producing enormous flares that are observed in both radio and optical emission (see Chapter 1). The flares may be 10^3 times brighter than the flares on the sun, on a star 10^{-3} times fainter, and in extreme cases may appear at a rate of more than one per hour. Their behaviour is so similar to the solar flare that a magnetic origin is generally attributed to them (see, for instance, Kunkel 1973; Mullan 1976). It would appear, then, that at least some of the dM subdwarfs have magnetic fields of greater strength and scope than the strongest fields observed in the sun.

The fields of 10^6–10^8 G in some of the white dwarfs (Kemp 1970, 1977; Angel 1977) and the inference that pulsars have polar fields of the general order of 10^{12} G (Pacini 1968; Gunn and Ostriker 1969) (see Chapter 1) suggest that these collapsed objects have retained their original stellar fields of some 10^3 G through the contraction to their present compact state. A white dwarf has a radius of the order of 10^{-2} the typical stellar radius, so that conservation of flux during collapse implies an increase in field by a factor of 10^4. A neutron star has a radius of 10^{-5} the typical stellar radius, implying an increase in field by a factor of 10^{10} in the process of becoming a pulsar. There may be more to the origin of the strong fields of white dwarfs and neutron stars, of course, but if so, it is obscured by the fact that the known contraction appears to offer an adequate explanation for the existing strong fields.

The gaseous disc of the galaxy contains a magnetic field of the order of a few μ G. The mean field is in the azimuthal direction around the disc (Hiltner 1949, 1956; Berge and Seielstad 1967; Manchester 1974). The strength of the field, approximately 3×10^{-6} G, is sufficiently large that the magnetic pressure $B^2/8\pi$ exerted outward on the gas is comparable to the mean interstellar gas pressure, so that the magnetic field contributes significantly to the support of the gas against the gravitational field perpendicular to the disc (Parker 1966, 1969). We may presume that the interstellar gas of other galaxies is permeated by comparable fields. Unfortunately, the human life span—indeed the entire span of existence of hominids on this planet—is not sufficient to observe whether galactic fields are stationary, or periodic, in character like so many stars.

Finally we should not fail to note that recent space flights show that all the planets from Mercury to Jupiter have intrinsic magnetic fields (Warwick 1961; Dolginov *et al.* 1973; Smith *et al.* 1974; Ness *et al.* 1976; Russell 1976). The four inner planets including Earth, are composed of silicate mantles enclosing liquid iron cores with a high electrical conductivity of $\sigma\cong10^{16}\,s^{-1}$. Jupiter is composed largely of an alloy of metallic hydrogen and helium, with an electric conductivity of $\sigma\cong10^{17}\,s^{-1}$. All planets rotate, although Venus and Mercury spin much more slowly than the others, with rotation rates 4×10^{-3} and $1{\cdot}7\times10^{-2}$, respectively, compared to Earth. The resistive decay time for the fields is of the general order of 10^4 years. Even for Jupiter the decay time is only 10^8 years. Hence we need a contemporary mechanism to account for the present-day fields (cf. Stevenson 1974).

Returning, then, to the historical development of ideas on the origin of magnetic fields, Hale's discovery of the sunspot magnetic fields (Hale 1908a, b) was an important step on the long intellectual journey to modern ideas on the origin of magnetic fields. Larmor (1919) noted the vortical designs in the H_α filaments around the sunspots, which he interpreted as stream lines of vortical motion of the gas. In effect, Larmor concluded that the sunspot is just the upper end of a tornado, or cyclonic hurricane, in the atmosphere beneath the visible surface. He suggested that the partially ionized gas swirling around the magnetic field carried electric currents and thereby caused the magnetic field. It was this suggestion that motivated Cowling (1934) to study the interaction of a conducting fluid with a magnetic field. The result of that work is known as *Cowling's theorem*, stating that a magnetic field with axial symmetry cannot be maintained, or generated, by fluid motions. Indeed, the statement can be generalized to assert that fluid motions cannot maintain any magnetic field in which there is anywhere a closed line of force (or a null line) around which the neighbouring lines of force circle. A magnetic field with axial symmetry is sufficient to illustrate the general principles of the theorem. The argument is very simple.

Consider a magnetic field with axial symmetry produced in a body of finite dimension λ. As in all physical circumstances the field strength is limited to finite values. The magnetic field vanishes at infinity at least as fast as the r^{-3} for a dipole. Place the z-axis of the coordinate system along the axis of symmetry and note, then, that the field decomposes naturally into an azimuthal field $B_\phi(\varpi, z)$ and a meridional field $\{B_\varpi(\varpi, z), B_z(\varpi, z)\}$ where ϖ is distance $(x^2+y^2)^{\frac{1}{2}}$ measured from the z-axis. The field is sketched in Fig. 18.1 for a simple dipole mode, although the conclusions are valid for any mode of higher order, or any combination of modes. The divergence of the azimuthal field alone is zero, because of the axial symmetry, $\partial/\partial\phi = 0$. Since the divergence of the total field is zero, it follows that the divergence of the meridional field vanishes and B_ϖ and B_z can be written in terms of the vector potential $A_\phi(\varpi, z)$ as $B_\varpi = -\partial A_\phi/\partial z$, $B_z = +\varpi^{-1}\,\partial\varpi A_\phi/\partial\varpi$. The lines of force of the meridional fields, sketched in the cross-section in Fig. 18.1, are given by

$$\varpi A_\phi = \textit{constant}.$$

Rotational symmetry about the z-axis requires that there be no lines of force crossing the z-axis, for if there were, they would define a preferred direction ϕ. The lines of force in each meridional plane are bounded by the z-axis on one side and fall off rapidly toward $\varpi = \infty$, $z = \pm\infty$ in the other directions. There is no magnetic flux extending to infinity because, as already pointed out, the field declines at least as fast as $r^{-3} = (\varpi^2 + z^2)^{-\frac{3}{2}}$. Hence, the lines of force form closed curves in the meridional plane, and there must be at least one neutral point of the meridional field in the meridional plane about which the lines circle as sketched in Fig. 18.1. Such a neutral point is called an O-type, as opposed to the X-type (see discussion in §14.8 and Fig. 14.8a, b). In higher order fields there are additional neutral points, appearing in pairs of one O and one X-type. The essential point is that there is at least one O-type, as shown in Fig. 18.1.

Note that the neutral points in each meridional plane collectively

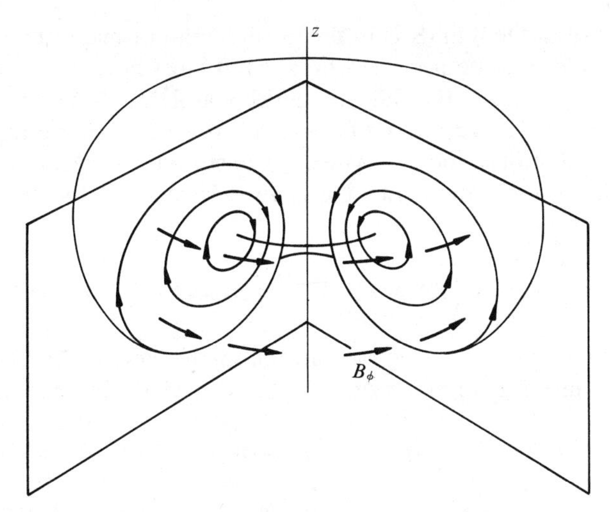

FIG. 18.1. A sketch of an axi-symmetric field configuration, made up of an azimuthal field, B_ϕ, shown by the heavy arrows, and a dipole meridional field B_ϖ, B_z indicated by the closed lines of force circling the neutral point in the meridional planes.

generate a circle concentric about the z-axis. The adjacent lines of force wind around the neutral circle. The basis for Cowling's theorem is that there is necessarily an electric current flowing azimuthally around the neutral circle, because the circulation of the neighbouring lines around the circle implies a non-vanishing azimuthal component of $\nabla \times \mathbf{B}$ along the neutral circle. On the other hand, the meridional field vanishes at the neutral point, so there is no way for the fluid motion to produce the necessary e.m.f. through the induction term $\mathbf{v}\times\mathbf{B}/c$ to drive the current. This is the contradiction that leads to Cowling's theorem.

To go through the formal argument step by step, note first that in the neighbourhood of the O-type neutral point the meridional field circles the neutral point and grows linearly with distance from the point. The uniform finite resistivity of the medium does not permit other than a linear increase. Hence $(\nabla\times\mathbf{B})_\phi = \partial B_\varpi/\partial z - \partial B_z/\partial\varpi$ is non-zero and the azimuthal current density, j_ϕ, is uniform and non-vanishing throughout the neighbourhood of the neutral point. The current density j_ϕ is related to the electric and magnetic fields by Ohm's law, in the form (4.7) and (4.8). Integrating j_ϕ around the azimuthal neutral line, at a radius $\varpi = a$, (see Fig. 18.1) it follows from the non-vanishing of j_ϕ that

$$\oint \mathrm{d}\mathbf{s}\,.\,\mathbf{j} = 2\pi a j_\phi \neq 0,$$

while from Ohm's law,

$$\oint \mathrm{d}\mathbf{s}\,.\,\mathbf{j} = \sigma\oint \mathrm{d}\mathbf{s}\,.(\mathbf{E}+\mathbf{v}\times\mathbf{B}/c).$$

The element of arc length $\mathbf{ds}$ is in the ϕ-direction, of magnitude $a\,\mathrm{d}\phi$, so that if $\mathbf{ds}\,.(\mathbf{v}\times\mathbf{B})$ is to be non-vanishing, $\mathbf{v}\times\mathbf{B}$ must have a non-vanishing ϕ-component, $v_z B_\varpi - v_\varpi B_z$. But the meridional field (B_ϖ, B_z) vanishes at the neutral point. Hence $\mathbf{ds}\,.(\mathbf{v}\times\mathbf{B})$ is zero. The remaining term $\mathbf{ds}\,.\mathbf{E}$ is just the usual Faraday induction term, transformed by Stokes theorem and the induction equation into the time derivative of the total flux Φ enclosed by the contour $\varpi = a$,

$$\oint \mathbf{ds}\,.\mathbf{j} = \sigma \int \mathbf{dS}\,.\nabla\times\mathbf{E} = -\frac{\sigma}{c}\int \mathbf{dS}\,.\partial\mathbf{B}/\partial t = -(\sigma/c)\partial\Phi/\partial t.$$

It follows, then, that the electric current in the neighbourhood of the neutral line can be maintained only by the decay of the total flux

$$\Phi = 2\pi \int_0^a \mathrm{d}\varpi\varpi B_z$$

through the contour defined by the neutral points. There is no way that the meridional portion of an axi-symmetric field can be maintained by fluid motions.

Suppose, then that there is no meridional field. It is readily shown from the hydromagnetic eqn (4.13) that an axi-symmetric azimuthal field in the absence of a meridional field cannot be maintained by fluid motions. The ϕ-component of (4.13) is

$$\partial B_\phi/\partial t + \partial v_\varpi B_\phi/\partial\varpi + \partial v_z B_\phi/\partial z = \eta(\nabla^2 - 1/\varpi^2)B_\phi \qquad (18.3)$$

which can be rewritten as

$$\frac{\mathrm{d}}{\mathrm{d}t}\frac{B_\phi}{\varpi\rho} = \frac{\eta}{\varpi\rho}\left\{\frac{\partial}{\partial\varpi}\left(\frac{1}{\varpi}\frac{\partial}{\partial\varpi}\varpi B_\phi\right) + \frac{\partial^2 B_\phi}{\partial z^2}\right\}. \qquad (18.4)$$

The left-hand side of (18.4) is just the Lagrangian derivative of $B_\phi/\varpi\rho$. The right-hand side is the vector diffusion term for B_ϕ, with a variable diffusion coefficient. There is no source term. The field can be manipulated but not generated. The quantity $B_\phi/\varpi\rho$ would be conserved except for the steady decay wrought by the resistive diffusion. It is a simple matter to integrate over a meridional plane ($\varpi \geqslant 0$) to show that the diffusion term can only cause the integral of $B_\phi/\varpi\rho$ to decline with time. The resistivity η leads only to spreading out of B_ϕ. The lines of force are destroyed by cancellation where opposite fields face each other across $\varpi = 0$, and where they escape through the outer boundary.

Cowling's theorem is the statement of the inability of the motion of a conducting fluid to produce or maintain either azimuthal or meridional magnetic fields with axial symmetry (for a more formal treatment see Backus and Chandrasekhar 1956). It is evident, from the topological nature of the proof that axial symmetry is not essential. Any field containing a closed line of force around which spiral the neighbouring lines of force can be subjected to the same analysis, with the same conclusion, that it cannot be maintained by fluid motion.

The implication of Cowling's theorem is that magnetic fields with the symmetric topology of the sunspot, or the dipole field of Earth, are not induced directly by the motion of a conducting fluid. The origin of such outwardly symmetric topologies must be more complicated. As a matter of fact, we now understand that the H_α filaments in the solar photosphere outline not the stream lines of the gas motion, but the magnetic lines of force. Thus, what Larmor thought to be evidence for a rotating vortex of fluid is instead a twisting of the spreading lines of force emerging from the sunspot. The twisting is concentrated in the upper expanded portion of the tube that extends above the surface (see §9.6). The result is the spiral H_α filaments along the spiral lines of force.

Subsequently it was shown (Bullard and Gellman 1954; Cowling 1957; Roberts 1967) that toroidal motions $v_\phi(\varpi, z)$ alone cannot maintain a magnetic field of any form. Altogether, then, the circular fluid motions postulated by Larmor cannot directly induce a magnetic field of any form, nor can the axi-symmetric field of the sunspot be created directly by fluid motions of any form. Nonetheless, Larmor's conjecture was in the right direction and sparked considerable theoretical progress.

A few years after Cowling put forth his theorem, Elsasser (1939, 1941) proposed that the dipole field of Earth might be explained by thermoelectric effects in the inhomogeneous composition and temperature distribution in the core and mantle. However, there was rapid progress in atomic and solid state physics throughout the decade of the thirties, and even during the period of the second World War, so that in 1945 Elsasser (1945, 1946, 1947, 1950a) was able to rule out any atomic process as the origin of the magnetic field of Earth. The developing quantum mechanical treatment of atoms and molecules made it clear that the thermoelectric effect is if anything diminished rather than enhanced by the high temperature and pressure in the interior of Earth, and hence is entirely inadequate to account for the geomagnetic field. Elsasser pointed out, therefore, that there is no mechanism available except induction of currents and fields by the motions of the liquid metal core. Hence he proposed to study the hydromagnetic induction equation in a systematic manner to elucidate the possibilities, from which the source of the geomagnetic dipole field could then be extracted (see the review by Elsasser 1950b, 1955, 1956b; Stevenson 1974).

The essential electromagnetic features of the planet Earth are the liquid metallic core occupying the inner half of the radius and the solid (plastic) silicate mantle occupying the outer half of the radius. The liquid core is of molten iron with an electrical conductivity which Elsasser estimated to be $\sigma \cong 10^{16}\,\mathrm{s}^{-1}$. The silicate mantle is a relatively poor conductor, which for the purposes at hand, Elsasser approximated as an insulator $\sigma = 0$. There is, evidently, a small solid, iron, inner core within the liquid core, whose presence does not change the physics in any essential way so that it was neglected in the theoretical considerations. The problem, then, was to formulate the electromagnetic equations (i.e. the hydromagnetic equation (4.13)) in a spherical volume of radius R surrounded by free space.

Any vector **B** may be written in terms of three scalar functions U, P, and T as

$$\mathbf{B} = -\nabla U + \nabla \times \mathbf{e}_\phi P + \nabla \times \nabla \times \mathbf{e}_\phi T.$$

Since $\nabla . \mathbf{B} = 0$, it follows that U is a solution of Laplace's equation. The second term represents the poloidal part of **B** and the third term the toroidal part. The field within the conducting sphere $r \leqslant R$ can be represented conveniently by P and T. In the non-conducting space outside $(r > R)$ there can be no electric currents. Hence $\nabla \times \mathbf{B} = 0$ and **B** is conveniently represented by U alone, with

$$U = \sum_{n=0}^{\infty} C_n \left(\frac{R}{r}\right)^{n+1} P_n^m(\cos\theta)\exp(\pm im\phi)$$

outside the conducting sphere. The mathematical treatment can then be cast into an elegant form by expressing P and T within the sphere in terms of tesseral harmonics $S_{nml} = j_n(k_{nl}r)P_n^m(\cos\theta)\exp(\pm im\phi)$ where j_n is the spherical Bessel function and the radial wave number is chosen so that $j_{n-1}(k_{nl}R) = 0$. Both P and T were expanded in the modes S_{nml}. The azimuthal field vanishes at $r = R$ while the poloidal field must fit smoothly onto the external vacuum field $-\nabla U$, i.e. B_r and B_θ must be continuous across $r = R$.

The velocity field of the fluid was similarly expanded, with the result that the induction term $\mathbf{v} \times \mathbf{B}$ involves the product of the harmonics in which **v** and **B** are expanded, so that the ith mode of the velocity interacts with the jth mode of the field to give additional modes (such as $i \pm j$) which in turn interact with the ith mode of the velocity to give additional modes, etc. Thus the induction term is a matrix, representing the interaction of the each velocity mode with the infinite series of magnetic modes. The idea was to choose a velocity represented by as few modes as possible—two or three—and compute the magnetic field.

Now quantitative magnetic observation of both the magnitude and direction of the geomagnetic field has been carried on from the time of Gauss (cf. Chapman and Bartels 1940: Elsasser 1950b, 1955, 1956a, b).

Worldwide surveys show a dozen or more identifiable local anomalies in the field intensity, i.e. local deviations from a dipole. The individual bulges in the field are changing with time, some increasing and others decreasing, with characteristic times of many centuries. There is a conspicuous tendency for the field inhomogeneities appearing on the magnetic maps to drift westward at a rate of about 0·18° per year. This worldwide effect suggests that the surface of the metal core, through which the fields emerge, rotates less rapidly than the overlying mantle, falling back some 0·18° per year. Since there is presumed to be no net drag between the core and the mantle, this slow rotation of the core surface at low and middle latitudes suggests a more rapid rotation of the core surface at high latitudes. That is to say, the angular velocity of the core appears to decrease with distance from the spin axis of Earth. The rate of 0·18° per year at the core radius of 3×10^3 km represents a westward flow of 0·3 mm s^{-1} relative to the mantle. It is not surprising, on elementary considerations, to find the outer part of the core rotating more slowly than the inner part. The conservation of angular momentum of the rising and sinking fluid (i.e. the coriolis force) in the convective cells should produce just such an effect. Whether, in fact, so simple a conclusion can be substantiated by a complete formal solution of the hydromagnetic equations for the convection fluid in the core remains to be seen (cf. Busse 1970a, b, giving dynamical solutions in which the onset of convection occurs in elongated rolls extending parallel to the axis of rotation, yielding a *higher* rotation rate at low latitudes, as observed in the sun). In any case, the slower rotation of the outer part of the core is suggested by observation and serves as a useful point of departure for exploration of the possibilities for the generation of magnetic field in convective fluid bodies. There is plenty of time, once the principles for the generation of field have been elucidated, to return to the question of the precise nature of the convection and circulation in the core of the Earth, in the sun, etc. Indeed, at the time of writing, some thirty years after Elsasser's pioneering work, we still have no unique and compelling answer for the form of the non-uniform rotation in either Earth or the sun. The last word on the questions seems yet some years away.

The first important fact to emerge from Elsasser's investigation was that the principal magnetic field in the core is an *azimuthal* field. The dipole field, observed on the surface of Earth to have a strength of 0·3 G at the equator and 0·6 G at the poles, extrapolates downward to the surface of the core ($r = R_c$) to a polar field strength of about 5 G. The lines of force of the dipole field extend through the core, from the north to the south with the usual sign convention, as sketched in Fig. 18.2. Consequently any non-uniform rotation of the core shears the lines of force of the dipole field drawing the lines out in the azimuthal direction.

The effect is described by the azimuthal component of (4.13)

$$\left\{\frac{\partial}{\partial t}-\eta\left(\nabla^2-\frac{1}{\varpi^2}\right)\right\}B_\phi = B_\varpi \varpi \frac{\partial}{\partial \varpi}\frac{v_\phi}{\varpi}+B_z\frac{\partial v_\phi}{\partial z},$$
$$= \varpi(B_\varpi\ \partial\Omega/\partial\varpi + B_z\ \partial\Omega/\partial z), \qquad (18.4)$$

where Ω is the angular velocity, $v_\phi = \varpi\Omega$. Figure 18.3 is a sketch of the lines of force stretched in a non-uniform rotation $\Omega(\varpi, z)$ in which the inner part of the core rotates more rapidly than the outer part, so that there is an eastward azimuthal field in the northern hemisphere and a westward azimuthal field in the southern hemisphere. The azimuthal field grows steadily with time until eventually it becomes so strong that either resistive decay balances the rate of production or the stresses halt the non-uniform rotation v_ϕ.

In the next section, there are some formal examples presented to illustrate the growth and final steady state of the azimuthal field at a fixed azimuthal velocity v_ϕ.

The characteristic resistive decay time for the azimuthal field is of the order of 10^4 years or more, during which time the non-uniform rotation progresses five times or more around the core. The azimuthal field B_ϕ, therefore, is many times stronger than the dipole field in the core. Indeed, B_ϕ must be so strong as to act as a brake on the non-uniform rotation. Bullard (1949a, b) suggested that Elsasser's azimuthal field grows until

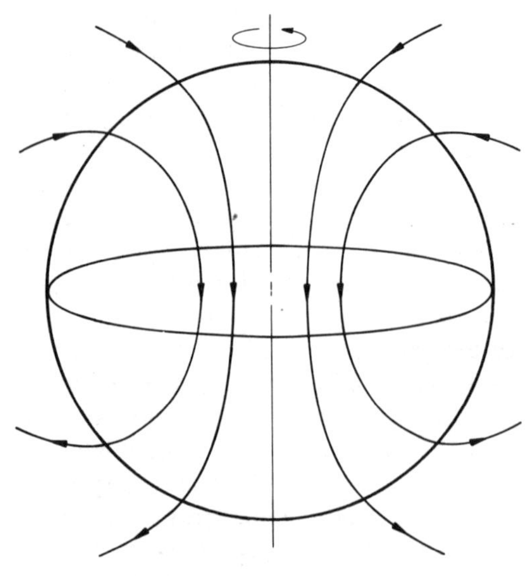

FIG. 18.2. A sketch of the lines of force of the dipole field of Earth as they extend through the liquid core.

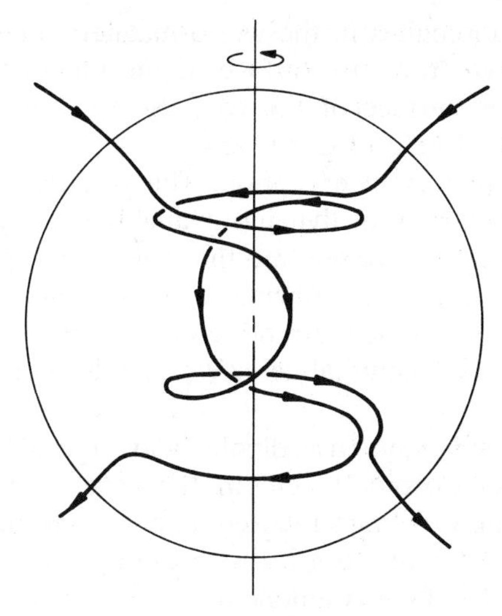

FIG. 18.3. Production of the azimuthal field B_ϕ from the lines of force of the dipole field as a consequence of the non-uniform rotation of the liquid core. The lines of force of the combined azimuthal and meridional field are sketched as though the non-uniform rotation, in which the angular velocity of the core decreases with distance from the axis, were switched on only a few centuries ago, so that the lines of the original purely meridional field have been carried only a little more than half a revolution around the core. Note that the resulting azimuthal field is eastward in the northern hemisphere and westward in the southern hemisphere.

the Lorentz force $B_\phi B_p/4\pi R_c$ (B_p is the meridional field of about 5 G) balances the coriolis forces of the convective motions, $B_\phi B_p \cong 4\pi\rho v\omega_E R_c$ where ω_E is the angular velocity of rotation of Earth ($\omega_E = 7\times10^{-5}\,\mathrm{s}^{-1}$), v is the convective velocity ($v \cong 10^{-2}\,\mathrm{cm\,s}^{-1}$) and ρ is the density, of the order of $10\,\mathrm{g\,cm}^{-3}$. On this basis B_ϕ may easily exceed 10^2 G. It is much stronger than the dipole field in the core. The azimuthal field cannot extend outside the core into the non-conducting mantle, of course, because it necessarily has meridional (poloidal) currents associated with it. Hence it is completely unobservable at the surface of Earth. We will never see the principal magnetic field of our planet. Only the secondary field—the dipole—is available to us, from which we infer the existence of the primary field in the core.

In stars, such as the sun, we see all the way in to the top of the convective zone where the fields are generated. Hence we see occasional local bulges of the azimuthal field extending out through the surface of the convection zone, in the form of the east–west bipolar magnetic regions. To this limited degree we are able to see the azimuthal field of the sun. In planets on the other hand, the convective core is overlain by thousands of km of opaque, non-conducting silicate, so that the observer is too far from the core to see the local bipolar regions. We see only shadowy magnetic anomalies in the world magnetic maps, which cannot be extrapolated down from the surface of the planet to obtain a clear magnetic map of the surface of the core. So we infer, but we do not observe, the azimuthal field of our planet.

The galaxy is an important exception. The gaseous disc of the galaxy comprises the one transparent dynamo available for observation. In our own, and in other galaxies, we see into the inner workings of the dynamo, and, as we might expect, the principal field is azimuthal. The poloidal components of the galactic field are not apparent, being lost in the strong local fluctuations of the azimuthal field. We shall have more to say on this in a later chapter.

So far, then, an axi-symmetric dipole field of Earth implies an axi-symmetric azimuthal (toroidal) field in the core as a consequence of non-uniform rotation. Cowling's theorem implies that the development of the theory has not yet come to grips with the dynamo mechanism. That requires an asymmetric field component.

Elsasser went ahead to work out the interaction matrix elements for simple asymmetric velocities, i.e. the lowest modes in the expansion of the poloidal velocity component. The plan was to truncate the expansion of **B** after a tractable number of terms. Bullard, modifying the radial functions in the Elsasser formalism, pushed the problem using numerical methods to invert the matrix and compute the coefficients for the field (Elsasser 1946, 1947, 1950a; Bullard 1949a, b; Bullard and Gellman 1954; Takeuchi and Elsasser 1954; Takeuchi and Shimazu 1954; Elsasser and Takeuchi 1955; Rikitake 1958, 1966). It became clear that the formal expansion (of the poloidal and toroidal components of the field) in terms of orthogonal surface harmonics converged very slowly, if at all. Even with the simplest non-symmetric velocity modes, the interaction of the velocity with each magnetic mode generates higher magnetic modes with amplitudes that are not small. The problem was restricted to the steady state, $\partial/\partial t = 0$, so that the calculation is an eigenvalue problem for the magnetic Reynolds number R_{m} of the fluid velocity to maintain a perfect balance between production $\nabla\times(\mathbf{v}\times\mathbf{B})$ and dissipation $\eta\nabla^2\mathbf{B}$. The slow convergence appears most clearly in the work of Bullard and Gellman (1954), where they show the strong dependence of R_{m} on the level of truncation of the expansion of **B**. For instance, keeping only the first and second order harmonics yields the eigenvalue $R_{\mathrm{m}} = 22$, while keeping terms to fourth order produces $R_{\mathrm{m}} = 67$ (see also Gibson and Roberts 1969; Lilley 1970; Gubbins 1973a for further exploration of the problem). The higher harmonics of the field are strong in spite of their higher rate of decay. Pekeris *et al.* (1973) recently presented a suitably idealized fluid motion for which the solution appears to converge, probably because the very strong helicity $\mathbf{v}\cdot\nabla\times\mathbf{v}$ in their fluid model makes a much more efficient dynamo, so that the magnitude of $\mathbf{v}$, and hence the magnitude of each successively higher order field harmonic, is smaller than in the rather inefficient fluid motions considered by Bullard. But we are getting ahead of the story.

The conspicuous absence of convergence led Bullard to speculate (Bullard 1950) that there might be some 'super Cowling's theorem' to the effect that fluid motions cannot maintain a steady magnetic field, i.e. there is no way to produce a detailed and precise balance of $\nabla\times(\mathbf{v}\times\mathbf{B})$ and $\eta\nabla^2\mathbf{B}$, whether the field is symmetric or not. In view of the later successful researches of Herzenberg (1958) and others we know now that Bullard's conjecture was wrong. But once again, a conjecture does not have to be correct to be useful. The conjecture suggested the idea that we should not remain preoccupied with the stationary case. The stationary case was the simplest place to begin, but if it becomes difficult and unproductive, then perhaps a time dependent dynamo effect should be considered.

This author took up the problem from the point of view of time dependent fluid motions[1]. First of all, it must be understood in what sense the dynamo is time dependent. The mean magnetic dipole field of Earth, for all of the local variation and secular drift, is basically a steady field. The dipole exists, at about its present strength, and alignment more or less along the axis of rotation, for periods of 10^5 years or more[2]. The non-uniform rotation of the core is presumably a steady phenomenon, and hence so is the dominant azimuthal field in the core. Thus, the entire field configuration sketched in Fig. 18.3 must be presumed to be more or less steady in time. The gap in the theory is the generation of the axi-symmetric dipole field. Unless that is regenerated in some way, it will decay and the azimuthal field derived from it will die away too. The dipole field represents a net circulation of magnetic field in the meridional plane. The question is what deformation of the azimuthal field gives a contribution to the circulation in the meridional plane. The answer proves to be simple enough (Parker 1955b). An azimuthal line of force, lying in a circle about the z-axis, may be deformed on a small scale into a spiral extending azimuthally around the z-axis, as sketched in Fig. 18.4. The spiral line of force contributes a net circulation of magnetic flux in each meridional plane.

A review of the fluid motions in the core brought to light the local convective cells. An upwelling of fluid to form a convective cell carries

[1] The author is deeply indebted to both Elsasser and Bullard for their comments and discussion of the problem as it stood at the time, in 1954.

[2] At that time the random occasional reversals at 10^5–10^7 year intervals were unknown so that, for all one knew, the field had always had its present strength and sense.

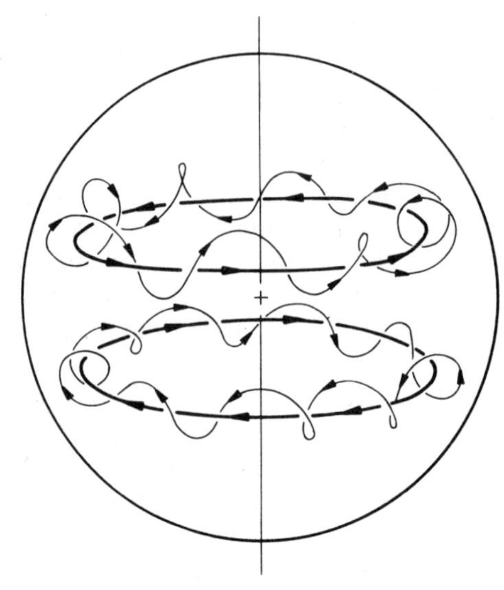

FIG. 18.4. A sketch of the azimuthal lines of force (heavy lines) in the northern and southern hemisphere carried into spirals by the local cyclonic convective cells. The azimuthal projection of the spiral on meridional planes contains a net magnetic circulation.

the lines of force of the azimuthal field upward with it, making a large upward dent in the field. The local field lines are shaped, then, like a capital Greek omega, Ω. It must be remembered, however, that a convective upwelling of fluid in a body spinning as rapidly as Earth is cyclonic, so that the fluid rotates as it rises, just as rising or sinking bodies of air in the atmosphere are cyclonic. The converging flow at the base of a cell causes the rising fluid to rotate more rapidly than its surroundings. The capital omega, then, is rotated so that the net result is one loop of a spiral in each convective cell, as sketched in Fig. 18.5. A number of convective cells throughout the core produce a general spiralling of the azimuthal lines of force, the final result being the original axi-symmetric azimuthal field plus a meridional component involving a net circulation of magnetic field, i.e. a contribution to the dipole field.

It is obvious that conservation of fluid implies as much downward as upward flow of fluid. If the downdrafts had the same form as the updrafts with the same sense of rotation, they would produce an opposite spiralling of the lines of force, thereby reducing the net effect to zero. But the non-linear nature of the hydrodynamic equations suggests that the downdrafts, formed by a convergence of fluid in the spacious outer part of the core and a divergence near the crowded centre, cannot have the same form as the updrafts, which begin near the centre and diverge in the outer core. For want of a more complete dynamical picture, we suppose merely that the downflow consists of a general subsidence throughout the broad region between the individual updrafts, without significant rotation. This simple picture is sufficient for the present kinematical exploration of the hydromagnetic origin of magnetic fields. Downdrafts can be included later when the precise form of the convection and non-uniform rotation is known from dynamical considerations.

When we turn to a formal mathematical representation of the average effect of cyclonic convective cells, it is immediately evident that the problem is very difficult if there are but a few large convective cells, whereas the problem is relatively easy in the limit of a large number of small convective cells. With small-scale convective motions the fields produced by the individual convective cells are of small scale (i.e. of very high order) decaying away rapidly so that only the mean contribution remains. There are two scales, then. One, denoted by λ, and comparable to the core radius R, refers to the large-scale quasi-stationary, axi-symmetric azimuthal and meridional fields. The other, denoted by L, refers to the small scale of the individual short-lived convective cells that interact with the large-scale field. In the limit where $L \ll \lambda$ the mathematics proceeds without difficulty. The expansions converge rapidly because

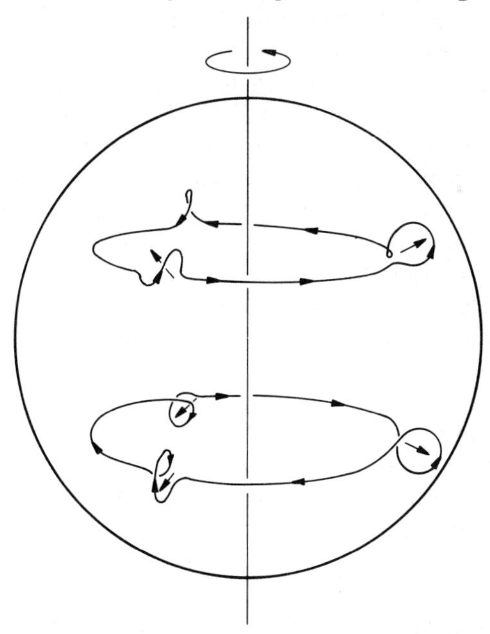

FIG. 18.5. A sketch of the deformation of azimuthal lines of force by rising rotating convective cells. The interaction of each rising cell with the azimuthal field produces a single spiral revolution, with non-vanishing circulation in the meridional plane. The direction of rotation of Earth is shown by the circular arrow at the top of the figure, indicating too the sense of rotation of the rising convective cell relative to its surroundings.

we treat the case opposite to that attempted by Elsasser and Bullard, who limited themselves to two or four convective cells, each of large dimension.

It is a straightforward matter to calculate the meridional magnetic circulation for a variety of fluid motions. To illustrate the basic principles with the simplest circumstances, suppose that we switch on the small convective cells for a brief period $\tau_1 \ll L^2/\eta$ with a characteristic velocity v that may be so large that the displacement $v\tau_1$ is as large as the dimensions L of the convective cell. The deformation of the magnetic field in the cell may be large, $\delta B \cong B_\phi$. The calculation of the deformation may be carried out neglecting resistive diffusion, because of the short life of the motions, i.e. the magnetic Reynolds number of individual cells Lv/η is large compared to one. Following the brief life τ_1 of the convective cell, the motion is switched off for a period τ_2 in excess of L^2/η, during which time the loops and spirals of the individual convective cells coalesce with their neighbours. The local magnetic circulation in the meridional plane produced by each convective cell merges with that of its neighbours so that there remains only the average overall circulation. The fields return to their original axi-symmetric state, with the meridional field augmented by the additional circulation gained from the loops and spirals of the individual convective cells. After a suitable time $\tau = \tau_1 + \tau_2$ a new round of convective cells is switched on for a time τ_1 to interact with the axi-symmetric fields. They are switched off and, again, a time τ_2 suitably in excess of L^2/η is allowed to elapse, during which the fields return to axial symmetry again, etc. This idealized form of motion serves to illustrate the basic physical effects while keeping the mathematics at a tractable level (Parker 1955b)[3]. The role of diffusion, represented by η, is essential in converting the separate loops of field produced by the individual cells into a contribution to the large-scale field (Parker 1955b; Parker and Krook 1956). The diffusion of the large-scale field is also essential and will appear naturally when we come to solve the dynamo equation (18.7).

The calculation of the fields produced within the local convective cell was carried out for a variety of forms for the fluid motion. The important point is that the magnetic field in the meridional plane is represented by the azimuthal vector potential $A_\phi(\varpi, z)$,

$$B_\varpi = -\partial A_\phi/\partial z, \qquad B_z = +(1/\varpi)\,\partial \varpi A_\phi/\partial \varpi. \tag{18.5}$$

Hence it is the mean production of azimuthal vector potential by the

[3] The reader interested in the a formal mathematical statement of the problem is referred to Backus (1958), who applied the technique to the Bullard–Gellman formalism, with immediate success (cf. Moffatt 1978).

small-scale convective cells that is the basis for the dynamo. For a given convective motion the magnetic circulation, and, hence, the vector potential, produced by the local convective cell is proportional to the large-scale field at that point. In the simplest case, where B_ϕ is the dominant field, one can write the local vector potential produced by the small, local convective cell as $B_\phi(x_k)f(\xi_k)$ for the cell centred on x_k, with local rectangular coordinates ξ_k. If, then, there are ν cells per unit volume and time, the mean rate of production of azimuthal vector potential is $B_\phi(x_k)\Gamma(x_k)$ with

$$\Gamma(x_k) = \nu \int d^3\xi\, f(\xi_k). \tag{18.6}$$

The hydromagnetic equation for the vector potential is, then,

$$\left\{\frac{\partial}{\partial t} - \eta\left(\nabla^2 - \frac{1}{\varpi}\right)\right\} A_\phi = \Gamma B_\phi. \tag{18.7}$$

This obvious result can be deduced formally by starting with (4.33), which can be written

$$\left(\frac{\partial}{\partial t} - \eta\nabla^2\right) A_i = -v_j\frac{\partial A_i}{\partial x_j} - A_j\frac{\partial v_j}{\partial x_i}$$

in rectangular coordinates, including the resistive dissipation term. Imagine, then, an ensemble of systems with identical large-scale fields $\langle A_i \rangle$ and with the same statistical distribution of convective cells, switched on at random positions for a brief time τ_1 and then off for a time τ_2 large compared to L^2/η, etc. Write

$$A_i = \langle A_i \rangle + a_i$$

where a_i represents the small-scale fields produced by each local convective cell. Then the ensemble average of the equation yields

$$\left(\frac{\partial}{\partial t} - \eta\nabla^2\right)\langle A_i \rangle = -\left\langle v_j\frac{\partial a_i}{\partial x_j} + a_j\frac{\partial v_j}{\partial x_i}\right\rangle,$$

where the right-hand side is just the mean rate of production of vector potential by the cyclonic convective cells. The time average production over a period τ in any given member of the ensemble is the same as the ensemble average, with the result (18.6) and (18.7). In our first calculations (Parker 1955b) we worked out the magnetic fields produced in each convective cell, using (4.31) to deduce the field after the convection from the field before, as well as a second order perturbation scheme to solve (4.16) directly. The resulting vector potential then follows from the fields. It is much easier to work directly with the vector potential, using (4.35) (Parker 1970b), and that will be the procedure used here when later, in §18.3, we deduce Γ for various forms of the convection.

Together (18.4), (18.5), and (18.7) make up a complete set of equations for the magnetic field (given the fluid velocity or Γ) usually referred to as the *dynamo equations*. A slightly more elaborate form of (18.7) is given in §18.3 where the interaction of the cyclonic motions with the meridional as well as the azimuthal field is included, and higher order effects are retained. The physical implications and the mathematical solutions of the dynamo equations are the subject of the next chapter. With the assumption of 'plausible' convective motions and non-uniform rotation in a sphere of conducting fluid, the dynamo equations lead to a hierarchy of solutions, the lowest mode being a dipole such as that possessed by Earth at the present time. There are higher modes, both stationary and periodic, giving the possibility for reversal of field, and generally periodic fields such as exhibited by the sun. Indeed, in a relatively thin shell, such as the convective zone of the sun, the only self-sustaining solutions are periodic in time. In a flat non-uniformly rotating disc, such as the gaseous disc of the galaxy, there are both stationary and periodic solutions, raising observational questions about the galactic field that the human race will not endure long enough to answer. The characteristic periods are of the order of 10^8 years.

To continue with the development of the basic ideas on hydromagnetic dynamos, the next important step in dynamo physics was taken by Herzenberg (1958), disproving the existence of a super-Cowling theorem. Herzenberg showed that two small conducting spheres spinning steadily in opposite directions about axes inclined to each other, and embedded in either a large spherical volume, or an infinite volume, of motionless conducting fluid, generate magnetic field. If the spheres are spun sufficiently rapidly, the field grows exponentially with time. Spun at just the right rate the field is stationary in time, and, of course, if spun more slowly, the field decays away because the regenerative effects are not sufficient to offset the resistive decay. Herzenberg's dynamo removes a major conceptual obstacle in the development of dynamo theory. Bullard's conjecture as to a super-Cowling's theorem was useful, like Larmor's conjecture before him, even if wrong in the end. Steady motions can maintain steady fields.

Indeed Lowes and Wilkinson (1963, 1968) have demonstrated Herzenberg's dynamo in the laboratory, taking advantage of the very high speeds with which a body with axial symmetry can be spun (see also Gailitis *et al.* 1972). What is more, Gailitis (1973) has been able to show that the steady rotation of the spheres in the Herzenberg dynamo can, under some circumstances, yield an alternating magnetic field. More recent theoretical work by Gibson (1968) and Roberts (1971) extends the Herzenberg dynamo to three and more spinning spheres, where the mathematical series converges more rapidly than with the original two spheres (see review in Moffatt, 1978).

As a matter of fact, the development of the regenerative effects of steady fluid motions has evolved into a rich subject of its own. Mathematical methods have been developed to handle the solution of the hydromagnetic equation

$$(\partial/\partial t - \eta\nabla^2)\mathbf{B} = \nabla\times(\mathbf{v}\times\mathbf{B})$$

for a variety of steady flows. This includes flows that are three-dimensional but invariant with respect to one coordinate, such as the azimuthal angle ϕ (Tverskoy 1965; Gailitis 1967, 1969, 1970) or one linear coordinate (Roberts 1972). The simple dynamos of Lortz (1968a, b, 1972) are particularly interesting and, given suitable steady helical motions, may be deformed to operate in a sphere such as the core of Earth (Benton 1975). Childress (1967a, b) has exhibited still another class of fluid motions capable of regenerating magnetic field. As a result of the many dynamo models that have been constructed with steady fluid motions, the remarkable fact emerges that almost all fluid motions, except some unlikely cases with too much symmetry, have magnetic regenerative effects as a consequence of their helicity (Roberts 1970). Not all the motions are efficient. The fluid may have to be run at very high velocity for the field to be self-sustaining. But almost all possess the power to regenerate magnetic field. The basic effect for the generation of field seems to be that in all but the most symmetric fluid motions, the rotation of the fluid about its direction of motion converts straight lines into spirals, thereby building up magnetic circulation in a plane perpendicular to the initial line. This can be done by steady fluid motions as well as time dependent motions. Resistive diffusion allows the magnetic circulation to escape from the region of its origin to be sheared and rotated further. In one way or another, the flows, somewhere in their active volume, produce the effect illustrated in Fig. 18.4. The conclusion is remarkable, and comes as a revelation after so many years of doubt and difficulty with the question of the hydromagnetic origin of magnetic fields.

The problem is, then, to decide what forms of field generation are the dominant ones in nature. What fluid motions do we expect, and what aspect of those motions is essential in the production of field? The first question is the most difficult because it is the dynamical question. What internal circulation and convection is expected in the interior of astrophysical bodies, from the cores of planets, to the convective zones of stars, to the gaseous discs of galaxies? As already noted, the dynamical problem is so difficult, particularly when the density varies rapidly with height (in the sun) or the energy source is not known (as in the core of Earth), that it has not yet been solved, nor is it likely to be solved very soon in any final way, in spite of the very hard thought and work that is presently directed to the task (cf. Busse 1970a, 1975a, b; Childress and

Soward 1972; Gilman 1969, 1972, 1974; Yoshimura 1972, 1973, 1975). Hence the best we can do is to rely on observations of fluid motions, and on any suggestive elementary dynamical principles as may seem appropriate. On this basis note that the gaseous disc of the galaxy is observed to rotate with an angular velocity that decreases outward from the centre, as one would expect from Newtonian mechanics and the concept of a massive galactic nucleus. The sun is the only star for which there is information available, and there the surface is observed to rotate non-uniformly, with the angular velocity of the polar regions only about two-thirds that of the equator. Sunspots rooted far below the surface of the sun move slightly faster than the surrounding photosphere, suggesting that the angular velocity increases downward for 10^4–10^5 km. But this is only superficial evidence. Various theoretical models, ignoring the enormous density stratification across the convective zone, produce an Ω that *declines* with depth, so that the polar regions are indicative of the angular velocity of the deep interior (Gilman 1972, 1974).

Earth is the only planet for which there is any observational information available. The westward drift of the magnetic anomalies, if they indicate the motion of the surface of the liquid core[4], implies that the equatorial regions rotate more slowly than the rest of the core. Altogether, then, the evidence is that there is steady non-uniform rotation in all convecting rotating astrophysical bodies, as we would expect from the coriolis forces on the convection. Indeed it would be remarkable to find a rotating, convecting fluid body that somehow managed to rotate uniformly. The rate of change of Ω with polar angle, $\partial\Omega/\partial\theta$, is known for the sun, and may be inferred for Earth. The radial gradient $\partial\Omega/\partial r$ is not known except for the galaxy.

The convection in any rotating body is cyclonic, so that altogether it appears that the universal motions are cyclonic convection of some form and non-uniform rotation. The degree to which these motions are time dependent is a complicated question. The non-uniform rotation of the galaxy may be presumed fixed. The non-uniform rotation of the sun, if it changes at all (see Eddy *et al.* 1976), does so only over periods of time that are long compared to the characteristic 22-year period of the solar magnetic cycle. The non-uniform rotation of the core of a planet may also vary, but presumably only very slowly in view of the quasi-steady character of the dipole of Earth.

The small-scale convective cells, on the other hand, appear to be less steady. The turbulence in the interstellar medium shows no signs of the

[4] It has been suggested that the westward drift may represent the westward propagation of Alfven waves in the core rather than a westward motion of the core itself.

geometrical order that one might expect in steady motions. The turbulence is driven, evidently, by the very irregular differential expansion of the gas by novae, supernovae, and the creation and passage of luminous stars. None of these effects would seem to lead to any but very unsteady motions. The convection in the sun, up to the scale of 3×10^4 km of the supergranules, is not steady. The convective patterns continually shift over periods of the order of 10^5 s, during which time they turn over only a fraction of a revolution. The magnetic anomalies in the field of Earth behave in such a way as to suggest convective cells that are far from permanent, presumably shifting and reforming over periods of a few centuries, during which time the fluid may turn through angles of the order of a radian.

On the basis of these facts and inferences, it is generally believed that most astrophysical magnetic fields are the result of a combination of non-uniform rotation and turbulent, i.e. time dependent, convection. Indeed the remainder of this chapter is devoted to an exploration and formulation of the dynamo equations on this basis. However, as already noted, the mathematical game is played first by taking the ultimate turbulent limit in which the convective cells are strong but their size and duration very small. The idealization illustrates the physical principles with the minimum of mathematical effort. It is not a precise quantitative mathematical model of the bulky, leisurely convective motions that produce the magnetic fields in the universe about us, of course. As in most circumstances the detailed, disorderly behaviour of the real world must be idealized if we are to illustrate the physical principles, rather than obscure them under a mass of formal mathematical and numerical calculation. Hence the successful idealization in other limits is an important complement to the theory of the turbulent dynamo. A number of authors have given serious consideration to the opposite extreme, in which the convective motions are of small amplitude, large-scale, and slowly varying, showing that the basic structure of the dynamo equations is the same and the general physical principles unaltered (cf. Braginskii 1964a, b).

Soward (1974) has pointed out that time varying motions are more effective than steady flows of the same structure in sustaining magnetic fields, in that smaller velocity amplitudes are sufficient. The poorer efficiency of the steady flows is evidently due to the simultaneity of the production of magnetic circulation in certain regions of the helical fluid motions and the escape of the magnetic circulation from those regions by resistive diffusion. It is more effective to shut off the flow temporarily while the magnetic circulation spreads without further interference from the fluid motion. So we will not pursue the fascinating subject of the stationary dynamo. However, the dynamo with motions that vary slowly and deviate but little from axial symmetry receives further discussion.

The nearly symmetric dynamo began with Braginskii (1964a, b, 1967; 1970, 1975). The approach was straightforward. If Cowling's theorem states that axially symmetric fields cannot be generated by fluid motions, then add small components of field and fluid motion that are not symmetric. Consider fluid motions that are azimuthal and axi-symmetric and strong, except for a small component that is not symmetric and which varies with time. Then the small non-symmetric velocity interacts with the strong azimuthal field to produce the small non-symmetric component of the magnetic field. That small non-symmetric component of field is then acted upon by the strong axi-symmetric azimuthal velocity, etc. The idealization that the field is dominated by the strong azimuthal component is presumably a reasonable representation of conditions within the core of Earth. It turns out, however, that in carrying out the mathematical manipulations leading to the equations for the mean fields, the strength of the meridional fluid motion and of the turbulent (non-symmetric) motion must be restricted to being small, $O(R^{-1})$ and $O(R^{-\frac{1}{2}})$, respectively, compared to the dominant azimuthal velocity v_ϕ, where R is the magnetic Reynolds number computed for v_ϕ. The scale L and correlation time τ of the turbulent motion v are also affected and must be restricted to $\tau \ll L^2/\eta$ and $\tau \leqslant \eta/v^2$ (i.e. $\tau \ll L/v$) so that the perturbations of the dominant azimuthal field remain small. The formulation is, then, effectively a combination of large magnetic Reynolds number for the axi-symmetric flow, with small magnetic Reynolds number for the turbulence. The proper ordering of the terms in the hydromagnetic equations leads to meridional and turbulent components of the magnetic field that are small compared to the dominant azimuthal field. Braginskii's approach, and the very elegant refinement by Soward (1972), are treated in §18.6. The formulation is interesting not only for its mathematical elegance in developing the dynamo equations, but also for the possibilities for extending it to the dynamical calculations (see Soward and Roberts 1976).

We are still faced with the fact, then, that no global theoretical connection has been established between the general statistical properties of the fluid motion and the production of magnetic field. It is clear that the cyclonic character of the fluid motions is essential, but there is no evident universal relation between the local helicity and the production of field. The relation is established for each special idealization. Of particular interest is the quasi-linear approximation first introduced by Steenbeck *et al.* (1966) to provide a broad treatment of the hydromagnetic dynamo effects of cyclonic turbulence with small magnetic Reynolds number ($vL\eta \ll 1$). The quasi-linear approach is one of the most productive idealizations, for, although it is restricted to small magnetic Reynolds number of the individual eddies, its simple analytical form permits the demonstration of higher order dynamo effects. In the quasi-linear approximation the vector potential generated by the interaction of cyclonic turbulence with the large-scale field **B** is proportional, to lowest order in L/λ, to the product of the helicity $\alpha \equiv \langle \mathbf{v} \cdot \nabla \times \mathbf{v} \rangle$ and the field **B**. The formal analysis of Steenbeck *et al.* (1966) and Rädler (1969) shows higher order terms contributing to the local electromotive force, as a consequence of gradients in the basic quantities of field and helicity (see also Gubbins 1974). Their results illustrate in a striking way the extraordinary variety of effects that contribute to the generation of magnetic field. Their work shows for turbulent velocities the fact, already pointed out by Roberts (1972) for steady velocities, that any break in the symmetry of the fluid motion produces at least some slight dynamo effect.

Unfortunately the construction of the higher order terms is based only on formal mathematical construction of combinations of the physical vectors with the proper geometrical properties for an electromotive force, such as $\mathbf{v} \times \mathbf{B}$. The scalar coefficients of the combinations are not known, so that the physical significance of the various effects cannot be judged. Hence in the development that follows we have gone back to the more powerful 'short–sudden' idealization (Parker 1955b) which leads to complete and explicit expressions for the effects at succeeding orders. The higher order contributions are presented in the formal expansions worked out in §18.3. We do not pursue them, however, beyond the second order contributions to the turbulent diffusion, producing the reduction, and possible reversal, of diffusion pointed out earlier by Kraichnan (1976a, b). The lowest order terms produce physical effects so complex as to merit our full attention at this stage of the development. Perhaps special circumstances, in which the lowest order effects vanish identically, will later motivate a study of the higher order terms neglected in the present writing. For the time being it is sufficient to limit the presentation to the zero order effect, wherein cyclonic convection deforms the local large-scale field into loops or spirals described by the dynamo equation (18.7). That effect alone takes us beyond the dynamical knowledge available on the fluid motions in astrophysical bodies.

As a matter of fact, theoretical studies of the dynamics of dynamos has come a long way in the last decades, even if the subject is still far from adequately developed. Braginskii (1964b, 1967, 1970) has worked out the wave properties of the fluid motions that make up his dynamo with near axial symmetry, noting the similarity to MAC waves and Alfven waves under geostrophic conditions. Moffatt (1970b, 1972) and Soward (1975) have treated a system of random inertial waves, solving the dynamical equations simultaneously with the induction equations to demonstrate a dynamo effect. Busse (1973) considered the convective dynamo including the reaction of the field back on the fluid motions (see also the work of Roberts and Stewartson 1974). Gilman (1969) has developed numerical models of the non-uniform rotation and local convective motion (in the form of Rossby waves) in the rotating sun including the Lorentz forces of the magnetic field generated by the motions. The resulting fields duplicate in remarkable degree the magnetic cycle of the sun (Parker 1955b, 1957, 1970a).

Dynamical effects can be included in a more elementary way in the calculations. Leighton (1969) pointed out that the magnetic buoyancy (Parker 1955a) provides a dynamical cut-off on the growth of the azimuthal magnetic field. He incorporated the effect into his numerical model for the solar dynamo where it plays an important regulatory role in the resulting magnetic cycle. Stix (1972), Deinzer and Stix (1971), and Yoshimura (1975) have also included the buoyant loss of the azimuthal field in the operation of the solar dynamo. Parker (1971) has shown that the magnetic buoyancy of the azimuthal field in both the sun and the gaseous disc of the galaxy leads to cyclonic convection and the generation of meridional field after the manner of (18.7). Thus, in principle, the only energy input need be the non-uniform rotation that generates the strong azimuthal field from the meridional field. Jepps (1975) has constructed a dynamo model that includes the suppression of the fluid motions by the growing magnetic field. Gilman (1972, 1974) has contributed to the theory of convection and non-uniform rotation in the sun in the absence of Lorentz forces. Unz and Walter (1969) have considered briefly the acceleration of the non-uniform rotation during the progress of the sunspot cycle, comparing the theoretical results with observations.

The development of the theoretical hydrodynamics of rotating, stratified, convective bodies continues at a rapid pace and is beyond the scope of this work. The reader is referred, therefore, to the concise but extensive mathematical reviews by Roberts and Soward (1975) and Soward and Roberts (1976) of the very difficult theoretical fluid dynamics of the hydromagnetic dynamo. Roberts and Soward (1975) demonstrate some of the difficulties encountered with the quasi-linear approximation when the calculations are extended to include the dynamics, showing the appearance of the fluid viscosity in the *denominator* of the effective rate of generation of field (cf. their eqn (4.26)) in the limiting case of *small* viscosity.

We pick up the generation of magnetic fields here at a much more elementary and mathematically secure level, demonstrating first (in §18.2) the generation of azimuthal field by the non-uniform rotation of a spherical body possessing a meridional field. This simple effect is responsible for the production of the principal magnetic field in planets, stars, and galaxies from the meridional field. We go on from there to demonstrate the generation of the meridional field (i.e. azimuthal vector potential) by the interaction of cyclonic turbulence with the azimuthal field. Sections §18.3–18.6 illustrate the effect in the various special cases that have so far been successfully formulated. The cases are restricted either to low or high magnetic Reynolds number for the turbulent (non-symmetric) component of the fluid motion. The formal analytical illustrations are limited in one way or another to low velocity in the individual eddies[5]. Moffatt (1974) has exhibited some of the formal difficulties with the convergence of certain integrals in the limit $t \to \infty$, that arise in the limit of large Reynolds number ($\eta = 0$). On the other hand Kraichnan's Monte Carlo calculations (based on $\eta = 0$), extending to $v_0 k_0 t = 4$, seem to show a well-behaved dynamo effect, with $\Gamma \cong 0{\cdot}6 v_0$. The calculations in §18.3 are also applicable to the high magnetic Reynolds number limit in a certain restricted sense described at the end of that section.

18.2. Generation of azimuthal magnetic field

In view of the fact that the principal magnetic field in a non-uniformly rotating body, such as Earth, the sun, or the galaxy, is the azimuthal field, it is worth while to consider in detail some simple illustrative examples of its generation. The azimuthal field cannot be observed in either the sun or the core of Earth, so our only knowledge of it is through the theory of its generation from the observed poloidal field. The situation is otherwise in the gaseous disc of the galaxy, where the azimuthal field is observed, but, unfortunately, the poloidal field from which it is generated, is not observed, so again a sound theoretical understanding of the basic principles is the best we can manage at the present time.

Consider, then, a poloidal field with rotational symmetry about the z-axis ($\partial/\partial\phi = 0$). In terms of the vector potential $\mathbf{A} = \mathbf{e}_\phi A = \mathbf{e}_\phi \Psi/\varpi = \mathbf{e}_\phi \Psi/r\sin(\theta)$ in cylindrical and spherical coordinates, the components of the poloidal field are

$$\varpi B_\varpi = -\partial\Psi/\partial z, \qquad \varpi B_z = +\partial\Psi/\partial\varpi$$

and

$$r\sin\theta B_r = +(1/r)\,\partial\Psi/\partial\theta, \qquad r\sin\theta B_\theta = -\partial\Psi/\partial r.$$

The lines of force are confined to the meridional plane and are given by $\Psi = constant$. Suppose, then, that at time $t = 0$ the azimuthal velocity $v_\phi(\varpi, z) = \varpi\Omega(\varpi, z)$ is switched on, for which the associated angular velocity $\Omega(\varpi, z)$ is not altogether uniform. The poloidal field is unaffected by v_ϕ, while the azimuthal component of the hydromagnetic equation (4.13) is

$$\left\{\frac{\partial}{\partial t} - \eta\left(\nabla^2 - \frac{1}{\varpi^2}\right)\right\} B_\phi = \frac{\partial(\Psi, \Omega)}{\partial(\varpi, z)} = \frac{1}{r}\frac{\partial(\Psi, \Omega)}{\partial(\theta, r)}. \tag{18.8}$$

[5] The exceptions are the effect pointed out by Soward and Roberts (1976) in their eqn (2.47b) and the fast dynamo of Vainshtein (1970) and Vainshtein and Zeldovich (1972).

Thus, insofar as the angular velocity Ω varies along the lines of force $\Psi = constant$, i.e. insofar as $\Omega \neq \Omega(\Psi)$, the non-uniform angular velocity produces an azimuthal field by shearing the poloidal field.

The initial growth of the azimuthal field is described by

$$B_\phi = \{\partial(\Psi, \Omega)/\partial(\varpi, z)\}t$$

until resistive dissipation becomes important. After a period of time of the general order of the resistive decay time λ^2/η of the region as a whole, the azimuthal field approaches an asymptotic state in which resistive dissipation balances the production of B_ϕ by the shearing of the poloidal field. Consider some simple examples to illustrate the resulting strength and distribution of B_ϕ.

18.2.1. Non-uniform rotation with cylindrical symmetry ($\Omega = \Omega(\varpi)$)

As a first example suppose that the angular velocity is a function only of the distance ϖ from the z-axis, as suggested by the researches of Busse (1970a, b) into the linear dynamics of convection in a rotating space. Then, for the simplest case, in the sphere $r < a$, let

$$\Omega = \Omega_0 - \Omega_1(\varpi/a)^m \tag{18.9}$$

where m is a positive number. If the fluid rotates less rapidly with distance ϖ from the axis, as we have imagined in some of the exposition, then Ω_1 is positive. Suppose that beyond the spherical surface the conductivity of the medium in negligible, as in the silicate mantle of Earth, or the medium is so tenuous as to offer no resistance to the escape of B_ϕ (see Chapter 13); then B_ϕ vanishes at $r = a$ and beyond. Suppose that the poloidal field has the simple form

$$\Psi = \tfrac{1}{2}B_p\varpi^2(1 - z^2/b^2),$$

$$B_\varpi = B_p\varpi z/b^2, \qquad B_z = B_p(1 - z^2/b^2),$$

essentially like that sketched in Fig. 18.2. Then (18.8) becomes

$$\left(\frac{\partial^2}{\partial\varpi^2} + \frac{1}{\varpi}\frac{\partial}{\partial\varpi} - \frac{1}{\varpi^2} + \frac{\partial^2}{\partial z^2}\right)B_\phi = -\frac{m\Omega_1 B_p a z}{b^2\eta}\left(\frac{\varpi}{a}\right)^{m+1}.$$

The solution to the inhomogeneous equation is readily shown to be

$$B_\phi = -\{mB_p\Omega_1 a^3/b\eta(m+2)(m+4)\}(z/b)(\varpi/a)^{m+3}.$$

To this must be added a solution of the homogeneous equation such that the total will vanish at $r = a$. In spherical coordinates the solution of the homogeneous equation that is finite throughout the interior of $r = a$ is

$$\sum_{n=0}^{\infty} C_n r^n P_n^1(\cos\theta).$$

Writing the solution to the inhomogeneous equation in spherical coordinates, the complete solution can be written

$$B_\phi = -\frac{mB_p\Omega_1 a^4}{(m+2)(m+4)b^2\eta}\left\{\left(\frac{r}{a}\right)^{m+4}\cos\theta\,\sin^{m+3}\theta + \sum_{n-1}^{\infty} D_n\left(\frac{r}{a}\right)^n P_n^1(\cos\theta)\right\}. \tag{18.10}$$

The coefficients D_n are to be determined from the boundary condition that $B_\phi = 0$ at $r = a$. Hence

$$0 = \cos\theta\,\sin^{m+3}\theta + \sum_{n-1}^{\infty} D_n P_n^1(\cos\theta).$$

The associated Legendre polynomials are orthogonal over the range $\theta = 0, \pi$, with

$$\int_{-1}^{+1} d\mu\{P_n^1(\mu)\}^2 = \frac{2n(n+1)}{2n+1}.$$

Hence

$$D_n = -\frac{2n+1}{2n(n+1)}\int_{-1}^{+1} d\mu\,\mu(1-\mu^2)^{\frac{1}{2}(m+3)}P_n^1(\mu).$$

The polynomials $P_n^1(\mu)$ are odd functions of μ when n is even, and even functions when n is odd. Hence the integral vanishes for odd n and the even-numbered coefficients are

$$D_{2n} = -\frac{4n+1}{4n(2n+1)}\int_{-1}^{+1} d\mu\,\mu(1-\mu^2)^{\frac{1}{2}(m+3)}P_{2n}^1(\mu). \tag{18.11}$$

Thus B_ϕ is an odd function of z (latitude)[6], for the obvious physical reason that the radial component B_ϖ, from which it is generated, is an odd function of z.

For $m = 1$ the non-uniform rotation $\partial\Omega/\partial\varpi$ is uniform across the radius. In this case the series converges only very slowly, with the first three coefficients given by

$$\begin{aligned} D_2 &= -25\pi/2^9 = -0{\cdot}1533, \\ D_4 &= -D_2\times 81/200 = -0{\cdot}4050D_2, \\ D_6 &= +D_2\times 13/320 = +0{\cdot}0406D_2. \end{aligned} \tag{18.12}$$

Concentrating the non-uniform rotation somewhat toward the outer boundary by choosing $m = 2$, the series terminates after three terms so

[6] Note that $P_2^1 = 3\mu(1-\mu^2)^{\frac{1}{2}}$; $P_4^1 = \frac{1}{2}\mu(35\mu^2 - 15)(1-\mu^2)^{\frac{1}{2}}$, $P_6^1 = (21\mu/8)(33\mu^4 - 30\mu^2 + 5)(1-\mu^2)^{\frac{1}{2}}$, etc.

that the solution is a polynomial, with

$$D_2 = -8/63,$$
$$D_4 = +24/385, \qquad (18.13)$$
$$D_6 = -8/693.$$

There are several points of interest. First of all, note that the final expression for B_ϕ is not separable into a product of a function of ϖ and a function of z, nor of r and θ. The reason is that the shear, producing B_ϕ, has cylindrical symmetry, while the outer boundary, through which the lines of force escape as a consequence of resistive diffusion, is spherical. The mixture of cylindrical and spherical geometries precludes a simple separable result in either cylindrical or spherical coordinates. It is evident, therefore that the solution should be simpler when the angular velocity is a function of r rather than of ϖ or of z. The circumstance $\Omega = \Omega(r)$ is treated in §18 2.3, where in simple cases, the final form for B_ϕ is seen to be separable in r and θ. We point out this clumsy aspect of the solutions in mixed cylindrical and spherical geometries because it arises again in Chapter 19 in solutions of the dynamo equations, of which the azimuthal equation (18.8) is an intimate part.

Note, too, the very slow convergence of the sum (18.12) $m = 1$ where the shear $\partial\Omega/\partial\varpi$ is spread uniformly over the radius. This is but one example of the general slow convergence of any expansion of B_ϕ in tesseral harmonics. Only for m equal to an even integer is the series terminated, after $m+1$ terms.

The radial dependence of B_ϕ is plotted in Fig 18.6 for $\theta = \pi/4$, shown by the solid curve labelled (18.13), in units of $-B_0\Omega_1 a^4/12b^2\eta$. Note the displacement of the maximum toward $r = a$, largely as a consequence of the active diffusive dissipation of the opposite azimuthal fields facing each across the z-axis. These effects are shown most dramatically by the examples given in §18.2.3. The angular distribution of B_ϕ is shown in Fig. 18.7 where the profile is plotted for $r = 0{\cdot}75a$.

18.2.2. Non-uniform rotation with plane symmetry $(\Omega = \Omega(z))$

As a second example suppose that the angular velocity is a function only of the distance z from the equatorial plane. Then, for the simplest case, let

$$\Omega = \Omega_0 + \Omega_1(z/a)^m \qquad (18.14)$$

for $r < a$. If Ω is to be an even function of z, as we would expect in most astrophysical bodies, then m is restricted to even integers. The parameter Ω_1 is positive if the fluid rotates more rapidly near the surface $r = a$ at high latitudes than at low latitudes, and negative otherwise. The simple

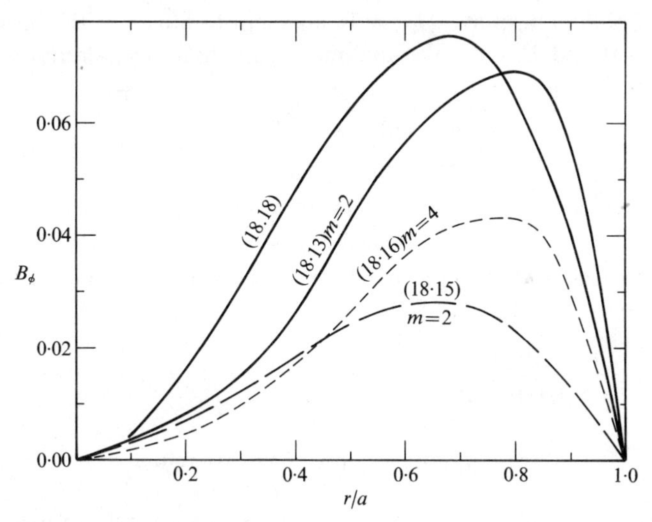

FIG. 18.6. The radial variation of B_ϕ at $\theta = \pi/4$. The solid curve labelled (18.13) represents B_ϕ in units of $-B_p\Omega_1 a^4/12b^2\eta$ from the coefficients (18.13) in (18.10). The dashed curve represents B_ϕ in units of $+B_p\Omega_1 a^2/3\eta$ plotted from (18.15) for $m=2$ in (18.14). The dotted curve represents B_ϕ in units of $+B_p\Omega_1 a^2/5\eta$ plotted from (18.16), for $m=4$ in (18.14). The solid curve labelled (18.18) represents B_ϕ in units of $-B_p\Omega_1 a^2/4\eta$ plotted from (18.18) for $m=1$ in (18.17).

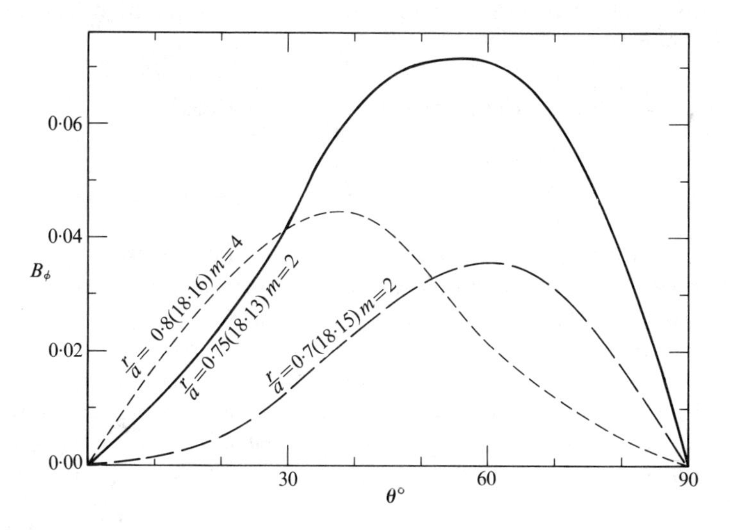

FIG. 18.7. The angular variation of B_ϕ at fixed r/a. The solid curve labelled (18.13) represents B_ϕ in units of $-B_p\Omega_1 a^4/12b^2\eta$ from the coefficients (18.13) in (18.10) at the fixed radius $r/a = 0{\cdot}75$. The dashed curve represents B_ϕ in units of $+B_p\Omega_1 a^2/3\eta$ plotted from (18.15), for $m=2$ in (18.14), at the fixed radius $r/a = 0{\cdot}7$. The dotted curve represents B_ϕ in units of $+B_p\Omega_1 a^2/5\eta$ plotted from (18.16) for $m=4$ in (18.14) at the fixed radius $r/a = 0{\cdot}8$.

uniform poloidal field $\Psi=\frac{1}{2}B_p\varpi^2$ is sufficient to illustrate the basic effects, with $B_\varpi=0$ and $B_z=B_p$ throughout $r\leqslant a$. Then (18.5) reduces to

$$\left(\frac{\partial^2}{\partial\varpi^2}+\frac{1}{\varpi}\frac{\partial}{\partial\varpi}-\frac{1}{\varpi^2}+\frac{\partial^2}{\partial z^2}\right)B_\phi=-\frac{B_p\Omega_1 m\varpi z^{m-1}}{\eta a^m}.$$

The solution to the inhomogeneous equation is

$$B_\phi=-\frac{B_p\Omega_1 a^2}{(m+1)\eta}\frac{\varpi}{a}\left(\frac{z}{a}\right)^{m+1}.$$

To this must be added solutions of the homogeneous equation so that B_ϕ vanishes at $r=a$. Thus

$$B_\phi=-\frac{B_p\Omega_1 a^2}{(m+1)\eta}\left\{\left(\frac{r}{a}\right)^{m+1}\sin\theta\cos^{m+1}\theta+\sum_{n=1}^{\infty}D_n\left(\frac{r}{a}\right)^n P_n^1(\cos\theta)\right\}$$

and

$$0=\sin\theta\cos^{m+1}\theta+\sum_{n=1}^{\infty}D_n P_n^1(\cos\theta),$$

as in the previous example. For $m=2$, it follows that $D_2=-1/7$, $D_4=-2/35$, with all other D_n equal to zero. The complete solution is, then,

$$B_\phi=+\frac{B_p\Omega_1 a^2}{3\eta}\left\{\frac{3}{7}-\frac{r}{a}\cos^2\theta+\frac{r^2}{a^2}(\cos^2\theta-\tfrac{3}{7})\right\}\left(\frac{r}{a}\right)^2\sin\theta\cos\theta. \tag{18.15}$$

For a non-uniform rotation more concentrated toward high latitude; put $m=4$. Then

$$B_\phi=+\frac{B_p\Omega_1 a^2}{5\eta}\left\{\tfrac{5}{21}+\tfrac{2}{77}(35\cos^2\theta-15)\left(\frac{r}{a}\right)^2-\left(\frac{r}{a}\right)^3\cos^4\theta\right.$$
$$\left.+\tfrac{1}{33}(33\cos^4\theta-30\cos^2\theta+5)\left(\frac{r}{a}\right)^4\right\}\left(\frac{r}{a}\right)^2\sin\theta\cos\theta. \tag{18.16}$$

Once again we note that B_ϕ does not separate into a function of r and a function of θ. The radial profile of B_ϕ is plotted in Fig. 18.6 for $\theta=\pi/4$ using both (18.15) and (18.16). The maximum of (18.16) lies at the larger radius because the source is concentrated toward larger radius. The angular profiles are shown in Fig. 18.7, from (18.15) for $r=0.7a$ and from (18.16) for $r=0.8a$. The profiles are different at other r, but these values suffice to show the general shape. The maximum B_ϕ lies toward smaller θ in (18.16) because the source is concentrated at large z.

18.2.3. Non-uniform rotation with spherical symmetry ($\Omega=\Omega(r)$)

Suppose that Ω is a function only of radial distance from the origin. Then, each layer $r=a$ rotates rigidly. For the simplest case

$$\Omega=\Omega_0+\Omega_1(r/a)^m,\qquad \Psi=\tfrac{1}{2}B_p r^2\sin^2\theta, \tag{18.17}$$

the hydromagnetic equation (18.8) for the steady state reduces to

$$\frac{1}{r^2}\left\{\frac{\partial}{\partial r}r^2\frac{\partial}{\partial r}+\frac{1}{\sin\theta}\frac{\partial}{\partial\theta}\sin\theta\frac{\partial}{\partial\theta}-\frac{1}{\sin^2\theta}\right\}B_\phi=-\frac{B_p m\Omega_1}{\eta}\left(\frac{r}{a}\right)^m\sin\theta\cos\theta.$$

The solution to the inhomogeneous equation is

$$B_\phi=-\frac{mB_p\Omega_1 a^2}{\eta(m^2+5m-2)}\left(\frac{r}{a}\right)^{m+2}\sin\theta\cos\theta.$$

Including the solution of the homogeneous equation so that B_ϕ vanishes at $r=a$, it follows that

$$B_\phi=-\frac{mB_p\Omega_1 a^2}{\eta(m^2+5m-2)}\left\{\left(\frac{r}{a}\right)^m-1\right\}\left(\frac{r}{a}\right)^2\sin\theta\cos\theta. \tag{18.18}$$

With the spherical symmetry of the non-uniform rotation, the resistive dissipation and the production by shear have the same form, so that B_ϕ has the same angular dependence as the shearing of the poloidal field and is separable in the coordinates r and θ. The radial profile of B_ϕ, in units of $-B_p\Omega_1 a^2/4\eta$, calculated from (18.18) is shown in Fig. 18.6 for $m=1$ at $\theta=\pi/4$ by the solid curve labelled (18.18). The angular dependence is just $\sin(2\theta)$ and is not plotted in Fig. 18.7.

As another example suppose that the fluid rotates with fixed angular velocity throughout $0<r<b<a$, and with a fixed value in $b<r<a$ that is larger by Ω_1. Then

$$\partial\Omega/\partial r=\Omega_1\delta(r-b). \tag{18.19}$$

The inhomogeneous term on the right-hand side of (18.8) is then

$$B_p\Omega_1 r\sin\theta\cos\theta\delta(r-b),$$

and the solution of (18.8) is the radial Green's function. The inhomogeneous term vanishes except at $r=b$. Hence, except at $r=b$, B_ϕ is given by a solution of the homogeneous equation, $r^nP_n^1(\cos\theta)$ or $r^{-n-1}P_n^1(\cos\theta)$. The solutions on either side of $r=b$ must be continuous across $r=b$. Integrating (18.8) over r from $b-\epsilon$ to $b+\epsilon$, it follows in the limit of $\epsilon\to 0$ that there is a jump in $\partial B_\phi/\partial r$ in the amount

$$\partial B_\phi/\partial r\Big|_{b-\epsilon}^{b+\epsilon}=-B_p\Omega b\delta(r-b)\sin\theta\cos\theta. \tag{18.20}$$

It is readily shown, then, that the desired solution in $r<b$ is

$$B_\phi=+B_p\frac{\Omega_1 a^2}{5\eta}\left(\frac{b}{a}\right)^5\left\{\left(\frac{a}{b}\right)^5-1\right\}\left(\frac{r}{a}\right)^2\sin\theta\cos\theta \tag{18.21}$$

while in $b<r<a$,

$$B_\phi = +\frac{B_p\Omega_1 a^2}{5\eta}\left(\frac{b}{a}\right)^5\left\{\left(\frac{a}{r}\right)^3 - \left(\frac{r}{a}\right)^2\right\}\sin\theta\cos\theta. \tag{18.22}$$

The angular dependence of B_ϕ is just $\sin(2\theta)$ again. The radial profile is plotted in Fig. 18.8 in units of $B_p\Omega_1 a^2/5\eta$ for various values of b/a. The peak B_ϕ occurs at the source $r=b$, and is

$$B_{\phi\,\text{peak}} = B_p(\Omega_1 a^2/5\eta)(b/a)^2\{1-(b/a)^5\}\sin\theta\cos\theta.$$

The field is strongest when the source is located so as to balance the losses at the outer boundary $r=a$ against diffusion across $\varpi=0$ into the opposite B_ϕ on the other side. This occurs for $b/a=(2/7)^{\frac{1}{5}}=0{\cdot}779$ and is shown in Fig. 18.8. Note how rapidly the field declines for a given $B_p\Omega_1 a^2/\eta$ as b/a becomes small, or approaches one.

These examples suffice to illustrate azimuthal fields produced by non-uniform rotation in the presence of poloidal (meridional) fields in simple form. The calculations appear again in the following chapter where simultaneous solution of (18.8) with the dynamo equation (18.7) is effected for a variety of cases (see also the examples given by Nakagawa and Swarztrauber 1969).

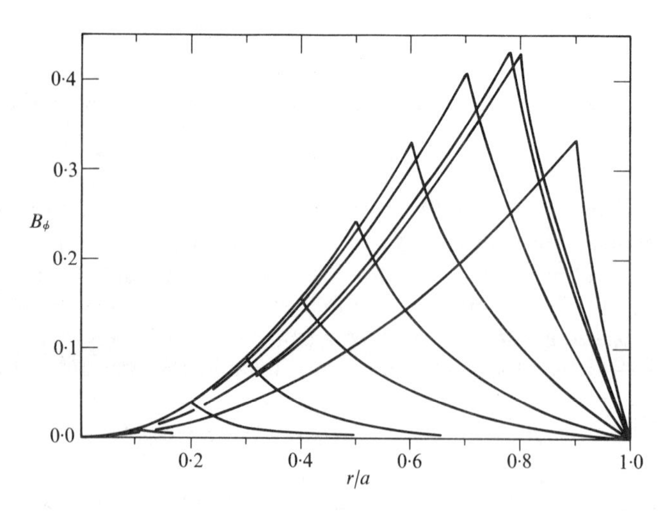

FIG. 18.8. The radial profile of B_ϕ in units of $+B_p(\Omega_1 a^2/5\eta)\sin\theta\cos\theta$ plotted from (18.21) and (18.22) for $b/a=0{\cdot}1, 0{\cdot}2, \ldots, 0{\cdot}9$. The strongest B_ϕ occurs for $b/a=0{\cdot}779$, shown by the curve with its peak at $r=0{\cdot}779a$.

18.3. Generation of meridional field by cyclonic turbulence

The generation of meridional field through spiralling of the azimuthal lines of force was described in §18.1 leading to the dynamo equations (18.6) and (18.7). The spiralling is produced by the rotation of indentations in the azimuthal field carried in cyclonic eddies (convective cells) in the fluid motion. The purpose of the present section is to derive the dynamo equations formally, giving explicit recipes for the calculation of Γ from the fluid motions. We will use the idealized fluid motion employed in the first paper on the generation of meridional field (Parker 1955b) wherein the cyclonic convective cells, of small scale (correlation length) L, are switched on for a short period $\tau_1 \ll L^2/\eta$ so that the local, strong deformation of the large-scale field can be computed using (4.32) and (4.34). The motion is then switched off for a period τ_2 large compared to the local diffusion time L^2/η so that the local deformations blend into their neighbours, leaving only the large-scale field when the next round of eddies is switched on again for the brief period τ_1. Rather than deducing the magnetic fields with the aid of (4.32) and extracting the vector potential from those fields (Parker 1955b), it is much easier to compute the vector potential directly from (4.34) (Parker 1970b), and that procedure will be used here.

18.3.1. The dynamo equations

The induction equation for the vector potential is readily obtained from (4.14) by writing $\mathbf{B} = \nabla \times \mathbf{A}$ and integrating out one curl operation, as was done in the development of (4.33). With the gauge $\phi = v_j A_j$ and the fact that the divergence of A_i is then not conserved, and, hence, generally not equal to zero, it is readily shown that

$$\left\{\delta_{in}\left(\frac{\partial}{\partial t} - \eta\nabla^2\right) + \eta\frac{\partial^2}{\partial x_i\,\partial x_n}\right\}A_n = -A_j\frac{\partial v_j}{\partial x_i} - v_j\frac{\partial A_i}{\partial x_j}. \qquad (18.23)$$

In the present circumstance the fluid velocity contains a large-scale velocity denoted by V_i, representing the axi-symmetric non-uniform rotation of most astrophysical bodies, and perhaps meridional circulation. The magnitude of V_i is denoted by V, and the scale is presumably comparable to the dimension λ of the entire body. The characteristic time T for variation of V_i is presumed to be not less than either λ/V or λ^2/η. If the small-scale turbulent velocity, switched on and off according to the prescription already noted, is represented by u_i, then the total fluid velocity v_i is

$$v_i = V_i + u_i. \qquad (18.24)$$

Consider an ensemble of systems, all with the same V_i, of course, and all with the same statistical properties for the small-scale velocity u_i, but with the location of the individual eddies uncorrelated among the members of the ensemble. Thus the ensemble average of u_i at any point is zero because there is no net displacement of fluid by the small-scale turbulence. Note that the turbulent displacement of fluid $\int dtu_i$ may be as large or larger than the eddy size L. Write the vector potential as $A_i(x_k, t)+a_i(x_k, t)$ where A_i denotes the large-scale ensemble average field and a_i represents the local fluctuations produced by the turbulent velocity u_i. The contribution a_i is defined to be that part of the vector potential with zero mean.

Now suppose that at time $t=t_0$ the local fluctuation a_i is zero in all members of the ensemble. A turbulent velocity is switched on simultaneously for a brief instant τ_1 in each ensemble, with u_i so large as to produce displacements of the general order of the correlation length L. The velocity fields are statistically identical but uncorrelated in detail among the members of the ensemble. In the limit of small τ_1 both the diffusion η and the large-scale velocity V_i have vanishing effect, $O(\tau_1)$, over the period τ_1, so that (18.23) reduces to

$$\frac{\partial}{\partial t}(A_i+a_i)=-(A_j+a_j)\frac{\partial u_j}{\partial x_i}-u_j\frac{\partial(A_i+a_i)}{\partial x_j}. \tag{18.25}$$

The solution is given by (4.34) as

$$A_i(x_k, t_0+\tau_1)+a_i(x_k, t_0+\tau_1)=A_j(X_k, t_0)\,\partial X_j/\partial x_i \tag{18.26}$$

where x_k is the coordinate at time $t_0+\tau_1$ of the fluid element at X_k at time t_0. The displacement of the fluid element during the time τ_1 is $\xi_k=x_k-X_k$, which may be $O(L)$ in magnitude.

In terms of the Lagrangian displacement ξ_i of the element of fluid write

$$A_i(X_k, t_0)=A_i(x_k, t_0)-\xi_j\,\partial A_i/\partial x_j +\frac{1}{2!}\xi_j\xi_k\frac{\partial^2 A_i}{\partial x_j\,\partial x_k}+\dots. \tag{18.27}$$

We retain the terms to all orders for the present, but note that the convergence of the series is rapid provided only that the displacement ξ_i is small compared to the characteristic scale of variation λ of the mean large-scale field A_i. It also follows that

$$\partial X_k/\partial x_j=\delta_{jk}-\partial\xi_k/\partial x_j. \tag{18.28}$$

Substituting into (18.26) the result is

$$A_i(x_k, t_0+\tau_1)+a_i(x_k, t_0+\tau_1)=(\delta_{ij}-\partial\xi_j/\partial x_i) \times\Big[A_j(x_k, t_0)-\xi_k\,\partial A_j/\partial x_k+\frac{1}{2!}\xi_k\xi_l\frac{\partial^2 A_i}{\partial x_k\,\partial x_l} -\frac{1}{3!}\xi_k\xi_l\xi_m\frac{\partial^3 A_i}{\partial x_k\,\partial x_l\,\partial x_m}+\dots \tag{18.29}$$

to all orders. The mean value of this relation yields the increment $\Delta A_i(x_k, t_0+\tau_1)$ in the large-scale field given by

$$\Delta A_i(x_k, t_0+\tau_1) \equiv A_i(x_k, t_0+\tau_1) - A_i(x_k, t_0) \tag{18.30}$$

$$= -A_j(x_k, t_0)\langle \partial\xi_i/\partial x_j\rangle + \{\langle \xi_k\, \partial\xi_j/\partial x_i\rangle - \delta_{ij}\langle \xi_k\rangle\}\, \partial A_j/\partial x_k$$

$$-\frac{1}{2!}\{\langle \xi_k\xi_l\, \partial\xi_j/\partial x_j\rangle - \delta_{ij}\langle \xi_k\xi_l\rangle\}\, \partial^2 A_j/\partial x_k\, \partial x_l + \ldots. \tag{18.31}$$

This is the jump in the mean field $A_i(x_k)$ over the brief period τ_1 of active turbulence.

We are concerned with turbulence for which there is no mean velocity $\langle u_i\rangle = 0$. Hence the mean displacement is zero and

$$\langle \partial\xi_j/\partial x_i\rangle = 0, \tag{18.32}$$

$$\langle \xi_i\xi_j\xi_k\rangle = 0. \tag{18.33}$$

Equation (18.31) reduces to

$$\Delta A_i(x_k, t_0+\tau_1) = \langle \xi_k\, \partial\xi_j/\partial x_i\rangle\, \partial A_j/\partial x_k$$

$$+\frac{1}{2!}\{\delta_{ij}\langle \xi_k\xi_l\rangle - \langle \xi_k\xi_l\, \partial\xi_j/\partial x_i\rangle\}\, \partial^2 A_j/\partial x_k\, \partial x_l$$

$$+\frac{1}{3!}\langle \xi_k\xi_l\xi_m\, \partial\xi_j/\partial x_i\rangle\, \partial^3 A_j/\partial x_k\, \partial x_l\, \partial x_m + \ldots. \tag{18.34}$$

Note that with ξ_k of the order of L the increment ΔA_i given by (18.34) is not larger in magnitude than $O(A_iL/\lambda)$. The increment is small even for strong displacement, $u\tau_1 \sim L$. For ξ_k comparable to L and varying over x_k with a characteristic scale L, the local variation b_i in the mean field B_i is comparable in magnitude to B_i. Since a_i is of the order of Lb_i and A_i is of the order of λB_i, it follows in this case that $a_i = O(A_iL/\lambda)$. Hence the mean increment ΔA_i is comparable to the local magnitude of a_i. Both are small compared to A_i.

Following the period of active turbulence the fluid is held motionless for a time τ_2 large compared to the characteristic diffusion time L^2/η for the small-scale field a_i, although short compared to the large-scale times λ/V and λ^2/η. During the period τ_2 the small-scale field a_i decays, so that at the end of that time, when the next round of eddies is switched on, there remains only a large-scale field A_i. The cycle is then repeated indefinitely. During the waiting period τ_2 the large-scale field follows from (18.23) as

$$\left\{\delta_{in}\left(\frac{\partial}{\partial t} - \eta\nabla^2\right) + \eta\frac{\partial^2}{\partial x_i\, \partial x_n}\right\}A_n = -A_j\frac{\partial V_j}{\partial x_i} - V_j\frac{\partial A_i}{\partial x_j}. \tag{18.35}$$

Over the entire period $\tau = \tau_1 + \tau_2$ the total change in the large-scale field is ΔA_i, given by (18.34), plus the large-scale evolution (18.35). The period τ is small compared to the times λ/V and λ^2/η over which A_i evolves, and the increment ΔA_i is small, $O(A_i L/\lambda)$. Hence the average evolution of A_i reduces to

$$\left\{\delta_{in}\left(\frac{\partial}{\partial t} - \eta\nabla^2\right) + \eta\frac{\partial^2}{\partial x_i\,\partial x_j}\right\}A_n = -V_j\frac{\partial A_i}{\partial x_j} - A_j\frac{\partial V_j}{\partial x_i} + \frac{\Delta A_i}{\tau} \tag{18.36}$$

in the limit of small L/λ. With the explicit form (18.34) for ΔA_i the equation can be written

$$\begin{aligned}\frac{\partial A_i}{\partial t} + \eta\frac{\partial}{\partial x_i}\frac{\partial A_n}{\partial x_n} = {} & -A_j\frac{\partial V_j}{\partial x_i} - V_j\frac{\partial A_i}{\partial x_j} + \frac{\langle \xi_k\,\partial\xi_j/\partial x_i\rangle}{\tau}\frac{\partial A_j}{\partial x_k}\\ & + \left\{\eta\,\delta_{kl} + \frac{\delta_{ij}\langle \xi_k\xi_l\rangle - \langle \xi_k\xi_l\,\partial\xi_j/\partial x_i\rangle}{2!\,\tau}\right\}\frac{\partial^2 A_j}{\partial x_k\,\partial x_l}\\ & + \frac{\langle \xi_k\xi_l\xi_m\,\partial\xi_j/\partial x_i\rangle}{3!\,\tau}\frac{\partial^3 A_j}{\partial x_k\,\partial x_l\,\partial x_m} + \ldots.\end{aligned} \tag{18.37}$$

We should not fail to point out that these same dynamo equations arise in the limit of $\eta = 0$. In that case there is no diffusion, so that (4.35) is applicable for all values of t. The time τ in the coefficient in (18.37) is then just the total length of time that the fluid is in motion to produce the mean displacement $\langle \xi_k\,\partial\xi_j/\partial x_i\rangle$ etc. The expansion (18.27) eventually becomes invalid, of course, when the displacement becomes a significant fraction of the scale λ of the large-scale field. Hence (18.37) is valid for times small compared to the long time λ^2/Lv in which an element of the turbulent fluid random walks a distance λ. It is the case already treated by Kraichnan (1967a, b) using Monte Carlo techniques, and the reader is referred to Fig. 1 of Kraichnan (1976b) for numerical values of the dynamo coefficient Γ.

18.3.2. The dynamo coefficients

Having derived the general dynamo equation, the next question is under what circumstances the coefficients on the right-hand side are non-vanishing. It is one thing to derive a general form, and another to demonstrate that it is non-trivial. The coefficient $\langle \xi_k\,\partial\xi_j/\partial x_i\rangle$ of $\partial A_j/\partial x_k$ represents the basic dynamo effect. The next term $\partial^2 A_j/\partial x_k\,\partial x_l$ represents turbulent diffusion, etc.

In isotropic turbulence with mirror symmetry, the lowest order coefficient $\langle \xi_i\,\partial\xi_j/\partial x_k\rangle$ vanishes. In that case the basic dynamo effect disappears, leaving turbulent diffusion as the lowest term. On the other hand, $\langle \xi_k\xi_l\,\partial\xi_j/\partial x_i\rangle$ need not vanish, as is demonstrated in 18.3.3. Kraichnan (1976b) has shown from Monte Carlo calculations the important fact that the turbulent diffusion coefficient may be negative in mirror-symmetric turbulence with long-lived eddies (see §17.4). The turbulent diffusion coefficient on the right-hand side of (18.37) is

$$\{\langle\xi_k\xi_l\rangle\delta_{ij}-\langle\xi_k\xi_l\,\partial\xi_j/\partial x_i\rangle\}/2\tau,$$

the first term of which is positive definite. Hence the coefficient is negative if and only if $\langle\xi_k\xi_l\,\partial\xi_j/\partial x_i\rangle$ is suitably large and positive. Examples are given in §18.4 to illustrate the effect of $\langle\xi_k\xi_l\,\partial\xi_j/\partial x_i\rangle$ in reducing the diffusion coefficient to negative values.

When the requirement of mirror symmetry is dropped, so that there is a net cyclonic rotation (helicity) of the eddies, the coefficient $\langle\xi_i\,\partial\xi_j/\partial x_k\rangle$ has a number of non-vanishing components, so that there is a net dynamo effect. The magnetic field is regenerated with any V_i containing significant shear, such as non-uniform rotation. Indeed, Krause and Steenbeck (1967) point out that V_i can be set equal to zero and the dynamo equations yield regenerative solutions on the basis of cyclonic turbulence alone. The dynamo action is then not as efficient as when there is strong non-uniform rotation to produce the azimuthal field, because the cyclonic eddies by themselves are extremely wasteful of field, producing so much small-scale resistive dissipation. But Krause and Steenbeck establish an important physical principle that large-scale motion is not necessary to produce large-scale field. Small-scale motions are sufficient in the presence of helicity. Then in the presence of mirror symmetry (vanishing net helicity) there is Kraichnan's point that turbulence can concentrate rather than disperse, the existing field, even if it cannot generate new flux. Finally, it should be noted that the dynamo action of steady fluid motions in the absence of small-scale turbulence, has been amply illustrated by Herzenberg (1958), Tverskoy (1965), Gailitis (1967, 1969, 1970), Childress (1967a, b), Lortz (1968a, b, 1972), and Roberts (1972). These points all bear on the old question of the net effect of turbulent fluid motion on magnetic field. It is clear that there is not a simple growth toward equipartition of energy between the velocity and magnetic field, as originally postulated by Alfven (1947) and Biermann and Schlüter (1951) (see also Biermann 1953; Chandrasekhar 1955). It is also clear that there is not necessarily a simple dispersal of the magnetic field by turbulent diffusion, as there is with a scalar field. Kraichnan (1976a, b) has emphasized that the problem is too complex for any of the present formal statistical approximations to produce a general mathematical theory, such as the direct interaction approximation is able to do for hydrodynamic turbulence. Hence, idealized examples, to illustrate the effects, are the most potent tool available at the present time.

Consider the lowest order term $(\partial A_j/\partial x_k)\langle\xi_k\,\partial\xi_j/\partial x_i\rangle$ in the dynamo equation (18.37). It is the basis for the dynamo effect produced by cyclonic convection and turbulence (Parker 1955b). Writing out the individual terms, it is readily seen that

$$\frac{\partial A_j}{\partial x_k}\left\langle \xi_k \frac{\partial \xi_j}{\partial x_i}\right\rangle = B_1\left\langle \xi_2 \frac{\partial \xi_3}{\partial x_i}\right\rangle + B_2\left\langle \xi_3 \frac{\partial \xi_1}{\partial x_i}\right\rangle + B_3\left\langle \xi_1 \frac{\partial \xi_2}{\partial x_i}\right\rangle$$
$$+\frac{1}{2}\left\{\frac{\partial A_1}{\partial x_1}\frac{\partial\langle \xi_1^2\rangle}{\partial x_i}+\frac{\partial A_2}{\partial x_2}\frac{\partial\langle \xi_2^2\rangle}{\partial x_i}+\frac{\partial A_3}{\partial x_3}\frac{\partial\langle \xi_3^2\rangle}{\partial x_i}\right\}$$
$$+\frac{\partial A_1}{\partial x_2}\frac{\partial}{\partial x_i}\langle \xi_1\xi_2\rangle+\frac{\partial A_2}{\partial x_3}\frac{\partial}{\partial x_i}\langle \xi_2\xi_3\rangle+\frac{\partial A_3}{\partial x_1}\frac{\partial}{\partial x_i}\langle \xi_3\xi_1\rangle, \quad (18.38)$$

where B_i is the large-scale field represented by the vector potential A_i, with $B_1 = \partial A_3/\partial x_2 - \partial A_2/\partial x_3$, etc. In mirror-symmetric, statistically homogeneous isotropic turbulence all the coefficients in (18.38) are zero, of course. Introducing a net helicity leads to the non-vanishing of $\langle \partial \xi_i\, \partial \xi_j/\partial x_k\rangle$ so that

$$\frac{\partial A_j}{\partial x_k}\left\langle \xi_k \frac{\partial \xi_j}{\partial x_i}\right\rangle = B_1\left\langle \xi_2 \frac{\partial \xi_3}{\partial x_i}\right\rangle + B_2\left\langle \xi_3 \frac{\partial \xi_1}{\partial x_i}\right\rangle + B_3\left\langle \xi_1 \frac{\partial \xi_2}{\partial x_i}\right\rangle. \quad (18.39)$$

When both mirror symmetry and homogeneity are given up, the general result is (18.38). The extra terms in (18.38), as compared to (18.39), represent gradients in the intensity of the turbulence and the helicity. Such coefficients as $\partial\langle \xi_1^2\rangle/\partial x_i$ are of the order $\langle \xi_1^2\rangle/\lambda$, where λ is the dimension of the convecting region and of the large-scale field produced therein. On the other hand, in the presence of strong helicity such coefficients as $\langle \xi_1\, \partial \xi_2/\partial x_i\rangle$ are of the order of $\langle \xi_1^2\rangle/L$, where L is the scale of the turbulence and is assumed to be small compared to λ. For this reason the form (18.39) is adequate for most purposes, the higher order terms $O(L/\lambda)$ being dropped in the present idealization of very small cyclonic cells.

The next step in the development, then, is to compute ξ_i and the various mean values $\langle \xi_i\, \partial \xi_j/\partial x_k\rangle$ etc. for specific cyclonic convective cells. A number of examples can be found in the literature (cf. Parker 1955b, 1970b). For the present purposes of exploration and illustration, it is convenient to restrict the discussion to idealized forms of cyclonic eddies to facilitate the calculation of various mean values.

In this most primitive form a cyclonic eddy, or convective cell, is composed of a rising (or sinking) body of fluid rotating about the local vertical direction and surrounded by the returning downward (upward) flow. There is no reason to make the rotation simultaneous with the vertical displacement of the fluid, nor for the rotation to overlap completely with the vertical motion in space. Therefore, to facilitate the computation, we separate for the moment the vertical and rotational components of the local cyclonic eddy in time.

18.3.3. *Illustrative models for dynamo coefficients*

Set up a local coordinate system (x, y, z) (having nothing to do with the global coordinate system employed in §18.1) so that the local cyclonic convective cell is oriented along the z-axis and rotates about the z-axis. We may as well assume rotational symmetry about the z-axis, so that the rotation can be described by an angular velocity $\Omega(\varpi, z)$ where $\varpi = (x^2+y^2)^{\frac{1}{2}}$ is distance measured from the axis of symmetry. It follows that the local rotational velocity is $v_\phi = \varpi\Omega$ and the total angular displacement from an initial azimuthal position ϕ_0 to a final position ϕ is

$$\phi = \phi_0 + \Omega(\varpi, z)t_2 \tag{18.40}$$

after a time t_2. Suppose that $\Omega(\varpi, z)$ is non-vanishing only in the cylindrical region $0 \leqslant \varpi < b$ and $-c < z < +c$ sketched in Fig. 18.9.

Describe the rising and sinking motions (the poloidal component of the fluid velocity) of the local convection in terms of the velocity components (v_ϖ, v_z), given by

$$v_\varpi = +\frac{1}{\varpi}\frac{\partial\psi}{\partial z}, \qquad v_z = -\frac{1}{\varpi}\frac{\partial\psi}{\partial\varpi} \tag{18.41}$$

for an incompressible flow. In keeping with the idealization of the localized convective cell, the stream lines $\psi = constant$ are presumed closed, in the planes $\phi = constant$ through the z-axis. It is sufficient that they close in $z > c$ and $z < -c$ above and below the region of rotation.

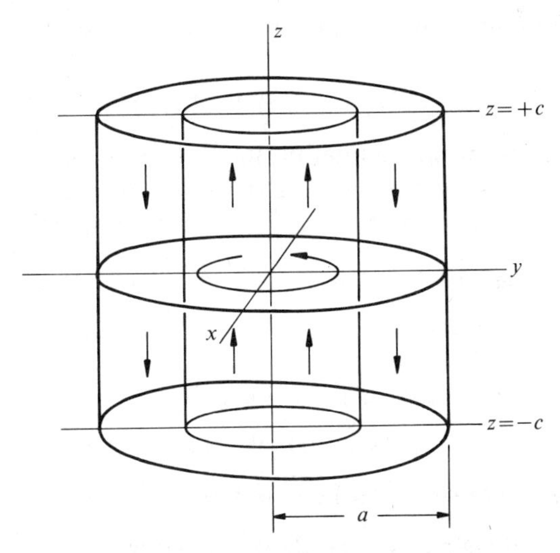

FIG. 18.9. A sketch of the portion of the cyclonic convective cell that is subject to rotation ($\varpi < a; -c < z < +c$). The vertical arrows indicate the poloidal circulation, while the circular arrow signifies the rotation.

Where the flow passes across the rotation, let

$$\psi = -\tfrac{1}{2}v_3\begin{cases}\varpi^2(1-\varpi/a)^2 & \text{for} \quad \varpi < a \\ 0 \quad \text{for} \quad \varpi > a\end{cases} \tag{18.42}$$

so that the velocity field is

$$v_\varpi = 0, \qquad v_z = v_3(1-\varpi/a)(1-2\varpi/a) \tag{18.43}$$

from the z-axis out to $\varpi = a$, and vanishes beyond. The rotation extends out to a radius b, which may be larger or smaller than a. If v_3 is taken as a positive quantity, then v_z is positive in $\varpi < \frac{1}{2}a$ with a peak value v_3 on the z-axis, and negative in $\frac{1}{2}a < \varpi < a$, where the minimum is $-\frac{1}{8}v_3$ at $\varpi = \frac{3}{4}a$. The net flow $2\pi\int \mathrm{d}\varpi \varpi v_z$ is zero. The fluid velocity vanishes at $\varpi = \frac{1}{2}a$ and a.

Suppose, then, that the cyclonic eddy consists of the local poloidal circulation (18.43) for a period t_3 followed by the rotation (18.40) for a period t_2, with $t_2 + t_3$ equal to the total life τ_1 of the cyclonic eddy employed in the discussion of §18.3.1. From all this is needed the Lagrangian displacement ξ_i of the element of fluid arriving at the position x_k after the eddy is finished, i.e. after the time $t_2 + t_3$, in order to compute the mean value $\langle \xi_i \, \partial\xi_j/\partial x_k \rangle$. The mean value is zero unless the element of fluid undergoes rotation. Hence the contribution to the mean comes only from those elements of fluid within the region of rotation $\varpi < b$, $-c < z < +c$. If we suppose that the flow (18.43) extends for a distance of at least $v_3 t_3$ outside the region of rotation, i.e. at least to $z = \pm(c + v_3 t_3)$, then all elements of fluid within the rotation region have undergone a displacement

$$\xi_z = v_3 t_3 (1-\varpi/a)(1-2\varpi/a) \tag{18.44}$$

in the z-direction from their initial position at $(\varpi, z-\xi_z)$. In addition to the displacement ξ_z, they have undergone the rotation (18.40) so that, if their initial position was $x_0 = \varpi \cos(\phi_0)$, $y_0 = \varpi \sin(\phi_0)$, their final position is

$$x = \varpi \cos(\phi_0 + \Omega t_2), \qquad y = \varpi \sin(\phi_0 + \Omega t_2),$$

and their displacements are

$$\begin{aligned}\xi_x &= x\{1-\cos(\Omega t_2)\} - y \sin(\Omega t_2) \\ &= \varpi[\cos\phi\{1-\cos(\Omega t_2)\} - \sin\phi \sin(\Omega t_2)],\end{aligned} \tag{18.45}$$

$$\begin{aligned}\xi_y &= y\{1-\cos(\Omega t_2)\} + x \sin(\Omega t_2) \\ &= \varpi[\sin\phi\{1-\cos(\Omega t_2)\} + \cos\phi \sin(\Omega t_2)].\end{aligned} \tag{18.46}$$

The derivatives $\partial\xi_i/\partial x_j$ are now readily computed, remembering that Ω is a function of both ϖ and z, falling to zero at $z = \pm c$. Thus

$$\begin{aligned}
\partial\xi_x/\partial x &= 1 - \cos(\Omega t_2) + \sin(\Omega t_2 - \phi)\varpi t_2 \cos\phi\, \partial\Omega/\partial\varpi,\\
\partial\xi_x/\partial y &= -\sin(\Omega t_2) + \sin(\Omega t_2 - \phi)\varpi t_2 \sin\phi\, \partial\Omega/\partial\varpi,\\
\partial\xi_x/\partial z &= \sin(\Omega t_2 - \phi)\varpi t_2\, \partial\Omega/\partial z,\\
\partial\xi_y/\partial x &= +\sin(\Omega t_2) + \cos(\Omega t_2 - \phi)\varpi t_2 \cos\phi\, \partial\Omega/\partial\varpi,\\
\partial\xi_y/\partial y &= 1 - \cos(\Omega t_2) + \cos(\Omega t_2 - \phi)\varpi t_2 \sin\phi\, \partial\Omega/\partial\varpi,\\
\partial\xi_y/\partial z &= \cos(\Omega t_2 - \phi)\varpi t_2\, \partial\Omega/\partial z,\\
\partial\xi_z/\partial x &= (v_3 t_3/a)\cos\phi(4\varpi/a - 3),\\
\partial\xi_z/\partial y &= (v_3 t_3/a)\sin\phi(4\varpi/a - 3),\\
\partial\xi_z/\partial z &= 0.
\end{aligned}$$

With this tabulation of ξ_i and $\partial\xi_j/\partial x_k$, it is a straightforward but tedious procedure to compute the mean values $\langle\xi_i\, \partial\xi_j/\partial x_k\rangle$. The contribution comes from the fluid that has experienced both rotation and displacement in the z-direction and hence arises entirely from the volume $\varpi < b$, $-c < z < +c$ subject to rotation. If there is, on the average, one cyclonic cell for a volume of space V, then, the spatial and/or ensemble average is

$$\left\langle \xi_i \frac{\partial\xi_j}{\partial x_k} \right\rangle = \frac{1}{V}\int_{-c}^{+c} \mathrm{d}z \int_0^b \mathrm{d}\varpi\varpi \int_0^{2\pi} \mathrm{d}\phi \xi_i \frac{\partial\xi_j}{\partial x_k}. \tag{18.47}$$

The volume V in the present context cannot be less than the minimum volume $2\pi a^2(c + v_3 t_3)$ occupied by the flow (18.43). For rectangular packing and $b \leqslant a$, the volume V per cell would be $8a^2(c + v_3 t_3)$.

Of the 27 quantities $\xi_i\, \partial\xi_j/\partial x_k$ it is evident that those for which $i = j$ can be written as $\frac{1}{2}\partial\xi^2/\partial x_k$, etc. Their mean value (volume integral) vanishes over an isolated eddy. Thus nine components can be neglected. They give no contribution because they do not combine rotation (ξ_x, ξ_y) with the rising and sinking motion ξ_z. Since the mean value of $\partial\xi_i\xi_j/\partial x_k$ vanishes over an isolated eddy, it follows that

$$\langle\xi_i\, \partial\xi_j/\partial x_k\rangle = -\langle\xi_j\, \partial\xi_i/\partial x_k\rangle, \tag{18.48}$$

so that of the remaining 18 components there are only nine distinct quantities to compute. Five of the nine vanish from symmetry in ϕ or because $\partial\xi_z/\partial z = 0$ or because $\Omega(\varpi, \pm c) = 0$. The only non-vanishing independent components are

$$\langle\xi_x\, \partial\xi_z/\partial x\rangle, \qquad \langle\xi_x\, \partial\xi_z/\partial y\rangle, \qquad \langle\xi_y\, \partial\xi_z/\partial x\rangle, \qquad \langle\xi_y\, \partial\xi_z/\partial y\rangle.$$

It is worthwhile giving a number of examples of these four.

Suppose that the angular velocity of rotation $\Omega(\varpi, z)$ is independent of z across the region $-c < z < c$, with a value $\Omega(\varpi)$ except at $z = \pm c$ where

Ω drops rapidly to zero. Then

$$\partial\Omega/\partial z = \Omega(\varpi)\{\delta(z+c)-\delta(z-c)\}.$$

The integration over z contributes only a factor $2c$.

To take the simplest case first, then, suppose that $\Omega(\varpi)$ has the uniform value Ω_3 from the z-axis out to $\varpi = a$, where it drops sharply to zero. Then $\partial\Omega/\partial\varpi = -\Omega_2\delta(\varpi - a)$. Note that the radial dependence of $\partial\xi_z/\partial x$ and $\partial\xi_z/\partial y$ is just $\partial v_z/\partial\varpi$, and that the mean values reduce to the radial integral

$$\int_0^a \mathrm{d}\varpi\varpi^2 \frac{\partial v_z}{\partial\varpi} = \int_0^a \mathrm{d}\varpi\left(\frac{\partial}{\partial\varpi}\varpi^2 v_z - 2\varpi v_z\right)$$
$$= \varpi^2 v_z \Big|_0^a - 2\int_0^a \mathrm{d}\varpi\varpi v_z$$

in each of the four terms. Since v_z vanishes at $\varpi = a$, the first term gives nothing. The integral $\int \mathrm{d}\varpi\varpi v_z$ vanishes because the eddy contains the return flow and represents no net displacement of fluid. All the components vanish then. The reason is simple enough. All elements of fluid experience the same uniform rotation in both the upward flow ($\varpi < \frac{1}{2}a$) and in the downward flow ($\frac{1}{2}a < \varpi < a$), so that there is no net helicity. The upward and the downward flows contribute equal and opposite spiralling of the lines of force of the large-scale field.

Suppose, then, that $\Omega(\varpi)$ has the uniform value Ω_3 from the z-axis out to $\varpi = \frac{1}{2}a$ where it drops sharply to zero. Then the rising fluid rotates while the sinking fluid does not. There should be a net spiralling of field produced by the cyclonic cell, in the manner illustrated in Figs. 18.4 and 18.5. It is readily shown that the eight non-vanishing components of $\langle\xi_i\,\partial\xi_j/\partial x_k\rangle$ are

$$\left\langle\xi_x\frac{\partial\xi_z}{\partial x}\right\rangle = -\left\langle\xi_z\frac{\partial\xi_x}{\partial x}\right\rangle = +\left\langle\xi_y\frac{\partial\xi_z}{\partial y}\right\rangle = -\left\langle\xi_z\frac{\partial\xi_y}{\partial y}\right\rangle$$
$$= -\frac{\pi c a^2 v_3 t_3}{V}F(\Phi) \tag{18.49}$$

for two equal indices, while for those in which all three indices are different

$$\left\langle\xi_x\frac{\partial\xi_z}{\partial y}\right\rangle = -\left\langle\xi_z\frac{\partial\xi_x}{\partial y}\right\rangle = -\left\langle\xi_y\frac{\partial\xi_z}{\partial x}\right\rangle = +\left\langle\xi_z\frac{\partial\xi_y}{\partial x}\right\rangle$$
$$= +\frac{\pi c a^2 v_3 t_3}{V}G(\Phi) \tag{18.50}$$

where Φ is the local rotation $\Omega_3 t_2$, and the functions F and G are

$$F(\Phi) = \tfrac{1}{8}(1-\cos\Phi) = \tfrac{1}{16}\Phi^2(1-\Phi^2/12+\ldots), \tag{18.51}$$
$$G(\Phi) = \tfrac{1}{8}\sin\Phi = \tfrac{1}{8}\Phi(1-\Phi^2/6+\ldots). \tag{18.52}$$

Thus the non-vanishing coefficients separate into two independent groups, described by the functions $F(\Phi)$ and $G(\Phi)$. The non-vanishing coefficients involve the cross-product of rotation (ξ_x, ξ_y) with convective displacement ξ_z along the axis of the rotation. The two groups of coefficients are distinguished by having two equal indices or no equal indices, the later class representing the helicity of the total displacement. For small Φ, then, the components with three unequal indices (representing the helicity) dominate

$$\langle \xi_x\, \partial\xi_z/\partial y\rangle = \ldots \cong (\pi c a^2 v_3 t_3/V)\tfrac{1}{8}\Phi. \tag{18.53}$$

The components with two equal indices, given by $F(\Phi)$, are smaller by the factor $\frac{1}{2}\Phi$. The physical implications of this have been noted briefly already (Parker 1970b) and will be taken up at the end of this subsection.

If, in some special circumstance, Φ is greater than one, then $F(\Phi)$ and $G(\Phi)$ are both of the order of unity. The coefficients with three unequal indices, given by $G(\Phi)$, oscillate about zero with undiminishing amplitude, while those with two equal indices oscillate about $+1$, shown by the light lines in Fig. 18.10. The undiminished oscillation with increasing Φ is a consequence of the rigid rotation of the fluid so that phase relations are preserved indefinitely.

In a more realistic situation $\Omega(\varpi)$ declines from a maximum value on the z-axis to smaller values at a distance, with the result that the inner

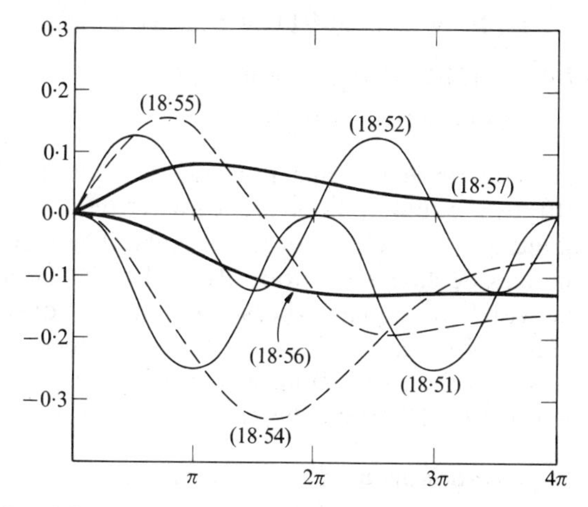

FIG. 18.10. A plot of the two independent coefficients $\langle \xi_x\, \partial\xi_z/\partial x\rangle = -(\pi a^2 c v_3 t_3/V)F(\Phi)$ and $\langle \xi_x\, \partial\xi_z/\partial y\rangle = +(\pi a^2 c v_3 t_3/V)G(\Phi)$ in units of $\pi a^2 c v_3 t_3/V$. The latter is positive as Φ increases from zero, while the former is negative. The equation numbers are written alongside the corresponding curves. Thus the light continuous curves represent the coefficients for uniform rotation in the rising fluid and none in the sinking fluid. The dashed curves are for rotation that declines linearly to zero across the rising and sinking columns, while the heavy lines are for rotation that declines to zero across the rising fluid and is zero in the sinking fluid beyond.

and outer part of the rotation soon get out of step with each other and mutually cancel. Consequently $\langle \xi_i \,\partial\xi_j/\partial x_k \rangle$ declines to zero with increasing Φ. To exhibit this, suppose that $\Omega(\varpi) = \Omega_3(1-\varpi/b)$, so that the angular velocity declines linearly to zero at $\varpi = b$ with $\partial\Omega/\partial\varpi = -\Omega_3/b$, and is zero everywhere beyond. Then with $b = a$, it follows that $F(\Phi)$ and $G(\Phi)$ in (18.49) and (18.50) are

$$F(\Phi) = (12/\Phi^4)\{\Phi(\Phi + \sin\Phi) - 4(1-\cos\Phi)\}$$
$$\cong (\Phi^2/30)(1-\Phi^2/14+\ldots), \qquad (18.54)$$

$$G(\Phi) = (2/\Phi^4)\{6\Phi(3+\cos\Phi) - (\Phi^3 + 24\sin\Phi)\}$$
$$\cong (\Phi/10)(1-3\Phi^2/14+\ldots). \qquad (18.55)$$

Again, for small Φ, the components with three unequal indices dominate. For Φ of the order of one, the non-vanishing components are comparable, while for large Φ the components with unequal indices diminish as Φ^{-1} while those with two equal indices diminish as Φ^{-2} and are again negligible. The variation with Φ is shown in Fig. 18.10 by the two dashed curves.

If the rotation is limited entirely to the rising column of fluid, putting $b = \frac{1}{2}a$, so that $\Omega = 0$ in the descending fluid in $\frac{1}{2}a < \varpi < a$ and beyond, it is a simple matter to show that $\partial\Omega/\partial\varpi = -2\Omega_0/a$ and

$$F(\Phi) = (1/8\Phi^4)\{\Phi(\Phi^3 + 12\sin\Phi) - 24(1-\cos\Phi)\}$$
$$\cong (\Phi^2/20)(1-3\Phi^2/112+\ldots), \qquad (18.56)$$

$$G(\Phi) = (1/4\Phi^4)\{6\Phi(1+\cos\Phi) + (\Phi^3 - 12\sin\Phi)\}$$
$$\cong (3\Phi/80)(1-5\Phi^2/16+\ldots). \qquad (18.57)$$

The components with three unequal indices are again $O(\Phi)$ for $\Phi \ll 1$, while those with two equal indices are only $O(\Phi^2)$. They are all of comparable magnitude for Φ of the order of unity. The components with three unequal indices decline asymptotically as Φ^{-1} for large Φ, whereas the components with two equal indices approach a constant value of $-\pi a^2 c v_3 t_3/8V$. The finite limit as $\Phi \to \infty$ is a consequence of the confinement of the rotation to the rising fluid, wherein $\int d\varpi\varpi v_z$ is not zero. The results are shown in Fig. 18.10 by the two heavy curves.

18.3.4. Physical properties of the dynamo coefficients

It is evident that $v_3 t_3 F(\Phi)$ is an even function of the rotation Φ and an odd function of the displacement $v_3 t_3$, while $v_3 t_3 G(\Phi)$ is an odd function of both. It follows that $F(\Phi)$ is non-vanishing even if there are equal numbers of eddies with opposite rotations. This is a dynamo effect, then, in turbulence with zero net helicity. It occurs along with Kraichnan's reduced, or negative, turbulent diffusion in rapidly rotating bodies and is explored briefly in §19.3. It has somewhat the same nature as the dynamo effects leading to the negative turbulent diffusion. It is different, however, in that $v_3 t_3 F$ is an odd function of v_3, so that equal numbers of eddies with opposite displacements give a mean value of zero, whereas the negative turbulent diffusion survives.

Equal numbers of eddies with opposite $v_3\Omega_3$ (i.e. helicity) yield $v_3 t_3 G = 0$, so that $\langle \xi_x \, \partial\xi_z/\partial y \rangle$ is non-vanishing only insofar as one sign dominates over the other. The expressions computed above may be considered to represent the residual after mutual cancellation of eddies with equal and opposite helicity.

This is perhaps the appropriate place to comment on the coefficients that would be computed if the rotation precedes, or is simultaneous with, the rising and sinking motions, rather than following as in the foregoing calculations. Suppose then that the rotation is switched on first for a time t_2 in the volume $-c < z < +c$. Following the rotating motions the rising and sinking motions are switched on for a time t_3 leading to the displacement $\xi_z(\varpi)$. Then the fluid which has undergone both rotation and convection (rising and sinking), so that it contributes to $\xi_i \, \partial\xi_j/\partial x_k$, is the fluid in $\varpi < a$, $-c + \xi_z(\varpi) < z < +c + \xi_z(\varpi)$. With both Φ and v_z independent of z across the region of rotation, the integration over z from $-c + \xi_z$ to $+c + \xi_z$ gives the same result as the integration over $-c$ from to $+c$ in (18.47). Thus the coefficients $\langle \xi_i \, \partial\xi_j/\partial x_k \rangle$ are the same as computed above.

If the rotation is simultaneous with the rising and sinking motion, and if the rotational motion Φ is carried along with the fluid velocity $v_z(\varpi)$, the result is also the same as computed above. Hence, the idealization by which the rotation precedes the rising and sinking motion is a computational convenience but does not affect the final result for $\langle \xi_i \, \partial\xi_j/\partial x_k \rangle$.

There are several conclusions to be drawn from the examples and the curves shown in Fig. 18.10. One is that both the magnitude and form of the coefficients $\langle \xi_x \, \partial\xi_z/\partial x \rangle$ and $\langle \xi_x \, \partial\xi_z/\partial y \rangle$ vary markedly depending upon the distribution of rotation across the rising and sinking fluid. The coefficients differ from zero only insofar as the rising and sinking regions undergo different rotation. The form of the coefficients varies strongly with the distribution of rotation, going asymptotically to zero, or to a finite value, with increasing rotation. These sharp differences show the precision with which the fluid velocity must be known from either observation or theory to give an accurate value for $\langle \xi_i \, \partial\xi_j/\partial x_k \rangle$. The mean value depends critically on the quantitative details of the distribution of strain rate $\partial v_j/\partial x_k$ across the velocity v_i.

In slowly rotating bodies, such as the sun or the galaxy, one expects the rotation during the life of an eddy to be small, $\Phi = \Omega t_2 \ll 1$, in which case the coefficients $\langle \xi_x \, \partial\xi_z/\partial y \rangle$ with unequal indices, representing the net helicity, are the dominant effect. Then the source term

$$\tau^{-1}(\partial A_j/\partial x_k)\langle \xi_k\, \partial\xi_j/\partial x_i\rangle$$

on the right-hand side of (18.37), shown in detail in (18.39), reduces to the vector $(\Gamma B_1, \Gamma B_2, 0)$ with

$$\Gamma = -\pi a^2 c v_3 t_3 G(\Phi)/V\tau.$$

This is, of course, just the result (18.7), in which the cyclonic convective cells interact with the horizontal field B_ϕ. The coefficient Γ is plotted, then, in Fig. 18.10 for three different forms of convective cell.

It is a simple matter, now, to estimate the numerical magnitude of the dynamo coefficient Γ in terms of the scale L and characteristic velocity v of the eddies and convective cells that appear in nature. A glance over the values of $G(\Phi)$ for the various configurations of cyclonic convection shows (see (18.52), (18.55), and (18.57)) that $G(\Phi) \cong \epsilon\Phi$ where $\epsilon = 0{\cdot}03$–$0{\cdot}1$. The volume V per eddy is not less than the volume $8a^2c$ occupied by the region of rotation, assuming rectangular close packing. The net translation $v_3 t_3$ is presumably comparable to the dimension L of the eddy. Hence, altogether,

$$\Gamma \cong \tfrac{1}{8}\pi\epsilon L\Phi/\tau.$$

It is readily shown that the mean cyclonic velocity $v_\phi = \varpi\Omega$ over the radius is

$$\begin{aligned}\langle v_\phi\rangle &\equiv \frac{1}{a}\int_0^a \mathrm{d}\varpi v_\phi\\ &= \mu a \Omega_3 t_2/\tau\\ &= \mu a\Phi/\tau\end{aligned}$$

where μ is a numerical coefficient, comparable to ϵ for each of the three configurations ($\epsilon = \frac{1}{8}, \frac{1}{10}, \frac{3}{80}$ while $\mu = \frac{1}{8}, \frac{1}{6}$, and $\frac{1}{24}$, respectively). Hence, with $a = L$ it follows that

$$\Gamma \cong 0{\cdot}4\langle v_\phi\rangle \qquad (18.58)$$

as an approximate rule of thumb for the dynamo coefficient Γ. The dynamo coefficient is essentially a direct measure of the cyclonic velocity in close packed cyclonic cells and eddies.

It is interesting to consider the dynamo effects in physical circumstances wherein the rotation goes beyond about one radian. One might expect that, in a body rotating as rapidly as Earth, completing some 10^5 rotations in the life span of a single eddy, the rotation proceeds to $\pi/2$ and beyond. There is the theoretical possibility that $\langle \xi_x\, \partial\xi_z/\partial y\rangle$ may become negative, when the raised dent in the lines of force is rotated beyond π so that its projection on the meridional plane (see Fig. 18.5) has the reverse sense of rotation. The light curve and the dashed curve (eqns (18.52) and (18.55)) in Fig. 18.10 show this reversal, with the corresponding change in the sign of Γ. The generation of the dipole field of Earth would then proceed as described in §18.1 if the angular velocity of the core *increased* outward from the axis of Earth, rather than decreased. Note too that if the principal cyclonic effects were associated with downdrafts rather than updrafts in the core, the generation of the dipole field with $\Phi < \pi$ would proceed with the angular velocity increasing outward, whereas if $\Phi > \pi$, it would proceed with an angular velocity decreasing outward. We are not yet in a position to assert, beyond the inference from the westward drift, the fluid motion in the liquid core of our planet. It remains for the future to decide these points, as emphasized in §18.1.

We should not take the reversal of Γ at $\Phi = \pi$ too seriously, however. The reversal is possible only in the idealized convective cell or eddy wherein all parts of the eddy maintain their rotational phase coherence for extended periods. There is no reason, of which we are aware, to think that this happens in nature. The circumstance where the angular velocity Ω varies linearly across the rising and sinking fluid is sufficient to destroy the coherence, with the result (18.57), shown by the upper heavy curve in Fig. 18.10, that Γ is positive for all Φ and declines only slowly to zero, as Φ^{-1}, with increasing Φ. There is always some part of the cell in which the rotation gives a positive contribution large enough to dominate the rest.

Thus the sign of $\langle \xi_x \, \partial\xi_z/\partial y \rangle$ is an open question if there are circumstances where the cyclonic eddies possess a *long-lived* rotation. There is in addition the fact that the coefficients $\langle \xi_x \, \partial\xi_z/\partial x \rangle$ with two equal indices, are no longer smaller than the coefficients with three unequal indices. All three terms in (18.39) must be retained and the dynamo equation does not reduce to the standard form (18.7), but has the more general form (18.37). This fact was pointed out some years ago (Parker 1955b).

The physical origin of the terms with two equal indices, e.g. $\xi_x \, \partial\xi_z/\partial x$ is elementary. A local convective eddy displaces the lines of force both upward and downward in equal amounts. The upward and downward indentations may be of different shape, but conservation of incompressible fluid means that they involve equal and opposite displacements, as sketched in the yz-plane in Fig. 18.11. If we think of each indentation as a local loop of field superimposed on the initial large-scale field, the two loops represent equal and opposite magnetic circulation, and hence no net circulation. Now suppose that the upward indentation is rotated about the vertical z-direction through any angle Φ. The projection of the upward indentation on the yz-plane is reduced by the factor $\cos \Phi$, while the downward indentation is unaffected. The net circulation in the yz-plane is then no longer zero, but is proportional to $1 - \cos \Phi$. It does not

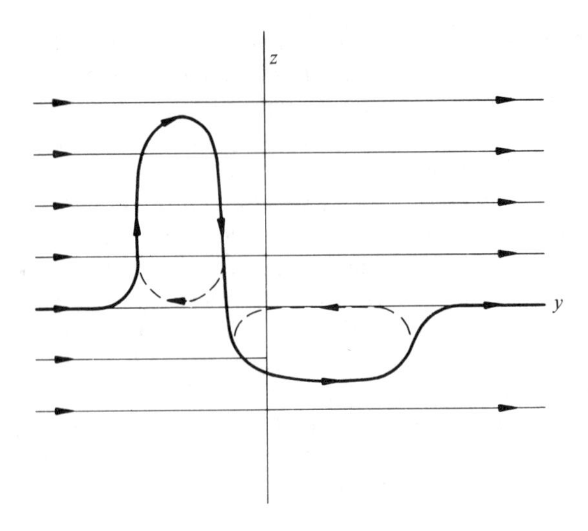

FIG. 18.11. A sketch of the upward and downward indentations in the lines of force produced by a convective displacement of fluid in the z-direction and the return flow.

matter whether Φ is positive or negative. Hence equal numbers of eddies with opposite rotation produce the same effect as if they all had the same rotation. The net generation of vector potential (in the x-direction) is independent of whether there is net helicity. Equal numbers of eddies with *opposite vertical* displacement patterns give zero effect, of course (see also Figs. 4(a), 4(b) in Parker 1955b).

The effect slips through the mathematical net cast by the formal theories of the hydromagnetic dynamo effect, reviewed in §18.5, because they deal with perturbations of small magnitude, equivalent to $\Phi \ll 1$. Altogether, then, the form (18.7) is adequate for the present purpose of a preliminary exploration of the theoretical possibilities for the generation of magnetic field in astrophysical bodies, such as Earth and the sun. But it must be kept in mind for the future that (18.7) is only the lowest order approximation for small Φ. Sooner or later it will be necessary to explore the solutions of the more general form (18.37) in which the non-vanishing of the coefficients with two equal indices is recognized.

18.3.5. Directional distributions of helical convective eddies

Up to this point the z-axis, about which the cyclonic eddy rotates, has been tacitly assumed to lie in the vertical (radial) direction. That is the sense of rotation of the cyclonic eddies in a stratified atmosphere. There may be circumstances in nature, however, in which there is net helicity in either or both of the horizontal directions. Indeed, in the formal mathematical theories developed in the last decade, the isotropic case is often treated, because of its formal simplicity, if not its physical applicability. We would expect cyclonic convection in an astronomical body to have axial symmetry about the vertical direction (as treated in the above examples) if it has any symmetry at all[7].

In any case it is instructive to look briefly at the source term of the dynamo equation for cyclonic eddies oriented in all three directions. The above calculations can be taken over in a very simple way to eddies with their axis of symmetry in, say, the x-direction. In that case the z-coordinate above becomes the x-coordinate. In order to maintain a right-handed coordinate system, then, x and y in the above calculations map into the y- and z-coordinates respectively. All subscripts and coordinates are changed accordingly. For eddies with their axes in the y-direction, z transforms to y while x and y map into z and x, respectively. Denote by v_1 and Ω_1 the velocity and rotation of the eddies with rotational symmetry about the x-direction and v_2, Ω_2 for the y-direction. Then write F_1 and G_1 for

$$(\pi ca^2 V)v_3 t_3 F(\Omega_1 t_2)$$

and

$$(\pi ca^2/V)v_3 t_3 G(\Omega_1 t_2),$$

respectively. Denote by F_2 and G_2 the same quantities for v_2 and Ω_2, and F_3, G_3 for v_3 and Ω_3. Then it is readily shown that the total effect is

$$\langle \xi_y \, \partial \xi_z/\partial z \rangle = \langle \xi_y \, \partial \xi_x/\partial x \rangle = F_1, \tag{18.59}$$

$$\langle \xi_x \, \partial \xi_y/\partial y \rangle = \langle \xi_x \, \partial \xi_z/\partial z \rangle = F_2, \tag{18.60}$$

$$\langle \xi_z \, \partial \xi_x/\partial x \rangle = \langle \xi_z \, \partial \xi_y/\partial y \rangle = F_3, \tag{18.61}$$

as the extension of (18.49), and

$$\langle \xi_x \, \partial \xi_z/\partial y \rangle = G_2 + G_3, \tag{18.62}$$

$$\langle \xi_y \, \partial \xi_x/\partial z \rangle = G_1 + G_2, \tag{18.63}$$

$$\langle \xi_z \, \partial \xi_y/\partial x \rangle = G_3 + G_1, \tag{18.64}$$

as the extension of (18.50). In view of (18.49), there are, then, 18 non-vanishing components of $\xi_i \, \partial \xi_j/\partial x_k$, described by six functions. For $i = j$ the components vanish, neglecting terms $O(1/\lambda)$. If we imagine that the turbulence is statistically isotropic, then $F_1 = F_2 = F_3$, $G_1 = G_2 = G_3$, and there are only two independent components.

[7] Moffatt (1976, 1978) has outlined some of the basic properties of axially symmetric turbulence in connection with the construction of formal dynamo theories.

18.4. Negative turbulent diffusion

18.4.1. Introductory discussion

This is the appropriate point to inquire into the nature of the negative turbulent diffusion discovered recently by Kraichnan (1976a, b) in model turbulence that is statistically mirror-symmetric (zero mean helicity). (See discussion in §§17.1 and 17.4). The essential feature of the model turbulence is the long persistence of the eddies. Each eddy was represented as a drifting cyclone, say of dimension l and internal velocity v. The individual cyclones live somewhat longer than the characteristic time l/v seen by a fixed observer. Each eddy, or cyclone, has strong internal helicity, but there are equal numbers with opposite helicity so that the total is zero. There is, then, no dynamo effect, of the character described in §18.3. There is, however, the curious property that the turbulence causes existing fields to bunch together rather than spread out. The effect has in common with the dynamo the ability to enhance the energy and order of the magnetic field.

Negative turbulent diffusion bears directly on the old questions of the effect of isotropic, mirror-symmetric turbulence on, say, a relatively weak ($B^2 \ll 4\pi\rho v^2$) initial magnetic field, and whether there is equipartition between the field and the large eddies, or the smaller eddies, etc.

The remarkable property of the magnetic fields of the sun, to bunch together—i.e. to break up—into isolated flux tubes widely separated from each other, comes to mind as a possible product of negative diffusion. However, the quantitative considerations on negative diffusion in a slowly rotating sun, (presented at the end of this section) make a dubious case for the idea. Krause and Roberts (1973) have treated turbulent diffusion in turbulence with small magnetic Reynolds number, using the quasi-linear approximation. They show, to that order, that there is no tendency for statistically stationary, isotropic, mirror-symmetric turbulence to sustain magnetic field. The only turbulent effect is the dispersal of the field by the eddy diffusivity just as a scalar field is dispersed. They emphasize, however, that their conclusions are valid only to the order of the quasi-linear calculation, and higher order approximations may yield different results. The direct interaction approximation also yields equality of the turbulent diffusion coefficients for vector and scalar fields. We have already seen one higher order result in the non-vanishing of $F(\Phi)$ in mirror-symmetric turbulence, as noted following eqn (18.56). Kraichnan's demonstration of negative turbulent diffusion of field, through direct numerical calculation of the fluid displacements and strains in strong eddies in an infinitely conducting fluid, was another step at higher order.

Consider, then, how negative turbulent diffusion arises. The turbulence is assumed to have equal numbers of eddies with opposite helicity so that the net helicity is zero. Yet the negative diffusion depends upon each individual eddy (with non-vanishing helicity) having a relatively long life, in excess of the characteristic time l/v, during which the rotation of the fluid in the eddy proceeds through some large angle, of the order of π. The rotation is an essential part of the effect. At first sight the effect appears paradoxical because whatever is accomplished by the rotation of one eddy would seem to be cancelled by its mirror image, with opposite rotation. The dynamo effect Γ cancels completely, as described in the foregoing sections, because the projections of the loops of field on the local plane *perpendicular* to the large-scale field have opposite signs of circulation, depending upon whether they are produced by fluid motion with positive or negative helicity. In the circumstance of negative diffusion, however, it is the projection of the loop on a plane that is *parallel*, rather than perpendicular, to the field. In that plane the sign of the circulation of the loop is *independent* of the *sign* of the rotation and of the helicity of the eddy that produced the loop. If there is no rotation, so that each eddy only dents the lines of force, the dents have a sense of circulation that produces the familiar positive turbulent diffusion of the large-scale field. If each dent is rotated through an angle $\pm\pi$, then its local circulation is reversed and the effect is negative diffusion. Eddies rotating equal amounts to the right and to the left accomplish the same, rather than opposite, effects on the field. The effect is a maximum for a rotation of π (or 3π etc.) and vanishes for 2π (or 4π etc.). Hence it arises for larger total rotation than the dynamo effect, which peaks at $\pi/2$.

18.4.2. Physical basis for negative diffusion

To illustrate the negative diffusion consider the interaction of four cyclonic eddies, each of small-scale l, with a large-scale field $B_y(z)$. The eddies each revolve, and their fluid is displaced in the z-direction as in Fig. 18.9. Their return flow is assumed to be broadly distributive and without significant rotation, so that each eddy has strong helicity. Two eddies rotate clockwise about the z-direction, while two rotate counterclockwise. The rotating portion of one of the clockwise eddies moves in the positive z-direction and the other in the negative z-direction. The same applies to the counterclockwise eddies, so that there is complete $\pm$ symmetry. The net rotation, displacement, and helicity are all identically zero. The distortion of a line of force by each of the four eddies is sketched in Fig. 18.12 for the simple case where the rotation is through an angle $\pm\pi$. The figure is drawn for $B_y(z)$ increasing linearly with z, the vector potential A_x being in the x-direction and increasing as z^2. Write

$$B_y = +\partial A_x/\partial z > 0, \qquad \partial B_y/\partial z = +\partial^2 A_x/\partial z^2 > 0$$

for the local large-scale field. Denote by ΔA_x the change in the vector

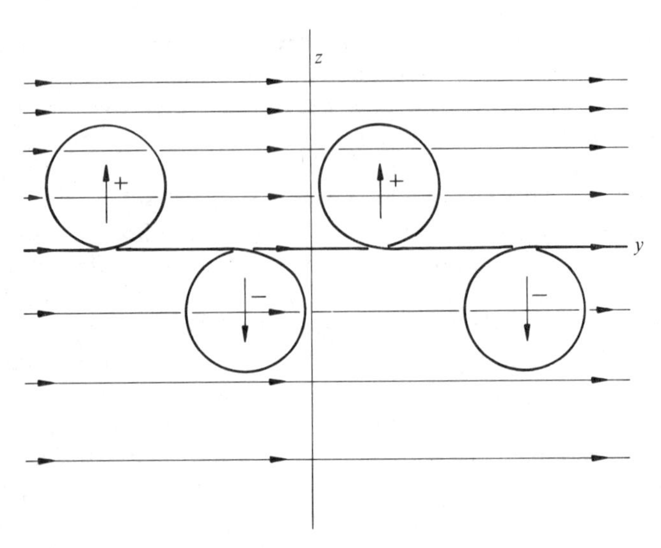

FIG. 18.12. A sketch of a line of force of the magnetic field $B_y(z)$ displaced from the y-axis by four eddies, two toward positive z and two toward negative z, one of each pair rotated about the direction of displacement through the angle $+\pi$ and the other through $-\pi$. Note that both loops displaced toward positive z have a counterclockwise rotation, indicated by the plus sign, while those displaced toward negative z have a clockwise rotation, indicated by the minus sign.

potential produced by the eddies. It may be seen from Fig. 18.12 that the lines of force displaced toward positive z represent a circulation of field around a local maximum in ΔA_x, indicated by the + sign in Fig. 18.12, whereas the lines of force displaced toward negative z are described by a local minimum in ΔA_x. Thus the first moment of ΔA_x is positive,

$$\int_{-\infty}^{+\infty} dy \int_{-\infty}^{+\infty} dz z \Delta A_x > 0.$$

The net effect of the four eddies is a displacement of vector potential in the positive z-direction, up the gradient in A_x. This is the negative turbulent diffusion produced by long-lived eddies, each with strong internal helicity. It is evident that the result is independent of whether all eddies have the same helicity, or whether there are equal numbers with positive and negative helicity.

The magnitude of the negative diffusion can be estimated from the fact that for the loops shown in Fig. 18.12 the magnitude of the inhomogeneity ΔA_x produced by each eddy of scale L is proportional to L times the strength of the field at the initial position of the line of force. Thus the magnitude of the positive ΔA_x transported to $z=0$ from $z=-L$ is proportional to $LB_y(-L)$ while the ΔA_x transported to $z=0$ from $z=+L$ is negative and proportional to $LB_y(+L)$. The net accumulation

δA_x at $z=0$ is of the order of

$$\begin{aligned}\delta A_x &\cong \Delta A_x(-L)+\Delta A_x(+L)\\ &\cong L\{B_y(-L)-B_y(+L)\}\\ &\cong -2L^2\,\partial B_y/\partial z.\end{aligned}$$

The accumulation δB_y of field is $\partial\delta A_x/\partial z$. Noting that half of the eddies displace the field in the positive z-direction and half in the negative direction in one eddy life τ, it follows that the mean rate of accumulation of field is

$$\frac{\partial B_y}{\partial t}\cong\frac{1}{2\tau}\frac{\partial\delta A_x}{\partial z}\cong-\frac{\partial}{\partial z}\left(\frac{L^2}{\tau}\frac{\partial B_y}{\partial z}\right).$$

The turbulent diffusion coefficient for magnetic field may be as negative, $-L^2/\tau$, as the turbulent diffusion coefficient for scalar fields is positive.

The effect is described formally by the term

$$\frac{1}{2\tau}\frac{\partial^2 A_x}{\partial z^2}\{\langle\xi_z^2\rangle-\langle\xi_z^2\,\partial\xi_x/\partial x\rangle\}$$

on the right-hand side of (18.37). The turbulent diffusion coefficient is, then,

$$\eta_{\mathrm{T}}=\frac{1}{2\tau}\{\langle\xi_z^2\rangle-\langle\xi_z^2\,\partial\xi_x/\partial x\rangle\}. \qquad (18.65)$$

The term $\langle\xi_z^2\rangle$ represents the usual turbulent diffusion of scalar and vector fields, discussed at length in Chapter 17. The additional effect, leading to negative diffusion, is $\langle\xi_z^2\,\partial\xi_x/\partial x\rangle$.

18.4.3. *Illustrative models*

To illustrate these two effects with the idealized eddies employed in §18.3 suppose that, in its own local coordinate system, the fluid is rotated by the amount $\Omega(\varpi, z)t_2$ throughout the cylindrical region $0<\varpi<b$, $-c<z<+c$ sketched in Fig. 18.9. That volume of fluid is then subjected to the translation (18.43) for a time t_3, producing the displacement

$$\xi_z=v_3t_3(1-\varpi/a)(1-2\varpi/a)$$

throughout $\varpi<a$. The fluid that has experienced both rotation and translation occupies the region

$$-c+\xi_z(\varpi)<z<+c+\xi_z(\varpi),$$

$\varpi<a$, so that

$$\begin{aligned}\langle\xi_z^2\,\partial\xi_x/\partial x\rangle &=\frac{1}{V}\int_{-c+\xi_z}^{+c+\xi_z}\mathrm{d}z\int_0^a \mathrm{d}\varpi\varpi\int_0^{2\pi}\mathrm{d}\phi\xi_z^2\,\partial\xi_x/\partial x\\ &=\frac{2c}{V}\int_0^a \mathrm{d}\varpi\varpi\int_0^{2\pi}\mathrm{d}\phi\xi_z^2\,\partial\xi_x/\partial x,\end{aligned}$$

where, as before, V represents the volume of space per eddy. The calculation of $\langle \xi_z^2 \rangle$ involves translation alone and depends, therefore, on how far the translation extends beyond the boundaries of the rotation. If v_z maintains the uniform value (18.43) from $z=-h$ to $z=+h$, then

$$\langle \xi_z^2 \rangle = \frac{1}{V} \int_{-h}^{+h} \mathrm{d}z \int_0^a \mathrm{d}\varpi \varpi \int_0^{2\pi} \mathrm{d}\phi \xi_z^2(\varpi).$$

At $z=\pm h$ there is presumably a thin sheet of radial flow transporting the fluid from the rising to the sinking region for the return flow. Hence, for (18.43)

$$\langle \xi_z^2 \rangle = \pi a^2 v_3^2 t_3^2 h / 15 V. \tag{18.66}$$

If the rotation is assumed to have the uniform value $\Omega_3 t_2$ from $\varpi = 0$ out to $\varpi = \frac{1}{2}a$, so that the rotation is confined to the rising column of fluid, then with (18.45) we obtain

$$\langle \xi_z^2 \, \partial \xi_x / \partial x \rangle = (\pi a^2 v_3^2 t_3^2 c / 15 V) \tfrac{13}{16} (1 - \cos \Phi)$$

where again $\Phi = \Omega_3 t_2$. Hence the turbulent diffusion coefficient (18.65) is

$$\eta_T = \frac{1}{2\tau} \frac{\pi a^2 v_3^2 t_3^2 c}{15 V} \left(\frac{h}{c} - \frac{13}{16} + \frac{13}{16} \cos \Phi \right). \tag{18.67}$$

If the rotation extends uniformly all the way to $\varpi = a$, then

$$\langle \xi_z^2 \, \partial \xi_x / \partial x \rangle = (\pi a^2 v_3^2 t_3^2 c / 15 V)(1 - \cos \Phi)$$

and

$$\eta_T = \frac{1}{2\tau} \frac{\pi a^2 v_3^2 t_3^2 c}{15 V} \left(\frac{h}{c} - 1 + \cos \Phi \right). \tag{18.68}$$

Note that the effect is an even function of the rotation Φ, so that the sign of rotation does not matter. An equal number of opposite rotations produces the same effect as rotation of all one sign.

Now h must be at least as large as $c + v_3 t_3$ to contain the region of rotation after distortion by ξ_z. But for purposes of comparing $\langle \xi_z^2 \rangle$ with $\langle \xi_z^2 \, \partial \xi_x / \partial x \rangle$ to determine the sign of η_T, it does not matter whether $v_3 t_3$ is large or small compared to c. To simplify the exposition, then, suppose that $v_3 t_3 \ll c$, so that $h \cong c$ and (18.67) and (18.68) reduce to

$$\eta_T = \frac{\pi a^2 c v_3^2 t_3^2}{30 V \tau} \frac{3 + 13 \cos \Phi}{16} \cong \frac{\pi a^2 c v_3^2 t_3^2}{30 V \tau} (1 - \tfrac{13}{32} \Phi^2 + \ldots), \tag{18.69}$$

$$\eta_T = \frac{\pi a^2 c v_3^2 t_3^2}{30 V \tau} \cos \Phi \cong \frac{\pi a^2 c v_3^2 t_3^2}{30 V \tau} (1 - \tfrac{1}{2} \Phi^2 + \ldots), \tag{18.70}$$

respectively. The first expression is negative for $103° 20'' < \Phi < 256° 40'$, while the second is negative for $90° < \Phi < 270°$. Note that the second case, with the rotation extending uniformly across both the rising and sinking columns of fluid gives the strongest effect. Yet in this case there is no net helicity. The coefficients $\langle \xi_x\, \partial\xi_z/\partial y\rangle$ etc. contributing to the dynamo effect vanish. For negative diffusion, on the other hand, the sign of the helicity does not matter. It is only the fact that, through either clockwise or counterclockwise rotation, the loops of field have been rotated. The rising column displaces its rotated loops in one direction and the sinking column in the other, as sketched in Fig. 18.12. All such loops contribute the same sign to the effect. The diffusion coefficients (18.69) and (18.70) are plotted in Fig. 18.13.

Suppose, then, that the rotation is not uniform but declines linearly to zero at $\varpi = \frac{1}{2}a$ and is zero beyond. Then $\langle \xi_z^2 \rangle$ is unaffected while $\Omega = \Omega_3(1-2\varpi/a)$ yields

$$\left\langle \xi_z^2 \frac{\partial \xi_x}{\partial x} \right\rangle = \frac{\pi c a^2 v_3^2 t_3^2}{15V} \left\{ \frac{3}{16} + \frac{15}{4\Phi^2} + \frac{135}{2\Phi^4} + \frac{900}{\Phi^6} + \frac{\sin \Phi}{\Phi}\left(\frac{45}{\Phi^2} - \frac{810}{\Phi^4}\right) + \cos \Phi \left(\frac{585}{2\Phi^4} - \frac{900}{\Phi^6}\right)\right\}$$

$$\cong \frac{\pi c a^2 v_3^2 t_3^2}{15V} \frac{3\Phi^2}{28}\{1 + O(\Phi)\}. \tag{18.71}$$

The resulting diffusion coefficient is plotted in Fig. 18.13. It barely goes negative. The inner portions of the eddy rotate more rapidly than the

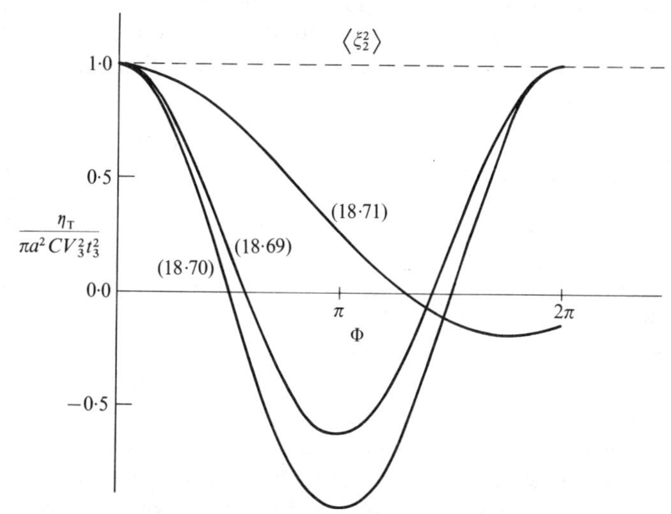

FIG. 18.13. A plot of the turbulent diffusion coefficient η_T in units of $\pi a^2 c v_3^2 t_3^2$ from eqns (18.69), (18.70), and (18.71), as a function of the angular rotation $\Phi = \Omega_3 t_2$ at the centre of the eddy.

outer portions, so that for any given $\Phi > \pi/2$, there are both positive and negative contributions, from regions that have rotated more or less than $\pi/2$. In the asymptotic limit of large Φ the rotation is so smeared across ϖ that the coefficient approaches the value

$$\langle \xi_z^2\, \partial \xi_x/\partial x\rangle \sim \frac{\pi a^2 c v_3^2 t_3^2}{15V}\left\{\frac{13}{16}+\frac{15}{4\Phi^2}+O\left(\frac{1}{\Phi^3}\right)\right\}. \tag{18.72}$$

The net diffusion coefficient approaches the constant, reduced value

$$\eta_{\mathrm{T}} \sim \frac{\pi a^2 c v_3^2 t_3^2}{30 V\tau}\left(\frac{3}{16}-\frac{15}{4\Phi^2}+\ldots\right). \tag{18.73}$$

This asymptotic limit is the same as the mean value of (18.69), because for large Φ the average of the rotation $\Omega_3(1-2\varpi/a)$ over $\varpi < \frac{1}{2}a$ in (18.71) gives the same effect as the average of (18.69) over many rotations. The diffusion coefficient is 3/16 its normal value for $\Phi = 0$.

The negative contribution of $\langle \xi_z^2\, \partial \xi_x/\partial x\rangle$ to the turbulent diffusion coefficient begins when the rotation Φ increases from zero. The growth from zero is quadratic in Φ, as may be seen from (18.69)–(18.71), and becomes significant only for $\Phi \gtrsim 1$. This brings us to the question of the circumstances in nature where the effect might be encountered in significant degree.

18.4.4. Negative diffusion in nature

In a slowly rotating body, in which the period of rotation T is long compared to the eddy life L/v, the net rotation of the individual eddies is small, with Φ of the order of $L/vT < 1$. Hence there is no evident reason to write η_{T} from (18.65) as other than

$$\eta_{\mathrm{T}} \cong \frac{1}{2\tau}\langle \xi_z^2\rangle(1-\epsilon\Phi^2+\ldots) \tag{18.74}$$

where in most astrophysical circumstances ϵ is a number less than 0·5 as in (18.69)–(18.71). When better information on the dynamics becomes available, this approximate value can be corrected to the complete expression (18.65).

On the other hand, in a body that turns through many revolutions during the life of a single eddy we might expect to find $\Phi \gtrsim 1$. The reduction of η_{T}, given by (18.65), is then large so that η_{T} is greatly reduced from (18.74) and may be negative. Of course, in such rapidly spinning objects there are not equal numbers of eddies with positive and negative rotations, but rather a strong domination of one sign over the other. That in no way affects $\langle \xi_z^2\, \partial \xi_x/\partial x\rangle$, which is an even function of Φ, but it means that there is a net helicity and an associated strong dynamo effect. The negative diffusion becomes important along with the dynamo effect when Φ is of the general order of unity, or larger. Hence Kraichnan's reduced, or negative, turbulent diffusion coefficient is primarily a concern for the strong hydromagnetic dynamo. The dynamo in rapidly rotating bodies must be treated with suitably reduced η_T. Just how much η_T is reduced in any special circumstance is hard to say, of course, without a quantitative knowledge of the dynamic motions.

In the sun the only convective cells positively identified so far by observation are the granules and supergranules, with characteristic lives of 10^3 and 10^5 s respectively. Both lives are short compared to the rotation period of 2×10^6 s of the sun as a whole. Hence we expect but little cyclonic rotation, $\Phi\lesssim0{\cdot}2$, and, in fact, none is observed ($\Phi\lesssim0{\cdot}2$). Hence $\Phi^2\ll1$ and η_T should be adequately represented by the first term in (18.74) (i.e. (17.71)). The same conclusion applies to most other stars, which rotate relatively slowly, with periods of 10^6 s or more. If there are longer-lived giant convective cells in the sun, then there is a possibility for significant rotation and, on some suitably large-scale, a reduced turbulent diffusion. But that is only a speculative possibility at the moment.

Similar conclusions apply to the gaseous disc of the galaxy, with a rotation period of the order of 2×10^8 years. The characteristic period of the internal disordered motions of the interstellar gas is of the order of 10^7 years so that the cyclonic rotation must be small, $\Phi^2\ll1$, and the reduction of η_T negligible.

On the other hand, there are among the hot stars many that rotate rapidly, with periods of 10^5 s or less. If they contain internal convection and circulation comparable to the supergranules on the sun, then we would expect turbulent diffusion of field with strong rotation of the eddies so that the turbulent diffusion is significantly reduced below the value (18.74). In such circumstances there is also a strong dynamo effect. The combination of a strong dynamo effect Γ and a reduced diffusion coefficient η_T produces strong magnetic fields, but perhaps in a high mode, i.e. distributed over the surface of the star in patterns of small wavelength. Hence, the mean field over the surface may be very small, even though the local fields are strong. This fact, together with the rapid rotation of the star, makes observational detection very difficult.

Planetary bodies rotate rapidly so that we might expect the turbulent diffusion to be greatly reduced from (18.74). Earth turns through more than 10^5 revolutions in the 10^3-year life of a convective cell in the core, suggesting the possibility of strong cyclonic rotation, with $\Phi>1$. With the expression (17.71) for the turbulent diffusion coefficient, and turbulent velocities of 10^{-3} cm s^{-1} with a life of 3×10^{10} s (10^3 years) we obtain $\eta_T=4\times10^4$ cm^2 s^{-1}, to be compared with the molecular resistive diffusion coefficient of $\eta\cong10^4$ cm^2 s^{-1}. Turbulent diffusion makes an important contribution to the decay of the large-scale fields in the core of Earth. The effective value of the diffusion coefficient may be significantly reduced, even to negative values, by the rotation of the convective cells. The effect should be included in any eventual quantitative model of the geomagnetic dynamo. At the present time we know so little about the fluid motions, not to mention η_T in (18.74) and its reduction by cyclonic rotation, that a quantitative model is out of the question.

If we estimate the reduced diffusion effect to be important in the core of Earth, then, it is doubly so in Jupiter where the planet spins faster and the resistive diffusion coefficient $\eta \cong 10^3\ \mathrm{cm^2\,s^{-1}}$ in the metallic hydrogen is smaller (cf. Chapter 20). Unfortunately we know nothing at all of the convective velocities of Jupiter, so it is not possible to say more. Altogether, then, the reduction of the diffusion coefficient from (18.74) occurs in strong dynamos, in rapidly rotating bodies. Elsewhere η_T from (18.74), i.e. the diffusion coefficient for scalar fields (17.71), is the best estimate of turbulent diffusion of magnetic fields available.

18.5. The dynamo effect at low magnetic Reynolds number

18.5.1. Quasi-linear approximation

Consider the production of vector potential by cyclonic eddies with small magnetic Reynolds number $R_m = Lv/\eta$ so that the quasi-linear approximation can be employed, as in §17.2. The equation for the mean field is given by (17.20). With net helicity, the correlation tensor $\langle v_i(x_k, t)v_j(x'_k, t')\rangle$ for isotropic (but not mirror-symmetric) turbulence is given by (A20) in Appendix A of Chapter 17. The correlations

$$\langle v_i(x_k, t)\, \partial v_j(x'_k, t')/\partial x'_n\rangle$$

are given by (A3) and (A4).

The first of the four terms on the right-hand side of (17.20) gives zero because of (A7) for an incompressible fluid. For the remaining three terms, expand $B_i(x'_n, t')$ in a Taylor series about x_n, with $\zeta_n = x'_n - x_n$. Keep all the terms through $\partial^2 B_i/\partial x_n\, \partial x_m$. Note that the Green's function G is a function only of the magnitude ζ of ζ_i, as are the correlation functions $A(\zeta)$, $B(\zeta)$, and $C(\zeta)$. Hence any terms with odd powers of ζ_k vanish upon integration over ζ_k. The only non-vanishing quantities, then, are the turbulent diffusion coefficient, yielding (17.22), and the two equal contributions from the second and third terms, yielding

$$\left(\frac{\partial}{\partial t} - \eta\nabla^2\right)\langle B_i\rangle = 2\frac{\partial B_j}{\partial x_k}\int_0^\infty \mathrm{d}s \int \mathrm{d}^3\mathbf{r}' G(\zeta, s) \times u^2\{\epsilon_{ikj}C(\zeta) + \epsilon_{ikn}C'(\zeta)\zeta_n\zeta_j/\zeta\} \quad (18.75)$$

where $s = t - t'$. Consider the $i = 1$ component. Noting that the integral of

$\zeta_n\zeta_j$ vanishes unless $n=j$, it is readily shown that

$$\left(\frac{\partial}{\partial t}-\eta\nabla^2\right)\langle B_1\rangle=2u^2\left(\frac{\partial B_3}{\partial x_2}-\frac{\partial B_2}{\partial x_3}\right)\int_0^\infty \mathrm{d}s\int \mathrm{d}^3\mathbf{r}'G(\zeta,s)(C+C'\zeta_3^2/\zeta),$$

where we have taken advantage of the fact that integrals over ζ_2^2 yield the same result as ζ_3^2. Next, do the angular integrations in $\mathrm{d}^3\mathbf{r}'$. Substituting in the explicit form (17.18) for the Green's function, the result reduces to

$$\left(\frac{\partial}{\partial t}-\eta\nabla^2\right)\langle B_i\rangle=\frac{u^2\epsilon_{ijk}}{3\pi^{\frac{1}{2}}\eta^{\frac{3}{2}}}\frac{\partial B_k}{\partial x_j}\int_0^\infty\frac{\mathrm{d}s}{s^{\frac{3}{2}}}\int_0^\infty \mathrm{d}\zeta\frac{\mathrm{d}}{\mathrm{d}\zeta}(\zeta^3C)\exp(-\zeta^2/4\eta s) \tag{18.76}$$

for the ith component. By comparison with the curl of (18.7) we see that the dynamo coefficient is the quantity

$$\Gamma=\frac{u^2}{3\pi^{\frac{1}{2}}\eta^{\frac{3}{2}}}\int_0^\infty\frac{\mathrm{d}s}{s^{\frac{3}{2}}}\int_0^\infty \mathrm{d}\zeta\frac{\mathrm{d}}{\mathrm{d}\zeta}(\zeta^3C)\exp\left(-\frac{\zeta^2}{4\eta s}\right). \tag{18.77}$$

To give an explicit value for Γ it is necessary to specify $C(\zeta,s)$. Suppose, for instance, that

$$C(\zeta,s)=-\frac{\alpha}{6u^2}\exp\left(-\frac{s^2}{\tau^2}\right)\exp\left(-\frac{\zeta^2}{L^2}\right) \tag{18.78}$$

where α is the mean helicity $\langle\mathbf{v}\cdot\nabla\times\mathbf{v}\rangle$. Do the integration over ζ first. The result is

$$\Gamma=-\tfrac{1}{3}\alpha\int_0^\infty\frac{\mathrm{d}s\,\exp(-s^2/\tau^2)}{(1+4\eta s/L^2)^{\frac{5}{2}}}\cdot \tag{18.79}$$

The quasi-linear approximation is valid for small magnetic Reynolds number for the individual eddies, $\tau\gg L^2/\eta$. Hence the Gaussian factor is essentially unity for any s for which $(1+4\eta s/L^2)^{-\frac{5}{2}}$ is non-negligible, with the result that

$$\Gamma=-\alpha L^2/18\eta. \tag{18.80}$$

This result was obtained first by Steenbeck *et al.* (1966) (see also Steenbeck and Krause 1966, 1967; Krause and Steenbeck 1967; Rädler 1968; Krause 1968; Krause and Roberts 1973; Roberts and Soward 1975).

18.5.2. Fourier transformed equations

Consider the calculation, of the production of vector potential by cyclonic eddies, working with the Fourier transformed equations presented in §17.3. Starting from (17.36)–(17.38), it is obvious that the first order term $b_i^{(1)}(\mathbf{k},t)$ has zero mean, because $\langle u_i\rangle=0$, while the second order term yields

$$\langle b_i^{(2)}(\mathbf{k},t)\rangle = -k_j\int_0^t \mathrm{d}t'\int_0^{t'}\mathrm{d}t''\int \mathrm{d}^3\mathbf{k}'\int \mathrm{d}^3\mathbf{k}''k_n'$$

$$\times\exp\{-\eta k^2(t-t')-\eta k'^2(t'-t'')-\eta k''^2 t''\}$$

$$\times[b_n(\mathbf{k}'',0)\{\langle u_i(\mathbf{k}-\mathbf{k}',t')u_j(\mathbf{k}'-\mathbf{k}'',t'')\rangle$$

$$-\langle u_j(\mathbf{k}-\mathbf{k}',t')u_i(\mathbf{k}'-\mathbf{k}'',t'')\rangle\}$$

$$-b_j(\mathbf{k}'',0)\langle u_i(\mathbf{k}-\mathbf{k}',t')u_n(\mathbf{k}'-\mathbf{k}'',t'')\rangle$$

$$+b_i(\mathbf{k}'',0)\langle u_j(\mathbf{k}-\mathbf{k}',t')u_n(\mathbf{k}'-\mathbf{k}'',t'')\rangle]. \qquad (18.81)$$

The correlation tensor $\langle u_i(\mathbf{k}-\mathbf{k}',t')u_j(\mathbf{k}'-\mathbf{k}'',t'')\rangle$ is given by (A23) and (A24) (Appendix A, Chapter 17). The symmetric part of the correlation tensor contributes nothing to the coefficient of $b_n(\mathbf{k}'',0)$ in the integrand, so that turbulent diffusion is derived entirely from the last two terms $b_i(\mathbf{k}'',0)$ and $b_j(\mathbf{k}'',0)$, as shown in (17.39). With non-vanishing helicity, then, the anti-symmetric part of $\langle u_iu_j\rangle$ is

$$-iu^2\delta(\mathbf{k}-\mathbf{k}')\gamma(k,s)\epsilon_{ijn}k_n.$$

This leads to the contribution $\beta_i(\mathbf{k},t)$ given by

$$\beta_i = iu^2\exp(-\eta k^2 t)\int_0^t \mathrm{d}t'\int_0^{t'}\mathrm{d}s\int \mathrm{d}^3\mathbf{k}'\exp\{-\eta(k'^2-k^2)s\}$$

$$\times\gamma(|\mathbf{k}-\mathbf{k}'|,s)[2b_n(\mathbf{k},0)k_jk_n'\epsilon_{ijm}(k_m-k_m')$$

$$+b_i(\mathbf{k},0)k_jk_n\epsilon_{jnm}(k_m-k_m')]$$

upon writing $s=t'-t''$, integrating over $\mathbf{k}''$, and noting that the divergence condition yields $k_jb_j(\mathbf{k},0)=0$. The simplest way to proceed from here is to write out the $(i=1)$-component of the quantity in brackets, noting that most of the terms cancel, leaving only

$$2b_n(\mathbf{k},0)k_n'(k_3k_2'-k_2k_3').$$

Since $k_3k_2'-k_2k_3'$ is just the $i=1$ component of $\epsilon_{ijk}k_j'k_k$, it follows that

$$\beta_i = 2iu^2\exp(-\eta k^2 t)\int_0^t \mathrm{d}t'\int_0^{t'}\mathrm{d}s\int \mathrm{d}^3\mathbf{k}'\exp\{-\eta(k'^2-k^2)s\}$$

$$\times\gamma(|\mathbf{k}-\mathbf{k}'|,s)b_n(\mathbf{k},0)k_n'\epsilon_{ijk}k_j'k_k. \qquad (18.82)$$

Note that the argument of γ is

$$|\mathbf{k}-\mathbf{k}'| = (k^2+k'^2)^{\frac{1}{2}}\{1-k_jk_j'/(k^2+k'^2)\}^{\frac{1}{2}},$$

so that, if γ is expanded in ascending powers of k_jk_j', there are terms both odd and even in $k_j k_j'$, while the coefficients depend only on k'^2. Altogether, then, we can write

$$\gamma(|\mathbf{k}-\mathbf{k}'|, s) = k_j k_j' I + J$$

where I and J are functions of $(k_j k_j')^2$, $k_j k_j$, and $k_j' k_j'$, so that they are independent of the sign of k_j'. This form for γ is multiplied by

$$b_n(\mathbf{k}, 0) k_n' \epsilon_{ijk} k_j' k_x$$

in the integral over $\mathrm{d}^3\mathbf{k}$. The $i = 1$ component is

$$(k_1' k_2' k_3 - k_1' k_3' k_2) b_1 + (k_2'^2 k_3 - k_2' k_3' k_2) b_2 + (k_3' k_2' k_3 - k_3'^2 k_2) b_3.$$

For those terms involving $k_2'^2$ and $k_3'^2$ it is evident that the integration over $\mathrm{d}^3\mathbf{k}'$ leads to nothing when multiplied by $k_j k_j' I$ because the factor k_j' is odd about the origin. For those terms involving $k_1' k_2'$, $k_2' k_3'$, or $k_3' k_1'$, it is evident that the integration over $\mathrm{d}^3\mathbf{k}'$ leads to nothing when multiplied by either $k_j k_j' I$ or J, because there is at least one odd power of k_i' in each term. There survives, then, only the term

$$b_2(\mathbf{k}, 0) k_3 k_2'^2 - b_3(\mathbf{k}, 0) k_2 k_3'^2.$$

The other factors in the integrand now are functions only of k'^2, so that $k_2'^2$ can be replaced by $k_3'^2$ and the volume element $\mathrm{d}^3\mathbf{k}$ reduces to

$$2\pi \,\mathrm{d}k k^2 \,\mathrm{d}\vartheta \sin\vartheta$$

where ϑ is the angle between $\mathbf{k}'$ and the $(i = 3)$-direction, $k_3' = k' \cos\vartheta$. Hence the $(i = 1)$-component is

$$\beta_1 = +i\{k_2 b_3(\mathbf{k}, 0) - k_3 b_2(\mathbf{k}, 0)\} tS(\mathbf{k}, t) \exp(-\eta k^2 t),$$

where

$$tS(\mathbf{k}, t) = -4\pi u^2 \exp(-\eta k^2 t) \int_0^t \mathrm{d}t' \int_0^{t'} \mathrm{d}s \int_0^\infty \mathrm{d}k' k'^4$$
$$\times \exp[-\eta(k'^2 - k^2)s] \int_0^\pi \mathrm{d}\vartheta \sin\vartheta \cos^2\vartheta \, J.$$

In general, then,

$$\beta(\mathbf{k}, t) = i\mathbf{k} \times \mathbf{b}(\mathbf{k}, 0) S(\mathbf{k}, t) t \exp(-\eta k^2 t). \tag{18.83}$$

To evaluate S note that the iteration procedure leading to (17.38) is valid if and only if η is large. In that limit the principal contribution to the integrals comes from k'^2 near k^2 and/or $s \cong 0$, in order that the exponential factor is not small. On the other hand the spectrum function $\gamma(k, s)$ peaks in the neighbourhood of some wave number k_0 of the general order of $1/L$, and declines rapidly toward both larger and small values of k. We are interested primarily in the large-scale mean field, for wave numbers k_i

of the order of $1/\lambda$, and, hence, wave numbers that are very small compared to the characteristic wave number k_0 for the turbulence. Thus, let $q_i = k_i' - k_i$. Then

$$\gamma(|\mathbf{k}' - \mathbf{k}|, s) = \gamma(q, s)$$

and

$$k'^2 - k^2 = q^2 + 2\mathbf{q} \cdot \mathbf{k}.$$

For those values of q for which $\gamma(q, s)$ is non-negligible, then, $k'^2 - k^2 \cong q^2$ and the exponential factor is $\exp(-\eta q^2 s)$, declining rapidly (with a characteristic time $1/\eta k_0^2$) to zero with increasing s. Over this short interval of time $J(q, s)$ can be adequately approximated by

$$J(q, 0) \cong \gamma(q, 0).$$

The integration over ϑ can be done immediately, producing a factor of 2/3. The integration over s can be carried out and the result is

$$\int_0^{t'} \mathrm{d}s \exp(-\eta q^2 s) = \{1 - \exp(-\eta q^2 t')\}/\eta q^2,$$

which reduces to $1/\eta q^2$ except over a very small interval of time

$$0 \leqslant t' \leqslant 1/\eta q^2 \cong 1/\eta k_0^2$$

which can be neglected. In a similar vein, $k'^4 \cong q^4$ to a sufficient approximation. The integral over t' yields now only a factor t, so that the final result is

$$S(\mathbf{k}, t) = -\frac{8\pi u^2}{3\eta} \int_0^\infty \mathrm{d}q q^2 \gamma(q, 0). \tag{18.84}$$

Note, then, that $S(\mathbf{k}, t)$ does not depend upon either $\mathbf{k}$ or t. It follows from (17.35) and (17.36) that

$$\mathbf{b}(\mathbf{k}, t) = \{\mathbf{b}(\mathbf{k}, 0) + i\mathbf{k} \times \mathbf{b}(\mathbf{k}, 0) S(\mathbf{k}, t) t\} \exp(-\eta k^2 t). \tag{18.85}$$

Consider, then, the Fourier transform of the curl of the dynamo equation (18.7),

$$\left(\frac{\partial}{\partial t} + \eta k^2\right) \mathbf{b}(\mathbf{k}, t) = i\Gamma \mathbf{k} \times \mathbf{b}(\mathbf{k}, t). \tag{18.86}$$

The solution for the initial value $\mathbf{b}(\mathbf{k}, 0)$ is

$$\mathbf{b}(\mathbf{k}, t) \cong \{\mathbf{b}(\mathbf{k}, 0) + i\Gamma t \mathbf{k} \times \mathbf{b}(\mathbf{k}, 0)\} \exp(-\eta k^2 t)$$

for values of t that are suitably small. This is precisely the result (18.85), with $S(\mathbf{k}, t) = \Gamma$. It follows from (18.84) that

$$\Gamma = -\frac{8\pi u^2}{3\eta} \int_0^\infty \mathrm{d}q q^2 \gamma(q, 0), \tag{18.87}$$

which is to be compared with (18.77). The two expressions are not identical, to be sure, but in the limit of large η, for which the correlation time $\tau < O(L/v)$ for the turbulence is large compared to the resistive dissipation time L^2/η, the results for Γ are the same. Consider the correlation function (18.78). It has a Fourier transform

$$c(q, s) = -\frac{\alpha \exp(-s^2/\tau^2)}{(2\pi)^3 6u^2} \int d^3\zeta \exp(-iq \cdot \zeta - \zeta^2/L^2)$$

$$= -\frac{\alpha L^3 \exp(-s^2/\tau^2)}{48\pi^{\frac{3}{2}} u^2} \exp(-q^2L^2/4).$$

Hence (see Appendix A, Chapter 17)

$$\gamma(q, s) = q^{-1} \, dc/dq$$

$$= \frac{\alpha L^5 \exp(-s^2/\tau^2)}{96\pi^{\frac{3}{2}} u^2} \exp(-q^2L^2/4), \tag{18.88}$$

and (18.87) yields (18.80) exactly.

18.5.3. Direct derivation

A particularly adroit extraction of Γ from the hydromagnetic equation has been performed by Moffatt (1970a, 1976, 1978) for low magnetic Reynolds number $R_m = Lv/\eta$. Starting from the position that η is large, so that the fluid flows readily across the field, each local eddy produces only a small and temporary perturbation in the field. The small local fluctuation δB_i produced by the turbulent flow v_i is balanced by diffusion at each instant in time. The diffusion $\eta \nabla^2 \delta B_i$ and the induction $B_j \, \partial v_i/\partial x_j$ are the principal terms of the hydromagnetic equation, so that

$$\eta \nabla^2 \delta B_i + B_j \, \partial v_i/\partial x_j \cong 0. \tag{18.89}$$

It is evident, then, that the local fluctuations δB_i have the same scale L as the turbulent eddies producing them, and δB_i is of the order of $B_j R_m$. The neglected term $\partial \delta B_i/\partial t$ is small, $O(R_m)$, compared to either of those retained, while the neglected term $v_j \, \partial B_i/\partial x_j$ is small $O(L/\lambda)$, where λ is the characteristic scale of B_i.

If the Fourier transform of $v_i(\mathbf{r}, t)$ is denoted by $u_i(\mathbf{k}, t)$ (see §17.3), while the transform of $\delta B_i(\mathbf{r}, t)$ is $b_i(\mathbf{k}, t)$, then the hydromagnetic equation (18.89) becomes

$$\eta k^2 b_i = i B_j k_j u_i. \tag{18.90}$$

To the same order the induction term $\mathbf{v} \times \delta \mathbf{B}$ in the equation (4.36) for the vector potential is

$$\epsilon_{ijk}v_j\delta B_k = \epsilon_{ijk}\int d^3\mathbf{k}\int d^3\mathbf{k}'u_j(\mathbf{k}', t)b_k(\mathbf{k}, t)\exp i(\mathbf{k}+\mathbf{k}')\cdot\mathbf{r}$$

$$= \frac{iB_n}{\eta}\epsilon_{ijk}\int d^3\mathbf{k}\int d^3\mathbf{k}'u_j(\mathbf{k}', t)u_k(\mathbf{k}, t)\frac{k_n}{k^2}\exp\{i(\mathbf{k}+\mathbf{k}')\cdot\mathbf{r}\}.$$

The mean value of the induction term involves the velocity correlation $\langle u_j u_k\rangle$ from A(23) and A(24) (Appendix A, Chapter 17). The sum $\epsilon_{ijk}\langle u_j u_k\rangle$ obliterates the symmetric part of $\langle u_j u_k\rangle$ and leaves only

$$\epsilon_{ijk}\langle v_j\delta B_k\rangle = -\frac{B_n u^2}{\eta}\epsilon_{ijk}\epsilon_{jkm}\int d^3\mathbf{k}\int d^3\mathbf{k}'\delta(\mathbf{k}'-\mathbf{k})$$

$$\times\gamma(k, 0)(k_m k_n/k^2)\exp i\{(\mathbf{k}-\mathbf{k}')\cdot\mathbf{r}\}.$$

Note that $\epsilon_{ijk}\epsilon_{jkm}k_m = +2k_i$. The integration over $\mathbf{k}'$ yields

$$\epsilon_{ijk}\langle v_j\delta B_k\rangle = -\frac{2B_n u^2}{\eta}\int d^3\mathbf{k}\frac{k_i k_n}{k^2}\gamma(k, 0). \tag{18.91}$$

The integration over $\mathbf{k}$ gives zero unless $i = n$, in which case the integration over angles gives $4\pi/3$ and

$$\epsilon_{ijk}\langle v_j\delta B_k\rangle = -\frac{8\pi u^2 B_i}{3\eta}\int_0^\infty dk k^2\gamma(k, 0). \tag{18.92}$$

Comparison with (4.36) and (18.7) yields (18.87).

18.5.4. Discussion

When the electrical conductivity of the fluid is small, so that the magnetic Reynolds number R_m of the individual eddy is small, the fluid motions produce only small, local perturbations δB_i in the large-scale field B_i. The quasi-linear approximation is applicable and provides a convenient formal procedure for deducing the turbulent diffusion coefficient η_T and the dynamo coefficient Γ. The resulting turbulent diffusion coefficient (cf. (17.25)) is small, $\eta_T = \eta O(R^{-2})$, and so is to be neglected compared to the 'zero order' diffusion coefficient η. The purpose in calculating η_T was to illustrate the physical principles rather than to establish a quantitative value. The dynamo coefficient Γ, with the dimensions of a velocity, is also small, $O(R^{-2})$. But Γ is the lowest order effect of its kind, and so is *not* negligible. It leads to a characteristic growth rate of the large-scale magnetic field of scale λ computed from (18.7) as Γ/λ. The growth rate Γ/λ is large compared to the resistive decay rate η/λ^2 in the limit of large λ in which we are working. Thus the limit of small magnetic Reynolds number is a non-trivial example of the dynamo effect, going beyond the illustration of physical principles to quantitative results for Γ that may perhaps have application in astronomy.

In this context note that Vainshtein (1970) has employed diagram techniques to sum the infinite sequence (17.35) when R_m is large but the correlation time τ is small, $\tau \ll L/v$. Then the time correlation is just $\tau\delta(s)$. Suitable rearrangement of the ordering of terms in the sum leads to the result that

$$\Gamma = \tfrac{1}{3}u^2 \int_0^\infty \mathrm{d}s \int_0^\infty \mathrm{d}q q^4 \gamma(q, s)$$

$$= \tfrac{1}{3}\tau u^2 \int_0^\infty \mathrm{d}q q^4 \gamma(q, 0). \tag{18.93}$$

This result is to be compared with (18.87), the ratio being of the order of

$$\eta\tau/L^2 = (v\tau/L)/R_m.$$

Both $v\tau/L$ and $1/R_m$ are small. Note that Vainshtein's result gives the simple scalar form for the dynamo coefficient, without the higher order terms computed in §18.3, because with $\tau \ll L/v$ the cyclonic rotation Φ is small and the mathematical methods capture only the terms that are first order in Φ, viz. the basic coefficient Γ appearing in (18.7).

18.6. The dynamo effect near axial symmetry

18.6.1. Braginskii's development of the dynamo effect

A different approach to the dynamo problem was developed by Braginskii (1964a, b, c), starting from the fact that the non-uniform rotation is the strongent fluid motion in most astrophysical circumstances. Consider a body of radius λ with large magnetic Reynolds number R, defined in terms of the maximum value $V_{\max}$ of $V_\phi(\varpi, z)$ as $R = \lambda V_{\max}/\eta$. With strong non-uniform rotation R may be made very large. The principal effect of the azimuthal velocity $V_\phi(\varpi, z)$ is the generation of a strong azimuthal field $B_\phi(\varpi, z)$ from whatever small meridional field may be present. Cowling's theorem states that the axi-symmetric velocity and magnetic field cannot be self-sustaining, so Braginskii introduced a small-scale velocity component $\mathbf{u}(\varpi, \phi, z)$ lacking axial symmetry, with the property that the average of each of the three components u_ϖ, u_ϕ, u_z over ϕ is *zero*. That is to say, the average is zero around any circle concentric about the z-axis. If angular brackets are used to indicate the ϕ-average, then

$$\langle u_\varpi \rangle = \frac{1}{2\pi} \int_0^{2\pi} \mathrm{d}\phi u_\varpi(\varpi, \phi, z) = 0$$

for any value of ϖ and z. Similarly $\langle u_\phi \rangle = \langle u_z \rangle = 0$. In this way Braginskii was able to set up a self-consistent set of equations yielding positive dynamo effects. What is more, he showed that a suitable redefinition of the vector potential reduced the equations to the simple form (18.7) and (18.8). Braginskii's dynamo works in the same general way as the dynamo with cyclonic eddies, pushing the azimuthal lines of force into a corkscrew pattern so as to generate the meridional field, portrayed in Fig. 18.4. From that general point of view the reduction to (18.7) is not unexpected. The mathematical manipulations, on the other hand, are so entirely different from the cyclonic convective dynamo, and the intermediate expressions so remarkably complicated, that the abrupt reduction comes as an algebraic surprise. Indeed the reduction to the simple form (18.7) led Braginskii to suggest that a more direct derivation of the final equations may be possible. Soward has provided that simplification based on the concept of closed stream lines, on which we will have more to say later. Braginskii (1967, 1970, 1975) and Tough and Roberts (1968) have gone on to show that the formulation is particularly amenable to simultaneous consideration of the dynamics. Braginskii's development of the dynamo equations provides a powerful tool for theoretical work on dynamical dynamos, in addition to its novel formulation and direct application to the terrestrial dynamo (Braginskii 1964a, b, c).

To follow the development write

$$\mathbf{v} = \mathbf{V}(\varpi, z) + \mathbf{u}(\varpi, \phi, z) \tag{18.94}$$

and

$$\mathbf{B} = \mathbf{B}(\varpi, z) + \mathbf{b}(\varpi, \phi, z), \tag{18.95}$$

where $\mathbf{V}(\varpi, z)$ and $\mathbf{B}(\varpi, z)$ are rotationally symmetric about the z-axis. The non-symmetric vectors $\mathbf{u}$ and $\mathbf{b}$ average to zero over ϕ. The axi-symmetric field $\mathbf{B}(\varpi, z)$ can be written

$$\mathbf{B}(\varpi, z) = \mathbf{e}_\phi B_\phi(\varpi, z) + \nabla \times \mathbf{e}_\phi A_\phi(\varpi, z) \tag{18.96}$$

in terms of the azimuthal (toroidal) and meridional (poloidal) components, with the meridional component expressed as the curl of the azimuthal vector potential A_ϕ. It follows that

$$\langle \mathbf{v} \times \mathbf{B} \rangle = \mathbf{V} \times \mathbf{B}(\varpi, z) + \langle \mathbf{v} \times \mathbf{b} \rangle$$

where the angular brackets again denote the average of the algebraic value of the individual components over ϕ.

The ϕ-component of the hydromagnetic equation (4.13) becomes

$$\frac{\partial B_\phi}{\partial t} + \varpi(\mathbf{V}_\mathrm{M} \cdot \nabla)\frac{B_\phi}{\varpi} = \left\{\nabla \frac{V_\phi}{\varpi} \times \nabla \varpi A_\phi\right\}_\phi + \{\nabla \times \langle \mathbf{u} \times \mathbf{b} \rangle\}_\phi + \eta(\nabla^2 - 1/\varpi^2)B_\phi \tag{18.97}$$

where $\mathbf{V}_\mathrm{M} \equiv (V_\varpi, 0, V_z)$ represents the meridional component of the axi-symmetric velocity. The azimuthal component V_ϕ appears in the non-uniform angular velocity V_ϕ/ϖ which shears the meridional field to produce B_ϕ. The vector potential A_ϕ follows from the azimuthal component of (4.36) as

$$\frac{\partial A_\phi}{\partial t}+\frac{1}{\varpi}(\mathbf{V}_\mathrm{M}\cdot\nabla)\varpi A_\phi=\langle\mathbf{u}\times\mathbf{b}\rangle_\phi+\eta(\nabla^2-1/\varpi^2)A_\phi. \qquad (18.98)$$

The interaction of the axi-symmetric velocity and magnetic fields cannot sustain the field, of course, so the dynamo effects must be contained in $\langle\mathbf{u}\times\mathbf{b}\rangle$. The equation for $\mathbf{b}$ is obtained by subtracting the ϕ-averaged equations from the complete equations, yielding

$$\partial\mathbf{b}/\partial t=\nabla\times\{\mathbf{u}\times\mathbf{B}+\mathbf{V}\times\mathbf{b}+\mathbf{u}\times\mathbf{b}-\langle\mathbf{u}\times\mathbf{b}\rangle\}+\eta\nabla^2\mathbf{b}.$$

Braginskii found that a self-consistent solution can be constructed if both the fluid velocity and magnetic field are developed in powers of $R^{-\frac{1}{2}}$. The non-symmetric part of the fluid velocity $\mathbf{u}$ must be made small $O(R^{-\frac{1}{2}})$ and the axi-symmetric meridional flow smaller yet, $O(R^{-1})$. Thus write

$$\mathbf{u}(\varpi,\phi,z)=R^{-\frac{1}{2}}\mathbf{w}(\varpi,\phi,z), \qquad (18.99)$$

$$\mathbf{V}_\mathrm{M}(\varpi,z)=R^{-1}\mathbf{U}_\mathrm{M}(\varpi,z). \qquad (18.100)$$

The velocity, then, is to be of the form

$$\mathbf{v}=\mathbf{e}_\phi V_\phi(\varpi,z)+R^{-\frac{1}{2}}\mathbf{w}(\varpi,\phi,z)+R^{-1}\mathbf{U}_\mathrm{M}(\varpi,z). \qquad (18.101)$$

With this ordering of magnitudes, the magnetic field has the same form,

$$\mathbf{B}=\mathbf{e}_\phi B_\phi(\varpi,z)+R^{-\frac{1}{2}}\mathbf{h}(\varpi,\phi,z)+R^{-1}\mathbf{H}_\mathrm{M}(\varpi,z)+O(R^{-\frac{3}{2}})B_\phi \qquad (18.102)$$

where $R^{-1}\mathbf{H}_\mathrm{M}$ is $\nabla\times\mathbf{e}_\phi A_\phi$, indicating that A_ϕ is small $O(R^{-1})$.

Altogether, then, the principal flow and magnetic field are axi-symmetric and azimuthal. The meridional parts are axi-symmetric and very small $O(R^{-1})$ while the unsymmetric parts are of intermediate magnitude $O(R^{-\frac{1}{2}})$.

Braginskii carried through the calculations, involving the complicated ϕ-averages, to obtain $\langle\mathbf{u}\times\mathbf{b}\rangle_\phi$ and $\{\nabla\times(\mathbf{u}\times\mathbf{b})\}_\phi$. The result is a rather complicated source term, but Braginskii showed the remarkable reduction of the equation by the introduction of the 'effective' vector potential and meridional velocity

$$A_\mathrm{e}=A_\phi+\chi B_\phi, \qquad \mathbf{U}_\mathrm{e}=\mathbf{U}_\mathrm{M}+\nabla\times\chi V_\phi \qquad (18.103)$$

where

$$\chi=\tfrac{1}{2}\varpi\langle\mathbf{w}_\mathrm{M}(\varpi,\phi,z)\times\int\mathrm{d}\phi\,\mathbf{w}_\mathrm{M}(\varpi,\phi,z)\rangle_\phi. \qquad (18.104)$$

The subscript ϕ denoting the meridional component, and the integration over ϕ is to be performed with the unit vector out in front of the integral.

Inspection of the terms in (18.103) shows that the quantity A_e is azimuthal and $\mathbf{U}_e$ meridional. Both contribute to the fields at $O(R^{-1})$. To write the final dynamo equations for B_ϕ and A_e, it is convenient to measure length in units of the overall dimension λ and time in units of the resistive decay time λ^2/η. It follows, then, that the ϕ-component is

$$\left\{\frac{\partial}{\partial t}-\left(\nabla^2-\frac{1}{\varpi^2}\right)\right\}B_\phi+\varpi(\mathbf{U}_e\,.\,\nabla)\frac{B_\phi}{\varpi}=\left\{\nabla\frac{V_\phi}{\varpi}\times\nabla\varpi RA_e\right\}_\phi. \tag{18.105}$$

The definition of the effective vector potential and meridional flow has incorporated the term $\{\nabla\times(\mathbf{u}\times\mathbf{b})\}_\phi$ in (18.97) into the other quantities, so that the form of the equation is exactly that for axi-symmetric fields and velocities. The equation for the effective vector potential A_e is

$$\left\{\frac{\partial}{\partial t}-\left(\nabla^2-\frac{1}{\varpi^2}\right)\right\}A_e+\frac{1}{\varpi}(\mathbf{U}_e\,.\,\nabla)\varpi A_e=\Gamma B_\phi, \tag{18.106}$$

where Γ is the dynamo coefficient, defined as

$$\varpi R\Gamma=\left\langle\mathbf{w}_M(\varpi,\phi,z)\times\int d\phi\,\mathbf{w}_M(\varpi,\phi,z)\right\rangle_\phi+\langle\mathbf{w}_M\times\partial\mathbf{w}_M/\partial\phi\rangle_\phi$$

$$+2\left\langle\nabla_M(\varpi w_\varpi+zw_z)\,.\,\nabla_M\int d\phi\,w_M(\varpi,\phi,z)\right\rangle. \tag{18.107}$$

The derivative $\partial\mathbf{w}_M/\partial\phi$ and the integration $\int d\phi\,\mathbf{w}_M$ are to be performed as though $\partial\mathbf{e}_\varpi/\partial\phi=0$. The subscript M denotes the meridional components of vectors and gradient operators. The dynamo coefficient Γ is small $O(R^{-1})$ so that the source term ΓB_ϕ is of the same magnitude as $\partial A_e/\partial t$ and $\nabla^2 A_e$.

The equations, then, have the basic axi-symmetric form (18.7), (18.8) with the dynamo effect appearing as the generation of net mean azimuthal vector potential by the average interaction of cyclonic fluid motions with the azimuthal field. Motivated by the surprising algebraic reduction of the equations to the basic form, Tough (1967) and Tough and Roberts (1968) carried the calculations to one higher order in $R^{-\frac{1}{2}}$ and found that, with suitable definition of the effective meridional velocity $\mathbf{U}_e$ and azimuthal vector potential A_e, the equations reduce again to the form (18.105) and (18.106), i.e. (18.7) and (18.8). The basic physical principle $\partial A/\partial t\propto\Gamma B_\phi$, that the dynamo effect arises from the cyclonic generation of azimuthal vector potential from the dominant azimuthal magnetic field, asserts itself in both the lowest order and in the next higher order.

It is apparent from inspection of (18.101) that the ordering of the velocity components is somewhat peculiar, and seriously restrictive. The non-uniform rotation is the leading term, as in the physical world, but why should the 'turbulent' component $\mathbf{u}$ adjust itself to be smaller than the non-uniform rotation by $R^{-\frac{1}{2}}$, and why should there have to be an axi-symmetric meridional circulation $O(R^{-1})$? It is doubtful that nature provides this artificial situation necessary for the systematic solution of the equations along the lines worked out by Braginskii.

The immediate retort is that no one has yet succeeded in reducing the hydromagnetic equations for cyclonic turbulence to a workable form without employing severe idealization, such as small magnetic Reynolds number for Steenbeck, Krause, and Rädler's use of the quasi-linear approximation or the abrupt 'switch on and off' eddies used by Parker, or the interlocking vortex rings employed by Kraichnan. The important fact is that Braginskii has found a scheme for treating another special case, based on strong non-uniform rotation and small-scale motions. The magnetic Reynolds number $R=\lambda V_{\max}/\eta$ of the non-uniform rotation is taken to be large, while the magnetic Reynolds number of the non-symmetric motions—the turbulence—is also large, $R_{\mathrm{m}}=O(R^{\frac{1}{2}})$, as will be shown later. Collectively the several different idealizations employed by various authors illustrate the basic physical principles over a wide range of circumstances. The range is so wide that the physical principles common to the entire range can be applied to the complex physical world with some confidence. It is to be hoped that further work will produce more general cases, of course, but Kraichnan's appraisal of the difficulties of the problem should not go unheeded. Turbulent diffusion of magnetic field, and the many effects of the internal helicity of turbulent eddies, depend upon quantitative details of the turbulent velocity field that are simply not provided by the gross statistical properties of the turbulence.

18.6.2. Soward's development of the Braginskii dynamo

Recent work by Soward (1971, 1972) has provided insight into the mathematical and physical origin of Braginskii's remarkable reduction of the hydromagnetic equations to the basic form (18.7). First of all, Soward considered the axi-symmetric azimuthal velocity $V_{\phi}(\varpi, z)$ with large magnetic Reynolds number $R=\lambda V_{\max}/\eta$ and a small irregular velocity $R^{-\frac{1}{2}}\mathbf{w}(\varpi, \phi, z)$ with vanishing average over ϕ. He showed that it is always possible to construct a small additional axi-symmetric velocity vector $R^{-1}\mathbf{W}(\varpi, z)$ such that the total velocity

$$\mathbf{v}_0(\varpi, \phi, z)=\mathbf{e}_{\phi}V_{\phi}(\varpi, z)+R^{-\frac{1}{2}}\mathbf{w}(\varpi, \phi, z)+R^{-1}\mathbf{W}(\varpi, z) \quad (18.108)$$

has closed stream lines. What is more, a velocity (18.101) can be written as

$$\mathbf{v}(\varpi, \phi, z)=\mathbf{v}_0(\varpi, \phi, z)+R^{-1}\mathbf{U}_{\mathrm{e}} \quad (18.109)$$

where $\mathbf{U}_e$ is the effective meridional velocity (18.103). Thus, it turns out that the total velocity $\mathbf{v}$ may be considered as made up of a principal velocity $\mathbf{v}_0$ with closed stream lines, plus a very small component $O(R^{-1})$ for which the stream lines are not closed. The concept of the closed stream line provides a wholly new approach to the problem, avoiding the complicated average over ϕ.

Soward noted that the deviation of the closed stream lines of $\mathbf{v}_0$ from circles concentric about the z-axis can be expressed by the displacement $\zeta_i(x_k, t)$ of the point x_i on the stream line from a corresponding nearby point x_i' on a concentric circle. With a suitable mapping $x_i = x_i(x_k')$ the velocity $\mathbf{v}_0(\varpi, \phi, z)$ can be transformed into $\mathbf{e}_\phi' v(\varpi', z')$ in terms of the new coordinates (ϖ', ϕ', z'). That is to say, expressed in suitable variables, the velocity is in the ϕ'-direction and independent of ϕ'. The entire velocity field becomes 'axi-symmetric' $(\partial/\partial\phi' = 0)$ in the new coordinate system. The magnetic field is also 'axi-symmetric', and the enormous complication of the average over ϕ drops out of the calculation. Any axi-symmetric field $v(\varpi', z')$ is automatically equal to its own ϕ'-average.

The execution of the idea requires that the hydromagnetic equations be expressed in the new x_i'-coordinate system, of course. Denoting x_i' by X_i, we follow Soward by introducing a coordinate transformation $X_i = X_i(x_k)$ that preserves volume and the field transformation

$$B_i^+(X_k) = B_j(x_k)\partial X_i/\partial x_j. \tag{18.110}$$

Since the field transformation (18.110) is just the Cauchy integral of the hydromagnetic equation (4.16) for vanishing resistive diffusion coefficient, it leaves invariant the Lagrangian equation

$$\mathrm{d}B_i/\mathrm{d}t = B_j\partial v_i/\partial x_j, \qquad v_i = \mathrm{d}X_i/\mathrm{d}t.$$

After some considerable calculation Soward showed that the complete hydromagnetic equation (4.13) transforms into

$$\frac{\partial B_i^+}{\partial t} + v_j\frac{\partial B_i^+}{\partial X_j} - B_j^+\frac{\partial v_i}{\partial X_j} = \eta\frac{\partial^2 B_i^+}{\partial X_j\partial X_j} + \eta\epsilon_{ijk}\frac{\partial \mathscr{E}_k^+}{\partial X_j} \tag{18.111}$$

where

$$\mathscr{E}_i^+ = \alpha_{ij}B_j^+ + \epsilon_{ijk}\beta_{kn}\,\partial B_j^+/\partial X_n, \tag{18.112}$$

$$\alpha_{ij} = \epsilon_{ikl}\frac{\partial X_k}{\partial x_n}\frac{\partial}{\partial X_j}\left(\frac{\partial X_l}{\partial x_n}\right), \tag{18.113}$$

$$\beta_{ij} = \frac{\partial X_i}{\partial x_n}\frac{\partial X_j}{\partial x_n} - \delta_{ij}. \tag{18.114}$$

The coordinate transformation takes the field equation into an axi-symmetric form, so that were it not for the term in $\mathscr{E}_i^+$ on the right-hand side of (18.111), Cowling's theorem would indicate no generation of field. The source of field, then, is the curl of $\mathscr{E}_i$ multiplied by the resistive diffusion coefficient η. If there were no diffusion, the source term for B_i^+ would vanish. This points out again the crucial role played by diffusion and reconnection of field in the generation of new magnetic flux. Without diffusion some components of the mean field can be increased by the fluid motions, as in the production of meridional field by the spiralling of the lines of force of the azimuthal field (see §18.1), but the production is reversible until diffusion has reconnected the lines of force.

To treat Braginskii's dynamo, Soward developed a transformation that makes the fluid velocity independent of the coordinate along the closed stream lines. The calculation in cylindrical coordinates has been given in detail by Soward (1971, 1972). For the present purposes, of illustrating the basic principles, we follow Moffatt (1976, 1978) who derives the result in cartesian coordinates, with significant reduction in algebraic complexity.

Consider, then, the rectangular coordinate system x_i with $y \equiv x_2$ corresponding to the azimuthal direction of ϕ. The fluid velocity (18.101) may be written

$$\mathbf{v}(x_k) = \mathbf{V}(x, z) + R^{-\frac{1}{2}}\mathbf{w}(x, y, z) \tag{18.115}$$

where now $\mathbf{V}(x, z)$ contains all that part of the velocity that is independent of y. The average over y is

$$\langle \mathbf{v} \rangle = \mathbf{V}(x, z), \qquad \langle \mathbf{w} \rangle = 0.$$

Equation (18.102) suggests that the associated magnetic field is to be written

$$\mathbf{B}(x_k) = \mathbf{B}(x, z) + R^{-\frac{1}{2}}\mathbf{b}(x, y, z). \tag{18.116}$$

The coordinate transformation from x_k to X_k is conveniently written

$$x_i = X_i + R^{-\frac{1}{2}}\zeta_i(X_k, t), \qquad \partial\zeta_j/\partial X_j = 0, \tag{18.117}$$

so that (18.110) becomes

$$\begin{aligned} B_i(x_k) &= B_j^+(X_k)\,\partial x_i/\partial X_j \\ &= B_j^+(X_k)(\delta_{ij} + R^{-\frac{1}{2}}\,\partial\zeta_i/\partial X_j). \end{aligned} \tag{18.118}$$

It is assumed that $\mathbf{b}$ and ζ_i vary over the same scale λ as the fields themselves, so that the relative ordering by $R^{\frac{1}{2}}$ is preserved by differentiation.

The purpose of the transformation is to eliminate $Y \equiv X_2$ from the fields, so that the fluid velocity can be written

$$v_i^+(X_k) = \delta_{i2}V^+(X, Z) + R^{-1}w_i^+(X, Z) \tag{18.119}$$

and the magnetic field can be written

$$B_i^+(X_k) = \delta_{i2} B^+(X, Z) + R^{-1} b_i^+(X, Z). \tag{18.120}$$

To work out the relation between $b_i^+(X, Z)$ and $b_i(x, y, z)$ compute $B_i(x_k)$ at the position X_k and compare it with (18.116). Note from (18.117) that

$$\partial X_i/\partial x_j = \delta_{ij} - R^{-\frac{1}{2}} \partial\zeta_i/\partial X_j + O(R^{-1}). \tag{18.121}$$

It follows from (18.118) that

$$\frac{\partial B_i}{\partial X_n} = \frac{\partial}{\partial X_n}\left\{B_j^+(X_k)\left(\delta_{ij} + R^{-\frac{1}{2}}\frac{\partial\zeta_i}{\partial X_j}\right)\right\} + O(R^{-1}).$$

Hence

$$\frac{\partial B_i}{\partial x_j} = \frac{\partial B_i}{\partial X_n}\frac{\partial X_n}{\partial x_j} = \frac{\partial B_i^+}{\partial X_j} + R^{-\frac{1}{2}}\left\{\frac{\partial}{\partial X_j}\left(B_n^+\frac{\partial\zeta_i}{\partial X_n}\right) - \frac{\partial B_i^+}{\partial X_n}\frac{\partial\zeta_n}{\partial X_j}\right\}. \tag{18.122}$$

Then expand $B_i(X_k)$ about the point x_k, so that

$$B_i(X_k) = B_i(x_k) + \frac{\partial B_i}{\partial x_n}(X_n - x_n) + \frac{1}{2}\frac{\partial^2 B_i}{\partial x_n \partial x_m}(X_n - x_n)(X_m - x_m) + \dots.$$

With (18.117)–(18.119) this can be written as

$$B_i(X_k) = B_i^+(X_k) + R^{-\frac{1}{2}} Q_i(X_k) + R^{-1} S_i(X_k) + O(R^{-\frac{3}{2}}) \tag{18.123}$$

where

$$Q_i(X_k) = B_j^+(X_k)\,\partial\zeta_i/\partial X_j - \zeta_j(X_k)\,\partial B_i^+/\partial X_j \tag{18.124}$$

and

$$S_i(X_k) = \zeta_n \frac{\partial}{\partial X_n}\left(\zeta_j \frac{\partial B_i}{\partial X_j} - B_j \frac{\partial\zeta_i}{\partial X_j}\right) - \tfrac{1}{2}\zeta_n\zeta_j \frac{\partial^2 B_i}{\partial X_n \partial X_j}. \tag{18.125}$$

Using the vector identity

$$\nabla \times (\mathbf{r} \times \mathbf{s}) = (\mathbf{s} \cdot \nabla)\mathbf{r} - (\mathbf{r} \cdot \nabla)\mathbf{s} + \mathbf{r}\nabla \cdot \mathbf{s} - \mathbf{s}\nabla \cdot \mathbf{r}$$

and noting that the divergences of B_i and ζ_i vanish, it follows that in vector form

$$\mathbf{Q} = \nabla \times (\boldsymbol{\zeta} \times \mathbf{B}^+). \tag{18.126}$$

The terms in S_i can be rearranged in such a way that

$$\begin{aligned} S_i &= S_i^{(1)} + S_i^{(2)} - \tfrac{1}{2}(\partial B_n^+/\partial X_j)(\partial\zeta_i\zeta_j/\partial X_n) \\ &\quad + \frac{1}{2}\frac{\partial}{\partial X_n}\left(\zeta_n\zeta_j \frac{\partial B_i^+}{\partial X_j}\right) - \tfrac{1}{2}B_i^+ \frac{\partial\zeta_n}{\partial X_j}\frac{\partial\zeta_j}{\partial X_n} \end{aligned} \tag{18.127}$$

where

$$S_i^{(1)} = \frac{1}{2}\left\{\left(B_n^+ \frac{\partial \zeta_j}{\partial X_n}\right)\frac{\partial \zeta_i}{\partial X_j} - \zeta_j \frac{\partial}{\partial X_j}\left(B_n^+ \frac{\partial \zeta_i}{\partial X_n}\right) + \zeta_i \frac{\partial}{\partial X_j}\left(B_n^+ \frac{\partial \zeta_j}{\partial X_n}\right)\right\},$$

$$S_i^{(2)} = \frac{1}{2}\left\{\left(\zeta_n \frac{\partial \zeta_j}{\partial X_n}\right)\frac{\partial B_i^+}{\partial X_j} - B_j^+ \frac{\partial}{\partial X_j}\left(\zeta_n \frac{\partial \zeta_i}{\partial X_n}\right) + B_i^+ \frac{\partial}{\partial X_j}\left(\zeta_n \frac{\partial \zeta_j}{\partial X_n}\right)\right\}.$$

In vector form

$$\mathbf{S}^{(1)} = \tfrac{1}{2}\nabla \times [\boldsymbol{\zeta} \times \{(\mathbf{B}^+ \cdot \nabla)\boldsymbol{\zeta}\}], \tag{18.128}$$

$$\mathbf{S}^{(2)} = \tfrac{1}{2}\nabla \times [\mathbf{B}^+ \times \{(\boldsymbol{\zeta} \cdot \nabla)\boldsymbol{\zeta}\}]. \tag{18.129}$$

Comparing (18.122) with (18.116) it is evident that the terms $O(R^{-\frac{1}{2}})$ are

$$\begin{aligned}\mathbf{b}(x_k) &= \mathbf{Q}(x_k)\\ &= \nabla \times (\boldsymbol{\zeta} \times \mathbf{e}_y B^+)\\ &= B^+\, \partial\boldsymbol{\zeta}/\partial Y - \mathbf{e}_y(\boldsymbol{\zeta} \cdot \nabla B^+)\end{aligned} \tag{18.130}$$

using (18.124) and noting from (18.120) that to lowest order B_i^+ is in the $i = 2$ direction.

Consider, then, the poloidal (meridional) component of $B_i(x_k)$. The poloidal component of $B_i(x_k)$ is denoted by

$$R^{-1}\mathbf{H}_M = \nabla \times (\mathbf{e}_y A_y)$$

in (18.102) with $\phi \to y$. It is $O(R^{-1})$ and expressible as the curl of a toroidal vector (in the y-direction). The term $O(R^{-1})$ in $\mathbf{B}$, as given by (18.123), is $R^{-1}S_i$. It is evident that, with $B_i^+ = \delta_{i2}B^+(X, Z)$, the terms involving B_i^+ on the right-hand side of (18.127) are vectors in the toroidal ($i = 2$) direction. Hence they contribute nothing to the poloidal ($i = 1, 3$) field. The other individual term reduces to $\frac{1}{2}(\partial B^+/\partial X_j)\, \partial\zeta_i\zeta_j/\partial Y$ where $Y \equiv X_2$, which has vanishing mean in the Y-direction and so produces no net poloidal component. The contribution $\mathbf{S}^{(2)}$ is given by (18.129) as the curl of the poloidal vector $B^+\mathbf{e}_y \times \{(\boldsymbol{\zeta} \cdot \nabla)\boldsymbol{\zeta}\}$. The curl is then not a poloidal vector. The only purely poloidal vector is that part of $\mathbf{S}^{(1)}$ arising from the curl of the *toroidal* component of $\boldsymbol{\zeta} \times [(\mathbf{B}^+ \cdot \nabla)\boldsymbol{\zeta}] = B^+\boldsymbol{\zeta} \times \partial\boldsymbol{\zeta}/\partial Y$. If $\boldsymbol{\zeta}_M$ denotes the meridional component of $\boldsymbol{\zeta}$, so that $\boldsymbol{\zeta}_M = (\zeta_X, 0, \zeta_Z)$, then the toroidal component is $B^+\boldsymbol{\zeta}_M \times \partial\boldsymbol{\zeta}_M/\partial Y$. Hence the effective toroidal vector potential, whose curl gives the poloidal field, is

$$\mathbf{A}_y = \tfrac{1}{2}R^{-1}B^+\langle\boldsymbol{\zeta}_M \times \partial\boldsymbol{\zeta}_M/\partial Y\rangle. \tag{18.131}$$

It follows that the poloidal component of $\mathbf{B}(x_k)$ is

$$\mathbf{b}_M = \tfrac{1}{2}\nabla \times \{B^+\langle\boldsymbol{\zeta}_M \times \partial\boldsymbol{\zeta}_M/\partial Y\rangle\} + \mathbf{b}_M^+ \tag{18.132}$$

where $\mathbf{b}_M^+$ is the poloidal component of $\mathbf{b}^+$ in (18.121). If we write

$$\mathbf{b}_M = \nabla \times \mathbf{e}_y a, \qquad \mathbf{b}_M^+ = \nabla \times \mathbf{e}_y a^+, \tag{18.133}$$

then

$$\mathbf{e}_y a = \tfrac{1}{2}B^+\langle \boldsymbol{\zeta}_M \times \partial\boldsymbol{\zeta}_M/\partial Y\rangle + \mathbf{e}_y a^+. \tag{18.134}$$

The calculation may be repeated for the velocity, with the result that $\mathbf{w}$ in (18.115) is given by

$$\mathbf{w} = \left(\frac{\partial}{\partial t} + V^+\frac{\partial}{\partial Y}\right)\boldsymbol{\zeta} - \mathbf{e}_y\boldsymbol{\zeta}\cdot\nabla V^+ \tag{18.135}$$

and the poloidal component of $\mathbf{v}(x_k)$ is

$$\mathbf{v}_M = \tfrac{1}{2}\nabla\times\left\{\boldsymbol{\zeta}_M \times \left(\frac{\partial}{\partial t} + V^+\frac{\partial}{\partial Y}\right)\boldsymbol{\zeta}_M\right\} + \mathbf{v}_M^+ \tag{18.136}$$

where $\mathbf{v}_M^+$ is the poloidal component of $\mathbf{v}^+$ in (18.119).

Consider the hydromagnetic equation (18.111) derived from the transformation. Note that the diffusion coefficient η is just $\lambda V_{\max}R^{-1}$. The toroidal Y-component of the equation involves the Y-component of $\nabla\times\mathscr{E}^+$, which depends upon $\mathscr{E}_X^+$ and $\mathscr{E}_Z^+$. Since B_j^+ has only a Y-component, it follows from (18.112) that $(\nabla\times\mathscr{E}^+)_Y$ is small $O(R^{-1})$ because the off-diagonal terms of α_{ij} and β_{ij}, defined in (18.113) and (18.114), are both second order in the transformation strain $\partial X_i/\partial x_j$, and hence $O(R^{-1})$. Consequently the term $\eta(\nabla\times\mathscr{E}^+)_Y$ is small $O(R^{-2})$ and can be neglected compared to the other terms in (18.111). The hydromagnetic equation (18.111) reduces, then, to the simple form

$$\partial B^+/\partial t = R^{-1}(\mathbf{b}_M\cdot\nabla V^+ - \mathbf{v}_M\cdot\nabla B^+ + \lambda V_{\max}\,\partial^2 B^+/\partial X_j\,\partial X_j) + O(R^{-\frac{3}{2}}). \tag{18.137}$$

With the neglect of $(\nabla\times\mathscr{E}^+)_Y$, the coordinate Y disappears from the equation and the form is just that for two-dimensional fields and motions, producing B^+ as a consequence of the shearing of $\mathbf{b}_M$ by V^+. The equation is just (18.4), with the addition of the small meridional velocity $\mathbf{v}_M^+(X, Z)$. The complexity caused by $\mathbf{w}(x, y, z)$ is removed by the coordinate transformation and the transformed field component $\mathbf{b}_M^+$.

The poloidal or meridional field $\mathbf{b}_M^+$ is conveniently written as $\nabla\times\mathbf{e}_Y a^*$. For the field (18.120) the meridional component of (18.111) is given by the curl of

$$R^{-1}\frac{\partial a^*}{\partial t} = (\mathbf{v}^+\times\mathbf{B}^+)_Y + R^{-1}\eta\frac{\partial^2 a^*}{\partial X_j\partial X_j} + \eta\mathscr{E}_Y^+.$$

The Y-component of $\mathbf{v}^+\times\mathbf{B}^+$ involves the meridional components of $\mathbf{v}^+$ and $\mathbf{B}^+$ from (18.119) and (18.120). It is readily shown to be $-R^{-2}\mathbf{w}_M\cdot\nabla a^+$. The Y-component of $\mathscr{E}_i^+$ follows from (18.112) as

$$\mathscr{E}_Y^+ = \langle\alpha_{22}\rangle B^+ + R^{-1}(\alpha_{2j}b_j^+ + \beta_{1j}\,\partial b_3^+/\partial X_j - \beta_{3j}\,\partial b_1^+/\partial X_j).$$

The leading term is $\langle\alpha_{22}\rangle B^{+}$, with the others smaller by the factor R^{-1}. It follows from (18.113) and (18.121) that

$$\alpha_{22}=R^{-\frac{1}{2}}\frac{\partial}{\partial X_2}\left(\frac{\partial\zeta_3}{\partial X_1}-\frac{\partial\zeta_1}{\partial X_3}\right)+R^{-1}\left\{\frac{\partial\zeta_3}{\partial X_n}\frac{\partial^2\zeta_1}{\partial X_n\partial X_2}-\frac{\partial\zeta_1}{\partial X_n}\frac{\partial^2\zeta_1}{\partial X_n\partial X_2}\right\}+O(R^{-\frac{3}{2}}),$$

so that the mean value over $X_2=Y$ is just

$$\langle\alpha_{22}\rangle=R^{-1}\left\langle\frac{\partial\zeta_3}{\partial X_n}\frac{\partial^2\zeta_1}{\partial X_n\partial X_2}-\frac{\partial\zeta_1}{\partial X_n}\frac{\partial^2\zeta_3}{\partial X_n\partial X_2}\right\rangle.$$

Altogether it follows that

$$R^{-1}\frac{\partial a^{+}}{\partial t}+R^{-2}\mathbf{w}_{\mathrm{M}}^{+}\cdot\nabla a^{+}=R^{-2}\lambda V_{\max}\,\partial^2 a^{+}/\partial X_j\partial X_j+\Gamma B^{+},\quad(18.138)$$

where

$$\Gamma\equiv\eta\langle\alpha_{22}\rangle=\lambda V_{\max}R^{-1}\langle\alpha_{22}\rangle$$

$$=\lambda V_{\max}R^{-2}\left\langle\frac{\partial\zeta_3}{\partial X_n}\frac{\partial^2\zeta_1}{\partial X_n\partial X_2}-\frac{\partial\zeta_1}{\partial X_n}\frac{\partial^2\zeta_3}{\partial X_n\partial X_2}\right\rangle.\quad(18.139)$$

The equation (18.137) for B^{+} is the cartesian analogue of (18.105) for the azimuthal field, while (18.138) corresponds to (18.106), with $A_e=R^{-1}a^{+}$. It is evident that, in Soward's development, the definition of the effective potential A_e comes as a direct consequence of the geometry of the fluid motion, rather than Braginskii's feat of serendipity. Soward has demonstrated that the choice of the effective potential is possible because the streamlines are essentially closed, permitting the averaging to be performed in a much simpler fashion (see also Gubbins 1973b).

The reader interested in further discussion of the mathematical properties of the Braginskii–Soward equations should consult the reviews by Soward and Roberts (1976) and by Moffatt (1976, 1978) as well as the original papers of Soward and Braginskii.

18.6.3. *Discussion of the Braginskii–Soward results*

The dynamo effect, of Braginskii's ordered solution near axial symmetry, is contained in the dynamo coefficient Γ, defined in (18.139). It is interesting to compare the dynamo coefficient for this case with the result (18.80) and (18.87) for small magnetic Reynolds number. The expressions for Γ are different algebraic form, of course, which is to be expected from the totally different physical circumstances for which they are derived. In the present case, of large overall magnetic Reynolds number $R=\lambda V_{\max}/\eta$, the dynamo effect is directly proportional to the diffusion coefficient η, while for small local magnetic Reynolds number $R_{\mathrm{M}}=Lv/\eta$, the dynamo effect is inversely proportional to η. On the other hand, in both circumstances the dynamo effect reduces to a single, scalar coefficient. This fact is to be attributed to the small amplitude of the magnetic effects of the individual fluctuations in the field in both cases. It is in contrast to the more complex dynamo effects of §18.3, arising from the large amplitude of the displacement of the field in each eddy. So long as the fluctuations are small the basic scalar form (18.7) prevails.

It is instructive to look briefly at an example of a simple fluid motion that leads to dynamo activity. First of all, it is evident from (18.139) that it is the fluid displacement (ζ_1, ζ_3) transverse to the principal field $B_2^+ = B^+(X, Z)$ that produces the dynamo effect, because the parallel displacement ζ_2 does not interact with B_2^+. Suppose, then, that

$$\zeta_i = L\{s_i \sin(kY) + c_i \cos(kY)\} \tag{18.140}$$

where L is the characteristic amplitude of the displacement and k is the wave number. The vectors s_i and c_i are fixed unit vectors in the meridional XZ-plane, with $s_2 = c_2 = 0$ so that the divergence of ζ_i vanishes, as required by (18.117). If the fluid velocity in the Y-direction has the local value $V^+(X, Z)$, as in (18.119), then the transformation displacement $R^{-\frac{1}{2}}\zeta_i$ involves the fluid velocity

$$\begin{aligned} R^{-\frac{1}{2}}\,\mathrm{d}\zeta_i/\mathrm{d}t &= R^{-\frac{1}{2}}V^+\,\partial\zeta_i/\partial Y \\ &= R^{-\frac{1}{2}}V^+kL\{s_i \cos(kY) - c_i \sin(kY)\} \end{aligned}$$

with the helicity

$$\alpha = R^{-1}V^{+2}k^3L^2(s_3c_1 - s_1c_3), \tag{18.141}$$

The dynamo coefficient is readily computed from (18.139) to be

$$\Gamma = \lambda V_{\max}R^{-2}k^3L^2(s_1c_3 - s_3c_1), \tag{18.142}$$

$$= -\lambda R^{-1}V_{\max}\alpha/V^{+2},$$

$$= -\alpha\eta/V^{+2}. \tag{18.143}$$

This result, it will be noted, is proportional to the helicity and *inversely* proportional to V^{+2}. The toroidal velocity enters directly into the dynamo coefficient, something it does not do in any of the other cases treated thus far. If the helicity is written in terms of the asymmetric 'turbulent' component of the velocity, appearing as $R^{-\frac{1}{2}}\mathbf{w}$ in (18.101) with a characteristic scale of the same order in R as the dimension λ of the mean field, then $\alpha = \epsilon(R^{-\frac{1}{2}}w)^2/\lambda$, where ϵ is a number of the order of unity and w is $|\mathbf{w}|$. The dynamo coefficient Γ becomes

$$\Gamma = -\epsilon(\eta/\lambda)R^{-1}(w/V^+)^2.$$

In order of magnitude, $w = V^+$ so that

$$\Gamma \cong -(\eta/\lambda)R^{-1} = -V^+R^{-2}.$$

The magnetic Reynolds number $R_M = (R^{-\frac{1}{2}}w)\lambda/\eta \cong R^{\frac{1}{2}}$ for the asymmetric motion is 'moderately' large, being of the order of the square root of the overall magnetic Reynolds number R.

As noted earlier, the immediate theoretical value of the Braginskii–Soward formulation of the hydromagnetic dynamo effect lies in its demonstration of the generality of the dynamo equation (18.7) in the extreme case of large magnetic Reynolds number in fluid motions that are nearly axi-symmetric. The form of the dynamo equation (18.138) shows again that the principal dynamo effect arises through the distortion of the toroidal lines of force into spirals, illustrated in Fig. 18.4, and, hence, is equivalent to the generation of a toroidal vector potential at a rate proportional to the mean helicity of the fluid motions.

Beyond this direct application, however, there is the intrinsic interest in the mathematical formulation, beginning with the general expression of the hydromagnetic equation in the form (18.111) in terms of the initial coordinates X_i of the Lagrangian transformation $x_i = x_i(X_k, t)$. Soward has emphasized that (18.111) is an exact and complete result, in no way restricted to the 'small' coordinate transformation (18.117) to which it was applied in the present calculations. Thus (18.111) gives the time evolution of a magnetic field in terms of the initial position X_k of each volume element of field and fluid. The potential for the theoretical treatment of the convection and diffusion of magnetic field in various physical circumstances is only beginning to be explored. Perhaps the most important consequence of the Braginskii–Soward formulation will prove to be its potential for treating dynamical problems (Braginskii 1964c, 1967, 1970; Roberts and Soward 1972; Soward and Roberts 1976; Moffatt 1978).

References

ALFVEN, H. (1947). *Mon. Not. Roy. Astron. Soc.* **107,** 211.
ANGEL, J. R. P. (1977). *Astrophys. J.* **216,** 1.
BABCOCK, H. D. (1959). *Astrophys. J.* **130,** 364.
BABCOCK, H. W. (1947). *Astrophys. J.* **105,** 105.
—— (1958). *Astrophys. J. Suppl.* **3,** (3) 141.
—— (1960a). *Stars and stellar systems* Vol. 6, *Stellar Atmospheres* (ed. J. L. GREENSTEIN) p. 282. University of Chicago Press.
—— (1960b). *Astrophys. J.* **132,** 521.
—— (1963). *Ann. Rev. Astron. Astrophys.* **1,** 41.
BABCOCK, H. W. and BABCOCK, H. D. (1955). *Astrophys. J.* **121,** 349.
BACKUS, G. E. (1958). *Ann. Phys.* **4,** 372.
—— and CHANDRASEKHAR, S. (1956). *Proc. Nat. Acad. Sci.* **42,** 105.
BENTON, E. R. (1975). *Geophys. J. Roy. Astron. Soc.* **42,** 385.
BERGE, G. L. and SEIELSTAD, G. A. (1967). *Astrophys. J.* **148,** 367.
BIERMANN, L. (1953). *Kosmische Strahlung* (ed. W. HEISENBERG). Springer-Verlag, Berlin.

BIERMANN, and SCHLÜTER, A. (1951). *Phys. Rev.* **82,** 863.
BLACKETT, P. M. S. (1947). *Nature* **159,** 658.
—— (1952). *Phil. Trans. Roy. Soc. Lond. A.* **245,** 309.
BRAGINSKII, S. I. (1964a). *Zh. Eksp. teor. Fiz.* **47,** 1084, 2178; *Sov. Phys. JETP* **20,** 726, 1462.
—— (1964b). *Geomagn. i. Aeronomiya* (USSR) **4,** 732; *Geomagn. and Aeron.* **4,** 572.
—— (1964c). ibid. **4,** 898; ibid. **4,** 702.
—— (1967). ibid. **7,** 1050; ibid. **7,** 851.
—— (1970). ibid. **10,** 3, 221; ibid. **10,** 1, 172.
—— (1975). ibid. **15,** 149.
BULLARD, E. C. (1949a). *Proc. Roy. Soc. Ser. A* **197,** 433.
—— (1949b). ibid. **199,** 413.
—— (1950). *Observatory* **70,** 139.
—— and GELLMAN, H. (1954). *Phil. Trans. Roy. Soc. Lond. A* **247,** 213.
BUSSE, F. H. (1970a). *J. Fluid Mech.* **44,** 441.
—— (1970b). *Astrophys. J.* **159,** 629.
—— (1973). *J. Fluid Mech.* **57,** 529.
—— (1975a). *Rev. Geophys. Space Res.* **13,** No. 3, 206.
—— (1975b). *J. Geophys. Res.* **80,** 278.
CHANDRASEKHAR, S. (1955). *Proc. Roy. Soc., Ser. A* **204,** 435.
CHAPMAN, S. and BARTELS, J. (1940). *Geomagnetism*, vol. I. Clarendon Press, Oxford.
CHILDRESS, (1967a). *Courant Inst. Math. Sci.* MF-53, MF-34. New York University, New York.
—— (1967b). *Geophysical Fluid Dynamics*, Ref. No. 67-54, vol. 1, p. 165. Woodshole Oceanographic Institute.
CHILDRESS, S. and SOWARD, A. M. (1972). *Phys. Rev. Letters* **29,** 837.
COWLING, T. G. (1934). *Mon. Notic. Roy. Astron. Soc.* **94,** 39.
—— (1957). *Quart. J. Mech. Appl. Math.* **10,** 129.
COX, A. and DALRYMPLE, G. B. (1967). *Earth Planet. Sci. Lett.* **3,** 173.
DEINZER, W. and STIX, M. (1971). *Astron. Astrophys.* **12,** 111.
DOELL, R. R. and COX, A. (1965). *J. Geophys. Res.* **70,** 3377.
DOLGINOV, SH. SH., YEROSHENKO, YE. G., and ZHUSGOV, L. N. (1973). *J. Geophys. Res.* **78,** 4779.
EDDY, J. A., GILMAN, P. A., and TROTTER, D. E. (1976). *Solar Phys.* **46,** 3.
ELSASSER, W. M. (1939). *Phys. Rev.* **55,** 489.
—— (1941). ibid. **60,** 876.
—— (1945). ibid. **69,** 106.
—— (1946). ibid. **70,** 202.
—— (1947). ibid. **72,** 821.
—— (1950a). ibid. **79,** 183.
—— (1950b). *Rev. Mod. Phys.* **22,** 1.
—— (1955). *Amer. J. Phys.* **23,** 590.
—— (1956a). *Rev. Mod. Phys.* **28,** 135.
—— (1956b). *Amer. J. Phys.* **24,** 85.
—— and TAKEUCHI, H. (1955). *Trans. Amer. Geophys. Union* **36,** 584.
GAILITIS, A. (1967). *Magnitnaya Gidrodinamika* **3,** 45.
—— (1969). ibid. **5,** 31.
—— (1970). ibid. **6,** 19.
—— (1973). ibid. **9,** 12.

—— and KLAWINA, A. P., LIELAUSIS, O. A., and TIMANIS, L. L. (1972). *Proc. 6th Riga Conf. Magnetohydrodynamics* **1,** 199.
GIBSON, R. D. (1968). *Quart. J. Mech. appl. Math.* **21,** 243, 257.
—— and ROBERTS, P. H. (1969). *The application of modern physics to the earth and planetary interiors* (ed. S. K. RUNCORN), p. 577. Wiley, London.
GILMAN, P. A. (1969). *Solar Phys.* **8,** 316; **9,** 3.
—— (1972). ibid. **27,** 3.
—— (1974). *Ann. Rev. Astron. Astrophys.* **12,** 47.
GUBBINS, D. (1973a). *Phil. Trans. Roy. Soc. Lond. A* **274,** 493.
—— (1973b). *Geophys. J. Roy. Astron. Soc.* **33,** 57.
—— (1974). *Studies in Appl. Math.*, (Mass. Inst. Tech.), **103,** 157.
GUNN, J. and OSTRIKER, J. P. (1969). *Nature.* **221,** 454.
HALE, G. E. (1908a). *Pub. Astron. Soc. Pacific* **20,** 220, 287.
—— (1908b). *Astrophys. J.* **28,** 315.
—— (1913). ibid. **38,** 27.
——SEARES, F. H., VAN MAANEN, A., and ELLERMAN, F. (1918). *Astrophys. J.* **47,** 206.
HERZENBERG, A. (1958). *Phil. Trans. Roy. Soc. Lond. A* **250,** 543.
HILTNER, W. A. (1949). *Astrophys. J.* **109,** 471.
—— (1956). *Astrophys. J. Suppl.* **2,** 389.
JEPPS, S. A. (1975). *J. Fluid Mech.* **67,** 625.
JOY, A. H. and HUMASON, M. L. (1949). *Pub. Astron. Soc. Pacific* **61,** 133.
KEMP, J. C. (1970). *Astrophys. J.* **162,** 169.
—— (1977). ibid. **213,** 794.
KRAICHNAN, R. H. (1976a). *J. Fluid Mech.* **75,** 657.
——(1976b). ibid. **77,** 753.
KRAUSE, F. (1968). *Z. Ang. Math. Mech.* **48,** 333.
—— and ROBERTS, P. H. (1973). *Astrophys. J.* **181,** 977.
—— and STEENBECK, M. (1967). *Z. Naturforsch. A* **22,** 761.
KUNKEL, E. E. (1973). *Astrophys. J. Suppl.* **25,** 1.
LARMOR, J. (1919). *Rep. Brit. Assoc. Adv. Sci.*, 159.
LEIGHTON, R. B. (1969). *Astrophys. J.* **156,** 1.
LILLEY, F. E. M. (1970). *Proc. Roy. Soc. Ser. A* **316,** 153.
LORTZ, D. (1968a). *Plasma Phys.* **10,** 967.
—— (1968b). *Phys. Fluids* **11,** 913.
—— (1972). *Z. Naturforsch. A* **27,** 1350.
LOVELL, B. (1971). *Quart. J. Roy. Astron. Soc.* **12,** 98.
LOWES, F. J. and WILKINSON, I. (1963). *Nature* **198,** 1158.
—— —— (1968). ibid. **219,** 717.
LUYTEN, W. J. (1949). *Astrophys. J.* **109,** 532.
MANCHESTER, R. N. (1974). *Astrophys. J.* **188,** 637.
MESTEL, L. and MOSS, D. L. (1977). *Mon. Notic. Roy. Astron. Soc.* **178,** 27.
MOFFATT, H. K. (1970a). *J. Fluid Mech.* **41,** 435.
—— (1970b). ibid. **44,** 705.
—— (1972). ibid. **53,** 385.
——(1974). ibid. **65,** 1.
——(1976). *Adv. Appl. Mech.* **16,** 119.
——(1978) *Magnetic field generation in electrically conducting fluids*, Cambridge University Press, Cambridge.
MOFFETT, T. J. (1975). *Ap. J. Suppl.* **29,** 1.
MOSS, D. L. (1977). *Mon. Notic. Roy. Astron. Soc.* **178,** 51, 61.

MULLAN, D. J. (1976). *Astrophys. J.* **210,** 702.
NAKAGAWA, Y. and SWARZTRAUBER, P. (1969). *Astrophys. J.* **155,** 295.
NESS, N. F. BEHANNON, K. W., LEPPING, R. P., and WHANG, Y. C. (1976). *Icarus* **28,** 479.
PACINI, F. (1968). *Nature* **219,** 145.
PARKER, E. N. (1955a). *Astrophys. J.* **121,** 491.
—— (1955b). ibid. **122,** 293.
—— (1957). *Proc. Nat. Acad. Sci.* **43,** 8.
—— (1966). *Astrophys. J.* **145,** 811.
—— (1969). *Space Sci. Rev.* **9,** 651.
—— (1970a). *Astrophys. J.* **160,** 383.
—— (1970b). ibid. **162,** 665.
—— (1971). ibid. **168,** 239.
—— and KROOK, M. (1956). *Astrophys. J.* **124,** 214.
PEKERIS, C. L., ACCAD, Y. and SHKOLLER, B. (1973). *Phil. Trans. Roy. Soc. Lond.* A **275,** 425.
PRESTON, G. W. (1967). *The magnetic and related stars* (ed. R. C. CAMERON), p. 3. Mono Book Corp., Baltimore.
—— (1971). *Astrophys. J.* **164,** 309.
RÄDLER, K. H. (1968). *Z. Naturforsch. A* **23a** 1841.
—— (1969). *Monat. Deutsch. Akad. Wiss.* **11,** 194, 272.
RIKITAKE, T. (1958). *Proc. Cambridge Phil. Soc.* **54,** 89.
—— (1966). *Electromagnetism and the earth's interior,* Elsevier, Amsterdam.
ROBERTS, G. O. (1970). *Phil Trans. Roy. Soc. Lond. A* **266,** 535.
—— (1972). ibid. **271,** 411.
ROBERTS, P. H. (1967). *An introduction to magnetohydrodynamics,* Longmans, London.
—— (1971), *Lectures in Applied Math.* (ed. W. H. REID), vol. 14, p. 129. American Math. Soc., Providence, Rhode Island.
—— and SOWARD, A. M. (1972). *Ann. Rev. Fluid Mech.* **4,** 117.
—— —— (1975). *Astron. Nachr.* **296,** 49.
—— and STEWARTSON, K. (1974). *Proc. Roy. Soc. Lond. A* **277,** 287.
RUNCORN, S. K. (1955). *Adv. Phys.* **4,** 244.
RUSSELL, C. T. (1976). *Geophys. Res. Lett.* **3,** 413, 589.
SMITH, E. J., DAVIS, L., JONES, D. E., COLEMAN, P. J., COLBURN, D. S., DYAL, P., SONETT, C. P., and FRANDSEN, A. M. A. (1974). *J. Geophys. Res.* **79,** 3501.
SOWARD, A. M. (1971). *J. Math. Phys.* **12,** 1900, 2052.
—— (1972). *Phil. Trans. Roy. Soc. Lond.* **272,** 431.
—— (1974). ibid. **275,** 611.
—— (1975). *J. Fluid Mech.* **69,** 145.
—— and ROBERTS, P. H. (1976). *Magnitnaya Gidrodinamika* **1,** 3.
STEENBECK, M. and KRAUSE, F. (1966). *Z. Naturforsch. A* **21,** 1285.
—— —— (1967). *Magnitnaya Gidrodinamika* **3,** 19.
—— —— and RÄDLER, K. H. (1966). *Z. Naturforsch. A* **21,** 369.
STENFLO, J. O. (1970). *Solar Phys.* **14,** 263.
STEVENSON, D. J. (1974). *Icarus* **22,** 403.
STIX, M. (1972). *Astron. Astrophys.* **13,** 203.
TAKEUCHI, H. and ELSASSER, W. M. (1954). *J. Phys. Earth* **2,** 39.
—— and SHIMAZU, Y. (1954). *J. Phys. Earth* **2,** 5.
TOUGH, J. G. (1967). *Geophs. J. Roy. Astr. Soc.* **13,** 393; **15,** 343.
—— and ROBERTS, P. H. (1968). *Phys. Earth Planet. Interiors* **1,** 288.

TVERSKOY, B. A. (1965). *Geomagn. Aeron.* **5,** 11.
UNZ, F. and WALTER K. (1969). *Solar Phys.* **8,** 310.
VAINSHTEIN, S. I. (1970). *Zh. Eksp. Teor. Fiz.* **58,** 153 (Soviet Phys. JETP **31,** 87).
—— and ZELDOVICH, YA. B. (1972). *Usp. Fiz. Nauk* **106,** 431 (Sov. Phys. Usp. **15,** 159).
WARWICK, J. A. (1961). *Astrophys. J.* **137,** 41.
WILSON, O. C. (1966). *Astrophys. J.* **144,** 695.
—— (1968). ibid. **153,** 221.
—— (1976). ibid. **205,** 823.
WILSON, R. L. (1966). *Geophys. J.* **10,** 413.
—— (1967). *Magnetism and the cosmos,* (ed. W. R. HINDMARSH, F. J. LOWES, P. H. ROBERTS, and S. K. RUNCORN), p. 79. Oliver and Boyd, London.
YOSHIMURA, H. (1972). *Astrophys. J.* **178,** 863.
—— (1973). *Solar Phys.* **33,** 131.
—— (1975). *Astrophys. J. Suppl.* **29,** 467.

19

THE PHYSICAL NATURE OF THE GENERATED FIELDS

19.1. The dynamo problem

THE generation of magnetic field by turbulent fluid motion was worked out and discussed at length in the foregoing chapter. The basic generation effect arises from the deformation of the magnetic lines of force from straight lines into spirals, illustrated schematically in Fig. 18.4. The spiralling represents a vector potential in the direction of the initial mean field, in the manner described by the basic dynamo equation (18.7) (Parker 1955). The question that confronts us now is the physical nature of the magnetic fields produced in this manner. In particular, the liquid core of Earth is a thick shell of convecting fluid and the external field it produces is a stationary dipole. The convective zone of the sun is a thin shell, and the field it produces varies periodically with time, on a 22-year time scale. The interstellar gas forms a thin disc and the field it produces is predominantly azimuthal; the time variation and the poloidal part are unknown.

The precise form and time behaviour of the magnetic field generated by a rotating, convecting fluid body depends, sometimes very sensitively, on the precise form of the fluid motions within the body. It has been emphasized that such detailed information is not yet available for any astronomical body, from either observation or dynamical theory. The basic *qualitative* features of the fluid motion are known, however, and permit deduction of the *qualitative* form of the magnetic fields. The general motions were discussed in §18.1 and their properties follow from the two facts that all astronomical bodies rotate, and most of them are subject to internal turbulence and convection. Consequently, the rotation is non-uniform and the convection is cyclonic. We can estimate or observe the magnitude of the non-uniform rotation for Earth and for the sun even if we cannot be sure of the direction of the gradient of the rotation. We can estimate the magnitude of the cyclonic rotation of the convective cells from the angular velocity of the parent body and the velocity of the convection. Hence the order of magnitude of the dimensionless numbers describing the dynamo effects can be estimated, giving, as we shall see, sound reason for believing that the magnetic fields of Earth, of the sun, and of the galaxy are generated by the fluid motions within. That is to say, the observed time behaviour and spatial form of the magnetic fields are encompassed by the formal solutions of the dynamo equations for the estimated values of the non-uniform rotation and cyclonic velocity of the convection. The general hydromagnetic theory for the origin of the magnetic fields in the astronomical universe proves entirely satisfactory within the framework of existing observational knowledge. It is to be hoped that present theoretical work on the very difficult problem of the fluid motions in convecting, rotating, stratified bodies will be successful in providing a more quantitative picture of the fluid motions. The dynamo theory can then be combined with the various dynamical models for the fluid motion and the two together compared with observation to build a more complete picture of the magnetic generation in the rotating astronomical body.

The present task is to develop a physical understanding of the general solutions of the dynamo equations and then to explore the special solutions that arise in the variety of circumstances that may be imagined in various astronomical bodies. The solution of even the simplest form (18.7) of the dynamo equations in a spherical body proves to be a complex undertaking, with complicated results. The discussion begins, therefore, on a much more elementary level. The solution of the dynamo equations (18.4) and (18.7) are made up of four basic wave modes. In rectangular coordinates in an infinite space the four modes, each with their own peculiar properties, can be expressed in simple terms. Then plane boundaries are introduced so that the reflection and refraction of the dynamo waves can be studied. Finally the waves are boxed in on all sides by boundaries (in §19.3) and the dynamo in a finite body is explored. This leads to simple rectangular models illustrating the general dynamo effects presumed to occur in the liquid core of Earth, in the convective zone of the sun, and in the gaseous disc of the galaxy. Only after this extensive study of the basic physical properties of dynamos and migratory dynamo waves do we venture on to treat the axi-symmetric dynamo in spherical and spheroidal bodies, in §19.4 and later sections. The spherical dynamo is essential, because it was, after all, a scientific curiosity about the magnetic field of Earth and the magnetic field of the sunspot that motivated the long history of development leading to the present theory. Fortunately a number of authors have devoted their efforts to the difficult solution of the dynamo equations (18.4) and (18.7) in spherical bodies, so that there is today an extensive catalogue of solutions illustrating the many possible configurations and time variations of the magnetic fields. The range of possibilities proves extensive. It will be seen from the first simple treatment of the plane dynamo waves that the behaviour of the field depends qualitatively on whether the rotation varies principally with radius r or with latitude θ. The sign and the latitude distribution of the cyclonic turbulence play a crucial role, as does the possibility for large-scale meridional circulation. Magnetic buoyancy has qualitative effects, and in the fast spinning core of Earth there is at least the theoretical possibility of negative diffusion. The occasional abrupt reversal of the magnetic field of Earth can be caused by an increase in the level of non-uniform rotation and convection in the core of Earth, a change in meridional circulation, or a sudden change in the distribution of any of these with no change in the general level. Indeed, the work carried out so far has served to open up, rather, than close off, the theoretical possibilities. The present chapter provides illustrative examples of the major physical effects that make up the turbulent dynamo.

19.2. The properties of the dynamo equation

19.2.1. The basic form of the equations

For the most part, the fluid motions in planets, stars, and galaxies consist of an axi-symmetric non-uniform rotation, $\Omega(\varpi, z)$ and an axi-symmetric statistical distribution of cyclonic convection. There may be giant circulation cells in the sun, or in the core of Earth, violating this simple generalization, but we know so little about non-symmetric circulation that we cannot include it in any meaningful way. Hence the discussion that follows is limited to the axi-symmetric case, $\partial/\partial\phi = 0$. The presentation begins in cartesian coordinates where the physical properties of the dynamo waves can be illustrated by simple mathematical examples. We adopt the convention that the y-axis corresponds to the azimuthal direction, of the ignorable coordinate ϕ. Hence $\partial/\partial y = 0$ and the non-uniform rotation is represented by a large-scale velocity $V(x, z)$ in the y-direction. For strong non-uniform rotation, i.e. large $\partial V/\partial x, \partial V/\partial z$, the dominant field, then is $B_y(x, z, t)$, corresponding to the azimuthal component in a spherical body.

We expect a local preferred direction for the rotation axis of the cyclonic eddies or convective cells. The direction is determined by the local large-scale shear ∇V and by the axis of rotation of the astronomical body. For the time being, working in an infinite space, it matters little whether that direction is thought of as 'vertical', or 'parallel' to the rotation axis, or whatever. In keeping with the convention of §18.3 the z-axis is oriented along the rotation axis of the local cyclones.

There are no natural circumstances known to us where there occurs the statistically isotropic orientation of cyclonic rotation explicitly assumed by many authors. However, as we shall see, if the discussion is limited to strong non-uniform rotation and to eddies with short correlation times or small magnetic Reynolds numbers, so that the net rotation Φ of the field in each eddy is small, the form of the dynamo equations is independent of the statistical orientation of the cyclonic eddies. Hence the assumption of statistical isotropy yields the relevant *form* of the equations whether the eddies are statistically isotropic or not. Thus the many illustrative examples to be found in the literature with the isotropy label are in fact directly applicable to the important physical circumstance of strong uniform rotation and $\Phi < 1$.

To get back to the general case, that the cyclonic rotation of an eddy is not small compared to one, suppose that the rotation of the cyclonic eddies is about the z-axis. Use (18.49) and (18.50) with $F(\Phi)$ and $G(\Phi)$ given by (18.51)–(18.57). Denote by Q and Γ the dynamo coefficients

$$Q = \frac{1}{\tau}\langle \xi_x \, \partial \xi_z / \partial x \rangle, \qquad \Gamma = \frac{1}{\tau}\langle \xi_z \, \partial \xi_x / \partial y \rangle, \tag{19.1}$$

so that from (18.49) and (18.50)

$$Q = -(\pi a^2 c v_3 t_3 / V\tau) F(\Phi), \qquad \Gamma = -(\pi a^2 c v_3 t_3 / V\tau) G(\Phi). \tag{19.2}$$

For convenience denote by G_x and G_z the components $\partial V/\partial x$ and $\partial V/\partial z$, respectively. Then (18.37) and (18.39) yield

$$\left(\frac{\partial}{\partial t} - \eta \nabla^2\right) A_x + \eta \frac{\partial}{\partial x}\frac{\partial A_j}{\partial x_j} = -G_x A_y + \Gamma B_x - Q B_y,$$

$$\left(\frac{\partial}{\partial t} - \eta \nabla^2\right) A_y = +Q B_x + \Gamma B_y$$

$$\left(\frac{\partial}{\partial t} - \eta \nabla^2\right) A_z + \eta \frac{\partial}{\partial z}\frac{\partial A_j}{\partial x_j} = -G_z A_y$$

where, with $\partial/\partial y = 0$, the field components are

$$B_x = -\frac{\partial A_y}{\partial z}, \qquad B_y = \frac{\partial A_x}{\partial z} - \frac{\partial A_z}{\partial x}, \qquad B_z = \frac{\partial A_y}{\partial x}. \tag{19.3}$$

The most convenient form of the equations is in terms of A_y and B_y alone, for which

$$\left(\frac{\partial}{\partial t} - \eta \nabla^2 + \frac{\partial}{\partial z} Q\right) B_y = -\frac{\partial}{\partial z}\Gamma \frac{\partial A_y}{\partial z} + \frac{\partial}{\partial x} G_z A_y - \frac{\partial}{\partial z} G_x A_y, \tag{19.4}$$

$$\left(\frac{\partial}{\partial t} - \eta \nabla^2 + \frac{\partial}{\partial z} Q\right) A_y = \Gamma B_y. \tag{19.5}$$

The magnetic field B_y is easily eliminated using (19.5), so that (19.4) reduces to

$$\left\{ D \frac{1}{\Gamma} D + \frac{\partial}{\partial z}\Gamma \frac{\partial}{\partial z} + \frac{\partial}{\partial z} G_x - \frac{\partial}{\partial x} G_z \right\} A_y = 0 \tag{19.6}$$

for the vector potential, where $D \equiv \partial/\partial t - \eta\nabla^2 + (\partial/\partial z)Q$. Equations (19.4)–(19.6) are particularly useful for deducing boundary conditions at discontinuities in Γ, G_x, etc.

Consider the case where Q and Γ are uniform in space and time. Equation (19.6) reduces to

$$\{(\partial/\partial t - \eta\nabla^2)^2 + 2Q(\partial/\partial t - \eta\nabla^2)\,\partial/\partial z + (Q^2 + \Gamma^2)\,\partial^2/\partial z^2 + G_x\,\partial/\partial z - G_z\,\partial/\partial x\}A_y = 0. \quad (19.7)$$

Writing the general solution in the form

$$A_y = \exp(t/\tau - ik_x x - ik_z z), \quad (19.8)$$

the dispersion relation is

$$(1/\tau + \eta k_x^2 + \eta k_z^2)^2 - 2Q(1/\tau + \eta k_x^2 + \eta k_z^2)\,ik_z - (Q^2 + \Gamma^2)k_z^2 + i(G_z k_x - G_x k_z) = 0. \quad (19.9)$$

The dispersion relation is second order in $1/\tau$, with the two roots

$$\begin{aligned} 1/\tau &= -\eta(k_x^2 + k_z^2) + ik_z Q \pm \Gamma k_z\{1 + i(G_x k_z - G_z k_x)/\Gamma k_z^2\}^{\frac{1}{2}} \\ &= -\eta(k_x^2 + k_z^2) \pm \Gamma k_z R \cos\mu + ik_z(Q \pm \Gamma R \sin\mu), \end{aligned} \quad (19.10)$$

where

$$R = \{1 + (G_x k_z - G_z k_x)^2/\Gamma^2 k_z^4\}^{\frac{1}{4}} \quad (19.11)$$

$$\tan(2\mu) = (G_x k_z - G_z k_x)/\Gamma k_z^2. \quad (19.12)$$

The solutions of the dynamo equations represent migratory waves, which Parker (1955) called *dynamo waves*. The propagation of the waves can be deflected, or blocked, by the boundaries of the region, giving an endless variety of form and behaviour to the resulting magnetic fields.

Consider the case where k_x and k_z are real, so that the solution is bounded throughout an infinite space. Amplification of the field requires $\mathrm{Re}(1/\tau) > 0$. No generality is lost if Γ is chosen to be positive. It follows from (18.49) and (18.50) that Q is then also positive, for the various cyclonic convective cells treated in §18.33. No generality is lost if k_z is taken to be positive[1]. In that case the regenerative mode arises when $\pm$ is taken to be $+$ and

$$\Gamma k_z RR \cos\mu > \eta(k_x^2 + k_z^2).$$

Note that

$$\cos^2\mu = \tfrac{1}{2}(1 + 1/R), \qquad \sin^2\mu = \tfrac{1}{2}(1 - 1/R).$$

It is convenient to define the dynamo coefficient Γ in units of the resistive

[1] Replacing (k_x, k_z) by $(-k_x, -k_z)$ also reverses the sign of $\mathrm{Im}(1/\tau)$, with no effect on the physical properties of the solution.

dissipation as

$$\gamma \equiv \Gamma k_z/\eta(k_x^2 + k_z^2), \tag{19.13}$$

and the large-scale shear in units of the dynamo coefficient as

$$g \equiv (G_x k_z - G_z k_x)/\Gamma k_z^2, \tag{19.14}$$

so that

$$R = (1+g^2)^{\frac{1}{2}}.$$

Hence the condition for $\mathrm{Re}(1/\tau) > 0$ becomes

$$R(R+1) \geqslant 2/\gamma^2$$

or

$$2R \geqslant (1 + 8/\gamma^2)^{\frac{1}{2}} - 1. \tag{19.15}$$

It is obvious that this condition is fulfilled for any suitably large g or suitably large γ. The general condition is

$$g^2 \geqslant \frac{1}{2}\left\{\frac{8}{\gamma^2}\left(1+\frac{1}{\gamma^2}\right) - 1 - \left(1+\frac{4}{\gamma^2}\right)\left(1+\frac{8}{\gamma^2}\right)\right\}. \tag{19.16}$$

When $\gamma^2 > 1$ the right-hand side of this inequality is negative, so that the requirement is satisfied even for $g^2 = 0$. The sufficient condition $\gamma^2 > 1$ is just the case where the cyclonic convection is so strong that it can maintain a field on its own. This is the α^2-dynamo, whose existence was first pointed out by Steenbeck *et al.* (1966). If $\gamma^2 < 1$, then the cyclonic convection must be supplemented by shear (non-uniform rotation) g to regenerate the field. This is the usual case in nature. Any object with enough rotation and convection to operate as an α^2-dynamo will presumably rotate so non-uniformly that the more efficient production of field by non-uniform rotation is the principal source of field. The generation of azimuthal field needs to be supplemented by the cyclonic effect Γ only to the extent necessary to evade Cowling's theorem. The cyclonic motions are then inhibited principally by the stresses in the azimuthal field, so that the net effect is $\gamma^2 < g^2$. If $\gamma^2 \ll 1$, then the criterion for regeneration reduces to

$$g^2\gamma^4 \geqslant 4.$$

It is interesting to note that the secondary dynamo coefficient Q, defined in (19.1), plays no role in the real part of $1/\tau$, i.e. in the basic regeneration of the field. The effect represented by Q, it will be recalled from §18.3.4, generates vector potential in the direction perpendicular to the principal field, rather than parallel to it as does Γ. The secondary coefficient Q contributes to the frequency $\omega = \mathrm{Im}(1/\tau)$ of the dynamo wave, for which

$$\omega = \Gamma k_z R \sin\mu + k_z Q,$$
$$= \pm\Gamma k_z\{\tfrac{1}{2}R(R-1)\}^{\frac{1}{2}} + k_z Q. \qquad (19.17)$$

The dynamo waves with real k_x and k_z are stationary only if $\omega = 0$, i.e. if

$$R(R-1) = 2Q^2/\Gamma^2$$

or

$$2R = 1 + (1 + 8Q^2/\Gamma^2)^{\frac{1}{2}}. \qquad (19.18)$$

Write R in terms of g. Then raise (19.18) to the fourth power. The result can be written

$$g^2 = \frac{1}{2}\left\{8\frac{Q^2}{\Gamma^2}\left(1+\frac{Q^2}{\Gamma^2}\right) - 1 + \left(1+\frac{4Q^2}{\Gamma^2}\right)\left(1+8\frac{Q^2}{\Gamma^2}\right)^{\frac{1}{2}}\right\}. \qquad (19.19)$$

If $Q = 0$, then $g = 0$ and the only solution for (19.16) for amplification of the field is $\gamma \geqslant 1$. For the special case where the field is completely stationary, $\text{Re}(1/\tau) = \text{Im}(1/\tau) = 0$, the equality sign obtains in (19.16) and we have the requirement

$$\frac{8}{\gamma^2}\left(1+\frac{1}{\gamma^2}\right) - \left(1+\frac{4}{\gamma^2}\right)\left(1+\frac{8}{\gamma^2}\right)^{\frac{1}{2}} = \frac{8Q^2}{\Gamma^2}\left(1+\frac{Q^2}{\Gamma^2}\right) + \left(1+4\frac{Q^2}{\Gamma^2}\right)\left(1+8\frac{Q^2}{\Gamma^2}\right)^{\frac{1}{2}}, \qquad (19.20)$$

relating γ to Q/Γ. If $Q = 0$, then $\gamma = 1$.

It is possible to go on with the general case, plotting families of curves for the real and imaginary parts of $1/\tau$ in terms of the physical parameters Q, Γ, G_x, and G_z. In preparation for the introduction of boundaries, the problem must be turned around too, computing the four roots of (19.9) for k_x or k_z for a given $1/\tau$. However, the purpose is not to display the numerical relations so much as it is to obtain an understanding of the physical properties of the dynamo. The best way to develop the physics is to consider a variety of special cases, illustrating the contributions of each physical parameter separately. Only when that task is accomplished can we hope to understand the physical significance of the general solution.

It is apparent that the secondary dynamo coefficient Q has quantitative effects on the generation of field. For if we set the principal dynamo coefficient Γ equal to zero, (19.5) and (19.8) give the dispersion relation

$$1/\tau = -\eta(k_x^2 + k_z^2) + ik_z Q.$$

If k_z is real, so that the solution is bounded for all z, then the real part of $1/\tau$ is just $-\eta(k_x^2 + k_z^2)$ and the fields undergo resistive decay. Note, however, that Q contributes an imaginary part to $1/\tau$, causing the waves to migrate in the positive z-direction with a phase velocity Q while they decay. A similar conclusion obtains if k_z is complex. For in that case write $k_z = q_z - iK_z$. It follows that

$$\mathrm{Re}(1/\tau) = K_z(Q + \eta K_z) - \eta(k_x^2 + q_z^2),$$

$$\mathrm{Im}(1/\tau = q_z(Q + 2\eta K_z).$$

It is evident that k_x is of secondary importance and may be put equal to zero. The real part of $1/\tau$ can be positive if K_z is positive and sufficiently large. Then $\mathrm{Im}(1/\tau) > 0$ and the wave migrates in the positive z-direction (whether q_z is positive or negative). Note that the wave amplitude varies as $\exp(-K_z z)$ so that the wave migrates from large amplitudes at negative z to small amplitude at positive z with a phase velocity $Q + 2\eta K$. Thus each wave crest declines as it migrates and the fields are clearly not amplified. Their existence is maintained only by some source at $z = -\infty$. A similar conclusion obtains if K_z is large and negative. Altogether, then, the secondary coefficient Q does not provide a source of field in the dynamo equations.

The coefficient Q may not be negligible in rapidly rotating bodies such as the Earth, of course, and its quantitative effects may have to be taken into account. The presence or absence of Q may mean the difference between a stationary or an oscillatory dynamo, for instance. In such circumstances the possibility of a reduced, or even a negative, diffusion coefficient for the turbulent transport of field cannot be excluded (see §18.4). For the present, however, our knowledge of the fluid motions is so rudimentary that we concentrate on understanding only the basic dynamo effects, produced by Γ and an isotropic turbulent diffusion coefficient η_T.

Consider again the point made by Steenbeck *et al.* (1966), that the dynamo coefficient Γ alone, without large-scale shear, is capable of generating magnetic field. Whether the idea is applicable to the rapidly spinning convective bodies in the astronomical universe or not, the point is of general theoretical interest to the physicist. The behaviour of such a dynamo is readily demonstrated from the dynamo equations (19.4) and (19.5). With $Q = 0$, and with $G_x = G_z = 0$ so that there is no large-scale shear, the dispersion relation (19.9) reduces to

$$1/\tau = -\eta(k_x^2 + k_z^2) \pm \Gamma k_z.$$

Thus, for real wave numbers, amplification occurs for the upper sign (assuming $\Gamma k_z > 0$) provided only that $\Gamma k_z > \eta(k_x^2 + k_z^2)$. Note that $1/\tau$ is real and there is no oscillation or migration of field. The dynamo is intrinsically stationary. Steenbeck and Krause (1969b) have noted that the stationary effect of Γ alone (called the α^2-dynamo) provides a natural explanation for the stationary dynamo of Earth. It is an interesting point, but in as much as there are other stationary dynamos, the question is open and must be decided by dynamical considerations on the strength of the non-uniform rotation. The basic point is that the Coriolis forces on the local convergence of fluid in the individual convective cells cause the cells to rotate, producing Γ, while the Coriolis forces on the overall fluid displacement within the cells cause the non-uniform rotation. Both effects are proportional to the angular velocity of the body as a whole. We are inclined to follow Elsasser and Braginskii in the view that the non-uniform rotation is generally the stronger of the two effects. But we must remember that the last word has not yet been said on the theoretical dynamics of the convection and circulation, and non-uniform rotation, in planetary cores and stellar convective zones. If the differential rotation in some special case should prove to be sufficiently weak, then the α^2-dynamo will share the centre of the stage with the dynamo driven by strong differential rotation.

To continue with the dynamo without large-scale shear, note from (19.5) that

$$B_y = k_z A_y$$

for the regenerative mode, so that

$$B_x = iB_y, \qquad B_z = -i(k_x/k_z)B_y$$

with

$$B_y = B_0 \exp\{t/\tau - ik_x x - ik_z z\}.$$

Steenbeck, Krause, and Rädler treat the case of isotropic cyclonic turbulence (i.e. isotropic but not mirror symmetry), rather than the present circumstance wherein the cyclonic rotation is only in the z-direction. For the isotropic case the dynamo equations are (see §18.3.5)

$$(\partial/\partial t - \eta\nabla^2)\mathbf{B} = \nabla\times\Gamma\mathbf{B},$$

yielding the dispersion relation

$$(1/\tau + \eta k^2)\{(1/\tau + \eta k^2)^2 - \Gamma^2 k^2\} = 0$$

where $k^2 \equiv k_x^2 + k_y^2 + k_z^2$. The regenerative mode is

$$1/\tau = \Gamma k - \eta k^2$$

assuming Γ to be positive. For real $1/\tau + \eta k^2$, it follows that the dynamo equation reduces to

$$\nabla\times\mathbf{B} = \{(1/\tau + \eta k^2)/\Gamma\}\mathbf{B}.$$

The fields are force-free, subject to torsion along the lines of force. The field components are related by

$$B_x = -B_y \frac{k_y k - ik_x k_z}{k_x k + ik_y k_z}, \qquad B_y = -B_z \frac{k_z k - ik_x k_y}{k_y k + ik_x k_z}.$$

If $k_y = 0$, as in the present circumstance, then

$$B_x = +B_y\, ik_z/k, \qquad B_y = +B_z\, ik/k_x.$$

A variety of models of this interesting dynamo have been worked out in the context of the spherical geometry appropriate to astronomical bodies (cf. Steenbeck and Krause 1969a, b; Roberts 1972). The effect should be kept in mind for the time when better information is available on the relative strengths of the large-scale shear and the dynamo coefficient Γ.

19.2.2. Dynamo waves for strong shear

19.2.2.1. *General relations.* One of the simplest circumstances to treat mathematically, and probably the one most commonly occurring in rotating convecting bodies, is $(k_zQ)^2 \ll (k_2\Gamma)^2 \ll (G_zk_x - G_xk_z)^2$. The first inequality obtains for small angular rotation Φ within each cyclonic eddy, because $Q \propto \Phi^2$ while $\Gamma \propto \Phi$, or where there are equal numbers of up- and downdrafts with the same Φ^2 (see §18.3.4). The second inequality obtains when the non-uniform rotation is strong, causing B_y to be the dominant field as a consequence of the rapid shear of B_x and/or B_z. In this limit, then, the general dynamo equations (19.4) and (19.5) reduce to (18.4) and (18.7)

$$(\partial/\partial t - \eta\nabla^2)B_y = G_z\, \partial A_y/\partial x - G_x\, \partial A_y/\partial z, \tag{19.21}$$

$$(\partial/\partial t - \eta\nabla^2)A_y = \Gamma B_y. \tag{19.22}$$

no longer depending upon whether the cyclonic eddies rotate about the z-axis, since Γ appears only as the algebraic coefficient B_y. Hence these equations may be applied with either x or z in the 'vertical' direction, and in fact x will be chosen to represent the vertical direction in many of the illustrative examples that follow. With solutions of the form (19.8) the dispersion relation is (Parker 1955)

$$\{1/\tau + \eta(k_x^2 + k_z^2)\}^2 = i\Gamma(G_xk_z - G_zk_x) \tag{19.23}$$

in place of (19.9). It follows that

$$1/\tau = -\eta(k_x^2 + k_z^2) \pm \{i\Gamma(G_xk_z - G_zk_x)\}^{\frac{1}{2}}.$$

For real wave numbers k_x and k_z, so that the dynamo waves are bounded throughout the infinite space, it follows that

$$\mathrm{Re}(1/\tau) = \pm|\tfrac{1}{2}\Gamma Gk|^{\frac{1}{2}} - \eta(k_x^2 + k_z^2) \tag{19.24}$$

where $Gk \equiv G_xk_z - G_zk_x$ and $k = (k_x^2 + k_z^2)^{\frac{1}{2}}$. If for our choice of (k_x, k_z) ΓG is positive then

$$\mathrm{Im}(1/\tau) = \pm\tfrac{1}{2}(\Gamma Gk)^{\frac{1}{2}}, \tag{19.25}$$

while if ΓGk is negative,

$$\mathrm{Im}(1/\tau) = \mp|\tfrac{1}{2}\Gamma Gk|^{\frac{1}{2}}.$$

The upper sign yields growing fields if

$$|\Gamma Gk| > 2\eta^2 k^4. \tag{19.26}$$

If $|\Gamma Gk| < 2\eta^2 k^4$, then the mode decays, but at a more leisurely pace than with resistivity alone. The lower sign represents a mode for which the fields are actively destroyed by the dynamo mechanism. The growth or destruction can be very rapid. If, for instance, $|\Gamma Gk|^{\frac{1}{2}}$ is large compared to the resistive decay $2\eta k^2$, then $\mathrm{Re}(1/\tau) = \mathrm{Im}(1/\tau) = \omega$, and the time dependence is $\exp[\pm(1+\mathrm{i})\omega t]$. One mode increases, and the other diminishes, by a factor of $\exp(2\pi) = 535$ in one period of oscillation. Thus, fields may be built up or destroyed in a time very short compared to the overall decay time $1/\eta k^2$. Such an active dynamo effect is presumably responsible for the occasional sudden reduction, reversal, and restoration of the geomagnetic field, all in a period of a few thousand years. On the other hand, the passive resistive decay of the geomagnetic field has a characteristic time estimated at the much longer span of 3×10^4 years.

There is a growing magnetic field, then, provided only that the product of the cyclonic convective strength Γ and the large-scale shear G is of suitably large magnitude to overcome the resistive dissipation. In terms of the dimensionless dynamo number

$$N_{\mathrm{D}} = |\Gamma G|/\eta^2 k^3$$

the requirement is merely $N_{\mathrm{D}} > 2$.

It is of some interest (Parker 1971c) to consider an isolated packet of dynamo waves. As an illustration suppose that $G_z = 0$ while $\Gamma G_x > 0$ so that the waves migrate in the z-direction. A group of waves, $B_y(z, t)$, can be expressed as

$$B_y(z, t) = \int_{-\infty}^{+\infty} \mathrm{d}k f(k) \exp\{t/\tau(k) - \mathrm{i}kz\}$$

with $\tau(k)$ given by (19.23) and $f(k)$ given by

$$f(k) = \frac{1}{2\pi}\int_{-\infty}^{+\infty} \mathrm{d}z B_y(z, 0) \exp(\mathrm{i}kz)$$

in terms of the initial field $B_y(z, 0)$. As an example suppose that the initial field is the wave packet

$$B_y(z, 0) = B_0 \exp(-z^2/\Lambda^2 + \mathrm{i}k_0 z)$$

where $\Lambda k_0 \gg 1$, so that the envelope is a broad Gaussian many wavelengths across. It follows that

$$f(k) = B_0(\Lambda/2\pi^{\frac{1}{2}}) \exp\{-\tfrac{1}{4}(k - k_0)^2 \Lambda^2\}$$

and

$$B_y(z,t)=\frac{B_0\Lambda}{2\pi^{\frac{1}{2}}}\int_{-\infty}^{+\infty}\mathrm{d}k\exp\{-\tfrac{1}{4}(k-k_0)^2\Lambda^2-\mathrm{i}kz-\eta k^2t+(\Gamma G_x/2)^{\frac{1}{2}}k^{\frac{1}{2}}(1+\mathrm{i})t\}$$

for the regenerative mode. For large Λ the Gaussian factor $\exp\{-\frac{1}{4}(k-k_0)^2\Lambda^2\}$ is negligible except for k very close to k_0. So let $k=k_0+\kappa$ and neglect all terms second order in κ in the other terms. Thus

$$B_y(z,t)=B_0(\Lambda/2\pi^{\frac{1}{2}})\exp(t/\tau_0-\mathrm{i}k_0z)$$

$$\times\int_{-\infty}^{+\infty}\mathrm{d}\kappa\exp[-\tfrac{1}{4}\Lambda^2\kappa^2+\kappa\{-\mathrm{i}z-2\eta k_0t+(\tfrac{1}{2}\Gamma G_xk_0)^{\frac{1}{2}}(1+\mathrm{i})t/2k_0\}]$$

$$=B_0\exp(t/\tau_0-\mathrm{i}k_0z)\exp\{[\{-2\eta k_0+(\tfrac{1}{2}\Gamma G_xk_0)^{\frac{1}{2}}(1+\mathrm{i})/2k_0\}t-\mathrm{i}z]^2/\Lambda^2\},$$

where the growth rate at $k=k_0$ is denoted by

$$1/\tau_0=-\eta k_0^2+(\tfrac{1}{2}\Gamma G_xk_0)^{\frac{1}{2}}(1+\mathrm{i}).$$

If, for instance, $\mathrm{Re}(1/\tau_0)=0$, so that the amplitude of the mode $k=k_0$ is steady in time, then $(\frac{1}{2}\Gamma G_xk_0)^{\frac{1}{2}}=\eta k_0^2$ and

$$B_y(z,t)=B_0\exp\{\mathrm{i}k_0(\eta k_0t-z)\}$$

$$\times\exp[\{(2\eta k_0t-z)(\eta k_0t+z)+\mathrm{i}\tfrac{3}{2}\eta k_0t(2z-\eta_0k_0t)\}/\Lambda^2].$$

The basic wave $\exp\{\mathrm{i}k_0(\eta k_0t-z)\}$ persists but is strongly modulated by the envelope $\exp\{(2\eta k_0t-z)(\eta k_0t+z)/\Lambda^2\}$ which is a Gaussian with the central peak of height $\exp\{(\frac{3}{2}\eta k_0t)^2/\Lambda^2\}$ at the moving position $z=\frac{1}{2}\eta k_0t$. On this pattern, then, is superimposed the additional factor $\exp\{\mathrm{i}3\eta k_0t(z-\frac{1}{2}\eta k_0t)/\Lambda^2\}$ which moves with the Gaussian envelope and involves corruscations of increasingly fine scale, the effective wave number $3\eta k_0t/\Lambda^2$ growing linearly with time. There are, then, the phase velocity ηk_0 and the group velocity $\frac{1}{2}\eta k_0$. It is clear that the wave is highly dispersive in the course of its migration. This depends very much on the relative magnitude of ΓG_x and η^2k^3, of course. For suppose that $(\frac{1}{2}\Gamma G_xk_0)^{\frac{1}{2}}$ is put equal to $4\eta k_0^2$ instead of ηk_0^2. The behaviour of the wave packet is then rather different, with

$$B_y(z,t)=B_0\exp\{\mathrm{i}k_0(4\eta k_0t-z)\}\exp\{3\eta k_0^2t-(z-2\eta k_0t)^2/\Lambda^2\}.$$

The basic wave $\exp\{\mathrm{i}k_0(4\eta k_0t-z)\}$ is now modulated only by the Gaussian envelope $\exp[-(z-2\eta k_0t)^2/\Lambda^2]$ with its growing maximum of $\exp(3\eta k_0^2t)$ at the position $z=2\eta k_0t$.

Altogether, then, the migratory dynamo wave has a complicated personality, even in the simplest of circumstances. The sections that follow are aimed at exploring the more elementary properties of the dynamo wave, to illustrate some of the effects before entering into the formal calculations in spherical (astronomical) bodies.

19.2.2.2. *Physical properties of periodic waves.* As noted above, the growth rate of the magnetic fields, given by $\mathrm{Re}(1/\tau)$ can be very large. Whatever the initial strength of the field, it is rapidly amplified to a level where the field stresses come into dynamic balance with the cyclonic eddies that produce the field, thereby inhibiting ΓG and reducing the growth rate toward zero. This is the common circumstance in which we find magnetic fields in nature today, with the fields of the sun as the most immediate example of oscillatory dynamo waves with a stable amplitude, $\mathrm{Re}(1/\tau)=0$, $\mathrm{Im}(1/\tau)\equiv\omega\neq 0$.

Suppose, then, that $\mathrm{Re}(1/\tau)=0$. It follows at once from (19.24) and (19.25) that

$$|\Gamma(G_x k_z - G_z k_x)| = 2\eta^2(k_x^2+k_z^2) = 2\omega^2.$$

Hence the frequency is equal to the dissipation rate,

$$\omega = \eta(k_x^2+k_z^2).$$

If the wavelength of the dynamo waves is Λ and the period is T, so that $k_x^2+k_z^2=4\pi^2/\Lambda^2$ and $\omega=2\pi/T$, then

$$T=\Lambda^2/2\pi\eta.$$

The period of oscillation, then, is directly connected to the diffusion coefficient η, being essentially the diffusion time over the wavelength Λ. The relation has immediate consequences for the sun because the oscillatory fields of the sun have a period of about 22 years (Parker 1955, 1957). The sign of B_y, i.e. the sign of the azimuthal field B_ϕ, reverses across the equator, as indicated by the opposite polarity of bipolar magnetic regions north and south of the equator. Bipolar regions appear within about 45° latitude on either side of the equator, indicating the half wavelength on either side to be some 5×10^5 km, or $\Lambda\cong10^6$ km. It follows from (19.27) that a period of 22 years and a wavelength of 10^6 km are compatible if the resistive diffusion coefficient is $\eta=3\times10^{12}\ \mathrm{cm^2\,s^{-1}}$.

Now it is possible to be a little more precise in formulating the relation between the period, the wave number, and the resistive diffusion coefficient. Later on in Chapter 21 quantitative models of the solar dynamo are presented and it will be interesting to see how they compare with the simple result derived here. We should, therefore, distinguish the vertical and horizontal dimension of the dynamo. Suppose, then, that k_x represents the vertical wave number, while $k_z=2\pi/\Lambda$ is the horizontal wave number. Note that a dynamo wave in a convective zone of depth $h=2\times10^5$ km cannot have a vertical wave number less than about π/h.

Hence for h rather less than Λ, it follows that $k_x^2+k_z^2=(\pi/h)^2(1+4h^2/\Lambda^2)\cong(\pi/h)^2$. Hence $\omega\cong\eta(\pi/h)^2$ and

$$T\cong 2h^2/\pi\eta. \tag{19.27}$$

It follows that the 22-year period requires $\eta=0.4\times10^{12}\,\mathrm{cm^2\,s^{-1}}$.

Now the resistive diffusion coefficient varies with depth in the manner prescribed by (4.39). At depths of 10^4 and 10^5 km the temperature is approximately 7×10^4 and 1.6×10^6 K, yielding $\eta=5\times10^5$ and $0.5\times10^4\,\mathrm{cm^2\,s^{-1}}$, respectively. Neither of these values is anywhere near the required figure of $10^{12}\,\mathrm{cm^2\,s^{-1}}$. Indeed, with η as small as $5\times10^5\,\mathrm{cm^2\,s^{-1}}$ a period of 22 years implies $\Lambda=400$ km, smaller than the granule size. Alternatively, $\Lambda=10^6$ km implies $T=10^8$ years. It is this glaring discrepancy in the solar dynamo that led Steenbeck and Krause (1966) and Leighton (1964, 1969) to point out the fundamental role played by turbulent diffusion in the generation of the magnetic fields of the sun. They note that the observed period and wavelength can be understood from the fact that turbulent eddies with a velocity v_0 and a mixing length l yield an effective turbulent diffusion coefficient $\eta_T\cong 0.3v_0 l$ (see (17.71)). The convective velocity in the sun can be estimated very roughly from the usual mixing length theory of heat transport in the convective zone (cf. Spruit 1974) putting the mixing length l equal to the local scale height. Thus, for instance, at a depth of 10^4 km, $v_0\cong10^4\,\mathrm{cm\,s^{-1}}$ with $l\cong3\times10^8$ cm, gives $\eta_T\cong10^{12}\,\mathrm{cm^2\,s^{-1}}$, while at a depth of 10^5 km, $v_0\cong3\times10^3\,\mathrm{cm\,s^{-1}}$ and $l=4\times10^9$ cm, giving $\eta_T=4\times10^{12}\,\mathrm{cm^2\,s^{-1}}$. It is immediately evident that the turbulent diffusion coefficient 1–$4\times10^{12}\,\mathrm{cm\,s^{-1}}$ is of the correct order of magnitude, $10^{12}\,\mathrm{cm^2\,s^{-1}}$, to reconcile the observed period and wavelength through (19.27). It is evident, too, that turbulent diffusion is an essential part of the operation of the solar dynamo. Without turbulent diffusion the fields of the sun would behave quite differently from what is observed.

Consider the direction of propagation of dynamo waves. If ΓG is so large and positive that the dissipation $\eta(k_x^2+k_z^2)$ can be ignored, then the solution is

$$B_y=\exp(\pm\omega t)\exp\mathrm{i}(\pm\omega t-k_x x-k_z z) \tag{19.28}$$

with ω given by (19.25) as $\pm(\frac{1}{2}\Gamma Gk)^{\frac{1}{2}}$. Choosing the upper sign, for growing waves, it is evident that the phase velocity of the wave is in the direction of the wave vector (k_x, k_z) at the speed

$$\begin{aligned} U_\phi &= \omega/(k_x^2+k_z^2)^{\frac{1}{2}} \\ &= \{\tfrac{1}{2}\Gamma(G_x k_z-G_z k_x)\}^{\frac{1}{2}}/k. \end{aligned} \tag{19.29}$$

The decaying mode propagates in the opposite direction of k_x, k_z, with the same speed U_ϕ. If, on the other hand, ΓG is negative rather than positive, then the direction of propagation is reversed, so that the growing mode propagates in the opposite direction of (k_x, k_z). The directions are the same when ΓG is small enough that the amplitude is constant $(\mathrm{Re}(1/\tau)=0)$. The solution is then

$$B_y = \exp(\pm i\omega t - ik_x x - ik_z z) \tag{19.30}$$

with ω again equal to $(\frac{1}{2}\Gamma Gk)^{\frac{1}{2}}$. Thus the formulae for the phase velocity are the same as for large ΓG. To be more specific, suppose that Γ, G_x, and k_z are positive, with $G_z = 0$ so that the large-scale shear is $\mathrm{d}V_y/\mathrm{d}x = G_x > 0$. Then the regenerative mode follows from the upper sign in (19.24), and

$$\mathrm{Re}(1/\tau) = (\tfrac{1}{2}\Gamma G_x k_z)^{\frac{1}{2}} - \eta(k_x^2 + k_z^2), \tag{19.31}$$

$$\mathrm{Im}(1/\tau) = (\tfrac{1}{2}\Gamma G_x k_z)^{\frac{1}{2}}. \tag{19.32}$$

The wave number k_x plays no essential role in the regeneration and migration, its only effect being a contribution to the dissipation. The essential wave number is k_z, so put $k_x = 0$. It follows from (19.8) that the dynamo waves migrate in the positive z-direction. If both Γ and G_x are negative, the direction of the migration is also in the positive z-direction, although the relative phases of the poloidal field B_x and toroidal field B_y are reversed. If one of Γ and G_x is negative while the other is positive, then the migration is in the negative z-direction.

Now suppose that k_x is not zero. The wave fronts are oblique and the phase velocity is also oblique, in the direction of (k_x, k_z) or the opposite. But the basic direction of migration of field is in the $\pm z$-direction, perpendicular to the direction of shear, G_x.

If, on the other hand, Γ, G_z, and k_x are positive while $G_x = 0$, then

$$\mathrm{Re}(1/\tau) = (\tfrac{1}{2}\Gamma G_z k_x)^{\frac{1}{2}} - \eta(k_x^2 + k_z^2), \tag{19.33}$$

$$\mathrm{Im}(1/\tau) = -(\tfrac{1}{2}\Gamma G_z k_x)^{\frac{1}{2}}. \tag{19.34}$$

The wave number k_z plays no essential role. Putting $k_z = 0$ it follows from (19.8) that the migration is in the negative x-direction. Changing the sign of either Γ or G_z causes the fields to migrate in the positive x-direction, and changing the sign of both puts the migration in the negative x-direction again.

In general the principal wave number and the direction of migration are perpendicular to the gradient of the large-scale velocity $V(x, z)$. In the axi-symmetric dynamo the migration is in the meridional plane perpendicular to the gradient of the angular velocity $\Omega(\varpi, z)$, i.e. the migration is along isorotation surfaces (Parker 1971c; Yoshimura 1975a).

The direction of propagation of dyamo waves is of particular interest for the sun, because the direction is known from the migration of the bipolar active regions through the sunspot cycle. Their phase velocity is towards the equator from either side. Suppose then, that the y-axis is pointed east (in solar coordinates) in the local azimuthal direction and the z-axis oriented in the direction of the local vertical (so that with a right-handed coordinate system the x-axis points south). Consider the effect of a vertical shear G_z and cyclonic convection rotating about a vertical axis. The direction of migration depends upon the sign of $\Gamma G_z k_x$, which is in no way affected by k_z. So put $k_z = 0$, with $k_x > 0$. Suppose that the angular velocity increases downward from the surface of the sun, as suggested by the observed fact that the sunspots rotate more rapidly than the visible surface of the photosphere. Then $G_z < 0$. Following Steenbeck *et al.* (1966) note that the rising convective cells generally expand as they move upward (and compress as they move downward) through the convective zone, as a consequence of the rapid decrease of pressure with height. Hence conservation of angular momentum suggests that the rising, expanding cells rotate less rapidly than the surrounding medium. That is to say, rising cells rotate backward, as seen by a local observer rotating with the sun. Consider, then, the northern hemisphere. As viewed from above the north pole, the sun rotates counterclockwise, as does Earth. Hence, a rising convective cell in the northern hemisphere, viewed from above, rotates clockwise, relative to the surrounding gas. This is the opposite of the rotation in (18.40) for $\Omega > 0$ in the examples in §18.3. Hence it follows from (19.1) and (18.50) that Γ is positive. The same sign for Γ follows for sinking cells, because both the translation and rotation are the reverse of the rising cell. Hence the product ΓG_z is negative. It follows that the regenerative mode migrates toward positive x. That is to say, the phase velocity in the northern hemisphere is southward toward the equator, in agreement with observation.

In the southern hemisphere Γ is reversed, so that ΓG_z is positive and the regenerative mode propagates in the negative x-direction, which is northward.

Altogether, then, the period and direction of migration of the principal field of the sun can be understood on the basis of the simple dynamo wave, as originally suggested by Parker (1955, 1957) and Yoshimura (1975c). It remains to be shown from dynamical studies of course, whether the angular velocity actually increases downward and whether the cyclonic rotation of the convective cells has the sign expected from elementary considerations on the Coriolis force. If the sign of both Γ and G_z are reversed, the direction of migration would be the same, of course, but the relative phase of the poloidal and toroidal fields is reversed, permitting an observational test of the signs of Γ and G_z separately.

Note that if the only non-uniform rotation in the sun were the $\partial\Omega/\partial\theta$ observed at the surface, then the gradient of the azimuthal velocity is in the horizontal north–south direction and $G_z=0$, $G_x\neq 0$. With the equator rotating more rapidly than the poles, and the x-axis pointing southward, it follows that G_x is positive in the northern hemisphere and negative in the southern. Then with Γ positive in the northern hemisphere, ΓG_x is positive there. The migration is in the positive z-direction, i.e. vertically upward. This circumstance leads to a more complicated dynamo. The phase velocity at the surface depends upon the manner in which the vertical propagation of the dynamo wave is deflected at the surface, and hence depends in detail on the distribution of Γ and G_x with latitude. The problem is taken up at some length in §19.3.

This is an appropriate place to inquire into the basic physics of the dynamo wave (Parker 1955). Suppose for instance that the large-scale shear is just $G_z=\mathrm{d}V_y/\mathrm{d}z$, with $G_x=0$, while the rising convective cells rotate clockwise as viewed from above (positive z), so that Γ is positive, as in the northern hemisphere of the sun. Suppose that there are bands of azimuthal field B_y of alternate sign, as sketched in Fig. 19.1(a), so that at time $t=0$

$$B_y=-\cos(k_x x).$$

The question is the subsequent effect of Γ and G_z on this initial field. The interaction of the cyclonic convective cells with B_y produces a large number of raised loops in the lines of force, all rotated clockwise so that they represent a net circulation in the meridional xz-plane, as sketched in Fig. 19.1(a). Subsequently the many small loops diffuse together leaving only the net magnetic circulation indicated by the closed loops drawn in the xz-plane. The closed meridional loops are subject to the shear $G_z=\partial V_y/\partial z$, of course, so they are drawn out in the y-direction as shown in Fig. 19.1(b). It may be seen from Fig. 19.1(a) that this generates new B_y in the gaps between the original bands of B_y. It is apparent from Fig. 19.1(b) that the new B_y has the same direction as the original B_y lying immediately toward positive x. That is to say, the original bands of B_y are augmented on their sides toward negative x and cancelled on their sides toward positive x. The centre of gravity of each band of B_y is shifted toward negative x. Repeating the process shifts the bands again toward negative x, etc. The direction of migration of the fields is toward negative x,

$$B_y=-\cos\{k_x(x+U_\phi t)\}.$$

If $\Gamma G_z k_x$ is larger than $2\eta^2 k_x^2$, there is a net gain in the total magnetic flux and the amplitude of B_y grows with time.

19.2.2.3. Blocked propagation and stationary fields. If the magnetic field of the sun is to be understood in terms of migrating dynamo waves, then what are we to think of the stationary field of Earth? The answer is

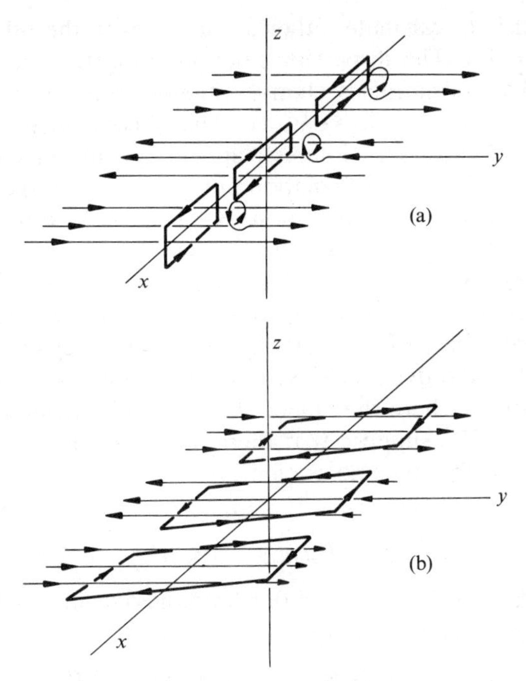

FIG. 19.1. A schematic drawing of a migratory dynamo wave composed of alternating bands of azimuthal magnetic field $B_y = -\cos(k_x x)$. (a) shows the loops in the lines of force produced by rising convective cells rotating clockwise (as viewed from above). The sense of the magnetic circulation in the meridional plane is indicated by the square loops. (b) shows the meridional circulation sheared by $\partial V_y/\partial z$, producing new B_y in the gaps between the initial bands of B_y.

that the migration can be blocked by the boundaries, by the secondary dynamo coefficient Q, and/or by meridional circulation, so that both the real and imaginary parts of $1/\tau$ are zero and the waves stand stationary in space. Whether the dynamo waves migrate or stand still depends upon the dynamo number and upon the geometry of the large-scale flow and the distribution of the cyclonic convection. When the propagation is blocked by boundaries, the wave numbers k_x and k_z are complex instead of real. The physical requirement that k_x and k_z be real applies only to an infinite space. There is no reason to expect k_x and k_z to be real in the liquid core of Earth. Indeed, we do not expect k_x and k_z to be entirely real in the convective zone of the sun, but, as will be shown later, their real parts are large enough in the thin shell of the convective zone that the simple relation (19.27) is applicable, at least in order of magnitude. For Earth (19.27) is not applicable, so that we must go back to (19.23) and begin the analysis anew. Consider then, dynamo waves in opposite 'hemispheres' blocking each other at the 'equator', $z = 0$.

The problem is to calculate either k_x or k_z given the other and given the growth rate $1/\tau$. The dispersion relation is fourth order in the wave numbers, so the calculation is clumsy. Suppose that $G_z = 0$ so that the large-scale shear is just $G_x = \mathrm{d}V_y/\mathrm{d}x$. The x-component of the wave vector is then of secondary interest, appearing only in the dissipation $\eta(k_x^2 + k_z^2)$. To facilitate the calculation put $k_x = 0$ so that the migration of the dynamo wave is in the z-direction. For ΓG_x positive the migration is in the positive z-direction. To illustrate a stationary solution of the dynamo equation, then, suppose that the large-scale shear has the uniform positive value $+G_x$ for all z, while the dynamo coefficient has the uniform positive values $+\Gamma$ in $z<0$ and the uniform negative value $-\Gamma$ in $z>0$. As a result the dynamo waves migrate from $z = \pm\infty$ toward $z = 0$ blocking each other where they meet at $z = 0$, and providing a stationary solution $(1/\tau = 0)$. We suppose that there are no fields at $z = \pm\infty$.

In $z>0$, where the dynamo coefficient is $-\Gamma$, the dynamo equations (19.21) and (19.22) reduce to

$$\mathrm{d}^2 B_y/\mathrm{d}z^2 = -G_x B_x/\eta, \qquad \mathrm{d}B_x/\mathrm{d}z = -\Gamma B_y/\eta \tag{19.35}$$

upon writing $B_x = -\mathrm{d}A_y/\mathrm{d}z$. In $z<0$, where the dynamo coefficient is $+\Gamma$ the dynamo equations reduce to

$$\mathrm{d}^2 B_y/\mathrm{d}z^2 = -G_x B_x/\eta, \qquad \mathrm{d}B_x/\mathrm{d}z = +\Gamma B_y/\eta. \tag{19.36}$$

Consider solutions of the form

$$B_y = C\exp(-\mathrm{i}qKz), \qquad B_y = D\exp(+\mathrm{i}qKz) \tag{19.37}$$

in $z>0$ and $z<0$, respectively, where K is the characteristic dynamo wave number defined by

$$K^3 = \Gamma G_x/\eta^2 \tag{19.38}$$

and C and D are arbitrary coefficients. It follows that

$$B_x = (\eta q^2 K^2/G_x)C\exp(-\mathrm{i}qKz), \qquad B_x = (\eta q^2 K^2/G_x)D\exp(+\mathrm{i}qKz) \tag{19.39}$$

in $z>0$ and $z<0$, respectively. Substituting these forms into the dynamo equations (19.35), it follows that $q^3 = -\mathrm{i}$ so that there are the three roots for q,

$$q_1 = +\exp(-\mathrm{i}\tfrac{1}{6}\pi), \qquad q_2 = -\exp(+\mathrm{i}\tfrac{1}{6}\pi), \qquad q_3 = +\mathrm{i}. \tag{19.40}$$

The root $q_3 = +\mathrm{i}$ yields $\exp(Kz)$ in $z>0$ and $\exp(-Kz)$ in $z<0$, both of which diverge at infinity. Hence the third root is without physical interest in the present problem. The fields are, then,

$$B_y = C_1\exp(-\mathrm{i}q_1Kz) + C_2\exp(-\mathrm{i}q_2Kz)$$
$$B_x = (\eta K^2/G_x)\{q_1^2 C_1\exp(-\mathrm{i}q_1Kz) + q_2^2 C_2\exp(-\mathrm{i}q_2Kz)\}$$

in $z>0$, while in $z<0$

$$B_y = D_1 \exp(+iq_1Kz) + D_2 \exp(+iq_2Kz),$$
$$B_x = (\eta K^2/G_x)\{q_1^2 D_1 \exp(+iq_1Kz) + q_2^2 D_2 \exp(iq_2Kz)\}.$$

The finite resistivity of the medium requires that B_x and B_y are continuous at $z=0$. Since G_xB_x/η is continuous and finite across $z=0$, it follows from (19.35) and (19.36) that dB_y/dz is also continuous. Since the dynamo coefficient jumps from $+\Gamma$ to $-\Gamma$ in passing from $z=-\epsilon$, to $z=+\epsilon$, while B_y is continuous, it follows from (19.35) and (19.36) that dB_x/dz reverses sign across $z=0$. Thus dB_x/dz at $z=+\epsilon$ is equal to the negative of dB_x/dz at $z=-\epsilon$. The first three of these boundary conditions yield

$$C_1 + C_2 = D_1 + D_2$$
$$q_1^2C_1 + q_2^2C_2 = q_1^2D_1 + q_2^2D_2$$
$$-q_1C_1 - q_2C_2 = +q_1D_1 + q_2D_2$$

for the continuity of B_y, B_x, and dB_y/dz, respectively. The reversal of sign of dB_x/dz leads to the same condition as for the continuity of B_y. Solving these equations for C_2, D_1, and D_2 in terms of C_1 gives

$$D_1 = C_1, \qquad D_2 = C_2 = -q_1C_1/q_2 = +C_1 \exp(-i\tfrac{1}{3}\pi). \qquad (19.41)$$

There is no dispersion relation because there are only three independent boundary conditions. With $B_0 \equiv 2C_1 \exp(-i\frac{1}{6}\pi)$, it follows that the fields in $z>0$ are

$$B_x = B_0(\eta K^2/G_x)\exp(-\tfrac{1}{2}Kz)\cos(\tfrac{1}{2}\sqrt{3}\,Kz + \tfrac{1}{6}\pi), \qquad (19.42)$$
$$B_y = B_0 \exp(-\tfrac{1}{2}Kz)\cos(\tfrac{1}{2}\sqrt{3}\,Kz - \tfrac{1}{6}\pi). \qquad (19.43)$$

In $z<0$ the fields are

$$B_x = B_0(\eta K^2/G_x)\exp(+\tfrac{1}{2}Kz)\cos(\tfrac{1}{2}\sqrt{3}\,Kz - \tfrac{1}{6}\pi), \qquad (19.44)$$
$$B_y = B_0 \exp(+\tfrac{1}{2}Kz)\cos(\tfrac{1}{2}\sqrt{3}\,Kz + \tfrac{1}{6}\pi). \qquad (19.45)$$

The fields are symmetric about $z=0$, representing two dynamo waves of equal and opposite strength deadlocked head-on across $z=0$. The fields are plotted in Fig. 19.2. Note that this is not an eigenvalue problem because there is no characteristic scale length in the environment from which to derive the dynamo number. The strength of the dynamo effect, characterized by $\Gamma G_x/\eta^2$, can be as large or as small as desired. Increasing $\Gamma G_x/\eta^2$ merely causes the fields to crowd closer together against $z=0$, as indicated by the increase of the characteristic dynamo wave number K given by (19.38). The scale $1/K$ of the fields adjusts so that the effective dynamo number, $\Gamma G_x/\eta^2K^3$, always equals one.

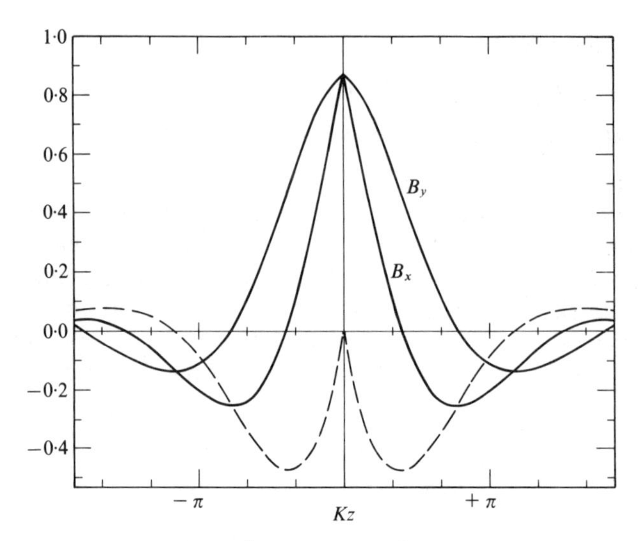

FIG. 19.2. The solid curves give B_x, in units of $B_0\eta K^2/G_x$ from (19.42) and (19.44), and a plot of B_y in units of B_0 from (19.43) and (19.45). The field B_x, in units of $B_0\eta K^2/G_x$ given by (19.47) and (19.49), is identical with B_y in units of B_0 from (19.43) and (19.45). The dashed curve is from (19.48) and (19.50).

The complementary problem is the circumstance wherein the dynamo coefficient has the same positive value $+\Gamma$ for all z while the large shear has the positive value $+G_x$ in $z<0$ and $-G_x$ in $z>0$. Then in $z>0$ the dynamo equations are

$$d^2B_y/dz^2 = +G_xB_x/\eta, \qquad dB_x/dz = +\Gamma B_y/\eta. \tag{19.46}$$

In $z<0$ they are given by (19.36). With the solutions (19.37) it follows that

$$B_x = -(\eta q^2K^2/G_x)\exp(-iqKz), \qquad B_x = +(\eta q^2K^2/G_x)D\exp(+iqKz)$$

in $z>0$ and $z<0$, respectively. The boundary conditions can be expressed in terms of the continuity of B_x, B_y, and dB_y/dz, from which it is readily shown that the coefficients are $D_2 = -D_1 = -C_2 = +C_1$. Hence in $z>0$ the fields are

$$B_x = +B_0(\eta K^2/G_x)\exp(-\tfrac{1}{2}Kz)\cos(\tfrac{1}{2}\sqrt{3}\,Kz - \tfrac{1}{6}\pi), \tag{19.47}$$

$$B_y = -B_0\exp(-\tfrac{1}{2}Kz)\sin(\tfrac{1}{2}\sqrt{3}\,Kz). \tag{19.48}$$

where now $B_0 = +2iC_1$. In $z<0$ the fields are

$$B_x = +B_0(\eta K^2/G_x)\exp(+\tfrac{1}{2}Kz)\cos(\tfrac{1}{2}\sqrt{3}\,Kz + \tfrac{1}{6}\pi), \tag{19.49}$$

$$B_y = -B_0\exp(-\tfrac{1}{2}Kz)\sin(\tfrac{1}{2}\sqrt{3}\,Kz). \tag{19.50}$$

It is apparent that B_x is symmetric about $z=0$, while B_y is antisymmetric. the x-component of the field, given by (19.47) and (19.49), has the same analytical form as the y-component given by (19.43) and (19.45). The field B_y is shown by the dashed curve in Fig. 19.2.

These two examples serve to illustrate the pile-up of oppositely moving dynamo waves when they meet head on, producing stationary magnetic fields from the otherwise oscillatory waves. The stronger the dynamo, the more closely the opposite waves crowd against each other. It should be noted that the oscillations can also be blocked by counter-streaming fluid. Roberts (1972) shows examples wherein an oscillatory dynamo is converted to a stationary dynamo by a meridional flow opposing the latitudinal migration of the dynamo wave. Indeed, the meridional flow is an intrinsic part of Braginskii's formal mathematical development of the dynamo equations (see §18.6), which led to the suggestion (Braginskii 1964) that meridional flow may be an essential part of the operation of the geomagnetic dynamo. It is an interesting possibility, but, as we shall see, it is not necessary for the operation of a stationary dynamo. Parker (1977a, b) has pointed out that the buoyancy of magnetic fields in a compressible medium can have the same effect as a meridional flow. And, of course, it was pointed out in Chapter 13 that the buoyancy is a powerful loss mechanism when the fields become very strong. On the other hand a downward streaming of fluid can counter the loss by the buoyant rise and greatly enhance the efficiency of the dynamo and/or allow the amplification to proceed to stronger fields. There is a large variety of effects that can be incorporated into the theoretical models, producing an unfortunate proliferation of undetermined parameters that are free to be adjusted to fit any observational result. Until firm knowledge of the fluid motions is available, there seems little to be gained by pursuing the many possibilities.

19.2.2.4. The four wave modes. When considering the migration of dynamo waves in the presence of boundaries, as already done for the stationary case in the subsection above, it is necessary to solve the dispersion relation (19.23) for the wave numbers. The equation is fourth order in the wave numbers, so there are four roots for a given $1/\tau$. Without loss of generality, the coordinate system may be rotated about the y-axis so that the large-scale shear is in the x-direction. Then $dV_y/dx \equiv G_x \equiv G$ and $G_z = 0$. It is convenient to write (19.23) in non-dimensional form, in terms of the characteristic dynamo wave number (19.38),

$$K^3 = \Gamma G/\eta^2$$

together with the dimensionless wave number $q = k_z/K$ and 'growth rate'

$$A = 1/\eta K^2\tau + k_x^2/K^2. \tag{19.51}$$

If the y-component of the field is written as

$$B_y = C \exp i(\omega t - k_x x - qKz), \tag{19.52}$$

then it follows from (19.22) that

$$B_x = +C(\eta K^2/G)(A+q^2)\exp i(\omega t - k_x x - qKz), \tag{19.53}$$

$$B_z = -C\{i\eta K k_x/G(A+q_n^2)\}\exp i(\omega t - k_x x - qKz). \tag{19.54}$$

The dispersion relation (19.23) becomes

$$(A+q^2)^2 - iq = 0. \tag{19.55}$$

Note again that $k_z = qK$ is the principal wave number, directly involved in the dynamo effect. The wave number k_x is secondary, contributing to the dissipation and producing B_z, but not being involved directly in the dynamo.

There is a formal algebraic solution to any quartic equation, of course, yielding the four roots $q_n (n = 1, 2, 3, 4)$ as algebraic functions of A. Unfortunately the expressions are too complicated to be of much use in the calculations that follow. Instead we note that the four roots of (19.55) satisfy

$$\sum_{n=1}^{4} q_n = 0$$

and content ourselves with the fact that they can be expanded in ascending powers of A, and in descending powers of A. If $|A| \ll 1$, it is readily shown that[2]

$$q_1 = -\exp(-i\tfrac{1}{6}\pi)\{1 - \tfrac{2}{3}A \exp(i\tfrac{1}{3}\pi) + \tfrac{1}{3}A^2 \exp(-i\tfrac{1}{3}\pi) + 28A^3/81 + \ldots\}, \tag{19.56}$$

$$q_2 = +\exp(+i\tfrac{1}{6}\pi)\{1 - \tfrac{2}{3}A \exp(-i\tfrac{1}{3}\pi) + \tfrac{1}{3}A^2 \exp(i\tfrac{1}{3}\pi) + 28A^3/81 + \ldots\}, \tag{19.57}$$

$$q_3 = -i(1 + \tfrac{2}{3}A - \tfrac{1}{3}A^2 + 28A^3/81 + \ldots), \tag{19.58}$$

$$q_4 = -iA^2(1 - 2A^3 + \ldots). \tag{19.59}$$

The expansions converge rapidly for $|A| \lesssim 0.5$. On the other hand, when $|A| \gg 1$, these same roots can be expanded in the asymptotic forms

$$q_1 = +iA^{\frac{1}{2}}\left(1 + \frac{i}{2A^{\frac{3}{4}}} + \frac{i}{64A^{\frac{9}{4}}} + \frac{1}{128A^3} - \frac{9i}{4096A^{\frac{15}{4}}} - \frac{55i}{131\,072A^{\frac{21}{4}}} + \ldots\right), \tag{19.60}$$

[2] The three roots q_1, q_2, q_3 in (19.40) for $A = 0$ were written for a dynamo coefficient $-\Gamma$, and so have the opposite sign of the roots defined here.

$$q_2 = +\mathrm{i}A^{\frac{1}{2}}\left(1 - \frac{\mathrm{i}}{2A^{\frac{3}{4}}} - \frac{\mathrm{i}}{64A^{\frac{9}{4}}} + \frac{1}{128A^3} + \frac{9\mathrm{i}}{4096A^{\frac{15}{4}}} + \frac{55\mathrm{i}}{131\,072A^{\frac{21}{4}}} + \ldots\right), \tag{19.61}$$

$$q_3 = -\mathrm{i}A^{\frac{1}{2}}\left(1 + \frac{1}{2A^{\frac{3}{4}}} - \frac{1}{64A^{\frac{9}{4}}} + \frac{1}{128A^3} - \frac{9}{4096A^{\frac{15}{4}}} + \frac{55}{131\,072A^{\frac{21}{4}}} + \ldots\right), \tag{19.62}$$

$$q_4 = -\mathrm{i}A^{\frac{1}{2}}\left(1 - \frac{1}{2A^{\frac{3}{4}}} + \frac{1}{64A^{\frac{9}{4}}} + \frac{1}{128A^3} + \frac{9}{4096A^{\frac{15}{4}}} - \frac{55}{131\,072A^{\frac{21}{4}}} + \ldots\right) \tag{19.63}$$

which converge rapidly for $|A| \gtrsim 1$. In the intermediate region $0.5 < |A| < 1$ the convergence is conditional, although useful if suitable care is taken.

To illustrate the character of the different modes put the secondary wave number k_x equal to zero and suppose that the magnetic fields have grown to a level where there is dynamical equilibrium, with the cyclonic motions and the large-scale shear inhibited to such a level that $\mathrm{Re}(1/\tau) = 0$ and the amplitude of the waves is fixed in time. Denoting the frequency by $\omega = \mathrm{Im}(1/\tau)$, it follows that $A = \mathrm{i}\omega/\eta K^2$. If the product ηK^2 is large so that $|A|$ is small, it follows from (19.56)–(19.59) that the wave numbers are

$$q_1 = -\exp(-\mathrm{i}\tfrac{1}{6}\pi)(1 + 2\omega/\eta K^2 + \ldots), \tag{19.64}$$

$$q_2 = +\exp(+\mathrm{i}\tfrac{1}{6}\pi)(1 - 2\omega/\eta K^2 + \ldots), \tag{19.65}$$

$$q_3 = -\mathrm{i} + \tfrac{2}{3}\omega/\eta K^2 + \ldots, \tag{19.66}$$

$$q_4 = +\mathrm{i}(\omega/\eta K^2)^2 - 2(\omega/\eta K^2)^5 + \ldots. \tag{19.67}$$

The phase velocity U_Φ is defined to be

$$U_\Phi = \omega/\mathrm{Re}(k_z) = \omega/K\,\mathrm{Re}(q).$$

Thus, for the four modes the phase velocities are

$$U_{\Phi 1} \cong -\frac{2}{\sqrt{3}}\frac{\omega}{K}, \qquad U_{\Phi 2} \cong +\frac{2}{\sqrt{3}}\frac{\omega}{K}, \tag{19.68}$$

$$U_{\Phi 3} \cong +\frac{3}{2}\frac{\eta K^2}{\omega}\frac{\omega}{K}, \qquad U_{\Phi 4} \cong -\frac{1}{2}\left(\frac{\eta K^2}{\omega}\right)^5\frac{\omega}{K}. \tag{19.69}$$

The amplitude of each mode varies as $\exp\{Kz\,\mathrm{Im}(q)\}$ along the wave train. Thus the first and fourth modes migrate in the negative z-direction ($U_\Phi < 0$), the amplitude of the first mode declining rapidly, as $\exp(+\frac{1}{2}Kz)$, and the amplitude of the fourth declining slowly, as $\exp\{+(\omega/\eta K^2)^2 Kz\}$, in the direction of migration for the present case that $\omega/\eta K^2 \ll 1$. The second and third modes migrate in the positive z-direction, the amplitude of the second increasing rapidly, as $\exp(+\frac{1}{2}Kz)$ and the third declining rapidly, as $\exp(-Kz)$, in the direction of migration. The phase velocity of the third mode is larger than ω/K, while the phase velocity of the fourth is very much larger than ω/K (for $\eta K^2 \gg \omega$). It is evident that for $|A| \ll 1$ the fourth mode differs from the other three in having a very large phase velocity.

If ΓG is negative rather than positive, K is replaced by $-|K|$ in the above expressions. The sense of migration and the direction of increasing amplitude are both reversed, so that they maintain the same relationship as for $\Gamma G > 0$.

19.2.2.5. Dynamo waves at an impenetrable boundary. To explore the interrelations of the four modes consider their interaction at a perfectly conducting boundary. Suppose that the space $z < 0$ is filled with a rigid, infinitely conducting material containing no magnetic field. The space $z > 0$ is filled with conducting fluid ($\eta > 0$) subject to the large-scale shear $\mathrm{d}V_y/\mathrm{d}x \equiv G$ and to the cyclonic convection characterized by $+\Gamma$. To establish the boundary conditions at $z = 0$ note that the dynamo equations (19.21) and (19.22) are written

$$\frac{\partial B_x}{\partial t} - \frac{\partial}{\partial z}\left\{\eta\left(\frac{\partial B_x}{\partial z} - \frac{\partial B_z}{\partial x}\right)\right\} = -\frac{\partial}{\partial z}(\Gamma B_y), \tag{19.70}$$

$$\frac{\partial B_y}{\partial t} - \frac{\partial}{\partial z}\left(\eta\frac{\partial B_y}{\partial z}\right) - \frac{\partial}{\partial x}\left(\eta\frac{\partial B_y}{\partial x}\right) = +GB_x, \tag{19.71}$$

$$\frac{\partial B_z}{\partial t} - \frac{\partial}{\partial x}\left\{\eta\left(\frac{\partial B_z}{\partial x} - \frac{\partial B_x}{\partial z}\right)\right\} = +\frac{\partial}{\partial x}(\Gamma B_y), \tag{19.72}$$

when η, Γ, and G are functions of x and z. Note first that in order to satisfy $\partial B_j/\partial x_j = 0$ it is necessary that B_z be continuous at $z = 0$, and hence equal to zero. Then integrate (19.70) and (19.71) from $z = -\epsilon$ to $z = +\epsilon$, with $\epsilon \to 0$. The terms $O(\epsilon)$ vanish, and there remain the two conditions

$$\eta\, \partial B_x/\partial z\, \Big|_{-\epsilon}^{+\epsilon} = \Gamma B_y\, \Big|_{-\epsilon}^{+\epsilon},$$

$$\eta\, \partial B_y/\partial z\, \Big|_{-\epsilon}^{+\epsilon} = 0.$$

Both the magnetic field and the coefficients η and Γ vanish at $z = -\epsilon$. Hence at $z = +\epsilon$,

$$\partial B_x/\partial z = \Gamma B_y/\eta, \qquad \partial B_y/\partial z = 0. \tag{19.73}$$

There are, then, three boundary conditions. Applying them to the solutions (19.52)–(19.54) leads to the three equations

$$\sum_{n=1}^{4} C_n/(A + q_n^2) = 0, \tag{19.74}$$

$$\sum_{n=1}^{4} C_n\{1+\mathrm{i}q_n(A+q_n^2)\}=0, \tag{19.75}$$

$$\sum_{n=1}^{\infty} C_n q_n=0, \tag{19.76}$$

for the vanishing of B_z, $\partial B_x/\partial z-\Gamma B_y/\eta$, and $\partial B_y/\partial z$, respectively. The quantity C_n is the amplitude C of the nth mode. The three equations specify any three of the C_n in terms of the fourth. It is readily shown that

$$\begin{aligned}
&C_1[(q_2-q_4)\{1-\mathrm{i}q_2q_4(q_2+q_4)\}/(A+q_1^2)\\
&\quad+(q_4-q_1)\{1-\mathrm{i}q_1q_4(q_4+q_1)\}/(A+q_2^2)\\
&\quad+(q_1-q_2)\{1-\mathrm{i}q_1q_2(q_1+q_2)\}/(A+q_4^2)]\\
&\qquad\qquad=C_3[(q_4-q_2)\{1-\mathrm{i}q_2q_4(q_2+q_4)\}/(A+q_3^2)\\
&\qquad\qquad\quad+(q_2-q_3)\{1-\mathrm{i}q_2q_3(q_2+q_3)\}/(A+q_4^2)\\
&\qquad\qquad\quad+(q_3-q_4)\{1-\mathrm{i}q_3q_4(q_3+q_4)\}/(A+q_2^2)]
\end{aligned} \tag{19.77}$$

and

$$\begin{aligned}
&C_2[(q_1-q_4)\{1-\mathrm{i}q_1q_4(q_1+q_4)\}/(A+q_2^2)\\
&\quad+(q_2-q_1)\{1-\mathrm{i}q_1q_2(q_1+q_2)\}/(A+q_4^2)\\
&\quad+(q_4-q_2)\{1-\mathrm{i}q_2q_4(q_4+q_2)\}/(A+q_1^2)]\\
&\qquad\qquad=C_3[(q_4-q_1)\{1-\mathrm{i}q_1q_4(q_1+q_4)\}/(A+q_3^2)\\
&\qquad\qquad\quad+(q_1-q_3)\{1-\mathrm{i}q_1q_3(q_1+q_3)\}/(A+q_4^2)\\
&\qquad\qquad\quad+(q_3-q_4)\{1-\mathrm{i}q_3q_4(q_3+q_4)\}/(A+q_1^2)].
\end{aligned} \tag{19.78}$$

The other four linear relations between pairs of coefficients follow by cyclic permutation of the indices.

In the limit of small A, the ratios of the amplitudes reduce to

$$C_4=C_3 \tag{19.79}$$

with C_1 and C_2 both small, $\mathrm{O}(A)$ compared to C_3. Thus, if ΓG is positive, the fourth mode migrates in the negative z-direction and is the 'incident' wave. The third mode then represents the 'reflected' wave, migrating away from the boundary in the positive z-direction, both declining in the direction of migration. Note that the first and second modes, with very small amplitudes at the boundary, both increase as $\exp(+\frac{1}{2}Kz)$ away from the boundary.

If ΓG is negative, on the other hand, then the third mode propagates in the negative z-direction and is the 'incident' wave, while the fourth is the 'reflected' wave. The amplitudes of both decline in the direction of propagation. The first and second modes continue to be small, $\mathrm{O}(A)$.

Note from (19.54) that if the wave fronts are precisely parallel to the boundary, i.e. if $k_x = 0$, then B_z is identically zero so that there are but two boundary conditions to be imposed. In this special case, then, any one of the four modes may vanish and there is still a consistent solution of the dynamo equations. The impenetrable boundary is capable of blocking any normally incident waves ($k_x = 0$) so that there are stationary solutions ($A = 0$) to the dynamo equations. The same cannot be said for oblique waves, however.

19.2.2.6. Dynamo waves at a vacuum boundary. If the region $z < 0$ is filled with a non-conducting medium ($\eta = \infty$), or filled with conducting medium so tenuous as to be pushed aside by the magnetic fields extending in from the dynamo region $z > 0$, then the magnetic field in $z < 0$ follows from Laplace's equation as

$$B_x = D \exp\{i(\omega t - k_x x) + k_x z\}, \tag{19.80}$$

$$B_y = 0, \tag{19.81}$$

$$B_z = iD \exp\{i(\omega t - k_x x) + k_x z\}, \tag{19.82}$$

where k_x is assumed to be positive. The fields (19.52)–(19.54) obtain in $z > 0$. The boundary conditions at $z = 0$ are readily specified because the conductivity is finite or zero, prohibiting current sheets and any discontinuities in B_i. Thus all three components of B_i are continuous across $z = 0$. Eliminating D between the equations for B_x and B_z, we are left with the two conditions from (19.52)–(19.54) that

$$\sum_{n=1}^{4} C_n = 0,$$

$$\sum_{n=1}^{4} C_n(-iq_n + k_x/K)/(A + q_n^2) = 0,$$

after some simplification of the coefficients with the aid of the dispersion relation (19.55). There are, then, but two restrictions on the four coefficients C_n. We are at liberty, therefore, to set any one of the four equal to zero. If, for instance, the first mode is absent ($C_1 = 0$), then

$$C_2 = C_4 \frac{(A+q_2^2)\{(iq_4 - k_x/K)(A+q_3^2) - (iq_3 - k_x/K)(A+q_4^2)\}}{(A+q_4^2)\{(iq_3 - k_x/K)(A+q_2^2) - (iq_2 - k_x/K)(A+q_3^2)\}}, \tag{19.83}$$

$$C_3 = C_4 \frac{(A+q_3^2)\{(iq_4 - k_x/K)(A+q_2^2) - (iq_2 - k_x/K)(A+q_4^2)\}}{(A+q_4^2)\{(iq_2 - k_x/K)(A+q_3^2) - (iq_3 - k_x/K)(A+q_2^2)\}}. \tag{19.84}$$

For small A these reduce to

$$C_2 = -C_3 = -C_4 \frac{k_x}{\sqrt{3}\,AK} \frac{(k_x/K)\exp(i\tfrac{1}{6}\pi) + \exp(-i\tfrac{1}{6}\pi)}{1 + k_x/K + k_x^2/K^2}. \tag{19.85}$$

The amplitude of the fourth mode is small, O(A), compared to C_2 and C_3. But note that when $K<0$ the fourth mode migrates in the negative z-direction and is the 'incident' wave. In that case the second and third modes represent the 'reflected' waves and have large amplitudes O(A^{-1}) compared to the 'incident' wave. In the limit of $k_x=0$ (normal incidence) C_2 and C_3 vanish and the fourth mode simply terminates at the boundary without any reflected wave (Parker 1971c).

We can put any one of the coefficients C_n equal to zero, obtaining any two of the coefficients in terms of the remaining one. The results can be written down directly by inspecting the ordering of the indices in (19.83) and (19.84) relative to the number of the excluded mode and the number of the coefficient in terms of which the other two are to be expressed. If $C_2=0$, then in the limit of small A,

$$C_1=-C_3,\ C_4=\surd 3\ AC_3\{\exp(\mathrm{i}\tfrac{1}{6}\pi)+(K/k_x)\exp(-\mathrm{i}\tfrac{1}{6}\pi)\}. \qquad (19.86)$$

In this instance one may think of C_3 as the 'incident' wave, producing principally the first mode as the reflected wave, with C_4 but weakly excited. But note that as $k_x\to 0$ (normal incidence) the fourth mode is more strongly excited. If we first set $k_x/K=0$ and then consider small A, the result is

$$C_1=C_3\exp(\mathrm{i}\tfrac{1}{3}\pi), \qquad C_4=C_3\exp(-\mathrm{i}\tfrac{1}{3}\pi). \qquad (19.87)$$

The fourth mode is excited equally with the first.

If, on the other hand, $C_3=0$, then for small A,

$$C_1=-C_2, \qquad C_4=\mathrm{i}\surd 3\ AC_2(1-K/k_x) \qquad (19.88)$$

where the second mode may now be considered as the 'incident' wave. The fourth mode is again but weakly excited except near normal incidence. In that case, putting $k_x/K=0$ and then considering small A,

$$C_1=+C_2\exp(-\mathrm{i}\tfrac{1}{3}\pi), \qquad C_4=-\surd 3\ C_2\exp(-\mathrm{i}\tfrac{1}{6}\pi), \qquad (19.89)$$

in which case the amplitude of the fast moving fourth mode is a little larger than the first.

Finally, note that if $C_4=0$, then for small A,

$$C_2=-C_1\frac{\exp(\mathrm{i}\tfrac{1}{3}\pi)-2(k_x/K)\exp(-\mathrm{i}\tfrac{1}{3}\pi)-(k_x/K)^2}{1+k_x/K+k_x^2/K^2},$$

$$C_3=-C_1\frac{(1-k_x/K)[\exp(\mathrm{i}\tfrac{1}{3}\pi)-(k_x/K)\exp(-\mathrm{i}\tfrac{1}{3}\pi)]}{1+k_x/K+k_x^2/K^2}.$$

In this case we may think of C_1 as the 'incident' wave in the circumstance that $K<0$. The second and third modes have comparable amplitudes. In the limit of normal incidence, $C_2=C_3=-C_1\exp(\mathrm{i}\tfrac{1}{3}\pi)$, representing the blocked (stationary) dynamo waves treated in §19.2.2.3.

It is a simple but tedious matter to treat the reflection and refraction of dynamo waves at interfaces where Γ, G, and/or η change discontinuously. The present examples suffice, however, to illustrate the general idea of the complex behaviour of the dynamo wave at a boundary. It is time to move on to specific examples of the generation of field in finite volumes of space.

19.3. Dynamos in rectangular boundaries

Consider what happens when the dynamo waves are penned up in a rectangular box. In particular we shall be interested in a box in which the dynamo coefficient Γ changes sign across the middle, simulating the reversal of sign of cyclonic motion across the equator of a rotating body. In the limit of strong shear the dynamo equations reduce to the simple form (19.21) and (19.22) (i.e. (18.4) and (18.7)) whether the cyclonic rotation is about the x- or z-direction, or any other direction in the xz-plane.

19.3.1. Dynamos with vertical shear

To stay with the conventions of the foregoing illustrations let $\partial V_y/\partial z \equiv G_z = 0$ again, and write simply $G_x \equiv \mathrm{d}V_y/\mathrm{d}x \equiv G$. The dispersion relation is given by (19.55) and the three components of the field by (19.52)–(19.54) for each of the four modes. The wave number k_x is secondary and appears only in the non-dimensional growth rate (19.51). The primary wave number k_z is given by (19.56)–(19.63) for each of the four modes.

Now in a rotating, convecting astronomical body the sense of rotation of the cyclonic motions, i.e. the helicity, changes sign across the equator. We expect that at low latitudes the non-uniform rotation is the result of a vertical gradient in angular velocity. To develop a rectangular dynamo with these same general properties along the lines discussed in §19.2, let the positive x-axis correspond to the vertical direction so that $G = \mathrm{d}V_y/\mathrm{d}x$ represents a 'vertical' shear. Let y represent the azimuthal direction pointing in the direction of rotation so that axi-symmetry becomes $\partial/\partial y = 0$, as in the discussion of §19.2. Then the z-axis of a right-handed coordinate system points north and the rectangular dynamo simulates in some degree the basic properties of the spherical dynamo in the manner sketched in Fig. 19.3, following Parker (1971b).

Let the dynamo coefficient have the uniform value $+\Gamma$ in $z>0$ and $-\Gamma$ in $z<0$, simulating the reversal of cyclonic motion (helicity) across the 'equator' $z=0$ of a rotating body. Let G be uniform throughout the entire dynamo region $-\lambda<z<+\lambda$.

The advantage of the artificial rectangular dynamo over a more realistic spherical one is the elementary analytical treatment, producing concise analytical expressions for the dynamo numbers and the magnetic fields of

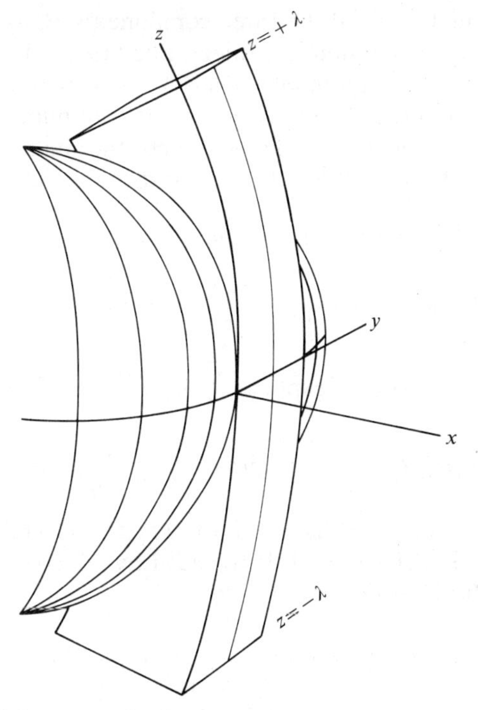

FIG. 19.3. Sketch of the rectangular dynamo region $-\lambda < z < +\lambda$ and the correspondence to the spherical geometry of Earth and of the sun.

the various stationary and periodic states. The calculations show the interesting fact that stationary modes can occur only in thick regions, where the dissipation of k_x is not large. For if the dynamo number $K\lambda$ is large, as it must be to overcome a large resistive dissipation ηk_x^2, the dynamo waves are oblique, of short length, and 'push' so hard that they cannot be blocked by the boundaries. Then only migratory (periodic) fields are generated. The stationary magnetic field is a special circumstance appearing only under favourable conditions for which modest values of $K\lambda$ are sufficient to maintain the field. The same general effects obtain within spherical boundaries too, with any large $K\lambda$ producing periodic fields.

The boundary conditions on the rectangular dynamo are easily prescribed. With $z = \pm\lambda$ corresponding to the north and south 'poles', only the vertical component B_x is non-vanishing there. So $B_y = B_z = 0$ at $z = \pm\lambda$. For simplicity suppose that the horizontal planes $x = \pm h = \pm\pi/2k_x$ represent free surfaces at which $B_y = 0$. The fields at the midplane $x = 0$ then satisfy the boundary conditions for an impenetrable boundary there, so that the solutions for free boundaries at $x = \pm h$ illustrate two cases at once.

Across the 'equator' $z=0$ the three components $B_x=-\partial A_y/\partial z$ B_y, and $B_z=+\partial A_y/\partial x$ must be continuous. Integrating (19.21) from $z=-\epsilon$ to $+\epsilon$ and noting that G_xB_x is continuous, leads to the conclusion that $\partial B_y/\partial z$ is continuous. On the other hand $\partial B_x/\partial z$ is not continuous. The continuity of the three components of the field and the jump in the dynamo coefficient from $-\Gamma$ in $z<0$ to $+\Gamma$ in $z>0$ dictates that $\partial B_x/\partial z$ jumps by $2\Gamma B_y(x,0)/\eta$ across $z=0$.

Following (19.52)–(19.54) denote the field in $z>0$ by

$$B_x=(K^2\eta/G)\cos(k_xx)\exp(t/\tau)\sum_{n=1}^{4}C_n(A+q_n^2)\exp(-iq_nKz), \quad (19.90)$$

$$B_y=\cos(k_xx)\exp(t/\tau)\sum_{n=1}^{4}C_n\exp(-iq_nKz), \quad (19.91)$$

$$B_z=-(k_xK\eta/G)\sin(k_xx)\exp(t/\tau)\sum_{n=1}^{4}\frac{C_n}{A+q_n^2}\exp(-iq_nKz) \quad (19.92)$$

with the four q_n representing the roots (19.56)–(19.63) of (19.55). In $z<0$ the dynamo coefficient $+\Gamma$ is replaced by $-\Gamma$ so that K is replaced by $-K$. Write the field in $z<0$ as

$$B_x=(K^2\eta/G)\cos(k_xx)\exp(t/\tau)\sum_{n=1}^{4}D_n(A+q_n^2)\exp(iq_nKz), \quad (19.93)$$

$$B_y=\cos(k_xx)\exp(t/\tau)\sum_{n=1}^{4}D_n\exp(iq_nKz), \quad (19.94)$$

$$B_z=+(k_xK\eta/G)\sin(k_xx)\exp(t/\tau)\sum_{n=1}^{4}\frac{D_n}{A+q_n^2}\exp(iq_nKz). \quad (19.95)$$

The continuity of B_x, B_y, B_z, and $\partial B_y/\partial z$ at $z=0$ leads to the four relations

$$\sum C_n(A+q_n^2)=+\sum D_n(A+q_n^2), \quad (19.96)$$

$$\sum C_n=+\sum D_n, \quad (19.97)$$

$$\sum C_n/(A+q_n^2)=-\sum D_n/(A+q_n^2), \quad (19.98)$$

$$\sum C_nq_n=-\sum D_nq_n, \quad (19.99)$$

respectively. The vanishing of B_y and B_z at $z=\pm\lambda$ yields the four relations

$$\sum C_n\exp(-iq_nK\lambda)=\sum D_n\exp(-iq_nK\lambda)=0, \quad (19.100)$$

$$\sum\frac{C_n}{A+q_n^2}\exp(-iq_nK\lambda)=\sum\frac{D_n}{A+q_n^2}\exp(-iq_nK\lambda)=0. \quad (19.101)$$

There are, then eight linear homogeneous equations for the eight coefficients C_n and D_n.

The substitution

$$E_n \equiv C_n + D_n, \qquad F_n \equiv C_n - D_n \tag{19.102}$$

breaks the set of eight equations into a set of four for E_n and another set of four for F_n. The coefficient E_n represents the modes (for $C_n = D_n$) for which the azimuthal and vertical fields B_y and B_x are even functions of z, while F_n represents the modes (for $C_n = -D_n$) for which B_y and B_x are odd functions of z.

For the even modes, then, the determinant of the coefficients of E_n in the four equations must vanish, leading to the dispersion relation

$$\begin{aligned} 0 = (q_2^2 - q_1^2)(q_4^2 - q_3^2)&\left[\frac{\exp\{-i(q_1+q_2)K\lambda\}}{(A+q_3^2)(A+q_4^2)} + \frac{\exp\{i(q_1+q_2)K_\lambda\}}{(A+q_1^2)(A+q_2^2)}\right] \\ +(q_2^2 - q_3^2)(q_1^2 - q_4^2)&\left[\frac{\exp\{-i(q_2+q_3)K\lambda\}}{(A+q_1^2)(A+q_4^2)} + \frac{\exp\{i(q_2+q_3)K\lambda\}}{(A+q_2^2)(A+q_3^2)}\right] \\ +(q_3^2 - q_1^2)(q_2^2 - q_4^2)&\left[\frac{\exp\{-i(q_1+q_3)K\lambda\}}{(A+q_2^2)(A+q_4^2)} + \frac{\exp\{i(q_1+q_3)K\lambda\}}{(A+q_1^2)(A+q_3^2)}\right], \end{aligned} \tag{19.103}$$

upon making use of the fact that $\sum q_n = 0$. For the odd modes the determinant of the F_n must vanish, yielding the same dispersion relation except that $+K\lambda$ is replaced by $-K\lambda$ throughout.

19.3.1.1. Stationary states. It will be seen that stationary states (real positive A) occur for modest values of the dynamo number $K\lambda$ and yield modest values of the generation rate $AK^2\eta$. Hence the expansion of q_n in ascending powers of A suggests itself (Parker 1971b). Using (19.56)–(19.59) it can be shown, after some tedious but elementary algebra, that the dispersion relation (19.103) (for the even modes) reduces to

$$0 = F(K\lambda) + AG(K\lambda) + O(A^2, A^2K\lambda, A^2K^2\lambda^2), \tag{19.104}$$

where

$$F(K\lambda) \equiv \exp(K\lambda) + 2\exp(-\tfrac{1}{2}K\lambda)\sin(\tfrac{1}{2}\sqrt{3}\,K\lambda - \tfrac{1}{6}\pi), \tag{19.105}$$

$$\begin{aligned} G(K\lambda) \equiv (1+\tfrac{2}{3}K\lambda)\exp(K\lambda) - \exp(-K\lambda) - 2\exp(\tfrac{1}{2}K\lambda)\cos(\tfrac{1}{2}\sqrt{3}\,K\lambda) \\ +2\{\cos(\tfrac{1}{2}\sqrt{3}\,K\lambda) - \tfrac{2}{3}K\lambda\sin(\tfrac{1}{2}\sqrt{3}\,K\lambda + \tfrac{1}{6}\pi)\}\exp(-\tfrac{1}{2}K\lambda). \end{aligned} \tag{19.106}$$

Note that, when $K\lambda$ is large and positive, F and G are both positive and increase monotonically. Hence there are no stationary or regenerative states ($A > 0$) for $K\lambda > 0$. Such states occur only for $K\lambda < 0$. Note that $A = 0$ (which is a stationary state in the limit of small k_x occurs at the zeros of $F(K\lambda)$ for the negative values

$$3^{\frac{1}{2}}K\lambda = -[2(n-\tfrac{1}{6})\pi + (-1)^n \exp\{-\sqrt{3}\,(n-\tfrac{1}{6})\} + \ldots], \quad (19.107)$$

where n is any positive integer. The exponential term, and the higher order terms not given, are all quite negligible, even for $n=1$. Hence $A=0$ for $K\lambda = -3{\cdot}01, -6{\cdot}63, -10{\cdot}25, -13{\cdot}9$. For $K\lambda < -3{\cdot}01$

$$A \cong 2\exp(\tfrac{1}{2}K\lambda)\sin(\tfrac{1}{2}\sqrt{3}\,K\lambda - \tfrac{1}{6}\pi). \quad (19.108)$$

The generation rate $K^2\lambda^2 A$ is plotted in Fig. 19.4 as a function of $|K\lambda|$. The growth rate A is positive for $3{\cdot}01 < -K < 6{\cdot}63$, with a maximum of $A = 0{\cdot}210$ at $K\lambda = -4{\cdot}22$. Indeed, there is a sequence of extrema

$$A = (-1)^{m+1}\sqrt{3}\exp\{-(m+\tfrac{1}{6})\pi/\sqrt{3}\} \quad (19.109)$$

at $K\lambda = -2(m+\tfrac{1}{6})\pi/3^{\frac{1}{2}}$, where $m = 1, 2, 3, \ldots$, but the exponential $\exp(\tfrac{1}{2}K\lambda)$ declines so rapidly with increasing m that only the first maximum, at $m=1$, is of physical interest. Note, then, that $A \ll 1$ so that the expansion of q_n in ascending powers of A converges rapidly and the results derived therefrom are accurate. Neglect of $(AK\lambda)^2$ in (19.104) is not so easily justified for the lowest mode but is clearly valid everywhere except in the immediate vicinity of the first maximum of A at $K\lambda = -4{\cdot}22$ where $AK\lambda = 0{\cdot}886$.

Now regenerative modes for negative dynamo number, $K\lambda < 0$, represent the even modes, for which (19.103) is the dispersion relation. The magnetic fields are even functions of z, representing the quadrupole field

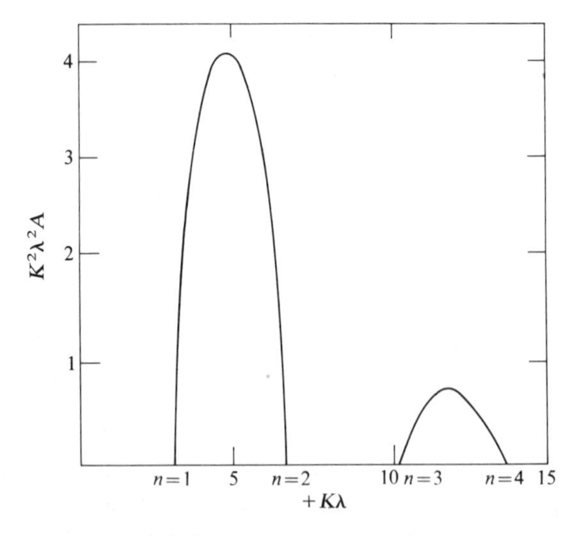

FIG. 19.4. The growth rate $K^2\lambda^2 A(1/\tau$ in units of $\eta\lambda^2)$ in the neighbourhood of the stationary states at $|K\lambda| = 3{\cdot}01, 6{\cdot}63, 10{\cdot}25, 13{\cdot}9$.

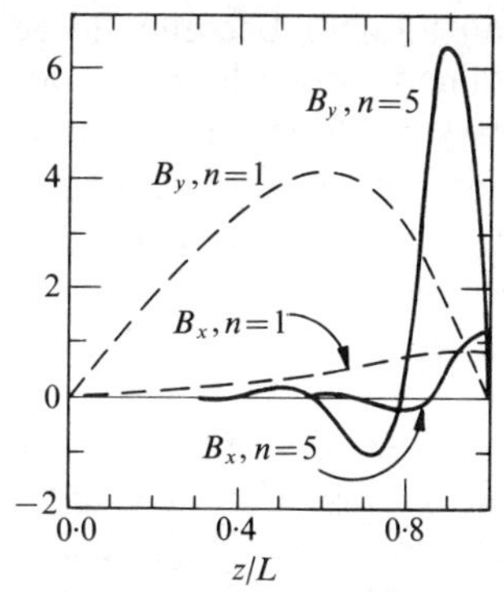

FIG. 19.5. The broken curves represent B_y in arbitrary units, and the associated B_x, for $\eta k^2 G = 0{\cdot}1$, for the first stationary odd mode ($n = 1$, $K\lambda = -3{\cdot}01$). Solid curves represent the stationary odd mode $n = 5$, $K\lambda = -17{\cdot}5$.

as the lowest mode, $n = 1$. The odd modes, for which the lowest is the dipole, are represented by (19.103) with $+K\lambda$ replaced by $-K\lambda$. Hence $+K$ is replaced by $-K$ in (19.105)–(19.109) and the regenerative modes occur for $K\lambda > 0$. The generation rate $K^2\lambda^2 A$ is plotted in Fig. 19.4, being the same function of $|K\lambda|$ as when $K\lambda < 0$. The idea that the outer layers of the liquid core of Earth rotate more slowly than the inner layers yields $G = \mathrm{d}V_y/\mathrm{d}x < 0$, while the more rapid rotation of rising convective cells yields $\Gamma < 0$. Altogether, then, $K\lambda > 0$, yielding the dipole mode as the most strongly generated field.

Computation of the coefficients for the odd mode from (19.96), (19.97), (19.100), and (19.101) yield

$$0 = C_1\{(q_4^2 - q_3^2)\exp(-iq_1K\lambda) + (q_1^2 - q_4^2)\exp(-iq_3K\lambda) + (q_3^2 - q_1^2)\exp(-iq_4K\lambda)\} + C_2\{(q_4^2 - q_2^2)\exp(-iq_2K\lambda) + (q_2^2 - q_4^2)\exp(-iq_3K\lambda) + (q_3^2 - q_2^2)\exp(-iq_4K\lambda)\}$$

with

$$0 = C_1(q_1^2 - q_4^2) + C_2(q_2^2 - q_4^2) + C_3(q_3^2 - q_4^2).$$

Then in the approximation that $K\lambda$ is large and positive, the fields reduce to

$$B_y \cong +\cos(k_x x)\exp(t/\tau)\{\exp(Kz) + 2\exp(-\tfrac{1}{2}Kz)\sin(\tfrac{1}{2}\sqrt{3}\,Kz - \tfrac{1}{6}\pi)\}, \tag{9.110}$$

$$B_x \cong -(\eta K^2/G)\cos(k_x x)\exp(t/\tau)\{\exp(-Kz) + 2\exp(\tfrac{1}{2}Kz)\sin(\tfrac{1}{2}\sqrt{3}\,Kz + \tfrac{1}{6}\pi)\}, \tag{19.111}$$

in $z > 0$ in the neighbourhood $A = 0$. The fields B_x and B_y are plotted in Fig. 19.5 for the dipole mode ($n = 1$). They are also plotted for $n = 5$ ($K\lambda = +17{\cdot}5$) to show the extreme crowding of the field against the poles

$z = \pm\lambda$ when the dynamo number becomes large (Parker 1971b). Such high modes can be sustained only when $k_x \ll K$, of course. The dipole mode is the only stationary mode strongly regenerated when $K\lambda > 0$.

There are a number of points of physical interest illustrated by the calculations. First of all, it follows from the definition (19.51) that the total generation rate $1/\tau + \eta k_x^2$ is given by

$$1/\tau + \eta k_x^2 = \eta K^2 A,$$

so that it is $K^2\lambda^2 A$, rather than A itself, that is a measure of the rate of generation of magnetic fields. Fig. 19.4 is a plot of $K^2\lambda^2 A$ as a function of $K\lambda$, showing the maximum generation rate $K^2\lambda^2 A \cong 4$ at $K\lambda \cong 5$. The maximum is capable of producing stationary fields ($1/\tau = 0$) for $k_x\lambda$ as large as 2. The limitation on k_x is readily understood from the fact that the open boundaries at $k_x x = \pm\frac{1}{2}\pi$ (where we have required $B_y = 0$) cause a rapid escape of magnetic flux. Hence if the total depth π/k_x of the dynamo region in the x-direction is smaller than the fraction $\frac{1}{4}\pi = 0{\cdot}785$ of the length 2λ in the z-direction, the dynamo effect throughout the interior cannot keep ahead of the losses from the surface. If one of the boundaries were impenetrable, of course, then the dynamo region could be half as deep. If for instance, the dynamo region is $0 < x < \pi/2k_x$, with no escape of B_y through $x = 0$ (so that $\partial B_y/\partial x = 0$ at $x = 0$), then the above solutions apply and the depth $\pi/2k_x$ can be as small as the fraction $\frac{1}{8}\pi = 0{\cdot}392$ of the length 2λ in the z-direction. On the basis of this simple rectangular model, then, it is not unreasonable to expect a stationary dynamo in the thick shell of convecting liquid iron, between the mantle and the solid inner core of Earth. It is clear, too, that it is not possible for a dynamo to operate in a shell as thin as the convective zone of the Sun. A thickness of only about 10^5 km $= 0.15 R_\odot$ is available in the lower convective zone of the Sun (because of the rapid buoyant losses from the upper regions of the convective zone), and 10^5 km is only about 10^{-1} of the north–south length of convective zone.

This brings us to the next question, why cannot the dynamo keep ahead of the losses out the surface if the dynamo number $K\lambda$ is made sufficiently large? A glance at Fig. 19.4 shows that the generation rate $K^2\lambda^2 A$ *declines* rapidly with increasing dynamo number. The answer to this question is shown in the next subsection to be the simple fact that the large dynamo number produces periodic rather than stationary fields. When the dynamo is vigorous, the dynamo waves resist being blocked by the boundaries. The waves are so strongly regenerative, and press forward so strongly, that they cycle around through the region, producing migratory dynamo waves and a periodic magnetic field. Only in the idealized symmetry of §19.2.2.3 can the waves be blocked so completely as to produce stationary fields in the limit of large dynamo number. Thus the stationary dynamo is a delicate creature, thriving only in the most favourable environments. The stationary dynamo is favoured by moderate dynamo number, in deep dynamo regions so that the losses through the surface are moderate. With large dynamo numbers, the natural tendency of the dynamo waves to migrate in the z-direction (for dV_y/dx) asserts itself and the fields are periodic in time.

Indeed, the occasional sudden (10^3 years) reversal of the geomagnetic field can be explained merely by an increase in dynamo number, causing the field to become periodic for a time. There are other explanations for the reversal too, such as a change in the distribution of the cyclonic eddies, but that is a topic for later consideration.

Examination of the principal maximum in Fig. 19.4 brings to light the 'unstable' character of the stationary dynamo (Levy 1974). It would seem that at the present time the terrestrial dynamo is in a quasi-stationary state with $1/\tau = 0$. There is presumably some dynamical balance between the hydrodynamic forces and the magnetic field such that, if the field grows stronger, the Lorentz forces suppress the cyclonic convection and non-uniform rotation, thereby reducing the dynamo number $K\lambda$ slightly. Hence, the field relaxes and equilibrium is restored. But it is evident from Fig. 19.4 that this simple picture of dynamical balance is possible only on the left side of each maximum. On the right side, toward larger $K\lambda$, an increase in the fields, causing a decline in the dynamo number $K\lambda$, *enhances* the generation of field, further depressing the dynamo number. The effect runs away until the dynamo number is pushed down below the maximum of the curve and the system resides on the left side of the maximum. On the other hand, suppose that the fields decline slightly while the system is operating on the right side of the maximum. If the reduction of the magnetic stresses permits stronger shear and cyclonic convection, then the dynamo number increases, and the generation rate $K^2\lambda^2A$ *declines*, leading to a further weakening of the field. What happens then depends upon what periodic modes are available, and is taken up in the next subsection. It is clear, however, that a dynamo, operating near the peak of the maximum in the generation rate $K^2\lambda^2A$, leads a precarious existence. A slight upward fluctuation of the dynamo number $K\lambda$ can carry the system over the maximum onto the declining side of the curve, with subsequent decay of the stationary field and the prospect of the onset of a periodic dynamo. It may be some such effect as this that occasionally leads to a reversal of the geomagnetic field.

The blocking of the dynamo waves to provide a stationary dynamo appears to be a tenuous effect (Parker 1971b; Lerche and Parker 1972), its success depending critically on the distribution of the non-uniform rotation and the cyclonic convection over depth and latitude (Parker 1969a; Levy 1972a, b). Roberts (1972) points out the important role that meridional circulation may have in producing stationary dynamos (by opposing the natural migration of the dynamo waves). Steenbeck and Krause (1966, 1967) employed numerical methods to treat the dynamo in spherical boundaries. With a dynamo coefficient Γ proportional to $\cos\theta$, where θ is the polar angle measured from the north pole, they found stationary solutions, whereas Braginskii, using the same non-uniform rotation but with $\Gamma \propto \sin^2\theta\cos\theta$ (so that the cyclones are concentrated at lower latitudes) failed to find stationary solutions. On the other hand, it is shown below that cyclones at high latitude alone fail to generate the dipole mode (Parker 1969a; Levy 1972a, b, c).

19.3.1.2. Periodic states. The occurrence of the stationary dynamo only at modest dynamo numbers suggests that strong dynamos (large dynamo numbers) produce periodic fields. The migration of their internal waves is not easily suppressed so that the fields are migratory, as observed on the sun. The search for stationary states in the section above was carried out expanding q_n in ascending powers of A and then supposing that A is so small that $\exp(AK\lambda)$ can be expanded in powers of $AK\lambda$. This second assumption will not do when we search for vigorous generation at large $K\lambda$. As a matter of fact, the expansion (19.104) of the exponential functions $\exp(AK\lambda)$ in ascending powers of $Ak\lambda$ seems to miss the periodic modes in a systematic way, because going to the terms second order in A yields no imaginary part to A. So again use (19.56)–(19.59) to express the q_n in terms of A. This time, however, do not expand the exponential functions in powers of $AK\lambda$. The dispersion relation for the even mode is

$$\begin{aligned}0 = {} & \mathrm{i}(1+A+\tfrac{1}{3}A^2)\exp\{(1+\tfrac{2}{3}A+\tfrac{2}{3}A^2)K\lambda\} \\ & -A(1+\tfrac{1}{3}A)\exp\{-(1+\tfrac{2}{3}A+\tfrac{2}{3}A^2)K\lambda\} \\ & -\{1-A\exp(\mathrm{i}\tfrac{1}{3}\pi)-\tfrac{1}{3}A^2\exp(-\mathrm{i}\tfrac{1}{3}\pi)\} \\ & \times\exp[\mathrm{i}\tfrac{2}{3}\pi-\{\exp(\mathrm{i}\tfrac{1}{3}\pi)+\tfrac{2}{3}A\exp(-\mathrm{i}\tfrac{1}{3}\pi)-\tfrac{2}{3}A^2\}K\lambda] \\ & -\mathrm{i}A\{1-\tfrac{5}{3}A\exp(\mathrm{i}\tfrac{1}{3}\pi)\}\exp[\{\exp(\mathrm{i}\tfrac{1}{3}\pi)+\tfrac{2}{3}A\exp(-\mathrm{i}\tfrac{1}{3}\pi)-\tfrac{2}{3}A^2\}K\lambda] \\ & -\mathrm{i}A\{1-\tfrac{5}{3}A\exp(-\mathrm{i}\tfrac{1}{3}\pi)\}\exp[\{\exp(-\mathrm{i}\tfrac{1}{3}\pi)+\tfrac{2}{3}A\exp(\mathrm{i}\tfrac{1}{3}\pi)-\tfrac{2}{3}A^2\}K\lambda] \\ & -\{1-A\exp(-\mathrm{i}\tfrac{1}{3}\pi)-\tfrac{1}{3}A^2\exp(\mathrm{i}\tfrac{1}{3}\pi)\} \\ & \times\exp[-\mathrm{i}\tfrac{1}{6}\pi-\{\exp(-\mathrm{i}\tfrac{1}{3}\pi)+\tfrac{2}{3}A\exp(\mathrm{i}\tfrac{1}{3}\pi)-\tfrac{2}{3}A^2\}K\lambda],\end{aligned} \tag{19.112}$$

keeping terms second order in A (Parker 1971b). Interest is centred on complex values of A, for complex $1/\tau$, so write

$$A = R\exp(\mathrm{i}\theta) \tag{19.113}$$

where R is a real positive number and $0<\theta<\pi$ (negative θ gives the same final expressions). In the limit of large, positive $RK\lambda$ any exponential function whose argument has a negative real part becomes very small.

Neglecting such terms there remain in (19.112) only the first, fourth, and fifth terms. The real part of the argument of the exponential in the fourth term is

$$\{\tfrac{1}{2}+\tfrac{2}{3}R\cos(\theta-\tfrac{1}{3}\pi)-\tfrac{2}{3}R^2\cos(2\theta)\}K\lambda, \tag{19.114}$$

and the real part of the fifth is

$$\{\tfrac{1}{2}+\tfrac{2}{3}R\cos(\theta+\tfrac{1}{3}\pi)-\tfrac{2}{3}R^2\cos(2\theta)\}K\lambda. \tag{19.115}$$

For $0<\theta<\pi$ it is readily shown that $\cos(\theta-\tfrac{1}{3}\pi)$ is greater than $\cos(\theta+\tfrac{1}{3}\pi)$, by the amount $3^{\frac{1}{2}}\sin\theta$. Hence for $K\lambda$ large and positive the exponential factor in the fifth term is very small, by the factor $\exp(+\tfrac{2}{3}RK\lambda 3^{\frac{1}{2}}\sin\theta)$, compared to the exponential in the fourth. The fifth term may be dropped (assuming $\theta\neq 0, \pi$); there remain the first and the fourth terms. The real part of the dispersion relation is, then,

$$(1+R\cos\theta+\tfrac{1}{3}R^2\cos 2\theta)\exp T=R\cos(S+\theta)-\tfrac{5}{3}R^2\cos(S+2\theta+\tfrac{1}{3}\pi) \tag{19.116}$$

and the imaginary part is

$$(\sin\theta+\tfrac{1}{3}R\sin 2\theta)\exp T=\sin(S+\theta)-\tfrac{5}{3}R\sin(S+2\theta+\tfrac{1}{3}\pi), \tag{19.117}$$

where T and S are the quantities

$$T\equiv[\tfrac{1}{2}+\tfrac{2}{3}R\{\cos\theta-\cos(\theta-\tfrac{1}{3}\pi)\}+\tfrac{4}{3}R^2\cos 2\theta]K\lambda, \tag{19.118}$$

$$S\equiv[\tfrac{1}{2}\sqrt{3}-\tfrac{2}{3}R\{\sin\theta-\sin(\theta-\tfrac{1}{3}\pi)\}-\tfrac{4}{3}R^2\sin 2\theta]K\lambda. \tag{19.119}$$

When $K\lambda$ is large and positive, the right-hand sides of (19.116) and (19.117) are of the order of $\exp T$ and of unity, respectively. The coefficients of $\exp T$ on the left-hand sides are also of the order of unity. However, it is obvious from (19.118) that for most values of θ, large positive $K\lambda$ implies a large positive T, so that $\exp T$ is very large indeed. Equations (19.116) and (19.117) cannot be satisfied in this case. They require that T be more or less of the order of unity, which requires that the coefficient of $K\lambda$ in (19.118) is small, $O(1/K\lambda)$. Write (19.118) as

$$R^2\cos 2\theta+\tfrac{1}{2}R\{\cos\theta-\cos(\theta-\tfrac{1}{3}\pi)\}+\tfrac{3}{8}\equiv\epsilon$$

where $\epsilon=T/K\lambda$ and is small $O(1/K\lambda)$. It follows that

$$R=\frac{1}{4\cos 2\theta}\{\cos(\theta-\tfrac{1}{3}\pi)-\cos\theta$$

$$\pm[\{\cos(\theta-\tfrac{1}{3}\pi)-\cos\theta\}^2-(6-16\epsilon)\cos 2\theta]^{\frac{1}{2}}\}. \tag{19.120}$$

Since $\cos(\theta-\tfrac{1}{3}\pi)-\cos\theta$ is positive for $0<\theta<\pi$, the only requirement to be placed on this expression is that the argument of the radical be positive, so that R is real and positive. Hence it must be required that

$$-(4\sqrt{39}-\sqrt{3})<27\tan\theta<4\sqrt{39}+\sqrt{3},$$

i.e.

$$0{\cdot}780<\theta<2{\cdot}43 \tag{19.121}$$

in order that there be a satisfactory solution. Here ϵ is neglected in the limit of large $K\lambda$.

Returning, then, to (19.116) and (19.117) with the compatibility conditions (19.120) and (19.121), eliminate $\exp T$ between the two equations. The result can be written

$$\sin(S+\theta)=R\{\tfrac{5}{3}\sin(S+2\theta+\tfrac{1}{3}\pi)-\sin(S)\} \tag{19.122}$$

neglecting terms second order in R. For small R, then, $\sin(S+\theta)$ must be small, $O(R)$, and have the same sign as the square brackets on the right-hand side. Hence

$$S+\theta=n\pi+r+O(R^2)$$

where r is a small correction

$$r=R\{\tfrac{5}{3}\sin(\theta+\tfrac{1}{3}\pi)+\sin\theta\}$$

and n is an integer, even or odd so as to make the signs consistent within (19.114) and (19.115). If (19.121) is substituted into (19.114) and (19.115) the result is the two relations

$$\sin\theta\exp(T)=(-1)^n\sin r,$$
$$\exp(T)=(-1)^nR\cos r,$$

to lowest order. For $0<\theta<\pi$ the left-hand sides are both positive, as are $\sin r$ and $\cos r$ for small r on the right. Hence n must be an even integer, and will henceforth be written $2n$. It follows that

$$S+\theta=2n\pi+r+O(R^2). \tag{19.123}$$

The general structure of the equations is now apparent from (19.119)–(19.123). Choose a value of θ and compute R from (19.120) neglecting the term $\epsilon=O(1/K\lambda)$ in the limit of large $K\lambda$. Then S is readily expressed in terms of $K\lambda$ from (19.119) and eliminated from (19.123) to obtain $K\lambda$ in terms of n, with an uncertainty in $K\lambda$ of the order of unity. That is to say, $K\lambda$ can then be computed to an accuracy that neglects terms $O(R^2)$ and $O(1/K\lambda)$ compared to one.

As an example illustrating the results in the limit of large $K\lambda$, let $\theta=\tfrac{1}{2}\pi-\Delta$ where Δ is small $O(1/RK\lambda)$. Then $\text{Re}(A)=R\Delta$ so that the regenerative power of the dynamo is

$$K^2\lambda^2\,\text{Re}(A)=RK^2\lambda^2\Delta=O(1)$$

and is sufficient to overcome any zero order dissipation ηk_x^2. With $\theta = \frac{1}{2}\pi$ (19.120) yields $R = \frac{1}{4}\sqrt{3} = 0{\cdot}433$, neglecting terms $O(1/RK^2\lambda^2)$ compared to one. It follows from (19.119) that $S = 5\sqrt{3}\,K\lambda/12$. The most direct and accurate calculation of S follows from (19.116) and (19.117). Put $R = \frac{1}{4}\sqrt{3}$ and $\theta = \frac{1}{2}\pi$, eliminate $\exp T$, and solve for S, obtaining

$$\tan S = -25/11\sqrt{3} = -1{\cdot}312. \tag{19.124}$$

Hence

$$\begin{aligned} S &= 2n\pi - 0{\cdot}92 \\ K\lambda &= (12/5\sqrt{3})(2n\pi - 0{\cdot}92) \\ &= +7{\cdot}43, +16{\cdot}1, +24{\cdot}8, +33{\cdot}5, \ldots . \end{aligned} \tag{19.125}$$

Even for the lowest mode $K\lambda$ is sufficiently large as to justify neglect of four of the six exponential factors in (19.112).

To show the degree of uncertainty in this result for $K\lambda$, put $R = \frac{1}{4}\sqrt{3}$ and $\theta = \frac{1}{2}\pi$ in (19.122), yielding $\tan S = -3\sqrt{3}/11$ and $S = 2n\pi - 0{\cdot}442$. On the other hand (19.123) yields $2n\pi - 0{\cdot}794$.

Finally note that $R = 0{\cdot}433$ is sufficiently large as to permit the calculation to be carried out with the asymptotic expansions (19.60)–(19.63) instead of (19.56)–(19.59). The result (Parker 1971b) for $\theta = \frac{1}{2}\pi$ is $R = (\frac{1}{2}\cos\frac{1}{8}\pi)^{\frac{4}{3}} = 0{\cdot}357$ (instead of $0{\cdot}433$) and $K\lambda = +11{\cdot}1, +23{\cdot}9, +36{\cdot}6$. etc. for the odd modes. Hence R is somewhat smaller and the dynamo numbers are about half again as large. The structure of the spectrum is identical, of course, the numerical differences resulting from the slow convergence of the expansions in both ascending and descending powers of A when $R \cong 0{\cdot}4$.

Now suppose that $K\lambda$ is large and negative. Then returning to (19.112) it is possible to drop the first, fourth, and fifth terms because of their exponential factors. The real parts of the arguments of the exponentials of the third and sixth terms are just the negatives of (19.114) and (19.115) respectively, so that the sixth term is small compared to the third, and may be neglected. There remain, then, only the second and third terms, whose real and imaginary parts are

$$\begin{aligned} &(R\sin\theta + \tfrac{5}{3}R^2\sin 2\theta)\exp(-T) \\ &\qquad = \cos(S - \tfrac{1}{6}\pi) - R\sin(S - \theta) - \tfrac{1}{3}R^2\cos(S - 2\theta + \tfrac{1}{6}\pi), \end{aligned} \tag{19.126}$$

and

$$\begin{aligned} &(R\cos\theta + \tfrac{5}{3}R^2\cos 2\theta)\exp(-T) \\ &\qquad = \sin(S - \tfrac{1}{6}\pi) + R\cos(S - \theta) - \tfrac{1}{3}R^2\sin(S - 2\theta + \tfrac{1}{6}\pi), \end{aligned} \tag{19.127}$$

respectively, where T and S are defined by (19.118) and (19.119). For large negative $K\lambda$ the consistency condition (19.120) must be satisfied again. Hence choosing θ yields R again. Eliminating $\exp(-T)$ between (19.126) and (19.127) yields S, and $K\lambda$ follows from (19.119).

To treat the case where $\theta = \frac{1}{2}\pi - \Delta$, where $\Delta = O(1/RK^2\lambda^2)$ in the limit of large negative $K\lambda$, it follows that $R = \frac{1}{4}\sqrt{3}$ again. Hence (19.126) and (19.127) yield

$$\tan S = -13\sqrt{3}/75 \tag{19.128}$$

so that

$$S = -2n\pi - 0{\cdot}29$$

where n is a positive integer. From (19.119)

$$\begin{aligned} K\lambda &= (5\sqrt{3}/12)S \\ &= -(12/5\sqrt{3})(2n\pi + 0{\cdot}29) \\ &= -9{\cdot}10, -17{\cdot}8, -26{\cdot}5, -35{\cdot}2, \ldots. \end{aligned} \tag{19.129}$$

Repeating the calculation with the expansion of q_n in descending powers of A yields $K\lambda = -16{\cdot}6, -29{\cdot}1, -41{\cdot}9$, etc. which are about half again as large in magnitude, but with the same structure.

To treat the odd modes it is necessary only to replace $K\lambda$ by $-K\lambda$ in the dispersion relation (19.112). Hence for $K\lambda$ positive the result (19.129) applies, with

$$K\lambda = +9{\cdot}10, +17{\cdot}8, +26{\cdot}5, +35{\cdot}2, \ldots \tag{19.130}$$

while for $K\lambda$ negative (19.125) applies, and

$$K\lambda = -7{\cdot}43, -16{\cdot}1, -24{\cdot}8, -33{\cdot}5, \ldots. \tag{19.131}$$

There are a number of important points that follow from these calculations (Parker 1971b). The first is, of course, that large dynamo numbers give strong regeneration of the field. The regenerative power is proportional to $(K\lambda)^2$, i.e. proportional to $(\Gamma G\lambda^3/\eta^2)^{\frac{2}{3}}$, and in the limit of large $K\lambda$ can overcome any resistive dissipation ηk_x^2. The second point is that large dynamo numbers produce only periodic fields. Hence the only possible field generated in a thin convective layer (thin in the direction of the shear, $\nabla\Omega$) is a periodic field, if the dynamo number is large enough to give any regeneration at all. It is no accident, then, that the magnetic field of the sun is periodic. If the principal shear is vertical, then the vertical wave number k_x is large, $k_x\lambda \sim 10$ and the dissipation ηk_x^2 is too large to permit a stationary state.

If a *very* thin convecting layer has a dynamo number so large as to overcome the dissipation ηk_x^2, then the large dynamo number produces a dynamo in a very high mode. The basic fact is that the horizontal wavelength cannot be more than a few times the vertical thickness π/k_x of the dynamo region. Hence the fields in a thin wide region necessarily break up into several separate waves (high modes). We know of no astrophysical object where this occurs, but the effect should be kept in mind. The dynamos of Earth and of the sun operate in regions sufficiently thick as to permit the lowest modes to appear.

One might conjecture that a dynamo automatically operates with a dynamo number at the lowest level consistent with net regeneration, because of the exponential growth of fields if the dynamo number is larger. For a larger dynamo number the magnetic field grows until the magnetic stresses became so strong as to cut back on Γ and G, reducing the dynamo number to the lowest level consistent with a self-sustaining field. Unfortunately this complicated dynamical picture still lies beyond the grasp of contemporary theory.

While we are on the subject of the lowest modes, note from (19.125) and (19.130) that, when $K\lambda > 0$, the lowest mode is the even mode with $K\lambda = 7{\cdot}43$ for $n = 1$, as opposed to the odd mode, for which the higher value $K\lambda = 9{\cdot}10$ is required. The difference between these two dynamo numbers is the sum of the S from (19.124) and the S from (19.128). Hence there is no question about the sign of the difference. Positive dynamo number favours the even mode. Negative dynamo number favours the odd mode, occurring for $K\lambda = -7{\cdot}43$, while the even mode requires $K\lambda = -9{\cdot}10$.

These theoretical facts have immediate implications for the sun. If we suppose that the angular velocity of the sun increases with depth ($\partial\Omega/\partial r < 0$), then the shear in the convective layer is equivalent to $G = \mathrm{d}V_y/\mathrm{d}x < 0$ in the present problem. If rising convective fluid is dominated by expansion, then it rotates less rapidly than its surroundings. A rising convective cell in northern latitudes rotates clockwise relative to its surroundings, as viewed from above the surface of the sun. It follows from (19.2) and the fact, from (18.51), that $G(\Phi)$ has the same sign as the rotation Φ, that $\Gamma > 0$. Hence the dynamo number $K\lambda = (\Gamma G\lambda^3/\eta^2)^{\frac{1}{3}}$ is negative. The lowest mode is the dipole. The fact that the solar field is essentially a dipole, then, follows directly from the idea that the dynamo number in the northern hemisphere is negative.

We should not fail to notice, however, that the lowest even mode (the quadrupole mode) lies nearby and may be excited intermittently by fluctuations in the convection and non-uniform rotation. Occasional excitation of the quadrupole mode would explain the observed sloppy odd symmetry of the solar dipole field about the equator. In the late 1960s and early 1970s, for instance, both poles had the same sign, although the positive pole remained more positive than the 'negative' pole. Leighton (1969) pointed out the tendency for the quadrupole mode to appear in his numerical experiments on the solar dynamo. The mode causes an erratic variation of the symmetry, much like that which is observed.

Present ignorance of the dynamical convection and circulation within planets and stars leaves open the possibility that there are stars with positive dynamo number $K\lambda$ in their northern hemispheres, and planets with negative $K\lambda$. Such stars would favour quadrupole fields as their lowest periodic modes. Stars of small mass (i.e. late-type stars) have a convective zone so thick as to raise the possibility of stationary magnetic fields. Evidently Earth has positive dynamo number $K\lambda$ in the northern hemisphere of the liquid core, giving rise to the stationary dipole. A planet with negative $K\lambda$ in the northern hemisphere would favour the stationary quadrupole as the lowest mode. It is interesting to note that if the positive dynamo number $K\lambda$ of Earth were to increase from its present value in the neighbourhood of 4 or 5 (defined in the context of the present rectangular dynamo) to somewhere in the neighbourhood of 7·5, the geomagnetic field would become an oscillating quadrupole. Raising the dynamo number to 9 or 10 would produce an oscillating dipole, and further increase in $K\lambda$ would give even higher modes. It should be noted however, that $K\lambda$ varies only as the cube root of ΓG so the increase in ΓG to bring about an oscillatory quadrupole is a factor of four.

Before going on to other models it is worth noting some of the properties of the magnetic fields themselves. An extensive discussion may be found in Parker (1971b). A brief treatment is adequate for the present purposes. Note first of all that for $R=\frac{1}{4}\sqrt{3}$ and $\theta=\frac{1}{2}\pi$ the frequency of oscillation of the fields is

$$\omega=\mathrm{Im}(1/\tau)=\tfrac{1}{4}\sqrt{3}\,\eta K^2$$

in the limit of large $K\lambda$. For $K\lambda=-7\cdot43$, favouring the dipole mode observed in the sun, this is $\omega=23.9\eta/\lambda^2$, to be compared with the elementary result worked out in §19.1.2.2, leading to (19.27). In place of (19.27) we have the more precisely defined period $T=0\cdot26\lambda^2/\eta$. The discussion following (19.27) is unaffected by this better value.

To illustrate the spatial form of the dynamo waves note that with $R=\frac{1}{4}\sqrt{3}$ and $\theta=\frac{1}{2}\pi$ the four roots (19.56)–(1959) reduce to

$$-iq_1=+0\cdot69+i1\cdot01, \qquad -iq_2=+0\cdot19-i0\cdot72$$

$$-iq_3=-1\cdot06-i0\cdot29, \qquad -iq_4=+0\cdot19+i0\cdot03$$

neglecting all terms $O(A^3)$ except in the fourth root. The dependence of the fields on q_n is given by (19.91) as $\exp\{-iq_n(K\lambda)z/\lambda\}$. With $K\lambda=-7\cdot43$, favouring the dipole mode observed in the sun, the amplitudes of the first and third modes vary extremely rapidly across $0<z<\lambda$, as $\exp\{K\lambda\ \mathrm{Im}(q_nz/\lambda)\}=\exp(-5\cdot13z/\lambda)$ and $\exp(+7\cdot88z/\lambda)$, respectively. Thus the first mode exists only near $z=0$ and the third mode near $z=\lambda$, where their amplitudes are comparable to the second and fourth modes. Their rapid exponential decline inward from $z=0$ and $z=\lambda$ means that they are negligible everywhere else in the region. The second and fourth modes are the principal modes spanning the dynamo region, the amplitudes of both varying moderately as $\exp(-1{\cdot}4z/\lambda)$. Both modes are stronger by the factor $\exp(+1{\cdot}4)=4$ at the 'equator' ($z=0$) than at the 'poles' ($z=\pm\lambda$). The wavelength of the fourth mode is $2\pi/\mathrm{Re}(q_4K)\cong 25\lambda$ so that the phase of the fourth mode is more or less uniform across the region. It oscillates up and down essentially as $\exp(i\omega t-1{\cdot}4z/\lambda)$ across the entire region. The second mode is the principal migratory mode. The wavelength is $2\pi/\mathrm{Re}(q_2K)=1{\cdot}17\lambda$ so that both the crest and the trough of the wave appear in $0<z<\lambda$. The phase velocity of the wave is $\mathrm{Im}(1/\tau)/\mathrm{Re}(q_2K)=\omega/0{\cdot}72K$ in the direction of negative z (for $K\lambda<0$), taking the wave from the 'pole' to the 'equator' in each period. The crests move from 'middle latitudes' $z=\frac{1}{2}\lambda$ to the equator in each half period, growing by about a factor of two as they progress. This is essentially the behaviour of the azimuthal field $B_\phi=B_y$ of the sun, as inferred from the observed appearance of bipolar sunspot groups at middle latitudes, and their subsequent migration and intensification toward the equator.

19.3.2. The dynamo with horizontal shear

Consider the operation of a dynamo in which the large-scale shear extends horizontally along the length of the dynamo, rather than vertically across its thickness (Lerche and Parker 1972). The direction of migration of the dynamo waves is then vertical instead of horizontal, with the result that fields behave in quite a different manner. Use the same rectangular geometry as in the subsection above, sketched in Fig. 19.1. Suppose that the dynamo lies between the two horizontal planes $x=0$ and $x=h=\pi/k_x$, and is bounded at its ends by $z=\pm\lambda$. Again the plane $z=0$ corresponds to the 'equator' and $z=\pm\lambda$ to the poles. The large-scale velocity is now $V_y(z)$ so that the shear is $dV_y/dz=G_z(x,z)$ with the vertical shear $\partial V_y/\partial x$ equal to zero. The dynamo coefficient $\Gamma(x,z)$ changes sign across the equator, $z=0$.

Recall again that, on the basis of elementary dynamical considerations (Steenbeck *et al.* 1966), one expects a rising convective cell to expand and rotate more slowly than the surrounding medium. Hence when we come later to discuss the physical implications of the mathematical results, it will be convenient, to fix ideas, to suppose that Γ is positive in the 'northern hemisphere', $0<z<\lambda$ and negative in the 'southern hemisphere', $-\lambda<z<0$. Similarly it is to be expected that the shear $G_z(x,z)$ changes sign across the equator $z=0$. The equatorial regions rotate more rapidly than the polar regions of the sun, in which case G_z is negative in $z>0$ and positive in $z<0$. The product ΓG_z is then negative throughout the entire dynamo region.

It should be noted that most models of the solar dynamo (Rädler 1968;

Steenbeck and Krause 1969a; Deinzer and Stix 1971; Parker 1971b) have employed vertical shear $\partial\Omega/\partial r$, although Babcock (1961) based his ideas on horizontal shear $\partial\Omega/\partial\theta$ alone, and Gilman (1969) and Yoshimura (1975c) employed horizontal shear. Leighton (1969) made the important quantitative point that the observed surface variation of Ω with θ is so weak as to require cyclonic convection stronger than that represented by the observed increasing tilt of the bipolar regions with time. The tilt is the observable surface effect, and the cyclonic motion may well be significantly stronger in the deep convective zone where the dynamo operates. Leighton's point must be kept in mind when treating the quantitative aspects of any solar dynamo based on horizontal shear. By contrast, a 10 or 20 per cent vertical change in Ω across the thin solar convective zone is a stronger shear than the observed non-uniform rotation of the surface at low latitudes (Yoshimura 1975a, b, c).

Whatever the total $\nabla\Omega$ in the sun may eventually prove to be, we should understand the contribution of $\partial\Omega/\partial\theta$ to the generation of fields. As noted already, the basic wave number is then the vertical wave number k_x, and the wave migration is vertical. With ΓG_z negative, as presumed for the sun, the migration is vertically upward, so that the dynamo waves are blocked by the surface of the sun. This is in the same direction of migration as that caused by magnetic buoyancy, about which more will be said later. The horizontal migration of the dynamo waves is caused by their deflection as they press up against the top of the convective zone and escape through the surface of the sun. Generally they migrate away from the latitude where they are most strongly produced. Hence their direction of migration at the blocking surface depends on the distribution of ΓG_z over latitude. The dynamo with horizontal shear alone is, then, a more complex creature, with such possibilities as a migration of field toward the equator at low latitudes and toward the poles at high latitudes, as high latitude magnetic prominences are observed to do (Lockyer 1931; Bumba and Howard 1965; Hansen *et al.* 1969).

We wish to study the behaviour of the magnetic fields generated by various distributions of Γ and G_z over latitude. For this we turn to the variational principle developed by Lerche (1972).

The dynamo equations are

$$(\partial/\partial t - \eta\nabla^2)B_x = -\partial\Gamma B_y/\partial z, \qquad (19.132)$$

$$(\partial/\partial t - \eta\nabla^2)B_y = G_z B_z \qquad (19.133)$$

$$(\partial/\partial t - \eta\nabla^2)B_z = +\partial\Gamma B_y/\partial x, \qquad (19.134)$$

where Γ and G_z are both fixed functions of position. The field B_x is secondary, with B_y and B_z the principal fields, interacting through

(19.133) and (19.134). We are interested in fields with the common time dependence $\exp(t/\tau)$. Then $\partial/\partial t$ is replaced by $1/\tau$ and the dynamo equations (19.133) and (19.134) follow from the extrema of the Lagrangian (Morse and Feshbach 1953).

$$\mathscr{L} = \int_0^h \mathrm{d}x \int_{-\lambda}^{+\lambda} \mathrm{d}z \Big(\eta \nabla B_y \,.\, \nabla B_y^\dagger + \frac{1}{\tau} B_y B_y^\dagger - G_z B_z B_y^\dagger + \eta \nabla B_z \,.\, \nabla B_z^\dagger + \frac{1}{\tau} B_z B_z^\dagger - B_z^\dagger \frac{\partial}{\partial x} \Gamma B_y \Big). \quad (19.135)$$

where the integration is over the volume of the dynamo. The fields $B_y^\dagger$ and $B_z^\dagger$ are the adjoint fields. If the field $B_y^\dagger$ is replaced by $B_y^\dagger + \delta B_y^\dagger$, where $\delta B_y^\dagger$ is an arbitrary variation, subject only to the condition that it vanish at the boundaries, then the Lagrangian is $\mathscr{L} + \delta\mathscr{L}$. Subtract (19.135) from the expression for $\mathscr{L} + \delta\mathscr{L}$ and set to zero the resulting expression for $\delta\mathscr{L}$ in terms of $\delta B_y^\dagger$ to obtain the conditions for the extremum. Integrate by parts, and note that for arbitrary $\delta B_y^\dagger$ the result can vanish only if the integrand is zero everywhere throughout the volume. There follows the dynamo equation (19.133). A similar variation $\delta B_z^\dagger$ yields (19.134). The variations δB_y and δB_z yield the adjoint equations

$$(1/\tau - \eta\nabla^2) B_y^\dagger = -\Gamma\, \partial B_z^\dagger/\partial x, \quad (19.136)$$

$$(1/\tau - \eta\nabla^2) B_z^\dagger = +G_z B_y^\dagger \quad (19.137)$$

for the adjoint fields $B_y^\dagger$ and $B_z^\dagger$. Comparison with the equations for B_y and B_z indicates that the dynamo equations are not self-adjoint. Hence, substituting trial functions for the fields and varying the parameters to make $\delta\mathscr{L} = 0$, yields an extremum of $\mathscr{L}$, rather than a minimum.

It is readily shown that the modes separate depending upon whether B_y is an even or an odd function of z. So let

$$B_y = \exp\left(\frac{t}{\tau}\right) \sum_{n,m=1}^{\infty} A_{nm} \sin\left(\frac{n\pi x}{h}\right) \sin\left(\frac{m\pi z}{\lambda}\right), \quad (19.138)$$

$$B_x = \exp\left(\frac{t}{\tau}\right) \sum_{n,m=1}^{\infty} B_{nm} \sin\left(\frac{n\pi x}{h}\right) \cos\left\{(m - \tfrac{1}{2}) \frac{\pi z}{\lambda}\right\} \quad (19.139)$$

for the odd modes and

$$B_y = \exp\left(\frac{t}{\tau}\right) \sum_{n,m=1}^{\infty} C_{nm} \sin\left(\frac{n\pi x}{h}\right) \cos\left\{(m - \tfrac{1}{2}) \frac{\pi z}{\lambda}\right\}, \quad (19.140)$$

$$B_x = \exp\left(\frac{t}{\tau}\right) \sum_{n,m=1}^{\infty} D_{nm} \sin\left(\frac{n\pi x}{h}\right) \sin\left(\frac{m\pi z}{\lambda}\right) \quad (19.141)$$

for the even modes throughout $-\lambda < z < +\kappa$, The sines and cosines form a complete set and automatically satisfy the boundary conditions that B_y and B_z vanish at the free surfaces $x=0, h$ and at the 'poles' $z=\pm\lambda$. To treat one free surface at $x=h$ and one impenetrable surface at $x=0$, we would use $\cos(n-\frac{1}{2})\pi x/h$ instead of $\sin(n\pi x/h)$, and for two impenetrable surfaces $\cos(n\pi x/h)$ etc.

Express the even and odd modes of the adjoint fields in the same way, with the coefficients E_{nm}, F_{nm}, G_{nm}, and H_{nm}, respectively, since they satisfy the same boundary conditions.

As a first example, consider the shear and dynamo coefficients.

$$G_z = G\sin(\pi z/2\lambda), \qquad \Gamma = \Gamma_0 \sin(\pi x/h)\sin(\pi z/2\lambda). \quad (19.142)$$

The minimum number of terms in the expansion (19.138)–(19.141) capable of exhibiting the migration of the dynamo waves is three, i.e. $n+m\leqslant 3$. It is a simple matter to compute $\mathscr{L}$ using the first three terms in the representation of each of the fields. To treat the odd modes, set equal to zero the derivatives of $\mathscr{L}$ with respect to the adjoint coefficients E_{nm} and F_{nm}. There arise a set of linear homogeneous equations. Equating the determinant of the coefficients to zero yields the dispersion relation

$$\psi(1,\tfrac{1}{2})\psi(1,1)\psi(1,\tfrac{3}{2})\psi(2,\tfrac{1}{2})\psi(2,1)+(\pi^2/32)(\Gamma_0 G/h)^2\{\psi(1,\tfrac{1}{2})+\psi(1,\tfrac{3}{2})\}=0 \quad (19.143)$$

where $\psi(n, m)$ is the algebraic function

$$\psi(n,m)=1/\tau+\eta\pi^2(n^2/h^2+m^2/\lambda^2), \quad (19.144)$$

corresponding to $A+q^2$ in the calculations of the foregoing sections. Note that $\Gamma_0 G$ appears to the second power, so that the sign of $\Gamma_0 G$ has no effect on the time variation $1/\tau$. The ratio of the coefficients for B_y is

$$A_{12}/A_{21}=+\pi\Gamma_0 G/8h\psi(1,2)\psi(1,\tfrac{3}{2}), \quad (19.145)$$

$$A_{21}/A_{11}=+\pi\Gamma_0 G/4h\psi(2,1)\psi(2,\tfrac{1}{2}). \quad (19.146)$$

Note here that the sign of the ratio A_{21}/A_{11}, giving the direction of the vertical migration, depends upon the sign of $\Gamma_0 G$. The sign of A_{12}/A_{11}, giving the direction of horizontal migration, depends upon the square of $\Gamma_0 G$ and so is independent of the sign of $\Gamma_0 G$. This demonstrates the point stated earlier, that the phase velocity of the dynamo waves observed at the surface depends upon the distribution of Γ and G over 'latitude' z, rather than on the sign of ΓG.

Consider a relatively thin dynamo layer, $h\ll\lambda$, so that to a first approximation

$$\psi(n,m)\cong 1/\tau+\eta\pi^2 n^2/h^2\equiv\phi(n) \quad (19.147)$$

for $\frac{1}{2}\leqslant m\leqslant\frac{3}{2}$. Diffusion across the length λ is negligible compared to the diffusive losses across h. The dispersion relation (19.143) becomes

$$\phi(1)\{\phi^2(1)\phi^2(2)+(\pi\Gamma_0 G/4h)^2\}=0.$$

The root $\phi(1)=0$ represents free decay and is without physical interest. It arises as the complete dispersion relation if only the lowest order term in the expansion is used. There remain, then, the roots

$$\phi(1)\phi(2)=\pm i\pi\Gamma_0 G/4h. \qquad (19.148)$$

Define the dynamo number

$$\mu\equiv+\Gamma_0 Gh^3/4\pi^3\eta$$

and the dimensions less growth rate $\delta=h^2/\pi^2\eta\tau$. Then (19.148) has the two roots

$$\sigma=\tfrac{1}{2}\{(9\pm4i\mu)^{\frac{1}{2}}-5\} \qquad (19.149)$$

which may have positive real parts. Steady operation of the dynamo arises when $\mathrm{Re}(\sigma)=0$. This occurs for $\mu=+10$ ($\Gamma_0 Gh^3/\eta^2=1240$) and yields $\sigma=\pm2i$ and $A_{21}=+\tfrac{1}{2}A_{11}\exp(\pm i\vartheta)$ where $\vartheta=\mathrm{Arc}\tan\tfrac{4}{3}=0{\cdot}927$. For $\mu=-10$, we obtain $\sigma=\mp2i$ and $A_{21}=-\tfrac{1}{2}A_{11}\exp\mp i\vartheta$. The real part of (19.138) yields

$$B_y=A_{11}\sin\left(\frac{\pi x}{h}\right)\sin\left(\frac{\pi z}{\lambda}\right)\left\{\cos\left(\frac{2\eta\pi^2 t}{h^2}\right)\left(1-\cos\left(\frac{\pi z}{\lambda}\right)\right)\mp\frac{3}{5}\cos\left(\frac{\pi x}{h}\right)\right.$$

$$\left.\pm\frac{4}{5}\sin\left(\frac{2\eta\pi^2 t}{h^2}\right)\cos\left(\frac{\pi x}{h}\right)\right\}, \qquad (19.150)$$

where the $\pm$ has the same sign as μ. It is evident that $x\to h+x$ transforms one case into the other.

The field B_y near the top ($x=h-\epsilon$) and bottom ($x=\epsilon$) of the dynamo region is plotted in Fig. 19.6 for $\mu=-10$, corresponding to the sign of the dynamo number in the northern hemisphere of the sun (based on $G_z<0$, $\Gamma_0>0$). Following the time sequence of the curves it is evident that the field migrates toward the equator at the top and toward the poles at the bottom of the convective zone. At the middle level $x=\tfrac{1}{2}h$ the field is a standing wave without migration. An observer looking at the upper surface of the dynamo sees a migration of the field toward the equator, as observed in the sun. For positive dynamo number the migration is reversed. The dynamo waves at the top behave for positive dynamo number as do the waves of the bottom for negative dynamo number.

To check on the reliability of the calculation using the three lowest terms in (19.138), the calculation was repeated (Lerche and Parker 1972) using the first six terms, $n+m\leqslant4$. The resulting dispersion relation provides roots for the higher modes allowed by the higher order terms,

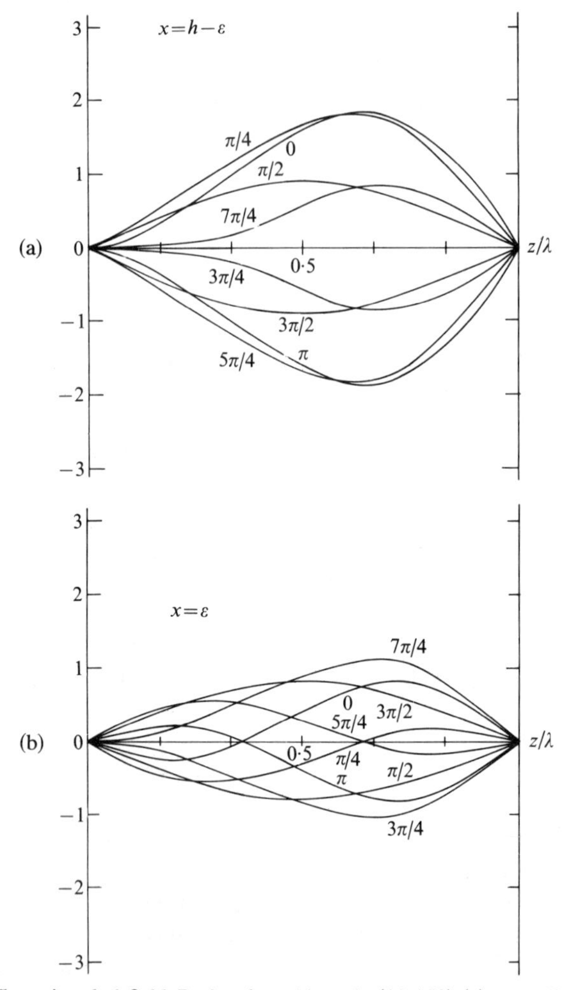

FIG. 19.6. The azimuthal field B_y for the odd mode (19.150) (a) a small distance ϵ below the top of the dynamo region and (b) a small distance ϵ above the bottom. The curves represent B_y at successive times $2\eta\pi^2 t/h^2 = 0, \frac{1}{4}\pi, \frac{1}{2}\pi$, etc.

but no essential change in the eigenvalue for the growth rate for the lowest mode (19.150). Eigenfunctions are always given less accurately then the eigenvalue, of course, so that if a precise description, rather than merely the general properties, of the field migration were required, the higher order terms should not be neglected. For the present purpose of qualitative exploration the terms $n+m \leqslant 3$ are sufficient.

The even modes are readily treated in the same fashion as the odd modes (Lerche and Parker 1972), with a steady amplitude for $\mu = \pm 10$ again. For $\mu = -10$ the migration of the fields at the upper surface of the dynamo, $x = h - \epsilon$ is toward the equator from both sides and away from the equator at the bottom $h = \epsilon$, as with the odd mode.

The purpose in exploring the dynamo with horizontal shear is to determine, among other things, the direction of migration of the fields at the observable upper surface. The example just presented is based on the dynamo distribution (19.142) with the simple result that the fields migrate toward the equator. The question now is how the direction of migration varies as the dynamo distribution ΓG_z is altered. The simple step function distribution

$$G_z = \begin{cases} +G & \text{for} \quad 0 < z < \lambda \\ -G & \text{for} \quad -\lambda < z < 0 \end{cases}, \qquad \Gamma = \begin{cases} +\Gamma_0 & \text{for} \quad 0 < z < \lambda \\ -\Gamma_0 & \text{for} \quad -\lambda < z < 0 \end{cases}, \tag{19.151}$$

is easily worked out. It yields steady amplitudes ($\mathrm{Re}(1/\tau) = 0$) for $\Gamma_0 G h^3/\eta = 442$ or $\mu = \pm 3{\cdot}49$ (with μ defined by (19.150)). In terms of the mean value of ΓG_z over $0 < z < \lambda$ this is $\langle \Gamma G_z \rangle = \Gamma_0 G = 442\eta^2/h^3$. In the previous example with the distribution (19.142) the mean is $\langle \Gamma_0 G_z \rangle = \Gamma_0 G/\pi = 395\eta^2/h^3$. Hence, in terms of the mean value $\langle \Gamma G_z \rangle$ there is little difference in the consequences of (19.142) and (19.151), because in both cases the dynamo is distributed broadly across each 'hemisphere'.

To see what happens when the shear is strongly localized, let the dynamo coefficient Γ be represented by the step function (19.151) but suppose that the shear is confined to thin sheets at $z = \pm b$, so that

$$G_z = G\lambda\{\delta(z - b) - \delta(z + b)\}.$$

Repeating the calculation of the Lagrangian (19.135), and the vanishing of its variation for the odd mode, the dispersion relation computed for the first three terms A_{11}, A_{12}, A_{21} is (Lerche and Parker 1972)

$$\psi(2,1)\psi(2,\tfrac{1}{2}) + \left(\frac{128\Gamma_0 G}{3\pi h}\right)^2 \sin s \cos(\tfrac{1}{2}s) \left\{\frac{\sin s}{3\psi(1,1)} + \frac{2\sin(2s)}{15\psi(1,2)}\right\}$$
$$\times \left\{\frac{\cos(\tfrac{1}{2}s)}{3\psi(1,\tfrac{1}{2})} - \frac{\cos(3s/2)}{5\psi(1,\tfrac{3}{2})}\right\} = 0$$

and

$$\frac{A_{12}}{A_{11}} = \frac{2\psi(1,1)}{\psi(1,2)} \cos s, \tag{19.152}$$

$$\frac{A_{11}}{A_{21}} = -\frac{128 G\Gamma_0 \sin s}{3\pi h\psi(1,1)} \left\{\frac{\cos(\tfrac{1}{2}s)}{3\psi(1,\tfrac{1}{2})} - \frac{\cos(3s/2)}{5\psi(1,\tfrac{3}{2})}\right\}, \tag{19.153}$$

where $s \equiv \pi b/\lambda$. For a thin layer, $h \ll \lambda$ the dispersion relation reduces to

$$(\sigma + 1)^2(\sigma + 4)^2 + \nu^2 = 0,$$

where now the dynamo number ν is defined as

$$\nu^2 = \tfrac{1}{3}\sin^2 s \cos(\tfrac{1}{2}s)\left(\frac{128K^3h^3}{3\pi^5}\right)^2(1+\tfrac{4}{5}\cos s)\{\tfrac{1}{3}\cos(\tfrac{1}{2}s)-\tfrac{1}{5}\cos(\tfrac{3}{2}s)\}. \tag{19.154}$$

With $0<s<\pi$, this definition for ν^2 is non-negative. The dispersion relation has the same structure as in the previous cases, with the growth rate $\sigma = h^2/\pi^2\eta\tau$ related to ν in exactly the same way it was related to μ in (19.149). Steady oscillations ($\mathrm{Re}(\sigma)=0$) occur for $\nu = \pm 10$, for which $\sigma = \pm 2\mathrm{i}$. In that case

$$A_{12} = 2A_{11}\cos s, \tag{19.155}$$

$$A_{21} = -A_{11}\exp(\pm \mathrm{i}\vartheta)\frac{\cos^{\frac{1}{2}}(\frac{1}{2}s)\{1+\frac{4}{5}\cos s\}^{\frac{1}{2}}}{2\sqrt{3}\,\{\frac{1}{3}\cos(\frac{1}{2}s)-\frac{1}{5}\cos(3s/2)\}^{\frac{1}{2}}}. \tag{19.156}$$

The important point is that in the present example A_{12} changes sign across $s = \frac{1}{2}\pi$, i.e. across the middle $b = \frac{1}{2}\lambda$ of each 'hemisphere', so that the direction of migration in the horizontal z-direction depends upon whether b is larger or smaller than $\frac{1}{2}\lambda$. The field is plotted in Fig. 19.7 for $b = \frac{1}{4}\lambda$ ($\nu = -10$, $K^3h^3 = 300$, $A_{12} = +2^{\frac{1}{2}}A_{11}$, $A_{21} = 0{\cdot}21A_{11}\exp(\pm\mathrm{i}\vartheta)$) and for $b = \frac{3}{4}\lambda$ ($\nu = -10$, $K^3h^3 = 770$, $A_{12} = -2^{\frac{1}{2}}A_{11}$, $A_{21} = 0{\cdot}21A_{11}\exp(\pm\mathrm{i}\vartheta)$). The principal migration is away from the source at $z = b$. Hence for $b = \frac{3}{4}\lambda$, placing the source near the 'poles', the migration is toward the equator. For $b = \frac{1}{4}\lambda$, placing the source near the equator, the migration is toward the poles. Examination of B_y at the bottom of the dynamo region ($x = \epsilon$) shows the expected counter migration. Thus, with $\Gamma_0 G$ the waves migrate upward, press against the upper boundary, and are deflected there toward whatever 'free' space is available.

To show the effects for concentrated cyclonic turbulence suppose that G_z is distributed uniformly over each hemisphere, as described by the step function (19.151), but now the dynamo coefficient Γ is confined to thin sheets at $z = \pm a$ so that

$$\Gamma = \Gamma_0\{\delta(z-a)-\delta(z+a)\}. \tag{19.157}$$

With $r = \pi a/\lambda$ the dispersion relation reduces to

$$(\sigma+1)^2(\sigma+4)^2+\rho^2 = 0 \tag{19.158}$$

for a thin layer ($h \ll \lambda$), where ρ has the same sign as $-\Gamma_0 G$ and

$$\rho^2 = \frac{1}{27}\left(\frac{128K^3h^3}{\pi^5}\right)^2\sin^2 r\cos(\tfrac{1}{2}r)$$
$$\times[\tfrac{1}{3}\cos(\tfrac{1}{2}r)-\tfrac{1}{5}\cos(\tfrac{3}{2}r)+4\cos r\,\{\tfrac{1}{15}\cos(\tfrac{1}{2}r)+\tfrac{1}{7}\cos(\tfrac{3}{2}r)\}]. \tag{19.159}$$

For steady operation of the dynamo, $\rho = \pm 10$ and $\sigma = \pm 2\mathrm{i}$, and the

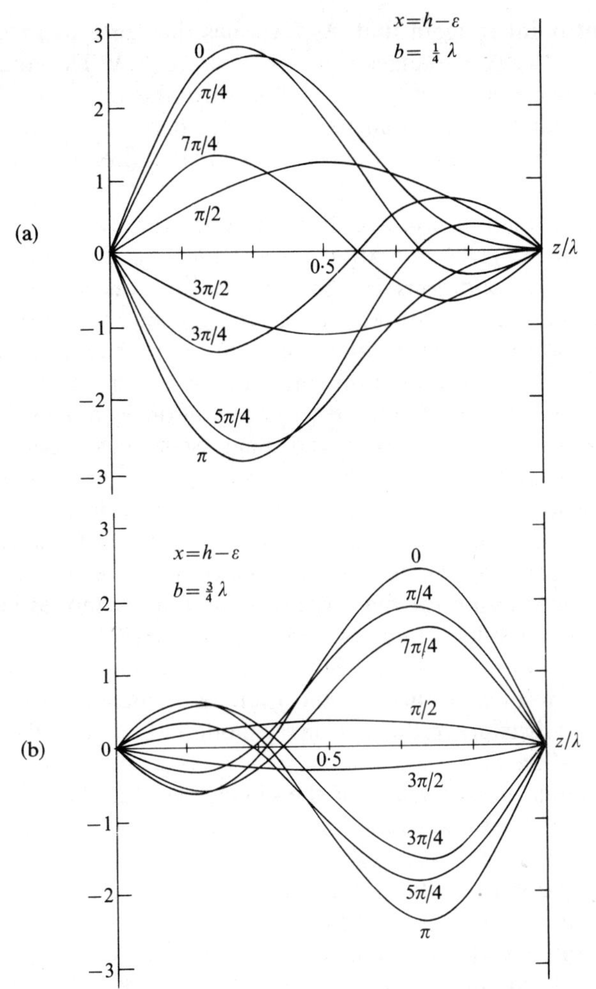

FIG. 19.7. *The azimuthal* field B_y for the odd mode (19.155), (19.156) a small distance ϵ below the top of the dynamo region $0<x<h$ (a) when the shear (19.151) is concentrated at $z=\frac{1}{4}\lambda$ and (b) when $z=\frac{3}{4}\lambda$. The curves represent B_y at successive times $2\eta\pi^2 t/h^2 = 0, \frac{1}{4}\pi, \frac{1}{2}\pi$, etc.

coefficients are

$$\frac{A_{12}}{A_{11}} = 2\frac{\frac{1}{15}\cos(\frac{1}{2}r)+\frac{1}{7}\cos(\frac{3}{2}r)}{\frac{1}{2}\cos(\frac{1}{2}r)-\frac{1}{5}\cos(\frac{3}{2}r)}, \tag{19.160}$$

$$\frac{A_{21}}{A_{11}} = -\frac{\cos^{\frac{1}{2}}(\frac{1}{2}r)}{2\sqrt{3}}\exp(\pm i\vartheta)$$
$$\times\frac{[\frac{1}{3}\cos(\frac{1}{2}r)-\frac{1}{5}\cos(\frac{3}{2}r)+4\cos(r)\{\frac{1}{15}\cos(\frac{1}{2}r)+\frac{1}{7}\cos(\frac{3}{2}v)\}]^{\frac{1}{2}}}{\frac{1}{2}\cos(\frac{1}{2}r)-\frac{1}{5}\cos(\frac{3}{2}r)} \tag{19.161}$$

The important point is again that A_{21}/A_{11} has the same sign for all r in $(0, \pi)$, whereas A_{12}/A_{11} changes sign across $a = 0{\cdot}41\lambda$. The migration is toward the poles when $a < 0{\cdot}41\lambda$ and toward the equator when $a > 0{\cdot}41\lambda$. It is the familiar rule that the waves are deflected at the blocking surface toward the largest free volume available (Lerche and Parker 1972).

To summarize, then, a uniform distribution of the dynamo effect ΓG_z for horizontal shear gives migration toward the equator. This seems to correspond to the conditions in the sun. Strong concentration of the dynamo effect at a given latitude causes the dynamo waves to migrate away from the source, toward either the pole or the equator, whichever provides the most free volume into which the wave can escape.

Lerche and Parker were unable to find any stationary modes, for which $1/\tau = 0$, for the dynamo with horizontal shear. Stationary modes exist for vertical shear (see (19.107)) in dynamo regions that are not too thin, $h > \frac{1}{2}\pi\lambda$, but evidently there are no stationary modes in a layer, either thick or thin, of horizontal shear. Thus, the introduction of small amounts of horizontal shear may destroy the stationary mode of a dynamo with vertical shear, accounting for the absence of stationary modes in various spherical dynamo models. For instance, Roberts (1972) could find no stationary modes without introducing meridional circulation. Both horizontal shear and concentration of the cyclonic turbulence toward high latitudes tend to discourage the stationary mode in favour of the periodic modes (Levy 1972a, c).

Finally, it should be noted that if the angular velocity of the sun or of the liquid core of Earth is a function principally of the distance ϖ from the axis of rotation (cf. Durney 1975), then the shear is vertical at the equator and horizontal in the polar regions. This, plus the fact that the geometry of the north–south meridians narrows toward the poles, makes it clear that the rectangular dynamo can illustrate the basic physical principles of the migration of dynamo waves (Parker 1955) but the detailed behaviour of any real dynamo can be treated only when spherical geometry with the precise non-uniform rotation and distribution of cyclonic convection are supplied. The analysis of the solar dynamo, for instance, based on the observed non-uniform rotation of the surface and the observed variation of the latitude of sunspots (as an indication of the underlying azimuthal field) has been studied at length in spherical geometry by Yoshimura (1975a, b, c) and Durney (1975) among others, with interesting results. The problem is taken up in §19.4 when we have finished the treatment of rectangular dynamos.

19.3.3. The galactic dynamo

As a final illustrative example of rectangular dynamos, consider a region with the characteristics of the thin gaseous disc of the galaxy. It was pointed out in §17.1 that the galactic field is a product of contemporary processes, presumably some sort of dynamo (Parker 1971a). The gaseous disc of the galaxy rotates non-uniformly, the angular velocity Ω depending strongly on the distance ϖ from the axis of rotation. The interstellar gas that makes up the disc is turbulent, with r.m.s. velocities of the order of $10\ \mathrm{km\,s^{-1}}$ over scales of 10^2 pc. In the neighbourhood of the Sun the characteristic thickness (scale height) of the disc is estimated to be about 200 pc (half width at half maximum) with broad wings that extend to a kpc (Badwar and Stephens 1977). The thickness of the disc is maintained by a dynamic balance between the inward force of gravity (toward the central plane of the disc) and the outward forces of the gas pressure, magnetic field pressure, and cosmic ray pressure (Parker 1966, 1969b) (see §§13.4 and 13.8). The turbulent velocities of $10\ \mathrm{km\,s^{-1}}$ indicate that individual eddies (interstellar clouds) in the disc move up and down through distances comparable to the scale height. Hence the rising gas is subject to expansion. Coriolis forces cause it to rotate less rapidly than the surrounding gas. Falling gas rotates more rapidly because it is contracting. The interstellar turbulence, then, is cyclonic and, together with the non-uniform rotation, the motions of the gaseous disc constitute a dynamo.

The geometry of the galactic dynamo is different from the geometry of either a planetary core or a stellar convective zone. The disc is broad and thin, with the large-scale velocity V_ϕ varying along the radius, while the dynamo coefficient has opposite signs on either side of the central plane of the disc. The disc is more than fifty times broader than it is thick. Hence the distant edges can have but little effect, and for a first look at the problem it is sufficient to treat an infinite slab. Following Parker (1971a), suppose that the y-axis of the local Cartesian coordinate system is oriented in the azimuthal direction, so that $\partial/\partial y = 0$ for the large-scale mean fields. The x-axis is pointed in the outward radial direction and the z-axis is perpendicular to the slab so that xyz forms a right-handed coordinate system. The xy-plane coincides with the midplane of the slab $z = 0$. The large-scale shear is then $G \equiv G_x = dV_y/dx$ with $G_z = 0$. We suppose that dV_y/dx extends uniformly across the thickness, $-h < z < +h$, of the slab. Denote the dynamo coefficient by $+\Gamma$ in $0 < z < h$ and by $-\Gamma$ in $-h < z < 0$. To keep the calculations as simple as possible Γ is assumed to be uniform within the slab, $z^2 < h^2$, and zero outside. The surfaces $z = \pm h$ are free surfaces, in the sense that the gas beyond is too tenuous to offer mechanical resistance to the fields escaping from the surface of the disc. Hence, beyond the surface the field is described in terms of the gradient of a solution of Laplace's equation. The dynamo equations in $0 < z < h$ follow from (19.22) and (19.23) as

$$(\partial/\partial t - \eta\nabla^2)B_x = -\partial(\Gamma B_y)/\partial z, \tag{19.162}$$

$$(\partial/\partial t - \eta\nabla^2)B_y = GB_x, \tag{19.163}$$

$$(\partial/\partial t - \eta\nabla^2)B_z = +\partial(\Gamma B_y)/\partial x. \tag{19.164}$$

The equations in $-h<z<0$ are the same with $+\Gamma$ replaced by $-\Gamma$.

The solutions of the dynamo equations can be written in the form (19.90)–(19.92) in $0<z<h$ with $\cos(k_x x)$ replaced by $\exp(-ik_x x)$ and $\sin(k_x x)$ replaced by $\mathrm{i}\exp(-ik_x x)$. Write

$$B_x = +E\exp\{t/\tau - \mathrm{i}k_x x - k_x(z-h)\}, \tag{19.165}$$

$$B_y = 0 \tag{19.166}$$

$$B_z = -\mathrm{i}E\exp\{t/\tau - \mathrm{i}k_x x - k_x(z-h)\} \tag{19.167}$$

in the infinite free space $z>+h$ outside. In $-h<z<0$, where the dynamo coefficient is $-\Gamma$, the appropriate solutions are (19.93)–(19.95), with $\cos(k_x x)$ and $\sin(k_x x)$ replaced by $\exp(-ik_x x)$ and $\mathrm{i}\exp(-ik_x x)$, respectively. Write

$$B_x = +F\exp\{t/\tau - \mathrm{i}k_x x + k_x(z+h)\}, \tag{19.168}$$

$$B_y = 0, \tag{19.169}$$

$$B_z = +\mathrm{i}F\exp\{t/\tau - \mathrm{i}k_x x + k_x(z+h)\}, \tag{19.170}$$

in the infinite space $z<-h$ outside. The dispersion relation is again (19.55), with the four roots (19.56)–(19.63) and $k_z = qK$.

The boundary conditions at $z=\pm h$ are the continuity of B_x, B_y, and B_z, because of the resistivity η. Hence $B_y=0$ at $z=\pm h$. At $z=0$ the fields are also continuous. It is readily shown by integrating (19.162) from $z=-\epsilon$ to $z=+\epsilon$, and letting $\epsilon\to 0$, that $\partial B_x/\partial z$ jumps by $2\Gamma B_y$ (evaluated at $z=0$), but the requirement is readily shown, with the aid of (19.55), to be equivalent to the continuity of B_z. Integrating (19.163) across $z=0$ leads to the requirement that $\partial B_y/\partial z$ is continuous. Continuity of $\partial B_z/\partial z$ follows from (19.164), but in view of the divergence condition $\nabla\cdot\mathbf{B}=0$, it is equivalent to the continuity of B_x and B_y. Altogether, then, the continuity of B_x, B_y, B_z, and $\partial B_y/\partial z$ at $z=0$ leads to the four basic relations

$$\sum C_n(A+q_n^2) = \sum D_n(A+q_n^2), \tag{19.171}$$

$$\sum C_n = \sum D_n \tag{19.172}$$

$$\sum C_n/(A+q_n^2) = \sum D_n/(A+q_n^2) \tag{19.173}$$

$$\sum C_n q_n = \sum D_n q_n, \tag{19.174}$$

respectively. At $z=+h$ the continuity of B_x and B_z gives

$$(K^2\eta/G)\sum C_n(A+q_n^2)\exp(-\mathrm{i}q_n Kh) = E, \tag{19.175}$$

$$(k_x K\eta/G)\sum\{C_n/(A+q_n^2)\}\exp(-\mathrm{i}q_n Kh) = E, \tag{19.176}$$

respectively. The vanishing of B_y requires

$$\sum C_n \exp(-iq_nKh) = 0. \tag{19.177}$$

At $z = -h$ the continuity of B_x and B_y gives

$$(K^2\eta/G) \sum D_n(A + q_n^2)\exp(-iq_nKh) = F, \tag{19.178}$$

$$(k_xK\eta/G) \sum [D_n/(A + q_n^2)]\exp(-iq_nKh) = F, \tag{19.179}$$

respectively. The vanishing of B_y requires

$$\sum D_n \exp(-iq_nKh) = 0. \tag{19.180}$$

There are, then, ten homogeneous equations relating the ten coefficients. It is sufficient for the present purpose to treat the even modes, for which B_x and B_y are even functions of the distance z from the midplane. In that case $C_n = D_n$ and $E = F$. The equations reduce to

$$\sum C_n/(A + q_n^2) = 0, \qquad \sum C_n q_n = 0, \tag{19.181}$$

$$\sum C_n \exp(-iq_nKh) = 0, \tag{19.182}$$

with (19.175), (19.176), (19.178), and (19.179) yielding

$$\sum C_n[(iq - k_x/K)/(A + q_n^2)]\exp(-iq_nKh) = 0 \tag{19.183}$$

upon elimination of E. With the aid of such identities as $\sum q_n = 0$ and

$$\frac{q_4}{A + q_1^2} - \frac{q_1}{A + q_4^2} = \frac{iA(q_1^2 - q_4^2)}{q_1 q_4},$$

the determinant of the coefficients can be reduced to

$$\begin{aligned}
0 = {} & (q_1^2 - q_2^2)(q_4^2 - q_3^2)[\{(A + q_1^2)(A + q_2^2) - (K/k_x)q_1q_2\}\exp\{+i(q_1 + q_2)Kh\} \\
& + \{(A + q_3^2)(A + q_4^2) - (K/k_x)q_3q_4\}\exp\{-i(q_1 + q_2)Kh\}] \\
& + (q_3^2 - q_1^2)(q_4^2 - q_2^2)[\{(A + q_1^2)(A + q_3^2) - (K/k_x)q_1q_3\}\exp\{+i(q_1 + q_3)Kh\} \\
& + \{(A + q_2^2)(A + q_4^2) - (K/k_x)q_2q_4]\exp\{-i(q_1 + q_3)Kh\}] \\
& + (q_2^2 - q_3^2)(q_4^2 - q_1^2)[\{(A + q_2^2)(A + q_3^2) - (K/k_x)q_2q_3\}\exp\{+i(q_2 + q_3)Kh\} \\
& + \{(A + q_1^2)(A + q_4^2) - (K/k_x)q_1q_4\}\exp\{-i(q_2 + q_3)Kh\}].
\end{aligned} \tag{19.184}$$

This dispersion relation gives the growth rate A in terms of the dynamo number Kh.

19.3.3.1. Stationary states. As we shall see, there are no regenerative states for k_x as large in magnitude as K, so suppose that $k_x \ll |K|$. Then the principal losses are through the surfaces $z = \pm h$. The quantity $(k_x/K)^2$ can be neglected compared to one so that $1/\tau \cong \eta K^2 A$. The steady state solutions arise for small or vanishing A. Hence use (19.56)–(19.59) to represent the roots q_n. Neglect terms second order in A. Then

(19.184) can be reduced to

$$M(K, k_x) - AN(K, k_x) = O\{A^2K^2h^2 \exp(\pm Kh)\}, \qquad (19.185)$$

where

$$\begin{aligned} M(K, k_x) \equiv\; & 2\exp(-\tfrac{1}{2}Kh)\{\cos(\tfrac{1}{2}\surd 3\, Kh + \tfrac{1}{3}\pi) - (K/k_x)\cos(\tfrac{1}{2}\surd 3\, Kh)\} \\ & -(1 + K/k_x)\exp(+Kh), \\ N(K, k_x) \equiv\; & 2\exp(-\tfrac{1}{2}Kh)\{\cos\tfrac{1}{2}\surd 3\, Kh - (2K/3k_x)(1 - k_x h)\cos(\tfrac{1}{2}\surd 3\, Kh - \tfrac{1}{3}\pi) \\ & -2Kh(K/k_x)\cos(\tfrac{1}{2}\surd 3\, Kh + \tfrac{1}{3}\pi)\} \\ & +2\{\sinh Kh + (K/3k_x)(1 - k_x h + Kh)\exp(Kh) \\ & \qquad - \exp(\tfrac{1}{2}Kh)\cos(\tfrac{1}{2}\surd 3\, Kh)\} \end{aligned}$$

The stationary states ($A = 0$) occur at the zeros of $M(K, k_x)$, where

$$(1 + K/k_x)\exp(\tfrac{3}{2}Kh) = (1 - 2K/k_x)\cos(\tfrac{1}{2}\surd 3\, Kh) - 3^{\frac{1}{2}}\sin(\tfrac{1}{2}\surd 3\, Kh).$$

For the present situation of small k_x this reduces to

$$\exp(\tfrac{3}{2}Kh) \cong -2\cos(\tfrac{1}{2}\surd 3\, Kh),$$

which has no roots for $Kh > 0$. For negative Kh there is a sequence of roots at $Kh \cong +(2n-1)\pi/\surd 3$ where $n = 1, 2, 3, \ldots$. To higher order in k_x and $\exp(Kh)$ the sequence is

$$\begin{aligned} Kh = & -(2n-1)\pi/\surd 3 - k_x h/(n - \tfrac{1}{2})\pi \\ & + \{(-1)^n/\surd 3\}\{1 - \tfrac{3}{2}k_x h/(n - \tfrac{1}{2})\pi\}\exp\{-\surd 3\,(n - \tfrac{1}{2})\pi\} \qquad (19.186) \end{aligned}$$

for the dynamo number giving $A = 0$. For $k_x h = 0$ this gives $Kh = -1{\cdot}857$, $-5{\cdot}43$, $-9{\cdot}06$, $-12{\cdot}70$. For large negative dynamo numbers, (19.185) reduces to

$$A \cong 2(K/k_x)\exp(\tfrac{1}{2}Kh)\cos(\tfrac{1}{2}\surd 3\, Kh)[1 + O(k_x/K) + O\{(K^2h/k_x)\exp(\tfrac{3}{2}Kh)\}].$$

This expression is valid only in the neighbourhood of the stationary states $A = 0$ at the dynamo numbers (19.186), of course, in order that the approximate form (19.185) be valid. It shows, however, that A is positive for $(4n-1)\pi < -\surd 3\, Kh < (4n+1)\pi$, with a maximum in the near vicinity of $\surd 3\, Kh = 4n\pi$. It follows from this expression for A that the growth rate is

$$1/\tau = \eta k_x^2\{2(K/k_x)^3\exp(\tfrac{1}{2}Kh)\cos(\tfrac{1}{2}\surd 3\, Kh) - 1\}. \qquad (19.187)$$

In order that the growth rate be non-negative it is necessary that $2(K/k_x)^3\exp(\tfrac{1}{2}Kh)$ be larger than one, which is satisfied for small k_x.

Note that if we first take the limit of $k_x \to 0$, then (19.185) reduces to

$$A = \frac{\cos(\tfrac{1}{2}\surd 3\, Kh) + \tfrac{1}{2}\exp(\tfrac{3}{2}Kh)}{2Kh\cos(\tfrac{1}{2}\surd 3\, Kh + \tfrac{1}{3}\pi) + \tfrac{2}{3}\cos(\tfrac{1}{2}\surd 3\, Kh - \tfrac{1}{3}\pi) - (1 + Kh)\exp(\tfrac{3}{2}Kh)},$$

so that, for large negative Kh, the growth rate is

$$A = \frac{\cos(\frac{1}{2}\sqrt{3}\,Kh)}{2Kh\cos(\frac{1}{2}\sqrt{3}\,Kh + \frac{1}{3}\pi)}, \qquad \frac{1}{\tau} = \frac{Kh}{2}\left(\frac{\eta}{h^2}\right)\frac{\cos(\frac{1}{2}\sqrt{3}\,Kh)}{\cos(\frac{1}{2}\sqrt{3}\,Kh + \frac{1}{3}\pi)},$$

valid in the neighbourhood of the zeros of $\cos(\frac{1}{2}\sqrt{3}\,Kh)$. Again the stationary states are at $Kh = -(2n-1)\pi/\sqrt{3}$. Altogether, then, the dynamo has regenerative modes for small k_x. In order of magnitude (19.187) requires

$$k_x h < |Kh| \exp(-\tfrac{1}{6}|Kh|)$$

for large negative Kh to regenerate the field. There are no regenerative solutions if k_x is as large as K. It follows, then, that the waves of field are very broad compared to the thickness $2h$ of the dynamo slab. For the lowest mode, $Kh \cong -1{\cdot}8$, we require $k_x h \lesssim 2{\cdot}5$.

The field distribution is readily computed from the coefficients C_n. It follows from (19.181)–(19.183) that, to lowest order,

$$C_1[\exp\{\sqrt{3}\,Kh\exp(\mathrm{i}\tfrac{1}{6}\pi)\} + \exp(\mathrm{i}\tfrac{1}{3}\pi)] \\ + C_2[\exp\{\sqrt{3}\,Kh\exp(-\mathrm{i}\tfrac{1}{6}\pi)\} + \exp(-\mathrm{i}\tfrac{1}{3}\pi)] = 0.$$

With $Kh \cong -(2n-1)\pi/\sqrt{3} \ll -1$ this reduces to

$$C_2 \cong C_1\exp(-\mathrm{i}\tfrac{1}{3}\pi)\{1 + \mathrm{i}(-1)^n\exp(\tfrac{3}{2}Kh)\}.$$

Then C_3 follows from $\sum C_n q_n = 0$ as

$$q_3 C_3 \cong -q_1 C_1 - q_2 C_2$$
$$C_3 \cong (-1)^n C_1 \exp(+\tfrac{3}{2}Kh - \mathrm{i}\tfrac{1}{6}\pi).$$

It follows from (19.182) that

$$C_4 \cong 3A(-1)^n C_1\exp(\tfrac{1}{2}Kh - \mathrm{i}\tfrac{1}{6}\pi).$$

The coefficient E follows from (19.176). Substituting these coefficients into (19.90)–(19.92), with $\cos(k_x x)$ and $\sin(k_x x)$ replaced by $\exp(-\mathrm{i}k_x x)$ and $\mathrm{i}\exp(-\mathrm{i}k_x x)$, respectively, it follows that

$$B_x \cong C_1(\eta K^2/G)[2\cos(\tfrac{1}{2}\sqrt{3}\,Kz - \tfrac{1}{6}\pi) + (-1)^{n-1}\exp\{\tfrac{3}{2}K(h-z)\}] \\ \times\exp(t/\tau - \mathrm{i}k_x x + \tfrac{1}{2}Kz - \mathrm{i}\tfrac{1}{6}\pi), \quad (19.188)$$

$$B_y \cong C_1\{2\cos(\tfrac{1}{2}\sqrt{3}\,Kz + \tfrac{1}{6}\pi) + (-1)^n\exp\tfrac{3}{2}K(z-h)\} \\ \times\exp(t/\tau - \mathrm{i}k_x x + \tfrac{1}{2}Kz - \mathrm{i}\tfrac{1}{6}\pi), \quad (19.189)$$

$$B_z = +C_1(\eta K k_x/G)\{2\sin(\tfrac{1}{2}\sqrt{3}\,Kz) + (-1)^{n-1}[3 - \exp\{\tfrac{3}{2}K(h-z)\}]\} \\ \times\exp(t/\tau - \mathrm{i}k_x x + \tfrac{1}{2}Kz + \mathrm{i}\tfrac{1}{3}\pi). \quad (19.190)$$

The interplay of the four basic wave modes is evident from these

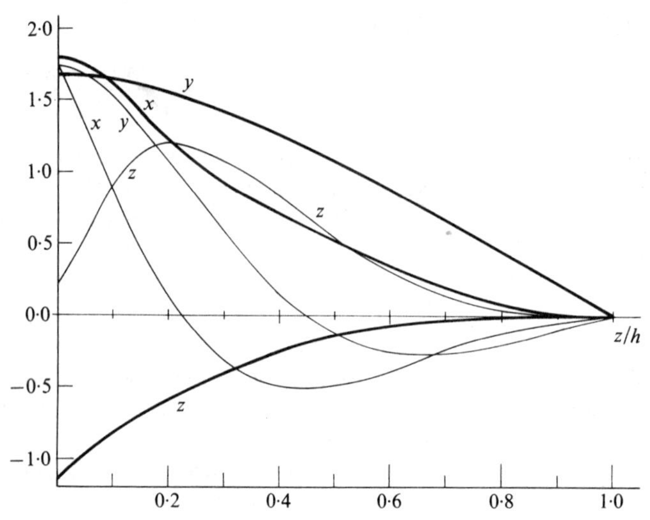

FIG. 19.8. A plot of the fields B_x, B_y, and B_z. The heavy lines are for the lowest even mode (19.188)–(19.190) ($n=1$, $Kh=-1{\cdot}85$), while the light lines represent the next higher mode, $n=2$. Curves labelled x represent B_x in units of $(\eta K^2/G)C_1 \exp(t/\tau+\frac{1}{2}Kz)\cos(k_x x-\frac{1}{6}\pi)$. The y-components are given in units of $C_1 \exp(t/\tau+\frac{1}{2}Kz)\cos(k_x x-\frac{1}{6}\pi)$ and the z-components in units of $(\eta Kk_x/G)C_1 \exp(t/\tau+\frac{1}{2}Kz)\sin(k_x x-\frac{1}{6}\pi)$.

expressions. The fields are plotted across $0<z<h$ in Fig. 19.8 for the lowest mode, $n=1(Kh=-1{\cdot}85)$ and for the next higher mode $n=2\,(Kh=-5{\cdot}43)$. Note that B_x and B_y are even functions of z, while B_z is an odd function.

It is worthwhile to consider the basic physics of the stationary dynamo in galactic geometry. The principal field is B_y of course, the x- and z-components being smaller by the factor $\eta K^2/G$ when the shear G is large. In the lowest mode the 'azimuthal' field B_y (which we take to be positive) has a simple maximum at the midplane $z=0$, produced by the shearing of B_x, and falls to zero at the open surfaces $z=\pm h$. The dynamo is regenerative when ΓG is negative in $0<z<h$. To see how the whole scheme fits into the structure of the galaxy, suppose that the negative x-axis points in the direction of the galactic centre, as sketched in Fig. 19.9, and the positive y-axis points in the direction of rotation. Hence, the galaxy rotates counterclockwise when viewed from large positive z. The rotational velocity of the galaxy declines outward, with increasing x, so that $G\equiv \mathrm{d}V_y/\mathrm{d}x$ is negative. A rising expanding eddy in $z>0$ rotates less rapidly than the ambient interstellar gas around it, so that it rotates clockwise relative to the ambient gas, as viewed from large positive z. Hence Γ is positive (based on the definition (19.2) and (18.51)). The product ΓG is negative in $z>0$, corresponding to the case $Kh<0$ treated here. Fig. 19.9 shows two twisted loops in the lines of force of the azimuthal field B_y, caused by rising cyclonic eddies on either side of the midplane $z=0$. The loops represent opposite circulation of magnetic flux in the xz-plane. The production of many such loops and the coalescence of the loops as a consequence of turbulent diffusion lead to a general circulation of field indicated by the broken lines. The circulation is such that the field near the midplane is in the negative x-direction. The field near the surfaces $z=\pm h$ is soon lost by diffusion into the empty space immediately outside, leaving only a net flux in the negative x-direction, which spreads out across the thickness of the disc from its place of origin in the vicinity of the midplane. The negative shear $\mathrm{d}V_y/\mathrm{d}x$ stretches the lines of force into the y-direction and regenerates B_y. From the hydromagnetic equation, the shear generates B_y at the rate

$$\partial B_y/\partial t = B_x\, \mathrm{d}V_y/\mathrm{d}x.$$

Since both $\mathrm{d}V_y/\mathrm{d}x$ and B_x are negative, it follows that $\partial B_y/\partial t$ is positive. It is in this way, then, that the galactic dynamo operates (Parker 1971a). The effect of the non-uniform rotation and the cyclonic turbulence is to generate an azimuthal field that is an even function of the distance z from the midplane of the galaxy (the odd modes, treated below, require $Kh>0$).

The free escape of lines of force from the surface is an essential part of the functioning of the galactic dynamo, because without it, the net flux in the x-direction, and hence the net production of B_y in the slab, would be zero. The only regenerative solutions are then migratory waves, leading

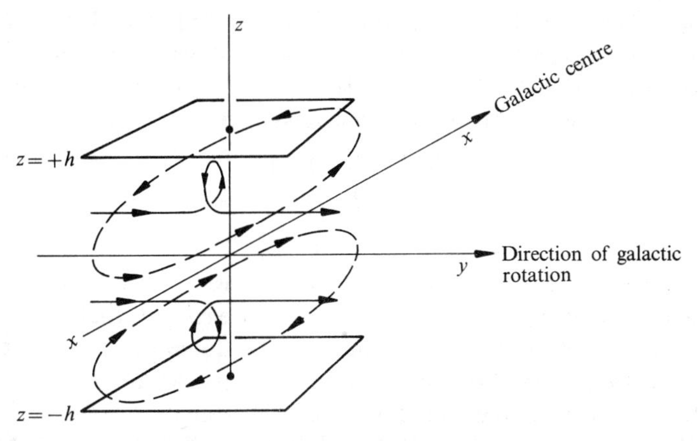

FIG. 19.9. A sketch of the internal fields in the generation of the lowest even mode $B_y(+z)=B_y(-z)$ in the galactic dynamo when $\mathrm{d}V_y/\mathrm{d}x<0$ across the slab $z^2<h^2$ and the dynamo coefficient is positive in $0<z<h$ and negative in $-h<z<0$, so that $Kh<0$. The 'azimuthal' field is shown by the continuous lines, in which there are shown the twisted loops produced by rising cells of gas. The large-scale circulation from the coalescence of many loops is indicated by the broken curves.

to periodic solutions. The escape of lines of force from the surface of a gaseous body is described in §13.7. We have shown by specific example (Parker 1971a) that the generation of stationary modes does not occur in the slab if the boundaries are closed by sheets of rigid, infinitely conducting material.

If the medium outside the slab has the same shear $G = \mathrm{d}V_y/\mathrm{d}x$ and diffusion coefficient η as the slab, but lacks only the cyclonic turbulence Γ, the dynamo is regenerative (Parker 1971a). The loss of field out of the slab is reduced by the external medium, so that the generation rate K^2h^2A is significantly reduced, as we would expect. But the rate is not zero. The free surface is a better approximation to the environment of the gaseous disc of the galaxy, in view of the magnetic buoyancy of the fields and the tenuous gas outside (Chapter 13), so the solutions worked out above appear to be appropriate. Interpretation of the observations of the galactic magnetic field (Jokipii and Lerche 1969; Jokipii *et al.* 1969; Manchester 1973) shows large local fluctuations in the field, as we would expect, together with the fact that the mean azimuthal field has the same sign across the entire thickness of the disc. Hence the galactic dynamo is evidently operating in the lowest even mode, for which $Kh \cong -1{\cdot}8$.

It is worthwhile noting the values of η, Γ, and G for the galactic disc. For Keplerian motion about a massive galactic nucleus, the angular velocity Ω varies as $\varpi^{-\frac{3}{2}}$ and the circular velocity is $V_y = V_\phi = \varpi\Omega \propto \varpi^{-\frac{1}{2}}$. The local shearing rate is then, $G = \varpi\,\mathrm{d}\Omega/\mathrm{d}\varpi = -\frac{3}{2}\Omega$. The characteristic cyclonic rotation of the convecting fluid, relative to the surrounding fluid, is of the order of $\Phi \cong \Omega\tau \cong \Omega L/v$ in an eddy of scale L, velocity v, and life τ. It was shown in the development of (18.58) that $\Gamma \cong \frac{1}{8}\pi\epsilon L\Phi/\tau \cong \frac{1}{8}\pi\epsilon L\Omega \cong 0{\cdot}04L\Omega$ since $\epsilon \cong 0{\cdot}1$. The turbulent diffusion coefficient may be estimated from (17.89) as $\eta_{\mathrm{T}} \cong 0{\cdot}2v^2\tau \cong 0{\cdot}2vL$. Altogether, then, the dynamo number in a slab of half thickness h is

$$\begin{aligned}(Kh)^3 &\equiv \Gamma Gh^3/\eta_{\mathrm{T}}^2\\ &\cong \tfrac{3}{2}h^3\Omega^2/v^2L\\ &\cong \tfrac{3}{2}(h/L)^3(\Omega\tau)^2.\end{aligned} \tag{19.191}$$

The period of rotation of the gaseous disc of the galaxy in the neighbourhood of the sun is $T = 2{\cdot}5\times10^8$ years, so that the angular velocity is $\Omega = 2\pi/T = 0{\cdot}8\times10^{-15}$ rad s^{-1}. The quantities h, v, and L are not well defined by observation, but convention dictates $L = 100$ pc and $v = 10$ km s^{-1}. Hence the characteristic eddy life is $\tau = 3\times10^{14}$ s $= 10^7$ years. On this basis $\Phi = \Omega\tau = 0{\cdot}24$ rad and is sufficiently small that the quadratic effects characterized by the secondary dynamo coefficient Q and the negative contributions to turbulent diffusion, are negligible. The galactic dynamo is dominated by the principal dynamo coefficient Γ and by conventional turbulent diffusion η_{T}.

The thickness h of the galactic disc is a particularly nebulous quantity. The recent work of Badwar and Stephens (1977) suggests a thickness of 600 pc at the half-density points of the gas and 1400 pc for the field. The thickness $2h$ employed in the present calculations represents the *full* thickness for the gas and field, which appears to be of the order of at least 1 kpc. On this basis then, write $h = 400$ pc as an estimate. The dynamo number is easily scaled up or down to fit other estimates of h. The numerical result is $(Kh)^3 = -6$ or $Kh = -1{\cdot}8$. This rough estimate is not less than the $Kh = -1{\cdot}8$ for the lowest even mode of the galactic dynamo, suggesting that the galactic field is indeed the offspring of the dynamo (Parker 1971a; Vainshtein 1972b; Stix 1975; White 1978). After Earth and the sun, then, we find another dynamo operating in its lowest mode. Presumably the field grows until its strength suppresses the cyclonic turbulence and reduces the dynamo number to the lowest value for a steady state. The comments at the end of §19.3.1.1 on the stability of the operation of the dynamo are appropriate in this context. Stable operation obtains when the dynamo is in its lowest steady state, with the generation rate K^2h^2A increasing with increasing dynamo number $|Kh|$.

The odd modes can be worked out by the same general methods employed for the even modes (Parker 1971a). They arise at somewhat higher dynamo numbers than the even modes, for the simple reason that there is a reversal of sign of B_y across the midplane $z = 0$, producing more loss by diffusion. In place of (19.186) the stationary modes for small k_x are given by

$$Kh \cong (2n - \tfrac{1}{3})\pi/\sqrt{3}$$

$$= +3{\cdot}02, +6{\cdot}65, +10{\cdot}28, +13{\cdot}81$$

in place of $Kh = -1{\cdot}86, -5{\cdot}43, -9{\cdot}06$, etc. for the even modes. Note that the stationary states of the odd modes are produced by positive dynamo numbers, while the stationary even modes are excited by negative dynamo numbers. We suggest (from the the simple dynamical picture where rising convective cells expand and rotate more slowly than their surroundings) that the dynamo number in the gaseous disc of the galaxy is negative and, hence, can excite only the even modes. Observations support the even mode idea. The odd modes are of interest in any circumstance where the dynamics produces a positive dynamo number. At the moment we cannot say where that might occur.

The physical nature of the odd modes is describable on the same elementary basis as the even modes. Fig. 19.10 for the odd mode is essentially a duplicate of Fig. 19.9 for the even mode, except that the sense of the 'azimuthal' field B_y is reversed in the lower half of the slab to give an odd mode. As a result, the loops produced by cyclonic convective cells on each side of the midplane have the *same* sense of rotation in the

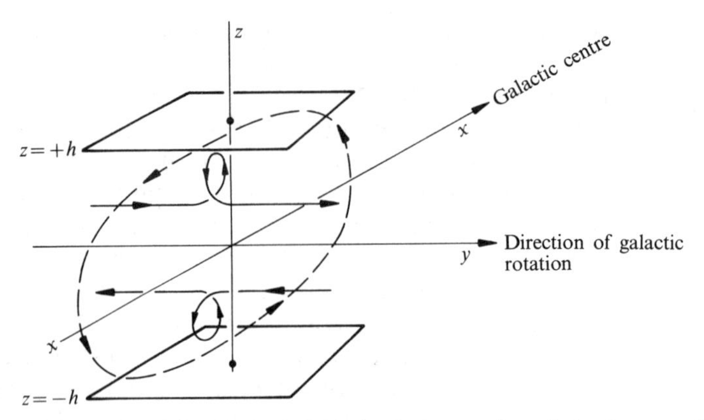

FIG. 19.10. A sketch of the internal fields in the generation of the lowest odd mode $B_y(+z) = -B_y(-z)$ in the galactic dynamo when $dV_y/dx > 0$ across the slab $z^2 < h^2$ and the dynamo coefficient is positive in $0 < z < h$ and negative in $-h < z < 0$, so that $Kh > 0$. The 'azimuthal' field B_y is shown by the continuous lines, in which there are shown the twisted loops produced by rising cells of gas. The large-scale circulation from the coalescence of many loops is indicated by the broken curve.

xz-plane. The coalescence of many such loops produces the large-scale magnetic circulation indicated by the large dashed curve. With $dV_y/dx > 0$ the lines of force of the circulating field are stretched out in the y-direction. The new B_y generated in this way adds to the initial field, so that there is a net generation. Note that there is no need here for lines of force to be lost out through the surfaces $z = \pm h$, as with the even mode. It is easy to show (Parker 1971a) that the dynamo functions efficiently in the odd mode when impervious boundaries are introduced at $z = \pm h$.

19.3.3.2. Periodic states. Consider the periodic states of the rectangular dynamo in the infinite slab (Parker 1971d). To keep the calculations as simple as possible, work in the limit $k_x/K \to 0$. Then (19.184) reduces to

$$\begin{aligned} 0 = &(q_1^2 - q_2^2)(q_4^2 - q_3^2)[q_1 q_2 \exp\{i(q_1 + q_2)Kh\} + q_3 q_4 \exp\{-i(q_1 + q_2)Kh\}] \\ &+ (q_3^2 - q_1^2)(q_4^2 - q_2^2)[q_1 q_3 \exp\{i(q_1 + q_3)Kh\} + q_2 q_4 \exp\{-i(q_1 + q_3)Kh\}] \\ &+ (q_2^2 - q_3^2)(q_4^2 - q_1^2)[q_2 q_3 \exp\{i(q_2 + q_3)Kh\} + q_1 q_4 \exp\{-i(q_2 + q_3)Kh\}] \end{aligned} \quad (19.192)$$

for the even modes. The odd mode in the same limit (Parker 1971a, d) gives

$$\begin{aligned} 0 = &(q_1^2 - q_2^2)(q_4^2 - q_3^2)\cosh\{(q_1 + q_2)Kh\} + (q_1^2 - q_3^2)(q_2^2 - q_4^2) \\ &\times \cosh\{(q_1 + q_3)Kh\} + (q_2^2 - q_3^2)(q_4^2 - q_1^2)\cosh\{(q_2 + q_3)Kh\}. \end{aligned} \quad (19.193)$$

The roots q_n may now be represented by the expansions in either ascending or descending powers of A, given by (19.56)–(19.63). The algebra then proceeds more or less along the lines of §19.3.1.2. Write $A = R\exp(i\theta)$ and look for solutions with $1/\tau = \eta K^2 A = i\eta K^2 R$ $(\theta = \frac{1}{2}\pi)$ for the real frequency $\omega = \eta K^2 R$. The results are given in Table 19.1 for the open boundary, i.e. $\eta = \infty$ outside the slab so that $B_y = 0$. The convergence of the expansions is conditional for θ in the neighbourhood of $\frac{1}{2}\pi$, so that the calculation was carried out with both ascending powers

TABLE 19.1

Dynamo numbers Kh *for the* nth *periodic mode* (19.192) *and* (19.193) *for* $Re(1/\tau)=0$, $Im(1/\tau)=\eta K^2 R$

(a) *Expansion in ascending powers of R*

Even modes, positive Kh $Kh=+(12{\cdot}6n-4{\cdot}9)$			Odd modes $Kh=\pm(12{\cdot}6n+1{\cdot}1)$		
n	Kh	R	n	Kh	R
1	7·7	0·362	1	±13·8	0·327
2	20·3	0·359	2	±26·4	0·342
3	32·9	0·358	3	±39·0	0·347

Even modes, negative Kh

$Kh=-(12{\cdot}6n-9{\cdot}5)$

n	Kh	R
1	−3·1	0·43
2	−15·7	0·36
3	−28·3	0·36

(b) *Expansion in descending powers of R*

Even modes, positive Kh $Kh=+(8{\cdot}7n-2{\cdot}5)$			Odd modes $Kh=\pm(8{\cdot}7n-0{\cdot}5)$		
n	Kh	R	n	Kh	R
1	6·2	0·53	1	±8·1	0·41
2	14·9	0·49	2	±17·0	0·42
3	23·6	0·47	3	±25·7	0·43

Even modes, negative Kh
$Kh=-(8{\cdot}4n+0{\cdot}1)$

n	Kh	R
1	no solution	—
2	−16·9	0·36
3	−25·3	0·39

Part (a) is calculated from the expansions (19.56)–(19.59) and (b) is calculated from the expansions (19.60)–(19.63).

of A, given in part (a) of the table, and in descending powers of A, given in part (b). The numerical values for the dimensionless frequency R differ by about 20 per cent between the two expansions, while the corresponding dynamo numbers differ by about one-third. We conclude therefore, that the results are approximately correct, and certainly illustrate the correct structure of the periodic modes of the dynamo.

In view of the sensitivity of the *stationary* even mode of the dynamo to the escape of magnetic flux from the dynamo, the calculation for the periodic mode was repeated for impenetrable boundaries, so that $\partial B_y/\partial z = 0$ at $z = \pm h$. Curiously, the dispersion relation for the even mode in the limit of $k_x \to 0$ turns out to be (19.193), already derived for the odd mode with open boundaries. The dispersion relation for the odd mode with impenetrable boundaries and $k_x \to 0$ is the same as (19.192) for the even mode with open boundaries if Kh in the exponentials is replaced by $-Kh$. The calculations show that the periodic modes are affected somewhat by the boundaries, but their regenerative powers are not crucially dependent upon whether flux is free to escape from the boundaries. This is not unexpected, because the migratory dynamo waves, making up the periodic dynamo, create and destroy their own fields locally. They do not depend upon the escape of field through the boundaries, as some of the stationary modes.

The tenuous gases outside the gaseous disc of the galaxy can impede the escape of field from the disc but little, whatever may be their electrical conductivity, so the free boundary is the case of principal physical interest. It is evident from Table 19.1 that the lowest oscillatory or periodic mode is for negative dynamo number, $Kh = -3{\cdot}1$. It is an even mode, and there is no fundamental objection, therefore, to the idea that galactic dynamo is presently operating in the lowest periodic, rather than stationary, mode (Parker 1971d). The period is of the order of 10^9 years. Hence, even with the broadest view, we are confined to theoretical, rather than observational conclusions. We note that the dynamo number Kh required to drive the periodic mode is more than half again as large as for the lowest stationary mode. The most conservative point of view is, then, that the galactic magnetic field is the product of the lowest mode of a stationary dynamo operating in the gaseous disc of the galaxy. The recent work of Stix (1975, 1978), White (1978), and Soward (1978) applying numerical techniques to an oblate spheroidal model of the gaseous disc, and exploring several distributions of the dynamo coefficient Γ across the thickness of the disc, provide a broad base for this conclusion. It should be kept in mind, however, that in other galaxies there may be circumstances, with thicker gaseous discs, for instance, in which the dynamo number is substantially larger, admitting the possibility of periodic fields in both the even and odd mode.

It is evident that we could go on from here to treat an endless variety of rectangular dynamos with various combinations of $\pm\Gamma$ and $\pm\partial V_y/\partial x$, $\pm\partial V_y/\partial z$, simulating an endless variety of physical circumstances, real or imagined. Our intent, however, is only to illustrate the basic physical principles of the captive, and sometimes stalled, dynamo waves that make up the periodic and stationary dynamos, respectively. Further examples do not show any outstanding new effects, and if there is one thing to be learned from the examples already given, it is the sensitivity of the behaviour of the dynamo to the boundary conditions and the distribution of Γ and G. Hence, at this point we turn to a consideration of the more complex, but more realistic, spherical or spheroidal geometry in which dynamos operate in the astronomical bodies of the universe. The rectangular dynamos have illustrated surprisingly well, as we shall see, the qualitative features of the stationary and periodic dynamos of Earth, the sun, and the galaxy. The results, therefore, are to be kept in mind as a guide throughout the numerical experiments with spherical geometries.

19.4. Axi-symmetric dynamos in spherical boundaries

19.4.1. Localized shear and cyclones

Consider the generation of magnetic fields by fluid motion with rotational symmetry confined to a spherical volume $r \leqslant R$. The fluid is subject to an axi-symmetric non-uniform angular velocity $\Omega(\varpi, z)$, where z is measured along the axis of rotation from the equatorial plane and $\varpi = (x^2+y^2)^{\frac{1}{2}}$ is distance from the axis of rotation. The dynamo coefficient Γ is also a function only of ϖ and z. The fluid has the uniform resistive diffusion coefficient η. Outside the sphere $r^2 = \varpi^2 + z^2 = R^2$ the medium, if any, is too tenous to resist being pushed aside by the field emerging from the dynamo region $r < R$. So, in view of the rapid reconnection (see §13.7.4), the external force-free field is a potential form. For all practical purposes it is the same as if $\eta = \infty$ outside.

If G_0 is the mean shear and Γ_0 the characteristic value of the dynamo coefficient, then the dynamo number is

$$N_D = G_0 \Gamma_0 R^3/\eta^2.$$

It was pointed out in (19.26) that dynamo waves with wave numbers k less than $(G_0\Gamma_0/2\eta^2)^{\frac{1}{3}}$ are amplified by the motions. The shortest wavelengths that will fit into the sphere are of the order of the diameter, $R = \pi/k$. Hence there is amplification if $N_D \gtrsim \pi^3$. In approximate terms the dynamo regenerates all waves for which $kR \leqslant N_D^{\frac{1}{3}}/\pi$. Thus, for instance, if N_D is a very large number, the spectrum of dynamo waves extends to small wavelength, the waves migrating in complicated ways throughout the region, as illustrated at length in §19.3. There are always periodic modes, for suitably high dynamo number. However, the more common circumstance, using Earth, the sun, and the galaxy as examples, appears to be a modest dynamo number sufficient to operate only the lower modes. Either the dynamo number is intrinsically low in such bodies, or the fields have grown to sufficient strength to suppress the fluid motions, and the dynamo number, to modest values. The stationary dynamo may be the lowest available mode, as in the thick core of Earth and in the galactic disc. A stationary mode may not always exist, of course, as in the thin convective zone of the sun, for which the lowest mode is periodic. We undertake first the problem of understanding the stationary modes.

The physics of the stationary mode can be expressed quite simply in terms of the Green's function of the dynamo equations. The complexity of the formal calculations, the diversity of the examples to be found in the literature, and the generalizations to which some authors have been tempted, prompt us to develop a firm physical understanding of the dynamo effect before venturing to review the extensive literature available on solutions of the dynamo equations. Using cylindrical coordinates (ϖ, ϕ, z) the dynamo equations can be written

$$\eta\nabla^2\mathbf{e}_\phi A_\phi = -\Gamma(\varpi, z)B_\phi(\varpi, z)\mathbf{e}_\phi, \tag{19.194}$$

$$\eta\nabla^2\mathbf{e}_\phi B_\phi = -\varpi(\nabla\Omega\,.\,\mathbf{B})\mathbf{e}_\phi \tag{19.195}$$

for stationary fields. The exposition of the physical properties begins by noting that the equation (19.194) for A_ϕ and the boundary conditions on A_ϕ are of exactly the same form as the familiar Poisson equation

$$\nabla^2\mathbf{e}_\phi A_\phi = -(4\pi/c)j_\phi\mathbf{e}_\phi \tag{19.196}$$

for the vector potential of an azimuthal current density j_ϕ. Hence the vector potential A_ϕ produced by a ring ($\varpi, z = constant$) of cyclonic turbulence has the same form as the well known vector potential of a ring of electric current circling the z-axis at the same position; the poloidal *field* produced by a ring of cyclones has exactly the same form as the field of a circular loop of current. The nearby lines of force circle closely around the electric current, while the more distant lines link through the current loop, taking the form of a dipole at large distances. The lines of force, $\varpi A_\phi = constant$, are sketched in Fig. 19.11. The basic point is that the ring of electric current

$$j_\phi(\varpi, z) = I\delta(\varpi - a)\delta(z - h)$$

of radius a in the plane $z = h$ has a magnetic dipole moment equal to I/c times the area πa^2 of the current loop. By direct analogy, then, a ring of cyclonic eddies

$$\Gamma(\varpi, z) = R^2\Gamma_0\delta(\varpi - a)\delta(z - h) \tag{19.197}$$

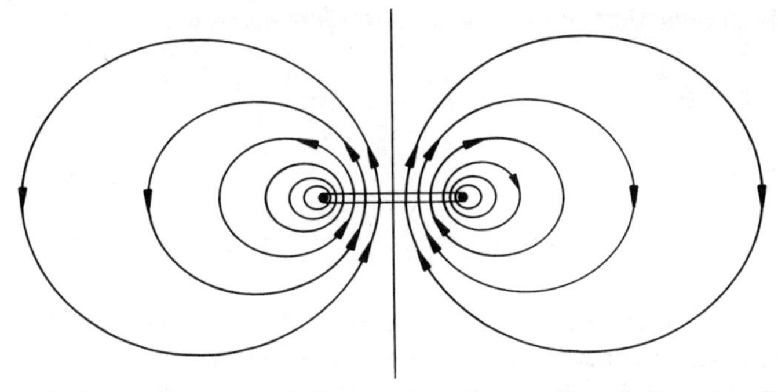

FIG. 19.11. A sketch of the magnetic lines of force in a meridional plane of a circular ring of electric current. The current flows in to the meridional plane on the right and out on the left.

leads to the same magnetic moment with $4\pi I/c$ replaced by $a^2\Gamma_0 B_\phi(a, h)/\eta$ i.e.

$$\mathcal{M} = \pi a^2\{R^2\Gamma_0 B_\phi(a, h)/4\pi\eta\}. \qquad (19.198)$$

At some distance outside the ring of cyclones, then, $A_\phi \sim \mathcal{M}\varpi/(\varpi^2+z^2)^{\frac{3}{2}}$, and the magnetic field has the familiar dipole form

$$B_\varpi \sim 3\mathcal{M}z\varpi/(\varpi^2+z^2)^{\frac{5}{2}}, \qquad B_z = \mathcal{M}(2z^2-\varpi^2)/(\varpi^2+z^2)^{\frac{5}{2}}. \quad (19.199)$$

The general expression for A_ϕ for a ring of current is well known (cf. Smythe 1936) and can be written in both cylindrical and spherical coordinates. It is convenient to use the Green's function $G_A(\varpi, z; a, h)$ defined as the solution of

$$(\nabla^2 - 1/\varpi^2)G_A = \delta(\varpi - a)\delta(z-h) = r_0^{-1}\delta(r-r_0)\delta(\theta-\theta_0)$$

with the property that $G_A \to 0$ at infinity and

$$a \equiv r_0 \sin(\theta_0), \qquad h = r_0 \cos(\theta_0).$$

Then the vector potential satisfying (19.194) for a ring of cyclones (19.197) is given by

$$A_\phi(\varpi, z) = -\{R^2\Gamma_0 B_\phi(a, h)/\eta\}G_A(\varpi, z; a, h). \qquad (19.200)$$

The Green's function can be written in a variety of forms, one of which is

$$G_A(\varpi, z; a, h) = (1/\pi p)(a/\varpi)^{\frac{1}{2}}\{E(p) - (1-\tfrac{1}{2}p^2)K(p)\}, \quad (19.201)$$

where $K(p)$ and $E(p)$ are the complete elliptic integrals of the first and second kind, and p is the modulus

$$p^2 = 4a\varpi/\{(\varpi+a)^2 + (z-h)^2\}. \qquad (19.202)$$

It will prove useful later to have G_A in spherical coordinates, for which it is readily shown that, with $a = r_0 \sin\theta_0$, $h = r_0 \cos\theta_0$,

$$G_A(r, \theta; r_0, \theta_0) = -\tfrac{1}{2}\sin\theta_0 \sum_{n=1}^{\infty} \left(\frac{r}{r_0}\right)^n \frac{P_n^1(\cos\theta_0)P_n^1(\cos\theta)}{n(n+1)} \tag{19.203}$$

for $r \leqslant r_0$ and

$$G_A(r, \theta; r_0, \theta_0) = -\tfrac{1}{2}\sin\theta_0 \sum_{n=1}^{\infty} \left(\frac{r_0}{r}\right)^{n+1} \frac{P_n^1(\cos\theta_0)P_n^1(\cos\theta)}{n(n+1)} \tag{19.204}$$

for $r \geqslant r_0$. The Green's function can, of course, be written in any other coordinate system in which the Laplacian operator is separable, but the two forms will suffice here. In any case, the important point is not the formal mathematics (to be used later to illustrate some simple aspects of the problem) but rather the well defined and familiar form of the field. The field represented by A_ϕ is essentially a dipole, with the form sketched in Fig. 19.11.

The solutions of (19.195) for a ring of shear of radius $\varpi = b$ at $z = l$ are equally simple in nature. Their mathematical form is well known. The form of the field $B_\phi(\varpi, z)$ is the same as the vector potential of a current ring in a spherical enclosure. Write

$$\nabla\Omega \cdot \mathbf{B} = W(b, l)\delta(\varpi - b)\delta(z - l) \tag{19.205}$$

where[3]

$$W(b, l) = b^2\{B_\varpi(b, l)\,\partial\Omega/\partial\varpi + B_z(b, l)\,\partial\Omega/\partial z\}.$$

Equation (19.195) becomes the Poisson equation

$$\eta\nabla^2\mathbf{e}_\phi B_\phi = -\mathbf{e}_\phi \varpi W(b, l)\delta(\varpi - b)\delta(z - l). \tag{19.206}$$

The form of the solution of this equation for the boundary condition $B_\phi = 0$ at the surface, $r = R$, is well known. The field B_ϕ has the same dependence on position (ϖ, z) as the vector potential of a current loop enclosed in an empty spherical cavity of radius R in a block of superconducting materials into which A_ϕ does not penetrate. This may be seen from the fact that $B_r = (r\sin(\theta))^{-1}\,\partial(\sin(\theta)A_\phi)/\partial\theta$ vanishes on $r = R$. Hence $\sin\theta A_\phi$ must be constant on $r = R$, implying $A_\phi = C/\sin\theta$ on $r = R$, where C is a constant. But A_ϕ is finite as one approaches the z-axis $(\theta = 0)$. Hence $C = 0$ and $A_\phi = 0$ on $r = R$. This is the required boundary condition for B_ϕ. The contours of constant B_ϕ are sketched in Fig. 19.12. The azimuthal field B_ϕ is very strong near the source ring and declines to zero toward the surface of the sphere.

[3] The vanishing of $\nabla\Omega \cdot \mathbf{B}$ elsewhere than at (b, l) means only that the surfaces of isorotation coincide with the magnetic lines of force, $\Omega = \Omega(\varpi A_\phi)$ as indicated by the right-hand side of (18.8). Thus (b, l) is the one location at which the surfaces of constant Ω and ϖA_ϕ are not parallel.

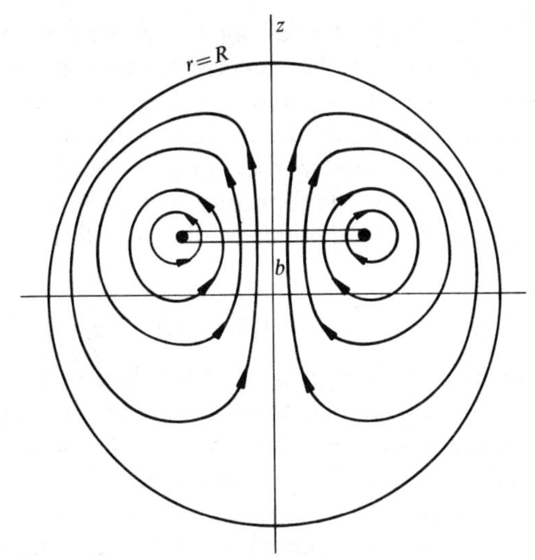

FIG. 19.12. A sketch of the magnetic lines of force in a meridional plane of a circular ring of electric current enclosed in a spherical cavity of radius R in an impenetrable medium.

The formal solution of (19.206) can be written as

$$B_\phi(\varpi, z) = -(b^2/\eta)\{B_\varpi(b, l)\,\partial\Omega/\partial\varpi + B_z(b, l)\,\partial\Omega/\partial z\}G_B(r, \theta; r_1, \theta_1) \tag{19.207}$$

in terms of the Green's function

$$\nabla^2 \mathbf{e}_\phi G_B = \mathbf{e}_\phi(\nabla^2 - 1/\varpi^2)G_B = \mathbf{e}_\phi r_1^{-1}\delta(r - r_1)\delta(\theta - \theta_1)$$

where $b \equiv r_1 \sin\theta_1$, $l = r_1 \cos\theta_1$. It is readily shown (Parker 1969a) that in the region $0 \leqslant r \leqslant r_1$

$$G_B(r, \theta; r_1, \theta_1) = \tfrac{1}{2}\sin(\theta_1)\sum_{n=1}^{\infty}\left\{\left(\frac{r_1}{R}\right)^{2n+1} - 1\right\}\left(\frac{r}{r_1}\right)^n \frac{P_n^1(\cos\theta_1)P_n^1(\cos\theta)}{n(n+1).} \tag{19.208}$$

while in $r_1 \leqslant r \leqslant R$

$$G_B(r, \theta; r_1, \theta_1) = \tfrac{1}{2}\sin(\theta_1)\sum_{n=1}^{\infty}\left\{\left(\frac{r_1}{R}\right)^{2n+1}\left(\frac{r}{r_1}\right)^n - \left(\frac{r_1}{r}\right)^{n+1}\right\} \times \frac{P_n^1(\cos\theta_1)P_n^1(\cos\theta)}{n(n+1)}. \tag{19.209}$$

But again, the important point is the simple form of the field illustrated in Fig. 19.12. The azimuthal field B_ϕ declines outward from its source at the ring to the surface.

Knowing the form of the dipole solutions for rings of shear and cyclones, sketched in Figs 19.11 and 19.12, it is a simple matter to develop in a rigorous way the physical properties of the solutions to

(19.194) and (19.195). In particular we are interested in the necessary and sufficient conditions for the generation of stationary dipole and quadrupole fields. As a first example, consider two axi-symmetric rings of cyclones situated above and below the equator at $z = \pm h$. If the dynamo coefficient of the ring in the northern hemisphere ($z = \pm h$) is $R^2\Gamma_0\delta(\varpi - a)\delta(z - h)$, then the opposite rotation of cyclones in the southern hemisphere yields the dynamo coefficient $-R^2\Gamma_0\delta(\varpi - a)\delta(z + h)$. Hence, altogether,

$$\Gamma(\varpi, z) = R^2\delta(\varpi - a)\{\delta(z - h) - \delta(z + h)\}. \tag{19.210}$$

The dynamo coefficient $\Gamma(\varpi, z)$ is an odd function of z. It interacts with the azimuthal field B_ϕ, which may be either an even or an odd function of z. As a first example, then, consider the poloidal field to be a dipole. Then B_ϕ is an odd function of z, as illustrated in §18.2. It follows that the source $\Gamma(\varpi, z)B_\phi(\varpi, z)$ is an even function of z. Both rings of cyclones make the same contribution to A_ϕ. Each ring is effectively a magnetic dipole, with moment $\mathcal{M}$, pointing along the z-axis at $z = \pm h$.

19.4.1.1. Central concentration of cyclones. Suppose that the two rings of cyclones have a small radius and lie close to the origin ($a, h \ll R$). Then everywhere throughout the dynamo region, $r \leq R$, except in the small neighbourhood of the rings, the poloidal field is that of a dipole of moment $2\mathcal{M}$, sketched in Fig. 19.13. It follows that B_ϖ is an odd function of z, while B_z is an even function.

Consider, then, the generation of B_ϕ by the shearing of B_ϖ and B_z by the non-uniform rotation $\Omega(\varpi, z)$. We expect $\Omega(\varpi, z)$ to be symmetric about the equatorial plane in most circumstances, so that $\partial\Omega/\partial\varpi$ is an even function, and $\partial\Omega/\partial z$ an odd function, of z. It follows that the generation of azimuthal field B_ϕ by both $B_\varpi\, \partial\Omega/\partial\varpi$ and $B_z\, \partial\Omega/\partial z$ is an odd function of z, consistent with the initial statement that the azimuthal field is an odd function of z, as described by (18.8) in §18.2. Thus the dipole form is self-consistent for the prescribed shear and dynamo coefficients. The next question is whether the signs work out to give generation or destruction of the dipole.

To facilitate the discussion suppose that Γ and B_ϕ are both positive in the northern hemisphere (and hence both negative in the southern hemisphere). Then $\mathcal{M}$ is positive in (19.198). Suppose that the angular velocity Ω increases with distance ϖ from the axis of rotation, as suggested by the equatorial acceleration of the sun. Then $\partial\Omega/\partial\varpi$ is positive and its interaction in the northern hemisphere with a positive B_ϖ makes a positive contribution to the azimuthal magnetic field there. The originally positive B_ϕ is reinforced and the dynamo cycle regenerates the field. It is obvious that the opposite combination, of negative $\partial\Omega/\partial\varpi$ and

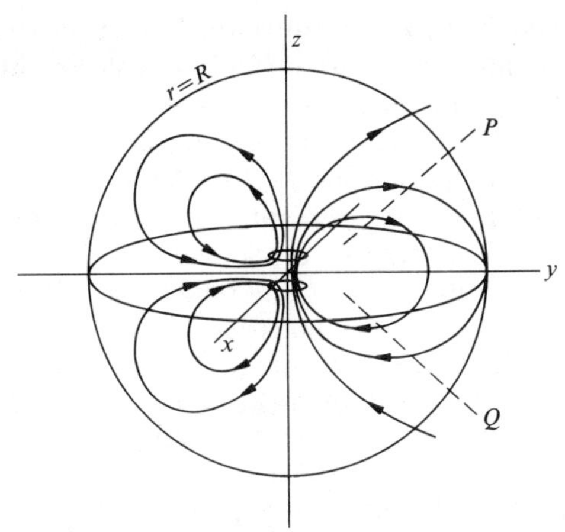

FIG. 19.13. A sketch of the magnetic lines of force in the meridional plane of two rings of cyclones near the origin, with dynamo coefficient $+\Gamma$ for the ring north of the equatorial plane and $-\Gamma$ for the southern ring. The lines on the right-hand half-plane are those of a dipole, for the case where B_ϕ is an odd function of latitude so that the azimuthal vector potential generated by ΓB_ϕ has the same sign for both rings. The left-hand half-plane shows the quadrupole field that arises when B_ϕ is an even function of latitude and ΓB_ϕ has opposite sign for the two rings.

negative Γ (as suggested for the core of Earth) is also regenerative. If one is negative and the other positive, however, then the dynamo actively destroys the magnetic field. Hence it is not possible to generate a dipole field with negative $\Gamma \partial\Omega/\partial\varpi$ when the cyclones are confined to the neighbourhood of the origin.

The situation is somewhat different if Ω is a function of z. For suppose that Ω declines with distance from the equatorial plane, so that $\partial\Omega/\partial z$ is negative in the northern hemisphere. The generation of azimuthal field is $B_z\, \partial\Omega/\partial z$, which must be positive to reinforce an initial positive B_ϕ. Hence, with negative $\partial\Omega/\partial z$ the azimuthal field B_ϕ is augmented only if B_z is negative at the location of the shear $\partial\Omega/\partial z$. If B_z is positive, then the shear actively destroys the azimuthal field and no stationary dipole mode exists. A glance at (19.199) shows that B_z is positive where $z^2 > \frac{1}{2}\varpi^2$ and negative in $z^2 < \frac{1}{2}\varpi^2$. The dynamo reinforces the field if the shear is located in the sector PQ in Fig. 19.13, representing $z^2 < \frac{1}{2}\varpi^2$ in which B_z is negative. Hence any shear in the PQ-sector is regenerative, reinforcing the original azimuthal field, whereas a shear located outside the sector PQ actively destroys B_ϕ. It is immediately evident, then, that a continuous distribution of shear extending across the boundary $z^2 = \frac{1}{2}\varpi^2$ into both regions produces both positive and negative B_ϕ. The net result is regenerative only if the contribution from within PQ outweighs the destructive contribution from outside. A careful quantitative calculation is required to determine the net effect.

On the other hand, if $\partial\Omega/\partial z$ is positive (or Γ is negative) in the northern hemisphere, the situation is reversed. Only the shear outside the sector PQ regenerates the magnetic field. It is evident at once why various authors, assuming different distributions of cyclones and shears, have come to a variety of conclusions as to whether positive or negative dynamo numbers do, or do not, generate stationary dipole fields. We can see, too, how a small change in location of the shear can change the dynamo from a regenerative to a degenerative state. Moving the location of the shear across the sector boundaries $z = \pm\varpi/\sqrt{2}$ in a regenerative dynamo leads to active destruction of the fields. It is simple changes of this nature that are presumed to be responsible (Parker 1969a; Levy 1972a, b) for the abrupt reversals (§18.1) of the geomagnetic field.

19.4.1.2. Low latitude cyclones. Suppose that the two rings of cyclones are of large radius a, lying at low latitudes immediately inside the surface (i.e. $R - a,\ h \ll R$) as sketched in Fig. 19.14. Then if B_ϕ and Γ are positive in the northern hemisphere, it follows that $\mathcal{M}$ is positive and both components (B_ϖ, B_z) are positive almost everywhere in the northern hemisphere. Hence if $\partial\Omega/\partial\varpi$ is positive, it interacts with B_ϖ to reinforce the initial positive B_ϕ. If $\partial\Omega/\partial z$ is positive, it interacts with positive B_z to reinforce B_ϕ. In both cases the cycle is regenerative. On the other hand if Γ is negative in the northern hemisphere, then $\partial\Omega/\partial\varpi$, and/or $\partial\Omega/\partial z$ must be negative to generate field. Altogether, then, Γ and $(\partial\Omega/\partial\varpi,\ \partial\Omega/\partial z)$ must have the same sign for amplification of the fields. Opposite signs

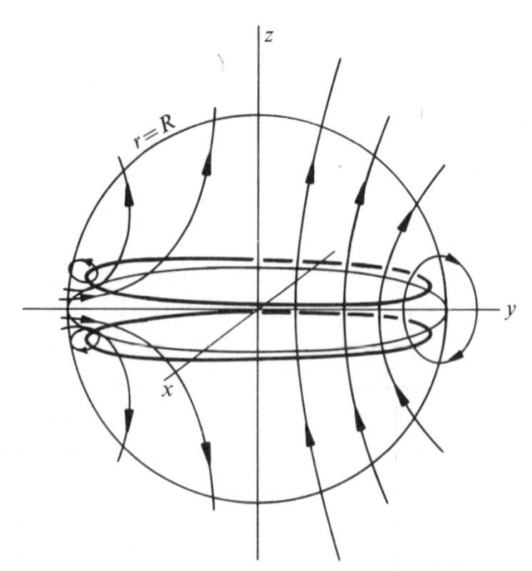

FIG. 19.14. A sketch of the magnetic lines of force in the meridional plane for two rings of cyclones near the surface at low latitudes. The right half-plane shows the lines when B_ϕ is an odd function of latitude, associated with a dipole field. The left half-plane shows the lines when B_ϕ is an even function, associated with a quadrupole field.

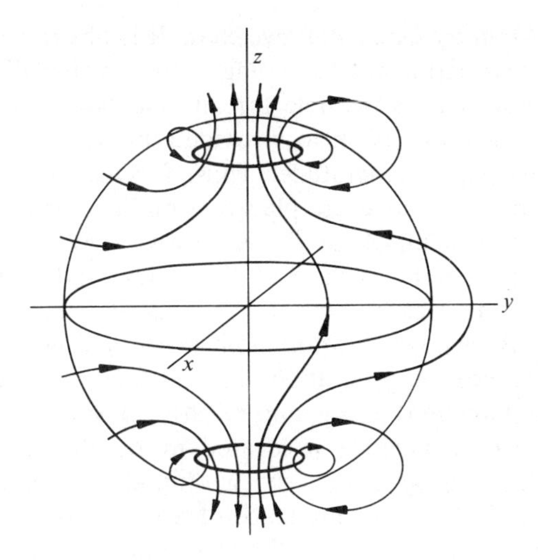

FIG. 19.15. A sketch of the magnetic lines of force in the meridional plane for two rings of cyclones near the surface at high latitudes. The right half-plane shows the lines when B_ϕ is an odd function of latitude, associated with a dipole field. The left half-plane shows the lines when B_ϕ is an even function, associated with a quadrupole field.

lead to active destruction of the dipole field by the dynamo process. The conclusion is independent of the location of the shear anywhere in $b < a$.

19.4.1.3. High latitude cyclones. Suppose that the two rings of cyclones are of small radius a, lying at high latitudes immediately beneath the surface at either pole, as sketched in Fig. 19.15. If B_ϕ and Γ are both positive in the northern hemisphere, it follows that $\mathcal{M}$ is positive, and it is obvious from Fig. 19.15 that B_z is positive but B_ϖ is negative almost everywhere within the dynamo. Hence a negative $\partial\Omega/\partial\varpi$ is required to interact with B_ϖ to produce positive B_ϕ and amplify the field. A positive $\partial\Omega/\partial z$ also suffices because B_z is positive. If Γ is negative, then a positive $\partial\Omega/\partial\varpi$, or a negative $\partial\Omega/\partial z$, is required in the northern hemisphere. If both Γ and $\partial\Omega/\partial\varpi$ are negative (as suggested by elementary dynamical considerations) in the northern hemisphere of the liquid core of Earth, then it is not possible to generate a stationary dipole. It is evident, then, that a dipole can be generated with negative Γ and $\partial\Omega/\partial\varpi$ only if the shear lies at some suitably larger distance from the equatorial plane than the cyclones. A shift in the location of the shear across the critical position can cause active destruction of the field. However, it should be pointed out that the most efficient dynamo places the shear and cyclones in the strongest parts of B_ϖ and B_ϕ, respectively, rather than in the weak fields near zeros. For such dynamos a strong shift would be needed to provide active destruction of the field.

19.4.1.4. Arbitrary location of cyclones. It is obvious from the physical principles illustrated in the foregoing examples that the generation of a stationary dipole field is possible for any location of the two rings of cyclones and the two rings of shear, provided the dynamo number has the right sign and sufficient magnitude. Thus, for instance, if the rings of cyclones of arbitrary radius a are placed at middle latitudes, it is evident that in the northern hemisphere B_ϖ is positive over most of the region between the ring and the surface at the poles, and negative over the major portion of the region between the rings and the equatorial plane. The component B_z is positive throughout $\varpi < a$ and negative throughout most of $\varpi > a$. Hence the generation of positive B_ϕ requires $\partial\Omega/\partial\varpi$ to be positive if located above the ring, and negative if located below. Alternatively $\partial\Omega/\partial z$ must be positive if located inside the radius $\varpi = a$ and negative if located outside. Then a dynamo number of suitable magnitude leads to amplification and maintenance of a steady dipole field. If these conditions are not satisfied, so that the production of B_ϕ is negative, then the field is actively destroyed.

19.4.1.5. A stationary dipole in a thin shell. Consider the question of the generation of a stationary dipole field in a thin shell, treated in §19.3.2. Suppose that the dynamo is confined to the thin spherical shell $R_1 < r < R$ $(R - R_1 \ll R)$ immediately inside the surface of the sphere $r = R$, as sketched in Fig. 19.16. The inner core $r < R_1$ is assumed to be fixed and impenetrable. The Green's function for a ring of cyclones is a

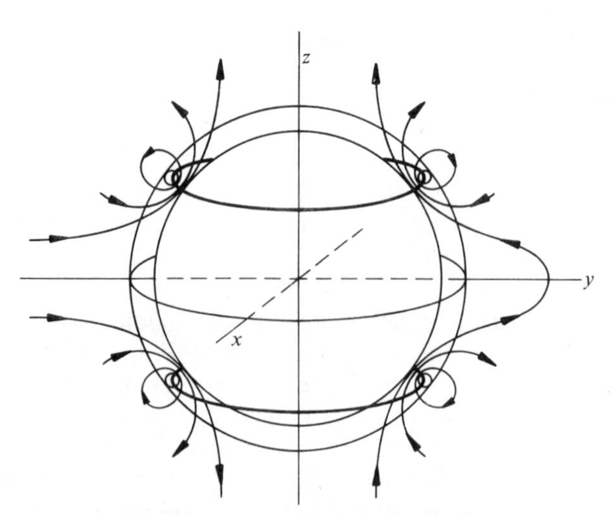

FIG. 19.16. A sketch of the magnetic lines of force in the meridional plane for two rings of cyclones in a thin shell surrounding an impenetrable core. The right half-plane shows the lines for a dipole field, and the left half-plane for a quadrupole field.

slight modification of (19.203) and (19.204) with $(r/r_0)^n$ in (19.203) replaced by $(r/r_0)^n-(R_1/r_0)^n(R_1/r)^{n+1}$ and the additional factor $\{1-(R_1/r_0)^{2n+1}\}$ inserted into (19.204) so that A_ϕ is continuous across r_0 and vanishes on $r=R_1$. The lines of force of the resulting poloidal field are sketched in Fig. 19.16 for the two rings of cyclones (19.210) of radius $\varpi=a$ $(R_1<a<R)$ at $z=\pm h$. The sense of the lines represents positive ΓB_ϕ. Fixing attention on the northern hemisphere, it is evident from Fig. 19.16 that, except for the immediate neighbourhood of the ring of cyclones, B_z is positive and B_ϖ is generally negative throughout the thin shell of the dynamo region. Hence the shear $\partial\Omega/\partial\varpi$ must be negative if $B_\varpi\ \partial\Omega/\partial\varpi$ is to reinforce the initially positive azimuthal field, while $\partial\Omega/\partial z$ must be positive. It should be noted, of course, that the poloidal field falls off rapidly away from the rings of cyclones, declining exponentially around the shell with a characteristic scale $R-R_1$, rather than R. Similarly the azimuthal field declines rapidly, with a scale $R-R_1$, away from the rings of shear. Hence, if the shear and the cyclones are widely separated, by distances of the order of R $(\gg R-R_1)$, the poloidal field at each ring of shear, and the azimuthal field at each ring of cyclones, are very weak. Consequently the dynamo is extremely inefficient, requiring an enormous dynamo number to maintain the fields. The important point is, however, that the dynamo exists. It is possible, with rings of cyclones and shear of suitable strength, to regenerate and amplify a dipole field in a thin spherical shell open to free space at its outer surface.

It will be recalled that the treatment of a dynamo in a thin slab in §19.3.2 led to the conclusion that the generation of field could not compensate for the rapid loss of field out of the sides of the slab. The necessary large dynamo numbers to compensate for the loss gave periodic rather than stationary fields. The migration of the dynamo waves cannot be blocked when the dynamo number is large. The explanation for the present result, that a stationary dynamo is possible for suitably high dynamo number, seems to be that dynamo waves depend for their existence (and migration) on a more or less continuous distribution of cyclones and shear. When the cyclones exist at only two points, and the shear at only two points, in the meridional plane, the dynamo waves are 'hooked' so that stationary fields are possible. Indeed, an obvious qualitative question is whether there are now periodic states of the dynamo. The answer would seem to be in the affirmative because there are propagating periodic solutions of the diffusion equation (without cyclones or shear) of the form $\exp\{iwt\pm ix(\omega/2\eta)^{\frac{1}{2}}\}\exp\{\pm x(\omega/2\eta)^{\frac{1}{2}}\}$ in rectangular coordinates so that the phase delay in the propagation of the poloidal field from the rings of cyclones to the rings of shear, and the phase delay of the toroidal field from the shear to the cyclones, can give a constructive effect for suitable (large) values of the dynamo number. However, because of the rapid attenuation of the fields with distance, the dynamo effect is very inefficient unless the cyclones and shear are located close to each other. The characteristic scale of attenuation is $(2\eta/\omega)^{\frac{1}{2}}$ which may be much smaller than the scale of attenuation $R-R_1$ for stationary fields. A related quantitative question, then, is the manner in which the dynamo waves become free to migrate as the separation of the discrete rings of cyclones and shear is decreased.

19.4.1.6. Formal solution of the dynamo equations. Now that the physics of the generation of a stationary field has been illustrated, it is interesting to note how the formal mathematical problem appears. For the two rings of cyclones (19.210) the formal solution (19.200) for the azimuthal vector potential becomes

$$A_\phi(\varpi, z) = -a^2(\Gamma_0/\eta)B_\phi(a, h)\{G_A(\varpi, z; a, h) + G_A(\varpi, z; a, -h)\} \tag{19.211}$$

upon noting that, in the present dipole problem, $B_\phi(a, h) = -B_\phi(a, -h)$. To write the formal solution for $B_\phi(\varpi, z)$, recall that a rotation $\Omega(\varpi, z)$ symmetric about the equatorial plane $z=0$ yields a shear $\partial\Omega/\partial z$ that is an odd function of z, while $\partial\Omega/\partial\varpi$ is an even function. For dipole symmetry B_ϖ is an odd function and B_z is an even function, so that $B_\varpi\,\partial\Omega/\partial\varpi + B_z\,\partial\Omega/\partial z$ is a purely odd function of z. This fact was noted in the development of the basic physics in the subsection above. It follows that rings of shear at $\varpi=b$, $z=\pm l$ yield

$$B_\phi(\varpi, z) = -(b^3/\eta\{B_\varpi(b, l)\,\partial\Omega/\partial\varpi + B_z(b, l)\,\partial\Omega/\partial z\}_{b,l} \times\{G_B(\varpi, z; b, l) - G_B(\varpi, z; b, -l)\}, \tag{19.212}$$

the first set of braces to be evaluated at $\varpi=b$, $z=l$. Note, then, that $B_\varpi = -\partial A_\phi/\partial z$ and $B_z = (1/\varpi)\,\partial(\varpi A_\phi)/\partial\varpi$. Differentiating (19.211) and substituting into (19.212) leads to

$$B_\phi(\varpi, z) = (\Gamma_0 a^2 b^3/\eta^2)B_\phi(a, h)\{G_B(\varpi, z; b, l) - G_B(\varpi, z; b, -l)\} \times\left[\frac{\partial\Omega}{\partial z}\frac{1}{\varpi}\frac{\partial}{\partial\varpi}\varpi\{G_A(\varpi, z; a, h) + G_A(\varpi, z; a, -h)\} - \frac{\partial\Omega}{\partial\varpi}\frac{\partial}{\partial z}[G_A(\varpi, z; a, h) + G_A(\varpi, z; a, -h)]\right]_{b,l}. \tag{19.213}$$

In particular, this relation must be satisfied at $\varpi=a$, $z=h$. If $B_\phi(a, h)\neq 0$, then, it can be cancelled from the equation, yielding the eigenvalue

$$\Gamma_0 a^2 b^3/\eta^2 = \{G_B(+) - G_B(-)\}^{-1} \times\left[\frac{\partial\Omega}{\partial z}\frac{1}{\varpi}\frac{\partial}{\partial\varpi}\varpi\{G_A(+) + G_A(-)\} - \frac{\partial\Omega}{\partial\varpi}\frac{\partial}{\partial z}\{G_A(+) + G_A(-)\}\right]_{b,l}^{-1} \tag{19.214}$$

for the dynamo number for stationary operation. We have written

$$G_A(\pm) = G_A(\varpi, z; a, \pm h), \qquad G_B(\pm) = G_B(a, h; b, \pm l),$$

and $G_A(\pm)$ is to be evaluated at $\varpi = b$, $z = l$ after carrying out the indicated differentiation with respect to ϖ and z. The Green's functions are to be expressed in terms of the expansions (19.203), (19.204) and (19.208), (19.209). The important point is that the Green's functions are real, and non-vanishing almost everywhere. Hence the eigenvalue exists and is real for almost all locations (a, h) and (b, l) of the cyclones and shear, respectively. The convergence of the expressions for the Green's function is uniform and absolute, so that there are no questions with the formal validity of (19.214). In practice, however, the convergence is sufficiently slow as to make the formal computation of the eigenvalues a tedious affair. Before going on to the general case, it is instructive to note one or two special cases. If, for instance, the shear and the cyclones are widely separated, then the equations reduce to a particularly simple form. If the cyclones are located close to the origin and the shear is not ($r_0^2 \ll r_1^2$), then for $r \gg a, h$

$$G_A(\varpi, z; a, h) \cong -\frac{1}{4}\left(\frac{a}{r}\right)^2 \sin\theta = -\frac{a^2\varpi}{4(\varpi^2 + z^2)^{\frac{3}{2}}}$$

and for $r \ll b, l$

$$G_B(\varpi, z; b, l) \cong \tfrac{1}{4}\sin\theta_1 \left[\left\{\left(\frac{r_1}{R}\right)^3 - 1\right\}\frac{r}{r_1}\sin\theta_1 \sin\theta \right.$$

$$\left. + 3\left\{\left(\frac{r_1}{R}\right)^5 - 1\right\}\left(\frac{r}{r_1}\right)^2 \sin\theta_1 \cos\theta_1 \sin\theta\cos\theta + \dots\right].$$

Hence

$$G_B(+) - G_B(-) = \frac{3}{2}\left\{\left(\frac{r_1}{R}\right)^5 - 1\right\}\frac{b^2 lah}{r_1^5}$$

to lowest order in r_0/r_1, where $r_0^2 = a^2 + h^2$, $r_1^2 = b^2 + l^2$. Note further that

$$\frac{1}{\varpi}\frac{\partial}{\partial\varpi}\varpi\{G_A(+) + G_A(-)\} = -\frac{a^2}{2r_1^3}(3\cos^2\theta_1 - 1),$$

$$\frac{\partial}{\partial z}\{G\ (+) + G_A(-)\} = +\frac{3a^2}{2r_1^3}\sin\theta_1\cos\theta_1$$

and are to be evaluated at $\varpi = r\sin\theta = b$, $z = r\cos\theta = l$. It follows from (19.214) that the dynamo number for stationary operation is

$$\frac{\Gamma_0 R^3}{\eta^2}\left\{\frac{3bl}{r_1}\frac{\partial\Omega}{\partial\varpi} + \frac{2l^2 - b^2}{r_1}\frac{\partial\Omega}{\partial z}\right\}_{b,l} = +\frac{4}{3}\left(\frac{r_1^6}{b^5 l}\right)\left(\frac{r_0^3}{a^2 h}\right)\left(\frac{r_1}{r_0}\right)^3 \frac{(R/a)^3}{1-(r_1/R)^5}. \tag{19.215}$$

The dynamo number is represented by the left-hand side of this equation. It is large, $O\{r_1R/r_0a)^3\}$, because of the location of the cyclones at small radius r_0, near both the z-axis and the equatorial plane, on which B_ϕ declines to zero.

A similar expression in closed form for the dynamo number is easily developed for the shear near the origin $r_1 \ll r_0$, or for either the cyclones or the shear circling closely about the z-axis. However, the above example is sufficient to illustrate the physical effects and we are more interested in the case where the cyclones and shear are near each other, producing an efficient dynamo, rather than widely separated.

A number of special cases are treated in the literature. Parker (1969a) worked out the result for cyclones confined to the neighbourhood of the origin and the sheet of shear $\partial\Omega/\partial\varpi = \Omega\delta(\varpi - b)$ just below the surface at the equator. The most complete results have been given by Levy (1972a, b) for the shell of non-uniform rotation $\partial\Omega/\partial r = \Omega\delta(r - r_1)$, summarized in §19.4.2 to illustrate the basic properties of the dynamo number as a function of location. The present subsection is concluded with a brief discussion of quadrupole fields.

19.4.1.7. Quadrupole fields. The treatment of the generation of dipole fields by two rings of cyclones with opposite signs at $z = \pm h$, and two rings of shear at $z = \pm l$, shows that stationary fields are possible provided that the dynamo number has the right sign. The sign depends upon the location of the cyclones and shear. When the dynamo number has the wrong sign, the dynamo actively destroys the dipole field. As it turns out, however, the sign is then correct to generate a stationary quadrupole field. For suppose that B_ϕ is an even function of z, rather than odd. Then, because Γ is an odd function of z, ΓB_ϕ is odd and the A_ϕ generated by ΓB_ϕ is then an odd function of z. That is to say, (B_ϖ, B_z) has the form of the field of a linear quadrupole. Suppose, then, that B_ϕ is everywhere positive with Γ positive in the northern hemisphere ($z > 0$) and negative in the southern. The lines of force of the quadrupole poloidal field of A_ϕ have the sense indicated in Figs 19.13 and 19.14. Field amplification follows if $B_\varpi\, \partial\Omega/\partial\varpi$ is positive, or if $B_z\, \partial\Omega/\partial z$ is positive, at the location of the shear. When the cyclones are close to the origin, as depicted in Fig. 19.13, the field is

$$B_\varpi \propto \frac{\varpi(4z^2 - \varpi^2)}{(\varpi^2 + z^2)^{\frac{7}{2}}}, \qquad B_z \propto \frac{z(2z^2 - 3\varpi^2)}{(\varpi^2 + z^2)^{\frac{7}{2}}}$$

almost everywhere, so that B_ϖ is positive for $z^2 > \frac{1}{4}\varpi^2$ and B_z is positive for $z^2 > \frac{3}{2}\varpi^2$. Hence positive $\partial\Omega/\partial\varpi$ amplifies the field if it is located in $z^2 > \frac{1}{4}\varpi^2$, while positive $\partial\Omega/\partial z$ amplifies the field if it is located in $z^2 > \frac{3}{2}\varpi^2$. Otherwise negative $\partial\Omega/\partial\varpi$ or $\partial\Omega/\partial z$ is required.

It will be recalled from §19.4.1.1. that regeneration of the dipole field by positive $\partial\Omega/\partial\varpi$ occurs without restriction on the location of the shear because B_ϖ was positive almost everywhere. On the other hand for $\partial\Omega/\partial z$ positive, amplification required $z^2 > \frac{1}{2}\varpi^2$. It is evident that the shear in any position for positive $\partial\Omega/\partial\varpi$ or $\partial\Omega/\partial z$ where it will amplify the quadrupole field is also suitably located to amplify the dipole field. On the other hand, positive $\partial\Omega/\partial\varpi$ in $z^2 < \frac{1}{4}\varpi^2$ amplifies a dipole field and destroys a quadrupole field; positive $\partial\Omega/\partial z$ in $\frac{1}{2}\varpi^2 < z^2 < \frac{3}{2}\varpi^2$ does the same; positive $\partial\Omega/\partial z$ in $z^2 < \frac{1}{2}\varpi^2$ destroys both. Finally, negative $\partial\Omega/\partial\varpi$ or $\partial\Omega/\partial z$ in $z^2 > \frac{1}{2}\varpi^2$ destroys both dipole and quadrupole fields; negative $\partial\Omega/\partial z$ in $z^2 < \frac{1}{2}\varpi^2$ amplifies both, etc.

We could go on to discuss the situation with the cyclones near the surface of the dynamo at low and high latitudes. The lines of force, with the sense indicated for positive Γ and B_ϕ in the northern hemisphere, are sketched in Figs 19.14 and 19.15, indicating the sense of B_ϖ and B_z. Amplification of B_ϕ requires that $\partial\Omega/\partial\varpi$ and $\partial\Omega/\partial z$ have the same sign as B_ϖ and B_z, respectively. The combination of signs and locations are endless. The purpose has been to make the physical principles clear.

The fields in a thin shell are sketched in the left half of Fig. 19.16. It is evident, from the fact that the fields in the northern and southern hemispheres are separated by distances large compared to the thickness of the shell, that there is very little interaction between them. Hence, unless the shear is concentrated in the immediate vicinity of the equator, the dynamo effects in the two hemispheres proceed independently of each other, so that the conditions for regeneration are essentially the same for both dipole and quadrupole symmetries.

Finally, it should be noted that under stationary conditions two rings of cyclones are limited to the generation of poloidal fields whose fundamental mode contains no more than two lobes. If ΓB_ϕ has the same sign for both rings, then the rings act together, producing only a single lobe, i.e. a dipole field. If ΓB_ϕ has opposite signs, then the rings oppose each other and each produces its own lobe, with a quadrupole field as the result. It is clear, then, that an octupole field can be produced with no less than three rings. Indeed, if we stick to the symmetry wherein Γ is an odd function of z and the rings are located symmetrically about the equatorial plane, four rings are the minimum to produce an octupole. A continuous distribution, then, can produce an octupole and higher modes under the right combination of signs, etc. But no fundamental mode above the quadrupole can be produced in the idealized circumstances of two rings.

19.4.2. Shear confined to a spherical surface

The stationary dynamo driven by two symmetric rings of cyclones and two symmetric rings of shear treated in §19.4.1 is the primitive spherical dynamo, illustrating the basic physical properties of the effect in spherical geometry (Parker 1969a). The dynamo in the spherical astronomical body with a continuous distribution of cyclones and shear is a superposition of rings of the primitive dynamo. The first step in that direction is the dynamo involving two symmetric rings of cyclones with the shear spread over a spherical shell. The dynamo coefficient is given by (19.210) again, but the axi-symmetric rotation is now $\Omega = \Omega_0$ for $0 < r < r_1$, and $\Omega = \Omega_0 + \Delta\Omega$ for $r_1 < r < R$. Then Ω is a function only of the distance r from the origin, and the shear is confined to a spherical surface of radius r_1, with

$$\partial\Omega/\partial r = \Delta\Omega\delta(r - r_1). \tag{19.216}$$

This case is tractable, if difficult, and the stationary states have been worked out in closed form by Levy (1972a, b), who extended the results (Levy 1972c) to include n symmetric pairs of rings of cyclones. The periodic states with constant amplitude, were then treated by Stix (1973). This synthesis of the primitive spherical dynamo provides both stationary and oscillatory dipole and quadrupole states. The effects are readily understood on the basis of the physical principles developed in the earlier sections of this chapter. It will be interesting to compare the solutions in §19.3 for the rectangular dynamo with the solutions for the spherical dynamo. We shall see that the rectangular dynamo captures the basic *qualitative* properties, with positive dynamo number producing the stationary dipole in regions of sufficient thickness and the oscillating quadrupole at higher dynamo number, while negative number gives the stationary quadrupole and the oscillating dipole. The implications for Earth and the sun, for which the dynamo numbers are believed to be positive and negative, respectively, are obvious. The primitive dynamo with two pairs of rings, and the dynamo with the shear distributed over a spherical shell, provide the basic demonstration of the dynamo in the core of Earth and in the convective zone of the sun.

19.4.2.1. Stationary fields. The dynamo equations for the stationary azimuthal vector potential is (19.194). With the dynamo coefficient given by (19.210), the solution is (19.211), so that for the cyclones at (r_0, θ_0) and $(r_0, \pi - \theta_0)$

$$A_\phi(r, \theta) = \frac{a^2\Gamma_0}{2\eta}\sin\theta_0 \sum_{n=1}^{\infty}\left(\frac{r}{r_0}\right)^n \frac{P_n^1(\cos\theta_0)}{n(n+1)} \times\{B_\phi(r_0, \theta_0)P_n^1(\cos\theta_0) - B_\phi(r_0, \pi-\theta_0)P_n^1(-\cos\theta_0)\} \tag{19.217}$$

for $r < r_0$, and the same expression with $(r/r_0)^n$ replaced by $(r_0/r)^{n+1}$ for $r_0 < r < R$. For odd (dipole) symmetry about the equatorial plane $(\theta = \frac{1}{2}\pi)$ B_ϕ is an odd function of $z = r\cos\theta$. Noting that $P_{2n}^1(x)$ is an odd function of x while $P^1_{2n+1}(x)$ is an even function of x, it follows from (19.217) for the dipole that

$$A_\phi(r, \theta) = \frac{a^2\Gamma_0}{2\eta} \sin\theta_0 B_\phi(r_0, \theta_0) \sum_{n=1}^{\infty} \frac{P^1_{2n-1}(\cos\theta_0)P^1_{2n-1}(\cos\theta)}{n(2n-1)} \left(\frac{r}{r_0}\right)^{2n-1}. \tag{19.218}$$

For even (quadrupole) symmetry about the equatorial plane B_ϕ is an even function of z and it follows from (19.217) that

$$A_\phi(r, \theta) = \frac{a^2\Gamma_0}{2\eta} \sin\theta_0 B_\phi(r_0, \theta_0) \sum_{n=1}^{\infty} \frac{P^1_{2n}(\cos\theta_0)P^1_{2n}(\cos\theta)}{n(2n+1)} \left(\frac{r}{r_0}\right)^{2n}. \tag{19.219}$$

The hydromagnetic equation for the azimuthal magnetic field is (19.195), which reduces to

$$\nabla^2 \mathbf{e}_\phi B_\phi = -\frac{\mathbf{e}_\phi}{\eta} B_r r \sin\theta \Delta\Omega \delta(r - r_1)$$

for the non-uniform rotation (19.216). In terms of the Green's function (19.208), then,

$$B_\phi(r, \theta) = -\tfrac{1}{2}\Delta\Omega r_1^2 \sum_{n=1}^{\infty} \left\{\left(\frac{r_1}{R}\right)^{2n+1} - 1\right\}\left(\frac{r}{r_1}\right)^n \frac{P^1_n(\cos\theta)}{n(n+1)}$$
$$\times \int_0^\pi d\theta_1 \sin^2\theta_1 P^1_n(\cos\theta_1) B_r(r_1, \theta_1) \tag{19.220}$$

for the region $0 < r < r_1$ inside the interior of the shell of velocity shear. The field B_r is

$$B_r = \frac{1}{r\sin\theta} \frac{\partial}{\partial\theta}(\sin\theta A_\phi).$$

Hence, for the modes with even (quadrupole) symmetry about the equatorial plane, it follows from (19.219) that, for $r > r_0$,

$$B_r(r, \theta) = \frac{a^2\Gamma \sin\theta_0}{2\eta r \sin\theta} B_\phi(r_0, \theta_0) \sum_{n=1}^{\infty} \left(\frac{r_0}{r}\right)^{2n+1} \frac{P^1_{2n}(\cos\theta_0)}{(2n+1)}$$
$$\times \{P^1_{2n+1}(\cos\theta) - \cos\theta_0 P^1_{2n}(\cos\theta)\}. \tag{19.221}$$

Note, then, that

$$\int_{-1}^{+1} d\mu \{P^1_{2n+1}(\mu) - \mu P^1_{2n}(\mu)\} P^1_m(\mu)$$
$$= \frac{4(2n+1)^2(n+1)}{(4n+3)(4n+1)} \delta_{m,2n+1} + \frac{4n(4n^2-1)}{16n^2-1} \delta_{m,2n-1}. \tag{19.222}$$

Substituting (19.211) into (19.220) and performing the integration over θ_1 with the aid of (19.222), it follows that

$$B_\phi(r, \theta) = -\frac{\Gamma_0 \Delta\Omega a^2 r_1}{\eta^2} B_\phi(r_0, \theta_0) \sin\theta_0 \times \sum_{n=1}^{\infty} \left[\frac{P^1_{2n+1}(\cos\theta) P^1_{2n}(\cos\theta_0)}{(4n+1)(4n+3)} \left\{ \left(\frac{r_1}{R}\right)^{4n+3} - 1 \right\} \frac{r_0^{2n+1} r^{2n+1}}{r_1^{4n+2}} - \frac{P^1_{2n-1}(\cos\theta) P^1_{2n}(\cos\theta_0)}{(4n+1)(4n-1)} \left\{ \left(\frac{r_1}{R}\right)^{4n-1} - 1 \right\} \frac{r_0^{2n+1} r^{2n-1}}{r_1^{4n}} \right]. \quad (19.223)$$

This relation must be satisfied by $B_\phi(r, \theta)$ at all positions in $r \leqslant r_1$. In particular, when the cyclones (r_0, θ_0) are located inside the shear $(r_0 < r_1)$ the relation must be valid at (r_0, θ_0). If $B_\phi(r_0, \theta_0)$ is not zero, then it may be cancelled from the relation, so that (19.223) reduces to the requirement

$$1 = +N_D \sin\theta_0 \sum_{n=1}^{\infty} \left[\frac{P^1_{2n+1}(\cos\theta_0) P^1_{2n}(\cos\theta_0)}{(4n+1)(4n+3)} \left\{ 1 - \left(\frac{r_1}{R}\right)^{4n+3} \right\} \left(\frac{r_0}{r_1}\right)^{4n+2} - \frac{P^1_{2n-1}(\cos\theta_0) P^1_{2n}(\cos\theta_0)}{(4n+1)(4n-1)} \left\{ 1 - \left(\frac{r_1}{R}\right)^{4n-1} \right\} \left(\frac{r_0}{r_1}\right)^{4n} \right], \quad (19.224)$$

for the generation of a stationary quadrupole field, where N_D is the dynamo number $\Gamma_0 \Delta\Omega R^2 r_1/\eta^2$. This relation determines the eigenvalue N_D for the stationary operation of the dynamo.

For odd (dipole) symmetry, it can be shown that the azimuthal field throughout $r < r_1$ is given by

$$B_\phi(r, \theta) = +N_D \sin\theta_0 B_\phi(r_0, \theta_0) \times \sum_{n=1}^{\infty} \left[\frac{P^1_{2n+1}(\cos\theta_0) P_{2n}(\cos\theta)}{(4n+1)(4n+3)} \left\{ \left(\frac{r_1}{R}\right)^{4n+1} - 1 \right\} \frac{r_0^{2n+2} r^{2n}}{r_1^{4n+2}} - \frac{P^1_{2n-1}(\cos\theta_0) P^1_{2n}(\cos\theta)}{(4n+1)(4n-1)} \left\{ \left(\frac{r_1}{R}\right)^{4n+1} - 1 \right\} \frac{r_0^{2n} r^{2n}}{r_1^{4n}} \right]. \quad (19.225)$$

Hence, if the cyclones are located inside the shell of shear $(r_0 < r_1)$ (19.225) must be satisfied at (r_0, θ_0), leading to the relation

$$1 = +N_D \sin\theta_0 \sum_{n=1}^{\infty} \left[\frac{P^1_{2n+1}(\cos\theta_0) P^1_{2n}(\cos\theta_0)}{(4n+1)(4n+3)} \left\{ \left(\frac{r_1}{R}\right)^{4n+1} - 1 \right\} \left(\frac{r_0}{r_1}\right)^{4n+2} - \frac{P^1_{2n-1}(\cos\theta_0) P^1_{2n}(\cos\theta_0)}{(4n+1)(4n-1)} \left\{ \left(\frac{r_1}{R}\right)^{4n+1} - 1 \right\} \left(\frac{r_0}{r_1}\right)^{4n} \right] \quad (19.226)$$

for the generation of a stationary dipole field.

Note that the series on the right-hand sides of (19.224) and (19.226) are real, and may be presumed to be non-vanishing for almost all locations (r_0, θ_0) of the cyclones. Hence there are eigenvalues N_D for almost all (r_0, θ_0). That is to say, the dipole field and the quadrupole field can be regenerated by suitable positive and negative N_D no matter where the cyclones are located. The question is then the sign and the magnitude of N_D to produce the generation. For Earth it is believed that N_D is positive, so we shall be particularly interested in the stationary fields for that circumstance.

Levy (1972a) went on to show that, if the shear lies immediately inside the surface of the region, $r_1 = R - \epsilon$ where $\epsilon \to 0$, then the series on the right-hand sides of (19.224) and (19.226) can be summed and expressed in closed form in terms of an integral of an algebraic function of trigonometric functions. The integral can be evaluated by conventional numerical techniques, giving the dynamo number as a function of the location (r_0, θ_0) of the cyclones. The summation technique is outlined in the original paper. The numerical results for the dynamo number $N_D = \Gamma_0 \Delta\Omega R^3/\eta^2$, equal to the reciprocals of the sums, are displayed in Figs. 19.17 and 19.18 for the dipole and quadrupole fields, (19.226) and (19.224), respectively, when N_D is positive. It is evident from Fig. 19.17

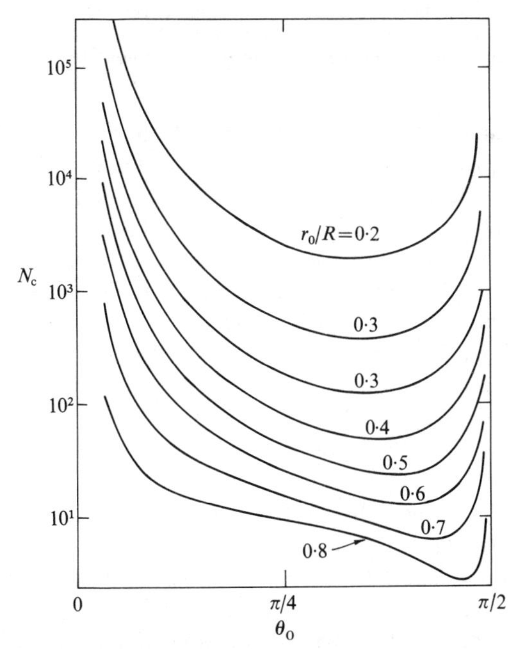

FIG. 19.17. The critical dynamo number $N_D > 0$ for the production of a stationary dipole, plotted as a function of the angular position θ_0 of the ring of cyclones for the radial positions r_0/R indicated on each curve (reproduced from Levy 1969a).

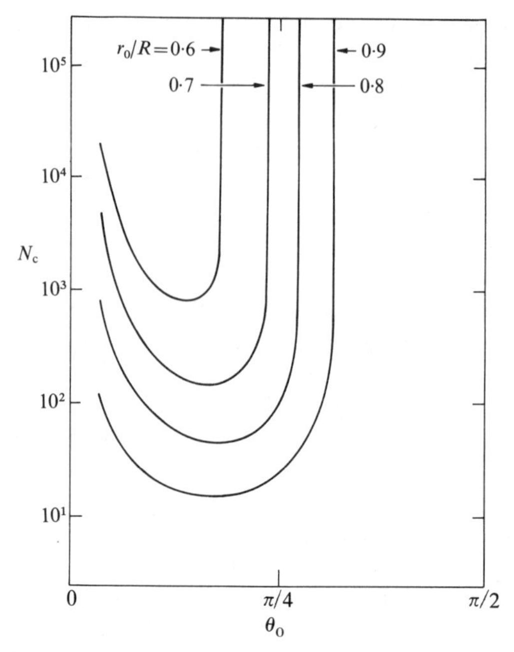

FIG. 19.18. The critical dynamo number $N_D > 0$ for the production of a stationary quadrupole. plotted as a function of the angular position θ_0 of the ring of cyclones for the radial positions r_0/R indicated in each curve (reproduced from Levy 1969a).

that the steady dipole field can be generated no matter what the location of the cyclones, provided only that N_D is large enough. It is evident from Fig. 19.18 that the quadrupole field can be generated only if the cyclones lie at sufficiently high latitudes, in the unshaded area shown in Fig. 19.19. At low latitudes the sense of the toroidal field is such that ΓB_ϕ opposes the existing A_ϕ and no regeneration is possible. That is to say, the series on the right-hand side of (19.224) sums to positive values only for sufficiently small θ_0. For larger θ_0 the sum of the series is negative so that no solution is possible for positive N_D. Comparison of Figs. 19.17 and 19.18 reveals that a higher dynamo number is required to generate the quadrupole field than the dipole field for any given location of the cyclones.

Since the series on the right-hand side of (19.226) for the dipole field is positive for all (r_0, θ_0), it follows that a negative dynamo number N_D cannot regenerate a dipole field no matter where the cyclones are located inside the shell of shear. On the other hand, the quadrupole field is regenerated for negative N_D where the sum of the series in (19.224) is negative. Thus the low latitude locations of the cyclones that could not generate the quadrupole field with $N_D > 0$ are the positions for which the

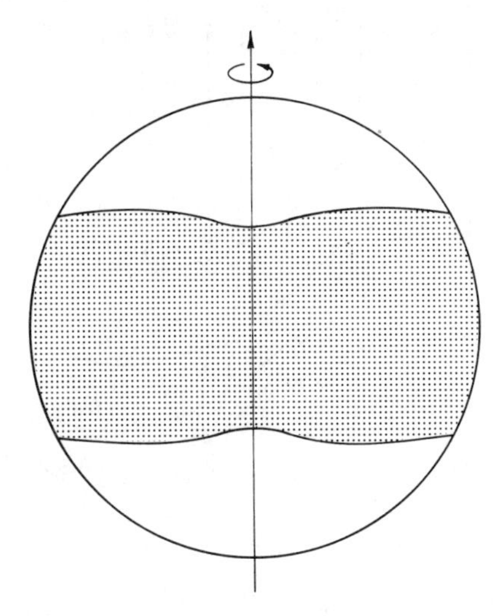

FIG. 19.19. The shaded area is the location of the ring of cyclones in the meridional plane for which $N_D>0$ cannot produce stationary quadrupole fields. It is the only region in which $N_D<0$ can produce stationary fields of quadrupole symmetry. A ring of cyclones in the unshaded area can generate both stationary dipole and quadrupole fields when $N_D>0$ and no stationary fields at all for $N_D<0$ (reproduced from Levy 1969a).

quadrupole field is generated by negative N_D. The dynamo number is given for the quadrupole field for $N_D<0$ in Fig. 19.20.

Altogether, the situation is only a little different from the two rings of shear treated in §19.4.1, because the basic principles are the same, i.e. that the azimuthal field B_ϕ at the location of the cyclones should yield ΓB_ϕ with the same sign as A_ϕ at that location. The generalization of the calculations to more than one symmetric pair of cyclones was carried out by Levy (1972c) in connection with questions on the reversal of the geomagnetic field.

19.4.2.2. Periodic fields. Stix (1973) has extended the calculations to the time-dependent dynamo equations

$$\left\{\frac{\partial}{\partial t}-\eta\left(\nabla^2-\frac{1}{r^2\sin^2\theta}\right)\right\}A_\phi=-\frac{R^2}{r_0}\Gamma_0B_\phi\{\delta(\theta-\theta_0)-\delta(\theta-\pi+\theta_0)\}\delta(r-r_0),$$

$$\left\{\frac{\partial}{\partial t}-\eta\left(\nabla^2-\frac{1}{r^2\sin^2\theta}\right)\right\}B_\phi=\Delta\Omega r_1B_r\delta(r-r_1)$$

for the shell (19.216) and rings of cyclones at $\varpi=a=r_0\sin\theta_0$, $z=\pm h=\pm r_0\cos\theta_0$. The time-dependent problem calls for an entirely different technique from the solution of the stationary equations (19.194) and

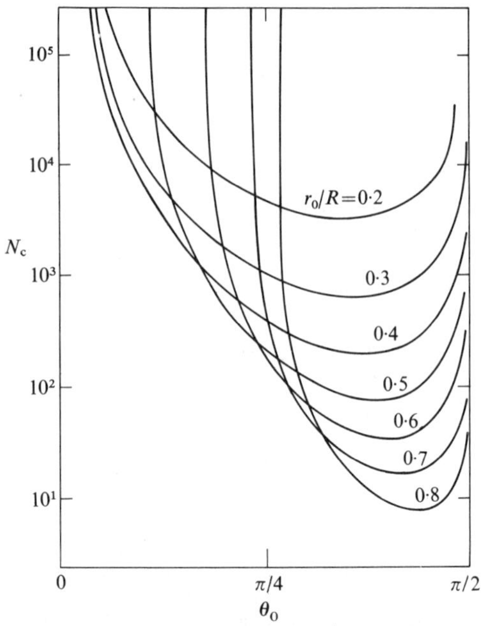

FIG. 19.20. The critical dynamo number $N_D<0$ for the production of a stationary quadrupole, plotted as a function of the angular position θ_0 of the ring of cyclones for the radial position r_0/R indicated on each curve (reproduced from Levy 1969a).

(19.195). Stix turned to the variational principle of Lerche (1972) (employed in §19.3.2 to treat the dynamo with horizontal shear, Lerche and Parker 1972).

Stix expanded the fields in terms of the functions $\exp(\lambda t) j_n(k_{pq}r) P_n^1(\cos\theta)$ where $q=n$, $n\pm1$, and $k_{nq}R$ is the qth zero of j_n so as to satisfy the boundary conditions at the surface of the sphere, more or less along the lines originally employed by Elsasser (1946, 1947, 1950, 1955, 1956). The calculations show clearly the relation of the stationary solutions to the periodic solutions with steady amplitudes. Stix expressed the results in terms of the parameter $R_c=\{|N_D|\,R^3/r_0(R-r_1)\}^{\frac{1}{2}}$, which is essentially the square root of the dynamo number N_D, employed by Levy (1972a). Figure 19.21 shows R_c for positive dynamo number as a function of the angular position θ_0 of the rings of cyclones, while Fig. 19.22 shows R_c for negative dynamo number. The plots are for cyclones at $r_0=\frac{1}{2}R$ and the shear at $r_1=0{\cdot}98R$. The stationary dipole and quadrupole are shown (see Fig. 19.17) so that the relative location of the lowest oscillatory state—the quadrupole—can be seen. The important point is that the stationary dipole is the lowest mode for positive dynamo number. The quadrupole lies somewhat higher (note that the vertical scale is linear in Figs 19.21 and 19.22 rather than logarithmic).

For negative dynamo number the situation is quite different. The stationary quadrupole can be generated only by cyclones above about 60° latitude ($\theta < 30°$), as pointed out by Levy. Stix showed that low latitude cyclones, that cannot produce a steady field of either symmetry, produce instead an oscillatory quadrupole, and at a substantially smaller (negative) dynamo number than the stationary quadrupole when the cyclones are located at high latitudes. Thus the lowest mode for negative N_D is not a stationary field but an oscillatory one, in the form of a quadrupole.

It is believed that the dynamo number in the core of Earth is positive. Hence the observed stationary dipole is the lowest mode. It is evident from Fig. 19.21 that the next higher mode for cyclones at low or middle latitudes is the periodic quadrupole, at about four times the dynamo number N_D necessary to produce the stationary quadrupole. Thus, if the core of Earth convected more vigorously, we might find ourselves residing in an alternating quadrupole field with a period of the general order of 10^3 years. The magnetic compass would be of little use in both low and high latitudes, and the rapid change of the field would require revision of the charts for navigational purposes every ten years or so. Field reversals would occur on a millenial basis. It was Braginskii (1964) who first made

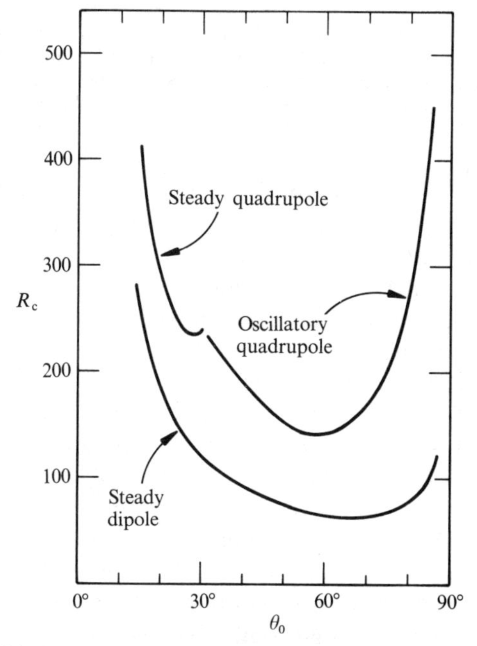

FIG. 19.21. The critical dynamo number R_c for positive N_D for the production of a periodic quadrupole (oscillatory with steady amplitude) plotted as a function of the angular position of the ring of cyclones for shear at $r_1/R = 0{\cdot}98$ and the cyclones at $r_0/R = 0{\cdot}5$ (reproduced from Stix, 1973).

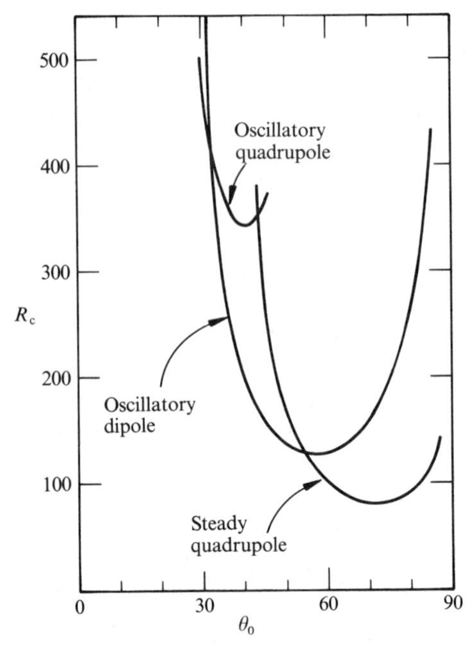

FIG. 19.22. The critical dynamo number R_c for negative N_D for the production of a stationary quadrupole and for the production of a periodic dipole and a periodic quadrupole (oscillatory with steady amplitude) plotted as a function of the angular position θ_0 of the ring of cyclones for the shear at $r_1/R = 0{\cdot}98$ and the cyclones at $r_0/R = 0{\cdot}5$ (reproduced from Stix 1973).

the point that the close proximity (in terms of N_D) of the stationary and oscillatory states provides a possibility for reversal of the geomagnetic field through fluctuations in the dynamo number.

It is believed that the dynamo number is negative in the convective zone of the sun. Hence the only stationary field is a quadrupole at the dynamo numbers already noted in Fig. 19.20. The stationary quadrupole is possible, however, only for cyclones at suitably low latitudes, as shown by Figs. 19.19 and 19.22 (for $r_0 = \frac{1}{2}R$). It was pointed out in §19.3.1 that the stationary states disappear, leaving only periodic states, when the dynamo is confined to a thin layer (Parker 1971b). For negative dynamo number the lowest state is then the oscillatory dipole. Hence the observed oscillatory field of the sun would appear to be the lowest available mode for the thin convective zone. Stix points out, however, that it is more than just the thickness of the dynamo shell that restricts the field to a periodic dipole. It is also the fact that both the cyclones and the shear are concentrated toward the surface of the sphere. In that location they produce the periodic dipole whether the fields are confined to a thin shell or have available the entire sphere. It should be noted, too, that the cyclonic rotation of the convection of a stratified atmosphere is proportional to the radial component of the angular velocity of the body as a whole, i.e. proportional to cos θ. Hence the cyclonic convection in the sun lies principally at higher latitudes and a stationary quadrupole might not be possible (see Fig. 19.22) even if the convective zone extended to greater depths. Hence if $\partial\Omega/\partial r$ lies near the surface, it would appear that only oscillatory fields are possible in the sun. Note then, that the oscillatory dipole lies at lower dynamo number than the oscillatory quadrupole unless the cyclones are above about 60° latitude, where the oscillatory quadrupole is favoured. Thus for cyclones at middle latitudes the lowest available mode is the oscillatory dipole. At high latitudes and large negative dynamo number, it is possible to excite an oscillatory quadrupole. Leighton (1969) was the first to point out this possibility, based on his numerical experiments with the solar dynamo. He noted the intermittent excitation of the oscillatory quadrupole, together with the oscillatory dipole, to produce an occasional strong north–south asymmetry, as is often observed in the fields of the sun.

Stix (1973) presents the critical value for the dynamo number R_c in a different fashion, shown in Fig. 19.23, where contours of constant positive dynamo number are plotted for the stationary dipole. Note the invariance of the result to interchange of r_0 and r_1, $R_c(r_0, r_1) = R_c(r_1, r_0)$. Stix points out that this symmetry of the eigenvalue follows directly from the symmetry of the equations, even though the fields are quite different upon interchange of r_0 and r_1 because of the asymmetry of the boundary conditions (B_ϕ vanishes at the surface $r = R$, but A_ϕ does not).

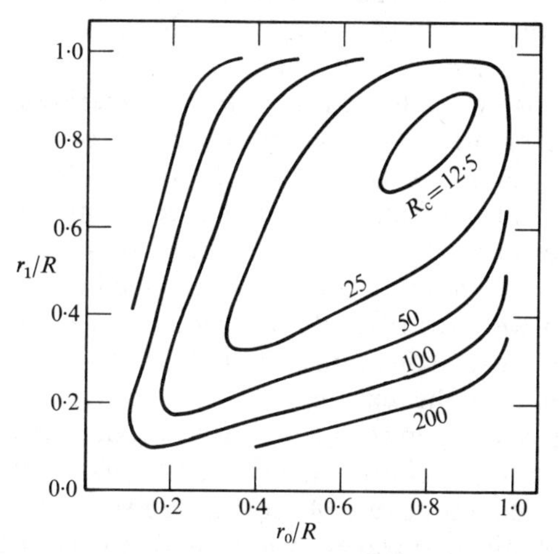

FIG. 19.23. Contours of constant critical dynamo number in the $(r_0/R, r_1/R)$ plane for the steady dipole field produced by $N_D > 0$ when the ring of cyclones is located at $\theta_0 = \frac{1}{4}\pi$ (reproduced from Stix 1973).

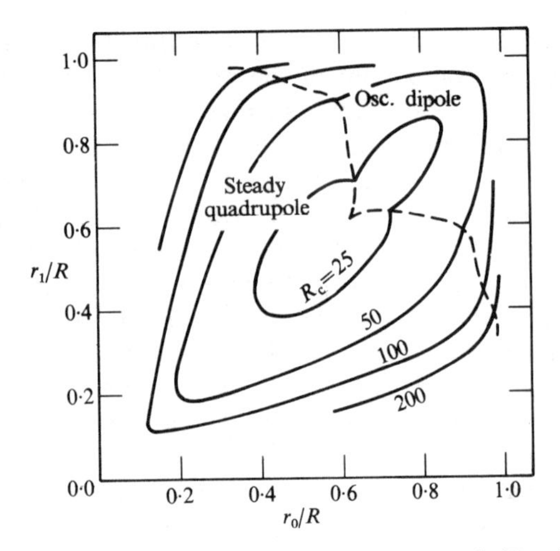

FIG. 19.24. Contours of constant critical dynamo number in the $(r_0/R, r_1/R)$ plane for the steady quadrupole field and the oscillatory dipole field produced by $N_D<0$ when the ring of cyclones is located at $\theta_0=\frac{1}{4}\pi$. The dashed curve separates the regions where the critical dynamo number is lower for the steady quadrupole or the periodic dipole.

When the dynamo number is negative, the contours of constant R_c are again symmetric about $r_0=r_1$, of course, but there are both stationary and oscillatory fields. The contours of constant R_c are plotted in Fig. 19.24, with the oscillatory dipole occurring when r_0 and r_1 are both greater than about 0·6. The steady quadrupole is most easily excited in the vicinity of $r_0=r_1=0{\cdot}55\ R$, while the periodic dipole of constant amplitude is most easily excited in the viciniy of $r_0=r_1=0{\cdot}75\ R$.

The question of periodic versus stationary solutions was first introduced by Braginskii (1964) who showed the effects of meridional circulation in stabilizing or destabilizing stationary solutions for both positive and negative dynamo numbers (Braginskii 1970a, b). Indeed, for positive dynamo number and a broad, rather than a concentrated, distribution of cyclones over the sphere Roberts (1972) found no stationary solutions. It is clear in terms of the physical principles described in §19.4.1 above, that Robert's cyclones were distributed throughout the azimuthal field in such a way that, under steady conditions, $\Gamma(r,\theta)B_\phi(r,\theta)$ generated a net A_ϕ of the opposite sense to the existing A_ϕ. Or, conversely, the shear was distributed relative to A_ϕ in such a way as to produce reversed B_ϕ. There were oscillatory states, of course, and Roberts found that the introduction of meridional circulation of suitable sign gave rise to stationary states. In terms of the basic dynamo waves, it is clear that part of the stabilization effect arose because the circulation opposed the migration of the waves sufficiently for the waves to be trapped by the geometry to produce stationary fields. Another facet of the circulation effect is that it redistributed both the azimuthal and meridional fields relative to the fixed location of the shear and the cyclones. Roberts noted that meridional circulation can enhance the efficiency of the dynamo if it transports B_ϕ from the site of production to the centre of the cyclonic action, without at the same time hindering the arrival of poloidal field, from which B_ϕ is produced, at the region of shear. Roberts suggested that the meridional circulation is a necessary part of a stationary dynamo such as the geomagnetic dynamo. The conjecture is not correct in general, in view of the many known stationary solutions of the dynamo equations (Braginskii 1964; Parker 1969a; Levy 1972a, b, c; Stix 1972), but, if interpreted in a narrower context, it makes an important physical point. Suppose, for instance, that the distribution of shear and cyclones in the core of Earth is such that they cannot generate a stationary dipole (Roberts's model is an example of such a distribution) but there is more or less coincidentally a meridional flow of such a nature that the cyclones and shear are coerced into generating a stationary dipole. Then suppose that the meridional flow is interrupted or altered for a brief time. During that period the dynamo may generate an oscillatory field so that, when the meridional circulation is restored, the new stationary field may be reversed relative to the earlier stationary field. It is one more of the many theoretical possibilities for an occasional reversal of the geomagnetic field.

While we are on the subject, note that the rectangular dynamo, treated in §19.3.1, reproduces the qualitative results of the dynamo with vertical (radial) shear in the spherical volume, including both the stationary dipole and quadrupole for negative and positive dynamo number, respectively, in a slab of sufficient thickness, and the preference for the oscillatory dipole mode when the dynamo number is negative. The similarity is gratifying and particularly interesting when we recall the great difference in the distribution of the cyclones and the shear within the dynamo region. When the dynamo number is positive, both the rectangular and the spherical dynamo clearly show the preference for the oscillatory quadrupole over the oscillatory dipole.

Historically, Braginskii's (1964) periodic solutions for the core of Earth were the first of their kind for the sphere, with the physical implications mentioned above. Steenbeck and Krause (1969a) were the first to cage the migratory dynamo waves (Parker 1955, 1957) within a spherical shell appropriate for the Sun. They used a continuous distribution of the dynamo coefficient, rather than shells or rings, and computed the complex radial wave number as a function of the dynamo number. Their pioneering work on periodic fields has been extended by Deinzer and Stix (1971) and by Stix (1973), as described above, and more recently by Deinzer *et al.* (1974) who confined the dynamo coefficient to a spherical shell, writing

$$\left\{\frac{\partial}{\partial t}-\eta\left(\nabla^2-\frac{1}{r^2\sin^2\theta}\right)\right\}A_\phi=\Gamma_0 R\delta(r-r_0)B_\phi\cos\theta$$

$$\left\{\frac{\partial}{\partial t}-\eta\left(\nabla^2-\frac{1}{r^2\sin^2\theta}\right)\right\}B_\phi=\Delta\Omega R\delta(r-r_1)\frac{\partial}{\partial\theta}A_\phi\sin\theta.$$

Their calculations were based on an expansion of the fields in terms of spherical Bessel functions and tesseral harmonics again. The results of the calculation show that cyclones and the shear distributed over spherical shells (of radius r_0 and r_1, respectively) have the same qualitative dynamo effects as when confined to rings. The only stationary modes have dipole symmetry when the dynamo number is positive and quadrupole symmetry when the dynamo number is negative. Deinzer *et al.* show that separating the two shells favours the stationary modes, with near juxtaposition of cyclones and shear leading to oscillatory modes. They treat a Gaussian radial distribution of cyclones and of shear to show that the effect is not a special consequence of the limitation to thin shells. Hence it is necessary to displace the location of the cyclones from the shear to obtain stationary solutions. This effect may explain in part why Roberts (1972) did not find stationary solutions, without a meridional flow to block the migration of the periodic dynamo waves. Deinzer *et al.* (1974) include a variety of graphs of critical dynamo numbers and, in particular, show extensive plots of both the poloidal lines of force and the contours of constant B_ϕ, thereby providing a graphic illustration of the form for the magnetic fields of a simple spherical dynamo (see also Bullard and Gubbins 1977).

There are a variety of papers providing examples with many different distributions of cyclones and shear. A study of their results, and particularly their extensive illustration of the poloidal lines of force and the distribution of B_ϕ for stationary dynamos, and at successive stages of the cycles of periodic dynamos, is an important part of the education of anyone with a professional interest in astronomical magnetic fields. Only the limitations of space in this volume prevent reproducing the many pages of graphic printout. Thus, for instance, Roberts and Stix (1972) employed the Elsasser–Bullard–Gellman formalism for treating the dynamo effect of stationary convective cells. They treated dynamos with cyclonic convection, with or without non-uniform rotation. They show, for instance, that the α^2-dynamo (cyclones without non-uniform rotation) in some circumstances has an asymmetric mode as its lowest state. They show that $\partial\Omega/\partial\theta$ is less effective than $\partial\Omega/\partial r$ in maintaining the solar magnetic fields, and that meridional circulation can play an important role in the time variation of the distribution of azimuthal field over latitude, i.e. in the butterfly diagram of the sunspot distribution.

The reader is referred to the original paper by Steenbeck *et al.* (1966) where the theoretical possibility of the α^2-dynamo was first pointed out. Busse (1975) has treated the α^2-dynamo from a semidynamical point of view recently. The calculations of α^2-dynamos show in general the very strong cyclonic convection that is necessary to maintain the field in the absence of non-uniform rotation.

Gubbins (1973a) has gone further with the Elsasser–Bullard–Gellman formalism, treating growing fields, $\exp(t/\tau)$. The extensive results, for various radial functions and orders (n, m) for the velocity provide a vivid portrayal of the fields produced by the dynamo action of steady convective motions with internal helicity. The exclusion of the field from the interior of the steady convective cells (see Chapter 16) is also illustrated by the graphic results presented in Gubbin's paper. Kumar and Roberts (1975) treat the dynamo effect of discrete upwelling cells of cyclonic fluid.

The important aspect of these calculations is their complementarity to the turbulent dynamo, outlined in the sections above. The dynamo effect of both stationary and rapidly fluctuating cyclonic motion is illustrated now with formal examples. The motions may deviate but little from the axi-symmetric non-uniform rotation of Braginskii's formulation, or there may be no axi-symmetric non-uniform rotation at all, as in Roberts and Stix (1972) and Gubbins (1973a). The breadth of the treatment shows the general character of the dynamo effect, wherever there is cyclonic fluid motion.

Braginskii (1964, 1970a, b) has provided a number of examples, with particular attention to the geomagnetic dynamo. Steenbeck and Krause (1969b) treat dynamos with steady fields. Braginskii (1970b), Roberts (1972), and Kumar and Roberts (1975) present numerical models of dynamos with meridional circulation, producing stationary fields. Braginskii (1970a), Stix (1972), and Jepps (1975) have worked out dynamos in which the dynamo coefficient is suppressed as the field increases. The result is, of course, a bounded growth of the field, with the dynamo coefficient sinking to a mean value just sufficient to maintain a stationary state or an oscillatory state with fixed amplitude. The solution of the equations for steady operation is no longer an eigenvalue problem, as it is in the purely kinematical circumstances treated above. Stix points out that the fluctuation of the dynamo coefficient Γ with the strength of the field can have the effect of increasing the frequency of oscillations of the field.

One of the most interesting results of all is the fact turned up by Krause (1971), Stix (1971), and Roberts and Stix (1972) that there are solutions of the dynamo equations for *axi-symmetric* distributions of cyclones and shear that are themselves without rotational symmetry about the axis of the dynamo. Some of the asymmetric solutions are *excited at lower dynamo number than any of the axi-symmetric solutions*, suggesting that there are circumstances in nature in which the asymmetric solutions may appear instead of the more convenient axi-symmetric fields. This raises the possibility that the very poor 'axial' symmetry of the observed solar magnetic fields may be more than just statistical fluctuations about an axi-symmetric mean. We do not propose to explore the problem here, but it is clear that it needs further attention.

Babcock (1961) and Nakagawa and Swarztrauber (1969) display specific examples of the growth and evolution of the azimuthal field derived from the interaction of the non-uniform rotation $\Omega(r, \theta)$ with the poloidal field. Köhler (1966) and Durney (1972) have provided examples of dynamos operating in spherical shells under conditions of interest for the sun, including meridional circulation and an inward increase of the angular velocity. Gilman (1969), Leighton (1969), Yoshimura (1972, 1975a, b), Durney (1975), and Köhler (1973) have provided theoretical examples of dynamos with an eye to direct application to the sun. Hence we defer their discussion to §21.2, where the magnetic properties of the sun are described in detail.

Extensive reviews of the generation of magnetic fields by various distributions of shear and cyclones are available (Elsasser 1950, 1955, 1956a, b; Rikitake 1966; Parker 1970; Roberts 1971; Weiss 1971; Roberts and Soward 1972, 1975; Vainshtein and Zeldovich 1972; Gubbins 1974; Moffatt 1976, 1978; Soward and Roberts 1976). The reviews of Roberts and Soward and of Moffatt in particular, go into the mathematical aspects of the dynamo equations, generally overlooked in the present discussion of the basic physics.

We emphasize again that the theory of the hydromagnetic dynamo has come through the exploratory stage. Collectively the work to date has established the generality of the dynamo effect in any fluid motion that is not free of helicity. This fundamental theoretical development explains the remarkably general occurrence of the magnetic fields in the rotating bodies that populate the astronomical universe. The physical principles by which the fluid motion generates the field are now understood. What remains to be done are the detailed models for each astronomical object, or class of objects. That task gets into the details of the fluid motions, which are presently not altogether known to us. Indeed, to appreciate the complexity of the problem it is instructive to review some of the alternatives that might have been pursued in the exposition of the physical principles thus far. To begin, we may ask ourselves whether the non-uniform rotation in the core of Earth is as strong as the elementary considerations on Coriolis forces would suggest, yielding strong azimuthal fields and reducing the dynamo equations to the simple form (18.4) and (18.7), (19.21) and (19.22). Or is the non-uniform rotation really not very strong, so that we should use the complete equations (18.37) with (18.38) and (18.39) reducing only to (19.4) and (19.5) in rectangular geometry? Or is the non-uniform rotation so weak that the shear terms $\partial V_i/\partial x_j$ can be dropped entirely from (18.37) and the dynamo is sustained by the small-scale cyclonic convection alone—the α^2-dynamo of Steenbeck *et al.* (1966)? What is the most realistic form for the small-scale cyclonic convection? The idealized rotating cells employed in §18.3.3 are clearly an oversimplification, to facilitate the calculation of $\xi_i\ \partial\xi_j/\partial x_k$ and to illustrate the basic physical principles. Is the convection in the core of Earth, in the sun, or in the gaseous disc of the galaxy, more of the nature of Rossby waves (Gilman 1969) or random inertial waves (Moffatt 1970; Soward 1975) or random turbulence (Vainshtein 1970, 1972a; Vainshtein and Vainshtein 1973)? Or is it more of the nature of acoustic turbulence (Vainshtein 1971)? Or are the convective cells really much too large to treat them statistically, as most authors do, requiring instead direct attention to the individual convective cell (Bullard and Gellman 1954; Gilman 1969; Gubbins 1973a; Kumar and Roberts 1975)? Or is the convection confined to cylindrical convective rolls with very little non-uniform rotation and azimuthal field (Busse 1973)? The azimuthal field, hidden away in the core and unobservable at the surface of Earth is, from the entirely pragmatic view, a waste of energy, and it will be shown in the next chapter that, with present estimates of the energy available to the core, there is not much to spare in accounting for the generation of the field. It would be a convenience if the azimuthal field were weak. But the azimuthal field may be strong nonetheless, so that the simple dynamo equations (18.4) and (18.7) (Parker 1955) or the Braginskii formulation (Braginskii 1964, 1970a, b; Soward 1972; Gubbins 1973b) is more appropriate. Certainly the azimuthal field in the gaseous disc of the galaxy is the dominant field. The poloidal field has not yet emerged from the noise in the data. The bipolar regions appearing at the surface of the sun indicate that the azimuthal field of the sun is also the dominant field. The azimuthal field of the sun must be of the order of 10^2 G or more to contain the flux coming up through the surface in one 11-year sunspot cycle, whereas the poloidal field of the sun is no more than 5 or 10 G. So the reduced dynamo equations would seem appropriate. On the other hand, the circumstances in the core of Earth are less well known, because we cannot say what the azimuthal field is. The poloidal field is evidently about 5 G there. These observational facts, together with the theoretical fact that the combination of non-uniform rotation and cyclonic convection is the lowest order dynamo effect (and hence the most efficient dynamo effect), have prompted this author (and others) to present the theory in the context of strong non-uniform rotation. We are inclined to let higher order terms lie sleeping until we have a specific need dictated by observations.

The statistical treatment of the individual eddies, waves, or convective cells, assumes that they are of small scale and short life compared to the large-scale fields. This is not an extremely accurate representation for the dozen convective cells identifiable in the core of Earth from maps of the surface field. Nor, on the other hand, does it seem to provide the opportunity for any interesting qualitative error. Hence, rather, than belabour the details, it is more interesting and useful to pass on to other questions.

The next step in the present writing is to consider the physics of the dynamo in the context of the individual planets, stars, and galaxies where the fields are produced and observed. To what extent can we infer the fluid motions hidden within from the fields observed extending through the surface? One might hope that one day the dynamical theory of convection within rotating spherical bodies will develop to where it can specify the precise fluid motions to be expected within each astronomical body. The ultimate achievement would be the deduction of the magnetic field from the theoretical dynamical flows, including the Lorentz forces of the field, predicting external magnetic fields resembling those observed. But that is more an ideal than a serious expectation for the foreseeable future. Not only are there serious questions as to the form of the energy sources that drive the internal meteorology of planets and stars, but the dynamical consequences of the energy are so sensitive to subtle aspects of the boundary conditions as to preclude unique results. Hydrodynamics at anything but the lowest Reynolds number is highly non-unique. But these questions lie far in the future. In the next chapter we summarize the present state of 'negative ignorance' of the magnetic fields of astronomical bodies. It is gratifying that so much can be stated. Progress has been rapid in the last decades and hopefully will continue at the same brisk pace for some time to come. Any failure to achieve the final idealistic goal of a completely deductive theory is less a consequence of the finite rate of advance as it is of the excessive distance to that goal.

References

Babcock, H. W. (1961). *Astrophys. J.* **133,** 572.
Badwar, G. D. and Stephens, S. A. (1977). *Astrophys. J.* **212,** 494.
Braginskii, S. I. (1964). *Geomagn. i. Aeronomiya* (USSR) **4,** 732; *Geomagn. and Aeron.* **4,** 572.
—— (1970a). ibid. **10,** 172.
—— (1970b). ibid. **10,** 221.
Bullard, E. C. and Gellman, H. (1954). *Phil. Trans. Roy. Soc. Lond. A* **247,** 213.
—— and Gubbins, D. (1977). *Geophys. Astrophys. Fluid Dynamics* **8,** 43
Bumba, V. and Howard, R. F. (1965). *Astrophys. J.* **141,** 1492.
Busse, F. H. (1973). *J. Fluid Mech.* **57,** 529.
—— (1975). *Geophys. J. Roy. Astron. Soc.* **42,** 281.
—— (1976). *Phys. Earth Planet. Inter.* **12,** 350.

Deinzer, W., v. Kusserow, H. U., and Stix, M. (1974). *Astron. Astrophys.* **36,** 69.
—— and Stix, M. (1971). *Astron. Astrophys.* **12,** 111.
Durney, B. R. (1972). *Solar Phys.* **26,** 3.
—— (1975). *Astrophys. J.* **199,** 761.
Elsasser, W. M. (1946). *Phys. Rev.* **69,** 106; **70,** 202.
—— (1947). ibid. **72,** 821.
—— (1950). *Rev. Mod. Phys.* **22,** 1.
—— (1955). *Amer. J. Phys.* **23,** 590.
—— (1956a). ibid. **24,** 85.
—— (1956b). *Rev. Mod. Phys.* **28,** 135.
Gilman, P. (1969). *Solar Phys.* **8,** 316; **9,** 3.
Gubbins, D. (1973a). *Phil. Trans. Roy. Soc. Lond.* A **274,** 493.
—— (1973b). *Geophys. J. Roy. Astron. Soc.* **33,** 57.
—— (1974). *Rev. Geophys. Space Sci.* **12,** 137.
Hansen, R. T., Garcia, C. J., and Hansen, S. F. (1969). *Solar Phys.* **1,** 417.
Jepps, S. A. (1975). *J. Fluid Mech.* **67,** 625.
Jokipii, J. R. and Lerche, I. (1969). *Astrophys. J.* **157,** 1137.
—— Lerche, I., and Schommer, R. A. (1969). *Astrophys. J. Lett.* **157,** L119.
Köhler, H. (1966). *Mitt. Astron. Ges.* **21,** 91.
—— (1973). *Astron. Astrophys.* **25,** 467.
Krause, F. (1971). *Astron. Nachr.* **293,** 187.
Kumar, S. and Roberts, P. H. (1975). *Proc. Roy. Soc., Ser. A* **344,** 235.
Leighton, R. B. (1964). *Astrophys. J.* **140,** 1559.
—— (1969). ibid. **156,** 1.
Lerche, I. (1972). *Astrophys. J.* **176,** 225.
Lerche, I. and Parker, E. N. (1972). *Astrophys. J.* **176,** 213.
Levy, E. H. (1972a). *Astrophys. J.* **171,** 621.
—— (1972b). ibid. **171,** 635.
—— (1972c). ibid. **175,** 573.
—— (1974). ibid. **187,** 361.
Lockyer, N. (1931). *Mon. Notic. Roy. Astron. Soc.* **91,** 797.
Manchester, R. N. (1973). *Astrophys. J.* **188,** 637.
Moffatt, H. K. (1970). *J. Fluid Mech.* **44,** 705.
—— (1976). *Adv. appl. Mech.* **16,** 120.
—— (1978). *Magnetic field generation in electrically conducting fluids*, Cambridge University Press, Cambridge.
Morse, P. M. and Feshbach, H. (1953). *Methods of theoretical physics*, pp. 1108–1109. McGraw-Hill, New York.
Nakagawa, Y. and Swarztrauber, P. (1969). *Astrophys. J.* **155,** 295.
Parker, E. N. (1955). *Astrophys. J.* **122,** 293.
—— (1957). *Proc. Nat. Acad. Sci.* **43,** 8.
—— (1966). *Astrophys. J.* **145,** 811.
—— (1969a). ibid. **158,** 815.
—— (1969b). *Space Sci. Rev.* **9,** 651.
—— (1970). *Astrophys. J.* **160,** 389.
—— (1971a). ibid. **163,** 255.
—— (1971b). ibid. **164,** 491.
—— (1971c). ibid. **165,** 139.
—— (1971d). ibid. **166,** 295.
—— (1977a). ibid, **215,** 370.
—— (1977b). *Ann. Rev. Astron. Astrophys.* **15,** 45.

Rädler, K. H. (1968). *Z. Naturforsch. A* **23,** 1841, 1851.
Rikitake, T. (1966). *Electromagnetism and the earth's interior.* Elsevier, Amsterdam
Roberts, P. H. (1971). *Lectures in appl. Math.* **14,** 129.
—— (1972). *Phil Trans. Roy. Soc. A* **272,** 663.
—— and Soward, A. M. (1972). *Ann. Rev. Fluid Mech.* **4,** 117.
—— —— (1975). *Astron. Nachr.* **296,** 49.
—— and Stix, M. (1972). *Astron. Astrophys.* **18,** 453.
Smythe, W. R. (1936). *Static and dynamic electricity.* McGraw-Hill, New York.
Soward, A. M. (1972). *Phil. Trans. Roy. Soc. Lond. A* **272,** 431.
—— (1975). *J. Fluid Mech.* **69,** 145.
—— (1978). *Astron. Nachr.* **299,** 26.
—— and Roberts, P. H. (1976). *Magnitnaya Gidrodinamika* **1,** 3.
Spruit, H. C. (1974). *Solar Phys.* **34,** 277.
Steenbeck, M. and Krause, F. (1966). *Z. Naturforsch.* **21,** 1285.
—— —— (1967). *Magnitnaya Gidrodinamika* **3,** 19.
—— —— (1969a). *Astron. Nachr.* **291,** 49.
—— —— (1969b). ibid. 271.
—— —— and Rädler, K. H. (1966). *Z. Naturforsch. A* **21,** 369.
Stix, M. (1971). *Astron. Astrophys.* **13,** 203.
—— (1972). ibid. **20,** 9.
—— (1973). ibid. **24,** 275.
—— (1975). ibid. **42,** 85.
—— (1978) ibid, **68,** 459.
Vainshtein, L. L. and Vainshtein, S. I. (1973). *Geomagn. Aeron.* **13,** 123.
Vainshtein, S. I. (1970). *Sov. Phys. JETP* **31,** 87.
—— (1971). *Sov. Phys. Dokl.* **15,** 1090.
—— (1972a). *Sov. Phys. JETP* **34,** 327.
—— (1972b). *Sov. Astron.–AJ.* **15,** 714.
—— and Zeldovich, Ya. B. (1972). *Usp. Fiz. Nauk* **106,** 431; *Sov. Phys. Usp.* **15,** 159.
Weiss, N. O. (1971). *Quart. J. Roy. Astron. Soc.* **12,** 432.
White, M. P. (1978). *Astron. Nachr.* **299,** 209.
Yoshimura, H. (1972). *Astrophys. J.* **178,** 863.
—— (1975a). ibid. **201,** 740.
—— (1975b). ibid. **202,** 285.
—— (1975c). *Astrophys. J. Suppl.* **29,** (294) 467.

20

THE MAGNETIC FIELDS OF PLANETS

20.1. General remarks

THE physics of the generation of large-scale magnetic fields by the motion of conducting fluids has been described in the preceding chapters. The exposition concentrated on amplification of magnetic field by the combined effect of non-uniform rotation and cyclonic turbulence. The next step is to apply the theory to the observed magnetic fields of the various astronomical bodies. The theory is kinematical, in that the fluid motions are prescribed and the field is deduced therefrom. So we choose what seem to be plausible fluid motions based on simple qualitative dynamical arguments and on observations of surface motions. The formal theoretical dynamics of the fluid motions within convecting, rotating, spherical bodies has been strongly exercised in the past few years, and is presently in a state of rapid development. Even so, the enormity of the dynamical problem decrees that the ultimate development of the theory, to provide unique models for the fluid motion and the associated magnetic fields, lies veiled in the mists of the future. So we review what is presently known of the fields, the fluid motions, and the sources of energy that cause the motions. As we shall see, the fields appear to be accounted for by the expected motions so that the physical principles stand firm. But no unique, detailed, quantitative models can be constructed, even for so familiar an object as Earth. Indeed, the review suggests that the final word on quantitative models may not be spoken in the foreseeable future. But that is a conclusion to be reached only after considering the facts.

In the absence of a complete dynamical theory of the fluid motions, one works backwards from the observed motions and external magnetic fields to ask what are the simplest internal fluid motions that explain the fields. At the most elementary level, for instance, it can be asserted that any planetary body with a significant intrinsic magnetic field has an electrically conducting, convecting fluid core. For stars the question is not quite so simple because a primordial field may be present, firmly clutched in the conductivity of the stable radiative core. But an active stellar magnetic cycle implies an active dynamo, i.e. internal convection and non-uniform rotation. A galaxy such as our own cannot retain a primordial field, so the existence of a galactic field implies turbulence and non-uniform rotation, both of which are readily observed, of course. Hence for the galaxy the argument runs in the other direction, with the conclusion that the fluid motions readily account for the magnetic field. For planets and stars, however, we do not observe their interiors. We infer convection and non-uniform rotation from the theory of stellar interiors, but there is no direct proof except the theoretical success in fitting the mass, radius, and luminosities, if not the neutrino emission, of a variety of stars. The existence of an active magnetic cycle in a star indicates convection and non-uniform rotation over a considerable depth.

We begin the review with the planets. They are close at hand and accessible to snooping spacecraft, so that their fields are known in greater detail than stellar fields, with the exception of the sun. It is now known from measurements *in situ* that all the planets out to Jupiter have intrinsic magnetic fields, with the possible exception of Venus and Mars. Recent detection of sporadic non-thermal radio emission from Saturn indicates that it too probably has a magnetic field, of about 1 G at the equator (Brown 1975; Kaiser and Stone 1975).

The field of Earth has been directly studied from the time of Gauss, so that the secular variation is known in addition to the present spatial configuration. Paleomagnetic studies extend the record back in time for some 10^9 years. So Earth is the showplace for studies of planetary magnetism, thanks to its hospitable climate and the curiosity of the creatures the climate has spawned.

20.2. General facts on planetary interiors and magnetic fields

20.2.1. Earth

The magnetic field at the surface of Earth ($r = R_E$) is principally a dipole, with a magnetic moment of $8{\cdot}2 \times 10^{25}$ G cm^3, pointing southward. To be more precise, the dipole moment is inclined some 11·5° to the axis of rotation and is offset from the centre of the planet by about 450 km, or 0·07 R_E. The mean field at the equator is 0·310 G and about twice that much at the north and south magnetic poles. The second harmonic has a magnitude of 0·043 G at the surface, and the third, 0·029 G (Chapman and Bartels 1940; Dolginov 1976). Extrapolated to the surface of the liquid core ($R_c = 3483$ km $= 0{\cdot}54\ R_E$) the dipole field strength B_p is 4·9 G and the second and third harmonics are 0·7 G and 1 G, respectively. Hence, the core produces a field that is basically a dipole, but with large distortions. The higher harmonics are evident at the surface in the form of a dozen or more large-scale inhomogeneities, visible on any magnetic map of the surface field (cf. Chapman and Bartels 1940; Elsasser 1950, 1955, 1956a,b). The silicate mantle overlying the liquid metal core has enough electrical conductivity to impede the escape of the small-scale inhomogeneities from the surface of the core, so we may presume that the actual field in the core is more chaotic than the simple extrapolation of a potential field from the surface would imply.

The geomagnetic field above the surface of Earth extends outward into space for a distance of some 10 R_E in the direction of the sun, where it is blocked by the impact pressure of the solar wind. The magnetic field has been extensively mapped by spacecraft (Zmuda 1973; Cain 1975). In the anti-solar direction the field is drawn out into an extended magnetic tail. The activity of this irregularly shaped magnetosphere and its long tail is fascinating, with direct consequences for our environment. Its dynamical behaviour, with storms, substorms, Van Allen belts, aurora, equatorial electrojets, etc., goes beyond the simple hydromagnetic theory into the complexities of the collisionless plasma, and is a whole topic in itself (cf. Akasofu 1968; Dyer 1972; Akasofu and Chapman 1972; Kennel 1973; Gendrin 1975; Gurevich *et al.* 1976; McCormac 1976).

The secular variation of the geomagnetic field is well known. Briefly, the dipole moment is presently diminishing by about one part in 2×10^3 per year. Paleomagnetic studies indicate that it went through a maximum some 10^3 years ago. The dipole axis is precessing westward at about 0·04°/year, suggesting a mean precession period of 9×10^3 years. The local inhomogeneities drift westward at a rate of 0·18°/year, indicating one revolution in about 2×10^3 years. This corresponds to a westward drift of 0·3 mm s^{-1} at the surface of the core. The strengths of the individual local inhomogeneities rise and fall over characteristic times of the order of 10^3 years.

Runcorn (1955) made the remarkable—and for a time, highly controversial—discovery some years ago that the residual magnetism of the crustal rocks, and hence the geomagnetic field at the time of their crystallization, reversed abruptly at random intervals of 10^5–10^7 years over the past geological ages. The field reversals, when they occur, go quickly, in only 10^3 years or a little more. Between reversals the dipole has had an intensity and inclination comparable to the present values for at least $2{\cdot}7\times10^9$ years (Smith 1967; Kobayashi 1968; McElhinny and Evans 1968; Schwartz and Symons 1969; Cox 1975). The sudden reversals, now used in much the same way as tree rings to date the spreading of the sea floor from the mid-ocean ridges, have been extensively studied (Van Zijl *et al.* 1962; Cox and Doell 1964; Doell and Cox 1965; Doell *et al.* 1966; Ninkovich *et al.* 1966; Wilson 1966; Cox and Dalrymple 1967; Burlatskaja *et al.* 1968; Heirtzler *et al.* 1968; Bullard 1968; Cox 1970, 1972, 1973, 1975; Kono 1972; Cox *et al.* 1975; Larson and Helsley 1975).

The westward drift of the inhomogeneities in the geomagnetic field is generally interpreted (Elsasser 1950; Bullard *et al.* 1950) as an indication of the non-uniform rotation of the liquid core, with the angular velocity increasing inward toward the axis by some 0·3°/year. Hide (1966) and Braginskii (1967) have pointed out, however, that the inhomogeneities may also be thought of as Alfven waves in the azimuthal field of the core, rather than bulk motion of the fluid, so that other interpretations are possible (see reviews by Roberts and Soward 1972; Acheson and Hide 1973).

The physical properties of the liquid metal core are known from the detailed seismic studies of the last hundred years and from theoretical extrapolation of the properties of iron and nickel, together with their oxides and sulfides, to the enormous pressures and temperatures in the core. Briefly, the core radius is $R_c = 3483 \pm 2$ km (Engdahl and Johnson 1974). The radius of the inner solid core is 1240 ± 10 km (Gilbert *et al.* 1973). It is composed of some crystalline iron–nickel alloy, presumably. The mean density of the core is 11 g cm^{-3} (Haddon and Bullen 1969) as a consequence of the high pressure (10^{12} dyn cm^{-1}). The temperature is estimated (Higgins and Kennedy 1971) to be about 4000 K in the core. The thermal conductivity is estimated from the Franz–Wiedemann law to be about 2.5×10^6 erg/cm s K, so that the total heat outflow from the core is at least 10^{19} erg s^{-1} (10^{12} W) in order merely to maintain the necessary melting point gradient. The electrical conductivity is estimated (Elsasser 1950; Gardiner and Stacey 1971; Jain and Evans 1972; Johnston and Strens 1973) to be $(0{\cdot}6 \pm 0{\cdot}2) \times 10^{16}$ s^{-1} [$(7 \pm 2) \times 10^5$ mho/m]. Hence the resistive diffusion coefficient is $\eta \cong 10^4$ cm^2 s^{-1}. The magnetic Reynolds number $R_c v/\eta$ based on the core radius and a westward drift of $v =$ 0·3 mm s^{-1} is, then, 10^3. The resistive decay time of the dipole field is about 3×10^4 years as a consequence of Joule dissipation alone. It was pointed out in §18.4.4 that the eddy diffusivity η_T may be somewhat larger, of the order of 4×10^4 cm^2 s^{-1} if not reduced by negative diffusion. The decay time may, therefore, be as little as 10^4 years.

The interpretation of these basic hydromagnetic properties has been discussed in §18.1 and need not be repeated here. The basic fact is that the non-uniform rotation (see the discussion in §20·2.2 below) is expected to produce a strong azimuthal field. Each revolution of the westward drift (in about 2×10^3 years) shears the lines of force of the dipole field B_p and adds some 20 G to the azimuthal field. If the azimuthal field B_ϕ grows so large that the Lorentz forces, of the order of $B_p B_\phi / 4\pi R_c$, balance the Coriolis forces $\rho \Omega v$ dyn cm^{-3}, then with $B_p \cong 5$ G and a convective velocity of $v \cong 10^{-3}$ cm s^{-1}, B_ϕ might be as large as several hundred G (but see discussion in §20.3). Consider, then, the energies involved in the magnetic field in the core. The energy density of the dipole field of 5 G is 1 erg cm^{-3} and the stresses which it exerts alone are $B_p^2/8\pi = 1$ dyn cm^{-2}. The energy density of an azimuthal field of 10^2 G would be 400 times greater, or 400 erg cm^{-3}. An azimuthal field of 30 G has an energy density 36 times greater, or about 36 erg cm^{-3}. Thus, the principal energy is presumed to lie in the azimuthal field B_ϕ. The volume of the core is $1{\cdot}7\times10^{26}$ cm^3, so that an r.m.s. azimuthal field of 10^2 G represents 7×10^{28} erg. The characteristic resistive decay time of the azimuthal field is only a little less than for the dipole field, of the order of 10^{12} s (turbulent diffusion may reduce this to something of the order of 3×10^{11} s). Hence an energy input of at least 7×10^{16} ergs^{-1} $=7\times10^9$ W is required to maintain a field of 10^2 G. It is an interesting coincidence that 7000 MW is equivalent to the total heat output of a single large nuclear power plant (with an electrical power output of 3000 MW). If the azimuthal field is only 30 G, of course, the power is only 700 MW, and any one of a number of nuclear or coal fired plants could handle the load. The ohmic dissipation of the geomagnetic field, then, releases 700–7000 MW ($7\times10^{15}-7\times10^{16}$ erg s^{-1}) of heat into the core, which is at most comparable only to the heat released into the atmosphere from a *single* big nuclear plant.

By comparison, the total heat flow out of the crust into the atmosphere of Earth is 3×10^{20} erg s^{-1} (Stacey 1969) comparable to the heat from 5000 of the biggest nuclear plants. All of these figures are miniscule when compared with the sunlight falling on Earth ($1{\cdot}4\times10^6$ erg cm^{-2} s^{-1} or 2×10^{24} erg s^{-1}). Altogether, then, the magnetic field of Earth involves a power input of the general order of 10^{16}–10^{17} erg s^{-1} or about 10^{-4} of the total heat flow through the crust.

Now the heat flow of 3×10^{20} erg s^{-1} through the crust of Earth is caused largely by the radioactivity in the crustal rocks. An unknown, and much debated, portion of the radioactivity resides in the core. Presumably it is the decay of K^{40}, in solution with iron sulfide in the core perhaps, that heats the core (Lewis 1971), but more will be said on this in §20.2.2. We go on to survey the moon and other planets to get an idea of the most crucial problems in planetary magnetism before discussing the energy problem.

The moon, as expected from its low density and low internal temperature, is constructed principally of silicates and so has no intrinsic magnetic field of its own. It is surprising to find, therefore, that the rocks on the surface of the moon show strong fossil magnetism, *implying* that they solidified in the presence of a steady (non-oscillatory) field of the order of a G. It is not clear that either Earth or the sun could have provided so strong a field. The effect, so far, defies explanation (cf. Coleman *et al.* 1972; Levy 1972c; Gose and Butler 1975; Dolginov 1976) and raises a variety of questions about conditions in the solar system some 3–4$\times10^9$ years ago when the present surface of the moon was formed. The question is taken up below, following a comparison of Earth and Mercury.

20.2.2. Venus

The sister planet of Earth may be presumed to have an internal structure closely resembling that of Earth. Venus has but little rotation however. The traditional figure for the period of the rotation of Venus, to be found in the standard text books until about a decade ago, was two weeks, with the same sense as Earth. The idea was based firstly on the supposition that Venus rotates with the same sense as Earth, there being no obvious reason to assume the contrary. The rate was based, then, on the absence of Doppler shifts of light from the rim, giving a *lower* limit of about two weeks for the period, and an absence of strong cooling of the night side of Venus, giving what was thought to be an *upper* limit of about two weeks. Hence, two weeks was a unique result. Radar studies of Venus show now that its rotation is retrograde, with a period of 244 days. The subsequent landings of spacecraft from the USSR have shown an atmosphere of CO_2 100 times more massive than the atmosphere of Earth, and with such strong winds as to keep the dark side amply supplied with heat. On the basis of present dynamo theory, in which both the non-uniform rotation and the strength of the cyclonic convection in the core are proportional to the angular velocity Ω of the planet as a whole, one would expect the magnetic field of Venus to be very weak compared to that of Earth.

The spacecraft, Mariner 2 in 1962 (Smith *et al.* 1965), Venera 4 (Dolginov *et al.* 1968), and Mariner 5 in 1967 (Dolginov *et al.* 1969), Mariner 10 in 1974 (Ness *et al.* 1974a), and Venera 9 and 10 in 1975 and 1976 (Dolginov *et al.* 1976) failed to find evidence of an intrinsic magnetic field in Venus. They found a shock wave in the solar wind, configured around Venus in a manner that can be explained if the solar wind collides directly with the top of the atmosphere of Venus, i.e. with the ionosphere. The turbulence and the small-scale magnetic fields behind the shock seem to be a natural consequence of the solar field carried in the wind and the violent flow of the wind around the planet.

On the other hand, Russell (1976) argues that a dipole field with a magnetic moment of about $1{\cdot}4\times10^{23}$ G cm^3, so that the field at the equatorial surface of the planet is perhaps 6×10^{-4} G and is just sufficient to push the wind back from the top of the atmosphere, gives a better description of the standing shock wave and the turbulence observed by the passing spacecraft. If this is correct, then, of course, the form of the shock should be highly variable. The solar wind impacts the sunlit ionosphere when blowing with above average strength, but is held off from the atmosphere on more ordinary days, such as the times when the spacecraft were passing by. In any case, the magnetic moment of the sister planet to Earth is at most 2×10^{-3} that of Earth. The rotation rate is smaller by 4×10^{-3}, so the small field, if indeed there is any at all, is understandable.

20.2.3. Mercury

The rotation period of Mercury is 59 days. The planet is small ($R_M = 2{\cdot}3\times10^8$ cm) and has a mean density, $\rho_M = 5{\cdot}44$ g cm^{-3}. The slow rotation rate suggests that it is unlikely to contain an operating dynamo. It was with no little surprise, therefore, that the passage of Mariner 10 near the planet in 1974 and again in 1975 showed a magnetic field (Ness *et al.* 1974b, 1975, 1976; Jackson and Beard 1977). The best fit to the spacecraft magnetometer curves yields a magnetic moment of $5{\cdot}2\times10^{22}$ G cm^3. The polarity is the same as that of Earth, with the dipole moment pointing southward. The dipole axis is inclined at about 12° to the axis of rotation. The field at the equatorial surface of the planet at the equator is 4×10^{-3} G. The magnetic moment is, then, smaller than the moment of Earth by a factor of about 1600, while the rotation rate is smaller by the factor of 59 (Ness 1978).

The metallic core of Mercury is estimated to be large in comparison to the size of the planet, with a radius $R_c = 1600$ km of just two-thirds the planetary radius of 2400 km (Peale 1972; Siegfreid and Solomon 1974; Toksoz and Johnston 1975). We may safely conclude from the existence of the dipole field that the core is at least partly liquid and in a state of convection. The small mean density of the planet, in spite of so large a core, is a consequence of the small gravitational field, so that the density of the iron in the core is more like 9 g cm^{-3} instead of 11 g cm^{-3} as in the core of Earth.

The radius of the core of Mercury is about half the radius of the core of Earth. There is every reason to think that the electrical conductivity is comparable. Hence the resistive decay time is one-fourth as long, of the order of 10^4 years or less. The surface field of $3{\cdot}5\times10^{-3}$ G extrapolates to about 7×10^{-3} G at the surface of the core. If, for instance, there were an azimuthal field some ten times stronger than the dipole field in the core, as has been suggested for Earth, the magnetic energy density would be about 2×10^{-4} erg cm^{-3}, so that the total magnetic energy might be as large as 3×10^{21} erg. A resistive decay time of 10^4 years $= 3\times10^{11}$ s leads to an expenditure of energy of a rate of 10^{10} erg s^{-1} = 1 kW. The resistive heating in the burner on a conventional electric stove involves more power than the resistive heating of the core of Mercury. A small petrol engine driving a generator could supply enough power to maintain the magnetic field of Mercury. The enormous breadth of the core of Mercury allows the tremendous total current, of the order of cB_M/R_M, to flow without much resistive loss. It would appear, then, that whatever the heat budget of the core of Mercury, the magnetic field represents only a microscopic fraction of it.

20.2.4. Mars

Turning our attention to Mars, there is evidence for a magnetic field, based on data from the passage of Mars 2 and 3 in 1972 and Mars 5 in 1974 (Dolginov *et al.* 1972, 1973, 1975; Dolginov *et al.* 1974). The magnetic moment is $2{\cdot}5\times10^{22}$ G cm^3, so that the field at the equatorial surface is $0{\cdot}6\times10^{-3}$ G. The magnetic moment is, then, smaller than the dipole moment of Earth by a factor of about 3×10^3. The polarity is opposite to that of Earth, so that the dipole moment of Mars points northward. The dipole is inclined about 15° to the axis of rotation.

On the other hand, Russell (1978) has re-analysed the data from Mars 3 and 5 and concludes that the observed fields are nothing more than the interplanetary magnetic field draped and compressed around the planet. Hence, the existence of an intrinsic field in Mars is presently an open question. The discussion that follows is based on the dipole moment suggested by Dolginov *et al.* If, in fact, the moment is effectively zero, as argued by Russell, the necessary modification of the discussion is obvious.

The rotation rate of Mars is very nearly equal to that of Earth (24 h 37 min period). The mean density of Mars is 0·72 times the mean density of Earth, or some 4 g cm^{-3}, implying only a small metallic core, if any at all (Ringwood and Clark 1971). Binder and Davis (1973) suggest that the mass of the core lies between 5 and 25 per cent of the total mass of the planet. The existence of the magnetic field requires that there be some conducting fluid core, certainly. If we suppose that the radius of the core is one-third of the 3×10^3-km radius of the planet, then, $R_c=10^3$ km. Assuming the same electrical conductivity as for the core of Earth, it follows that the resistive decay time is 3×10^3 years $=10^{11}$ s. This is long enough for convective motions and non-uniform rotation of 10^{-2} cm s^{-1} to make several circuits around the core, the magnetic Reynolds number $R_c v/\eta$ being of the order of 10^2.

The dipole field extrapolates to 2×10^{-2} G at the surface of the core. The azimuthal field is presumably somewhat stronger. If it is five times stronger, i.e. 10^{-1} G, the total magnetic energy in the core would be about 2×10^{21} erg. The resistive dissipation would be 2×10^{10} erg s$^{-1}=$ 2 kW, comparable to the resistive heating of the core of Mercury. It can hardly represent a significant perturbation in the heat budget of the core.

20.2.5. Jupiter

The magnetic field of Jupiter was first pointed out by Warwick (1961) on the basis of the intermittent polarized non-thermal decimetric and decametric radio emission. The simplest explanation for the decametric radiation is Cerenkov radiation near the cyclotron frequency from relativistic electrons trapped in a magnetic field. The emission comes from regions about 3 radii ($R_J = 7 \times 10^{11}$ km) from the centre of the planet, where the field is presumably about 0·4 G. Hence a magnetic dipole moment of the general order of 4×10^{36} G cm^3 is required. The field at the equatorial surface is then about 10 G. The magnetic dipole is offset from the centre of the planet somewhat, and inclined to the rotation axis by some 10° (Warwick 1967; Hide, 1974).

The passage of Pioneer 10 and Pioneer 11 near Jupiter in 1974 provided direct observations of the field, showing it to be profoundly distorted beyond a few times R_J by the interaction with the solar wind and by the pressure of the trapped particles. The best extrapolation from the magnetometer data from Pioneer 10, which approached to within a radial distance $r = 2{\cdot}8\ R_J$, and Pioneer 11 which approached to $r = 1{\cdot}6\ R_J$ yields a mean dipole equatorial surface field of about 4·1 G (polar fields of about 8 G) corresponding to a dipole moment of $1{\cdot}4 \times 10^{36}$ G cm^3, inclined about 10° to the axis of rotation and with the same polarity as Mars (Smith *et al.* 1974). Dolginov (1976) raises the question of whether the different periods of rotation of Jupiter derived from the radio emission from the dipole field and from System III (which may represent the inner atmosphere of Jupiter) are a consequence of the precession (westward drift) of the dipole moment relative to the planet itself. The rotation of Jupiter is very rapid, with a period of about 9 h 50 min, based on determinations using different reference systems on Jupiter. The difference between the rotation period of the radio emission and the frame of reference III is 0·4 s, or about 6 min per terrestrial year. The difference of 6 min represents 3·6° of rotation of Jupiter. Hence the precession would be extremely rapid by terrestrial standards, where it is 0·04°/year for the dipole moment and 0·18°/year for the small-scale inhomogeneities.

The interior of Jupiter is believed to be made up of an alloy of metallic hydrogen and helium (Hubbard and Smoluchowski 1973; Hide 1974). The alloy is presumably a liquid throughout much of the core, with an electrical conductivity of the general order of $\sigma = 10^{17}$ s^{-1} (Stevenson and Ashcraft 1974; see also Salpeter 1973), which is about the same as that of copper or silver at room temperature. Hence $\eta \cong 10^3$ cm^2 s^{-1} and the characteristic decay time for the field is $R_J^2/4\eta \cong 10^{16}$ s $= 3 \times 10^8$ years. The existence of the strong dipole field implies that the interior constitutes a dynamo, and hence is in a state of convection. Hence a turbulent eddy diffusivity η_T may be more appropriate than the simple ohmic diffusivity η. However, in our present state of ignorance there is not much that can be done to estimate the magnitude of the effects.

Suppose, by analogy with Earth that there is an azimuthal field ten times stronger than the dipole field, throughout the volume of Jupiter. Say $B_\phi \cong 10^2$ G. Then the total magnetic energy is $0{\cdot}6\times10^{33}$ erg (although this is nothing more than a guess on which to base a discussion). With the ohmic decay rate quoted above, the rate of dissipation is 6×10^{16} erg s^{-1} and hence is only comparable to the dissipation in the core of Earth. This remarkable fact is largely the result of the much larger radius and lower electrical resistivity of the interior of Jupiter. If there is significant turbulent diffusion, the dissipation rate could be much larger, of course. But whatever it is, it seems to be a truly negligible fraction of the total heat budget of the interior of Jupiter, estimated to be 8×10^{24} erg s^{-1} (Armstrong *et al.* 1972; Chase *et al.* 1974). It is an interesting coincidence that the output of heat from the interior of Jupiter is presently estimated to be comparable to the sunlight falling on the planet. For Earth the internal heat source is only 10^{-4} of the incident sunlight.

20.3. The source of planetary magnetic fields

At least three of the five inner planets have strong dipole fields, and none of these fields can be accounted as primordial. Therefore, each planet has within it a conducting, convecting, fluid core. It is not known for any case but Earth whether the field is steady or oscillatory.

The physics of the generation of both stationary and periodic fields is the subject of the previous chapter. A variety of kinematical models of the geomagnetic dynamo were presented in §19.4, and more are available in the literature, exploring a variety of forms of non-uniform rotation, and/or distributions of cyclonic convection (cf. Braginskii 1964b, 1970; Steenbeck and Krause 1969; Deinzer and Stix 1971; Roberts 1972; Roberts and Stix; 1972; Roberts and Soward 1972; Stix 1973; Gubbins 1973, 1974; Deinzer *et al.* 1974; Busse 1975b; Kumar and Roberts 1975; Soward and Roberts 1976; Busse and Carrigan 1976; Busse and Cuong 1977). The basic fact is that dynamical considerations suggest that the dynamo number is positive in the northern hemisphere and negative in the southern hemisphere of the liquid core of a rotating planet, producing fields with dipole (odd) symmetry (see discussion in §§19.3.1 and 19.4). The dipole is the lowest mode, so it appears that the planetary fields are a manifestation of the simple fact of convection and non-uniform rotation in the planetary interiors.

The energy that drives the convection and produces the magnetic field is negligible compared to the total heat budget of the planetary core, except perhaps for Earth. When the thermodynamic efficiency of the convective heat engine is taken into account, as well as the turbulent diffusion and dissipation of field, the necessary energy source in the core of Earth comes close to the estimated radioactive heating. It is fortunate that Earth provides the critical case, where the most direct information is available to guide the inquiry.

To lay out the entire picture for Earth we note again that the secular variation of the dozen or so large-scale inhomogeneities observed in the geomagnetic field of the surface of Earth indicate that the core is convecting. As already noted, the westward drift is most simply viewed as a direct indication of non-uniform rotation (see §20.2.1 above) (Elsasser 1946, 1950, 1955, 1956a,b; Bullard *et al.* 1950; Inglis 1955) driven by the Coriolis force on the radial convective motions. On the other hand, it has been suggested that the westward drift may represent the progress of Alfven waves along the geomagnetic field, having nothing whatever to do with non-uniform rotation (Braginskii 1964a,b, 1967; Hide 1966; Acheson 1972; Roberts and Stix 1972). In Braginskii's picture the Alfven waves may be the cyclonic convection responsible for the dynamo coefficient. The reader is referred to the reviews by Roberts and Soward (1972) and Acheson and Hide (1973) for a discussion of the general problem. It seems to this author that neither non-uniform rotation nor the propagation of the local inhomogeneities in the azimuthal field at the Alfven speed can be avoided. But how much each contributes to the observed westward drift is an open question. For instance, the westward drift of 0·18° per year corresponds to 3×10^{-2} cm s^{-1}, while the Alfven speed in the liquid metal ($\rho\cong 11$ g cm^{-3}) is three times larger in a field of only 1 G. But the field may be 30–100 G, yielding Alfven speeds far in excess of the observed drift. It is clear, then, that the problem is complicated, with the fluid pressure, the magnetic field and the Coriolis forces nearly balanced (Busse 1973, 1976a,b; Roberts and Soward 1972). This point of magnetogeostrophic balance comes up again in the theoretical estimation of the dynamo number in the paragraphs below.

A direct balance of Lorentz and Coriolis forces yields something of the order of $B_\phi B_p/4\pi\lambda=\rho\Omega v$ (Bullard 1949; Bullard and Gellman 1954) where λ is the scale of variation of B_ϕ, as already noted. Convective motions of $v\cong10^{-3}$ cm s^{-1}, a poloidal field of $B_p=5$ G, a characteristic scale $\lambda=10^8$ cm in a fluid of density $\rho=11$ g cm^{-3} with an overall rotation $\Omega=7\times10^{-5}$ s^{-1} yields $B_\phi=200$ G. Braginskii (1963) estimates 200 G from considerations on the secular variations. If, on the other hand the Coriolis forces are balanced largely by the fluid pressure, then B_ϕ may be very much less than 200 G (Busse 1971, 1973, 1977).

If the azimuthal field in the core of Earth has an r.m.s. value of 10^2 G, then the dissipation rate is 7×10^{16} erg s^{-1} as a consequence of the resistivity of the liquid metal. If turbulent diffusion is included (see §20.2.1) then the dissipation is perhaps four times larger, or 3×10^{17} erg s^{-1}. If the field is only 30 G on the average, then the dissipation is 3×10^{16} erg s^{-1}, including the effects of turbulent diffusion.[1] For the purposes of discussion, suppose that the rate is 10^{17} erg s^{-1}.

Now the latent heat of fusion and the growth of the solid inner core may contribute a significant amount of heat (Verhoogen 1961). The radioactive decay of the K^{40} concentrated in the core, perhaps in solution with iron sulfide, liberates heat at a rate estimated to be 2×10^{11}–2×10^{12} cal s$^{-1}\cong 10^{19}$–10^{20} erg s^{-1} (Lewis 1971; Goettel, 1972, 1976; Verhoogen 1973). It is generally assumed that the outflow of heat is sufficient to produce a superadiabatic temperature gradient (Bullard 1950; Lewis 1971) causing the fluid to overturn (Braginskii 1964b, 1965; Kovetz 1969; Busse 1971, 1973, 1975b), but there has been some controversy on this point. Higgins and Kennedy (1971) have argued that the core is stably stratified, so that there is no convective overturning. More recently they have modified their view somewhat, admitting the possibility that the lower part of the core may convect (Kennedy and Higgins 1973). Busse (1972, 1975a), Verhoogen (1973), and Jacobs (1973) have expressed quite different views. The fundamental difficulty is that the composition of the liquid core is poorly known and the properties of any of the possible compositions are poorly known at the pressures (several megabars) and the temperatures (4000 K) in the liquid core.

Braginskii (1963), Busse (1972), Elsasser (1972), and Malkus (1973) have suggested that the settling of heavy components in the core may, in some way, drive the convection directly. The total energy requirement for 10^{17} erg s^{-1} over the 5×10^{9} year age of Earth is $1{\cdot}5\times 10^{34}$ erg, which is a very small fraction of the gravitational energy, 2×10^{38} erg, of the core. Very little differentiation in density would be needed. But of course the settling could not cause overturning of the fluid if the core is stably stratified.

We cannot avoid the basic fact that the dipole field exists, and the local inhomogeneities in the field vary with time. The conviction that the core is convecting is hard to escape. The convective motions are evidently $v\cong 10^{-3}$–10^{-2} cm s^{-1}.

Malkus (1963, 1968) and others (Barret 1970; Malkus and Proctor 1975) have pointed out that the precession of the axis of rotation of Earth feeds energy into the fluid motions in the core, but Rochester *et al.* (1975)

[1] Pekeris *et al.* (1973) have worked out a model of the geomagnetic dynamo, based on closed vortex rings, which requires only a tenth as much energy because it lacks an azimuthal field. Busse (1971, 1973) treats cylindrical convection rolls without overall non-uniform rotation so that B_ϕ is small and the energy requirements are greatly reduced (see discussion in Soward and Roberts 1976; Moffatt 1976, 1978).

estimate that less than 10^{15} erg s^{-1} is available to the convection in this way.

The evidence is, then, that the heat supply to the core is some 10^{19}–10^{20} erg s^{-1}. The fraction of this available to drive the convection is limited by the usual thermodynamic considerations on heat engines. For ideal gases no more than the fraction $(T_2-T_1)/T_2$ of the thermal energy can be converted into mechanical work when the convection operates between a low temperature of T_1 and a high temperature of T_2. In practice, of course, the efficiency is usually very much less than the ideal value $(T_2-T_1)/T_2$ because the heat transfer is not carried out reversibly, and friction converts some of the work back to heat, etc. But it makes no difference whether the working substance is an ideal gas (employed in so many textbook illustrations), a liquid, or a solid. For suppose that the working material absorbs (reversibly) an amount of energy Q_2 at the fixed temperature T_2. The entropy of the material increases by $S_2=Q_2/T_2$. Then the material is expanded reversibly and adiabatically to a lower temperature T_1, for which there is no change in entropy. Then the material is compressed while being held at T_1 and gives up (reversibly) an amount of heat, Q_1, into some convenient reservoir at T_1. The isothermal compression is stopped when the material reaches a volume such that reversible adiabatic compression restores it to its initial volume at the higher temperature T_2. The net loss of entropy from the system at the temperature T_1 is $S_1=Q_1/T_1$. The adiabatic compression restores the system to its original state. Restoration of the system to its original thermodynamic state also restores the original entropy, of course. Hence, the total change of entropy is zero and $S_1=S_2$. The work W is equal to the missing thermal energy Q_2-Q_1. Hence the fraction of the input heat Q_2 converted into mechanical work is

$$W/Q_2=1-Q_1/Q_2=1-T_1/T_2.$$

This relation applies to any working substance, as already emphasized. The limitation imposed by the working material is the restriction of the temperature difference T_2-T_1: a gas can be expanded adiabatically from T_2 down to any lower temperature T_1, but a non-volatile liquid can be cooled by only a limited amount as the pressure falls to lower values. The enormous compression of the liquid iron and the modest temperature difference $(T_2-T_1)/T_2 \lesssim 0{\cdot}1$ in the core of Earth would seem to impose no restriction, however.

Higgins and Kennedy (1971) suggest $T \cong 4000$ K and $T_2-T_1 \equiv \Delta T=$ 500 K for the liquid core of Earth. On this basis the maximum efficiency would be 1/8. If the convection produces only 4 per cent of the mechanical work that is thermodynamically available, then 1/200 of the energy output goes into mechanical work, for a total of $0{\cdot}5$–$5{\cdot}0\times10^{17}$ erg s^{-1}

This number is very uncertain, then, but it would seem to cover the 10^{17} erg s^{-1} for the dynamo. Indeed, until a more quantitative picture of the superadiabatic temperature gradient in the core can be worked out, and the dynamical problem gives a general and unique theoretical solution, there is little more that can be said on the adequacy of energy sources.

Consider, then, the dynamo number for the core of Earth based on the fluid motions inferred from the secular variations of the inhomogeneities in the geomagnetic field, as outlined in §20·2.1 above. The dynamo coefficient is given by (18·58) in terms of the cyclonic component v_ϕ of the individual convective cell, $\Gamma \cong 0{\cdot}4 v_\phi$. The convective velocities are of the order of 10^{-3}–10^{-2} cm s^{-1}. If we suppose that the cyclonic (rotational) component of this motion is $v_\phi = 10^{-3}$ cm s^{-1}, then $\Gamma \cong -0{\cdot}4 \times 10^{-3}$ cm s^{-1}. The westward drift of $R_c \Delta\Omega \cong 0{\cdot}03$ cm s^{-1} (0·18° per year) across the depth h of the core, of some 2×10^3 km, implies a vertical shear rate $G = R_c \Delta\Omega / h \cong -1{\cdot}5 \times 10^{-10}$ s^{-1}. The length λ of the dynamo region from the equatorial plane around to the axis of rotation is of the order of the core radius R_c. Hence the dynamo number is defined in terms of R_c as $N_D = \Gamma G R_c^3 / \eta^2 \equiv K^3 \lambda^3$. With $R_c = 3 \times 10^8$ cm and $\eta \cong 4 \times 10^4$ cm^2 s^{-1} for turbulent diffusion, it follows that $N_D \cong +10^3$, $K\lambda = 10\cdot$

It was shown in §19.3.1 that the stationary dipole (odd) mode is produced by positive dynamo numbers in the vicinity of $K\lambda \cong +4.4$. The very rough estimate of $K\lambda = +10$ for the core of Earth is a comparable figure, indicating the plausibility of the idea that the magnetic field of Earth is produced by cyclonic fluid motion in the core (see Braginskii 1964a,b, 1967; Roberts and Soward 1972; Soward and Roberts 1976 for a detailed discussion).

What, then, can be said of the other planets? The internal structure of Jupiter is so different from the terrestrial planets that we can do no more than point out that internal convection and non-uniform rotation provide a tentative explanation for the magnetic field. We may presume, however, that Mercury, Venus, and Mars have liquid metal cores with much the same qualitative properties as the core of Earth. Mercury has a dipole field, and perhaps Mars and Venus too (see review by Stix 1977).

Whether the dipoles of Mercury and Mars are stationary or periodic matters little (see (19.130)). The requirement on the dynamo number is roughly the same. They must have dynamo numbers $K\lambda$ of the general order of 5 or 10 so as to excite the lowest dipole mode, as in Earth. The startling fact is that Mercury, with a core radius about half as large, and a rotation rate Ω one-fiftieth as large, as Earth, has a dynamo number approximately equal to that of Earth. Since it is the Coriolis force of the rotation that produces both the cyclonic component of the convection, to give Γ, and the non-uniform rotation, to give G, we would expect

$N_D \propto \Omega^2 R_c^3$, $K\lambda \propto \Omega^{2/3} R_c$, suggesting that $K\lambda$ is smaller in Mercury than in Earth by a factor of about 30. Mars has a rotation rate comparable to Earth, but a core radius believed to be about one-third that of Earth. So $K\lambda$ should be only about a third that of Earth. Yet each planet seems to have a dipole field.

This puzzle drives home the point noted earlier that the fluid motion in the core of Earth must be near some dynamical magnetogeostrophic balance, as pointed out by Braginskii (1946b, 1967) and by Busse (1971, 1976a, b), among others, so that the fluid motions in the core of Earth are strongly suppressed by the Lorentz forces. Then presumably in Mercury and Mars, where the fields are weak, they are not suppressed, and the dynamo numbers are as large as for Earth in spite of the slow rotation or the small size of the core. It is the cyclonic component of the fluid motion in the core of Earth that is remarkable for its low level.

Evidently, then, the fluid is freer to circulate in a more nearly hydrodynamic fashion in Mercury, Venus, and Mars, without such complete suppression by strong Lorentz forces. To explore this simple possibility, suppose that the core of Mercury is subject to convective motions of 10^{-3} cm s^{-1}, as in Earth's core, with characteristic dimensions of $L =$ 500 km. The angular velocity of Mercury is $\Omega_M \cong 10^{-6}$ s^{-1}, so that conservation of angular momentum of an element of fluid displaced a distance L implies cyclonic velocities of the order of $\Omega L \cong 50$ cm s^{-1}. Thus, from a naive point of view we might expect such velocities to appear as nonuniform rotation or as rotation of an individual convective cell. Such motions yield dynamo numbers $K\lambda$ too large by a factor of at least 10^2. We are forced to conclude, then, that even in a slowly rotating planet like Mercury the fluid motions are near some stationary dynamical balance between fluid pressure, Coriolis force, viscosity (perhaps eddy viscosity), Lorentz forces, etc. for if the fluid motions were running free they would produce magnetic fields of very high order, whereas only the dipole mode is observed. The situation in the core of Earth must be even more closely constrained. This emphasizes the fact that the kinematic dynamo provides an excellent vehicle for illustrating the general principles of the hydromagnetic dynamo, but only a very tightly formulated treatment of the *dynamical* equations can give a quantitative picture of the precise form of the production of the fields in the various planets. The occasional abrupt reversal of the geomagnetic field may, then, be understood as a consequence of a momentary imbalance between enormously powerful, opposing, and finely balanced forces. Presumably the dynamical system is not strongly stable so that a small disturbance of some sort leads to a large tremor (if not total disorder, with fluid velocities of centimetres per second) that upsets the dynamo and actively destroys the field. Just how tranquillity is restored with fluid velocities of only 10^{-3}–10^{-1} cm s^{-1}, is an interesting and very difficult question. Braginskii (1964b) pointed out that the proximity of the dynamo numbers of the stationary dipole and the oscillatory quadrupole provide a ready mechanism for reversing the field. Even a modest increase in the dynamo number for a few hundred years would set the field into a state of oscillation and repeated reversal. Parker (1969) pointed out that a sudden shift in the location of the cyclonic convection can cause an abrupt reversal. Levy (1972a,b) has shown how extremely sensitive the dynamo is to the location of the cyclonic convection and has given examples showing reversal as a consequence of changes as small as 20 per cent in the location and strength of the cyclonic convection. The observed excursions of the strength of the geomagnetic dipole and the continual evolution of the higher harmonics imply that the dynamical system of field and fluid in the core teeters back and forth by substantial amounts and yet remains in balance. Indeed, it seems remarkable that the liquid core ever settles into a dynamical state so finally balanced as to suppress the fluid motions so much as to generate the lowest (dipole) mode. That the field is a stationary dipole most of the time shows the *naiveté* of the elementary dynamical considerations.

It should be noted again (see §19.2.1) that there may, in some special circumstances, be so little differential rotation that the field is generated principally through the dynamo coefficient Γ alone. This is the so-called α^2-dynamo, pointed out by Krause and Steenbeck (1967), Steenbeck and Krause (1969), and Roberts and Stix 1972), about which we have said relatively little, except for the plane dynamo waves treated in §19.2.1, on the presumption that the non-uniform rotation is sufficient to produce a strong azimuthal field in most situations. Hopefully the rapid advance of the theoretical dynamics of the convection and circulation in planetary cores will soon establish the degree of non-uniform rotation and the relative importance of the α^2-dynamo.

In contrast to the balance of fluid and field forces in the core of Earth, the fluid motions in the sun and in the galaxy appear to be more free-running. Thus, for instance, $\Omega \cong 2\times10^{-6}\,\mathrm{s}^{-1}$ for the sun, so that circulation over the depth $h = 2\times10^5$ km of the convective zone suggests non-uniform rotation of the order of $\Omega h \cong 0{\cdot}4\,\mathrm{km\,s}^{-1}$. This is more or less what is observed, with the equatorial surface velocity of the sun equal to $2\,\mathrm{km\,s}^{-1}$, and the polar rotation velocities about one-third smaller than the equatorial value suggesting $h\Delta\Omega \cong 0{\cdot}7\,\mathrm{km\,s}^{-1}$. On this same basis the cyclonic velocity $v_\phi \cong L\Omega$ for the supergranules ($L \cong 2\times10^9$ cm) is expected to be some $4\times10^3\,\mathrm{cm\,s}^{-1}$ with $\Gamma = 0{\cdot}4v_\phi = +1{\cdot}6\times10^3\,\mathrm{cm\,s}^{-1}$, according to (18.58). If $\Delta\Omega$ appears across the depth of the convective zone, then $G \cong -\Delta\Omega \cong -0{\cdot}7\times10^{-6}\,\mathrm{s}^{-1}$. The turbulent diffusion coefficient was estimated in §19.2.2 to be of the general order of $\eta_T \cong 3\times10^{12}\,\mathrm{cm}^2\,\mathrm{s}^{-1}$ in the convective zone. Hence with the length λ of the dynamo region comparable to one solar radius ($R_\odot = 0{\cdot}7 \times 10^{11}$ cm), the dynamo number is $N_D = \Gamma G R_\odot^3/\eta_T^2 \equiv K^3\lambda^3$. Numerically, then, $N_D \cong -4\times10^4$ and $K\lambda = -34$. The theoretical model requires (19.131), or $K\lambda = -7{\cdot}43$ or $-11{\cdot}1$ (for expansion in ascending and descending powers of the growth rate, respectively). Thus the simple numerical estimate of $K\lambda$ for the sun is of the same general magnitude as that required by the rectangular model (see §21.2 for a more detailed discussion). A similar result was obtained in §19.3.3.1 for the galactic dynamo. There is, then, no evidence for more than nomimal hindrance of the convection and non-uniform rotation by the magnetic field in either the sun or the galaxy. The situation is quite different from the peculiar case of nearly complete balance of forces and the extraordinary suppression of convection and non-uniform rotation below the 'elementary' values in the cores of planets.

Another way of making the comparison is to note that the Lorentz force exerted by the general magnetic field of the sun is weak compared to the Coriolis forces, while in Earth the Lorentz force closely balances the Coriolis force (Soward and Roberts 1976). At the same time the period in which a convective cell (say, a supergranule cell) in the sun turns over is somewhat less than the period of rotation of the sun. The same is true for the turbulence in the gaseous disc of the galaxy. In contrast to the sun or the galaxy, the turn over time of a convective cell in the core of Earth, or other terrestrial planet, is long compared to the rotation period. So there are distinct differences between the dynamics of the planetary dynamo on the one hand and the stellar or galactic dynamo on the other.

The dynamics of the magnetogeostrophic balance in the core of Earth is subject to formal mathematical analysis within the framework of the nearly axi-symmetric dynamo of Braginskii (see discussion in Moffatt 1976, 1978; Soward and Roberts 1976). Indeed, one might consider the dynamics in terms of stationary fluid motions, generating magnetic fields along the lines treated by Lortz (1968a,b, 1972) (see discussion by Benton 1975). But it is difficult to attack the dynamics of core convection without being sure of the magnitude and form of the heat source that drives the convection. Altogether, it is no wonder that the strength of the fields of the various planets are not related by some simple scaling law involving angular velocity, core radius, etc.

As a final point, consider the puzzling fact of the remnant magnetism of the moon noted in §20·2.1. The residual magnetization of the surface rocks of the moon (see review by Dyal *et al.* 1974; Dolginov 1976) implies that the rocks solidified (some 3–4×10^9 years ago) in a magnetic field of the order of 0·05–1 G. The obvious possibilities are that the field responsible for the residual magnetism was the field of the moon itself (Runcorn and Urey 1973, Strangway and Sharpe 1974), the field of Earth (Alfven and Linberg 1974), or the field of the sun (Nagata *et al.* 1974) at that early time. It is difficult to understand how a field of the sun could produce 0·1 G at the orbit of Earth. The moon was much closer to Earth in those days, so that 10–50 G of the surface of Earth might produce 0·1 G at a distance of 6 or 7 times the radius of Earth. It would be remarkable indeed if Earth ever had such strong convection in its core as to produce a field 10^2 times stronger than the present one. Consequently Levy (1972c) looked into the question of whether the moon itself might once have generated its own field. He compared the lunar dynamo to the present dynamo in the core of Earth using the simple scaling law that the non-uniform rotation is proportional to the rotation Ω and the dynamo coefficient Γ is proportional to $R_c\Omega$ in a core of radius R_c. Hence the dynamo number scales as $R_c^4\Omega^2$. The low density of the moon implies a core radius no more than one-tenth the radius of the core of Earth (i.e. $R_c = 350$ km). Hence, the same dynamo number as Earth, to produce the dipole mode, requires an angular velocity of the moon 10^2 times greater than Earth has today (i.e. a rotation period of about 15 min). Unfortunately the moon cannot rotate more than about 20 times faster (90-min period) without bursting. Hence Levy suggested that the moon has never had a dynamo field of its own. He points out, however, that his argument neglects the possible suppression of the fluid motions in the core of Earth by the Lorentz forces. The arguments given in the paragraphs above, based on a comparison of Earth and Mercury, indicate that the suppression of fluid motion by the field is extreme in the core of Earth, and perhaps not negligible even in Mercury. On this basis, it would seem now more appropriate to apply Levy's arguments to a comparison of the moon with Mercury. Then the lunar core radius is smaller by a factor of about 4·5, requiring an angular velocity larger than Mercury by $4{\cdot}5^2 \cong 20$, i.e., a rotation period of about 50 h. The moon may well have rotated much faster than this at an early stage in history. Therefore, if in fact the moon has ever had a convecting, electrically conducting, metal core, it may well have generated its own field at one time. Whether the Coriolis forces were strong enough to produce the necessary field of at least 0·1 G at the surface of the moon (five times the radius of the core) is an open question, of course. That would require a dipole field of at least 12 G at the surface of the core. The requirement has to be weighed against the alternative that the early Earth had a *surface* field of, say, 10–50 G so as to produce 0·1 G or more at the nearby moon. Unfortunately the question is more accessible to conjecture than to formal theory.

20.4. The theoretical problems posed by planetary magnetic fields

Given that a rotating planet possesses a convecting, conducting fluid core, the theoretical principles of magnetohydrodynamics lead to the conclusion that the planet has a magnetic field, generated in the core. The dynamo effects of the motions are so strong that it could hardly be otherwise. What is remarkable is the fact that the fields of the planets are all principally dipolar. The enormous variation of dynamo number that one might expect from the differences in size, convective velocity, and rate of rotation would suggest instead a whole spectrum of fields ranging from none at all (as perhaps for Mars and Venus) through the stationary dipole to high order multipoles that oscillate in time. Earth and Jupiter would be the prime candidates for the higher order oscillatory fields. The universal appearance of the dipole field indicates a close dynamical balance between field and fluid so that the effective dynamo number $K\lambda$ is of the order of 10 for all of the planets, in spite of their different physical characteristics. The answer to this riddle must lie in a general theoretical treatment of the fluid dynamics of the planetary core. Indeed, the dynamics of the planetary core is the next step in the development of the theory of planetary magnetic fields. The subject is complex and has yet to reach the point where unique models of the convection can be constructed.

Ideally, a prescription for the energy input would lead to an *unambiguous* prediction of the fluid circulation (with or without the neglect of the Lorentz forces of the magnetic field generated by the circulation), but, as a matter of fact, the non-uniqueness of turbulent flow poses severe limits on the possibilities for quantitative prediction. Numerical experiments will play a role, but certainly have their limitations. And finally, it is doubtful that a hard statement on the energy source of the core will ever be possible. So the theoretical research of the future will continue in the same spirit as in the past, groping with illustrative examples to understand the basic physical principles, as has already been done for the kinematical dynamo. The problem will remain open-ended for the foreseeable future.

The dynamical behaviour of the external field of the planets is another facet of planetary magnetism that has not been pursued here. It enters into the labyrinthian complexity of plasma physics, and, for the more interesting effects, into plasma turbulence and suprathermal particles as noted in §20.1. Together the ionospheric effects and the activity of the external magnetic field appear to play some role in the continual caprice of the terrestrial weather, although the connections are presently based on statistical correlations rather than sound theory, and are warmly debated (see, for instance, Wilcox 1972; Roberts and Olsen 1973a,b; Hines and Halevy 1975; Wilcox *et al.* 1975, 1976; Leith 1975).

The plasma effects in the magnetosphere, and particularly the plasma turbulence and acceleration of particles, resemble on a small scale some of the gigantic and extreme magnetic activity observed throughout the cosmos in the many active stars, nebulae, and galaxies. The geomagnetic fields, then, and the surrounding solar wind, and the sun that drives the whole system, form the one laboratory where direct experiments and observations of the plasma turbulence and particle acceleration can be carried out to point the way for the development of physical theory (see §21.5).

The distant x-ray source, the distant pulsar, and the distant quasar are fascinating objects to observe. Their remarkable outward behaviour suggests such exotic effects as relativistic accretion discs whirling about black holes, magnetic fields of 10^{12} G, and the gravitational collapse of galactic masses. Their contemplation has spurred the development of quantum electrodynamics in the Kerr metric, and quantum electrodynamics and nuclear physics in strong magnetic fields. The future may hold even more exciting observational and theoretical discoveries. But, however stimulating and fashionable these exotic studies may be, their completion is blocked by their remoteness. We cannot see the details to show that the theoretical ideas are the correct or relevant ones. We know from detailed studies of the 'ordinary' activity of the geomagnetic field and the fields of the sun how wrong the naive application of theoretical concepts can be. To put the matter simply, the x-ray source, the active star, the pulsar, the active galaxy and the quasar are 'turbulent', like all other interesting objects; they are so complex as to lie outside the realm of unique theoretical prediction. Hence, they must be observed in detail. The details lie below the limit of resolution, and the objects are often too transparent or too opaque, and always too complicated, to determine the internal structure. The solid aspects of the science of magnetic–plasma–particle activity are restricted to direct studies of the magnetic fields of Earth (and other planets), the solar wind, and the sun, through detailed observations and the direct passage of instruments through the active regions. The local phenomena available for such definitive studies are limited, of course. Only rarely are relativistic nuclei produced. Gravitational accretion and collapse are essentially non-existent. Only the distant dim objects can tantalize us with the extreme conditions. But only the nearby plasma effects can be developed into a hard science. Their role in the terrestrial environment and their role as a foundation for the more extreme effects in distant objects places them at the heart of the science of astronomy.

References

Acheson, D. J. (1972). *J. Fluid Mech.* **52,** 529.
——, and Hide, R. (1973). *Rep. Progr. Phys.* **36,** 159.
Akasofu, S. I. (1968). *Polar and magnetospheric substorms.* Reidel, Dordrecht.
——, and Chapman, S. (1972). *Solar–terrestrial physics.* Clarendon Press, Oxford.
Alfven, H. and Linberg, L. (1974). *The Moon* **10,** 323.

Armstrong, K. R., Harper, D. A., and Low, F. J. (1972). *Astrophys. J. Lett.* **178,** L89.
Barret, K. E. (1970). *Phys. Earth Planet. Int.* **4,** 32, 322.
Benton, E. R. (1975). *Geophys. J. Roy. Astron. Soc.* **42,** 385.
Binder, A. B. and Davis, D. R. (1973). *Phys. Earth Planet Int.* **7,** 477.
Braginskii, S. I. (1963). *Dokl. Acad. Nauk. SSSR* **145,** 1311.
——, (1964a). *Sov. Phys. JETP* **20,** 776, 1462.
——, (1964b). *Geomag. Aeronom.* **4,** 732.
——, (1965). *Sov. Phys. JETP* **20,** 726, 1462.
——, (1967). *Geomag. Aeronomi.* **7,** 851.
——, (1970). ibid. **10,** 172, 221.
Brown, L. W. (1975). *Astrophys. J. Lett.* **198,** L89.
Burlatskaja, S. P., Nechaeva, T. B., Petrova, G. I., and Izvestija, A. N. (1968). *SSSR Fiz. Zemla* **12,** 62.
Bullard, E. C. (1949). *Proc. Roy. Soc. Ser. A* **197,** 481.
——, (1950). *Mon. Notic. Roy. Astron. Soc. Geophys. Suppl.* **6,** 36.
——, (1968). *Phil. Trans. Roy. Soc. Lond. A* **263,** 481.
——, Freeman, C., Gellman, H., and Nixon, J. (1950). *Phil Trans. Roy. Soc. Lond. A* **243,** 67.
——, and Gellman, H. (1954). *Phil. Trans. Roy. Soc. Lond. A* **247,** 213.
Busse, F. H. (1971). *Z. Geophys.* **37,** 153.
——, (1972). *J. Geophys. Res.* **77,** 1589.
——, (1973). *J. Fluid Mech.* **57,** 529.
——, (1975a). *J. Geophys. Res.* **80,** 278.
——, (1975b). *Geophys. J. Roy. Astron. Soc.* **42,** 281.
——, (1976a). *Phys. Earth Planet. Int.* **12,** 350.
——, (1976b). *J. Geophys.* **43,** 441.
——, and Carrigan, C. R. (1976). *Science* **191,** 81.
——, and Cuong, P. G. (1977). *Geophys. Astrophys. Fluid Dynamics* **8,** 17.
Cain, J. C. (1975). *Res. Geophys. Space Sci.* **13,** No. 3, 203.
Chapman, S. and Bartels, J. (1940). *Geomagnetism*, vol. I. Clarendon Press, Oxford.
Chase, S. C., Ruiz, R. D., Nunch, G., Neugebauer, G., Schroeder, M., and Trafton, L. M. (1974). *Science* **183,** 315.
Coleman, P. J., Russell, C. T., Sharp, L. R., and Schubert, G. (1972). *Phys. Earth Planet Int.* **6,** 167.
Cox, A. (1970). *J. Geophys. Res.* **75,** 7501.
——, (1972). ibid. **77,** 4339, 6459, 7065.
——, (1973). ibid. **78,** 6977.
——, (1975). *Rev. Geophys. Space Sci.* **13,** 35.
——, and Dalrymple, G. B. (1967). *Earth Planet Sci. Lett.* **3,** 173.
——, and Doell, R. R. (1964). *Bull. Seismol. Soc. Amer.* **54,** 2243.
——, Hillhouse, J., Fuller, M. (1975). *Rev. Geophys. Space Phys.* **13,** (3) 195.
Deinzer, W., v. Kusserrow, H. U., and Stix, M. (1974). *Astron. Astrophys.* **36,** 65.
——, and Stix, M. (1971). *Astron Astrophys.* **12,** 111.
Doell, R. R. and Cox, A. (1965). *J. Geophys. Res.* **70,** 3377.
——, Dalrymple, G. B. and Cox, A. (1966). *J. Geophys. Res.* **71,** 531.
Dolginov, Sh. Sh. (1976). *Planetary magnetism and dynamo mechanism problem.* Preprint N. 15a, Inst. Terr. Mag. Ionosphere, and Radio Wave Prep., USSR Acad. Sci.

Dolginov, Sh. Sh., Yeroshenko, Ye. G., and Davis, L. (1969). *Kosmicheskije Issledovanija* **7,** 747.
—— —— ——, (1975). ibid. **13,** 108.
—— ——, Zhuzgov. L. N. (1968). *Kosmicheskije Issledovanija* **6,** 561.
—— —— ——, (1972). *Dokl. Acad. Nauk SSSR* **207,** No. 6.
—— —— ——, (1973). *J. Geophys. Res.* **78,** 4779.
—— —— ——, (1975). *Kosmicheskije Issledovanija* **13,** 108.
—— —— ——, Buzin, V. B., and Sharova, V. A. (1976). *Pisma v. Astronomicheskij zhurnal* **2,** No. 2.
—— —— ——, Sharova, V. A. (1974). *Dokl. Acad. Nauk. SSSR* **218,** No. 4.
Dyal, P., Parkin, C., and Daly, W. (1974). *Rev. Geophys. Space Phys.* **12,** 56.
Dyer, E. R. (1972). *Solar terrestrial physics* (ed. E. R. Dyer). Reidel Co. Dordrecht.
Elsasser, W. M. (1946). *Phys. Rev.* **69,** 106; **70,** 202.
——, (1950). *Rev. Mod. Phys.* **22,** 4.
——, (1955). *Amer. J. Phys.* **23,** 590.
——, (1956a). ibid. **24,** 85.
——, (1956b). *Rev. Mod. Phys.* **28,** 135.
——, (1972). *Trans Amer. Geophys. Union* **53,** 605.
Engdahl, E. R. and Johnson, L. E. (1974). *Geophys. J. Roy. Astron. Soc.* **39,** 435.
Gardiner, R. B. and Stacey, F. D. (1971). *Phys. Earth Planet. Int.* **4,** 406.
Gendrin, R. (1975). *Space Sci. Rev.* **18,** 145.
Gilbert, F., Dziewonski, A. M., and Brune, J. (1973). *Proc. Nat. Acad. Sci. USA* **70,** 1410.
Goettel, K. A. (1972). *Phys. Earth Planet. Int.* **6,** 161.
——, (1976). *Geophys. Survey* **2,** 369.
Gose, W. A. and Butler, R. F. (1975). *Rev. Geophys. Space Phys.* **13,** 189.
Gubbins, D. (1973). *Phil. Trans. Roy. Soc. Lond. A* **274,** 493.
—, (1974). *Rev. Geophys. Space Sci.* **12,** 137.
Gurevich, A. V., Krylov, A. L., and Tsedilina, E. E. (1976). *Space Sci. Rev.* **19,** 59.
Haddon, R. A. and Bullen, K. E. (1969). *Phys. Earth Planet. Int.* **2,** 35.
Heirtzler, J. R., Dickson, G. O., Herron, E. M., Pitman, W. C., and LePichon, X. (1968). *J. Geophys. Res.* **73,** 2119.
Hide, R. (1966). *Phil. Trans. Roy. Soc. Lond. A* **259,** 615.
——, (1974). *Proc. Roy. Soc. Ser. A* **336,** 63.
Higgins, G. and Kennedy, G. C. (1971). *J. Geophys. Res.* **76,** 1870.
Hines, C. O. and Halevy, I. (1975). *Nature* **258,** 313.
Hubbard, W. B. and Smoluchowski, R. (1973). *Space Sci. Res.* **14,** 599.
Inglis, D. E. (1955). *Rev. Mod. Phys.* **27,** 212.
Jackson, D. J. and Beard, D. B. (1977). *J. Geophys. Res.* **82,** 2828.
Jacobs, J. A. (1973). *Nature Phys. Sci.* **243,** 113.
Jain, A. and Evans, R. (1972). *Nature Phys. Sci.* **235,** 165.
Johnston, M. J. S. and Strens. R. G. J. (1973). *Phys. Earth Planet. Int.* **7,** 217.
Kaiser, M. L. and Stone, R. G. (1975). *Science* **189,** 285.
Kennedy, G. C. and Higgins, G. (1973). *J. Geophys. Res.* **78,** 900.
Kennel, C. F. (1973). *Space Sci. Res.* **14,** 511.
Kobayashi, K. (1968). *Phys. Earth Planet. Int.* **1,** 387.
Kono, M. (1972). *Phys. Earth Planet. Int.* **5,** 140.
Kovetz, A. (1969). *Phys. Earth Planet. Int.* **2,** 83.

Krause, F. and Steenbeck, M. (1967). *Z. Naturforsch. A* **22,** 671.
Kumar, S. and Roberts, P. H. (1975). *Proc. Roy. Soc. Lond. A* **344,** 235.
Larson, R. L. and Helsley, C. E. (1975). *Rev. Geophys. Space Phys.* **13,** (3) 174.
Leith, C. E. (1975). *Rev. Geophys. Space Sci.* **13,** 681.
Levy, E. H. (1972a). *Astrophys. J.* **171,** 635.
——, (1972b). ibid. **175,** 573.
——, (1972c). *Science* **178,** 52.
Lewis, J. S. (1971). *Earth Planet Sci. Lett.* **11,** 130.
Lortz, D. (1968A). *Plasma Phys.* **10,** 967.
——, (1968b). *Phys. Fluids* **11,** 913.
——, (1972). *Z. Naturforsch. A* **27,** 1350.
Malkus, W. V. R. (1963). *J. Geophys. Res.* **68,** 289.
——, (1968). *Science* **160,** 259.
——, (1973). *Geophys. Fluid Dyn.* **4,** 267.
——, and Proctor, M. R. E. (1975). *J. Fluid Mech.* **67,** 417.
McCormac, B. M. (1976). *Magnetospheric particles and fields* (ed. B. M. McCormac). Reidel, Dordrecht.
McElhinny, M. W. and Evans, M. E. (1968). *Phys. Earth Planet. Int.* **1,** 485.
Moffatt, H. K. (1976). *Adv. Appl. Mech.* **16,** 119.
——, (1978). *Magnetic field generation in electrically conducting fluids.* Cambridge University Press, Cambridge.
Nagata, T., Fisher, R. M., Schwerer, F. C., Fuller, M. D. and Dunn, I. R. (1974). *Geochem. Cosmochem. Acta. Suppl.* **36,** (3) 2423.
Ness, N. F. (1978). *Space Sci. Rev.* **21,** 527.
Ness, N. F., Behannon, K. W., Lepping, R. P. and Whang, Y. C. (1975). *J. Geophys. Res.* **80,** 2708.
—— —— —— ——, (1976). *Icarus* **28,** 479.
—— —— —— ——, Schatten, K. H. (1974a). *Science* **183,** 1301.
—— —— —— —— ——, (1974b). ibid. **185,** 151.
Ninkovich, D., Opdyke, N., Heezen, B. C. and Foster, J. H. (1966). *Earth Planet. Sci. Lett.* **1,** 476.
Parker, E. N. (1969). *Astrophys. J.* **158,** 815.
Peale, S. J. (1972). *Icarus* **17,** 168.
Pekeris, C. L., Accad, Y., and Shkoller, B. (1973). *Phil. Trans. Roy. Soc. Lond.* **275,** 425.
Ringwood, A. E. and Clark, S. P. (1971). *Nature* **234,** 89.
Roberts, P. H. (1972). *Phil. Trans. Roy Soc. Lond. A.* **272,** 663.
——, and Soward, A. M. (1972). *Ann. Rev. Fluid Mech.* **4,** 117.
——, and Stix, M. (1972). *Astron. Astrophys.* **18,** 453.
Roberts, W. O. and Olson, R. H. (1973a). *J. Atmos. Sci.* **30,** 135.
—— ——, (1973b). *Rev. Geophys. Space Sci.* **11,** 731.
Rochester, M. G., Jacobs, J. A., Smylie, D. E. and Chang, K. F. (1975). *Geophys. J. Roy. Astron. Soc.* **43,** 661.
Runcorn, S. K. (1955). *Adv. Phys.* **4,** 244.
Runcorn, S. K. and Urey, H. C. (1973). *Science* **180,** 636.
Russell, C. T. (1976). *Geophys. Res. Lett.* **3,** 413, 589.
——, (1978). *ibid* **5,** 81, 85.
Salpeter, E. E. (1973). *Astrophys. J. Lett.* **181** L83.
Schwartz, E. J. and Symons, D. T. A. (1969). *Phys. Earth Planet. Int.* **2,** 11.
Siegfried, R. W. and Solomon, S. C. (1974). *Icarus* **23,** 192.

Smith, P. J. (1967). *Geophys. J.* **12,** 321.
Smith, E., Davis, L., Coleman, P., and Sonett, C. (1965). *J. Geophys. Res.* **70,** 1571.
—— ——, Jones, D. E., Coleman, P. J., Colburn, D. S., Dyal, P., Sonett, P. C. P., and Frandsen, A. M. R. (1974). *J. Geophys. Res.* **79,** 3501.
Soward, A. M. and Roberts, P. H. (1976). *Magnitnaya Gidrodinamika,* **1,** 3.
Stacey, F. D. (1969). *Physics of the earth.* John Wiley and Sons, New York.
Steenbeck, M. and Krause, F. (1969). *Astron. Nachr.* **291,** 271.
Stevenson, D. J. and Ashcraft, N. W. (1974). *Phys. Rev. A* **9,** 782.
Strangway, D. W. and Sharpe, H. A. (1974). *Nature* **249,** 227.
Stix, M. (1973). *Astron. Astrophys.* **24,** 275.
——, (1977). *Z. Geophys.,* **43,** 695.
Toksoz, M. N. and Johnston, D. H. (1975). *Proc. Soviet–American conf. moon and planets,* Akademiia nauk SSSR, National Aeronautics and Space Adm.
Van Zijl, J. S. V., Graham, K. W. T., and Hales, A. L. (1962). *Geophys. J. Roy. Astron. Soc.* **7,** 169.
Verhoogen, J. (1961), *Geophys. J.* **4,** 276.
——, (1973). *Phys Earth Planet. Int.* **7,** 47.
Warwick, J. A. (1961). *Astrophys. J.* **137,** 41.
——, (1967). *Space Sci. Rev.* **6,** 841.
Wilcox, J. M. (1972). *Rev. Geophys. Space Sci.* **10,** 1003.
——, Svalgaard, L., and Scherrer, P. H. (1975). *Nature* **255,** 539.
—— —— ——, (1976). *J. Atmos. Sci.* **33,** 1113.
Wilson, R. L. (1966). *Geophys. J. Roy. Astron. Soc.* **10,** 413.
Zmuda, A. J. (1973). *Geomagn. Aeronomy.* **13,** 929.

21

THE MAGNETIC FIELD OF THE SUN AND STARS

21.1. The magnetic field of the sun

THE sun is the only star whose magnetic field can be resolved, to give the distribution of field over the surface. For other stars the observer is restricted to measuring the total integrated Zeeman effect, representing, more or less, the algebraic mean (over the visible hemisphere) of the component of field in the line of sight (see discussion in §18.1). In such cases the mean field must be 10^2 G or more to be detectable even in the most favourable circumstances. For the sun, on the other hand, with its large angular size and intense light, the threshold for observation of the mean field is a fraction of a gauss.

The sun is also the only star whose surface motions can be observed, including the non-uniform rotation, circulation, and convection. So it is to the sun, rather than to the more spectacular magnetic stars, that the physicist turns first to explore the origin and the activity of the magnetic fields of stars. The display of magnetic activity on the sun is spectacular, even if its magnitude does not rival some of the more distant stars. The magnetic fields of the sun display effects that no one anticipated. Indeed, there are effects remaining outside our understanding even after years of close observation and theoretical study. It is curious that the sun, being such a mediocre, middle-aged, and generally conservative star (for otherwise we would not be nearby to study it), yet displays so much vigour and variety in its magnetic effects.

The sun appears to be an average dwarf main-sequence star. Hence we infer that most lower main-sequence stars produce similar magnetic effects, too distant for us to detect. Indeed the u.v. Ceti flare stars among the little dM-dwarfs are much more active than the sun (see §18.1). However the sun provides quite enough in the way of well-resolved puzzles to occupy the scientist. We leave it to the philosopher to contemplate the unresolved.

The sun has been studied for millenia by the Chinese astronomer–astrologers so that sunspots are known to have been present on the sun for at least two thousand years. The invention of the telescope in 1610 introduced sunspots onto the European stage and made possible their quantitative study. Oddly, two centuries of telescopic observations elapsed before the 11-year sunspot cycle was pointed out by Schwabe in 1842 (von Humboldt 1851). Carrington (1858, 1863) and Spörer (1894) noted the tendency for the region of formation of sunspots to begin at middle latitudes and drift toward the equator. The general irregularity of the amplitude and period of the sunspot cycle became well known, with the conspicuous failure to find any means of forecasting the time and strength of distant future cycles. There is some tendency for alternate sunspot cycles to be related in strength. A strong (weak) sunspot cycle implies that the cycle 22 years later is likely to be stronger (weaker) than average. That is to say, the 22-year period sometimes manifests itself in the sunspot number as well as in the magnetic polarity. Waldmeier (1957) noted that the more rapid the onset of the spots of a new cycle, and the higher the latitude at which the spots appear, the stronger is the new cycle when it finally reaches its maximum. These characteristics are evident in retrospect from inspection of Wolf's sunspot number plotted for each year, and from inspection of Maunder's butterfly diagrams, showing the evolution of the distribution of sunspots with latitude with the passage of time (Maunder 1904; Spencer Jones 1955; Becker 1955).

It was not till the end of the nineteenth century that Clerke (1894) and Maunder (1890, 1894a,b, 1922) pointed out from the records the curious fact that the sunspots, and, indeed, all outward signs of solar activity, were absent over the last half of the 17th century. And another eighty years elapsed before this curious and portentous fact was taken up seriously, by Eddy (1976, 1977). Eddy discovered from the C^{14} record that solar activity has been absent in a number of earlier centuries too. He went on to point out the remarkable coincidence of cold centuries in the northern temperate zone of Earth with the centuries of inactivity at the sun.

It remained for Hale to detect the Zeeman splitting of the spectral lines from sunspots and thus to show that the 11-year sunspot cycle is, in fact, a magnetic cycle with a period of 22-years. The spots in opposite hemispheres and of each successive sunspot cycle have opposite polarity (Hale 1908a,b, 1913; Hale and Nicholson 1938). It has gradually come to be realized that all the irregularity and activity on, and above, the surface of the sun is a manifestation of the magnetic fields. Were it not for magnetic fields the sun would be a star in the classical mould—placid, serene, and, once the internal energy source is understood and the theoretical neutrino emission squared with observations, not particularly interesting to the professional scientist. Instead the surface of the sun is disfigured by the cool sunspot with its field of 3000 G, distorted by quiescent prominences, disturbed by the frequent coronal eruptions and by flares and subflares, mottled by plages, and enormously pimpled by spicules. Everywhere over the surface of the sun the magnetic field exhibits the remarkable property of self-concentration into isolated flux tubes of 1500 G or more. The corona is intensified by the strong fields of the bipolar magnetic regions. On the other hand, in the weak fields of coronal holes the corona is particularly tenuous and transparent, expanding rapidly to give the faster streams in the solar wind. Everywhere over the surface of the sun there is enhanced heating of the tenuous upper atmosphere (the chromosphere and corona) associated with magnetic fields. The effect is presumed to be a consequence of the dissipation of Alfvén waves generated in the field where the lines of force pass down through the convective zone and/or the dissipation of field energy as a consequence of the wrapping of the lines about each other by the subsurface convection (see the discussion of topological dissipation in Chapter 14; Parker 1972; Glencross 1975). It is this universal dissipative and eruptive aspect of the magnetic field in astronomical circumstances that has motivated the inquiry into the general non-equilibrium of magnetic fields.

The origin of the magnetic fields that drive the activity of the sun is the other side of the question, and the purpose to which the first sections of this chapter are directed. The fields continually erupt through the surface, their creation occurring somewhere below, hidden from view. Indeed the solar magnetic fields are known only insofar as they emerge through the surface of the sun. It was pointed out in §8.8 that there could be a primordial field of as much 10^6 G hidden in the core of the sun, for all we know at the present time. But the general periodic poloidal surface fields represent the poloidal field of the hydromagnetic dynamo below the surface. The fields of the active regions on the surface are made up of sections of the toroidal field that float up to form the familiar east–west bipolar magnetic regions (Parker 1955a, 1975c) as sketched in Fig. 21.1.

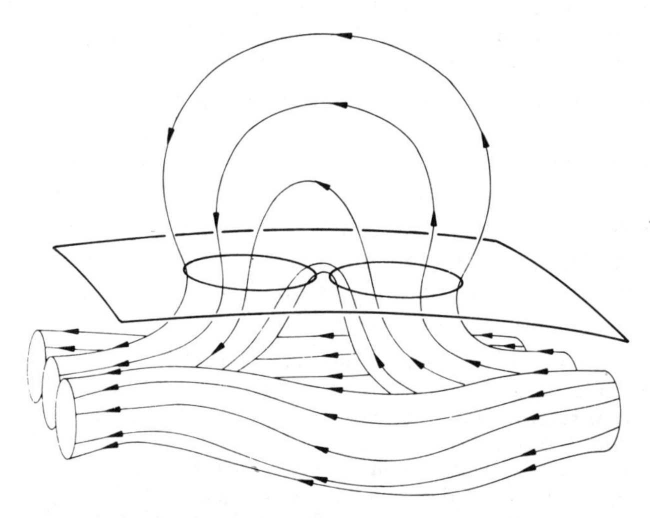

FIG. 21.1. A sketch of the magnetic configuration of a bipolar magnetic region on the surface of the sun. The subsurface configuration is inferred from the continual emergence of toroidal tubes of flux to form the bipolar regions.

The total azimuthal magnetic flux, appearing as bipolar regions on either side of the equator in any 11-year period, is of the general order of 10^{23} Mx. We conclude from this that the azimuthal field in the lower convective zone must be at least 10^2 G, so as to contain as much flux as is observed. On the other hand, the azimuthal field must lie at a depth of at least 10^5 km and cannot be much stronger than 10^2 G if it is to remain submerged for the period of several years necessary for its generation (see discussion of magnetic buoyancy and the rate of rise of flux tubes in §§8.7 and 8.8 and the rate of generation of azimuthal field in §18.2).

It is generally assumed, then, that the bipolar regions in general, and the sunspots in particular, provide a direct indication of the location and relative strength of the underlying azimuthal magnetic field of the sun[1]. With this hypothesis, the works of Schwabe and Hale tell us that the azimuthal field beneath the surface has opposite signs in the northern and southern hemispheres and reverses sign at 11-year intervals.

The work of Carrington indicates that the azimuthal field first develops in middle latitudes and subsequently migrates toward the equator. The strength and/or the depth of the azimuthal field evidently varies from one 11-year cycle to the next, with a tendency for the relative magnitudes of two consecutive 11-year cycles to be repeated on a 22-year basis. The fields in the northern and southern hemispheres are often not well matched because in many 11-year periods the great majority of sunspots are confined to one or the other hemisphere. Waldmeier's work indicates that in those periods when the field grows particularly rapidly, it also appears first at higher latitudes and reaches a higher maximum than the average cycle. Bumba *et al.* (1968) have pointed out that the magnetic regions of the new sunspot cycle often emerge in the midst of magnetic regions still lingering on from the old cycle.

Altogether, then, the azimuthal field first appears in strength at about 40° latitude, with opposite signs in the northern and southern hemispheres. From there it migrates toward the equator over an 11-year period, increasing in strength all the way to about 10° latitude, and then declining toward the equator as it meets the opposite field from the other hemisphere. The reversed azimuthal fields of the new half cycle make their appearance at +40° latitude at about the same time that the old fields are disappearing at the equator.

[1] It is not difficult to think of complications in this simple interpretation, of course. Large-scale circulation of 10 m s^{-1} could displace the flux tubes horizontally by as much as a solar radius during their rise, or block the vertical rise in some regions altogether, while accelerating it in others. The rise may be as much the migration of a dynamo wave as it is the buoyancy of the field (Parker 1977a) (see §19.3.2). But we may as well stick to the simplest interpretation until one or more of these complications is thrust upon us. A discussion of the problem based on a specific model is given by Yoshimura (1976b).

Turning now to the poloidal (meridional) field, the effects are directly visible in the coronal streamers from high latitudes, observed when the disc of the sun is eclipsed by the moon. The poleward migration of the polar prominences (D'Azambuja and D'Azambuja 1948; Anathakrishnan 1954) shows the evolution of the poloidal field in more detail. The positive identification of these effects with the poloidal field and a more or less quantitative measurement of the strength of the poloidal field had to wait for the development of the Babcock solar magnetograph, permitting systematic mapping of the fields at the surface of sun on a regular basis (Babcock and Babcock 1955). With the Babcock instrument, and the many refinements since, the general nature of the dipolar magnetic regions emerged, with the regions of intense field (10–10^2 G) outlined by the plages. The poloidal field appears in a pure form in the polar regions, above latitudes of about 55°. The work of Babcock (1959) and others (Howard 1965, 1972; Gillespie *et al.* 1973) shows that the poloidal and toroidal magnetic fields of the sun are coupled, presumably all part of the same dynamo effect, because they vary in step together throughout the sunspot cycle, the polar fields reversing at about the time of sunspot maximum and the polar fields strongest near sunspot minimum. It was shown that at the onset of each new sunspot cycle the preceding spots of the bipolar magnetic region have the same polarity as the polar field of the same hemisphere. That is to say, the poles resemble the preceding spots of the same hemisphere up to the time of reversal at sunspot maximum.

It is principally the poloidal field of the sun that is extended by the solar wind throughout interplanetary space (Parker 1958, 1963). The extended lines of force come mainly from the middle and high latitudes (Wilcox and Howard 1968; Yoshimura 1977a,b) and make up the observed interplanetary field. The interplanetary field, as observed near the equatorial plane of the sun, is divided into sectors of inward and outward pointing field (Ness and Wilcox 1966, 1967; Wilcox and Ness 1967). There are usually four (but sometimes two) sectors of alternating sign observed in the equatorial plane of the sun.

The axi-symmetric poloidal component of the magnetic field of the sun is most apparent in the magnetograph poleward of about 55° latitude. The mean strength is estimated to be of the order of 5–10 G, although there is considerable uncertainty because the field is so inhomogeneous on a small scale and the observations give only the component in the line of sight, which becomes perpendicular to the mean field direction over the poles (Howard 1977). The polar fields reversed in 1958–1959 at about the maximum of the activity (and azimuthal field strength) at low latitudes, and again in 1969–1971 (Babcock 1959, 1961; Howard 1974b). The south polar field appeared to reverse before the north polar field. On the other hand, Altschuler *et al.* (1974) point out that the polar fields form only a small part of the total north–south axi-symmetric dipole moment of the sun. Most of the total north–south dipole moment at the surface of the sun, in fact, is contributed by the large patches of strong field at middle and low latitudes. Hence the polar fields alone are not necessarily the best criterion for the reversal of the poloidal field.

It is possible to trace the magnetic cycle of the poloidal field in the coronal activity, as indicated by the coronal green line at 5303 Å. The brightness of the line is a direct indication of magnetic activity and heating in the corona. The line brightens first at latitudes of 50° to 70° one to four years before the minimum of the magnetic activity at low latitudes (Trellis 1957). The brightening of the line drifts slowly towards the poles over a period of 4–6 years, arriving at about the time of the peak of the next sunspot maximum, and then disappearing (Müller 1955). The activity of the high latitude prominences also starts up at about the minimum of the low latitude activity and moves poleward (de Jager 1959; Gnevyshev 1967; Yoshimura 1976a, 1977a). These two effects indicate enhancement of the poloidal component of the field at high latitudes during the minimum of the low latitude activity. Altogether then, the poloidal field is a maximum at the time when the toroidal field is a minimum and a minimum when the toroidal field is at its maximum.

Magnetographic studies of the sun (Howard 1974a) suggest the fortieth parallels of latitude as convenient boundaries for the equatorial zone, with a polar zone at each end. About 95 per cent of the magnetic lines of force crossing the surface of the sun do so between latitudes 40° N and 40° S. This broad equatorial zone is the site of the bipolar magnetic regions, where the most intense magnetic fields and the most intense magnetic activity are observed. It is here too that the unipolar magnetic regions emerge (Babcock and Babcock 1955; Bumba and Howard 1965; Wilcox 1971) giving the very strong asymmetry of the magnetic fields, so that one face of the sun is predominantly positive and the other side negative. This asymmetry arises from the asymmetric emergence of the *toroidal* field. Hence it outweighs the much weaker axi-symmetric poloidal field and, as pointed out by Altschuler *et al.* (1974), represents the principal dipole component of the sun, with its axis in the equatorial plane, perpendicular to the axis of rotation. This component is prominent during the maximum activity of each cycle, when the emergence of the toroidal field is most pronounced. It is interesting to note that at sunspot minimum a quadrupole component seems to make a brief appearance (Stenflo 1970, 1971; Howard 1972). The general drift of the large-scale sectors of field relative to the visible surface (Wilcox and Howard 1968; Wilcox *et al.* 1970) has been interpreted variously, as the propagation of hydromagnetic waves (Suess 1971) or as migratory dynamo waves

(Roberts and Stix 1972). It shows the ambiguity of the implications of the general slow motion of sunspots relative to the surrounding photosphere.

Bumba and Howard (1965) and Ward (1965b) have pointed out that the motions of individual magnetic features on the surface of the sun indicate large-scale convection and circulation (see discússion in Simon and Weiss 1968 and the review by Howard 1967). The strongest circulation is the non-uniform rotation of the sun (Carrington 1863; Ward 1965a, b), based on spectroscopic studies and on the motion of sunspots, in which the equatorial regions rotate about 50 per cent faster than the polar regions. There are indications (Eddy *et al.* 1976) that the non-uniform rotation changes with the disappearance of the normal sunspots for a century at a time. The most fundamental unknown aspect of the non-uniform rotation, so far as the solar dynamo is concerned, is the vertical gradient of the angular velocity beneath the visible surface. Unfortunately the available evidence on the question is indirect and not without ambiguity. For instance, the strong magnetic fields of sunspots extend downward from the surface for a distance comparable at least to the radius of the spot, which may be as large as $1{\cdot}5\text{–}2\times10^4$ km. Hence a sunspot might be expected to rotate at a rate representative of the dense gas far below the surface, thereby giving the difference in angular velocity between the surface and the depths. A careful analysis of the rotation rates of sunspots by Ward (1966) discloses slightly different rotation rates for large spots and small spots, and elongated and circular spots. The large long-lived circular spots, for instance, rotate as much as 2 per cent more slowly than smaller or elongated spots. Ward points out that the uncertainties in the measurements are such as to preclude establishing any difference between the rotation at depth and the rotation of surface features (plages, flocculi, etc.). The small backward tilt of solar fields (Howard 1974c), so that the lines of force trail as they extend outward from the surface, suggests that the angular velocity increases downward ($\partial\Omega/\partial r<0$).

There are a variety of theoretical possibilities for the vertical gradient $\partial\Omega/\partial r$. Theoretical models in which the observed non-uniform rotation of the visible surface is presumed to be a consequence of the anisotropy of the eddy viscosity of the convective zone (Kippenhahn 1963; Cocke 1967; Köhler 1970) lead to the conclusion that the angular velocity *decreases* downwards through the convective zone, with the rotation more or less confined to cylindrical surfaces, $\Omega=\Omega(\varpi)$ with $\partial\Omega/\partial\varpi>0$. On the other hand, models in which the observed non-uniform rotation of the visible surface is presumed to be a consequence of meridional circulation (Gierasch 1974; Durney 1974, 1975) lead to the conclusion that the angular velocity *increases* downwards through the convective zone. Other models, involving large-scale Rossby waves have been treated as well. A review and discussion of the problem may be found in Gilman (1974) and in Durney (1976).

21.2. The solar dynamo

21.2.1. Exploratory models and physical principles

If the magnetic field of the sun is continually replenished by the dynamo effect of the non-uniform rotation and cyclonic convection (Parker 1955b, 1957b, 1970), then there are a number of constraints and requirements that can be placed on the fluid motions beneath the surface of the sun. Steenbeck and Krause (1969) have solved the dynamo equations (19.21) and (19.22) in spherical geometry with axial symmetry, giving a formal demonstration that the qualitative behaviour of the theoretical dynamo resembles the behaviour of the fields in the sun, with a toroidal field that forms in middle latitudes and migrates toward the equator, coupled to a poloidal field that reverses in step with the migratory waves of toroidal field. Their model locates the regions of cyclonic convection and non-uniform rotation in separate spherical shells, so that turbulent diffusion is the only communication between shells. They emphasized the fundamental role of turbulent diffusion in the operation of the hydromagnetic dynamo.

Leighton (1969) assembled the dynamo principles into a numerical model confined to a spherical surface. He included a critical value of the toroidal field for the onset of buoyancy and the appearance of sunspots. He built the model around the explicit assumption that the tilt of the bipolar sunspot groups is the sole cyclonic convective effect producing the poloidal field more or less along the lines described by Babcock (1961). He assumed that the horizontal random walk of the supergranules (Leighton 1969) is the sole diffusion. He found that, if the angular velocity of rotation of the sun increases downward across the convective zone by about 10 per cent, in addition to the observed latitude dependence, the behaviour of the calculated magnetic fields bears a striking resemblance to the observed behaviour of the field at the surface of the sun, including the proper butterfly diagrams, and the variation from one half cycle to the next. He found that the oscillatory quadrupole mode, for which the dynamo number is only slightly larger than the oscillatory dipole mode (see §19.3.1.2), was occasionally strongly excited, giving the fields and the sunspot number an extraordinary asymmetry about the equator, much like the asymmetry sometimes observed. He found also the strong fluctuation in the strength of the fields that is observed from one cycle to the next. The numerical experiment, being confined to a surface, does not recognize the role of vertical diffusion except in the loss of azimuthal field through the eruption of bipolar magnetic regions through the surface of the sun. The experiment also does not recognize the vertical migration of the dynamo waves as a consequence of the horizontal shear $\partial\Omega/\partial\theta$ (see §19.3.2), so it is not clear that all of the features are either quantitatively or qualitatively correct. Nonetheless the model is a vivid portrayal of the behaviour of the basic dynamo effect, and represents an important exploratory step. The most interesting aspect of the model is the irregularity of the cycles, so like the observed sunspot numbers and so conspicuously absent from other numerical models to be found in the literature.

Exploring in another direction Gilman (1969) constructed a numerical experiment on the solar dynamo, in which the dynamics of the convection is included with the dynamo equations. To make the problem tractable he idealized the convective zone to a thin ring of rectangular cross-section, equivalent to the rectangular geometry treated in §19.3. He included, too, a modest variation of density with depth in the quasi-Boussinesq approximation. In place of the small-scale cyclonic turbulence leading to (19.2) and (19.22) he worked directly with the hydromagnetic equations (4.12), introducing Rossby waves of large dimensions (comparable to the depth and width of each 'hemisphere' of the dynamo region) and driven by a small temperature gradient between the equator and the poles along the lines proposed earlier by Ward (1064, 1965a). The Rossby waves, then, give a local lift and a twist to the lines of force of the azimuthal magnetic field to produce the poloidal field. The loops are followed in detail by the computer, rather than treating the net average contribution to the azimuthal vector potential as in Chapters 18 and 19. In Gilman's dynamo the diffusion soon spreads out the loops, of course, and merges them to form the overall poloidal field. The Lorentz forces of the field on the fluid were included in the calculation, so that, starting with weak fields, the fields grow rapidly and then level off as the convective and inertial forces come into balance with the magnetic forces. In view of the complexity of the calculation, several simplifications were introduced, based on what is known of the behaviour of Rossby waves under other circumstances; the numerical scheme divided the region vertically into four layers, separated by five equally spaced grid points (planes) so the numerical accuracy is not high. But vertical diffusion is included and the vertical migration of dynamo waves is permitted. The results are that the fields are oscillatory, with a strong toroidal field appearing at middle latitudes and migrating toward the equator if the angular velocity of rotation of the sun increases downward through the convective zone—the usual condition. The variations of the field settle into a periodic state, with a period of about 2 years when the non-uniform rotation and turbulent diffusion were adjusted to reproduce the form of Maunder's butterfly diagrams. Gilman pointed out that the simplified numerical scheme contains so few degrees of freedom that the regularity is not surprising. He showed, too, the curious fact that, as the diffusion is reduced, the oscillations of the field become increasingly irregular, with variations in the amplitude of succeeding cycles by a factor of the order of two. This point does not turn up in any of the kinematical numerical experiments of which we are aware, so it may be the dynamical effect of the stronger small-scale fields that arise when the diffusion is reduced. It would be interesting to explore the effect further. The period of the oscillations is not greatly changed by the decreasing diffusivity, although the period is reduced somewhat as the magnetic forces begin to inhibit the fluid motions.

21.2.2. Detailed models

In addition to the exploratory studies in spherical geometry, there have been a number of investigations into the quantitative aspects of specific models for the kinematic solar dynamo. The purpose, besides providing an illustration of the solar dynamo, is to deduce the non-uniform rotation rate and the dynamo coefficient (which are neither subject to direct observation nor yet available from the formal dynamical theory) from the observed variation of the surface fields in space and time.

21.2.2.1. Köhler's model. Köhler (1973) constructed an elaborate numerical experiment in which the convective zone of the sun, representing a spherical shell of thickness h, is subject to the observed surface variation of angular velocity Ω with latitude, plus a linear variation of Ω with radius r. The linear variation is expressed in terms of the total change $\Delta\Omega$ across the depth $h = 0{\cdot}3225\ R_\odot$ of the convective zone, denoted by the fraction g of the angular velocity Ω_s at the equatorial surface ($\Omega_s \cong 3\times10^{-6}$ rad s^{-1}), so that $\Delta\Omega = g\Omega_s$. Köhler supposes that both the turbulent diffusion η_T and the dynamo coefficient $\Gamma \equiv \alpha$ vanish at the top and bottom of the convective zone and have a parabolic profile between, with the peak values, at a depth $\frac{1}{2}h$, denoted by $\eta_* \equiv c_\eta^{-1}\ 10^{13}$ cm s^{-1} and $\alpha^* \cos\theta \equiv c_\alpha \cos\theta \times 10$ cm s^{-1}, respectively. Thus Köhler takes the dynamo coefficient to vary as $\cos\theta$, in proportion to the strength of the Coriolis forces on a rising expanding convective cell. The factors c_η and c_α are free parameters, and the dynamo number is proportional to $gc_\eta^2 c_\alpha$. He denotes by P the combination $c_\eta^2 c_\alpha$. Köhler makes the usual assumption that the shear g is sufficiently strong that the toroidal field B_ϕ is large compared to the poloidal field $\nabla\times\mathbf{e}_\phi A_\phi$, so that the dynamo equations reduce to (19.21) and (19.22) (expressed in spherical coordinates, of course). He takes the lower boundary of the convective zone to be impenetrable, so that $B_r = 0$ there, and the upper boundary to be free so that B_ϕ vanishes there and the poloidal field outside is of potential form. He expands the azimuthal field and vector potential in the axi-symmetric form $\Sigma C_n(r)P_n^1(\cos\theta)\exp(i\omega t)$ and proceeds to compute the radial functions $C_n(r)$ by numerical methods[2]. The angular frequency ω is complex, of course, the magnitude of the real and imaginary parts depending upon the parameters g and $P = c_\eta^2 c_\alpha$. Since the radial gradient of the angular velocity Ω is unknown and parametrized by g, Köhler was particularly interested in studying its effect on the observable fields at the surface of the sun. He begins, then, with a study of the case $c_\alpha > 0$ and $g = 0$, so that the only shear is the observed non-uniform rotation $\partial\Omega/\partial\theta$. The computations show that the azimuthal field appears first, and begins to grow at the beginning of each new magnetic cycle, at the middle latitude $\theta \cong 45°$. The field spreads out from there in both directions, toward the pole and the equator. The maximum of the field remains at $\theta \cong 45°$, as illustrated by Fig. 21.2, reproduced from Köhler's paper. This is different from the result obtained by Leighton (1969) for the same circumstances, where the field migrated toward the equator instead of standing still while spreading out in both directions. To understand the difference, note from §19.3.2 that the migration of the dynamo waves is vertically upward when the angular velocity is independent of r and varies only with latitude. Hence the horizontal phase velocity of the dynamo waves at the upper surface of the dynamo region is a question of the deflection of the dynamo waves as they reach the surface, and that depends upon the detailed distribution of the dynamo coefficient and the field dissipation. So the qualitative difference between Leighton's and Köhler's models is not surprising. The numerical models are themselves qualitatively different, with the vertical dimension absent from Leighton's model but included in Köhler's.

When vertical shear is introduced, the azimuthal field migrates as a whole, rather than merely spreading out equally in both directions. As pointed out in §19.1.2.2 the direction of migration depends upon the sign of the shear g. The migration is toward the equator if the angular velocity increases downward through the convective zone $(g > 0)$. Hence the observed behaviour of sunspots requires a non-zero and positive value for g.

With increasing positive g Köhler finds that the azimuthal field begins its growth at higher and higher latitudes. For instance, for $g = 0{\cdot}2$ the azimuthal field first appears at $\pm 65°$ latitude, as may be seen from Fig. 21.3, reproduced from Köhler's paper. The field grows as it migrates toward the equator, reaching a maximum strength at about $\pm 20°$ latitude and then dying out closer to the equator as the dynamo coefficient (proportional to $\cos\theta$) goes to zero and as the field mixes with the

[2] Köhler writes $\exp(i\Omega t)$ instead of $\exp(i\omega t)$, but we have used Ω to describe the angular velocity of the sun. To avoid confusion in the present text, ω is used here for the dynamo frequency.

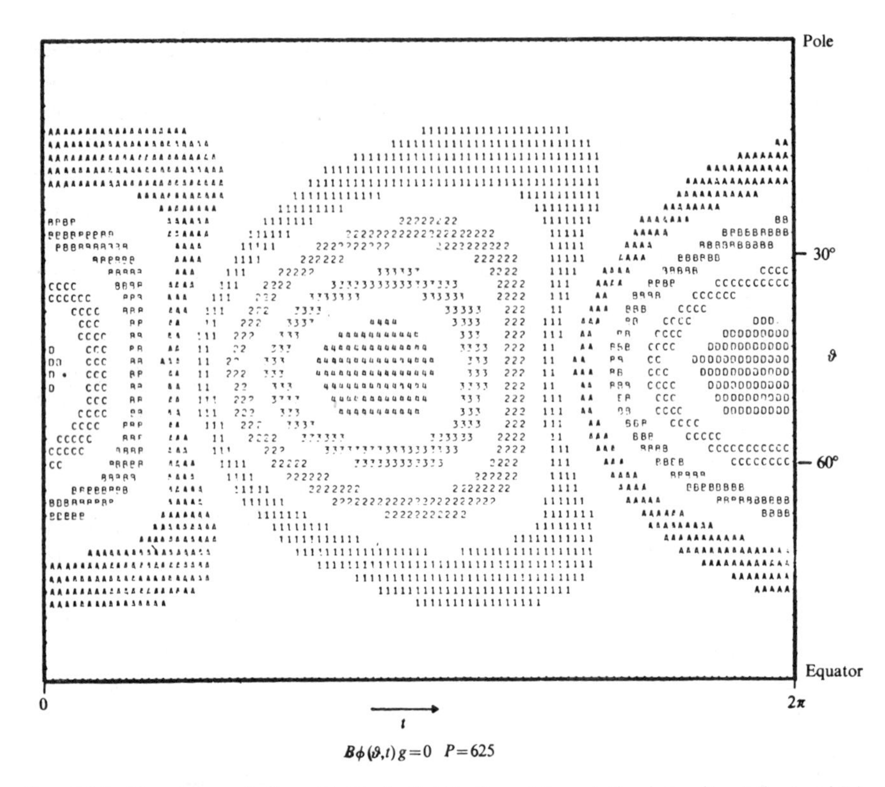

FIG. 21.2. The time evolution of latitude distribution of the relative intensity of the toroidal magnetic field immediately below the upper surface of Köhler's dynamo model with vanishing radial shear ($g=0, P=625$). The numbers represent the relative intensity of one polarity and the letters the other polarity (reproduced from Köhler 1973).

opposite field from the other hemisphere. If the number and size of the sunspots observed at the surface of the sun are assumed to be a direct measure of the azimuthal field beneath the surface, then the model would have an occasional sunspot appearing at latitudes above 60°, contrary to observation. But if it is assumed, after Babcock (1961) and Leighton (1969), that the azimuthal field must exceed some threshold before magnetic buoyancy brings the fields to the surface to form spots, then the first appearance of spots at 40° latitude is readily explained by suitable adjustment of the threshold.

Köhler points out that the larger the diffusion coefficient, the faster the diffusion takes place and the shorter the period of the oscillation of the fields, assuming that g and c_α are made sufficiently large to compensate for the increased diffusion. The basic relation is (19.27) for constant amplitude, for which the period of the oscillation is inversely proportional to the diffusion coefficient (Fig. 21.3 was computed for $g = 0{\cdot}2$, $P = 125$, $\mathrm{Im}(\omega)/\mathrm{Re}(\omega) = 3{\cdot}62 \times 10^{-3}$, so it represents nearly constant amplitude).

Köhler gives a plot of the real and imaginary parts of ω for various values of g and P. The most interesting case is for constant amplitude ($\mathrm{Im}(\omega) = 0$), a period of about 22 years ($\mathrm{Re}(\omega) = 0{\cdot}91 \times 10^{-8}\,\mathrm{s}^{-1}$), and a migration of the azimuthal field toward the equator at a mean rate of about 30° per 11 years, given by the slope of the Maunder butterfly diagrams. Köhler expresses ω as $\tilde{\omega}\eta^*/R_\odot^2$ where $R_\odot = 0{\cdot}7 \times 10^{11}$ cm is the radius of the sun, and η^* is the peak value of η, at the middle of the convective zone. It follows from his plot of the real and imaginary parts of ω that for steady amplitude ($\mathrm{Im}(\omega) = 0$) and a migration rate of 30° per 11 years, the model dynamo requires $g \cong 0{\cdot}24$ and $P \cong 110$, yielding $\mathrm{Re}(\tilde{\omega}) = 67$. From the observed value of $\mathrm{Re}(\omega)$, it follows that $\eta^* \cong 0{\cdot}66 \times 10^{12}\,\mathrm{cm}^2\,\mathrm{s}^{-1}$. The mean value of η across the convective zone is, then, $\frac{2}{3}\eta^* = 0{\cdot}44 \times 10^{12}\,\mathrm{cm}^2\,\mathrm{s}^{-1}$. It follows from the value of η^* that the

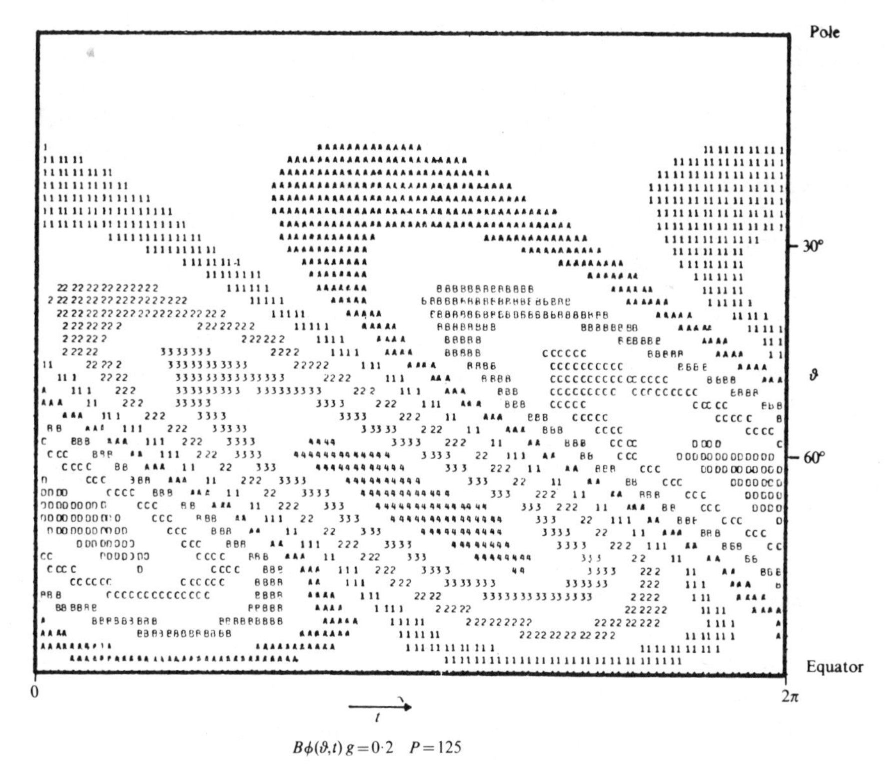

$B\phi(\vartheta, t)\ g = 0{\cdot}2 \quad P = 125$

FIG. 21.3. The time evolution of the latitude distribution of th relative intensity of the toroidal magnetic field immediately below the upper surface of Köhler's dynamo model with moderate radial shear ($g = 0{\cdot}2$, $P = 125$). The numbers represent the relative intensity of one polarity and the letters the other polarity (reproduction from Köhler 1973).

parameter c_η has the value 15·1. With $P=110$, then, $c_\alpha \cong 0{\cdot}5$. Hence the maximum value of the dynamo coefficient, at the middle of the convective zone, is $\Gamma = 5\cos\theta\ \mathrm{cm\,s^{-1}}$. From the value $g=0{\cdot}24$ it follows that the angular velocity increases by $\Delta\Omega = 0{\cdot}24\Omega_s = 0{\cdot}7\times10^{-6}\ \mathrm{rad\,s^{-1}}$ from the surface at the equator to the bottom of the convective zone.

In §19.2.2.2 it was estimated, from the basic properties of the plane dynamo wave, that a constant amplitude and a period of 22 years in a convective zone with a depth of 2×10^5 km require a diffusion coefficient of the general order of $0{\cdot}4\times10^{12}\ \mathrm{cm^2\,s^{-1}}$, remarkably close to Köhler's mean value. Hence Köhler's numerical results are readily understood in terms of the general properties of the free dynamo wave.

It is interesting, then, to go on to compare the rectangular dynamo results with Köhler's quantitative conclusions. Note that the maximum values of Γ and η used by Köhler give a dynamo number $(Kh)^3 \equiv \Gamma G h^3/\eta^{*2}$, where G is the shear rate $r\partial\Omega/\partial r \cong -\Delta\Omega R_\odot/h = -2\times 10^{-6}\ \mathrm{rad\,s^{-1}}$ at the equator for $h=0{\cdot}32R_\odot = 2{\cdot}2\times10^{10}$ cm. It follows that $Kh=-6{\cdot}2$. In terms of the pole–equator length λ of the convective zone, note that $\lambda = \frac{1}{2}\pi R_\odot \cong 5h$, so that $K\lambda = -30$ based on the maximum values. The mean value averaged over the surface area is $-30/\pi^{\frac{1}{3}} \cong -20$. Turning back to (19.131) we see that the dynamo number for the dipole mode was computed as $K\lambda = -7{\cdot}43$ and $-11{\cdot}1$ by the two expansion schemes. In view of the different geometries and the different spatial distributions of Γ used in the rectangular model and Köhler's model, the agreement is as close as can be expected.

21.2.2.2. Yoshimura's model. A more elaborate numerical model of the solar dynamo has been put together by Yoshimura (1971, 1972a,b, 1973, 1975,b) with the final kinematical model based on considerations of the dynamics of the non-uniform rotation and associated meridional circulation of the sun. In the preliminary papers Yoshimura considers a variety of effects including a large-scale non-axi-symmetric convection and circulation, somewhat along the lines of Gilman's (1969) Rossby waves. Yoshimura suggests that the helicity of the convection has opposite signs in the upper and lower convective zone. The idea is that the expansion and consequent *retrograde* rotation of rising fluid currents dominate the upper convective zone, so that the dynamo coefficient Γ (as defined in (19.1)) is positive in the upper levels of the convective zone in the northern hemisphere and negative in the upper levels in the southern hemisphere. The converging flows feeding the rising currents lead to *direct* rotation of the rising fluid in the lower convective zone, so that Γ is negative in the northern and positive in the southern hemispheres. An extensive discussion of that dynamical picture has been given by Durney (1975) (see also Köhler 1966; Durney 1972, 1974) showing the apparent consistency of the overall idea that the non-uniform rotation is driven by a general meridional circulation with $\partial\Omega/\partial r<0$ (Gilman 1977; Belvedere and Paterno 1977). The global picture is that fluid rises at the poles, flows toward lower latitudes and sinks in the general vicinity of the equator. The characteristic meridional velocity is 1–2 m s^{-1} while the rising and sinking currents are of the order of 0·1 m s^{-1}. Unfortunately such motions lie below the observational limits of about 10 m s^{-1}, so the ideas cannot be directly checked.

In any case Yoshimura (1975b) uses the non-uniform angular velocity

$$\Omega(r,\theta)=\Omega_s(1+G_{\theta 1}\sin^2\theta+G_{\theta z}\sin^4\theta)\{1-G_r(r_s-r)/r_s\},$$

where Ω_s is the angular velocity at the equatorial surface, r_s is the radius of sun, and $(G_{\theta 1}, G_{\theta z}, G_r)$ are dimensionless constants describing the variation of the angular velocity with latitude ($G_{\theta 1}\cong -0{\cdot}2$, $G_{\theta z}\cong 0$, Newton and Nunn 1955). Lacking a detailed model for the radial variation of Ω Yoshimura adopts the simplest form, employing a linear variation with r. Since $G_{\theta 1}$, $G_{\theta z}$, and G_r are small compared to one, it is evident that Yoshimura and Köhler are using essentially the same form for $\Omega(r,\theta)$.

In the representation of the dynamo coefficient $\Gamma(r,\theta)$ Yoshimura denotes by r_c the lower boundary of the convective zone and writes $r=r_b$ for the spherical surface within the convective zone across which the dynamo coefficient changes sign. Thus in the upper convective zone, between r_b and the surface r_s in the northern hemisphere he writes

$$\Gamma(r,\theta)=+\Gamma_0 N_u(\sin\theta)^{R_{\theta 1}}(\cos\theta)^{R_{\theta 2}}\exp(R_{r1}r/r_s)\sin\left\{\pi\left(\frac{r-r_b}{r_s-r_b}\right)\right\} \quad (21.1)$$

where N_u, $R_{\theta 1}$, $R_{\theta 2}$, and R_{r1} are all dimensionless parameters. In the lower convective zone between r_c and r_b in the northern hemisphere he writes

$$\Gamma(r,\theta)=-\Gamma_0 N_L(\sin\theta)^{R_{\theta 1}}(\cos\theta)^{R_{\theta 2}}\exp(R_{r1}r/r_s)\sin\left\{\pi\left(\frac{r-r_c}{r_b-r_c}\right)\right\}. \quad (21.2)$$

The free radius r_b is expressed in terms of r_c by the parameter R_{r2}, so that $r_b=r_c+R_{r2}(r_s-r_c)$. Hence R_{r2} represents the fraction of the thickness of the convective zone committed to the reverse Γ. The same forms, with opposite signs, are used for the southern hemisphere.

For boundary conditions, Yoshimura takes the lower surface of the convective zone to be impenetrable ($\eta=0$) so that the normal component B_r vanishes there. At the outer surface r_s of the convective zone he supposes that $B_\theta=0$ for reasons of mathematical simplicity. (Köhler (1973), it will be recalled, assumed $\eta=\infty$ outside the sun, so that the external field is a potential field fitting smoothly onto the interior field). It is not clear whether Yoshimura's simplification to $B_\theta = 0$ has any essential effects on the operation of the dynamo.

Now the kinematical dynamo equations are linear and homogeneous so that the task of computing a steady oscillation of fixed amplitude is an eigenvalue problem for the dynamo number. On the other hand, if there is introduced into the equations some effect which seriously limits the growth of the fields above some critical value, then it is possible to assume any value for the dynamo number, provided only that it is suitably large, and to proceed with the calculation, obtaining a steady oscillation after the passage of a few cycles. To this end Yoshimura subtracts out a portion of the azimuthal field B_ϕ and azimuthal vector potential A_ϕ whenever the magnitude of the toroidal field exceeds a critical value B_c. The arithmetical subtraction is presumed to represent a loss of field as a consequence of the magnetic buoyancy, more or less along the lines used earlier by Leighton (1969) in his representation of the dynamo. Thus, if the maximum value of rB_ϕ, denoted by Φ_{max}, exceeds some critical value $B_c r_s$, then a fraction E_r of both B_ϕ and A_ϕ is removed, where

$$E_r \equiv (1 - B_c r_s/\Phi_{max})^{E_x}$$

where $E_x (\ll 1)$ is a number to be chosen later. Otherwise the dissipation is limited to the uniform resistive diffusion coefficient η.

Yoshimura (1975b) explores a number of cases, defined by the several parameters. The fundamental obstacle to a unique solution is the fact that the number of free parameters in the model exceeds the number of observable quantities to which the model can be fitted. There are, altogether, the six free parameters $R_{\theta 1}$, $R_{\theta 2}$, R_{r1}, R_{r2}, B_c, and E_x, in addition to the usual specification of the dynamo coefficient Γ_0 and the vertical and horizontal shear, and the diffusion coefficient η. The coefficients $G_{\theta 1}$ and $G_{\theta 2}$ are prescribed by observation of the surface of the sun, of course. Yoshimura uses the observed 22-year period of the sunspot cycle and the usual assumption that the butterfly diagram for the sunspots marks the location of the azimuthal field beneath and, hence, gives the rate of migration of the azimuthal field. He notes, in addition, that the migration of poloidal field from middle latitudes toward the poles (inferred from high latitude prominences and coronal activity, and beginning just prior to the onset of each new sunspot cycle), provides a number of new constraints on the theoretical model. First of all it requires that there be a secondary dynamo wave migrating poleward, originating at middle latitudes at about the time of sunspot minimum and developing to a strength somewhat less than the main toroidal field moving toward the equator. Further, the relative magnitudes of the general poloidal and toroidal fields of the sun determine the relative strengths of Γ_0 and $G_{\theta 1}$, G_r.

Yoshimura states a number of general conclusions based on the case studies. First of all, he finds that both vertical and horizontal shear, G_r and $G_{\theta 1}$, are necessary to give a complete representation of the observed behaviour of the solar dynamo. For $G_{\theta 1}$ (i.e. $\partial\Omega/\partial\theta$) alone the dynamo waves have two branches, starting at middle latitudes and migrating poleward and equatorward. The toroidal fields of the two branches are about equally strong, as in Köhler's example for $\partial\Omega/\partial\theta$ alone (see Fig. 21.2). As in Köhler's model, it is necessary to add a negative $\partial\Omega/\partial r$, introducing a migration toward the equator, to pull the field away from the poles and reduce the toroidal field of the polar branch of the dynamo wave to the small magnitude that is observed. It should be noted, however, that the radial gradient reduces the toroidal field but enhances the poloidal field, automatically accounting for radial fields somewhat stronger and more active in the poleward branch than in the equatorward branch, in accordance with observation. If the variation of Ω with latitude is dropped altogether ($G_\theta = 0$), then there is no poleward branch. The fields originate at high latitude and move toward the equator. It would appear, then, that the magnetic activity originating at high latitudes a year or so before the appearance of the first spots of a new cycle is a direct product of the equatorial acceleration of the sun, $\partial\Omega/\partial\theta$. The radial gradient $\partial\Omega/\partial r$ partially suppresses the development of this poleward branch. On the other hand, the principal fields, arising at middle and high latitudes, from where they migrate toward the equator, are mainly the product of the radial gradient $\partial\Omega/\partial r$. In this way, then, the details of the subsurface rotation of the sun are directly reflected in the behaviour of the surface fields.

Yoshimura points out that the set of values

$$G_{\theta 1} = -0{\cdot}2, \quad G_{\theta 2} = 0, \quad G_r = -0{\cdot}1, \quad R_{\theta 1} = 2, \quad R_{\theta 2} = 3,$$
$$R_{r1} = 1, \quad R_{r2} = 0{\cdot}75, \quad r_c = \tfrac{1}{2} r_s, \quad r_b = \tfrac{7}{8} r_s, \quad E_x = 10^{-3}$$

gives a good representation of the observed surface fields of the sun, although there are other combinations that do just as well. If the fields are expressed in units of B_0, equal to the critical value for the present example, the results are as illustrated in Figs. 21.4 and 21.5 (both reproduced from Yoshimura 1975b). The calculations were performed for the fixed value $\Gamma_0\Omega_s r_s^3/\eta_0^2 = 2\times 10^5$ for the dynamo number, with Γ_0 and η_0 then adjusted to the values $\Gamma_0 = 4{\cdot}8$ cm s^{-1} and $\eta_0 = 0{\cdot}16\times 10^{12}$ cm^2 s^{-1} to bring the period of the dynamo to 22 years. The procedure was to specify an initial, relatively weak field that is anti-symmetric about the equatorial plane. The time dependent equations were then integrated step by step. The field grows rapidly for a few cycles until the growth is limited by the non-linear loss, whereupon it settles into an oscillation with

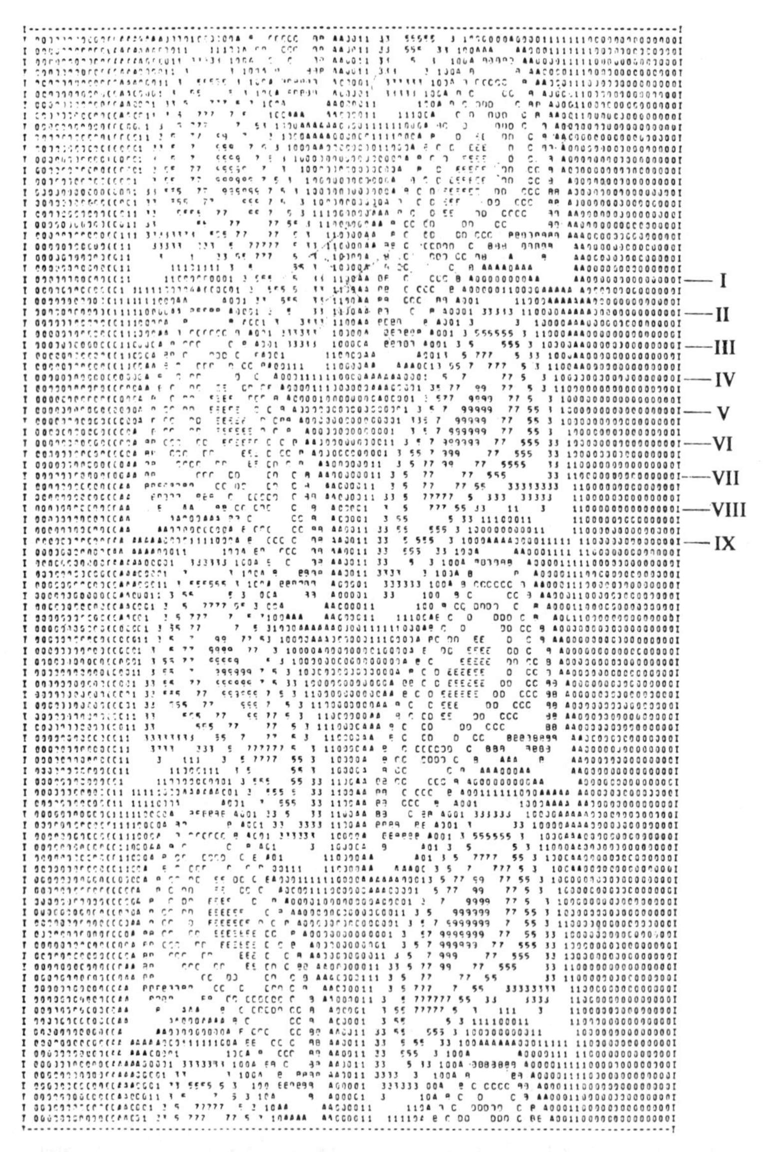

FIG. 21.4. The time evolution of the latitude distribution of the relative intensity of the toroidal magnetic field immediately below the upper surface of Yoshimura's standard dynamo model. The abscissa is sin θ with the south pole at the left and the north pole at the right. The time runs vertically and corresponds to a total period of 48 years for the sun. The odd numbers represent the relative strength of one polarity and the letters the other polarity (reproduced from Yoshimura 1975b).

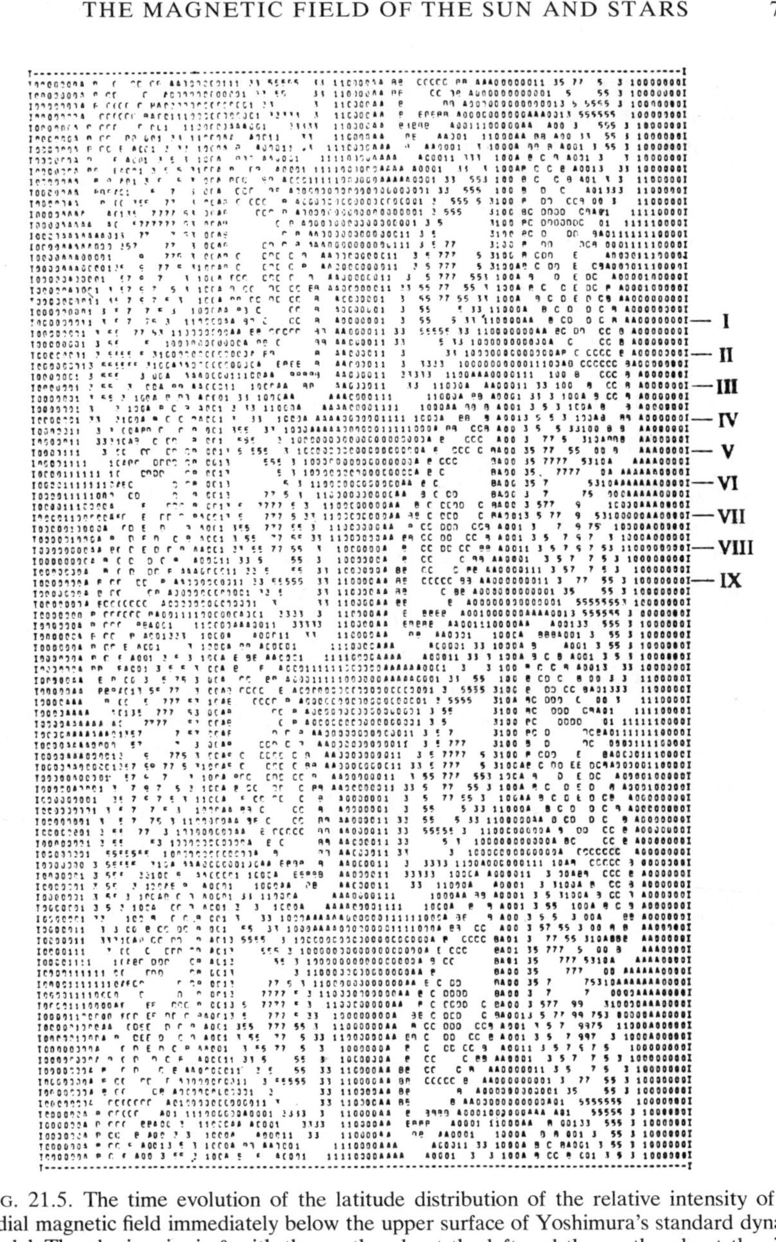

FIG. 21.5. The time evolution of the latitude distribution of the relative intensity of the radial magnetic field immediately below the upper surface of Yoshimura's standard dynamo model. The abscissa is sin θ with the south pole at the left and the north pole at the right. Time runs vertically and corresponds to a total period of 48 years for the sun. The odd numbers represent the relative strength of one polarity and the letters the other polarity (reproduced from Yoshimura 1975b).

essentially constant amplitude. Figure 21.4, shows the variation of the toroidal field in the steady oscillations, showing the small poleward branch. Figure 21.5 shows the radial component of the field. The polar branch of the radial field is clearly separated, and is somewhat stronger, than the branch that migrates toward the equator in association with the dominant toroidal field.

Comparison of the toroidal field strength shown in Fig. 21.4 with the Maunder butterfly diagrams of the sun requires some assumption as to the threshold for the field strength to produce sunspots. If, for instance, the threshold for the production of spots is taken to be half the maximum value, then only the equatorward branch of the toroidal field produces sunspots. If the threshold is one-third, then a few spots appear in the poleward wave. Becker (1958) suggests that such spots occasionally appear. On the other hand, if the threshold is larger than the maximum reached by the toroidal field, no sunspots appear at all. Yoshimura raises the question of whether it is something of this nature that led to the 70-year Maunder minimum, with a fluctuation of some sort dropping B_ϕ below the threshold for the production of sunspots. Yoshimura's model is too stable to produce any significant fluctuations in field strength, but Leighton's (1969) model is notable for the variation in amplitude from one cycle to the next. It should be noted, however, that if the subsidiary historical evidence from solar eclipses, pointed out by Eddy (1976, 1977), is correct, there were no active coronal regions and hence no biploar magnetic regions, with or without sunspots, during much of the Maunder minimum. This would appear to be a more stringent requirement than the mere absence of fully developed sunspots.

Yoshimura (1975b) gives extensive graphic displays of the poloidal field lines and the distribution of the toroidal field in the meridional planes throughout the dynamo cycle, to which the interested reader is referred. He also shows the variation of the fields in a dynamo in which $\partial\Omega/\partial r = 0$, and in a dynamo in which the upper portion of the convective zone, $r_b < r < r_s$ (in which the dynamo coefficient is given by (21.1)) is deeper, extending from the surface down to $0{\cdot}775\ r_s$. He points out that there is a tendency for the principal toroidal field to reach its maximum at lower latitudes for deeper convective zones (i.e., smaller r_c).

21.2.2.3. Stix's model. More recently Stix (1976a,b) has carried out numerical experiments to study dynamos in which the rotation rate Ω is constant on cylindrical surfaces, $\Omega = \Omega(\varpi)$, and to contrast their behaviour with that of dynamos in which Ω is constant on spherical surfaces, $\Omega = \Omega(r)$, as described above. He has given particular attention to the question of the phase relationships of the poloidal (radial) and azimuthal field components in order to determine the sign of $\varpi\partial\Omega/\partial\varpi$ separately from the dynamo number (involving the product $\Gamma \partial\Omega/\partial\varpi$). He points out that the phase relations inferred from observation bring one face to face with the question of the time lag between the generation of the azimuthal field (at depths of 10^5 km, see §§8.7 and 8.8) and the arrival of the buoyant flux tubes at the surface to produce the bipolar magnetic regions. Köhler and Yoshimura placed the upper surface of their dynamos near the surface of the sun and compared the calculated fields of the upper surface of the dynamo with the observed Maunder butterfly diagram of the erupting sunspots. Stix incorporates the fact that the generation of azimuthal field lies mainly at depths of 10^5 km or more, as a consequence of the rapid rate of rise at any higher level (see discussion in §21.2.3). Consequently, Stix uses the maximum azimuthal field, arising at some depth, for comparison with the Maunder butterfly diagrams.

Stix chooses the non-uniform rotation to be of the form $\varpi\partial\Omega/\partial\varpi \propto \varpi^2$. Then he writes the dynamo coefficient as $\Gamma = \alpha_0 G(r)\cos\theta$ and considers several different forms for the radial variation $G(r)$. In particular he uses

$$G(r) = \pm 1,$$

$$G(r) = \pm\tfrac{1}{2}[1 + \Phi\{(r - r_1)/d\}],$$

$$G(r) = \begin{cases} 0 & \text{for} \quad r < r_c \\ -\exp\{r - \tfrac{1}{2}(r_c + r_b\}\sin\left(\pi \dfrac{(r - r_c)}{r_b - r_c}\right) & \text{for} \quad r_c \leq r \leq r_b \\ +\exp\{r - \tfrac{1}{2}(r_c + r_b)\}\sin\left(\pi \dfrac{(r - r_b)}{1 - r_b}\right) & \text{for} \quad r_b \leq r \leq 1, \end{cases}$$

$$G(r) = -\sin\left(2\pi \frac{(r - r_c)}{1 - r_c}\right),$$

where Φ is the error function. The third example for G resembles Yoshimura's model, with the dynamo coefficient of opposite sign in the upper and lower levels of the convective zone. The outer radius is normalized to $r = 1$. Stix fits an external potential field to the dynamo field at the surface $r = 1$. He calculates the critical dynamo numbers for the oscillatory solutions for which B_ϕ is either symmetric or antisymmetric about the equatorial plane. The first three numerical experiments, using the first three functions $G(r)$, are conducted for the dynamo extending throughout the entire sphere. The calculations are repeated for the case where the inner half ($r < \tfrac{1}{2}$) of the sphere has a resistive diffusion coefficient η so small that the oscillatory fields are unable to penetrate into it, thereby simulating a stable core beneath the convective zone (confined to $r < \tfrac{1}{2}$). The results show that cycindrical rotation $\Omega(\varpi)$ leads to the anti-symmetric oscillatory mode at somewhat lower dynamo numbers than the symmetric mode when the product $\Gamma\partial\Omega/\partial\varpi$ is negative (or mainly negative) in the northern hemisphere and positive in the southern hemisphere. On the other hand, positive $\Gamma\partial\Omega/\partial\varpi$ in the northern hemisphere favors symmetric B_ϕ. The azimuthal field is observed to be an odd function of latitude (i.e. anti-symmetric), suggesting once again that $\Gamma\partial\Omega/\partial\varpi$ is negative over the major part of the northern hemisphere of the convective zone. This is the same result as in the spherical models with $\Omega = \Omega(r)$ treated by Köhler and Yoshimura (see also the earlier models of Roberts 1972; Roberts and Stix 1972; Deinzer *et al.* 1974) and in the rectangular models presented in §19.3. The anti-symmetric nature of B_ϕ is closely related, then, to the migration of the fields toward the equator at low latitudes.

Stix points out that if Ω is assumed to be a function only of the distance ϖ from the axis of rotation, rather than a function of radius r, then the observed non-uniform rotation of the surface decrees that $\partial\Omega/\partial\varpi > 0$ He goes on, then, to consider the phase relation between the poloidal and toroidal fields to check the sign of $\partial\Omega/\partial\varpi$.

The radial component of the magnetic field is directly observed at middle and low latitudes with the magnetograph. The facts are that B_r, observed directly, and B_ϕ, inferred from the polarity of the bipolar magnetic regions, followed each other closely through the 1965–1975 sunspot cycle, with B_r negative at low latitudes in the northern hemisphere where B_ϕ was positive. That is to say, the two were π out of phase. Stix points out that the phase difference between the radial and azimuthal fields in the propagating dynamo wave depends upon the sign of $\partial\Omega/\partial r$, as well as upon the sign of the product $\Gamma\partial\Omega/\partial r$, as is evident from (19.37) and (19.39). This property is reflected in Stix's present numerical experiments (see also Steenbeck and Krause 1969; Deinzer and Stix 1971; Stix 1976a), which require that $\partial\Omega/\partial r < 0$ if B_r and B_ϕ are to be approximately π out of phase with each other. Hence, a purely cylindrical model, for which $\partial\Omega/\partial\varpi$ is *constrained* to *positive* values by the observed equatorial acceleration of the solar surface, seems to be ruled out. The phase difference, plus the fact that B_ϕ is observed to be an odd function of latitude, dictate the $\Gamma > 0$ with $\partial\Omega/\partial r < 0$ over the major part of the convective zone in the northern hemisphere, and $\Gamma < 0$, $\partial\Omega/\partial r < 0$ over the major part of the southern hemisphere. This conclusion is particularly interesting because of the theoretical dynamical questions about the non-uniform rotation and the dynamical *possibility* that the angular velocity Ω is constant on cylindrical, as opposed to spherical, surfaces (see, for instance, Köhler 1970; Gilman 1974, 1977). The same dynamical question arises in connection with the convecting core of Earth, where the boundary conditions are quite different, of course (see Busse and Cuong 1977).

21.2.3. *Discussion*

Altogether the numerical experiments provide an instructive display of the dynamo effects of the convective zone of the sun, or other star. The use of two levels in Yoshimura's model, with opposite dynamo coefficient, is novel; other authors model the dynamo with only one level. The calculations of Köhler and Yoshimura, for all of their arbitrary features, indicate that both $\partial\Omega/\partial r$ and $\partial\Omega/\partial\theta$ contribute to the operation of the solar dynamo. It is clear, too, that the surface behaviour of the solar magnetic fields is explained most simply by an angular velocity that increases downwards ($\partial\Omega/\partial r<0$). Only a very modest change in Ω is required across the convective zone. For Yoshimura's value $G_r=-0{\cdot}1$ there is a difference $\Delta\Omega$ of only $0{\cdot}05\,\Omega_s$ across the dynamo region ($r_c=0{\cdot}5\,r_s$). Note that Köhler's model operates with a somewhat larger difference, $\Delta\Omega=0{\cdot}24\,\Omega_s$ and a diffusion coefficient about four times larger, in a thinner convective zone $r_c=0{\cdot}62\,r_s$. The magnitudes of the dynamo coefficients are essentially the same for the two models, with somewhat different distributions, of course. The upper level of Yoshimura's model is less than half as thick as the whole convective zone (presumed to have a depth of 2×10^5 km) while the lower level in which Γ is reversed, extends down to $3{\cdot}5\times10^5$ km and is much deeper than the conventional convective zone. It would be interesting to know how the numerical model behaves when the depth of the dynamo is only as large as the conventional depth of 2×10^5 km.

Note that Yoshimura's dynamo number, defined in terms of the radius of the sun, is 2×10^5 while Köhler's is 4×10^3. The higher value of Yoshimura is to be understood from the fact that he curtails the otherwise exponential growth of the dynamo fields with a non-linear dissipation effect, whereas Köhler adjusts the dynamo number downward to the minimum value that will sustain the field. By comparison, note again that the periodic dynamo in rectangular coordinates yields the lower dynamo numbers of 410 and 1300, depending upon the formal expansion employed in the calculation.

While on the subject of dynamo numbers, Yoshimura points out that when the initial field is chosen to be symmetric about the solar equator rather than anti-symmetric, the symmetry is maintained during the growth phase and continues into the final asymptotic state of steady oscillation, so that the dynamo produces a quadrupole rather than a dipole field. Leighton (1969) noted the occasional appearance of a quadrupole component in his numerical model. The explanation is simply the point made in §19.3.1.2, that for negative dynamo number in the northern hemisphere and positive in the southern, the quadrupole mode is generated at a dynamo number that is only slightly higher than that required for the dipole mode, $K\lambda=-9{\cdot}10$ instead of $-7{\cdot}43$. That is to say, $(K\lambda)^3 = -750$ instead of -410. Yoshimura's dynamo number of -2×10^5 is more than adequate to excite either of these two modes, so the model runs in whichever symmetry the initial conditions place it.

The large fluctuations of the solar dynamo from one cycle to the next, and over periods of a century, giving the Maunder minimum in the 17th century, are evidently to be understood as consequences of changes in the mode of convection and circulation in the sun, much as occurs over short periods in the atmosphere of Earth, and as has been postulated to occur over long periods in association with the major climatic changes. The extent to which the changes in the convection and circulation in the sun cause changes in the solar luminosity must await the solution of the dynamics of the solar convective envelope. It is clear, however, that the enhanced e.u.v. and x-ray emissions (mainly from active regions) vary enormously. Leighton (1969) studied the fluctuations of the solar cycle by introducing random changes in the rate of emergence of sunspots, with results bearing a striking resemblance to the irregularities in the sunspot cycle as recorded by the annual mean sunspot numbers. Indeed Leighton notes that, with a threshold for the eruption of bipolar regions, which are his sole cyclonic effect, it is possible for the dynamo to die out entirely. The physical significance of the numerical exercises is not clear, of course, but its similarity to the behaviour of the real sunspot cycle is remarkable.

As a final exercise in the solar dynamo, consider the usual dynamical estimate of the dynamo coefficient, based on mixing length theory. The characteristic cyclonic rotation of the convecting fluid, relative to the surrounding fluid, is given by mixing length theory to be of the order of $\Phi \cong \Omega\tau = \Omega L/v$ in an eddy of scale L and velocity v in a body with angular velocity Ω. Hence it follows from the development of (18.58) that $\Gamma \cong \frac{1}{8}\pi\epsilon L\Phi/\tau \cong 0{\cdot}04 L\Omega$ for $\epsilon = 0{\cdot}1$. In the standard mixing length models of the convective zone the mixing length is taken to be comparable to the scale height $\Lambda = p/\rho g$, where g is the gravitational acceleration, equal to about 3×10^4 cm s^{-2} in the solar convective zone. At a depth of 10^4 km in the sun, the temperature is 7×10^4 K so that the scale height is 3×10^3 km. With $\Omega = \Omega_s = 3\times 10^{-6}$ rad s^{-1}, it follows that $\Gamma = 40$ cm s^{-1}. The turbulent velocity at this depth is estimated (Spruit 1974) to be of the general order of 10^4 cm s^{-1}, so that the eddy diffusivity is $\eta_T = 0{\cdot}2vL \cong 10^{12}$ cm^2 s^{-1}. Thus the mixing length theory predicts a value of Γ that is eight times larger than the 5 cm s^{-1} value used by Köhler and Yoshimura to duplicate the solar cycle. The mixing length theory predicts an eddy diffusivity that is half again as large as the value used by Köhler and six times larger than that used by Yoshimura. Deeper in the convective zone the scale height is larger so that the Γ and η_T estimated from mixing length theory are even larger.

In view of this difference, one might ask, then, whether the azimuthal magnetic field in the sun might suppress the turbulent convection and thereby reduce the effective values of Γ and η_T to the levels that seem to make the dynamo models work best. The answer depends upon the strength of the azimuthal field, of course. We have estimated in §8.7, that the rate of buoyant rise of a flux tube through the solar convective zone is so high that fields of no more than a few hundred G can be expected to reside there. At a depth of 10^5 km the turbulent velocity forces $\frac{1}{2}\rho v^2$ are comparable to the stresses in a field of the general order of 3×10^3 G (at 10^4 km it is 10^2 G). Hence, the Maxwell stresses exerted by a field of a few hundred G deep in the convective zone are small compared to the Reynolds stresses of the convective turbulence, suggesting that the field has but little effect on the convective velocities. Hence we have to live with the fact that the mixing length theory gives a dynamo coefficient and eddy diffusion rather larger than required by the numerical models of the solar dynamo. It should be noted, however, that the two quantities Γ and η_T appear as Γ/η_T^2 in the dynamo number, so the excess value Γ is largely compensated by the excess η_T with little effect on the dynamo number. The real problem appears in adjusting the period (expressed in units of $R_\odot^2/\eta_T$) to the observed 22 years without jeopardizing other quantitative aspects of the dynamo model. And of course it must not be forgotten that the application of mixing length theory to the convective zones of stars is based on convention rather than deduction. We use the mixing length theory because it is a familiar dialect and the only one we know, with no assurance that the dialect speaks the ultimate truth.

There are several other blind spots in the treatment of the solar dynamo, as well. One particularly obvious problem is the depth from which the fields rise to the surface of the sun to produce the bipolar magnetic regions. Most authors assume that the bipolar regions at the surface reflect the toroidal fields not far below (cf. Leighton 1969; Köhler 1973; Yoshimura 1975a,b, 1976a,b). It was pointed out, however, that the toroidal field cannot be less than about 10^2 G; yet even so weak a field as 10^2 G rises rapidly through the upper convective zone (§8.7). It can reside for a period of years only in the lower half of the convective zone. The toroidal field is produced from the poloidal field by the non-uniform rotation in a characteristic time of several years. Hence the production of a toroidal field of 10^2 G cannot occur much above a depth of 10^5 km. The solar dynamo is concentrated in the lower, rather than the upper, half of the convective zone, (Parker 1975a, 1977a). Hence it is the non-uniform rotation $\partial\Omega/\partial r$ and the dynamo coefficient Γ in the middle and lower convective zone that are responsible for the generation of field and the migration of the field toward the equator. This leads to the question of the way in which the upper part of the convective zone should be included in models of the dynamo. It is a complex problem and only the most primitive theoretical studies have been made (Parker 1977a) so far. What level in the convective zone is responsible for the Maunder butterfly diagrams at the surface?

Turning attention to Yoshimura's model of the solar dynamo, note that the negative Γ in the lower region causes the migration of the toroidal field to be away from the equator, rather than toward the equator, there. Hence the proper migration (toward the equator) would require that $\partial\Omega/\partial r$ be positive, of course, with the interior of the sun rotating at about the same rate as the polar regions at the surface. Indeed, these are the simplest forms for solar rotation, with the angular velocity increasing simply with distance ϖ from the axis ol rotation.

Once again we end with questions on the dynamics of the convection, circulation, and non-uniform rotation in the convective zone. To these standard questions there is now to be added the dynamical problem of the buoyant rise of the fields (see §§8.7, 8.8 and Chapter 13). The buoyant rise impinges directly on the depth of the dynamo in the sun and the interpretation of the Maunder butterfly diagrams of the toroidal field emerging at the surface.

21.3. The magnetic fields of stars

If the magnetic field of the sun is to be understood as a consequence of the dynamo effect of the convective zone, what can be said about the other stars? Their fields cannot be resolved and can be detected only when the algebraic mean of the line of sight component over the visible hemisphere exceeds 10^2 G in the most favourable cases of sharp lines and slow rotation. Only a handful of freak stars have fields so strong as to produce a detectable signal (see Chapter 1 and §18.1). For all other stars there is no direct indication of magnetic fields, so we can only infer that they have magnetic fields by analogy to the sun. In this connection Wilson has shown that the dwarf main-sequence stars have strong chromospheric lines (of Ca II) in their youth, fading out by middle age (Wilson 1963, 1966; Wilson and Skumanich 1964). To interpret this result, note that the localized Ca II emission observed in the solar chromosphere (detectable only because of the proximity of the sun and the high spatial resolution) is roughly proportional to the strength of local magnetic field, and becomes detectable wherever the field exceeds some 10 or 20 G (Howard 1959; Leighton 1959). If the relation between Ca II emission and field strength in the aging sun applies to younger dwarf main-sequence stars, then it can be inferred that the dwarf main-sequence stars have general poloidal fields of the order of 10^2 G for the first 10^8 years or so (Wilson 1963). By contrast, in middle age the general poloidal field of the sun is about 5 G, far below the level of detection in any distant star.

Wilson (1971) finds from repeated measurement of the Ca II emission that the strength of the emission varies with a period of several years or more, just as the total Ca II emission from the sun goes up and down with the 11-year sunspot cycle. Then noting that the dwarf main-sequence stars rotate and have vigorous convective zones like the sun, it follows that there is a dynamo effect and we would expect a periodic variation of the fields with periods of several years. Putting together all these observational facts and theoretical principles, we infer that the dwarf main-sequence stars are magnetic, in much the same way that the sun is magnetic. It is to be presumed that, if we could get close to any of them, we would see the magnetic effects in the form of plages, prominences, starspots, flares, x-ray bright spots, etc., and coronal streamers and stellar winds. The very young dwarf main-sequence stars evidently have much stronger fields than the sun, so that a close look would show a high level of magnetic activity. The young stars must seethe with activity, but their magnetic virility declines over a period of 10^8 years.

It is interesting to note, therefore, that some of the very faint dM *subdwarf* main-sequence stars have sporadic transient brightenings that resemble solar flares but involve 10^3 times more energy (see description and references in Chapter 1 and §18.1). Between the transient flare outbursts it can be seen that there is a significant periodic variation of the luminosity, indicating a strong variation of surface brightness around the star. The obvious possible causes for variation of surface brightness are magnetic inhibition of heat transport to the surface, or enhanced plages, or the outright formation of huge starspots (in analogy to the sunspot) (cf. Mullan 1975). The general magnetic fields of the flaring dM stars must be very strong indeed. Precisely how strong is hard to say.

Not all dM stars are observed to be flare stars, however, raising the question of whether it is an activity confined to their youth, or a special ability of certain individuals. The mass of a dM dwarf may be only $0{\cdot}1$–$0{\cdot}2\ M_\odot$, and their luminosities as low as 10^{-5} times the luminosity of the sun. Their life on the main sequence is, then, of the general order of 10^{14} years. From that point of view, viz. the evolution on the hydrogen-burning main sequence, all dwarfs are youths. But there may be other facets of their structure, such as the rotation rate, that decline relatively rapidly, in 10^8–10^9 years at the beginning of their life. So the question remains. So far there is no indication that the flare stars are members of young populations, suggesting that the enormous activity is a special feature of certain individuals.

Now we recall that most of the stars in the galaxy are to be found on or near the lower main sequence, between the sun and the dM dwarfs. We are led to conclude, then, that most of the stars in the galaxy have magnetic fields. The fields may vary periodically, as in the sun, but it should be kept in mind that the cooler main-sequence stars have deep convective zones, so they may produce stationary, rather than periodic, magnetic fields (see §19.3.1.1 and 19.3.1.2).

Levy and Rose (1974) consider the generation and the evolutionary effects of stellar magnetic fields in massive main sequence stars ($M_* > 1.5\ M_\odot$) with particular attention to the redistribution of angular momentum during the red giant stage and the slow rotation of the final white dwarf. More recently Schüssler and Pähler (1978) treat the diffusion of the fields from the core to the surface of such stars during their life on the main sequence. Schatzman (1962) has pointed out the close connection between the existence of a convection zone and the slow rate of rotation in stars less massive than the middle of type F, suggesting that the strong magnetic field generated by the convective zone in the vigorous youth of each star may play a role in the loss of mass and angular momentum from the star. Theoretical estimates of the rate of loss of angular momentum (Dicke 1964, 1972; Modisette 1967; Weber and Davis 1967; Alfonso–Faus 1967; Durney and Latour 1978) suggest that the idea may be correct. Havnes and Conti (1971) consider the opposite possibility, treating the selective accretion of heavy atoms from the interstellar gas to account for the anomalous atomic abundances of the magnetic stars.

It is a curious fact, then, that there are a few anomalous hot stars (mostly A-stars), evidently without convective zones, that have magnetic fields so strong ($>10^2$ G) as to show a detectable Zeeman effect. The discovery of these *magnetic* stars was made by Babcock. The star 78 Virginis (see discussion in §18.1) was the first to be identified, with a field of the order of 10^3 G. There are now over a hundred known (Preston 1967a). They appear to form a class, or perhaps several classes, distinct from the normal main sequence stars. Most of the magnetic stars are to be found among the peculiar A-stars—the Ap-stars. The Ap-stars are identified by their apparent overabundance of such metals as Si, Mn, Sr, and some of the rare earths. As a matter of fact, magnetic fields are now found among the peculiar stars from B-stars all the way down to F-stars, as cool as 8000 K. Essentially all of the Ap-stars whose lines are not broadened by rotation show magnetic fields. The most common polar field strengths are $2–3\times10^3$ G, with some individuals having field strengths as high as 10^4 G and more (Preston 1971). There appears to be an anti-correlation between the angular velocity and the strength of the field (Landstreet *et al.* 1975), with the most rapidly rotating Ap-stars showing no signs of a field. The slowly rotating Ap-stars have mean fields ranging from about 10^2 G (the minimum detectable field) up to 3×10^4 G for HD 215441 (Babcock 1960b). When we remember that these numbers represent the mean field in the line of sight, it means that the peak fields are several times greater (see discussion in §18.1). They are magnetic stars indeed!

There are some metallic line stars—the Am-stars, distinguished by their weak Ca II K-lines—that are magnetic, with fields of hundreds of G, but only about one in ten of the magnetic stars is of this type. One in twenty of the magnetic stars is a late-type star. The slowly rotating Ap-star seems to be the principal magnetic star (see the review by Bidelman 1967 on the properties of the Ap- and Am-stars.)

The periodic time variation of the mean field in the line of sight, and the associated spectral variations, indicate that the fields rotate rigidly with the star (Stibbs 1950; Deutsch 1954, 1958a,b). There are patches of enhanced emission and anomalous chemical abundances associated with the field, representing active magnetic regions on the surface of the stars (see review by Preston 1967a and the more recent observational studies, Preston 1967b, 1969, 1971, and references therein). To a first approximation the magnetic field is a dipole, with its axis strongly inclined to the axis of rotation of the star. Indeed the magnetic dipole is perpendicular to the rotation axis in so many cases as to suggest a systematic effect. Detailed analysis of the time variation of the field and emission patterns shows higher harmonics of significant strength, of course (Steinitz 1964, 1967; Kodaira and Unno 1969; Pyper 1969; Cohen 1970; Deutsch 1970; Monaghan 1973). In practice, then, the field of the magnetic star is a distorted and offset dipole. The periods of rotation extend upward from 5 days to several months, which is to be compared with the rotation period of a day for the normal main-sequence A-stars, which show no magnetic effects. The stronger fields appear in the more slowly rotating stars.

There are two evident possibilities for the origin of the fields of the magnetic Ap-stars. One is that there is a dynamo operating in the interior of the star. There are a number of difficulties with this idea, starting with the fact that an A-star is not expected to have a convective zone beneath its photosphere. There is a possibility of an internal convective core, perhaps, but the extension of the fields through the stable shell above poses interesting questions (see discussion in Schüssler and Pähler 1977).

The simplest explanation of the magnetic field might then seem to be in terms of fossil fields, i.e., primordial fields, caught up in the collapsing gas that formed the star, compressed to 10^3 G or more, and thereafter clutched in the stable core of the star (Cowling 1953). There the field remains for the life of the star, with the hair of its head showing out one side and the soles of its feet out the other. Eventual senility of the star, with the dwindling hydrogen in the core, leads to collapse of the core to form a white dwarf or neutron star. The collapse compresses the field to the 10^7 G, observed in some white dwarfs (see review by Angel 1977) and to the 10^{12} G inferred from the behaviour of the spinning neutron star. Mestel (1965a,b, 1967) has considered the mechanics of the trapping of interstellar magnetic fields in the formation of stars from gas clouds.

Roxburgh (1967) has reviewed the magnetic effects of the fields within the star once it is formed (see discussion and references in Chapter 1). The braking of the rotation of the magnetic star through the interaction of its strong dipole field with the surrounding interstellar material has been examined by Mestel and Selley (1970). The braking is evidently not large unless the star resides in relatively dense interstellar gas clouds.

Consider, then, the arguments that the magnetic stars show fossilized primordial fields. The anti-correlation of field strength and angular velocity seems to argue against the dynamo, and hence in favour of a trapped primordial field. In all dynamo theories, and in the planets and the young main-sequence dwarfs, the field strength is generally stronger in the more rapidly rotating cases. The field strength and the rotation rate are positively, rather than negatively, correlated.

The large obliquity of the rigid rotating dipole field of the magnetic star poses an interesting question that has received considerable attention. First of all, Rädler (1975) has demonstrated an oblique field produced by the dynamo effect of fluid motions with symmetry about the axis of rotation. Hence the mere fact of obliquity cannot by itself be used as an argument for fossil fields. But the preference for perpendicularity is a stronger condition that argues in favor of fossil fields. That effect does not arise in the dynamo, so far as anyone is aware, but can be understood, evidently, for primordial fields. Moss (1977a,b) and others (Mestel and Moss 1977 and references therein) has considered the Eddington–Sweet circulation (Eddington 1929; Sweet 1950)[3] in which fluid rises at the poles and sinks at the equator. The effect is to increase the obliquity of any dipole that is not precisely aligned with the axis of rotation at the outset. The circulation is slow but Moss estimates that it can carry the dipole to obliquities near $\frac{1}{2}\pi$ in the life of the star. Mestel and Moss (1977) and Moss (1977a) (see also Mestel and Selley 1970) point out that if the Ap-stars began their lives on the main sequence with fossil fields of arbitrary strength, then the rotation of those with the strongest fields would be braked the most, producing the observed inverse correlation of B and Ω. What is more, they show that the stronger Eddington–Sweet circulation of the fast rotators tends to bury the magnetic field beneath the surface, the fluid motion pushing the field down out of sight. So there are at least two contributions to the inverse relation between B and Ω.

[3] In a rotating star with a heat flux outward from its centre the isothermal surfaces do not coincide precisely with the level surfaces. Hence there is no static equilibrium (Von Zeipel 1924). There is instead a dynamical equilibrium including the Biermann battery effect (see §21.1) and meridional circulation of the fluid. The circulation is very slow in a stable stellar interior because the vertical motion is opposed by the convective stability. The flow velocity is limited to something of the order of 10^{-6}–10^{-8} cm s^{-1} by the necessity for cooling (or warming) by radiative transfer, in the same manner as the rising flux tube treated in §8.8.

The circulation also has the effect of burying any dynamo fields, leading to the conclusion that, whatever may be generated in a convective core, the surface fields are evidently fossil in nature[4].

On the other hand, the ease with which the primordial field concept can be argued for the magnetic stars should not blind us to one fundamental question that has not yet been answered. That is the problem of preserving the primordial field through the early Hayashi phase when the star is believed to convect all the way from the centre to the surface. This point was mentioned in §17.1 where it was noted that any initial magnetic fields are quickly lost by mixing to the surface, from where they escape into space. The global convection of the Hayashi phase may be expected to work as a dynamo, of course, generating fields throughout the star. But it seems that these fields would also disappear as the convection comes to an end at the centre of the star and a non-convecting core begins to grow. The magnetic buoyancy alone would provide rapid escape as the interior of the star passes slowly through neutral convective stability. It is clear, then, that the problem needs careful quantitative study if we are to establish a firm explanation for the fields of the magnetic stars. So far we have only an extensive but unassembled set of ideas.

Bidelman (1967) has emphasized that the whole peculiar and metallic line A-star phenomenon, with their strange chemical abundances, is puzzling and must be considered in any final deliberations on the origin of the magnetic fields. Why should a star with strange chemical abundances have magnetic fields when other stars do not? It may be that the high metallic abundances are only at the surface, produced by the extensive acceleration of fast particles and resulting nuclear transmutations in some sort of magnetic activity. But the Ap-stars show no such explosive activity at the present time, and the flare stars that show gigantic activity do not show such anomalous abundances. So the peculiar abundances may reflect peculiar conditions of origin, such as the debris of a supernova, rather than being caused by the strong magnetic field. It is difficult to say what relation exists between the fields and the abundances.

21.4. The activity of stellar magnetic fields

Several varieties of stars (and indeed many galaxies) exhibit activity of one form or another, much of it presumably magnetic in origin. Unfortunately only the grossest features, as they are represented by the variations in the integrated light from the stars, can be studied, so that it is difficult to progress beyond conjecture as to the nature of the activity. The most plausible conjectures in fact, are based on analogies to the activity

[4] It is interesting to contemplate how a convective core would be shared by its own dynamo field and by a fossil field.

observed on the sun. For this reason, then, we deal with the remarkable activity occurring on and above the surface of the sun (see review by Kiepenheuer 1953). The sun becomes the prototype of all magnetic activity, and serves as a guide for the theoretical discussion. Consequently many of the earlier sections and chapters have begun with physical descriptions of the various aspects of solar activity, in order to define the problems to be treated. We do not attempt to repeat all the many physical details here but rather to pull together the overall picture of solar activity with a few brief statements, and references to the earlier chapters, on how the activity is to be understood. A general understanding of the basic active effects in the sun, and the anticipation of the active effects in other stars, and in galaxies, is the goal of this writing. We shall see now in what ways the previous chapters have contributed to progress toward that goal.

First of all, on the broadest terms, magnetic fields are generated in the convective zone of the sun at depths of the order of 10^5 km (see §§21.2, 21.3). The fields rise to the surface as a consequence of their buoyancy (§8.7, Chapter 12), where we observe them. The twisting of the magnetic flux tubes, and any other topological complexities to be found along the lines of force, rise to the top of the field and concentrate in the expanded portion of the field above the surface (§§9.5 and 9.6). Hence the general non-equilibrium of magnetic fields (Chapters 14 and 15) in the sun, or another star, arises principally in that part of the field that extends above the surface. The activity in the fields above the surface is fostered by both the concentration of topological complexity, which causes the non-equilibrium, and the high Alfven speed (because of the low gas density) causing the rapid dissipation of that complexity. This, then, is the general cause of the continual magnetic activity of the sun and stars.

The activity takes on a variety of forms. At its lowest level the rapid reconnection occurs more or less steadily in the topological variations of the fields, producing a general heating and uplifting of the chromosphere and corona in the regions of strong field (cf. the observations and discussion by Leighton 1959; Howard 1959; Vaiana, Davis, *et al.* 1973). The heating may be augmented, and perhaps dominated, by dissipation of gravitational and hydromagnetic waves generated in the convective zone beneath the visible surface (Schwartz and Stein 1975). Transient bursts of heating and particle acceleration are not uncommon in active regions where new magnetic flux is coming up through the surface. Sporadic flaring seems to be part of the picture. Altogether, the magnetic fields imply continuing suprathermal effects.

It was surprising, therefore to discover that the fast solar wind streams originate from the hydrodynamic expansion of the quiet corona of the coronal holes (Krieger *et al.* 1973; Nolte *et al.* 1976). The effect is to be understood, evidently, as the direct consequence of the rapid divergence of the expanding corona as it streams outward in the void between the strong localized magnetic fields on either side (Kopp and Holzer 1976).

The general magnetic field of the sun exhibits the remarkable property of spontaneous self-concentration into isolated, intense flux tubes everywhere over the surface of the sun. Essentially all of the magnetic flux through the surface of the sun is gathered into isolated tubes. In quiet regions, where the mean field is of the order of 5 G, the field gathers itself into separate flux tubes of about 10^{18} Mx, with an intensity of 1500 G over a diameter of 300 km. The evidence is that the spicule phenomenon (Lynch *et al.* 1973) is associated with the individual concentrated flux tubes (Parker 1974a) and with fine H_α fibrils to be seen against the disc (Beckers 1963, 1968; Veeder and Zirin 1970; Zirin 1972). Evidently, then, the H_α fibrils represent the cores of individual flux tubes.

In active regions the field separates into tubes of about 10^{19} Mx and about 1500 G over a diameter of 1000 km. The tiny magnetic region where the individual tube comes up through the surface of the sun is called a *magnetic knot*, first clearly identified by Beckers and Schröter (1968). It is the gathering together of flux into bundles (magnetic knots) of increasing size that leads to the growth of sunspots. A magnetic region first appears dark on the surface of the sun when two or three flux tubes in an active region coalesce to give a total flux of 2×10^{19} Mx, with about 1500 G over a diameter of 1500 km. Most such *pores* break apart (in a few hours) into two or three separate tubes again. But sometimes they grow through the accumulation of more flux tubes and through merging with other pores, developing into full fledged sunspots, with a field strength of about 3000 G and a surface temperature reduced below 4000 K (Gokhale and Zwaan 1972) when the total flux reaches 4×10^{20} Mx across a diameter of 4000 km. The sunspot appears black against the surrounding photosphere, the total radiation being reduced to about 20 per cent of the normal value of 6×10^{10} erg cm^{-2} s^{-1} = 6 kW cm^{-2}, and the visible radiation to less than 10 per cent, by the lower temperature. The accumulation of additional knots and pores may continue, the spot increasing in size and developing the remarkable filamentary penumbral region around the periphery of the dark umbra (see the striking photographs in Danielson 1961a,b). The field strength in the sunspot generally does not increase beyond about 3000 G as the spot grows in size (although fields up to 4500 G have been observed on occasion). The average spot grows to some 10^{21} Mx at the peak of its development after a time of 5–10 days. On rare occasions the growth may progress to fluxes in excess of 2×10^{22} Mx, across a diameter of 3×10^4 km, or more. Then, in most cases, disintegration of the sunspot begins when the accumulation of new flux ceases, after a time of 5–15 days. The total flux and the size then decline, the spot breaking into smaller spots and decomposing back into magnetic knots over periods of 10–50 days. Thus a large spot may represent the accumulated flux of some 2×10^3 knots, which, for some reason, have come together and coalesced into a whole. The spot breaks up again into about 2×10^3 knots at the end of its life. Indeed, a magnetic knot of 10^{19} Mx is a convenient unit of flux for discussing sunspots. It is only fear of confusion with nautical terminology that prevents us from specifying sunspot sizes in knots, a three-knot spot being a small pore, and a kiloknot spot a rather large sunspot.

Sunspots develop within large bipolar magnetic regions, where a segment of a large flux tube has come up through the surface of the sun (Fig. 21.1). Consequently sunspots are themselves often bipolar, with the spots forming in the preceding magnetic region (in the sense of the solar rotation) generally having the same magnetic polarity as the region, and the spots in the following region having the opposite polarity, to match the following magnetic region. Spots form more copiously in the preceding region, gathering together on the average about three times as much flux as the spots in the following region. The spots in the preceding region usually last longer than in the following region. The bipolar active magnetic regions are made up of a collection of magnetic knots. It is evident from Fig. 21.6 that the bipolar regions move apart as the flux tubes rise through the surface. The rise ceases when the tubes extend straight up through the surface from their anchor points deep in the convective zone. At this stage the separation of the bipolar regions at the surface ceases to increase. It is an observed fact that the sunspots begin to

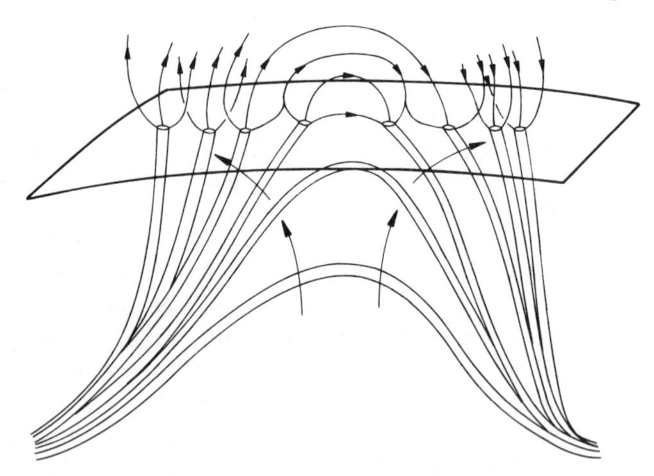

FIG. 21.6. A schematic drawing of the magnetic flux tubes of a bipolar magnetic region emerging through the surface of the sun. The expansion of the fields to fill all the available volume above the surface of the sun is indicated by the lines of force sketched above the surface. The separate arrows indicate the general upward motion of the field and fluid.

break up and disperse when the separation ceases to increase. The spots are soon reduced to magnetic knots again which spread out over the broad area that makes up the bipolar active region. On some rare occasions a large spot in the preceding region gathers itself into a coordinated circular form with a field-free 'moat' around it, surviving alone as a unipolar spot for several solar rotations (see description by Beckers and Schröter 1969; Bray and Loughhead 1964).

Zwaan (1978) gives a detailed description of the manner in which magnetic knots and pores merge to produce the growth of a sunspot, and the breakup of a sunspot into pores and ultimately into magnetic knots (Beckers and Schröter 1968) at the end of its life. The bipolar sunspot group grows by the addition of new flux emerging in the central region between the two regions of opposite polarity. The flux arrives at the surface in small (10^{19} Mx) intense (1500 G) flux tubes, and these individual knots then move toward the sunspots to which they finally add their flux. Zirin (1972) provides a vivid description of the reconnection of the magnetic fields of the little bipolar regions as each separate flux tube comes to the surface. The knots may coalesce with other knots on the way to the sunspot, often following a common path across the surface of the sun on the way to the spot. Zwaan makes the point that, so far as one can tell from the intersection of the flux tubes with the surface of the sun, the large flux tube producing the general bipolar region is composed of many separate strands. The separate strands rise through the surface as a scattered group, depicted in Fig. 21.6. The observer, then, sees a scattered group of magnetic knots which move toward each other as the tubes rise up from below.

The problem is to understand why the general magnetic field of a star behaves in this peculiar fashion. It is not, as we have remarked before, a behaviour that could have been anticipated from the elementary properties of magnetic fields in conducting fluids. On an elementary basis the behaviour of the magnetic fields appearing at the surface of the sun is largely baffling. It is remarkable, therefore, that the observed behaviour of the fields has been accepted so widely in the published literature without critical comment. The most superficial and irrelevant explanations often serve to allay the feeble curiousity of the reader. Yet the observed cohesion and stability of a sunspot have not been demonstrated from basic principles (Parker 1974b, 1975d; but see the recent work of Meyer *et al.* 1977). The cooling of a sunspot, in a manner consistent with its cohesion has not been demonstrated (Parker 1977b). The internal circulation and the general character of the convection in the sun are not known, although, recognizing the fundamental importance of the problem, there are several theoreticians around the world exploring this very difficult theoretical question. The formation of the solar prominence is not entirely clear, although there has been serious work on the problem over the years (see §§6.6 and 6.7). The theory of the acceleration of particles to high energy in a solar flare has received wide attention for at least two decades, but is far from resolved at the present time. The remarkable efficiency, of the order of 10–50 per cent, with which magnetic energy is converted into the kinetic energy of fast particles, has yet to be established by any self-consistent theoretical illustrative example. There is a widespread feeling that the problem has been solved, when in fact some major hurdles remain (cf. Parker 1976; Eichler 1977). Altogether, much work remains to be done to develop an understanding of solar activity.

But there has been some progress, and many effects can be understood from the basic principles. To treat the principles in some detail, then, the concept of the flux tube (Chapter 8) confined by a pressure increment $\Delta p = B^2/8\pi$ at its surface was established thirty years ago and has been used effectively since that time. The recent observational discovery that the broad fields of the sun are fragmented into isolated flux tubes has placed the concept in a central position in the theory of solar magnetic fields. It seems in retrospect that the fragmentation of a broad flux tube beneath the surface of the sun into a thousand or more small, isolated, tubes of 10^{18}–10^{19} Mx each, concentrated to the remarkable strength of 1500 G or more, is the end result of a sequence of effects. The breakup begins with the instabilities of the upper surface of any broad field that may be buried deep in the convective zone. The calculations show that lengths of the field, longer than several scale heights, rise up from the surface of the broad field and float upward through the convective zone of the sun (see §§13.2–13.4). These omega-shaped re-entrant loops rise up through the surface of the sun, so that the apex of each loop can be seen by the terrestrial observer. This provides the evidence of the general azimuthal field buried deep in the convective zone of the sun (see Fig. 21.1 and the discussion in §21.2). It is a curious feature of the flux tube (see the theoretical presentation in §§9.5–9.7) that the twisting and braiding of the lines of force are concentrated by the buoyancy into the apex of each tube, above the surface of the sun. The resulting non-equilibrium progresses rapidly, because of the high Alfven speed (see Chapters 14 and 15), in the transparent chromosphere and corona where it can be seen by the terrestrial observer.

Now the breakup of the upper surface of the broad azimuthal field in the convective zone proceeds with wavelengths along the field that are several scale heights long (i.e., $\gtrsim 10^5$ km). On the other hand the wavelength of the breakup in the transverse direction across the field is relatively short. It is shown in §13.4 that the transverse scale is generally less than the scale of decline of the magnetic field at its upper surface, and indeed may be very much less than that scale, limited only by the viscosity and resistivity of the fluid. The general turbulence of the convective zone precludes any quantitative statements, unfortunately. Thus the broad field may spawn clusters of small tubes directly from its upper surface. Certainly any broad tube rising from the surface of the submerged field is unstable against further breakup (see discussion in §§9.7 and 9.8) unless it is strongly twisted for some reason. The evidence is that the broad tubes are not strongly twisted, because the broad flux tubes extending through the surface of the sun (actually a cluster of many small tubes unless coagulated into a sunspot) show only moderate twisting at the most, and often none at all. Yet we expect that the twisting is concentrated in the field above the surface of the sun where we can measure it (see §§9.5 and 9.7). This implies but little twisting of the tubes beneath the surface, and hence but little resistance to decomposition into small, separate flux tubes. Thus, flux tubes of various diameters free themselves from the upper surface of the broad fields deep in the convective zone and rise to the surface of the sun, breaking progressively into smaller tubes as they rise. The individual strands of field, then, are further broken apart and pushed around and organized by the granules and supergranules so that they appear at the surface only in the converging flows and downdrafts at the boundaries of the upwelling convective cells (§10.2). It is not clear why the individual tubes end up at 10^{19} Mx in active regions and 10^{18} Mx in quiet regions. That question must have something to do with the dynamical stability of the tubes, and is a subject that is but little understood. But the existence of a breakup of the broad field into many small flux tubes can be understood in a straightforward manner from the general physical principles expounded in Chapters 8, 9, and 13. Their general activity above the surface of the sun seems to follow from the principles presented in Chapters 9, 11, 12, 14, and 15—the general non-equilibrium and rapid reconnection of magnetic fields with complex topologies concentrated above the surface of the sun.

The most surprising aspects of the individual isolated flux tubes is their concentration to the enormous intensity of 1500 G. The effect cannot be a consequence of the twisting of the tubes, even if we imagine that the small tubes are strongly twisted. The effect is not produced by the direct dynamical effects of the convective, turbulent fluid motions, because such effects, reviewed at length in Chapter 10, cannot, under ordinary circumstances, push the field stresses $B^2/4\pi$ much beyond the Reynolds stresses ρv^2. Thus fluid velocities in excess of 5 km s^{-1} are required at the surface of the sun to produce fields of 1500 G. We are reminded again, of course, that the spicules, with velocities of the order of 20 km s^{-1} in the chromosphere, appear to be associated with the flux tubes, but there is, so far, no firm observation of the necessary 5–8 km s^{-1} at the photosphere. It is evident, then, that the flux tubes are a dynamical entity, as indicated by the spicules, but the dynamics does not seem to be the basic cause of the intense field.

Instead, the intense field appears to be caused by cooling from the *superadiabatic effect*, stemming from the general downdraft of the gas in and around the flux tube and/or by the emission of Alfven waves (Roberts 1976). The effect is illustrated in §10.10, and may be presumed to arise in other dwarf main-sequence stars like the sun. The effect suggests that the gas motions within, without, and across the magnetic fields above the surface of the sun are very complicated and not yet understood (see discussion in §10.10).

If the individual flux tubes are strongly concentrated, so that the Alfven speed (7 km s^{-1}) within the tubes is larger than any turbulent velocities outside, it follows that the individual flux tubes are extremely buoyant. Their rate of rise is of the general order of the Alfven speed (i.e., km s^{-1}) (§8.7). Hence neighbouring flux tubes are strongly attracted toward each other by the hydrodynamic forces in spite of the mutual repulsion of the fields of like polarity above the surface of the sun. The mutual attraction is described in §§8.9 and 8.10. The observed tendency for neighbouring magnetic knots (flux tubes) to attract each other, and to coalesce into pores and sunspots, may be understood as a direct consequence of the hydrodynamic forces on the rising tubes. The observed tendency for magnetic knots to follow a common path into, or out of, a sunspot may perhaps be explained by the wake of each passing tube being maintained by the convective forces to provide an easy path for the next tube to come that way (see §8.10).

Now if the mutual attraction of magnetic knots is to be understood as a consequence of their buoyant rise through the surface of the sun, then it follows that the attraction ceases when the tubes are fully emerged and standing more or less vertically beneath the surface of the sun (see Fig. 21.6). This seems to be in agreement with the general observation that in most cases a sunspot begins to disintegrate into pores and magnetic knots, and the knots lose their mutual attraction, when no more new flux emerges through the surface of the sun.

It is a curious fact, then, that the buoyant rise of a flux tube produces a mutual hydrodynamic attraction of neighbouring flux tubes, causing them to coalesce. This is just the opposite effect of the buoyant Rayleigh–Taylor instability that caused the initial breakup into separate strands of field. The decomposition back into elemental flux tubes of 10^{19} Mx begins again when the buoyant rise ceases.

Now, if it seems that the remarkable properties of mutual attraction of the rising flux tubes can be explained, it must be pointed out, on the other hand, that the cooling and cohesion of the many flux tubes that have been assembled into a sunspot has yet to be demonstrated from basic physical principles. The compression of the field to 3000 G in the sunspot may be understood from the observed cooling, as put forth in §§8.3 and 10.10, whereby the gas is pulled down out of the vertical magnetic field by gravity, causing the field to be compressed by the surrounding gas (Parker 1955a). But while the cooling may provide an equilibrium, that equilibrium by itself is strongly unstable to the hydromagnetic exchange instability (Parker 1975d; Meyer *et al.* 1977) so that we would expect the sunspot to come apart in a matter of hours. In fact the sunspot comes apart slowly in the expected way at the end of its life after weeks or months. But why does it not do so in its formative stages? The sunspot flux tube is not rising, so far as one can tell. Only the small knots and pores that coalesce to form the spot seem to be rising.

Then the cause of the cooling has yet to be demonstrated. Suitably copious generation of Alfven waves by the convective overstability of the field beneath the surface provides one theoretical possibility (Parker 1974b; Boruta 1977). But it is necessary that nearly three-quarters of the heat flow be diverted into waves. Thus far there is no observational support for the Alfven waves that one might expect above the sunspot (Beckers 1976; Beckers and Schneeberger 1977). The fact that the strong magnetic field inhibits the turbulent heat transport (Biermann 1941) produces a cooling at the surface, but a buildup of heat below. By itself it does not produce cooling in a form which concentrates the field (§10.10). It may be a combination of high turbulent heat transport at depth beneath the sunspot, with a strongly reduced heat transport near the surface, that permits a net cooling and compression. But the dynamical formation of a sunspot has not yet been illustrated in any self-consistent way, so we have no basis for understanding it. Meyer *et al.* (1974) address themselves to the possibility of the dynamical stabilization of the long-lived unipolar spot by a convective flow set up by the continued presence of the sunspot. Perhaps some such convective dynamical effects operate earlier in the life of a sunspot.

Concentrated bipolar active regions appear in all sizes, down to the little ephemeral active regions (Harvey and Martin 1973; Harvey *et al.* 1975) only $5\text{–}10 \times 10^3$ km across, with a mean life of about twelve hours. Evidently even such small flux tubes can maintain their coherence as portions of them rise up to the surface to produce a bipolar magnetic region, although the east west alignment of the raised portion of the small tubes (the ephemeral active regions) is lost. Some 500 or so ephemeral active regions are present on the surface of the sun at any one time. They frequently give rise to tiny flares, at the end of their brief lives. In particular the ephemeral active regions produce the x-ray bright points (Vaiana *et al.* 1973; Golub *et al.* 1974; Golub *et al.* 1977). The x-ray bright points are produced by a steady input of energy to the enhanced coronal plasma over the region, presumably as a consequence of rapid reconnection (current sheets) in twisted flux tubes (Glencross 1975; Parker 1975b; see §9.7 and Chapter 15). The enhanced x-ray emission from the corona above the large bipolar regions is well known.

Indeed the enhanced emission of x-ray, u.v., H_α, and various coronal lines is the outstanding feature of the atmosphere above active regions both large and small. Superimposed on the steady enhanced emission is a frequent winking and sputtering of tiny points. At intervals larger areas may brighten and fade, sometimes appearing with explosive suddenness (in 10^2 s) and lasting for some time thereafter (10^3 s or more). Occasional enormous explosive outbursts (up to 10^{32} erg over 2×10^3 s) are observed involving strong particle acceleration and the associated nonthermal radio, x-ray, γ-ray, and 'solar cosmic ray' emission. All of these transient phenomena come under the heading of the *solar flare*, although it is clear from the observations that they differ in many features besides their magnitude.

Traditionally the solar flare was observed in the enhanced H_α emission, but the x-rays and e.u.v. are produced by the extreme temperatures in the core of the flare, so that they give a more intimate view of the source of the flare. The H_α emission seems to be a peripheral effect, wherever the intense energy of the flare spills into the dense and relatively cool hydrogen of the lower chromosphere. One of the most remarkable features of the larger flares appearing in a given active region is the occasional similarity of the irregularities of the form in both space and time of successive flares (Waldmeier 1938; Ellison *et al.* 1960). The effect is best shown in the H_α pictures, suggesting that the energy spills down the lines of force of the fixed magnetic pattern of the region at the time of each successive flare.

The largest flares produce a blast that drives off a major portion of the corona overhead, appearing as a blast wave in the solar wind at the orbit of Earth (Parker 1963; Hundhausen 1972). Even the smallest seem to throw off eruptive loops of matter entrained in magnetic flux tubes. Indeed, it was the remarkable sequence of expanding loops from minor explosions on the sun that was one of the spectacular discoveries of *Skylab* (Brueckner 1976; Kahler *et al.* 1975; Cheng and Widing 1975; Widing 1975). Rust *et al.* (1975) have shown from coordinated studies of one flare in the x-ray and e.u.v. emission, together with the H_α and magnetic pictures, that the eruptions of the flare originated in the vicinity of a very high temperature kernel on the activated filament. Although the flare developed only slowly and was classified as a 1B flare, they estimate the total energy to have been 10^{31} erg. Petrasso *et al.* (1975) report a rapid sequence of high-resolution x-ray pictures of a later flare, in which the centre of the flare was clearly seen to be a tiny kernel, in the form of a ribbon about 3000 km wide, located at the brightest point of a loop structure over the active region (see also the earlier observations of Vaiana and Giacconi 1969; Krieger *et al.* 1970; Krieger *et al.* 1971; Neupert *et al.* 1974). The larger flares, then, appear to be an extraordinary energy release in a remarkably localized region on a pre-existing magnetic loop.

Flares occur when loops of flux—i.e., portions of magnetic flux tubes—emerge through the surface of the sun into the established fields of active regions (cf. Priest and Heyvaerts 1974). The more complex is the magnetic pattern of positive and negative polarities of the established fields and the energing loops, the more frequent and violent are the flares. In old, well stabilized regions the flares are usually only small sporadic bursts of energy here and there throughout the region, whereas in complex regions in the formative stages the colossal two-ribbon flare is likely to occur.

Flares have been studied intently for half a century or more, with rapid progress in the past twenty. Present knowledge of solar flares of all kinds and sizes makes it clear that the flare is made up of many effects that appear in varying proportions from one flare to the next. Indeed the flare is so complex that the extensive morphological studies have, for many years, merely broadened the bewildering array of observational facts on the variety of individual flares without providing a direct idea of the detailed inner working of the flare itself. It was known that the total energy output of the largest flares might be 2×10^{32} erg and that such flares always occur in the strong fields around and between sunspots (which might have a total field energy of 2×10^{33} erg above the surface of the sun). No source of energy besides the magnetic energy is adequate to supply the flare (Parker 1957a; Sweet 1969). Hence the energy of the flare may be presumed to be accumulated in the distortion of the magnetic field from the simple potential form $\mathbf{B}=-\nabla\psi$ that represents the minimum energy for a given flux distribution at the surface of the sun. The sudden conversion of magnetic energy into accelerated particles and extraordinarily hot gases (3×10^7 K) to make the flare is presumed to be through some form of rapid reconnection (See Chapter 15) enhanced and complicated by anomalous resistivity (caused by plasma turbulence), the resistive tearing mode instability, the hydromagnetic exchange instability, etc. The solar flare, and particularly the many tiny flares that continually wink in an evolving active region, represent the jerks and starts in the evolution of the complex topology of the twisted and distorted flux tubes emerging through the surface of the sun to make the active region. It was pointed out in Chapter 9 that all the twists and cross-overs in a magnetic field propagate upward along the flux tubes to the most expanded portion, above the surface of the sun. As a consequence of the low gas density and consequently large Alfven speed in the chromosphere and corona, it is possible for the intermittent readjustments to proceed rapidly, reducing the topology to simple form. Zirin (1972) and Sheeley *et al.* (1975) have remarked on how closely the observed magnetic connections between sunspots etc. soon reduce to the potential form, however complicated may be the topological connections as the fields first emerge through the surface of the sun.

With the challenge of observing solar flares and solar activity from spacecraft above the atmosphere of Earth, the art of high resolution x-ray and e.u.v. telescopic observations has been developed to where the time evolution of flares can be followed with a resolution of about 2000 km (Kahler *et al.* 1975; Cheng and Widing 1975; Thomas 1975; Brueckner 1976; Vorpahl *et al.* 1976). At the same time, the ground-based instruments in visible and infrared (with a resolution of about 700 km) have been improved and coordinated to carry out simultaneous observations in several different wavelengths. The time evolution of a flare is studied in H_α, the calcium K line, etc. together with high resolution magnetograms and Dopplergrams to show the simultaneous behaviour of the magnetic field and the gas velocities (Dunn *et al.* 1974). This formidable combination of instrumentation is presently brought to bear on the flare problem, with immediate results. The high resolution ground-based observations give a detailed picture of the fields and the gas densities, temperatures, and motions in the general flare region, while the e.u.v. and x-ray observations see the tiny intense kernels that are the site of the energy release, with temperatures of 10^7–10^8 K. At the same time the accelerated particles escaping from the flare are studied in space, with careful separation of atomic number and masses. It has been shown, for instance, that some flares, of modest size in the periphery of active centres, accelerate He^3 in numbers exceeding the normal He^4, and without detectable H^2 and H^3 (Anglin 1975; Hurford *et al.* 1975; Serlemitsos and Balasubrahmanyan 1975).

Excellent reviews of the background in which flares occur, and of the general feature of the flare, may be found in the monographs by Smith and Smith (1963) and Tandberg–Hanssen (1967) (see also the review by Sweet 1969). An extensive but concise review of the extraordinarily complex and varied properties of the solar flares has been compiled and organized in a recent volume by Svestka (1976). Svestka provides not only a summary of the radio, optical, u.v., x-ray, γ-ray, and particle observations up to 1975, but also a systematic review of the various hypothetical scenarios that have been proposed to account for the flare by field annihilation and infalling gas (see also the review volume edited by de Jager and Svestka 1976). The fact that an entire volume is required for a concise summary and discussion of flare observation and theory is evidence of the enormous complexity of the flare phenomenon. It is clear that a flare is made up of a host of effects. The flare embodies the hydromagnetic effects expounded in Chapters 9, 14, and 15, and an array of plasma effects besides. The extensive theoretical studies to be found in the literature exhibit the rich variety of forms the effects may take in any one flare. A comprehensive theoretical treatment of the flare would be premature at the moment. It is to be hoped, however, that the rapid improvement and extension of observations to all frequencies of the electromagnetic spectrum, together with studies of the particle emission, may suitably define the flares so that a more concrete picture can be developed in the next few years. The solar flare is not only the most violent form of solar activity, with direct consequences in the space environment and upper atmospheres of the planets, but it is the prototype, evidently, of the enormous outbursts of the u.v. Ceti stars (Chapter 1) and perhaps of some of the outbursts in distant galaxies. Indeed, the magnetic flare is the ultimate manifestation of the activity of the magnetic field.

21.5. Basic theoretical ideas on solar flares

For all the complexity of the flare, there are, however, a few basic theoretical ideas that have grown up with the observations and which seem to be fundamental to the flare mechanism. As already noted, flaring occurs when loops of flux intrude from below the surface of the sun up into fields of magnetic regions already established. The more complex the topology of the established and the emerging fields, the more frequent and more violent the flares. Observations cited above show that the flare consists of rapid heating at some point near the top of the emerging loops, where they press most strongly into the fields above. The x-ray and e.u.v. observations show the thermal phase of the flare to consist of a growing kernel of intensely hot gas (3×10^7 K) spreading along a magnetic loop from a point somewhere near the apex, and then jumping to neighbouring loops, like a brush fire spreading to the nearby bushes and trees. The *non-thermal* x-ray and microwave radio emissions, and the rapid brightening and widening of the H_α emission line are first detected at about the time the high temperature kernel appears, increasing thereafter with the passage of time. A concise review of the detailed progress of the flare may be found in Van Hoven (1976) and in Heyvaerts *et al.* (1977) (see also Rust and Bridges 1975; Rust 1976; Dere *et al.* 1977; and the very interesting description of the activity in flare loops by Jeffries and Orrall 1965). Svestka (1976) provides a review in depth.

Now it was pointed out some years ago (Sweet 1964; Dungey 1964) that the high intensity and short life of the solar flare indicates that it is caused by a non-linear instability. In some way the distortion of the magnetic field from a potential form builds up to a high level before the instability turns on (Barnes and Sturrock 1972). Once turned on, the instability proceeds rapidly to devour the distortion, reducing it to a level well below that which is necessary to initiate the instability. A linear effect cannot accomplish this. For if there were a linear dissipation effect, arising when the distortion of the field exceeds some threshold, then the rate of dissipation would simply keep the distortion limited to a level just a small amount above the threshold. Presumably rapid reconnection in the simple hydromagnetic form (Chapter 15) would, of itself, have this property, proceeding at some fraction of the Alfven speed in the distorted component of the field. The result is a steady release of magnetic energy, giving a general enhancement of the density and temperature of the gas, and of the thermal x-ray emission. It is some such effect that leads to the steady glow of the chromosphere and corona above an active region. But it cannot, so far as we can see, produce the transient explosive energy release that is the flare, where the clock spring is first wound up tight and then suddenly released so that a very large amount of energy is available in a short period of time. The field is left at a much lower energy level than that at which it began. In recognition of this requirement, and given that rapid reconnection goes on continually in the complex non-equilibrium topology above an evolving active region, it has been suggested that a flare occurs when somewhere the curl of the field in the neighbourhood of a neutral point becomes sufficiently large, where the gas is sufficiently tenuous, that the electron conduction velocity approaches the ion sound speed. Several plasma instabilities may then arise, the first being the ion–acoustic mode. The growth of the ion–acoustic mode proceeds rapidly and soon produces strong, non-linear plasma turbulence. The consequent scattering of the conduction electrons greatly enhances the effective resistivity. The enhanced resistivity hastens the rapid reconnection, causing a thickening of the region of dissipation. It may also give rise to the resistive tearing mode instability, greatly increasing the rate of dissipation and reconnection (Furth *et al.* 1963; Jaggi 1964; Coppi 1964; Sturrock 1968; Coppi and Friedland 1971; Van Hoven and Cross 1971, 1973; Rutherford 1973).

Van Hoven (1976) has recently reviewed the enhancement of rapid reconnection in solar flares, comparing Petschek's rapid reconnection and the consequences of the resistive tearing mode instability in the context of recent observations of flares and of laboratory experiments on neutral point reconnection. The point made is that the sequence of instabilities is triggered only after the field is strongly distorted. It is argued that, once initiated, the enhanced dissipation is self-sustaining, so that the distortion of the field is grossly reduced before the dissipation disappears. The scenario is not implausible, nor has it been firmly demonstrated. Indeed the entire question of the development of the resistive tearing mode instability into its non-linear stages, and the associated large-scale reconnection, is a very murky and difficult question. It has been studied extensively in connection with the Tokamak plasma confinement geometry, but much work remains to be done in connection with the solar flare.

Heyvaerts *et al.* (1977) have proposed an interesting interpretation of the flare, based on a theoretical interpretation of the observed properties of the magnetic loops and the intense x-ray kernels that appear when the loops push upward into the established overhead field of an active region. They suggest, as have others before them, that when the current sheet involved in the neutral-point rapid reconnection is pushed by the rising loops to a sufficiently high altitude (generally somewhere in the upper chromosphere) thermal instability leads to runaway temperatures (Parker 1953; Syrovatsky 1976; Heyvaerts and Priest 1976). The gas density becomes so small that the electron conduction speed approaches the ion sound speed, so that the ion–acoustic mode is suddenly and strongly excited, enhancing the resistivity (by a factor of the order of 10^4, or more). The increased resistivity causes the current sheet to lengthen and thicken, the reconnection rate increasing by a factor of about four, and the electric field increasing to the point of producing runaway electrons. This gives rise, they suggest, to the impulsive and flash phases of the flare. The runaway particles are accelerated to high energies, and, upon streaming down along the magnetic lines of force, impact the dense gases in the low chromosphere to produce the classical H_α flare. Those accelerated electrons which stream upward along the magnetic lines of force produce the type-III radio bursts. They suggest that the flash phase ends, and the prolonged main phase of the solar flare begins, when the current sheet readjusts to a quasi-equilibrium state that is marginally unstable to plasma turbulence. In this state the enhanced resistivity and the associated enhanced width of the current sheet, are maintained in balance. The point is that the total current per unit length in the current sheet depends entirely on the change in field ΔB across the sheet. For a given field configuration, then, the current density, and the electron conduction velocity, are inversely proportional to the thickness of the current sheet. On the other hand, the thickness of the current sheet increases with the resistivity. If, then, the plasma turbulence declines, the current sheet narrows and the electron conduction velocity increases, thereby exciting more plasma turbulence, etc. It seems plausible that a more or less stable balance is struck, which persists until so much of the field is consumed that it is no longer possible to excite plasma turbulence. The main phase of the flare continues for some time, then, until 'combustion' falls below the critical level and the 'fire' goes out.

Heyvaerts *et al.* suggest that this scenario accounts for the small flares, up to perhaps 10^{30} erg. It does not provide for a very large flare, with nearly a thousand times as much energy. It is to account for the large flares that other authors have invoked the resistive tearing mode instability, which arises when the current sheet becomes sufficiently thin. On the other hand, Heyvaerts *et al.* suggest that the large flares are produced when the established field is very strongly twisted prior to the intrusion of the magnetic loops from below. This provides an entirely new 'twist' to the theory, for the role of the intruding fields is then simply a knife, cutting across the distorted lines of force of the established field so that the large energy of their twisting is released. The cutting involves but little energy in itself—only enough to make a small flare. It is the twang of the newly severed lines of force of the established field that makes the large flare. The observed field topology is sketched in Fig. 21.7 and has the general form of an athletic fieldhouse—a vaulted magnetic dome. Imagine some magnetic loops, then, emerging through the surface of the sun somewhere under the middle of the dome. The emerging loops are not aligned with the fields above, so there is rapid reconnection where they press against the overlying magnetic dome. The lines of the rising loops become connected to the lines forming the dome, thereby releasing the large excess energy of the lines of the dome. The rapid reconnection of the lines takes place up in the dome structure so that the accelerated particles escape downward along the lines of force of the dome to give a bright H_α ribbon in the low chromosphere on each side. This is the origin, presumably, of the big two-ribbon H_α flares.

Spicer (1976, 1977) and Colgate (1977) have gone a step further, with the suggestion that a twisted rope of flux, of suitable dimension and field

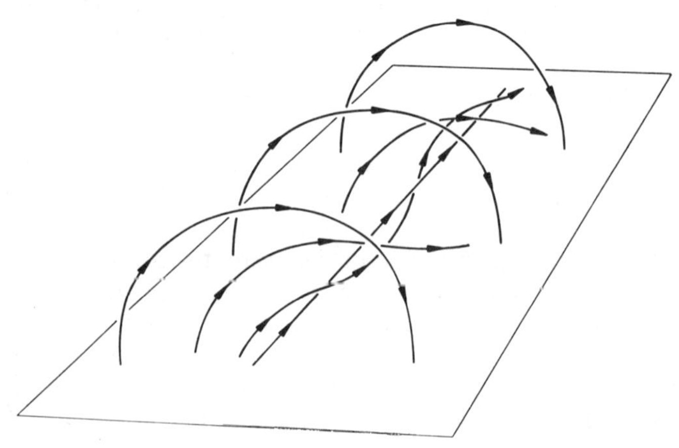

FIG. 21.7. A sketch of the strongly twisted field forming the magnetic dome structure in which two-ribbon H_α flares occur.

strength, produces the solar flare by itself. Spicer argues that the resistive tearing mode instability is endemic throughout any region of strongly twisted field, so that the field is subject to rapid dissipation across its entire volume. The idea arises from the well known fact (§14.8) that the projection of the magnetic field $B_i(\mathbf{r})$ onto the plane perpendicular to $B_i(\mathbf{r}_0)$ has a neutral point at $\mathbf{r}_0$. In a twisted flux rope the neutral point in the projected field may be an x-type neutral point for nearly all $\mathbf{r}_0$. Hence Spicer argues that the entire flux rope is subject to the resistive tearing mode instability. The result is the destruction of the twisted component of the field. The dissipation may be increased by the enhanced resistivity of plasma turbulence excited by the heavy localized electric current densities flowing along the large-scale twisted field.

Colgate (1977) has explored the idea further and finds a remarkable, and apparently natural, coincidence of the theoretical plasma temperatures and densities with those inferred from x-ray observations of real flares Using the comprehensive observations of the large flare of 4 August 1972, Colgate notes that the total energy release was of the order of 10^{31} erg over a region extending some 2×10^4 km along the local magnetic field. Assuming that the flare arose in a flux tube with a diameter of 2×10^3 km, twisted to a pitch of the order of $\pi/4$, the field strength must have been about 10^3 G in order to release 10^{31} erg. The duration of the flare was about 10^3 s. The current density along the twisted tube was $j_{\|}=2\times10^4$ e.s.u. It follows from the induction eqn (4.2) that the dissipation of a major part of the magnetic field $B=10^3$ G in a time $t=10^3$ s across a radius $r=10^8$ cm has an electric field of the order of $E=Br/ct\cong 1\text{ V cm}^{-1}$ associated with it. The total potential difference between the ends of the tube is then of the order of 10^9 V. The dissipation rate is $Ej_{\|}=10^2\text{ erg cm}^{-3}\text{ s}^{-1}$. Over the entire volume, of about 10^{26} cm^3, the power is $10^{28}\text{ erg s}^{-1}=10^{21}$ W, yielding 10^{31} erg in the 10^3 s life of the flare. Colgate works out the heating of the ambient plasma, including thermal conduction along the lines of force into the photosphere, and arrives at a temperature of 2×10^7 K over most of the tube. The idea is that some $10^2\text{ erg cm}^{-3}\text{ s}^{-1}$ is converted to thermal energy, which escapes either as x.u.v. radiation or by conduction along the tube to the lower 'ends' in the chromosphere and photosphere. Thermal conduction is proportional to $T^{\frac{5}{2}}$, with the result that the temperature is nearly uniform along the tube, dropping abruptly at the ends. For a tube with uniform energy input extending in the x-direction with uniform cross-section, and a maximum temperature T_m at $x=0$, the temperature is described by $T(x)=T_m(1-x^2/L^2)^{\frac{2}{7}}$. The temperature goes steeply to zero at the two ends. Pressure balance along the lines of force from the photosphere gives a pressure of $2N_ekT\cong6\times10^3\text{ dyn cm}^{-2}$ along the tube, so that $N_e\cong10^{12}\text{ cm}^{-3}$. These results are in good agreement with values inferred from both the observed x-ray continuum and the observed x-ray line intensities (cf. the models of Shapiro and Moore 1977; Phillips and Zirker 1977; Dere *et al.* 1977). The gas pressure is small compared to the magnetic pressure $B^2/8\pi = 40 \times 10^3$ dyn cm^{-2}, so that the field remains nearly force-free. The heat conducted downward along the lines of force into the low chromosphere produces the classical H_α flare. Colgate goes on to argue that the hard non-thermal x-rays (5–100 keV) are a consequence of runaway electrons in the field E in the current filaments presumed to develop as the flare progresses (cf. Smith 1977). The electric field of 1 V cm^{-1} is so much stronger than that required for runaway electrons that most of the electric current must be carried by an electron stream constrained mainly by the enhanced resistivity of the plasma turbulence which the current excites.

Altogether the model has many unique attractive features, suggesting it is part of the flare phenomenon. It is to be hoped that further theoretical analysis and close comparison with the high resolution x-ray and x.u.v. studies of future flares from spacecraft will sharpen the picture and allow a more definitive statement of the many effects that contribute to the various flares occurring on the sun. Flares appear on so many different scales and in so many different forms that a variety of mechanisms must be involved.

The variety of scenarios proposed for the solar flare is reviewed in concise form by Svestka (1976). The challenge presented by the solar flare to both the observer and the theoretician has played a major role in solar physics for a decade. The flare is the central figure in planning for the NASA Solar Maximum Mission for 1980 and 1981, and, has been a major concern to a variety of space missions in the past, such as the Orbiting Solar Observatories, the Skylabs, and a variety of Explorer and Discoverer spacecraft, the Interplanetary Monitoring Platforms, the Orbiting Geophysical Observatories and the Polar Orbiting Geophysical Observatories, to mention only a few. These latter spacecraft were designed to study the effects of energetic flare particles and the blast waves from flares on the magnetic field and upper atmosphere of Earth. They gave direct attention to the flare particles, and most of what we know of the flare acceleration mechanism comes from their observations of the energetic flare particles reaching the orbit of Earth.

As a matter of fact, one of the most interesting and important discoveries of the spacecraft exploring the terrestrial magnetosphere is the general non-equilibrium and flare activity of the magnetosphere itself. It was known in the nineteenth century that the magnetosphere is active, but the theories that began to emerge in the late nineteenth and early twentieth centuries viewed the magnetosphere only as a passive structure that was shaken, and often penetrated, by the particles from the sun, without playing an active role itself (see historical review by Dessler 1967). Thus, for instance, Störmer suggested early in the century (see Störmer 1955) that the aurora are produced directly by mono-energetic protons from the sun (with an energy of 10–10^3 MeV) which enter directly into the magnetosphere where they are deflected into the curved patterns so often assumed by the auroral curtains. His calculations of the auroral patterns based on his erroneous assumption show striking similarities to the many features of the aurora. The idea was nagged by the fact that protons, with the energies necessary to penetrate through the geomagnetic field to the auroral zone and lower latitudes, would penetrate far more deeply into the atmosphere than the aurora. But the remarkable similarity of the theoretical and observed shapes was too seductive, so the idea held out for many decades after it might have been put to rest.

The magnetic storm theory of Chapman and Ferraro (1931, 1933, 1940) (see review in Chapman and Bartels 1940; Chapman 1964) viewed the distortion of the geomagnetic field as a consequence of the direct intrusion of solar plasma into the field. Dessler and Parker (1959) and Dessler *et al.* 1961) used the same point of departure in their proposal that the main phase of the storm is a consequence largely of 10 keV protons trapped in the geomagnetic field at a distance of 4–6 Earth's radii. The existence of the particles and of a strong electron component has been verified now by direct detection (Frank 1967; Frank and Owens 1970) of the magnetic storm protons. The amusing fact is, however, that the general observational study of the magnetosphere has shown no evidence that the storm particles are introduced directly from the sun. Most of the acceleration takes place within the magnetosphere. In particular, the magnetic substorm phenomenon has much the character of a flare in the neutral sheet of the geomagnetic tail (Williams and Mead 1969; Vasyliunas and Wolf 1973; Frank *et al.* 1976). The magnetospheric flare is invisible to observations from the ground because the gas density in the tail is so low that there is no significant electromagnetic radiation produced. But the fast particles generated in and around the neutral sheet in the geomagnetic tail produce a variety of low-frequency electromagnetic plasma waves within the magnetosphere, a variety of fluctuations of the geomagnetic field itself, and the electromagnetic emission in the upper atmosphere that makes up the visible aurora. The hydromagnetic conditions in the magnetosphere are complicated by the fact of the non-conducting atmosphere at the foot of the magnetic lines of force. The magnetic lines of force are not prevented from convecting, and, indeed, the convection plays a fundamental role in the activity of the magnetosphere (Axford and Hines 1961). The activity is more complex than the basic hydromagnetic theory in a uniformly highly conducting fluid, to which the present volume is limited. The theory works more directly with the motions of the individual particles. The electric currents are channelled along the magnetic lines of force so that the magnetosphere is a complex exercise in plasma physics rather than hydromagnetic theory. (cf. Akasofu and Chapman 1972; Kennel and Coroniti 1975; Paulikas 1975; Fairfield 1976; Kivelson 1976; Stern 1977).

The geomagnetic field, then, provides a second environment for the study of magnetic activity. The magnetic activity on the sun is visible to telescopic observation because of the strength of the fields and the high density of the gas. It is not subject to study by direct measurements with instruments placed inside the region of activity, for obvious reasons. The activity of the terrestrial magnetosphere, while invisible to telescopic observation, can be studied by instruments placed directly into the site of the activity. Hence the solar and terrestrial magnetic activity provide the physicist with complementary opportunities to study the magnetic flare phenomenon. The flare is the ultimate form of magnetic activity, occurring on planetary, stellar, and, evidently, galactic scales. It cannot be emphasized too strongly that the development of a solid understanding of the magnetic activity, occurring in so many forms in so many circumstances in the astronomical universe, can be achieved only by coordinated study of the various forms of activity that are accessible to quantitative observation in the solar system. The sun, the solar wind, and the magnetosphere of Earth and of the other planets provide a variety of activities, each partially observable. The entire array of activity must be studied if we are to see the many faces of the magnetic activity of the sun and planets as an introduction to the study of the magnetic activity elsewhere (Kennel *et al.* 1978).

References

Akasofu, S. I. and Chapman, S. (1972). *Solar terrestrial physics.* Clarendon Press, Oxford.

Alfonso-Faus, A. (1967). *J. Geophys. Res.* **72,** 5576.

Altschuler, M. D., Trotter, D. E., Newkirk, G., and Howard, R. (1974). *Solar Phys.* **39,** 3.

Anathakrishnan, R. (1954). *Proc. Ind. Acad. Sci.* **40,** 72.

Angel, J. R. P. (1977). *Astrophys. J.* **216,** 1.

Anglin, D. (1975). *Astrophys. J.* **98,** 733.

Axford, W. I. and Hines, C. O. (1961). *Can. J. Phys.* **39,** 1433.

Babcock, H. D. (1959). *Astrophys. J.* **130,** 364.

——, (1960b). ibid. **132,** 521.

——, (1961). ibid. **133,** 572.

Babcock, H. W. and Babcock,, H. D. (1955). *Astrophys. J.* **121,** 349.

Barnes, C. W. and Sturrock, P. (1972). *Astrophys. J.* **174,** 659.

Becker, J. M. *Z. Astrophys.* **37,** 47.

——, (1958). unpublished, quoted by de Jager (1959), p. 327.

Beckers, J. M. (1963). *Astrophys. J.* **138,** 648.

BECKERS, J. M. (1968). *Solar Phys.* **3,** 367.
——, (1976). *Astrophys. J.* **203,** 739.
——, and SCHNEEBERGER, T. J. (1977). *Astrophys. J.* **215,** 356.
——, and SCHRÖTER, E. H. (1968). *Solar Phys.* **4,** 192.
——, (1969). ibid. **10,** 384.
BELVEDERE, G. and PATERNO, L. (1977). *Solar Phys.* **54,** 289.
BIDELMAN, W. P. (1967). *The magnetic and related stars* (ed. R. C. CAMERON), P. 29, Mono Book Corp., Baltimore.
BIERMANN, L. (1941). *Vierteljahrsschr. Astron. Gesselsch.* **76,** 194.
BORUTA, N. (1977). *Astrophys. J.* **215,** 364.
BRAY, R. J. and LOUGHHEAD, R. E. (1964). *Sunspots.* Chapman and Hall, London.
BRUECKNER, G. E. (1976). *Phil Trans. Roy. Soc. Lond. A* **281,** 443.
BUMBA, V. and HOWARD, R. F. (1965). *Astrophys. J.* **141,** 1502.
—— ——, MARTRES, M. J., and SORU-ISOVICI, I. (1968). *Structure and development of solar active regions, IAU Symposium No. 35* (ed. K. O. KIEPENHAHN), p. 13. Reidel, Dordrecht.
BUSSE, F. H. and CUONG, P. G. (1977). *Geophys. Astrophys. Fluid Dynamics* **8,** 17.
CARRINGTON, R. C. (1858). *Mon. Notic. Roy. Astron. Soc.* **19,** 1.
——, (1863). *Observations on the spots of the sun from Nov. 9, 1853 to March 24, 1861 made at Redhill,* Williams and Norgate, London.
CHAPMAN, S. (1964). *Solar plasma, geomagnetism and aurora.* Gordon and Breach, New York.
——, and BARTELS, (1940). *Geomagnetism.* Clarendon Press, Oxford.
——, and FERRARO, V. C. A. (1931). *Terr. Magn. Atmos. Elect.* **36,** 77, 171.
—— ——, (1933). ibid. **38,** 79.
—— ——, (1940). ibid. **45,** 245.
CHENG, C. C. and WIDING, K. G. (1975). *Astrophys. J.* **201,** 735.
CLERKE, A. M. (1894). *Knowledge* **17,** 206.
COCKE, W. J. (1967). *Astrophys. J.* **150,** 1041.
COHEN, J. G. (1970). *Astrophys. J.* **159,** 473.
COLGATE, S. A. (1977). *Astrophys. J.* **221,** 1068.
COPPI, B. (1964). *Ann. Phys.* **30,** 178.
——, and FREIDLAND, A. B. (1971). *Astrophys. J.* **169,** 379.
COWLING, T. G. (1953). *The sun,* (ed. G. P. KUIPER), chap. 8. University of Chicago Press.
DANIELSON, R. E. (1961a). *Astrophys. J.* **134,** 275.
——, (1961b). ibid. 289.
D'AZAMBUJA, L. and D'AZAMBUJA, M. (1948). *Ann. Obs. Meudon* **6,** 52.
DEINZER, W. and STIX, M. (1971). *Astron. Astrophys.* **12,** 111.
DEINZER, W., v. KUSSEROW, H. U., and STIX, M. (1974). *Astron. Astrophys.* **36,** 69.
DE JAGER, C. (1959). *Handbuch d. Physik* **52,** 322.
DE JAGER, C. and SVESTKA, Z. (1976). *Solar Phys.* **47,** No. 1.
DERE, K. P., HORAN, D. M., and KREPLIN, R. W. (1977). *Astrophys. J.* **217,** 976.
DESSLER, A. J. (1967). *Rev. Geophys. Space Phys.* **5,** 1.
——, HANSON, W. B., and PARKER, E. N. (1961). *J. Geophys. Res.* **66,** 3631.
——, and PARKER, E. N. (1959). *J. Geophys. Res.* **64,** 2239.
DEUTSCH, A. J. (1954). *Trans. IAU* Cambridge University Press, Cambridge. **8,** 801.

DEUTSCH, A. J. (1958a). *Electromagnetic phenomena in cosmical physics, IAU Symp. No. 6* (ed. B. LEHNERT), p. 209. Cambridge University Press.
——, (1958b). *Handbuch d. Phys.* **51,** 689.
——, (1970). *Astrophys. J.* **159,** 985.
DICKE, R. H. (1964). *Nature* **202,** 432.
——, (1972). *Astrophys. J.* **171,** 331.
DUNGEY, J. W. (1964). *Proc. AAS–NASA Symp. on the physics of solar flares,* NASA SP–50, p. 415. National Aeronautics and Space Administration, Washington, *D.C.*
DUNN, R. B., RUST, D. M., and SPENCE, G. E. (1974). *Proc. Soc. Photo-Optical Inst. Eng.* **44,** 109.
DURNEY, B. (1972). *Solar Phys.* **26,** 3.
——, (1974). *Astrophys. J.* **190,** 211.
——, (1975). ibid. **199,** 761.
——, (1976). in *Basic mechanisms of solar activity* (ed. Bumba, V. and Kleczek, J.) I.A.U. Symposijm No. 71. Kleczek.
——, and LATOUR, J. (1977). *Geophys. Astrophys. Fluid Dynamics* **9,** 241.
EDDINGTON, A. S. (1929). *Mon. Notic. Roy. Astron. Soc.* **90,** 54.
EDDY, J. A. (1976). *Science* **192,** 1189.
——, (1977). *Scientific American* **236,** (5) 80.
——, GILMAN, P. A., and TROTTER, D. E. (1976). *Solar Phys.* **46,** 3.
EICHLER, D. (1977). *Astrophys. J.* **216,** 174.
ELLISON, M. A., MCKENNA, S. M. P., and REID, J. H. (1960). *Dunsink Obs. Publ.* **1,** 1.
FAIRFIELD, D. H. (1976). *Rev. Geophys. Space Phys.* **14,** 117.
FRANK, L. A. (1967). *J. Geophys. Res.* **72,** 3753.
——, ACKERSON, K. L., and LEPPING, R. P. (1976). *J. Geophys. Res.* **81,** 5859.
——, and OWENS, H. D. (1970). *J. Geophys. Res.* **75,** 1269.
FURTH, H. P., KILLEEN, J., and ROSENBLUTH, M. N. (1963). *Phys. Fluids* **16,** 1054.
GIERASCH, P. J. (1974). *Astrophys. J.* **190,** 199.
GILMAN, P. A. (1969). *Solar Phys.* **8,** 316; **9,** 3.
——, (1974). *Ann. Rev. Astron. Astrophys.* **12,** 47.
——, (1977). *Geophys. Astrophys. Fluid Dynamics* **8,** 101.
GILLESPIE, B., HARVEY, J., and LIVINGSTON, W. (1973). *Astrophys. J. Lett.* **186,** L85.
GLENCROSS, W. M. (1975). *Astrophys. J. Lett.* **199,** L53.
GOKHALE, M. H. and ZWAAN, C. (1972). *Solar Phys.* **26,** 52.
GOLUB, L., KRIEGER, A. S., SILK, J. K., TIMOTHY, A. F., and VAIANA, G. S. (1974). *Astrophys. J. Lett.* **189,** L93.
—— ——, and VAIANA, G. S. (1977). *Solar Phys.* **49,** 79.
GNEVYSHEV, M. N. (1967). *Solar Phys.* **1,** 107.
HALE, G. E. (1908a). *Pub. Astron Soc. Pac.* **20,** 220, 287.
——, (1908b). *Astrophys. J.* **28,** 100, 315.
——, (1913). ibid. **38,** 27.
——, and NICHOLSON, S. B. (1938). *Magnetic observations of sunspots,* 1917–1924, Part I, Pub. Carnegie Inst. No. 498.
HARVEY, K. L., HARVEY, J. W., and MARTIN, S. F. (1975). *Solar Phys.* **40,** 87.
——, and MARTIN, S. F. (1973). *Solar Phys.* **32,** 389.
HAVNES, O. and CONTI, P. (1971). *Astron. Astrophys.* **14,** 1.
HEYVAERTS, J. and PRIEST, E. R. (1976). *Solar Phys.* **47,** 223.

HEYVAERTS, J., PRIEST, E. R., and RUST, D. M. (1977). *Astrophys. J.* **216,** 123.
HOWARD, R. (1959). *Astrophys. J.* **130,** 193.
——, (1965). *Stellar and magnetic fields, IAU Symp. No.* 22, (ed. R. LÜST), p. 129. North-Holland, Amsterdam.
——, (1967). *Ann. Rev. Astron. Astrophys.* **5,** 1.
——, (1972). *Solar Phys.* **25,** 5.
——, (1974a). ibid. **38,** 59.
——, (1974b). **38,** 283.
——, (1974c). ibid. **39,** 275.
——, (1977). ibid. **52,** 243.
HUNDHAUSEN, A. J. (1972). *Solar wind and coronal expansion.* Springer–Verlag, Berlin.
HURFORD, G. J., MEWALDT, R. A., STONE, E. C., and VOGT, (1975). *Astrophys. J. Lett.* **201,** L95.
JAGGI, R. K. (1964). *AAS–NASA symposium on the physics of solar flares* NASA Spec. Publ. SP-50 (ed. W. N. HESS), p. 425. National Aeronautics and Space Administration, Washington, D. C.
JEFFRIES, J. T. and ORRALL, F. Q. (1965). *Astrophys. J.* **141,** 505.
KAHLER, S. W., KRIEGER, A. S., and VAIANA, G. S. (1975). *Astrophys. J. Lett.* **199,** L57.
KENNEL, C. F. and CORONITI, F. V. (1975). *Space Sci. Rev.* **17,** 857.
——, LANZEROTTI, L. J., and PARKER, E. N. (1978). (editors) *Solar system plasma physics.* North–Holland, Amsterdam.
KIEPENHEUER, K. O. (1953). *The sun* (ed. G. P. KUIPER), chap. 6. University of Chicago Press.
KIPPENHAHN, R. (1963). *Astrophys. J.* **317,** 664.
KIVELSON, M. G. (1976). *Rev. Geophys. Space Phys.* **14,** 189.
KODAIRA, K. and UNNO, W. (1969). *Astrophys. J.* **157,** 769.
KÖHLER, H. (1966). *Mitt. Astron. Ges.* **21,** 91.
——, (1970). *Solar Phys.* **13,** 3.
——, (1973). *Astron. Astrophys.* **25,** 467.
KOPP, R. A. and HOLZER. T. E. (1976). *Solar Phys.* **49,** 43.
KRIEGER, A. S., TIMOTHY, A. F., and ROELOF, E. C. (1973). *Solar Phys.* **29,** 505.
KRIEGER, A. S., VAIANA, G. S., and VAN SPEYBROECK, L. P. (1971). *Solar magnetic fields, IAU symposium No. 43* (ed. R. HOWARD), p. 397. Reidel, Dordrecht.
KRIEGER, A. S., VAIANA, G. S., VAN SPEYBROECK, L. P., ZEHNPFENNIG, T., and ZOMBECK, M. (1970). *Bull. Amer. phys. Soc.* **15,** 612.
LANDSTREET, J. D., BORRO, E. F., ANGEL, J. R. P., and ILING, R. M. E. (1975). *Astrophys. J.* **201,** 624.
LEIGHTON, R. B. (1959). *Astrophys. J.* **130,** 366.
——, (1969). ibid. **156,** 1.
LEVY, E. H. and ROSE, W. K. (1974). *Astrophys. J.* **193,** 419.
LYNCH, D. K., BECKERS, J. M., and DUNN, R. B. (1973). *Solar Phys.* **30,** 63.
MAUNDER, E. W. (1890). *Mon. Notic. Roy. Astron. Soc.* **50,** 251.
——, (1894a). *J. Brit. Astron. Soc.* **5,** 47.
——, (1894b). *Knowledge* **17,** 173.
——, (1904). *Mon. Notic. Roy. Astron. Soc.* **64,** 760.
——, (1922). *J. Brit. Astron. Soc.* **32,** 140.
MESTEL, L. (1965A). *Stellar and solar magnetic fields, IAU Symp. No. 22* (ed. R. LÜST), p. 420. North–Holland, Amsterdam.

MESTEL, L. (1965b). *Quart. J. Roy. Astron Soc.* **6,** 265.
——, (1967). *The magnetic and related stars* (ed. R. C. CAMERON), p. 101. Mono Book Corp., Baltimore.
——, and MOSS, D. L. (1977). *Mon. Notic. Roy. Astron. Soc.* **178,** 27.
——, and SELLEY, C. S. (1970). *Mon. Notic. Roy. Astron. Soc.* **149,** 197.
MEYER, F., SCHMIDT, H. U., and WEISS, N. O. (1977). *Mon. Notic. Roy. Astron. Soc.* **179,** 741.
—— —— ——, and WILSON, P. R. (1974). *Mon. Notic. Roy. Astron. Soc.* **169,** 35.
MODISETTE, J. L. (1967). *J. Geophys. Res.* **72,** 1521.
MONAGHAN, J. J. (1973). *Ap. J.* **186,** 631.
MOSS, D. L. (1977a). *Mon. Notic. Roy. Astron. Soc.* **178,** 51.
——, (1977b). ibid. **178,** 61.
MULLAN, D. J. (1975). *Astrophys. J.* **198,** 563; **200,** 641.
MÜLLER, R. (1955). *Z. Astrophys.* **38,** 212.
NESS, N. F. and WILCOX, J. M. (1966). *Astrophys. J.* **143,** 23.
——, (1967). *Solar Phys.* **2,** 351.
NEUPERT, W. M., THOMAS, R. J., and CHAPMAN, R. D. (1974). *Solar Phys.* **34,** 349.
NEWTON, H. W. and NUNN, M. L. (1955). *Mon. Notic Roy. Astron. Soc.* **111,** 413.
NOLTE, J. T., KRIEGER, A. S., TIMOTHY, A. F., GOLD, R. E., ROELOF, E. C., VAIANA, G. S., LAZARUS, A. J., SULLIVAN, J. D., and MCINTOSH, P. S. (1976). *Solar Phys.* **46,** 303.
PARKER, E. N. (1953). *Astrophys. J.* **117,** 431.
——, (1955a). ibid. **121,** 491.
——, (1955b). ibid. **122,** 293.
——, (1957a). *Phys. Rev.* **107,** 830.
——, (1957b). *Proc. Nat. Acad. Sci.* **43,** 8.
——, (1958). *Astrophys. J.* **128,** 664.
——, (1963). *Interplanetary dynamical processes.* Wiley–Interscience, New York.
——, (1970). *Astrophys. J.* **162,** 665.
——, (1972). ibid. **174,** 499.
——, (1974a). ibid. **190,** 429.
——, (1974b). *Solar Phys.* **36,** 249; **37,** 127.
——, (1975a). *Astrophys. J.* **198,** 205.
——, (1975b). ibid. **201,** 494.
——, (1975c). ibid. **202,** 523.
——, (1975d). *Solar Phys.* **40,** 291.
——, (1976). *The physics of nonthermal radio sources* (ed. G. SETTI), p. 137. Reidel, Dordrecht.
——, (1977a). *Astrophys. J.* **215,** 370.
——, (1977b). *Mon. Notic. Roy. Astron. Soc.* **179,** 93P.
PAULIKAS, G. A. (1975). *Rev. Geophys. Space Phys.* **13,** 709.
PETRASSO, R. D., KAHLER, S. W., KRIEGER, A. S., SILK, J. K., and VAIANA, G. S. (1975). *Astrophys. J. Lett.* **199,** L127.
PHILLIPS, K. J. H. and ZIRKER, J. B. (1977). *Solar Phys.* **53,** 41.
PRESTON, G. W. (1967A). *The magnetic and related stars* (ed. R. C. CAMERON), p. 3. Mono Book Corp., Baltimore.
——, (1967b). *Astrophys. J.* **147,** 804; **150,** 547.
——, (1969). ibid. **156,** 967; **158,** 243, 251.
——, (1971). ibid. **164,** 309.
PRIEST, E. R. and HEYVAERTS, J. (1974). *Solar Phys.* **36,** 433.

PYPER, J. M. (1969). *Astrophys. J. Suppl.* **18,** 347.
RÄDLER, K. H. (1975). *Mem. Soc. R. Sci. Liege* **6,** 109.
ROBERTS, B. (1976). *Astrophys. J.* **204,** 268.
ROBERTS, P. H. (1972). *Phil. Trans. Roy. Soc. Lond. A* **272,** 663.
——, and STIX, M. (1972). *Astron. Astrophys.* **18,** 453.
ROXBURGH, I. W. (1967). *The magnetic and related stars* (ed. R. C. CAMERON), p. 45. Mono Book Corp., Baltimore.
RUST, D. M. (1976). *Phil. Trans. Roy. Soc. Lond. A* **281,** 427.
——, and BRIDGES, C. A. (1975). *Solar Phys.* **43,** 129.
RUST, D. M., NAKAGAWA, Y., and NEUPERT, W. M. (1975). *Solar Phys.* **41,** 397.
RUTHERFORD, P. H. (1973). *Phys. Fluids* **16,** 1903.
SCHATZMAN, E. (1962). *Ann. Astrophys.* **25,** 1.
SCHWARTZ, R. A. and STEIN, R. F. (1975). *Astrophys. J.* **200,** 499.
SCHÜSSLER, M. and PÄHLER, A. (1978). *Astron. Astrophys.* in press.
SERLEMITSOS, A. T. and BALASUBRAHMANYAN, V. K. (1975). *Astrophys. J.* **198,** 195.
SHAPIRO, P. R. and MOOE, R. T. (1977). *Astrophys. J.* **217,** 621.
SHEELEY, N. R., BOHLIN, J. D., BRUECKNER, G. E., PURCELL, J. D., SCHERRER, V., and TOUSEY, R. (1975). *Solar Phys.* **40,** 103.
SIMON, G. and WEISS, N. (1968). *Structure and development of solar active regions, IAU Symp. No. 35*, p. 108. Reidel, Dordrecht.
SMITH, D. F. (1977). *Astrophys. J.* **217,** 644.
SIMTH, H. J. and SMITH, E. V. P. (1963). *Solar flares*, Macmillan, New York.
SPENCER JONES, H. (1955). *Sunspot and geomagnetic storm data derived from Greenwich observations* 1874–1954. Her Majesty's Stationery Office, London.
SPICER, D. S. (1976). *Solar Phys.* **53,** 305.
——, (1977). ibid. **54,** 379.
SPÖRER, G. (1894). *Pub. Potsdam Observ.* **10,** (1) 144.
SPRUIT, H. D. (1974). *Solar Phys.* **34,** 277.
STEENBECK, M. and KRAUSE, F. (1969). *Astron. Nachr.* **291,** 49, 271.
STEINITZ, R. (1964). *Bull. Astron. Inst. Netherlands* **17,** 504.
——, (1967). *The magnetic and related stars* (ed. R. C. CAMERON), p. 83. Mono Book Corp., Baltimore.
STENFLO, J. O. (1970). *Solar Phys.* **13,** 43.
——, (1971). *Solar magnetic fields* (ed. R. HOWARD), p. 714. Reidel, Dordrecht.
STERN, D. P. (1977). *Rev. Geophys. Space Phys.* **15,** 156.
STIBBS, D. W. N. (1950). *Mon. Notic. Roy. Astron. Soc.* **110,** 395.
STIX, M. (1976a). *Astron. Astrophys.* **47,** 243.
——, (1976b). *Basic mechanisms of solar activity, IAU Symposium No. 71* (ed. J. KLECZEK), D. Reidel, Dordrecht.
STÖRMER, C. (1955). *The polar aurora.* Clarendon Press, Oxford.
STURROCK, P. A. (1968). *Structure and development of solar active regions, IAU Symposium No. 35* (ed. K. O. KIEPENHEUER), p. 471. Reidel, Dordrecht.
SUESS, S. T. (1971). *Solar Phys.* **18,** 172.
SVESTKA, Z. (1976). *Solar flares.* Reidel, Dordrecht.
SWEET, P. A. (1950). *Mon. Notic. Roy. Astron. Soc.* **110,** 548.
——, (1964). *Proc. AAS–NASA Symp. on the physics of solar flares*, NASA SP–50, p. 409. National Aeronautics and Space Administration, Washington, D. C.
——, (1969). *Ann. Rev. Astron. Astrophys.* **7,** 149.
SYROVATSKY, S. I. (1976). *Sov. Astron. Lett.* **2,** 1.

Tandberg–Hanssen, E. (1967). *Solar activity*. Blaisdell, Waltham, Mass.
Thomas, R. J. (1975). *Solar gamma, x-, and e.u.v. radiation, IAU Symp. No. 68* (ed. S. R. Kane), p. 25. Reidel, Dordrecht.
Trellis, N. (1957). *Ann. d. Ap. Suppl.* **5.**
Vaiana, G. S., Davis, J. M., Giacconi, R., Krieger, A. S., Silk, J. K., Timothy, A. F., and Zombeck, M. (1973). *Astrophys, J. Lett.* **185,** L47.
——, Krieger, A. S., and Timothy, A.F. (1973). *Solar Phys.* **32,** 81.
——, and Giacconi, R. (1969). *Plasma instabilities in astrophysics*, (ed. D. G. Wentzel and D. A. Tidman), p. 91. Gordon and Breach, New York.
Van Hoven, G. (1976). *Solar phys.* **49,** 95.
——, and Cross, M. A. (1971). *Phys. Fluids* **14,** 1141.
——, (1973). *Phys. Rev. A* **7,** 1347.
Vasyliunas, V. M. and Wolf, R. A. (1973). *Rev. Geophys. Space Phys.* **11,** 181.
Veeder, G. J. and Zirin, H. (1970). *Solar Phys.* **12,** 391.
Von Humboldt, A. (1851). *Cosmos* **3.**
Von Zeipel, H. (1924). *Mon. Notic. Roy. Astron. Soc.* **84,** 665.
Vorpahl, J. A., Gibson, E. G., Landecker, P. B., McKenzie, D. L., and Uderwood, J. H. (1976). *Solar Phys.* **45,** 199.
Waldmeier, M. (1938). *Z. Astrophys.* **15,** 299.
——, (1957). *Z. Astrophys.* **43,** 29.
Ward, F. (1964). *J. Pure Appl. Geophys.* **58,** 157.
——, (1965a) ibid. **60,** 126.
——, (1965b). *Astrophys. J.* **141,** 534.
——, (1966). ibid. **145,** 416.
Weber, E. J. and Davis, L. (1967). *Astrophys. J.* **148,** 217.
Widing, K. G. (1975). *Solar gamma, x-, and e.u.v. radiation, IAU Symp. No. 68* (ed. S. R. Kane), p. 153. Reidel, Dordrecht.
Wilcox, J. M. (1971). *Solar magnetic fields, IAU Symp. No. 43* (ed. R. F. Howard), p. 744. Reidel, Dordrecht.
——, and Howard, R. (1968). *Solar Phys.* **5,** 564.
——, and Ness, N. F. (1967). *Solar Phys.* **1,** 347.
——, Schatten, K., Tannenbaum, A., and Howard, R. (1970). *Solar Phys.* **14,** 255.
Williams, D. J. and Mead, (editors) (1969). *Rev. Geophys. Space Phys.* **7,** (1,2).
Wilson, O. C. (1963). *Astrophys. J.* **138,** 832.
——, (1966). ibid. **149,** 695.
——, (1971). *Proc. Asilomar Solar Wind Conf.* 21–26 April, Pacific Grove, California.
——, and Skumanich, A. (1964). *Astrophys. J.* **140,** 1401.
Yoshimura, H. (1971). *Solar phys.* **18,** 417.
——, (1972a). ibid. **22,** 20.
——, (1972b). *Astrophys. J.* **178,** 863.
——, (1973). *Solar Phys.* **33,** 131.
——, (1975a). *Astrophys. J.* **201,** 740.
——, (1975b). *Astrophys. J. Suppl.* **29,** 467 (No. 294).
——, (1976a). *Solar Pkys.* **47,** 581.
——, (1976b). ibid. **50,** 3.
——, (1977a). ibid. **52,** 41.
——, (1977b). ibid. **54,** 229.
Zirin, H. (1972). *Solar Phys.* **22,** 34.
Zwaan, C. (1978). *Solar Phys.* in press.

22

THE MAGNETIC FIELD OF THE GALAXY

22.1. The gas–field–cosmic ray disc of the galaxy

22.1.1. General properties

THE magnetic field of the galaxy can be observed only through its integrated effects over long distances. Consequently its form and its activity cannot be studied in detail, as we are able to do with the magnetic field of Earth and other planets, and of the sun. But it is clear that the magnetic field of the galaxy is active, in its own way and in its own time. The stellar winds of the individual stars interact with their local galactic field (Parker 1960, 1963; Yu, 1974). Expanding supernova remnants interact strongly with the field to produce various radio spurs and active objects. On a grander scale the galactic magnetic field is active because it possesses no stable equilibrium. It is a participating dynamical constituent of the gaseous disc of the galaxy, contributing strongly to the inhomogeneity and activity of the disc. The interstellar gas is the visible component of the disc, of course, but the two invisible constituents—the magnetic field and the cosmic rays—are equal dynamical partners with the gas and drive the instability that prevents the system from reaching equilibrium. The interstellar gas and the cosmic ray gas are both tightly coupled to the magnetic field, forming a single active dynamical system (Parker 1957, 1958a, 1966b, 1967a,b 1969; see §13.4–13.9).

Altogether, the gas–field–cosmic ray system is confined by the weight of the interstellar gas in the gravitational potential well of the galaxy. The weight of the gas is the only confining force. The gas is inflated by the magnetic field and the cosmic rays, whose pressures are balanced against the weight of the gas to determine the thickness of the gaseous disc (Parker 1966b). The whole system is dynamically unstable and vigorously disturbed by the active stars (Parker 1969).

In the neighbourhood of the sun the scale height Λ of the gaseous disc is inferred from observations to be of the general order of 150–200 pc (cf. Baker and Burton 1975). The inferred scale height is consistent with present estimates of the mean gas density of 1–2 hydrogen atoms/cm^3 in the disc, a mean azimuthal field of about 3×10^{-6} G, and the cosmic ray pressure of $0{\cdot}5\times10^{-12}$ dyn cm^{-2} measured at the position of the sun. The simple local barometric equilibrium of the disc is strongly unstable (Parker 1966b, 1969) along the lines described in §§13.4, 13.7, and 13.8.

The magnetic field and cosmic rays are essentially weightless, so they are free to bulge upward wherever the system is suitably perturbed. The gas drains downward from the upward bulge of the lines of force, unloading the field and cosmic rays, and permitting them to bulge further. The accumulation of gas in the low places along the lines of force weighs those places down further. The characteristic growth time of the effect is comparable to the free-fall time $(\Lambda/g)^{\frac{1}{2}}$, or Alfven transit time Λ/V_A, over the scale height Λ. With $V_A \cong 5\ \text{km s}^{-1}$, the characteristic time is of the order of 3×10^7 years, or about 10^{-1} of the rotation period of the galaxy, and perhaps 2×10^{-3} of the age of the galaxy.

The galactic magnetic field is the central element in the dynamical instability of the disc. It is the means by which the cosmic rays are coupled into the system and it also contributes directly to the activity itself (see §§13.4 and 13.7) through magnetic buoyancy. It is frustrating that the gas–field–cosmic ray disc of the galaxy is so transparent that only the gross features can be observed. The system must be extremely active, yet we cannot directly see much of the activity. It is ironic that the sun is too opaque to see the activity of the fields beneath the surface, while the galaxy is too transparent.

22.1.2. The interstellar gas

The observations of the interstellar absorption lines in the spectra of distant stars show immediately the extreme inhomogeneity and violent agitation of the interstellar gas (Adams 1943, 1948; Greenstein 1946; Stromgren 1948; Heiles and Habing 1974). The strong clumping of the gas is a combination of thermal instability (Parker 1953; Zanstra 1955; Weymann 1960; Field 1969) and the mutual attraction caused by the aforementioned interaction of the gas with the galactic magnetic and gravitational fields (Parker 1966b, 1967a). The thermal instability causes the gas to prefer either a hot ($T > 10^3$ K) or a cold ($T \cong 10^2$ K) phase, so that the pressure compresses the cold phase to densities sometimes as high as 10–10^2 hydrogen atoms/cm^3 to form the dense interstellar clouds, while the hot phase expands, dropping the density below $0{\cdot}1\ \text{cm}^{-3}$ (Greenstein 1946; Stromgren 1948). The interaction of the gas with the galactic gravitational and magnetic fields causes a strong mutual attraction. Formally it can be shown (Parker 1967a) that the mutual attraction of two gas clouds, suspended on the horizontal lines of force of the galactic magnetic field B in the gravitational field g perpendicular to the plane of the disc, is stronger than their mutual gravitational attraction by the factor g^2/GB^2. The factor is equal to 15 for the typical values $g = 3\times10^{-9}\ \text{cm s}^{-2}$ (at a distance of 150 pc from the central plane) and $B = 3\times10^{-6}$ G. The point is that the gas is suspended on the magnetic lines of force, so that it collects into clumps which sag downward in the field as sketched in Fig. 13.3. It is downhill in the gravitational acceleration g of the galactic disc into each clump, so that the effective self-attraction of the gas is much stronger than self-gravitation except in the most massive or concentrated clouds whose local gravitational fields are as large as that of the galaxy.

Appenzeller (1971) has shown the striking resemblance of the magnetic field configuration in the α-Persei cluster to the expected configuration around a massive gas cloud, with the lines of force arching up on either side, as sketched in Fig. 13.3, and helping to support the weight of the gas cloud (see also the recent detailed study of Markkanen 1977).

This hybrid magneto–gravitational interaction (see §13.8) is largely responsible for the collection of the interstellar gas into the giant cloud complexes and H_{II} regions at irregular intervals of 1 kpc (Parker 1966b; Mouschovias 1974; Shu 1974; Mouschovias *et al.* 1974; Asseo, *et al.* 1978). Mouschovias (1976) (See also Mouschovias and Spitzer 1976) has treated the equilibrium of individual, strongly concentrated gas clouds under self-gravitation when the condensation has become extreme.

The violent agitation of the gas is evidently a consequence of supernova explosions (including the possibility of the explosive production of cosmic rays) and the explosive formation and lighting up of massive luminous stars (Oort and Spitzer, 1955; Savedoff 1956; Khan and Dyson 1965; Kaplan 1966; Münch 1968; Pikelner 1968; Spitzer 1968; Parker 1968b; see review in Field 1970). The total input of such mechanical energy to the turbulence of the interstellar gas is estimated to be of the order of $0{\cdot}5\times10^{-2}$ erg g^{-1} s^{-1}. For a mean hydrogen density of 1 atom/cm^3 this is $0{\cdot}8\times10^{-26}$ erg cm^{-3} s^{-1}, comparable to the energy input of cosmic rays on the basis of the short 3×10^6 year life hypothesis. Hence, the combination of the mechanical effects of the birth and death of the massive stars (i.e. the creation of new O, B, and A stars, and their expiration as supernovae) seems to be adequate to account for the observed turbulence.

On the other hand, the thermodynamic state of the gas, and the heat balance that maintains that thermodynamic state, are not at all clear at the present time. Based on optical absorption and emission lines of atomic hydrogen, calcium, and sodium, the traditional interpretation was that the interstellar gas exists in a cold and a hot phase—the interstellar clouds and the hot intercloud medium (see the excellent reviews by Kaplan 1966; Spitzer 1968; Münch 1968; Pikelner 1968). The intercloud medium, at a temperature of the order of 10^3–10^4 K, is assumed to be in approximate pressure balance with the cold dense clouds at 10^2 K and 10 atoms cm^{-3} and in radiative heat balance with the ambient starlight (cf. Savedoff and Spitzer 1950; Spitzer 1954; Savedoff 1956).

Radio measurements of the 21 cm line of atomic hydrogen seem to confirm this picture. The observations of the profiles of the 21 cm absorption and emission lines are interpreted as requiring a general background of hot (10^3–10^4 K), tenuous (0·1–0·7 atoms cm^{-3}), and largely neutral (1–20 per cent ionization) hydrogen (Kerr 1968; Hughes *et al.* 1971; Radhakrishnan *et al.* 1972; Baker and Burton 1975; Dickey *et al.* 1977). The cold dense clouds are to be found here and there, embedded in the hot background plasma. The existence of molecular hydrogen as an important component of the cold dense clouds was suggested on theoretical grounds (Salpeter and Gould 1963; Salpeter *et al.* 1963) but could not be observed because its principal absorption lines lie in the ultraviolet. More recently the extensive observations of the u.v. spectra of early-type stars by the Copernicus satellite have provided a general view of the interstellar absorption of molecular hydrogen, with the mean densities of 0·9 hydrogen atoms/cm^3 and 0·14 hydrogen molecules/cm^3 over the volume of space within 500 pc of the sun.

As so often happens when a new field of observation is opened up, the u.v. observations showed a surprising result, viz, widespread absorption lines of O VI, suggesting the widespread occurrence of a superhot plasma. The mean density of the superhot plasma is deduced to be very low (0·003 atoms/cm^3) and the temperature ranges from 3×10^5 to 7×10^5 K (Jenkins and Meloy 1974; York 1974, 1977), suggesting an approximate pressure balance with the cool clouds and with many of the H_{II} emission regions. What is more, Williamson *et al.* (1964) discovered a background of soft x-rays that is more or less what would be expected from just such a superhot plasma. McKee and Ostriker (1977) suggest that the superhot gas is produced by the eruption of supernovae. The u.v. observations fail to detect any evidence for the traditional hot phase (10^3–10^4 K) with its many neutral atoms, leading Spitzer and Jenkins (1975) to suggest that there may be only the superhot component, above 10^5 K. On their interpretation, the hot gas (10^4 K), such as it is, would be limited to the coronas, or transition regions, surrounding cold clouds.

On the other hand Castor *et al.* (1977, 1978) suggest that the O VI lines arise in the circumstellar shells of the very hot and luminous stars that the Copernicus spacecraft has been directed to observe. It should be noted, too, that the evidence for O VI involves observations of stars in many cases far above the gaseous disc of the galaxy in the halo region (cf. Shull 1977; Shull and York 1977; York 1977) so it is not clear that the results, based on the integrated effects along the line of sight to the stars, pertain to the gas in the disc. Gas clouds are known to exist in the halo, well above the gasous disc (Münch and Zirin 1961; Cohen 1974) where the hot intercloud component may be quite different from what it is in the disc. Recent optical observations of sodium lines in the disc provide evidence for a hot phase in the range of 10^3–10^4 K (Hobbs 1976b) in some directions from the sun, but not in others. Hence a final verdict on the relative importance of the hot phase and the super hot phase is difficult (Hobbs 1978). It is not impossible that the sodium lines are produced in circumstellar regions.

One of the fundamental difficulties in interpreting the observations is the lack of a basic theory for the heating of the interstellar gas. It was suggested by Hayakawa *et al.* (1961) that the heating may be accomplished by an abundance of low energy cosmic rays (below 10^2 MeV per nucleon) which ionize the gas through which they are passing. The existence of such low energy cosmic rays cannot be tested by direct observation in the inner solar system because they are excluded by the outward sweep of the solar wind (cf. Parker 1963). The theory was developed at length by Field *et al.* (1969) and predicted a hot phase at several thousand K. Hobbs (1974, 1976a) has shown, however, with observations of the resonant absorption lines K I, that the number of free electrons and the ionization rate are so small as to exclude a significant contribution to the heating by either cosmic rays or a soft x-ray background (Shull *et al.* 1977). O'Donnell and Watson (1974) employed a complex analysis of the chemical reaction involved in the formation of HD and arrived at a similar conclusion.

Fortunately the interaction of the interstellar gas with the galactic magnetic field depends in large part only on the existence of the gas, its mean density (1–2 H atoms/cm^3), and the dimensions (10^2–10^3 pc) and velocities (10 km s^{-1}) of the dominant inhomogeneities, which are readily available from observations (cf. Heiles and Habing 1974). Hence, in spite of the many questions concerning the remarkable state of the gas, it is possible to describe its large-scale dynamical instability, as was done in §§13.4, 13.7, and 13.8, and its regenerative effects on the field, described in §19.3.3.

22.1.3 *The cosmic rays*

Any attempt to work out the basic dynamical state of the gas and magnetic field comes face to face with the question of the origin of cosmic rays. The energy density U of the cosmic rays at the position of the solar system is about $1\ \text{eV cm}^{-3} = 1{\cdot}6\times10^{-12}\ \text{erg cm}^{-3}$ (Ginzburg and Syrovatsky 1964; Parker 1968c). The pressure P exerted by the cosmic rays is $P=\frac{1}{3}U$ (for an extreme relativistic gas), or about $0{\cdot}5\times 10^{-12}\ \text{dyn cm}^{-2}$. Evidently the sources of cosmic rays—whatever they may be—in the gaseous disc raise the pressure of the cosmic rays until the cosmic rays are able to push their way out of the field and escape from the galaxy (Parker 1966b, 1969; Jokipii and Parker 1969), there being no other effective means of escape (see examples of escaping magnetic field in §13.7). Hence we expect P to be of the same general order as $B^2/8\pi$, suggesting that B is of the order of $3\text{–}5\times10^{-6}$ G. But we will come back to this question later. The rate at which cosmic rays inflate the magnetic field determines in large degree the overall activity of the field in the disc, beyond the dynamical instability treated in §§13.4, 13.7, and 13.8. Unfortunately the basic sources, and their rate of production of cosmic rays, are as poorly known for the cosmic rays as for the interstellar gas. Some estimates of the rate of production of cosmic rays are as high as 2×10^{-26} erg cm^{-3} s^{-1} throughout the disc of the galaxy, based on an inferred cosmic ray life of $\tau=3\times10^{6}$ years before pushing their way out of the magnetic field of the disc. If this estimate is correct, then the cosmic rays are one of the principal sources of energy that drive the dynamical motions of the disc (Parker 1958a, 1969). Their total production rate over the volume of the gaseous disc would be $10^{40}\text{–}10^{41}$ erg s^{-1}. A significant fraction of the total energy ($10^{42}\text{–}10^{43}$ erg s^{-1}) expended by supernovae would go into cosmic rays, if indeed the supernova is their source. Such rapid production of cosmic rays inflates the upward bulges of the galactic magnetic field at a speed $\Lambda/\tau\cong50$ km s^{-1}. The upward bulge would be pushed outward from both faces of the disc at a lively rate (Parker 1965, 1969; see Figs 13.3 and 13.5). On this basis we would expect an active layer of magnetic field and cosmic ray bubbles extending outward from the disc. It may be some such halo of magnetic bulges, inflated with cosmic rays, that gives the broad distribution of non-thermal radio emission extending outward a distance of the order of 1 kpc from either side of the disc of the galaxy (cf. Badwar and Stephens 1977; Badwar *et al.* 1977; Stecker and Jones 1977). It may also be such an effect that produces the thin radio halo observed around some other galaxies (Ekers and Sancisi 1977; Allen *et al.* 1977).

In the opposite extreme, Peters and Westergaard (1977) have argued that the observed properties of cosmic rays can be explained just as well on the assumption that there is no escape of cosmic rays from the galactic disc, i.e., no progressive inflation of the field, with the effective life τ more like 10^{8} years. The requirements for the production of cosmic rays are greatly reduced, obviously, and the cosmic rays produce no active radio halo extending outward from the two surfaces of the gaseous disc. The only outward bulges of field would be those produced by the magnetic and cosmic ray instability developed in §§13.4, 13.7, and 13.8. There would be no additional rapid inflation by cosmic rays.

A middle course may be had from the conventional interpretation of the H^2, He^3, Li, Be, and B nuclei among the cosmic rays as fragments knocked off heavier cosmic ray nuclei, mainly C, N, and O. The relative abundances of the fragments and the heavier nuclei indicate that the heavier nuclei have passed through about 5 g cm^{-2} of hydrogen in order to suffer the necessary fragmentation (Silberberg 1966; Shapiro and

Silberberg 1970; Garcia–Munoz *et al.* 1975b). Thus if the heavy nuclei move with a velocity w through the interstellar hydrogen with a mean number density N, the necessary path length is $(3\times 10^{24}/N)$ cm, traversed in a time $10^{14}(c/w)/N$ s. The canonical value for the mean interstellar gas density in the disc of the galaxy, $N = 1/\text{cm}^3$, combined with $w \cong c$, yields 3×10^6 years. However, the mean interstellar gas density in the local neighbourhood of the sun, where we measure the fragmentation, is of the order of 10^{-1} atoms cm^{-3} (Münch and Unsöld 1962; Bertaux and Blamont 1971; Thomas and Krassa 1971) or less (Savage *et al.* 1977). Hence, if the cosmic rays spend most of their life within a kpc (in the azimuthal direction) of the sun, the estimated cosmic ray life is 3×10^7 years.

Now a direct determination of the age of the cosmic ray particles is theoretically possible from the abundance of the unstable isotope Be^{10} among the cosmic rays (Hayakawa *et al.* 1958). The Be^{10} is unstable with a half life of $1{\cdot}5\times 10^6$ years, decaying into B^{10}. The recent observations of Garcia–Munoz *et al.* (1975a, 1977) find the ratio Be_e^{10}/Be^9 to be so low compared to the ratio produced by the fragmentation of C, N, O as to require an age of $1{\cdot}5\times 10^7$ years or more (see also Jokipii 1976). Thus, if the passage through matter during that life is limited to $5\ \text{g cm}^{-2}$, the mean density of the gas traversed by the cosmic rays is not more than 0·2 hydrogen atoms cm^{-3}. This may reflect the low mean density in the solar neighbourhood, or it may reflect a large portion of the life spent in magnetic bubbles extending out from the gaseous disc.

It is clear, then, that the cosmic ray picture is less than clear. The cosmic rays may, or may not, inflate the galactic field at a high rate ($\sim 50\ \text{km s}^{-1}$) to produce an active halo of a kpc in thickness on each side of the disc (see, for instance, the empirical model of Badwar and Stephens 1977). It may, or may not, be that supernovae are the principal sources of cosmic rays (Ginzburg and Syrovatsky 1964). If supernovae are the principal source, it remains to be shown how the cosmic rays can be accelerated with efficiencies as high as a few per cent and then deftly injected into the gaseous disc of the galaxy without being decelerated again in the expanding supernova shell or blasting their way out of the disc altogether. Cosmic rays are extremely mobile, but their bulk streaming is limited to a few times the local Alfven speed (computed for the ionized component of the thermal gas) by their interaction with the short wavelength Alfven waves excited by their passage (Lerche 1966, 1967b; Wentzel 1968, 1974; Tademaru 1969; Kulsrud and Pearce 1969; Lee 1972). The kinetic properties of cosmic rays are simple (Parker 1966a; Lerche and Parker 1966) only if their collective interaction with the background plasma (Lerche 1976a, c, 1969; Wentzel, 1974) is ignored.

The fascinating properties of the galactic cosmic ray gas are beyond the scope of this work (see reviews by Ginzburg and Syrovatsky 1964; Parker

1968c, 1969; Lerche 1969; Daniel and Stephens 1975). Some of their dynamical effects were noted in §13.8, where they contribute to the general support of the interstellar gas, and to the dynamical instability, much as does the magnetic field of the galaxy. It is the global properties of the cosmic rays that most concern us here.

It is not possible at the present time, then, to state how active the cosmic ray component of the gaseous disc of the galaxy is. It is known that the pressure of the cosmic ray gas has, generally, over the last 10^9 years, been near its present value at the position of the sun (see review in Parker 1968c). Recent gamma ray observations (Fichtel *et al.* 1975; Bignami *et al.* 1975; Stecker *et al.* 1975) show the extension of the cosmic rays throughout the gaseous disc[1]. The continued existence and broad extent of the cosmic rays in the gaseous disc are sufficient to establish that they contribute to the support and the dynamical instability of the gaseous disc, even when they are not sufficient to determine the rate of inflation of the upward bulges of magnetic field (see §§13.4 and 13.8).

22.1.4. *The galactic magnetic field*

Consider, then, the magnetic field of the galaxy. Fermi (1949, 1954) was the first to propose that the disc of the galaxy contains a general large-scale magnetic field. The basis for the proposal was the existence of cosmic rays, with individual energies typically 1–100 GeV per nucleon. It is not possible to confine such energetic particles to the solar system, say with a dipole field at the sun, nor is it an attractive idea to suppose that the entire expanding universe is maintained full of cosmic rays. The only hypothesis anywhere between these two extremes is that the cosmic rays are confined to the galaxy by a galactic magnetic field. Fermi suggested field strengths of the general order of 10^{-6}–10^{-5} G as being adequate but not extravagant.

In the same year Hiltner (1949) and Hall (1949) discovered the systematic polarization of the light of distant stars. The degree of polarization increased with the degree of reddening of the light by passage through interstellar dust (Hiltner 1949, 1951, 1956) indicating that the polarization is caused by the selective scattering from aligned grains. The polarization was found to be strong and aligned more or less along the plane of the Milky Way for stars lying in the general direction of the galactic centre ($l^{\text{I}} = 325°$, $l^{\text{II}} = 0$) or anti-centre ($l^{\text{I}} = 145°$, $l^{\text{II}} = 180°$). Figure 22.1 shows a sample of Hiltner's data (reproduced from Hiltner 1956), looking in the anti-centre direction. The lines in the figure indicate

[1] It is to be hoped that before long measurements of the local anisotropy of the cosmic rays will give an indication of the local bulk streaming velocity, providing an indication of the direction to the local sources of cosmic rays and an indication of their strength.

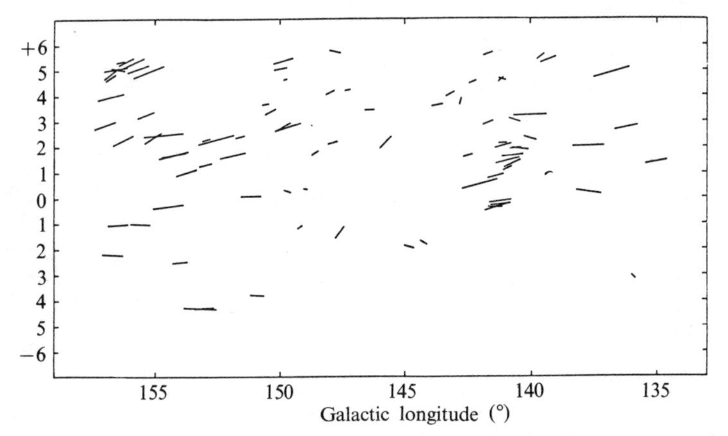

FIG. 22.1. Plane of polarization of stars in low galactic longitude in the general direction of the anti-centre ($l^{\mathrm{I}} = 145°$) of the galaxy (reproduced from Hiltner 1956). The length of the lines indicates the degree of polarization and the direction indicates the orientation of the electric vector of the polarized component.

the plane of vibration of the electric vector. In contrast to the situation when looking in the radial direction, the plane of polarization of the light from stars lying in the azimuthal directions ($l^{\mathrm{I}} = 55°, 235°$, $l^{\mathrm{II}} = 90°, 270°$) is scattered, with almost random orientation. Figure 22.2 shows a sample of Hiltner's data (reproduced from Hiltner 1956), looking in the azimuthal direction ($l^{\mathrm{I}} = 55°$, $l^{\mathrm{II}} = 90°$). As a matter of fact, the greatest confusion arises in the general range $l^{\mathrm{I}} = 350°–45°$, so that the direction is only poorly defined, presumably as a consequence of local fluctuations. The difficulty is our restriction to observations only of the local neighbourhood of the sun. A broader view is needed for an unambiguous determination of the field configuration. Segalovitz *et al.* (1976) and Tosa and Fujimoto (1978) have looked at the Faraday rotation measure of the radio emission and the plane of polarization of the optical emission from sources in the spiral arms of M51. Their assumption that the total number of electrons in the line of sight is independent of whether the line of sight intersects an arm or an interarm region leads them to the conclusion that the field lies along the arms. They suggest that the field is primordial, but the dynamical instability of the field and the rapid escape would seem to exclude this possibility. So the question remains open. Coming back, then, to the polarization of starlight in our own galaxy, we would expect to find a nearly random orientation of the plane of polarization when looking parallel to the local field (in the local azimuthal direction). Figure 22.2 shows the effect very nicely.

The fluctuation of the plane of polarization from one star to the next indicates local variations ΔB_i in the field. The polarization, of course,

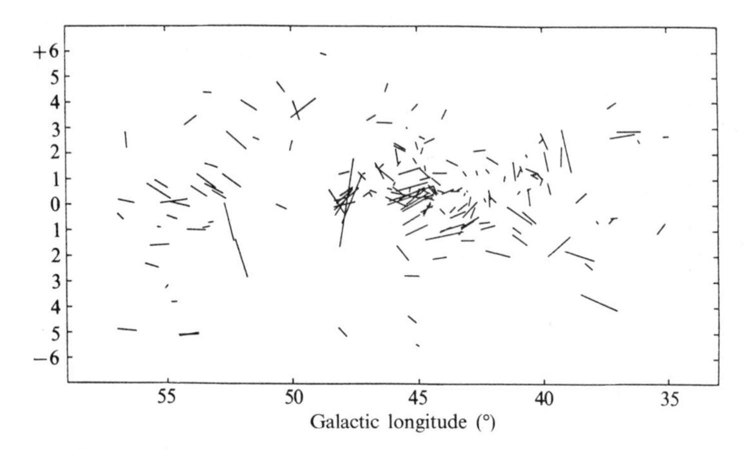

FIG. 22.2. Plane of polarization of stars in low galactic longitude in the azimuthal direction ($l^{\mathrm{I}} = 55°$) (reproduced from Hiltner 1956).

averages the many local fluctuations along the line of sight, so that the local values of $\Delta B_i/B$ are larger than the fluctuations in the directions of polarization to be seen in Fig. 22.1. If these fluctuations are considered to be largely the deflections of the mean field B by the disordered motions v of the interstellar gas of density ρ, then $v \cong \Delta B/(4\pi\rho)^{\frac{1}{2}}$ (see §7.3). On this basis Davis (1951) pointed out that for $\Delta B \cong 0{\cdot}2\ B$, $\rho = 10^{-23}$ g cm^{-3} (6 H-atom cm^{-3}) and $v = 20$ km s^{-1}, the field must be of the general order of $B = 10^{-4}$ G. Repeating the calculation later with the better values $\rho = 2\times 10^{-24}$ g cm^{-3} and $v = 5$ km s^{-1} available from observations of the 21 cm line, Chandrasekhar and Fermi (1953) and Pikelner (1956) suggested $B = 7\times 10^{-6}$ G. Parker (1951) pointed out that the fine curved dust striations in the Pleiades can be understood only if the individual grains are tied to magnetic lines of force of a general field. Otherwise the striations would be smeared by the interstellar winds there. If the charge on each grain is one electron, a field not less than 2–3×10^{-6} G is required to preserve the observed striations (see development in parker 1958b).

Spitzer and Tukey (1951) worked out the alignment of ferromagnetic dust grains in a large-scale magnetic field, while Davis and Greenstein (1951) did the same for spinning paramagnetic grains. It become clear that the observed polarization of starlight is, indeed, a manifestation of the general field of the galaxy, but that the strength of the field cannot be determined from quantitative measurements of the degree of polarization unless the precise shape and composition of the grains are known. (See additional references and review in van de Hulst 1967; Davis and Berge 1968). Since that time Greenberg (1968) and Purcell (1969a,b, 197?) (see also Purcell and Spitzer 1971; Purcell and Pennypacker 1973; Purcell and Shapiro 1977) have gone into the physics of the interstellar dust grain, the thermal and photoelectric effects that spin the grains, the alignment of spinning grains in a magnetic field, and the scattering of radiation by aligned grains. The work has shown that the field strength can be determined if the composition and structure of the grains are precisely specified, but that there is no general quantitative statement that can be made at the present time.

The direction, if not the strength of the galactic field, is provided by the orientation of the polarization, as already noted. Hiltner's observations (Hiltner 1951, 1956) show that the field is generally in the azimuthal direction around the gaseous disc of the galaxy, with large local fluctuations. Analysis of the fluctuations ΔB (Jokipii and Lerche 1969; Jokipii *et al.* 1969; Appenzeller 1969) show that $\Delta B \cong B$ and that the fluctuations in the field have a characteristic scale of about 10^2 pc. This is entirely compatible with the observed motions of the interstellar gas if B is a few μG, as noted above.

Turning to other effects of the galactic magnetic field, the background non-thermal radio emission from the disc of the galaxy is believed to be the result of synchrotron emission from supernova remnants and from the electrons among the cosmic rays in the general galactic field (Ginzburg and Syrovatski 1964; see review by Davis and Berge 1968). Models based on the electron intensities presently observed at the position of the solar system, and based on the assumption that all the radio emission is from cosmic ray electrons in the galactic magnetic field, require a general field of the galaxy as strong as $5\text{–}6 \times 10^{-6}$ G across a disc with characteristic thickness of 2 kpc (Badwar and Stephens 1977).

The Faraday rotation measure of the radio emission from extragalactic sources provides a direct value for the integral of the product of the electron density N_e and the component of the magnetic field along the line of sight, $\int d\mathbf{S} \cdot \mathbf{B}\, N_e$ (see discussion in Davis and Berge 1968). If one knows N_e and its correlation with B along the line of sight, then the Faraday rotation measure gives the mean value of the field in the line of sight. Unfortunately N_e is not well known, although recent high resolution observations of the weak resonant line of potassium by Hobbs (1974, 1978) have permitted a direct determination of the mean electron density in certain clouds. Using conventional numbers for the electron density Gardner *et al.* (1969) and Gardner and Whiteoak (1969) suggest fields of the order of 3×10^{-6} G (see review by Heiles 1976). The dispersion measure obtained for several pulsars gives the columnar electron density $\int ds\, N_e$. Assuming that both N_e and $\mathbf{B}$ are uniform along the line of sight, Manchester (1972, 1974) has combined the dispersion and rotation measures for a number of pulsars and comes up with $2\text{–}3 \times 10^{-6}$ G for the mean galactic magnetic field within about 1 kpc of the sun. The mean field density derived from the Faraday rotation measure on the assumption of a uniform distribution along the line of sight would be too low if the field and electron distributions were strongly non-uniform and anti-correlated for some reason, and too large if non-uniform and correlated (note the complications pointed out by Elmergreen 1975).

The number 3×10^{-6} G seems not unreasonable, in that it is compatible with the observed cosmic ray pressure, for which it yields $P\cong B^2/8\pi$, and it is compatible with the observed scale height in the gaseous disc of the galaxy. This point merits brief elaboration because it is often overlooked in speculation on the strength of the galactic magnetic field in connection with the non-thermal radio emission. Fields stronger than 3×10^{-6} G over scale heights in excess of 200 pc are required if the non-thermal radio emission is to be the result of a cosmic ray electron density throughout the disc that is no more than that presently observed at the position of the solar system. A field of 10×10^{-6} G would be adequate, or a scale height of 1 kpc, or some compromise combination. But 3×10^{-6} G and 200 pc are not adequate. Hence it is sometimes suggested that the field is 10×10^{-6} G. However, a field of 10 μG has too large a pressure to be confined to 200 pc by the weight of the observed interstellar gas of 1 atom cm^{-3}. The field strength, the gas density, and the scale height are not independent quantities and cannot be assigned arbitrary values without attention to the consequences (see discussion in §§13.4 and 13.8). The scale height is

$$\Lambda \equiv (\rho u^2 + B^2/8\pi + P)/\rho\langle g\rangle = (1+\alpha+\beta)u^2/\langle g\rangle \qquad (22.1)$$

where α and β are the ratios of the magnetic and cosmic ray pressures to the gas pressure ρu^2, and where $\langle g\rangle$ is the mean value of the gravitational acceleration over the scale height ($\langle g\rangle \sim 2\times 10^{-9}$ cm s^{-2} for Λ of the order of 200 pc, with $\langle g\rangle \propto \Lambda$ for both larger and smaller Λ) (Oort 1959; Hill 1960). The conventional numbers $B=3\times 10^{-6}$ G ($B^2/8\pi = 0{\cdot}4\times 10^{-12}$ dyn cm^{-2}), $P=0{\cdot}5\times 10^{-12}$ dyn cm^{-3}, $\rho = 1{\cdot}6\times 10^{-24}$ g cm^{-3} (1 H-atom cm^{-3}), and $u=7$ km yield a dynamical pressure $\rho u^2 = 0{\cdot}4\times 10^{-12}$ dyn cm^{-2}. Hence $\alpha\cong\beta\cong 1$ and $\Lambda = 7\times 10^{20}$ cm $\cong$ 200 pc, in agreement with the observational estimates (cf. Baker and Burton 1975). However, if we put $B=10\times 10^{-6}$ G, then $\alpha = 10$, $\langle g\rangle \cong 4\times 10^{-9}$, and we deduce the scale height to be 500 pc, rather larger than the conventional value of 150–200 pc inferred from observations.

Various adjustments and compromises in the numbers are possible, but it is not clear which way to turn. Badwar and Stephens (1977) provide a self-consistent compromise, combining a field strength of $5\text{–}6\times 10^{-6}$ G with the first scale height above the central plane of the disc equal to 400 pc, and a broad 'upper atmosphere' above (for which $\Lambda\cong 1$ kpc). The non-thermal radio emission then follows from the cosmic ray electron density observed at the position of the sun. It cannot be ruled out, however, that old supernova remnants may contribute a major fraction of the galactic non-thermal radio background, so that there is no need to assume fields in excess of 3 or 4 μG over scale heights in excess of 200 pc (Parker 1968c). The possibility of a general superhot component of the interstellar gas (for which $u \cong 10^2$ km s^{-1}) further complicates the picture.

It should be remembered, of course, that the simple relation $\Lambda = (1+\alpha+\beta)u^2/g$ applies to a horizontal magnetic field in an isothermal gas whereas the galactic field is composed of magnetic arches bulging upward between regions of dense interstellar gas, and the gas consists of a cool, a hot, and a superhot phase. The simple relation for the scale height is imprecise in the real world. The definition of the mean scale height is complicated (cf. Parker 1969) and any general inference from the observations from our single location in the galactic disc becomes problematical. So far the question has not received either the theoretical or the observational attention that it deserves.

Altogether, then, the galactic magnetic field is an extensive, strong, and active magnetic field. The activity is inferred from the theoretical dynamical instability and non-equilibrium (§§13.4, 13.7, and 13.8). The dynamical instability and the general turbulent state of the interstellar gas permit the field to escape from the gaseous disc in characteristic times of the order of 10^8 years. The existence of the galactic field in this late epoch in time suggests, then, that the gaseous disc continually generates more field to make up for the continual loss.

22.2. Generation of the galactic magnetic field

22.2.1. Energy requirements

Consider the means by which magnetic field is generated in the gaseous disc of the galaxy. The basic facts are discussed in the foregoing section and may be summarized as follows. The gaseous disc of the galaxy contains a general azimuthal field with a mean strength of $3\text{–}4\times10^{-6}$ G in the neighbourhood of the sun, based on Faraday rotation measures, and on dispersion measures and other estimates of the mean electron density. This value for the mean field is consistent with the estimated scale height of 200 pc for the gaseous disc and the mean density of 1–2 hydrogen atom cm^{-3} inferred from radio, optical, and ultraviolet observations of the interstellar gas. The field is subject to large fluctuations ΔB, comparable in magnitude to the mean field itself. Hence the field direction fluctuates locally by ±1 rad. The scale of the fluctuations is estimated to be of the order of 10^2 pc. The fluctuations are expected from the fact that the r.m.s. turbulent velocities in the interstellar gas are of the order of $v = 10$ km s^{-1}, so that in any one dimension $u = v/3^{\frac{1}{2}} \cong 6$ km s^{-1} and $\frac{1}{2}\rho u^2$ is comparable to $B^2/8\pi$.

It is believed, from the theory of the hydromagnetic dynamo, that the azimuthal field of the galaxy is accompanied by a meridional or poloidal field. Unfortunately there seems to be little prospect for observing the poloidal component of the field. In contrast to the circumstances for all other astronomical bodies, we observe the galactic magnetic field from deep inside the dynamo. The observations are swamped with the local fluctuations of the dominant azimuthal field. The situation is the reverse of that for planets where the inner workings of the dynamo are hidden from view and only the poloidal field penetrates out through the surface to be observed. The best that we can hope is to estimate the poloidal field of the galaxy by applying the dynamo theory to the observed azimuthal field (see §22.2.3).

Now the characteristic decay time of the mean field in a disc of thickness $2h$ is $t = h^2/4\eta$. The dominant diffusion effect in the galaxy is turbulent diffusion, given by (17.89) as $\eta = \eta_T = 0{\cdot}2vL$. Hence if τ is the characteristic eddy time $\tau = L/v$, it follows that the decay time is

$$t = 1{\cdot}25\tau(h/L)^2.$$

If the thickness h is 400 pc (scale height of 200 pc) while $L = 100$ pc, then $t = 20\tau$. For interstellar turbulent velocities of the order of $v = 10\ \mathrm{km\ s^{-1}}$, it follows that the characteristic eddy life is something of the order of $\tau = 3\times 10^{14}\ \mathrm{s} = 10^7$ years so that the characteristic decay time for the galactic field is $t = 2\times 10^8$ years. This number is very uncertain, of course, because the interstellar gas is so extremely inhomogeneous that the simple mixing length concepts, on which this estimate is based, may need some modification. For instance, the coefficient $0{\cdot}2$ in (17.89) is obtained from numerical experiments with homogeneous fluids. Is it the appropriate value for anything so lumpy as the interstellar gas? Perhaps more important is the neglect of the magnetic buoyancy and dynamical instability of the galactic magnetic field, for which the characteristic growth time is closer to 10^7 years. We suggest that 10^8 years represents the order of magnitude of the decay time of the galactic field. The important fact is that the decay time is small compared to the 10^{10}-year age of the galaxy, so that we are obliged to conclude that the present-day galactic field is not primordial but is maintained by some form of dynamo activity in the gaseous disc.

The energy density of a field of 3×10^{-6} G is $0{\cdot}4\times 10^{-12}\ \mathrm{erg\ cm^{-3}}$. If the characteristic decay time is 10^8 years, the dissipation rate is $1{\cdot}3\times 10^{-28}\ \mathrm{erg\ cm^{-3}}$. This is to be compared to the estimated input of mechanical energy to the interstellar turbulence of $80\times 10^{-28}\ \mathrm{erg\ cm^{-3}}$, mentioned in §22.1.2. As a matter of fact, the azimuthal field is produced from the poloidal component of the galactic magnetic field by the differential rotation of the gaseous disc, so it is more meaningful perhaps to compare the dissipation rate with the rotational energy. At the position of the solar system the rotational velocity is 250 km s^{-1} (corresponding to a radius of 10 kpc and a rotation period of $2{\cdot}5\times10^8$ years, $\Omega\cong10^{-15}$ rad s^{-1}). Hence the kinetic energy density of the rotation is 3×10^{19} erg g^{-1}. For a mean interstellar gas density of 1 H-atom cm^{-3} this is 5×10^{-10} erg cm^{-3}. The dissipation rate of $1{\cdot}3\times10^{-28}$ erg cm^{-3} over the 10^{10}-year age of the galaxy consumes $0{\cdot}4\times10^{-10}$ erg cm^{-3}, or about one-tenth of the rotational energy. It is of course the *differential* rotation $\varpi\partial\Omega/\partial\varpi$ that is consumed in the generation of the azimuthal field, rather than the total rotation. It would appear, then, that the galactic magnetic field may have introduced some significant readjustment of the radial distribution of the gas in the disc, by distances of the general order of a kpc, over the life of the galaxy.

It is important in this connection to note that a galactic magnetic field greatly in excess of 3×10^{-6} G has important long-term effects on the galactic disc. The 10×10^{-6} G, sometimes postulated to account for the non-thermal radio emission (see discussion in §22.1), has an energy density ten times larger. For a given dissipation time t, then, the dissipation rate is ten times larger, amounting to 13×10^{-28} erg cm^{-3} s^{-1} in the present epoch. The accumulated effect of this over a period of 10^{10} years is the enormous energy of 4×10^{-10} erg cm^{-3}, comparable to the rotational energy of the gaseous disc. If fields as strong as 10 μG are an actuality, then their evolutionary effect on the gaseous disc has been profound. The effect on the observed scale height of the gaseous disc would also be profound (as noted in §22.1), suggesting a serious readjustment of the mean gas density and scale height inferred from observation. Hence, once again, we come back to 3×10^{-6} G as the best available estimate, attributing much of the galactic non-thermal radio noise to old supernova remnants, etc.

22.2.2. Slab models

The simple slab model of the galactic magnetic field was worked out in §19.3.3 to illustrate the basic dynamo properties of the gaseous disc of the galaxy. The example was chosen with the dynamo coefficient Γ an odd function of distance z from the central plane of the disc. In fact Γ was chosen to have a fixed value $+\Gamma_0$ from $z=0$ to $+h$, and $-\Gamma_0$ in $-h<z<0$, with $\Gamma=0$ in the region $z^2>h^2$ outside. The direction of positive z is defined such that a right-handed screw with the rotation of the galaxy moves in the direction of positive z, i.e. viewed from the positive z-axis, the galaxy rotates counter-clockwise. It follows that the galactic rotation Ω is positive, and, choosing the x-coordinate to represent the local radial direction, we have $\partial\Omega/\partial\varpi=\partial\Omega/\partial x<0$. The resulting azimuthal field B_y for the lowest mode (see (19.188)–(19.190)) is then an even or an odd function of z, depending upon whether the dynamo coefficient Γ_0 is positive or negative, so that the dynamo number $Kh = (G\Gamma_0 h^3/\eta_T^2)^{\frac{1}{3}}$ is negative or positive. On the assumption that a rising cell of gas is expanding, and therefore rotating less rapidly than its surroundings, Γ_0 is positive, Kh is negative, and the azimuthal field is an even function of z, in the lowest mode. This is in contrast to the lowest modes of the planets and of the sun, where the azimuthal field is an odd function of distance from the equatorial plane. The accumulated observational information indicates that the mean field, for all the local fluctuations, is an even function of z and does not reverse sign across the mid-plane. Hence the idea of a cyclonic dynamo in the galactic disc leads to the right qualitative features.

The dynamo number is given by (19.186) to be $Kh = -1{\cdot}86$ for the lowest mode with $k_x h \ll 1$. Equation (19.191) gives an order of magnitude estimate of Kh based on elementary dynamical considerations. With $h = 400$ pc for the half-thickness of the gaseous disc and $L = 100$ pc for the characteristic scale of the turbulent eddies within the disc, we obtained the estimate $Kh = -1{\cdot}8$. Hence the estimated Kh is equal to the value needed to produce the lowest mode. It is curious that the dynamo numbers estimated on simple dynamical considerations for the cores of planets are so much larger than the eigenvalues for the observed stationary dipole modes, whereas for the sun and the galaxy the elementary dynamical estimates are much closer to the calculated eigenvalues. It suggests that the motions in planetary cores are greatly restricted in some magnetogeostrophic balance, already noted in §20.3. Thus in planetary cores the Reynolds stress $\frac{1}{2}\rho v^2$ is generally small compared to the Maxwell stress $B_\phi^2/8\pi$, whereas in the sun and the galaxy $\frac{1}{2}\rho v^2$ is comparable to $B_\phi^2/8\pi$.

Now, to come back to the question of the poloidal field of the galaxy, note from (19.188)–(19.190) that for $k_x h \ll 1$ the principal poloidal field is the radial component B_x. The ratio of the amplitude of B_x to the amplitude of the azimuthal field B_y is, in order of magnitude,

$$B_x/B_y \cong \eta K^2/G \cong (\Gamma^2/\eta G)^{\frac{1}{3}}.$$

With the same expressions for Γ, G, and η as were used in the development of (19.191), it follows that

$$\frac{B_x}{B_y} = \tfrac{1}{5}(\tfrac{2}{3}\Omega\tau)^{\frac{1}{3}}$$
$$\cong 0{\cdot}1$$

for $\Omega\tau = 0{\cdot}24$. Thus the radial component of the galactic field is only about one-tenth of the azimuthal field. There is no prospect for observing it, so far as we can see. It is buried in the noise inside the dynamo where we observe the fields.

The generation of the weak poloidal field from the azimuthal field is illustrated in Fig. 19.9, the azimuthal field then being generated by the very strong shearing of the weak poloidal field B_x extending radially across the differential rotation of the disc.

It was pointed out in §19.3.2 that the galactic dynamo possesses oscillatory, as well as stationary, states. The calculations show, however, that the lowest self-sustaining, oscillatory mode for negative Kh arises for $Kh = -3{\cdot}1$ ($N_D \cong 30$). This is substantially larger than the $Kh = -1{\cdot}8$($N_D \cong 6$) for the stationary mode and substantially larger than the simple estimates of the dynamo number from elementary dynamical considerations. Hence, the evidence is that the galactic field is stationary, rather than periodic. If oscillatory, the period would be 10^8 years, so the oscillation is not directly observable.

22.2.3. *Spheroidal models*

The gaseous disc of the galaxy is extremely thin compared to the radius a, with $h/a \sim 0{\cdot}04$. Hence the illustrative model based on a slab of infinite breadth and width captures the basic physics. The distant edges of the galaxy cannot possibly have much effect on the local state of the field unless we imagine some critical situation where the weight of a hair might somehow be sufficient to tip the balance. Nonetheless it is clear that the slab model cannot give any clear idea of the distribution of field across the disc because the disc is, in fact, defined by its edges. Therefore, several authors have recently taken up the more difficult problem of solving the dynamo equations for axi-symmetric fields in an oblate spheroid. An oblate spheroid, well flattened, should capture the basic physics of the edge effects, and so provide an illustration of the location and distribution of the field across the disc.

Stix (1975) developed the first such model, using the standard oblate spheriodal coordinates

$$\varpi = a(1-\eta^2)^{\frac{1}{2}}(1+\xi^2)^{\frac{1}{2}},$$

$$z = a\eta\xi.$$

The surfaces $\xi = constant$ are oblate spheroids with semi-major axis $R = a(1+\xi^2)$, the distance a being the radial distance to the focus. The spheroids become smaller, with increasing eccentricity, with decreasing ξ, shrinking to a thin disc ($z = 0$), of radius a as $\xi \to 0$. On that disc we have $\varpi = a(1-\eta^2)^{\frac{1}{2}}$ or $\eta = (1-\varpi^2/a^2)^{\frac{1}{2}}$, which is a useful approximation for many purposes in a highly flattened spheroid. The spheroids become less eccentric as ξ increases, with $r^2 = \varpi^2 + z^2 \cong a^2\xi^2$ and $\eta \cong \cos\theta$ as $\xi \to \infty$.

The surfaces $\eta = constant$ are hyperboloids of revolution. As $\eta \to 0$ the hyperboloids fold in from both sides against the $z=0$ plane, the two leaves ($z \lessgtr 0$) becoming a single sheet outside $\varpi = a$. As $\eta \to \pm 1$ the hyperboloid shrinks onto the z-axis (cf. Morse and Feshbach 1953).

Stix assumes a vacuum outside the spheroid $\xi = \xi_0$, so that the toroidal field is identically zero and the poloidal field has a potential form, vanishing at infinity. Within the spheroid the diffusion coefficient is finite and uniform, and the dynamo coefficient and differential rotation are non-vanishing so that the region is a dynamo. The field is continuous across the boundary $\xi = \xi_0$. Stix uses the form $\alpha_0 z/b$ for the dynamo coefficient Γ, where α_0 is a constant and b is the semi-minor axis of the spheroid ξ_0, so that $b = a\xi_0$. Thus Γ passes linearly through zero across the central plane of the spheroid and is a maximum at the surface. For the non-uniform rotation stix writes $G \equiv \varpi\, \mathrm{d}\Omega/\mathrm{d}\varpi = \omega_0 \varpi^2/R^2$, so that the shear vanishes on the axis of the region and increases quadratically with radius ϖ to the maximum value ω_0 at the outer rim.

The dynamo equations are solved by expanding the fields in terms of $P_n^1(\eta)$, the integration over ξ then being done numerically. The eigenvalues are worked out for the lowest modes for positive and negative dynamo numbers, for a variety of eccentricities. The solutions show B_ϕ an odd function of z for positive dynamo number and an even function for negative dynamo number, of course. Stix defines the dynamo number as

$$P \equiv \alpha_0 \omega_0 R^3/\eta^2$$
$$= (\alpha_0 \omega_0 b^3/\eta^2)(R/b)^3.$$

In order to compare this with the numerical results of the slab model, we write it in terms of the mean values of the dynamo coefficient Γ and the shear G. The simple linear mean of Γ is $\frac{1}{2}\alpha_0$ and the linear mean of G is $\frac{1}{3}\omega_0$. For the slab model, where Γ and G are uniform, and hence represent the mean values, we wrote $N_\mathrm{D} = \Gamma G h^3/\eta^2$. Based on the average values for Stix's model, then, we would write $N_\mathrm{D} = \frac{1}{6}\alpha_0\omega_0 b^3/\eta^2$, as nearly as the two geometries can be compared. To proceed with Stix's results, the strongly flattened case $b/R = 1/30$ yields $P = +1{\cdot}73 \times 10^7$ for the lowest stationary mode[2] for which B_ϕ is an odd function of z. It follows that $N_\mathrm{D} = +107$ and $Kh = 4{\cdot}7$. This is to be compared with $Kh = +3{\cdot}02$ for the slab model. The lowest stationary mode for which B_ϕ is an even function of z occurs for $P = -2{\cdot}80 \times 10^6$. It follows that $N_\mathrm{D} = -17{\cdot}1$ and $Kh = -2{\cdot}6$.

[2] It was pointed out by White (1977) that a factor of R/b is missing from the values of P tabulated by Stix. Hence the value of P quoted here is 30 times the value in Table 1 of Stix (1975).

Quite recently Soward (1978) has worked out the asymptotic expression for the dynamo number for Stix's model in the limit of a flattened oblate spheroid ($b/R \to 0$). For the stationary even mode he obtained

$$P \sim -84{\cdot}8(R/b)^3\{1+1{\cdot}64(b/R)^{\frac{2}{3}}+O(b^{\frac{4}{3}}/R^{\frac{4}{3}})\}$$

Hence for $b/R = 1/30$, the result is $P = -99{\cdot}2$, $N_D = -16{\cdot}5$, and $Kh = -2{\cdot}54$, in agreement with result $Kh = -2{\cdot}6$ obtained earlier by Stix. These numbers are to be compared with $Kh = -1{\cdot}86$ for the slab model. Altogether, then, it is apparent that the dynamo number depends upon the distribution of the cyclonic turbulence and shear, as we would expect, but the dependence is not critical. The same general values for the dynamo number are obtained for several distributions.

The most important aspect of Stix's calculations is their illustration of the distribution of field throughout the disc. We are interested in negative dynamo number and B_ϕ an even function of z, appropriate to the galaxy. Stix plots the lines of force of the poloidal field and the distribution of the azimuthal field for the spheroid $b/R = 2/15$. The plots show that the poloidal field is a flattened linear quadrupole, with the radial component changing sign across the central plane of the spheroid. The radial component is the principal component of the poloidal field because of the highly flattened geometry. Both the radial field and the azimuthal field are weak over the inner third of the radius of the spheroid, because the shear is so small there. They are both confined largely to the outer two-thirds of the spheroid, falling to zero in the vicinity of the rim at $\varpi = R$.

Stix (1975) worked out the oscillatory modes and, for $b/R = 1/15$, found $P = -1{\cdot}21 \times 10^7$ for the lowest mode in which B_ϕ is an even function of z. This corresponds to $N_D = -6 \times 10^3$ and $Kh = -18{\cdot}2$. The dynamo number necessary to produce the oscillatory mode is again larger than the dynamo number $N_D \cong -6$ estimated in §19.3.3.1 from elementary dynamical considerations on the gaseous disc of the galaxy. It seems doubtful, therefore, that the field of the galaxy is oscillatory rather than stationary.

More recently White (1978) has worked out the behaviour of a dynamo inside the spheroidal region $\xi = \xi_0$, with vacuum outside, for the dynamo coefficient Γ of the form $\alpha_0\eta$. Thus, in a highly flattened disc ($\xi_0 \ll 1$) the dynamo coefficient in the upper half ($z > 0$) of the spheroid is

$$\Gamma \cong \alpha_0\{1-(\varpi/a)^2\}^{\frac{1}{2}},$$

so that Γ is a maximum at the center ($\varpi = 0$) and declines to zero at the outer rim. In the lower half the same form is used but with a negative sign in front (because η is negative in $z < 0$). He supposes that $G \equiv \varpi\, \mathrm{d}\Omega/\mathrm{d}\varpi = \omega_0\varpi/a$, so that the shear increases linearly, rather than quadratically, with radial distance. For negative dynamo number, so that B_ϕ is an even function of z, White finds that the lowest stationary mode is maintained for a dynamo number $P = -4 \times 10^5$ when $b/R = 1/15$. This is the same value obtained by Stix. White gives extensive plots of the fields.

White takes up the question of the oscillatory dynamo, showing the field configuration at intervals of $\frac{1}{6}\pi$ throughout the cycle. He goes on to explore the effects of the distribution of the dynamo coefficient Γ across the disc. In particular he employs the continuous distribution $\Gamma = 1{\cdot}5686\ \alpha_0 \exp(-|z/b|)\sin(\pi z/b)$ where the numerical factor is chosen so that α_0 represents the maximum value of Γ, at $z/b = \frac{1}{4}\pi$. Following Jepps (1975), White has also experimented with a non-linear cut-off, in which the dynamo coefficient is diminished by the factor $\{1 + (B_\phi/B_0)^2\}$ as the field increases. The effect is to give stationary fields for any dynamo number P in excess of the eigenvalue for the stationary state when the dynamo coefficient is not suppressed. Examples are worked out to illustrate the consequences of a periodic variation of the diffusion coefficient with radius ϖ, to simulate the spiral structure of the galaxy. An illustrative model is constructed based on the observed galactic rotation curve instead of the idealized form $G \propto \varpi$. The result is a modest improvement in efficiency, as indicated by the calculated eigenvalue P for the lowest stationary mode.

It was pointed out by Parker (1971) (see remarks in §19.3.3.1) that the generation of the galactic magnetic field, along the lines illustrated in Fig. 19.9, requires the escape of the reverse flux generated near the surface of the disc or slab. It was shown that an impenetrable boundary at $z = \pm h$, which chokes off the escape, blocks the mode for B_ϕ an even function of z. The dynamo does not function because reverse radial field formed immediately inside the upper and lower surfaces is confined to the disc and cannot escape. Hence the total radial flux, including the main radial field near the central plane, is zero. The differential rotation shears the radial field, then, producing an azimuthal field with vanishing net flux. There may be some higher order mode produced, or an oscillatory mode, if the dynamo number is large enough. But the lowest stationary even mode cannot function. Its existence requires that the reverse field near the surface be allowed to escape so that only the radial field near the central plane remains, with a non-vanishing radial flux.

An illustrative example was worked out, too, in which the medium outside was constrained to have the same shear G and the same diffusion coefficient η_T as inside the slab, but the dynamo coefficient was zero outside. The calculations show an increase in the dynamo number necessary to maintain the lowest stationary state for which B_ϕ is an even function of z (negative dynamo number). On the other hand, White (1977) works out an example for the spheroidal model in which the gas outside the spheroid is constrained to be *motionless* and has the same diffusion coefficient as inside. He found a modest *reduction* in the dynamo number for the lowest stationary even mode. This result is puzzling. It may be that there is some initial gain in partial closing of the surface of the galactic disc to the escape of field. The losses to the azimuthal field, which is the principal component, are reduced and this may help so long as some of the reversed radial field can still get out. Ultimately, when $\eta_T \to 0$ outside, the generation of the lowest even mode is halted.

It must be remembered, of course, that the purpose of these models is to explore the physical principles of the galactic dynamo, because, in fact, the external gas is so tenuous that no matter how high its conductivity and how low its effective diffusion coefficient η_T, it cannot block the dynamical escape of magnetic field from the disc of the galaxy. So for constructing illustrative models of the galaxy itself, one puts $\eta_1 = \infty$ outside, equivalent to a vacuum.

22.3. Summary

Altogether, with the simple slab model and the several numerical examples provided by Stix and White for highly flattened oblate spheroidal regions, the basic physical principles of the operation of the kinematical dynamo in the gaseous disc of the galaxy are fairly well illustrated. It seems that we know how the thing works, along the lines sketched in Fig. 19.9. The outstanding questions now are two-fold. On the observational side there is difficult work yet to be done to determine in more detail, and with more confidence, the general form and strength of the field in the disc. The observational picture of the interstellar gas, that comprises the dynamo, still leaves many basic aspects of the motions but poorly defined. The difficulty is that the observational information is limited to signal strength as a function of radial velocity at each point in the sky (cf. Heiles and Habing 1974), whereas it would be most illuminating to see the total velocity and the gas density as a function of radial distance. It is not obvious how this can be accomplished, there being no way to determine the radial distance to an element of gas responsible for the absorption at a particular frequency in the radio signal received from a distant source.

On the theoretical side, there remains, of course, the problem posed by the dynamics of a turbulent gas with an embedded field whose Maxwell stresses are comparable to the Reynolds stresses of the fluid motions. This is the general theoretical problem now for all dynamos, in planets, stars, and galaxies. The problem is easiest in the planetary core where the fluid density is uniform and the Mach number so low that sound waves are unimportant. The problem is more difficult in the convective zone of a star, because of the high Reynolds number and the extreme variation of the density from top to bottom. The most difficult problem of all is that of the gaseous disc of the galaxy, where there is a strong variation of the density, a Mach number of the order of unity, and a hot and a cold phase.

The activity of the galactic field is an obscure subject. In view of the dynamical instability of a magnetic field confined to a gaseous disc by the weight of the gas, we can say only that the field is intrinsically active on a large scale, over times of 10^7–10^8 years (§8.7; Chapters 13, 14, and 15). The field is active on a small scale because of the activity of the internal stars, such as supernovae. Unfortunately we cannot see the large-scale features of the field with any clarity, nor does the transient human journey through time permit us to follow the temporal changes of the features. Many of the peculiar galaxies to be seen at great distance in the universe represent the extreme cases of magnetic activity (see review edited by Setti 1976). In the extreme cases the magnetic activity is driven by some colossal energy source, such as gravitational collapse, within the galaxy, with evidence from the high material velocities and intense synchrotron emission that there is magnetic field strongly inflated with relativistic particles, i.e, cosmic rays. Unfortunately, these fascinating objects are so remote as to make it difficult to discern their internal machinations.

References

ADAMS, W. S. (1943). *Astrophys. J.* **97,** 105.
—— (1948). ibid. **109,** 354.
ALLEN, R. J., BALDWIN, J. E., and SANCISI, R. (1978). *Astron. Astrophys.*, **62,** 397.
APPENZELLER, I. (1969). *Astrophys. J.*, **151,** 907.
—— (1971). *Astron. Astrophys.* **12,** 313.
ASSEO, E., CESARSKY, C. J., LACHIEZE-REY, M. and PELLAT, R. (1978) *Astrophys. J. Lett.* **225,** L21.
BADWAR, G. D., DANIEL, R. R., and STEVENS, S. A. (1977). *Astrophys. Space Sci.* **49,** 133.
—— and STEPHENS, S. A. (1977). *Astrophys. J.* **212,** 494.
BAKER, P. L. and BURTON, W. B. (1975). *Astrophys. J.* **198,** 281.
BERTAUX, J. L. and BLAMONT, J. E. (1971). *Astron. Astrophys.* **11,** 200.
BIGNAMI, G. F., FICHTEL, C. E., KNIFFEN, D. A., and THOMPSON, D. J. (1975). *Astrophys. J.* **199,** 54.
CASTOR, J., MCCRAY, R., and WEAVER, R. (1977). *Astrophys. J. Lett.* **200,** L107.
CHANDRASEKHAR, S. and FERMI, E. (1953). *Astrophys. J.* **118,** 113.
COHEN, J. G. (1974). *Astrophys. J.* **194,** 37.
DANIEL, R. R. and STEPHENS, S. A. (1975). *Space Sci. Rev.* **17,** 45.
DAVIS, L. (1951). *Phys. Rev.* **81,** 890.
—— and BERGE, G. L. (1968). *Nebulae and interstellar matter*, vol. VII, *Stars and stellar systems* ed. MIDDLEHURST B. M. and ALLER, L. H. Chap. XV. University of Chicago Press.
DAVIS, L. and GREENSTEIN, J. (1951). *Astrophys. J.* **114,** 206.
DICKEY, J. M., SALPETER, E. E., and TERZIAN, Y. (1977). *Astrophys. J. Lett.* **211,** L77.
EKERS, R. D. and SANCISI, R. (1977). *Astron. Astrophys.* **54,** 973.
ELMERGREEN, B. G. (1975). *Astrophys. J. Lett.* **198,** L31.
FERMI, E. (1949). *Phys. Rev.* **75,** 1169.
—— (1954). *Astrophys. J.* **119,** 1.

FICHTEL, C. E., HARTMAN, R. C., KNIFFEN, D. A., THOMPSON, D. J., BIGNAMI, G. F., ÖGELMAN, H., ÖZEL, M. E., and TÜMER, T. (1975). *Astrophys. J.* **198,** 163.
FIELD, G. B. (1969). *Astrophys. J.* **142,** 531.
—— (1970). *Interstellar gas dynamics IAU symposium No. 39* (ed. HABING, H. J.), p. 51. Reidel, Dordrecht.
—— GOLDSMITH, D. W., and HABING, H. J. (1969). *Astrophys. J. Lett.* **188,** L107.
GARCIA-MUNOZ, M., MASON, G. M., and SIMPSON, J. A. (1975a). *Astrophys. J. Lett.* **201,** L141.
—— (1975b). ibid. **201,** L145.
—— (1977). *Astrophys. J.* **217,** 859.
GARDNER, F. F., MORRIS, D., and WHITEOAK, J. B. (1969). *Aust. J. Phys.* **22,** 79, 813.
—— and WHITEOAK, J. B. (1969). *Aust. J. Phys.* **22,** 107.
GINZBURG, V. L. and SYROVATSKY, S. I. (1964). *Origin of cosmic rays.* Pergamon Press, New York.
GREENBERG, J. M. (1968). *Nebulae and interstellar matter,* vol. VII, *Stars and stellar systems* (ed. MIDDLEHURST B. M. and ALLER L. H.), chap. VI. University of Chicago Press.
GREENSTEIN, J. L. (1946). *Astrophys. J.* **104,** 414.
HALL, J. S. (1949). *Science,* **109,** 166.
HAYAKAWA, S., ITO, K., and TERASHIMA, Y. (1958). *Progr. Theoret. Phys.* **8,** (*Suppl.* 6) 1.
—— NISHIMURA, S., and TAKAYANAGI, K. (1961), *Pub. Astron. Soc. J.* **13,** 184.
HEILES, C. (1976). *Ann. Rev. Astron. Astrophys.* **14,** 1.
—— and HABING, H. J. (1974). *Astron. Astrophys. Suppl.* **14,** 1.
HILL, E. R. (1960). *Bull. Astron. Inst. Netherlands* **15,** 1.
HILTNER, W. A. (1949). *Astrophys. J.* **109,** 471.
—— (1951). ibid. **114,** 241.
—— (1956). *Astrophys. J. Suppl.* **2,** 389.
HOBBS, L. M. (1974). *Astrophys. J. Lett.* **188,** L107.
—— (1976a). *Astrophys. J.* **203,** 143.
—— (1976b). *Astrophys. J. Lett.* **206,** L117.
—— (1978). *Astrophys. J.,* **222,** 491.
HUGHES, M. P., THOMPSON, A. R., and COLVIN, R. S. (1971). *Astrophys. J. Suppl.* **23,** 323.
JENKINS, E. B. and MELOY, D. A. (1974). *Astrophys. J. Lett.* **193,** L121.
JEPPS, S. A. (1975). *J. Fluid Mech.* **67,** 625.
JOKIPII, J. R. (1976). *Astrophys. J.* **208,** 900.
—— and LERCHE, I. (1969). *Astrophys. J.* **157,** 1137.
—— —— and SCHOMMER, R. A. (1969). *Astrophys. J. Lett.* **157,** L119.
—— and PARKER, E. N. (1969). *Astrophys. J.* **155,** 777, 799.
KAHN, F. D. and DYSON, J. E. (1965). *Ann. Rev. Astron. Astrophys.* **3,** 65.
KAPLAN, S. A. (1966). *Interstellar gas dynamics,* Pergamon Press, London and New York.
KERR, F. J. (1968). *Nebulae and interstellar matter,* vol. VII, *Stars and Stellar systems* (ed. MIDDLEHURST B. M. and ALLER L. H.), chap. X. University of Chicago Press.
KULSRUD, R. and PEARCE, W. P. (1969). *Astrophys. J.* **156,** 446.
LEE, M. A. (1972). *Astrophys. J.* **178,** 837.
LERCHE, I. (1966). *Phys. Fluids* **9,** 1073.

LERCHE, I. (1967a). *Astrophys. J.* **147,** 681.
—— (1967b). ibid. 689.
—— (1967c). ibid. **150,** 651.
—— (1969). *Wave phenomena in the interstellar plasma, advances in plasma physics* (ed. THOMPSON W. B. and SIMON A.), vol. II, p. 47. Wiley-Interscience, New York.
—— and PARKER, E. N. (1966). *Astrophys. J.* **145,** 106.
MANCHESTER, R. N. (1972). *Astrophys. J.* **172,** 43.
—— (1974). ibid. **188,** 637.
MARKKANEN, T. (1977). *Report No. 1*, Observatory and Astrophys. Lab. University of Helsinki.
MCKEE, C. F. and OSTRIKER, J. P. (1977). *Astrophys. J.*, **218,** 148.
MORSE, P. M. and FESHBACH, H. (1953). *Methods of theoretical physics*, vol. I, p. 662. McGraw-Hill New York.
MOUSCHOVIAS, T. Ch. (1974). *Astrophys. J.* **192,** 37.
—— (1976). ibid. **206,** 753; **207,** 141.
—— SHU, F. H., and WOODWARD, P. R. (1974). *Astron. Astrophys.* **33,** 73.
—— and SPITZER, L. (1976). *Astrophys. J.* **210,** 326.
MÜNCH, G. (1968). *Nebulae and interstellar matter*, vol. VII, *Stars and stellar systems* (ed. MIDDLEHURST B. M. and ALLER L. H.) chap. VII. University of Chicago Press.
MÜNCH, G. and UNSÖLD, A. (1962). *Astrophys. J.* **135,** 711.
—— and ZIRIN, H. (1961). *Astrophys. J.* **133,** 11.
O'DONNELL, E. J. and WATSON, W. D. (1974). *Astrophys. J.* **191,** 89.
OORT, J. H. (1959). *Bull. Astron. Inst. Netherlands* **15,** 45.
—— and SPITZER, L. (1955). *Astrophys. J.* **121,** 6.
PARKER, E. N. (1951). Ph. D. Thesis, Dept. Physics, California Institute of Technology.
—— (1953). *Astrophys. J.* **117,** 169, 431.
—— (1957). *Astrophys. J. Suppl.* **3,** 51.
—— (1958a). *Phys. Rev.* **109,** 1328.
—— (1958b). *Rev. Mod. Phys.* **30,** 955.
—— (1960). *Astrophys. J.* **132,** 821.
—— (1963). *Interplanetary dynamical processes*, chap. IX, XV. Wiley-Interscience, New York.
—— (1965). *Astrophys. J.* **142,** 584.
—— (1966a). ibid. **144,** 916.
—— (1966b). ibid. **145,** 811.
—— (1967a). ibid. **149,** 517.
—— (1967b). ibid. 535.
—— (1968a). ibid. **154,** 57.
—— (1968b). ibid. **154,** 875.
—— (1968c). *Nebulae and interstellar matter*, vol. VII, *Stars and stellar systems* (ed. MIDDLEHURST B. M. and ALLER L. H.), chap. XIV. University of Chicago Press.
—— (1969). *Space Sci. Rev.* **9,** 651.
—— (1971). *Astrophys. J.* **163,** 255; **166,** 295.
PETERS, B. and WESTERGAARD, N. J. (1977). *Astrophys. Space Sci.* **48,** 21.
PIKELNER, S. B. (1956). *Usp. Fiz. Nauk.* **58,** 285.
—— (1968). *Ann. Rev. Astron. Astrophys.* **6,** 165.
PURCELL, E. M. (1969a). *Physica*, **41,** 100.

Purcell, E. M. (1969b). *Astrophys. J.* **158,** 433.
—— (1973). *Interstellar grains as pinwheels*, Whipple Symposium Harvard.
—— and Pennypacker, C. R. (1973). *Astrophys. J.* **186,** 705.
—— and Shapiro, P. R. (1977). *Astrophys. J.* **214,** 92.
—— and Spitzer, L. (1971). *Astrophys. J.* **167,** 31.
Radhakrishnan, V., Murray, J. D., Lockhart, P., and Whipple, R. P. J. (1972). *Astrophys. J. Suppl.* **24,** 15.
Salpeter, E. E. and Gould, R. J. (1963). *Astrophys. J.* **138,** 393.
—— —— and Gold, T. (1963). *Astrophys. J.* **138,** 408.
Savage, B. D., Bohlin, R. C., Drake, J. F., and Budich, W. (1977). *Astrophys. J.* **216,** 291.
Savedoff, M. (1956). *Astrophys. J.* **124,** 533.
—— and Spitzer, L. (1950). *Astrophys. J.* **111,** 593.
Segalovitz, A., Shane, W. W., and de Bruyn, A. G. (1976). *Nature* **264,** 222.
Setti, G. (1976). *The physics of non-thermal radio sources* (Proc. NATO Advanced Study Inst., Urbino, June, 1975) (ed. Setti, G.) Reidel, Dordrecht.
Shapiro, M. M. and Silberberg, R. (1970). *Ann. Rev. Nucl. Sci.* **20,** 323.
Shu, F. H. (1974). *Astron. Astrophys.* **33,** 55.
Shull, J. M. (1977). *Astrophys. J.* **212,** 102.
—— —— and York, D. G. (1977). *Astrophys. J.* **211,** 803.
—— —— and Hobbs, L. M. (1977). *Astrophys. J.* **211,** L139.
Silberberg, R. (1966). *Phys. Rev.* **148,** 1247.
Soward, A. M. (1978). *Astron. Nachr.* **229,** 26.
Spitzer, L. (1954). *Astrophys. J.* **120,** 1.
—— (1968). *Nebulae and interstellar matter*, vol. VII, *Stars and stellar systems* (ed. Middlehurst B. M. and Aller L. H.), chap I. University of Chicago Press.
—— and Jenkins, E. B. (1975). *Ann. Rev. Astron. Astrophys.* **13,** 133.
—— and Tukey, J. W. (1951). *Astrophys. J.* **114,** 186.
Stecker, F. W. and Jones, F. C. (1977). *Astrophys. J.* **217,** 843.
—— Solomon, P. M., Scoville, N. Z., and Ryter, C. E. (1975). *Astrophys. J.* **201,** 90.
Stix, M. (1975). *Astron. Astrophys.* **42,** 85.
Stromgren, B. (1948). *Astrophys. J.* **108,** 242.
Tademaru, E. (1969). *Astrophys. J.* **158,** 959.
Thomas, G. E. and Krassa, R. F. (1971). *Astron. Astrophys.* **11,** 218.
Tosa, M. and Fujimoto, M. (1978). *Publ. Astron. Soc. Japan* **30,** (in press).
Van de Hulst, H. C. (1967). *Ann. Rev. Astron. Astrophys.* **5,** 167.
Weaver, R., McCray, R., Castor J., Shapiro, P. and Moore, R. (1977). *Astrophys. J.* **218,** 377.
——, (1978). Ibid. **220,** 742.
Wentzel, D. (1968). *Astrophys. J.* **152,** 987.
—— (1974). *Ann. Rev. Astron. Astrophys.* **12,** 71.
Weymann, R. (1960). *Astrophys. J.* **132,** 452.
White, M. P. (1978) *Astron. Nachr.* **299,** 209.
Williamson, F. O., Sanders, W. T., Kraushaar, W. L., McCammon, D., Borken, R., and Bunner, A. N. (1974). *Astrophys. J. lett.* **193,** L133.
York, D. G. (1974). *Astrophys. J. Lett.* **193,** L127.
—— (1977). *Astrophys. J.* **213,** 43.
Yu, G. (1974). *Astrophys. J.* **194,** 187.
Zanstra, H. (1955). *Gas dynamics of cosmic clouds* (ed. Burgers J. M. and van de Hulst, H. C)., North-Holland, Amsterdam.

23

MAGNETIC FIELDS AND MAGNETIC ACTIVITY

IT was remarked in the opening chapter that the magnetic field appears wherever there is a rotating astronomical body with an internal energy source. Even the slight radioactivity, internal chemical inhomogeneity, or differential precession in a slowly rotating planet is sufficient to generate a magnetic field. Stars, of course, have an energy source so vast that their production of field is, in most cases, but a minor diversion in their main business of illuminating the space around them. It is the thermal energy of stars combined with galactic rotation that produces the magnetic field in the gaseous disc of the galaxy, and it is presumed to be stars in their final convulsions that inflate that field with cosmic rays.

The hydromagnetic theory of large-scale fields in conducting fluids provides a comprehensive picture of the origin and the active non-equilibrium of the magnetic fields. The magnetic fields are spontaneously amplified within the body and grow to considerable strength before escaping into space. It is their departure through the surface into space that produces the suprathermal pyrotechnic effects that sputter and flare in so many astronomical bodies. On the other hand, while the general properties of the origin and activity of magnetic fields may now be in hand, there remain a variety of associated problems that are still very much unsolved. For instance, the dynamics of the internal circulation and convection of the sun, the formation of sunspots, the nature of coronal heating, the structure and activity of the galactic field, the acceleration of fast particles, the origin of cosmic rays, and many of the complex plasma effects of solar activity and planetary magnetospheres, to name only a few, are still in a preliminary state of development.

It was pointed out in the introductory chapter that the magnetic field plays an unusual role in astronomy, in many respects like a viable organism—indeed, an infectious organism—in a well defined ecological niche in the mechanical and electrodynamical operation of the astronomical body. From this point of view the magnetic field is an exceedingly simple organism, whose spontaneous creation is assured by the thermal battery effects in any rotating body. The growth and continued existence of the field is described in terms of the fluid motion by the linear electromagnetic equations. The spontaneous generation of field can be demonstrated from the theory under the most austere and idealized conditions. None of the complex circumstances and special environment necessary for the spontaneous development of biological life forms are needed. Magnetic fields thrive almost everywhere in the universe, whereas there is no evidence, either theoretical or observational, for the common occurrence of biological life. Biology is restricted by its loosely bound complexity to the occasional cool planet with just the right temperature and atmosphere. Thus, the ubiquitous magnetic field exhibits no progressive evolution, whereas biological organisms evolve through an endless sequence of individual forms, each with their own novel and peculiar aspects. Indeed, in some special cases their evolutionary survival reflexes have developed to such a degree that they detect, record, and contemplate each other, although where that particular evolutionary quirk will lead remains to be seen. But if biological forms are notable for their complexity, the magnetic forms are notable for their grand displays over astronomical dimensions. The impact of magnetic forms on their environment far exceeds the impact of biological forms. Magnetic forms produce activity and violence in the otherwise serene thermal degradation of the cosmic landscape. Nuclear and gravitational forces are the basic sources of energy and sometimes produce explosive events, but, without resident magnetic fields, the explosions would be limited to shock waves and thermal radiation. It is the magnetic fields that produce the suprathermal effects and the relativistic particles, and consequently the enhanced radio, visible, u.v., x-ray, and γ-ray emissions that are so conspicuous in even the gentlest circumstances. Even the lethargic sun is roused to explosive suprathermal events by the magnetic fields spawned beneath its surface.

The magnetic field is created from the convection and turbulence within each rotating astronomical body. The mechanics of the turbulence in the rotating frame of the body produces the cyclonic properties essential for the generation of magnetic field. The magnetic field rotates with the body while being generated and amplified. The field appears deep within the body, in the *material* manifestation of a slight density reduction. The reduced density is buoyant and rises out of the interior to the surface, carrying its magnetic field with it. In passing through the surface into the tenuous atmosphere above, the partial density reductions become extreme. The internal stresses of the field dominate, producing non-equilibrium and rapid reconnection. Dissipation through rapid reconnection, escape, and expansion into space is the ultimate fate of most of the magnetic flux that is generated in stars and galaxies, providing the remarkable activity that appears in so many forms to the astronomical observer (cf. Setti, 1976).

Planetary fields have a somewhat less spectacular history. They are essentially non-buoyant and extend out through the surface only as a consequence of resistive, and perhaps turbulent, diffusion. Their ultimate fate is an intrinsic part of their diffusion. The massive fluid in which they are created and anchored inhibits any more spectacular route to oblivion. The vigorous and complex activity of the field above the surface of the planet is driven by the solar wind, itself a product of activity of the sun. Jupiter rotates fast enough to contribute additional activity.

The supernova remnants, radio galaxies, quasars, and x-ray sources are dependent upon their magnetic fields for their extreme suprathermal activity. The cosmic rays and radio halos of galaxies are but another manifestation of magnetic fields. The chromosphere, corona, and solar wind, are largely a consequence of the magnetic flux extending up from the interior of the sun. The quiescent prominence, the sunspot, the spicule, the flare, the surge, the eruptive prominences and arches, the radio bursts, and the solar cosmic rays, are all the death rattle of the magnetic fields from the cyclonic motions beneath the surface of the sun. The activity of magnetic fields is evident on all scales from planets to galaxies.

Indeed, the suggestion has been made that there is a large-scale extra-galactic magnetic field, perhaps of cosmological extent (Sofue *et al.* 1968; Kawabata *et al.* 1969; Reinhardt and Thiel 1970; Fujimoto *et al.* 1971; Reinhardt 1971). The suggestion is based on the global asymmetry in the Faraday rotation measures of extra-galactic radio sources, that would be explained by a homogeneous field of about 10^{-9} G extending from $l^{\text{II}} \approx 280°$, $b^{\text{II}} \approx +30°$ to $l^{\text{II}} \approx 100°$, $b^{\text{II}} \approx -30°$. The energy density of such a field is small compared to the rest energy of the matter and the universal 3° black-body radiation, so it is not obvious that it has any profound implications for cosmology (see Doroshkevich 1965; Zeldovich 1965, 1969; Zeldovich and Novikov 1967). If it is assumed that the field is expanding isotropically with the universe as a whole, then the magnetic energy density declines in proportion to the four-thirds power of the density of matter, the same as the radiation field. It follows that, if the magnetic energy density is small compared to the energy density of the background thermal radiation at the present time, then it has always been, and always will be, small compared to the radiation. The origin of a weak cosmological field, if indeed there is such a field, may be presumed to lie in the very early stages of the universe. We are aware of no dynamo effect to be expected in a primordial fireball. The question lies outside the scope of the present work.

One of the most remarkable realizations of the last decade is the fact, noted in §21.1, that the conditions at the surface of Earth, which so critically affect the biological forms of the planet, are strongly influenced by the activity of the sun. There are now identified, from the C^{14} production rate over the last seven thousand years, ten different centuries in which activity was largely absent from the sun and eight centuries in which the activity was enormously enhanced above present-day levels (Eddy 1977). The historical information, as well as the geophysical evidence, show that the climate in the northern temperate zone was cold, with the advance of glaciers and ice sheets, in those ten different centuries when the magnetic activity of the sun was absent, while the climate was warm, with the retreat of glaciers, etc. in those eight separate centuries when the magnetic activity of the sun was extraordinarily high. The correspondence is one to one, with no exceptions, over the period of 7000 years for which information is available (see also the short-term variations pointed out by King 1975 and Wilcox 1975). The historical records show the economic hardship and political unrest in the most recent (15th and 17th) centuries of low activity and cold climate (cf. McNeill 1976).

The cause of the climate effect is not known, but there are several possibilities. For instance the u.v. and x-rays (Smith and Gottlieb 1974), the solar wind, and the solar cosmic rays vary greatly with the coming and going of magnetic activity at the sun (Parker 1963; Hundhausen 1972; Akasofu and Chapman 1972; Williams 1976). It is these hard radiations, and the associated polar cap absorption, ionospheric disturbances, aurorae, and magnetic storms and substorms that control conditions in the upper atmosphere of our planet (Ratcliffe 1960, 1970; Chapman 1964; Davies 1965; Hines *et al.* 1965; Akasofu and Chapman 1972; Williams 1976). The assertion of the upper atmosphere upon the dynamics of the troposphere has remained in doubt for many years, but it now appears that a close coupling exists (Bates 1977). The north–south heat transport by the ultra-long wavelength motions in the troposphere is strongly enhanced by heating of the stratosphere.

There is another possibility for the relation of climate and solar activity, too, in the fact that the coming and going of magnetic field at the sun implies changes in the convection and circulation that produce the fields. It is also the convection and circulation that deliver the heat to the surface of the sun. Is it possible, then, to alter the circulation so much as to destroy or enhance the appearance of magnetic fields at the surface of the sun without altering the luminosity of the sun by some small amount, say one per cent?

Whatever the connection between solar activity and terrestrial climate, it is difficult to avoid the fact that the historical records of climate, and the geological indicators of climate, have kept in step with the general level of solar activity for 18 strides at irregular intervals of time. A simple coincidence of two unrelated phenomena is too improbable to consider. It appears, then, that magnetic activity has a longer arm and a heavier hand than anyone could have imagined.

References

AKASOFU, S. I. and CHAPMAN, S. (1972). *Solar–terrestrial physics.* Clarendon Press, Oxford.

BATES, J. R. (1977). *Quart. J. Roy. Meterolog. Soc.* **103,** 397.

CHAPMAN, S. (1964). *Solar plasma, geomagnetism and aurora.* Gordon and Breach, New York.

DAVIES, K. (1965). *Ionospheric radio propagation.* Constable, London. Dover, New York.

DOROSHKEVICH, A. G. (1965). *Astrofizika* **1,** 255.

EDDY. J. A. (1977). *Climatic Change* **1,** 173.

FUJIMOTO, M., KAWABATA, K., and SOFUE, Y. (1971). *Prog. theor. Phys. Suppl.* (49) 181.

HINES, C. O., PAGHIS, I., HARTZ, T. R., and FEJER, J. A. (1965). *Physics of the earth's upper atmosphere.* Prentice Hall, Englewood Cliffs. New Jersey.

HUNDHAUSEN, A. J. (1972). *Coronal expansion and solar wind.* Springer-Verlag, Berlin.

KAWABATA, K., FUJIMOTO, M., SOFUE, Y., and FUKUI, M. (1969). *Pub. Astron. Soc. Jpn.* **21,** 293.

KING, J. W. (1975). *Aeronautics and Astronautics* **13,** (4) 10.

MCNEILL, W. H. (1976). *Plagues and peoples.* Anchor Press, Garden City, New York.

PARKER, E. N. (1963). *Interplanetary dynamical processes.* Wiley-Interscience, New York.

RATCLIFFE, J. A. (1960). *Physics of the upper atmosphere.* Academic Press, New York. (1970). *Sun, earth, and radio.* McGraw-Hill, New York.

REINHARDT, M. (1971). *Astrophys. Lett.* **8,** 181.

—— and THIEL, M. A. F. (1970). *Astrophys. Lett.* **7,** 101.

SETTI, G. (1976). *The physics of non-thermal radio sources.* Reidel, Dordrecht.

SMITH, E. V. P. and GOTTLIEB, D. M. (1974). *Space Sci. Rev.* **16,** 771.

SOFUE, Y., FUJIMOTO, M., and KAWABAtA, K. (1968). *Pub. Astron. Soc. Jpn.* **20,** 368.

WILCOX, J. M. (1975). *J. Atmos. Terrestr. Phys.* **37,** 237.

WILLIAMS, D. J. (1976). *Physics of solar planetary environments,* vols. I and II. American Geophysical Union.

ZELDOVICH, Ya. B. (1965). *Zh. eksp teor. Fiz.* **48,** 986.

—— (1969). *Zh. eksp. teor. Fiz.* **56,** L6.

—— and NOVIKOV, I. D. (1967), *Relativistkaya astrofizika,* chap. 21. Izdatel'stvo Nauka, Moscow.

INDEX OF COMMON SYMBOLS

Page references indicate first appearance of symbol.

a — half thickness of column of field, 161; of slab, 444; vector potential of x- and y-components of magnetic field, 365; radius, 448; semi-major axis of spheroid, 812.

a_i — vector potential, 551.

A — cross-sectional area of flux tube, 124; vector potential for x- and y-components of magnetic field, 65; correlation function, 525; dimensionless growth rate, 637.

A_i, $\mathbf{A}$ — vector potential, 39.

α — one component of the Euler potential (α, β), 36; constant in Newton's law of cooling, 147; free parameter, 154, 321; dimensionless function of time, 236; ratio of magnetic to gas pressure, 326; angle between standing Alfven wave front and x-axis, 413; function of position along magnetic line of force, 475; $2\tau_1\mu$, 494; $d\gamma/dt$, 496.

b — semi-minor axis of spheroid, 812.

$\mathrm{ber}_n(x)$, $\mathrm{bei}_n(x)$ — real and imaginary parts of $J_n(i^{\frac{3}{2}}x)$, 448

$\mathbf{b}$ — magnetic field, 167; Fourier transform of $\mathbf{B}$, 484; non-axisymmetric part of magnetic field, 600.

B — magnitude of magnetic field, $|\mathbf{B}|$, 23; correlation function, 525.

B_i, $\mathbf{B}$ — magnetic field, 17; axisymmetric part of magnetic field, 600.

$\mathbf{B}'$ — magnetic field in moving frame of the fluid, 17.

β — one component of the Euler potential (α, β), 36; dimensionless parameter, 322; ratio of cosmic ray pressure to gas pressure, 353; initial angle of inclination of lines of force to x-axis, 413; exponent, 419.

c — speed of light, 17.

C — correlation function, 527.

C_D — aerodynamic drag coefficient, 142.

γ — relativistic factor, $(1-w^2/c^2)^{-\frac{1}{2}}$, 50; ratio of specific heats, 114; Euler's constant, 0.577, 144; final inclination of lines of force, 414; contraction of γ_{ijk}, 495; dimensionless dissipation, 621.

γ_{ijk} — Lagrangian correlation function, 495.

Γ — dimensionless constant, 89; gamma function, 140; dynamo coefficient, 551, 580, 619.

D — diffusion coefficient for scalar field, 529; differential operator, 620.

ξ	elliptical coordinate, 200; oblate spheroidal coordinate, 811.
ξ_i	integral of turbulent velocity at fixed point, 483; total Lagrangian displacement of a point, 490.
ξ_{ijkl}	Lagrangian correlation function, 495.
p	fluid pressure, 53.
p_e	ambient fluid pressure outside flux tube, 124.
p_i	fluid pressure inside flux tube, 124.
P	function representing axisymmetric poloidal field, 69; total fluid and magnetic pressure, 220; cosmic ray pressure, 353.
$P_n^m(x)$	associated Legendre polynomial, 560.
ρ	fluid density, 52.
ρ_e	ambient fluid density outside flux tube, 124.
ρ_i	fluid density inside flux tube, 124.
ϖ	radial distance $(x^2+y^2)^{\frac{1}{2}}$ from the z-axis, 67.
Π	radial distance $(X^2+Y^2)^{\frac{1}{2}}$ from the Z-axis, 176.
q	electric charge on a particle, 23; arbitrary parameter, 85, 185; fractional reduction of heat transport, 257; pressure perturbation, 364.
q_i	dimensionless wave number, $k_i\Lambda$, 327; k_z/K, 637.
Q	dimensionless constant, $\ln(4/N_R+\frac{1}{2}-\gamma)$, 144; dimensionless wave number $k_z\Lambda+\frac{1}{2}(s-1)$, 327; secondary dynamo coefficient, 619.
Q_1, Q_2	heat transferred during isothermal legs of Carnot cycle, 727.
Q_i, $\mathbf{Q}$	Poynting vector, 51.
dQ/dt	rate of heat loss, 147.
r	radial distance $(x^2+y^2+z^2)^{\frac{1}{2}}$ from origin.
R	cyclotron radius of particle, 23; effective mass of moving cylinder, 153; vector potential in resistive region, 402; magnitude of dimensionless growth rate, 652; characteristic radius, 812.
R_λ	magnetic Reynolds number over small scale λ, 418.
R_m	magnetic Reynolds number, 43.
R_{ij}	correlation function, 284; Reynolds stress tensor, 468.
R_{ijkl}	correlation function, 280.
s	arc length, 34.
S	coordinate orthogonal to lines of force in the xy-plane, 371; special function, 405.
S_{ij}	correlation function, 289.
σ	electrical conductivity, 19.
t	time, 17; relaxation time of magnetic field, 808.
t_2	duration of rotation in convective cell, 574.
t_3	duration of longitudinal velocity v_3 in convective cell, 574.
T	temperature, 19; trace of T_{ij}, 57; function representing axisymmetric toroidal field, 69.

T_e ambient temperature of gas outside flux tube, 124.
T_i temperature of gas inside flux tube, 124.
$\Delta T \equiv T_e - T_i$, 127.
T_{ij} kinetic tensor, 57.
$\mathcal{T}$ total tension along flux tube, 137; tension per unit area, 162.
τ characteristic dissipation time, 117; growth time, 216; eddy correlation time, 482.
τ_1 velocity correlation time, 494.
τ_2 vorticity correlation time, 494.
u rate of rise of buoyant flux tube, 142; x-component of fluid velocity, 153; ϖ^2/L^2, 176; fluid velocity along tube, 221; rms thermal velocity in any one direction, 326; ion thermal velocity, 429.
u_i Fourier transform of v_i, 484.
u_T mean upward flow velocity of thermal energy, 149.
$\mathbf{u}$ non axi-symmetric component of fluid motion, 600.
U thermal energy density, 114; magnetic potential, 542.
U_ϕ phase velocity, 629.
v y-component of fluid velocity, 153.
v_0 characteristic velocity of dominant eddies, 477.
v_3 longitudinal velocity in cyclonic convective cell, 574.
$\mathbf{v}$ fluid velocity, 17.
V equipartition velocity, 237; volume per cyclonic cell, 575.
V_A Alfven speed, 108.
V_i large-scale, slowly varying component of fluid velocity, 567.
$\mathbf{V}$ large-scale, axisymmetric component of fluid velocity, 600.
ϕ electrostatic potential, 22; azimuthal angle measured around z-axis from the x-axis, 67; velocity potential, 229.
Φ total magnetic flux, 33; probability distribution, 282; potential energy, 339; total rotation, $\Omega_3 t_2$, 576.
w thermal velocity, 45; drift velocity, 217; velocity of magnetic merging, 393.
$w_\parallel$ particle velocity parallel to the magnetic field, 23.
$w_\perp$ particle velocity perpendicular to the magnetic field, 23.
$\mathbf{w}$ particle velocity, 23.
W particle energy per unit volume, 54; magnetic potential, 129; potential function, 350; total work done by Carnot cycle, 727.
ψ gravitational potential per unit mass, 63; dimensionless function, 178; relative magnetic field density, 242; probability distribution, 281; stream function, 396; polynomial, 662.
Ψ stream function, 310; velocity potential, 316; field pressure gradient, 369; probability distribution of position of origin of Lagrangian point, 491; vector potential, 559.

GENERAL INDEX